COLD SPRING HARBOR SYMPOSIA ON QUANTITATIVE BIOLOGY

VOLUME LXVII

The Cardiovascular System

Meeting Organized by Bruce Stillman and David Stewart
COLD SPRING HARBOR LABORATORY PRESS
2002

COLD SPRING HARBOR SYMPOSIA ON QUANTITATIVE BIOLOGY VOLUME LXVII

©2002 by Cold Spring Harbor Laboratory Press
Cold Spring Harbor, New York
International Standard Book Number 0-87969-678-8 (cloth)
International Standard Book Number 0-87969-679-6 (paper)
International Standard Serial Number 0091-7451
Library of Congress Catalog Card Number 34-8174

Printed in the United States of America
All rights reserved

COLD SPRING HARBOR SYMPOSIA ON QUANTITATIVE BIOLOGY
Founded in 1933 by
REGINALD G. HARRIS
Director of the Biological Laboratory 1924 to 1936
Previous Symposia Volumes

I (1933) Surface Phenomena
II (1934) Aspects of Growth
III (1935) Photochemical Reactions
IV (1936) Excitation Phenomena
V (1937) Internal Secretions
VI (1938) Protein Chemistry
VII (1939) Biological Oxidations
VIII (1940) Permeability and the Nature of Cell Membranes
IX (1941) Genes and Chromosomes: Structure and Organization
X (1942) The Relation of Hormones to Development
XI (1946) Heredity and Variation in Microorganisms
XII (1947) Nucleic Acids and Nucleoproteins
XIII (1948) Biological Applications of Tracer Elements
XIV (1949) Amino Acids and Proteins
XV (1950) Origin and Evolution of Man
XVI (1951) Genes and Mutations
XVII (1952) The Neuron
XVIII (1953) Viruses
XIX (1954) The Mammalian Fetus: Physiological Aspects of Development
XX (1955) Population Genetics: The Nature and Causes of Genetic Variability in Population
XXI (1956) Genetic Mechanisms: Structure and Function
XXII (1957) Population Studies: Animal Ecology and Demography
XXIII (1958) Exchange of Genetic Material: Mechanism and Consequences
XXIV (1959) Genetics and Twentieth Century Darwinism
XXV (1960) Biological Clocks
XXVI (1961) Cellular Regulatory Mechanisms
XXVII (1962) Basic Mechanisms in Animal Virus Biology
XXVIII (1963) Synthesis and Structure of Macromolecules
XXIX (1964) Human Genetics
XXX (1965) Sensory Receptors
XXXI (1966) The Genetic Code
XXXII (1967) Antibodies
XXXIII (1968) Replication of DNA in Microorganisms
XXXIV (1969) The Mechanism of Protein Synthesis
XXXV (1970) Transcription of Genetic Material
XXXVI (1971) Structure and Function of Proteins at the Three-dimensional Level
XXXVII (1972) The Mechanism of Muscle Contraction
XXXVIII (1973) Chromosome Structure and Function
XXXIX (1974) Tumor Viruses
XL (1975) The Synapse
XLI (1976) Origins of Lymphocyte Diversity
XLII (1977) Chromatin
XLIII (1978) DNA: Replication and Recombination
XLIV (1979) Viral Oncogenes
XLV (1980) Movable Genetic Elements
XLVI (1981) Organization of the Cytoplasm
XLVII (1982) Structures of DNA
XLVIII (1983) Molecular Neurobiology
XLIX (1984) Recombination at the DNA Level
L (1985) Molecular Biology of Development
LI (1986) Molecular Biology of *Homo sapiens*
LII (1987) Evolution of Catalytic Function
LIII (1988) Molecular Biology of Signal Transduction
LIV (1989) Immunological Recognition
LV (1990) The Brain
LVI (1991) The Cell Cycle
LVII (1992) The Cell Surface
LVIII (1993) DNA and Chromosomes
LIX (1994) The Molecular Genetics of Cancer
LX (1995) Protein Kinesis: The Dynamics of Protein Trafficking and Stability
LXI (1996) Function & Dysfunction in the Nervous System
LXII (1997) Pattern Formation during Development
LXIII (1998) Mechanisms of Transcription
LXIV (1999) Signaling and Gene Expression in the Immune System
LXV (2000) Biological Responses to DNA Damage
LXVI (2001) The Ribosome

Front Cover (*Paperback*): Composite image composed of lateral views of the vascular anatomy of the zebrafish at 1.5, 3, 5, and 7 days post fertilization. Image courtesy of Sumio Isogai and Brant Weinstein, National Institutes of Health.

Authorization to photocopy items for internal or personal use, or the internal or personal use of specific clients, is granted by Cold Spring Harbor Laboratory Press, provided that the appropriate fee is paid directly to the Copyright Clearance Center (CCC). Write or call CCC at 222 Rosewood Drive, Danvers, MA 01923 (508-750-8400) for information about fees and regulations. Prior to photocopying items for educational classroom use, contact CCC at the above address. Additional information on CCC can be obtained at CCC Online at http://www.copyright.com/

All Cold Spring Harbor Laboratory Press publications may be ordered directly from Cold Spring Harbor Laboratory Press, 500 Sunnyside Boulevard, Woodbury, NY 11797-2924. Phone: 1-800-843-4388 in Continental U.S. and Canada. All other locations: (516) 422-4100. FAX: (516) 422-4097. E-mail: cshpress@cshl.org. For a complete catalog of all Cold Spring Harbor Laboratory Press publications, visit our World Wide Web Site http://www.cshlpress.com/

Symposium Participants

ABMAN, STEVEN, Dept. of Pediatrics, The Children's Hospital, University of Colorado, Denver
ACEVEDO-BOLTON, GABRIEL, Dept. of Bioengineering, California Institute of Technology, Pasadena, California
ALBRECHT, BARBARA, PH-R CVII, Bayer AG, Wuppertal, Germany
ALITALO, KARI, Lab. of Molecular and Cancer Biology, University of Helsinki, Helsinki, Finland
ALLEN, CLAIRE, Developmental Genetics Programme, Biomedical Science, University of Sheffield, Sheffield, United Kingdom
APTEL, HERVE, Dept. of Pharmacy and Pharmacology, Bath University, Bath, United Kingdom
ARAD, MICHAEL, Dept. of Genetics, Harvard Medical School, Boston, Massachusetts
ARBEIT, JEFFREY, Cancer Genetics Program, University of California, San Francisco
ARNETT, DONNA, Dept. of Epidemiology, University of Minnesota, Minneapolis
AUSONI, SIMONETTA, Dept. of Biomedical Sciences, University of Padua, Padova, Italy
AZENE, EZANA, Institute of Molecular Cardiobiology, Johns Hopkins University School of Medicine, Baltimore, Maryland
BALDINI, ANTONIO, Dept. of Pediatrics and Cardiology, Baylor College of Medicine, Houston, Texas
BANG, MARIE-LOUISE, Institute of Molecular Medicine, University of California at San Diego, La Jolla
BARRANS, DAVID, Dept. of Medicine, Harvard Medical School, Brigham and Women's Hospital, Boston, Massachusetts
BARRON, MATTHEW, Dept. of Molecular and Cellular Biology, Baylor College of Medicine, Houston, Texas
BELAGAJE, RAMA, Dept. of Cardiovascular Research, Eli Lilly and Co., Indianapolis, Indiana
BENEZRA, ROBERT, Dept. of Cell Biology, Memorial Sloan-Kettering Cancer Center, New York, New York
BENJAMIN, LAURA, Dept. of Pathology, Beth Israel Deaconess Medical Center, Harvard Medical School, Boston, Massachusetts
BERGERS, GABRIELLE, Dept. of Neurological Surgery, University of California, San Francisco
BLASIOLE, BRIAN, Dept. of Pharmacology, Pennsylvania State College of Medicine, Hershey, Pennsylvania
BLAU, HELEN, Dept. of Microbiology and Immunology, Baxter Laboratory, Stanford University School of Medicine, Stanford, California
BREITSCHOPF, KRISTIN, Div. of Cardiovascular Diseases, Aventis Pharma AG, Frankfurt, Germany
BROECKEL, ULRICH, Human and Molecular Genetics Center, Medical College of Wisconsin, Milwaukee
BURGON, PATRICK, Dept. of Genetics, Howard Hughes Medical Institute, Harvard Medical School, Boston, Massachusetts
CAMPIONE, MARINA, Dept. of Biomedical Sciences, University of Padua, Padova, Italy
CHANG, DAVID, Dept. of Cardiovascular Sciences, Baylor College of Medicine, Houston, Texas
CHEAH, KATHRYN, Dept. of Biochemistry, University of Hong Kong, Hong Kong, China
CHEN, HAIFENG, Dept. of Advanced Technology, Avigen, Inc., Alameda, California
CHEN, HSIAO-LIN, Institute of Biomedical Sciences, Academia Sinica, Taipei, Taiwan
CHERESH, DAVID, Dept. of Immunology and Vascular Biology, The Scripps Research Institute, La Jolla, California
CHI, XUAN, Dept. of Cellular and Molecular Biology, Baylor College of Medicine, Houston, Texas
CHIEN, KENNETH, Institute of Molecular Medicine, University of California at San Diego, La Jolla
CHOU, MIN-YUAN, Biomedical Engineering Center, Industrial Technology Research Institute, Chutung, Taiwan
CLOUTHIER, DAVID, Dept. of Molecular, Cellular and Craniofacial Biology, University of Louisville, Louisville, Kentucky
COFFMAN, THOMAS, Dept. of Medicine, Duke University Medical Center, Durham, North Carolina
COHEN, JONATHAN, Dept. of Internal Medicine, Southwestern Medical Center, University of Texas, Dallas
CONLON, FRANK, Dept. of Genetics, University of North Carolina, Chapel Hill
COUGHLIN, SHAUN, Cardiovascular Research Institute, University of California, San Francisco
COWAN, CHAD, Center for Developmental Biology, Southwestern Medical Center, University of Texas, Dallas
CROISSANT, JEFFREY, Anterogen Co., Roxbury, Massachusetts
CUI, YINGJIE, Dept. of Thoracic Surgery, First Hospital of Beijing University, Beijing, China
CZIROK, ANDRAS, Dept. of Anatomy and Cell Biology, Medical Center, University of Kansas, Kansas City
DAVEY, MEGAN, Div. of Cell and Developmental Biology, School of Life Sciences, University of Dundee, Dundee, Scotland, United Kingdom
DEATON, REBECCA, Cardiovascular Research Institute, Health Science Center, University of North Texas, Fort Worth, Texas
DE LANEROLLE, PRIMAL, Dept. of Physiology and Biophysics, University of Illinois, Chicago

DELOT, EMMANUELE, Dept. of Orthopaedic Surgery, School of Medicine, University of California, Los Angeles
DESCHEPPER, CHRISTIAN, Dept. of Cardiovascular Biology, Montréal Clinical Research Institute, Montréal, Québec, Canada
DOMINGUEZ MACIAS, JORGE, Dept. of Experimental Biology, University of Jaén, Jaén, Spain
DOWNES, MEREDITH, Div. of Cellular and Developmental Biology, Institute for Molecular Bioscience, Brisbane, Queensland, Australia
DVORAK, HAROLD, Research Pathology, Beth Israel Deaconess Medical Center, Boston, Massachusetts
EBERHARD, DANIEL, Dept. of Developmental Biology and Molecular Pathology, University of Bielefeld, Bielefeld, Germany
ELDAR, MICHAEL, Sheba Medical Center, Neufeld Cardiac Research Institute, Tel Hashomer, Israel
ENGEL, FELIX, Dept. of Cardiology, Children's Hospital, Boston, Massachusetts
EPSTEIN, JONATHAN, Dept. of Molecular Cardiology, Hospital of the University of Pennsylvania, Philadelphia
EPSTEIN, NEAL, Dept. of Cardiology, National Heart, Lung, and Blood Institute, National Institutes of Health, Bethesda, Maryland
FARRANCE, IAIN, Dept. of Biochemistry, School of Medicine, University of Maryland, Baltimore
FERRARA, NAPOLEONE, Dept. of Molecular Oncology, Genentech, Inc., South San Francisco, California
FISCHER, ANDREAS, Dept. of Medicine, Children's Hospital, Boston, Massachusetts
FISCHER, JAN, Klinik Balgrist, University of Zurich, Zurich, Switzerland
FISHMAN, GLENN, Division of Cardiology, New York University School of Medicine, New York, New York
FISHMAN, MARK, Dept. of Cardiovascular Research, Cardiology Division, Massachusetts General Hospital, Harvard Medical School, Boston, Massachusetts
FRAIDENRAICH, DIEGO, Dept. of Cell Biology, Memorial Sloan-Kettering Institute for Cancer Research, New York, New York
FRANCO, DIEGO, Dept. of Experimental Biology, University of Jaén, Jaén, Spain
FRASCH, MANFRED, Dept. of Biochemistry and Molecular Biology, Mount Sinai School of Medicine, New York, New York
FUJIWARA, KEIGI, Dept. of Cardiology, Center for Cardiovascular Research, University of Rochester, Rochester, New York
FUKAI, TOHRU, Dept. of Cardiology, Emory University, Atlanta, Georgia
FULLER, GERALDINE, Dept. of Veterinary Biosciences, Ohio State University, Columbus, Ohio
GE, RUOWEN, Dept. of Biological Sciences, National University of Singapore, Singapore
GESSLER, MANFRED, Biocenter, PC I, University of Würzburg, Würzburg, Germany
GOLDBERG, ITZHAK, Dept. of Radiation Oncology, North Shore–Long Island Jewish Health System, Manhasset, New York
GOLFETTI, ROSELI, Dept. of Cell Biology and Neuroscience, Rutgers University, Highland Park, New Jersey
GONG, XIAOHUA, Dept. of Cell Biology, The Scripps Research Institute, La Jolla, California
GOYAL, LAKSHMI, Editorial Office, Cell Press, Cambridge, Massachusetts
GRANT, STEPHEN, Cardiovascular Research Institute, Health Science Center, University of North Texas, Fort Worth, Texas
GRANVILLE, DAVID, Dept. of Molecular and Experimental Medicine, The Scripps Research Institute, La Jolla, California
GRAZIANO, MICHAEL, Dept. of Cardiovascular and Endocrine Biology, Schering-Plough Research Institute, Kenilworth, New Jersey
GRIDLEY, THOMAS, The Jackson Laboratory, Bar Harbor, Maine
GU, XING HUA, Cardiovascular System Research, Zensun Science and Technology Ltd., Pudong, Shanghai, China
GUERRIERO, ANASTASIA, Dept. of Pathology, Abramson Family Cancer Research Institute, University of Pennsylvania, Philadelphia
HALL, GENTZON, Dept. of Biochemistry and Molecular Biology, School of Medicine, University of Maryland, Baltimore
HARVEY, RICHARD, Developmental Biology Unit, Victor Chang Cardiac Research Institute, Darlinghurst, Australia
HEICKLEN, ALICE, Dept. of Developmental and Molecular Biology, Albert Einstein College of Medicine, Bronx, New York
HENKEMEYER, MARK, Center for Developmental Biology, Southwestern Medical Center, University of Texas, Dallas
HICK, ELIZABETH, Dept. of Genetics, Harvard Medical School, Boston, Massachusetts
HONER, CHRISTIAN, Dept. of Metabolic and Cardiovascular Diseases, Novartis Pharmaceuticals Corp., Summit, New Jersey
HORTON, JAY, Dept. of Internal Medicine and Molecular Genetics, Southwestern Medical Center, University of Texas, Dallas
HOSOKAWA, HIROSHI, Dept. of Cardiology, Center for Cardiovascular Research, University of Rochester, Rochester, New York
HOVE, JAY, Dept. of Bioengineering, California Institute of Technology, Pasadena, California
HU, BING, Dept. of Research and Development, Centocor, Inc., Malvern, Pennsylvania
HU, CHENG-JUN, Dept. of Cell and Developmental Biology, University of Pennsylvania, Philadelphia
HU, ERDING, Dept. of Vascular Biology, GlaxoSmithKline, King of Prussia, Pennsylvania
HUBNER, NORBERT, Dept. of Cardiovascular Genetics, Max-Delbrück-Center for Molecular Medicine, Berlin, Germany
HUSSAIN, M. MAHMOOD, Dept. of Anatomy and Cell Biology, Downstate Medical Center, State University of New York, Brooklyn

HWANG, PAUL, National Heart, Lung, and Blood Institute, National Institutes of Health, Bethesda, Maryland
HYNES, RICHARD, Dept. of Biology, Center for Cancer Research, Massachusetts Institute of Technology, Cambridge, Massachusetts
IIZUKA, MASAKI, Dept. of Biology, Molecular and Cellular Biology, Nippon Boehringer Ingelheim Co., Ltd., Kawanishi, Hyogo, Japan
IZUMO, SEIGO, Dept. of Biology, Cardiovascular Research, Beth Israel Deaconess Medical Center, Harvard University, Boston, Massachusetts
JAIN, RAKESH, Dept. of Radiation Oncology, Massachusetts General Hospital, Boston, Massachusetts
JAISSER, FREDERIC, Hôpital Xavier Bichat, INSERM, Paris, France
JENG, ARCO, Dept. of Biology, Research, Novartis Pharmaceuticals Corp., Summit, New Jersey
JOHNSON, LYNNE, Dept. of Biology, Cardiology, Rhode Island Hospital, Brown University, Providence, Rhode Island
JONES, ELIZABETH, Dept. of Biology, California Institute of Technology, Pasadena, California
KALLURI, RAGHU, Dept. of Biology, Medicine, Beth Israel Deaconess Medical Center, Boston, Massachusetts
KANSE, SANDIP, Institute for Biochemistry, University of Giessen, Giessen, Germany
KELLY, DANIEL, Dept. of Medicine and Molecular Biology, Center for Cardiovascular Research, Washington University School of Medicine, St. Louis, Missouri
KIBERSTIS, PAULA, Editorial Office, *Science* Magazine, Washington, D.C.
KIM, YONGSOK, National Heart, Lung, and Blood Institute, National Institutes of Health, Bethesda, Maryland
KIOUSSI, CHRISSA, Dept. of Medicine, Howard Hughes Medical Institute, University of California at San Diego, La Jolla, California
KISHI, HIROKO, Lab. of Molecular Cardiology, National Heart, Lung, and Blood Institute, National Institutes of Health, Bethesda, Maryland
KITAJIMA, SATOSHI, Dept. of Cellular and Molecular Toxicology, National Insititute of Health Sciences, Tokyo, Japan
KOICHIRO, KUWAHARA, Dept. of Medicine and Clinical Science, Kyoto University Graduate School of Medicine, Kyoto, Japan
KRANZ, ANDREA, Dept. of Internal Medicine, Southwestern Medical Center, University of Texas, Dallas
KRASNOW, MARK, Dept. of Biochemistry, Stanford University School of Medicine, Stanford, California
KRIEG, PAUL, Dept. of Cell Biology and Anatomy Life Sciences, College of Medicine, University of Arizona, Tucson
KROLL, JENS, Dept. of Internal Medicine, Southwestern Medical Center, University of Texas, Dallas
KRUPINSKI, JOHN, Dept. of Cardiovascular Research, Bristol-Myers Squibb, Pennington, New Jersey
LAM, JASON, Institute of Molecular Medicine, University of California at San Diego, La Jolla
LAMBERT, DAN, Dept. of Biochemistry and Molecular Biology, University of Leeds, Leeds, United Kingdom
LAPING, NICHOLAS, Dept. of Renal and Urology Research, GlaxoSmithKline, King of Prussia, Pennsylvania
LASSAR, ANDREW, Dept. of Biological Chemistry and Molecular Pharmacology, Harvard Medical School, Boston, Massachusetts
LATINKIC, BRANKO, Dept. of Developmental Biology, National Institute for Medical Research, London, United Kingdom
LAU, LESTER, Dept. of Molecular Genetics, University of Illinois, Chicago
LAW, SIMON, Dept. of Biology, Linden Technologies, Inc., Woburn, Massachusetts
LAWSON, NATHAN, Lab. of Molecular Genetics, National Institute of Child Health Development, National Institutes of Health, Bethesda, Maryland
LE BLANCQ, SYLVIE, Doris Duke Charitable Foundation, New York, New York
LEINWAND, LESLIE, Dept. of Molecular, Cellular and Developmental Biology, University of Colorado, Boulder
LI, QING, Dept. of Cell Biology, Harvard Medical School, Boston, Massachusetts
LI, QUANYI, Dept. of Cardiology, The Children's Hospital of Philadelphia, Philadelphia, Pennsylvania
LI, XURI, Center for Transgene Technology and Gene Therapy, Catholic University of Leuven, Leuven, Belgium
LIEW, CHOONG-CHIN, Dept. of Medicine, Harvard Medical School, Brigham and Women's Hospital, Boston, Massachusetts
LIFTON, RICHARD, Boyer Center for Molecular Medicine, Yale University School of Medicine, New Haven, Connecticut
LIN, JIING-HUEY, Dept. of Cardiovascular Research, Berlex Biosciences Inc., Richmond, California
LIN, JIUANN-HUEY, Dept. of Cardiovascular Sciences, Baylor College of Medicine, Houston, Texas
LIU, ZHI-PING, Dept. of Molecular Biology, Southwestern Medical Center, University of Texas, Dallas
LOPEZ-PEREZ, ELVIRA, Dept. de Biologie, Glaxo Smith Kline, Les Ulls, France
MACK, FIONA, Dept. of Cell Growth and Cancer, University of Pennsylvania, Philadelphia
MAHONEY, WILLIAM, Dept. of Biochemistry and Molecular Biology, University of Maryland, Baltimore
MANN, DOUGLAS, Dept. of Medicine and Cardiology, Baylor College of Medicine, Houston, Texas
MANSFIELD, KYLE, Abramson Family Cancer Research Institute, University of Pennsylvania, Philadelphia
MARBÁN, EDUARDO, Institute of Molecular Cardiobiology, Johns Hopkins University, Baltimore, Maryland
MARKS, ANDREW, Center for Molecular Cardiology, Columbia University, New York, New York
MARVIN, MARTHA, Dept. of Cell Biology, Harvard Medical School, Boston, Massachusetts
MASCARENO, EDUARDO, Dept. of Anatomy and Cell Biology, Downstate Medical Center, State University of New York, Brooklyn
MAY, SCOTT, Dept. of Neurology, Ernest Gallo Clinic and Research Center, University of California, San Francisco

MCGARRY, THOMAS, Dept. of Medicine and Signal Transduction, Beth Israel Deaconess Medical Center, Boston, Massachusetts
MCKINNON, DAVID, Dept. of Neurobiology and Behavior, State University of New York, Stony Brook
MCNALLY, ELIZABETH, Dept. of Medicine, Cardiology Section, University of Chicago, Chicago, Illinois
MERCOLA, MARK, Dept. of Cell Biology, Harvard Medical School, Boston, Massachusetts
MERKI, ESTHER, Institute of Molecular Medicine, University of California at San Diego, La Jolla
MIANO, JOSEPH, Dept. of Medicine and Cardiology, University of Rochester Medical Center, Rochester, New York
MITTAL, VIVEK, Dept. of Cancer Genomics, Genome Center, Cold Spring Harbor Laboratory, Woodbury, New York
MIURA, GRANT, Dept. of Developmental Genetics, New York University School of Medicine, New York, New York
MIURA, NAOYUKI, Dept. of Biochemistry, Hamamatsu University School of Medicine, Hamamatsu, Japan
MOCKRIN, STEPHEN, Div. of Heart and Vascular Diseases, National Heart, Lung, and Blood Institute, National Institutes of Health, Bethesda, Maryland
MOHUN, TIMOTHY, Dept. of Molecular Biology, National Institute for Medical Research, London, United Kingdom
MOLDOVAN, NICANOR, Davis Heart and Lung Research Institute, Ohio State University, Columbus, Ohio
MONTI, JAN, Dept. of Cardiovascular Genetics, Max-Delbrück-Center for Molecular Medicine, Berlin, Germany
MORRISON, JOHN, Dept. of Medicine, North Shore University Hospital–New York University, Manhasset, New York
MOSKOWITZ, IVAN, Dept. of Genetics, Harvard Medical School, Boston, Massachusetts
MOULTON, KAREN, Dept. of Surgical Research, Children's Hospital, Boston, Massachusetts
MUKHOPADHYAY, DEBABRATA, Dept. of Pathology, Beth Israel Deaconess Medical Center, Harvard Medical School, Boston, Massachusetts
MUNDEL, PETER, Dept. of Medicine, Div. of Nephrology, Albert Einstein College of Medicine, Bronx, New York
MURRY, CHARLES, Dept. of Pathology, School of Medicine, University of Washington, Seattle
NABEL, ELIZABETH, Clinical Research Programs, Cardiovascular Branch, National Heart, Lung, and Blood Institute, National Institutes of Health, Bethesda, Maryland
NAKAGAWA, YASUAKI, Dept. of Medicine and Clinical Science, Kyoto Graduate School of Medicine, Kyoto, Japan
NAKAMURA, TOMOYUKI, Institute of Molecular Medicine, University of California at San Diego, La Jolla
NASEVICIUS, AIDAS, Discovery Genomics, Inc., Minneapolis, Minnesota
NHAN, THOMAS, Dept. of Pathology, University of Washington, Seattle
NORBY, PEDER, Dept. of Cloning Technology and Immunology, Novo Nordisk, Bagsvaerd, Denmark
OLSON, ERIC, Dept. of Molecular Biology, Southwestern Medical Center, University of Texas, Dallas
O'ROURKE, JAMES, Dept. of Pathology, Health Center, University of Connecticut, Farmington
OUVRARD-PASCAUD, ANTOINE, Medicine Faculty, INSERM U478, Paris, France
PABON, LIL, Dept. of Pathology, University of Washington, Seattle
PAN, YI, Dept. of Cellular and Developmental Biology, School of Medicine, University of Pennsylvania, Philadelphia
PARK, WOO JIN, Dept. of Life Science, Kwangju Institute of Science and Technology, Kwangju, Korea
PASQUALINI, RENATA, Dept. of Medicine and Cancer Biology, M.D. Anderson Cancer Center, University of Texas, Houston
PAXTON, CHRISTIAN, Dept. of Animal Science, Iowa State University, Ames, Iowa
PEALE, JR., FRANKLIN, Dept. of Research Pathology, Genentech, Inc., South San Francisco, California
PHILIP, MOHAN, RHeoGene, Charlottesville, Virginia
PHOON, COLIN, Pediatric Cardiology Program, New York University School of Medicine, New York, New York
PURDY, RALPH, Dept. of Pharmacology, University of California, Irvine
RADICE, GLENN, Center for Research on Reproduction and Women's Health, University of Pennsylvania, Philadelphia
RAMÍREZ-BERGERON, DIANA, Abramson Family Cancer Research Institute, Howard Hughes Medical Institute, University of Pennsylvania, Philadelphia
RANADE, KOUSTUBH, Dept. of Clinical Discovery, Bristol-Myers Squibb, Pennington, New Jersey
REBAGLIATI, MICHAEL, Dept. of Anatomy and Cell Biology, University of Iowa, Iowa City
REECY, JAMES, Dept. of Animal Science, Iowa State University, Ames, Iowa
RITTER, ARTHUR, Dept. of Pharmacology and Physiology, New Jersey Medical School-UMDNJ, Newark, New Jersey
RIVERA-FELICIANO, JOSE, Dept. of Genetics, Harvard Medical School, Boston, Massachusetts
ROBERTS, WILMER, Dept. of Molecular and Cellular Biology, Baylor College of Medicine, Houston, Texas
ROCKMAN, HOWARD, Dept. of Cardiology, Duke University Medical Center, Durham, North Carolina
ROMAN, RICHARD, Dept. of Physiology, Medical College of Wisconsin, Milwaukee, Wisconsin
ROSATI, BARBARA, Dept. of Physiology and Biophysics, State University of New York, Stony Brook, Stony Brook
ROSENFELD, MICHAEL, Howard Hughes Medical Institute, School of Medicine, University of California at San Diego, La Jolla
ROSENTHAL, NADIA, Mouse Biology Programme, European Molecular Biology Laboratory-Monterotondo, Monterotondo-Scala, Italy
RUAS, JORGE, Dept. of Cell and Molecular Biology, Karolinska Institutet, Medical Nobel Institute, Stockholm, Sweden
RUIZ-LOZANO, PILAR, Institute of Molecular Medicine, University of California at San Diego, La Jolla

RUPNICK, MARIA, Dept. of Cardiovascular Medicine, Brigham and Women's Hospital, Harvard Medical School, Boston, Massachusetts
RUPP, PAUL, Dept. of Anatomy and Cell Biology, Medical Center, University of Kansas, Kansas City
SAAD, YASSER, Dept. of Molecular Cardiology, The Cleveland Clinic Foundation, Cleveland, Ohio
SAGA, YUMIKO, Dept. of Mammalian Development, National Institute of Genetics, Mishima, Japan
SATO, THOMAS, Dept. of Cell Biology, Southwestern Medical Center, University of Texas, Dallas
SAVLA, USHMA, *Nature Medicine*, Nature Publishing Group, New York, New York
SCHACHTNER, SUSAN, Dept. of Pediatric Cardiology, Children's Hospital, University of Pennsylvania, Philadelphia
SCHLAEGER, THORSTEN, Dept. of Medicine, Children's Hospital, Boston, Massachusetts
SCHNEIDER, MICHAEL, Center for Cardiovascular Development, Baylor College of Medicine, Houston, Texas
SCHOENEBECK, JEFFREY, Dept. of Developmental Genetics, New York University, New York, New York
SCHULTHEISS, THOMAS, Dept. of Molecular Medicine, Beth Israel Deaconess Medical Center, Boston, Massachusetts
SCHWARZ, KARIN, Division of Nephrology, AECOM, Albert Einstein College of Medicine, Bronx, New York
SEHNERT, AMY, Dept. of Pediatrics, University of California, San Francisco
SEIDMAN, CHRISTINE, Dept. of Genetics, Harvard Medical School, Brigham and Women's Hospital, Boston, Massachusetts
SEIDMAN, JONATHAN, Dept. of Genetics, Harvard Medical School, Howard Hughes Medical Institute, Boston, Massachusetts
SHAUL, PHILIP, Dept. of Pediatrics, Southwestern Medical Center, University of Texas, Dallas
SIMON, M. CELESTE, Abramson Family Cancer Research Institute, Howard Hughes Medical Institute, Medical School, University of Pennsylvania, Philadelphia
SINGH, JAIPAL, Dept. of Cardiovascular Research, Eli Lilly and Co., Indianapolis, Indiana
SKOPICKI, HAL, Dept. of Medicine and Circulatory Physiology, Columbia Presbyterian Medical Center, New York, New York
SPEE, PIETER, Dept. of Cloning Technology and Immunology, Novo Nordisk A/S, Bagsværd, Denmark
SRIVASTAVA, DEEPAK, Dept. of Pediatrics and Molecular Biology, Southwestern Medical Center, University of Texas, Dallas
STAINIER, DIDIER, Dept. of Biochemistry, University of California, San Francisco
STARR, LOIS, Center for Human Molecular Genetics, Medical Center, University of Nebraska, Omaha
STENBIT, ANTINE, Dept. of Medicine, Div. of Cardiology, University of California at San Diego, La Jolla
STRUMAN, INGRID, Dept. of Molecular Biology and Genetic Engineering, University of Liège, Liège, Belgium
SUN, BING, Dept. of Vascular Biology, Otsuka Maryland Research Institute, Rockville, Maryland
SUNDARAM, NAMBIRAJAN, Div. of Developmental Biology, Children's Hospital Medical Center, Cincinnati, Ohio
SUNDARAVADIVEL, BALASUBRAMANIAN, Dept. of Medicine, Gazes Cardiac Research Institute, Medical University of South Carolina, Charleston
TACHIBANA, KAZUNOBU, Mount Sinai Hospital, Samuel Lunenfeld Research Institute, Toronto, Canada
TAKEDA, KAZUYO, Lab. of Molecular Cardiology, National Heart, Lung, and Blood Institite, National Institutes of Health, Bethesda, Maryland
TAKEZAKO, TAKANOBU, Dept. of Molecular Cardiology, The Cleveland Clinic Foundation, Cleveland, Ohio
TALLQUIST, MICHELLE, Dept. of Molecular Biology, Southwestern Medical Center, University of Texas, Dallas
TANG, YI, Dept. of Physiology and Functional Genomics, University of Florida, Gainesville
THOMSEN, GERALD, Dept. of Biochemistry and Cell Biology, State University of New York, Stony Brook
TON, CHRISTOPHER, Dept. of Medicine, Harvard Medical School, Brigham and Women's Hospital, Boston, Massachusetts
TSAI, HUAI-JEN, Institute of Fisheries Science, National Taiwan University, Taipei, Taiwan
TU, CHI-TANG, Institute of Fishcries Science, National Taiwan University, Taipei, Taiwan
UPALAKALIN, JAN, Dept. of Pathology, Beth Israel Deaconess Medical Center, Boston, Massachusetts
USHIO-FUKAI, MASUKO, Division of Cardiology, Emory University, Atlanta, Georgia
VOGEL, ANDREAS, Dept. of Vascular Biology, Exelixis Deutschland GmbH, Tuebingen, Germany
VON DEGENFELD, Georges, Stanford University Medical Center, Baxter Laboratories for Genetical Pharmacology, Palo Alto, California
VON DER AHE, DIETMAR, Dept. of Vascular Genomics, Kerckhoff-Clinic, Bad Nauheim, Germany
WAGNER, MICHAEL, Dept. of Anatomy and Cell Biology, Downstate Medical Center, State University of New York, Brooklyn
WANG, YIBIN, Dept. of Physiology, School of Medicine, University of Maryland, Baltimore
WANG, ZHIGAO, Dept. of Molecular Biology, Southwestern Medical Center, University of Texas, Dallas
WEHRENS, XANDER, Center for Molecular Cardiology, Columbia University, New York, New York
WEI, LEI, Dept. of Medicine, Baylor College of Medicine, Houston, Texas
WEINSTEIN, BRANT, Lab. of Molecular Genetics, National Institute of Child Health and Human Development, National Institutes of Health, Bethesda, Maryland
WELIKSON, ROBERT, Dept. of Biochemistry, University of Washington, Seattle
WHITNEY, MARSHA, Dept. of Pathology, University of Washington, Seattle
WILLIAMS, R. SANDERS, Dept. of Medicine and Pharmacology, Duke University Medical Center, Durham, North Carolina
WILSON, EMILY, Dept. of Medical Physiology, Texas A&M Health Science Center, College Station, Texas
WINITSKY, STEVE, Lab. of Molecular Cardiology, National

Heart, Lung, and Blood Institute, National Institutes of Health, Bethesda, Maryland

YANCOPOULOS, GEORGE, Regeneron Pharmaceuticals, Inc., Tarrytown, New York

YANG, JING, Dept. of Vascular Research, Centocor, Inc., Malvern, Pennsylvania

YAO, YIHONG, Dept. of Molecular and Cellular Biology, Abbott Bioresearch Center, Worcester, Massachusetts

YASHIRO, KENTA, Institute for Molecular and Cellular Biology, Osaka University, Suita, Osaka, Japan

YASUNO, SHINJI, Dept. of Medicine and Clinical Science, Kyoto Graduate School of Medicine, Kyoto, Japan

YELON, DEBORAH, Developmental Genetics Program, Skirball Institute, New York University School of Medicine, New York, New York

YOON, YOUNG-SUP, Dept. of Cardiovascular Research, St. Elizabeth's Medical Center, Boston, Massachusetts

YOST, H. JOSEPH, Huntsman Cancer Institute, Center for Children, University of Utah, Salt Lake City

ZHONG, TAO, Dept. of Medicine and Cell Biology, Vanderbilt University Medical Center, Nashville, Tennessee

ZHOU, MINGDONG, Cardiovascular System Research, Zensun Science and Technology Ltd., Pudong, Shanghai, China

First row: A. Jeng, A. Banfi, G. von Degenfeld; M. Nemer; J. Reecy, C. Paxton, C. Kioussi
Second row: D. Srivastava, E. Delot, J. Epstein; E. Azene, X. Wehrens; M. Frasch, D. Stainier
Third row: R. Jain, U. Savla, S. Mockrin, M. Schneider; R. Benezra, D. Fraidenraich, E. Stillwell
Fourth row: M. Tallquist; G. Yancopoulos; B. Latinkic, F. Conlon

First row: S. Winitsky, C. Murry; J.P. Singh, E. McNally
Second row: M.C. Simon, A. Lassar; M. Krasnow; J. Dominguez, D. Franco, L. Pabon
Third row: R.S. Williams, E. McNally, M. Schneider; D. McKinnon, B. Rosati, D. Franco
Fourth row: A. Ritter, B. Sundaravadivel; Symposium picnic

First row: M. Gessler, M. Fishman; L. Starr; R. Kalluri
Second row: D. McKinnon, N. Epstein; B. Stillman, R. Lifton
Third row: F. Mack, D. Ramirez-Bergen, Y. Pan, K. Mansfield; B. Albrecht, K. Breitschopf

First row: T. Arrhenius, L. Starr; L. Johnson, E. Wilson
Second row: M. Gessler, A. Fischer, A. Gossler; P. Kiberstis; G. Bergers, L. Komuves
Third row: R. Harvey, N. Rosenthal; T. Mohun, B. Latinkic, F. Conlon

First row: H.-L. Chen, M.-Y. Chou; D. Cheresh, R.S. Williams; P. Norby, P. Spee
Second row: R. Hynes, D. Stewart; Outside Grace, meeting for dinner parties
Third row: R. Harvey, P. Krieg, M. Nemer; C. Phoon, P. Hwang

First row: R. Deaton, R. Purdy; R. Lifton, J. Cohen; L. Goyal, E. Olson
Second row: I. Struman, A. Nasevicius; K. Fujiwara, N. Takeda; U. Savla, N. Rosenthal
Third row: B. Stillman, R. Hynes, C. Seidman, R. Benezra; R. Belagaje, J.P. Singh
Fourth row: R. Kalluri, L. Goyal; T. Chen, X. Li; C. Seidman

Foreword

The Cold Spring Harbor Symposium on Quantitiative Biology has rarely been devoted to a single organ. One notable exception was the successful Symposium in 1990 on the brain, which marked in part President Bush's proclamation of the 1990s as the decade of the brain, and also the beginning of the current Cold Spring Harbor initiative in neuroscience. There has been no presidential pronouncement about the heart, and Cold Spring Harbor has not focused research on it, but there have been remarkable developments in molecular understanding of the cardiovascular system over the last decade. It seemed fitting, therefore, that this year's Symposium should focus on such an important biological topic.

Development of the cardiovascular system during embryogenesis is a major area of research, particularly the patterning that forms the branches of the peripheral blood and lymphatic systems and development of the heart itself. The Symposium made clear that there is now a good general understanding of how the system develops. An equally important area of inquiry was the signaling pathways that permit the cardiovascular system to maintain its function. And cardiovascular disease, the western world's most common cause of death, also received appropriate attention.

Help in organizing this Symposium came from many sources, particularly my co-organizer Dr. David Stewart, Executive Director of our Meetings and Courses program, who played a key role in identifying speakers and topics. I thank Shaun Coughlin, Richard Lifton, Eric Olson, Christine Seidman, Celeste Simon, and Sandy Williams for their generous advice about a field with details that were rather foreign to me.

The meeting ran for five days and included 257 participants with 68 oral presentations and 108 poster presentations. I thank Dr. Richard Lifton for his superb presentation at the Dorcas Cummings Memorial public lecture on the genetics of hypertension providing a clear example of how the Human Genome Project is having a large impact on understanding human disease. I also thank the first-night speakers, Drs. Richard Harvey, Mark Fishman, Jonathan Seidman, and Rakesh Jain for their superb overview presentations. I am particularly grateful to Dr. Christine Seidman for agreeing to summarize the meeting.

As always, I greatly appreciate the efficiency of the Meetings and Courses staff under the leadership of David Stewart and Terri Grodzicker. I thank Joan Ebert, Patricia Barker, and Danny deBruin in the Cold Spring Harbor Laboratory Press for again putting together an important volume in the Cold Spring Harbor Symposium series. Finally, I am pleased to acknowledge the funding from companies, foundations, and the federal government, listed on the following page, without which the meeting would not have been possible.

Bruce Stillman
March 2003

SPONSORS

This meeting was funded in part by the **National Cancer Institute**.
Contributions from the following companies provide core support for the Cold Spring Harbor meetings program.

Corporate Benefactors

Aventis Pharma AG
Bristol-Myers Squibb Company
Eli Lilly and Company
GlaxoSmithKline
Novartis Pharma AG
Pfizer Inc.

Corporate Sponsors

Abbott Laboratories
Amersham Biosciences, Inc.
Amgen Inc.
Applied Biosystems
BioVentures, Inc.
Chugai Pharmaceutical Co., Ltd.
Cogene BioTech Ventures, Ltd.
Diagnostic Products Corporation
Forest Laboratories
Genentech, Inc.
Hoffmann-La Roche Inc.
Johnson & Johnson Pharmaceutical Research & Development, L.L.C.
Kyowa Hakko Kogyo Co., Ltd
Merck Research Laboratories
New England BioLabs, Inc.
OSI Pharmaceuticals, Inc.
Pall Corporation
Pharmacia Corporation
ResGen, Inc.
Schering-Plough Research Institute
Wyeth Genetics Institute

Plant Corporate Associates

MeadWestvaco Corporation
Monsanto Company
Pioneer Hi-Bred International, Inc

Corporate Associates

Affymetrix, Inc.
Ceptyr, Inc.

Corporate Contributors

Biogen, Inc.
Epicentre Technology
KeyGene
ImmunoRx
Integrated DNA Technologies, Inc.
Invitrogen
Lexicon Genetics, Inc.
Prolinx, Inc.
Qbiogene
ZymoGenetics, Inc.

Foundations

Albert B. Sabin Vaccine Institute, Inc.

Contents

Symposium Participants   v
Foreword   xvii

Development of the Cardiovascular System

Cardiogenesis in the *Drosophila* Model: Control Mechanisms during Early Induction and
 Diversification of Cardiac Progenitors *S. Zaffran, X. Xu, P.C.H. Lo, H.-H. Lee, and M. Frasch*   1
Regulation of Cardiac Muscle Differentiation in *Xenopus laevis* Embryos *T. Mohun, B. Latinkic,
 N. Towers, and S. Kotecha*   13
Genetic Regulation of Cardiac Patterning in Zebrafish *D. Yelon, J.L. Feldman, and B.R. Keegan*   19
Zebrafish Hearts and Minds: Nodal Signaling in Cardiac and Neural Left–Right Asymmetry
 S. Long, N. Ahmad, and M. Rebagliati   27
Cardiac Left–Right Development: Are the Early Steps Conserved? *K.L. Kramer and H.J. Yost*   37
Erythropoietin and Retinoic Acid Signaling in the Epicardium Is Required for Cardiac Myocyte
 Proliferation *I. Stuckmann and A.B. Lassar*   45
Endocardial Cushion Formation in Zebrafish *D.Y.R. Stainier, D. Beis, B. Jungblut, and T. Bartman*   49
Neural Crest Migration and Mouse Models of Congenital Heart Disease *A.D. Gitler, C.B. Brown,
 L. Kochilas, J. Li, and J.A. Epstein*   57

Transcription Factors in Cardiovascular Development

Hey bHLH Factors in Cardiovascular Development *A. Fischer, C. Leimeister, C. Winkler,
 N. Schumacher, B. Klamt, H. Elmasri, C. Steidl, M. Maier, K.-P. Knobeloch, K. Amann,
 A. Helisch, M. Sendtner, and M. Gessler*   63
Molecular Mechanisms of Chamber-specific Myocardial Gene Expression: Transgenic Analysis
 of the ANF Promoter *E.M. Small and P.A. Krieg*   71
Pitx Genes during Cardiovascular Development *C. Kioussi, P. Briata, S.H. Baek,
 A. Wynshaw-Boris, D.W. Rose, and M.G. Rosenfeld*   81
Pitx2 and Cardiac Development: A Molecular Link between Left/Right Signaling and
 Congenital Heart Disease *M. Campione, L. Acosta, S. Martínez, J.M. Icardo, A. Aránega,
 and D. Franco*   89
Regulation of Cardiac Growth and Development by SRF and Its Cofactors *D. Wang, R. Passier,
 Z.-P. Liu, C.H. Shin, Z. Wang, S. Li, L.B. Sutherland, E. Small, P.A. Krieg, and E.N. Olson*   97
Homeodomain Factor Nkx2-5 in Heart Development and Disease *R.P. Harvey, D. Lai, D. Elliott,
 C. Biben, M. Solloway, O. Prall, F. Stennard, A. Schindeler, N. Groves, L. Lavulo, C. Hyun,
 T. Yeoh, M. Costa, M. Furtado, and E. Kirk*   107
Causes of Clinical Diversity in Human *TBX5* Mutations *T. Huang, J.E. Lock, A.C. Marshall,
 C. Basson, J.G. Seidman, and C.E. Seidman*   115
Molecular Mechanisms of Ventricular Hypoplasia *D. Srivastava, P.D. Gottlieb, and E.N. Olson*   121
Hypoxia, HIFs, and Cardiovascular Development *M.C. Simon, D. Ramirez-Bergeron, F. Mack,
 C.-J. Hu, Y. Pan, and K. Mansfield*   127
Quiescent Hypervascularity Mediated by Gain of HIF-1α Function *J.M. Arbeit*   133

Vascular Biology

The Diverse Roles of Integrins and Their Ligands in Angiogenesis *R.O. Hynes, J.C. Lively,
 J.H. McCarty, D. Taverna, S.E. Francis, K. Hodivala-Dilke, and Q. Xiao*   143
Arteries, Veins, Notch, and VEGF *B.M. Weinstein and N.D. Lawson*   155
Cell Cycle Signaling and Cardiovascular Disease *E.G. Nabel, M. Boehm, L.M. Akyurek,
 T. Yoshimoto, M.F. Crook, M. Olive, H. San, and X. Qu*   163
Selective Functions of Angiopoietins and Vascular Endothelial Growth Factor on
 Blood Vessels: The Concept of "Vascular Domain" *T.N. Sato, S. Loughna, E.C.
 Davis, R.P. Visconti, and C.D. Richardson*   171

Survival Mechanisms of VEGF and PlGF during Microvascular Remodeling *J.N. Upalakalin, I. Hemo, C. Dehio, E. Keshet, and L.E. Benjamin* — 181

Molecular Mechanisms of Lymphangiogenesis *T. Mäkinen and K. Alitalo* — 189

Protease-activated Receptors in the Cardiovascular System *S.R. Coughlin* — 197

Migration of Monocytes/Macrophages In Vitro and In Vivo Is Accompanied by MMP12-dependent Tunnel Formation and by Neovascularization *M. Anghelina, A. Schmeisser, P. Krishnan, L. Moldovan, R.H. Strasser, and N.I. Moldovan* — 209

The Role of EG-VEGF in the Regulation of Angiogenesis in Endocrine Glands *J. LeCouter, R. Lin, and N. Ferrara* — 217

Profiling the Molecular Diversity of Blood Vessels *R. Pasqualini and W. Arap* — 223

Tumor Angiogenesis

VEGF-A Induces Angiogenesis, Arteriogenesis, Lymphangiogenesis, and Vascular Malformations *J.A. Nagy, E. Vasile, D. Feng, C. Sundberg, L.F. Brown, E.J. Manseau, A.M. Dvorak, and H.F. Dvorak* — 227

Angiogenesis and Lymphangiogenesis in Tumors: Insights from Intravital Microscopy *R.K. Jain* — 239

A Genetic Approach to Understanding Tumor Angiogenesis *R. Benezra, P. de Candia, H. Li, E. Romero, D. Lyden, S. Rafii, and M. Ruzinova* — 249

Discovery of Type IV Collagen Non-collagenous Domains as Novel Integrin Ligands and Endogenous Inhibitors of Angiogenesis *R. Kalluri* — 255

Complementary and Coordinated Roles of the VEGFs and Angiopoietins during Normal and Pathologic Vascular Formation *N.W. Gale, G. Thurston, S. Davis, S.J. Wiegand, J. Holash, J.S. Rudge, and G.D. Yancopoulos* — 267

Involvement of G Proteins in Vascular Permeability Factor/Vascular Endothelial Growth Factor Signaling *D. Mukhopadhyay and H. Zeng* — 275

Targeted Delivery of Mutant Raf Kinase to Neovessels Causes Tumor Regression *J.D. Hood and D.A. Cheresh* — 285

Combining Antiangiogenic Agents with Metronomic Chemotherapy Enhances Efficacy against Late-stage Pancreatic Islet Carcinomas in Mice *G. Bergers and D. Hanahan* — 293

Genetics and Genomics of the Cardiovascular System

Zebrafish: The Complete Cardiovascular Compendium *C.A. MacRae and M.C. Fishman* — 301

Consomic Rats for the Identification of Genes and Pathways Underlying Cardiovascular Disease *R.J. Roman, A.W. Cowley, Jr., A. Greene, A.E. Kwitek, P.J. Tonellato, and H.J. Jacob* — 309

A Mouse Model of Congenital Heart Disease: Cardiac Arrhythmias and Atrial Septal Defect Caused by Haploinsufficiency of the Cardiac Transcription Factor Csx/Nkx2.5 *M. Tanaka, C.I. Berul, M. Ishii, P.Y. Jay, H. Wakimoto, P. Douglas, N. Yamasaki, T. Kawamoto, J. Gehrmann, C.T. Maguire, M. Schinke, C.E. Seidman, J.G. Seidman, Y. Kurachi, and S. Izumo* — 317

Genetic Dissection of the DiGeorge Syndrome Phenotype *F. Vitelli, E.A. Lindsay, and A. Baldini* — 327

A Missense Mutation in a Highly Conserved Region of CASQ2 Is Associated with Autosomal Recessive Catecholamine-induced Polymorphic Ventricular Tachycardia in Bedouin Families from Israel *M. Eldar, E. Pras, and H. Lahat* — 333

Regulation of the Heart

Calcium-dependent Gene Regulation in Myocyte Hypertrophy and Remodeling *R.S. Williams and P. Rosenberg* — 339

A Gradient of Myosin Regulatory Light-chain Phosphorylation across the Ventricular Wall Supports Cardiac Torsion *J.S. Davis, S. Hassanzadeh, S. Winitsky, H. Wen, A. Aletras, and N.D. Epstein* — 345

Molecular and Functional Maturation of the Murine Cardiac Conduction System *S. Rentschler, G.E. Morley, and G.I. Fishman* — 353

The Yin/Yang of Innate Stress Responses in the Heart *D.L. Mann* — 363

Regulatory Networks Controlling Mitochondrial Energy Production in the Developing, Hypertrophied, and Diabetic Heart *B.N. Finck, J.J. Lehman, P.M. Barger, and D.P. Kelly* — 371

Heart Disease and Failure

Molecular Epidemiology of Hypertrophic Cardiomyopathy *H. Morita, S.R. DePalma, M. Arad, B. McDonough, S. Barr, C. Duffy, B.J. Maron, C.E. Seidman, and J.G. Seidman*	383
The Sarcoglycan Complex in Striated and Vascular Smooth Muscle *M.T. Wheeler, M.J. Allikian, A. Heydemann, and E.M. McNally*	389
The MLP Family of Cytoskeletal Z Disc Proteins and Dilated Cardiomyopathy: A Stress Pathway Model for Heart Failure Progression *M. Hoshijima, M. Pashmforoush, R. Knöll, and K.R. Chien*	399
From Sarcomeric Mutations to Heart Disease: Understanding Familial Hypertrophic Cardiomyopathy *A. Maass, J.P. Konhilas, B.L. Stauffer, and L.A. Leinwand*	409
Vascular Endothelium in Tissue Remodeling: Implications for Heart Failure *S.M. Dallabrida and M.A. Rupnick*	417
Using a Gene-switch Transgenic Approach to Dissect Distinct Roles of MAP Kinases in Heart Failure *B.G. Petrich, P. Liao, and Y. Wang*	429
G-Protein-coupled Receptor Function in Heart Failure *S.V. Naga Prasad, J. Nienaber, and H.A. Rockman*	439

Hypertension, Atherosclerosis, and Lipid Homeostasis

Salt and Blood Pressure: New Insight from Human Genetic Studies *R.P. Lifton, F.H. Wilson, K.A. Choate, and D.S. Geller*	445
Gene-targeting Studies of the Renin–Angiotensin System: Mechanisms of Hypertension and Cardiovascular Disease *S.B. Gurley, T.H. Le, and T.M. Coffman*	451
Modulation of Endothelial NO Production by High-density Lipoprotein *C. Mineo and P.W. Shaul*	459
Plaque Angiogenesis: Its Functions and Regulation *K.S. Moulton*	471
Extracellular Superoxide Dismutase, Uric Acid, and Atherosclerosis *H.U. Hink and T. Fukai*	483
SREBPs: Transcriptional Mediators of Lipid Homeostasis *J.D. Horton, J.L. Goldstein, and M.S. Brown*	491
Genetic Defenses against Hypercholesterolemia *H.H. Hobbs, G.A. Graf, L. Yu, K.R. Wilund, and J.C. Cohen*	499

Repairing the Cardiovascular System

Insulin-like Growth Factor Isoforms in Skeletal Muscle Aging, Regeneration, and Disease *N. Winn, A. Paul, A. Musarò, and N. Rosenthal*	507
Cellular Therapies for Myocardial Infarct Repair *C.E. Murry, M.L. Whitney, M.A. Laflamme, H. Reinecke, and L.J. Field*	519
Gene Therapy for Cardiac Arrhythmias *E. Marbán, H.B. Nuss, and J.K. Donahue*	527
Myocardial Disease in Failing Hearts: Defective Excitation–Contraction Coupling *X.H.T. Wehrens and A.R. Marks*	533
Summary: The Coming of Age of Cardiovascular Science *C.E. Seidman and J.G. Seidman*	543

Author Index 551

Subject Index 553

Cardiogenesis in the *Drosophila* Model: Control Mechanisms during Early Induction and Diversification of Cardiac Progenitors

S. ZAFFRAN,* X. XU,† P.C.H. LO, H.-H. LEE, AND M. FRASCH
Brookdale Department of Molecular, Cell, and Developmental Biology, Mount Sinai School of Medicine, New York, New York 10029

The dorsal vessel of *Drosophila* displays developmental, functional, and morphological similarities to the primitive linear heart tube of early vertebrate embryos. Because these similarities extend to the genetic and molecular level, *Drosophila* has become a fruitful model to study control mechanisms of early heart development. Herein we summarize recently obtained insights into control mechanisms during early induction and diversification of cardiac progenitors in *Drosophila*. We also show that induction of *tinman*, a key cardiogenic gene, in the dorsal mesoderm by Dpp (*Drosophila* BMP) involves protein/protein interactions between Tinman and the Smad proteins Mad and Medea, in addition to their DNA-binding activities to specific *tinman* enhancer sequences. Furthermore, we present evidence that binding of a high-mobility-group protein, HMG-D, to the Dpp-responsive enhancer of *tinman* as well as to the Tinman protein may be involved in the formation of a fully active enhancer complex.

MORPHOLOGY AND PATTERNING OF THE DORSAL VESSEL

The dorsal vessel consists of a linear tube extending along the dorsal midline of the *Drosophila* larva (Fig. 1A) and late-stage embryo (Fig. 1B). This organ is composed of an inner tube formed by contractile cardiomyocytes (termed cardioblasts; Fig. 1A,B) that are surrounded by non-contractile pericardial cells (Fig. 1C,D). In late-stage embryos, the cardiomyocytes are arranged in two bilateral rows of cells that enclose the lumen of the dorsal vessel (Fig. 1B) (Rugendorff et al. 1994). Peripheral components of the dorsal vessel include the segmentally arranged alary muscles, which attach the dorsal vessel to the body wall (Fig. 1A). Like the vertebrate heart tube, the *Drosophila* dorsal vessel has an anterior/posterior polarity and pumps the hemolymph from posterior to anterior. This polarity is also reflected in the subdivision of the dorsal vessel into two chamber-like portions, an anterior one with a narrower lumen, termed "aorta," and a posterior one with a wider diameter, termed "heart" (Fig. 1A,B). In addition, the heart portion features larger alary muscles (Fig. 1A) and three bilateral pairs of inflow tracts, termed "ostia" (Fig. 1B). Although more simple, this type of anteroposterior subdivision is reminiscent of the anteroposterior subdivision of the linear vertebrate heart tube into presumptive heart chambers.

In addition to the broad anteroposterior polarity, the dorsal vessel also displays an intrasegmental polarity along its anteroposterior extent. Indeed, expression analysis of various molecular markers has revealed that, analogous to the remainder of the insect body, the dorsal vessel is composed of segmentally repeated units. For example, the NK homeobox gene *tinman* (*tin*), which has an essential early function in cardiogenesis (see below), is expressed in only four of the six bilateral cardioblasts in each segment of the mature dorsal vessel (Fig. 1B–E). On the basis of the differential expression of the sulfonylurea receptor (a potassium channel subunit; Fig. 1E) as well as other differentiation markers, it appears that these four cardioblasts acquire different physiological characteristics from the two *tin*-negative cells. The specific expression of the homeobox gene *ladybird* (*lb*) in two bilateral, anterior pairs of *tin*-positive cardioblasts within each segment provides an additional level of diversity among these cardioblasts (Jagla et al. 1997). Similarly, the *tin*-negative pair of cardioblasts in each hemisegment is characterized by the expression of other transcription factors. Notably, these include *seven-up* (*svp*; Fig. 1C), which encodes a nuclear orphan receptor that is homologous to vertebrate COUP-TF, and the three related *Dorsocross* genes (*Doc1* aka *Tb66F2*, *Doc2*, *Doc3*; Fig. 1D) that are closest to the Tbx4, 5, and 6 subgroup of vertebrate T-box genes. In the "heart" portion of the dorsal vessel, the *tin*-negative/*svp*- and *Doc*-positive cells form the ostia (inflow tracts), demonstrating that the intrasegmental diversification of cardioblasts leads to physiological as well as functional and morphological differences among these cells (Fig. 1B) (Molina and Cripps 2001).

Genetic studies have uncovered some of the mechanisms that control intrasegmental cell fate diversification within the dorsal vessel (Gajewski et al. 2000; Lo and Frasch 2001). Specifically, it has been shown that *svp* acts as an upstream regulator to repress *tin* and activate *Doc* in two of the six bilateral cardioblasts in each segment (Fig.

Present addresses: *Department de Biologie du Developpement, Institut Pasteur, 75015 Paris, France; †Cardiovascular Research Center, Massachusetts General Hospital, Harvard Medical School, Charlestown, Massachusetts 02120.

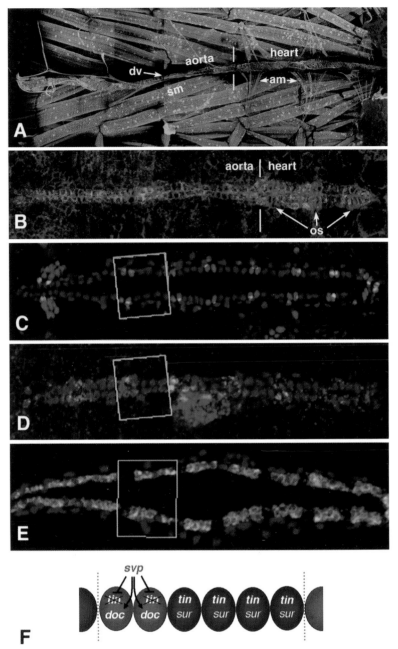

Figure 1. Morphology of the dorsal vessel and diversification of cell types. (*A*) Dorsal area of third-instar larva stained with phalloidin (*green*) and anti-MEF2 (*red*), which shows the dorsal vessel (dv) including heart and aorta, alary muscles (am), and somatic muscles (sm) (length ~2 mm). (*B*) Dorsal vessel of late-stage embryo stained with anti-spectrin (*green*) and anti-Tinman (*red*) (length ~300 µm). Larger, non-Tin-stained cardiomyocytes in the heart portion form the ostia (os). (*C*) Developing dorsal vessel in late-stage *svp-lacZ* embryo stained for Tin (*red*) and βGal (*green*). One segment is highlighted in *C–E*. (*D*) Dorsal vessel in late-stage embryo stained for Tin (*red*) and *Doc1* mRNA (*green*). (*E*) Developing dorsal vessel stained for Tin (*red*) and *sulfonylurea receptor* (*sur*) mRNA (*green*). (*F*) Summary of genetic interactions within one hemisegment of the dorsal vessel.

1F). The sulfonylurea receptor-encoding gene, *sur*, may be activated by *tin* in the remaining four bilateral cardioblasts (Nasonkin et al. 1999), although its potential direct repression by *svp* and/or *Doc* cannot be ruled out. In the "heart" portion, *svp* is required for specifying the identity of the cells forming the inflow tracts. Interestingly, a vertebrate homolog of *svp*, *COUP-TFII*, has also been shown to function in the development of the inflow tracts (sinus venosus) and atria of the early mouse heart (Pereira et al. 1999). Moreover, expression of vertebrate *Tbx5* becomes rapidly restricted to areas of the prospective atria and sinus venosus (as well as the left ventricle), and functional knockouts of *Tbx5* in frog and mouse cause strong disruptions of heart development, particularly the sinoatrial structures (Horb and Thomsen 1999; Bruneau et al. 2001). Thus, it appears that some of the functions of *svp*, *doc*, and *tin*-related genes have been evolutionarily conserved in vertebrates, and it is conceivable that these similarities extend to the regulatory interactions among these genes that have been defined in the *Drosophila* system.

SIGNALING PROCESSES DURING EARLY INDUCTION OF BILATERAL CARDIAC FIELDS

The cardiac progenitors arise from bilateral fields of cells in the lateral (i.e., dorsal) mesoderm in the early *Drosophila* embryo, which are analogous to the bilateral heart fields that are formed in the early vertebrate embryo. In subsequent stages, the bilateral rows of cardiac cells move toward the dorsal midline, where they merge to form the dorsal vessel (for late phases of this process, see Fig. 1C,D).

It has been shown that the formation of the bilateral heart fields is triggered by inductive signals from the dorsal ectoderm. A key signaling factor mediating this process was identified as the *Drosophila* BMP homolog Dpp, which is expressed in a spatially restricted domain within the dorsal ectoderm (Fig. 2A). Moreover, the expression of the NK homeobox gene *tinman* (*tin*) was found to be induced in response to Dpp signaling in dorsal mesodermal cells that are located directly underneath the Dpp-expressing ectodermal cells (Fig. 2A) (Frasch 1995). The activity of *tin* in turn is critical for the formation of all types of heart progenitors, as well as for the formation of visceral and somatic muscle progenitors that arise within this same area of the dorsal mesoderm (Azpiazu and Frasch 1993; Bodmer 1993).

Genetic and molecular studies have identified Smad proteins, which in *Drosophila* are represented by Mad (Smad1/5) and Medea (Smad4), as intracellular effectors of BMP signals (for review, see Raftery and Sutherland 1999). To further investigate the spatial pattern of Dpp signaling activity in the early *Drosophila* embryo, we performed stainings with an anti-phospho-Smad1 antibody that specifically detects the receptor-activated form of Mad. As predicted, nuclear P-Mad was specifically detected in the dorsal ectoderm as well as in the underlying dorsal mesoderm. The coincidence between the mesodermal domain of P-Smad with the domain of *tin* induction (Fig. 2A,B and data not shown) provides additional support for the proposed mechanism of dorsal *tin* induction by Dpp signals. As evidenced by the lack of *tin* induction in the dorsal ectoderm, Dpp signaling alone is not sufficient to elicit induction of *tin*. Indeed, it was found that the activity of *tin* itself is also required, suggesting a mechanism in which Dpp signals are required in combination with *tin* autoactivation to induce *tin* (Fig. 2C). The main source of the autoregulatory pool of Tin protein is likely derived from the transient and mesoderm-autonomous activation of *tin* by the bHLH protein Twist in the whole mesoderm at earlier stages of development (Yin et al. 1997).

MOLECULAR MECHANISMS OF *tinman* INDUCTION BY Dpp SIGNALS

Induction of *tin* by Dpp in the dorsal mesoderm is mediated by a 350-bp enhancer element, which is located downstream of the *tin* gene and is distinct from a *twist*-responsive and a dorsal vessel-specific enhancer (Yin et al. 1997). Functional dissection of this Dpp-responsive element, termed tinD, in vivo and in vitro has provided important information on molecular mechanisms in cardiac induction (Xu et al. 1998). These studies identified three short sequence elements, D1 (which is present in two identical copies), D3, and D6, that were essential and sufficient for induction by Dpp in vivo (see Fig. 3) (Xu et al. 1998). Consequently, the D1 and D3 sequences were used as baits in yeast one-hybrid screens to isolate cDNAs encoding regulatory factors that specifically bind to these sequences. The results of these screens were highly instructive, as they yielded Tinman as a D1-binding factor and Medea (Smad4) as a D3-binding factor. The presence of a Tinman/Nkx2-5 optimal binding site motif, TCAAGTG (Chen and Schwartz 1995), in both copies of D1 is consistent with this result. Likewise, D3 contains two GC-rich sequence motifs, which can bind Smad proteins, and two additional Smad-binding sites of this type were identified in D6 (Xu et al. 1998; see also Kim et al. 1997). In addition, D3a (Fig. 3) contains two AGAC motifs that constitute the core of a second type of Smad site (Yingling et al. 1997). The presence of at least six Smad-binding sites within TinD could potentially allow binding of two heterotrimeric Medea/Mad complexes (Kawabata et al. 1998). Taken together, the unbiased identification of Tinman and Smads as specific binding factors of the Dpp-responsive enhancer of *tin* has provided a molecular explanation of the genetic data described above (Fig. 2). Together with additional experiments, these data have suggested a model in which Tin binding makes this enhancer competent to respond to Dpp, and the combinatorial binding of Tin and Dpp-activated Smad complexes provides the synergy that is needed for the induction of *tin* in the dorsal mesoderm.

In related experiments, it has been shown recently that similar molecular pathways are operative during the in-

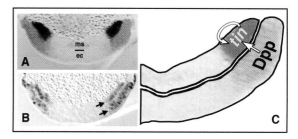

Figure 2. Induction of *tinman* expression by Dpp in the dorsal mesoderm. (*A*) Cross-sectioned germ band (ec, ectoderm; ms, mesoderm) of stage-10 embryo that carries a *dpp-lacZ* construct and was stained for *tin* mRNA (*purple*) and βGal (*yellow*). (*B*) Embryo section as in *A* stained with an anti-phospho-Smad1 antibody. Arrows point to ectodermal and mesodermal P-Mad, respectively. (*C*) Summary of interactions during dorsal mesodermal induction of *tinman*.

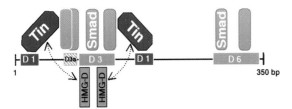

Figure 3. Summary of protein/DNA and protein/protein interactions at the Dpp-responsive enhancer tinD from the *tinman* gene that are discussed in this paper.

duction of the tinman homolog *Nkx2-5* in the cardiac crescent during early vertebrate cardiogenesis. In particular, *Nkx2-5* induction in the chick and frog systems was shown to require BMP signaling (Schultheiss et al. 1997; Andrée et al. 1998; Shi et al. 2000), and enhancer elements of mouse *Nkx2-5* were found to contain functionally important binding sites for Smad proteins (Liberatore et al. 2002; Lien et al. 2002). Hence, it appears that at least some aspects of the molecular pathways in cardiac induction have been conserved between invertebrates and vertebrates.

In this paper, we describe additional molecular events that contribute to the activation of the Dpp-responsive enhancer of *tin* in the dorsal mesoderm. We demonstrate that, in addition to their combinatorial DNA-binding activities, Tinman and Smad proteins also undergo mutual protein/protein interactions, and we define the protein domains that are required for these interactions. Furthermore, we show that a high-mobility-group protein, HMG-D, binds to tinD enhancer sequences as well as to the Tin protein, and these interactions appear to potentiate binding of Tin to its target sites. Finally, we discuss that the protein/DNA and protein/protein interactions that have been defined for the Dpp-responsive enhancer of *tin* are also crucial for the induction of other target genes in the cardiac mesoderm that depend on Dpp and *tin* activities in combination with additional signals.

METHODS

Yeast two-hybrid and yeast one-hybrid screens. Yeast strain CG1945 carrying both His and lacZ reporter genes was used in the MATCHMAKER two-hybrid system (Clontech). A full-length *tinman* (*XmnI-PstI*) fragment was added in-frame to the GAL4 DNA-binding domain encoded in the pGBT9 vector and used as bait to screen ~750,000 colonies of a 0–16-hour embryonic cDNA library in pGAD10 (Clontech), which resulted in the isolation of three independent *HMG-D* clones. The *Medea* and *Mad* constructs in this study were previously described by Xu et al. (1998). The tin^{1-321}, tin^{1-56}, and $tin^{300-416}$ constructs were obtained by PCR amplification and cloned into pGBT9 vector. (Primer sequences will be provided upon request.) In a yeast one-hybrid screen with the above-described fusion library (for details, see Xu et al. 1998), one clone encoding full-length HMG-D was obtained.

Yeast assays for lacZ activities. Fresh colonies of the yeast strain SFY526 (Clontech) carrying a lacZ reporter gene and the respective interaction constructs were grown in 3 ml of selective medium at 30°C until saturation (OD_{600}~1.0) and used to grow new cultures with a starting OD_{600} of 0.1 and a final OD_{600} of ~0.4. Of the chilled culture, 1 ml was pelleted, resuspended in 0.2 ml of 0.1 M Tris (pH 7.5), 0.05% Triton X-100, and immediately frozen on dry ice. For the assay, cells were thawed on ice, resuspended in 1 ml of ONPG solution (0.8 mg/ml ONPG, 1.25 mM DTT, 0.0062% SDS, 40 mM β-mercaptoethanol in Z-Buffer (60 mM Na_2HPO_4, 40 mM NaH_2PO_4, 10 mM KCl, 1 mM $MgSO_4$), and incubated at 30°C. When a medium-dark yellow appearance was reached, color reaction times were recorded, stopped with 0.5 ml of 1 M Na_2CO_3, and centrifuged. OD_{420} values were measured from the supernatants and β-galactosidase units were calculated with the formula: OD_{420}/OD_{600} × 1000/(1 ml × minutes reaction time). The lacZ activities from five independent colonies were averaged for each construct.

In vitro protein-binding assay. For full-length GST-Tinman, the *XmnI/PstI* fragment of the pNB40-*tin* plasmid was subcloned into pGEX2T (Pharmacia). The *Medea* and *Mad* constructs used in the GST pull-down assay were described by Xu et al. (1998). Deletion of amino acids 301 to 364 of Tinman via site-directed in vitro mutagenesis was done by inverse PCR on *tin* cDNA to replace the HD domain with a *BamHI* site. $tin^{301-416}$ (see above) was used in pBluescript KS for the in vitro protein-binding assays.

[^{35}S]Methionine-labeled proteins were produced using the TNT-coupled lysate system (Promega). Glutathione S-transferase (GST) fusion proteins were expressed in *Escherichia coli* BL21 and purified on glutathione-agarose (Sigma). For in vitro binding assays, 30 µl (~3 µg) of purified GST or GST fusion proteins was bound to glutathione-agarose beads, incubated at 4°C for 30 minutes with 15 µl of lysate containing ^{35}S-labeled proteins, diluted with 150 µl of 0.5% NP-40/PBS for another 30 minutes, after which the beads were washed five times for 30 minutes in 0.5% Tween-20/PBS.

UAS-tinΔ42-301 and UAS-tinΔ42-124 transformants and other Drosophila strains. The $tin^{\Delta 42-301}$ and $tin^{\Delta 42-124}$ were generated by recircularizing *EcoRV/BstEII* and *EcoRV/PvuII*-digested *tin* cDNA in pBluescript KS. The resulting constructs were subcloned into the pUAST vector and introduced into yw^-. Other *Drosophila* lines used in this study are *P(en2.4-GAL4)en* (gift from A. Brand), *UAS-tin* (Yin and Frasch 1998), and *tin-D7-25/lacZ* (Xu et al. 1998). Embryo stainings were done as described by Azpiazu and Frasch (1993) and Xu et al. (1998).

RESULTS

Tinman and Smad Proteins Undergo Specific Protein/Protein Interactions

The observed synergism between Tinman and Smad proteins as well as the localization of two Tinman-binding sites in proximity to several Smad-binding sites within the Dpp-responsive enhancer of *tin* suggested the possibility that protein–protein interactions between Tinman and Smads could play an additional role in the formation of an active enhancer complex. To test whether Mad and Medea proteins can physically interact with Tinman, we used the yeast two-hybrid system as an interaction assay. As shown in Figure 4A, full-length Tinman protein interacts with both Mad and Medea proteins in this assay. In order to define the domains involved in these protein–protein interactions, various internally deleted and carboxy-terminally truncated versions of Tinman were tested. These experiments showed that the amino-terminal 124 amino acids of Tinman are sufficient for the interaction with Smads (Fig. 4A; Tin^{1-416}, Tin^{1-321}, and Tin^{1-124}). The TN-domain does not appear to be required and, interestingly, deletion of residues 1–50 improves the interaction (Fig. 4A, $Tin^{\Delta TN-124}$). In contrast, carboxy-terminal deletions beyond residue 56 (Tin^{1-56}, Tin^{50-130}), amino-terminal deletion of residues 1–300 ($Tin^{300-416}$), and internal deletion of residues 42–124 ($Tin^{\Delta 42-124}$) prevent the interaction. Taken together, the

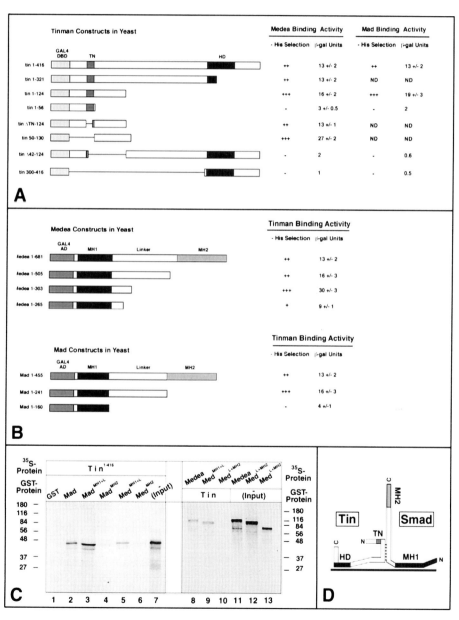

Figure 4. Protein interactions between Tinman and the Smad proteins Mad and Medea. (*A*) Interactions between Tinman derivatives (fused with the GAL4 DNA-binding domain) and full-length Medea and Mad, respectively (each fused with the GAL4 activation domain). Interactions were tested by assaying for colony growth ("–His selection") and activation of a *lacZ* reporter ("βgal Units") in yeast two-hybrid assays. (*B*) Interactions between full-length Tinman and various derivatives of Medea and Mad, respectively, which were tested in yeast two-hybrid assays as described in *A*. (*C*) In vitro protein interactions between full-length Tinman and different versions of Mad and Medea using "GST pull-down" assays. For the gel to the left, various Mad and Medea polypeptides fused to GST were used as indicated, and for the gel to the right, full-length Tin was used as a GST-fusion protein to bind ^{35}S-labeled proteins. (*D*) Schematic drawing of protein interactions between DNA-bound Tin and Smad proteins (MH, Mad homology domains; HD, homeodomain; TN, Tin-homology domain).

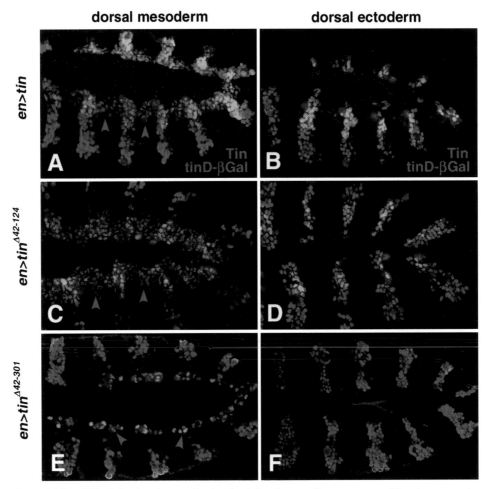

Figure 5. In vivo activities of Tin derivatives in co-activating the tinD enhancer. Shown are stage-10 embryos carrying a *tinD-lacZ* reporter construct together with an *en-GAL4* driver construct and different *UAS-tin* derivatives. Embryos were stained for βGal (*red*) and Tin (*green*). Confocal scans on the left show mesodermal layers with endogenous Tin expression (*arrowheads*) as well as ectopic Tin stripes in the ventral ectoderm, whereas scans to the right show only ectodermal layers with ectopic Tin and βGal stripes in dorsal areas. (*A*) Ectopic expression of full-length Tin. (*B*) Ectopic expression of Tin$^{\Delta 42-124}$. (*C*) Ectopic expression of Tin$^{\Delta 42-301}$.

residues between ~56 and 124 in the amino-terminal portion of Tinman appear to be necessary and sufficient for the interaction with both Smads proteins, Mad and Medea.

As has been shown for other BMP-dependent Smad proteins, the MH1 domains of Medea and Mad are required for DNA binding (Xu et al. 1998), whereas the MH2 domains are required for heteromer formation (S. Zaffran and M. Frasch, unpubl.). To identify the regions of the Smad proteins that are required for the interaction with Tinman protein, we tested various derivatives of these proteins in the yeast two-hybrid assay. As shown in Figure 4B, full-length Medea and Mad and versions with deletions of their MH2 domains support interaction with Tinman (compare Medea$^{1-681}$ with Mad$^{1-455}$ and Medea$^{1-505}$ with Mad$^{1-241}$, respectively). Additional deletion of the carboxy-terminal ~2/3 of the linker domain of Medea (Medea$^{1-303}$) does also not abolish Tin binding. In contrast, additional deletion of residues 266–303 in the amino-terminal portion of the Medea linker region (Medea$^{1-265}$) and deletion of the entire Mad linker region (Mad$^{1-160}$) severely reduce or abolish Tin binding (Fig. 4B). Together, these data demonstrate that the Tin interacting domains are located within the linker regions of these Smad proteins.

For an independent confirmation, we performed in vitro binding assays with GST-fusions of different versions of Mad and Medea proteins. As predicted from the yeast two-hybrid data, both GST-Mad and GST-Mad^{MH1+L} bind ^{35}S-labeled Tinman protein, whereas GST-Mad^{MH2} is unable to pull down Tin (Fig. 4C, lanes 2–4). Similar to Mad, the MH1-L portion of Medea binds to the Tinman protein (Fig. 4C, lanes 5 and 9), whereas the MH2 domain does not (lane 6). Notably, we were unable to identify a physical interaction between Tinman and dSmad2 protein, suggesting that the interaction with Tin is specific to the Smads of the BMP pathway (data not shown). The observed interactions between Tin and these Smad proteins are schematically shown in Figure 4D.

We used an ectopic expression assay based on the GAL4/UAS system (Brand and Perrimon 1993) to examine the significance of the Smad interaction domain of

Tin, particularly in the context of the Dpp-responsive TinD enhancer activity, in vivo. Previous data have shown that ectopic expression of *tinman* in the ectoderm results in the activation of tin-D specifically within the Dpp domain (Xu et al. 1998). In particular, ectopic expression of full-length Tinman in a segmental pattern driven by the *engrailed* enhancer in the ectoderm activates *tinD-lacZ* specifically in the domains where Tin and Dpp signaling intersect (Fig. 5A,B) (Xu et al. 1998). In an analogous experiment using a version of Tin with a deletion of Smad-interacting domain (Tin$^{\Delta 42-174}$), ectopic activation of tinD in the ectoderm is still observed, although apparently at slightly reduced levels (Fig. 5C,D). Interestingly, when the Smad-interaction domain is deleted together with additional residues amino-terminally to the homeodomain (Tin$^{\Delta 42-301}$), ectopic activation of tinD is no longer supported (Fig. 5E,F). In summary, whereas the minimal Smad interaction domain of Tin is not absolutely required in this assay, a larger region in the amino-terminal portion of Tin that includes the Smad interaction domain appears essential for the synergistic activation of Dpp response element by Tinman and activated Smad complexes.

HMG-D Associates Both with the Tinman Protein and with Sequences of the Dpp-responsive Enhancer of *tinman*

We used yeast two-hybrid screens to address the question of whether there exist yet unknown protein partners of Tinman in addition to Mad and Medea. A *Drosophila* library of hybrid cDNAs from 0–16-hour embryos was screened using GAL4/Tinman fusion protein as a bait. Among 50 positive isolates, three clones encoded the *High Mobility Group-D* gene (*HMG-D*; Wagner et al. 1992) and contained the complete *HMG-D* ORF in frame with the yeast GAL4 activating domain. In vitro protein-binding assays using immobilized GST fusions of HMG-D and ^{35}S-labeled in vitro-translated proteins of Tin confirm that Tinman binds to full-length HMG-D (Fig. 6A). Additional experiments with truncated derivatives of Tin show that the carboxy-terminal portion of Tinman including the homeodomain interacts with HMG-D (Fig. 6A, Tin$^{\Delta 301-416}$). Interestingly, a derivative of Tin lacking only the homeodomain (Tin$^{\Delta 301-364}$) fails to bind HMG-D in this assay, suggesting that the Tinman/HMG-D interaction requires either a portion or the entire homeodomain. The observation that *HMG-D* is preferentially expressed in the mesoderm during the stages when *tin* is induced by Dpp and promotes early cardiogenesis (Fig. 6B) is compatible with a role of HMG-D/Tin interactions in vivo.

Gel mobility shift assays were used to examine whether the association of HMG-D influences the DNA-binding activity of Tinman. In these experiments, Tinman and GST-HMG-D were incubated either individually or in combination with a ^{32}P-labeled probe consisting of a dimer of the Tin-binding D1 subsequence from the tinD enhancer. Consistent with the presence of two Tin-binding sites, high and intermediate concentrations of full-length Tin alone produce two retarded bands (Fig. 6C). Presumably, the weaker band with higher mobility (B4) has only one Tin site occupied, whereas the stronger one with lower mobility (B3) is associated with two Tin molecules. At lower concentrations of Tin, the intensity of both bands, and in particular of B3, is strongly reduced (lane 3). However, addition of increasing concentrations of HMG-D to low Tin produces intensified bands and an increasing preference for B3 and higher-molecular-weight forms (Fig. 6C, lanes 4–6). At the highest concentration, HMG-D binds to D1 on its own (lanes 6,7), and in combination with low Tin there is an additional band (B5) that may correspond to D1 associated with both Tin and HMG-D (lane 6). Similar experiments with the amino-terminally truncated derivative Tin$^{301-416}$, which contains the HMG-D-interaction domain (see above), produces analogous results (Fig. 6, lanes 8–14; B7, one Tin site; B6, both Tin sites occupied). Together, these data suggest that interactions between HMG-D and Tin (as well as tinD, see below), which are transient at lower concentrations and stable at higher concentration of HMG-D, enhance the binding of Tin protein to its target sites.

In addition to the isolation of HMG-D with Tinman as a bait, we also obtained HMG-D in yeast one-hybrid screenings using the D3 sequence of the tinD enhancer as a bait. Gel mobility shift experiments were performed to verify the specific binding of HMG-D to D3 sequences. As shown in Figure 6D, a GST-HMG-D fusion protein generates a retarded band with a D3 probe (lane 15) that can be competed by excess amounts of unlabeled D3 (lane 16) and, more efficiently, by a D3 derivative in which the GC-rich Smad-binding sites are mutated (lane 18, d3g..c) (Xu et al. 1998). In contrast, mutation of both copies of a CAATGT motif adjacent to the Smad-binding sites disrupts the ability of D3 to compete (lane 17, d3c..t). These data indicate that HMG-D associates specifically with the CAATGT motif(s) within D3, which is consistent with previous findings that this protein preferentially binds to CA dinucleotide sequences that are embedded in A/T-rich sequences (Churchill et al. 1995). Furthermore, mutation of the CAATGT motifs within the context of a minimal Dpp-response element disrupts its inducibility by Dpp+Tin in vivo (Xu et al. 1998). Although it is possible that HMG-D shares this binding site with other essential factors, the available data suggest that the association of HMG-D with D3 sequences and with the D1-associated Tin protein is important for the formation of a functional Dpp-responsive enhancer complex.

DISCUSSION

Protein/DNA and Protein/Protein Complexes Transmitting Dpp Signals to the *tinman* Gene

The relatively low DNA-binding affinities and degenerate target sequences of Smad proteins raise the question of how Dpp/BMP signals are able to induce highly selective responses during cardiac induction and other processes. Data provided herein indicate that targeting of these signals to specific cardiogenic genes is achieved through two additional molecular features; first, the clustering of Smad-binding sites on Dpp-dependent en-

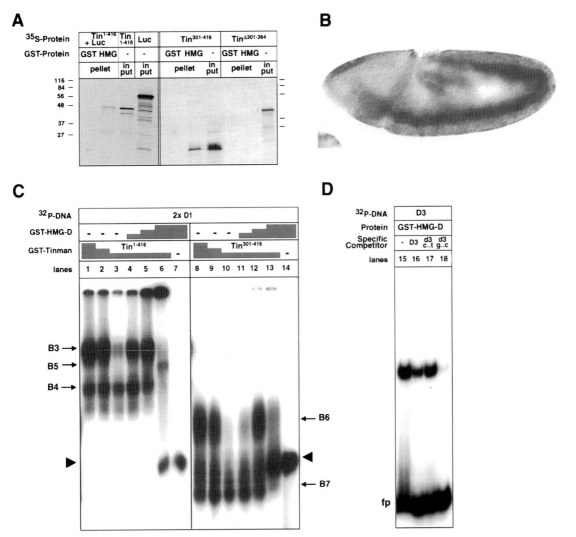

Figure 6. Interactions of HMG-D with Tinman protein and tinD enhancer sequences. (*A*) "GST pull-down" assays with GST-HMG-D fusion protein and wild type as well as truncated versions of Tin. Labeled Luciferase (Luc) was used as a negative control. (*B*) *HMG-D* mRNA expression is predominant in the mesoderm of a stage-9/10 embryo. (*C*) Enhancement of Tin binding to D1 by HMG-D. Shown are gel mobility shift experiments with GST-Tin derivatives, GST-HMG-D, and mixtures of Tin and HMG-D fusion proteins using tandemly duplicated D1 DNA as a probe (free probes are not shown). Protein concentrations were 150 ng (lanes *1, 8*), 75 ng (lanes *2, 9*), 20 ng (lanes *3–6, 10–13*) of GST-Tin and GST-Tin$^{301-416}$, respectively, and 50 ng (lanes *4, 11*), 100 ng (lanes *5, 12*), and 200 ng (lanes *6, 7, 13, 14*) of GST-HMG-D. At high concentrations HMG is able to bind by itself, presumably to the Tin-binding site which contains an HMG-binding motif (TCAAT; see text), to produce a retarded band (lanes *6, 7*, and *13, 14, arrowhead*). (*D*) Gel mobility shift experiments with GST-HMG-D and the D3 sequence from the tinD enhancer as a probe (fp, free probe; lane *16*, wild type D3; lanes *17, 18*, mutated versions of D3).

hancers, and second, specific protein/protein interactions with additional DNA-binding factors. Specifically, we have shown that both Mad and Medea are able to bind via a portion of their linker domains to a distinct domain within the amino-terminal portion of the Tin protein. This type of protein/protein interaction should be favored in vivo at the tinD enhancer, which contains interspersed Tin- and Smad-binding sites (Fig. 3). Similar interactions are likely to occur at yet-unidentified enhancers of other Dpp/Tin-dependent cardiogenic target genes. In addition to targeting of Dpp signals to specific enhancers, the Tin/DNA and Tin/Smad interactions also serve as a mechanism to target the Dpp signals specifically to the mesoderm, since only this germ layer contains Tin proteins that are derived from the transient early activation of the *tin* gene by *twist*.

Ectopic expression of Tin in the ectoderm can activate the Dpp-responsive enhancer of *tin* in this germ layer, where it is normally silent. Under our current experimental conditions, the Smad-interaction domain of Tin does not appear to be essential for the ectopic induction of this enhancer, although the efficiency of induction may be reduced. It is conceivable that the higher levels of ectopic Tin in the ectoderm as compared to the endogenous levels in the mesoderm can overcome the requirement for Tin/Smad interactions. In the normal situation, Tin/Smad protein interactions may be required for full levels but not for on/off switches of Dpp/Tin-dependent enhancers.

Nevertheless, deletion of Smad-interaction sequences together with carboxy-terminally adjacent sequences in Tin leads to a complete abolishment of ectopic tinD enhancer induction, indicating that the Smad interaction domain may be critical if a second interaction domain with a yet-unknown factor(s) is also missing.

In summary, Tinman represents a new example of specific DNA-binding factors that enhance the target specificity of Dpp/BMP signals via protein interactions with Smad proteins. Previously reported factors acting in an analogous fashion in various BMP pathways include a Hox protein (Yang et al. 2000), the zinc finger proteins Schnurri and OAZ (Dai et al. 2000; Hata et al. 2000; Udagawa et al. 2000), and Runt-domain proteins (Hanai et al. 1999). The identification of HMG-D in both yeast two-hybrid screening for Tin-interacting proteins and yeast one-hybrid screenings for TinD enhancer-binding factors strongly supports the notion that HMG-D is a genuine component of the *tin* induction process in vivo. On the basis of our DNA-binding studies, we propose that the interaction of HMG-D with Tin, which occurs through the homeodomain of Tin, serves to strengthen the interaction of Tin with its target DNA sequences. Analogous findings have previously been reported for a vertebrate homolog of HMG-D, HMG-1, which was shown to enhance sequence-specific binding of Hox proteins by interactions with their homeodomains (Zappavigna et al. 1996). Conversely, the interaction of HMG-D with Tin may also serve to recruit HMG-D, which itself has a relatively low binding specificity, to the Dpp-responsive *tin* enhancer. This activity of Tin would be comparable to the recruitment of Smads via protein/protein interactions, as discussed above. Because of the known activity of HMG-D as well as other HMG proteins in inducing DNA bending (for review, see Thomas and Travers 2001), it is very likely that HMG-D-binding to the Dpp-response element has an architectural role in addition to the enhancement of Tin binding. We speculate that HMG-D-induced DNA bending may generate a proper tertiary structure of an "enhanceosome" that facilitates protein/protein interactions among bound factors that define the active state of the enhancer (Fig. 3) (for a discussion of related activities of HMG-1, see Mitsouras et al. 2002).

Spatial and Molecular Integration of Dpp and Wingless Signals in Cardiac Induction

The combinatorial action of Dpp and Tin is required not only for the induction of *tin* itself in the dorsal mesoderm, but also for the induction of other key genes that control the development of dorsal mesodermal derivatives, including the dorsal vessel. Although the expression of *tin* is induced within the entire domain of the dorsal mesoderm, most if not all of the known downstream genes of *tin*+*dpp* are induced in segmentally restricted sub-areas within this domain. As a result, cardiogenic and visceral muscle-generating genes become expressed in alternating, mutually exclusive domains along the antero-posterior embryo axis. Well-characterized examples include the homeobox gene *even-skipped* (*eve*), which is required for normal differentiation of pericardial cells (Su et al. 1999), and the NK homeobox gene *bagpipe* (*bap*), as well as the forkhead domain gene *biniou* (*bin*), both of which are essential for the formation of visceral muscles (Fig.7) (Azpiazu and Frasch 1993; Zaffran et al. 2001).

The observation that only one-half of each segment shows a cardiogenic response to *dpp*+*tin* suggests the involvement of additional, spatially restricted cues in cardiac induction. Indeed, several studies have demonstrated that the Wnt family member Wingless (Wg) is required in combination with Tin and Dpp to elicit cardiogenic responses in the dorsal mesoderm (Wu et al. 1995; Azpiazu et al. 1996; Park et al. 1996; Carmena et al. 1998). Wg is expressed in transverse stripes in the ectoderm and is required for the induction of cardiogenic genes such as *eve* in the underlying cells of the dorsal mesoderm (Fig. 7). Additional studies showed that the cardiogenic activity of Wg signals acts through at least two pathways. One pathway involves the striped induction of the forkhead domain encoding gene *sloppy paired* (*slp*) in the mesoderm, whereas the other is *slp*-independent (Lee and Frasch 2000). It was further demonstrated that *slp* is required for the repression of *bin* within the cardiogenic domain. *bin* is a positive regulator of visceral mesoderm development while it represses cardiogenesis (Fig. 7) (Zaffran et al. 2001). Therefore, one important activity of Wg signaling is to prevent the induction of negative regulators of cardiogenesis by Dpp/Tin in the cardiogenic domain.

Two recent studies have provided important insight into the question of how Dpp/Tin activities and Wg signals are integrated at the molecular level (Halfon et al. 2000; Knirr and Frasch 2001). These studies defined short and overlapping enhancer elements located downstream from the *eve* gene that are able to confer a full response to these cues in cardiogenic (as well as dorsal somatic muscle) progenitors. The *eve* enhancer shares interesting similarities with the tinD enhancer, as it contains multiple and interspersed binding sites for Tin and Smad proteins that have essential (although partially re-

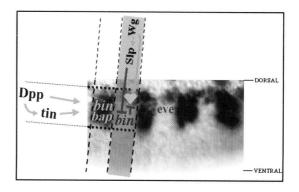

Figure 7. Summary of intersecting pathways during cardiac progenitor specification. Shown is a lateral view of the mesoderm of a stage-10 embryo that was stained for Even-skipped (Eve) protein (cardiac/dorsal muscle progenitors, *brown*) and *bagpipe* (*bap*) mRNA (*purple*). *biniou* (*bin*) is coexpressed with *bap* in visceral muscle progenitors. One of the four segments shown is highlighted (*left*).

dundant) roles in the induction of the enhancer activity (Fig. 8A). On the basis of the data presented herein, we suggest that in addition to their DNA-binding activities, protein/protein interactions between Tin and Smads contribute to the full activation of this *eve* enhancer as well. A potential role of HMG-D has not been examined in this context.

The Wg inputs via the *slp*-independent route (Fig.7) into this enhancer appear to involve the canonical Wg pathway, which is mediated through the effectors Armadillo (Arm = *Drosophila* β-catenin) and the HMG protein dTCF (also known as Pangolin). The mesodermal *eve* enhancer includes several binding sites for dTCF that confer Wg inputs. As expected, inactivation of subsets of these dTCF-binding sites results in decreased enhancer activity (Halfon et al. 2000; Knirr and Frasch 2001). Surprisingly, simultaneous inactivation of all dTCF sites causes a strong expansion of enhancer activity into dorsal mesodermal areas in which *eve* is not expressed (Knirr and Frasch 2001). We conclude that dTCF has an important repressing function in the areas of the mesoderm that do not receive Wg signals. Such a role of dTCF is consistent with previous findings that dTCF/TCF/Lef-1 proteins can associate with Groucho co-repressors and that this repressing activity is relieved upon Wg/Wnt signaling by the binding of Armadillo/β-catenin (for review, see Korswagen and Clevers 1999). From this, we conclude that in the absence of Wg signals, binding of dTCF/co-repressor complexes to the mesodermal *eve* enhancer negates the activating function of bound Tin+Smad protein/DNA and protein/protein complexes, which prevents *eve* from being induced by Dpp (Fig. 8B). In contrast, in cells of the cardiogenic areas, which receive Wg signals, this repression is abolished and

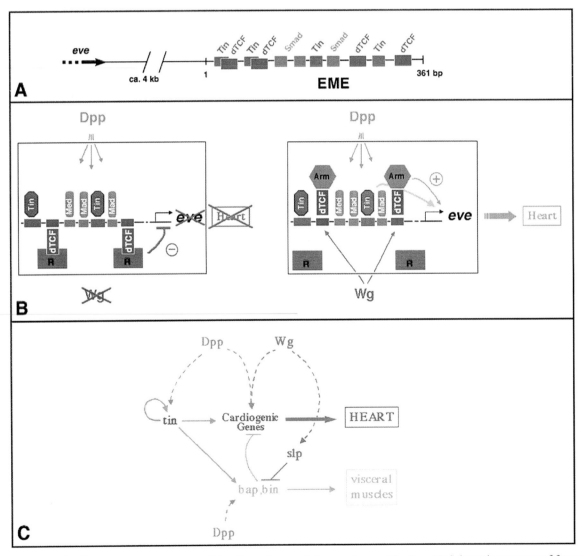

Figure 8. Molecular and genetic integration of signaling pathways during cardiac specification. (*A*) Schematic summary of factor-binding sites within the 361-bp EME enhancer of *even-skipped* (*eve*). (*B*) Model of de-repression of the Dpp-responsive *eve* EME enhancer upon Wg signaling (R: co-repressor[s] of dTCF, which include Groucho). (*C*) Summary of signaling pathways during the activation of cardiogenic genes and the repression of visceral mesoderm regulators within the cardiogenic domains of the dorsal mesoderm. Note that, based on the model in *B*, the Wg inputs are thought to act via inactivation of dTCF/corepressor complexes.

Tin+Smad complexes (together with additional positive inputs that involve the Ras pathway; see Halfon et al. 2000) are able to activate *eve* expression (Fig. 8B). Based on the observed reduction in the levels of reporter gene expression upon mutation of dTCF sites, Armadillo also appears to have a direct positive effect, albeit minor, on the enhancer activity (Fig. 8B). Preliminary data from our laboratory (not shown) indicate that other regulatory genes in the cardiogenic areas are controlled through an analogous Wg-induced switch from the repressed to the activated state.

CONCLUSIONS

On the basis of data presented in this study and other published work, we conclude that the combination of ectodermally derived Dpp (BMP) signals and mesodermal Tinman provides a key activity for the induction of cardiogenic genes in the dorsal mesoderm. We have shown that, in molecular terms, this activity is mediated via combinatorial binding of Tin and activated Smad proteins to enhancer sequences of cardiogenic genes, which also involves Tin/Smad protein interactions. In addition, HMG-D is likely to play an auxiliary role in the formation of an active "enhanceosome" through protein/protein interactions with Tinman as well as DNA bending upon HMG-D binding to *tin* enhancer sequences. We have discussed that Dpp acts at multiple steps in the cardiogenic pathway through this type of mechanism, initially to induce *tinman* within the dorsal mesoderm, and subsequently, to induce other regulatory genes specifically in segmental cardiogenic areas of the dorsal mesoderm (Fig. 8C). These latter induction events require Wingless signaling in addition to Dpp+Tin. We have proposed a mechanism in which the main function of Wingless is to inactivate cardiogenic repressors. Wg exerts this de-repression through at least two pathways; first, the (indirect) transcriptional repression of negative regulators such as the FoxF gene *biniou* in the cardiogenic domains (Fig. 8C), and second, through post-transcriptional events of the Wg signaling cascade that remove co-repressors such as Groucho from dTCF proteins which are bound to enhancers of cardiogenic genes.

ACKNOWLEDGMENT

This work was supported by National Institutes of Health grants HD-30832 and DK-59406.

REFERENCES

Andrée B., Duprez D., Vorbusch B., Arnold H.-H., and Brand T. 1998. BMP-2 induces ectopic expression of cardiac lineage markers and interferes with somite formation in chicken embryos. *Mech. Dev.* **70:** 119.

Azpiazu N. and Frasch M. 1993. *tinman* and *bagpipe*: Two homeo box genes that determine cell fates in the dorsal mesoderm of *Drosophila*. *Genes Dev.* **7:** 1325.

Azpiazu N., Lawrence P., Vincent J.-P., and Frasch M. 1996. Segmentation and specification of the *Drosophila* mesoderm. *Genes Dev.* **10:** 3183.

Bodmer R. 1993. The gene *tinman* is required for specification of the heart and visceral muscles in *Drosophila*. *Development* **118:** 719.

Brand A.H. and Perrimon N. 1993. Targeted gene expression as a means of altering cell fates and generating dominant phenotypes. *Development* **118:** 401.

Bruneau B.G., Nemer G., Schmitt J.P., Charron F., Robitaille L., Caron S., Conner D.A., Gessler M., Nemer M., Seidman C.E., and Seidman J.G. 2001. A murine model of Holt-Oram syndrome defines roles of the T-box transcription factor Tbx5 in cardiogenesis and disease. *Cell* **106:** 709.

Carmena A., Gisselbrecht S., Harrison J., Jimenez F., and Michelson A. 1998. Combinatorial signaling codes for the progressive determination of cell fates in the *Drosophila* embryonic mesoderm. *Genes Dev.* **15:** 3910.

Chen C. and Schwartz R. 1995. Identification of novel DNA binding targets and regulatory domains of a murine *tinman* homeodomain factor, Nkx-2.5. *J. Biol. Chem.* **270:** 15628.

Churchill M.E., Jones D.N., Glaser T., Hefner H., Searles M.A., and Travers A.A. 1995. HMG-D is an architecture-specific protein that preferentially binds to DNA containing the dinucleotide TG. *EMBO J.* **14:** 1264.

Dai H., Hogan C., Gopalakrishnan B., Torres-Vazquez J., Nguyen M., Park S., Raftery L., Warrior R., and Arora K. 2000. The zinc finger protein schnurri acts as a Smad partner in mediating the transcriptional response to decapentaplegic. *Dev. Biol.* **227:** 373.

Frasch M. 1995. Induction of visceral and cardiac mesoderm by ectodermal Dpp in the early *Drosophila* embryo. *Nature* **374:** 464.

Gajewski K., Choi C., Kim Y. and Schulz R. 2000. Genetically distinct cardial cells within the *Drosophila* heart. *Genesis* **28:** 36.

Halfon M., Carmena A., Gisselbrecht S., Sackerson C., Jimenez F., Baylies M. and Michelson A. 2000. Ras pathway specificity is determined by the integration of multiple signal-activated and tissue-restricted transcription factors. *Cell* **103:** 63.

Hanai J., Chen L., Kanno T., Ohtani-Fujita N., Kim W., Guo W., Imamura T., Ishidou Y., Fukuchi M., Shi M., Stavnezer J., Kawabata M., Miyazono K., and Ito Y. 1999. Interaction and functional cooperation of PEBP2/CBF with Smads. Synergistic induction of the immunoglobulin germline *Calpha* promoter. *J. Biol. Chem.* **274:** 31577.

Hata A., Seoane J., Lagna G., Montalvo E., Hemmati-Brivanlou A. and Massagué J. 2000. OAZ uses distinct DNA- and protein-binding zinc fingers in separate BMP-Smad and Olf signaling pathways. *Cell* **100:** 229.

Horb M.E. and Thomsen G.H. 1999. Tbx5 is essential for heart development. *Development* **126:** 1739.

Jagla K., Frasch M., Jagla T., Dretzen G., Bellard F., and Bellard M. 1997. *ladybird*, a new component of the cardiogenic pathway in *Drosophila* required for diversification of heart precursors. *Development* **124:** 3471.

Kawabata M., Inoue H., Hanyu A., Imamura T., and Miyazono K. 1998. Smad proteins exist as monomers in vivo and undergo homo- and hetero-oligomerization upon activation by serine/threonine kinase receptors. *EMBO J.* **17:** 4056.

Kim J., Johnson K., Chen H., Carroll S., and Laughon A. 1997. *Drosophila* Mad binds to DNA and directly mediates activation of *vestigial* by Decapentaplegic. *Nature* **388:** 304.

Knirr S. and Frasch M. 2001. Molecular integration of inductive and mesoderm-intrinsic inputs governs *even-skipped* enhancer activity in a subset of pericardial and dorsal muscle progenitors. *Dev. Biol.* **238:** 13.

Korswagen H. and Clevers H. 1999. Activation and repression of Wingless/Wnt target genes by the TCF/LEF-1 family of transcription factors. *Cold Spring Harbor Symp. Quant. Biol.* **64:** 141.

Lee H. and Frasch M. 2000. Wingless effects mesoderm patterning and ectoderm segmentation events via induction of its downstream target *sloppy paired*. *Development* **127:** 5497.

Liberatore C., Searcy-Schrick R., Vincent E., and Yutzey K. 2002. *Nkx-2.5* gene induction in mice is mediated by a Smad consensus regulatory region. *Dev. Biol.* **244:** 243.

Lien C., McAnally J., Richardson J. and Olson E. 2002. Cardiac-specific activity of an *Nkx2-5* enhancer requires an evolutionarily conserved Smad binding site. *Dev. Biol.* **244:** 257.

Lo P.C. and Frasch M. 2001. A role for the *COUP-TF*-related gene *seven-up* in the diversification of cardioblast identities in the dorsal vessel of *Drosophila*. *Mech. Dev.* **104:** 49.

Mitsouras K., Wong B., Arayata C., Johnson R., and Carey M. 2002. The DNA architectural protein HMGB1 displays two distinct modes of action that promote enhanceosome assembly. *Mol. Cell. Biol.* **22:** 4390.

Molina M. and Cripps R. 2001. Ostia, the inflow tracts of the *Drosophila* heart, develop from a genetically distinct subset of cardial cells. *Mech. Dev.* **109:** 51.

Nasonkin I., Alikasifoglu A., Ambrose C., Cahill P., Cheng M., Sarniak A., Egan M., and Thomas P. 1999. A novel sulfonylurea receptor family member expressed in the embryonic *Drosophila* dorsal vessel and tracheal system. *J. Biol. Chem.* **274:** 29420.

Park M., Wu X., Golden K., Axelrod J.D., and Bodmer R. 1996. The Wingless signaling pathway is directly involved in *Drosophila* heart development. *Dev. Biol.* **177:** 104.

Pereira F., Qiu Y., Zhou G., Tsai M. and Tsai S. 1999. The orphan nuclear receptor COUP-TFII is required for angiogenesis and heart development. *Genes Dev.* **13:** 1037.

Raftery L. and Sutherland D. 1999. TGF-beta family signal transduction in *Drosophila* development: From Mad to Smads. *Dev. Biol.* **210:** 251.

Rugendorff A., Younossi-Hartenstein A., and Hartenstein V. 1994. Embryonic origin and differentiation of the *Drosophila* heart. *Roux's Arch. Dev. Biol.* **203:** 266.

Schultheiss T., Burch J., and Lassar A. 1997. A role for bone morphogenetic proteins in the induction of cardiac myogenesis. *Genes Dev.* **11:** 451.

Shi Y., Katsev S., Cai C., and Evans S. 2000. BMP signaling is required for heart formation in vertebrates. *Dev. Biol.* **224:** 226.

Su M., Fujioka M., Goto T., and Bodmer R. 1999. The *Drosophila* homeobox genes *zfh-1* and *even-skipped* are required for cardiac-specific differentiation of a *numb*-dependent lineage decision. *Development* **126:** 3241.

Thomas J. and Travers A. 2001. HMG1 and 2, and related 'architectural' DNA-binding proteins. *Trends Biochem. Sci.* **26:** 167.

Udagawa Y., Hanai J., Tada K., Grieder N., Momoeda M., Taketani Y., Affolter M., Kawabata M., and Miyazono K. 2000. Schnurri interacts with Mad in a Dpp-dependent manner. *Genes Cells* **5:** 359.

Wagner C., Hamana K., and Elgin S. 1992. A high-mobility-group protein and its cDNAs from *Drosophila melanogaster*. *Mol. Cell. Biol.* **12:** 1915.

Wu X., Golden K., and Bodmer R. 1995. Heart development in *Drosophila* requires the segment polarity gene *wingless*. *Dev. Biol.* **169:** 619.

Xu X., Yin Z., Hudson J., Ferguson E., and Frasch M. 1998. Smad proteins act in combination with synergistic and antagonistic regulators to target Dpp responses to the *Drosophila* mesoderm. *Genes Dev.* **12:** 2354.

Yang X., Ji X., Shi X., and Cao X. 2000. Smad1 domains interacting with Hoxc-8 induce osteoblast differentiation. *J. Biol. Chem.* **275:** 1065.

Yin Z. and Frasch M. 1998. Regulation and function of *tinman* during dorsal mesoderm induction and heart specification in *Drosophila*. *Dev. Genet.* **22:** 187.

Yin Z., Xu X.-L. and Frasch M. 1997. Regulation of the twist target gene *tinman* by modular *cis*-regulatory elements during early mesoderm development. *Development* **124:** 4971.

Yingling J., Datto M., Wong C., Frederick J., Liberati N., and Wang X.-F. 1997. Tumor suppressor Smad4 is a transforming growth factor β-inducible DNA binding protein. *Mol. Cell. Biol.* **17:** 7019.

Zaffran S., Küchler A., Lee H.H., and Frasch M. 2001. *biniou* (*FoxF*), a central component in a regulatory network controlling visceral mesoderm development and midgut morphogenesis in *Drosophila*. *Genes Dev.* **15:** 2900.

Zappavigna V., Falciola L., Helmer-Citterich M., Mavilio F., and Bianchi M. 1996. HMG1 interacts with HOX proteins and enhances their DNA binding and transcriptional activation. *EMBO J.* **15:** 4981.

Regulation of Cardiac Muscle Differentiation in *Xenopus laevis* Embryos

T. MOHUN, B. LATINKIC, N. TOWERS, AND S. KOTECHA
Division of Developmental Biology, National Institute for Medical Research, London, United Kingdom

Amphibians offer an attractive model for studying vertebrate heart formation because their embryos are relatively big, they are available in large numbers, and their development is rapid. Within a day and a half of egg fertilization, differentiation of cardiac tissue commences, and within 3 days, the tadpole heart is fully formed. The process of heart formation in amphibians appears very similar to that found in mammals. The tadpole heart arises from bilateral patches of mesodermal tissue that are first formed during gastrulation and whose fate is apparently specified by cell–cell interactions. This tissue eventually fuses along the ventral midline, forming a contractile myocardial tube that encompasses the endocardium. Looping of the heart tube is rapidly followed by differentiation of distinct atrial and ventricular chambers, giving rise to a mature tadpole heart comprising two atria and a common ventricle. Because eggs are laid in water, cardiac tissue and its progenitors are readily accessible at all stages of development for experimental study. The size and robust character of frog embryos facilitates the use of embryological methods such as tissue explantation, grafting, and extirpation. In addition, their capacity to tolerate the injection of synthetic oligonucleotide or nucleic acid sequences allows the roles of individual gene products in cardiac differentiation and morphogenesis to be examined. A significant limitation for such studies is the relatively poor targeting of individual tissue types that can be achieved by injection of embryonic cells. Within a few divisions after fertilization, it is no longer feasible to inject individual blastomeres, and there is also considerable variation in the precise cleavage pattern. As a result, it is only possible to target injected material to broad regions of the embryo, and such difficulties are increased when the targeted tissue, such as the heart, forms relatively late in embryo development.

TRANSGENESIS IN AMPHIBIANS

With the discovery of a method to create transgenic frog embryos (Kroll and Amaya 1996), there is now a way to overcome this limitation in targeting and to fully exploit the potential advantages of amphibian embryos for cardiac research. Using appropriate promoters, it is possible to drive the expression of reporter genes, candidate regulator genes, or their mutated counterparts to investigate individual steps in cardiogenesis. Crucially, the technique is rapid, allowing a significant number of transgenic embryos to be produced for each transgene, and their effects on cardiogenesis to be monitored within a few days of egg fertilization. Because the transgene is integrated into the genome of the sperm prior to (or during) fertilization, each transgenic embryo is a potential founder for a transgenic line.

As with transgenesis in other organisms, the method facilitates three distinct types of experiments. First, by using an appropriate reporter, it is possible to create transgenic frog lines in which the cardiac progenitors or particular tissues of the developing heart are labeled. Embryos from such lines are of great potential value, since simple observation of reporter expression (such as fluorescence from the GFP family of proteins) can replace more complex assays of gene expression as a means of identifying cardiac tissue and following the onset of its differentiation. A second use of transgenic embryos is in the mapping of regulatory sequences necessary for expression of a particular gene within cardiac tissue. The relative simplicity of the procedure allows a wide variety of promoter modifications to be quickly tested and compared (see, e.g., Latinkic et al. 2002; Polli and Amaya 2002). Finally, the procedure also allows the contribution of individual gene products to normal cardiogenesis to be tested (Breckenridge et al. 2001). Using appropriate regionally and temporally restricted expression conferred by individual promoters, it is possible to target transgene expression and to assess the impact of particular gene products or mutated versions on normal cardiogenesis. Such an approach can be used to investigate regional differentiation within the heart, regulation of heart morphogenesis, or the specific signaling pathways necessary for differentiation of cardiac tissues.

To appreciate the current feasibility of such experiments, it is necessary to emphasize several features of current amphibian transgenesis procedures and to outline their limitations. The present method is based on the random introduction of breaks into the DNA of sperm to allow the integration of transgenes into the sperm genome. This can be achieved either by the use of limited restriction enzyme digestion of swollen sperm nuclei (Amaya and Kroll 1999) or by taking advantage of the damage caused to sperm nuclei by cycles of freezing and thawing (Sparrow et al. 2000a). In either case, the method is difficult to reproduce in a consistent manner, either because of its inherent complexity or because of an apparent variability between batches of recipient eggs. Whereas variations in transgenesis efficiency are commonly encountered and generally attributed to differences in egg "quality," we do not yet understand the basis for this, and we can neither quantify nor control it. In the face of such formidable problems, it may seem remarkable that transgenesis can be achieved at all. The fact that it remains a relatively simple and reliable procedure can be explained by the prodigious number of recipient eggs that are available and the relative ease and speed with which several thousand can be injected.

The efficiency with which transgenic embryos can be obtained is affected by several other factors besides egg batch. Empirically, it is clear that the efficiency of transgenesis is (within limits) directly related to the concentration of DNA used in the sperm treatment. However, it is also evident that the viability of transgenic embryos is inversely related to the DNA concentration, with the result that there is an inevitable trade-off between the frequency with which transgenic embryos are obtained and their subsequent survival. Other aspects of the technique also affect frequency of success. The modified sperm nuclei are injected into unfertilized eggs in a manner that attempts to maximize successful injection while minimizing the frequency with which eggs receive multiple nuclei. As a result, the procedure is inherently inefficient, the majority of injected eggs failing subsequently to show normal patterns of cell cleavage. Where sperm nuclei have been swollen to facilitate transgene integration, a relatively large-bore needle is used for egg injection in order to minimize further damage to the nuclei from shearing forces. Injection with such needles causes significant damage to the unfertilized egg and can further reduce the proportion of fertilized eggs that survive.

Because transgenesis currently depends on randomly induced breaks in sperm DNA, the associated damage is difficult to control and can have a severe impact on normal development of the transgenic embryo. This appears to have a progressive and cumulative effect on embryo development, with the result that older transgenic embryos are more frequently abnormal, and survival through to metamorphosis is often very poor. This effect may be of little consequence if transgene expression is only required early in development. However, since heart formation is a relatively late event in amphibian embryos, it can have a severe effect on studies of cardiogenesis and can hamper attempts to establish transgenic lines.

Integration of transgenes occurs randomly, with the result that both the location and copy number within the genome will be different in each transgenic embryo. This may produce variable and unexpected effects on expression of the transgene and prevents any simple quantitative comparisons in transgene expression. Empirically, it appears that although expression levels may vary considerably between individual embryos possessing the same transgene, when tens or (preferably) hundreds of transgenic embryos are compared, the level of expression is broadly consistent (albeit not readily quantifiable). Furthermore, when tissue- or region-restricted promoters are used, inappropriate or ectopic transgene expression is relatively rare, suggesting that any effects of integration site are either too subtle to detect or so profound that they simply block transgene expression.

A final practical limitation that affects the use of transgenesis to study heart formation is the finding that a relatively common malformation seen with amphibian embryos affects the antero-ventral region. This manifests itself as a large, possibly edematous, expansion of tissue into a vesicle-like structure adjacent to or encompassing the developing heart, arising both spontaneously and as an apparent consequence of damage from the transgenesis procedure. Whatever its precise cause, it is nonspecific with respect to transgene but can hamper the study of heart formation in F_0 generation transgenic embryos.

A common thread linking all the limitations associated with transgenesis is the damaging and random nature of the current procedure. If this can be replaced by an alternative method, we may expect greater efficiencies, better survival, and a reduction in developmental abnormalities. Two candidate approaches are currently being investigated with reports of preliminary success. One involves the use of engineered transposons (see, e.g., Ivics et al. 1997) to produce highly efficient, random integration of transgenes into the genome of the fertilized egg. The second uses targeted recombination via Frt or Cre recombinases (Werdien et al. 2001). This will require the prior establishment of a recipient transgenic line bearing a target flp or loxP recombination site, but it has the potential advantage of eliminating variations in transgene integration site or copy number, thereby facilitating proper, quantitative comparison of transgene expression levels.

CARDIAC-SPECIFIC TRANSGENIC LINES

The current limitation on establishment of cardiac-specific transgenic reporter lines reflects the dearth of well-characterized promoters in *Xenopus*, rather than difficulties inherent in the transgenesis technique. From our own studies, we have identified regions of the Nkx2-5, cardiac actin, and myosin light-chain 2a (MLC2a) promoters that confer the appropriate patterns of transgene expression, and we are using these to construct transgenic GFP reporter lines in *Xenopus laevis* and *Xenopus tropicalis*. Of these, the best characterized is the *X. laevis* cardiac actin line which utilizes 570 bp of promoter sequence and is currently bred to the F_2 generation (Latinkic et al. 2002). Like its endogenous counterpart, the GFP reporter is expressed in both skeletal and cardiac muscle of embryos from the onset of differentiation (Fig. 1). Interestingly, little or no expression is detected in adult frogs, indicating that additional regulatory elements are necessary to maintain expression after metamorphosis. The MLC2a transgene is expressed exclusively in the myocardial layer of the developing heart, whereas the XNkx2-5.GFP transgene is expressed within the cardiac mesoderm prior to the onset of terminal differentiation and also within the underlying endoderm (Sparrow et al. 2000b). No promoters are yet characterized from *Xenopus* that can be used to drive expression specifically within the endocardium, al-

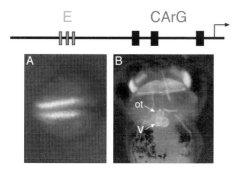

Figure 1. Expression of a cardiac actin.GFP transgene. A 570-bp portion of the proximal promoter containing conserved CArG box and E box motifs (*diagram*) is first expressed in the differentiating skeletal muscle of the myotomes in the neurula embryo (*A*). In the feeding tadpole (*B*, ventral view), expression is now also evident in musculature of the pharynx and in the heart. (ot) Outflow tract; (*V*) ventricle.

though this might be possible using a portion of the SMAD 3 promoter (Howell et al. 2001). Endothelial markers might provide an alternative, and the Fli.GFP reporter used so successfully to map development of the vasculature in zebrafish embryos (Lawson and Weinstein 2002) also shows similar expression in tadpoles (T. Mohun, unpubl.). From studies in chick and the mouse, we might expect that the IRX4 and ANF promoters would provide a means of targeting transgene expression to the ventricular and atrial myocardium, respectively (Bao et al. 1999; Christoffels et al. 2000). However, in *Xenopus*, both genes are actually expressed throughout the myocardium during the period of heart chamber formation (Small and Krieg 2000; Garriock et al. 2001) and, at least in the case of the ANF promoter, atrial restriction is only established later.

TRANSGENIC STUDIES OF THE CARDIAC ACTIN PROMOTER

Since only 570 bp of the cardiac actin promoter is sufficient to drive expression of a transgene reporter within both skeletal and cardiac muscle of the tadpole, we have used the transgenesis assay to examine both the common and distinct sequences utilized in these different tissues. Our results demonstrate a remarkable complexity for such a short promoter sequence (Latinkic et al. 2002). The most striking observation is that mutation of the proximal CArG box/SRE motif selectively abolishes transgene expression in the myocardium while having little effect on expression in skeletal muscle of the myotomes. This indicates that serum response factor (SRF) is an important regulator of cardiac-specific expression from this promoter. Interestingly, mutation of four nucleotides downstream from the CArG box/SRE has a similar effect, and gel-shift studies indicate that this may be an indirect effect, since such mutations reduce the affinity of SRF for the adjacent SRE in vitro. Consistent with this, mutations within the SRE motif that directly reduce SRF affinity also reduce or block transgene expression in the myocardium. Within the 570-bp region of promoter, there are in fact two other CArG/SRE motifs (Mohun et al. 1989), and the potential importance of this arrangement is suggested by the observation that multiple CArG/SRE motifs are conserved in the cardiac and skeletal actin genes across all vertebrates studied. In the *Xenopus* cardiac actin gene promoter, these distal sites have a lower affinity for SRF than that of the proximal site, but their presence nevertheless appears to be essential. Their precise sequence is unimportant, since myocardial expression is maintained with transgenes in which all three have been replaced by multimerized copies of the cFos SRE.

SRF is expressed in a wide range of tissues throughout the vertebrate embryo and has been identified as a regulator of both muscle-specific and growth factor-regulated transcription. It therefore seems unlikely that SRF alone mediates myocardial expression of the cardiac actin promoter. Previous studies have demonstrated that SRF can interact with several other transcriptional regulators, and the protein is perhaps best viewed as a "platform" that facilitates the formation of a transcriptional regulatory complex with cell- or tissue-specific factors (Chen and Schwartz 1996; Belaguli et al. 2000; Morin et al. 2001). In myocardial muscle cells, one such factor is myocardin, a powerful transcriptional activator that can interact with SRF and whose transcripts are found in cardiac and smooth muscle cells (Wang et al. 2001). Interestingly, in cell transfection studies, myocardin-mediated transcriptional activation of an SM22a promoter requires the presence of two SREs. If this proves to be a general requirement for myocardin-mediated gene activation, it may account for the importance of multiple CArG/SREs for myocardial expression from the cardiac actin promoter.

A surprising finding from our mutagenesis experiments is that several regions within the 570-bp promoter are apparently necessary to establish correct expression of the cardiac actin.GFP transgene in skeletal muscle of the tadpole. The clearest distinction can be made between a distal portion of the promoter (that encompasses previously identified myogenic factor-binding sites) and a more proximal region encompassing the CArG/SRE motifs. Each of these is redundant to the extent that expression is retained at some level within the tadpole myotomes in their absence, but only their combination results in uniform, high-level expression that is appropriately restricted to muscle tissue (Fig. 2).

TRANSGENIC STUDIES OF THE MYOSIN LIGHT-CHAIN 2A PROMOTER

Although the *Xenopus* cardiac actin promoter is active in all striated muscle, the MLC2a gene is expressed solely within cardiac muscle. Its transcripts are present in the heart myocardium and the myocardial muscle of the pulmonary vasculature. To investigate whether cardiac-specific expression from the two promoters results from common regulatory mechanisms, we have used transgenesis to map the MLC2a regulatory sequences. These studies demonstrate that myocardial expression of this promoter is dependent on quite different *cis*-binding sites

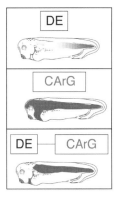

Figure 2. The distal enhancer (DE) of the cardiac actin promoter drives graded expression of a GFP reporter transgene in the tadpole myotomes while the proximal CArG box-containing region (CArG) drives widespread expression throughout the myotomes, heart, and head region. Correct expression requires both regions (DE-CArG).

from those in the cardiac actin promoter and is therefore achieved by a distinct regulatory mechanism.

Early studies suggested that MEF2 proteins played an important role in myocardial expression of the *Xenopus* MLC2a gene, since ectopic expression of MEF2D in explants of prospective ectodermal tissue resulted in robust MLC2a transcription (Chambers et al. 1994). Several potential MEF2-binding sites are indeed present within 3 kb of the gene promoter, but selective mutation of these proved to have little or no effect on transcription of the transgene in tadpoles. Through a combination of deletion analysis and selective mutagenesis, we have established that a series of binding sites for members of the GATA family of zinc finger transcription factors are important for cardiac-specific transgene expression. These show some functional redundancy, but at least two such sites are required along with an adjacent CArG box-like motif that overlaps with a consensus binding site for the YY1 transcriptional regulator. Indeed, a 93-bp sequence encompassing the two most proximal GATA sites and the intervening CArG/YY1 motifs is capable of conferring myocardial-restricted expression on a transgene reporter driven by a heterologous minimal promoter (Fig. 3).

What can we infer about the regulatory interactions from these studies? As with the cardiac actin promoter, we envisage that the core proximal regulatory region we

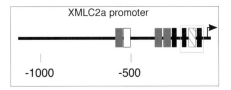

Figure 3. The proximal 1 kb of the *Xenopus* MLC2a promoter contains binding sites for GATA factors (*black*), MEF2 proteins (*gray*), and SRF (*white*). In addition, there is a YY1-binding site that overlaps a weak SRF-binding site (*hatched*). Heart-specific expression of a GFP transgene can be driven by a minimal fragment (*boxed*) that contains the YY1/SRE-like sequence and the flanking two GATA motifs.

have identified allows the formation of a multiprotein complex that drives myocardial transcription. Although the CarG-like motif is a relatively poor binding site for SRF, mutagenesis demonstrates that it is essential. The adjacent GATA motifs provide binding sites for GATA4 in vitro, and GATAs 4–6 are all expressed in cardiac mesoderm (Jiang and Evans 1996; Gove et al. 1997). GATA4 has been shown to interact with SRF on DNA (Morin et al. 2001), and it is likely that such a complex yields synergistic activation of the MLC2a promoter. Since the promoter also contains more distally located CArG/SRE motifs, we cannot exclude the possibility that myocardin may also contribute to the regulatory complex, but if so, this is unlikely to be an indispensable component, since the minimal cardiac-specific promoter sequences contain only a single potential SRF-binding site. Finally, an important component of the regulatory complex appears to be the transcriptional regulator, YY1 (Thomas and Seto 1999). Mutations that selectively abolish YY1 binding while retaining or enhancing the overlapping SRF-binding site can result in extensive ectopic expression of the transgene. These findings suggest that YY1 may act as a negative regulator of the MLC2a promoter in several tissues.

INDUCTION OF CARDIAC TISSUE IN EMBRYONIC TISSUE

Although promoter mapping can provide some evidence for the identity of regulatory factors that drive cardiac-specific transcription, it is necessarily indirect. An alternative approach is to establish whether candidate transcription factors, acting either alone or in combination, can trigger cardiac muscle-specific transcription. Frog embryos are well suited for this type of assay, since ectopic expression is simple to achieve by microinjection of synthetic RNA into the fertilized egg, and the activity of endogenous genes or introduced transgenes can then be monitored in subsequent embryonic tissue. By using animal pole explants from blastula-stage embryos, the assay can be restricted to cells that would normally contribute to neural or epidermal tissue. Furthermore, by following the time course of cardiac marker gene induction and the range of markers that are activated, it is possible to distinguish between apparently direct transcriptional activation of particular gene targets and the induction of novel cardiac tissue in the explants.

We have used this approach to test several of the transcription factor families that have been implicated as regulators of cardiac muscle-specific transcription, including members of the NKX2 family, MADS family proteins (SRF and MEF2 factors), and GATA transcription factors. Initial experiments demonstrated that combined ectopic expression of these factors triggered expression of the endogenous MLC2a gene in blastula animal pole explants that were cultured until tadpole stages (Fig. 4). Subsequent studies confirmed that additional myocardial-specific markers (MHCα and cardiac troponin I) were also activated and that explants showed spontaneous beating, consistent with the formation of myocardial muscle tissue. Remarkably, this de novo induction of

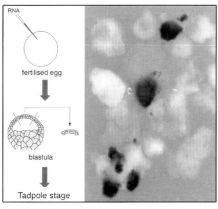

Figure 4. Ectopic expression of GATA4 in explants from the animal pole of early embryos can trigger cardiac muscle differentiation. GATA4 RNA is injected into fertilized eggs, then explants are isolated at the blastula stage and cultured until sibling embryos reach swimming tadpole stages. Myocardial muscle differentiation can be detected by whole-mount in situ hybridization for the cardiac-specific marker, MLC2a.

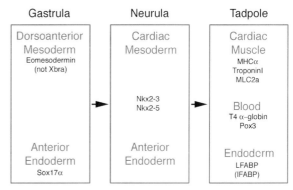

Figure 5. Model for the induction of cardiac tissue in animal pole explants by ectopic expression of GATA factors. GATA4 and GATA5 appear to trigger differentiation of both dorsoanterior mesoderm and anterior endoderm in the equivalent of gastrula stage. This results in the formation of cardiac mesoderm (as well as other dorsoanterior mesodermal derivatives). Explants cultured until sibling embryos reach tadpole stages contain cardiac muscle, blood, and anterior endodermal derivatives, as judged by expression of cell-type-specific molecular markers.

cardiac tissue from explant tissue that would otherwise form atypical epidermis can be achieved using ectopic expression of only GATA4 or GATA5.

Using a range of other markers for both early and late embryonic tissue types, we find that the GATA factors initially induce both mesodermal and endodermal cell types. Interestingly, this proves to be a far more restricted set than those induced in similar explants by members of the TGFβ growth factor family (Kimelman and Griffin 2000). The GATA factors do not, for example, induce any skeletal muscle tissue, as judged by assays for the myogenic regulator, MyoD, or the contractile protein myosin light chain1. The early mesoderm marker eomesodermin (Ryan et al. 1996) is detectable in GATA-expressing explants at early stages, whereas the related posterior mesoderm marker, brachyury (Smith et al. 1991), is not. Induction of endodermal tissue is suggested by the expression of Sox17α from early stages onward. In older explants, we detect expression of anterior endodermal markers such as liver fatty acid binding protein (LFABP) and sometimes more posterior markers such as its intestinal counterpart (IFABP). Together, these results suggest that ectopic expression of GATA4 or GATA5 induces a restricted set of anterior mesendodermal tissues, a later consequence of which is the differentiation of functional cardiac muscle tissue (Fig. 5).

That GATA4 and GATA5 can have such a profound effect on the fate of early embryonic blastomeres is a remarkable and surprising result. Expression in newly formed anterior endoderm of the frog embryo has been reported (Weber et al. 2000), but there is little evidence to date for their role in normal mesoderm formation. However, it is interesting to note that a similarly profound effect of ectopic GATA5 expression has been noted in the zebrafish embryo (Reiter et al. 1999), again apparently resulting in the formation of ectopic cardiac muscle tissue. Together, these observations raise the possibility of a hitherto unidentified role of GATA factors in the earliest events of embryo cell fate determination. If there proves

to be no such role, then ectopic expression of GATA4 or GATA5 must, in some manner, trigger pathways leading to mesendoderm formation, and this may prove a useful tool for investigating their control. One intriguing observation, for example, is that cardiac tissue induction can be triggered remarkably late in development. Using ectopic expression of an inducible GATA4-GR construct, we have found that activation of the fusion protein in explants several hours after the onset of explant culture still results in the formation of cardiac muscle foci. These can be readily seen using eggs from transgenic reporter lines that provide fluorescent markers for cardiac muscle formation (Fig. 6). Indeed, even if induction of GATA4-GR is delayed until uninduced explants have undergone extensive differentiation into ectodermal tissue derivatives, cardiac differentiation is still subsequently detected. Under these conditions, induction of cardiac tissue results either from the trans-differentiation of such ectodermal cells or from the presence of a limited population within the explant that remains pluripotent and therefore susceptible to the effects of ectopic GATA factor expression.

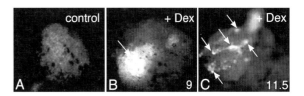

Figure 6. Using RNA encoding a synthetic GATA4-GR fusion protein, it is possible to induce cardiac differentiation surprisingly late in development. Explants from the cardiac actin.GFP transgenic frog line show GFP fluorescence if muscle differentiation occurs. Uninduced explants show no fluorescence at tadpole stages (*A*); induction of GATA4-GR activity by dexamethasone treatment immediately after explant isolation (stage 9 blastula) results in high levels of fluorescence at the tadpole stage (*B*); induction several hours later (stage 11.5 late gastrula) results in punctate fluorescence at the tadpole stage (*C, arrows*).

For studies of cardiac tissue formation, the ability to induce cardiac muscle formation from naïve embryonic tissue is potentially very useful. In normal embryogenesis, the inductive tissue interactions thought to result in cardiac mesoderm specification are intimately linked with the cell movements associated with gastrulation and are therefore inherently difficult to study. By facilitating cardiac tissue induction from explant tissue, the complications of gastrulation movements are eliminated and the role played by particular tissue interactions can be examined more easily.

CONCLUSIONS

Transgenesis in *Xenopus* provides a rapid and simple method that can aid the study of amphibian cardiogenesis. The use of transgenic reporter lines can greatly assist in tracking differentiation of cardiac tissue, and with cardiac tissue-restricted promoters, it is possible to target any transgene with precision. Sufficient numbers of F_0 transgenic embryos can be readily generated to allow screening of promoter mutations. We have begun to use this approach to delineate the mechanisms responsible for expression of individual genes within the differentiating tadpole heart. Our results reveal that distinct regulatory controls not only account for expression of the same gene (cardiac actin) within different striated muscle types. In addition, apparently identical expression profiles (shared by cardiac actin and MLC2a genes) within differentiating cardiac muscle are achieved by equally diverse mechanisms.

REFERENCES

Amaya E. and Kroll K.L. 1999. A method for generating transgenic frog embryos. *Methods Mol. Biol.* **97:** 393.

Bao Z.Z., Bruneau B.G., Seidman J.G., Seidman C.E., and Cepko C.L. 1999. Regulation of chamber-specific gene expression in the developing heart by Irx4. *Science* **283:** 1161.

Belaguli N.S., Sepulveda J.L., Nigam V., Charron F., Nemer M., and Schwartz R.J. 2000. Cardiac tissue enriched factors serum response factor and GATA-4 are mutual coregulators. *Mol. Cell. Biol.* **20:** 7550.

Breckenridge R.A., Mohun T.J., and Amaya E. 2001. A role for BMP signalling in heart looping morphogenesis in *Xenopus*. *Dev. Biol.* **232:** 191.

Chambers A.E., Logan M., Kotecha S., Towers N., Sparrow D., and Mohun T.J. 1994. The RSRF/MEF2 protein SL1 regulates cardiac muscle-specific transcription of a myosin light-chain gene in *Xenopus* embryos. *Genes Dev.* **8:** 1324.

Chen C.Y. and Schwartz R.J. 1996. Recruitment of the tinman homolog Nkx-2.5 by serum response factor activates cardiac alpha-actin gene transcription. *Mol. Cell. Biol.* **16:** 6372.

Christoffels V.M., Habets P.E., Franco D., Campione M., de Jong F., Lamers W.H., Bao Z.Z., Palmer S., Biben C., Harvey R.P., and Moorman A.F. 2000. Chamber formation and morphogenesis in the developing mammalian heart. *Dev. Biol.* **223:** 266.

Garriock R.J., Vokes S.A., Small E.M., Larson R., and Krieg P.A. 2001. Developmental expression of the *Xenopus* Iroquois-family homeobox genes, Irx4 and Irx5. *Dev. Genes Evol.* **211:** 257.

Gove C., Walmsley M., Nijjar S., Bertwistle D., Guille M., Partington G., Bomford A., and Patient R. 1997. Over-expression of GATA-6 in *Xenopus* embryos blocks differentiation of heart precursors. *EMBO J.* **16:** 355.

Howell M., Mohun T.J., and Hill C.S. 2001. *Xenopus* Smad3 is specifically expressed in the chordoneural hinge, notochord and in the endocardium of the developing heart. *Mech. Dev.* **104:** 147.

Ivics Z., Hackett P.B., Plasterk R.H., and Izsvak Z. 1997. Molecular reconstruction of Sleeping Beauty, a Tc1-like transposon from fish, and its transposition in human cells. *Cell* **91:** 501.

Jiang Y. and Evans T. 1996. The *Xenopus* GATA-4/5/6 genes are associated with cardiac specification and can regulate cardiac-specific transcription during embryogenesis. *Dev. Biol.* **174:** 258.

Kimelman D. and Griffin K.J. 2000. Vertebrate mesendoderm induction and patterning. *Curr. Opin. Genet. Dev.* **10:** 350.

Kroll K.L. and Amaya E. 1996. Transgenic *Xenopus* embryos from sperm nuclear transplantations reveal FGF signaling requirements during gastrulation. *Development* **122:** 3173.

Latinkic B.V., Cooper B., Towers N., Sparrow D., Kotecha S., and Mohun T.J. 2002. Distinct enhancers regulate skeletal and cardiac muscle-specific expression programs of the cardiac alpha-actin gene in *Xenopus* embryos. *Dev. Biol.* **245:** 57.

Lawson N.D. and Weinstein B.M. 2002. In vivo imaging of embryonic vascular development using transgenic zebrafish. *Dev. Biol.* **248:** 307.

Mohun T.J., Taylor M.V., Garrett N., and Gurdon J.B. 1989. The CArG promoter sequence is necessary for muscle-specific transcription of the cardiac actin gene in *Xenopus* embryos. *EMBO J.* **8:** 1153.

Morin S., Paradis P., Aries A., and Nemer M. 2001. Serum response factor-GATA ternary complex required for nuclear signaling by a G-protein-coupled receptor. *Mol. Cell. Biol.* **21:** 1036.

Polli M. and Amaya E. 2002. A study of mesoderm patterning through the analysis of the regulation of Xmyf-5 expression. *Development* **129:** 2917.

Reiter J.F., Alexander J., Rodaway A., Yelon D., Patient R., Holder N., and Stainier D.Y. 1999. Gata5 is required for the development of the heart and endoderm in zebrafish. *Genes Dev.* **13:** 2983.

Ryan K., Garrett N., Mitchell A., and Gurdon J.B. 1996. Eomesodermin, a key early gene in *Xenopus* mesoderm differentiation. *Cell* **87:** 989.

Small E.M. and Krieg P.A. 2000. Expression of atrial natriuretic factor (ANF) during *Xenopus* cardiac development. *Dev. Genes Evol.* **210:** 638.

Smith J.C., Price B.M., Green J.B., Weigel D., and Herrmann B.G. 1991. Expression of a *Xenopus* homolog of Brachyury (T) is an immediate-early response to mesoderm induction. *Cell* **67:** 79.

Sparrow D.B., Latinkic B., and Mohun T.J. 2000a. A simplified method of generating transgenic *Xenopus*. *Nucleic Acids Res.* **28:** E12.

Sparrow D.B., Cai C., Kotecha S., Latinkic B., Cooper B., Towers N., Evans S.M., and Mohun T.J. 2000b. Regulation of the tinman homologues in *Xenopus* embryos. *Dev. Biol.* **227:** 65.

Thomas M.J. and Seto E. 1999. Unlocking the mechanisms of transcription factor YY1: Are chromatin modifying enzymes the key? *Gene* **236:** 197.

Wang D., Chang P.S., Wang Z., Sutherland L., Richardson J.A., Small E., Krieg P.A., and Olson E.N. 2001. Activation of cardiac gene expression by myocardin, a transcriptional cofactor for serum response factor. *Cell* **105:** 851.

Weber H., Symes C.E., Walmsley M.E., Rodaway A.R., and Patient R.K. 2000. A role for GATA5 in *Xenopus* endoderm specification. *Development* **127:** 4345.

Werdien D., Peiler G., and Ryffel G.U. 2001. FLP and Cre recombinase function in *Xenopus* embryos. *Nucleic Acids Res.* **29:** E53.

Genetic Regulation of Cardiac Patterning in Zebrafish

D. YELON, J.L. FELDMAN, AND B.R. KEEGAN

Developmental Genetics Program and Department of Cell Biology, Skirball Institute of Biomolecular Medicine, New York University School of Medicine, New York, New York 10016

The embryonic vertebrate heart is composed of two major chambers: a ventricle and an atrium. To achieve this form, it is first necessary to select specific numbers of ventricular and atrial precursors from among a field of cells possessing cardiac potential. The molecular mechanisms that establish and refine cardiac potential in the early embryo remain poorly understood. The recent emergence of the zebrafish as a model organism provides excellent opportunities for classic genetic analyses of cardiac patterning. Through studies of relevant zebrafish mutations, it is possible to identify critical genes and their precise roles in regulating cardiac cell fate decisions. Here, we review recent insights regarding cardiac patterning in zebrafish.

DEVELOPMENTAL POTENTIAL OF ORGAN FIELDS

The heart, like other organs, arises from an organ field—a region of the embryo with the developmental potential to form a specific structure. Classic embryological manipulations have demonstrated that organ fields are typically larger than the number of cells that are actually fated to contribute to the organ (Huxley and deBeer 1934; Jacobson and Sater 1988; Fishman and Chien 1997). However, organ fields are thought to be highly dynamic: At first, a field may be relatively large, but as development proceeds, the dimensions of the field will gradually be refined. Interestingly, some organ fields, including the heart field, appear to be regulative, such that if definitive precursors are damaged or removed at early stages, cells from other regions of the field can compensate for the loss. These traits of organ fields are of great interest, but the molecular mechanisms that establish and refine developmental potential within a field are not well understood.

The mechanisms responsible for patterning the heart field are particularly intriguing. In vertebrates, cardiac precursors are found within the anterior lateral plate mesoderm (ALPM) (Fishman and Chien 1997; Harvey 2002), such that there are two primary heart fields, one on either side of the embryo (Fig. 1). Myocardial precursors arise from within these fields, and these precursors are further divided into ventricular and atrial subsets, each with distinct chamber-specific gene expression programs. All myocardial precursors migrate toward the embryonic midline, where they reorganize to form an embryonic heart tube with ventricular cells at its anterior end and atrial cells at its posterior end. Thus, formation of a properly patterned heart tube depends on the initial patterning decisions made within the heart fields.

An appealing model for cardiac patterning divides the process into three general phases (Yelon 2001). First, a broad region of cardiac potential is established within the ALPM. Next, a combination of inductive and repressive influences refines the boundaries of cardiac potential, resulting in the selection of the definitive myocardial precursors. Finally, additional fate specification signals define separate populations of ventricular and atrial precursors. Although this model is oversimplified, it is nevertheless valuable to search for factors that could regulate each of these three proposed steps. One goal of our laboratory's research is to identify genes that are critical for establishing and refining the heart field, using the zebrafish as a model organism.

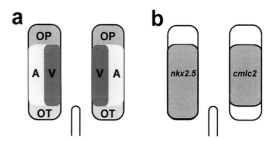

Figure 1. Zebrafish heart fields. (*a*) Arrangement of cell fate domains within the zebrafish heart fields at mid-somitogenesis stages (Goldstein and Fishman 1998; Serbedzija et al. 1998; Yelon et al. 1999). Dorsal view, anterior to the top. Heart fields are depicted as two rectangles, one on either side of the embryo. Notochord is depicted as a white rod, with its rostral tip adjacent to the posterior portions of the heart fields. The approximate boundaries of adjacent fate domains are indicated, including domains for optic mesenchyme (OP, *green*, anterior), otic mesenchyme (OT, *blue*, posterior), ventricular myocardium (V, *red*), and atrial myocardium (A, *yellow*). (*b*) Comparison of gene expression patterns (*purple*) of *nkx2.5* and *cmlc2* within the zebrafish heart fields at mid-somitogenesis stages (Yelon et al. 1999). Both genes are normally expressed bilaterally; for the purposes of comparison, *nkx2.5* is shown on the left and *cmlc2* is shown on the right. *cmlc2* expression is only found in the *nkx2.5*-expressing cells that are anterior to the rostral tip of the notochord.

THE ZEBRAFISH AS A MODEL ORGANISM FOR ANALYSIS OF CARDIAC PATTERNING

Many features of the zebrafish make it a desirable model organism for the analysis of cardiac patterning. First, the externally fertilized and transparent zebrafish embryo allows extensive inspection and manipulation of cardiac cells throughout their development (Stainier and Fishman 1994). Not only is the contracting heart tube easy to view, but it is also extremely convenient to use Nomarski optics and/or molecular markers to achieve single-cell resolution of the myocardial precursors within the bilateral heart fields. Furthermore, the transparency and manipulability of the zebrafish embryo make possible a variety of fate mapping, transplantation, and ablation techniques that are useful for the analysis of pattern formation (Westerfield 1995). Finally, the zebrafish presents excellent opportunities for large-scale genetic screens, primarily due to its small size, fecundity, and brief generation time (Westerfield 1995). A classic genetic approach, in which mutagenized organisms are screened for phenotypes relevant to cardiac patterning, is likely to reveal novel players, thereby complementing and extending the observations made with targeted mouse mutations in known genes.

Several groups have conducted genetic screens focused on the identification of zebrafish mutations that disrupt cardiogenesis. Some of these screens have focused on finding mutations that cause visible defects in cardiac morphology (Chen et al. 1996; Stainier et al. 1996), other screens have concentrated on mutations that alter cardiac function (Warren et al. 2000), and still other screens have employed cardiac molecular markers for the sensitive detection of abnormalities in gene expression (Chen et al. 1997; Alexander et al. 1998). Depending on the screening criteria and on the developmental stages examined, a variety of mutant phenotypes can be detected, ranging from reduction of precardiac mesoderm at early stages to abnormal contractile rhythms at later stages. Our laboratory has employed both morphological and molecular criteria to identify a panel of more than 40 mutations that affect myocardial specification, differentiation, and morphogenesis (Alexander et al. 1998; D. Yelon, unpubl.). A number of these mutations disrupt early steps in cardiac patterning and are therefore relevant to the analysis of the zebrafish heart field (Table 1).

THE ZEBRAFISH HEART FIELD: LIVING UP TO ITS POTENTIAL

To analyze the regulation of cardiac patterning in zebrafish, it is first necessary to define the dimensions of the zebrafish heart fields in wild-type embryos. Fate mapping experiments at mid-somitogenesis stages have indicated that the myocardial precursors are located within the ALPM, just anterior to the level of the rostral tip of the notochord (Fig. 1) (Serbedzija et al. 1998). Furthermore, gene expression patterns suggest that the ventricular precursors are located more medially than the atrial precursors (Fig. 1) (Yelon et al. 1999). The immediate neighbors of the myocardial precursors appear fated to become head mesenchyme: The region anterior to the myocardial precursors contributes to optic mesenchyme, and the region posterior to the myocardial precursors contributes to otic mesenchyme (Fig. 1) (Serbedzija et al. 1998). Interestingly, it seems that both of these flanking areas fall within the heart fields: Cells from these regions can form myocardium under specific experimental conditions. For example, following notochord ablation, cells normally fated to become otic mesenchyme can contribute to the heart (Goldstein and Fishman 1998). In addition, ablation of definitive myocardial precursors can induce cells from the optic mesenchyme domain to become myocardial (Serbedzija et al. 1998). These observations suggest that the zebrafish heart fields extend over a broad area of the ALPM, and only limited portions of these fields normally realize their cardiac potential.

The arrangement of fates within the zebrafish heart fields raises a number of interesting questions about the molecular mechanisms that allocate developmental potential of the ALPM. For instance, which genetic pathways establish myocardial potential? Additionally, how are myocardial fates promoted in the centers of the fields

Table 1. Zebrafish Mutations That Disrupt Cardiac Patterning

Mutation	Gene	Type of gene product	Cardiac patterning phenotype	Reference
faust (*fau*)	*gata5*	GATA transcription factor	reduced *nkx2.5* and myocardial/ventricular differentiation	Reiter et al. (1999)
acerebellar (*ace*)	*fgf8*	growth factor	reduced *nkx2.5* and myocardial/ventricular differentiation	Reifers et al. (2000)
one-eyed pinhead (*oep*)	*oep*	Nodal signaling cofactor	reduced *nkx2.5* and myocardial/ventricular differentiation	Reiter et al. (2001)
swirl (*swr*)	*bmp2b*	growth factor	reduced *nkx2.5* and myocardial differentiation	Reiter et al. (2001)
hands off (*han*)	*hand2*	bHLH transcription factor	reduced myocardial/ventricular differentiation	Yelon et al. (2000)
pandora (*pan*)	*spt6*	transcription elongation factor	reduced myocardial/ventricular differentiation	Keegan et al. (2002)
foggy (*fog*)	*spt5*	transcription elongation factor	reduced myocardial differentiation	Keegan et al. (2002)

and repressed on the anterior and posterior ends? Furthermore, are there signals that differentially affect the medial and lateral portions of the fields, thereby establishing ventricular and atrial identities? The answers remain largely mysterious. For example, although experimental evidence suggests that the notochord somehow represses the assumption of myocardial fates (Goldstein and Fishman 1998), the molecular nature of the proposed negative signal produced by the notochord is not yet clear. Additionally, it is not yet known how the anterior portion of the heart field can sense injury to the definitive myocardial precursors (Serbedzija et al. 1998). However, analyses of zebrafish mutations that affect the number, position, and organization of myocardial precursors within the heart fields are beginning to clarify some of the critical steps in cardiac patterning. Recent studies have focused on seven of these mutations, each with its own essential role in promoting the formation of cardiac precursors (Table 1).

ESTABLISHING THE ZEBRAFISH HEART FIELDS

Mutations that affect cardiac patterning can generally be divided into two classes: those that affect the initial designation of the heart fields and those that affect the refinement of these fields. Several zebrafish mutations (*faust* [*fau*], *acerebellar* [*ace*], *one-eyed pinhead* [*oep*], and *swirl* [*swr*]) appear to belong to the former class (Table 1) (Reiter et al. 1999, 2001; Reifers et al. 2000). In each of these cases, mutant embryos exhibit reduced expression of the homeodomain transcription factor gene *nkx2.5* at early stages. Although *nkx2.5* does not precisely mark the zebrafish heart fields, defects in its expression surely represent defects within the field, since it is expressed in both the myocardial precursor domain and the otic mesenchyme domain (Fig. 1) (Goldstein and Fishman 1998; Serbedzija et al. 1998). Identification of the genes disrupted by the *fau*, *ace*, *oep*, and *swr* mutations and analysis of their relationships have suggested a hypothetical pathway for the initial establishment of myocardial potential in zebrafish (Yelon 2001). Specifically, Nodal signaling, mediated by the Nodal cofactor Oep (Gritsman et al. 1999), is essential to establish mesendodermal expression of *fau/gata5* during gastrulation stages (Reiter et al. 2001). Additionally, during the same stages, Swr/Bmp2b is critical for the assignment of ventrolateral, including cardiac, fates (Kishimoto et al. 1997; Nguyen et al. 1998); *swr/bmp2b* mutants also exhibit reduced *fau/gata5* expression (Reiter et al. 2001). Fau/Gata5 function is then required in the ALPM to promote *nkx2.5* expression; in fact, overexpression of *fau/gata5* is sufficient to induce *nkx2.5* expression in ectopic locations (Reiter et al. 1999). Like Fau/Gata5, the growth factor Ace/Fgf8 is also necessary for the normal expression of *nkx2.5* (Reifers et al. 2000). *ace/fgf8* expression can be detected in and near the ALPM both during and after gastrulation (Reifers et al. 2000), suggesting that Fgf8 signaling may cooperate with Fau/Gata5 activity to define regions of cardiac potential.

REFINING THE ZEBRAFISH HEART FIELDS

After the heart fields are established, the selection of the definitive myocardial precursors is clearly indicated by their expression of myocardial differentiation markers, like *cmlc2* (Fig. 1) (Yelon et al. 1999). Several zebrafish mutations interfere with this selection process (Table 1). As predicted by their early heart field defects,

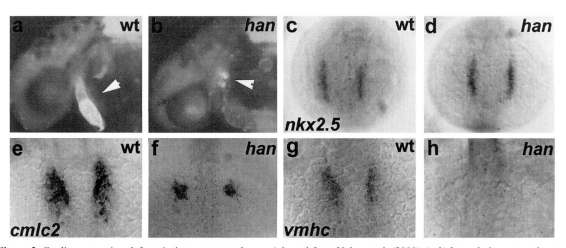

Figure 2. Cardiac patterning defects in *han* mutant embryos. Adapted from Yelon et al. (2000). (*a,b*) Lateral views, anterior to the left, at 36 hours postfertilization. MF20/S46 immunofluorescence (Alexander et al. 1998) labels ventricular tissue (*red*) and atrial tissue (*yellow*) in the wild-type (wt) heart (*a*, *arrowhead*). Only small groups of atrial cells (*b*, *yellow*, *arrowhead*) are evident in *han* mutant embryos. (*c,d*) Dorsal views, anterior to the top, of in situ hybridizations at the 10-somite stage. *nkx2.5* expression is normal in *han* mutant embryos. (*e–h*) Dorsal views, anterior to the top, of in situ hybridizations at the 15-somite stage. *han* mutant embryos have very few *cmlc2*-expressing cells and no cells expressing the ventricular marker *vmhc* (Yelon et al. 1999).

fau, *ace*, *oep*, and *swr* mutants all exhibit severe reductions in the expression of *cmlc2* and other differentiation markers (Reiter et al. 1999, 2001; D. Yelon, unpubl.). These deficiencies may be secondary to a reduction in Nkx2.5 activity and/or may reflect additional roles of each mutated gene. The effects of *nkx2.5* loss of function have not yet been studied in zebrafish, but *nkx2.5* overexpression can induce the ectopic expression of some myocardial markers (Chen and Fishman 1996). However, it is clear that *nkx2.5* expression is not the sole requirement for the selection of myocardial precursors, since the most posterior *nkx2.5*-expressing cells normally become otic mesenchyme rather than myocardium (Fig. 1) (Goldstein and Fishman 1998; Serbedzija et al. 1998). Furthermore, some zebrafish mutations affect *cmlc2* expression without disrupting *nkx2.5* expression, suggesting additional genetic regulation of myocardial specification downstream from or parallel to *nkx2.5*. For example, we have shown that mutation of the *hands off/hand2* (*han*) locus results in a dramatic myocardial deficiency without affecting the initial expression of *nkx2.5* (Fig. 2) (Yelon et al. 2000). Therefore, the transcription factor Hand2 is dispensable for *nkx2.5* induction, but essential for myocardial differentiation. It is interesting to note that *han*, *fau*, *oep*, and *ace* mutants all exhibit particularly striking ventricular deficiencies, with atrial tissue less dramatically affected; evidently, each of these factors is especially important for promoting the selection of ventricular precursors within the heart field (Reiter et al. 1999, 2001; Reifers et al. 2000; Yelon et al. 2000). *hand2*, like *gata5*, is expressed throughout the ALPM (Yelon et al. 2000), and so it is not yet clear how its activity regulates the selection of myocardial precursors in general and ventricular precursors in particular. It is therefore important to identify other factors with roles similar to that of Hand2 in order to clarify the process by which competence for myocardial differentiation is refined downstream from heart field induction.

pandora AND *foggy*: THE IMPORTANCE OF TRANSCRIPTIONAL EFFICIENCY DURING MYOCARDIAL DIFFERENTIATION

In search of additional players that influence the selection of myocardial precursors, we identified two zebrafish mutations that cause cardiac phenotypes reminiscent of the *han* mutant phenotype. Like *han*, both the *pandora* (*pan*) and *foggy* (*fog*) mutations severely inhibit myocardial differentiation, but neither mutation significantly affects *nkx2.5* induction (Fig. 3) (Keegan et al. 2002). *pan* mutants are ultimately able to generate a small amount of myocardial tissue, which is primarily atrial (Yelon et al. 1999), again reminiscent of *han* mutants (Fig. 2) (Yelon et al. 2000). *pan* and *fog* mutants share a number of additional, noncardiac phenotypes, including defects in pigmentation, tail length, and ear formation (Fig. 4) (Keegan et al. 2002). The overall similarity of the *pan* and *fog* mutant phenotypes suggested that the *pan* and *fog* gene products might play comparable roles in a number of embryonic tissues.

Intrigued by the similarities between *pan* and *fog*, we proceeded to clone both mutated genes. We found that *pan* is tightly linked to a zebrafish *spt6* gene and that *fog* is tightly linked to a zebrafish *spt5* gene; *spt6* and *spt5* had previously been proposed to perform related roles in yeast (see below). Molecular analysis of one *pan* allele and two *fog* alleles demonstrated that *pan^{m313}* and *fog^{sk8}* are point mutations that cause missplicing of *spt6* and *spt5* mRNA, respectively, resulting in severely truncated Spt6 and Spt5 proteins, and that *fog^{s30}* is a deletion that eliminates the entire *spt5* gene (Keegan et al. 2002). Fur-

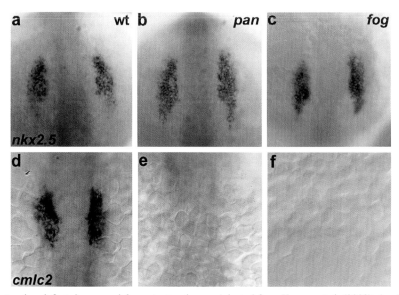

Figure 3. Cardiac patterning defects in *pan* and *fog* mutant embryos. Adapted from Keegan et al. (2002). (*a–c*) Dorsal views, anterior to the top, of in situ hybridizations at the 8-somite stage. *nkx2.5* expression is normal in *pan* and *fog* mutant embryos. (*d–f*) Dorsal views, anterior to the top, of in situ hybridizations at the 15-somite stage. *pan* and *fog* mutant embryos do not have any *cmlc2*-expressing cells at this stage.

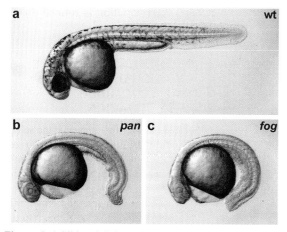

Figure 4. Additional defects in *pan* and *fog* mutant embryos. Adapted from Keegan et al. (2002). (*a–c*) Lateral views, anterior to the left, of live embryos at 36 hours postfertilization. Compared to wild-type (wt) embryos, *pan* and *fog* mutant embryos exhibit reduced pigmentation, short tails, small ears, and pericardial edema.

thermore, injection of wild-type *spt6* or *spt5* synthetic mRNA into *pan* or *fog* mutant embryos can rescue the respective mutant phenotypes, confirming the identities of *pan* and *fog* (Keegan et al. 2002).

Why might mutation of *spt5* and *spt6* cause similar embryonic phenotypes? Genetic studies in yeast have previously suggested that both Spt5 and Spt6 regulate transcription elongation (Winston et al. 1984; Swanson and Winston 1992; Bortvin and Winston 1996; Hartzog et al. 1998), which can be a rate-limiting step in transcript synthesis (Lis 1998). For example, heat shock response genes, such as *hsp70*, are constitutively occupied by a RNA polymerase II (Pol II) complex that is paused adjacent to the promoter after transcription initiation (Rougvie and Lis 1988; Rasmussen and Lis 1993). Transcription elongation is inhibited until heat shock stimulation occurs, at which time the paused Pol II becomes hyperphosphorylated and transcript synthesis proceeds. Biochemical studies have demonstrated that Spt5, together with another factor (Spt4), forms a physical complex known as DSIF (DRB sensitivity inducing factor) that directly interacts with Pol II (Hartzog et al. 1998; Wada et al. 1998). The DSIF complex has been implicated in both stimulation and repression of transcription elongation in vitro (Wada et al. 1998). In fact, elimination of only the repressive activity of Spt5 in zebrafish results in a reduction of dopaminergic neurons and an excess of serotonergic neurons in the embryonic brain (Guo et al. 2000). Like Spt4 and Spt5, Spt6 may stimulate transcription elongation: in *Drosophila*, Spt6 colocalizes with Spt5 at sites of active transcription and is recruited to heat shock loci upon heat shock stimulation (Andrulis et al. 2000; Kaplan et al. 2000). However, the biochemical properties of Spt6 are distinct from those of Spt4 and Spt5. Spt6 is not tightly bound to the DSIF complex (Hartzog et al. 1998), and, unlike Spt4 and Spt5, Spt6 binds to histones, particularly histone H3 (Bortvin and Winston 1996), suggesting that Spt6 could regulate transcription by modulating chromatin assembly. Altogether, Spt5 and Spt6 are likely to each contribute independently to the regulation of transcription elongation in vivo.

On the basis of the previous studies of Spt5 and Spt6 in other systems, we suspected that Spt5 and Spt6 could also function as transcription elongation factors in the zebrafish embryo. To examine the efficiency of transcription elongation in *pan* and *fog* mutants, we devised a real-time RT-PCR assay to quantify *hsp70* transcript synthesis, since the *hsp70* locus is known to be regulated at the level of elongation (Andrulis et al. 2000; Kaplan et al. 2000). Comparison of the rates of *hsp70* induction in wild-type and mutant embryos demonstrated that Spt5 and Spt6 are each independently required to promote efficient kinetics of embryonic transcript production (Fig. 5) (Keegan et al. 2002). Extrapolating from our analysis of *hsp70* transcripts, we propose that Spt6 and Spt5 promote transcription elongation of many loci during development, including a number of genes that are essential during myocardial differentiation. However, it is still not clear whether Spt5 and Spt6 are required for the majority of embryonic transcripts or simply for a select subset of critical genes. Resolution of these issues will require high-throughput analysis of the transcripts affected in *fog* and *pan* mutant embryos.

It is important to note that *spt5* and *spt6* are both maternally and zygotically supplied (Keegan et al. 2002). Therefore, zygotic *pan* and *fog* mutants contain some normal Spt5 and Spt6 at early stages. This calls into question our initial interpretation that *pan* and *fog* are not required to establish the heart fields (Fig. 3). Instead, maternal supplies may compensate for the absence of zygotic *spt5* or *spt6* during early phases of cardiac patterning. Transcriptional inefficiency would then become apparent only after maternal stores are depleted, perhaps around the time that myocardial precursors are selected within the heart fields. However, even if Spt5 and Spt6 are required at multiple stages during cardiac patterning, the *pan* and *fog* zygotic mutants still provide valuable tools for future studies of the transcriptional program associated with myocardial differentiation.

CONCLUSIONS

It is clear that genetic analysis of cardiac patterning in the zebrafish embryo can reveal a variety of relevant factors—transcription factors and signaling molecules, as well as relatively specific factors and more broadly employed factors. Future studies of additional mutations that affect cardiac patterning will surely add to the arsenal of genes known to affect the establishment and refinement of the heart fields. It will be especially exciting to complement our understanding of the factors that promote myocardial differentiation by studying the other side of the coin—the factors that inhibit myocardial differentiation on the edges of the heart fields. In this regard, important progress will surely come from the analysis of newly identified mutations that result in the formation of an excessive number of myocardial precursors (J.L. Feldman et al., unpubl.). By coupling these studies with lin-

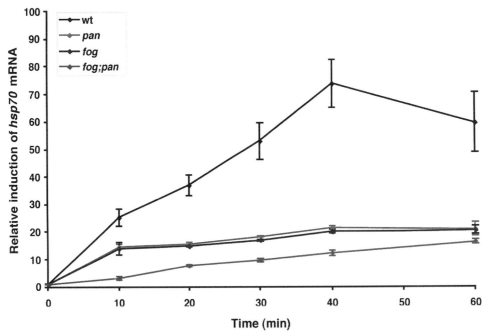

Figure 5. Induction of endogenous *hsp70* expression during an hour-long heat shock, detected by real-time RT-PCR. Adapted from Keegan et al. (2002). Wild-type (wt) embryos (*black*) accumulate significantly higher levels of *hsp70* more rapidly than *pan* mutant embryos (*green*) or *fog* mutant embryos (*blue*). *hsp70* induction is further inhibited in *fog;pan* double mutant embryos (*red*). Each data point represents the average degree of induction of *hsp70* expression at a particular time point, relative to the low, but detectable, levels of *hsp70* expression at time zero. Each experiment was performed in triplicate; standard deviation from the mean is indicated by error bars. To account for small variances in RNA extraction and cDNA synthesis, levels of *hsp70* expression were normalized relative to levels of stable β-*actin* expression.

eage tracing in wild-type and mutant embryos (B.R. Keegan and D. Yelon, unpubl.), we will be able to determine how repressive forces influence cell fate decisions in myocardial and head mesenchyme lineages. Undoubtedly, future studies of cardiac patterning in zebrafish will illuminate the balance between positive and negative influences that results in the selection of the proper numbers of ventricular and atrial myocardial precursors within the heart field.

ACKNOWLEDGMENTS

We are grateful to N. Glickman, E. Berdougo, J. Schoenebeck, and A. Schier for many helpful discussions. Research in the Yelon laboratory is supported by the Burroughs Wellcome Fund, the New York City Council Speaker's Fund, the American Heart Association, the National Institutes of Health, and a Whitehead Fellowship.

REFERENCES

Alexander J., Stainier D.Y., and Yelon D. 1998. Screening mosaic F1 females for mutations affecting zebrafish heart induction and patterning. *Dev. Genet.* **22**: 288.

Andrulis E.D., Guzman E., Doring P., Werner J., and Lis J.T. 2000. High-resolution localization of Drosophila Spt5 and Spt6 at heat shock genes in vivo: Roles in promoter proximal pausing and transcription elongation. *Genes Dev.* **14**: 2635.

Bortvin A. and Winston F. 1996. Evidence that Spt6p controls chromatin structure by a direct interaction with histones. *Science* **272**: 1473.

Chen J.-N. and Fishman M.C. 1996. Zebrafish *tinman* homolog demarcates the heart field and initiates myocardial differentiation. *Development* **122**: 3809.

Chen J.-N., Haffter P., Odenthal J., Vogelsang E., Brand M., van Eeden F.J., Furutani-Seiki M., Granato M., Hammerschmidt M., Heisenberg C.P., Jiang Y.-J., Kane D.A., Kelsh R.N., Mullins M.C., and Nüsslein-Volhard C. 1996. Mutations affecting the cardiovascular system and other internal organs in zebrafish. *Development* **123**: 293.

Chen J.N., van Eeden F.J., Warren K.S., Chin A., Nusslein-Volhard C., Haffter P., and Fishman M.C. 1997. Left-right pattern of cardiac BMP4 may drive asymmetry of the heart in zebrafish. *Development* **124**: 4373.

Fishman M.C. and Chien K.R. 1997. Fashioning the vertebrate heart: Earliest embryonic decisions. *Development* **124**: 2099.

Goldstein A.M. and Fishman M.C. 1998. Notochord regulates cardiac lineage in zebrafish embryos. *Dev. Biol.* **201**: 247.

Gritsman K., Zhang J., Cheng S., Heckscher E., Talbot W.S., and Schier A.F. 1999. The EGF-CFC protein one-eyed pinhead is essential for nodal signaling. *Cell* **97**: 121.

Guo S., Yamaguchi Y., Schilbach S., Wada T., Lee J., Goddard A., French D., Handa H., and Rosenthal A. 2000. A regulator of transcriptional elongation controls vertebrate neuronal development. *Nature* **408**: 366.

Hartzog G.A., Wada T., Handa H., and Winston F. 1998. Evidence that Spt4, Spt5, and Spt6 control transcription elongation by RNA polymerase II in Saccharomyces cerevisiae. *Genes Dev.* **12**: 357.

Harvey R.P. 2002. Patterning the vertebrate heart. *Nat. Rev. Genet.* **3**: 544.

Huxley J.S. and deBeer G.R. 1934. The mosaic style of differentiation. In: *The Elements of Experimental Embryology* (ed. J. Barcroft and J.T. Saunders), pp. 195-270. Cambridge University Press, London.

Jacobson A.G. and Sater A.K. 1988. Features of embryonic induction. *Development* **104**: 341.

Kaplan C.D., Morris J.R., Wu C., and Winston F. 2000. Spt5 and

Spt6 are associated with active transcription and have characteristics of general elongation factors in *D. melanogaster*. *Genes Dev.* **14:** 2623.

Keegan B.R., Feldman J.L., Lee D.H., Koos D.S., Ho R.K., Stainier D.Y.R., and Yelon D. 2002. The elongation factors Pandora/Spt6 and Foggy/Spt5 promote transcription in the zebrafish embryo. *Development* **129:** 1623.

Kishimoto Y., Lee K.H., Zon L., Hammerschmidt M., and Schulte-Merker S. 1997. The molecular nature of zebrafish *swirl*: BMP2 function is essential during early dorsoventral patterning. *Development* **124:** 4457.

Lis J. 1998. Promoter-associated pausing in promoter architecture and postinitiation transcriptional regulation. *Cold Spring Harb Symp. Quant. Biol.* **63:** 347.

Nguyen V.H., Schmid B., Trout J., Connors S.A., Ekker M., and Mullins M.C. 1998. Ventral and lateral regions of the zebrafish gastrula, including the neural crest progenitors, are established by a *bmp2b/swirl* pathway of genes. *Dev. Biol.* **199:** 93.

Rasmussen E.B. and Lis J.T. 1993. *In vivo* transcriptional pausing and cap formation on three Drosophila heat shock genes. *Proc. Natl. Acad. Sci.* **90:** 7923.

Reifers F., Walsh E.C., Léger S., Stainier D.Y.R., and Brand M. 2000. Induction and differentiation of the zebrafish heart requires fibroblast growth factor 8 (*fgf8/acerebellar*). *Development* **127:** 225.

Reiter J.F., Alexander J., Rodaway A., Yelon D., Patient R., Holder N., and Stainier D.Y.R. 1999. Gata5 is required for the development of the heart and endoderm in zebrafish. *Genes Dev.* **13:** 2983.

Reiter J.F., Verkade H., and Stainier D.Y. 2001. Bmp2b and Oep promote early myocardial differentiation through their regulation of *gata5*. *Dev. Biol.* **234:** 330.

Rougvie A.E. and Lis J.T. 1988. The RNA polymerase II molecule at the 5' end of the uninduced *hsp70* gene of D. melanogaster is transcriptionally engaged. *Cell* **54:** 795.

Serbedzija G.N., Chen J.N., and Fishman M.C. 1998. Regulation in the heart field of zebrafish. *Development* **125:** 1095.

Stainier D.Y.R. and Fishman M.C. 1994. The zebrafish as a model system to study cardiovascular development. *Trends Cardiovasc. Med.* **4:** 207.

Stainier D.Y.R., Fouquet B., Chen J.N., Warren K.S., Weinstein B.M., Meiler S.E., Mohideen M.A., Neuhauss S.C., Solnica-Krezel L., Schier A.F., Zwartkruis F., Stemple D.L., Malicki J., Driever W., and Fishman M.C. 1996. Mutations affecting the formation and function of the cardiovascular system in the zebrafish embryo. *Development* **123:** 285.

Swanson M.S. and Winston F. 1992. SPT4, SPT5 and SPT6 interactions: Effects on transcription and viability in *Saccharomyces cerevisiae*. *Genetics* **132:** 325.

Wada T., Takagi T., Yamaguchi Y., Ferdous A., Imai T., Hirose S., Sugimoto S., Yano K., Hartzog G.A., Winston F., Buratowski S., and Handa H. 1998. DSIF, a novel transcription elongation factor that regulates RNA polymerase II processivity, is composed of human Spt4 and Spt5 homologs. *Genes Dev.* **12:** 343.

Warren K.S., Wu J.C., Pinet F., and Fishman M.C. 2000. The genetic basis of cardiac function: dissection by zebrafish (*Danio rerio*) screens. *Philos. Trans. R. Soc. Lond. B. Biol. Sci.* **355:** 939.

Westerfield M. 1995. *The Zebrafish Book*. Univ. of Oregon Press, Eugene, Oregon.

Winston F., Chaleff D.T., Valent B., and Fink G.R. 1984. Mutations affecting Ty-mediated expression of the *HIS4* gene of *Saccharomyces cerevisiae*. *Genetics* **107:** 179.

Yelon D. 2001. Cardiac patterning and morphogenesis in zebrafish. *Dev. Dyn.* **222:** 552.

Yelon D., Horne S.A., and Stainier D.Y. 1999. Restricted expression of cardiac myosin genes reveals regulated aspects of heart tube assembly in zebrafish. *Dev. Biol.* **214:** 23.

Yelon D., Ticho B., Halpern M.E., Ruvinsky I., Ho R.K., Silver L.M., and Stainier D.Y.R. 2000. The bHLH transcription factor Hand2 plays parallel roles in zebrafish heart and pectoral fin development. *Development* **127:** 2573.

Zebrafish Hearts and Minds: Nodal Signaling in Cardiac and Neural Left–Right Asymmetry

S. LONG,* N. AHMAD, AND M. REBAGLIATI

Department of Anatomy and Cell Biology, Roy J. and Lucille A. Carver College of Medicine, University of Iowa, Iowa City, Iowa 52242

The establishment of the left–right (L-R) axis is essential for the correct development of the cardiovascular system; for instance, for heart looping and for the asymmetric remodeling of the aortic arches (for review, see Capdevila et al. 2000). Consequently, defects in L-R patterning can give rise to congenital defects of the heart and the vasculature (Ferencz et al. 1997). Because of its rapid development, transparency, and genetic tractability, the zebrafish has proven to be a useful model system in which to study cardiovascular development. Many early aspects of heart formation appear to be similar between the zebrafish and other vertebrates. An embryonic zebrafish heart is analogous in key respects to the heart of a human embryo at about 3 weeks postimplantation (for review, see Driever and Fishman 1996). The embryonic zebrafish heart also resembles other hearts in that it undergoes the same rightward looping (D-looping) as is seen for the hearts of mammals and other vertebrates. Thus, studies in the zebrafish can provide insight into the early L-R patterning processes that are involved in the development of all vertebrate hearts. In addition, zebrafish is the only model system in which the development of diencephalic and visceral organ asymmetry can be analyzed concurrently. This makes the zebrafish system especially well-suited to addressing the issue of whether and how brain and cardiac/visceral L-R asymmetry are coordinated or linked. In this paper, we summarize progress made in analyzing cardiac L-R asymmetry in the zebrafish as well as the broader issue of how L-R asymmetry in the cardiovascular system is coordinated with the L-R morphogenesis of other organs.

LEFT–RIGHT ASYMMETRIES IN THE ZEBRAFISH

Zebrafish embryos display many L-R asymmetries (Table 1). The earliest sign of L-R asymmetry is the onset of left-side expression of several genes within the trunk or tail. During mid-somitogenesis (17–22 somites), four markers (*cyclops*, *lefty1*, *lefty2*, and *pitx2*) are expressed asymmetrically within the lateral plate mesoderm (LPM), with expression being more extensive in or exclusive to the left LPM, in domains that overlap the heart field at their anterior limit (Fig. 1D) (Rebagliati et al. 1998b; Sampath et al. 1998; Bisgrove et al. 1999; Campione et al. 1999; Thisse and Thisse 1999; Essner et al. 2000; Faucourt et al. 2001). The heart field, the region with cardiogenic potential, is by definition slightly larger than the heart tube (which comprises the actual cardiac precursors); the heart tube itself can be visualized as a domain of *nkx2.5*-expressing cells (Fig. 2A). At 22 somites, BMP4 is also expressed asymmetrically and shows much stronger expression on the left side of the heart tube (Chen et al. 1997). At about the same stages, *cyc*, *lft1*, and *pitx2* are also expressed on the left side of the dorsal diencephalon in the epiphyseal/habenular region (Fig. 1C) (Rebagliati et al. 1998a; Sampath et al. 1998; Bisgrove et al. 1999; Thisse and Thisse 1999; Essner et al. 2000; Liang et al. 2000). Many of these genes show a similar pattern of asymmetric expression in the LPM in other vertebrates, both amniotes and non-amniotes (see next section). However, the diencephalic expression is unique to the zebrafish. Slightly later, morphological aspects of L-R asymmetry become apparent. By 24 hours postfertilization, the zebrafish heart tube has acquired a leftward displacement from the midline that is referred to as a left cardiac jog (Fig. 2A); the displacement is greatest for the atrial end of the heart tube. "Jogging" is the first visible morphological sign of L-R asymmetry in zebrafish embryos (Chen et al. 1997). From 36–72 hours postfertilization, the heart tube acquires its rightward or D-loop (rightward bending of the ventricle) and assumes its final

Table 1. Left–Right Asymmetries in the Zebrafish Embryo

Visceral:
  Cardiac jogging
  Cardiac looping
  Organ looping (intestines)
  Organ placement (e.g., pancreas)
  Expression of *lefty1*, *lefty2*, *pitx2*, *cyclops*, and *southpaw* in left LPM
Neural:
  Projections of left versus right habenular nuclei
  Left-side position of the parapineal gland
  Early: Left-side expression of *lefty1*, *pitx2*, and *cyclops* in the diencephalon
  Late: Expression of *otx5* and *crx* in the parapineal gland

LPM: lateral plate mesoderm

---

*Present address: Animal Behavior Graduate Group, University of California, Davis, California 95616.

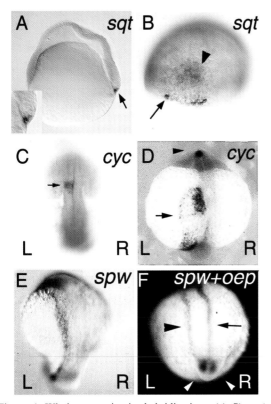

Figure 1. Whole-mount in situ hybridizations. (*A, B*) *squint* RNA, shield stage. (*A*) Lateral view, dorsal to right. (*B*) Dorsal view. (*Arrow*) Expression in dorsal forerunner cells. (*Arrowhead*) Low-level *squint* RNA in anterior part of shield. (*C,D*) *cyclops* RNA, 20-somite stage. (*C*) Dorsal view of head, left-side *cyc* expression in diencephalon (*arrow*). (*D*) Dorsal view of anterior trunk, left-side *cyc* expression in lateral plate mesoderm (*arrow*). (*Arrowhead*) Left-side diencephalic expression of *cyc*. (*E, F*) *southpaw* RNA, 13- to 14-somite stage. (*E*) Dorsal view of trunk, *spw* RNA in left LPM. (*F*) Dorsal view of tail, *spw* and *oep* expression. *spw* RNA in left LPM (*black arrowhead*) or in bilateral zones flanking tailbud (*white arrowheads*). *oep* RNA in notochord (*arrow*). (LPM) Lateral plate mesoderm. (*L*) Left; (*R*) right. (*A–C*, Adapted, with permission, from Rebagliati et al. 1998b [copyright Academic Press].).

is straight, respectively (Chen et al. 1997). In zebrafish and other vertebrates, the morphological defects are usually foreshadowed by misregulation of the early asymmetrically expressed genes, i.e., of *lefty1/2*, *pitx2*, and nodal-related genes (for review, see Bisgrove et al. 2000; Burdine and Schier 2000). Because of this correlation, much of the work on L-R patterning has focused on the question of whether the asymmetric expression of the aforementioned genes is actually required for heart L-R asymmetry.

THE Nodal–Lefty–Pitx2 CASSETTE

As indicated, the focus on nodal-related genes stems from the fact that these genes display a conserved pattern of asymmetry that correlates with organ asymmetry. Nodal expression in the lateral plate mesoderm is restricted largely to the left LPM in mice, chickens, zebrafish, and frogs. The expression pattern of the human *nodal-related* gene(s) is not known, but other studies support a role for nodal proteins in human laterality (Bam-

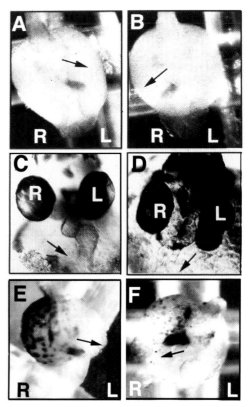

Figure 2. Cardiac jogging and looping in zebrafish embryos. (*A, B*) *nkx2.5* RNA expression in heart tube, 24 hours postfertilization, dorsal view, anterior at bottom. (*A*) Embryo with a left jog of the heart tube (*arrow*). (*B*) Embryo with right jog (*arrow*). (*C, D*) Ventral view, myosin antibody MF20 staining of heart, ~2-day embryos. Arrow highlights looping polarity. (*C*) Embryo with D-looping. (*D*) Embryo with L-looping. (*E, F*) 28 hours postfertilization, dorsal view with anterior to bottom. (*Purple*) *nkx2.5* RNA in heart tube. Arrow denotes jogging direction. (*Blue*) Nuclear β-gal (lacZ) staining. (*E*) Injected with 35 pg of *pCS2(+)nucLacZ*. (*F*) Coinjected with 10 pg of *pCS2(+)nucLacZ* plus 25 pg of *pCS2(+)Squint*. (R and L) Right and left.

A-P orientation (Fig. 2C) (for review, see Chin et al. 2000). Typically, 90–100% of the embryos in a wild-type clutch will have a leftward jogging heart tube that then develops into a D-looped heart (Chen et al. 1997). During the same period, morphological L-R asymmetries will develop in the other visceral organs as well as in the brain. For the viscera, these include the rotation of the gut and the rightward positioning of the pancreas; for the brain, these include the leftward displacement of the parapineal gland and the elaboration of differences between right and left habenular projections (Yan et al. 1999; Concha et al. 2000).

Defects in the L-R patterning process will lead to mispositioning of organs with respect to the midline, to isomerisms within an organ, or to alterations in chirality. In the case of the zebrafish heart, defective L-R patterning results in reversals of jogging (right jog) and looping (L-loop) (Fig. 2B, D) or to "no jog" or "no loop" phenotypes, where the heart tube is either positioned on the midline or

ford et al. 2000). Consistent with the expression patterns, studies in chicken, mouse, and *Xenopus* embryos have implicated the nodal signaling pathway directly in the establishment of heart and visceral organ L-R asymmetry. These studies also place the *lefty* and *pitx2* genes downstream from asymmetric nodal signaling (for review, see Burdine and Schier 2000; Wright 2001). In the mouse, complete loss of Nodal function leads to early embryonic lethality. To uncover later phenotypes, investigators have analyzed embryos in which early Nodal function is present, but in which later Nodal signaling is reduced or absent in the left LPM. This was done by producing embryos that are heterozygous for a null and either a hypomorphic or a down-regulated *nodal* allele (Lowe et al. 2001; Brennan et al. 2002; Norris et al. 2002). In these genotypes, one sees defects in heart looping, septation defects, and abnormalities in the aorta and pulmonary trunk, including transposition of the great vessels. These are accompanied by down-regulation of normally left-side genes like *lefty2* and *pitx2*. Likewise, ectopic expression of nodal-class proteins on the right side leads to a reversal or randomization of cardiac looping in both chicken and amphibian embryos and to up-regulation of *pitx2* on the right side (Levin et al. 1997; Sampath et al. 1997; Campione et al. 1999). The *lefty* genes act to limit the spatial expression and influence of nodal-related genes (for review, see Hamada et al. 2002). *Pitx2* is a homeodomain-class transcription factor that is itself implicated in cardiac development and visceral organ L-R asymmetry (Logan et al. 1998; Chijen et al. 1999). Taken together, these studies have led to the view that Nodal functions as part of a conserved vertebrate L-R "cassette," whereby asymmetric nodal signaling induces both conserved downstream antagonists (Lefty1, Lefty2) and conserved downstream effectors (Pitx2) (for review, see Hamada et al. 2002), ultimately driving correct cardiac and visceral L-R development. Molecules like Flectin may be regulated by this cassette to mediate specific downstream events (Tsuda et al. 1996). In general, however, very little is known about what molecules would lie farther downstream and would actually mediate the morphological and mechanical aspects of L-R morphogenesis. Herein lies one of the potential advantages of zebrafish, namely its suitability for large-scale genetic screens that could in principle identify novel molecules with conserved roles in cardiac looping or other aspects of L-R asymmetry. As a prelude to such screens, we and other workers have investigated the extent to which the L-R patterning functions of Nodal signaling are conserved in the zebrafish.

RESULTS AND DISCUSSION

Does the Nodal–Lefty–Pitx2 Cassette Have a Conserved Role in Zebrafish L-R Asymmetry?

In the zebrafish, the nodal-related gene *cyclops* exhibits the same conserved pattern of expression in the left lateral plate (Fig. 1D) (Rebagliati et al. 1998a; Sampath et al. 1998). This suggested that nodal signaling may also organize cardiac L-R asymmetry in the zebrafish. Surprisingly, we and others have found that, despite its asymmetric expression, *cyclops* appears to have, at best, only a small influence on visceral organ L-R asymmetry (Chen et al. 1997; Chin et al. 2000; Bisgrove and Yost 2001; M. Rebagliati, unpubl.). For instance, in the case of embryos that are homozygous for the point mutation cyc^{tf219}, one finds that nearly all the embryos exhibit a normal left cardiac jog (Fig. 3) (Chen et al. 1997). The lesion in cyc^{tf219} alters the start codon of the open reading frame and is ei-

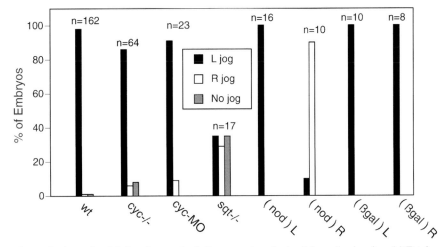

Figure 3. Effects of perturbations of nodal signaling on the L-R asymmetry of zebrafish cardiac jogging. (*n*) Total number of embryos scored. (Y axis) Percent of the *n* total embryos that have a given jogging pattern. (L jog) Left jog of heart tube (*black bars*). (R jog) Right jog of heart tube (*white bars*). (No jog) Heart tube remains on midline and doesn't jog (*gray bars*). (X axis) Embryos of different genotypes or injected with various reagents, analyzed for jogging polarity. (wt) Wild-type embryos. ($cyc^{-/-}$) Homozygous for cyc^{tf219} allele. (Cyc-MO) Injected with 8 ng of *cyclops* morpholino. ($sqt^{-/-}$) Homozygous for sqt^{cz35} allele. (nod) L: Embryos having a higher level of *pCS2(+)Squint* plus *pCS2(+)nucLacZ* on their left side. (nod) R: Embryos having a higher level of *pCS2(+)Squint* plus *pCS2(+)nucLacZ* on their right side. (β-gal) L: *pCS2(+)nucLacZ*-injected embryos with higher β-gal activity on their left side, no *pCS2(+)Squint*. (β-gal) R: *pCS2(+)nucLacZ*-injected embryos with higher β-gal activity on their right side, no *pCS2(+)Squint*.

ther a null mutation or a strong hypomorph (Rebagliati et al. 1998a). Incrosses using other *cyclops* loss-of-function alleles give the same result. In contrast, two other lesions that block nodal signaling in the zebrafish, the *one-eyed-pinhead (oep)* and *schmalspur (sur)* mutations, do disrupt both cardiac and visceral L-R asymmetry (Chen et al. 1997; Yan et al. 1999; Bisgrove et al. 2000). *oep* mutations perturb a membrane-associated protein, possibly a coreceptor, that is required for activation of the nodal signaling pathway (Zhang et al. 1998; Gritsman et al. 1999). *sur* mutations inactivate the Fast1/FoxH1 protein, a transcription factor that interacts with Smad class proteins to regulate the transcription of specific genes in response to nodal signals (Pogoda et al. 2000; Sirotkin et al. 2000).

The lack of agreement between the *cyc* and *oep/sur* phenotypes prompts the question as to whether the role of nodal signaling in the left LPM is conserved in zebrafish compared to other vertebrates. One alternative interpretation of the cyc^{tf219} data is that there is sufficient residual function in cyc^{tf219} (or other *cyc* alleles) to allow normal cardiac jogging and looping. To evaluate loss of *cyc* function by alternative means, we used a *cyclops* morpholino that should effectively block translation of both zygotic and maternal *cyclops* mRNAs. The efficacy and specificity of this morpholino had been established in prior studies (Karlen and Rebagliati 2001). Like the point mutations, injection of the *cyclops* MO induces the characteristic "cyclopic" phenotype. However, it too has little effect on the asymmetry of cardiac jogging (Fig. 3).

The genetic and morpholino-based studies appear to rule out a major requirement for *cyc* within the left LPM for cardiac L-R morphogenesis. *oep* and *sur* mutants suffer generalized defects in nodal signaling that would be expected to block signaling by any and all zebrafish nodal-class ligands. In light of this, one might wonder whether reductions in the activity of the other published zebrafish nodal-related gene, *squint (sqt)*, affect cardiovascular L-R asymmetry. Loss of *squint* function does randomize the polarity of cardiac jogging, giving rise with nearly the same frequency to embryos with a left jog, no jog, or a right jog phenotype (Fig. 3). However, late *squint* expression is restricted to a subset of cells in the dorsal organizer (shield) (Fig. 1A,B). *squint* is never expressed within the left LPM (Feldman et al. 1998; Rebagliati et al. 1998b), nor is it up-regulated in the LPM in embryos lacking *cyc* activity (M. Rebagliati, unpubl.).

The *cyc* and *sqt* phenotypes raise the possibility that nodal signaling is not needed in the zebrafish LPM for cardiac and visceral asymmetry. Perhaps nodal signaling is only required in the dorsal organizer for L-R patterning, and this would then explain the *oep/sur* phenotypes. To determine whether the zebrafish lateral plate is indeed responsive to nodal ligands as L-R cues, we manipulated the relative levels of nodal activity on the right and left sides of the embryos by injecting an expression vector driving a nodal-related cDNA. A *squint* expression vector was chosen as a convenient way to enhance nodal signaling. Unlike the situation with *Xenopus* embryos, the left and right sides of a zebrafish embryo cannot be predicted prior to somitogenesis. However, it has been observed that if one injects a single blastomere at the 8- to 16-cell stage, a fraction of the embryos will have the injected material deposited predominantly on the left or the right side (Hammerschmidt et al. 1996). By coinjecting the *squint* expression construct with a *lacZ* (β-gal) expression vector, we were able to to identify embryos in which levels of the expression plasmids—and by inference overall nodal activity—was stronger on the left or right flank (Fig. 2E,F). Misexpression of β-*gal* itself preferentially on the left or right side has no effect on cardiac L-R morphogenesis (Figs. 2E and 3, bars marked [β-gal]L and [β-gal]R). Under these conditions, cardiac jogging polarity was dictated by the relative levels of nodal plasmid on the left and right sides. When ectopic *squint* was higher on the left, the heart tube jogged to the left, whereas when *squint* expression was stronger on the right, heart jogging polarity was almost always reversed (Fig. 3, bars marked [nod]L and [nod]R). Figure 2F shows one such example of a *squint*-injected embryo: As inferred from the punctate nuclear β-gal stain, the amount of *squint* expression vector was higher on the right side, and the heart tube (purple stain, *nkx2.5* probe) shows a reversed (right) jog. One caveat is that the expression vectors become transcriptionally active at the mid-blastula stage. As a result, one cannot rule out the possibility that the laterality defects are a consequence of ectopic Squint protein acting prior to somitogenesis. However, the injected embryos were not overtly hyperdorsalized, which suggests that ectopic Squint activity is low at pre-somite stages. Overall, these experiments are consistent with the idea that the zebrafish lateral plate can respond to nodal cues to generate a bias in heart L-R asymmetry. Since neither Cyclops nor Squint serves as the primary cue in embryos, the results in Figure 3 suggest, ultimately, that there are other nodal-related genes which generate asymmetric nodal activity within the zebrafish LPM.

New Zebrafish Nodal-related Genes: The *southpaw* Gene

Motivated by the aforementioned experimental results, we undertook a molecular screen for new zebrafish nodal-related genes. This screen identified a new zebrafish nodal-related gene, *southpaw* (S. Long et al., in prep.). As the name implies, the *southpaw (spw)* locus is expressed within the left LPM as well as showing bilateral expression in the tailbud (Fig 1E, F). Left-side expression commences at the 10- to 12-somite stage, well before the onset of asymmetric expression of *cyc*, *lft1*, *lft2*, and *pitx2*. This makes *southpaw* the earliest molecular marker of L-R asymmetry in the zebrafish. The pattern of expression within the lateral plate is itself highly dynamic (S. Long et al., in prep.). The spatial restriction of *southpaw* to the left LPM suggests that *spw* may supply the endogenous nodal signal to which the lateral plate responds during zebrafish embryogenesis. We resorted to surrogate genetic approaches to address the issue of function, specifically, morpholino oligonucleotide-mediated knockdown of Spw activity. Although morpholinos are generally highly specific in zebrafish, some nonspecific effects have been observed; in particular, generalized necrosis in the embryonic central nervous system. To control for nonspecific morpholino effects, we used two *spw* morpholinos that would block either translation or splicing. The results of these studies are detailed elsewhere (S. Long et al., in prep.), and salient results are only summarized here.

Injection of either *southpaw* morpholino caused a strong disruption of the normal leftward bias of cardiac jogging. The frequency of "reversed jogging" or of "no jogging" embryos was strongly increased relative to what is observed in uninjected embryos or embryos injected with control morpholinos (Fig. 4A). Likewise, we also observed increases in the frequency of hearts with reversed looping or no looping (Fig. 4B; only Spw-MO1 was tested for looping effects). However, for any given embryo, there was no correlation between cardiac jogging polarity and looping polarity.

The *southpaw* morpholino phenotype resembles that seen in a subset of the known zebrafish L-R asymmetry mutants (Chin et al. 2000). This class of zebrafish L-R mutations (mom^{th211}, boz^{m168}, flh^{n1}, ntl^{b160}) perturbs both jogging and looping and uncouples the normal linkage between the two events. On the basis of these "uncoupling" mutants, it was proposed that heart placement (relative to the midline) and heart looping are established in the zebrafish by two separable mechanisms: Early L-R asymmetry cues within the LPM would bias heart placement (jogging), and a later process would somehow translate this initial heart position into a heart looping chirality (Chin et al. 2000). Judging by the morpholino phenotype, left-side *southpaw* expression appears to be required for both processes.

Left-side *spw* expression precedes the left-side expression of *cyc*, *lft1*, *lft2*, and *pitx2* within the LPM. This suggests that Spw may induce the expression of these other L-R patterning genes. The morpholino knockdown experiments confirm this. In most Spw-MO-injected embryos, there is no detectable expression of *cyc*, *lft1*, *lft2*, and *pitx2* within the left LPM (S. Long et al., in prep.). These results show that *spw* also regulates other key L-R patterning genes that are expressed asymmetrically in the LPM. Consistent with these effects, we find that the spw morpholinos also disrupt the left-right asymmetry of the pancreas.

In summary, both "ectopic expression" (Fig. 3) and "loss of function" (Fig. 4) (S. Long et al., in prep.) experiments lead to the same conclusion. These results all suggest that nodal signaling within the zebrafish left LPM is required for normal cardiac and visceral L-R asymmetry, similar to what is seen in other vertebrates. The crucial nodal signal in this regard is provided by a new nodal-related gene, *southpaw*.

BMP Antagonism by *cyclops*: A Role in L-R Patterning?

The aforementioned experiments point to a crucial role for *southpaw* in zebrafish heart L-R asymmetry. What is to be made, then, of the fact that *cyclops* is also expressed asymmetrically in the left lateral plate? One possibility is that *cyclops* may be critical for L-R patterning of other visceral organs besides the heart. Another possibility, not

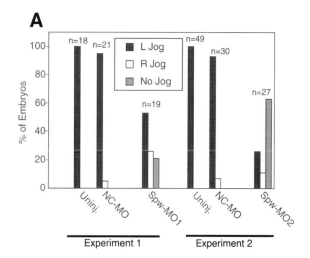

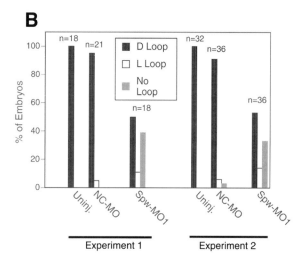

Figure 4. Effects of *southpaw* morpholinos on cardiac jogging and looping. (*n*) Total number of embryos scored per uninjected control or per injection. (*A*) (Y axis) Percent of the *n* total embryos that have a given jogging pattern. (L jog) Left jog of heart tube (*black bars*). (R jog) Right jog (*white bars*). (No jog) Heart tube remains on midline and doesn't jog (*gray bars*). (X axis) 2 injection experiments. (*B*) (Y axis) Percent of the *n* total embryos that have a given looping pattern. (D-Loop) Normal looping (*black bars*). (L-Loop) Reversed looping (*white bars*). (No Loop) Straight heart, unlooped (*gray bars*). (X axis) 2 injection experiments. For each experiment, embryos were separated into 3 groups and either not injected (Uninj.), injected with 10 ng of a negative control morpholino (NC-MO), or injected with 10 ng of a translation blocking (Spw-MO1) or splice-blocking (Spw-MO2) *spw* morpholino. Graph based on data from S. Long et al. (in prep.).

mutually exclusive with the first, is that *cyclops* may have a subtle role in heart jogging, looping, or heart differentiation—a role that may only be evident at the molecular level with specific probes. Lending credence to this possibility, *Xenopus* induction assays suggest that the Cyclops protein is functionally distinct from Southpaw or Squint.

The *Xenopus* animal cap explant assay is one of the standard assays that is used to characterize the inductive properties of peptide growth factors. This assay is particularly useful because it detects functional distinctions that would be missed in other assays. For instance, nodal-related proteins all produce the same effect, dorsalization of the mesoderm, after microinjection into intact zebrafish embryos. However, nodal-class ligands can demonstrate markedly different inductive properties in *Xenopus* explants. The assay itself is simple. mRNA encoding the protein of interest is injected into fertilized *Xenopus* eggs. At the blastula stage, the ectoderm around the animal pole is then excised and cultured in vitro. Naive, i.e., untreated, ectoderm normally develops into an atypical epidermis, whereas ectoderm that is exposed to the inductive influence of a protein will be diverted from this fate and express other molecular markers. For example, treatments that increase the phosphorylation of Smad2/3 drive the ectodermal cap to primarily a dorsal mesodermal fate, as shown by induction of *brachyury* (*Xbra*), *goosecoid* (*gsc*), and other markers. A small amount of neuroectoderm is also induced, but this is a secondary induction mediated by the newly formed dorsal mesoderm of the explant. In contrast, treatments that inhibit the bone morphogenic protein (BMP) signaling pathway result in explants containing large amounts of neural tissue but little or no mesoderm. When Squint is assayed in the animal cap assay, it behaves as a mesoderm inducer; i.e., mesoderm markers are strongly expressed and neuroectodermal markers are detected at only low levels (Fig. 5) (Rebagliati et al. 1998b). Southpaw, which is most similar to Squint, has not been tested yet but would be predicted to act in the same way by virtue of its close structural similarity to Squint. Injection of *cyclops* mRNA, on the other hand, leads to a very different result. Cyclops protein induces high levels of *nrp-1*, a pan-neural marker, but induces little, if any, expression of mesodermal markers (Fig. 5) (Rebagliati et al. 1998b). Thus, Cyclops behaves as a strong neural inducer in the *Xenopus* cap assay, the implication being therefore that it antagonizes the BMP pathway. This antagonism does not occur through the induction of any of the standard BMP antagonists that are thought to play an endogenous role in neural induction (Noggin, Chordin, or Follistatin). Although the *Xenopus* assays reveal two different inductive effects comparing Squint and Cyclops, these two inductive abilities are not mutually exclusive. In analogous studies, mouse Nodal was shown to have the ability to activate Smad2 phosphorylation in a receptor-dependent fashion, as well as the ability to antagonize BMPs by formation of heterodimers between Nodal and BMP proteins, a receptor-independent event (Yeo and Whitman 2001). However, the results with Squint and Cyclops imply that, in vivo, a given nodal-related ligand may be better at activating the Smad2/3 pathway than in heterodimerizing with BMPs or vice versa. If so, then it is possible that Southpaw may provide the main receptor-dependent cue for asymmetry within the left LPM, with Cyclops having an accessory role in modulating or damping the influence of nearby sources of BMPs. Studies in other species have suggested that the balance of BMP and nodal signals is important for normal L-R patterning. In other vertebrates, BMP antagonists like Cerberus are expressed asymmetrically, and several studies have reported that changes in BMP activity in vivo do alter L-R asymmetry. However, some studies suggested that BMP signaling antagonizes left-side nodal signaling, thereby weakening the cues that drive "left-sidedness," whereas other studies have suggested the opposite, that BMP activity is necessary for full induction of left-side expression of nodal-related genes (Esteban et al. 1999; Yokouchi et al. 1999; Zhu et al. 1999; Fujiwara et al. 2002; Piedra and Ros 2002; Schlange et al. 2002). Future studies will be needed to resolve whether and how Cyclops and BMPs interact during zebrafish development and if such interactions modulate Southpaw-dependent L-R asymmetry, in particular, heart jogging and looping.

Conservation of Nodal Expression and Regulation in Vertebrates

On the basis of comparisons in the zebrafish, chicken, frog, and mouse, a set of common features now emerges for the nodal-related genes that mediate cardiac/visceral organ asymmetry. Although there are also species-specific differences, the embryos of all four species share a similar late pattern of nodal-related gene expression domains (comparing mouse *nodal*, chicken *cnr-1*, frog *xnr-1*, and zebrafish *spw*) (Levin et al. 1995; Lowe et al. 1996; S. Long et al., in prep.). This pattern consists initially of a pair of bilateral *nodal* zones (domain I in Fig. 6) flanking a cluster of monociliated cells (Essner et al.

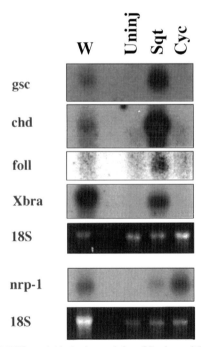

Figure 5. Differential inducing activity of Squint and Cyclops in *Xenopus* animal caps. RNAs (listed on top) encoding Sqt (400 pg) or Cyc (400 pg) were injected into the animal region, and explants cultured until equivalent stage 11.5 for the assay of *goosecoid (gsc), chordin (chd), Xenopus brachyury (Xbra)*, and *follistatin (foll)*, and to equivalent stage 25–27 for assay of *nrp-1* by Northern blotting. Stained 18S RNA is shown as a loading control. (W) Whole untreated embryo. (Adapted, with permission, from Rebagliati et al. 1998b [copyright Academic Press].)

2002). These ciliated cells also express a specific dynein, left-right dynein (Lrd); in all four species, the lrd^+ ciliated cells appear before the onset of any asymmetric nodal expression. Another common feature is that the monociliated cells either lie within a region with dorsal organizing activity or are descended from cells that were originally resident within a dorsal organizer (Essner et al. 2002). Proximity to the dorsal organizer may be required to ensure linkage of L-R asymmetry to the anterior–posterior and dorsal–ventral axes. The nodal-expressing cells of domain I will contribute to axial or paraxial mesoderm of posterior character in the mouse, zebrafish, and *Xenopus*, and possibly also in the chicken (Levin et al. 1995; Lowe et al. 1996; Brennan et al. 2002; S. Long et al., in prep.). In chicken and mouse embryos, the left component of the bilateral nodal zones is transiently stronger than its right counterpart (Levin et al. 1995; Lowe et al. 1996). The significance of this transient "domain I" asymmetry is unclear, as it is not required in the mouse for generating L-R asymmetry in the LPM, nor is this transient asymmetry ever seen in *Xenopus* or zebrafish

(Norris et al. 2002). Chicken, mouse, frog, and zebrafish embryos also display a second common nodal domain (domain II in Fig. 6), which is, of course, the stripe of strong expression present along much of the anterior–posterior extent of the left LPM (Levin et al. 1995; Lowe et al. 1996; S. Long et al., in prep.).

One challenge for the future is to determine whether the common expression pattern reflects a common regulatory mechanism for the spatial control of nodal-related genes. Studies in the mouse support the idea that the monociliated, lrd^+ cells in the ventral layer of the node have a critical role in activating *nodal* transcription in the left LPM. According to this nodal flow hypothesis, the cilia may generate a leftward flux that leads to the asymmetric accumulation of a diffusible morphogen, which could be Nodal protein itself or another factor (for review, see Nonaka et al. 1998, 2002; Hamada et al. 2002). Because this morphogen would be present at a higher concentration on the left side, it would then induce *nodal* expression in the left LPM. This model is tantalizing in light of the fact that the cells in domain I abut a group of ciliated cells in all four species. However, one simple argument against this model in the zebrafish is that the ciliated cells here simply line the internal surface of an epithelium that forms a closed sac, Kupfer's vesicle (Melby et al. 1996). It is hard to see how this topological arrangement could lead to a polarized flow of a morphogen toward the left. Another significant difference between zebrafish and mouse embryos is the necessity or dispensability of domain I for the induction of nodal transcription in the LPM. When domain I is lost in mouse embryos, *nodal* does not become expressed in the lateral plate (Brennan et al. 2002). However, in zebrafish, when domain I is eliminated genetically (in embryos that are homozygous for a null allele of *no tail* [S. Long et al., in prep.]), *southpaw* becomes expressed in both the right and left LPM, instead of being eliminated from the LPM. This result is also incompatible with the idea that a leftward flux of Spw protein from domain I is required for the induction of domain II. Comparative studies of nodal regulation in zebrafish, frogs, chickens, and mice should help resolve these discrepancies and clarify what aspects of the upstream regulatory apparatus are conserved or species-specific.

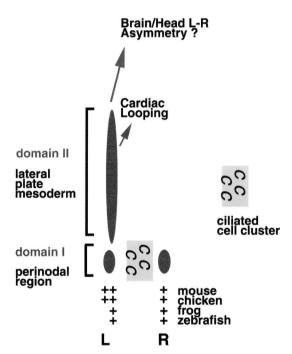

Figure 6. Late nodal-related gene expression domains that are conserved in chicken (*cnr-1* gene), mouse (*nodal* gene), frog (*xnr-1* gene), and zebrafish (*spw* gene) embryos. (domain I) Bilateral zones of nodal-related gene expression (*red*) flanking a cluster of ciliated cells (*blue*). Since the ciliated cells lie within a region of (or are descended from) the dorsal organizer (mouse or chicken node, zebrafish shield), the flanking nodal-related expression zones comprising domain I are designated as being "perinodal," in keeping with the mouse terminology (Brennan et al. 2002). (++) denotes that expression is transiently stronger in the left perinodal region in mouse and chicken embryos. (domain II) Nodal-related gene expression in the left lateral plate mesoderm (*red*). Domain II is implicated in inducing cardiac D-looping in all four species (*red arrow*). Studies in zebrafish (*spw*) and chicken (*cnr-1*) also suggest a possible role for domain II in inducing brain/head L-R asymmetry (*blue arrow*). (L) Left side, (R) right side.

Linkage of Cardiac L-R Asymmetry and the L-R Asymmetry of the Central Nervous System

Studies of L-R asymmetry have only begun to look at the larger question of whether and how cardiac/visceral organ asymmetry is coordinated with asymmetries in the head and CNS. On the face of it, this issue seems a bit esoteric, but the answer to this question could have important clinical implications. For example, if there were an obligate linkage between neural and visceral L-R asymmetry, one would have to consider the possibility that patients with congenital heart defects might also be at risk for specific neurological defects, with both lesions arising from abnormal L-R patterning. The analysis of this question has been hampered by the scarcity of early embryonic markers of head and CNS asymmetry. However, three examples of neural or cephalic L-R asymmetry have been identified to date. The first is the expression of mouse and chicken *lefty1* within the left half of the floor plate, the structure that forms the ventral midline of the embryonic spinal cord (Meno et al. 1997; Schlange et al. 2001). The second is the observation of left-side-specific expression of the *cerberus* (*cCer*) gene in head mesenchyme in chicken embryos (Zhu et al. 1999). Finally, the third example, which has been mentioned before, is the existence of molecular and morphological asymmetries within the epiphysis and habenular nuclei of the diencephalon. To date, the molecular asymmetries of the diencephalon (left-side expression of *cyclops*, *pitx2*, and *lefty1*, and later, *otx5* and *crx*) have only been seen in the zebrafish (Rebagliati et al. 1998b; Sampath et al. 1998; Bisgrove et al. 1999; Campione et al. 1999; Thisse and Thisse 1999; Essner et al. 2000; Liang et al. 2000; Gamse et al. 2002). The pronounced morphological asymmetries between the habenular nuclei have been observed in other animals as well, including certain birds, reptiles, fish, and amphibians (Von Wollwaerth 1950; Miralto and Kemali 1980). The Swiss embryologist Von Wollaerth, for instance, looked at newts in the wild and reported striking differences in the size and morphology of the left and right habenular nuclei. Interestingly, he also found that spontaneous situs inversus of the organs was usually accompanied by reversals of habenular L-R asymmetry (Von Wollwaerth 1950).

In all three of the aforementioned cases of neural or cephalic asymmetry, the asymmetries do not develop if nodal signaling is blocked. The experiments in the chicken and the zebrafish are particularly informative regarding the question of linkage. In stage 9 chick embryos, *cerberus* (*cCer*) is expressed in the left lateral plate and in left-side head mesenchyme. *cCer* expression in the head depends on a chick nodal-related gene (*cnr-1*) with a similar expression pattern. Left-side expression of both *cCer* and *cnr-1* is first seen within the LPM. Moreover, implantation of a Nodal pellet in the right flank of a chicken embryo induces ectopic *cCer* expression in the head (Zhu et al. 1999). In the zebrafish, the *oep* and *sur* mutations were used to block late zygotic nodal signaling. This resulted in the randomization of embryonic L-R asymmetry within both the heart and diencephalon (Yan et al. 1999; Bisgrove et al. 2000; Concha et al. 2000; Liang et al. 2000). However, since *oep* and *sur* mutations block signaling by most if not all nodal-class proteins, it could not be determined whether it was the loss of a common nodal signal that affected both head and visceral asymmetry. Our experiments with *spw* morpholinos have shown that it is a single nodal signal, Spw, that is required for organizing both brain and cardiac L-R asymmetry in zebrafish embryos (S. Long et al., in prep.). When *southpaw* is inhibited, there is no asymmetric neural expression of *cyclops*, *lefty1*, and *pitx2*. For all three genes, the left-side expression domains are lost, *both* in the diencephalon and in the LPM. Moreover, as described previously, heart jogging and looping are also disrupted by reductions in Spw activity. In sum, brain and heart/visceral organ asymmetry are closely linked in the zebrafish through their dependence on a single polarity cue, *southpaw.*

Because we have never detected *spw* expression in the head, the simplest model for these data is that left-side *spw* expression just rostral to the developing heart is responsible for establishing diencephalic L-R asymmetry (Fig. 6). This could occur either through direct diffusion of Southpaw from the anterior LPM or through another protein(s) that is induced by a Spw-dependent signal relay. One could invoke a similar explanation for the induction of *cCer* in left-side head mesenchyme in chicken embryos (Zhu et al. 1999). Nodal proteins behave as short-range signals in some contexts and as long-range morphogens in other situations (Jones et al. 1996; Chen and Schier 2001). Thus, direct diffusion and a signal relay are both plausible mechanisms. This model provides a simple explanation for linkage or coordination of diencephalic and cardiac left-right asymmetry. The same spatial cue, *spw* expression in anterior left LPM, would trigger left-side-specific differentiation in both the head and heart. Two goals for the future are (1) to rigorously test this model in the zebrafish and (2) to identify early molecular markers of brain L-R asymmetry in mammals, thereby opening the way for tests of this model in other vertebrates.

Over a short span of time, a remarkable amount of progress has been made in understanding the molecular basis of vertebrate L-R asymmetry. This reflects the fact that investigators have taken advantage of the unique strengths of several different animal models to attack this problem. Based on studies in the zebrafish, mouse, *Xenopus,* and chicken, it is clear that the nodal signaling pathway is a key, conserved component of the genetic program that establishes cardiac L-R asymmetry. Studies of *southpaw* expression and function show in addition that the same nodal cue can organize both brain and cardiovascular L-R asymmetry. The source of this Spw signal may be the left anterior LPM, but this proposition needs to be tested directly. A second challenge for the future is to understand how Spw exerts its polarizing influence on both the heart and brain—in terms of the downstream effector genes that alter cell shape and behavior. Zebrafish genetic screens can be used to find mutations that affect heart jogging and looping, that disrupt diencephalic asymmetry, or that perturb the regulation of *southpaw*.

It is hoped that these experimental approaches will contribute to an understanding of the molecules and mechanisms used by all vertebrates to establish cardiovascular and cephalic L-R asymmetry.

ACKNOWLEDGMENTS

We thank Igor Dawid and the Laboratory of Molecular Genetics-National Institute of Child Health and Human Development, National Institutes of Health for supporting the initial stages of this work. Additional support was provided by a Pilot Award from the University of Iowa Diabetes and Endocrinology Research Center, National Institutes of Health DK25295, and a Biosciences Pilot Grant from the state of Iowa.

REFERENCES

Bamford R.N., Roessler E., Burdine R.D., Saplakoglu U., dela Cruz J., Splitt M., Goodship J.A., Towbin J., Bowers P., Ferrero G.B., Marino B., Schier A.F., Shen M.M., Muenke M., and Casey B. 2000. Loss-of-function mutations in the EGF-CFC gene CFC1 are associated with human left-right laterality defects. *Nat. Genet.* **26:** 365.

Bisgrove B.W. and Yost H.J. 2001. Classification of left-right patterning defects in zebrafish, mice, and humans. *Am. J. Med. Genet.* **101:** 315.

Bisgrove B.W., Essner J.J., and Yost H.J. 1999. Regulation of midline development by antagonism of lefty and nodal signaling. *Development* **126:** 3253.

———. 2000. Multiple pathways in the midline regulate concordant brain, heart and gut left-right asymmetry. *Development* **127:** 3567.

Brennan J., Norris D.P., and Robertson E.J. 2002. Nodal activity in the node governs left-right asymmetry. *Genes Dev.* **16:** 2339.

Burdine R.D. and Schier A.F. 2000. Conserved and divergent mechanisms in left-right axis formation. *Genes Dev.* **14:** 763.

Campione M., Steinbeisser H., Schweickert A., Deissler K., van Bebber F., Lowe L.A., Nowotschin S., Viebahn C., Haffter P., Kuehn M.R., and Blum M. 1999. The homeobox gene Pitx2: Mediator of asymmetric left-right signaling in vertebrate heart and gut looping. *Development* **126:** 1225.

Capdevila J., Vogan K.J., Tabin C.J., and Izpisua Belmonte J.C. 2000. Mechanisms of left-right determination in vertebrates. *Cell* **101:** 9.

Chen J.N., van Eeden F.J., Warren K.S., Chin A., Nüsslein-Volhard C., P. Haffter P., and Fishman M.C. 1997. Left-right pattern of cardiac BMP4 may drive asymmetry of the heart in zebrafish. *Development* **124:** 4373.

Chen Y. and Schier A.F. 2001. The zebrafish Nodal signal Squint functions as a morphogen. *Nature* **411:** 607.

Chijen R., Kioussi C., O'Connell S., Briata P., Szeto D., Forrest L., Izpisua-Belmonte J., and Rosenfeld M. 1999. Pitx2 regulates lung asymmetry, cardiac positioning and pituitary and tooth morphogenesis. *Nature* **401:** 279.

Chin A.J., Tsang M., and Weinberg E.S. 2000. Heart and gut chiralities are controlled independently from initial heart position in the developing zebrafish. *Dev. Biol.* **227:** 403.

Concha M.L., Burdine R.D., Russell C., Schier A.F., and Wilson S.W. 2000. A nodal signaling pathway regulates the laterality of neuroanatomical asymmetries in the zebrafish forebrain. *Neuron* **28:** 399.

Driever W. and Fishman M.C. 1996. The zebrafish: Heritable disorders in transparent embryos. *J. Clin. Invest.* **97:** 1788.

Essner J.J., Branford W.W., Zhang J., and Yost H.J. 2000. Mesendoderm and left-right brain, heart and gut development are differentially regulated by pitx2 isoforms. *Development* **127:** 1081.

Essner J.J., Vogan K.J., Wagner M.K., Tabin C.J., Yost H.J., and Brueckner M. 2002. Conserved function for embryonic nodal cilia. *Nature* **418:** 37.

Esteban C., Capdevila J., Economides A.N., Pascual J., Ortiz A., and Izpisua-Belmonte J. 1999. The novel Cer-like protein Caronte mediates the establishment of embryonic left-right asymmetry. *Nature* **401:** 243.

Faucourt M., Houliston E., Besnardeau L., Kimelman D., and Lepage T. 2001. The pitx2 homeobox protein is required early for endoderm formation and nodal signaling. *Dev. Biol.* **229:** 287.

Feldman B., Gates M.A., Egan E.S., Dougan S.T., Rennebeck G., Sirotkin H.I., Schier A.F., and Talbot W.S. 1998. Zebrafish organizer development and germ-layer formation require nodal-related signals. *Nature* **395:** 181.

Ferencz C., Loffredo C., Correa-Villasenor A., and Wilson P. 1997. *Genetic and environmental risk factors of major cardiovascular malformations: The Baltimore-Washington Infant Study 1981–1989.* Futura Publishing, Armonk, New York.

Fujiwara T., Dehart D.B., Sulik K.K., and Hogan B.L. 2002. Distinct requirements for extra-embryonic and embryonic bone morphogenetic protein 4 in the formation of the node and primitive streak and coordination of left-right asymmetry in the mouse. *Development* **129:** 4685.

Gamse J.T., Shen Y.C., Thisse C., Thisse B., Raymond P.A., Halpern M.E., and Liang J.O. 2002. Otx5 regulates genes that show circadian expression in the zebrafish pineal complex. *Nat. Genet.* **30:** 117.

Gritsman K., Zhang J., Cheng S., Heckscher E., Talbot W.S., and Schier A.F. 1999. The EGF-CFC protein one-eyed pinhead is essential for nodal signaling. *Cell* **97:** 121.

Hamada H., Meno C., Watanabe D., and Saijoh Y. 2002. Establishment of vertebrate left-right asymmetry. *Nat. Rev. Genet.* **3:** 103.

Hammerschmidt M., Bitgood M.J., and McMahon A.P. 1996. Protein kinase A is a common negative regulator of Hedgehog signaling in the vertebrate embryo. *Genes Dev.* **10:** 647.

Jones C.M., Armes N., and Smith J.C. 1996. Signalling by TGF-beta family members: Short-range effects of Xnr-2 and BMP-4 contrast with the long-range effects of activin. *Curr. Biol.* **6:** 1468.

Karlen S. and Rebagliati M. 2001. A morpholino phenocopy of the cyclops mutation. *Genesis* **30:** 126.

Levin M., Johnson R.L., Stern C.D., Kuehn M., and Tabin C. 1995. A molecular pathway determining left-right asymmetry in chick embryogenesis. *Cell* **82:** 803.

Levin M., Pagan S., Roberts D.J., Cooke J., Kuehn M.R., and Tabin C.J. 1997. Left/right patterning signals and the independent regulation of different aspects of situs in the chick embryo. *Dev. Biol.* **189:** 57.

Liang J.O., Etheridge A., Hantsoo L., Rubinstein A.L., Nowak S.J., Izpisua Belmonte J.C., and Halpern M.E. 2000. Asymmetric nodal signaling in the zebrafish diencephalon positions the pineal organ. *Development* **127:** 5101.

Logan M., Pagan-Westphal S.M., Smith D.M., Paganessi L., and Tabin C.J. 1998. The transcription factor Pitx2 mediates situs-specific morphogenesis in response to left-right asymmetric signals. *Cell* **94:** 307.

Lowe L.A., Yamada S., and Kuehn M.R. 2001. Genetic dissection of nodal function in patterning the mouse embryo. *Development* **128:** 1831.

Lowe L.A., Supp D.M., Sampath K., Yokoyama T., Wright C.V., Potter S.S., Overbeek P., and Kuehn M.R. 1996. Conserved left-right asymmetry of nodal expression and alterations in murine situs inversus (comments). *Nature* **381:** 158.

Melby A.E., Warga R.M., and Kimmel C.B. 1996. Specification of cell fates at the dorsal margin of the zebrafish gastrula. *Development* **122:** 2225.

Meno C., Ito Y., Saijoh Y., Matsuda Y., Tashiro K., Kuhara S., and Hamada H. 1997. Two closely-related left-right asymmetrically expressed genes, lefty-1 and lefty-2: Their distinct expression domains, chromosomal linkage and direct neuralizing activity in *Xenopus* embryos. *Genes Cells* **2:** 513.

Miralto A. and Kemali M. 1980. Asymmetry of the habenulae in the elasmobranch "Scyllium stellare". II. Electron microscopy. *Z. Mikrosk. Anat. Forsch.* **94:** 801.

Nonaka S., Shiratori H., Saijoh Y., and Hamada H. 2002. Determination of left-right patterning of the mouse embryo by artificial nodal flow. *Nature* **418:** 96.

Nonaka S., Tanaka Y., Okada Y., Takeda S., Harada A., Kanai Y., Kido M., and Hirokawa N. 1998. Randomization of left-right asymmetry due to loss of nodal cilia generating leftward flow of extraembryonic fluid in mice lacking KIF3B motor protein. *Cell* **95:** 829.

Norris D.P., Brennan J., Bikoff E.K., and Robertson E.J. 2002. The Foxh1-dependent autoregulatory enhancer controls the level of Nodal signals in the mouse embryo. *Development* **129:** 3455.

Piedra M.E. and Ros M.A. 2002. BMP signaling positively regulates Nodal expression during left right specification in the chick embryo. *Development* **129:** 3431.

Pogoda H.M., Solnica-Krezel L., Driever W., and Meyer D. 2000. The zebrafish forkhead transcription factor FoxH1/Fast1 is a modulator of nodal signaling required for organizer formation. *Curr. Biol.* **10:** 1041.

Rebagliati M.R., Toyama R., Haffter P., and Dawid I.B. 1998a. cyclops encodes a nodal-related factor involved in midline signaling. *Proc. Natl. Acad. Sci.* **95:** 9932.

Rebagliati M.R., Toyama R., Fricke C., Haffter P., and Dawid I.B. 1998b. Zebrafish nodal-related genes are implicated in axial patterning and establishing left-right asymmetry. *Dev. Biol.* **199:** 261.

Sampath K., Cheng A.M., Frisch A., and Wright C.V. 1997. Functional differences among *Xenopus* nodal-related genes in left-right axis determination. *Development* **124:** 3293.

Sampath K., Rubinstein A.L., Cheng A.M., Liang J.O., Fekany K., Solnica-Krezel L., Korzh V., Halpern M.E., and Wright C.V. 1998. Induction of the zebrafish ventral brain and floorplate requires cyclops/nodal signalling. *Nature* **395:** 185.

Schlange T., Arnold H.H., and Brand T. 2002. BMP2 is a positive regulator of Nodal signaling during left-right axis formation in the chicken embryo. *Development* **129:** 3421.

Schlange T., Schnipkoweit I., Andree B., Ebert A., Zile M.H., Arnold H.H., and Brand T. 2001. Chick CFC controls Lefty1 expression in the embryonic midline and nodal expression in the lateral plate. *Dev. Biol.* **234:** 376.

Sirotkin H.I., Gates M.A., Kelly P.D., Schier A.F., and Talbot W.S. 2000. Fast1 is required for the development of dorsal axial structures in zebrafish. *Curr. Biol.* **10:** 1051.

Thisse C. and Thisse B. 1999. Antivin, a novel and divergent member of the TGFbeta superfamily, negatively regulates mesoderm induction. *Development* **126:** 229.

Tsuda T., Philp N., Zile M.H., and Linask K.K. 1996. Left-right asymmetric localization of flectin in the extracellular matrix during heart looping. *Dev. Biol.* **173:** 39.

Von Wollwaerth C. 1950. Experimentelle Untersuchungen über den Situs Inversus der Eingeweide und der Habenula des Zwischenhirns bei Amphibien. *Roux's Arch. Entw. Mech. Organ.* **144:** 178.

Wright C.V. 2001. Mechanisms of left-right asymmetry: What's right and what's left? *Dev. Cell.* **1:** 179.

Yan Y.T., Gritsman K., Ding J., Burdine R.D., Corrales J.D., Price S.M., Talbot W.S., Schier A.F., and Shen M.M. 1999. Conserved requirement for EGF-CFC genes in vertebrate left-right axis formation. *Genes Dev.* **13:** 2527.

Yeo C. and Whitman M. 2001. Nodal signals to Smads through Cripto-dependent and Cripto-independent mechanisms. *Mol. Cell* **7:** 949.

Yokouchi Y., Vogan K.J., Pearse R.V., II, and Tabin C.J. 1999. Antagonistic signaling by Caronte, a novel Cerberus-related gene, establishes left-right asymmetric gene expression. *Cell* **98:** 573.

Zhang, J., Talbot W.S., and Schier A.F. 1998. Positional cloning identifies zebrafish one-eyed pinhead as a permissive EGF-related ligand required during gastrulation. *Cell* **92:** 241.

Zhu L., Marvin M.J., Gardiner A., Lassar A.B., Mercola M., Stern C.D., and Levin M. 1999. Cerberus regulates left-right asymmetry of the embryonic head and heart. *Curr. Biol.* **9:** 931.

Cardiac Left–Right Development: Are the Early Steps Conserved?

K.L. KRAMER AND H.J. YOST

*Center for Children, Huntsman Cancer Institute, Department of Oncological Sciences,
University of Utah, Salt Lake City, Utah 84112*

The external features of the vertebrate body plan are bilaterally symmetric. In contrast, many of the internal organs, including the heart, gut, and regions of the brain, are organized in left–right (LR) asymmetric patterns that are consistently organized with respect to the dorsal–ventral (DV) and rostral–caudal axes (in adult anatomy; the anterior–posterior or AP axis in embryonic anatomy). Looping of the cardiac tube along the embryonic LR axis is a critical step in the development of the cardiovascular system. The correct direction and extent of cardiac tube looping allows subsequent events, such as septation, valve formation, and outflow tract development, to occur normally. Defects in this pathway lead to a wide range of cardiovascular and other developmental defects, from phenotypes traditionally associated with situs anomalies, such as heterotaxy and situs inversus, to more frequent forms of complex congenital heart defects. Collectively, laterality defects form a significant portion of the morbidity and mortality of congenital heart malformations.

INTERMEDIATE STEPS IN LR PATHWAYS ARE CONSERVED AMONG VERTEBRATES

The general mechanisms of cardiac tube looping along the embryonic LR axis are highly conserved in vertebrates, so it is reasonable to expect that the underlying genetic pathways and developmental mechanisms that control LR development are also highly conserved. In the past 7 years, the field of LR development has seen an explosion of information, with more than 50 genes implicated in LR development, and the list is growing rapidly (Bisgrove et al. 2003). From studies in several vertebrate model systems, it is clear that a complex genetic pathway in the early embryo establishes LR axis information well before LR organ morphogenisis begins (Burdine and Schier 2000; Capdevila et al. 2000; Hamada et al. 2002).

The field at this point could be described as a series of isolated islands; some of the footpaths on each island are intensely studied, but the routes between the islands are poorly charted. For example, most of the heart myocardium and gut mesoderm are derived from primordia that arise in the lateral plate mesoderm (LPM) during neurulation. In the left LPM, it is clear that the path involves Nodal (a TGFβ-family member) and Lefty (an inhibitory TGFβ-family member) working with an EGF-CFC coreceptor (epidermal growth factor-Crypto, FRL-1, Cryptic; *one-eyed pinhead* in zebrafish) to activate an ALK transmembrane receptor complex and signal through SMAD and FAST1 proteins to activate transcription of *pitx2* (a bicoid-family transcription factor) via its asymmetric regulatory enhancer elements (Burdine and Schier 2000; Capdevila et al. 2000; Hamada et al. 2002). The left-sided LPM expression of *nodal* and *lefty2* is transient, disappearing before the LPM gives rise to the cardiac tube at the ventral midline of the embryo. *pitx2* expression persists into organogenesis. Very little is known about genes that are asymmetrically expressed in response to *pitx2* expression, or the downstream cellular mechanics that drive LR morphogenesis (i.e., what drives the bending, twisting, and looping of the cardiac tube). Nonetheless, it is clear that genetic, molecular, or embryological manipulations that perturb the asymmetric expression of *nodal*, *lefty2*, and *pitx2* can also alter at least some components of LR cardiac morphogenesis.

It is important to note that a molecule in the LR pathway could function asymmetrically, as a LR instructive factor, or symmetrically, as a bilaterally permissive factor. In a well-characterized example, the EGF-CFC proteins have been implicated in LR development in several vertebrates (Gaio et al. 1999; Yan et al. 1999; Bamford et al. 2000; Shen and Schier 2000), but they are symmetrically expressed LR permissive factors, not LR instructive factors. EGF-CFCs are cofactors that are essential for signaling by the TGFβ ligand Nodal (Gritsman et al. 1999), and it is asymmetric *nodal* expression that instructs normal LR development in LPM (Gaio et al. 1999; Yan et al. 1999; Bamford et al. 2000). In comparison, the TGFβ transmembrane receptors and components of the downstream intracellular signaling pathways are symmetrically expressed and should be considered permissive, not instructive, factors in LR development. Although genetics and molecular embryology are uncovering a rapidly growing number of genes implicated in LR development, rigorous distinctions between permissive and instructive developmental factors have not been applied in this field.

IF AN INTERMEDIATE DEVELOPMENTAL STEP IS CONSERVED, MUST THE PRECEDING STEPS BE CONSERVED?

One of the principal unanswered questions in the field of LR development is whether the molecular mechanisms preceding asymmetric expression of *nodal*, *lefty2*, and *pitx2* are the same in all classes of vertebrates (Wright 2001; Hamada et al. 2002). Although it is tempting to assume that all steps that developmentally precede a highly conserved step must also be highly conserved, this is not necessarily the case. A clear illustration is gastrulation, the process in which mesodermal and endodermal cells move from the surface of the embryo to the interior.

The process of gastrulation is distinct in different classes of vertebrates, in part because distinct geometrical constraints are placed upon each gastrula in order to provide nutrients and gaseous exchange before the advent of the cardiovascular circulatory system. For example, holoblastic cell cleavages divide early amphibian embryos so that large quantities of yolk are stored in each cell. The spherical gastrula must then redistribute a relatively small number of large yolk-laden cells. In contrast, the yolk that provides nutrition to the zebrafish embryo is extraembryonic; the dome-shaped embryo envelops the yolk cell during epiboly and gastrulation, essentially internalizing the nonembryonic yolk cell. Similarly, the yolk cell remains extraembryonic in chick gastrula, allowing relatively large numbers of yolk-free cells to divide and migrate rapidly within a flat disk on top of the yolk cell during gastrulation. The mammalian embryo is not provided with yolk and must rapidly generate extraembryonic membranes in order to implant into the uterine wall and thereby obtain nutrients and exchange gases. Consequently, the embryonic cells of the cylindrical mouse gastrula are able to divide and migrate rapidly.

Although the mechanics of gastrulation appear strikingly distinct between vertebrate embryos, all vertebrate post-gastrula embryos form a highly conserved AP axis and have relatively similar external morphology during neurula and tailbud stages. The expression patterns of homoeobox gene family members are also remarkably conserved along the AP axis. The developmental period during which disparate forms of gastrula converge upon a highly similar post-gastrula form is called the phylotypic period (Arthur 2002). Morphologies and gene expression patterns that precede the phylotypic period are much less constrained evolutionarily (Galis and Metz 2001). We propose that a similar developmental convergence from disparate forms to a highly conserved pattern of *nodal*, *lefty*, and *pitx2* expression in LPM might occur in LR development. This opens the possibility that the earliest biomechanical and molecular LR asymmetries, preceding the highly conserved asymmetric expression of genes in the LPM during neurula and tailbud stages, are distinct among different classes of vertebrates.

Alternatively, the earliest steps in establishing LR asymmetries could be conserved but not well described in different classes of vertebrates. This suggestion in LR development is paralleled by the observation in gastrulation that the principal cell mechanics and movements are remarkably conserved among vertebrates (Schoenwolf and Smith 2000). For example, cell convergence during vertebrate gastrulation constricts the three germ layers mediolaterally while cell extension lengthens the embryo along the AP axis. In either gastrulation or LR development, the mechanisms are not sufficiently understood in multiple classes of vertebrates to allow one to conclude whether the developmental pathways are convergent or conserved. The predominant problem is that a given developmental pathway is discovered in one class of vertebrates at a time, and not yet examined in multiple vertebrate classes used in experimental embryology.

MONOCILIATED NODE CELLS ARE PRESENT IN SEVERAL VERTEBRATE MODEL SYSTEMS

Spemann's organizer is a small region in the amphibian embryo that is essential for gastrulation and formation of the bilateral midline and AP axis. When organizer cells are transplanted to other areas of the embryo, the transplanted cells induce adjacent cells to form a secondary embryonic axis, resulting in conjoined twins (Oelgeschlager et al. 2003). Cells with similar functions and patterns of gene expression are collectively called the shield in zebrafish and the node in chick and mouse. The node has a conserved role in AP axis formation, and work described below indicates that it is important in LR development as well. When the node is surgically reoriented in chick (Pagan-Westphal and Tabin 1998), or the organizer is altered by ectopic gene expression in *Xenopus* (Danos and Yost 1995), LR development is disrupted, suggesting that the node or organizer cells have a conserved role in LR development.

In late gastrula-stage mice embryos, a single monocilium projects from the surface of each node cell. These monocilium are motile, and fluorescent beads placed in the node move predominantly from the right to the left side of the node. These observations led to an elegant model in which extracellular fluid flow in the node, driven by monocilia, breaks left–right asymmetry across the midline (Nonaka et al. 2002). It is hypothesized that the leftward nodal flow moves or concentrates an extracellular molecule to the left side of the node. This extracellular molecule, perhaps a peptide growth factor, then activates left-sided signaling pathways. Alternatively, the nodal flow could differentially activate a biomechanical signal across the LR axis. Several genes have been implicated in the formation and function of monocilia in node cells, including members of kinesin and dynein families of the microtubule-dependent motor complexes. Mutations in these genes result in altered formation or movement of monocilia, altered nodal flow, changes in LPM asymmetric gene expression patterns, and aberrant LR heart morphogenesis.

At least some components of the nodal flow hypothesis have been demonstrated in other classes of vertebrates. A gene related to the *left-right dynein* (*lrd*) in mice is expressed in a small group of cells in chick, frog, and zebrafish embryos (Essner et al. 2002). These cells form monocilia during late gastrulation, suggesting that the nodal flow mechanism is conserved. However, the functions of these monociliated cells have not been reported. Below, we summarize recent evidence that monociliated cells function in LR development in zebrafish, and that several steps in LR development precede the appearance of monociliated cells in *Xenopus*. Together, this suggests that monociliated cells have a role in LR development, but not as the first step in establishing LR asymmetry.

MONOCILIATED CELLS FUNCTION IN LR DEVELOPMENT IN ZEBRAFISH

In zebrafish, *lrd* is expressed in dorsal forerunner cells (DFCs). DFCs are a small group (~25 cells) at the leading edge of the zebrafish shield (Essner et al. 2002), and their function was previously unknown (Alexander et al. 1999). DFCs do not involute with other cells during gastrulation, but remain within the organizing center of the zebrafish shield, giving rise after gastrulation to a specialized epithelial structure in the tailbud of the embryo known as Kupffer's vesicle (Cooper and D'Amico 1996). Monocilia develop on the DFC progeny within Kupffer's vesicle (Essner et al. 2002).

On the basis of the expression of *lrd* in DFCs and the appearance of monocilia in Kupffer's vesicle, we used three approaches to assess how these cells might function in LR development (J. Essner et al., unpubl.). First, laser-mediated ablation of DFCs and microsurgical ablation of Kupffer's vesicle altered the expression patterns of LPM markers, as well as the LR orientation of the heart. These treatments did not alter the morphology or gene expression patterns in the embryonic midline (notochord and neural floorplate), indicating that the effects on LR development were not due to perturbations of midline development.

Second, several zebrafish mutants that have LR developmental defects (Bisgrove et al. 2003) were examined for defects in *lrd* expression in DFCs. Mutations in the gene encoding the T-box transcription factor *no tail* (a homolog of mouse *brachyury*) and mutations in genes that encode components of the *nodal* signaling pathway result in a loss of *lrd* expression in DFCs. This indicates that two parallel pathways, *no tail* transcriptional activity and *nodal* signaling, are upstream of *lrd* expression in DFCs. These results suggest that the absence of *lrd* in DFCs results in LR developmental defects. However, it should be noted that these mutants have later defects in midline formation, which are probably not due to DFC defects (since DFC ablation has no effects on the midline) but which might contribute to altered LPM gene expression patterns and subsequent heart LR defects.

Third, since a mutation in *lrd* is not currently available in zebrafish, we used antisense morpholinos to block *lrd* expression. Injection of antisense morpholinos against *lrd* results in LR defects, directly demonstrating that *lrd* has a role in LR development in zebrafish. Together, these results indicate that monociliated cells, and *lrd* expression within these monociliated cells, have roles in LR development that are conserved from mice to zebrafish.

ASYMMETRIES OCCUR BEFORE THE APPEARANCE OF MONOCILIA IN *XENOPUS*

Mesodermally derived monociliated node cells exist in several classes of vertebrates and function in mice and zebrafish LR development (at least), indicating that this early step in LR development is conserved. Nonetheless, it has recently become clear that LR asymmetries exist and are required in *Xenopus* embryos before the developmental appearance of monocilia. Figure 1 is a relative time line of these early steps, summarizing a variety of embryological, pharmacological, and molecular manipulations that alter LR development (each step is discussed in the following sections). It is important to point out that we do not know whether any step is contingent upon a step that precedes it in the time line, with the exception of protein kinase Cγ (PKCγ) and Syndecan-2, which are epistatic in the right embryonic ectoderm.

Results from embryological manipulations suggested that the animal cap ectoderm in *Xenopus* gastrula-stage embryos directly patterns the underlying mesoderm with LR axis information (Yost 1992). To understand how transient cell interactions between ectoderm and mesoderm are involved in LR development, a brief overview of the geometric relationships in the *Xenopus* blastula and gastrula is important (Fig. 2). Before gastrulation, the embryo is a ball of cells with an internal extracellular space, called the blastocoel. The cells at the top (animal pole) are fated to form ectoderm. The basal surfaces of these "animal cap" ectoderm cells line the "roof" of the blastocoel space. The cells at the equator are fated to form mesoderm (red), and the cells at the base (vegetal pole), including cells that make the "floor" of the blastocoel, are fated to form endoderm. During the process of gastrulation, mesoderm cells converge toward the dorsal side of the embryo (the region containing the Spemann orga-

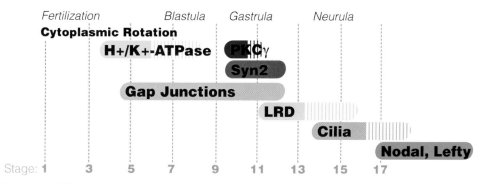

Figure 1. Time line of *Xenopus* LR development. Each step is listed according to when it is susceptible to perturbations (H+/K+-ATPase, gap junctions, PKCγ, and Syndecan-2) or when it appears in development (asymmetric *H+/K+-ATPase* α-subunit RNA, asymmetric phosphorylation states of Syndecan-2, *LR dynein* (*lrd*), node monocilia, and asymmetric expression of *nodal* and *lefty* in LPM). See text for references. (Adapted from Kramer et al. 2002.)

nizer) and involute toward the animal pole, migrating (indicated by arrows) on the basal surface of ectoderm cells.

SYNDECANS IN THE ECTODERM SIGNAL TO MIGRATING MESODERM

Interestingly, the heparan sulfate proteoglycans (HSPGs) s*yndecan-1* and *syndecan-2* are specifically expressed in the deep layer of ectoderm cells that interacts with the mesoderm during gastrulation (Fig. 2, blue) (Teel and Yost 1996), indicating that mesodermal cells contact Syndecan ectodomains during gastrulation. Embryological and biochemical results indicated that HSPGs are important for LR development (Yost 1990, 1992), so we used several complementary approaches to determine whether Syndecans regulate LR development. Perturbation of Syndecan-2 function specifically disrupts the asymmetric expression of *nodal*, *lefty*, and *pitx2* in the mesoderm (Kramer and Yost 2002). Furthermore, Syndecan-2 selectively binds and mediates Vg1 (a TGFβ cell signaling molecule), and asymmetric overexpression of an active Vg1 ligand or its activated receptor in the mesoderm rescues the normal LR axis in embryos where Syndecan-2 function is disrupted in the ectoderm. These results suggest that Syndecan-2 transmits LR information to migrating mesoderm by functioning as a *trans*-acting, cell nonautonomous cofactor of Vg1.

But how does Syndecan-2 function in LR development? Syndecan-2 is phosphorylated in the right ectoderm during early gastrulation, concurrent with the period during which Syndecan-2 and PKCγ activity are required in LR development (Fig. 3) (Kramer et al. 2002). During this period, left mesendoderm migrates across ectoderm that expresses nonphosphorylated Syndecan-2 while right mesendoderm migrates across ectoderm that expresses phosphorylated Syndecan-2 (Fig. 4). Interestingly, both phosphorylated Syndecan-2 in the right ectoderm and nonphosphorylated Syndecan-2 in the left ectoderm have roles in LR development. PKCγ phosphorylates Syndecan-2, and loss of PKCγ activity can be rescued by expression of phospho-mimetic Syndecan-2, indicating that PKCγ is upstream of Syndecan-2 in the same instructive LR developmental pathway. Collectively, these results suggest that Syndecan-2 directly patterns the mesoderm (Fig. 5) and that the mesoderm, including the mesodermally derived node, does not receive LR patterning information until gastrulation. Several results indicate that LR patterning of the ectoderm occurs even earlier in development.

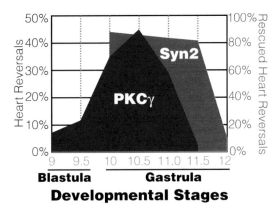

Figure 3. PKCγ and Syndecan-2 function during overlapping gastrula stages. Percent heart reversals (*blue*) in embryos treated with conventional PKC inhibitors (blocking PKCγ) at various blastula and gastrula stages of *Xenopus* development. For Syndecan-2 (Syn2, *red*), embryos were first injected with a dominant negative *syndecan-2* at the 32-cell stage. Dominant negative-treated embryos were then treated at several gastrula stages with an enzyme that selectively sheds the dominant negative from the cell surface and rescues normal LR development. Syndecan-2 rescue data are expressed as the percent rescued heart reversals to best illustrate the functional overlap of PKCγ and Syndecan-2 during gastrulation. (Data from Kramer and Yost 2002; Kramer et al. 2002.).

IS LR ESTABLISHED DURING THE FIRST CELL CYCLE AFTER FERTILIZATION?

In *Xenopus*, the cytoplasm of the fertilized egg rotates during the first cell cycle to establish bilateral symmetry

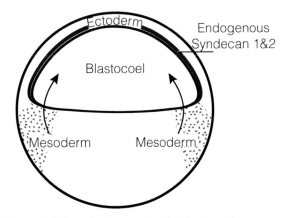

Figure 2. Schematic diagram of a blastula-stage *Xenopus* embryo. The cells at the equator are fated to form mesoderm (*red*), and the cells at the base (vegetal pole), including cells that make the "floor" of the blastocoel, are fated to form endoderm. During gastrulation, the leading edges of mesoderm cells (indicated by *arrows*) migrate on the basal surface of ectoderm cells, contacting Syndecan-1 and Syndecan-2 (*blue*). (Adapted from Kramer and Yost 2002.)

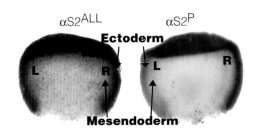

Figure 4. Endogenous Syndecan-2 is asymmetrically phosphorylated. Syndecan-2 protein is present in both left and right ectoderm, as detected by an antibody that recognizes all forms of Syndecan-2, regardless of phosphorylation (*left*). However, antibodies that recognize only phosphorylated forms of Syndecan-2 (*right*) indicate that Syndecan-2 is phosphorylated in right-side ectoderm (R) but not in left-side ectoderm (L).

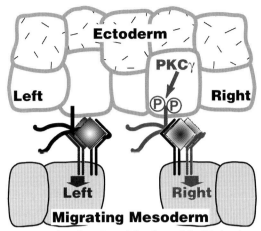

Figure 5. Roles of PKCγ and Syndecan-2 in LR development. Schematic illustrates inside-out transmission of LR information from ectoderm to migrating mesoderm during gastrulation. The function of PKCγ is necessary in ectoderm cells on the right, but not on the left, for normal LR development. Each phosphorylation state (off in the left cells, on in the right cells) changes the structure or function of Syndecan-2, which then regulates whether migrating mesoderm receives a "left" or "right" signal. The LR information transduced through Syndecan-2 leads to asymmetric patterning of the mesoderm where *nodal*, *lefty*, and *pitx2* are expressed only in the left mesoderm.

and the dorsoanterior body axes (for review, see Gerhart et al. 1989; Yost 1995). Normally, microtubule-dependent motors drive this rotation. The rotation can be blocked with a variety of inhibitors of microtubule assembly, including UV irradiation of the vegetal pole. In embryos that have been UV-irradiated during the first cell cycle, the cytoplasmic rotation does not occur, and development proceeds to give rise to "ventralized," radially symmetric embryos. During the first cell cycle, these UV-irradiated embryos can be tilted with respect to gravity, which drives the internal cytoplasm to rotate (the vegetal cytoplasm is heavier than the animal pole cytoplasm) and "rescues" the embryos back to normal dorsoanterior development (Gerhart et al. 1989). These results indicate that dorsoanterior development requires the cytoplasmic rotation to move factors from the vegetal pole to the prospective dorsal side. Furthermore, as long as the factors are moved (by gravity or transplantation) in the absence of microtubules, normal dorsoanterior development can occur (Holowacz and Elinson 1995; Miller et al. 1999).

In contrast, we observed that these otherwise normal "rescued" embryos have altered cardiac and visceral left–right development, suggesting that the microtubules in the first cell cycle are necessary for normal LR development (Yost 1991; Danos and Yost 1995). This suggests that LR asymmetry is established by events in the first cell cycle, but it is possible that this early manipulation has effects that do not ramify until late in development. Because these observations were made before the advent of LR molecular markers, we do not know whether events in the first cell cycle control the expression of *lrd* in the organizer or the asymmetric expression of *nodal*, *lefty*, and *pitx-2* in LPM.

H+/K+–ATPASE IN ECTODERM CELLS

Temporally, the next event (Fig. 1) is the transient asymmetric distribution of a maternal RNA encoding the α-subunit of *H+/K+-ATPase* (Levin et al. 2002). This RNA is diminished in the left ventral animal cap cell compared to the right ventral animal cap cell in 4-cell embryos. At midblastula transition (MBT), the zygotic genome is transcriptionally activated and the α-subunit of *H+/K+-ATPase* RNA again becomes symmetrically distributed. Pharmacological inhibitor studies indicate that normal LR development requires H+/K+-ATPase activity from the 4-cell stage (stage 2) to approximately stage 6 to 8 (MBT).

Co-injections of the α- and β-subunits of *H+/K+-ATPase* as well as *Kir4.1* (an outward K+ current channel) into the 1-cell embryo alter LR development (Levin et al. 2002). All three RNAs have to be coexpressed in order to alter LR development; overexpression of α-subunit of *H+/K+-ATPase* alone has no effect. Asymmetric injections of these three RNAs (i.e., into just the left side in 4-cell embryos) did not alter LR development, perhaps because of a delay in protein synthesis. In contrast to our studies that show a strong LR asymmetry in PKCγ function (Kramer et al. 2002), it is not clear whether H+/K+-ATPase α-subunit protein (antibodies are not currently available) or activity is asymmetric in early LR development, although the transient asymmetric distribution of the α-subunit RNA suggests that it is asymmetric.

GAP JUNCTIONS

In the frog 32-cell embryo, gap junction communication appears to be asymmetric along the DV axis, such that dorsal cells are more coupled than ventral cells when assessed by dye transfer (Olson et al. 1991). This analysis indicates that the two cells on the ventral side are junctionally isolated across the ventral midline. Gap junctions have been strongly implicated in LR development. Injections of connexin RNAs (*Cx26*, *Cx37*, or *Cx43*) into both ventral cells to increase gap junction communication, or dorsal injection of a dominant negative connexin (*H7* construct) that blocks gap junction communication, result in altered LR development (Levin and Mercola 1998). To alter LR development by connexin overexpression, a cell lineage on each side of the ventral midline (i.e., both a left and a right cell) must be injected with *connexin* RNA, probably because both cells need to synthesize connexin in order to form functional junctions across the ventral midline. This observation emphasizes the importance of junctional isolation across the ventral midline for normal LR development. Pharmacological agents that alter gap junction communication can result in perturbed LR development, if applied to embryos during a period ranging from stage 5 (blastula) through stage 12 (late gastrula) (Levin and Mercola 1998). Thus, the developmental period during which gap junctions is fairly long, both preceding and following the more specific stages during which PKCγ and Syndecan-2 are required and extending into the period of *lrd* expression in the organizer (Fig. 1).

DO EARLY STEPS EXIST IN OTHER VERTEBRATES?

We do not know enough about early LR development in multiple organisms to establish whether the early steps identified in *Xenopus* are conserved among vertebrates (Table 1). Although it is clear in *Xenopus* and other species that the DV and AP axes are patterned by a fertilization-driven cortical rotation in the 1-cell embryo (Vincent and Gerhart 1987), only recently has evidence suggested that this mechanism occurs in mouse (Piotrowska and Zernicka-Goetz 2002). If the DV and AP axes are established during the first cell cycle in mouse, the key question becomes, What would be the advantage of not initiating LR axis formation in mouse until much later in development, near the end of gastrulation? One possibility is that late LR axis specification around the node evolved because cells near the chick and mouse node proliferate more rapidly than in other vertebrates (Mathis and Nicolas 2002); it might be difficult to maintain earlier LR specification within a large group of cells that are contributing to the embryonic axes. Another possibility is that the distinct geometries of vertebrate gastrulation (as described above) preclude efficient initiation of LR development before gastrulation. Alternatively, but not exclusively, specification of the LR axis in a short period more proximal to the phylotypic period would reduce vulnerability to perturbations during the initial phase of embryonic development (Galis and Metz 2001). The unique asymmetric expression of several genes in the chick node might therefore be viewed as an advantageous developmental reprogramming that makes earlier steps unnecessary. However, asymmetric gene expression in the chick node is patterned by adjacent tissue (Pagan-Westphal and Tabin 1998), suggesting that earlier LR developmental steps occur in chick. Consequently, steps in LR development that precede monocilium formation and link LR axis formation with the generation of the DV and AP axes are likely to exist in other vertebrates. If the LR axis is established during the first cell cycle in all vertebrates, monocilia would be relegated to an intermediate step required for the maintenance, but not initiation, of LR asymmetry.

CONCLUSIONS: A MODEL THAT LINKS THE EARLY STEPS

With the exception of the connection of PKCγ and Syndecan-2 asymmetric function, we do not know whether any of the aforementioned events (first cell cycle rotation; H+/K+-ATPase activity, gap junction communication, and monocilium function) occur in the same LR developmental pathway or in parallel pathways. Thus, the field currently has a series of isolated islands of knowledge about the earliest steps in LR development that precede monocilium function. We propose a model that charts the routes between the islands. Although it is possible that each of these steps occurs in independent, parallel pathways, each of which is necessary for normal LR development, we propose that these steps are components of a unified LR pathway.

The advantage of studying frog embryos is that specific cell lineages can be readily identified, and results from manipulating the function of a molecule in specific cell lineages can provide hints that multiple molecules are linked in a developmental pathway. Two striking observations lead us to suggest that these disparate steps in left–right development are in the same pathway. First, both the high accumulation of *H+/K+-ATPase* α-subunit RNA during the 4-cell stage and the PKCγ-dependent phosphorylation of Syndecan-2 during early gastrulation occur in the same cell lineage, derived from the animal cap ectoderm cell on the ventral right side. Second, junctionally isolated cell lineages (i.e., cells on the right side vs. left side of the ventral midline) respond very differently to MO knockdown of PKCγ function or to the expression of Syndecan-2 phospho-state mutants (Kramer et al. 2002). In contrast, the junctionally connected cells have similar LR responses to PKCγ or Syndecan-2 manipulations. Together, these observations suggest that establishment or maintenance of the distinct Syndecan-2 phosphorylation states in left and right lineages requires junctional isolation.

Putting these cell lineage correlations together, we suggest that the cytoplasmic rotation in the first cell cycle drives the asymmetric accumulation of H+/K+-ATPase RNA in ventral right-side ectoderm. This cytoplasmic rotation also drives DV axis formation (Gerhart et al. 1989), which we propose forms junctional isolation across the ventral midline. This would explain how the mechanisms that establish the LR axis and the DV axis are linked in the first cell cycle after fertilization (Yost 1995). Once the left and right cell lineages are junctionally isolated in the early blastula stages, the asymmetric H+/K+-ATPase activity could then drive a PKCγ activator in the right cell lineages. PKCγ then phosphorylates Syndecan-2 in right cell lineages, and the absence of a PKCγ activator in the junctionally isolated left cell lineages prevents the phosphorylation of Syndecan-2 in left cell lineages. This results in phosphorylated Syndecan-2 in right-side ectoderm and nonphosphorylated Syndecan-2 in left-side ectoderm. The phosphorylation states of Syndecan-2 con-

Table 1. Mechanisms That Precede Phylotypic LR Asymmetric Gene Expression in LPM

	1st Cell cycle	H+/K+ ATPase	PKCγ	Syn2	Gap junction	Nodal cilia	Node asymmetry	Midline	Asymmetric TGFβ signaling
Fish	?					+	–	+	
Frog	+	+	+	+	+	+	–	+	+
Chick	?	+			?	+	+	–	+
Mouse	?					+	–	+	

trol differential cell nonautonomous signaling to migrating mesoderm during gastrulation (Kramer and Yost 2002; Kramer et al. 2002). The migrating mesoderm receives this asymmetric information and utilizes it to activate the left-sided *nodal-lefty-pitx2* pathway in left LPM and a distinct pathway in right LPM. The LR asymmetric information in LPM is transmitted to the precardiac mesoderm before cardiac tube formation and is utilized during LR cardiac looping. Although steps that precede monocilium formation in LR development likely exist in other vertebrates, the early steps might not be conserved because they occur during a period in vertebrate development that is less conserved evolutionarily.

ACKNOWLEDGMENTS

We thank Jeff Amack, Brent Bisgrove, Jeff Essner, Erin Harris, and Molly Wagner in our lab for sharing unpublished observations. K.L.K. was supported by a NRSA from National Institutes of Health/NIHLB. Core facilities are supported by National Cancer Institute CCSG. This work was supported by grants to H.J.Y. from an NIH/NIHLB SCOR program and the Huntsman Cancer Foundation.

REFERENCES

Alexander J., Rothenberg M., Henry G.L., and Stainier D.Y. 1999. casanova plays an early and essential role in endoderm formation in zebrafish. *Dev. Biol.* **215:** 343.

Arthur W. 2002. The emerging conceptual framework of evolutionary developmental biology. *Nature* **415:** 757.

Bamford R.N., Roessler E., Burdine R.D., Saplakoglu U., dela Cruz J., Splitt M., Towbin J., Bowers P., Marino B., Schier A.F., Shen M.M., Muenke M., and Casey B. 2000. Loss-of-function mutations in the EGF-CFC gene CFC1 are associated with human left-right laterality defects. *Nat. Genet.* **26:** 365.

Bisgrove B.W., Morelli S.H., and Yost H.J. 2003. Genetics of human laterality disorders: Insights from vertebrate model systems. *Annu. Rev. Genomics Hum. Genet.* **3:** (in press).

Burdine R.D. and Schier A.F. 2000. Conserved and divergent mechanisms in left-right axis formation. *Genes Dev.* **14:** 763.

Capdevila J., Vogan K.J., Tabin C.J., and Izpisua Belmonte J.C. 2000. Mechanisms of left-right determination in vertebrates. *Cell* **101:** 9.

Cooper M.S. and D'Amico L.A. 1996. A cluster of noninvoluting endocytic cells at the margin of the zebrafish blastoderm marks the site of embryonic shield formation. *Dev. Biol.* **180:** 184.

Danos M.C. and Yost H.J. 1995. Linkage of cardiac left-right asymmetry and dorsal-anterior development in *Xenopus*. *Development* **121:** 1467.

Essner J.J., Vogan K.J., Wagner M.K., Tabin C.J., Yost H.J., and Brueckner M. 2002. Conserved function for embryonic nodal cilia. *Nature* **418:** 37.

Gaio U., Schweickert A., Fischer A., Garratt A.N., Muller T., Ozcelik C., Lankes W., Strehle M., Britsch S., Blum M., and Birchmeier C. 1999. A role of the cryptic gene in the correct establishment of the left-right axis. *Curr. Biol.* **9:** 1339.

Galis F. and Metz J.A. 2001. Testing the vulnerability of the phylotypic stage: On modularity and evolutionary conservation. *J. Exp. Zool.* **291:** 195.

Gerhart J., Danilchik M., Doniach T., Roberts S., Rowning B., and Stewart R. 1989. Cortical rotation of the *Xenopus* egg: Consequences for the anteroposterior pattern of embryonic dorsal development. *Development* **107:** 37.

Gritsman K., Zhang J., Cheng S., Heckscher E., Talbot W.S., and Schier A.F. 1999. The EGF-CFC protein one-eyed pinhead is essential for nodal signaling. *Cell* **97:** 121.

Hamada H., Meno C., Watanabe D., and Saijoh Y. 2002. Establishment of vertebrate left-right asymmetry. *Nat. Rev. Genet.* **3:** 103.

Holowacz T. and Elinson R.P. 1995. Properties of the dorsal activity found in the vegetal cortical cytoplasm of *Xenopus* eggs. *Development* **121:** 2789.

Kramer K.L. and Yost H.J. 2002. Ectodermal syndecan-2 mediates left-right axis formation in migrating mesoderm as a cell-nonautonomous Vg1 cofactor. *Dev. Cell* **2:** 115.

Kramer K.L., Barnette J.E., and Yost H.J. 2002. PKCgamma regulates syndecan-2 inside-out signaling during *Xenopus* left-right development. *Cell* **111:** 981.

Levin M. and Mercola M. 1998. Gap junctions are involved in the early generation of left-right asymmetry. *Dev. Biol.* **203:** 90.

Levin M., Thorlin T., Robinson K., Nogi T., and Mercola M. 2002. Asymmetries in H(+)/K(+)-ATPase and cell membrane potentials comprise a very early step in left-right patterning. *Cell* **111:** 77.

Mathis L. and Nicolas J.F. 2002. Cellular patterning of the vertebrate embryo. *Trends Genet.* **18:** 627.

Miller J.R., Rowning B.A., Larabell C.A., Yang-Snyder J.A., Bates R.L., and Moon R.T. 1999. Establishment of the dorsal-ventral axis in *Xenopus* embryos coincides with the dorsal enrichment of dishevelled that is dependent on cortical rotation. *J. Cell Biol.* **146:** 427.

Nonaka S., Shiratori H., Saijoh Y., and Hamada H. 2002. Determination of left right patterning of the mouse embryo by artificial nodal flow. *Nature* **418:** 96.

Oelgeschlager M., Kuroda H., Reversade B., and De Robertis E.M. 2003. Chordin is required for the Spemann organizer transplantation phenomenon in *Xenopus* embryos. *Dev. Cell* **4:** 219.

Olson D.J., Christian J.L., and Moon R.T. 1991. Effect of wnt-1 and related proteins on gap junctional communication in *Xenopus* embryos. *Science* **252:** 1173.

Pagan-Westphal S.M. and Tabin C.J. 1998. The transfer of left-right positional information during chick embryogenesis. *Cell* **93:** 25.

Piotrowska K. and Zernicka-Goetz M. 2002. Early patterning of the mouse embryo: Contributions of sperm and egg. *Development* **129:** 5803.

Schoenwolf G.C. and Smith J.L. 2000. Gastrulation and early mesodermal patterning in vertebrates. *Methods Mol. Biol.* **135:** 113.

Shen M.M. and Schier A.F. 2000. The EGF-CFC gene family in vertebrate development. *Trends Genet.* **16:** 303.

Teel A.L. and Yost H.J. 1996. Embryonic expression patterns of *Xenopus* syndecans. *Mech. Dev.* **59:** 115.

Vincent J.P. and Gerhart J.C. 1987. Subcortical rotation in *Xenopus* eggs: An early step in embryonic axis specification. *Dev. Biol.* **123:** 526.

Wright C.V. 2001. Mechanisms of left-right asymmetry: What's right and what's left? *Dev. Cell* **1:** 179.

Yan Y.T., Gritsman K., Ding J., Burdine R.D., Corrales J.D., Price S.M., Talbot W.S., Schier A.F., and Shen M.M. 1999. Conserved requirement for EGF-CFC genes in vertebrate left-right axis formation. *Genes Dev.* **13:** 2527.

Yost H.J. 1990. Inhibition of proteoglycan synthesis eliminates left-right asymmetry in *Xenopus laevis* cardiac looping. *Development* **110:** 865.

———. 1991. Development of the left-right axis in amphibians. *Ciba Found. Symp.* **162:** 165.

———. 1992. Regulation of vertebrate left-right asymmetries by extracellular matrix. *Nature* **357:** 158.

———. 1995. Vertebrate left-right development. *Cell* **82:** 689.

Erythropoietin and Retinoic Acid Signaling in the Epicardium Is Required for Cardiac Myocyte Proliferation

I. STUCKMANN AND A.B. LASSAR

Department of Biological Chemistry and Molecular Pharmacology, Harvard Medical School, Boston, Massachusetts 02115

Cardiac myocytes proliferate until around the time of birth in the mouse (Soonpaa et al. 1996) and the rat (Li et al. 1996) or the time of hatching in the chick (Li et al. 1997). The major proliferation-driven increase in thickness of the ventricular wall of the heart takes place from E11.5 to E14.5 in the mouse (Erokhina 1968) and from E8 to E14 in the chick (Rychterova 1971, 1978). The maintained proliferation of cardiac myocytes at these embryonic stages results in the expansion of the ventricular wall. The region of the heart that shows the highest rate of proliferation lies adjacent to the epicardium (Tokuyasu 1990) and is termed the compact zone. Postnatally, the further increase in the size of the heart is generally attributed to hypertrophy of cardiac myocytes (Li et al. 1996, 1997; Soonpaa et al. 1996). The precise mechanisms that regulate cardiac myocyte cell proliferation are largely unknown. By analyzing the clonal progeny of chick heart cells infected with viruses encoding dominant-negative fetal growth factor (FGF) receptors, Mikawa and coworkers have demonstrated an FGF-dependent, early phase and an FGF-independent, late phase for cardiac myocyte proliferation (Mima et al. 1995). Whereas cardiac myocytes infected at E3 with a retrovirus encoding a dominant-negative FGF receptor gave rise to small colonies of cells relative to a control retrovirus infection, cardiac myocytes infected at E7, during the time of ventricular wall expansion, gave rise to normal-size colonies. This finding suggested that only an early phase of cardiac myocyte proliferation or survival requires FGF signals (Mima et al. 1995).

In addition to FGF, a number of other molecules have been implicated as playing a role in cardiac myocyte proliferation. Several knockout mice have been described that show a hypoplastic cardiac phenotype associated with a thin ventricular wall, or reduced trabeculation, suggesting a possible role in cardiac myocyte proliferation for such diverse molecules as RARα (Kastner et al. 1997), RXRα (Sucov et al. 1994), VCAM-1 (Kwee et al. 1995), α-integrin (Yang et al. 1995), erythropoietin (Wu et al. 1999), erythropoietin receptor (Wu et al. 1999), neuregulin (Meyer and Birchmeier 1995), erbB2 (Lee et al. 1995), erbB4 (Gassmann et al. 1995), N-myc (Moens et al. 1993), TEF-1 (Chen et al. 1994), WT-1 (Kreidberg et al. 1993), gp 130 (Yoshida et al. 1996), Jak2 (Neubauer et al. 1998), β-ARK (Jaber et al. 1996), p300 (Yao et al. 1998), Pax-3 (Li et al. 1999), and FOG-2 (Tevosian et al. 2000). However, it is unclear whether the affected gene products in these mutant mice directly control cardiac wall expansion per se, because it is possible that indirect mechanisms such as a disturbance of angiogenesis or an upstream obstruction of blood flow may result in a hypoplastic ventricular phenotype as well. In the case of VCAM-1 (Kwee et al. 1995), α-integrin (Yang et al. 1995), or FOG-2 (Tevosian et al. 2000) deficient animals, it was noted that the interaction between the epicardium and the myocardium was partially disrupted. The resultant thin ventricular wall in these animals (Kwee et al. 1995; Yang et al. 1995; Tevosian et al. 2000) is consistent with the notion that communication between the epicardium and myocardium may play a role in the expansion of the ventricular wall. Interestingly, the requirement for RXRα (Sucov et al. 1994), gp 130 (Yoshida et al. 1996), and the erythropoietin receptor (Wu et al. 1999) to promote the expansion of the compact zone is non-cell-autonomous with respect to the cardiac myocyte lineage (Chen et al. 1998; Tran and Sucov 1998; Hirota et al. 1999; Wu et al. 1999), suggesting that these signaling molecules may be required in cells other than cardiac myocytes to promote proliferation of this cell type.

RA AND EPO INDIRECTLY CONTROL CARDIAC MYOCYTE CELL PROLIFERATION

By employing a novel heart slice culture system, we have demonstrated that signals from the epicardium are necessary and sufficient for myocardial cell proliferation (Stuckmann et al. 2003). One such signal is apparently retinoic acid (RA), as this substance is produced by the epicardium (Moss et al. 1998), and administration of the RA-antagonist, Ro415253, to intact heart slices can block myocardial cell proliferation. In addition, we have found that administration of anti-epo receptor antisera can similarly block cardiac myocyte proliferation in chick heart slice cultures (Stuckmann et al. 2003). These results indicate that RA and epo signaling are necessary for cardiac myocyte proliferation in chick heart slice cultures, and are consistent with data from knockout mice embryos which suggested that these two signaling pathways might be necessary for the expansion of the ventricular wall (Sucov et al. 1994; Wu et al. 1999). Interestingly, administration of RA and epo, either alone or in combination, to heart slices cultured in the absence of the epicardium failed to rescue cardiac myocyte proliferation in these cultures (Stuckmann et al. 2003). Thus, although these signals are necessary to support proliferation of the myocardium, they are insufficient to directly drive cardiac myocyte proliferation in the absence of the epicardium. A resolution to this paradox has emerged from experiments employing medium conditioned by epicardial cells. We have found that either RA or epo induces the synthesis or activity of a cardiac myocyte mitogen in epicardial cells which is secreted into epicardial cell conditioned medium (Stuckmann et al. 2003). That RA and epo fail to directly drive myocardial cell proliferation, but rather promote the expression of a cardiac myocyte mitogen in the epicardium, is consistent with prior work indicating that both RXRα (Sucov et al. 1994) and the erythropoietin receptor (Wu et al. 1999) may be required in cells other than cardiac myocytes to promote expansion of the ventricular wall (Chen et al. 1998; Tran and Sucov 1998; Hirota et al. 1999; Wu et al. 1999). We think it is likely that RA and epo induce the synthesis of a cardiac myocyte mitogen in epicardial cells via parallel pathways (outlined schematically in Fig. 1), as exogenous epo administration can rescue myocardial proliferation in heart slices exposed to the RA antagonist, Ro415253, and exogenous RA administration can similarly rescue cardiac myocyte proliferation in heart slices exposed to anti-epo receptor antisera (Stuckmann et al. 2003).

CARDIAC MYOCYTE PROLIFERATION REQUIRES PHYSICAL CONTACT BETWEEN THE EPICARDIUM AND MYOCARDIUM

The involvement of the epicardium in modulating cardiac myocyte proliferation has been suggested by studies in mice engineered to lack either V-cam (Kwee et al. 1995) or α-integrin (Yang et al. 1995). In these genetically altered mice, there is a correlation between a loss of contact between the epicardium and the myocardium and the subsequent development of a hypoplastic ventricular wall around midgestation. However, it was not clear from any of these studies employing knockout or transgenic mice whether signals from the epicardium directly regulated ventricular wall expansion per se, or whether the ensuing cardiomyopathy resulted from a disturbance of cardiac vascularization, which originates from the epicardium (Komiyama et al. 1987), and from the resultant decreased blood flow through the heart, or whether the cardiomyopathy resulted from an upstream obstruction of blood flow. Using the heart slice culture technique, we have been able to demonstrate that soluble signals from the epicardium, induced by RA and epo, are required for cellular proliferation in the myocardium. In contrast, removal of the endocardium and trabeculae from the heart slice cultures failed to affect myocardial cell proliferation (Stuckmann et al. 2003).

EPICARDIAL RETINOIC ACID IS A PLEIOTROPIC REGULATOR OF CARDIAC MYOCYTE DIFFERENTIATION, SURVIVAL, AND PROLIFERATION

Recently, the epicardium has been recognized to be the principal source of RA in the developing heart, as RALDH2, the rate-limiting enzyme in RA synthesis, is highly expressed in the epicardium, but not in cardiac myocytes (Moss et al. 1998). Although RA is solely produced in the epicardium, expression of RA-responsive reporter constructs in transgenic mice has indicated that both epicardial and myocardial cells in the heart respond to RA signaling during development (Moss et al. 1998). We have demonstrated that whereas RA signaling in the epicardium is necessary for cardiac myocyte proliferation, this is not because of a direct mitogenic effect of epicardial RA on cardiac myocytes. Although the RA-antagonist, Ro415253, dramatically reduced proliferation and the subsequent survival of cardiac myocytes in slice culture in the presence of an intact epicardium, RA by itself could not substitute for the epicardium (Stuckmann et al. 2003). This finding suggests that RA secreted by the epicardium either works in concert with another epicardial factor(s) to promote myocardial cell proliferation, or that RA induces the synthesis of a distinct cardiac myocyte mitogen within the epicardium itself. The idea that RA signaling in non-cardiac myocytes is necessary to promote myocardial cell proliferation is supported by two recent studies: The Chien and Sucov laboratories have recently demonstrated that mice with a cardiac myocyte-specific deletion of RXRα fail to develop hypoplastic hearts (Chen et al. 1998) and that cardiomyocytes deficient in RXRα develop normally and contribute to the ventricular chamber wall in chimeric mice embryos composed of mutant and wild-type cells (Tran and Sucov 1998). Together with prior work indicating that RXRα signaling is necessary for ventricular compact zone expansion (Sucov et al. 1994), these studies plus our new data indicate that RA signals work in non-cardiac myocytes to promote proliferation of this cell type. In addition to indirectly modulating cardiac myocyte proliferation and survival, RA may also directly affect cardiac myocyte differentiation (Zhou et al. 1995; Kastner et al. 1997). Thus, RA signaling may indirectly control myocardial cell proliferation and affect cardiac myocyte differentiation.

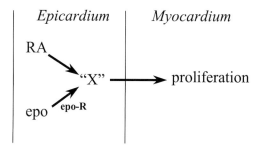

Figure 1. Retinoic acid and erythropoietin work in parallel to induce the expression or activity of a soluble cardiac mitogen in epicardial cells. RA and epo signaling in the epicardium work in parallel to induce the secretion of a soluble mitogen ("X") from the epicardium which stimulates cardiac myocyte proliferation in the adjacent myocardium.

EPO SIGNALING IN THE EPICARDIUM INDUCES THE SYNTHESIS OR EXPRESSION OF A CARDIAC MYOCYTE MITOGEN IN EPICARDIAL CELLS

Genetically engineered mice that lack either epo or the epo receptor (epo-R) have recently been shown to develop a hypoplastic heart phenotype in addition to defects in erythropoiesis at midgestation (Wu et al. 1999). Interestingly, epo-R is expressed in the epicardium, but not in cardiac myocytes (Wu et al. 1999), suggesting that epo signaling in cells other than cardiac myocytes is necessary for normal cardiac growth. In heart slice cultures, we found that administration of epo failed to rescue cardiac myocyte proliferation following removal of the epicardium, suggesting that epo does not directly induce myocardial cell proliferation. In contrast, conditioned medium from epo-treated epicardial cells, but not from non-epo-treated epicardial cells, was able to rescue myocardial cell proliferation in heart slices that lacked the epicardium (Stuckmann et al. 2003). In light of these findings, we propose that the observed hypoplastic heart phenotype in epo$^{-/-}$ and epo-R$^{-/-}$ mice may be due to the requirement for epo signaling in the epicardium to promote the expression or activity of a cardiac myocyte mitogen in the epicardial cells themselves.

Our results suggest that the control of myocardial growth is regulated by the activity of both RA and epo signaling pathways in the epicardium, which in turn controls the expression or activity of a myocardial mitogen that is secreted by the epicardium. That the control of myocardial cell proliferation is mediated by such a complex chain of molecular intermediaries is perhaps not surprising in view of the pleiotropic effects of RA and epo signaling, and the need for both spatially and temporally precise control of proliferation in the developing heart. In support of this hypothesis, it is known that epo-R is expressed in the mouse epicardium from E10.5 to E13.5 (Wu et al. 1999), coincident with the relatively massive growth of the ventricular wall from E11.5 in mouse (Erokhina 1968). In contrast, the overall responsiveness to RA signaling in the heart, as judged by expression of RA receptors, is maintained for a significantly longer period of time (Dolle et al. 1990). Thus, by constraining the time of expression of epo-R in the epicardium, the consequence of RA signaling may have far different outcomes at differing times of development within the same tissue. Although our results indicate that a signal(s) from the epicardium is necessary for both the proliferation and survival of the adjacent myocardium, it is unclear whether such an epicardial signal is similarly required for survival of adult myocardial cells, or for that matter, whether adult myocardial cells, which are postmitotic, can respond to the mitogenic activity of the embryonic epicardium.

The utilization of epo, an erythropoietic growth and differentiation factor, for regulating heart size is potentially intriguing from a physiologic and perhaps an evolutionary standpoint. We speculate that it might be advantageous to modulate the heart size of an animal according to the oxygenation status during fetal development. Since it is known that epo in the liver is stimulated by hypoxia-inducible factor, HIF-1α (Ebert and Bunn 1999), it is possible that a systemic increase in epo levels, following oxygen deprivation during fetal development, could both increase rates of erythropoiesis and increase myocardial cell proliferation. Both outcomes of epo signaling would act to improve the oxygenation status of the developing tissues, by increased erythropoiesis and by increased tissue perfusion, respectively. An understanding of how epo signaling independently regulates blood cell development or heart growth awaits further characterization of the molecular pathways regulating each of these effects.

CONCLUSIONS

To directly study the influence of the epicardium or endocardium on cardiac myocyte proliferation, we have developed a novel heart slice culture system that allows us to mechanically separate the epicardium or endocardium/trabeculae from the ventricular wall (Stuckmann et al. 2003). We have found that a signal(s) from the epicardium, but not from the endocardium, is required for cardiac myocyte proliferation. We have investigated whether retinoic acid or erythropoietin signaling contributes to the mitogenic action of the epicardium on ventricular myocytes. We found that administration of either the RAR antagonist, Ro-415253, or an anti-epo receptor antibody to heart slice cultures mimics the effect of removal of the epicardium and blocks cardiac myocyte proliferation. Furthermore, we demonstrated that inhibition of cardiac myocyte proliferation by Ro-415253 can be overcome by the addition of erythropoietin (epo), and that inhibition of cardiac myocyte proliferation by the anti-epo receptor antibody can be overcome by addition of exogenous RA. Finally, we showed that epo and RA do not directly induce proliferation of cardiac myocytes but rather induce the secretion of a soluble cardiac myocyte mitogen from the epicardium. Together, our findings support a model in which a network of retinoic acid and epo signals in the epicardium induce the expression (or activity) of a mitogen that directly regulates cardiac myocyte proliferation.

ACKNOWLEDGMENTS

We thank Dr. A. Pignatello (Hoffman-LaRoche) for the kind gift of Ro-415253. This work was supported by grants to A.B.L. from the National Institutes of Health. This work was done during the tenure of an established investigatorship from the American Heart Association to A.B.L. I.S. was supported by a fellowship from the Deutsche Forschungsgemeinschaft.

REFERENCES

Chen J., Kubalak S.W., and Chien K.R. 1998. Ventricular muscle-restricted targeting of the RXRalpha gene reveals a non-cell-autonomous requirement in cardiac chamber morphogenesis. *Development* **125**: 1943.

Chen Z., Friedrich G.A., and Soriano P. 1994. Transcriptional enhancer factor 1 disruption by a retroviral gene trap leads to

heart defects and embryonic lethality in mice. *Genes Dev.* **8:** 2293.
Dolle P., Ruberte E., Leroy P., Morriss-Kay G., and Chambon P. 1990. Retinoic acid receptors and cellular retinoid binding proteins. I. A systematic study of their differential pattern of transcription during mouse organogenesis. *Development* **110:** 1133.
Ebert B.L. and Bunn H.F. 1999. Regulation of the erythropoietin gene. *Blood* **94:** 1864.
Erokhina E.L. 1968. Proliferation dynamics of cellular elements in the differentiating mouse myocardium (Russian). *Tsitologiia* **10:** 1391.
Gassmann M., Casagranda F., Orioli D., Simon H., Lai C., Klein R., and Lemke G. 1995. Aberrant neural and cardiac development in mice lacking the ErbB4 neuregulin receptor (comments). *Nature* **378:** 390.
Hirota H., Chen J., Betz U.A., Rajewsky K., Gu Y., Ross J., Muller W., and Chien K.R. 1999. Loss of a gp130 cardiac muscle cell survival pathway is a critical event in the onset of heart failure during biomechanical stress. *Cell* **97:** 189.
Jaber M., Koch W.J., Rockman H., Smith B., Bond R.A., Sulik K.K., Ross J., Lefkowitz R.J., Caron M.G., and Giros B. 1996. Essential role of beta-adrenergic receptor kinase 1 in cardiac development and function. *Proc. Natl. Acad. Sci.* **93:** 12974.
Kastner P., Messaddeq N., Mark M., Wendling O., Grondona J.M., Ward S., Ghyselinck N., and Chambon P. 1997. Vitamin A deficiency and mutations of RXRalpha, RXRbeta and RARalpha lead to early differentiation of embryonic ventricular cardiomyocytes. *Development* **124:** 4749.
Komiyama M., Ito K., and Shimada Y. 1987. Origin and development of the epicardium in the mouse embryo. *Anat. Embryol.* **176:** 183.
Kreidberg J.A., Sariola H., Loring J.M., Maeda M., Pelletier J., Housman D., and Jaenisch R. 1993. WT-1 is required for early kidney development. *Cell* **74:** 679.
Kwee L., Baldwin H.S., Shen H.M., Stewart C.L., Buck C., Buck C.A., and Labow M.A. 1995. Defective development of the embryonic and extraembryonic circulatory systems in vascular cell adhesion molecule (VCAM-1) deficient mice. *Development* **121:** 489.
Lee K.F., Simon H., Chen H., Bates B., Hung M.C., and Hauser C. 1995. Requirement for neuregulin receptor erbB2 in neural and cardiac development (comments). *Nature* **378:** 394.
Li F., McNelis M.R., Lustig K., and Gerdes A.M. 1997. Hyperplasia and hypertrophy of chicken cardiac myocytes during posthatching development. *Am. J. Physiol.* **273:** R518.
Li F., Wang X., Capasso J.M., and Gerdes A.M. 1996. Rapid transition of cardiac myocytes from hyperplasia to hypertrophy during postnatal development. *J. Mol. Cell. Cardiol.* **28:** 1737.
Li J., Liu K.C., Jin F., Lu M.M., and Epstein J.A. 1999. Transgenic rescue of congenital heart disease and spina bifida in Splotch mice. *Development* **126:** 2495.
Meyer D. and Birchmeier C. 1995. Multiple essential functions of neuregulin in development (comments) (erratum in *Nature* [1995] **378:** 753). *Nature* **378:** 386.
Mima T., Ueno H., Fischman D.A., Williams L.T., and Mikawa T. 1995. Fibroblast growth factor receptor is required for in vivo cardiac myocyte proliferation at early embryonic stages of heart development. *Proc. Natl. Acad. Sci.* **92:** 467.
Moens C.B., Stanton B.R., Parada L.F., and Rossant J. 1993. Defects in heart and lung development in compound heterozygotes for two different targeted mutations at the N-myc locus. *Development* **119:** 485.
Moss J.B., Xavier-Neto J., Shapiro M.D., Nayeem S.M., McCaffery P., Drager U.C., and Rosenthal N. 1998. Dynamic patterns of retinoic acid synthesis and response in the developing mammalian heart. *Dev. Biol.* **199:** 55.
Neubauer H., Cumano A., Muller M., Wu H., Huffstadt U., and Pfeffer K. 1998. Jak2 deficiency defines an essential developmental checkpoint in definitive hematopoiesis. *Cell* **93:** 397.
Rychterova V. 1971. Principle of growth in thickness of the heart ventricular wall in the chick embryo. *Folia Morphol.* **19:** 262.
———. 1978. Development of proliferation structure of the ventricular heart wall in the chick embryo between the 6th and 14th day of embryogenesis. *Folia Morphol.* **26:** 131.
Soonpaa M.H., Kim K.K., Pajak L., Franklin M., and Field L.J. 1996. Cardiomyocyte DNA synthesis and binucleation during murine development. *Am. J. Physiol.* **271:** H2183.
Stuckmann I., Evans S., and Lassar A.B. 2003. Erythropoietin and retinoic acid, secreted by the epicardium, are required for cardiac myocyte proliferation. *Dev. Biol.* (in press).
Sucov H.M., Dyson E., Gumeringer C.L., Price J., Chien K.R., and Evans R.M. 1994. RXR alpha mutant mice establish a genetic basis for vitamin A signaling in heart morphogenesis. *Genes Dev.* **8:** 1007.
Tevosian S.G., Deconinck A.E., Tanaka M., Schinke M., Litovsky S.H., Izumo S., Fujiwara Y., and Orkin S.H. 2000. FOG-2, a cofactor for GATA transcription factors, is essential for heart morphogenesis and development of coronary vessels from epicardium. *Cell* **101:** 729.
Tokuyasu K.T. 1990. Co-development of embryonic myocardium and myocardial circulation. In *Developmental cardiology: Morphogenesis and function* (ed. E.B. Clark and A. Takao), p. 205. Futura Publishing, Mount Kisco, New York.
Tran C.M. and Sucov H.M. 1998. The RXRalpha gene functions in a non-cell-autonomous manner during mouse cardiac morphogenesis. *Development* **125:** 1951.
Wu H., Lee S.H., Gao J., Liu X., and Iruela-Arispe M.L. 1999. Inactivation of erythropoietin leads to defects in cardiac morphogenesis. *Development* **126:** 3597.
Yang J.T., Rayburn H., and Hynes R.O. 1995. Cell adhesion events mediated by alpha 4 integrins are essential in placental and cardiac development. *Development* **121:** 549.
Yao T.P., Oh S.P., Fuchs M., Zhou N.D., Ch'ng L.E., Newsome D., Bronson R.T., Li E., Livingston D.M., and Eckner R. 1998. Gene dosage-dependent embryonic development and proliferation defects in mice lacking the transcriptional integrator p300. *Cell* **93:** 361.
Yoshida K., Taga T., Saito M., Suematsu S., Kumanogoh A., Tanaka T., Fujiwara H., Hirata M., Yamagami T., Nakahata T., Hirabayashi T., Yoneda Y., Tanaka K., Wang W.Z., Mori C., Shiota K., Yoshida N., and Kishimoto T. 1996. Targeted disruption of gp130, a common signal transducer for the interleukin 6 family of cytokines, leads to myocardial and hematological disorders. *Proc. Natl. Acad. Sci.* **93:** 407.
Zhou M.D., Sucov H.M., Evans R.M., and Chien K.R. 1995. Retinoid-dependent pathways suppress myocardial cell hypertrophy. *Proc. Natl. Acad. Sci.* **92:** 7391.

Endocardial Cushion Formation in Zebrafish

D.Y.R. STAINIER, D. BEIS,* B. JUNGBLUT,* AND T. BARTMAN*

Department of Biochemistry and Biophysics, Programs in Developmental Biology, Genetics and Human Genetics, University of California, San Francisco, California 94143-0448

Cardiac valves form at three sites along the anteroposterior (AP) axis of the developing heart: the outflow tract, atrioventricular (AV) boundary, and sinus venosus. Their function is to ensure unidirectional blood flow through the heart. Cardiac valves derive from endocardial cells that first form endocardial cushions, which are later remodeled into valves and septae. Prior to valve formation, the cushions themselves may be critical in controlling intracardiac flow; therefore, derangement of endocardial cushion formation can lead to lethality from early embryonic stages. In mammals, disrupted endocardial cushion formation leads to various heart defects, including particular types of atrial septal defects (ASDs), ventricular septal defects (VSDs), and, in the most severe form, AV canal defects (aka AV septal or endocardial cushion defects). AV canal defects account for 5% of cases of human congenital heart disease (CHD) (i.e., 1600 cases in the U.S. per year), making it the fourth most common congenital heart malformation, and the fifth most common cause of death due to congenital heart disease (Emmanouilides et al. 1998). AV canal defects are also common in Down syndrome, where 16% of patients exhibit complete AV canal defects with all four chambers of the heart in open communication. AV canal defects are also common in isomerism syndromes (especially right isomerism) and in DiGeorge and Ellis-van Creveld syndromes (both of which are associated with incomplete AV canal defects). Septal defects (VSDs and ASDs), which are much more common than AV canal defects, can result from minor defects in endocardial cushion formation.

Cardiac valves derive from the endocardium, a layer of endothelial cells that lines the lumen of the heart tube and is continuous with the endothelium of blood vessels. Genetic studies in mouse and zebrafish indicate that endothelial and endocardial cell differentiation utilizes many common genetic pathways (Shalaby et al. 1995; Stainier et al. 1995; for review, see Jin et al. 2002). Fate-mapping studies in zebrafish indicate, however, that the endocardial cell progenitors originate from a specific region of the blastula, namely the region that gives rise to myocardial cells, or muscle cells of the heart (Lee et al. 1994). In a clonal analysis, ~52% of ventral marginal blastomeres gave rise to both endocardial and myocardial cells, indicating that in zebrafish the cells of the two cardiac layers share a common progenitor (Lee et al. 1994). These data are important because they suggest that interactions between endocardial and myocardial cells may start prior to the onset of gastrulation. Although the myocardial progenitors have been carefully fate-mapped in mouse and chick embryos, further studies are required to determine the spatial relationship between the endocardial and myocardial progenitors in these organisms.

After formation of the primitive heart tube in mouse and chick, the cardiac jelly (or extracellular matrix [ECM] between the endocardial and myocardial layers) expands within the AV canal and outflow tract regions. These swellings are the first morphological manifestations of endocardial cushion formation and may also function as valves to direct blood flow at early stages of development (for review, see Eisenberg and Markwald 1995). Subsequently, endocardial cells lining the AV canal and outflow tract undergo an epithelial to mesenchymal transformation (EMT) and invade these swellings (Krug et al. 1987). Studies in chick embryos have shown that myocardial cells from the AV boundary, but not those from the ventricle, were able to induce endocardial cells to undergo EMT (Mjaatvedt et al. 1987). Likewise, only endocardial cells from the AV boundary, but not those from the ventricle, can undergo EMT in response to a myocardial signal (Runyan and Markwald 1983). Therefore, the endocardial and myocardial cells at the AV boundary have distinct properties in comparison to other cardiac cells. This specialization of the AV boundary cells is likely to be a consequence of the AP patterning of the myocardium, which is also responsible for the formation of the cardiac chambers. One candidate gene required for AV boundary formation is *Irx4*, which is expressed in the ventricle and AV canal and has been shown to regulate chamber-specific gene expression in the developing chick and mouse heart (Bao et al. 1999; Christoffels et al. 2000; Bruneau et al. 2001). Tbx2 is also expressed in the mouse AV canal at the time of endocardial cushion formation and appears to play a critical role in regulating gene expression in this region of the embryo (Habets et al. 2002).

The process of EMT leading to endocardial cushion formation has been studied extensively in chick and mouse embryos (for review, see Eisenberg and Markwald 1995; Nakajima et al. 2000). For example, it has been shown that extracts of cardiac ECM can induce EMT in cell culture (Krug et al. 1985, 1987). Molecules secreted

---

* These authors contributed equally to this paper.

by the AV myocardium associated with EMT include Bmp2, TGFβ1, 2, and 3, as well as fibronectin, transferrin, ES130, and hLAMP-1 (for review, see Nakajima et al. 2000). The role of TGFβs during endocardial cushion formation has been extensively studied in the chick. Anti-TGFβ2 antibodies specifically inhibit the initial steps of EMT in AV canal explants, including endocardial cell–cell separation (Boyer et al. 1999b), whereas anti-TGFβ3 antibodies inhibit later steps of EMT such as cell migration (Boyer et al. 1999b). Experiments using three-dimensional collagen gel cultures have indicated that inductive signals from the myocardium up-regulate the expression of TGFβ3 in the AV endocardial cells at the onset of EMT, that the expression of TGFβ3 by the endocardial cells is required for the induction of the initial phenotypic changes of EMT, and that myocardial Bmp2 acts synergistically with TGFβ3 for the migration into the ECM (Yamagishi et al. 1999). Additional molecules important for EMT include Slug, a member of the Snail family of transcription factors (Romano and Runyan 1999, 2000), as well as Gi (Runyan et al. 1990; Boyer et al. 1999a). According to one current model, TGFβ2 and Gi signaling are required together with Slug to initiate EMT (Romano and Runyan 2000), whereas TGFβ3 and Bmp signaling are important for later steps of EMT (Boyer et al. 1999a).

In mouse, several mutations affect AV valve formation. The *hdf (heart defect)* mutation causes ventricular differentiation defects, reduced cardiac jelly production, and a failure in cellular invasion at the AV boundary (Yamamura et al. 1997). The *hdf* locus has been shown to encode Versican, a chondroitin sulfate proteoglycan. Hyaluronic acid synthase 2 (Has2), the earliest-functioning of the three mammalian HA synthases, is another protein involved in the regulation of EMT. HA is a high-molecular-weight glucosaminoglycan polymer that binds water, thereby creating a voluminous, highly structured gel. HAs not only have an architectural role, but are also involved in cell adhesion and migration (for review, see Bourguignon 2001). Mutation of murine *Has2* leads to a lack of cardiac jelly production, ventricular trabeculation, and endocardial cushion formation (Camenisch et al. 2000). A third gene involved in the regulation of EMT is *Bmp4*. Although the majority of *Bmp4* mutant embryos die before E10.5 (Winnier et al. 1995), the few that do progress past the egg cylinder stage have severe defects in endocardial cushion formation. In these embryos, although the ECM seems to be deposited in normal amounts, the cushions remain acellular. It is unclear at this point whether these defects are caused by a requirement for Bmp4 in EMT induction, or whether these embryos are only delayed in their development and unhealthy. Bmp6 and 7 and TGFβ2 have also been implicated in mouse endocardial cushion formation (Bartram et al. 2001; Kim et al. 2001).

Endocardial cells at the AV boundary are induced by myocardial cells to transform into mesenchymal cells and migrate into the cardiac jelly. Proteins expressed in activated endocardial and invasive mesenchymal cells include α smooth muscle actin (Nakajima et al. 1997), cell surface β 1-4 galactosyltransferase (Loeber and Runyan 1990), fibrillin (Wunsch et al. 1994), Msx-1 (Chan-Thomas et al. 1993), Mox-1 (Candia et al. 1992), type I pro-collagen (Sinning et al. 1988), and urokinase (McGuire and Orkin 1992). Although these and other molecules are associated with the migration of activated endocardial cells into the cardiac jelly, only vinculin has been shown to play a key role. Inactivation of *Vinculin* in mouse leads to defects in valve formation and death around E10 (Xu et al. 1998). In contrast to *Versican* and *Has2* mutants, the ECM at the AV boundary is present in *Vinculin* mutants. Since *Vinculin* has been implicated in the formation of focal adhesion plaques, it is likely to play a role in the migration of endocardial cells into the cardiac jelly at the AV boundary.

Once formed, endocardial cushions undergo complex morphogenesis. The AV cushions fuse to form the mitral (left) and tricuspid (right) AV valves as well as the septae in four-chambered hearts. Studies in mouse have shown that ALK3/BMPR-IA, a type I BMP receptor, is required for endocardial cushion morphogenesis including the expression of TGFβ2 during this process (Gaussin et al. 2002). In the absence of ALK3 function, endocardial cushions appear to form normally, but fail to fuse, and are reduced in size and misaligned. Ca^{++} signaling, via its activation of the transcription factor NFAT, has also been implicated in late aspects of valve morphogenesis. The NFATc1 transcription factor is nuclearly localized (and thus activated) in the endocardial cells that will form the AV and outflow tract valves. *NFATc1*-deficient embryos exhibit defects in valve morphology, arrest in development, and die around E14.5 (de la Pompa et al. 1998; Ranger et al. 1998). *Connexin 45*-deficient embryos exhibit a phenotype similar to that seen in *NFATc1*-deficient embryos, and NFATc1 fails to localize into the nucleus of the valve-forming cells in *Connexin 45*-deficient embryos (Kumai et al. 2000).

Thus, although several genes have been implicated in endocardial cushion formation, many more remain to be identified and their mechanism of action further explored. The zebrafish is a well-established model system for studies of vertebrate heart formation (for review, see Stainier 2001). It offers several distinct advantages for genetic and embryological studies, including the external fertilization, rapid development, and optical clarity of its embryos. In addition, because of their small size, zebrafish embryos are not completely dependent on a functional cardiovascular system. Even in the absence of blood circulation, they receive enough oxygen by passive diffusion to survive and continue to develop in a relatively normal fashion for several days, thereby allowing a detailed analysis of animals with severe cardiovascular defects. In terms of tools, in addition to its proven track record for gene discovery through forward genetics, the zebrafish embryo is now also amenable to reverse genetics through morpholino knock-down experiments. Finally, the relative simplicity of the zebrafish heart, which does not septate and remains two-chambered, is also an asset when dissecting the complex morphogenetic events leading to endocardial cushion formation.

RESULTS AND DISCUSSION

Time Line of Endocardial Cushion Development in Zebrafish

The zebrafish heart begins to beat at 22 hours post-fertilization (hpf) and loops by 36 hpf. Whereas the first morphological differences between cardiac chambers can be observed after the formation of the linear heart tube, molecular differences are apparent much earlier (for review, see Yelon and Stainier 1999). The first molecular indication of AV boundary formation in zebrafish occurs at approximately 37 hpf with the restriction of *bmp4* and *versican* expression to the AV myocardium (Walsh and Stainier 2001). At ~45 hpf, *notch1b* transcripts become similarly restricted to the AV endocardium (Westin and Lardelli 1997; Walsh and Stainier 2001). At 48 hpf, the expression of the ECM glycoprotein gene *fibulin-1* marks the differentiation of the AV boundary myocardium (Zhang et al. 1997). In zebrafish, the developing valve prevents retrograde blood flow starting at ~48 hpf, although endocardial cushions are only visible by 96 hpf (see below and also Hu et al. 2000). Several questions arise at this point. Is there at the AV boundary in zebrafish a local swelling of the ECM analogous to the acellular cushions in chick and mouse, and does this structure function to prevent retrograde blood flow? Do endocardial cells at the AV boundary and outflow tract undergo EMT, and do they subsequently migrate into the underlying ECM? What is the composition of the endocardial cushions in zebrafish? Do they form both at the AV boundary and in the outflow tract? What are the morphogenetic processes responsible for the remodeling of zebrafish endocardial cushions into valve leaflets? In order to start answering these questions, we have generated and utilized a GFP transgenic line (tie2::GFP; Motoike et al. 2000) that allows the identification and detailed analysis of endothelial and endocardial cells in living embryos. We have complemented these studies with histological observations at several time points. Preliminary data from this work suggest that the endocardial cells at the AV boundary appear to undergo EMT between 60 and 72 hpf, and that they have formed actual cushions by 96 hpf (Fig. 1C, F). Clearly, additional studies will be required to give a more detailed picture of the various morphogenetic events leading to endocardial cushion formation at the AV boundary. Currently, we are trying to develop techniques that will allow us to examine, with high resolution, the critical steps of this process as they happen in the living embryo.

Mutations Affecting Endocardial Cushion Development in Zebrafish

Given the relative ease of visualizing blood flow through the heart in the zebrafish embryo, it should be fairly straightforward to identify mutations that cause retrograde blood flow after 48 hpf, either a toggling phenotype where all the blood moves back and forth between the atrium and ventricle, or more subtle defects where only part of the blood moves in a retrograde direction across the AV boundary. (The assumption here is that some of these mutations affect the formation of endocardial cushions.) Indeed, many mutants leading to retrograde blood flow were identified in the initial large-scale screens, but unfortunately, few were kept (Chen et al. 1996; Stainier et al. 1996).

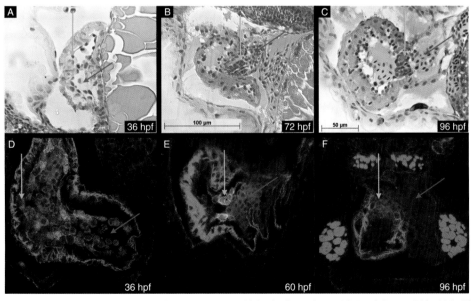

Figure 1. Atrioventricular (AV) boundary formation. (*A–C*) 5-μm-thick plastic sections, H&E staining, at 36 hpf (*A*), 72 hpf (*B*), and 96 hpf (*C*). (*D–F*) Confocal images of vibratome sections of tie2::GFP embryos at 36 hpf (*D*), 60 hpf (*E*), and 96 hpf (*F*); red stain is phalloidin. (*Blue arrows*) Atrium; (*green arrows*) endocardial cells or endocardial cushions. At 36 hpf (*A, D*), a single layer of endocardial cells lines the lumen of the heart. By 60 hpf (*E*), EMT appears to be initiated at the AV boundary, and at 72 hpf (*B*), the endocardial cushions appear to be about 2 cells thick. By 96 hpf (*C, F*), well-developed endocardial cushions are apparent.

jekyll. The first toggling mutant from the Boston screen that was analyzed in detail is *jekyll^{m151}* (Stainier et al. 1996; Walsh and Stainier 2001). *jekyll* mutant embryos display toggling between the atrium and ventricle so that by 48 hpf no blood leaves the ventricle. In addition, histological analyses showed that *jekyll* mutants lack valve tissue at 96 hpf (Stainier et al. 1996). Gene expression analyses further show that *bmp4*, *versican*, and *notch1b* fail to become restricted at the AV boundary of *jekyll* mutants (Fig. 2A–L). Moreover, tie2::GFP expression fails to be up-regulated in the AV boundary (Fig. 2M–P). On the basis of these data, our current working model positions Jekyll early in the pathway that leads to AV boundary establishment. Jekyll encodes uridine 5´-diphosphate-glucose dehydrogenase (Ugdh) (Walsh and Stainier 2001), a homolog of *Drosophila* Sugarless (Häcker et al. 1997). Ugdh is required for the conversion of UDP-glucose into UDP-glucuronate, which is a basic building block of hyaluronic acid as well as heparan and chondroitin sulfate glycosaminoglycans (for review, see Esko and Selleck 2002). To date, three proteins involved in proteoglycan biosynthesis: Sugarless, Sulfateless (a *Drosophila* homolog of heparan sulfate N-deacetylase/N-sulfotransferase), and the heparin sulfate proteoglycan Dally, have been shown to be involved in Dpp (TGFβ) (Jackson et al. 1997), Fgf (Lin et al. 1999), and Wg (Wnt) (Häcker et al. 1997; Tsuda et al. 1999) signaling. In zebrafish the mutations identified thus far in these signaling pathways or in proteoglycan biosynthesis do not exhibit a cardiac valve phenotype, with the exception of *jekyll*. However, two mutations, *knypek/glypican* (Topcsewski et al. 2001) and *pipetail/wnt5a* (Rauch et al. 1997), show similar defects to *jekyll* in jaw formation. These data suggest that during jaw formation, Jekyll modifies Glypican, which functions to regulate Wnt5a signaling.

We are currently investigating further the role of *jekyll* in AV boundary formation using both forward and reverse genetics. We are screening for genetic interactors of *jekyll* by crossing heterozygous *jekyll* females to F$_1$ males from mutagenized stocks. Through this screen we hope to identify additional components of the Jekyll pathway or genes that, when compromised and in combination with a reduction in Jekyll function, lead to a toggling phenotype. In addition, we are trying to determine in which cells (myocardial or endocardial) Jekyll is required. These cell-autonomy studies will be greatly facilitated by the cell-type-specific GAL4 driver transgenic lines under construction in the lab (see below). In summary, the cloning of *jekyll* was clearly an important step toward understanding endocardial cushion formation in zebrafish; however, the nature of the Jekyll protein strongly suggests that there is a signaling molecule involved in this process, and that its modification and/or distribution requires *jekyll*. One of our goals is therefore to identify this signaling molecule.

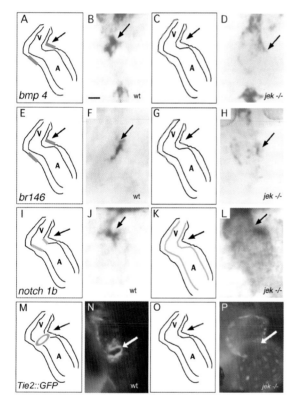

Figure 2. Atrioventricular (AV) boundary formation in wild-type and *jekyll* mutant embryos. Markers for the AV boundary region fail to be restricted in *jekyll* mutant embryos, suggesting early defects in valve development. Schematized representations are shown to the left of the actual data. *bmp4*, *br146* (*versican*), and *notch1b* are initially expressed throughout the AP axis of the heart, and they become restricted to the AV boundary prior to the formation of recognizable endocardial cushions. At 48 hpf, *bmp4* (*A, B*) and *br146* (*E, F*) are expressed in the myocardium and *notch1b* (*I, J*) in the endocardium. RNA in situ hybridization for these genes reveals defects in *jekyll* mutant hearts (*D, H, L*). Expression of *bmp4* (*D*) and *br146* (*H*) becomes restricted from the atrium and largely from the ventricle, but there is no heightened expression at the valve-forming region as in wild type (*B, F*). In the endocardium, *notch1b* expression (*L*) is not increased at the AV boundary and not restricted from the ventricle or atrium, unlike what is seen in wild type (*J*). Tie2::GFP transgenes show heightened expression in the AV boundary endocardial cells at 43 hpf (*arrow* in *N*), in wild-type but not in *jekyll* mutant hearts (*P*). (A) Atrium; (V) ventricle; (wt) wild type. Bar, 20 µm.

cardiofunk. Another mutant we have been analyzing is *cardiofunk^{s11}*, a mutant identified in one of our cardiac screens (Alexander et al. 1998). Like *jekyll*, *cardiofunk* abrogates endocardial cushion formation and leads to toggling of the blood between the atrium and ventricle with resulting loss of blood flow in the embryo. At the light microscopy level, there are no other primary phenotypes caused by the mutation. However, *cardiofunk* has a number of intriguing genetic characteristics. First, it exhibits low penetrance: Clutches from heterozygous adults, raised at the usual 28°C, have a mean percentage of affected embryos around 10% (e.g., 40% penetrance). Second, the penetrance is highly dependent on background. A specific pair of heterozygous adults when crossed on a weekly basis has a similar penetrance from week to week, but the penetrance of different pairs ranges from 20% to 80%. Third, the mutation is temperature-sensitive. Raising embryos at 32–33°C brings the mean penetrance closer to 100%. Fourth, during the positional cloning of the *cardiofunk* gene, we noticed that a number of embryos heterozygous at the locus exhibit the phenotype. About 10% of the phenotypically abnormal embryos collected for the mapping project are heterozygous for *cardiofunk*. These data suggest that the mutation can function in a dominant fashion, although at low penetrance. Thus, in a typical clutch of 100 embryos with 10 showing the phenotype, 9 will be homozygous and 1 heterozygous. Since there are 25 homozygous and 50 heterozygous embryos in the clutch, the penetrance (at 28°C) for homozygous embryos is about 36% (9/25) and for heterozygous embryos about 2% (1/50). This situation is rather unusual in studies of zebrafish mutations that affect heart development, but may be more analogous to mutations that have been found to affect heart development in the human population. Indeed, many human mutations cause heart defects at low penetrance, and/or variable expressivity, in heterozygous individuals. It will be most interesting to isolate the *cardiofunk* gene and explore the molecular basis for the peculiar genetic behavior of this allele.

Additional mutations. In addition to the screen for *jekyll* interactors, we are currently conducting large-scale F_2 genetic screens to identify new regulators of endocardial cushion development. These screens are done according to a breeding scheme similar to the one used in the initial large-scale screens done in Boston and Tübingen (Driever et al. 1996; Haffter et al. 1996). To identify embryos with defects in endocardial cushion formation, we are screening intercrosses of F_2 families between 50 and 60 hpf by visual inspection of live embryos for retrograde blood flow through the heart. Embryos are also being screened on day 5 post-fertilization to find potential defects in later steps of valve development. In ~900 families screened so far, we have identified 20 putative mutations that cause retrograde blood flow in the heart. As shown above, endocardial cells are easily visualized in tie2::GFP transgenic zebrafish. The formation of endocardial cushions can be observed in great detail and in vivo in this line. Therefore, in order to analyze rapidly the cellular phenotype in newly identified mutant embryos, the heterozygous F_2 fish are being outcrossed to the tie2::GFP line. So far six mutations that cause specific cellular defects in endocardial cushion formation have been recovered after outcrossing. Visual inspection of the GFP-expressing endocardial cells under fluorescence, as well as with Nomarski optics, gives us a relatively detailed picture of the cellular phenotype of the developing endocardial cushions in these mutants and allows us to classify them into phenotypic groups. Mutants that fall into the same phenotypic group will be tested by complementation analysis. Figure 3 illustrates the phenotypes of three new mutants that represent three different phenotypic groups. In wild-type embryos at 60 hpf, we can observe a ring formed by a single layer of endocardial cells—the endocardial ring—lining the AV canal (Fig. 3A). In embryos homozygous for the 342.1 mutation, this ring of differentiated endocardial cells, the precursors of the endocardial cushions, appears to be missing (Fig. 3B). In embryos homozygous for the 330.11 mutation (Fig.

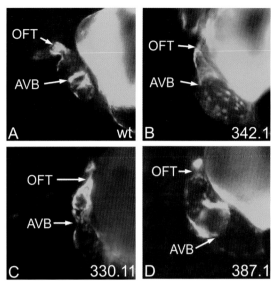

Figure 3. Endocardial cushion mutants. GFP fluorescence micrographs showing the endocardium of wild-type (*A*), 342.1 mutant (*B*), 330.11 mutant (*C*), and 387.1 mutant (*D*) embryos at 60 hpf in a tie2::GFP genetic background. In *A* only the ventricle is visible because the atrium is out of focus. The two heart chambers communicate through the AV canal, which is surrounded by a ring of endocardial cells with a heightened GFP expression; these endocardial cells are believed to give rise to the endocardial cushions. Similarly, endocardial cells located at the outflow tract (OFT) form a smaller ring of cells with heightened GFP expression. Embryos homozygous for the 342.1 mutation lack the endocardial ring of differentiated cells both at the AV boundary (AVB) and the OFT (*B*). In embryos homozygous for the 330.11 mutation, an endocardial ring forms within the ventricular compartment and not at the AV boundary (*C*). In embryos homozygous for the 387.1 mutation, the endocardial cells showing heightened GFP expression do not form a ring, but two rods oriented perpendicular to the AV boundary (*D*).

3C), heart looping is defective, and the endocardial ring is absent from the AV canal. An ectopic arch of round cells is visible on the ventricular surface. A defect in chamber differentiation or in AV boundary formation could account for the defect observed in 330.11 mutants. In embryos homozygous for the 387.1 mutation, endocardial cells differentiate in the vicinity of the AV boundary (Fig. 3D). We can distinguish these endocardial cells by their morphology and their up-regulation of GFP expression. Instead of forming a ring around the AV canal, these cells form two rows oriented perpendicularly to the AV canal. Intense blood regurgitation in these mutants indicates a lack of valve function, which suggests that in wild-type embryos the endocardial ring functions as a valve.

To identify the primary defect during endocardial cushion formation in the newly identified mutants, further analysis with molecular markers will address the questions of whether AP patterning of the heart chambers and the formation of the AV boundary is affected. Molecular cloning of the mutated genes will be greatly facilitated by the ongoing sequencing of the zebrafish genome and the availability of increasingly better genetic and physical maps. Functional analysis of the identified genes should greatly enhance our understanding of the mechanisms regulating endocardial cushion formation.

CONCLUSIONS

Endocardial cushion formation is a multistep process, and we are interested in questions such as how the AP patterning of the heart is connected to the formation of the endocardial cushions, what ensures the spatio-temporal control of endocardial cell differentiation at the AV boundary, and what is the role of the ECM in this process. Our ongoing descriptive analyses of the events associated with endocardial cushion formation and subsequent valve remodeling in wild-type embryos should identify the critical time points for these processes. We have also identified many mutations that appear to affect endocardial cushion formation, and their further molecular and cellular analyses will help us set priorities for the cloning of the corresponding genes.

In zebrafish, forward genetics has been very effectively used as an unbiased approach to identify genes involved in a certain process. Cloning the loci identified in our screens should provide novel players and elucidate mechanisms underlying valve development. However, a significant contribution to our understanding of this process has already been made through reverse genetics in mouse, and we have implicated signaling molecules such as members of the BMP family (Kim et al. 2001) and the Type III TGFβ receptor (Brown et al. 1999). To test the role of these molecules and others in endocardial cushion formation in zebrafish, we are utilizing the GAL4/UAS binary expression system (Fig. 4). The goal here is to activate or block specific signaling pathways, for example, Bmp, Notch, or Ras, during endocardial cushion formation and to unravel the role of these molecules in this process. This role could be masked by earlier functions for these molecules, and therefore it needs a more targeted approach than mRNA overexpression throughout the embryo. Generating valve-specific driver lines will provide a powerful tool to test these signaling molecules in vivo for their involvement in the making of a valve. In the future, forward genetic screens looking for molecules which if activated from our driver lines, give a valve phenotype, could be designed, using retroviral vectors (Amsterdam et al. 1999) and the UAS minimal promoter. Moreover, the availability of tissue-specific (endocardial or myocardial) and temporally restricted driver lines will be an important tool to test the cell-autonomous requirements of the genes isolated in our ongoing genetic screens.

In summary, we are taking multiple approaches to analyze endocardial cushion formation in zebrafish and hope to identify new genes and investigate molecular mechanisms involved in this process. Ultimately, the comparative analysis of endocardial cushion development in zebrafish, chick, and mouse will lead to a better understanding of the etiology of several common congenital heart disorders in humans.

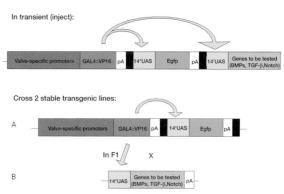

Figure 4. Testing gene function. Signaling molecules can be tested for their involvement in endocardial cushion and valve development using the GAL4/UAS binary expression system. (*1*) Valve-specific promoter elements driving expression of dominant negative or constitutively active forms of signaling molecules or receptors can be tested in transients by injecting DNA into the one-cell-stage embryo. Their expression will be driven in cells where GFP is also expressed. (*2*) Alternatively, to bypass the possibility that the above described transgenes exhibit a lethal phenotype or to overcome the mosaicism that is observed in transients, stable lines can be generated. In this scheme, the valve-specific driver line (*A*) is crossed to a line transgenic for a signaling molecule under the control of UAS elements (*B*). In the F_1 progeny of such a cross, GAL4 is driving expression of both the gene tested and GFP in the same cells.

ACKNOWLEDGMENTS

We thank the following granting agencies for their support of this work: Howard Hughes Medical Institute (T.B.), the Max Planck Society and the American Heart Association (B.J.), Human Frontier Science Program (D.B.), and the National Institutes of Health (D.Y.R.S.) and acknowledge the pioneering contributions of Emily Walsh in many aspects of this work.

REFERENCES

Alexander J., Stainier D.Y.R., and Yelon D. 1998. Screening mosaic F1 females for mutations affecting zebrafish heart induction and patterning. *Dev. Genet.* **22:** 288.

Amsterdam A., Burgess S., Golling G., Chen W., Sun Z., Townsend K., Farrington S., Haldi M., and Hopkins N. 1999. A large-scale insertional mutagenesis screen in zebrafish. *Genes Dev.* **13:** 2713.

Bao Z.Z., Bruneau B.G., Seidman J.G., Seidman C.E., and Cepko C.L. 1999. Regulation of chamber-specific gene expression in the developing heart by Irx4. *Science* **283:** 1161.

Bartram U., Molin D.G., Wisse L.J., Mohamad A., Sanford L.P., Doetschman T., Speer C.P., Poelmann R.E., and Gittenberger-de Groot A.C. 2001. Double-outlet right ventricle and overriding tricuspid valve reflect disturbances of looping, myocardialization, endocardial cushion differentiation, and apoptosis in TGF-β2-knockout mice. *Circulation* **103:** 2745.

Bourguignon L.Y. 2001. CD44-mediated oncogenic signaling and cytoskeleton activation during mammary tumor progression. *J. Mammary Gland Biol. Neoplasia* **6:** 287.

Boyer A.S., Erickson C.P., and Runyan R.B. 1999a. Epithelial-mesenchymal transformation in the embryonic heart is mediated through distinct pertussis toxin-sensitive and TGFβ signal transduction mechanisms. *Dev. Dyn.* **214:** 81.

Boyer A.S., Ayerinskas II, Vincent E.B., McKinney L.A., Weeks D.L., and Runyan R.B. 1999b. TGFβ2 and TGFβ3 have separate and sequential activities during epithelial-mes-

enchymal cell transformation in the embryonic heart. *Dev. Biol.* **208**: 530.
Brown C.B., Boyer A.S., Runyan R.B., and Barnett J.V. 1999. Requirement of type III TGF-β receptor for endocardial cell transformation in the heart. *Science* **283**: 2080.
Bruneau B.G., Bao Z.Z., Fatkin D., Xavier-Neto J., Georgakopoulos D., Maguire C.T., Berul C.I., Kass D.A., Kuroski-de Bold M.L., de Bold A.J., Conner D.A., Rosenthal N., Cepko C.L., Seidman C.E., and Seidman J.G. 2001. Cardiomyopathy in Irx4-deficient mice is preceded by abnormal ventricular gene expression. *Mol. Cell. Biol.* **21**: 1730.
Camenisch T.D., Spicer A.P., Brehm-Gibson T., Biesterfeldt J., Augustine M.L., Calabro A., Jr, Kubalak S., Klewer S.E., and McDonald J.A. 2000. Disruption of hyaluronan synthase-2 abrogates normal cardiac morphogenesis and hyaluronan-mediated transformation of epithelium to mesenchyme (comments). *J. Clin. Invest.* **106**: 349.
Candia A.F., Hu J., Crosby J., Lalley P.A., Noden D., Nadeau J.H., and Wright C.V. 1992. Mox-1 and Mox-2 define a novel homeobox gene subfamily and are differentially expressed during early mesodermal patterning in mouse embryos. *Development* **116**: 1123.
Chan-Thomas P.S., Thompson R.P., Robert B., Yacoub M.H., and Barton P.J. 1993. Expression of homeobox genes *Msx-1* (*Hox-7*) and *Msx-2* (*Hox-8*) during cardiac development in the chick. *Dev. Dyn.* **197**: 203.
Chen J.-N., Haffter P., Odenthal J., Vogelsang E., Brand M., van Eeden F.J., Furutani-Seiki M., Granato M., Hammerschmidt M., Heisenberg C.P., Jiang Y.-J., Kane D.A., Kelsh R.N., Mullins M.C., and Nüsslein-Volhard C. 1996. Mutations affecting the cardiovascular system and other internal organs in zebrafish. *Development* **123**: 293.
Christoffels V.M., Habets P.E., Franco D., Campione M., de Jong F., Lamers W.H., Bao Z.Z., Palmer S., Biben C., Harvey R.P., and Moorman A.F. 2000. Chamber formation and morphogenesis in the developing mammalian heart. *Dev. Biol.* **223**: 266.
de la Pompa J.L., Timmerman L.A., Takimoto H., Yoshida H., Elia A.J., Samper E., Potter J., Wakeham A., Marengere L., Langille B.L., Crabtree G.R., and Mak T.W. 1998. Role of the NF-ATc transcription factor in morphogenesis of cardiac valves and septum. *Nature* **392**: 182.
Driever W., Solnica-Krezel L., Schier A.F., Neuhauss S.C., Malicki J., Stemple D.L., Stainier D.Y., Zwartkruis F., Abdelilah S., Rangini Z., Belak J., and Boggs C. 1996. A genetic screen for mutations affecting embryogenesis in zebrafish. *Development* **123**: 37.
Eisenberg L.M. and Markwald R.R. 1995. Molecular regulation of atrioventricular valvuloseptal morphogenesis. *Circ. Res.* **77**: 1.
Emmanouilides G.C., Allen H.D., Riemenschneider T.A., and Gutgesell H.P. 1998. *Clinical synopsis of Moss and Adams' heart disease in infants, children, and adolescents*. Williams and Wilkins, Baltimore, Maryland.
Esko J.D. and Selleck S.B. 2002. Order out of chaos: Assembly of ligand binding sites in heparan sulfate. *Annu. Rev. Biochem.* **71**: 435.
Gaussin V., Van De Putte T., Mishina Y., Hanks M.C., Zwijsen A., Huylebroeck D., Behringer R.R., and Schneider M.D. 2002. Endocardial cushion and myocardial defects after cardiac myocyte-specific conditional deletion of the bone morphogenetic protein receptor ALK3. *Proc. Natl. Acad. Sci.* **99**: 2878.
Habets P.E., Moorman A.F., Clout D.E., van Roon M.A., Lingbeek M., van Lohuizen M., Campione M., and Christoffels V.M. 2002. Cooperative action of Tbx2 and Nkx2.5 inhibits ANF expression in the atrioventricular canal: Implications for cardiac chamber formation. *Genes Dev.* **16**: 1234.
Häcker U., Lin X., and Perrimon N. 1997. The *Drosophila* sugarless gene modulates Wingless signaling and encodes an enzyme involved in polysaccharide biosynthesis. *Development* **124**: 3565.

Haffter P., Granato M., Brand M., Mullins M.C., Hammerschmidt M., Kane D.A., Odenthal J., van Eeden F.J., Jiang Y.J., Heisenberg C.P., Kelsh R.N., Furutani-Seiki M., Vogelsang E., Beuchle D., Schach U., Fabian C., and Nüsslein-Volhard C. 1996. The identification of genes with unique and essential functions in the development of the zebrafish, *Danio rerio*. *Development* **123**: 1.
Hu N., Sedmera D., Yost H.J., and Clark E.B. 2000. Structure and function of the developing zebrafish heart. *Anat. Rec.* **260**: 148.
Jackson S.M., Nakato H., Sugiura M., Jannuzi A., Oakes R., Kaluza V., Golden C., and Selleck S.B. 1997. Dally, a *Drosophila* glypican, controls cellular responses to the TGFβ-related morphogen, Dpp. *Development* **124**: 4113.
Jin S.-W., Jungblut B., and Stainier D.Y.R. 2002. Angiogenesis during zebrafish development. In *Genetics of angiogenesis* (ed. J.B. Hoying), p. 101. Bios, Oxford, United Kingdom.
Kim R.Y., Robertson E.J., and Solloway M.J. 2001. Bmp6 and bmp7 are required for cushion formation and septation in the developing mouse heart. *Dev. Biol.* **235**: 449.
Krug E.L., Mjaatvedt C.H., and Markwald R.R. 1987. Extracellular matrix from embryonic myocardium elicits an early morphogenetic event in cardiac endothelial differentiation. *Dev. Biol.* **120**: 348.
Krug E.L., Runyan R.B., and Markwald R.R. 1985. Protein extracts from early embryonic hearts initiate cardiac endothelial cytodifferentiation. *Dev. Biol.* **112**: 414.
Kumai M., Nishii K., Nakamura K., Takeda N., Suzuki M., and Shibata Y. 2000. Loss of connexin45 causes a cushion defect in early cardiogenesis. *Development* **127**: 3501.
Lee R.K., Stainier D.Y., Weinstein B.M., and Fishman M.C. 1994. Cardiovascular development in the zebrafish. II. Endocardial progenitors are sequestered within the heart field. *Development* **120**: 3361.
Lin X., Buff E.M., Perrimon N., and Michelson A.M. 1999. Heparan sulfate proteoglycans are essential for FGF receptor signaling during *Drosophila* embryonic development. *Development* **126** 3715.
Loeber C.P. and Runyan R.B. 1990. A comparison of fibronectin, laminin, and galactosyltransferase adhesion mechanisms during embryonic cardiac mesenchymal cell migration in vitro. *Dev. Biol.* **140**: 401.
McGuire P.G. and Orkin R.W. 1992. Urokinase activity in the developing avian heart: A spatial and temporal analysis. *Dev. Dyn.* **193**: 24.
Mjaatvedt C.H., Lepera R.C., and Markwald R.R. 1987. Myocardial specificity for initiating endothelial-mesenchymal cell transition in embryonic chick heart correlates with a particulate distribution of fibronectin. *Dev. Biol.* **119**: 59.
Motoike T., Loughna S., Perens E., Roman B.L., Liao W., Chau T.C., Richardson C.D., Kawate T., Kuno J., Weinstein B.M., Stainier D.Y., and Sato T.N. 2000. Universal GFP reporter for the study of vascular development. *Genesis* **28**: 75.
Nakajima Y., Yamagishi T., Hokari S., and Nakamura H. 2000. Mechanisms involved in valvuloseptal endocardial cushion formation in early cardiogenesis: Roles of transforming growth factor (TGF)- β and bone morphogenetic protein (BMP). *Anat. Rec.* **258**: 119.
Nakajima Y., Miyazono K., Kato M., Takase M., Yamagishi T., and Nakamura H. 1997. Extracellular fibrillar structure of latent TGFβ binding protein-1: Role in TGFβ-dependent endothelial-mesenchymal transformation during endocardial cushion tissue formation in mouse embryonic heart. *J. Cell Biol.* **136**: 193.
Ranger A.M., Grusby M.J., Hodge M.R., Gravallese E.M., de la Brousse F.C., Hoey T., Mickanin C., Baldwin H.S., and Glimcher L.H. 1998. The transcription factor NF-ATc is essential for cardiac valve formation. *Nature* **392**: 186.
Rauch G.J., Hammerschmidt M., Blader P., Schauerte H.E., Strahle U., Ingham P.W., McMahon A.P., and Haffter P. 1997. Wnt5 is required for tail formation in the zebrafish embryo. *Cold Spring Harbor Symp. Quant. Biol.* **62**: 227.

Romano L.A. and Runyan R.B. 1999. Slug is a mediator of epithelial-mesenchymal cell transformation in the developing chicken heart. *Dev. Biol.* **212:** 243.

———. 2000. Slug is an essential target of TGFβ2 signaling in the developing chicken heart. *Dev. Biol.* **223:** 91.

Runyan R.B. and Markwald R.R. 1983. Invasion of mesenchyme into three-dimensional collagen gels: A regional and temporal analysis of interaction in embryonic heart tissue. *Dev. Biol.* **95:** 108.

Runyan R.B., Potts J.D., Sharma R.V., Loeber C.P., Chiang J.J., and Bhalla R.C. 1990. Signal transduction of a tissue interaction during embryonic heart development. *Cell Regul.* **1:** 301.

Shalaby F., Rossant J., Yamaguchi T.P., Gertsenstein M., Wu X.F., Breitman M.L., and Schuh A.C. 1995. Failure of blood-island formation and vasculogenesis in Flk-1-deficient mice. *Nature* **376:** 62.

Sinning A.R., Lepera R.C., and Markwald R.R. 1988. Initial expression of type I procollagen in chick cardiac mesenchyme is dependent upon myocardial stimulation. *Dev. Biol.* **130:** 167.

Stainier D.Y. 2001. Zebrafish genetics and vertebrate heart formation. *Nat. Rev. Genet.* **2:** 39.

Stainier D.Y.R., Weinstein B.M., Detrich H.W., Zon L.I., and Fishman M.C. 1995. *cloche*, an early acting zebrafish gene, is required by both the endothelial and hematopoietic lineages. *Development* **121:** 3141.

Stainier D.Y.R., Fouquet B., Chen J.N., Warren K.S., Weinstein B.M., Meiler S.E., Mohideen M.A., Neuhauss S.C., Solnica-Krezel L., Schier A.F., Zwartkruis F., Stemple D.L., Malicki J., Driever W., and Fishman M.C. 1996. Mutations affecting the formation and function of the cardiovascular system in the zebrafish embryo. *Development* **123:** 285.

Topczewski J., Sepich D.S., Myers D.C., Walker C., Amores A., Lele Z., Hammerschmidt M., Postlethwait J., and Solnica-Krezel L. 2001. The zebrafish glypican Knypek controls cell polarity during gastrulation movements of convergent extension. *Dev. Cell* **1:** 251.

Tsuda M., Kamimura K., Nakato H., Archer M., Staatz W., Fox B., Humphrey M., Olson S., Futch T., Kaluza V., Siegfried E., Stam L., and Selleck S.B. 1999. The cell-surface proteoglycan Dally regulates Wingless signalling in *Drosophila*. *Nature* **400:** 276.

Walsh E.C. and Stainier D.Y. 2001. UDP-glucose dehydrogenase required for cardiac valve formation in zebrafish. *Science* **293:** 1670.

Westin J. and Lardelli M. 1997. Three novel Notch genes in zebrafish: Implications for vertebrate Notch gene evolution and function. *Dev. Genes Evol.* **207:** 51.

Winnier G., Blessing M., Labosky P.A., and Hogan B.L. 1995. Bone morphogenetic protein-4 is required for mesoderm formation and patterning in the mouse. *Genes Dev.* **9:** 2105.

Wunsch A.M., Little C.D., and Markwald R.R. 1994. Cardiac endothelial heterogeneity defines valvular development as demonstrated by the diverse expression of JB3, an antigen of the endocardial cushion tissue. *Dev. Biol.* **165:** 585.

Xu W., Baribault H., and Adamson E.D. 1998. Vinculin knockout results in heart and brain defects during embryonic development. *Development* **125:** 327.

Yamagishi T., Nakajima Y., and Nakamura H. 1999. Expression of TGFβ3 RNA during chick embryogenesis: A possible important role in cardiovascular development. *Cell Tissue Res.* **298:** 85.

Yamamura H., Zhang M., Markwald R.R., and Mjaatvedt C.H. 1997. A heart segmental defect in the anterior-posterior axis of a transgenic mutant mouse. *Dev. Biol.* **186:** 58.

Yelon D. and Stainier D.Y.R. 1999. Patterning during organogenesis: Genetic analysis of cardiac chamber formation. *Semin. Cell Dev. Biol.* **10:** 93.

Zhang H.-Y., Lardelli M., and Ekblom P. 1997. Sequence of zebrafish *fibulin-1* and its expression in developing heart and other embryonic organs. *Dev. Genes Evol.* **207:** 340.

Neural Crest Migration and Mouse Models of Congenital Heart Disease

A.D. Gitler, C.B. Brown, L. Kochilas, J. Li, and J.A. Epstein
Department of Medicine, University of Pennsylvania Health System, Philadelphia, Pennsylvania 19104

Congenital heart disease is common, affecting nearly 1% of live births. Perhaps one-third of the cases of congenital heart disease involve abnormal morphogenesis of the outflow tract of the heart and/or abnormal patterning of the great vessels (Driscoll 1994; Goldmuntz et al. 1998). The development of these structures is influenced by migrating neural crest cells that arise from the dorsal neural tube and populate the pharyngeal arches and the heart (Kirby et al. 1983). Ablation studies in chick embryos suggest that neural crest migration defects can result in congenital heart disease. Recently, however, analyses of a series of mouse models with genetic forms of congenital heart disease that resemble human disease suggest that migration defects are uncommon. Newly developed molecular markers and fate-mapping techniques suggest that functional defects of post-migratory neural crest cause some forms of common structural cardiovascular disorders. These defects can be cell autonomous or can arise from defects in cells surrounding the neural crest migration pathways.

CARDIAC NEURAL CREST

During late stages of mammalian and avian cardiogenesis, the looped heart tube undergoes a series of septation events that result in the establishment of parallel pulmonary and systemic circulations (Fishman and Chien 1997). At the same time, the aortic arch arteries, which arise as paired sets of bilaterally symmetric vessels, undergo dramatic remodeling in order to produce the asymmetric adult vascular system (Epstein and Buck 2000). Landmark studies performed in the 1980s identified a critical requirement for neural crest cells in these processes (Kirby et al. 1983; Kirby 1988).

Neural crest cells arise from the dorsal neural tube and migrate throughout the developing embryo. They function as pluripotent progenitor cells and can differentiate into most mesodermal cell types (Hall 1999; Le Douarin and Kalcheim 1999). A subset of neural crest cells migrates from the level of the first three somites and contributes significant cell mass to the developing branchial arches. DiI labeling studies and quail–chick chimera analyses indicate that some of these migrating cells populate the outflow tract of the heart. Most significantly, ablation of these cells in chick embryos before they emerge from the neural tube results in predictable forms of congenital heart disease, including persistent truncus arteriosus, in which the aorta and pulmonary arteries fail to arise as distinct vessels due to the absence of outflow tract septation (Kirby et al. 1983). Other defects include double-outlet right ventricle and interruption of the aortic arch. These important experiments suggest that genetic defects which result in the absence of neural crest migration in mammals may cause congenital heart disease.

In mammals, however, studies to track neural crest migration or to perform ablation studies have been more difficult than in avian species due to technical considerations and to the relative dearth of reliable molecular markers. However, a number of mouse models of congenital heart disease involving the outflow tract of the heart have arisen either spontaneously or via targeted mutation of specific genes (Epstein 1996, 2001; Epstein and Buck 2000). Our laboratory and others have developed new markers that are specific for migrating neural crest cells and that permit the accurate identification of these cells in normal embryos and in mutant strains (Waldo et al. 1999; Yamauchi et al. 1999; Jiang et al. 2000; Li et al. 2000; Brown et al. 2001). The surprising finding that is emerging from the analysis of a series of murine models with outflow tract and aortic arch defects is that neural crest migration is generally preserved. None of these models is caused by a complete loss of neural crest migration that would be analogous to the chick ablation models.

Pax3 FUNCTION AND CARDIAC DEFECTS IN *Splotch* mice

Homozygous mutations in the gene encoding the transcription factor Pax3 result in congenital heart disease which closely resembles that seen in chick embryos after neural crest ablation (Franz 1989; Epstein 1996). $Pax3^{-/-}$ embryos succumb at embryonic day 13.5 (E13.5) with persistent truncus arteriosus and also show defects in other neural-crest-derived structures, including peripheral ganglia. *Pax3* is expressed in the dorsal neural tube beginning as early as E8.5, before cardiac neural crest cells have delaminated and initiated migration toward the branchial arches (Goulding et al. 1991). Expression is extinguished shortly after migration has initiated (Epstein et al. 2000). Hence, it was attractive to assume that Pax3 was required for cardiac neural crest migration (Moase and Trasler 1991, 1992) and that lack of migration resulted in cardiovascular defects. Some early studies indicated abnormal migratory behavior of cardiac neural crest cells derived from *Splotch* embryos (Moase and Trasler 1990), and later studies appeared to confirm these results (Conway et al. 1997). However, subsequent reanalysis of these latter studies suggests that the migratory population dependent on Pax3 is actually a pool of hypoglossal and hypaxial muscle progenitors rather than cardiac neural crest (Epstein et al. 2000).

We reexamined neural crest migratory behavior in *Splotch* embryos using a transgenic mouse line to aid in the identification of neural crest cells. A portion of the connexin 43 (Cx43) gene upstream regulatory region was used to drive expression of *lacZ* and was found to direct expression to neural crest cells in developing embryos (Waldo et al. 1999). We crossed Cx43-lacZ mice with *Splotch* mice and examined neural crest patterning at E12.5 in wild-type and homozygous *Splotch* embryos. Interestingly, we found significant numbers of labeled cells that had migrated throughout the branchial arches, encased the aortic arch arteries, and invaded the outflow tract of the heart in both wild-type and mutant embryos (Epstein et al. 2000). The precise patterning of neural crest derivatives in the heart was not identical since the outflow tract septum was poorly formed or absent in mutant embryos, and there appeared qualitatively to be fewer labeled cells at the most distal zones of migration in *Splotch* embryos. In addition, in some embryos, labeled cells were not as tightly clustered in the region of the endocardial cushion in mutant embryos when compared to wild type. Nevertheless, the unequivocal findings from these studies were that neural crest migration is not entirely dependent on Pax3 function, and the *Splotch* cardiac phenotype including persistent truncus arteriosus is not due to absence of neural crest cells in the heart. Rather, the defect is more subtle, stochastic, or related to postmigratory neural crest function.

Cre-lox APPROACHES TO FATE-MAP NEURAL CREST

We have taken an alternative approach to follow the fate of Pax3-expressing neural crest cells as they migrate throughout the embryos and undergo differentiation. We have identified a portion of the *Pax3* upstream genomic region that is sufficient to recapitulate *Pax3* expression in the dorsal neural tube while omitting expression in other domains of endogenous *Pax3* expression such as the somite (Li et al. 1999). Like *Pax3* itself, expression of a *lacZ* transgene driven by this regulatory region declines shortly after migration of neural crest cells initiates. However, using Cre-lox approaches, we have been able to indelibly label neural crest derivatives that expressed the *Pax3* transgene earlier during development (Li et al. 2000). We characterized Pax3-Cre mice and demonstrated expression of Cre recombinase in the dorsal neural tube, although the lines of mice characterized to date exhibit ectopic expression in the caudal regions in mesenchymal tissue adjacent to the neural tube. By crossing Pax3-Cre mice with Cre reporter mice, such as R26R mice in which *lacZ* expression is initiated by Cre-mediated recombination, cells that express Pax3-Cre are labeled by *lacZ* expression. Since *lacZ* is driven by a ubiquitous promoter in R26R mice, once *lacZ* expression is activated, it will continue to be expressed for the lifetime of that cell and by all daughter cells derived from that cell. Hence, *Pax3*-expressing neural crest precursors are fate-mapped, and their derivatives can be identified.

We have used this system to fate-map *Pax3*-expressing neural crest precursors in *Splotch* embryos. These studies confirm the observation that neural crest migrates through the branchial arches and populates the heart in both wild-type and mutant embryos (Fig. 1). Moreover, Pax3-expressing precursors give rise to smooth muscle cells in the aortic arches and ductus arteriosus, and they compose the vast majority of the aorto-pulmonary septum in the outflow tract (Epstein and Buck 2000).

Sema3C DEFICIENCY AND NEURAL-CREST-RELATED CARDIAC DEFECTS

Semaphorins represent a family of related cell-surface and secreted molecules that play important roles in mediating repulsive guidance cues during central nervous system axonal migration (Yu and Kolodkin 1999; Tamagnone and Comoglio 2000). Certain secreted semaphorin molecules mediate growth cone collapse and hence direct patterning of neuronal circuitry. Inactivation of *Semaphorin 3C* (*Sema3C*) in the mouse unexpectedly resulted in aortic arch and cardiac outflow tract defects that included interruption of the aortic arch and persistent truncus arteriosus (Feiner et al. 2001). Sema3C is a secreted semaphorin and is presumed to bind to a heterodimeric receptor composed of a plexin subunit and a neuropilin subunit. *Sema3C* is expressed in the non-neural-crest-derived mesenchyme of the branchial arches and is likely to have a non-cell-autonomous function that results in neural crest abnormalities and cardiovascular defects (Feiner et al. 2001).

We sought to identify a potential Sema3C receptor expressed by migrating neural crest cells. Although a number of molecules have been suggested as markers of migrating cardiac neural crest in mammals, none has been definitively proven to label these cells. Hence, we were aided considerably by the availability of the Cre-lox fate-mapping system described above in order to identify postmigratory neural crest cells while costaining for potential semaphorin receptors. This approach allowed the identification of PlexinA2 expression by migrating cardiac neural crest (Brown et al. 2001). PlexinA2 is expressed by two streams of cells invading the cardiac outflow tract and forming the aorto-pulmonary septum. This pattern appears identical to that of fate-mapped Pax3-expressing neural crest cells in the heart.

Because semaphorins are closely linked with regulation of migration, we examined the patterning of cardiac neural crest cells in wild-type and Sema3C-deficient embryos. Despite the absence of Sema3C, cardiac neural crest was able to find the proper migratory routes to the

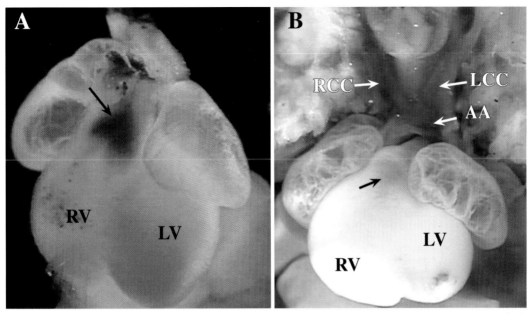

Figure 1. Cardiac neural crest cells populate the cardiac outflow tract and great vessels. (*A*) Fate mapping was performed using mice that express Cre recombinase in Pax3-expressing neural crest precursors and R26R reporter mice. Neural crest derivatives are labeled blue. E11.5 heart shows neural crest cells in the cardiac outflow tract (*arrow*). (*B*) At E14.5, neural crest cells can be seen encasing the great vessels including the right common carotid (RCC), left common carotid (LCC), and aortic arch (AA) arteries. Neural crest cells in the cardiac outflow tract (*black arrow*) are partially obscured by myocardium. (LV) Left ventricle. (RV) Right ventricle.

cardiac outflow tract, and the overall patterning of cardiac neural crest was unchanged. We did note subtle deficiencies in the maximal distance that cardiac neural crest migrated into the outflow tract, suggesting that subtle migratory defects may occur (Brown et al. 2001; Feiner et al. 2001). Migration in the region of the aortic arch arteries within the branchial arches was grossly unaffected despite the fact that aortic arch interruption developed in these embryos. These results suggest that defects of neural crest derivatives, other than migration defects, may account for some or all of the phenotypes seen in Sema3C-deficient mice. Perhaps semaphorin signaling affects differentiation or survival of this cell population.

TYPE I NEUROFIBROMATOSIS GENE AND DOUBLE-OUTLET RIGHT VENTRICLE

The type I neurofibromatosis gene (*NF1*) encodes a large intracellular protein capable of down-regulating Ras signaling (Cichowski and Jacks 2001). Mutation of this tumor suppressor gene in humans leads to type I neurofibromatosis, which is characterized by benign and malignant tumors of neural crest origin. Patients are heterozygotes and sustain a "second hit" in affected somatic tissues. In mice, heterozygous mutation of *Nf1* also leads to a predisposition to cancer (Brannan et al. 1994; Jacks et al., 1994). Homozygous mutation leads to embryonic lethality, and affected embryos display congenital heart disease. Heart defects include double-outlet right ventricle, in which both the pulmonary artery and the aorta arise from the right ventricle (Brannan et al. 1994; Jacks et al.

1994; Lakkis and Epstein 1998). This abnormality is associated with a ventricular septal defect allowing blood to exit the left ventricle. Double-outlet right ventricle is also seen in chick embryos after ablation of premigratory neural crest cells.

The association of type I neurofibromatosis with neural crest tumors, and the association of double-outlet right ventricle with neural crest ablation, suggested that congenital heart disease in $Nf1^{-/-}$ embryos was due to a defect in cardiac neural crest cells (Brannan et al. 1994; Jacks et al. 1994). Therefore, we examined neural crest migration in wild type and $Nf1^{-/-}$ embryos using molecular markers and Cre-lox fate-mapping systems. We also examined affected embryos for other signs of neural-crest-related cardiovascular defects.

In 1 of 20 $Nf1^{-/-}$ E13.5 embryos examined, we identified abnormal remodeling of the aortic arch arteries resulting in retro-esophageal right sublclavian artery, a defect associated with inappropriate regression of the right fourth aortic arch segment (Fig. 2). Interestingly, we identified a low incidence of the identical defect in *Nf1* embryos in which the *Nf1* gene was deleted only in neural crest cells using a Cre-lox approach. However, we did not detect any instances of more severe aortic arch or neural crest defects such as persistent truncus arteriosus or interruption of the aortic arch. Nevertheless, all mutant embryos exhibited double-outlet right ventricle.

Our preliminary results also suggest that neural crest migration in *Nf1* mutant embryos is intact. Using Cre-lox approaches described above, we followed neural crest cells throughout their migration along the aortic arch seg-

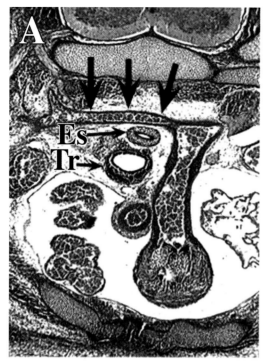 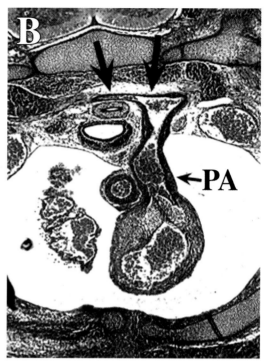

Figure 2. Retroesophageal right subclavian artery (*black arrows*) is present in a minority of Nfl$^{-/-}$ E13.5 embryos (*A*) and in an embryo lacking Nfl only in neural crest cells produced by a Cre-lox approach (*B*). Transverse sections stained with H&E at the level of the cardiac outflow tract are shown. Ventral is at the top; dorsal at the bottom. The right subclavian artery normally travels dorsal to the esophagus (Es) and trachea (Tr). These sections are at the level of the pulmonary valve and pulmonary artery (PA) and the ductus arteriosus is patent, allowing continuity between the PA and the right subclavian which is originating ectopically from the descending aorta near the junction of the ductus arteriosus and the aorta.

ments and into the cardiac outflow tract. In both wild type and *Nf1* mutants, abundant neural crest cells were identified in the appropriate locations. Of note, double-outlet right ventricle is completely penetrant in *Nf1* mutant embryos, suggesting that, although this defect can be caused by an absence of neural crest migration, it can also occur through other mechanisms.

MOUSE MODELS OF DiGEORGE SYNDROME AND Tbx1

DiGeorge syndrome is a common human genetic disorder occurring in 1 of 4000 live births (Driscoll 1994; Epstein and Buck 2000; Epstein 2001). The syndrome is characterized by cardiac outflow tract defects, parathyroid deficiency, thymus and thyroid abnormalities, and craniofacial defects. Learning difficulties can also be associated. Many of the affected organs have neural crest contributions, suggesting that DiGeorge syndrome may be related to abnormal neural crest development. In many cases, deletions on human chromosome 22q11 have been associated with familial and sporadic cases (Goldmuntz et al. 1998).

Recently, a series of mouse models of DiGeorge syndrome have been created that take advantage of the fact that the commonly deleted region of human chromosome 22 is homologous to a conserved region of mouse chromosome 11 (Lindsay et al. 1999, 2001; Epstein 2001; Merscher et al. 2001; Schinke and Izumo 2001). Heterozygous deletion of a ~1.5-Mb region, including about 20 genes, results in mice with cardiac outflow tract defects, including interruption of the aortic arch, retroesophageal subclavian artery, and related defects (Lindsay et al. 1999, 2001; Merscher et al. 2001). These mice also have parathyroid and thymus defects reminiscent of human patients. Hence, these mice provide a reasonable animal model for at least some aspects of DiGeorge syndrome.

A series of elegant complementation experiments using additional mouse lines that contained extra copies of some of the deleted genes allowed researchers to focus on a small group of genes within the deleted region that were responsible for the cardiovascular abnormalities (Lindsay et al. 2001; Merscher et al. 2001). Subsequent analysis identified Tbx1 as the critical gene, and heterozygous targeted deletion of Tbx1 results in similar cardiovascular abnormalities (Jerome and Papaioannou 2001; Lindsay et al. 2001; Merscher et al. 2001). Interestingly, Tbx1 is apparently not expressed by cardiac neural crest cells, but rather is expressed by the core mesenchyme of the pharyngeal arches through which cardiac neural crest migrates.

We asked whether Tbx1 haploinsufficiency affected neural crest migration. We examined the expression of PlexinA2, and we used Cre-lox strategies to examine neural crest patterning in wild-type and Tbx1$^{+/-}$ littermates. We were unable to detect any differences in neural crest migration or location or gene expression in affected embryos, despite the evidence for aortic arch artery abnormalities. Hence, we conclude that wild-type expression of Tbx1 is not required for normal cardiac neural crest migration, and the cardiovascular defects seen in these mouse models of DiGeorge syndrome are not related to absence of neural crest cells in the pharyngeal region or the heart.

Additional data from our lab and others, however, suggest that neural crest differentiation is abnormal in mouse models of DiGeorge syndrome. Cardiac neural crest cells encasing the aortic arch arteries invest the medial layer of these vessels and differentiate into smooth muscle. This differentiation process is deficient in DiGeorge models, and early markers of smooth muscle are reduced or absent in specific aortic arch segments including the 4th aortic arch artery. Since Tbx1 is expressed by adjacent mesenchyme, these results suggest a model in which Tbx1 induces the secretion of a critical growth factor that is required for proper survival or differentiation of nearby neural crest derivatives.

CONCLUSIONS

In summary, recent advances in genetics and cell biology have allowed the identification of new markers for cardiac neural crest cells and for genetic mechanisms for performing fate-mapping studies in the mouse. These tools have permitted the further evaluation of a series of mouse models of congenital heart disease that are associated with neural crest defects. Despite the fact that earlier studies in chick embryos suggested that neural crest migration defects would be associated with congenital heart disease, the surprising finding that is emerging from ongoing studies is that neural crest migration is relatively preserved in several diverse models of outflow tract and aortic arch disorders. It now appears likely that gross migration defects are relatively uncommon, whereas functional defects of postmigratory neural crest cells may be more common. These defects may involve survival, proliferation, and/or differentiation, and are likely to depend on critical interactions with non-neural-crest-derived neighboring cells.

ACKNOWLEDGMENTS

This work was supported by grants from the National Institutes of Health to J.A.E., C.B.B., and L.K. and from the W.W. Smith Foundation (J.A.E.) and the American Heart Association (J.A.E.). We are grateful to our collaborators Jonathan Raper, Luis Parada, and Cecilia Lo.

REFERENCES

Brannan C.I., Perkins A.S., Vogel K.S., Ratner N., Nordlund M.L., Reid S.W., Buchberg A.M., Jenkins N.A., Parada L.F., and Copeland N.G. 1994. Targeted disruption of the neurofibromatosis type-1 gene leads to developmental abnormalities in heart and various neural crest-derived tissues. *Genes Dev.* **8:** 1019.

Brown C.B., Feiner L., Lu M.M., Li J., Ma X., Webber A.L., Jia L., Raper J.A., and Epstein J.A. 2001. PlexinA2 and semaphorin signaling during cardiac neural crest development. *Development* **128:** 3071.

Cichowski K. and Jacks T. 2001. NF1 tumor suppressor gene function: Narrowing the GAP. *Cell* **104:** 593.

Conway S., Henderson D., and Copp A. 1997. Pax3 is required for cardiac neural crest migration in the mouse: Evidence from the *splotch (Sp^{2H})* mutant. *Development* **124:** 505.

Driscoll D.A. 1994. Genetic basis of DiGeorge and velocardiofacial syndromes. *Curr. Opin. Pediatr.* **6:** 702.

Epstein J.A. 1996. Pax3, neural crest and cardiovascular development. *Trends Cardiovasc. Med.* **6:** 255.

———. 2001. Developing models of DiGeorge syndrome. *Trends Genet.* **17:** S13.

Epstein J.A. and Buck C.A. 2000. Transcriptional regulation of cardiac development: Implications for congenital heart disease and DiGeorge syndrome. *Pediatr. Res.* **48:** 717.

Epstein J.A., Li J., Lang D., Chen F., Brown C.B., Jin F., Lu M.M., Thomas M., Liu E., Wessels A., and Lo C.W. 2000. Migration of cardiac neural crest cells in Splotch embryos. *Development* **127:** 1869.

Feiner L., Webber A.L., Brown C.B., Lu M.M., Jia L., Feinstein P., Mombaerts P., Epstein J.A., and Raper J.A. 2001. Targeted disruption of semaphorin 3C leads to persistent truncus arteriosus and aortic arch interruption. *Development* **128:** 3061.

Fishman M.C. and Chien K.R. 1997. Fashioning the vertebrate heart: Earliest embryonic decisions. *Development* **124:** 2099.

Franz T. 1989. Persistent truncus arteriosus in the Splotch mutant mouse. *Anat. Embryol.* **180:** 457.

Goldmuntz E., Clark B.J., Mitchell L.E., Jawad A.F., Cuneo B.F., Reed L., McDonald-McGinn D., Chien P., Feuer J., Zackai E.H., Emanuel B.S., and Driscoll D.A. 1998. Frequency of 22q11 deletions in patients with conotruncal defects. *J. Am. Coll. Cardiol.* **32:** 492.

Goulding M.D., Chalepakis G., Deutsch U., Erselius J.R., and Gruss P. 1991. Pax-3, a novel murine DNA binding protein expressed during early neurogenesis. *EMBO J.* **10:** 1135.

Hall B.K. 1999. *The neural crest in development and evolution.* Springer, New York.

Jacks T., Shih T.S., Schmitt E.M., Bronson R.T., Bernards A., and Weinberg R.A. 1994. Tumour predisposition in mice heterozygous for a targeted mutation in Nf1. *Nat. Genet.* **7:** 353.

Jerome L.A. and Papaioannou V.E. 2001. DiGeorge syndrome phenotype in mice mutant for the T-box gene, Tbx1. *Nat. Genet.* **27:** 286.

Jiang X., Rowitch D.H., Soriano P., McMahon A.P., and Sucov H.M. 2000. Fate of the mammalian cardiac neural crest. *Development* **127:** 1607.

Kirby M.L. 1988. Nodose placode contributes autonomic neurons to the heart in the absence of cardiac neural crest. *J. Neurosci.* **8:** 1089.

Kirby M.L., Gale T.F., and Stewart D.E. 1983. Neural crest cells contribute to normal aorticopulmonary septation. *Science* **220:** 1059.

Lakkis M.M. and Epstein J.A. 1998. Neurofibromin modulation of ras activity is required for normal endocardial-mesenchymal transformation in the developing heart. *Development* **125:** 4359.

Le Douarin N. and Kalcheim C. 1999. *The neural crest*, 2nd edition. Cambridge University Press, Cambridge, United Kingdom.

Li J., Chen F., and Epstein J.A. 2000. Neural crest expression of Cre recombinase directed by the proximal Pax3 promoter in transgenic mice. *Genesis* **26:** 162.

Li J., Liu K.C., Jin F., Lu M.M., and Epstein J.A. 1999. Transgenic rescue of congenital heart disease and spina bifida in Splotch mice. *Development* **126:** 2495.

Lindsay E.A., Botta A., Jurecic V., Carattini-Rivera S., Cheah Y.C., Rosenblatt H.M., Bradley A., and Baldini A. 1999. Congenital heart disease in mice deficient for the DiGeorge syndrome region. *Nature* **401:** 379.

Lindsay E.A., Vitelli F., Su H., Morishima M., Huynh T., Pramparo T., Jurecic V., Ogunrinu G., Sutherland H.F., Scambler

P.J., Bradley A., and Baldini A. 2001. Tbx1 haploinsufficieny in the DiGeorge syndrome region causes aortic arch defects in mice. *Nature* **410:** 97.

Merscher S., Funke B., Epstein J.A., Heyer J., Puech A., Lu M.M., Xavier R.J., Demay M.B., Russell R.G., Factor S., Tokooya K., Jore B.S., Lopez M., Pandita R.K., Lia M., Carrion D., Xu H., Schorle H., Kobler J.B., Scambler P., Wynshaw-Boris A., Skoultchi A.I., Morrow B.E., and Kucherlapati R. 2001. TBX1 is responsible for cardiovascular defects in velo-cardio-facial/DiGeorge syndrome. *Cell* **104:** 619.

Moase C.E. and Trasler D.G. 1990. Delayed neural crest cell emigration from Sp and Spd mouse neural tube explants. *Teratology* **42:** 171.

———. 1991. N-CAM alterations in splotch neural tube defect mouse embryos. *Development* **113:** 1049.

———. 1992. Splotch locus mouse mutants: Models for neural tube defects and Waardenburg syndrome type I in humans. *J. Med. Genet.* **29:** 145.

Schinke M. and Izumo S. 2001. Deconstructing DiGeorge syndrome. *Nat. Genet.* **27:** 238.

Tamagnone L. and Comoglio P.M. 2000. Signalling by semaphorin receptors: Cell guidance and beyond. *Trends Cell Biol.* **10:** 377.

Waldo K.L., Lo C.W., and Kirby M.L. 1999. Connexin 43 expression reflects neural crest patterns during cardiovascular development. *Dev. Biol.* **208:** 307.

Yamauchi Y., Abe K., Mantani A., Hitoshi Y., Suzuki M., Osuzu F., Kuratani S., and Yamamura K. 1999. A novel transgenic technique that allows specific marking of the neural crest cell lineage in mice. *Dev. Biol.* **212:** 191.

Yu H.H. and Kolodkin A.L. 1999. Semaphorin signaling: A little less per-plexin. *Neuron* **22:** 11.

Hey bHLH Factors in Cardiovascular Development

A. FISCHER,* C. LEIMEISTER,* C. WINKLER,* N. SCHUMACHER,* B. KLAMT,*
H. ELMASRI,* C. STEIDL,* M. MAIER,* K.-P. KNOBELOCH,† K. AMANN,‡
A. HELISCH,¶ M. SENDTNER,** AND M. GESSLER*

*Theodor-Boveri-Institute (Biocenter), Physiological Chemistry I, Am Hubland, 97074 Wuerzburg, Germany;
†Institute for Molecular Pharmacology, 13125 Berlin, Germany; ‡Institute for Pathology, University of
Erlangen, 91054 Erlangen, Germany; ¶Max-Planck-Institute for Physiological and Clinical Research,
61231 Bad Nauheim, Germany; **Institute for Clinical Neurobiology, University of Wuerzburg,
97078 Wuerzburg, Germany

Notch signal transduction is involved in a broad variety of developmental decisions throughout the animal kingdom. Nevertheless, there are only a small number of downstream effectors, most notably the *Hes* family of basic helix-loop-helix (bHLH) proteins that are known to mediate such signals (Davis and Turner 2001). The *Hey* gene family, which encodes a new subclass of *hairy/Enhancer-of-split*-related bHLH transcriptional regulators, represents a novel set of mediators for *Notch* signaling events. *Hey* genes are dynamically expressed during mouse development with an emphasis on somitogenesis, neurogenesis, and cardiac and vascular morphogenesis. *Hey* proteins can dimerize with multiple partners, and they show specific binding to a subset of E-box sequences. The most prominent features of *Hey* proteins, bHLH and orange domains and a characteristic carboxyl terminus, are depicted schematically in Figure 1. Several lines of evidence suggest that they are intimately involved in several aspects of cardiovascular development.

A KIDNEY DEVELOPMENT SCREEN IDENTIFIES *Hey1*

We have used the kidney organ culture system to devise a screen which should yield immediate response genes that participate in and orchestrate the initial steps of kidney organogenesis (Leimeister et al. 1999a). Metanephric mesenchyme, which is part of the intermediate mesoderm, gets induced by the ingrowing ureteric bud to condense and transform into epithelial precursors. Initially, vesicles are generated that elongate to form comma- and S-shaped structures, which later turn into major parts of the secretory and resorptive parts of the nephrons. In a transfilter culture system, we were able to generate pure populations of induced mesenchyme and to compare the transcriptional program with uninduced cells by differential display PCR. One of the induced transcripts, named *Hey1*, turned out to be related to *hairy* and the *Enhancer-of-split* genes in *Drosophila*. *Hey1* is strongly up-regulated upon condensation of mesenchymal precursors and especially in early epithelial comma- and S-shaped bodies (Leimeister et al. 1999b). Database searches revealed the presence of two additional genes related to *Hey1* that were cloned and named *Hey2* and *HeyL* (Leimeister et al. 1999b; Steidl et al. 2000). These genes have been described independently by several other laboratories since, and they were named either *Hesr1/2*, *Hrt1/2/3*, *CHF1/2*, or *Herp1/2* (see Table 1).

Hey GENES: A NEW SUBFAMILY OF *hairy*-RELATED bHLH FACTORS

Hey genes form a novel subfamily of *hairy/Enhancer-of-split*-related genes that are highly conserved throughout evolution. There is even an ancestral *Drosophila* gene, *dHey*, which has not been studied in detail up to now. It is possible that *dHey* is identical to the *HLH44A* gene (Frise et al. 1997), but again, there is no published information available. Three mammalian *Hey* genes were identified in the human and mouse genome (Steidl et al. 2000). In chicken we could detect *Hey1* and *Hey2* ho-

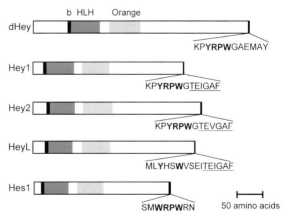

Figure 1. *Hey* proteins: *hairy/Enhancer-of-split*-related with YRPW motif. *Hey1* and *Hey2* show identical domain organization, whereas *HeyL* lacks the full KPYRPWG motif, which resembles the characteristic carboxy-terminal WRPW motif of *hairy/Enhancer-of-split* and *Hes* proteins. The conserved terminal TEIGAF motif present in all three mammalian *Hey* proteins has not been found in other protein sequences before, and its significance is unclear. In the amino-terminal half, the basic domain immediately precedes the helix-loop-helix domain, which is in turn followed by the orange domain. The latter is capable of forming two additional helices.

Table 1. Nomenclature Grid of *Hey* Genes

Leimeister et al. (1999b); Steidl et al. (2000)	Nakagawa et al. (1999)	Kokubo et al. (1999)	Chin et al. (2000); Sun et al. (2001)	Iso et al. (2001)
Hey1	*Hrt1*	*Hesr1*	*CHF2*	*Herp2*
Hey2	*Hrt2*		*CHF1*	*Herp1*
HeyL	*Hrt3*			

mologs but have failed to identify a *HeyL* counterpart as yet (Fig. 2) (Leimeister et al. 2000b). There are GenBank entries or ESTs for *Hey* genes in rat, dog, frog, and several other species, but notably not in *Caenorhabditis elegans*.

In a more systematic approach, we have detected *Hey* family members in different fish species by PCR with degenerate primers and, more recently, by screening available genomic sequence assemblies. In zebrafish the same three types of *Hey* genes are detected (Fig. 2). Not unexpectedly, we found evidence for additional duplicates of *Hey1*, but only in Fugu and Tetraodon—not in zebrafish (C. Winkler et al., in prep.). On the other hand, *Hey2* appears to be absent in Tetraodon and highly divergent in Fugu. This suggests that a reshuffling of tasks for individual *Hey* gene family members must have taken place during evolution, as discussed below.

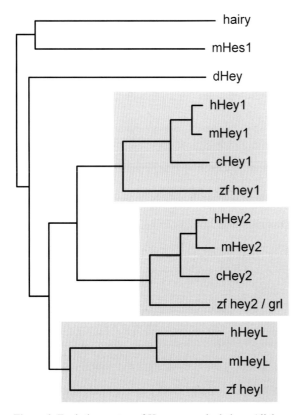

Figure 2. Evolutionary tree of *Hey* genes and relatives. All three types of vertebrate *Hey* genes and the *Drosophila* ancestral *dHey* gene form a separate subgroup of *hairy*-related genes. Sequence alignment unequivocally identifies the respective vertebrate homologs. This is also supported by chromosomal synteny in many cases.

INTERACTION PARTNERS OF HEY PROTEINS

hairy-related bHLH factors frequently function as transcriptional repressors. These factors form dimers through their HLH domains and bind to genomic target sequences via the basic domain. The orange domain, which appears capable of forming two additional helices, is less well characterized. We have utilized two approaches, GST-pulldown assays and yeast two-hybrid analysis, to characterize the interaction properties of *Hey* proteins (Leimeister et al. 2000b). In both assays, *Hey* proteins form homodimers and heterodimers via their HLH domain. They can also interact with *Hes* family proteins like *c-hairy1*, the myogenic factor *myoD*, and even the Class I E-proteins *E2-2* and *E2-5*. This has also been shown by others using coimmunoprecipitation of *Hey1* (i.e., *CHF2*) and *myoD* from transfected cells (Sun et al. 2000). Failure of *Hey1/2* to interact with *E12* or *myf5* shows that there is partial restriction to the choice of partners for *Hey* proteins (Leimeister et al. 2000b). Furthermore, quantification of interaction strength in yeast suggests a graded preference for different partner molecules. It is interesting to note that the orange domain appears to significantly improve the interaction of *Hey1* and *Hey2* or *Hey2* and *c-hairy1*, suggesting that HLH and orange domain may form an extended binding interface for dimerization (Fig. 3a).

Especially *hairy*, the *Enhancer-of-split*, and *Hes* proteins depend on an interaction with the *groucho* corepressor (*TLE/grg* in human and mouse) to exert their negative transcriptional effect on target genes. This interaction is mediated through a conserved carboxy-terminal WRPW motif (Fisher et al. 1996). Interestingly, the related VWRPY motif of *runt* family proteins and the FRPW motif of the *Drosophila huckebein* protein can also bind *groucho* proteins (Aronson et al. 1997; Goldstein et al. 1999). Contrary to expectations, the YRPW motif of *Hey* proteins shows little or no interaction with *TLE1* or *TLE2* proteins, the human homologs of *groucho*. When GST fusion proteins were generated containing only the carboxy-terminal 13 amino acids of mouse *Hey1*, there was either no or only negligible binding of *TLE* proteins in pulldown assays (Fig. 3b). A mutant *Hey1* carboxyl terminus with a single Y→W exchange to generate a WRPW motif displayed efficient binding of *TLE1* and *TLE2*, however. This clearly showed that the assay was functional and the context of surrounding *Hey1* sequence was not responsible for drastic loss of *TLE* binding in wild-type *Hey1*. The fact that a WRPY sequence within the *Hey1* context is similarly compromised in *TLE* binding suggests that neighboring amino acids

Hey GENES IN CARDIOVASCULAR DEVELOPMENT

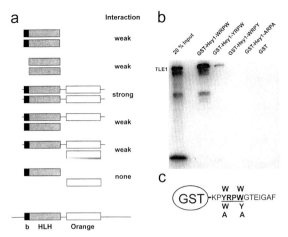

Figure 3. *Hey* protein interaction analysis. (*a*) *Hey* protein dimerization is improved by the orange domain. Although both HLH and orange domains of either *Hey1* or *Hey2* are capable of dimer formation, this process is much more efficient if both domains can synergize. There is no interaction between HLH and orange domains, however. (*b*) The carboxy-terminal YRPW motif of *Hey1* fails to efficiently recruit *TLE* proteins. GST fusion proteins containing only the terminal 13 amino acids of *Hey1* in wild-type or mutant form (see *c*) were used in pulldown experiments with radiolabeled *TLE* proteins to demonstrate efficient interaction of the mutant WRPW form, but very weak, and in some experiments even no, interaction with YRPW or other related sequence motifs.

like the preceding valine in VWRPY may well be relevant under certain circumstances.

DNA BINDING OF *Hey* PROTEINS

Specific residues like an invariant proline residue in the basic domain are thought to determine specific DNA-binding properties of *hairy* and *Hes* proteins; i.e., preferential binding to so-called N-box sequences (CACNAG). To determine potential target sites of *Hey* proteins, we employed a binding site selection scheme. GST-*Hey1* fusion proteins (amino acids 39–122 of mouse *Hey1*) were used to enrich potential binding sites from a pool of random $(N)_{14}$ oligonucleotides. After four rounds of selection on glutathione–Sepharose columns and reamplification, the preferred sites were cloned and sequenced. Alignment of 35 sequences revealed an E-box motif with the following characteristics (Table 2): The core-binding site was CACGTG in most cases. Although the central nucleotides tend to be more variable for other E-proteins (CANNTG), it is the center of the half-site, namely nucleotides 2 and 5 of the hexamer core, that appear to be under less stringent selection, although in most cases the purine in position 2 is represented by A. In addition to the central hexamer site, there was a weaker, but clear, preference for flanking nucleotides, which extends the optimal target site to a tggCACGTGcca sequence.

The preferred GST-*Hey1*-binding site is also recognized efficiently by in-vitro-translated full-length *Hey1*, *Hey2*, and *HeyL* proteins in electrophoretic mobility shift assays. Complex formation can be efficiently competed by excess cold oligonucleotides (Fig. 4). The classic tandem E-box motif of the MCK enhancer (CACCTG) that is recognized by myogenic factors does not compete at all using the same competitor oligonucleotide concentration. Thus, the consensus sequence obtained for *Hey* proteins is highly specific.

Interestingly, the same consensus sequence has been identified by Jennings et al. (1999) in experiments aimed at determining the preferred target for *Drosophila Enhancer-of-split* proteins. In this case, an identical core-binding sequence and even the same preferences for flanking sequences were selected.

Hey GENE EXPRESSION PATTERNS IN THE MOUSE

Analysis of expression patterns revealed a highly dynamic pattern for all three *Hey* genes in multiple processes of embryonic development. Examples are neurogenesis, somitogenesis, vasculogenesis, and cardiogenesis, as well as formation of limb buds, branchial arches, and the ear (Fig. 5). Detailed descriptions have been published elsewhere (Kokubo et al. 1999; Leimeister et al. 1999b, 2000a; Nakagawa et al. 1999; Chin et al. 2000). One of the most interesting aspects is the cycling expression of *Hey2* in the presomitic mesoderm, where periodic waves of *Hey2* transcription may actually be part of the

Table 2: Binding Site Selection for *Hey1* Identifies a tggCACGTGcca Consensus

	Position within the left half-site of the E-box motif					
	−6	−5	−4	−3	−2	−1
G	16 23%	**41** **59%**	**34** **49%**	0 0%	5 7%	0 0%
A	13 19%	11 16%	20 29%	0 0%	65 93%	0 0%
T	**28** **40%**	9 13%	6 9%	0 0%	0 0%	0 0%
C	13 19%	9 13%	10 14%	70 100%	0 0%	70 100%
consensus	t	g	g	**C**	**A**	**C**

Only the half-site of the symmetrical consensus sequence is shown here. Numbering is from the center of the E-box motif.

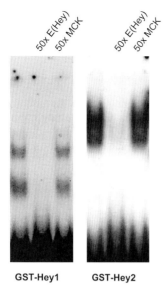

Figure 4. *Hey* proteins bind to specific E-box sequences. GST-fusion proteins were used here to demonstrate electrophoretic mobility shifts using the ^{32}P-labeled, optimized tggCACGT-Gcca oligonucleotide E(Hey). Protein–DNA complex formation was inhibited by incubation with 50-fold excess of unlabeled E(Hey) oligonucleotide, but not by the related E-box sequence (CACCTG) from the MCK enhancer at the same concentration.

molecular clock to coordinate somite formation and thus generation of a metameric body plan (Leimeister et al. 2000b).

All *Hey* genes are expressed in cardiac and vascular precursors at the earliest developmental time points, albeit in different compartments. At E8.5, *Hey1* is detected in precursors of the dorsal aorta and the sinus venosus, the future atria of the heart. *Hey2* is seen in complementary patterns in ventricular precursors and, furthermore, in the allantois, which harbors large numbers of vascular precursor cells. At later stages, *Hey1* and *Hey2* continue to demarcate the atrial and ventricular compartments of the developing heart, but there is no expression of *HeyL*. In the vasculature, *Hey1* is only weakly expressed, predominantly in arterial endothelial cells, but *Hey2* and *HeyL* are expressed more strongly, including, or even restricted to, the smooth muscle cell layer. For *Hey1*, a specific role in endothelial proliferation and tube formation has been proposed based on in vitro assay systems (Henderson et al. 2001).

EXPRESSION OF ZEBRAFISH *hey* GENES

In zebrafish, *hey* genes show a generally more restricted mode of expression during embryonic development when compared to the mouse. Although mouse *Hey* genes are frequently expressed in overlapping patterns, the only domain where we found two of three *hey* paralogs, *hey1* and *hey2*, expressed simultaneously in fish is the telencephalon. Other than that, zebrafish *hey* genes show very specific domains of expression. *hey1* is the only zebrafish gene expressed in somites and presomitic mesoderm. Similar to mouse *Hey1* and *HeyL*, it is

strongly and dynamically expressed in the anterior PSM (Fig. 6a), but it is also rhythmically expressed in the posterior PSM like mouse *Hey2* (C. Winkler et al., in prep.). It is therefore tempting to speculate that during evolution of higher vertebrates, several members of this gene family have taken over subfunctions that are combined into one gene function in fish. Alternatively, it is likewise possible that during evolution of the fish lineage some family members lost certain control regions or functions that were originally present in a common ancestor.

In the cardiovascular system, the situation again appears more restricted in zebrafish. Whereas all three mouse genes are expressed in the aorta, and *Hey1* and *Hey2* in complementary patterns in the developing heart, *hey2/grl* is the only member in zebrafish that is expressed in these structures from earliest stages on. Transcription in the precardiac field and angioblast precursors is evident from early somitogenesis stages (C. Winkler et al., in prep.) and persists at high levels until these organs have completed their morphogenetic process (Fig. 6b,c). In contrast to the mouse, zebrafish *heyl* is only detected at low levels in single cells of the ventral ectoderm (Fig. 6d) and in the floor plate. Interestingly, some of the prominent *Hey* expression domains in the mouse, like limb buds, retina, otic vesicles, and the nasal placodes, are absent in fish.

The two hallmarks of *hey* expression in zebrafish are transcription in earliest precursor cells of, e.g., somites, heart, and dorsal aorta, and specific expression of paralogs in nonoverlapping domains, which differs considerably from the situation in mouse. Whereas *Hey* gene expression profiles suggest certain degrees of functional redundancy in mouse, this seems to be unlikely in zebrafish. These features offer the opportunity to analyze the function of each single *Hey* transcription factor in fish using gene knock-down analyses, as has already been successfully demonstrated for *hey2* (Zhong et al. 2001).

Hey GENES ARE DIRECT TARGETS OF *Notch* SIGNALS

The *Notch* pathway is known for its multifaceted role and probably ubiquitous involvement in diverse aspects of embryonic development and subsequent maintenance. *Delta, Serrate,* or *Jagged* ligands activate *Notch* receptors, which are cleaved within the transmembrane domain to release the intracellular transcriptional activator domain (NICD, Notch intracellular domain). The *Drosophila suppressor of hairless (su(H))* or vertebrate *RBPJ-k* proteins, although they are transcriptional repressors, can be turned into activator complexes upon binding of NICD. In *Drosophila* the *hairy/Enhancer-of-split* gene family has been defined as the predominant class of direct intracellular targets. Until recently, the homologous mammalian *Hes* gene family represented the only vertebrate gene family to mediate *Notch* signals.

Inspection of the mammalian *Hey* promoter sequences revealed the presence of several potential target sites for *RBPJ-k*, the mammalian counterpart of the *Drosophila su(H)* protein (Fig. 7). It is interesting to note that a com-

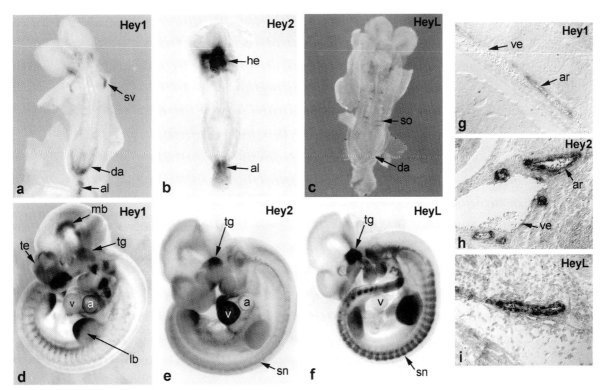

Figure 5. Expression of *Hey* genes during mouse embryogenesis. (*a–c*) Whole-mount in situ hybridization of E8.5 embryos reveals expression of all three *Hey* genes in the cardiovascular system. (*a*) *Hey1* is expressed in the dorsal aorta (da), the allantois (al), and the sinus venosus (sv). (*b*) *Hey2* transcripts are found in the allantois and the heart (he). (*c*) A very weak expression of *HeyL* was observed in the dorsal aorta and the somites. (*d–f*) At E10.5, *Hey* genes are expressed in several tissues. (*d*) *Hey1* transcripts are found in the telencephalon (te), along the ventral neural tube, and also in the ventral part of the midbrain (mb), the trigeminal ganglion (tg), and the limb buds. In the developing heart, *Hey1* expression is restricted to the atrium. (*e*) *Hey2* is expressed in the peripheral nervous system including the trigeminal ganglion and the spinal nerves (sn). Unlike *Hey1*, *Hey2* is detected in the heart ventricle and the outflow tract. (*f*) *HeyL* shows a partly overlapping expression with *Hey1* and *Hey2* in the trigeminal ganglion and the spinal nerves. *HeyL* is not expressed in the embryonic heart, however. (*g, h*) At later developmental stages, all *Hey* genes are found in the arteries. (*g*) E14.5: *Hey1* expression is restricted to the endothelial cells of the arteries. (*h*) E17.5: *Hey2* is likewise found in arteries, but here in smooth muscle precursors. (*i*) E17.5: *HeyL* is similarly expressed in the smooth muscle precursors of the arteries.

putational approach to identify *Notch* target genes in the completely sequenced *Drosophila* genome identified *dHey* as one of the ten prime candidates with eight potential *su(H)*-binding sites within or close to the promoter (Rebeiz et al. 2002).

Transient transfection assays with reporter constructs containing the mouse *Hey1*, *Hey2*, and *HeyL* promoters showed clear activation by NICD, comparable to that seen with the *Hes1* promoter (Fig. 7) (Maier and Gessler 2000). The transcriptional response was dependent on the presence of *RBPJ-k* target sites. Similar data have since been reported by several other laboratories for *Hey1* and *Hey2* (Nakagawa et al. 2000; Iso et al. 2001).

These in vitro data are corroborated by *Hey* expression analysis in *Notch1* and *Dll1* knockout mice that revealed a partial loss of certain *Hey* gene expression domains (Leimeister et al. 2000a,b). On the other hand, this suggests that regulation of *Hey* gene expression may well be dependent on additional *Notch*-independent signaling pathways. Direct proof for an in vivo activation of an endogenous *Hey* gene came from the analysis of transgenic mice expressing NICD specifically in the cortical layer of the hair roots. Here, a strong induction of the endogenous *HeyL* gene was detected (Lin et al. 2000). Since then, several studies have further supported the role of *Hey* genes in *Notch* signaling. In the vascular smooth muscle cell line A10, both *Hey1* and *Hey2* could be activated upon stimulation of the *Notch* pathway (Iso et al. 2002).

ZEBRAFISH *grl* (*Hey2*) IS NEEDED TO BUILD THE AORTA

The *Hey* gene family attracted enormous attention with the discovery that the zebrafish *gridlock* mutant is due to a point mutation in the *grl* gene, the zebrafish *Hey2* homolog (Zhong et al. 2000). Mutant fish embryos show an aortic maturation defect of the anterior-most part of the aorta, the bifurcation. This leads to a lack of posterior blood flow, and it has been hypothesized that *gridlock* resembles human aortic coarctation. In *gridlock* mutants, the *grl* gene has a point mutation affecting the stop codon, which leads to a transcriptional read-through, extending the protein by another 44 amino acids. Although not formally proven, the *gridlock* mutation likely represents a hypomorphic allele.

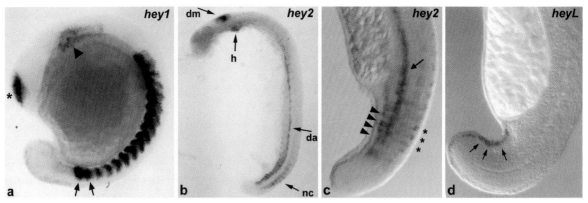

Figure 6. *Hey* genes are expressed in nonoverlapping domains during zebrafish embryogenesis. (*a*) Expression of *hey1* in presomitic mesoderm (PSM), somites, and head region of an embryo at the 15-somite stage (16hpf). Transcription in the two anterior-most somitomeres of the PSM (indicated by *arrows*), head neural crest (*arrowhead*), and the telencephalon (*asterisk*) is shown. (*b, c*) Expression of *hey2* in the dorsal midbrain (dm), heart (h), dorsal aorta (da), and neural crest (nc) of an embryo at 22hpf, lateral views. In *c*, a higher magnification view shows *hey2* transcription in the dorsal aorta (*arrow*) and aorta precursor cells (*arrowheads*). Asterisks demarcate *hey2* expression in cells that appear to be neural crest cells, based on their location lateral to the neural tube (data not shown). (*d*) Expression of *heyL* is restricted to the ventral ectoderm of the tailbud (indicated by *arrows*) and floor plate cells (not shown).

Experiments using graded *grl* inhibition by morpholinos showed that loss of *grl* abolishes arterial fate determination and differentiation of angioblast precursors (Zhong et al. 2001). It has yet to be shown how much of this arterial versus venous fate decision is due to *grl* alone or due to *Notch* signaling acting through additional target molecules besides *grl*. However, it is clear that in zebrafish *grl* is an essential component to build the aorta.

Hey2 LOSS IN THE MOUSE MAINLY AFFECTS THE HEART

A complete knockout allele for the mouse *Hey2* gene was generated by replacing exons 2 and 3 in-frame with the lacZ gene. Heterozygous mice do not show any developmental defects, and the expression of β-galactosidase faithfully mirrors the activity of *Hey2* in most places.

Intercrossing of heterozygous mice leads to a greatly reduced number of knockout mice at weaning. Immediately after birth, all genotypes are seen in Mendelian ratios, suggesting that there is no embryonic lethality in $Hey2^{-/-}$ mice. However, ~80% of the knockout animals die, mostly within the first week of life. $Hey2^{-/-}$ mice generally show severe growth retardation, but survivors catch up after several weeks to reach normal size and life expectancy compared to controls (Gessler et al. 2002).

Angiography performed at 3 days or up to 4 weeks of age did not reveal any evidence of aortic stenosis, and there was no sign of increased collateral blood flow (Fig. 8a). This effectively rules out the assumption that loss of *Hey2* may lead to aortic coarctation in mammals as hypothesized from the analysis of the zebrafish *gridlock* mutant. The most striking difference between *Hey2* knockout and control mice was an enormously enlarged heart, which was clearly evident soon after birth (Fig. 8b). This was not progressive, but heart shape changes remained visible throughout later life in survivors. Our entire initial set of either fully or partially sectioned hearts at different developmental or postnatal stages did not show any evidence for a gross structural anomaly in these mice. Ultrastructural examination of hearts from young mice revealed disorganized and reduced contractile elements, activated capillary endothelia, atypical mitoses, and an overall myoblast-like appearance (Fig. 8c,d). The poor cardiomyocyte development suggested that the resulting functionally compromised heart would not be able to cope efficiently with the load of postnatal circulation. The massive cardiac enlargement was paralleled by robust and again transient induction of classic markers of hypertrophic response, like *ANF, BNP*, or *CARP*.

It is interesting to note that a similar knockout of *Hey2* by M.T. Chin and colleagues (pers. comm.) resulted in a very high incidence of ventricular septum defects on a mixed 129SV/C57BL/6 mouse background. We have

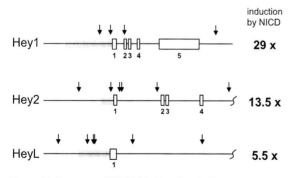

Figure 7. Consensus *RBPJ-k*-binding sites in *Hey* gene promoters and transcriptional activation by NICD. A DNA stretch from –3000 to +5000 nucleotides around the start site of human *Hey* genes was analyzed for consensus *RBPJ-k* binding sites (*arrows*) using the MatInspector (www.genomatix.de) software. Less than one site would be expected at random within the sequence used. Induction of these promoters by activated NICD was assayed by transient cotransfection using luciferase-reporter constructs containing the promoter fragments shaded in gray. Average induction values are given on the right. For details, see Maier and Gessler (2000).

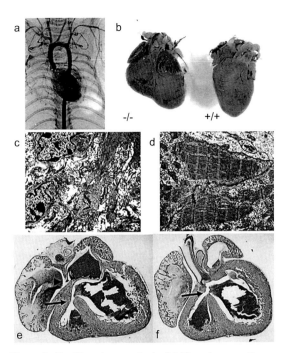

Figure 8. *Hey2* knockout analysis. (*a*) There is no aortic coarctation or induction of collateral vessels in a 4-week-old *Hey2* knockout mouse as visualized by angiography. (*b*) In all cases the knockout hearts show massive biventricular enlargement. (*c*) Ultrastructural analysis of *Hey2*$^{-/-}$ myocardium reveals sparse and disorganized contractile elements compared to controls (*d*). (*e,f*) Cross-sections of hearts from *Hey2*$^{-/-}$ mice that died within two days after birth exhibiting enlarged atria and ventricles, membranous ventricular septum defects (*arrows*), and an atrial septum defect (*arrowhead*).

since examined additional mice from our colony and found that some of our *Hey2* knockout mice also have ventricular and atrial septum defects, although this was clearly absent in the initial series of histopathological analyses upon reexamination (Fig. 8e,f). It remains to be seen how frequent septal defects or other structural malformations occur in *Hey2* knockout mice on different mouse backgrounds. The most parsimonious explanation would be that lack of *Hey2* results in a cardiomyocyte developmental defect with pleiotropic effects that may manifest either as septal defects or as a globally insufficient working myocardium.

Hey1 KNOCKOUT MICE EXHIBIT NO OBVIOUS DEFECTS

The clear distinction of atrial *Hey1* and ventricular *Hey2* expression suggested that a loss of *Hey1* may lead to a different type of cardiac maldevelopment, affecting the atria in this case. A complete knockout of *Hey1* was generated using the same lacZ insertion strategy. Unexpectedly, homozygous animals are viable, fertile, and do not show any cardiac problems yet. Thus, *Hey1* does not appear to fulfill essential functions that cannot be compensated for by other genes, at least under laboratory conditions.

Hey GENES MAY COMPENSATE FOR EACH OTHER IN MAMMALS

Knockout mice lacking either *Dll1* or *Notch1* die with embryonic hemorrhage around midgestation, showing that *Notch* pathway components are indeed crucial for vessel formation and integrity in mice (Gridley 2001). During evolution, complexity and demands on the vascular system have clearly increased. Examples are the addition of a placental circulation and the need for a regulated high-pressure arterial system in land-living animals. It is conceivable that this could only be accomplished through inclusion of additional genetic components. This may explain the lack of a *gridlock*-like phenotype in *Hey2*$^{-/-}$ mice. It may have become necessary to recruit additional *Hey* genes to ensure proper control of formation and function of blood vessels and to provide at least some redundancy. Nevertheless, even such an improved control system of higher vertebrates can still be destabilized by combined inactivation of *Hey* genes. Initial results from intercrosses of *Hey1* and *Hey2* knockout mice suggest that a loss of both genes leads to striking vascular complications (M. Gessler et al., unpubl.). Compound heterozygous/knockout mice are severely underrepresented, and most of them apparently die in utero. Although phenotypic characterization is not yet complete, it appears as if a combined *Hey1/Hey2* knockout may at least partly resemble the loss of *hey2/grl* in zebrafish, as seen in either the *gridlock* hypomorphic mutants or the *grl* morphant embryos (Zhong et al. 2000, 2001). This is in line with the restricted expression of *hey* genes, namely just *grl/hey2*, in the developing blood vessels in zebrafish, but the transcription of all three *Hey* genes in partly overlapping compartments of mouse blood vessels.

CONCLUSIONS

The *Hey* gene family has emerged as an essential component of the *Notch* signal transduction cascade in vertebrates, although additional regulatory upstream signals will certainly be uncovered in future. Although the related *Hes* gene family apparently serves nonredundant functions during segmentation, neurogenesis, and pancreatic organogenesis, the *Hey* genes seem to have unique roles in cardiovascular development. Despite clear differences in expression patterns and currently known phenotypes in fish and mouse models, it will be very interesting to perform analyses of loss and gain of function for these factors in both species to elucidate their functional roles. This may be of relevance for the elucidation of human congenital cardiovascular malformations with septum defects as a leading component, as well as for understanding subsequent vascular maintenance as highlighted by the *Notch* pathway disruption in CADASIL patients (Joutel et al. 1996).

ACKNOWLEDGMENTS

This work was supported by grants from the Deutsche Forschungsgemeinschaft Ge539/9 and SFB 465 (TP A4) to M.G.

REFERENCES

Aronson B.D., Fisher A.L., Blechman K., Caudy M. and Gergen J.P. 1997. Groucho-dependent and -independent repression activities of Runt domain proteins. *Mol. Cell. Biol.* **17:** 5581.

Chin M.T., Maemura K., Fukumoto S., Jain M.K., Layne M.D., Watanabe M., Hsieh C.M. and Lee M.E. 2000. Cardiovascular basic helix loop helix factor 1, a novel transcriptional repressor expressed preferentially in the developing and adult cardiovascular system. *J. Biol. Chem.* **275:** 6381.

Davis R.L. and Turner D.L. 2001. Vertebrate hairy and Enhancer of split related proteins: Transcriptional repressors regulating cellular differentiation and embryonic patterning. *Oncogene* **20:** 8342.

Fisher A.L., Ohsako S. and Caudy M. 1996. The WRPW motif of the hairy-related basic helix-loop-helix repressor proteins acts as a 4-amino-acid transcription repression and protein-protein interaction domain. *Mol. Cell. Biol.* **16:** 2670.

Frise E., Feger G., Feder J., Jan L.Y. and Jan Y.N. 1997. Identification of a novel bHLH gene which is transiently expressed in newborn neurons of the central nervous system. In *Abstracts from the 38th Annual* Drosophila *Research Conference* (Chicago), Genetics Society of America, Bethesda, Maryland, p. 191B.

Gessler M., Knobeloch K.P., Helisch A., Amann K., Schumacher N., Rohde E., Fischer A. and Leimeister C. 2002. Mouse gridlock: No aortic coarctation or deficiency but fatal cardiomyopathy in Hey2 $^{-/-}$ mice. *Curr. Biol.* **12:** 1601.

Goldstein R.E., Jimenez G., Cook O., Gur D. and Paroush Z. 1999. Huckebein repressor activity in *Drosophila* terminal patterning is mediated by Groucho. *Development.* **126:** 3747.

Gridley T. 2001. Notch signaling during vascular development. *Proc. Natl. Acad. Sci.* **98:** 5377.

Henderson A.M., Wang S.J., Taylor A.C., Aitkenhead M., and Hughes C.C. 2001. The basic helix-loop-helix transcription factor HESR1 regulates endothelial cell tube formation. *J. Biol. Chem.* **276:** 6169.

Iso T., Chung G., Hamamori Y., and Kedes L. 2002. HERP1 is a cell type-specific primary target of Notch. *J. Biol. Chem.* **277:** 6598.

Iso T., Sartorelli V., Chung G., Shichinohe T., Kedes L., and Hamamori Y. 2001. HERP, a new primary target of Notch regulated by ligand binding. *Mol. Cell. Biol.* **21:** 6071.

Jennings B.H., Tyler D.M., and Bray S.J. 1999. Target specificities of *Drosophila* enhancer of split basic helix-loop-helix proteins. *Mol. Cell. Biol.* **19:** 4600.

Joutel A., Corpechot C., Ducros A., Vahedi K., Chabriat H., Mouton P., Alamowitch S., Domenga V., Cecillion M., Marechal E., Maciazek J., Vayssiere C., Cruaud C., Cabanis E.A., Ruchoux M.M., Weissenbach J., Bach J.F., Bousser M.G., and Tournier-Lasserve E. 1996. Notch3 mutations in CADASIL, a hereditary adult-onset condition causing stroke and dementia. *Nature* **383:** 707.

Kokubo H., Lun Y., and Johnson R.L. 1999. Identification and expression of a novel family of bHLH cDNAs related to *Drosophila* hairy and enhancer of split. *Biochem. Biophys. Res. Commun.* **260:** 459.

Leimeister C., Bach A., Woolf A. and Gessler M. 1999a. Screen for genes regulated during early kidney morphogenesis. *Dev. Genet.* **24:** 273.

Leimeister C., Externbrink A., Klamt B., and Gessler M. 1999b. Hey genes: A novel subfamily of hairy- and Enhancer of split related genes specifically expressed during mouse embryogenesis. *Mech. Dev.* **85:** 173.

Leimeister C., Schumacher N., Steidl C., and Gessler M. 2000a. Analysis of HeyL expression in wild-type and Notch pathway mutant mouse embryos. *Mech. Dev.* **98:** 175.

Leimeister C., Dale K., Fischer A., Klamt B., Hrabe de Angelis M., Radtke F., McGrew M.J., Pourquie O., and Gessler M. 2000b. Oscillating expression of c-Hey2 in the presomitic mesoderm suggests that the segmentation clock may use combinatorial signaling through multiple interacting bHLH factors. *Dev. Biol.* **227:** 91.

Lin M., Leimeister C., Gessler M. and Kopan R. 2000. Activation of the notch pathway in the hair cortex leads to aberrant differentiation of the adjacent hair-shaft layers. *Development.* **127:** 2421.

Maier M.M. and Gessler M. 2000. Comparative analysis of the human and mouse Hey1 promoter: Hey genes are new Notch target genes. *Biochem. Biophys. Res. Commun.* **275:** 652.

Nakagawa O., Nakagawa M., Richardson J.A., Olson E.N., and Srivastava D. 1999. HRT1, HRT2, and HRT3: A new subclass of bHLH transcription factors marking specific cardiac, somitic, and pharyngeal arch segments. *Dev. Biol.* **216:** 72.

Nakagawa O., McFadden D.G., Nakagawa M., Yanagisawa H., Hu T., Srivastava D., and Olson E.N. 2000. Members of the HRT family of basic helix-loop-helix proteins act as transcriptional repressors downstream of Notch signaling. *Proc. Natl. Acad. Sci.* **97:** 13655.

Rebeiz M., Reeves N.L., and Posakony J.W. 2002. SCORE: A computational approach to the identification of *cis*-regulatory modules and target genes in whole-genome sequence data. *Proc. Natl. Acad. Sci.* **99:** 9888.

Steidl C., Leimeister C., Klamt B., Maier M., Nanda I., Dixon M., Clarke R., Schmid M., and Gessler M. 2000. Characterization of the human and mouse HEY1, HEY2, and HEYL genes: Cloning, mapping, and mutation screening of a new bHLH gene family. *Genomics.* **66:** 195.

Sun J., Kamei C.N., Layne M.D., Jain M.K., Liao J.K., Lee M.E., and Chin M.T. 2001. Regulation of myogenic terminal differentiation by the hairy-related transcription factor CHF2. *J. Biol. Chem.* **276:** 18591.

Zhong T.P., Childs S., Leu J.P., and Fishman M.C. 2001. Gridlock signalling pathway fashions the first embryonic artery. *Nature* **414:** 216.

Zhong T.P., Rosenberg M., Mohideen M.A., Weinstein B., and Fishman M.C. 2000. *gridlock*, an HLH gene required for assembly of the aorta in zebrafish *Science* **287:** 1820.

Molecular Mechanisms of Chamber-specific Myocardial Gene Expression: Transgenic Analysis of the *ANF* Promoter

E.M. SMALL AND P.A. KRIEG
Department of Cell Biology and Anatomy, The University of Arizona Health Sciences Center, Tucson, Arizona 85724

The mature vertebrate heart is a complex organ containing numerous different tissue types. Within the myocardial layer alone, the atrial and ventricular chambers and the specialized myocardium of the conduction system show distinct morphological, contractile, connective, and electrical properties. The molecular genetic pathways leading to the development of these different types of myocardial cells are largely unknown, although recent promoter studies in transgenic animals have begun to reveal the first details of the DNA sequences necessary for correct chamber- or region-specific gene expression. Much less is known about the transcription factors that bind to the DNA sequence to mediate region-specific expression in the heart.

In recent years, development of the insect heart-like organ, the dorsal vessel, has been used as a genetically accessible model for understanding development of the more complex vertebrate heart. This approach has been very successful, and it is clear that there is a high degree of conservation in the regulatory pathways leading to initial specification and differentiation of the myocardial tissues in invertebrates and vertebrates (Bodmer and Venkatesh 1998). The mature dorsal vessel consists of a simple linear tube of contractile muscle, but this tube is subdivided into at least two morphologically distinct domains, the aorta and the heart proper, that appear to serve different physiological functions. During early cardiogenesis, the vertebrate heart also goes through a brief period when it consists of a simple linear tube; however, subsequent looping and morphogenetic movements result in realignment of different heart regions and formation of the complex multichambered organ. During the looping process, portions of the myocardial layer "balloon" out from the outer curvature of the tube, forming morphologically distinct atrial and ventricular chambers. The inner curvature retains many characteristics of the linear heart tube and becomes the inflow tract, atrio-ventricular canal, and outflow tract (Christoffels et al. 2000). The looped tube then becomes septated, resulting in the creation of the morphologically mature multichambered heart.

The acquisition of the gene expression patterns that define the mature atrial and ventricular myocardial tissues appears to occur over an extended period of early heart development. Expression of some genes is limited to distinct precursors of the future atrial or ventricular chambers, even before formation of the linear tube. On the other hand, some genes that are atrium- or ventricle-specific in the adult heart do not assume this expression pattern until long after overt morphological development of the chambers is complete. Expression of still other genes becomes atrium- or ventricle-specific at some time between these two extremes. On the basis of these distinct expression profiles, it seems highly unlikely that a single global regulatory mechanism is responsible for controlling chamber-specific expression. Therefore, it will be necessary to study numerous different genes in order to learn the different mechanisms by which chamber-specific and region-specific expression is achieved.

THE *DROSOPHILA* DORSAL VESSEL AS A MODEL FOR REGION-SPECIFIC GENE EXPRESSION

Different *Drosophila* Cardiac Cell/Tissue Types Have Different Properties

The *Drosophila* dorsal vessel serves a function approximately equivalent to that of the vertebrate heart. The dorsal vessel is a muscular organ which undergoes regular contractions that result in movement of hemolymph within the insect body. The cells that give rise to the dorsal vessel originate in a reiterated pattern in the embryo. These cells are originally located in two rows, in a medial, dorsal position within each segment of the trunk (T2-A7) (Fig. 1A). The heart precursors on the left and right side of the embryo migrate to the midline and merge during dorsal closure to form a muscular tube (Bodmer and Frasch 1999). Although outwardly the dorsal vessel appears to be a simple linear tube, closer inspection reveals that it contains a number of specialized cell types. The cells that form the dorsal vessel are arranged into an inner layer of 104 contractile myocardial cells, also called cardioblasts, and an outer layer of pericardial cells (Rugendorff et al. 1994; Zaffran et al. 1995; Bodmer and Frasch 1999). Further specialization within each of these tissue layers is revealed by different patterns of gene expression and different physiological functions. Differentiation of the various cell types is orchestrated by a combination of genetic factors resulting from overlapping domains of instructive molecules (Gajewski et al. 2000; Lockwood and Bodmer 2002), which are discussed below.

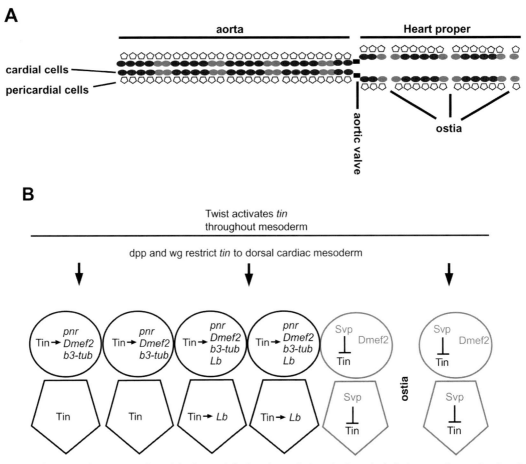

Figure 1. (*A*) Diagrammatic representation of the *Drosophila* dorsal vessel. Anterior is to the left. An outer layer of pericardial cells (*pentagons*) surrounds an inner layer of contractile myocardial cells (*ovals*). The myocardial tube consists of seven repeating segments of four *tin*-positive cells (*black ovals*) and two *svp*-positive cells (*gray ovals*). Two morphologically and functionally distinct regions are separated by the aortic valve. The anterior four segments, and a portion of the fifth, comprise the "aorta," and the posterior two and a half segments form the heart proper. Within the heart proper, three pairs of ostia, or valves, are flanked by the two *svp*-positive cells per segment. (*B*) Simplified model of the transcriptional regulation within one segment of the *Drosophila* dorsal vessel. Inductive signals from twist, and maintenance signals from dpp and wg, activate *tin* in all cardial cells of the segment. Tin expression is directly inhibited in the posterior two cells per segment by svp. Tin directly activates the regulatory factors *pnr*, *Dmef2*, and *lb*, and the differentiation product *b3 tubulin*. Dmef2 is activated in the svp-positive cells by a Tin-independent mechanism.

Myocardial cells are defined by their morphology and by their expression of a number of muscle-specific proteins including the MADS box transcription factor, *Dmef2* (Bour et al. 1995; Lilly et al. 1995) and the muscle differentiation products, *muscle myosin* (Kiehart and Feghali 1986) and *b3 tubulin* (Leiss et al. 1988; Kimble et al. 1989).

Surrounding the myocardial cells is an outer layer consisting of several types of non-myogenic pericardial cells. Pericardial cells do not express muscle genes, and the overall function of this tissue layer is unclear, although it may filter the hemolymph as it enters the dorsal vessel. The pericardial cells can be grouped into at least five different subtypes based on their AP position in the embryo and on their pattern of gene expression within each segment. More information on pericardial cell gene expression and function is available in a recent review (Cripps and Olson 2002).

Within the myocardial layer of the dorsal vessel there are a number of different cell types and different morphological domains. Although these domains do not appear to be analogous to the chambers of the vertebrate heart, they do represent regions of the myocardium with distinct functional properties. For example, the four anterior segments (and a portion of the fifth segment) of the dorsal vessel form a narrow lumen and comprise a domain that has been called the "aorta." The three posterior segments form a tube with a wider lumen and are described as the heart proper. Molecular studies have shown that the aorta and heart proper regions correspond to expression domains of the AbdA and Ubx homeobox genes, respectively (Karch et al. 1990; Bate 1993). As illustrated in Figure 1A, each of the three segments of the heart proper contains a pair of valves called ostia, and the heart proper is also physically separated from the aorta by the aortic valve (Rizki 1978). When the heart expands, the ostia open, allowing hemolymph from the body cavity to enter the heart proper. During this phase, the aortic valve is closed, and therefore hemolymph accumulates in the heart. Contraction of the heart proper then causes the ostia to close and the aortic valve to open, forcing hemolymph through the aorta and into the open circulatory system (Bodmer and Frasch 1999). These few specialized cardiac cell types are crucial to the proper functioning of the dorsal vessel, and the mechanisms surrounding the differentiation of these cells are beginning to be elucidated.

Region-specific Transcriptional Regulation in the *Drosophila* Dorsal Vessel

The differentiation of the different cell types comprising the dorsal vessel is controlled by a number of regulatory molecules that have distinct spatial and temporal expression patterns. Among the most closely studied factors is the homeodomain transcription factor, Tinman. *Tinman* expression is dependent on the activity of the mesoderm inducer *twist*, and *tinman* transcripts are initially present throughout the trunk mesoderm. Later during development, signaling by the maintenance factors dpp and wg results in restriction of *tinman* expression to cardiac precursors (Bodmer et al. 1990; for review, see Bodmer and Frasch 1999; Lockwood and Bodmer 2002). As the heart tube forms, *tinman* expression becomes further restricted to the four anterior-most cardial and pericardial cells of each segment (Jagla et al. 1997). Within the myocardial layer, *tinman* is responsible for direct transcriptional activation of the muscle genes *Dmef2* (Gajewski et al. 1997) and *b3 tubulin* (Kremser et al. 1999). As indicated in Figure 1B, tinman activity, in conjunction with wg signaling, activates expression of the homeobox gene *ladybird* (*lb*) in two of the four *tin*-positive cardioblasts per segment (Wu et al. 1995; Jagla et al. 1997). As a result of the overlapping and nonoverlapping activity of the *tin* and *lb* genes alone, three different domains of heart cells can be defined within the cardioblast layer of the dorsal vessel. This combinatorial regulation represents a good illustration of a mechanism by which myocardial cell diversity can be generated.

The remaining two *tin*-negative cells per segment express the gene *seven-up* (*svp*) (Gajewski et al. 2000). This gene appears to be the ortholog of the vertebrate *COUP-TFI* and *TFII* orphan nuclear receptors, which are also expressed in the developing heart, but which do not show region-specific expression (Pereira et al. 1999). *svp* is important for regulating cell fate by repressing expression of *tin* and thus the expression of *tin*-dependent genes (Gajewski et al. 2000; Lo and Frasch 2001). Although the repression of *tin* by *svp* has been shown to be direct, binding of the svp protein to the *tin* enhancer has not been observed (Lo and Frasch 2001). Possible repressor mechanisms could include protein–protein interactions that alter the function of the transcriptional machinery, or that result in titration out of transcriptional activators. An analogous activity has been reported in vertebrate cell culture studies where Nkx2-5 function appears to be repressed by COUP-TFI, by competition for the Nkx2-5-binding site (Guo et al. 2001). Ultimately, *tin* repression by *svp* results in a further diversification of cell types within the heart tube. These mechanisms may be conserved in vertebrate cardiac chamber formation and are discussed in more detail below.

Another gene exhibiting a conserved role in regulation of heart development in flies and vertebrates is the zinc finger transcription factor *pannier* (*pnr*). Pnr is a *Drosophila* homolog of the vertebrate *GATA* factors. Pnr has recently been shown to be a direct transcriptional target of the Tin protein (Gajewski et al. 2001), just as *GATA-6* is directly regulated by Nkx2-5 in mice (Davis et al. 2000; Molkentin et al. 2000). Furthermore, the tin and pnr proteins are able to cooperate in the activation of the muscle transcription factor gene *Dmef2* (Gajewski et al. 2001). In vertebrates, the Nkx2-5 and GATA-4 proteins have been shown to physically interact and to synergistically activate transcription from the *ANF* and *Cardiac α actin* promoters (Durocher et al. 1997; Durocher and Nemer 1998; Lee et al. 1998; Sepulveda et al. 1998). Additional levels of *Dmef2* regulation must exist in the dorsal vessel, however. Although tin and pnr are able to activate *Dmef2* directly, expression of *Dmef2* is also observed in the two *tin*-negative cardioblasts. A tin-independent enhancer of *Dmef2* has recently been isolated, and this is sufficient to drive expression in the *tin*-negative cells (Gajewski et al. 2000), but the factors binding to this enhancer have not yet been identified. Taken together, the tin-dependent and tin-independent enhancers drive expression of the *Dmef2* gene in every myocardial cell. This finding is evidence for modular regulation of cardiac promoters, an effect that has also been observed in vertebrates for a number of different myocardial regulatory and differentiation genes, including *Nkx2-5* (for review, see Schwartz and Olson 1999), *GATA-5* (MacNeill et al. 2000), *GATA-6* (He and Burch 1997), *CARP* (Kuo et al. 1999), and *MLC3F* (Franco et al. 1997; Kelley et al. 1999).

Within the cardioblast cells of the *Drosophila* dorsal vessel, therefore, there exists a regulatory network consisting of both activating and repressing mechanisms that restrict expression of genes to particular domains. In at least some cases, the vertebrate homologs of the *Drosophila* dorsal vessel regulatory genes seem to have similar functions, suggesting that mechanisms for achieving region-specific gene expression in the heart may be evolutionarily conserved.

VERTEBRATE CHAMBER FORMATION AND REGION-SPECIFIC REGULATION

Regulation of Regionalized Gene Expression in the Vertebrate Heart

The process of vertebrate heart morphogenesis, and differentiation of the myocardium into specialized tissue types, is a much more complex process than *Drosophila* dorsal vessel formation. However, several aspects of the early stages of heart development seem to be conserved, both in terms of the general cellular movements involved and in the molecular genetic program leading to cardiac differentiation. In the vertebrate embryo, pre-cardiac regions in the anterior mesoderm migrate to the ventral midline where they fuse, forming a linear, rhythmically contracting heart tube that is in some respects reminiscent of the *Drosophila* dorsal vessel. Analogous to the various cell types of the muscle layer of the dorsal vessel, the vertebrate myocardial tissue contains specialized cell types. These include the distinct contractile musculature of the atrial and ventricular chambers, the outflow tract, inflow tract, and the electrically active conduction system (De-Haan 1965; Icardo and Manasek 1992; Lyons 1994). Based on expression of molecular markers, there is evidence that specialized tissue types are at least partially defined, well in advance of overt morphological differentiation of the tissues.

Within the myocardial layer of the vertebrate heart, chamber- and region-specific gene expression has been described for a large number of cardiac differentiation products including contractile proteins, cell adhesion molecules, peptide hormones, and gap junction proteins. Although there are some exceptions, most genes that are expressed in a chamber- or region-specific manner in the adult heart are initially expressed throughout the embryonic heart tube. Expression of these genes becomes restricted to specific regions of the myocardium during later cardiogenesis. Examples include the mouse α*MHC*, *MLC1A*, and *MLC2A* genes, the quail *MyHC3* gene, and the peptide hormone *atrial natriuretic factor* (*ANF*) (Zeller et al. 1987; Lyons et al. 1990; Seidman et al. 1991; Kubalak et al. 1994; Wang et al. 1996; Kelly et al. 1999; Small and Krieg 2000). The expression of the Gap junction gene *Cx40*, which is an integral component of the conduction system, is also initiated throughout the atria and ventricles. Later in development, *Cx40* expression is down-regulated in the working ventricular myocardium, only remaining in the atrium and future conductive tissues (Van Kempen et al. 1996). It is probably significant that these different genes all show a distinct temporal pattern of restriction to the specialized tissue. For example, *MLC2a*, α*MHC*, and *ANF* become enriched in the atria at progressively later timepoints during heart development. *MLC2a* is the earliest gene to become restricted to the atria, with ventricular down-regulation beginning as early as E9 and becoming atrium-specific by E12. The α*MHC* gene is initially expressed in a graded fashion, with higher levels observed in the atrial pole. Down-regulation of α*MHC* in the ventricles begins at E10.5, and the transcripts are nearly atrium-specific by E16.5 (Lyons et al. 1990). *ANF* mRNA, on the other hand, does not become restricted to the atria until approximately the time of birth, and low levels of transcript remain in the trabeculations of the ventricles into adulthood (Zeller et al. 1987; Argentin et al. 1994). These observations imply that different mechanisms are responsible for regulating chamber-specific expression of different genes. Studies of various promoters that shed light on the possible regulatory mechanisms of chamber restriction are outlined below.

A relatively small number of differentiation products exhibit region- or chamber-specific expression from the onset of transcription. These include the chicken *AMHC1*, mouse *MLC2v*, and zebrafish atrial *cmlc2* and ventricular *vmhc* genes (O'Brien et al. 1993; Yutzey et al. 1994; Yelon et al. 1999). In addition to these myocardial differentiation products, regionalized expression has also been reported for a small number of regulatory factors, including *Irx4*, *Tbx5*, *eHAND*, *dHAND*, and *Pitx2* (Biben and Harvey 1997; Srivastava et al. 1997; Bao et al. 1999; Bruneau et al. 1999; Campione et al. 2001; Garriock et al. 2001). Although the specific mechanisms leading to region-specific expression of these transcription factors have not been elucidated, it is likely that they involve the same general principles implicated in *Drosophila* dorsal vessel specification; i.e., combinatorial activation, modular regulation of promoters, and cross-repression.

Perhaps the best-studied example of transcription factor-dependent chamber specification is the role that *Irx4* plays in ventricular differentiation. *Irx4* is a homeodomain transcription factor that is expressed exclusively in the ventricular region of the heart from the time that its transcription is initiated (Bao et al. 1999; Garriock et al. 2001). Misexpression of chicken *Irx4* throughout the heart results in abnormal expression of the ventricular marker, *VMHC1*, in the atria, with a corresponding loss of *AMHC1* expression (Bao et al. 1999). Furthermore, mice that lack a functional *Irx4* gene have reduced levels of *eHAND* in the ventricle, and misexpress *ANF* in the ventricles (Bruneau et al. 2001). Ultimately, *Irx4*-deficient mice exhibit adult-onset cardiac hypertrophy; however, ventricular morphology is not appreciably altered, suggesting that *Irx4* is not the single factor responsible for ventricle formation. The mechanism by which *Irx4* is able to affect both atrial and ventricle-specific gene expression is not clear; however, it appears that it may involve interactions with other regulatory proteins.

The promoters of several chamber-specific genes have been analyzed using transgenic methods, and this has resulted in the identification of regulatory regions and transcription factor-binding sites that are required for chamber-specific gene expression. The most extensively studied promoter is that of the ventricle-specific *MLC-2v* gene (Nguyên-Trân et al. 1999). *MLC-2v* is one of the subset of ventricular genes that is expressed in the ventricle from the onset of transcription (O'Brien et al. 1993; Nguyên-Trân et al. 1999). The regulatory elements in the *MLC-2v* promoter that confer ventricle specificity have been localized to a 250-bp fragment (Lee et al. 1992; Nguyên-Trân et al. 1999). Furthermore, this ventricle specificity is dependent on the 28-bp tandem HF-1a and HF-1b/MEF-2 regulatory elements, a single copy of which can confer cardiac-specific expression in primary cardiomyocytes (Zhu et al. 1991). In transgenics, two tandem copies of the 28-bp element are required for ventricular expression (Ross et al. 1996). It is interesting to note that the factors known to bind the 28-bp *cis* regulatory element are not ventricle-specific, and so the precise mechanism by which this promoter region confers chamber-restricted expression is not clear. It may be worth considering the possibility that different factors, perhaps exhibiting chamber-specific expression, also interact with this promoter region to regulate region-specific expression.

Extensive studies have also been carried out on the regulation of the avian *slow myosin heavy chain 3* gene (*slow MyHC3*) (Wang and Stockdale 1999). This gene is initially expressed throughout the tubular heart and later becomes restricted to the atrium (Wang et al. 1996). Analysis of the 5' regulatory sequences identified an atrial regulatory domain (ARD1) that is able to drive atrium-specific gene expression. Introduction of an ARD1-driven CAT reporter into chick embryos by replication-competent retrovirus results in atrium-specific CAT expression (Wang et al. 1996). The ARD1 contains a vitamin D receptor-binding element (VDRE) that is required for inhibition of slow MyHC3 expression in the ventricle (Wang et al. 1996). This ventricular inhibition of slow MyHC3 has recently been shown to involve the homeodomain protein, Irx4. In the proposed model, Irx4 is included in an inhibitory complex with the vitamin D and retinoid X receptors, which binds to the MyHC3 promoter via the VDRE (G.F. Wang et al. 2001). Irx4 does not bind directly to the VDR-binding domain but has been shown to interact with the VDR/RXR complex by coimmunoprecipitation and by yeast two-hybrid assays (G.F. Wang et al. 2001). These results indicate that atrium-specific expression of slow MyHC3 is achieved by an inhibitory mechanism, where a ventricle-specific transcriptional repressor acts to restrict gene expression to the atrium.

Atrial Restriction of *ANF* Expression Requires the GATA- and Nkx2-5-binding Sites

To further examine the molecular mechanisms underlying chamber-specific expression, we have used the *Xenopus* transgenesis procedure to study in vivo regulation of the *ANF* promoter. Characterization of the *Xenopus ANF* 5' regulatory sequences reveals significant conservation of regulatory elements between mammals and frog. In fact, the *Xenopus ANF* promoter contains an Nkx2-5-binding site (NKE), two SRF-binding sites (SRE), and two GATA sites that are nearly identical to the mammalian promoters in both sequence and spacing. The high level of sequence similarity between these evolutionarily distant organisms suggests that the molecular mechanisms regulating *ANF* expression are conserved. A 625-bp *ANF* promoter fragment, when driving GFP in transgenic *Xenopus* embryos, faithfully recapitulates the expression of the endogenous *ANF* gene, including initial expression throughout the heart tube followed by restriction to the atrial compartment. We have carried out an analysis of the *ANF* promoter using constructions containing mutations in conserved regulatory elements (Fig. 2A). The efficiency of expression from the different mutated promoters in transgenic embryos is illustrated in Figure 2B. First, we find that mutation of either the proximal or the distal SRE results in a dramatic reduction in overall promoter activity. This is in agreement with results suggesting that the *ANF* promoter may be regulated by the potent transcription activator, myocardin, which requires interactions with two SRF proteins for maximal transcription activation in the heart (D. Wang et al. 2001). Similarly, we find that the distal GATA element (GATAd) is required for efficient transcription from the *ANF* promoter.

Second, we observe that a relatively high level of promoter activity is retained when the NKE sequence, the proximal GATA site (GATAp), or both the NKE and GATAp in combination, are mutated. This is somewhat

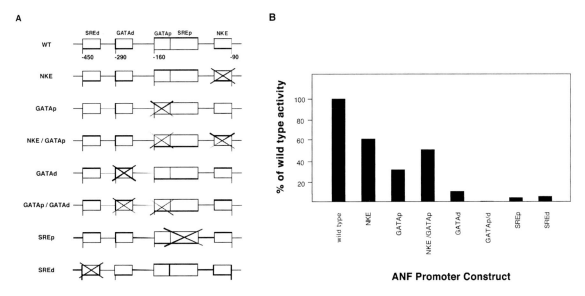

Figure 2. (*A*) *ANF* promoter constructions used in *Xenopus* transgenesis studies. The locations of the conserved Nkx2-5-binding site (NKE), two GATA-binding sites, and two SRF-binding sites (SRE) are indicated; the numbers refer to the approximate location of the elements relative to the transcription start site. Mutated binding elements in each *ANF* promoter construction are depicted with an X. (*B*) Graph showing the activity of the mutated promoter constructions, relative to the wild-type *ANF* promoter. Mutations of the NKE, proximal GATA, or both the NKE and proximal GATA, result in relatively minor reductions in transcriptional activity. Mutations of the distal GATA, both proximal and distal GATA, and either the proximal or distal SRE, greatly reduce *ANF* promoter activity relative to the wild-type construction.

surprising because mouse knockout experiments have shown that *ANF* expression is severely reduced or absent in animals lacking Nkx2-5 function (Tanaka et al. 1999). However, several different studies have indicated that protein–protein interactions with either SRF or GATA-4 may be sufficient to tether Nkx2-5 to the promoter, in the absence of DNA binding (Chen and Schwartz 1996; Durocher et al. 1997; Morin et al. 2001).

Surprisingly, mutation of either the NKE or the proximal GATA site resulted in failure of *ANF* expression to become restricted to the atria (Fig. 3D and data not shown). In transgenics carrying these promoter mutations, reporter gene activity persisted at extremely high levels throughout the ventricle, long after restriction of expression of the endogenous gene, or the wild-type transgene, to the atrial compartment (Fig. 3, cf. A and C with D). This result indicates that presence of the NKE and the GATA element is required for regulation of atrium-specific expression of *ANF*. Mutation of the GATA and NKE sites also resulted in ectopic *ANF* expression in a number of non-cardiac tissues, including the kidneys and facial muscles (data not shown). Several studies have indicated that Nkx2-5 and GATA are expressed in both the atrial and ventricular myocardium (Kasahara et al. 1998; Parmacek and Leiden 1999; our unpublished data), and so localization of these regulators is unlikely to be involved in atrial restriction of *ANF* promoter activity. Although a number of different models can be proposed (perhaps involving chamber-specific posttranslational modifications to Nkx2-5 or GATA-4), the most economical hypothesis is that the NKE and the proximal GATA site are involved in binding a repressor protein that inhibits *ANF* expression in the ventricle (Fig. 4). In this model, the repressor protein would compete with Nkx2-5 for binding to the NKE, resulting in reduction in *ANF* promoter activity in the ventricle. Expression in the atria would remain dependent on Nkx2-5 activity. This proposed mechanism is, in broad terms, equivalent to that proposed for regulation of slow *MyHC3*, where a ventricular repressor (in this case proposed to be Irx4) results in chamber-specific inhibition of promoter activity. Unlike the slow *MyHC3* promoter, however, it appears that direct binding of the repressor protein to the DNA is necessary for inhibition of *ANF* expression. A repressor protein model is diagrammed in Figure 4 that depicts an

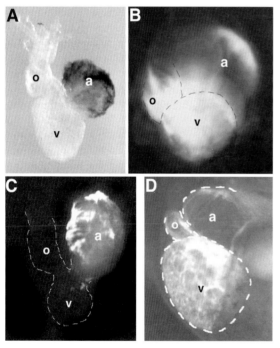

Figure 3. Transgenic expression of the *ANF* promoter-driven GFP reporter. (*A*) In situ hybridization on a dissected stage-49 *Xenopus* heart showing that endogenous *ANF* expression is restricted entirely to the atrium (Small and Krieg 2000). (*B*) GFP fluorescence in a living stage-46 transgenic tadpole carrying the wild-type *ANF*-driven GFP construction showing correct expression throughout the atrium, ventricle, and outflow tract. (*C*) By stage 49, GFP is restricted entirely to the atrium of wild-type *ANF* transgenic tadpoles, recapitulating the endogenous *ANF* expression pattern. A dashed line outlines the ventricle and outflow tract, which are not visible in this fluorescence image. (*D*) Transgenic embryos with a mutation of the NKE fail to show restriction of GFP expression to the atrium. GFP is strongly expressed throughout the atrium, ventricle, and outflow tract of a living stage-51 NKE-mutant tadpole, a full week after endogenous *ANF* and the wild-type *ANF* promoter-driven GFP are restricted to the atrium. (a) Atrium; (v) ventricle; (o) outflow tract.

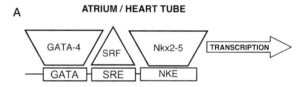

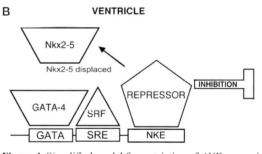

Figure 4. Simplified model for restriction of *ANF* expression to the atrium by active inhibition. Our study indicates that binding sites for Nkx2-5 and GATA are required for atrial restriction of *ANF* expression late in development. (*A*) In the embryonic linear heart tube and late-stage atrium, Nkx2-5 and GATA-4 proteins assemble into a transcription complex to activate *ANF* transcription. Mutation of the NKE or GATA sites results in only slight reduction in expression, perhaps due to protein–protein interactions that tether GATA and Nkx2-5 proteins to the promoter. (*B*) In the ventricle and outflow tract, we propose that a repressor protein binds to the NKE and displaces Nkx2-5, inhibiting the transcription of *ANF*. A corollary of this model is that the repressor protein must bind the promoter directly, as mutation of the NKE sequence in the DNA abolishes the ventricular inhibition. Similar repressor mechanisms may serve to regulate chamber-specific expression of other cardiac genes.

active inhibition by a ventricle-specific protein that can outcompete Nkx2-5 for binding to the NKE. Somewhat surprisingly, a number of candidate repressor proteins that might compete with Nkx2-5 for binding to the NKE have already been identified. These include members of the Nkx3-1 family, Hmx1, and COUP-TFI, all of which may possess repressor activity, and which appear to show affinity for a binding sequence related to the NKE (Tsai and Tsai 1997; Amendt et al. 1999; Steadman et al. 2000; Guo et al. 2001). In a variant of the model, the ventricle-specific repressor protein could bind to Nkx2-5 or GATA-4 in such a way as to abolish their interaction with SRF. Presumably this would then make binding to the *ANF* promoter completely dependent on the DNA-binding activity of the Nkx2-5 or GATA proteins, rather than on their presence in a transcription complex. Since Nkx2-5 is absolutely required for *ANF* expression (Biben et al. 1997; Tanaka et al. 1999), this would eliminate *ANF* expression in the ventricle. In either case, the putative repressor protein would either need to be expressed exclusively in the ventricle, or would need to possess ventricle-specific repressor activity, perhaps regulated by posttranslational modification of the protein.

Although there are still many unanswered questions regarding restriction of gene expression within the heart, recent studies in vertebrates and *Drosophila* have significantly increased our understanding of the molecular mechanisms underlying region- or chamber-specific expression. Comparison of the differentiation and diversification of the various cell types within the vertebrate heart or *Drosophila* dorsal vessel suggests these evolutionarily distant organisms may utilize similar regulatory mechanisms. Modular regulation of promoters via combinatorial interactions of *trans*-acting factors appears to be the theme for many cardiac genes. The modular layout of promoters results in independent regulation of a single gene in various transcriptional domains. This greatly increases the potential intricacy of the regulatory mechanisms involved and also greatly increases our challenge to understand the molecular and genetic details of the process. Elegant studies in both flies and vertebrates have suggested that chamber, or cell-type, restriction of cardiac gene expression is often achieved through active inhibition. Although different mechanisms are probably involved in regulation of expression in different regions and chambers of the heart, it appears that studies of the *Drosophila* dorsal vessel may, once again, provide a useful starting point for understanding vertebrate heart development.

ACKNOWLEDGMENTS

We thank Rob Garriock and Steve Vokes for critical reading of the manuscript. P.A.K. is the Allan C. Hudson and Helen Lovaas Endowed Professor of the Sarver Heart Center at the University of Arizona College of Medicine. This work was supported by the Sarver Heart Center and by the National Heart, Lung, and Blood Institute of the National Institutes of Health, grant HL-74763 to P.A.K.

REFERENCES

Amendt B.A., Sutherland L.B., and Russo A.F. 1999. Transcriptional antagonism between Hmx1 and Nkx2.5 for a shared DNA-binding site. *J. Biol. Chem.* **274:** 11635.

Argentin S., Ardati A., Tremblay S., Lihrmann I., Robitaille L., Drouin J., and Nemer M. 1994. Developmental stage-specific regulation of atrial natriuretic factor gene transcription in cardiac cells. *Mol. Cell. Biol.* **14:** 777.

Bao Z.-Z., Bruneau B.G., Seidman J.G., Seidman C.E., and Cepko C.L. 1999. Regulation of chamber-specific gene expression in the developing heart by Irx4. *Science* **283:** 1161.

Bate M. 1993. The mesoderm and its derivatives. In *The development of* Drosophila melanogaster (ed. M. Bate and A. Martinez-Arias), p. 1013. Cold Spring Harbor Laboratory Press, Cold Spring Harbor, New York.

Biben C. and Harvey R.P. 1997. Homeodomain factor Nkx2-5 controls left/right asymmetric expression of bHLH gene *eHand* during murine heart development. *Genes Dev.* **11:** 1357.

Biben C., Palmer S., Elliott D.A., and Harvey R.P. 1997. Homeobox genes and heart development. *Cold Spring Harbor Symp. Quant. Biol.* **62:** 395.

Bodmer R. and Frasch M. 1999. Genetic determination of *Drosophila* heart development. In *Heart development* (ed. R.P. Harvey and N. Rosenthal), p. 65. Academic Press, New York.

Bodmer R. and Venkatesh T.V. 1998. Heart development in *Drosophila* and vertebrates: Conservation of molecular mechanisms. *Dev. Genet.* **22:** 181.

Bodmer R., Jan L.Y., and Jan Y.N. 1990. A new homeobox-containing gene, *msh-2* (*tinman*), is transiently expressed early during mesoderm formation in *Drosophila*. *Development* **110:** 661.

Bour B.A., O'Brien M.A., Lockwood W.L., Goldstein E.S., Bodmer R., Tagheart P.H., Abmayr S.M., and Nguyen H.T. 1995. *Drosophila* MEF2, a transcription factor that is essential for myogenesis. *Genes Dev.* **9:** 730.

Bruneau B.G., Logan M., Davis N., Levi T., Tabin C.J., Seidman J.G., and Seidman C.E. 1999. Chamber-specific cardiac expression of Tbx5 and heart defects in Holt-Oram syndrome. *Dev. Biol.* **211:** 100.

Bruneau B.G., Bao Z.-Z., Fatkin D., Xavier-Neto J., Georgakopoulos D., Maguire C.T., Berul C.I., Kass D.A., Kuroski-deBold M.L., deBold A.J., Conner D.A., Rosenthal N., Cepko C.L., Seidman C.E., and Seidman J.G. 2001. Cardiomyopathy in Irx4-deficient mice is preceded by abnormal ventricular gene expression. *Mol. Cell. Biol.* **21:** 1730.

Campione M., Ros M.A., Icardo J.M., Piedra E., Christoffels V.M., Schweickert A., Blum M., Franco D., and Moorman A.F.M. 2001. Pitx2 expression defines a left cardiac lineage of cells: Evidence for atrial and ventricular molecular isomerism in the *iv/iv* mice. *Dev. Biol.* **231:** 252.

Chen C.Y. and Schwartz R.J. 1996. Recruitment of the Tinman homolog Nkx-2.5 by serum response factor activates cardiac α-actin gene transcription. *Mol. Cell. Biol.* **16:** 6372.

Christoffels V.M., Habets P.E.M.H., Franco D., Campione M., de Jong F., Lamers W.H., Bai Z.-Z., Palmer S., Biben C., Harvey R.P., and Moorman A.F. 2000. Chamber formation and morphogenesis in the developing mammalian heart. *Dev. Biol.* **223:** 266.

Cripps R.M. and Olson E.N. 2002. Control of cardiac development by an evolutionarily conserved transcriptional network. *Dev. Biol.* **246:** 14.

Davis D.L., Wessels A., and Burch J.B.E. 2000. An Nkx-dependent enhancer regulates *cGATA-6* gene expression during early stages of heart development. *Dev. Biol.* **217:** 310.

DeHaan R.L. 1965. Morphogenesis of the vertebrate heart. In *Organogenesis* (ed. R.L. DeHaan and H. Ursprung), p.377. Holt, Rinehart and Winston, New York.

Durocher D. and Nemer M. 1998. Combinatorial interactions regulating cardiac transcription. *Dev. Genet.* **22:** 250.

Durocher D., Charron F., Warren R., Schwartz, R.J., and Nemer M. 1997. The cardiac transcription factors Nkx2-5 and

GATA-4 are mutual cofactors. *EMBO J.* **16:** 5687.

Franco D., Kelly R., Lamers W., Buckingham M., and Moorman A.F. 1997. Regionalized transcriptional domains of myosin light chain 3f transgenes in the embryonic mouse heart: Morphogenetic implications. *Dev. Biol.* **188:** 17.

Gajewski K., Choi C.Y., Kim Y., and Schulz R.A. 2000. Genetically distinct cardial cells within the *Drosophila* heart. *Genesis* **28:** 36.

Gajewski K., Kim Y., Lee Y., Olson E., and Schulz R. 1997. D-mef2 is a target for Tinman activation during *Drosophila* heart development. *EMBO J.* **16:** 515.

Gajewski K., Zhang Q., Choi C.Y., Fosset N., Dang A., Kim Y.H., Kim Y., and Schulz R.A. 2001. *Pannier* is a transcriptional target and partner of Tinman during *Drosophila* cardiogenesis. *Dev. Biol.* **233:** 425.

Garriock R.J., Vokes S.A., Small E.M., Larson R., and Krieg P.A. 2001. Developmental expression of the *Xenopus* Iroquois-family homeobox genes, *Irx4* and *Irx5*. *Dev. Genes Evol.* **211:** 257.

Guo L., Lynch J., Nakamura K., Fliegel L., Kasahara H., Izumo S., Komuro I., Agellon L.B., and Michalak M. 2001. COUP-TF1 antagonizes Nkx2.5-mediated activation of the *calreticulin* gene during cardiac development. *J. Biol. Chem.* **276:** 2797.

He C.-Z. and Burch J.B.E. 1997. The chicken *GATA-6* locus contains multiple control regions that confer distinct patterns of heart region-specific expression in transgenic mouse embryos. *J. Biol. Chem.* **272:** 28550.

Icardo J.M. and Manasek F.J. 1992. Cardiogenesis: Development mechanisms and embryology. In *The heart and cardiovascular system* (ed. H.A. Fozzard et al.), p.1563. Raven Press, New York.

Jagla K., Frasch M., Jagla T., Dretzen G., Bellard F., and Bellard M. 1997. *ladybird*, a new component of the cardiogenic pathway in *Drosophila* required for diversification of heart precursors. *Development* **124:** 3471.

Karch F., Bender W., and Weiffenbach B. 1990. *abdA* expression in *Drosophila* embryos. *Genes Dev.* **4:** 1573.

Kasahara H., Bartunkov S., Schinke M., Tanaka M., and Izumo S. 1998. Cardiac and extracardiac expression of Csx/Nkx2.5 homeodomain protein. *Circ. Res.* **82:** 936.

Kelly R.G., Franco D., Moorman F.M., and Buckingham M. 1999. Regionalization of transcriptional potential in the myocardium. In *Heart development* (ed. R.P. Harvey and N. Rosenthal), p.333. Academic Press, New York.

Kiehart D.P. and Ferghali R. 1986. Cytoplasmic myosin from *Drosophila melanogaster*. *J. Cell Biol.* **103:** 1517.

Kimble M., Incardona J., and Raff E.C. 1989. A variant beta-tubulin isoform of *Drosophila melanogaster* (beta-3) is expressed primarily in tissues of mesodermal origin in embryos and pupae, and is utilized in populations of transient microtubules. *Dev. Biol.* **131:** 415.

Kremser T., Gajewski K., Schulz R.A., and Renkawitz-Pohl R. 1999. Tinman regulates the transcription of the β3 tubulin gene in the dorsal vessel of *Drosophila*. *Dev. Biol.* **216:** 327.

Kubalak S.W., Miller-Hance W.C., O'Brien T.X., Dyson E., and Chien K.R. 1994. Chamber specification of atrial myosin light chain-2 expression precedes septation during murine cardiogenesis. *J. Biol. Chem.* **269:** 16961.

Kuo H.-C., Chen J., Ruiz-Lozano P., Zou Y., Nemer M., and Chien K.R. 1999. Control of segmental expression of the cardiac-restricted ankyrin repeat protein gene by distinct regulatory pathways in murine cardiogenesis. *Development* **126:** 4223.

Lee K.J., Ross R.S., Rockman H.A., Harris A.N., O'Brien T.X., van Bilsen M., Shubeita H.E., Kandolf R., Brem G., Price J., and Chien K.R. 1992. Myosin light chain-2 luciferase transgenic mice reveal distinct regulatory programs for cardiac and skeletal muscle-specific expression of a single contractile protein gene. *J. Biol. Chem.* **267:** 15875.

Lee Y., Shioi T., Kasahara H., Jobe S.M., Wiese R.J., Markham B.E., and Izumo S. 1998. The cardiac tissue-restricted homeobox protein Csx/Nkx2.5 physically associates with the zinc finger protein GATA4 and cooperatively activates atrial natriuretic factor gene expression. *Mol. Cell. Biol.* **18:** 3120.

Leiss D., Hinz U., Gasch A., Mertz R., and Renkawitz-Pohl R. 1988. β3-tubulin expression characterizes the differentiating mesodermal germ layer during *Drosophila* embryogenesis. *Development* **104:** 525.

Lilly B., Zhao B., Ranganayakulu G., Patterson B.M., Schulz R.A., and Olson E.N. 1995. Requirement of MADS domain transcription factor D-MEF2 for muscle formation in *Drosophila*. *Science* **267:** 688.

Lo P.C.H. and Frasch M. 2001. A role for the COUP-TF-related gene seven-up in the diversification of the cardioblast identities in the dorsal vessel of *Drosophila*. *Mech. Dev.* **104:** 49.

Lockwood W.K. and Bodmer R. 2002. The patterns of wingless, decapentaplegic, and tinman position the *Drosophila* heart. *Mech. Dev.* **114:** 13.

Lyons G.E. 1994. In situ analysis of the cardiac muscle gene program during embryogenesis. *Trends Cardiovasc. Med.* **4:** 70.

Lyons G.E., Schiaffino S., Sassoon D., Barton P., and Buckingham M. 1990. Developmental regulation of myosin gene expression in mouse cardiac muscle. *J. Cell Biol.* **111:** 2427.

MacNeill C., French R., Evans T., Wessels A., and Burch J.B.E. 2000. Modular regulation of cGATA-5 gene expression in the developing heart and gut. *Dev. Biol.* **217:** 62.

Molkentin J.D., Antos C., Mercer B., Taigen T., Miano J.M., and Olson E.N. 2000. Direct activation of a *GATA6* cardiac enhancer by Nkx2.5: Evidence for a reinforcing regulatory network of Nkx2.5 and GATA transcription factors in the developing heart. *Dev. Biol.* **217:** 301.

Morin S., Paradis P., Aries A., and Nemer M. 2001. Serum response factor-GATA ternary complex required for nuclear signaling by a G-protein-coupled receptor. *Mol. Cell. Biol.* **21:** 1036.

Nguyêñ-Trân V.T.B., Chen J., Ruiz-Lozano P., and Chien K.R. 1999. The MLC-2 paradigm for ventricular heart chamber specification, maturation, and morphogenesis. In *Heart development* (ed. R.P. Harvey and N. Rosenthal), p.255. Academic Press, New York.

O'Brien T.X., Lee K.J., and Chien K.R. 1993. Positional specification of ventricular myosin light chain 2 expression in the primitive murine heart tube. *Proc. Natl. Acad. Sci.* **90:** 5157.

Parmacek M.S. and Leiden J.M. 1999. GATA transcription factors and cardiac development. In *Heart development* (ed. R.P. Harvey and N. Rosenthal), p. 291. Academic Press, New York.

Pereira F.A., Qiu Y., Zhou G., Tsai M.J., and Tsai S.Y. 1999. The orphan nuclear receptor COUP-TFII is required for angiogenesis and heart development. *Genes Dev.* **13:** 1037.

Rizki T.M. 1978. The circulatory system and associated cells and tissues. In *The genetics and biology of* Drosophila (ed. M. Ashburner and T.R.F. Wright), p.397. Academic Press, London.

Ross R.S., Navansakasattusas S., Harvey R.P., and Chien K.R. 1996. An HF-1a/HF-1b/MEF-2 combinatorial element confers cardiac ventricular specificity and establishes an anterior-posterior gradient of expression. *Development* **122:** 1799.

Rugendorff A., Younossi-Hartenstein A., and Hartenstein V. 1994. Embryonic origin and differentiation of the *Drosophila* heart. *Roux's Arch. Dev. Biol.* **203:** 266.

Schwartz R.J. and Olson E.N. 1999. Building the heart piece by piece: Modularity of *cis*-elements regulating *Nkx2-5* transcription. *Development* **126:** 4187.

Seidman C.E., Schmidt E.V., and Seidman J.G. 1991. *cis*-dominance of rat *atrial natriuretic factor* gene regulatory sequences in transgenic mice. *Can. J. Physiol. Pharmacol.* **69:** 1486.

Sepulveda J.L., Belaguli N., Nigam V., Chen C.-Y., Nemer M., and Schwartz R.J. 1998. GATA-4 and Nkx-2.5 coactivate Nkx-2 DNA binding targets: Role for regulating early cardiac gene expression. *Mol. Cell. Biol.* **18:** 3405.

Small E.M. and Krieg P.A. 2000. Expression of atrial natriuretic factor (ANF) during *Xenopus* cardiac development. *Dev. Genes Evol.* **210:** 638.

Srivastava D., Thomas T., Lin Q., Kirby M.L., Brown D., and Olson E.N. 1997. Regulation of cardiac mesodermal and neural crest development by the bHLH transcription factor, dHAND. *Nat. Genet.* **16:** 154.

Steadman D.J., Giuffrida D., and Gelmann E.P. 2000. DNA-binding sequence of the human prostate-specific homeodomain protein NKX3.1. *Nucleic Acids Res.* **28:** 2389.

Tanaka M., Chen Z., Bartunkova S., Yamasaki N., and Izumo S. 1999. The cardiac homeobox gene Csx/Nkx2.5 lies genetically upstream of multiple genes essential for heart development. *Development* **126:** 1269.

Tsai S.Y. and Tsai M.J. 1997. Chick ovalbumin upstream promoter-transcription factors (COUP-TFs): Coming of age. *Endocr. Rev.* **18:** 229.

Van Kempen M.J.A., Vermeulen J.L.M., Moorman A.F.M., Gros D., Paul D.L., and Lamers W.H. 1996. Developmental changes of connexin40 and connexin43 mRNA distribution patterns in the rat heart. *Cardiovasc. Res.* **32:** 886.

Wang G.F. and Stockdale F.E. 1999. Chamber-specific gene expression and regulation during heart development. In *Heart development* (ed. R.P. Harvey and N. Rosenthal), p. 357. Academic Press, New York.

Wang G.F., Nikovits W., Jr., Bao Z.-Z., and Stockdale F.E. 2001. Irx4 forms an inhibitory complex with the vitamin D and retinoic X receptors to regulate cardiac chamber-specific slow MyHC3 expression. *J. Biol. Chem.* **276:** 28835.

Wang G.F., Nikovits W., Jr., Schleinitz M., and Stockdale F.E. 1996. Atrial chamber-specific expression of the slow myosin heavy chain 3 gene in the embryonic heart. *J. Biol. Chem.* **271:** 19836.

Wang D., Chang P.S., Wang Z., Sutherland L., Richardson J.A., Small E., Krieg P.A., and Olson E.N. 2001. Activation of cardiac gene expression by myocardin, a transcriptional cofactor for serum response factor. *Cell* **105:** 851.

Wu X., Golden K., and Bodmer R. 1995. Heart development in *Drosophila* requires the segment polarity gene *wingless*. *Dev. Biol.* **169:** 619.

Yelon D., Horne S.A., and Stainier D.Y. 1999. Restricted expression of cardiac myosin genes reveals regulated aspects of heart tube assembly in zebrafish. *Dev. Biol.* **214:** 23.

Yutzey K.E., Rhee J.T., and Bader D. 1994. Expression of the atrial-specific myosin heavy chain AMHC1 and the establishment of anteroposterior polarity in the developing chicken heart. *Development* **120:** 871.

Zaffran S., Astier M., Gratecos D., Guillen A., and Semeriva M. 1995. Cellular interactions during heart morphogenesis in the *Drosophila* embryo. *Biol. Cell* **84:** 13.

Zeller R., Bloch K.D., Williams B.S., Arceci R.J., and Seidman C.E. 1987. Localized expression of the atrial natriuretic factor gene during cardiac embryogenesis. *Genes Dev.* **1:** 693.

Zhu H., Garcia S., Ross R.S., Evans S.M., and Chien K.R. 1991. A conserved 28 bp element (HF-1) within the rat cardiac myosin light chain-2 gene confers cardiac specific and α-adrenergic inducible expression in cultured neonatal rat myocardial cells. *Mol. Cell. Biol.* **13:** 4432.

Pitx Genes during Cardiovascular Development

C. Kioussi,*§ P. Briata,‡ S.H. Baek,* A. Wynshaw-Boris,¶ D.W. Rose,† and M.G. Rosenfeld*

*Howard Hughes Medical Institute, †Department of Medicine, ¶Department of Pediatrics and UCSD Cancer Center, University of California, San Diego, La Jolla, California 92093; §Department of Biochemistry and Biophysics, Oregon State University, 2041 ALS, Corvallis, Oregon 97331; ‡ Istituto Nazionale per la Ricerca sul Cancro, 16132 Genova, Italy

The evolution of multicellularity and complex body plans is accompanied by a simultaneous development of a cardiovascular system to serve cellular nutrition and physiological homeostasis. The heart is the first functional organ to develop in the vertebrate embryo, initially forming prior to and during gastrulation. Mesodermal cells invaginate the primitive streak and migrate through the cranial mesoderm to the midline to form the cardiac crescent, which is followed by the appearance of a single heart tube. The looped heart tube undergoes septation to form the mature four-chamber multifunctional structure. Cardiac development involves an endogenous self-differentiation mechanism and the morphogenetic effects of exogenous cells, with the cardiac neural crest cells participating in conotruncal septation of the developing outflow tract. Neural crest cells follow well-defined paths and migrate from the posterior rhombencephalon to the third, fourth, and sixth branchial arches where they contribute to the vascular development of the great arteries. Proper migration of these cells is mediated by a variety of molecular cues, and non-proper development causes numerous human conotruncal cardiac malformations, including persistent truncus arteriosus (PTA), double outlet from the right ventricle (DORV), and aortic arch anomalies. A number of signaling molecules, receptors, and transcription factors are involved in this dynamic pathway. Signaling molecules, like Wnts, can control proliferation and differentiation events in the heart by activating the Dvl2/β-catenin/Pitx2 pathway (Kioussi et al. 2002).

Pitx2, highly homologous to *Pitx1* (Gage and Camper 1997), was initially identified as one of the genes responsible for the human Rieger syndrome, an autosomal dominant condition characterized by ocular anterior chamber abnormalities causing glaucoma, dental hypoplasia, craniofacial dysmorphism, and umbilical stump defects, as well as occasional abnormal cardiac and pituitary development (Semina et al. 1996). *Pitx2* is asymmetrically expressed in the left lateral-plate mesoderm, and mutant mice with laterality defects often exhibited altered patterns of *Pitx2* expression that correlate with changes in the visceral symmetry (*situs*). Ectopic expression of *Pitx2* in the right lateral-plate mesoderm causes inversion of cardiac looping and gut and reverses body rotation in chick and *Xenopus* embryos (Logan et al. 1998; Meno et al. 1998; Piedra et al. 1998; Ryan et al. 1998; Yoshioka et al. 1998; Campione et al. 1999). *Pitx2*-deficient mice are characterized by a series of defects, including failure of body-wall closure, arrest in turning, ocular defects, right pulmonary isomerism, altered cardiac position with valvular and atrial septation defects, and a block in early determination events in anterior pituitary gland and tooth organogenesis (Gage et al. 1999; Kitamura et al. 1999; Lin et al. 1999; Lu et al. 1999), partially resembling humans affected by Rieger syndrome.

Pitx1 and *Pitx2* are expressed in diverse tissues, regulated by distinct tissue-selective enhancers (Szeto et al. 1999; Smidt et al. 2000), and likely therefore to be under the control of distinct regulating mechanisms in each tissue. *Pitx1* is a critical transcriptional component of limb development, anterior pituitary, and derivatives of the first branchial arch. *Pitx1*-gene-deleted mice exhibit striking abnormalities in morphogenesis and growth of hindlimb structural changes in the tibia fibia, in addition to the reciprocal abnormalities of the pituitary cell types. The ventral gonadotropes and thyrotropes are diminished, whereas the dorsal corticotropes are increased in both number and expression level of ACTH (Szeto et al. 1999).

In several tissues, *Pitx2*-gene-deleted mice exhibit an arrest of organ growth progression. Based on the pleiotropic effects of the *Pitx1* and *Pitx2* gene deletions, it was of particular interest to begin to understand the molecular events controlled by these factors that mediate the cardiac phenotypes. In examining these events genetically, an intriguing link was revealed in the development of the cardiac outflow tract.

Another group of genes that play a role in neural crest determination involves the endothelins and their receptors. $ET_A^{-/-}$ mice suffer from craniofacial and cardiovascular defects (Clouthier et al. 1998), including malformation in ventricular septation and outflow tract development. ET_A is down-regulated in $Pitx1^{+/-}/Pitx2^{+/-}$ mice, which suggests that it might be a downstream gene of the *Pitx* gene family.

METHODS

Genetic manipulations. $Pitx2^{+/-}$ mice (Lin et al. 1999) of 129/Sv background were crossed with $Pitx1^{+/-}$ mice of 129/Sv background (Szeto et al. 1999) to generate double heterozygous animals.

Antibodies, immunohistochemistry, and in situ hybridization. The mouse Pitx1 and Pitx2 proteins were produced in the baculovirus system according to standard procedures (Briata et al. 1999). To produce αPitx2, IgG polyclonal antisera guinea pigs were injected four times with 1 mg of each protein, and antibody was used at 1:500 dilution. Pax3 (Gross et al. 2000) rat polyclonal antibody was used at 1:400 dilution.

The following commercially available antibodies were used: β-galactosidase, rabbit polyclonal (Cappel 1:100), horseradish peroxidase-conjugated (Chemicon), or Cy-coupled (Jackson). Immunohistochemistry was done on 14-μm cryosections. Sections were incubated with primary antibodies overnight at 4°C. Hybridization with ^{35}S-labeled antisense RNA probes was performed as described previously (Simmons et al. 1990) on 18-μm cryosections.

Chromatin immunoprecipitation assays. For the chromatin immunoprecipitation assay, αT$_3$-1, a murine pituitary cell line, was utilized, and LiCl (10 mM) was added for 1 hour prior to harvest. Cells were washed twice with PBS and cross-linked with 1% formaldehyde for 10 minutes at room temperature. Cross-linked cells were treated as described previously (Hecht and Grunstein 1999; Scully et al. 2000; Shang et al. 2000). Cells were then resuspended in 0.3 ml of lysis buffer (1% SDS, 10 mM EDTA, 50 mM Tris-HCl [pH 8.1], protease inhibitors) and sonicated three times for 10 seconds followed by centrifugation for 10 minutes. Cross-linked materials were resuspended in TE buffer and chromatin sheared by sonicating three times for 12 minutes each in a Branson 250 sonicator at the power setting of 1.5 and 100% duty cycle, with 20-minute, 4°C incubations between pulses. Average size of sheared fragments was expected to be about 300–500 bp. Immunoprecipitates were eluted three times with 1% SDS, 0.1 M NaHCO$_3$. Eluates were pooled and heated at 65°C for 6 hours to reverse the formaldehyde cross-linking. DNA fragments were purified with a QIAquick spin kit (QIAGEN).

Single-cell nuclear microinjection assays. Affinity-purified αPitx2 IgG was used. Each experiment was performed on three independent coverslips consisting of 1000 cells. Where no experimental antibody was used, preimmune IgGs were coinjected, allowing the unambiguous identification of injected cells, in addition to serving as a preimmune control.

RESULTS

Pitx1 and Pitx2 during Embryogenesis and Organogenesis

Pitx1 was originally discovered through its ability to interact with the transcriptional domain of the pituitary-specific POU domain protein, Pit1 (Lamonerie et al. 1996; Szeto et al. 1996). *Pitx2* was discoverd by positional cloning for the Rieger syndrome gene (Gage and Camper 1997). All *Pitx2* genes share a conserved structural homology in the 60-amino-acid DNA-binding homeodomain with other members of homeobox genes (Fig.1A,C). *Pit* genes bind the TAATCC bicoid site (Driever et al 1989; Amendt et al. 1998) and transactivating reporter constructs under promoters of several genes (Szeto et al. 1996; C. Kioussi et al., in prep.).

Pitx1 and *Pitx2* are expressed from embryonic day 8 (e8) onward in specific mesenchymes and ectodermal priomordium. *Pitx1* (Lanctot et al. 1999; Szeto et al. 1999) exhibit altered length of femur, tibia, and fibula. In addition, the patella is absent, with loss of the Zucker nodes proximally (Fig. 1A), and the hindlimb-specific marker *Tbx4* is reduced. *Pitx1* is also expressed in a small population in the e13 heart in ring-shape cells of the developing pulmonary artery (Fig. 1B).

Pitx2$^{-/-}$ mice are characterized by failure of the body wall closure, a counterclockwise bending of the anterior–posterior axis, and right pulmonary isomerism. Tooth and pituitary development proceeds through the initial signaling and determination phases, but the emergence, migration, and expansion of distinct cell types in the developing ectoderm fail to progress past e11 (Fig. 1C) (Gage et al. 1999; Kitamura et al. 1999; Lin et al. 1999; Lu et al. 1999). *Pitx2* is also expressed in the developing heart at the left portions of the primary heart tube, whereas after looping, it will become the ventral portion of the right ventricle and atrium.

Pitx2 can serve as a Wnt-induced transcription factor required for normal proliferation of *Pitx2*-expressing cells. To subserve its proliferative effects, *Pitx2* must bind to its cognate DNA sites and requires an amino-terminal activation domain, but not the carboxyl terminus, consistent with a previous report of inhibitory effects of the Pitx2 carboxyl terminus. Three independent events underlie Pitx2-dependent activation of cell-type-specific proliferation. Wnt-dependent activation of Pitx2, in addition to Wnt and growth-factor-dependent relief of Pitx2 repression function, and serial recruitment of a series of specific coactivator complexes, includes a MYST family member, Tip60. β-Catenin interacts with several transcription factors, including Pitx2, to dismiss HDAC1-containing complexes on specific promoters, analogous to actions on LEF1. Although it is possible that Pitx2 may influence the cell cycle at more than one stage, we have shown that Pitx2 exerts key actions in early to mid G$_1$ and stimulates expression of specific growth-control genes regulating this phase of the cell cycle, including genes that, intriguingly, have not been found to harbor functional E2F sites. Activation of the Wnt pathway results in rapid recruitment of the *Pitx2* gene and binding of Pitx2 to promoters of specific growth-control genes. Pitx2 thus provides a direct nuclear target for synergy between the Wnt and growth factor pathways in mediating cell-type-specific regulation of growth control gene expression. A series of coactivators are sequentially and transiently recruited, all but one of which is required in a fashion reminiscent of the regulation of the HO locus in yeast.

Pitx1/Pitx2 in Cardiac Neural Crest

Cardiac outflow tract abnormalities correspond to 30% of all cardiovascular malformations in humans (Chien

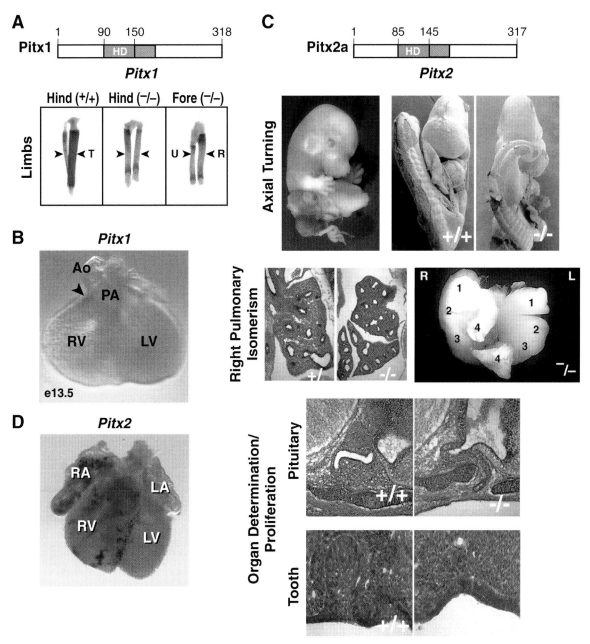

Figure 1. *Pitx1-* and *Pitx2*-deficient mice. (*A*) *Pitx1*$^{-/-}$ mice exhibit hindlimb malformations. (*B*) *Pitx1* is expressed in the developing heart. Whole-mount *lacZ* staining of a e13.5 *Pitx1*$^{+/-}$ embryonic heart showing β-gal-expressing cells in the wall of the proximal pulmonary artery (*arrowhead*). (*C*) *Pitx2*$^{-/-}$ mice characterized by visceral and axial turning abnormalities, right pulmonary isomerism, and pituitary and tooth determination and further proliferation. (*D*) *Pitx2* is asymmetrically expressed in the developing heart.

2000; Srivastava and Olson 2000). Cardiac neural crest cells are essential for normal development of the outflow tract. These cells originate from the caudal hindbrain and migrate into the caudal pharyngeal arches (third, fourth, and sixth), and a subset continues to migrate into the cardiac outflow tract where it will organize the outflow septum. If cardiac neural crest cells are removed prior to migration, several predictable outflow tract phenotypes are observed after development of the heart and great arteries is complete (Kirby and Waldo 1995; Creazzo et al. 1998).

In addition to previously described non-septated atrium and valvular deficiencies (Gage et al. 1999; Kitamura et al. 1999; Lin et al. 1999; Lu et al. 1999), we noted that all the *Pitx2*$^{-/-}$ mice that survive up to e14 invariantly exhibit major cardiac outflow tract abnormalities, including DORV (30%) and PTA (70%) (Fig. 2A). We next investigated a potential genetic linkage between *Pitx1* and *Pitx2*. Although most *Pitx1*$^{+/-}$/*Pitx2*$^{+/-}$ mice die at birth due to cleft palate (data not shown), >50% of the double heterozygotes exhibit severe cardiac outflow tract mal-

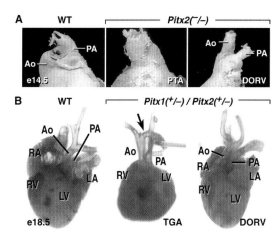

Figure 2. Genetic evidence of *Pitx2* and *Pitx1* roles in cardiac outflow tract development. (*A*) Scanning electron microscopy (SEM) of hearts from e14.5 showed PTA and DORV in *Pitx2*−/− mice. (*B*) Osmium tetroxide staining of hearts from *Pitx1*+/−/*Pitx2*+/− e18.5 showed TGA and DORV. The arrow points to the aortic arch. (Ao) Aorta; (PA) pulmonary artery; (RA) right atrium; (LA) left atrium; (LV) left ventricle; (RV) right ventricle.

formations including DORV (60%) and transposition of the great arteries (TGA) (40%), with sporadic PTA, that basically phenocopy the cardiac outflow tract defects in *Pitx2*−/− mice (Fig. 2B and data not shown).

The cardiac outflow tract deficiencies in *Pitx2*−/− mice appear to be due to a failure of the appropriate appearance of cardiac neural crest cells in the outflow tract. The disposition of *Pitx2*-expressing presumptive cardiac neural crest cells could be followed in the *Pitx2*+/− and *Pitx2*−/− mice using the β-gal knock-in allele. At e9.5, β-gal-expressing cells normally invade the presumptive outflow tract of the developing cardiac tube after migration through the third, fourth, and sixth branchial arches in *Pitx2*+/− mice. Significantly, β-gal-positive presumptive cardiac neural crest cells were barely detectable in the developing cardiac outflow tract of *Pitx2*−/− mice, and the absence of β-gal staining in the cranial nerves IX, X, and XI (glossopharyngeal, vagus, and accessory, respectively) was also observed (Fig. 3A). Thus, *Pitx2*-expressing presumptive neural crest cells fail to populate the cardiac outflow tract of *Pitx2*−/− mice.

To investigate Pitx2 expression in cells of neural crest origin, double-labeling immunofluorescence was performed. In *Pitx2*+/− mice, *Pitx2* was coexpressed with neural crest cell markers, tyrosine hydroxylase (TH) (Fig. 3B, a–c) (Groves et al. 1995), or Pax3 (Fig. 3B, d–f) (Goulding et al. 1991), in distinct subpopulations of neural crest cells migrating from the neural tube at e9.5 (Fig. 3B, a–f). Most importantly, a subpopulation of *Pitx2*-positive cells located in the presumptive cardiac outflow tract at e10.5 were Pax3-positive (Fig. 3B, g–i), indicating that the Pitx2-expressing cells display cardiac neural crest cellular properties (Conway et al. 1997).

Although *Pitx1*−/− mice never exhibited cardiac malformations and *Pitx1* could not be detected in migrating neural crest cells (not shown), *Pitx1* was expressed late in

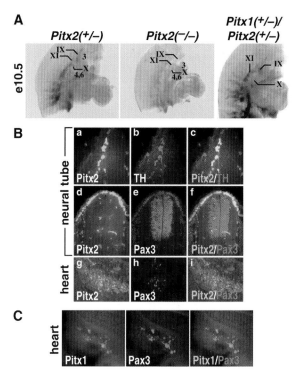

Figure 3. *Pitx2* and *Pitx1* expression in cardiac neural crest cells and cardiac outflow tract. (*A*) Whole-mount β-gal staining of e10.5 *Pitx2*+/−, *Pitx2*−/−, and *Pitx2*+/− / *Pitx1*+/− embryos. At e10.5, β-gal staining in the cranial nerves IX, X, and XI is absent in *Pitx2*−/− embryos. Whole-mount *LacZ* staining of e10.5 *Pitx1*+/−/*Pitx2*+/− embryos revealed an apparently normal β-gal-positive cell population invading the cardiac outflow tract from the fourth and sixth branchial arches and normal staining in the cranial nerves IX, X, and XI. (*B*) Coexpression of Pitx2 and tyrosine hydroxylase (TH) (*a–c*) and of Pitx2 and Pax3 (*d–i*) neural crest cells migrating from the neural tube (*a–f*) and in a subpopulation of heart outflow tract cells (*g–i*). Immunofluorescence of cross-sections from e9.5 (*a–f*) and e10.5 (*g–i*) *Pitx2*+/− embryos is shown. Anti β-gal antibody was used to mark Pitx2-positive cells in *Pitx2*+/− embryos. (*C*) Coexpression of Pitx1 and Pax3 in a population of heart outflow tract cells assessed by immunofluorescence using an anti β-gal antibody to mark *Pitx1*-positive cells in e13.5 *Pitx1*+/− mice.

embryonic cardiac development, at e13.5, in a very restricted cell population in the primordium of the proximal outflow tract. We suspect that *Pitx1* is expressed in an earlier time in the cNCC that populate the outflow tract. The location of β-gal-positive *Pitx2*-expressing cardiac neural crest cells was apparently normal in branchial arches, and the cardiac outflow tract of the double mutant *Pitx1*+/−/*Pitx2*+/− mice (Fig. 3C). At e13.5, double-labeling immunofluorescence experiments revealed that Pitx1 and Pax3 were coexpressed in cells located in the proximal outflow tract (Fig. 3C).

Endothelin Receptor A as a *Pitx1/Pitx2* Target

Consistent with the observation that cardiac neural crest cells were able to traverse the branchial arches and enter the outflow tract in the *Pitx1*+/−/*Pitx2*+/− mice, *endothelin converting enzyme-1* (*ECE-1*) (Yanagisawa et al. 1988), *connexin 43* (Waldo et al. 1999) (not shown),

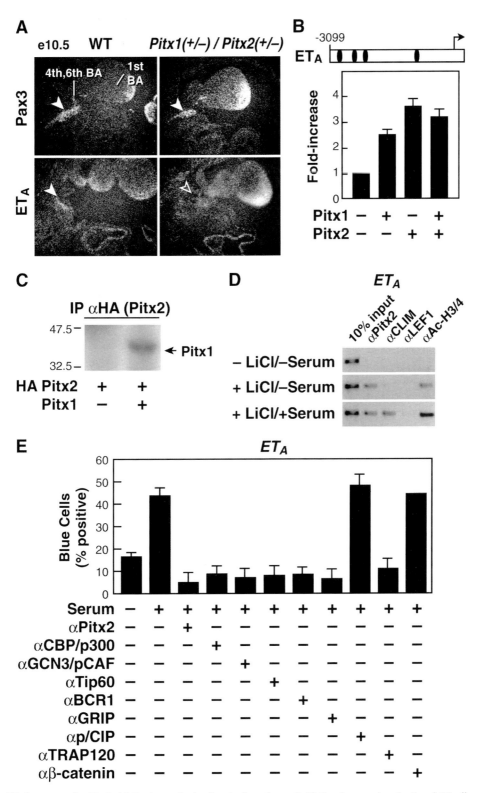

Figure 4. ET_A is a target for *Pitx2*. (*A*) In situ analysis of sagittal sections of e10.5 embryos using *Pax3* and ET_A riboprobes showed that *Pax3* is expressed, while ET_A was not expressed, in the cardiac neural crest cells of the fourth and sixth branchial arches in $Pitx1^{+/-}/Pitx2^{+/-}$ mice. (*B*) Both *Pitx1* and *Pitx2* activate human ET_A 5′ flanking region. Transcriptional activity was evaluated in transiently transfected 293 cells using the reporter (luciferase) under control of 3 kb of human ET_A 5′ flanking sequences, and CMV-*Pitx1* or CMV-*Pitx2* expression vectors (mean ± S.E.M.). Boxes indicate Pitx1/2 binding elements. (*C*) Pitx2 co-immunoprecipitates with Pitx1 in extracts from CV-1 cells transiently transfected with the indicated CMV expression vectors encoding Pitx2 or Pitx1. (*D*) ChiP assay was performed using Pitx2, CLIM, LEF1, β-catenin, and acetylated H3/H4 antibodies on C2C12 cell extracts. (*E*) Role of Pitx2 and coactivator complexes on ET_A gene activation. C2C12 cells were subjected to single microinjection with a Pitx2 IgG and other specific IgGs and an ET_A–*LacZ* reporter construct.

and *Pax3* (Fig. 4A)-positive cells were detected at e10.5, indicating that the expression of these genes is unaffected in the double heterozygotes. In contrast, no expression of *endothelin-A receptor* (ET_A) was detected in the fourth and sixth branchial arches (Fig. 4A). Consistent with the complete lack of cardiac neural crest cells in development of the cardiac outflow tract of *Pitx2*$^{-/-}$ mutant mice, neither ET_A, Pax3, ECE-1, or *connexin 43* was expressed (data not shown). $ET_A^{-/-}$ mice (Clouthier et al. 1998) display similar cardiac outflow tract abnormalities, consistent with the hypothesis that *Pitx1*, *Pitx2*, and ET_A participate in a common pathway. Indeed, the 5´ regulatory sequences of the human ET_A gene harbors multiple consensus binding sites for *Pitx1* and *Pitx2*, and either can stimulate the ET_A transcriptional activity in transient transfection assays in heterologous cells (Fig. 3G). These data argue that the ET_A is a functionally important direct Pitx1 and Pitx2 target gene. Finally, *Pitx1* and *Pitx2* proteins were capable of physically interacting, as indicated by co-immunoprecipitation assay in CV1 cells expressing Pitx1 and HA-tagged Pitx2 proteins (Fig. 3C). Thus, *Pitx1* and *Pitx2* are dosage-dependent components of a genetic pathway upstream of ET_A.

The next issue was to determine whether ET_A promoter recruits Pitx2. Thus, using ChiP analysis in mouse murine C2C12 cells, we noted that Pitx2 was present on the ET_A promoter after induction with LiCl and LiCl and serum together. The coactivator CLIM was present only after induction with serum and LiCl, as well as with the acetylated H3 and H4. No LEF was recruited on this promoter with or without induction (Fig. 4D).

Expression of a reporter under the control of mouse ET_A promoter was specifically blocked in a single cell nuclear microinjection assay in C2C12 cells with anti-Pitx2 IgG, further supporting the direct recruitment of Pitx2 in ET_A gene activation (Fig. 4E). The same effect was observed by microinjecting specific IgGs against the coactivators CBP/p300, GCN3/pCAF, Tip60, BCR1 and GRIP, and TRAP120. No block of promoter activity was observed after microinjecting IgGs against p/CIP and β-catenin (Fig. 4E). Therefore, there is a specific recruitment of specific coactivator complexes by Pitx2, each of which is required for activation of the ET_A gene.

DISCUSSION

Pitx1/Pitx2 in Cardiac Outflow Tract Development

In this paper, we have probed the molecular mechanisms which underlie the regulation of tissue-restricted factors that modulate specific aspects of general signaling pathways in specific cell types, in this case the cardiac neural crest cells that are required for development of the cardiac outflow tract. *Pitx* genes have provided an ideal model for investigating this issue because their deletion causes a failure of cell-type-specific proliferation at specific stages of development. We have documented a genetic linkage between *Pitx2* and *Pitx1*. Although *Pitx2* appears to be required for effective appearance of the cardiac neural crest cells in the nascent outflow tract primordium of the cardiac tube, the phenotype of the *Pitx1*$^{+/-}$/*Pitx2*$^{+/-}$ mice reveals that these factors also exert critical dose-dependent and redundant activities during the transient period of their coexpression in the cardiac neural crest cells that populate the area of the developing outflow tract.

Pitx genes are clearly under the control of multiple enhancers, implying that distinct pathways regulate the spatial and temporal patterns of *Pitx* gene expression (Lamonerie et al. 1996; Szeto et al. 1996, 1999; Gage et al. 1999; Kitamura et al. 1999; Lanctot et al. 1999; Lin et al. 1999; Lu et al. 1999). *Pitx2* executes the asymmetric morphogenesis of primordia of visceral organs acting via distinct regulatory regions (Shiratori et al. 2001) downstream of the nodal signaling pathway (for review, see Capdevila et al. 2000). The *Pitx2* gene encodes three isoforms, *Pitx2a* and *Pitx2b*, which are generated by alternative splicing mechanisms, and *Pitx2c*, which uses an alternative promoter located upstream of exon 4 (Kitamura et al. 1999). The cardiac atria require low *Pitx2c* levels, whereas the duodenum and lungs need higher *Pitx2c* dosage (Liu et al. 2001). In the case of cardiac outflow tract development, *Pitx2* itself appears to be upstream of an additional regulatory component of the pathway, the gene encoding the endothelin A receptor (ET_A), apparently required for proper proliferation and maturation of cardiac neural crest cells in the outflow tract. There is a failure of ET_A expression in *Pitx1*$^{+/-}$/*Pitx2*$^{+/-}$ mice, and $ET_A^{-/-}$ mice exhibit a similar defect in outflow tract development (Clouthier et al. 1998). These data are consistent with the linkage of endothelin signaling to growth of other neural crest-derived cells (Opdecamp et al. 1998).

REFERENCES

Amendt B.A., Sutherland L.B., Semina E.V., and Russo A.F. 1998. The molecular basis of Rieger syndrome. Analysis of Pitx2 homeodomain protein activities. *J. Biol. Chem.* **273:** 20066.

Briata P., Ilengo C., Bobola N., and Corte G. 1999. Binding properties of the human homeodomain protein OTX2 to a DNA target sequence. *FEBS Lett.* **445:** 160.

Campione M., Steinbeisser H., Schweickert A., Deissler K., van Bebber F., Lowe L.A., Nowotschin S., Viebahn C., Haffter P., Kuehn M.R., and Blum M. 1999. The homeobox gene Pitx2: Mediator of asymmetric left-right signaling in vertebrate heart and gut looping. *Development* **126:** 1225.

Capdevila J., Vogan K.J., Tabin C.J., and Izpisua-Belmonte J.C. 2000. Mechanisms of left-right determination in vertebrates. *Cell* **101:** 9.

Chien K.R. 2000. Genomic circuits and the integrative biology of cardiac diseases. *Nature* **407:** 227.

Clouthier D.E., Hosoda K., Richardson J.A., Williams S.C., Yanagisawa H., Kuwaki T., Kumada M., Hammer R.E., and Yanagisawa M. 1998. Cranial and cardiac neural crest defects in endothelin-A receptor-deficient mice. *Development* **125:** 813.

Conway S.J., Henderson D.J., and Copp A.J. 1997. Pax3 is required for cardiac neural crest migration in the mouse: Evidence from the splotch (Sp2H) mutant. *Development* **124:** 505.

Creazzo T.L., Godt R.E., Leatherbury L., Conway S.J., and Kirby M.L. 1998. Role of cardiac neural crest cells in cardiovascular development. *Annu. Rev. Physiol.* **60:** 267.

Driever W., Thoma G., and Nüsslein-Volhard C. 1989. Determination of spatial domains of zygotic gene expression in the *Drosophila* embryo by the affinity of binding sites for the bicoid morphogen. *Nature* **340:** 363.

Gage P.J. and Camper S.A. 1997. Pituitary homeobox 2, a novel member of the bicoid-related family of homeobox genes, is a potential regulator of anterior structure formation. *Hum. Mol. Genet.* **6:** 457.

Gage P.J., Suh H., and Camper S.A. 1999. Dosage requirement of Pitx2 for development of multiple organs. *Development* **126:** 4643.

Goulding M.D., Chalepakis G., Deutsch U., Erselius J.R., and Gruss P. 1991. Pax-3, a novel murine DNA binding protein expressed during early neurogenesis. *EMBO J.* **10:** 1135.

Gross M.K., Moran-Rivard L., Velasquez T., Nakatsu M.N., Jagla K., and Goulding M. 2000. Lbx1 is required for muscle precursor migration along a lateral pathway into the limb. *Development* **127:** 413.

Groves A.K., George K.M., Tissier-Seta J.P., Engel J.D., Brunet J.F., and Anderson D.J. 1995. Differential regulation of transcription factor gene expression and phenotypic markers in developing sympathetic neurons. *Development* **121:** 887.

Hecht A. and Grunstein M 1999. Mapping DNA interaction sites of chromosomal proteins using immunoprecipitation and polymerase chain reaction. *Methods Enzymol.* **304:** 399.

Kioussi C., Briata P., Baek S.H., Rose D.W., Hamblet N.S., Herman T., Otigi K.A., Lin C., Gleiberman A., Wang J., Brault V., Ruiz-Lozano P., Nguyen H.D., Kessler R., Glass C.K., Wynshaw-Borris A., and Rosenfeld M.G. 2002. Identification of a wnt/Dvl/β-catenin→Pitx2 pathway mediating cell-type specific proliferation during development. *Cell* **111:** 673.

Kirby M.L. and Waldo K.L. 1995. Neural crest and cardiovascular patterning. *Circ. Res.* **77:** 211.

Kitamura K., Miura H., Miyagawa-Tomita S., Yanazawa M., Katoh-Fukui Y., Suzuki R., Ohuchi H., Suehiro A., Motegi Y., Nakahara Y., Kondo S., and Yokoyama M. 1999. Mouse Pitx2 deficiency leads to anomalies of the ventral body wall, heart, extra- and periocular mesoderm and right pulmonary isomerism. *Development* **126:** 5749.

Lamonerie T., Tremblay J.J., Lanctot C., Therrien M., Gauthier Y., and Drouin J. 1996. Ptx1, a bicoid-related homeo box transcription factor involved in transcription of the pro-opiomelanocortin gene. *Genes Dev.* **10:** 1284.

Lanctot C., Moreau A., Chamberland M., Tremblay M.L., and Drouin J. 1999. Hindlimb patterning and mandible development require the Ptx1 gene. *Development* **126:** 1805.

Lin C.R., Kioussi C., O'Connell S.M., Briata P., Szeto D., Liu F., Izpisua-Belmonte J.C., and Rosenfeld M.G. 1999. Pitx2 regulates lung asymmetry, cardiac positioning and pituitary and tooth morphogenesis. *Nature* **401:** 279.

Liu C., Liu W., Lu M.F., Brown N.A. and Martin J. 2001. Regulation of left-right asymmetry by thresholds of Pitx2c activity. *Development* **128:** 2039.

Logan M., Pagan-Westphal S.M., Smith D.M., Paganessi L., and Tabin C.J. 1998. The transcription factor Pitx2 mediates situs-specific morphogenesis in response to left-right asymmetric signals. *Cell* **94:** 307.

Lu M.F., Pressman C., Dyer R., Johnson R.L., and Martin J.F. 1999. Function of Rieger syndrome gene in left-right asymmetry and craniofacial development. *Nature* **401:** 276.

Meno C., Shimono A., Saijoh Y., Yashiro K., Mochida K., Ohishi S., Noji S., Kondoh H., and Hamada H. 1998. lefty-1 is required for left-right determination as a regulator of lefty-2 and nodal. *Cell* **94:** 287.

Opdecamp K., Kos L., Arnheiter H., and Pavan W.J. 1998. Endothelin signalling in the development of neural crest-derived melanocytes. *Biochem. Cell Biol.* **76:** 1093.

Piedra M.E., Icardo J.M., Albajar M., Rodriguez-Rey J.C., and Ros M.A. 1998. Pitx2 participates in the late phase of the pathway controlling left-right asymmetry. *Cell* **94:** 319.

Ryan A.K., Blumberg B., Rodriguez-Esteban C., Yonei-Tamura S., Tamura K., Tsukui T., de la Pena J., Sabbagh W., Greenwald J., Choe S., Norris D.P., Robertson E.J., Evans R.M., Rosenfeld M.G., and Izpisua-Belmonte J.C. 1998. Pitx2 determines left-right asymmetry of internal organs in vertebrates. *Nature* **394:** 545.

Scully K.M., Jacobson E.M., Jepsen K., Lunyak V., Viadiu H., Carriere C., Rose D.W., Hooshmand F., Aggarwal A.K., and Rosenfeld M.G. 2000. Allosteric effects of Pit-1 DNA sites on long-term repression in cell type specification. *Science* **290:** 1127.

Semina E.V., Reiter R., Leysens N.J., Alward W.L., Small K.W., Datson N.A., Siegel-Bartelt J., Bierke-Nelson D., Bitoun P., Zabel B.U., Carey J.C., and Murray J.C. 1996. Cloning and characterization of a novel bicoid-related homeobox transcription factor gene, RIEG, involved in Rieger syndrome. *Nat. Genet.* **14:** 392.

Shang Y., Hu. X., DiRenzo J., Lazar M.A., and Brown M. 2000. Cofactor dynamics and sufficiency in estrogen receptor-regulated transcription. *Cell* **102:** 843.

Shiratori H., Sakuma R., Watanabe M., Hashiguchi H., Mochida K., Sakai Y., Nishino J., Saijoh Y., Whitman M., and Hamada H. 2001. Two-step regulation of left-right asymmetric expression of Pitx2: Initiation by nodal signaling and maintenance by Nkx2. *Mol. Cell* **7:** 137.

Simmons D.M. Voss J.W., Ingraham H.A., Holloway J.M., Broide R.S., Rosenfeld M.G., and Swanson L.W. 1990. Pituitary cell phenotypes involve cell-specific Pit-1 mRNLA translation and synergistic interactions with other classes of transcription factors. *Genes Dev.* **4:** 695.

Smidt M.P., Cox J.J., van Schaick H.S., Coolen M., Schepers J., van der Kleij A.M., and Burbach J.P. 2000. Analysis of three Ptx2 splice variants on transcriptional activity and differential expression pattern in the brain. *J. Neurochem.* **75:** 1818.

Srivastava D. and Olson E.N. 2000. A genetic blueprint for cardiac development. *Nature* **407:** 221.

Szeto D.P., Ryan A.K., O'Connell S.M., and Rosenfeld M.G. 1996. P-OTX: A PIT-1-interacting homeodomain factor expressed during anterior pituitary gland development. *Proc. Natl. Acad. Sci.* **93:** 7706.

Szeto D.P., Rodriguez-Esteban C., Ryan A.K., O'Connell S.M., Liu F., Kioussi C., Gleiberman A.S., Izpisua-Belmonte J.C., and Rosenfeld M.G. 1999. Role of the Bicoid-related homeodomain factor Pitx1 in specifying hindlimb morphogenesis and pituitary development. *Genes Dev.* **13:** 484.

Waldo K.L., Lo C.W., and Kirby M.L. 1999. Connexin 43 expression reflects neural crest patterns during cardiovascular development. *Dev. Biol.* **208:** 307.

Yanagisawa M., Inoue A., Ishikawa T., Kasuya Y., Kimura S., Kumagaye S., Nakajima K., Watanabe T.X., Sakakibara S., and Goto K., et al. 1988. Primary structure, synthesis, and biological activity of rat endothelin, an endothelium-derived vasoconstrictor peptide. *Proc. Natl. Acad. Sci.* **85:** 6964.

Yoshioka H., Meno C., Koshiba K., Sugihara M., Itoh H., Ishimaru Y., Inoue T., Ohuchi H., Semina E.V., Murray J.C., Hamada H., and Noji S. 1998. Pitx2, a bicoid-type homeobox gene, is involved in a lefty-signaling pathway in determination of left-right asymmetry. *Cell* **94:** 299.

Pitx2 and Cardiac Development: A Molecular Link between Left/Right Signaling and Congenital Heart Disease

M. Campione,† L. Acosta,* S. Martínez,* J.M. Icardo,‡ A. Aránega,* and D. Franco*

Department of Experimental Biology, University of Jaén, Jaén, Spain; †National Research Council, Center for Muscle Biology and Physiopathology, Department of Biomedical Sciences, University of Padova, Padova, Italy; ‡Department of Anatomy and Cell Biology, University of Cantabria, Santander, Spain

Cardiac development is a complex and dynamic process (Fishman and Chien 1997). The heart develops from a bilateral set of promyocardial cells located at either side of the embryonic midline. The left and right sides of the cardiac crescent eventually fuse along the midline of the embryo, leading to a straight cardiac tube. This straight tube comprises an outer myocardial layer lined on the inner side by an endocardial layer (Manasek 1968). Soon after, the heart tube bends rightward, displaying the first morphological sign of left/right asymmetry in the developing embryo (Stalsberg 1969). With further development, atrial and ventricular chamber-specific gene expression is activated in discrete regions of the looped heart, delineating the embryonic heart (Christoffels et al. 2000). At this stage, five different regions can be observed in the developing heart: inflow tract, atrial myocardium, atrioventricular canal, ventricular myocardium, and outflow tract. These regions display different gene expression profiles and have different functional properties (Franco et al. 1998, 2000; Moorman et al. 1998). In mammals, each of these regions divides into two components to generate a double circulatory system. Interestingly, myocardial differentiation continues at the inflow tract of the heart, allowing the pulmonary and caval vessels to acquire distinct myocardial components. We have recently documented in mice that the gene expression profiles of the caval and pulmonary vessels diverge from each other, as well as from the surrounding atrial myocardium, supporting the notion that multiple myocardial domains are present in the developing and fully formed inflow tract (Franco et al. 2000). In essence, the developing heart exhibits extensive transcriptional diversity, suggesting that at least seven different domains can be delineated.

The heart is the first organ to display morphological left/right asymmetry. However, several genes such as *nodal*, *lefty-1*, and *lefty-2*, among other members of the TGFβ superfamily, display left/right asymmetric expression during early gastrulation (Collignon et al. 1996; Lowe et al. 1996; Meno et al. 1996, 1998). Side-specific gene signaling cascades, which originate in Hensen's node, trigger asymmetric gene expression in the lateral plate mesoderm (LPM) (Yost 1999; Burdine and Schier 2000; Wright 2001). Such differential gene expression confers side identity to the left and right LPM. Although the components of the cascade and the mode of activation differ between species, the sequence *nodal/Pitx2* appears to be highly conserved. *nodal* is expressed in the left LPM at the midgastrulation period and induces the expression of *Pitx2*. *Pitx2* is a bicoid-related homeobox transcription factor with restricted expression to the left LPM (Logan et al. 1998; Piedra et al. 1998; Ryan et al. 1998; Yoshioka et al. 1998; Campione et al. 1999). Of greater interest is the finding that the left-side asymmetric expression of *Pitx2* persists during subsequent development and can be observed during the morphogenesis of handed organs such as the heart and gut (Campione et al. 2001).

The distinct expression pattern of *Pitx2* and the results of several functional studies in different experimental models suggested an important role for this transcription factor in the determination of the direction of cardiac looping (Logan et al. 1998; Piedra et al. 1998; Ryan et al. 1998; Yoshioka et al. 1998; Campione et al. 1999). However, null *Pitx2* mutants revealed normal rightward looping, casting doubts as to the specific role of *Pitx2* in this process (Gage et al. 1999; Kitamura et al. 1999; Lu et al. 1999; Liu et al. 2001). Molecular studies have demonstrated that three different *Pitx2* isoforms (*Pitx2a*, *Pitx2b*, and *Pitx2c*) are expressed through development in mice, generated by alternative splicing and promoter usage. Recently, a fourth *Pitx2* isoform (*Pitx2d*) has been described, although it appears to occur only in humans (Cox et al. 2002). Among all isoforms, only *Pitx2c* is expressed asymmetrically within the LPM and the developing heart (Schweickert et al. 2000).

The discrete expression profile of *Pitx2* in the left LPM indicates that this transcription factor is involved in left/right signaling in the early embryo. However, the role of the persistent expression of *Pitx2* during heart development has been poorly understood (Schweickert et al. 2000). Over the last years, our laboratories have become interested in the putative role of *Pitx2* during cardiogenesis. Our detailed study of the *Pitx2* expression profile during normal and abnormal cardiac development suggested that *Pitx2* might play key roles during cardiac septation. In this paper, we describe our recent findings concerning *Pitx2* expression during cardiac morphogenesis and our working hypothesis as to how *Pitx2* might exert its function during cardiac septation.

EXPERIMENTAL EVIDENCE

Pitx2 in the Developing Heart

The expression of *Pitx2* is already confined to the left cardiac crescent at the cardiac crescent stage (Fig. 1a) (mouse embryonic day [E] E7.5; chick Hamburger and Hamilton (1951) stage [HH] HH8). As the primitive heart tube is formed, the expression of *Pitx2* remains confined to the left part of the forming cardiac tube, both in the mouse (E8.0) and the chick (HH10) (Fig. 1b). With further development, the cardiac tube bends to the right, configuring the prospective ventricular and atrial chambers (mouse E8.5; chicken HH12). Interestingly, *Pitx2* transcript expression follows the torsion of the looping heart. Such an effect is prominently observed on the ventricular side, becoming associated with an anterior-ventral expression domain in the prospective ventricular chambers, but not on the atrial side (Fig. 1c), where it remains confined to the left domain. Fate-mapping experiments in chick embryos have demonstrated that *Pitx2* expression can be used as a molecular marker to follow the fate of cells derived from the left portion of the cardiac tube (Campione et al. 2001). In these experiments, we observed that *Pitx2*-expressing cells emerging from the left side of the linear heart tube are located in the most ventral region of the ventricular chambers, whereas the cells from the right side of the tube are located to the dorsal aspect of these chambers. Second, and most importantly, the expression of *Pitx2* remains confined to the left atrial component during cardiac looping. Three main conclusions can be derived from these experiments: (1) The ventricular chambers develop in series, each definitive ventricle having both left (ventral position) and right (dorsal position) components of the linear heart tube; (2) the atrial chambers develop in parallel, the left and right atria originating from their respective half-tube hearts; (3) *Pitx2* is a molecular marker that permits visualization of the fate of the left portion of the primitive heart tube.

By the end of looping, the heart has acquired five distinct cardiac regions (mouse E10.5; chicken HH16). At this stage, *Pitx2* expression remains essentially similar to that observed during cardiac looping. The outflow tract, the atrioventricular canal, and the inflow tract are now clearly distinguishable. At this stage, *Pitx2* expression is confined to the left part of the inflow tract and atrioventricular canal, whereas in the outflow tract *Pitx2* expression smoothly shifts from a ventral to a left position, extending from the ventricular junction to the aortic arch (Fig. 1d) (Campione et al. 2001). During cardiac septation (mouse E12.5–E16.5; chicken HH24–HH34), expression of *Pitx2* decreases in the atrioventricular canal, the ventricular chambers, and the outflow tract. However, high expression levels are maintained in the atrial appendages and developing pulmonary and caval vessels. Detailed analyses of the expression pattern of *Pitx2* in these structures have revealed that both caval and pulmonary vein myocardial domains have distinct left and right components (Franco et al. 2000). Interestingly, within the caval vein myocardium, the left superior caval vein expresses *Pitx2* transcripts and the right superior caval vein does not. On the contrary, almost the entire myocardial component of the pulmonary veins displays *Pitx2* expression, including both primary and secondary atrial septa, except for a small portion spanning from the secondary atrial septum to the entrance of the right superior caval vein (Fig. 1e, f). The expression of *Pitx2* is not observed beyond E16.5, nor in the adult heart.

Pitx2 and Left/Right Cardiac Asymmetry

The possibility of using *Pitx2* as a bona fide molecular marker to follow the fate of cardiac cells of left origin prompted us to investigate its expression in *iv* mutant mice, a model of abnormal laterality (Hummel and Chapman 1959; Supp et al. 1997).

Several groups have independently reported that the expression of *nodal/Pitx2* in the LPM is randomized in *iv* mutant mice; that is, it can be left-restricted, right-restricted, bilateral, or absent in the LPM (Piedra et al. 1998; Campione et al. 1999). Regardless of such diverse profiles of *nodal/Pitx2* expression in the LPM, cardiac looping always develops in the *iv* mutant mice. We have observed a number of interesting features in the expression profile of *Pitx2* in *iv* mutant mice during embryonic and fetal stages of cardiogenesis. Consistent with previous reports on left/right differential expression of endogenous genes and transgenes (Thomas et al. 1998; Franco et al. 2001), *Pitx2* expression was found either to be normally expressed or to appear in a mirror-image arrangement. In other cases, and regardless of the direction of cardiac looping, the atria may show bilateral or absent

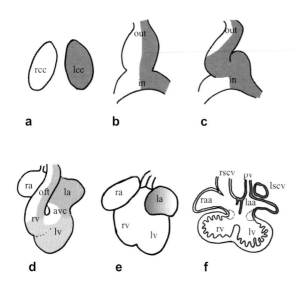

Figure 1. Schematic representation of *Pitx2* expression during different stages of cardiac development. (*a*) Cardiac crescent, (*b*) straight cardiac tube, (*c*) looping heart, (*d*) embryonic heart, (*e*) fetal heart, and (*f*) fetal heart (venous pole). (rcc) Right side of the cardiac crescent, (lcc) left side of the cardiac crescent, (in) inflow region, (out) outflow region, (ra) right atrium, (la) left atrium, (oft) outflow tract, (avc) atrioventricular canal, (rv) right ventricle, (lv) left ventricle, (raa) right atrial appendage, (laa) left atrial appendage, (rscv) right superior caval vein, (pv) pulmonary veins, (lscv) left superior caval vein.

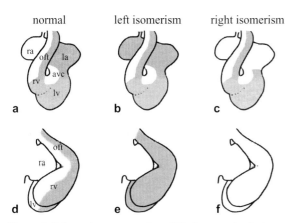

Figure 2. Schematic representation of *Pitx2* expression in *iv* mutant mice with normal (*a,d*), left isomerism (*b,e*), and right isomerism (*c,f*) within the atrial (*a–c*) and ventricular (*d–f*) domains. Thus, panel *b* represents left atrial isomerism and panel *c* represents right atrial isomerism. Similarly, panel *e* represents left ventricular isomerism and panel *f* represents right ventricular isomerism. (ra) Right atrium, (*la*) left atrium, (oft) outflow tract, (avc) atrioventricular canal, (rv) right ventricle, (lv) left ventricle.

Pitx2 expression (Fig. 2a–c). Analysis of the specimens with aberrant atrial expression shows the presence of atrial isomerism, a situation in which the two atria are morphologically identical. The fact that *Pitx2* expression in the atria may be aberrant, yet normal in the ventricular chambers, reveals for the first time a modular nature of left/right signaling during cardiogenesis.

Given that left/right embryonic signals are converted into dorsoventral signals in the developing ventricles, ventricular isomerism would be expected to perturb dorsoventral *Pitx2* expression instead of left/right (Fig. 2d–f). Careful analysis of *iv* mouse hearts has revealed aberrant expression of *Pitx2* in the ventricular chambers. A total of 11 specimens (11/23; 48%) showed *Pitx2* expression in both dorsal and ventral aspects of the ventricles, constituting the first evidence, to our knowledge, of ventricular isomerism detectable at the molecular level (Campione et al. 2001). More importantly, abnormal *Pitx2* expression in the atrial and ventricular chambers correlated well with abnormal outflow tract development, in particular with double-outlet right ventricle (9/11; 82%) (Campione et al. 2001).

TOWARD A MECHANISM OF ACTION

The acquisition of left/right asymmetry observed during development always follows the same pattern, positioning different organs such as the spleen, the gut, and the heart within the same orientation, a condition dubbed situs solitus. Rarely, the entire body plan is inverted, a condition dubbed situs inversus, which is fully compatible with normal development and subsequently with normal life. However, in other cases, some organs show normal development while others display the mirror-image arrangement, a condition termed situs ambiguus (heterotaxia). In the heterotaxia syndromes, the heart displays a wide spectrum of anomalies ranging from atrial or ventricular septal defects to life-threatening malformations such as double-outlet right ventricle (DORV), common atrioventricular canal (CAVC), or double-inlet left ventricle (DILV). These malformations may appear singly or in combination (Icardo and Sanchez de la Vega 1991). For example, CAVC is often accompanied by DORV, atrial isomerism, persistence of the sinus venosus, and anomalous venous return (Seo et al. 1992). At present there is little understanding of the morphological and molecular mechanisms underlying the spectrum of cardiac abnormalities associated with heterotaxia.

Misexpression studies in chick and *Xenopus* embryos have suggested a key role for *Pitx2* in the establishment of rightward cardiac looping (Logan et al. 1998; Piedra et al. 1998; Ryan et al. 1998; Campione et al. 1999). However, more recently, *Pitx2* null mutant mice have been shown to undergo normal heart looping, raising doubts over the role of this transcription factor in this process. Interestingly, the hearts of *Pitx2* null mice exhibit several other cardiac morphological abnormalities, such as right atrial isomerism and abnormal ventriculo-arterial connections (Gage et al. 1999; Kitamura et al. 1999; Lu et al. 1999; Lui et al. 2001).

Pitx2 Domains during Cardiac Development

Fate-mapping experiments in the mouse (Franco et al. 2001; Kelly et al. 2001) and the chick (de la Cruz et al. 1989; de la Cruz and Markwald 2001) have revealed that the linear heart tube contributes mostly to the ventricular regions. However, the expression of *Pitx2* in the embryonic heart extends to all myocardial compartments, including the most anterior outflow tract and most posterior inflow tract regions. Although *Pitx2* expression at early stages is not restricted to the myocardial boundaries, that is to say, it extends throughout the adjacent mesenchyme and epithelium, it is not known how *Pitx2* expression becomes activated in the atrial/inflow tract and outflow tract myocardium. Kelly et al. (2001) have recently reported, using a transgenic reporter mouse model and DiI experiments, that the outflow tract and right ventricular myocardium are derived from a medial region of the splanchnic mesoderm, dubbed the *anterior* (or *secondary*) *heart field*. Interestingly, asymmetrical expression of genes such as *nodal*, *lefty-2,* and *Pitx2* is observed in the left portion of this region (Kelly and Buckingham 2002), supporting the notion that left/right identity is conferred to the developing right ventricle/outflow tract by the canonical *nodal/lefty/Pitx2* signaling cascade (Fig. 3). Similarly, activation of *Pitx2* in the most caudal sinoatrial region could be operating throughout a similar mechanism. It remains to be elucidated whether a *posterior heart field* could give rise to the sinoatrial region and whether this putative *posterior heart field* has received left/right signaling before or after becoming committed to the myocardial fate.

Interestingly, the putative subdivision of the developing heart in three different sets of myocardial progenitor cells, the *anterior heart field*, the *classical heart field*

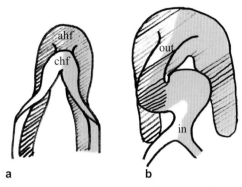

Figure 3. Schematic representation of the *classical heart field* (chf) and *anterior heart field* (ahf; *dashed area*) development and the *Pitx2* expression. Note that *Pitx2* is expressed not only in the left side of the *classical heart field* but also in the developing *anterior heart field* at an early stage of development. (in) Inflow region, (out) outflow region.

(straight myocardial tube), and the *posterior heart field* is concordant with the idea that left/right signaling operates independently within three distinct modules of the developing heart (outflow tract/ventricles, atrioventricular canal/atria, inflow tract). It remains to be established whether the myocardial boundaries of these domains correspond to different derivates of the three heart fields.

Pitx2 and Cardiac Isomerism: Evidence for Atrial and Ventricular Molecular Isomerism

Studies on *Pitx2* expression, in conjunction with fate-mapping experiments in the chick, have demonstrated that *Pitx2* represents a bona fide left cardiac lineage marker. We have observed that, during cardiac looping, the contribution of the right and left sides of the cardiac crescent to the developing ventricular chambers becomes relocated to the dorsal and ventral aspects of each ventricle. In contrast, the left and right atrial chambers, the atrioventricular canal, and the inflow and outflow tracts maintain their initial left/right regionalization (Campione et al. 2001). These observations are in agreement with the fact that morphological atrial isomerism is observed in patients with heterotaxia, while no pulmonary/systemic ventricular isomerism can be detected. However, in the search for ventricular isomerism, there are the dorsal and ventral aspects of each ventricle that should be compared, not the right and left ventricular chambers themselves. The lack of any anatomical hallmark to identify ventricular isomerism, together with the absence of any easily recognizable dorsoventral asymmetry in the normal developing heart, explains why no morphological ventricular isomerism has ever been reported (Brown and Anderson 1999).

Pitx2 and Abnormal Venous Return: A Left/Right Signaling Defect

We have documented that *Pitx2* expression in the fetal heart is restricted mainly to the venous component. Comparison between the expression patterns of *Pitx2* and other molecular markers in the developing inflow tract has revealed that each myocardial region (atria, atrioventricular canal, atrial appendages, cava and pulmonary veins), has distinct left and right components (Franco et al. 2000). Precise boundaries can be delineated within these myocardial domains. The embryonic vena cava system can be separated into left and right (superior and inferior) components, and only the myocardial component of the left veins shows *Pitx2* expression. In contrast, most of the pulmonary vein myocardium is left-derived and, consequently, shows *Pitx2* expression. The atrial septal complex (primary and secondary atrial septa) also expresses *Pitx2*. The atrial myocardium extending from the secondary atrial septum to the entrance of the left superior vena cava is right-derived and lacks *Pitx2* expression.

The intricate development of the inflow tract, together with modular left/right signaling, may be at the root of the wide phenotypic variation observed in patients with heterotaxia (Fig. 4a–d). Patients with right atrial isomerism normally display bilateral caval vein drainage and abnormal pulmonary vein development (Fig. 4b). In other cases, there is normal vena cava drainage with or without pulmonary vein formation (Fig. 4c). In patients with left atrial isomerism, the pulmonary veins develop normally in most cases, but the development of the caval system is

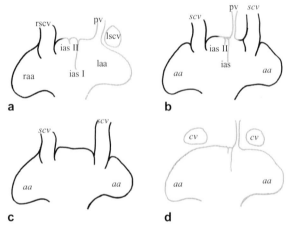

Figure 4. Schematic representation of a model for signaling by *Pitx2*, in normal (*a*) and abnormal venous (*b–d*) return. Light coloration corresponds to the *Pitx2*-expressing myocardial domains. Note that, in normal conditions, *Pitx2* is found within the entire left side of the venous pole (panel *a*). In right isomerism, if only a subset of the venous pole receives left (*Pitx2*) signals, normal pulmonary vein drainage will result (panel *b*). If none of the myocardial domains receives these signals, no pulmonary veins will drain into the atrial chambers (panel *c*). In contrast, if all the venous pole develops with left identity, the atrial appendages will be left-isomeric, with normal pulmonary veins but impaired caval vein drainage (panel *d*). (raa) Right atrial appendage, (laa) left atrial appendage, (rscv) right superior caval vein, (pv) pulmonary veins, (lscv) left superior caval vein, (iasI) primary atrial septum, (iasII) secondary atrial septum. Italic wording such as (*cv*), caval vein, (*scv*), superior caval vein, (*aa*) atrial appendage, reflects impaired left/right identity of these morphological structures.

abnormal (Fig. 4d). It must be underscored that the developing caval and pulmonary veins open into the embryonic sinus venosus. The sinus venosus is incorporated into the dorsal part of the developing atria, carrying with it the vein trunks. Thus, the veins reach their definitive connections. Defective incorporation of the sinus venosus results in anomalous venous return, explaining many of the venous anomalies.

Analyses of *Pitx2* transcripts and other molecular markers allowing differentiation of the distinct atrial myocardial components (Franco et al. 2000) have revealed that specification of discrete atrial regions is impaired in mice with laterality defects. Thus, in some cases, either the caval vein myocardial component or the pulmonary vein myocardium does not form. Impaired development of either of these regions in the presence or absence of left signaling molecules (i.e., *Pitx2*) provides an explanation for the degree of variability observed in heterotaxia patients. Nonetheless, it remains to be determined why certain phenotypes are more frequent than others.

Pitx2 and Congenital Heart Disease

We have previously reported that *Pitx2* represents a left cardiac lineage marker. Thus, *Pitx2* should be either bilaterally expressed in the atrial appendages in cases of left atrial isomerism or not expressed in either atrial appendage in cases of right atrial isomerism. These patterns of expression were observed in an analysis of *Pitx2* transcript distribution in *iv* mutant mice (Campione et al. 2001). Thus, the embryos are likely to represent genuine examples of atrial isomerism.

Similarly, if ventricular molecular isomerism does occur, either *Pitx2* will be expressed on both the dorsal and the ventral aspects of the developing left and right ventricular chambers, representing a left-derived ventricular molecular isomerism, or else no ventricular expression will be observed, representing a right-derived ventricular molecular isomerism. Detailed analysis of *Pitx2* expression in the ventricular components of *iv* mutant mice revealed a considerable number of specimens displaying both dorsal and ventral expression of *Pitx2* but failed to reveal any cases lacking *Pitx2* expression in the ventricular chambers.

Of particular interest is our observation that 80% of the hearts with abnormal atrial and/or ventricular *Pitx2* expression show DORV. This correlation is statistically significant and suggests a cause–effect relationship. We think that abnormal myocardial specification of the inner heart curvature is at the origin of the DORV (Fig. 5). This hypothesis is supported by the fact that abnormal *Pitx2* expression in the atrial region also affects the atrioventricular canal. Similarly, abnormal ventricular expression of *Pitx2* involves the inner curvature of the developing heart. On the other hand, remodeling and division of the atrioventricular canal, and particularly its inner contour, is crucial for normal atrioventricular septation (Kim et al. 2001). In fact, DORV in *iv* mice is often accompanied by CAVC (Icardo and Sanchez de la Vega 1991). It appears likely that abnormal remodeling of the inner curvature of the heart, due to abnormal *Pitx2* signaling, underlies atrioventricular canal malformations and abnormal ventriculo-arterial connections. Similarly, defective *Pitx2* signaling may modify the normal developmental patterns

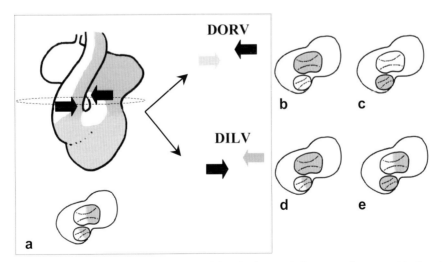

Figure 5. Model of abnormal development of ventriculo-arterial and atrio-ventricular connections caused by impaired expression of *Pitx2* in the inner curvature of the heart. Panel *a* represents the normal expression profile of *Pitx2* in the embryonic heart (ventral view and cranial view of the ventricles alone) and the morphological movements of the outflow tract (OFT) and atrio-ventricular canal (AVC) that lead to normal cardiac development. If there is an abnormal growth of the OFT, DORV will result, whereas if there is an abnormal growth of the AVC, DILV will develop. Panels *b–e* represent different combinations of abnormal atrial/AVC and ventricular/OFT *Pitx2* expression. Note that atrial and ventricular misexpression of *Pitx2* converge in the inner curvature of the heart. Therefore, an excess of left signaling (*Pitx2*) in this region can result in DILV due to abnormal rightward growth of the AVC (see panels *b* and *d*). Similarly, an excess of *Pitx2* expression in the OFT can result in impaired leftward displacement of this region, leading to DORV (see panels *c* and *e*). The multiple combination of AVC/OFT misexpression pattern of *Pitx2* in the inner curvature could therefore be at the root of the great number of atrio-ventricular and ventriculo-arterial abnormalities observed in the *iv* mutant mice.

at the sinoatrial region, resulting in anomalous venous return, atrial septal malformations, and abnormal atrioventricular connections (i.e., DILV). Current research in our laboratories is aimed at testing these hypotheses.

CONCLUSIONS

Our present studies reveal an intricate pattern of left/right signaling within the developing heart, permitting us to discern a segmental modulation in distinct myocardial domains; the outflow/ventricles, the atrioventricular canal/atria, and the inflow tract (including herein both the caval vein and pulmonary vein myocardium). We suggest that the abnormal expression of *Pitx2* is responsible for the wide heterogeneity of cardiac phenotypes observed in heterotaxia patients. Furthermore, we have convincingly documented that left and right sides of the cardiac tube contribute to the dorsal and ventral aspects of both left and right developing ventricles, whereas they maintained their left and right configuration in the developing atria. These observations demonstrate that each embryonic ventricle obtains a contribution of both sides of the cardiac crescent, whereas each embryonic atrium has a predominant contribution of a single side of the cardiac crescent. Thus, abnormal left/right signaling will equally affect the two developing ventricles, resulting in dorsoventral molecular isomerism. No signs of molecular isomerism or of morphological symmetries will be found if the pulmonary and systemic ventricles are compared to each other. On the other hand, each atrium derives from the corresponding side of the primitive cardiac tube. Abnormal left/right signaling will therefore affect only one atrium, resulting in atrial isomerism.

Finally, the fact that abnormal expression of *Pitx2* in different cardiac compartments correlates with a high incidence of DORV in abnormal laterality mutant mice provides new insights into the relationship between the left/right signaling pathway and impaired cardiac septation. Our laboratory has recently started several approaches to elucidate the molecular mechanisms underlying the role of *Pitx2* during cardiogenesis and its putative influence on the development of common congenital malformations such as DORV and DILV.

ACKNOWLEDGMENTS

We thank Robert Kelly for critical reading of the manuscript. This work is supported by a grant from the Ministry of Science and Technology of the Spanish Government to D.F., A.A., and J.M.I. (BCM2000-0118-C02-01 and BCM2000-0118-C02-02). S.M. is supported by a fellowship from the Ministry of Science and Technology of the Spanish Government (BCM2000-0118-C02-02). M.C. is supported by the Telethon Foundation (452/bi).

REFERENCES

Brown N.A. and Anderson R.H. 1999. Symmetry and laterality in the human heart: Developmental implications. In *Heart development* (ed. R.P. Harvey and N. Rosenthal), p. 447. Academic Press, San Diego, California.

Burdine R.D. and Schier A. 2000. Conserved and divergent mechanisms in left-right axis formation. *Genes Dev.* **14:** 763.

Campione M., Ros M.A., Icardo J.M., Piedra E., Christoffels V.M., Schweichert A., Blum M., Franco D., and Moorman A.F.M. 2001. *Pitx2* expression defines a left cardiac lineage of cells: Evidence for atrial and ventricular molecular isomerism in the iv/iv mice. *Dev. Biol.* **231:** 252.

Campione M., Steinbeisser H., Schweickert A., Deissler K., van Bebber F., Lowe L.A., Nowotschin S., Viebahn C., Haffter P., Kuehn M.R., and Blum M. 1999. The homeobox gene *Pitx2*: Mediator of asymmetric left-right signalling in vertebrate heart and gut looping. *Development* **126:** 1225.

Christoffels V.M., Habets P.E., Franco D., Campione M., de Jong F., Lamers W.H., Bao Z.-Z., Palmer S., Biben C., Harvey R.P., and Moorman A.F.M. 2000. Chamber formation and morphogenesis in the developing mammalian heart. *Dev. Biol.* **223:** 266.

Collignon J., Varlet I., and Robertson E.J. 1996. Relationship between asymmetric nodal expression and the direction of embryonic turning. *Nature* **381:** 155.

Cox C.J., Espinoza H.M., McWilliams B., Chappell K., Morton L., Hjalt T.A., Semina E.V., and Amendt B.A. 2002. Differential regulation of gene expression by PITX2 isoforms. *J. Biol. Chem.* **277:** 25001.

de la Cruz M.V. and Markwald R.R., Eds. 2001. *Living morphogenesis of the heart*. Birkhäuser, Boston, Massachusetts.

de la Cruz M.V., Sánchez-Gómez C., and Palomino M.A. 1989. The primitive cardiac regions in the straight tube heart (Stage 9⁻) and their anatomical expression in the mature heart: An experimental study in the chick embryo. *J. Anat.* **165:** 121.

Fishman M.C. and Chien K.R. 1997. Fashioning the vertebrate heart: Earliest embryonic decisions. *Development* **124:** 2099.

Franco D., Lamers W.H., and Moorman A.F.M. 1998. Patterns of gene expression in the developing myocardium: Towards a morphologically integrated transcriptional model. *Cardiovasc. Res.* **38:** 25.

Franco D., Kelly R., Moorman A.F.M., Lamers W.H., Buckingham M., and Brown N.A. 2001. MLC3F transgene expression in iv mutant mice reveals the importance of left-right signaling pathways for the acquisition of left and right atrial but not ventricular compartment identity. *Dev. Dyn.* **221:** 206.

Franco D., Campione M., Kelly R., Zammit P.S., Buckingham M., Lamers W.H., and Moorman A.F.M. 2000. Four transcriptional domains, with a distinct left-right component, are distinguished in the developing mouse cardiac atria. *Circ. Res.* **87:** 984.

Gage P.J., Suh H., and Camper S.A. 1999. Dosage requirement of *Pitx2* for development of multiple organs. *Development* **126:** 4643.

Hamburger V. and Hamilton H. 1951. A series of normal states in the development of the chick embryo. *J. Morphol.* **88:** 49.

Hummel K.P. and Chapman D.B. 1959. Visceral inversion and associated anomalies in the mouse. *J. Hered.* **50:** 9.

Icardo J.M. and Sanchez de la Vega M.J. 1991. Spectrum of heart malformations in mice with *situs inversus*, *situs solitus* and associated visceral heterotaxy. *Circulation* **84:** 2547.

Kelly R.G. and Buckingham M.E. 2002. The anterior heart-forming field: Voyage to the arterial pole of the heart. *Trends Genet.* **18:** 210.

Kelly R.G., Brown N.A., and Buckingham M.E. 2001. The arterial pole of the mouse heart forms from FGF10-expressing cells in pharyngeal mesoderm. *Dev. Cell* **1:** 435.

Kim J.-S., Viragh S., Moorman A.F.M., Anderson R.H., and Lamers W.H. 2001. Development of the myocardium of the atrioventricular canal and the vestibular spine in the human heart. *Circ. Res.* **88:** 395.

Kitamura K., Miura H., Miyagawa-Tomita S., Yanazawa M., Katoh-Fukui Y., Suzuki R., Ohuchi H., Suehiro A., Motegi Y., Nakahara Y., Kondo S., and Yokoyama M. 1999. Mouse *Pitx2* deficiency leads to anomalies of the ventral body wall, heart, extra- and periocular mesoderm and right pulmonary isomerism. *Development* **126:** 5749.

Logan M., Págan-Westphal S.M., Smith D.M., Paganessi L., and Tabin C.J. 1998. The transcription factor *Pitx2* mediates situs-specific morphogenesis in response to left-right asymmetric signals. *Cell* **94:** 307.

Lowe L.A., Supp D.M., Sampath K., Yokoyama T., Wright C.V.E., Potter S.S., Overbeek P., and Kuehn M.R. 1996. Conserved left-right asymmetry of nodal expression and alterations in murine situs inversus. *Nature* **381:** 158.

Lu M.-F., Pressman C., Dyer R., Johnson J.L., and Martin J.F. 1999. Function of Rieger syndrome gene in left-right asymmetry and craniofacial development. *Nature* **401:** 276.

Lui C., Lui W., Lu M.-F., Brown N.A., and Martín J.M. 2001. Regulation of left-right asymmetry by thresholds of *Pitx2c* activity. *Development* **128:** 2039.

Manasek F.J. 1968. Embryonic development of the heart. I. A light and electron microscopic study of myocardial development in the early chick embryo. *J. Morphol.* **125:** 329.

Meno C., Saijoh Y., Fujii H., Ikeda M., Yokoyama T., Yokoyama M., Toyoda Y., and Hamada H. 1996. Left-right asymmetric expression of the TGF-β-family member lefty in mouse embryos. *Nature* **381:** 151.

Meno C., Shimono A., Saijoh Y., Yashiro K., Mochida K., Ohishi S., Noji S., Kondoh H., and Hamada H. 1998. Lefty-1 is required for left-right determination as regulator of lefty-2 and nodal. *Cell* **94:** 287.

Moorman A.F.M., de Jong F., Denyn M.M., and Lamers W.H. 1998. Development of the cardiac conduction system. *Circ. Res.* **82:** 629.

Piedra M.E., Icardo J.M., Albajar M., Rodriguez-Rey J.C., and Ros M.A. 1998. *Pitx2* participates in the late phase of the pathway controlling left-right asymmetry. *Cell* **94:** 319.

Ryan A.K., Blumberg B., Rodriguez-Esteban C., Yonei-Tamura S., Tamura K., Tsukui T., de la Peña J., Sabbagh W., Greenwald J., Choe S., Norris D.P., Robertson E.J., Evans R.M., Rosenfeld M.G., and Izpisúa-Belmonte J.C. 1998. *Pitx2* determines left-right asymmetry of internal organs in vertebrates. *Nature* **394:** 545.

Schweickert A., Campione M., Steinbeisser H., and Blum M. 2000. *Pitx2* isoforms: Involvement of *Pitx2*c but not *Pitx2*a or *Pitx2*b in vertebrate left-right asymmetry. *Mech. Dev.* **90:** 41.

Seo J.W., Brown N.A., Ho S.Y., and Anderson R.H. 1992. Abnormal laterality and congenital cardiac anomalies. *Circulation* **86:** 642.

Stalsberg H. 1969. The origin of heart asymmetry: Right and left contributions to the early chick embryo heart. *Dev. Biol.* **19:** 109.

Supp D.M., Witte D.P., Potter S.S., and Brueckner M. 1997. Mutation of an axonemal dynein affects left-right asymmetry in inversus viscerum mice. *Nature* **389:** 963.

Thomas T., Yamagishi H., Overbeek P.A., Olson E.N., and Srivastava D. 1998. The bHLH factors, dHAND and eHAND, specify pulmonary and systemic cardiac ventricles independent of left-right sidedness. *Dev. Biol.* **196:** 228.

Wright C.V.E. 2001. Mechanisms of left-right asymmetry: What´s right and what´s left? *Dev. Cell* **1:** 179.

Yoshioka H., Meno C., Koshiba K., Sugihara M., Itoh H., Ishimaru Y., Inoue T., Ohuchi H., Semina E.V., Murray J.C., Hamada H., and Noji S. 1998. *Pitx2*, a bicoid-type homeobox gene, is involved in a lefty-signaling pathway in determination of left-right asymmetry. *Cell* **94:** 199.

Yost H.J. 1999. Diverse initiation in a conserved left-right pathway? *Curr. Opin. Genet. Dev.* **9:** 422.

Regulation of Cardiac Growth and Development by SRF and Its Cofactors

D. WANG,*† R. PASSIER,*¶ Z.-P. LIU,* C.H. SHIN,* Z. WANG,* S. LI,* L.B. SUTHERLAND,*
E. SMALL,‡ P.A. KRIEG,‡ AND E.N. OLSON*

*Departments of Molecular Biology, University of Texas, Southwestern Medical Center at Dallas, Dallas, Texas 75390-9148; ‡Department of Cell Biology and Anatomy, University of Arizona College of Medicine, Tucson, Arizona 85724; ¶Hubrecht Laboratory, 3584CT Utrecht, The Netherlands

The generation of specialized cell types during development is dependent on combinatorial interactions between widely expressed and cell-type-restricted transcription factors that establish cell identity and couple extracellular signals to specific programs of gene expression. The MADS (MCM1, Agamous, Deficiens, serum response factor) box family of transcription factors is especially illustrative of the importance of combinatorial control in gene regulation during development (Shore and Sharrocks 1995).

There are five MADS box proteins in vertebrates: serum response factor (SRF) and the four myocyte enhancer factor-2 (MEF2) factors—MEF2A, B, C, and D (Black and Olson 1998). These factors are expressed in a wide range of cell types where they associate with a plethora of positive and negative cofactors and serve as intranuclear targets for signal transduction pathways that control cell growth and differentiation. SRF and MEF2 are required for the opposing processes of cell proliferation and muscle cell differentiation, which is seemingly paradoxical since myoblast differentiation is coupled to withdrawal from the cell cycle, and signals that induce cell proliferation interfere with myogenesis. How these factors discriminate between different target genes in the pathways for cell proliferation and myogenesis is a fundamental question in the field. Here, we describe two transcriptional cofactors that act in an antagonistic manner to modulate the activity of SRF in the developing heart, and we consider how combinatorial interactions of SRF with positive and negative partners dictate the transcriptional readout of striated muscle cells. Although we focus in this paper on SRF, many of the same principles apply to the regulation of MEF2 activity, as reviewed recently (McKinsey et al. 2002).

MADS BOX PROTEINS

The structures of SRF and MEF2 factors are schematized in Figure 1A. The MADS box, which mediates dimerization and DNA binding, contains conserved residues in SRF and MEF2, whereas the region immediately carboxy-terminal to the MADS box, which influences dimerization and interactions with other transcriptional cofactors, is divergent (Fig. 1B). The crystal structures of SRF and MEF2 have revealed commonalities in their modes for DNA binding (Pellegrini et al. 1995; Santelli and Richmond 2000). The conserved amino-terminal portions of the MADS boxes of these proteins comprise α-helices that are oriented in an antiparallel manner to form a bipartite DNA-binding domain. An unstructured amino-terminal extension makes critical base contacts in the minor groove that stabilize DNA binding, and two β-sheets at the carboxy-terminal end of the α-helical DNA-binding region in each monomer extend away from the DNA-binding site, thereby permitting homodimerization and association with accessory factors.

The conservation in the structure of the DNA-binding domains of SRF and MEF2 is reflected in the similarity in sequence of their DNA-binding sites (Fig. 1C). Both proteins recognize an A/T-rich core sequence with different flanking nucleotides. Although SRF and MEF2 regulate different sets of genes, both factors control the expression of muscle structural genes and growth-factor-regulated genes. In addition to their modulation by cofactors that associate directly with their DNA-binding domains, SRF and MEF2 cooperate with other transcription factors that bind sites adjacent to their DNA-binding sites.

REGULATION OF CELL GROWTH AND MUSCLE DIFFERENTIATION BY SRF

The essential roles of SRF in cell proliferation and muscle cell differentiation have been documented by gain- and loss-of-function experiments in cultured cells. SRF is required for phosphatidylinositol-3-kinase (PI3K)-regulated cell proliferation, and expression of constitutively active SRF is sufficient for cell cycle reentry (Poser et al. 2000). Microinjection of anti-SRF antibodies also blocks cell cycle progression in serum-stimulated fibroblasts and causes growth arrest in the G$_1$ phase (Gauthier-Rouviere et al. 1991). Conversely, microinjection of anti-SRF antibodies or expression of dominant negative SRF mutants or SRF antisense transcripts in

---

†Present address: Carolina Cardiovascular Biology Center and Department of Cell and Developmental Biology, University of North Carolina, Chapel Hill, North Carolina 27599-7126.

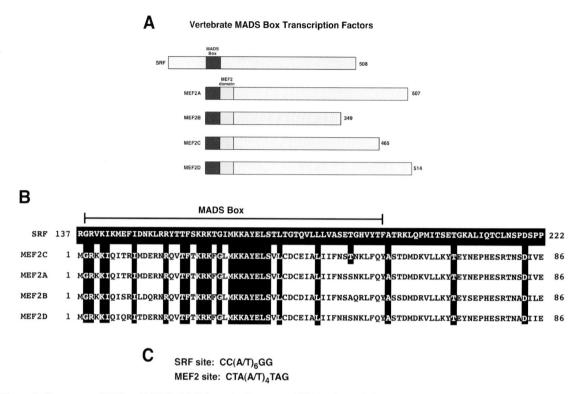

Figure 1. Structures of SRF and MEF2. (*A*) Schematic diagrams of SRF and MEF2 factors. Amino acid positions are indicated. (*B*) Sequence homologies between the DNA-binding domains of SRF and MEF2 factors. (*C*) DNA-binding sites of SRF and MEF2 factors.

skeletal muscle cells prevents differentiation (Vandromme et al. 1992; Croissant et al. 1996; Soulez et al. 1996). SRF is also required for mesoderm formation during gastrulation, as shown by the lack of mesoderm formation in mice lacking SRF (Arsenian et al. 1998). Embryonic stem (ES) cells lacking SRF are defective in activation of immediate early genes in response to serum stimulation, but they are able to proliferate (Schratt et al. 2001). SRF-null ES cells are also impaired in the ability to form mesodermal derivatives. However, in the presence of wild-type cells or specific extracellular factors, such as retinoic acid (Weinhold et al. 2000), some mesodermal genes can be expressed, including skeletal muscle α-actin. The potential to activate cardiac or smooth muscle genes in SRF-null ES cells has not been described. This is an especially interesting issue in light of studies described below on the requirement of the SRF cofactor myocardin for cardiac gene expression in vivo. Together, studies of ES cells lacking SRF indicate that at least a portion of the requirement of SRF for mesoderm formation in vivo reflects a non-cell-autonomous function. The specific SRF target genes responsible for mesoderm induction in vivo by SRF remain to be identified.

The SRF-binding site, known as a CArG box or serum response element (SRE) (Treisman 1986; Norman et al. 1988; Pellegrini et al. 1995), is found in the control regions of numerous sarcomeric genes expressed in cardiac and skeletal muscle and is required for expression of nearly all smooth-muscle-specific genes. In contrast to growth-factor-inducible genes, like *c-fos*, which generally contain single CArG boxes, most muscle-specific genes contain multiple CArG boxes that act cooperatively to induce transcription.

In serum-stimulated cells, SRF associates with a family of Ets domain transcription factors, which serve as targets for MAP kinase signaling (for review, see Treisman 1994). This family of ternary complex factors (TCFs) includes the ubiquitously expressed proteins SAP-1, SAP-2/Net (Dalton and Treisman 1992; Giovane et al. 1994), and elk-1 (Hipskind et al. 1991). In addition to binding to the MADS domain of SRF, the TCF proteins can also directly bind to the Ets domain-binding consensus motif GGA(A/T).

The MADS box of SRF also recruits several homeodomain proteins. The paired-type protein Prx1 interacts with SRF and enhances SRF DNA binding (Grueneberg et al. 1992). Similarly, SRF associates with the Nkx2.5, resulting in synergistic activation of cardiac genes (Chen and Schwartz 1996; Sepulveda et al. 2002). The Barx2 homeodomain protein also stimulates SRF-DNA binding, but the physiological significance of this interaction has not been determined (Herring et al. 2001). Numerous other SRF accessory factors have been described (Treisman 1994).

MYOCARDIN

In an effort to discover novel regulators of cardiogenesis, we have taken a bioinformatics approach to identify genes expressed specifically in the cardiac lineage. One of the genes identified by this approach encodes a novel transcription factor called myocardin, which is expressed at a high level in the myocardium and is necessary and sufficient for activation of myocardial gene expression (Wang et al. 2001; for review, see Hauschka 2001).

Myocardin belongs to the SAP (SAF-A/B, Acinus, PIAS) domain family of transcription factors and stimulates SRF-dependent transcription by interacting with the SRF MADS box. The SAP domain is a conserved 35-amino-acid motif comprising two amphipathic α-helices that resemble helices 1 and 2 of the homeodomain (Aravind and Koonin 2000). SAP domains are found in a variety of nuclear proteins, including the nuclear matrix attachment factors SAF-A and -B (Gohring et al. 1997; Kipp et al. 2000); Acinus, a target for caspase cleavage that participates in chromatin degradation during apoptosis (Sahara et al. 1999); and PIAS (*P*rotein *I*nhibitor of *A*ctivated *STAT*), a transcriptional repressor that associates with a variety of transcription factors (Liu et al. 1998). The SAP domains of SAF-A and PIAS have been shown to mediate association with matrix attachment regions (MARs), which has been proposed to potentiate transcription by forming transcriptionally active domains of chromatin (Kipp et al. 2000; Sachdev et al. 2001). Myocardin also contains a basic region, glutamine (Q)-rich region, and a transactivation domain, which are not found in other SAP domain proteins (Fig. 2).

CARDIAC AND SMOOTH MUSCLE-SPECIFIC EXPRESSION OF MYOCARDIN

During mouse embryogenesis, myocardin is initially expressed in the cardiac crescent concomitant with the cardiac-restricted homeobox gene Nkx2.5 (Lints et al. 1993; Wang et al. 2001). Together, these genes represent the earliest known markers of the cardiac lineage. Myocardin expression is maintained in the atrial and ventricular chambers of the heart throughout embryogenesis to adulthood (Fig. 3). In addition, myocardin is expressed in a subset of vascular and visceral smooth muscle cell types (Wang et al. 2001). The timing of its expression in developing cardiac and smooth muscle cells suggests its involvement in myocyte differentiation. Intriguingly, however, myocardin is expressed in only a subset of smooth muscle cell types, consistent with the known diversity of smooth muscle cells (Li et al. 1996; Kim et al. 1997) and suggesting that there may be differences in the mechanisms for smooth muscle gene transcription or that other related factors are expressed in those smooth muscle types in which myocardin is not expressed. Myocardin is

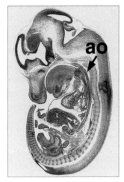 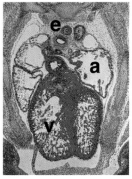

Figure 3. In situ hybridization of myocardin. Myocardin transcripts were detected by in situ hybridization to a sagittal section (*left*) and abdominal section (*right*) of E13.5 and 15.5 mouse embryos, respectively. (a) Atrium; (ao) aorta; (e) esophagus; (v) ventricle. Silver grains were pseudo-colored red using Adobe Photoshop and subsequently superimposed on identical brightfield images.

not expressed in skeletal muscle cells, despite the essential role of CArG boxes in expression of numerous skeletal muscle structural genes and the requirement of SRF for differentiation of skeletal muscle cells in culture (Vandromme et al. 1992; Croissant et al. 1996).

We have identified two myocardin-related transcription factors (MRTFs), -A and -B, which share homology with the SAP domain, as well as the basic and glutamine-rich domains, and the amino-terminal domain (NTD) of myocardin (Fig. 2). Unlike myocardin, MRTF-A and -B are widely expressed in embryonic and adult tissues. The human *MRTF-A* gene (also called *MAL* or *MKL*), which is located on chromosome 22, is translocated to chromosome 1 in acute megakaryocytic leukemia (AML) (Ma et al. 2001; Mercher et al. 2001). This translocation creates a fusion protein in which a novel protein called one-twenty-two (OTT)/RBM15 (RNA-binding motif protein-15) is fused to the amino terminus of MRTF-A. Little is known about the biological function of OTT, but based on our finding that MRTF-A is a transcriptional coactivator of SRF, it will be interesting to determine whether OTT-MAL induces aberrant growth via SRF.

ACTIVATION OF SRF-DEPENDENT TRANSCRIPTION BY MYOCARDIN

Myocardin potently activates transcription of reporter genes containing two or more CArG boxes. The promoters of the SM22 and smooth muscle α-actin genes, which

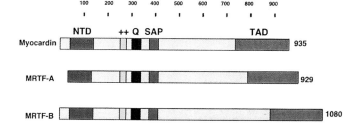

Figure 2. Myocardin and MRTFs. Schematic diagrams of myocardin, MRTF-A, and MRTF-B proteins. (++) Basic region; (NTD) amino-terminal domain; (Q) glutamine-rich domain; (TAD) transcription activation domain. The number of amino acids in each protein is shown to the right.

are transiently expressed in the early heart and are later specific to smooth muscle cells, are among the most sensitive target genes to myocardin identified to date (Fig. 4A) (D.Wang and E.N. Olson, unpubl.). The atrial natriuretic factor (ANF) gene, which is expressed in the embryonic heart and is induced in the adult myocardium in response to stress signals (Hines et al. 1999), is also activated by myocardin (Fig. 4A). All of these promoters contain a pair of CArG boxes that is essential for responsiveness to myocardin. In contrast, myocardin is unable to activate promoters, such as the c-fos promoter, which contain a single CArG box. These findings suggest that myocardin acts cooperatively. Recent studies suggest that this cooperativity is mediated by oligomerization of myocardin (Z. Wang and E.N. Olson, unpubl.).

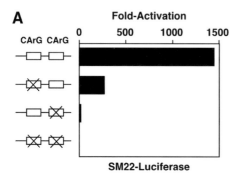

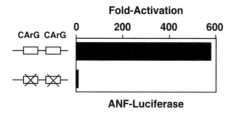

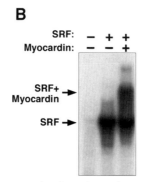

Figure 4. Ternary complex formation and transcriptional synergy by SRF and myocardin. (*A*) COS cells were transiently transfected with luciferase reporters linked to the SM22 or ANF promoters, as indicated. Values are expressed as relative luciferase activity compared to the maximal level. Transfections included 0.1 µg of myocardin. (*B*) Gel mobility shift assays were performed with a CArG box probe and in vitro translation products of SRF and myocardin, as indicated. Only the region of the gel containing the shifted probe is shown. Myocardin alone does not bind the probe, but in the presence of SRF, it gives rise to a ternary complex.

Transcriptional activation by myocardin requires a transcription activation domain (TAD) near the carboxyl terminus. This TAD is extremely powerful and is similar to the viral coactivator VP16 in its potency. The TAD of myocardin does not appear to convey transcriptional specificity to SRF, since it can be replaced with VP16. MRTF-A and -B also stimulate SRF activity, although MRTF-B is a relatively weak activator, due to its lower affinity for SRF.

Myocardin does not bind DNA alone, but forms a stable ternary complex with SRF bound to DNA (Fig. 4B). Mutational analysis of myocardin and SRF has shown that ternary complex formation is dependent on the basic and Q-rich domains of myocardin (Wang et al. 2001). Despite the amino acid homology between the MADS box regions of SRF and MEF2, myocardin does not interact with MEF2 factors. The specific amino acids within the SRF MADS box that are recognized by myocardin remain to be defined.

Myocardin can also bind MARs through its SAP domain (Z. Wang and E.N. Olson, unpubl.). Interestingly, a myocardin mutant lacking the SAP domain retains its ability to transactivate the SM22 promoter but is unable to activate the ANF promoter. These findings demonstrate that the SAP domain can discriminate between different SRF target genes, which may begin to explain the different expression patterns of SRF-dependent muscle genes. Whether this differential requirement of the SAP domain of myocardin for transcriptional specificity reflects a role of MAR binding or association of myocardin with other cofactors that regulate a subset of target genes remains to be determined.

The transcriptional cooperativity between SRF and myocardin is remarkably sensitive to the level of SRF expression (Wang et al. 2001). In transfected cells, elevation of SRF expression enhances the activity of myocardin over a very narrow range of SRF levels. However, exceeding this narrow range results in strong repression of myocardin activity by SRF. We presume this repressive effect is due to "squelching" in which excess SRF competes with DNA-bound SRF for binding to myocardin. Consistent with this notion, a mutant form of SRF that cannot bind DNA is as effective as wild-type SRF in repressing myocardin activity. Our results indicate that all of the transcriptional activity of myocardin in transfection assays with SRF-dependent reporter genes reflects the association of myocardin with endogenous SRF, since mutation of SRF-binding sites eliminates all myocardin activity. This conclusion is supported by the inability of myocardin to activate SRF-dependent reporters in SRF-null ES cells (D. Wang and E.N. Olson, unpubl.).

REGULATION OF CARDIAC GENE EXPRESSION IN FROG EMBRYOS BY MYOCARDIN

To assess the function of myocardin in vivo, we have injected mRNAs encoding wild-type and mutant forms of the protein into the ventral blastomeres of 8-cell *Xenopus* embryos and have assayed for cardiac gene expression in embryos at later stages. Injection of dominant negative forms of myocardin that can associate with SRF, but lack the TAD, eliminates cardiac gene expression at the tadpole stage (Fig. 5). Intriguingly, CArG box-dependent genes, such as α-cardiac actin, are expressed normally in skeletal muscle cells within the somites of injected embryos. Previous studies have shown that expression of α-cardiac actin in the somites requires the CArG boxes in the promoter of the gene (Taylor et al. 1989). Since the injected blastomeres contribute to the skeletal muscle lineage, it is unclear why dominant negative myocardin does not block actin gene expression in this lineage. One possibility is that other factors or signals in the skeletal muscle lineage override the inhibitory activity of the myocardin dominant negative mutant.

An obvious question is whether dominant negative myocardin eliminates the entire cardiac lineage or simply suppresses expression of cardiac genes in cells that have become committed to a cardiac cell fate. This question has been difficult to answer because we have not yet identified cardiac genes that are expressed in the presumptive heart-forming region of embryos injected with dominant negative myocardin. Even Nkx2.5 expression is downregulated in these embryos, which indicates that the mutant protein blocks a very early step in the cardiac pathway. Histological analysis of injected embryos has revealed disorganized cells in the presumptive heart-forming region, but their precise identity is currently unclear.

It will be particularly interesting to determine the phenotype of myocardin-null mice. Based on the phenotype of frog embryos expressing dominant negative myocardin, it might be predicted that such mutant mice would lack a heart. However, forced overexpression of a dominant negative mutant in a frog embryo need not phenocopy a loss-of-function mutation, particularly since MRTFs, which can mimic myocardin in transfection assays, are inactivated by dominant negative myocardin, but not by targeted inactivation of myocardin. Thus far, no mouse mutants in any cardiogenic regulatory gene have been found to prevent specification of the cardiac lineage, which suggests that this process is controlled by redundant regulators (Srivastava and Olson 2000).

In preliminary experiments, we have also examined whether forced expression of myocardin can induce ectopic cardiac gene expression in *Xenopus* embryos and isolated animal caps. In both assays, we have observed the misexpression of cardiac sarcomeric genes in the presence of ectopic myocardin. Injected embryos are highly sensitive to the level of myocardin expression; injection of wild-type transcripts in amounts equal to the amount of dominant negative transcripts that prevents heart formation results in severe developmental abnormalities. We presume these general abnormalities arise from the high potency of the myocardin TAD, which may interfere nonspecifically with general transcription. Although forced expression of myocardin can activate ectopic cardiac gene expression in frog embryos and animal caps, myocardin does not seem to act like a cardiac MyoD, in that it cannot induce ectopic expression of cardiac genes in transfection assays in cultured cells. Its ability to do so in injected frog embryos suggests that other factors provided by the embryo cooperate with myocardin in activating the cardiac gene program. Importantly, forced expression of myocardin does not induce the expression of skeletal muscle genes that are dependent on SRF. This specificity is consistent with the failure of dominant negative myocardin to affect skeletal muscle development in injected frog embryos. Whether myocardin also induces ectopic smooth muscle gene expression remains to be determined.

FUTURE QUESTIONS

In the future, it will be especially interesting to determine whether myocardin has partners in addition to SRF and to determine whether myocardin antagonizes cell proliferation by interfering with the interaction of TCFs with SRF. The roles of MRTF-A and -B in vivo also remain to be defined. Considering their widespread expression, it will be of interest to determine whether they modulate SRF activity in non-muscle cell types. How extracellular stimuli might regulate the functions of the myocardin family of transcription factors also warrants investigation.

INHIBITION OF SRF ACTIVITY BY HOP

The activity of SRF is also modulated by inhibitory factors. One such inhibitory factor is a novel homeodomain protein called homeodomain-only protein (HOP) that we identified by searching genomic databases with a homeodomain consensus sequence. We initially called this protein Cameo (Cardiac homeodomain), and Jon Epstein's lab, which identified it independently, called it Toto (Chen et al. 2002; Shin et al. 2002).

HOP contains only 73 amino acids, and its only recognizable protein motif is a 60-amino-acid homeodomain.

Dominant Negative Myocardin Blocks Cardiac Gene Expression

 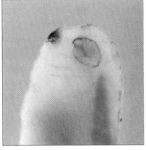

Control / Dominant Negative Mutant

Figure 5. Inhibition of cardiac gene expression in injected *Xenopus* embryos by myocardin. *Xenopus* embryos at the 8-cell stage were injected with transcripts encoding a myocardin deletion mutant lacking the TAD. At stage 28, α-cardiac actin transcripts were detected by whole-mount in situ hybridization. Expression of α-cardiac actin in the somites is unaffected, but expression in the heart is eliminated by dominant negative myocardin.

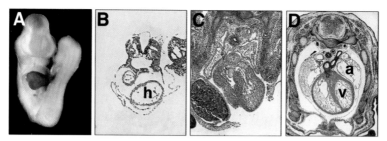

Figure 6. Expression of HOP in the developing heart. HOP transcripts were detected by whole-mount in situ hybridization at (*A*) E11.5, or in situ hybridization to sections of mouse embryos at (*B*) E8.5, (*C*) E12.5, and (*D*) E14.5. (a) Atrium; (h) heart; (v) ventricle. Silver grains in *B–D* were pseudo-colored red using Adobe Photoshop and subsequently superimposed on identical bright-field images.

The HOP homeodomain contains numerous residues that are conserved throughout the homeodomain superfamily, but it also contains divergent residues at several positions that are essential for DNA binding. Accordingly, HOP is unable to bind DNA, although the protein is localized predominantly to the nucleus.

During mouse embryogenesis, HOP is expressed in the lateral regions of the cardiac crescent at E7.75 and throughout the heart thereafter (Fig. 6). After E11.5, HOP expression is highest in the trabecular zone, where proliferation is diminished relative to the adjacent compact zone in the ventricular wall. In addition to its cardiac expression, HOP is expressed in the branchial arches, as well as the lung, liver, and neural tube. Cardiac expression of HOP is dramatically down-regulated in Nkx2.5 null embryos, suggesting that HOP acts downstream from Nkx2.5 in a cascade from cardiogenic transcription factors.

CARDIAC ABNORMALITIES IN HOP MUTANT MICE

Targeted inactivation of the HOP gene in mice results in cardiac abnormalities, characterized by an excess or deficiency of cardiomyocytes (Chen et al. 2002; Shin et al. 2002). A minor subset of homozygous mutants in an isogenic 129Sv background dies at E11.5 from apparent cardiac insufficiency. The hearts of these mutant embryos show a thin-walled myocardium and a lack of trabeculation, suggesting that cardiomyocyte growth is impaired, with resulting lethality due to an inability to sustain hemodynamic load during the rapid phase of embryonic growth.

The majority of HOP mutants survive to adulthood. Hearts from these animals show an ~30% excess of cardiomyocytes, which can be attributed to delayed exit from the cell cycle during the first few days after birth. These mutant hearts show increased ventricular wall thickness without evidence of myocyte hypertrophy at young age. However, at advanced ages (>6 months) a subset of these mice develops severe cardiomyopathy with myocyte hypertrophy and ventricular fibrosis.

The cardiomyocyte hyperplasia during the perinatal period and eventual hypertrophy at adulthood seen in HOP mutant mice are remarkably similar to the phenotypes of mice that overexpress telomerase in the heart (Oh et al. 2001). Whether HOP might act to repress telomerase expression or activity, such that in the absence of HOP, telomerase function is enhanced, remains to be determined.

ACTIVATION OF SRF-DEPENDENT GENE EXPRESSION IN HOP MUTANT HEARTS

In an effort to determine the molecular basis for the cardiac abnormalities in HOP mutant mice, we performed microarray analysis to compare the gene expression profiles of wild-type and mutant hearts at postnatal day 1 (P1). These analyses revealed the up-regulation of numerous genes associated with cell proliferation and a cardiac stress response. In addition, several smooth muscle structural genes were expressed at elevated levels in the mutant hearts. A remarkable number of the genes that were up-regulated in HOP mutant hearts were known targets for SRF, which prompted us to explore the possibility that HOP might act as an antagonist of SRF activity in vivo, such that in its absence, SRF activity would be enhanced.

ANTAGONISM OF SRF ACTIVITY BY HOP

As shown in Figure 7A, the DNA-binding activity of SRF is diminished in the presence of increasing amounts of HOP. This inhibitory activity of HOP is mediated by its direct interaction with the MADS box of SRF. As expected from the ability of HOP to interfere with SRF DNA binding, the ability of SRF to transactivate a variety of SRF-dependent promoters is reduced in the presence of HOP in a dose-dependent manner (Fig. 7B). HOP also interferes with the cooperativity between myocardin and SRF. We presume this reflects the inhibition of SRF DNA binding by HOP, but it is also possible that HOP competes with myocardin for interaction with SRF, since both proteins associate with the MADS box of SRF.

Although HOP can diminish SRF activity in transfection and DNA-binding assays, it does not completely abolish SRF activity. Typically, we observe up to about 50% inhibition of SRF activity in the presence of HOP, and further increasing the relative ratio of HOP to SRF does not result in further inhibition. The partial inhibitory effect of HOP on SRF may account for the partially pen-

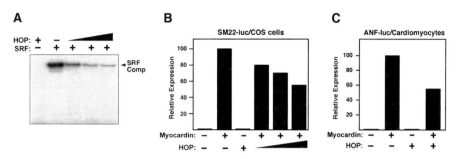

Figure 7. Inhibition of SRF DNA binding and transcriptional activity by HOP. (*A*) SRF (0.2 μg of cDNA) and increasing amounts of HOP (0.2, 0.4, and 0.6 μg of cDNA) were translated in vitro and used in gel mobility shift assays with a labeled CArG box probe. HOP interferes with SRF DNA binding. COS cells (*B*) or rat cardiomyocytes (*C*) were transiently transfected with luciferase reporters linked to the SM22 or ANF promoters, as indicated. Values are expressed as relative luciferase activity compared to the maximal level. Transfections included 0.2 μg of myocardin expression vector and 0.1, 0.4, and 0.8 μg of HOP (*B*) or 0.5 μg of HOP (*C*).

etrant phenotypes of HOP mutant mice, and suggests that HOP subtly alters SRF activity but does not have an all-or-none effect on SRF functions.

The lack of DNA-binding activity of HOP and its ability to interfere with DNA binding of SRF is similar to the properties of I-POU, a member of the POU-homeodomain family that cannot bind DNA and inhibits the activity of other POU-domain transcription factors by forming inactive heterodimeric complexes (Treacy et al. 1991). The HLH protein Id, which lacks a basic region and forms inactive heterodimers with bHLH proteins, acts in an analogous manner (Benezra et al. 1990).

ANTITHETICAL PHENOTYPES OF CARDIOMYOCYTES IN HOP MUTANT MICE

The segregation of HOP mutant mice into two classes characterized by an excess or deficiency of cardiac myocytes seems paradoxical. How can these findings be explained? One explanation is that they reflect the dual role of SRF as a regulator of cell proliferation and differentiation and the dysregulation of SRF activity in the absence of HOP. Prior to E11.5, when the heart must rapidly grow to meet hemodynamic demands of the embryo, SRF might be required for cardiomyocyte proliferation. In this setting, HOP is expressed at highest levels in the trabecular zone, which is highly proliferative, and may antagonize the differentiation-promoting function of SRF, thereby favoring cell proliferation. Conversely, later during perinatal development, SRF may be required primarily for myofibrillogenesis and terminal differentiation. In this case, the lack of HOP could tip the activity of SRF toward a cell proliferation function with a resulting increase in number of cardiomyocytes. The ultimate phenotype at each of these time points in development would be influenced by the spectrum of other SRF cofactors, as well as by growth signals.

With regard to the hypertrophic phenotype of HOP mutants at adulthood, previous studies have reported that overexpression of SRF in the heart results in hypertrophy (Zhang et al. 2001). SRF has also been shown to be a target for stress signaling to fetal cardiac genes that are induced in the adult myocardium in response to hypertrophic signals (Paradis et al. 1996).

Although the opposing phenotypes of HOP mutant mice can be explained by the antithetical roles of SRF in the control of cell proliferation and myocyte differentiation, our results do not unequivocally prove that the excess and deficiency of cardiomyocytes in the mutants is solely due to the release of SRF from the inhibitory influence of HOP. It remains possible that HOP has additional partners in the cardiac lineage that may be dysregulated in HOP mutant mice. In this regard, homeodomain proteins have been shown to associate with GATA and bHLH transcription factors (Lee et al. 1998; Sun et al. 2001; Zhou et al. 2001), which play key roles in cardiac myogenesis and morphogenesis (Laverierre et al. 1994; Srivastava et al. 1995; Sepulveda et al. 2002). Thus, the absence of HOP may alter the activities of these families of transcription factors in the developing heart, with consequent abnormalities in growth and differentiation of cardiomyocytes. It is also formally possible that HOP has transcription-activating functions, in addition to its repressive effects on SRF. Of note, HOP displays a high degree of transcriptional activity in yeast when fused to the GAL4 DNA-binding domain (R. Passier and E. Olson, unpubl.).

A POTENTIAL FEEDBACK LOOP BETWEEN Nkx2.5 AND HOP

The coexpression of Nkx2.5 and HOP in the cardiac crescent beginning at E7.75, and the loss of HOP expression in Nkx2.5 mutant embryos, suggest that HOP is regulated, either directly or indirectly, by Nkx2.5. In this regard, we have identified a 1.2-kb cardiac enhancer upstream of the HOP gene that contains multiple binding sites for Nkx2.5 and is activated by Nkx2.5 in transfection assays (C. Shin, Z.-P. Liu, and E. Olson, unpubl.).

Since Nkx2.5 cooperates with SRF to activate cardiac gene expression, and HOP interferes with SRF activity, HOP can be viewed as a component of a negative feedback loop for modulation of Nkx2.5 activity. Such a mechanism raises the possibility that some abnormalities in HOP mutant mice may reflect enhanced activity of

Nkx2.5. Previous studies have shown that overexpression of Nkx2.5 in mouse, frog, or zebrafish embryos results in cardiac enlargement (Chen and Fishman 1996; Cleaver et al. 1996; Kasahara et al. 2001). Haploinsufficiency or missense mutations of Nkx2.5 in mice and humans also result in structural abnormalities in the heart that include atrial-septal, ventricular-septal, and outflow tract defects, as well as atrioventricular conduction defects (Schott et al. 1998; Benson et al. 1999; Biben et al. 2000). Thus, HOP is likely to influence cardiac development via its indirect effects on Nkx2.5/SRF synergy.

ISSUES FOR THE FUTURE

A simplified schematic diagram of the antagonistic roles of myocardin and HOP in modulating SRF activity in the cardiac lineage is shown in Figure 8. The regulation of SRF activity is extraordinarily complex and involves dynamic interactions with numerous additional cofactors that are influenced by intracellular signals (for review, see Reecy et al. 1998). The phenotypes of HOP mutant mice exemplify the complexity of SRF regulation, with elimination of a single negative regulator of SRF resulting in opposing phenotypes. Given the multitude of signals that influence muscle cell growth, differentiation, and remodeling, the apparent complexity in regulation of gene expression during these processes is perhaps not surprising. Many of the key regulators of these processes are now known, but much remains to be learned about the mechanisms whereby these factors cooperate to coordinately regulate the spectrum of genes expressed by a muscle cell at a specific time and place. Modulating SRF activity through its association with positive and negative cofactors like myocardin and HOP is likely to have important roles in congenital and acquired cardiac disease.

ACKNOWLEDGMENTS

Work in E. Olson's laboratory has been supported by grants from the National Institutes of Health, The Donald W. Reynolds Cardiovascular Clinical Research Center, The Muscular Dystrophy Association, The McGowan Charitable Fund, and The Robert A. Welch Foundation.

REFERENCES

Aravind L. and Koonin E.V. 2000. SAP–a putative DNA-binding motif involved in chromosomal organization. *Trends Biochem. Sci.* **25:** 112.

Arsenian S., Weinhold B., Oelgeschlager M., Ruther U., and Nordheim A. 1998. Serum response factor is essential for mesoderm formation during mouse embryogenesis. *EMBO J.* **17:** 6289.

Benson D.W., Silberbach G.M., Kavanaugh-McHugh A., Cottrill C., Zhang Y., Riggs S., Smalls O., Johnson M.C., Watson M.S., Seidman J.G., Seidman C.E., Plowden J., and Kugler J.D. 1999. Mutations in the cardiac transcription factor NKX2.5 affect diverse cardiac developmental pathways. *J. Clin. Invest.* **104:** 1567.

Benezra R., Davis R., Lockshon D., Turner L., and Weintraub H. 1990. The protein Id: A negative regulator of helix-loop-helix DNA binding proteins. *Cell* **61:** 49.

Biben C., Weber R., Kesteven S., Stanley E., McDonald L., Elliott D.A., Barnett L., Koentgen F., Robb L., Feneley M., and Harvey R.P. 2000. Cardiac septal and valvular dysmorphogenesis in mice heterozygous for mutations in the homeobox gene Nkx2-5. *Circ. Res.* **87:** 888.

Black B.L. and Olson E.N. 1998. Transcriptional control of muscle development by myocyte enhancer factor-2 (MEF2) proteins. *Annu. Rev. Cell Dev. Biol.* **14:** 167.

Chen C.Y. and Schwartz R.J. 1996. Recruitment of the tinman homolog Nkx-2.5 by serum response factor activates cardiac α-actin gene transcription. *Mol. Cell. Biol.* **16:** 6372.

Chen F., Kook H., Milewski R., Gitler A.D., Lu M.M., Nazarian R., Schnepp R., Jen K., Liang H., Li J., Biben C., Runke G., Mackay J., Novotny J., Harvey R.P., Mullins M., and Epstein J.A. 2002. *Hop* is a novel type of homeobox gene that modulates cardiac development. *Cell* **110:** 713.

Chen J. and Fishman M. 1996. Zebrafish tinman homolog demarcates the heart field and initiates myocardial differentiation. *Development* **122:** 3809.

Cleaver O., Patterson K., and Krieg P. 1996. Overexpression of the tinman-related genes XNkx2.5 and XNkx2.3 in *Xenopus* embryos results in myocardial hyperplasia. *Development* **122:** 3549.

Croissant J., Kim J., Eichele G., Goering L., Lough J., Prywes R., and Schwartz R. 1996. Avian serum response factor expression restricted primarily to muscle cell lineages is required for α-actin gene transcription. *Dev. Biol.* **177:** 250.

Dalton S. and Treisman R. 1992. Characterization of SAP-1, a protein recruited by serum response factor to the c-fos serum response element. *Cell* **68:** 597.

Gauthier-Rouviere C., Basset M., Blanchard J., Cavadore J., Fernandez A., and Lamb N. 1991. Casein kinase II induces c-fos expression via the serum response factor. *Science* **257:** 1089.

Giovane A., Pintzas A., Maira S.M., Sobieszczuk P., and Wasylyk B. 1994. Net, a new ets transcription factor that is activated by Ras. *Genes Dev.* **8:** 1502.

Gohring F., Schwab B.L., Nicotera P., Leist M., and Fackelmayer F.O. 1997. A novel SAR binding domain of scaffold attachment factor A (SAF-A) is a target in apoptotic nuclear breakdown. *EMBO J.* **16:** 7361.

Grueneberg D.A., Natesan S., Alexandre C., and Gilman M.Z. 1992. Human and *Drosophila* homeodomain proteins that enhance the DNA-binding activity of serum response factor. *Science* **275:** 1089.

Hauschka S. 2001. Myocardin, a novel potentiator of SRF-mediated gene transcription in cardiac muscle. *Mol. Cell* **8:** 1.

Herring B.P., Kriegel A.M., and Hoggatt A.M. 2001. Identification of Barx2b, a serum response factor-associated homeodomain protein. *J. Biol. Chem.* **276:** 14482.

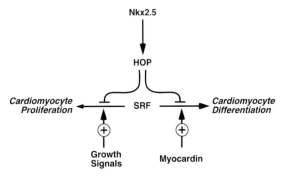

Figure 8. Model of the role of myocardin and HOP in the control of SRF-dependent transcription. SRF influences the balance between cardiomyocyte proliferation and differentiation. The activity of SRF is influenced by extracellular signaling and positive and negative cofactors. Overexpression of myocardin can tip the balance of SRF activity toward myocyte differentiation. Removal of HOP can stimulate or inhibit proliferation, depending on the stage of development and other influences on SRF activity.

Hines W.A., Thorburn J., and Thorburn A. 1999. A low-affinity serum response element allows other transcription factors to activate inducible gene expression in cardiac myocytes. *Mol. Cell. Biol.* **19:** 1841.

Hipkind R.A., Rao V.N., Mueller C.G., Reddy E.S., and Nordheim A. 1991. Ets-related protein Elk-1 is homologous to the c-fos regulatory factor p62TCF. *Nature* **354:** 531.

Kasahara H., Wakimoto H., Liu M., Maguire C.T., Converso K.L., Shioi T., Huang W.Y., Manning W.J., Paul D., Lawitts J., Berul C.I., and Izumo S. 2001. Progressive atrioventricular conduction defects and heart failure in mice expressing a mutant Csx/Nkx2.5 homeoprotein. *J. Clin. Invest.* **108:** 189.

Kim S., Ip H.S., Lu M.M., Clendenin C., and Parmacek M.S. 1997. A serum response factor-dependent transcriptional regulatory program identifies distinct smooth muscle cell sublineages. *Mol. Cell. Biol.* **17:** 2266.

Kipp M., Gohring F., Ostendorp T., Van Drunen C.M., Van Driel R., Przbylski M., and Fackelmayer F.O. 2000. SAF-box, a conserved protein domain that specifically recognizes scaffold attachment region DNA. *Mol. Cell. Biol.* **20:** 7480.

Laverriere A., MacNeill C., Mueller C., Poelmann R.E., Burch J.B., and Evans T. 1994. GATA-4/5/6, a subfamily of three transcription factors transcribed in developing heart and gut. *J. Biol. Chem.* **269:** 23177.

Lee Y., Shioi T., Kasahara H., Jobe S., Wiese R., Markham B., and Izumo, S. 1998. The cardiac tissue-restricted homeobox protein Csx/Nkx2.5 physically associates with the zinc finger protein GATA4 and cooperatively activates atrial natriuretic factor gene expression. *Mol. Cell. Biol.* **18:** 3120.

Li L., Miano J.M., Cserjesi P., and Olson E.N. 1996. SM22α, a marker of adult smooth muscle, is expressed in multiple myogenic lineages during embryogenesis. *Circ. Res.* **78:** 188.

Lints T., Parsons L., Hartley L., Lyons I., and Harvey R. 1993. Nkx2.5: A novel murine homeobox gene expressed in early heart progenitor cells and their myogenic descendants. *Development* **119:** 969.

Liu B., Liao J., Rao X., Kushner S.A., Chung C.D., Chang D.D., and Shuai K. 1998. Inhibition of Stat1-mediated gene activation by PIAS1. *Proc. Natl. Acad. Sci.* **95:** 10626.

Ma Z., Morris S.W., Valentine V., Li M., Herbrick J.A., Cui X., Bouman D., Li Y., Mehta P.K., Nizetic D., Kaneko Y., Chan G.C., Chan L.C., Squire J., Scherer S.W., and Hitzler J.K. 2001. Fusion of two novel genes, RBM15 and MKL1 in the t(1;22) (p13;q13) of acute megakaryoblastic leukemia. *Nat. Genet.* **28:** 220.

McKinsey T.A., Zhang C.L., and Olson E.N. 2002. MEF2: A calcium-dependent regulator of cell division, differentiation and death. *Trends Biochem. Sci.* **27:** 40.

Mercher T., Coniat M.B., Monni R., Mauchauffe M., Khac F.N., Gressin L., Mugneret F., Leblanc T., Dastugue N., Berger R., and Bernard O.A. 2001. Involvement of a human gene related to the *Drosophila* spen gene in the recurrent t(1;22) translocation of acute megakaryocytic leukemia. *Proc. Natl. Acad. Sci.* **98:** 5776.

Norman C., Runswick M., Pollock R., and Treisman R. 1988. Isolation of properties of cDNA clones encoding SRF, a transcription factor that binds to the c-fos serum response element. *Cell* **55:** 989.

Oh H., Taffet G.E., Youker K.A., Entman M.L., Overbeek P.H., Michael L.H., and Schneider M.D. 2001. Telomerase reverse transcriptase promotes cardiac muscle cell proliferation, hypertrophy, and survival. *Proc. Natl. Acad. Sci.* **98:** 10308.

Paradis P., MacLellan W.R., Belaguli N.S., Schwartz R.J., and Schneider M.D. 1996. Serum response factor mediates AP-1 dependent induction of the skeletal alpha-actin promoter in ventricular myocytes. *J. Biol. Chem.* **271:** 10827.

Pellegrini L., Tan S., and Richmond T.J. 1995. Structure of serum response factor core bound to DNA. *Nature* **376:** 490.

Poser S., Impey S., Trinh K., Xia Z., and Storm D. 2000. SRF-dependent gene expression is required for PI3-kinase-regulated cell proliferation. *EMBO J.* **19:** 4955.

Reecy J., Belaguli N., and Schwartz R. 1998. SRF/homeobox protein interactions. In *Heart development* (ed. R. Harvey and N. Rosenthal), p. 273. Academic Press, San Diego, California.

Sachdev S., Bruhn L., Sieber H., Pichler A., Melchior F., and Grosschedl R. 2001. PIASy, a nuclear matrix-associated SUMO E3 ligase, represses LEF1 activity by sequestration into nuclear bodies. *Genes Dev.* **15:** 3088.

Sahara S., Aoto M., Eguchi Y., Imamoto N., Yoneda Y., and Tsujimoto Y. 1999. Acinus is a caspase-3-activated protein required for apoptotic chromatin condensation. *Nature* **401:** 168.

Santelli E. and Richmond T.J. 2000. Crystal structure of MEF2A core bound to DNA at 1.5 Å resolution. *J. Mol. Biol.* **297:** 437.

Schott J.J., Benson D.W., Basson C.T., Pease W., Silberbach G.M., Moak J.P., Maron B.J., Seidman C.E., and Seidman J.G. 1998. Congenital heart disease caused by mutations in the transcription factor NKX2-5. *Science* **281:** 108.

Schratt G., Weinhold B., Lundberg A.S., Schuck S., Berger J., Schwartz H., Weinberg R.A., Ruther U., and Nordheim A. 2001. Serum response factor is required for immediate early gene activation yet is dispensable for proliferation of embryonic stem cells. *Mol. Cell. Biol.* **21:** 2933.

Sepulveda J.L., Vlahopoulos S., Iyer D., Belaguli N., and Schwartz R.J. 2002. Combinatorial expression of GATA4, Nkx2-5 and serum response factor directs early cardiac gene activity. *J. Biol. Chem.* **30:** 30.

Shin C., Liu Z.-P., Passier R., Zhang C.L., Wang D., Harris T.M., Yamagishi H., Richardson J.A., Childs G., and Olson E.N. 2002. Modulation of cardiac growth and development by HOP, an unusual homeodomain protein. *Cell* **110:** 725.

Shore P. and Sharrocks A.D. 1995. The MADS-box family of transcription factors. *Eur. J. Biochem.* **229:** 1.

Soulez M., Rouviere C., Chafey P., Hentzen D., Vandromme M., Lautredou N., Lamb N., Kahn A., and Tuil D. 1996. Growth and differentiation of C2 myogenic cells are dependent on serum response factor. *Mol. Cell. Biol.* **16:** 6065.

Srivastava D. and Olson E.N. 2000. A genetic blueprint for cardiac development. *Nature* **407:** 226.

Srivastava D., Cserjesi P., and Olson E.N. 1995. A new subclass of bHLH proteins required for cardiac morphogenesis. *Science* **270:** 1995.

Sun T., Echelard Y., Lu R., Yuk D., Kaing S., Stiles C., and Rowitch D. 2001. Oligo bHLH proteins interact with homeodomain proteins to regulate cell fate acquisition in progenitors of the ventral neural tube. *Curr. Biol.* **11:** 1413.

Taylor M., Treisman R., Garrett N., and Mohun T. 1989. Muscle-specific (CArG) and serum-responsive (SRE) promoter elements are functionally interchangeable in *Xenopus* embryos and mouse fibroblasts. *Development* **106:** 67.

Treacy M.N., He X., and Rosenfeld M.G. 1991. I-POU: A POU-domain protein that inhibits neuron-specific gene activation. *Nature* **350:** 577.

Treisman R. 1986. Identification of a protein-binding site that mediates transcriptional response of the c-fos gene to serum factors. *Cell* **46:** 567.

———. 1994. Ternary complex factors: Growth factor regulated transcriptional activators. *Curr. Opin. Genet. Dev.* **4:** 96.

Vandromme M., Gauthier-Rouviere C., Carnac G., Lamb N., and Fernandez A. 1992. Serum response factor p67 SRF is expressed and required during myogenic differentiation of both mouse C2 and rat L6 muscle cell lines. *J. Cell Biol.* **118:** 1489.

Wang D., Chang P.S., Wang Z., Sutherland L., Richardson J.A., Small E., Krieg P.A., and Olson E.N. 2001. Activation of cardiac gene expression by myocardin, a transcriptional cofactor for serum response factor. *Cell* **105:** 851.

Weinhold B., Schratt G., Arsenian S., Berger J., Kamino K., Schwartz H., Ruther U., and Nordheim A. 2000. Srf-/- ES cells display non-cell autonomous impairment in mesodermal differentiation. *EMBO J.* **19:** 5835.

Zhang X., Azhar G., Chai J., Sheridan P., Nagano K., Brown T., Yang J., Khrapko K., Borras A.M., Lawitts J., Misra R.P., and Wei J.Y. 2001. Cardiomyopathy in transgenic mice with cardiac-specific overexpression of serum response factor. *Am. J. Physiol. Heart Circ. Physiol.* **280:** H1782.

Zhou Q., Choi G., and Anderson D. 2001. The bHLH transcription factor Olig2 promotes oligodendrocyte differentiation in collaboration with Nkx2.2. *Neuron* **31:** 791.

Homeodomain Factor Nkx2-5 in Heart Development and Disease

R.P. Harvey,*† D. Lai,* D. Elliott,* C. Biben,* M. Solloway,* O. Prall,*
F. Stennard,* A. Schindeler,* N. Groves,* L. Lavulo,* C. Hyun,* T. Yeoh,*
M. Costa,* M. Furtado,* and E. Kirk*‡

*Victor Chang Cardiac Research Institute, Darlinghurst 2010, New South Wales, Australia; †Faculties of Medicine and Life Science, University of New South Wales, Kensington 2051, New South Wales, Australia; ‡Sydney Children's Hosptial, Randwick 2031, New South Wales, Australia

The mammalian heart is a highly modified muscular vessel (Fig.1). During embryonic development, it is the first organ to form, and contraction begins as soon as a primitive tubular structure and vasculature are in evidence. The subsequent morphogenetic progressions that give rise to the four-chambered organ are the product of both embryonic and local patterning systems (Harvey 2002b) and a feed-forward circuitry that links form with function (Sedmera et al. 1999). The result is a structure in which flow into and between chambers occurs in an elegant and efficient sling-like fashion (Kilner et al. 2000).

The genetic circuitry that guides heart development and function is being dissected with increased vigor (Cripps and Olson 2002; Harvey 2002b; Zaffran and Frasch 2002). A detailed description of these processes is central to our understanding of congenital heart disease, normal adaptive mechanisms such as myocyte hypertrophy, as well as pathological hypertrophy and heart failure. The recent suggestion that the mammalian heart has capacity for regeneration (Orlic et al. 2001; Hughes 2002) opens many new issues in developmental biology, as well as great clinical potential.

Key steps in the elucidation of the genetic basis for heart development are (1) the identification of the developmental building blocks and (2) the isolation and characterization of the important regulatory and signaling molecules. In this paper, we briefly review the morphogenesis of the heart and discuss findings from the perspective of our analyses of the cardiac homeodomain factor Nkx2-5.

OVERVIEW OF HEART MORPHOGENESIS

During gastrulation, heart progenitor cells move through the node and organizer region, and primitive streak (Kinder et al. 2001), and take a lateral migratory path toward the anterior and anterior-lateral parts of the embryo (Tam and Schoenwolf 1999). Here, they form the epithelial cardiac crescent (Fig. 2a,b). Progenitors then begin to express cardiac transcription factors such as GATA4/5/6, Nkx2-5, Mef2B/C, and Tbx5/20, and a primitive myogenic program is activated (Harvey 2002a). After formation of the intraembryonic coelom, which separates heart progenitors from dorsal mesocardial and pericardial tissues (Fig. 2b), heart progenitors move ventrally and fuse to form the linear heart tube (Fig. 2c,d). Migration occurs in concert with ingression of the foregut pocket and requires correct differentiation of foregut endoderm (Narita et al. 1997). The linear heart tube is a transient structure which comprises an inner endocardial layer shrouded by a myogenic layer that beats in a peristaltic fashion (Moorman et al. 2000). The heart tube then rapidly elongates and begins the process of looping morphogenesis, adopting a spiral conformation with the ventricular and outflow regions bending rightward before joining the aortic sac at the midline (Fig. 2e,f). Some aspects of heart form, including the direction of ventricular bending and the morphological identity of the atrial appendages, are guided by the left/right embryonic axial system (Hamada et al. 2002). During looping, primitive ventricular chambers become evident at the outer curvature (see below), and the internal relief of the heart undergoes considerable remodeling (Mjaatvedt et al. 1999). In the atrioventricular canal, opposing tissue masses called endocardial cushions are laid down (Fig. 2f,h). These are formed from local endocardial cells that migrate into the extracellular matrix located between the endocardium and myocardium (the cardiac jelly), where they proliferate (Mjaatvedt et al. 1999). The cushions form the precursors of the mitral and tricuspid valves and become muscularized to varying degrees in different species (van den Hoff et al. 2001). Endocardial cushions are also prominent in the outflow region of the heart, and these give rise to the aortic and pulmonary valves, as well as to the aorticopulmonary septum. Further remodeling of the heart leads to septation of the atrial and ventricular chambers, formation of valve leaflets, and the elaboration of a myogenic trabecular network on the inner surface of the ventricles and, to a lesser extent, the atria (Fig. 2g,h). During looping, the left and right atria move dorsal and anterior, such that they are positioned in register with their cognate ventricles. On the ventral side, the outflow region becomes positioned between the left and right atria and ventricles. With this arrangement, the embryonic heart has achieved a basic form that resembles that of the adult heart (Figs.1 and 2g,h).

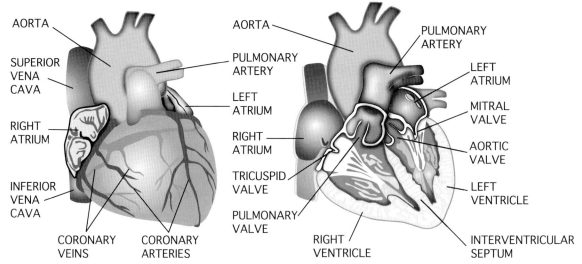

Figure 1. Diagram showing external (*left*) and internal (*right*) anatomical features of the mature human heart. (Adapted from Harvey 2002b.)

THE SECONDARY HEART FIELD AND EXTRA-CARDIAC LINEAGES

The complex form of the heart is generated through dynamic interactions between the myogenic and endocardial progenitors that arise in the primary heart field, the global embryonic patterning systems to which they are exposed, and the staged deployment of extracardiac lineages (Epstein 2001; Harvey 2002a; Reese et al. 2002). A recent finding is that heart progenitors are already compartmentalized into fields that are deployed at different times of development. The lateral cells of the cardiac crescent constitute the primary heart field, which forms the heart tube and its primitive ventricle and atrium. The more medial cells of the crescent express cardiac transcription factor genes such as *Gata4* and *Nkx2-5*, and inductive growth factor genes such as *Fgf8* and *Fgf10*, but do not contribute to the heart tube until about a day later (in the mouse), when they migrate into the cranial part of the heart tube to build the outflow tract and possibly right ventricle (Kelly et al. 2001; Mjaatvedt et al. 2001; Waldo et al. 2001). These so-called secondary (or anterior) heart field cells may be analogous to heart field cells in *Xenopus laevis* that play a regenerative role if definitive heart progenitors are removed (Raffin et al. 2000).

Other extracardiac lineages contribute to the heart. Cranial neural crest cells enter the cranial region of the heart and play a critical role in septation of the outflow tract (Epstein 2001). There is likely to be an active dialogue between the cardiac neural crest cells and those of the secondary heart field which form the outflow tract into which neural crest cells migrate. Another population of neural crest cells has been described which enter the caudal aspect of the chick heart to become transiently associated with elements of the conduction system (Poelmann and Gittenberger-de Groot 1999). Mesenchymal cells adjacent to the primitive atrium of the heart, which may also be part of the secondary heart field (Stanley et al. 2002), are drawn into the atria during its septation (Webb et al. 1998). Furthermore, the septum transversum gives rise to clusters of cells at the base of the ventricles called the pro-epicardial organ, and these structures seed the outer epicardial layer of the heart, which gives rise to the coronary circulation, its smooth muscle component, and interstitial fibroblasts.

BUILDING BLOCKS OF THE HEART

The heart is often considered to have a segmental origin, although there is little evidence for a true metameric structure. There are, however, several key processes that can be identified which contribute to the basic ground plan of the heart in the early looping stages. The first is the graded functional activity (beat rate) and graded expression of transcription factor genes that likely betray the graded activity of morphogens in early heart progenitors (Harvey 2002a). The second, mentioned above, is the delayed deployment of the secondary heart field cells to the forming heart. These cells contribute the outflow tract and quite possibly right ventricle (Kelly et al. 2001; Mjaatvedt et al. 2001; Waldo et al. 2001). Region-specific expression patterns of endogenous genes and numerous transgenes suggest that many cardiac genes will be regulated differentially in the primary and secondary heart field lineages. This "modularity" in cardiac gene regulation (Firulli and Olson 1997) may reflect the progressive addition of new anatomical modules or lineage specializations to the heart during the course of vertebrate evolution (Harvey 2002a).

Formation of the caudal (sinuatrial) aspect of the heart requires retinoic acid signaling. Inhibition of retinoid synthesis or receptor signaling leads to hearts that are severely hypoplastic in this area (Rosenthal and Xavier-Neto 2000; Niederreither et al. 2001). Furthermore, excess retinoic acid "atrializes" the ventricular region. It is likely that the caudal region of the heart progenitor zone is exposed to retinoids expressed in the node and primitive streak during gastrulation, and that the timing and

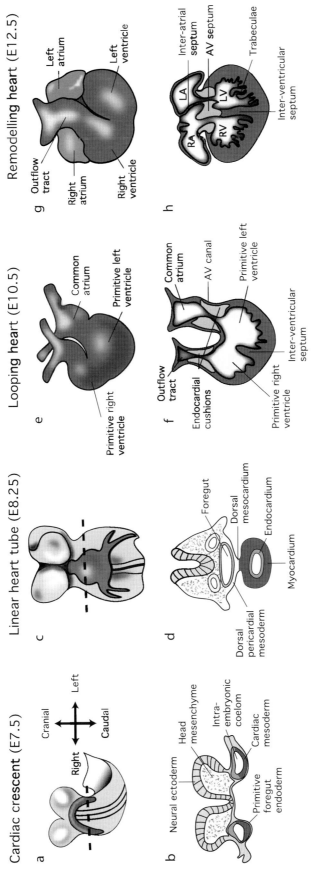

Figure 2. Diagramatic representation of the key stages in mammalian heart development. Figure shows whole embryos (*a,c*) or hearts (*e,g*) (*top* panels) and representative sections (*b,d,f,h*; *bottom* panels). Myocardium is dark-shaded. The dashed line in *a* and *c* represents the approximate plane of section in *b* and *d*, respectively. The compass in *a* represents the approximate axes of the embryos shown. (Adapted from Harvey 2002b.)

spatial dynamics of ingression of heart progenitors through the node and streak restrict the initial domain of retinoid influence to the caudal heart progenitors (Rosenthal and Xavier-Neto 2000). This important aspect of heart development likely occurs prior to heart tube formation and the initiation of myogenesis. Retinoids may also play other broader roles in heart development (Kastner et al. 1997).

A further aspect of early heart development relates to the formation of the muscles of the heart chambers, often called working myocardium. There is now evidence that chamber myocardium arises as a specialization of the more primitive muscle type that constitutes the primary heart tube (Harvey 1999; Christoffels et al. 2000). Expression patterns for genes encoding the hormone atrial natriuretic factor (ANF) and the cytoskeletal protein Chisel show that specification of chamber myocardium occurs in discrete zones at the outer curvature of the looping heart tube (Christoffels et al. 2000; Palmer et al. 2001). Whether these zones are specified by intrinsic patterning processes or by form/function relationships, or both, remains to be determined. The notion of zones of specialized working myocardium on the outer curvature of the looping heart tube (ballooning model of heart chamber formation) is generally incompatible with a segmental view of heart patterning and chamber formation (Christoffels et al. 2000). This radically different view of heart patterning is likely to help in interpretation of mutant phenotypes in mice and congenital heart defects in humans (Harvey 1999; Christoffels et al. 2000).

Nkx2-5 AND CARDIAC TRANSCRIPTION FACTOR PATHWAYS

Nkx2-5 (also called Csx) is a homeodomain transcription factor of the NK2-class that is expressed from the earliest times in the developing mammalian heart (Harvey 1996). The *Nkx2-5* gene was first cloned by virtue of its homology with the *Drosophila* gene *tinman*, required for specification of the heart-like vessel present in the fly larva (Komuro and Izumo 1993; Lints et al. 1993). The recognition that NK2 homeobox genes function in the hearts of both flies and man has led to a revision of the textbook view that vertebrate and invertebrate hearts arose through convergent evolution (Erwin and Davidson 2002). Considerable evidence now suggests that the two structures share a homologous core genetic pathway that utilizes NK2-class homeobox genes and many other cardiac transcription factors (Cripps and Olson 2002; Harvey 2002a; Zaffran and Frasch 2002). However, there is still debate as to whether this conserved pathway represents an ancient pathway for a cardiac-like myogenic program, or whether it incorporates aspects of heart morphogenesis (Erwin and Davidson 2002). What is clear, however, is that for each species, transcription factors acting in cardiomyogenic program have been co-opted for the control of morphogenesis (Harvey 2002a). Trans-species gene-rescue experiments failed to prove strict homology between Nkx2-5 and tinman with respect to their cardiogenic functions (Park et al. 1998; Ranganayakulu et al. 1998). However, we now know that the domain structure of other invertebrate NK2 proteins conforms very closely to their mammalian counterparts and therefore that the structure and function of tinman in insects has diverged (R.P. Harvey et al., unpubl.).

Both *tinman* and *Nkx2-5* sit high in the genetic hierarchies that govern heart development in their respective species (Lyons et al. 1995; Harvey 1996). Nkx2-5 is expressed from early cardiac crescent stages in a zone that encompasses the precursors of the ventricles and, later, the atria (Lints et al. 1993; Redkar et al. 2001). In situ hybridization and lineage studies utilizing Cre recombinase suggest that *Nkx2-5* is also expressed in secondary heart field cells and mesenchymal cells at the caudal aspect of the heart that are drawn into the atria during its septation (Stanley et al. 2002). Nkx2-5 directly interacts and synergizes with several other cardiac-expressed transcription factors and guides positive regulation of some cardiac genes. Nkx2-5-interacting factors include SRF, GATA4, and Tbx5 (Chen and Schwartz 1996; Lee et al. 1998; Hiroi et al. 2000). We have also found that Nkx2-5 interacts with another cardiac T-box gene, Tbx20 (R.P. Harvey and F. Stennard, unpubl.). Thus, Nkx2-5 is one component of transcriptional complexes that positively regulate the cardiac program. Negative regulators of this program include the minimal homeodomain protein, HOP, and the ankyrin-repeat protein, CARP; these appear to be part of Nkx2-5 and GATA factor-dependent negative feedback loops (Jayaseelan et al. 1997; Zou et al. 1997; Chen et al. 2002).

GENETIC DISSECTION OF *Nkx2-5* FUNCTION IN MICE

Null alleles of *Nkx2-5* have been created in mice and extensively studied (Lyons et al. 1995; Biben and Harvey 1997; Tanaka et al. 1999, 2001; Biben et al. 2000; Yamagishi et al. 2001). Homozygous null embryos are able to form a heart tube, but this does not progress past the early stages of cardiac looping, and embryos die around E10 (Lyons et al. 1995). Mutant hearts do not undergo ventricular bending (Harvey et al. 1999) and have no trabeculae or endocardial cushions (Lyons et al. 1995). A number of cardiac genes, including those encoding the transcription factors Hand1, Irx4, CITED1, and Nmyc, and myogenic or cytoskeletal proteins such as MLC2V, SM22, ANF, and Chisel, are down-regulated in *Nkx2-5* mutant hearts (Lyons et al. 1995; Biben et al. 1997; Tanaka et al. 1999; Palmer et al. 2001). The patterns of dysregulation suggest that the primary myogenic program is largely unaffected. The main defect is a failure of the heart tube to specify and differentiate the specialized working myocardium of the heart chambers (Harvey 1999). The gene expression signature in the mutant hearts is similar to that of the primary heart tube. Although there has been some discussion as to whether the left ventricular chamber was specifically missing in this mutant (Biben et al. 1997), our recent studies support the view that a primitive ventricle (neither right nor left in identity) is correctly patterned along the anterior/posterior axis but is blocked from myogenic specialization and further differentiation. It is also likely that secondary heart field cells do not migrate into the heart in this model. Thus, the mutant heart tubes have arrested in their most primitive form. These studies demonstrate that the more primitive (primary) and specialized chamber myogenic programs can be discriminated genetically.

Hand1 and Hand2 are members of the basic helix-loop-helix family of transcription factors and individually have also been shown to play key roles in chamber formation in the early heart (Srivastava et al. 1997; Firulli et al. 1998; Riley et al. 1998, 2000). Targeted deletion of *Hand1* appears to lead to loss of the left ventricle, whereas deletion of *Hand2* leads to loss of the right ventricle, a reciprocity that is reflected in their predominant sites of expression (Harvey 2002a). In *Nkx2-5* homozygous mutants, *Hand1* (although not *Hand2*) is down-regulated (Biben and Harvey 1997). *Nkx2-5/Hand2* double mutant mice have also been made (Yamagishi et al. 2001). The double mutant embryos lack normal expression of both *Hand* genes, and both left and right ventricles undergo apoptosis after their initial formation, leaving a rudimentary heart composed mostly of atrial tissue. Thus, the *Nkx2-5* gene, acting in part through regulation of *Hand1*, plays an essential role in differentiation, growth, and survival of chamber myocardium.

REGULATORS OF CHAMBER MYOCARDIUM

As discussed above, specification and/or differentiation of specialized chamber myocardium requires *Nkx2-5*. The spatial specificity of chamber formation is guided, in part, by transcriptional repression within the myocardium. The T-box family transcriptional repressor, Tbx2, is expressed in an evolving pattern in the forming heart tube that is mutually exclusive of the zones that become chamber myocardium (Habets et al. 2002). Tbx2 acts interdependently with Nkx2-5 on a repressive element in the proximal promoter of the ANF gene (a marker of chamber myocardium). Repression through this site results in down-regulation of ANF gene expression in non-chamber myocardium.

The endocardial layer of the heart is also an important signaling tissue for formation of chamber muscle. Trabeculae are the spongiform layer of myocytes that form on the inner surface of the developing chambers adjacent to the endocardial layer. Trabecular myocytes are ultrastructurally more differentiated and less proliferative than other layers of the heart (Pasumarthi and Field 2002) and may be the force-generating component of the embryonic ventricle. Trabeculae also have privileged conduction properties and, at least in mice, appear to be the precursors of the Purkinje fibers of the conduction system (Moorman et al. 2000; Rentschler et al. 2002). In mouse, deletion of genes encoding the EGF-related signaling molecule neuregulin-1, or its myocardial receptors ErbB2 and ErbB4, leads to elimination of trabecular myocardium (Gassmann et al. 1995; Lee et al. 1995; Meyer and Birchmeier 1995). Neuregulin-1 is expressed in endocardium, and its ErbB receptors in myocardium. Exogenous neuregulin-1 induces excessive trabeculation and up-regulation of a transgenic marker of the Purkinje system (Hertig et al. 1999; Rentschler et al. 2002). Our recent studies show that neuregulin-1 acts by maintaining expression of *Nkx2-5* and, in fact, a whole cardiac transcription factor gene network in the myocardium (R.P. Harvey and D. Lai, unpubl.). One implication of this finding is that the neuregulin-1 pathway augments and extends the myogenic pathways established in the cardiac crescent. Neuregulin-1 is also a key homeostatic factor in the adult heart (Crone et al. 2002).

Nkx2-5 AND CONGENITAL HEART DISEASE

Congenital heart disease (CHD) is an important cause of morbidity and mortality in human populations. It is estimated that some 0.8% of live births are associated with cardiac malformations and that cardiac defects are overrepresented in stillborn babies by 10-fold (Hoffman 1995a,b). Genetic etiologies have long been suspected in many of these cases and, indeed, specific mutations are now being identified, providing clues to mechanisms that control both normal and pathological heart development. Heterozygous mutations in NKX2-5 were described in families exhibiting progressive atrioventricular (AV) conduction disease and secundum atrial septal defect (ASD) as dominant features, after the discovery of genetic linkage between these particular defects and polymorphic DNA markers located at chromosome 5q35 (Benson et al. 1998, 1999; Schott et al. 1998). NKX2-5 mutations are also associated in individuals within mutant families with a wide range of other cardiac congenital abnormalites including tetralogy of Fallot (TOF), ventricular septal defect (VSD), Ebstein's anomaly, subvalvular aortic stenosis, and tricuspid valve anomaly, some of these without conduction defects. Indeed, mutations can be associated with all of the major classes of common CHD. Heterozygous mutations have also been associated with some 4% of cases of TOF that are unassociated with deletions in chromosome 22q11, which can also cause this condition (Goldmuntz et al. 2001). Thirteen different mutations have now been described, many within the homeodomain. Although the numbers are small, there is a possible genotype/phenotype correlation between two non-homeodomain mutations and TOF unassociated with conduction disease (Goldmuntz et al. 2001). Some mutations have been tested in vitro and are suggested to have dominant-negative effects in that they can inhibit the activity of normal NKX2-5 protein (Kasahara et al. 2000). Overexpression of one mutation in transgenic mice leads to severe conduction disease and heart failure postnatally (Kasahara et al. 2001b). These effects may be due to the fact that NKX2-5 can form homodimers as well as heterodimers with other transcription factors, and may require phosphorylation for its proper action (Kasahara and Izumo 1999; Kasahara et al.

2001a). Mice overexpressing a mutant NKX2-5 that develop AV conduction block also show decreased expression of gap junction proteins connexin 40 and 43, which are essential for propagation of the cardiac action potential (Kasahara et al. 2001b). These proteins are likely to be direct targets of NKX2-5 (R.P. Harvey and M. Costa, unpubl.). *Nkx2-5* is also expressed at high levels in developing Purkinje fibers in chick embryos (Kasahara et al. 2001b) and may therefore play a direct role in development of components of the specialized conduction system. Diminishment of this function may contribute to the AV conduction block in human patients and transgenic mice. The studies of Goldmuntz on TOF patients (Goldmuntz et al. 2001), together with our analysis of ASD patients unselected for familial history or association with conduction disease (Elliott et al. 2003), suggest that one to a few percent of humans with CHD may carry *NKX2-5* mutations.

An important role played by NKX2-5 in atrial septation and conduction system development and/or homeostasis was demonstrated directly in *Nkx2-5* knockout mice. Heterozygosity for null mutations was found to confer ASD and milder forms of atrial septal dysmorphogenesis (patent foramen ovale [PFO] and septal aneurysm), as well as a mild conduction delay (longer PR interval in females) (Biben et al. 2000). Genetic background played a major role in manifestation of the septal phenotypes, suggesting a role for modifier genes affecting septal morphogenesis in susceptible mouse strains, and also in the human disease (Biben et al. 2000). We found that the length of the atrial septal flap valve (septum primum) in mature mice correlates very well with the mean frequency of PFO in a particular mouse strain (with or without heterozygosity for *Nkx2-5* mutation). PFO frequency varies from 0 to 94%, depending on the strain. We are exploiting this finding to map the mouse strain-specific gene variants that underlie septal abnormalities and that are known to interact with the *Nkx2-5* pathway (Biben et al. 2000).

The Nkx2-5 protein can interact directly with the T-box transcription factor Tbx5, and these factors can synergistically collaborate to activate expression of cardiac genes such as ANF and connexin 40 (Hiroi et al. 2000; Bruneau et al. 2001). Humans with heterozygous mutations in TBX5 have Holt-Oram syndrome, characterized by forelimb abnormalities and abnormal cardiac electrophysiology, in particular AV conduction block (Basson et al. 1997; Li et al. 1997). Other congenital abnormalities can also be present, including ASD, VSD, and TOF. This spectrum of defects is reminiscent of those in patients with NKX2-5 mutations. Structure–function studies of various Holt-Oram TBX5 mutations show that some are associated with a predominance of cardiac over limb abnormalities, whereas others show the reverse association (Basson et al. 1999). A cardiac-type mutation in TBX5 did not activate the ANF promoter and did not show synergistic activation with NKX2-5, whereas a limb-type mutation was able to activate the ANF promoter to the same extent as wild-type TBX5 (Basson et al. 1999). The TBX5 mutations are predicted to produce null or nonfunctioning proteins. Indeed, heterozygous null mutations in *Tbx5* in mice show a Holt-Oram-like phentoype, with first or second degree AV conduction block, ASD, VSD, and decreased expression of ANF and connexin 40 genes (Bruneau et al. 2001). This provides strong evidence that an interdependent Nkx2-5/Tbx5 pathway is disrupted in both NKX2-5 disease and Holt-Oram syndrome.

CONCLUSIONS

The location, migration, induction, patterning, and morphogenesis of vertebrate heart progenitors can now be described in some detail, and many key regulatory genes have been identified. However, there is much to learn about these processes and how they relate to cardiac congenital disease. We must complete the definition of the cardiac building blocks and define the points of integration between the molecular genetic systems impinging on the heart field and forming heart tube. The *Nkx2-5* gene appears to regulate several different aspects of heart development, and its further investigation should reveal new insights into molecular and cellular mechanisms underpinning heart development. A new frontier is to define the relationships between form and function in the developing heart. Furthermore, the recent demonstration of intrinsic regenerative potential in the mammalian heart opens many new challenges in cardiac developmental biology that will continue to make this a vibrant field in the decades to come.

ACKNOWLEDGMENTS

We thank A. Tung and N. Rosenthal for help with artwork.

REFERENCES

Basson C.T., Bachinsky D.R., Lin R.C., Levi T., Elkins J.A., Soults J., Grayzel D., Kroumpouzou E., Traill T.A., Leblanc-Straceski J., Renault B., Kucherlapati R., Seidman J.G., and Seidman C.E. 1997. Mutations in human Tbx5 cause limb and cardiac malformation in Holt-Oram syndrome. *Nat. Genet.* **15:** 30.

Basson C.T., Huang T., Lin R.C., Bachinsky D.R., Weremowicz S., Vaglio A., Bruzzone R., Quadrelli R., Lerone M., Romeo G., Silengo M., Pereira A., Krieger J., Mesquita S.F., Kamisago M., Morton C.C., Pierpont M.E.M., Muller C.W., Seidman J.G., and Seidman C.E. 1999. Different TBX5 interactions in heart and limb defined by Holt-Oram syndrome mutations. *Proc. Natl. Acad. Sci.* **96:** 2919.

Benson D.W., Sharkey A., Fatkin D., Lang P., Basson C.T., McDonough B., Strauss A.W., Seidman J.G., and Seidman C.E. 1998. Reduced penetrance, variable expressivity, and genetic heterogeneity of familial atrial septal defects. *Circulation* **97:** 2043.

Benson D.W., Silberbach G.M., Kavanaugh-McHugh A., Cottrill C., Zhang Y., Riggs S., Smalls O., Johnson M.C., Watson M.S., Seidman J.G., Seidman C.E., Plowden J., and Kugler J.D. 1999. Mutations in the cardiac transcription factor *Nkx2-5* affect diverse cardiac developmental pathways. *J. Clin. Invest.* **104:** 1567.

Biben C. and Harvey R.P. 1997. Homeodomain factor Nkx2-5 controls left-right asymmetric expression of bHLH *eHand* during murine heart development. *Genes Dev.* **11:** 1357.

Biben C., Palmer D.A., Elliott D.A., and Harvey R.P. 1997. Homeobox genes and heart development. *Cold Spring Harbor Symp. Quant. Biol.* **62:** 395.

Biben C., Weber R., Kesteven S., Stanley E., McDonald L., Elliott D.A., Barnett L., Koentgen F., Robb L., Feneley M., and Harvey R.P. 2000. Cardiac septal and valvular dysmorphogenesis in mice heterozygous for mutations in the homeobox gene *Nkx2-5*. *Circ. Res.* **87:** 888.

Bruneau B.G., Nemer G., Schmitt J.P., Charron F., Robitaille L., Caron S., Conner D.A., Gessler M., Nemer M., Seidman C.E., and Seidman J.G. 2001. A murine model of Holt-Oram syndrome defines roles of the T-box transcription factor Tbx5 in cardiogenesis and disease. *Cell* **106:** 709.

Chen C.Y. and Schwartz R.J. 1996. Recruitment of the tinman homologue Nkx-2.5 by serum response factor activates cardiac α-actin gene transcription. *Mol. Cell. Biol.* **16:** 6372.

Chen F., Kook H., Milewski R., Gitler A., Lu M., Li J., Nazarian R., Schnepp R., Jen K., Biben C., Runke G., Mackay J., Novotny J., Schwartz R., Harvey R., Mullins M., and Epstein J. 2002. Hop is an unusual homeobox gene that modulates cardiac development. *Cell* **110:** 713.

Christoffels V.M., Habets P.E.M.H., Franco D., Campione M., de Jong F., Lamers W.H., Bao Z.-Z., Palmer S., Biben C., Harvey R.P., and Moorman A.F.M. 2000. Chamber formation and morphogenesis in the developing mammalian heart. *Dev. Biol.* **223:** 266.

Cripps R.M. and Olson E. 2002. Control of cardiac development by an evolutionarily conserved transcriptional network. *Dev. Biol.* **246:** 14.

Crone S.A., Zhao Y.-Y., Fan L., Gu Y., Minamisawa S., Liu Y., Peterson K.L., Chen J., Kahn R., Condorelli G., Ross J., Jr., Chien K.R., and Lee K.-F. 2002. ErbB2 is essential in the prevention of dilated cardiomyopathy. *Nat. Med.* **8:** 459.

Elliot D.A., Kirk E., Yeoh T., Chander S., McKenzie F., Taylor P., Grossfeld P., Fatkin D., Jones O., Hayes P., Feneley M., and Harvey R.P. 2003. NKX2-5 mutations and congenital heart disease: Associations with ASD and hyperplastic left heart syndrome. *J. Am. Coll. Cardiol.* (in press).

Epstein J. 2001. Developing models of DiGeorge syndrome. *Trends Genet.* **17:** S13.

Erwin D.H. and Davidson E.H. 2002. The last common bilaterian ancestor. *Development* **129:** 3021.

Firulli A.B. and Olson E. 1997. Modular regulation of muscle gene transcription: A mechanism for muscle cell diversity. *Trends Genet.* **13:** 364.

Firulli A.B., McFadden D.G., Lin Q., Srivastava D., and Olson E.N. 1998. Heart and extra-embryonic mesoderm defects in mouse embryos lacking the bHLH transcription factor Hand1. *Nat. Genet.* **18:** 266.

Gassmann M., Casagranda F., Orioli D., Simon H., Lai C., Klein R., and Lemke G. 1995. Aberrant neural and cardiac development in mice lacking the ErbB4 neuregulin receptor. *Nature* **378:** 390.

Goldmuntz F., Geiger B., and Benson D.W. 2001. NKX2.5 mutations in patients with tetralogy of fallot. *Circulation* **104:** 2565.

Habets P.E., Moorman A.F., Clout D.E., van Roon M.A., Lingbeek M., van Lohuizen M., Campione M.. and Christoffels V.M. 2002. Cooperative action of Tbx2 and Nkx2.5 inhibits ANF expression in the atrioventricular canal: Implications for cardiac chamber formation. *Genes Dev.* **16:** 1234.

Hamada H., Meno C., Watanabe M., and Saijoh Y. 2002. Establishment of vertebrate left/right asymmetry. *Nat. Rev. Genet.* **3:** 103.

Harvey R.P. 1996. NK-2 homeobox genes and heart development. *Dev. Biol.* **178:** 203.

———. 1999. Seeking a regulatory roadmap for heart morphogenesis. *Semin. Cell Dev. Biol.* **10:** 99.

———. 2002a. Molecular determinants of cardiac development and congenital disease. In *Mouse development: Patterning, morphogenesis, and organogenesis* (ed. J. Rossant and P.P.L. Tam), p. 331. Academic Press, San Diego, California.

———. 2002b. Patterning the vertebrate heart. *Nat. Rev. Genet.* **3:** 544.

Harvey R.P., Biben C., and Elliott D.A. 1999. Transcriptional control and pattern formation in the developing vertebrate heart: Studies on NK-2 class homeodomain factors. In *Heart development* (ed. R.P. Harvey and N. Rosenthal), p. 111. Academic Press, San Diego, California.

Hertig C.M., Kubalak S.W., Wang Y., and Chien K.R. 1999. Synergistic roles of neuregulin-1 and insulin-like growth factor-I in activation of the phosphatidylinositol 3-kinase pathway and cardiac chamber morphogenesis. *J. Biol. Chem.* **274:** 37362.

Hiroi Y., Kudoh S., Oka T., Monzen K., Ikeda Y., Hosoda T., Niu P., Nagai R., and Komuro I. 2000. Tbx5 associates with CSX/NKX2.5 and synergistically activates atrial natriuretic peptide (ANP) promoter. *Circulation (suppl. II) vol.* **102**.

Hoffman J.I. 1995a. Incidence of congenital heart disease. I. Postnatal incidence. *Pediatr. Cardiol.* **16:** 103.

———. 1995b. Incidence of congenital heart disease. II. Prenatal incidence. *Pediatr. Cardiol.* **16:** 155.

Hughes S. 2002. Cardiac stem cells. *J. Pathol.* **197:** 468.

Jayaseelan R., Poizat C., Baker R.K., Abdishoo S., Isterabadi L.B., Lyons G.E., and Kedes L. 1997. A novel cardiac-restricted target for doxorubicin, CARP, a nuclear modulator of gene expression in cardiac progenitor cells and cardiomyocytes. *J. Biol. Chem.* **272:** 22800.

Kasahara H. and Izumo S. 1999. Identification of the in vitro casein kinase II phosphorylation site within the homeodomain of the cardiac tissue-specifying homeobox gene product Csx/Nkx2.5. *Mol. Cell. Biol.* **19:** 526.

Kasahara H., Usheva A., Ueyama T., Aoki H., Horikoshi N., and Izumo S. 2001a. Characterization of homo- and heterodimerization of cardiac Csx/Nkx2.5 homeoprotein. *J. Biol. Chem.* **276:** 4570.

Kasahara H., Lee B., Schott J.-J., Benson D.W., Seidman J.G., Seidman C.E., and Izumo S. 2000. Loss of function and inhibitory effects of human CSX/NKX2.5 homeoprotein mutations associated with congenital disease. *J. Clin. Invest.* **106:** 299.

Kasahara H., Wakimoto H., Liu M., Maguire C.T., Converso K.L., Shioi T., Huang W.Y., Manning W.J., Paul D., Lawitts J., Berul C.I., and Izumo S. 2001b. Progressive atrioventricular conduction defects and heart failure in mice expressing a mutant Csx/Nkx2.5 homeoprotein. *J. Clin. Invest.* **108:** 189.

Kastner P., Messaddeq N., Mark M., Wendling O., Grondona J.M., Ward S., Ghyselinck N., and Chambon P. 1997. Vitamin A deficiency and mutations of RXRα, RXRβ and RARα lead to early differentiation of embryonic ventricular cardiomyocytes. *Development* **124:** 4749.

Kelly R.G., Brown N.A., and Buckingham M.E. 2001. The arterial pole of the mouse heart forms from Fgf10-expressing cells in pharyngeal mesoderm. *Dev. Cell* **1:** 435.

Kilner P.J., Yang G.-Z., Wilkes A.J., Mohiaddin R.H., Firmin D.N., and Yacoub M.H. 2000. Asymmetric redirection of flow through the heart. *Nature* **404:** 759.

Kinder S.J., Tsang T.E., Wakamiya M., Sasaki H., Behringer R.R., Nagy A., and Tam P.L. 2001. The organiser of the mouse gastrula is composed of a dynamic population of progenitor cells for the axial mesoderm. *Development* **128:** 3623.

Komuro I. and Izumo S. 1993. *Csx*: A murine homeobox-containing gene specifically expressed in the developing heart. *Proc. Natl. Acad. Sci.* **90:** 8145.

Lee K.-F., Simon H., Chen H., Bates B., Hung M.-C., and Hauser C. 1995. Requirement for neuregulin receptor erbB2 in neural and cardiac development. *Nature* **378:** 394.

Lee Y., Shioi T., Kasahara H., Jobe S.M., Wiese R.J., Markham B.E., and Izumo S. 1998. The cardiac tissue-restricted homeobox protein Csx/Nkx2.5 physically associates with the zinc finger protein GATA4 and cooperatively activates atrial natriuretic factor gene expression. *Mol. Cell. Biol.* **18:** 3120.

Li Q.Y., Newbury-Ecob R.A., Terrett J.A., Wilson D.I., Curtis A.R.J., Yi C.H., Gebuhr T., Bullen P.J., Robson S.C., Strachan T., Bonnet D., Lyonnet S., Young I.D., Raeburn A., Buckler A.J., Law D.J., and Brook J.D. 1997. Holt-Oram syndrome is caused by mutations in *TBX5*, a member of the *Brachyury* (T) gene family. *Nat. Genet.* **15:** 21.

Lints T.J., Parsons L.M., Hartley L., Lyons I., and Harvey R.P. 1993. *Nkx-2.5:* A novel murine homeobox gene expressed in early heart progenitor cells and their myogenic descendants. *Development* **119:** 419.

Lyons I., Parsons L.M., Hartley L., Li R., Andrews J.E., Robb L., and Harvey R.P. 1995. Myogenic and morphogenetic defects in the heart tubes of murine embryos lacking the homeobox gene *Nkx2-5*. *Genes Dev.* **9:** 1654.

Meyer D. and Birchmeier C. 1995. Multiple essential functions of neuregulin in development. *Nature* **378:** 386.

Mjaatvedt C.H., Yamamura H., Wessels A., Ramsdell A., Turner D., and Markwald R.R. 1999. Mechanisms of segmentation, septation, and remodeling of the tubular heart: Endocardial cushion fate and cardiac looping. In *Heart development* (ed. R.P. Harvey and N. Rosenthal), p. 159. Academic Press, San Diego, California.

Mjaatvedt C.H., Nakaoka T., Moreno-Rodriquez R., Norris R.A., Kern M.J., Eisenberg C.A., Turner D., and Markwald R.R. 2001. The outflow tract of the heart is recruited from a novel heart-forming field. *Dev. Biol.* **238:** 97.

Moorman A.F.M., Schumacher C.A., de Boer P.A.J., Hagoort J., Bezstarosti K., van den Hoff M.J.B., Wagenaar G.T.M., Lamers J.M.J., Wuytack F., Christoffels V.M., and Fiolet J.W.T. 2000. Presence of functional sarcoplasmic reticulum in the developing heart and its confinement to chamber myocardium. *Dev. Biol.* **223:** 279.

Narita N., Bielinska M., and Wilson D.B. 1997. Wild-type endoderm abrogates the ventral developmental defects associated with GATA-4 deficiency in the mouse. *Dev. Biol.* **189:** 270.

Niederreither K., Vermot J., Messaddeq N., Schuhbaur B., Chambon P., and Dolle P. 2001. Embryonic retinoic acid synthesis is essential for heart morphogenesis in the mouse. *Development* **128:** 1019.

Orlic D., Kajstura J., Chimenti S., Limana F., Jakoniuk I., Quaini F., Nadal-Ginard B., Bodine D.M., Leri A., and Anversa P. 2001. Mobilized bone marrow cells repair the infarcted heart, improving function and survival. *Proc. Natl. Acad. Sci.* **98:** 10344.

Palmer S., Groves N., Schindeler A., Yeoh T., Biben C., Wang C.-C., Sparrow D.B., Barnett L., Jenkins N., Copeland N., Koentgen F., Mohun T., and Harvey R.P. 2001. The small muscle-specific protein Csl modifies cell shape and promotes myocyte fusion in an insulin-like growth factor 1-dependent manner. *J. Cell Biol.* **153:** 985.

Park M., Lewis C., Turbay D., Chung A., Chen J.N., Evans S., Breitbart R.E., Fishman M.C., Izumo S., and Bodmer R. 1998. Differential rescue of visceral and cardiac defects in *Drosophila* by vertebrate tinman-related genes. *Proc. Natl. Acad. Sci.* **95:** 9366.

Pasumarthi K.B. and Field L.J. 2002. Cardiomyocyte cell cycle regulation. *Circ. Res.* **90:** 1044.

Poelmann R.E. and Gittenberger-de Groot A.C. 1999. A subpopulation of apoptosis-prone cardiac neural crest cells targets to the venous pole: Multiple functions in heart development? *Dev. Biol.* **207:** 271.

Raffin M., Leong L.M., Rones M.S., Sparrow D., Mohun T., and Mercola M. 2000. Subdivision of the cardiac Nkx2.5 expression domain into myogenic and nonmyogenic compartments. *Dev. Biol.* **218:** 326.

Ranganayakulu G., Elliott D.A., Harvey R.P., and Olson E.N. 1998. Divergent roles for *NK-2* class homeobox genes in cardiogenesis in flies and mice. *Development* **125:** 3037.

Redkar A., Mongomery M., and Litvin J. 2001. Fate map of early avian cardiac progenitor cells. *Development* **128:** 2269.

Reese D.E., Mikawa T., and Bader D.M. 2002. Development of the coronary vessel system. *Circ. Res.* **91:** 761.

Rentschler S., Zander J., Meyers K., France D., Levine R., Porter G., Rivkees S.A., Morley G.E., and Fishman G.I. 2002. Neuregulin-1 promotes formation of the murine cardiac conduction system. *Proc. Natl. Acad. Sci.* **99:** 10464.

Riley P., Anson-Cartwright L., and Cross J.C. 1998. The Hand1 bHLH transcription factor is essential for placentation and cardiac morphogenesis. *Nat. Genet.* **18:** 271.

Riley P.R., Gertsenstein M., Dawson K., and Cross J.C. 2000. Early exclusion of Hand1-deficient cells from distinct regions of the left ventricular myocardium in chimeric mouse embryos. *Dev. Biol.* **227:** 156.

Rosenthal N. and Xavier-Neto J. 2000. From the bottom of the heart: Anteroposterior decisions in cardiac muscle differentiation. *Curr. Opin. Cell Biol.* **12:** 742.

Schott J.-J., Benson D.W., Basson C.T., Pease W., Silberach G.M., Moak J.P., Maron B.J., Seidman C.E., and Seidman J.G. 1998. Congenital heart disease caused by mutations in the transcription factor *NKX2-5*. *Science* **281:** 108.

Sedmera D., Pexieder T., Rychterova V., Hu N., and Clark E.B. 1999. Remodelling of chick embryonic ventricular myoarchitecture under experimentally changed loading conditions. *Anat. Rec.* **254:** 238.

Srivastava D., Thomas T., Lin Q., Brown D., and Olson E.N. 1997. Regulation of cardiac mesodermal and neural crest development by the bHLH transcription factor, dHAND. *Nat. Genet.* **16:** 154.

Stanley E.G., Biben C., Elefanty A., Barnett L., Koentgen F., Robb L., and Harvey R.P. 2002. Efficient Cre-mediated deletion in cardiac progenitor cells conferred by a 3´UTR-ires-Cre allele of the homeobox gene Nkx2-5. *Int. J. Dev. Biol.* **46:** 431.

Tam P.L. and Schoenwolf G.C. 1999. Cardiac fate maps: Lineage allocation, morphogenetic movement, and cell commitment. In *Heart development* (ed. R.P. Harvey and N. Rosenthal), p. 3. Academic Press, San Diego, California.

Tanaka M., Chen Z., Bartunkova S., Yamasaki N., and Izumo S. 1999. The cardiac homeobox gene Csx/Nkx2.5 lies genetically upstream of multiple genes essential for heart development. *Development* **126:** 1269.

Tanaka M., Schinke M., Liao H.-S., Yamasaki N., and Izumo S. 2001. Nkx2.5 and Nkx2.6, homologues of *Drosophila tinman*, are required for development of the pharynx. *Mol. Cell. Biol.* **21:** 4391.

van den Hoff M.J., Kruithof B.P., Moorman A.F., Markwald R.R., and Wessels A. 2001. Formation of myocardium after the initial development of the linear heart tube. *Dev. Biol.* **240:** 61.

Waldo K.L., Kumiski D.H., Wallis K.T., Stadt H.A., Hutson M.R., Platt D.H., and Kirby M.L. 2001. Conotruncal myocardium arises from a secondary heart field. *Development* **128:** 3179.

Webb S., Brown N.A., and Anderson R.H. 1998. Formation of the atrioventricular septal structures in the normal mouse. *Circ. Res.* **82:** 645.

Yamagishi H., Yamagishi C., Nakagawa O., Harvey R.P., Olson E.N., and Srivastava D. 2001. The combined activities of Nkx2-5 and dHand are essential for cardiac ventricle formation. *Dev. Biol.* **239:** 190.

Zaffran S. and Frasch M. 2002. Early signals in cardiac development. *Circ. Res.* **91:** 457.

Zou Y., Evans S., Chen J., Kou H.-C., Harvey R.P., and Chien K.R. 1997. CARP, a cardiac ankyrin repeat protein, is downstream in the *Nkx2-5* homeobox gene pathway. *Development* **124:** 793.

Causes of Clinical Diversity in Human *TBX5* Mutations

T. HUANG,*† J.E LOCK,‡ A.C MARSHALL,‡ C. BASSON,¶ J.G. SEIDMAN,† AND C.E. SEIDMAN†

*Division of Genetics and ‡Department of Cardiology, Children's Hospital, Boston, Massachusetts 02115;
†Department of Genetics and Howard Hughes Medical Institute, Harvard Medical School, Boston,
Massachusetts 02115; ¶Cardiology Division, Department of Medicine and Department of Cell Biology,
Weill Medical College of Cornell University, New York, New York 10021*

Discovery of the many molecular signals that orchestrate cardiac development, a critical and complex embryologic process requiring integration of cell commitment, growth, looping, septation, and chamber specification, has predominantly come from studies in lower species. The study of human inherited cardiovascular malformations complements these investigations by providing abundant and unparalleled detail about the consequences of cardiovascular development gone awry. Five percent of human live births and 10% of stillbirths have congenital heart disease (D'Alton and DeCherney 1993). Although the cause for these prevalent malformations is often unknown, in rare instances cardiovascular malformations arise as a consequence of single gene mutations. Holt-Oram syndrome (HOS), an autosomal dominant disorder characterized by congenital heart defects and upper limb anomalies is caused by mutations of *TBX5*, a member of the T-box gene family of transcription factors. Although a variety of different human *TBX5* mutations have been identified located throughout the gene, most of these are predicted to encode a truncated protein, thereby suggesting the predominant mechanism by which HOS occurs is haploinsufficiency of *TBX5*. Remarkably, the clinical phenotypes caused by *TBX5* deficiency vary considerably, even in individuals with identical mutations, an observation that both alludes to the critical importance of TBX5 in many cardiovascular developmental processes and indicates the influence of other factors in modulating phenotype. Identification of the molecular basis for HOS in concert with meticulous clinical evaluations that preface surgical intervention to correct human congenital heart disease has provided the opportunity to assess the role of modifying genes and environment on the molecular program of cardiac embryogenesis.

TBX5 MUTATIONS

Human genetics studies of HOS defined a single disease locus on chromosome 12q12 (Basson et al. 1994) where the T-box transcription factor *TBX5* is encoded. Multiple distinct *TBX5* mutations have been reported (Basson et al. 1997, 1999; Li et al. 1997) that most often are private, or occurring in only one family. The range of *TBX5* mutations identified in HOS includes missense codons, insertions, deletions, and a chromosomal translocation that sever regulatory elements from protein-encoding sequences. The consequence of most of these mutations is premature termination of translation, so that the mutant *TBX5* allele cannot produce a functional protein. Haploinsufficiency or a reduction by half in the normal amount of *TBX5* throughout development therefore appears to be the mechanism responsible for human disease.

Studies of the pattern of expression of *Tbx5* in a variety of species indicate that this transcription factor might be important for development of the forelimb, heart, eye, trachea, and pulmonary vasculature (Gibson-Brown et al. 1998; Bruneau et al. 1999). However, neither deficits in vision nor in pulmonary function have been observed, and the clinical consequences of *TBX5* haploinsufficiency in humans are primarily restricted to the forelimb and heart, albeit sometimes with thoracic skeletal malformations (Holt and Oram 1960; Poznanski et al. 1970; Basson et al. 1994). The reason that a reduction in physiologic levels of Tbx5 perturbs development of some structures, but not other organs that also express this T-box protein, presumably reflects molecular redundancy and other unknown compensatory mechanisms. More broadly, such data indicate that the response to transcription factor insufficiency can be modified.

CLINICAL DIVERSITY OF HOS

The cardiac manifestations of HOS vary considerably in type and severity. Septation defects of the atria (ostium secundum) and the muscular interventricular septum comprise the most common heart malformation in HOS (Newbury-Ecob et al. 1996; Sletten and Pierpont 1996; Bruneau et al. 1999). Not infrequently, septation defects are accompanied by conduction system disease, but electrophysiologic defects also occur in the absence of structural malformations. In addition to these simple malformations, multiple septation defects and more complex cardiovascular lesions have also been observed in HOS (Smith et al. 1979; Ruzic et al. 1981; Sahn et al. 1981; Glauser et al. 1989). Tetralogy of Fallot, anomalous pulmonary venous return, atrioventricular canal, hypoplastic left heart syndrome, truncus arteriosus, coarctation of the aorta, patent ductus arteriosus, valvular lesions (severe aortic stenosis, mitral valve prolapse, or tricuspid atresia), and total anomalous pulmonary venous return in HOS collectively indicate that multiple components of the cardiac development are compromised by *TBX5* haploinsufficiency.

Skeletal abnormalities of the upper limbs are always present in HOS and may be the only clinical evidence of a *TBX5* mutation (Poznanski et al. 1970; Temtamy and McKusick 1978). Affected forelimb structures are usually derivatives of the embryonic radial rays. The thumb is frequently involved and may be triphalangeal (digit-like) and nonopposable, hypoplastic, or even completely absent. Carpal bone malformations and bone fusion are common skeletal findings in HOS. As with the congenital heart defects, the severity of limb defects varies considerably from mild anomalies detected only by radiography, to markedly severe deformities such as phocomelia (Basson et al. 1994). Upper limb malformations can be remarkably asymmetric, and clinical reports have suggested more severe involvement of the left extremity.

MODIFYING FACTORS OF *TBX5* MUTATIONS

Those factors that influence which of the many potential malformations will arise in HOS are poorly understood. Although allelic variation might account for some of these differences, this explanation is difficult to reconcile with the presumed common pathogenetic mechanism for disease, *TBX5* haploinsufficiency. Furthermore, the considerable variation observed between individuals with different *TBX5* mutations is also seen in affected family members with the same disease-causing mutation, suggesting that factors other than the inciting gene defect contribute to phenotype. Several distinct factors can be envisioned that might influence the response to a *TBX5* mutation: in the heart, hemodynamic parameters, and in the heart and limbs, environmental factors and modifier genes. Although the interplay between pump functions of the embryonic heart and cardiac development remains largely unexplored, it is conceivable that some structural deficits could affect broader developmental programs, specifically because of adverse consequences on embryonic cardiovascular hemodynamics. Delineating potential flow-mediated effects from the primary consequences of *TBX5* haploinsufficiency remains, however, a daunting challenge. In contrast, the relative influence of environmental factors and/or modifier genes in regulating the developmental response to a *TBX5* mutation can be more readily addressed from human studies. To do so, we have recently analyzed monozygotic twins with HOS and compared their cardiac and skeletal phenotypes in the context of identical *TBX5* mutations.

Three females born by Caesarean section at 36 weeks of gestation due to placenta insufficiency were noted to have bilateral thumb abnormalities and heart murmurs at birth. The mother's pregnancy was uneventful and without known exposures to teratogens. Birth weight for the three infants ranged between the 50th and 75th percentile. One infant had separate chorionic sacs with fused placenta. One triplet died at day 4 post-birth from necrotizing enterocolitis. The surviving infants (subsequently identified as Twin A and Twin B) (see Fig. 1) underwent cardiac and skeletal evaluations (Table 1), resulting in a clinical diagnosis of HOS. Both twins failed to thrive and were referred for further clinical evaluation in anticipation of cardiac surgery. After informed consent in accordance with institutional guidelines, they also participated in genetic studies to elucidate the genetic basis for the twins' malformations.

Cardiac echocardiography demonstrated multiple ventricular septal defects and pulmonary artery narrowing in each child. In addition, Twin B had a patent ductus arteriosus (Fig. 2D), which presumably helped to relieve the effects of the pulmonic artery stenosis. Cardiac catheterization and direct visualization during open-heart surgery further defined the spectrum of cardiac malformations in each child (Table 1 and Fig. 2). Each twin had an identical large membranous ventricular septal defect as well as multiple muscular ventricular septal defects, referred to as a "Swiss cheese" septum (Fig. 2C). The location of the largest of these lesions was atypical for most ventricular septal defects, and occurred anterior to the cardiac apex. In addition, each twin also had a secundum atrial septal defect (Fig. 2A,B) and mild tricuspid regurgitation. The locations of the secundum atrial septal defects differed. In Twin A, the atrial defect was near the tricuspid valve, and the left superior vena cava drained to the coronary sinus with a small connecting vein. In Twin B, the secundum atrial septal defect was nearer to the superior vena cava. The twins underwent patch closure of the membranous ventricular septal defects, suture closure of their atrial septal defects, and repair of the pulmonary artery stenoses with a pericardial patch. During their surgery, both twins developed atrioventricular block.

Skeletal evaluations of the twins also revealed multiple forelimb defects (Table 1 and Fig. 3), including hypoplastic radial bones, delayed carpal ossification, and abnormal thumb development. A hooked clavicle was present in each. In addition to these shared skeletal malformations, there were some differences. Twin A had hypoplastic thumbs (Fig. 3A,B), which on the right was so severe that the proximal metacarpal was missing and only remnants of the proximal and distal phalanges were present, adjacent to the second distal and middle phalange. The left proximal phalanx and distal phalanx were

Table 1. Phenotypes in Identical Twins with Holt-Oram Syndrome

Organ	Twin A	Twin B
Heart		
Atrium	septal defect	septal defect
Ventricle	membranous septal defect	membranous septal defect
	multiple muscular septal defects ("Swiss cheese")	multiple muscular septal defects ("Swiss cheese")
Vasculature		
Pulmonary artery	stenotic	stenotic
Ductus arteriosus	closed	patent
Limb		
Thumb	left: hypoplastic	left: absent
	right: remnant	right: absent
Radius	hypoplastic	hypoplastic
Clavicle	hooked	hooked

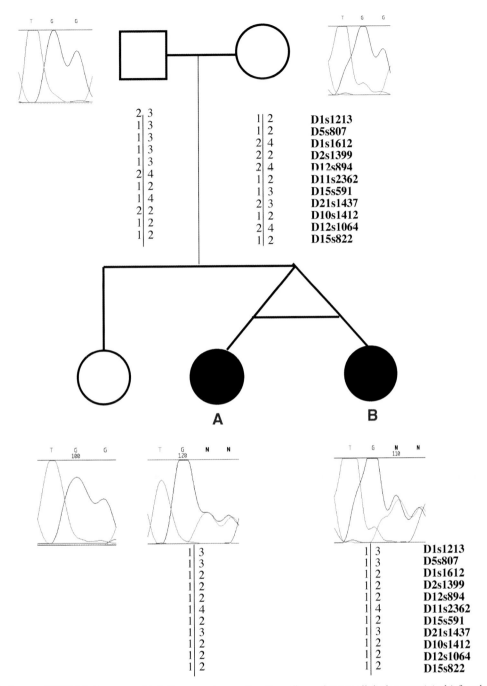

Figure 1. A de novo *TBX5* deletion causes Holt-Oram syndrome. Family pedigree denotes clinical status: (*circle*) female; (*square*) male; (*filled symbol*) affected; (*open symbol*) unaffected. DNA sequence traces defined a point deletion in exon 7, ΔGlu243Fster, as the cause of disease in this family. Haplotypes demonstrated genetic identity of twins.

also hypoplastic, and both were decreased in width and length. Twin B lacked thumbs on both hands (Fig. 3 C). In addition, Twin A, but not Twin B, had a bifid rib (data not shown). Because rib abnormalities are not usually reported in HOS, the relatedness of this abnormality to *TBX5* mutation is uncertain.

To identify the precise genetic cause of HOS in these children, DNA was purified from peripheral blood samples, and *TBX5* coding sequences (exons 2–9) were amplified and sequenced as described previously (Basson et al. 1997, 1999), and data were analyzed using the DNAstar program. A single guanine deletion (residue 727) was detected in exon 7 in DNA samples derived from both Twin A and Twin B (Fig. 1). As a consequence of this deletion the normal glutamic acid encoded at residue 243 is lost and followed by a frameshift in the *TBX5* coding sequences with premature termination at amino acid residue 263 (designated ΔGlu243FSter). The sequence variant was confirmed by oligonucleotide hybridization using wild-type probe (5´-ATGACATGGAGCTGCAC-

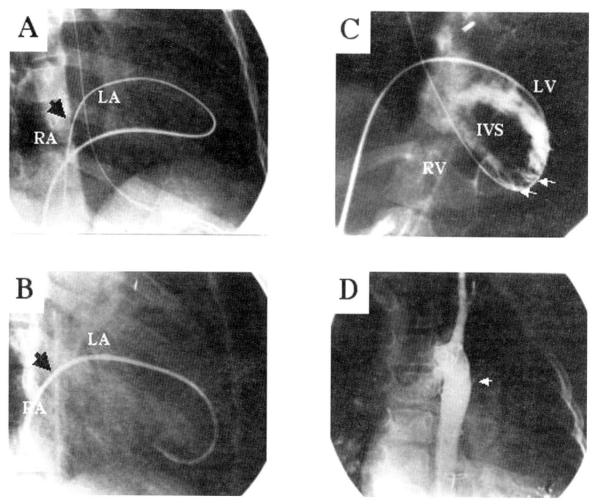

Figure 2. Congenital heart malformations caused by a *TBX5* deletion in identical twins. Cardiac catheters passed from right atrium (RA) to the left atrium (LA) through a secundum atrial septal defect (*arrows*) in Twin A and Twin B (panels *A* and *B*, respectively). (*C*) Injection of contrast into the left ventricular (LV) catheter revealed multiple defects in the interventricular muscular septum (IVS) (termed a "Swiss cheese septum") (*arrows*) with contrast evident in the right ventricle (RV). (*D*) A jet of dye in the aortic arch enters the main pulmonary artery through a patent ductus arteriosus (*arrow*) in Twin B.

3′) and mutant probe (5′-ATGACATGAGCTGCAC-3′). The sequence variant was not identified in DNA samples derived from either parent or in over 200 normal chromosomes. To exclude the possibility of genetic mosaicism in the parents, *TBX5* exon 7 sequences were amplified from parental DNA samples, cloned into a bacterial plasmid vector and characterized by sequence analysis. ΔGlu243FSter was not found in any of 100 cloned copies studied (data not shown).

To establish the genetic relationship between the twins and their parents, genotypes were determined at 11 highly polymorphic loci (D10S1213, D5S807, D1S1612, D2S1399, D13S894, D11S2362, D19S591, D21S1437, D10S1412, D12S1064, D15S822; obtained from Research Genetics, Huntsville, Alabama). At each of the eleven polymorphic loci, the affected twins had identical alleles shared by each parent (data not shown). The probability that dizygotic twins would share all of these alleles was calculated to equal $1/2^{11}$, indicating a very high likelihood that the twins were monozygotic.

Like previously described HOS defects, these genetic data indicate the cause of disease to be a mutant *TBX5* allele which encodes a severely truncated protein. The consequence of the point deletion is truncation of the T-box or DNA-binding domain, which presumably renders the transcription factor dysfunctional. Genetic studies further indicated that ΔGlu243FSter is a de novo mutation. The *TBX5* mutation and the considerable allele sharing of both twins strongly suggest their genetic identicalness.

Although there are a few subtle differences in the malformations observed in these twins' cardiac and limb malformations, the uniformity of their phenotypes is remarkable. Shared malformations of the atria and ventricular septa, including the unusual number and location of these defects, and identical involvement of the radial and carpal bones were observed. What might account for the observed differences? The patent ductus arteriosus found in one twin provides evidence of the important effect that flow can have on cardiac development. In utero, the ductus arteriosus plays an important physiologic role by con-

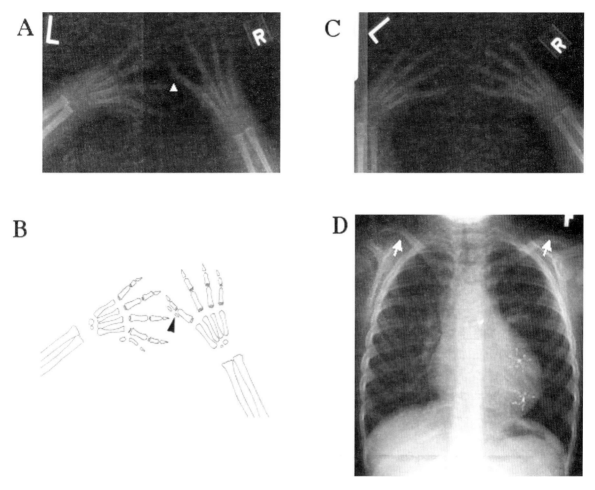

Figure 3: Comparison of skeletal abnormalities in HOS twins. (*A*) Severe thumb hypoplasia in Twin A. The proximate metacarpal is not seen and only remnants of the distal and proximate phalanges remain (*arrowhead*), which are adjacent to the second distal and middle phalanges and hypoplastic with decreased width and length. The radius is also hypoplastic, and delayed carpal ossification is evident. (*B*) Schematic of bone malformations found in Twin A. (*C*) Absence of bilateral thumbs in Twin B with a hypoplastic radial bone and delayed carpal ossification. (*D*) Chest x-ray reveals hooked clavicles in both twins (*arrow*).

necting the main pulmonary trunk with the descending aorta, so as to divert blood flow from the high-resistance pulmonary circulation into the systemic circulation. When the lungs inflate at birth, pulmonary pressures fall and blood flow through the pulmonary vasculature increases oxygen tension and decreases prostaglandins, factors that precipitate closure of this vascular shunt. We presume that persistent patency of the ductus arteriosus in one twin with HOS probably reflects persistently elevated pulmonary pressures, secondary to the pulmonary artery stenosis. Rather than indicating a role for TBX5 in closure of the ductus arteriosus, the patent ductus arteriosus more likely reflects pathologic hemodynamics. The finding of patent ductus arteriosus in association with congenital heart lesions that alter hemodynamics, but are not due to *TBX5* haploinsufficiency, further supports the importance that flow can have in adapting the malformed heart, but in so doing compounding congenital heart lesions.

The role of environment may have been important for the differences observed between the thumb structures of the twins. Evidence for the importance of intrauterine environment in the development of digits comes from children born with asymmetric finger malformations due to digit entrapment within the uterus with resultant hypoplasia and/or truncation. Since Twin A had rudimentary thumb structures in both hands, whereas the left and right thumbs were entirely absent in Twin B, we suspect that the failure to develop even primitive thumb structures may indicate a superimposition of intrauterine environment (i.e., position and pressure) on the underlying *TBX5* mutation.

CONCLUSIONS

The remarkable identity of the congenital malformations found in these twins indicates that although environmental factors and hemodynamics have some influence in the phenotypic response to a *TBX5* mutation, genotype has a far greater effect. A corollary to this conclusion is that the considerable variability in Holt-Oram syndrome produced by identical *TBX5* mutations (Basson et al. 1994) reflects the influence of modifying background genes. Although the direct identification of genetic modifiers of *TBX5* mutations in man will be very difficult, the development of a mouse engineered to lack one Tbx5 allele (Bruneau et al. 2001) provides a useful tool for this investigation. Analyses of genes dysregulated in this murine model of HOS led to the demonstration that *Tbx5* and *Nkx2.5* synergistically activate target sequences, a result that suggests that these transcription factors also function as genetic modifiers. Assessment of whether *NKX2.5* sequences effect the phenotype of HOS should formally test this hypothesis. Regardless of the success or failure of those studies, these data provide strong evidence for the fundamental importance that genotype has in determining congenital malformations associated with *TBX5* mutations.

ACKNOWLEDGMENTS

We are grateful to family members and clinicians who assisted in these studies, especially Barbara McDonough, R.N., and Barbara Weissman, M.D., of the Radiology Department, Brigham and Women's Hospital. This work was supported in part by grants from the National Institutes of Health, NIH/NIGMS grant 3M01RR02172-16S, and the Howard Hughes Medical Institute.

REFERENCES

Basson C.T., Cowley G.S., Solomon S.D., Weissman B., Poznanski A.K., Traill T.A., Seidman J.G., and Seidman C.E. 1994. The clinical and genetic spectrum of the Holt-Oram syndrome (heart-hand syndrome). *N. Engl. J. Med.* **330:** 885.

Basson C.T., Bachinsky D.R., Lin R.C., Levi T., Elkins J.A., Soults J., Grayzel D., Kroumpouzou E., Traill T.A., Leblanc-Straceski J., Renault B., Kucherlapati R., Seidman J.G., and Seidman C.E. 1997. Mutations in human *TBX5* gene cause limb and cardiac malformation in Holt-Oram syndrome. *Nat. Genet.* **15:** 30.

Basson C.T., Huang T., Lin R.C., Bachinsky D.R., Weremowicz S., Vaglio A., Bruzzone R., Quadrelli R., Lerone M., Romeo G., Silengo M., Pereira A., Krieger J., Mesquita S.F., Kamisago M., Morton C.C., Pierpont M.E., Muller C.W., Seidman J.G., and Seidman C.E. 1999. Different TBX5 interactions in heart and limb defined by Holt-Oram syndrome mutations. *Proc. Natl. Acad. Sci.* **96:** 2919.

Bruneau B.G., Logan M., Davis N., Levi T., Tabin C.J., Seidman J.G., and Seidman C.E. 1999. Chamber-specific cardiac expression of *Tbx5* heart defects in Holt-Oram syndrome. *Dev. Biol.* **211:** 100.

Bruneau B.G., Nemer G., Schmitt J.P., Charron F., Robitaille L., Caron S., Conner D.A., Gessler M., Nemer M., Seidman C.E., and Seidman J.G. 2001. A murine model of Holt-Oram syndrome defines roles of the T-box transcription factor *Tbx5* in cardiogenesis and disease. *Cell* **106:** 709.

D'Alton M.E. and DeCherney A.H. 1993. Prenatal diagnosis. *N. Engl. J. Med.* **328:** 114.

Gibson-Brown J.J., Agulink S.I., Silver L.M., and Papaioannou V.E. 1998. Expression of T-box genes *Tbx2-Tbx5* during chick organogenesis. *Mech. Dev.* **74:** 165.

Glauser T.A., Zackai E., Weinberg P., and Clancy R. 1989. Holt-Oram syndrome associated with the hypoplastic left heart syndrome. *Clin. Genet.* **36:** 69.

Holt M. and Oram S. 1960. Familial heart disease with skeletal malformations. *Br. Heart J.* **22:** 236.

Li Q.Y., Newbury-Ecob R.A., Terrett J.A., Wilson D.I., Curtis A.R.J., Yi C.H., Gebuhr T., Bullen P.J., Robson S.C., Strachan T., Bonnet D., Lyonnet S., Young I.D., Raeburn J.A., Buckler A.J., Law D.J., and Brook J.D. 1997. Holt-Oram syndrome is caused by mutations in *TBX5*, a member of the Brachyury (T) gene family. *Nat. Genet.* **5:** 21.

Newbury-Ecob R.A., Leanage R., Raeburn J.A., and Young I.D. 1996. Holt-Oram syndrome: A clinical genetic study. *J. Med. Genet.* **33:** 300.

Poznanski A.K., Gall J.C., Jr., and Stern A.M. 1970. Skeletal manifestations of the Holt-Oram syndrome. *Radiology* **94:** 45.

Ruzic B., Bosnar B., and Beleznay O. 1981. An unusual type of congenital heart disease associated with the Holt-Oram-syndrome *Radiologe* **21:** 296.

Sahn D.J., Goldberg S.J., Allen H.D., and Canale J.M. 1981. Cross-sectional echocardiographic imaging of supracardiac total anomalous pulmonary venous drainage to a vertical vein in a patient with Holt-Oram syndrome. *Chest* **79:** 113.

Sletten L.J. and Pierpont M.E.M. 1996. Variation in severity of cardiac disease in Holt-Oram syndrome. *Am. J. Med. Genet.* **65:** 128.

Smith A.T., Sack G.H., and Tylor G.J. 1979. Holt-Oram syndrome. *J. Pediatr.* **95:** 538.

Temtamy S.A. and McKusick V.A. 1978. The genetics of hand malformations. *Birth Defects Orig. Artic. Ser.* **14:** 1.

Molecular Mechanisms of Ventricular Hypoplasia

D. SRIVASTAVA,*† P.D. GOTTLIEB,‡ AND E.N. OLSON†

*Departments of *Pediatrics and †Molecular Biology, University of Texas Southwestern Medical Center, Dallas, Texas 75390-9148; ‡Section of Molecular Genetics and Microbiology, University of Texas, Austin, Texas 78712*

Hypoplasia of either the right or left ventricle is compatible with intrauterine life but has lethal consequences in the newborn period. Although they are rare, the severity of hypoplastic ventricular conditions results in the single largest contribution to childhood mortality from congenital heart disease. Understanding the molecular and physiologic cues that initiate ventricular specification, growth, and enlargement is essential for future diagnostic, preventive, and therapeutic approaches.

Ventricular cells become specified as early as the cardiac crescent stage (E7.75 in mouse), at which time subpopulations of ventricular and atrial cells can be identified along the anterior-posterior (AP) axis (Fig. 1) (for review, see Srivastava and Olson 2000). Right ventricular and left ventricular cells are indistinguishable at this stage, but within hours they become segregated to specific segments of the straight heart tube with the right ventricular precursors lying anterior to cells fated to form the left ventricle. A dorsal-ventral polarity of gene expression in the straight heart tube distinguishes ventral cells as more proliferative compared to the dorsal surface of the heart tube. As the heart tube loops and twists, the ventral surface becomes the outer curvature of the heart, which is thought to balloon ventrally by virtue of the increased proliferation compared to the inner curvature. This results in expansion of the right and left ventricles, which achieve their relative positioning along the left–right (LR) axis by rightward looping of the heart tube. The cells of the inner curvature possess a distinct pattern of gene expression and ultimately evacuate the region, allowing appropriate remodeling of the folding heart tube necessary for alignment of the atrioventricular and ventriculo-arterial connections.

Hemodynamic forces related to pumping functions of the heart are intimately connected with morphogenetic processes. As a result, disruptions in cardiac development

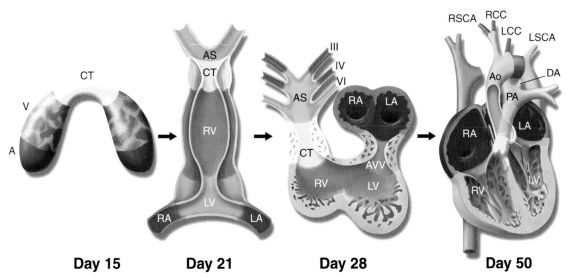

Figure 1. Schematic of cardiac morphogenesis. Illustrations depict cardiac development with color coding of morphologically related regions, seen from a ventral view. Cardiogenic precursors form a crescent (*left-most* panel) that is specified to form specific segments of the linear heart tube, which is patterned along the AP axis to form the various regions and chambers of the looped and mature heart. Each cardiac chamber balloons out from the outer curvature of the looped heart tube in a segmental fashion. Neural crest cells populate the bilaterally symmetric aortic arch arteries (III, IV, and VI) and aortic sac (AS) that together contribute to specific segments of the mature aortic arch, also color-coded. Mesenchymal cells form the cardiac valves from the conotruncal (CT) and atrioventricular valve (AVV) segments. Corresponding days of human embryonic development are indicated. (RV) Right ventricle; (LV) left ventricle; (RA) right atrium; (LA) left atrium; (PA) pulmonary artery; (Ao) aorta; (DA) ductus arteriosus; (RSCA) right subclavian artery; (RCC) right common carotid; (LCC) left common carotid; (LSCA) left subclavian artery. (Reprinted, with permission, from Srivastava and Olson 2000 [copyright Nature Publishing Group].)

often result in non-cell-autonomous defects in morphogenesis of other regions of the heart. For example, although a subset of children born with hypoplasia of the right or left ventricles may have a primary defect of ventricular growth, another population likely develops hypoplasia secondary to decreased or absence of blood flow that occurs in the setting of outflow tract obstruction. Such outflow tract defects, either on the aortic or pulmonary side, are typically caused by abnormalities in neural-crest-derived cells, yet they have a profound effect on mesodermally derived ventricular cardiomyocytes.

The discovery of numerous cardiac-specific transcription factors has begun to establish a network of events that regulate ventricular morphogenesis. These studies have revealed genetic and epigenetic events that regulate ventricular development and provide evidence for primary defects in expansion of cardiac chambers as a potential etiology for subsets of hypoplastic right or left ventricular conditions.

VENTRICULAR-SPECIFIC DNA-BINDING TRANSCRIPTION FACTORS

The first entry into identification of chamber-specific transcription factors came several years ago with the isolation of two proteins belonging to the basic helix-loop-helix (bHLH) family of transcription factors, dHAND and eHAND. Although these two proteins are co-expressed throughout the linear heart tube in the chick and appear to have genetic redundancy (Srivastava et al. 1995), they have chamber-specific expression patterns in the mouse. Both are co-expressed in the cardiac crescent, but *dHAND* gradually becomes restricted mostly to the right ventricle and outflow tract with lower expression in the left ventricle. In contrast, *eHAND* is expressed in the outflow tract and left ventricular segments but is excluded from the right ventricular precursors as early as the straight heart tube stage (Biben and Harvey 1997; Srivastava et al. 1997). Correspondingly, hearts of mice lacking *dHAND* form normally until the straight heart tube stage, at which time *eHAND* becomes down-regulated from the right ventricle. Soon thereafter, the right ventricular precursors undergo programmed cell death, presumably because this is the domain lacking any eHAND protein, resulting in a hypoplastic right ventricle (Srivastava et al. 1997; Yamagishi et al. 2001). eHAND also appears to be necessary for development of left ventricular cells (Firulli et al. 1998; Riley et al. 1998, 2000) and is absent in mice lacking the cardiac homeobox gene, Nkx2.5, which have only one ventricle (Biben and Harvey 1997). Mice lacking *dHAND* and *Nkx2.5* are null for both *HAND* genes and fail to form any ventricular chamber (Yamagishi et al. 2001), consistent with the notion of genetic redundancy between *dHAND* and *eHAND*. A heart is still distinguishable, but it consists only of an atrial chamber (Fig. 2), as defined by atrial-specific markers. Interestingly, a small group of ventricular cells are specified and congregate on the ventral surface of the atrium, but fail to expand ventrally to form a chamber, supporting the "ballooning" model of ventriculogenesis (Christoffels et al. 2000). dHAND and Nkx2.5 are together necessary for expression of the ventricular-specific homeobox gene, *Irx4*, in this subdomain, consistent with the partial down-regulation of *Irx4* in each single

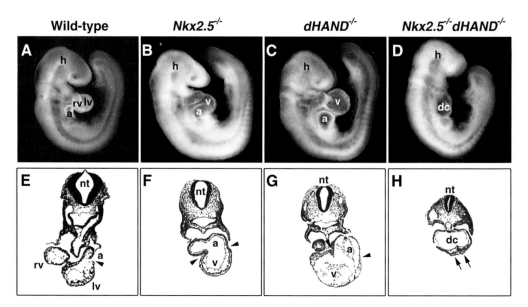

Figure 2. Absence of ventricles in embryos lacking *dHAND* and *Nkx2.5*. Embryonic day (E) 9.25 embryos of wild-type (*A*), *Nkx2.5*[−/−] (*B*), *dHAND*[−/−] (*C*), and *Nkx2.5*[−/−]*dHAND*[−/−] (*D*) embryos are shown in right lateral views. Transverse sections through the heart of wild type (*E*), *Nkx2.5*[−/−] (*F*), *dHAND*[−/−] (*G*), and *Nkx2.5*[−/−]*dHAND*[−/−] (*H*) embryos at E9.25. The segmentation of chambers into atria (a) and ventricles (v) is demarcated by arrowheads. There was only a dorsal chamber (dc) apparent in the *Nkx2.5*[−/−]*dHAND*[−/−] mutant. This chamber expressed atrial markers. An accumulation of cells was observed in the ventral region of the heart (*H, arrowheads*). (lv) Left ventricle; (rv) right ventricle; (nt) neural tube; (h) head.

mutant (Bruneau et al. 2000). In chick, Irx4 is sufficient to activate ventricular-specific gene expression and suppress atrial-specific genes (Bao et al. 1999), suggesting that regulation of *Irx4* may in part contribute to defects in ventriculogenesis.

The transcriptional network necessary for ventricular development appears tightly conserved across species. Large-scale mutagenesis of zebrafish, which have only a single ventricular chamber and a single atrial chamber, yielded a mutant line of fish that was deficient in ventricular precursors (*hands-off*), similar to mice lacking expression of both *HAND* genes. Positional cloning of the affected locus revealed point mutations in *dHAND* (Yelon et al. 2000), the one *HAND* gene present in zebrafish (Angelo et al. 2000), consistent with the role of *HAND* genes defined in mice.

EPIGENETIC FACTORS REGULATING VENTRICULAR DEVELOPMENT

In addition to the DNA-binding proteins described above, there is growing evidence of epigenetic factors that are critical for ventricular morphogenesis by virtue of their ability to regulate gene expression through chromatin remodeling events. Covalent modification of the amino-terminal tails of histones, particularly H3 and H4, regulates higher-order chromatin structure and gene expression (Fig. 3). Modifications include acetylation of specific lysine residues (e.g., lysine^9 of histone H3) by histone acetyl transferases (HATs), deacetylation by histone deacetylases (HDACs), phosphorylation (serine^{10} of histone H3) by kinases and, most recently, methylation of lysine^9 of histone H3 by histone methyl transferases (HMTs) (Cheung et al. 2000; Khochbin et al. 2001; Marmorstein and Roth 2001).

mBop is one such factor that is expressed specifically in cardiac and skeletal muscle during development and contains two interesting domains that promote condensation of heterochromatin, resulting in transcriptional silencing (Gottlieb et al. 2002). The mBop protein contains a MYND domain most similar to that of the ETO protein whose fusion with the AML1 protein in chronic myelogenous leukemia converts AML1, normally a transcriptional activator, into a transcriptional repressor (Lutterbach et al. 1998a, b). The MYND domain of ETO is essential for this conversion and appears to function by recruiting the nuclear co-repressor, N-CoR, which in turn recruits the Sin3/HDAC complex to DNA sites specified by AML1 binding. mBop also recruits HDACs through the MYND domain and functions as a transcriptional repressor in part through this mechanism (Gottlieb et al. 2002).

It is unique that mBop also contains a SET domain that, in other proteins, contains the catalytic domain necessary for HMT activity (Rea et al. 2000). Most of the essential residues for HMT activity are conserved in mBop, suggesting that it plays a role through regulation of the methylation state of histones. It is interesting that the lysine residues of histone tails that get methylated must first be deacetylated, raising the possibility that mBop is able

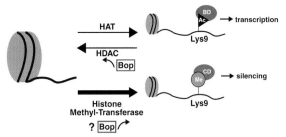

Figure 3. Schematic of epigenetic events regulating chromatin structure and transcription. Acetylation of specific lysine residues in tails of histone H3 or H4 by histone acetyl-transferases (HATs) results in relaxation of chromatin structure, making target DNA more accessible to DNA-binding transcription factors. The reverse reaction is catalyzed by histone deacetylases (HDACs) and results in condensation of chromatin into a transcriptionally silent state. Deacetylated residues can be methylated by histone methyl-transferases (HMTs), causing a more permanent state of transcriptional silencing. Acetylated (Ac) or methylated (Me) residues are recognized by bromodomain (BD)- or chromodomain (CD)-containing proteins, respectively.

to both recruit HDACS to "prepare" specific residues, and subsequently methylate those residues.

Investigation of the in vivo function of mBop was undertaken by targeted disruption in mice (Gottlieb et al. 2002). Mouse embryos lacking mBop displayed right ventricular hypoplasia and immature ventricular cardiomyocytes (Fig. 4); surprisingly, atrial cardiomyocytes appeared to differentiate normally. This phenotype was similar but more severe than that observed in mice lacking *dHAND*. Consistent with this, mBop was required for *dHAND* expression in the precardiac mesoderm, well before right ventricular formation, suggesting that regulation of *dHAND* may contribute to the right ventricular hypoplasia in *Bop* mutants (Fig. 4). Consistent with mBop's effects on dHAND, *Irx4* was also down-regulated in *Bop* mutants. Because mBop likely functions in vivo as a repressor of transcription, it is probable that there is an intermediate protein regulated by mBop that subsequently affects *dHAND* transcription and further downstream events. Identification of the molecular steps leading to mBop regulation of dHAND may yield insights into the precise targets to which mBop is recruited.

Other DNA-binding transcription factors also interact with HDACs and may regulate ventricular development through this mechanism. One member of the Mef2 family of transcriptional regulators, Mef2c, is essential for formation of the right and left ventricles in mice (Lin et al. 1997). Silencing of Mef2-dependent transcription through interaction with HDACs is necessary for regulation of hypertrophic growth of postnatal cardiomyocytes (McKinsey et al. 2000; Zhang et al. 2002). Whether a similar mechanism is involved in embryonic development of the heart remains unknown.

Finally, the hairy related transcription factors, Hrt1, Hrt2, and Hrt3, are expressed abundantly in the developing cardiovascular system and are transcriptional repressors that mediate events downstream of signaling by the

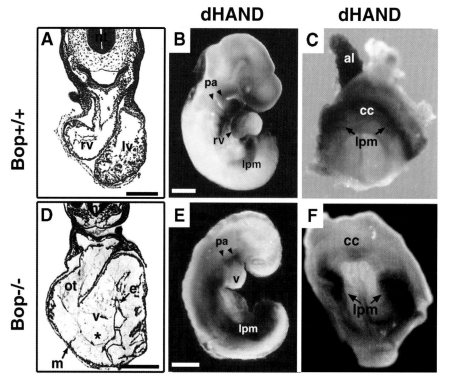

Figure 4. *dHAND* is down-regulated in *Bop*-null cardiac precursors. Transverse sections of $Bop^{-/-}$ or wild-type embryos at E9.25 reveal a single left-sided ventricle (v) that abruptly connects to an outflow tract (ot) in the mutant (*A, D*). *dHAND* expression is down-regulated in E9.0 *Bop*-null embryos specifically in the heart compared to wild type (*B, E*). Lateral plate mesoderm (lpm) and pharyngeal arch (pa) expression is unaffected. At E7.75, *dHAND* is barely detectable in the cardiac crescent (cc) but is expressed normally in the bilateral lateral plate mesoderm compared to wild type (*C, F*). (e) Endocardium, (m) myocardium, (rv) right ventricle, (lv) left ventircle, (al) allantois.

transmembrane receptor, Notch (Nakagawa et al. 1999, 2000). The zebrafish ortholog of *Hrt2*, *gridlock*, regulates the sorting of endothelial precursors into arterial or venous endothelial cells by mediating Notch signals (Zhong et al. 2001). In the mouse, *Hrt2* transcripts are present specifically in the ventricles but not atria, suggesting a ventricular-specific role for this gene in mammals (Nakagawa et al. 1999). Recent evidence suggests that the transcriptional repression by Hrt2 is mediated in part by recruitment of HDACs (Iso et al. 2001), although the impact of ventricular-specific histone modifications mediated by Hrt2 remains unknown.

SUMMARY

We have established the beginnings of a road map to understand how ventricular cells become specified, differentiate, and expand into a functional cardiac chamber (Fig. 5). The transcriptional networks described here provide clear evidence that disruption of pathways affecting ventricular growth could be the underlying etiology in a subset of children born with malformation of the right or left ventricle. As we learn details of the precise mechanisms through which the critical factors function, the challenge will lie in devising innovative methods to augment or modify the effects of gene mutations on ventricular development. Because most congenital heart disease likely occurs in a setting of heterozygous, predisposing mutations of one or more genes, modulation of activity of critical pathways in a preventive fashion may be useful in averting disease in genetically susceptible individuals.

ACKNOWLEDGMENTS

This work was supported by grants from the NHLBI/National Institutes of Health and the Donald W. Reynolds Clinical Cardiovascular Center to D.S. and E.N.O.; from the March of Dimes to D.S.; and from the National Institutes of Health to P.D.G.

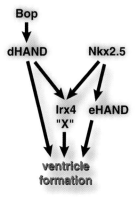

Figure 5. Molecular pathway for ventriculogenesis. A model for some known genes involved in ventricular development.

REFERENCES

Angelo S., Lohr J., Lee K.H., Ticho B.S., Breitbart R.E., Hill S., Yost H.J., and Srivastava D. 2000. Conservation of sequence and expression of *Xenopus* and zebrafish dHAND during cardiac, branchial arch and lateral mesoderm development. *Mech. Dev.* **95:** 231.

Bao Z.Z., Bruneau B.G., Seidman J.G., Seidman C.E., and Cepko C.L. 1999. Regulation of chamber-specific gene expression in the developing heart by Irx4. *Science* **283:** 161.

Biben C. and Harvey R.P. 1997. Homeodomain factor Nkx2-5 controls left/right asymmetric expression of bHLH gene eHAND during murine heart development. *Genes Dev.* **11:** 1357.

Bruneau B.G., Bao Z.Z., Tanaka M., Schott J.J., Izumo S., Cepko C.L., Seidman J.G., and Seidman C.E. 2000. Cardiac expression of the ventricle-specific homeobox gene Irx4 is modulated by Nkx2-5 and dHAND. *Dev. Biol.* **217:** 266.

Cheung P., Allis C.D., and Sassone-Corsi P. 2000. Signaling to chromatin through histone modifications. *Cell* **103:** 263.

Christoffels V.M., Habets P.E., Franco D., Campione M., de Jong F., Lamers W.H., Bao Z.Z., Palmer S., Biben C., Harvey R.P., and Moorman A.F. 2000. Chamber formation and morphogenesis in the developing mammalian heart. *Dev. Biol.* **223:** 266.

Firulli A.B., McFadden D.G., Lin Q., Srivastava D., and Olson E.N. 1998. Heart and extra-embryonic mesodermal defects in mouse embryos lacking the bHLH transcription factor Hand1. *Nat. Genet.* **18:** 266.

Gottlieb P.D., Pierce S.A., Sims R.J., Yamagishi H., Weihe E.K., Harriss J.V., Maika S.D., Kuziel W.A., King H.L., Olson E.N., Nakagawa O., and Srivastava D. 2002. Bop encodes a muscle-restricted protein containing MYND and SET domains and is essential for cardiac differentiation and morphogenesis. *Nat. Genet.* **31:** 25.

Iso T., Sartorelli V., Poizat C., Iezzi S., Wu H.Y., Chung G., Kedes L., and Hamamori Y. 2001. HERP, a novel heterodimer partner of HES/E(spl) in Notch signaling. *Mol. Cell. Biol.* **21:** 6080.

Khochbin S., Verdel A., Lemercier C., and Seigneurin-Berny D. 2001. Functional significance of histone deacetylase diversity. *Curr. Opin. Genet. Dev.* **11:** 162.

Lin Q., Schwarz J., Bucana C., and Olson E.N. 1997. Control of mouse cardiac morphogenesis and myogenesis by transcription factor MEF2C. *Science* **276:** 1404.

Lutterbach B., Sun D., Schuetz J., and Hiebert S.W. 1998a. The MYND motif is required for repression of basal transcription from the multidrug resistance 1 promoter by the t(8;21) fusion protein. *Mol. Cell. Biol.* **18:** 3604.

Lutterbach B., Westendorf J.J., Linggi B., Patten A., Moniwa M., Davie J.R., Huynh K.D., Bardwell V.J., Lavinsky R.M., Rosenfeld M.G., Glass C., Seto E., and Hiebert S.W. 1998b. ETO, a target of t(8;21) in acute leukemia, interacts with the N-CoR and mSin3 corepressors. *Mol. Cell. Biol.* **18:** 7176.

Marmorstein R. and Roth S.Y. 2001. Histone acetyltransferases: Function, structure, and catalysis. *Curr. Opin. Genet. Dev.* **11:** 155.

McKinsey T.A., Zhang C.L., Lu J., and Olson E.N. 2000. Signal-dependent nuclear export of a histone deacetylase regulates muscle differentiation. *Nature* **408:** 106.

Nakagawa O., Nakagawa M., Richardson J., Olson E.N., and Srivastava D. 1999. HRT1, HRT2 and HRT3: A new family of bHLH transcription factors marking specific cardiac, somitic and branchial arch segments. *Dev. Biol.* **216:** 72.

Nakagawa O., McFadden D.G., Nakagawa M., Yanagisawa H., Hu T., Srivastava D., and Olson E.N. 2000. Members of the HRT family of basic helix-loop-helix proteins act as transcriptional repressors downstream of Notch signaling. *Proc. Natl. Acad. Sci.* **97:** 13655.

Rea S., Eisenhaber F., O'Carroll D., Strahl B.D., Sun Z.W., Schmid M., Opravil S., Mechtler K., Ponting C.P., Allis C.D., and Jenuwein T. 2000. Regulation of chromatin structure by site-specific histone H3 methyltransferases. *Nature* **406:** 593.

Riley P., Anson-Cartwright L., and Cross J.C. 1998. The Hand1 bHLH transcription factor is essential for placentation and cardiac morphogenesis. *Nat. Genet.* **18:** 271.

Riley P.R., Gertsenstein M., Dawson K., and Cross J.C. 2000. Early exclusion of hand1-deficient cells from distinct regions of the left ventricular myocardium in chimeric mouse embryos. *Dev. Biol.* **227:** 156.

Srivastava D. and Olson E.N. 2000. A genetic blueprint for cardiac development: Implications for human heart disease. *Nature* **407:** 221.

Srivastava D., Cserjesi P., and Olson E.N. 1995. A subclass of bHLH proteins required for cardiac morphogenesis. *Science* **270:** 1995.

Srivastava D., Thomas T., Lin Q., Kirby M.L., Brown D., and Olson E.N. 1997. Regulation of cardiac mesodermal and neural crest development by the bHLH transcription factor, dHAND. *Nat. Genet.* **16:** 154.

Yamagishi H., Yamagishi C., Nakagawa O., Harvey R.P., Olson E.N., and Srivastava D. 2001. The combinatorial activities of Nkx2.5 and dHAND are essential for cardiac ventricle formation. *Dev. Biol.* **239:** 190.

Yelon D., Ticho B., Halpern M.E., Ruvinsky I., Ho R.K., Silver L.M., and Stainier D.Y. 2000. The bHLH transcription factor hand2 plays parallel roles in zebrafish heart and pectoral fin development. *Development* **127:** 2573.

Zhang C., McKinsey T., Chang S., Antos C., Hill J., and Olson E. 2002. Class II histone deacetylases act as signal-responsive repressors of cardiac hypertrophy. *Cell* **110:** 479.

Zhong T.P., Childs S., Leu J.P., and Fishman M.C. 2001. Gridlock signalling pathway fashions the first embryonic artery. *Nature* **414:** 216.

Hypoxia, HIFs, and Cardiovascular Development

M.C. Simon,*† D. Ramirez-Bergeron,*† F. Mack,†‡ C.-J. Hu,† Y. Pan,*†
and K. Mansfield†‡

*Howard Hughes Medical Institute, †Abramson Family Cancer Research Institute, ‡Cell and Molecular Biology Graduate Program, University of Pennsylvania School of Medicine, Philadelphia, Pennsylvania 19104

Proper development and function of the cardiovascular system requires complex signaling between hematopoietic cells, vascular endothelial cells, and various support cells, including smooth muscle and pericytes. The specification of hematopoietic and endothelial cell fates from common precursors and establishment of a differentiated vascular network is driven by a genetically controlled pathway, as indicated by loss-of-function mutations in numerous genes encoding specific growth factors (vascular endothelial growth factor [VEGF], angiopoietin-1, angiopoietin-2, etc.) and their receptors (Fong et al. 1995; Sato et al. 1995; Shalaby et al. 1995; Ferrara et al. 1996). However, this genetic program appears to be regulated in turn by metabolic cues such as oxygen deprivation (hypoxia). Diffusion of oxygen (O_2) in the embryo is limited by its size shortly after gastrulation (Ramirez-Bergeron and Simon 2001). During later embryogenesis, rapid cellular proliferation and organogenesis create localized regions of hypoxia. We have demonstrated that an inability to modulate gene expression in response to low O_2 levels has profound effects on development of all components of the embryonic cardiovascular system (blood, vessels, heart, and placenta). Our results suggest that O_2 levels, and O_2 gradients, function as critical developmental signals in many aspects of cardiovascular differentiation. This is reminiscent of the hypoxic zones observed in solid tumors and ischemic or infected tissues in adults (Semenza 2000). Therefore, a systematic examination of cellular and molecular responses to developmental hypoxia will provide important insights into the treatment of various disease states including tumor growth, diabetic retinopathy, preeclampsia, wound healing, and ischemia.

HYPOXIA-MEDIATED SIGNAL TRANSDUCTION

Activation of the hypoxia inducible factor (HIF) transcriptional complex represents the primary molecular mechanism by which O_2 regulates gene expression (Semenza 1999). HIF complex accumulation is inversely related to O_2 tension: It becomes detectable in cells grown at 8–10% O_2, and levels increase almost linearly as O_2 concentration drops (Jiang et al. 1996). Interaction of the HIF complex with its consensus DNA-binding site is required for hypoxia-induced target gene expression. The majority of genes induced by HIF are involved in cellular, tissue, and systemic responses to hypoxia (Semenza 1999). At a cellular level, O_2-starved cells respond by switching from oxidative phosphorylation to glycolytic production of ATP. At a localized or tissue level, affected tissues respond by improving the vascular density of a local region by stimulating the production of angiogenetic growth factors. At a systemic level, O_2-deprived organisms respond by stimulating the expression of erythropoietin (which increases red blood cell mass) in cells that reside within the kidney or the liver. Approximately 100 HIF target genes have been identified that encode a wide array of proteins associated with cellular metabolism and survival, including glycolytic enzymes, glucose transporters, and paracrine growth factors (VEGF and erythropoietin) (Semenza 1999; C.-J. Hu et al., in prep.).

HIF typically consists of a heterodimer of two basic helix-loop-helix (bHLH)-PAS proteins, HIF-1α and aryl hydrocarbon receptor nuclear translocator (ARNT) (Wang et al. 1995). These bHLH-PAS transcription factors induce gene expression by binding to a 50-base pair hypoxia response element (HRE) containing a core 5´-ACGTG-3´ sequence (Pugh et al. 1991; Semenza et al. 1991). Other bHLH-PAS proteins regulate circadian rhythms, neurogenesis, and toxin metabolism. Three bHLH-PAS proteins in vertebrates respond to hypoxia: HIF-1α, HIF-2α (EPAS), and HIF-3α. These factors heterodimerize with ARNT or ARNT2. Both HIF-1α and ARNT are widely expressed. Therefore, this complex is the master regulator of O_2 homeostasis and induces a network of genes involved in the cellular and systemic responses to O_2 deprivation. HIF-2α is more restricted in its expression and appears at high levels in endothelial cells (Ema et al. 1997; Flamme et al. 1997; Tian et al. 1997), whereas HIF-3α is expressed at high levels in the thymus, GI tract, and lung (Gu et al. 1998).

As shown in Figure 1, in the presence of O_2, HIFα subunits are ubiquinated by the von Hippel-Lindau protein (pVHL) and degraded via the proteosome (Kallio et al. 1999; Maxwell et al. 1999; Ohh et al. 2000; Cockman et al. 2000). Under hypoxic conditions, HIFα is instead stabilized, translocated to the nucleus, and complexed with ARNT to promote HRE-driven transcription of O_2-regulated genes. Considerable progress has been made in understanding the transcription factors activated during hypoxia, but the underlying mechanisms of O_2 sensing by cells are incompletely understood. Some evidence indi-

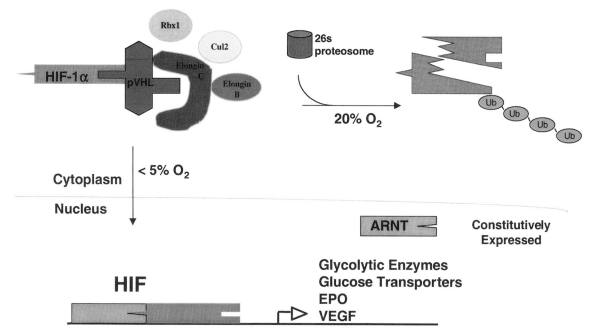

Figure 1. Model of HIF regulation by oxygen. In the presence of high levels of O_2 (20%), HIFα subunits are ubiquinated by pVHL and degraded by the 26s proteosome. The HIF-1α subunit is shown as an example. Under hypoxic conditions (less than 5% O_2), HIFα subunits are translocated to the nucleus and dimerize with ARNT, generating the HIF transcriptional activator of genes containing HREs (e.g., glycolytic enzymes, glucose transporters, EPO, and VEGF).

cates that hypoxia activates HIF via a mitochondria-dependent signaling process involving increased reactive O_2 species (ROS) (Chandel et al. 1998, 2000). More recent experiments have shown that prolyl hydroxylases modify HIFα subunits on proline residues within the O_2-dependent degradation domain (ODD) of HIF-1α, HIF-2α, and HIF-3α (Ivan et al. 2001; Jaakkola et al. 2001; Yu et al. 2001). Although prolyl hydroxylation of the ODDs is necessary and sufficient to allow HIFα to bind pVHL, causing subsequent ubiquination, it remains a formal possibility that ROS act upstream of prolyl hydroxylase enzymatic activity. We are currently investigating a model whereby mitochondrial function is essential for prolyl hydroxylation on all HIFα subunits during normoxic culture conditions.

HIF AND OXYGEN HOMEOSTASIS IN THE EMBRYO

By creating targeted mutations in murine HIF subunit genes, we have begun to assess the developmental consequences of impaired O_2 signaling. Our results demonstrate an absolute requirement for HIF function in hematopoietic, vascular, cardiac, and placental development. These findings underscore the importance of O_2 signaling in mammalian ontogeny and offer potential targets for hematopoietic and angiogenic therapies.

To genetically assess the function of HIF in mice, we have created $Arnt^{-/-}$ embryonic stem (ES) cells (Maltepe et al. 1997). $Arnt^{-/-}$ ES cells are devoid of HIF activity and fail to activate genes normally regulated by low O_2 (e.g., PGK1, PFLK, ALDA, GAPDH, GLUT1, and VEGF). These results demonstrate that HIF is essential for O_2-regulated gene expression. Of note, ~100 genes regulated by O_2 deprivation are not induced in ES cells deficient in either ARNT or HIF-1α (C.-J. Hu et al., in prep.). Given that the diffusion of O_2 in the embryo is limited by its size shortly after gastrulation, embryos defective in HIF activity should display severe developmental defects. Importantly, multiple growth factors essential for normal blood vessel development such as VEGF, angiopoietin-1, angiopoietin-2, bFGF, and PDGFβ are HIF target genes (Semenza 1999). We found that $Arnt^{-/-}$ embryos exhibit an early lethality associated with defective vascularization of the yolk sac, placenta, branchial arches, and cranium (Maltepe et al. 1997; Adelman et al. 2000). Vascular development is thought to take place in multiple steps; the first step is vasculogenesis, during which angioblasts expand, extend, project, and interconnect to form the endothelial cells of the primary vascular plexus (Risau 1997). Further remodeling of primary vessels via angiogenesis results in branching, sprouting, migration, and proliferation of preexisting cells, deposition of extracellular matrix (ECM) components, and recruitment of stromal cells for structural support. Although the initial development of vascular beds remains intact, vessel remodeling throughout the $Arnt^{-/-}$ embryo is abnormal. Of note, VEGF is expressed at low levels in E8.5 mutant embryos relative to wild type. These findings were largely confirmed by targeted mutation of the Hif-$1α$ gene (Iyer et al. 1998; Ryan et al. 1998). Mutant embryos derived from Hif-$2α^{-/-}$ ES cells develop severe vascular defects in the yolk sac, and the embryo proper also exhibits angiogenic defects (Peng et al. 2000). Collectively, these results indicate that HIF dimers consisting of both HIF-1α and HIF-2α play an important role at postvasculogenesis stages and are required for the remodeling of a primary network into mature vascular patterns.

HIF AND HEMATOPOIESIS

Given the spatial and functional relationship between primitive endothelial cells (ECs) and hematopoietic stem cells (HSCs), we assessed blood cell development in mice lacking HIF. We determined that $Arnt^{-/-}$ embryos produce decreased numbers of hematopoietic progenitors within the yolk sac (Adelman et al. 1999). The yolk sac progenitor defect is recapitulated by in vitro differentiation of $Arnt^{-/-}$ ES cells into embryoid bodies (EBs), which fail to exhibit hypoxia-mediated progenitor proliferation. Mutant progenitors exhibit a decrease in HIF-dependent VEGF expression, and progenitor numbers are increased by exogenous VEGF. Furthermore, the progenitor defect is cell extrinsic, as determined by generating chimeric animals consisting of $Arnt^{+/+}$ and $Arnt^{-/-}$ cells throughout a number of tissues. Here, progenitors cultured from yolk sac, fetal liver, and bone marrow reveal appropriate contribution of $Arnt^{-/-}$ ES cells to CFU numbers. Our data indicate that the physiologic hypoxia encountered by the E9.5 embryo is essential for the proliferation and/or survival of hematopoietic precursors within the yolk sac.

Hypoxia plays an important role in later stages of hematopoietic development as well. For example, we expect to observe similar defects in the para-aortic region within the early embryo. Our preliminary experiments indicate that indeed para-aortic hematopoietic progenitors and endothelial precursors are also deficient in $Arnt^{-/-}$ embryos. At later stages of development, blood cells are exposed to low O_2 tensions as they migrate between blood and different tissues. Recent studies have shown that $HIF-1α^{-/-}$ lymphocytes exhibit defects in differentiation and function, and HIF-1α deficiency in $Rag2^{-/-}$ chimeras results in an elevation in the number of peritoneal B1-like lymphocytes that express high levels of B220 on their surface (Kojima et al. 2002). Such animals display autoimmunity with the accumulation of anti-dsDNA antibodies and rheumatoid factor in serum and deposits of IgG and IgM in the kidney. In conclusion, hypoxia regulates hematopoietic tissue homeostasis in a HIF-dependent manner. It will be important to determine the mechanism by which hypoxia and HIF regulate the function of mature blood cells.

Given that $Arnt^{-/-}$ embryos exhibit both vascular and hematopoietic defects, we are studying the development of hemangioblasts (precursors to endothelial and blood cells) in $Arnt^{-/-}$, $Hif-1α^{-/-}$, and $Hif-2α^{-/-}$ embryos, and in vitro cultured ES cells. We propose that HIF coordinates angiogenic and hematopoietic differentiation by promoting VEGF expression. We have learned that $Arnt^{-/-}$ EBs produce greatly reduced numbers of hemangioblasts in vitro (D.L. Ramirez-Bergeron et al., in prep.). Similarly, $Vegf^{-/-}$ EBs also exhibit hemangioblastic defects. It will be critical to learn whether other HIF targets (bFGF, angiopoietin-1, angiopoietin-2) participate in this early developmental process.

OXYGEN-MEDIATED DEVELOPMENT OF THE PLACENTA AND HEART

Placental development is greatly influenced by O_2 tension. Human cytotrophoblasts proliferate under low O_2 in vitro but differentiate as O_2 levels rise, mimicking the developmental transition they undergo as they invade the maternal tissue of the endometrium (Genbacev et al. 1997). As stated previously, the embryo resides in a low O_2 environment prior to E9.5 and relies primarily on glycolysis to satisfy its metabolic needs. Establishment of placental circulation by E9.0 to E9.5 permits O_2 and nutrient delivery through the fetal vasculature to the rapidly growing embryo. Given the requirement for HIF and transcriptional responses to hypoxia, we reasoned that HIF may be critical to placentation. We demonstrated that placental cell fates are regulated by HIF-mediated hypoxia responses (Adelman et al. 2000). Specifically, $Arnt^{-/-}$ placentas show aberrant architecture due to reduced labyrinthine and spongiotrophoblast layers and increased numbers of giant cells. Importantly, trophoblast stem (TS) cells cultured in low O_2 preferentially differentiate into spongiotrophoblasts at the expense of giant cells in a HIF-dependent manner. Therefore, O_2 levels regulate cell fate determination in vivo, a model proposed using human placental explants.

In addition to yolk sac hematopoietic and endothelial defects, $Arnt$ mutant embryos exhibit hypoplastic tissues within the developing heart (Adelman et al. 2000). The ventricular myocardium is significantly reduced in size, while endocardial cushions that ultimately line the atrioventricular (AV) canal and develop into leaflets of the AV valves are severely underdeveloped. This produces an unusually large AV canal and impaired ventricular function. The result is poor cardiac output and dilated vessels throughout the embryo and yolk sac. Yolk sac vascularization depends on hydrostatic pressure and associated shear forces for vascular remodeling. In conclusion, "physiologic" hypoxia encountered by the developing embryo and HIF activation are critically important for the proper differentiation of specific cell types within the cardiovascular system. In fact, the entire cardiovascular system is abnormal in HIF-deficient embryos: vessels, blood, placenta, and heart.

UNIQUE AND OVERLAPPING ROLES FOR HIFα AND HIFβ SUBUNITS

The HIF protein family in mammals is complex in that three genes encode HIFα subunits and three genes encode ARNT (HIFβ) subunits in mice. ARNT has been shown to dimerize with HIF-1α as well as HIF-2α and HIF-3α under hypoxic conditions. Similarly, ARNT exhibits a high degree of sequence homology with ARNT2 and ARNT3. To reveal potential redundancy between these proteins, we have created both $Arnt2^{-/-}$ and $Arnt3^{-/-}$ mice. We demonstrated that ARNT2 forms functional

HIF complexes and regulates hypoxia-responsive genes (Maltepe et al. 2000; Keith et al. 2001). In addition, ARNT2 interacts with another bHLH-PAS protein, SIM1, to regulate neuroendocrine cell differentiation within the hypothalamus, demonstrating that individual bHLH-PAS proteins can fulfill multiple functions (Michaud et al. 2000; Keith et al. 2001). We have recently shown that $Arnt^{-/-}Arnt2^{-/-}$ embryos die at E7.5 and hypothesize that such mutants exhibit no hypoxia responses, given the severity of this phenotype relative to the $Arnt^{-/-}$, $Hif-1\alpha^{-/-}$, and $Hif-2\alpha^{-/-}$ single mutant animals. Unlike ARNT and ARNT2, ARNT3 (MOP3) does not effectively form HIF complexes and exhibits no role in hypoxic gene induction in vivo (Cowden and Simon 2002). Instead, ARNT3 is a nonredundant partner of CLOCK, an essential component of the circadian pacemaker in mammals (Bunger et al. 2000). Future studies will exploit the availability of targeted mutations in all members of the hypoxia complex ($Hif-1\alpha^{-/-}$, $Hif-2\alpha^{-/-}$, $Arnt^{-/-}$, and $Arnt2^{-/-}$ mice) to discern unique roles played by individual complexes (HIF-1α/ARNT versus HIF-2α/ARNT, etc.) during cardiovascular development.

We have recently determined via microarray analyses that many HIF-1α target genes (e.g., glycolytic enzymes) are not regulated by HIF-2α (C.-J. Hu et al., in prep.). Instead, HIF-2α regulates a subset of HIF-1α targets (e.g., VEGF, glucose transporters) in conjunction with HIF-1α. Why some HREs are exclusively responsive to HIF-1α while others can be regulated by both HIF-1α and HIF-2α is currently unclear. We are using both $Hif-1\alpha^{-/-}$ and $Hif-2\alpha^{-/-}$ ES and endothelial cells to identify HIF-2α-specific target genes. We have learned that although ES cells express high levels of HIF-2α, it is inert. $Hif-1\alpha^{-/-}$ endothelial cells should allow the identification of genes regulated by HIF-2α only. These experiments are critical to understanding *which* complex participates in both cellular differentiation and neoplastic transformation. Certain malignancies could express HIF-1α whereas others express HIF-2α. We wish to determine whether inappropriate activation of HIF-1α versus HIF-2α leads to different kinds of malignancy.

HIF-3α is unique in that it appears to lack a transcriptional activation domain (Gu et al. 1998). We have shown that HIF-3α expression is regulated by HIF-1α and potentially provides inhibition of HIF-1α and HIF-2α function via a negative feedback loop. We have created 293 clones, which inducibly express nondegradable HIF-1α, HIF-2α, and HIF-3α to molecularly dissect unique and overlapping roles provided by these three HIFα subunits.

HIF, VON HIPPEL-LINDAU DISEASE, AND OTHER NEOPLASMS

von Hippel-Lindau (VHL) is a hereditary cancer syndrome, where patients develop a variety of highly vascularized tumors, such as renal clear-cell carcinoma, retinal angioma, CNS hemangioblastoma, and pheochromocytoma (Kondo and Kaelin 2001). pVHL, the protein product of the *VHL* gene, interacts through its α domain with elongin B, elongin C, Cul2, and Rbx1 to become an E3 ubiquitin ligase (Krek 2000). As stated above, this multiprotein complex targets HIFα subunits for proteosome-mediated degradation. Mutations within the β domain of pVHL disrupt its interaction with HIFα subunits, leading to constitutive α-subunit stabilization and activity in VHL tumors. Renal tumor cell lines deficient in pVHL exhibit constitutive HIF function under normoxic conditions and high levels of expression of HIF target genes such as erythropoietin, VEGF, glucose transporters, and glycolytic enzymes.

Thus far, studies correlating loss of pVHL function with HIF dysregulation have been conducted in tumor-derived cell lines with a number of genetic abnormalities. Although reintroduction of pVHL into renal carcinoma cells eliminates their tumorigenic phenotype, this occurs within a cellular background including a number of genetic lesions. Therefore, we have eliminated murine *Vhl* from mouse ES cells, a primary cell line. We demonstrated that loss of pVHL is sufficient to completely deregulate HIF stabilization, DNA-binding activity, target gene activation, and nuclear localization, conferring a hypoxic phenotype to normoxic $Vhl^{-/-}$ ES cells (F. Mack et al., in prep.). Surprisingly, teratomas derived from $Vhl^{-/-}$ ES cells injected into immunocompromised mice exhibit smaller volume as compared to $Vhl^{+/-}$ control tumors. These results suggest that other genetic changes must occur to facilitate tumorigenesis in the absence of *Vhl*. It has been hypothesized that HIF protein stabilization, HIFα transit to the nucleus, interaction with transcriptional co-activators, and DNA binding are all regulated by O_2 availability. Our results argue that if this is the case, all of these steps must be downstream of pVHL. It will be important in the future to determine what other genetic lesions are necessary to confer a fully transformed neoplastic phenotype to pVHL-deficient cells.

Other tumor suppressor genes have been reported to regulate HIF function. For example, high levels of HIF-1α protein have been detected in glioblastoma and prostate cancer cells with mutations in the tumor suppressor PTEN (Zhong et al. 2000; Zundel et al. 2000) and colon cancer cells deficient in p53 (Ravi et al. 2000). When challenged by hypoxic stress, cells frequently undergo cell cycle arrest and/or apoptosis (Graeber et al. 1996; Gardner et al. 2001). It has been proposed that hypoxia can stimulate p53 protein stabilization (Graeber et al. 1994). In fact, this may occur by direct association with the HIF-1α protein (An et al. 1998). These results suggest a model whereby hypoxia induces p53 which ultimately down-regulates HIF-1α by recruiting mdm-2-associated ubiquitinization machinery (Ravi et al. 2000). However, we have closely examined multiple human cancer cell and ES cell lines and found that hypoxia alone is not sufficient to elevate p53 protein levels (Y. Pan and M.C. Simon, in prep.). Furthermore, the expression of p53 decreases under anoxic conditions, presumably due to a general inhibition of translation in cells starved for O_2. Our studies suggest that hypoxia per se is not a bona fide stimulus of p53 stabilization. We propose that the p53 pathway is activated in hypoxic regions of tumors via secondary effects brought by hypoxia and lack of proper vascularization. Further investigation of the mechanisms leading to p53 accumulation in hypoxic tumors should help provide a framework for understanding its role in regulating tumor angiogenesis.

CONCLUSIONS AND FUTURE DIRECTIONS

We provide evidence here that physiologic parameters, such as O_2 concentration, affect the expression of genes essential for vascular, hematopoietic, placental, and cardiac cell differentiation. Furthermore, production of an intact cardiovascular system is dependent on HIF, a master regulator of O_2 homeostasis during both fetal and postnatal life. Postnatally, HIF is required for a variety of responses to chronic hypoxia such as that encountered during tissue ischemia, stroke, and neoplasia. Therefore, HIF plays a pivotal role in embryogenesis, cellular and systemic physiology, and pathophysiology. Future experiments will address the critical question of functional redundancy between HIF subunit family members in these complex events. We hope to define unique and overlapping roles for all HIF proteins involved in O_2 homeostasis: HIF-1α, HIF-2α, HIF-3α, ARNT, and ARNT2. Furthermore, it will be important to define the scope of the response in different hypoxic tissues and to delineate tissue-specific HIF target genes. It will also be necessary to understand how the HIF signaling pathway is ultimately down-regulated once cells become re-oxygenated. We want to understand the molecular mechanisms by which HIF complexes mediate O_2-regulated signal transduction during development and disease.

ACKNOWLEDGMENTS

We acknowledge all members of our laboratory past and present for hard work, stimulating discussions, and critical reading of this manuscript. The research cited here was supported by the National Institutes of Health (grants 52094 and 66310), the Abramson Family Cancer Research Institute, and the Howard Hughes Medical Institute.

REFERENCES

Adelman D.M., Maltepe E., and Simon M.C. 1999. Multilineage embryonic hematopoiesis requires hypoxic ARNT activity. *Genes Dev.* **13:** 2478.

Adelman D.M., Gertsenstein M., Nagy A., Simon M.C., and Maltepe E. 2000. Placental cell fates are regulated in vivo by HIF-mediated hypoxia responses. *Genes Dev.* **14:** 3191.

An W.G., Kanekal M., Simon M.C., Maltepe E., Blagosklonny M.V., and Neckers L.M. 1998. Stabilization of wild-type p53 by hypoxia-inducible factor 1alpha. *Nature* **392:** 405.

Bunger M.K., Wilsbacher L.D., Moran S.M., Clendenin C., Radcliffe L.A., Hogenesch J.B., Simon M.C., Takahashi J.S., and Bradfield C.A. 2000. Mop3 is an essential component of the master circadian pacemaker in mammals. *Cell* **103:** 1009.

Chandel N.S., Maltepe E., Goldwasser E., Mathieu C.E., Simon M.C., and Schumacker P.T. 1998. Mitochondrial reactive oxygen species trigger hypoxia-induced transcription. *Proc. Natl. Acad. Sci.* **95:** 11715.

Chandel N.S., McClintock D.S., Feliciano C.E., Wood T.M., Melendez J.A., Rodriguez A.M., and Schumacker P.T. 2000. Reactive oxygen species generated at mitochondrial complex III stabilize hypoxia-inducible factor-1alpha during hypoxia: A mechanism of O_2 sensing. *J. Biol. Chem.* **275:** 25130.

Cockman M.E., Masson N., Mole D.R., Jaakkola P., Chang G.W., Clifford S.C., Maher E.R., Pugh C.W., Ratcliffe P.J., and Maxwell P.H. 2000. Hypoxia inducible factor-alpha binding and ubiquitylation by the von Hippel-Lindau tumor suppressor protein. *J. Biol. Chem.* **275:** 25733.

Cowden K.D. and Simon M.C. 2002. The bHLH/PAS factor MOP3 does not participate in hypoxia responses. *Biochem. Biophys. Res. Commun.* **290:** 1228.

Ema M., Taya S., Yokotani N., Sogawa K., Matsuda Y., and Fujii-Kuriyama Y. 1997. A novel bHLH-PAS factor with close sequence similarity to hypoxia-inducible factor 1alpha regulates the VEGF expression and is potentially involved in lung and vascular development. *Proc. Natl. Acad. Sci.* **94:** 4273.

Ferrara N., Carver-Moore K., Chen H., Dowd M., Lu L., O'Shea K.S., Powell-Braxton L., Hillan K.J., and Moore M.W. 1996. Heterozygous embryonic lethality induced by targeted inactivation of the VEGF gene. *Nature* **380:** 439.

Flamme I., Frohlich T., von Reutern M., Kappel A., Damert A., and Risau W. 1997. HRF, a putative basic helix-loop-helix-PAS-domain transcription factor is closely related to hypoxia-inducible factor-1 alpha and developmentally expressed in blood vessels. *Mech. Dev.* **63:** 51.

Fong G.H., Rossant J., Gertsenstein M., and Breitman M.L. 1995. Role of the Flt-1 receptor tyrosine kinase in regulating the assembly of vascular endothelium. *Nature* **376:** 66.

Gardner L.B., Li Q., Park M.S., Flanagan W.M., Semenza G.L., and Dang C.V. 2001. Hypoxia inhibits G1/S transition through regulation of p27 expression. *J. Biol. Chem.* **276:** 7919.

Genbacev O., Zhou Y., Ludlow J.W., and Fisher S.J. 1997. Regulation of human placental development by oxygen tension. *Science* **277:** 1669.

Graeber T.G., Peterson J.F., Tsai M., Monica K., Fornace A.J., Jr., and Giaccia A.J. 1994. Hypoxia induces accumulation of p53 protein, but activation of a G1-phase checkpoint by low-oxygen conditions is independent of p53 status. *Mol. Cell. Biol.* **14:** 6264.

Graeber T.G., Osmanian C., Jacks T., Housman D.E., Koch C.J., Lowe S.W., and Giaccia A.J. 1996. Hypoxia-mediated selection of cells with diminished apoptotic potential in solid tumours (comments). *Nature* **379:** 88.

Gu Y.Z., Moran S.M., Hogenesch J.B., Wartman L., and Bradfield C.A. 1998. Molecular characterization and chromosomal localization of a third alpha-class hypoxia inducible factor subunit, HIF3alpha. *Gene Expr.* **7:** 205.

Ivan M., Kondo K., Yang H., Kim W., Valiando J., Ohh M., Salic A., Asara J.M., Lane W.S., and Kaelin W.G., Jr. 2001. HIFalpha targeted for VHL-mediated destruction by proline hydroxylation: Implications for O_2 sensing. *Science* **292:** 464.

Iyer N.V., Kotch L.E., Agani F., Leung S.W., Laughner E., Wenger R.H., Gassmann M., Gearhart J.D., Lawler A.M., Yu A.Y., and Semenza G.L. 1998. Cellular and developmental control of O_2 homeostasis by hypoxia-inducible factor 1 alpha. *Genes Dev.* **12:** 149.

Jaakkola P., Mole D.R., Tian Y.M., Wilson M.I., Gielbert J., Gaskell S.J., Kriegsheim A., Hebestreit H.F., Mukherji M., Schofield C.J., Maxwell P.H., Pugh C.W., and Ratcliffe P.J. 2001. Targeting of HIF-alpha to the von Hippel-Lindau ubiquitylation complex by O_2-regulated prolyl hydroxylation. *Science* **292:** 468.

Jiang B.H., Semenza G.L., Bauer C., and Marti H.H. 1996. Hypoxia-inducible factor 1 levels vary exponentially over a physiologically relevant range of O_2 tension. *Am. J. Physiol.* **271:** C1172.

Kallio P.J., Wilson W.J., O'Brien S., Makino Y., and Poellinger L. 1999. Regulation of the hypoxia-inducible transcription factor 1alpha by the ubiquitin-proteasome pathway. *J. Biol. Chem.* **274:** 6519.

Keith B., Adelman D.M., and Simon M.C. 2001. Targeted mutation of the murine arylhydrocarbon receptor nuclear translocator 2 (Arnt2) gene reveals partial redundancy with Arnt. *Proc. Natl. Acad. Sci.* **98:** 6692.

Kojima H., Gu H., Nomura S., Caldwell C.C., Kobata T., Carmeliet P., Semenza G.L., and Sitkovsky M.V. 2002. Abnormal B lymphocyte development and autoimmunity in hypoxia-inducible factor 1alpha-deficient chimeric mice. *Proc. Natl. Acad. Sci.* **99:** 2170.

Kondo K. and Kaelin W.G., Jr. 2001. The von Hippel-Lindau tumor suppressor gene. *Exp. Cell Res.* **264:** 117-25.

Krek W. 2000. VHL takes HIF's breath away. *Nat. Cell. Biol.* **2:** E1.3.

Maltepe E., Keith B., Arsham A., Brorson J., and Simon M.C. 2000. The role of ARNT2 in tumor angiogenesis and the neural response to hypoxia. *Biochem. Biophys. Res. Commun.* **273:** 231.

Maltepe E., Schmidt J.V., Baunoch D., Bradfield C.A., and Simon M.C. 1997. Abnormal angiogenesis and responses to glucose and oxygen deprivation in mice lacking the protein ARNT. *Nature* **386:** 403.

Maxwell P.H., Wiesener M.S., Chang G.W., Clifford S.C., Vaux E.C., Cockman M.E., Wykoff C.C., Pugh C.W., Maher E.R., and Ratcliffe P.J. 1999. The tumour suppressor protein VHL targets hypoxia-inducible factors for oxygen-dependent proteolysis. *Nature* **399:** 271.

Michaud J.L., DeRossi C., May N.R., Holdener B.C., and Fan C. 2000. ARNT2 acts as the dimerization partner of SIM1 for the development of the hypothalamus. *Mech. Dev.* **90:** 253.

Ohh M., Park C.W., Ivan M., Hoffman M.A., Kim T.Y., Huang L.E., Pavletich N., Chau V., and Kaelin W.G. 2000. Ubiquitination of hypoxia-inducible factor requires direct binding to the beta-domain of the von Hippel-Lindau protein. *Nat. Cell Biol.* **2:** 423.

Peng J., Zhang L., Drysdale L., and Fong G.H. 2000. The transcription factor EPAS-1/hypoxia-inducible factor 2alpha plays an important role in vascular remodeling. *Proc. Natl. Acad. Sci.* **97:** 8386.

Pugh C.W., Tan C.C., Jones R.W., and Ratcliffe P.J. 1991. Functional analysis of an oxygen-regulated transcriptional enhancer lying 3′ to the mouse erythropoietin gene. *Proc. Natl. Acad. Sci. USA* **88:** 10553-7.

Ramirez-Bergeron D.L. and Simon M.C. 2001. Hypoxia-inducible factor and the development of stem cells of the cardiovascular system. *Stem Cells* **19:** 279.

Ravi R., Mookerjee B., Bhujwalla Z.M., Sutter C.H., Artemov D., Zeng Q., Dillehay L.E., Madan A., Semenza G.L., and Bedi A. 2000. Regulation of tumor angiogenesis by p53-induced degradation of hypoxia- inducible factor 1alpha. *Genes Dev.* **14:** 34.

Risau W. 1997. Mechanisms of angiogenesis. *Nature* **386:** 671.

Ryan H.E., Lo J., and Johnson R.S. 1998. HIF-1α is required for solid tumor formation and embryonic vascularization. *EMBO J.* **17:** 3005.

Sato T.N., Tozawa Y., Deutsch U., Wolburg-Buchholz K., Fujiwara Y., Gendron-Maguire M., Gridley T., Wolburg H., Risau W., and Qin Y. 1995. Distinct roles of the receptor tyrosine kinases Tie-1 and Tie-2 in blood vessel formation. *Nature* **376:** 70.

Semenza G.L. 2000. HIF-1 and human disease: One highly involved factor. *Genes Dev.* **14:** 1983.

Semenza G.L., Nejfelt M.K., Chi S.M., and Antonarakis S.E. 1991. Hypoxia-inducible nuclear factors bind to an enhancer element located 3′ to the human erythropoietin gene. *Proc. Natl. Acad. Sci.* **88:** 5680.

———. 1999. Regulation of mammalian O_2 homeostasis by hypoxia-inducible factor 1. *Annu. Rev. Cell Dev. Biol.* **15:** 551.

Shalaby F., Rossant J., Yamaguchi T.P., Gertsenstein M., Wu X.F., Breitman M.L., and Schuh A. 1995. Failure of blood-island formation and vasculogenesis in *Flk*-1 deficient mice. *Nature* **376:** 62.

Tian H., McKnight S.L., and Russell D.W. 1997. Endothelial PAS domain protein 1 (EPAS1), a transcription factor selectively expressed in endothelial cells. *Genes Dev.* **11:** 72.

Wang G.L., Jiang B.-H., Rue E.A., and Semenza G.L. 1995. Hypoxia-inducible factor 1 is a basic-helix-loop-helix-PAS heterodimer regulated by cellular O_2 tension. *Proc. Natl. Acad. Sci.* **92:** 5510.

Yu F., White S.B., Zhao Q., and Lee F.S. 2001. HIF-1alpha binding to VHL is regulated by stimulus-sensitive proline hydroxylation. *Proc. Natl. Acad. Sci.* **98:** 9630.

Zhong H., Chiles K., Feldser D., Laughner E., Hanrahan C., Georgescu M.M., Simons J.W., and Semenza G.L. 2000. Modulation of hypoxia-inducible factor 1alpha expression by the epidermal growth factor/phosphatidylinositol 3-kinase/PTEN/AKT/FRAP pathway in human prostate cancer cells: Implications for tumor angiogenesis and therapeutics. *Cancer Res.* **60:** 1541.

Zundel W., Schindler C., Haas-Kogan D., Koong A., Kaper F., Chen E., Gottschalk A.R., Ryan H.E., Johnson R.S., Jefferson A.B., Stokoe D., and Giaccia A.J. 2000. Loss of PTEN facilitates HIF-1-mediated gene expression. *Genes Dev.* **14:** 391.

Quiescent Hypervascularity Mediated by Gain of HIF-1α Function

J.M. ARBEIT

UCSF Comprehensive Cancer Center and Department of Surgery, University of California, San Francisco, California 94143-0808

CONCEPT AND CHALLENGE OF THERAPEUTIC ANGIOGENESIS

The goal of therapeutic angiogenesis is an increase in perfusion of ischemic tissues such as the heart, limbs, and brain. We now understand a great deal about the molecular control of angiogenesis in the context of both cell culture and animal models, and clinical trials using angiogenic growth factors or vascular precursor/stem cells have been implemented. A novel concept is modulation of transcription factor function to affect therapeutic angiogenesis. A potential bonus of this approach is that downstream genetic networks regulated by the transcription factor may tailor additional cellular functions, such as metabolism, which may facilitate ischemic adaptation and survival. Hypoxia-inducible factor-1 alpha (HIF-1α) is one transcription factor that may fulfill this promise. Not only has the molecular and cellular biology of HIF-1α been extensively characterized, but molecules and signaling pathways regulating HIF-1α protein stability and transcriptional activity have been identified (Semenza 2002). This body of data offers the opportunity to tailor manipulation of HIF-1α function to a particular disease or condition. Recent data demonstrate that HIF-1α overexpression can produce an increase in tissue vascularity and maintain these vessels with normal morphology and permeability. As such, manipulation of HIF-1α function is one mechanism to achieve therapeutic angiogenesis.

MOLECULAR INDUCTION OF ANGIOGENESIS, AND ALTERATIONS OF MICROVESSEL PERMEABILITY AND ARCHITECTURE

The most extensively characterized molecules inducing adult angiogenesis concomitantly increase microvessel permeability and produce an abnormal neovascular architecture. Both of these features are potentially disadvantageous in therapeutic angiogenesis. Increases in permeability produce tissue edema, which can elevate tissue interstitial pressure and force microvessel closure (Boucher et al. 1996; Carmeliet and Jain 2000). Abnormal architecture can impair intravascular flow and hence tissue perfusion (Jain 2001). An example of this paradigm is angiogenesis induced by VEGF-A. VEGF-A is one of the most potent angiogenic factors (Dvorak 2000; Yancopoulos et al. 2000; Ferrara 2001). Transgenic mouse models have been engineered to constitutively express either the 164 or the 120 isoforms of murine VEGF-A in the epidermal keratinocytes using the keratin-14 or the keratin-5 enhancer/promoters (Detmar et al. 1998; Larcher et al. 1998; Thurston et al. 1999). In both of these models, a marked increase in microvessel number compared to nontransgenic controls was elicited in skin. However, the microvessels in both models were tortuous and leaky to small molecules (Evans blue dye, 9100 kD), and the skin was inflamed. Parallel studies using adenoviral vectors have confirmed the marked alterations in vessel architecture consequent to overexpression of VEGF 164 in skin (Sundberg et al. 2001). VEGF-A-mediated increase in permeability is due to a combination of opening of fenestrae and induction of transcellular vesicle transport by caveolae within cells, and creation of gaps between cells (Feng et al. 1999; Dvorak and Feng 2001; Ferrara 2002). Gap formation is likely due to a pathway of VEGF-A integrin signaling, c-src and Rac activation in endothelial cells (Eliceiri et al. 1999). Induction of inflammation is due to up-regulation of intercellular adhesion molecules and selectin-mediated increased leukocyte rolling (Detmar et al. 1998). Induction of a microvessel network of abnormal architecture, tissue edema, and inflammation was also seen following transplantation of myoblasts stably transduced with retroviruses expressing VEGF164 (Springer et al. 1998). The abnormal microvessel architecture induced by VEGF164 was progressive in both the transgenic mice and myoblast models, such that vascular malformations ulcerating the overlying skin occurred in the transgenic mice (Thurston et al. 1999) and angiomas developed in the mice implanted with VEGF164 transduced myoblasts (Lee et al. 2000). Complications such as edema have also been seen in clinical trials of VEGF164 gene therapy for ischemic limb disease (Epstein et al. 2001a).

Angiopoietin-1 (Ang-1) has also been shown to affect microvascular number and permeability. Ang-1 has been shown to increase adhesion between microvessel pericytes and endothelial cells (Yancopoulos et al. 2000). This interaction stabilizes nascent vessels, ensures their survival, and decreases their permeability (Davis and Yancopoulos 1999). Transgenic mice engineered to overexpress Ang-1 in skin demonstrated an increase in dermal microvessels that were large, dilated, and markedly tortuous (Suri et al. 1998). Ang-1/VEGF164 double transgenic mice also displayed a dilated and highly tortuous microvasculature, morphologically a combination of each single transgenic mouse, but the Ang-1-mediated leakage resistance following Evans blue dye injection was dominant over the pronounced leak induced by VEGF. An interesting extension of this work was the effect of systemic Ang-1 delivery produced by intravenous Ang-1 adenoviral injection followed by hepatic secretion (Thurston et al. 2000). Here, neither angiogenesis nor alterations in microvasculature morphology were induced in adult mice, but microvascular leak was significantly reduced in response to acute topical application of either VEGF or a generic inflammatory challenge such as mustard oil. An outstanding question is whether the enhanced vascular leak accompanying VEGF-A-induced angiogenesis in adults can be ameliorated by Ang-1 treatment (Koh et al. 2002).

Additional molecules regulate microvessel permeability in the context of loss function. Mouse germ-line knockouts deficient in the cytoplasmic tyrosine kinase c-src have leakage-resistant vasculature in response to tissue inflammation, ischemia, and intradermal VEGF-A administration (Eliceiri et al. 1999; Paul et al. 2001). Deficiency of c-src has been shown to diminish VEGF-A signaling via cross-talk to integrin receptors, which itself appears to alter the endothelial cytoskeleton and effect gap formation and leakage (Eliceiri et al. 2002). Gene deletion of placental growth factor (PlGF) has also been shown to induce leakage resistance both to exogenous VEGF-A and to inflammatory stimuli (Carmeliet et al. 2001). However, PlGF knockout mice were also deficient in adult angiogenesis and failed to heal wounds or to collateralize ischemic myocardium. Conversely, PlGF overexpression in epidermal keratinocytes of transgenic mice was strikingly angiogenic, but these vessels were leaky compared to nontransgenic controls (Odorisio et al. 2002). Mechanisms for enhancement of VEGF angiogenesis and vascular leak by PlGF appear to be multifactorial, and include passive PlGF occupation of VEGFR1 limiting its ability to function as a VEGF-A "sink," active signaling via VEGFR-1 induced by PlGF that cooperatively enhances VEGF signaling via VEGFR-2 (Carmeliet et al. 2001), an additional signal for recruitment of bone marrow progenitor cells to sites of angiogenesis (Hattori et al. 2002). Similar to Ang-1, the hypothesis that the VEGF-induced microvasculature can be "normalized" without leakage by negative titration of PlGF signaling using anti-PlGF antibodies remains to be tested (Luttun et al. 2002a).

ENHANCEMENT OF ANGIOGENESIS USING VASCULAR AND HEMATOPOIETIC PRECURSORS AND STEM CELLS

An emerging concept is that microvessel formation in the adult engages vasculogenesis, the formation of new blood vessels from individual precursor or stem cell progenitors in conjunction with "classical" angiogenesis, which is microvessel sprouting from a preexisting vasculature. In this scenario, angiogenic factors and proteases produced locally in response to such stimuli as tissue injury, ischemia, tumor formation, or inflammation enter the circulation, are delivered to the bone marrow, and activate cognate receptors on immature, vascular, and hematopoietic precursors and stem cells (Rafii 2000; Gill et al. 2001; Rafii et al. 2002b). These activated precursors and stem cells travel back to the initial site of stimulation where they locally release additional angiogenic factors and proteases, and incorporate into new microvessels (Rafii et al. 2002a). The differentiation of these precursors into endothelium is mediated both by angiogenic stimuli locally and by cell–cell contacts with adjacent endothelial cells. Studies using genetically marked endothelial cells have documented the existence of this "adult vasculogenesis." Both VEGF-A and PlGF signaling via VEGFR-1 have been shown to augment recruitment of bone marrow precursor cells to sites of angiogenesis (Hattori et al. 2002; Luttun et al. 2002b). Gene knockout studies have identified additional molecules regulating this process, such as matrix metalloproteinase-9 (Heissig et al. 2003), the c-kit receptor and its ligand (Heissig et al. 2002), and the Id family of transcriptional repressors (Benezra et al. 2001). Importantly, vascular and hematopoietic precursors can be sorted from peripheral blood, cord blood, or bone marrow using cell surface markers (VEGFR-1, CD133), expanded in culture, and autologously readministered to recipients with tissue (cardiac or limb) ischemia (Yurasov et al. 1996; Peichev et al. 2000; Kawamoto et al. 2001; Rafii et al. 2002b). Studies in rodent models have documented incorporation of precursor cells at sites of ischemia, and increases in blood flow to the ischemic organ or limb (Freedman and Isner 2001). Increases in microvessel number and the biological characteristics of these new vessels have been less well demonstrated.

HYPOXIA, ANGIOGENESIS, AND HIF TRANSCRIPTION FACTORS

Cells within hypoxic tissues are potent inducers of a neovasculature that attempts to offset the hypoxic insult (Brown and Giaccia 1998). The biology and function of this microvasculature depend on the condition producing hypoxia, the extent to which the hypoxic insult is offset, and the time interval between hypoxic onset and vascular analysis. The transcription factors in the hypoxia-inducible factor (HIF) family are important components in the cellular response and adaptation to hypoxia, and to hypoxic angiogenesis (Iyer et al. 1998; Ryan et al. 1998; Grunstein et al. 2000). HIF transcription factors activate a diverse collection of target genes that can be categorized in five functional groups: genes controlling glucose metabolism, and molecules regulating tissue oxygenation, tissue perfusion, invasion, and apoptosis (Fig. 1) (Semenza 2000). Most importantly, the particular repertoire of HIF target genes is controlled by both cell type and tissue context.

HIF transcription factors are heterodimers of α- and β-subunits. α-Subunits are under multi-molecular regulation, including hypoxia, proliferation, and survival path-

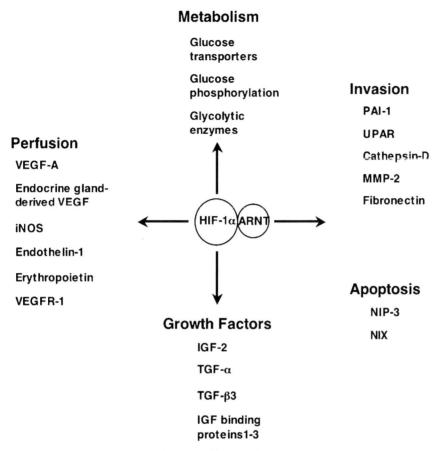

Figure 1. Functional classification of HIF-1 target genes.

way signaling (Semenza 2002). Hypoxia stabilizes HIF-α proteins whereas, in most instances, the molecules are rapidly degraded in normoxic, unstimulated cells. In contrast, the β subunit (also designated aryl hydrocarbon receptor nuclear translocator, ARNT) is constitutively expressed regardless of oxygen tension. Both HIF-α and HIF-β molecules are basic helix-loop-helix (bHLH) PERiod-ARNT-SIMple minded (PAS) domain proteins (Fig. 2). The bHLH domain both binds to DNA at a canonical A/GCGTG hypoxia response element (HRE) and contributes to heterodimer stabilization. The PAS domain is both a heterodimerization domain and the target for regulatory protein (HSP90) binding (Isaacs et al. 2002). In addition to the bHLH and PAS domain, HIF-α molecules contain amino and carboxyl *trans*-activation domains ($NAD_{531-575}$ and $CAD_{786-826}$, respectively [Jiang et al. 1997]; amino acid positions apply to HIF-1α), and an "oxygen-dependent degradation domain" (ODD), within which both the NAD and the adjacent inhibitory domain are located (Fig. 2). Three HIF-α family members encoded by different genes (HIF-1, 2, and 3a) have been identified. Similarly, there are separate genes encoding ARNT family members. HIF-α proteins are obligate HIF-β partners, HIF-α molecules do not bind DNA as homodimers, and ARNT knockout mice phenocopy HIF-1α knockouts (Maltepe et al. 1997). HIF-1α and HIF-2α are similarly regulated by hypoxia and are positive effectors of target gene expression, whereas HIF-3α lacks a *trans*-activation domain and appears to function as a dominant-negative transcriptional repressor (Hara et al. 2001). While under similar oxygen regulation, the tissue distribution and functions of HIF-1α and HIF-2α are distinct (Talks et al. 2000). This dichotomy was underscored by mouse knockouts. HIF-1α knockout mice die at embryonic day 9.5–10.5 of a failure of the neural plate microvasculature to grow and arborize into the developing brain (Iyer et al. 1998; Ryan et al. 1998). In addition, HIF-1α mice display cardiac defects (see Simon et al., this volume). In contrast, HIF-2α knockout mice die either at day 16–18 antepartum due to a failure of development of the neuroendocrine organ of Zuckerandl, the critical source of fetal catecholamines, or soon after birth, because of failure to mature the lungs (Tian et al. 1998). These data demonstrate that HIF-α molecules are not redundant and cannot compensate for their counterpart's constitutional deficiency in development. More recently, tissue-specific HIF-1α gene deletion also demonstrates lack of HIF-2α compensation when HIF-1α is deficient in cartilage or in macrophages (Schipani et al. 2001). Loss of HIF-1α function produces widespread apoptotic death of fetal cartilage cells and a failure to vascularize the growth plates of long bones and sternum (Schipani et

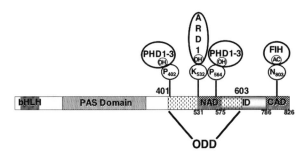

Figure 2. Domains and sites of regulation within the HIF-1α molecule. Heterodimerization of hypoxia-inducible factor-1α (HIF-1α) with HIF-1β (Aryl hydrocarbon nuclear translocator-ARNT) comprises the transcription factor HIF-1. Binding to the hypoxia response element (HRE) is mediated by the basic helix-loop-helix (bHLH) domain, which also provides additional stability for the HIF-1 heterodimer. Heterodimerization is mediated by the PAS domain (*P*eriod, *A*RNT, *S*imple-minded). The 200-amino acid ODD (*o*xygen-*d*ependent *d*egradation domain) contains four amino acids whose modifications target HIF-1α for ubiquitination and 26S proteasomal degradation. The prolyl hydroxylase family members (PHD-1, -2, and -3), the asparaginyl hydroxylase (FIH- factor inhibiting HIF), and the acetylating enzyme, ARD-1, are depicted as are their amino acid targets. (Pro-402 and Pro-564 are hydroxylated at the same atomic position; the mirror image is presented here for clarity.) Proline hydroxylation is the signal for pVHL binding and ultimate ubiquitination of HIF-1α. Three transcription domains are present in HIF-1α, two activating domains (NAD: amino-terminal activating domain and CAD; carboxy-terminal activating domain), and an inhibitory domain (ID). Asparaginyl hydroxylation abrogates p300 binding to the CAD and HIF-1α transcriptional activity. HIF-2α has the same domain structure and is regulated by the same enzymes, except the amino acid positions are different.

al. 2001). In other studies, transplanted HIF-1α$^{-/-}$ ES cells form smaller teratocarcinoma tumors significantly later than wild-type counterparts. The HIF-1α$^{-/-}$ tumors are also poorly vascularized (Ryan et al. 1998, 2000). Collectively, these data suggest that HIF-1α may be an important determinant of hypoxic angiogenesis.

REGULATION OF HIF-α MOLECULES

HIF transcription factors have two levels of regulation, protein stability and transcriptional activity, controlled by discrete molecules, which are themselves regulated by changes in tissue oxygen level, or growth factor stimulation. Hypoxic regulation of HIF-α protein stability is mediated by the ODD (Fig. 2) (Salceda and Caro 1997; Huang et al. 1998). The ODD is a 200-amino-acid long, regulatory subunit that confers normoxic protein instability when fused to non-HIF-α chimeric molecules such as Gal4 (Huang et al. 1998). Conversely, deletion of the ODD from HIF-1α produces a constitutively expressed and transcriptionally active molecule in normoxia. The ODD is the site of binding of the von Hippel-Lindau tumor suppressor protein (pVHL) to HIF-1α and -2α (Maxwell et al. 1999; Cockman et al. 2000; Ohh et al. 2000). HIF-α–pVHL interactions were suggested by the clinical findings that renal cancers in VHL patients were hypervascular, and that renal carcinoma cell lines lacking the VHL gene expressed high levels of VEGF-A (Kaelin and Maher 1998; Yang and Kaelin 2001). VHL is a component of a multiprotein E3 ubiquitin ligase (Kim and Kaelin 2003), and VHL-mediated polyubiquitination of HIF-1α directs the molecule to the 26S proteosome and subsequent degradation. The signals for pVHL binding to the HIF-1/2α ODDs are regulated by hydroxylation of two prolines, Pro-402 and Pro-564 (Fig. 2) (Ivan et al. 2001; Jaakkola et al. 2001). X-ray crystallography reveals that VHL binds to the ODD on a surface groove of the molecule, with the hydroxylated Pro-564 residue extending into a pocket stabilized by extensive hydrogen bonding (Min et al. 2002). Recently, a family of mammalian (and a single *Caenorhabditis elegans*) prolyl hydroxlases (PHDs 1–3) were cloned, each with differential expression levels and HIF-α ODD-binding affinities (Bruick and McKnight 2001; Epstein et al. 2001b). The HIF-α PHDs have important functional characteristics for oxygen sensors: They bind molecular oxygen and ferrous iron, and they use the tricarboxylic acid cycle intermediate α-ketoglutarate as a cofactor to facilitate transfer of a hydroxyl group to proline (McNeill et al. 2002).

Two other proteins covalently modify HIF-α to effect transcriptional activity. Factor inhibiting HIF (FIH) was initially cloned by the yeast two-hybrid technique (Mahon et al. 2001). FIH was subsequently shown to be a hydroxylase similar in structure and cofactor requirements to the PHDs, but targeting asparagine instead (Lando et al. 2002). In particular, asparagine at position 803, in the CAD (amino acid position in HIF-1α), is the specific FIH target (Fig. 2). This asparagine hydroxylation prevents HIF-α interaction with p300 and as such, inhibits transcription in normoxia. Co-crystals containing HIF-α CADs bound to either p300 or FIH have been obtained and demonstrate that Asn-803 is at the tip of a finger-like structure protruding into p300 (Freedman et al. 2002). Asparagine hydroxylation sterically prevents the intrusion of the CAD into the p300 binding core (Elkins et al. 2003; Lee et al. 2003). An additional enzyme regulating HIF-1α has been identified. Acetylation of a lysine at position 532 (HIF-1α) by ARD-1 (Fig. 2), its gene cloned by the yeast two-hybrid technique using the ODD as bait, both inhibits HIF-1α transcriptional activity in normoxia and accelerates HIF-1α protein degradation in both normoxia and hypoxia (Jeong et al. 2002). Importantly, the precise contribution of modifying molecules such as FIH and ARD-1 to HIF-1α biology in vivo remains to be determined. For instance, our work with HIF-1αΔODD in the skin of transgenic mice (see below), and the work of others (P. Ratcliff, pers. comm.) suggests that the activity of these enzymes varies according to cell type and tissue context.

In addition to the HIF-1α modifying enzymes, additional molecules have been shown to posttranslationally regulate HIF-1α. Increases in growth factor signaling by IGF-1, insulin, EGF, and FGF have all been shown both to induce HIF-1α function in normoxia and to augment transcriptional activity in hypoxia (Fig. 3). Loss of the tumor suppressor PTEN produces a similar significant transcriptional up-regulation of HIF-1α and a modest stabilizing effect in normoxia (Zundel et al. 2000). Further work has demonstrated that signaling via the MAP kinase

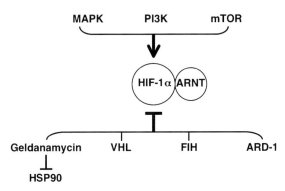

Figure 3. Summary of positive and negative regulation of HIF-1α. HIF-1α is regulated at the level of protein stability and transcriptional activation (see text for details). Independent pathways of HIF-1α regulation are depicted. The von Hippel-Lindau protein binds to hydroxylated prolines 564 and 402 and ubiquitinates HIF-1α, targeting the molecule for destruction in the 26S proteosome. ARD-1 also regulates HIF-1α protein stability by acetylation of lysine 532. At present, it is unclear whether ARD-1 functions independent of VHL. Geldanamycin inhibits binding of HIF-1α to heat shock protein 90 (HSP90) and targets HIF-1α for proteosomal degradation independent of VHL. FIH hydroxylation of aparagine 803 in the CAD inhibits HIF-1α transcriptional activity. Conversely, signaling via both the MAP kinase and PI3 kinase pathways stimulates HIF-1α transcriptional activity and contributes to protein stability in both normoxia and hypoxia. The mammalian target of rapamycin (mTOR) similarly stimulates HIF-1α target gene transcription and protein stability, particularly in hypoxia. mTOR is both a component of the PI3 kinase pathway and with additional extrinsic signaling inputs.

or PI3 kinase pathways are both independently responsible for producing functional HIF-1α in normoxic cells (for review, see Semenza 2002). Signaling via the cytoplasmic signal tyrosine kinase c-src also increases both HIF-1α mRNA and protein via a global massive increase in protein translation (Karni et al. 2002). The mammalian target of rapamycin can modestly increase HIF-1α levels and transcriptional activity in normoxia, with a pronounced potentiation of hypoxia response which is mediated by the ODD (Hudson et al. 2002). Finally, HSP90 has been shown to be a required molecular chaperone for HIF-1α (Isaacs et al. 2002). Inhibition of HSP90–HIF-1α binding by geldanamycin, an inhibitor of HSP90 nucleotide binding, produces HIF-1α degradation in a VHL-independent mechanism that overrides hypoxic stabilization (Isaacs et al. 2002).

HIF-1α IN DISEASE

Despite genetically informative data obtained from studies of constitutional and tissue-specific HIF-1α gene deletion, the pathophysiology of HIF-1α is gain of function. HIF-1α or its signature target genes have been shown to be up-regulated in the brain following stroke, in the myocardium following acute infarction, in normal wound healing, and in most epithelial cancers. Patterns of HIF-1α expression in tissues reflect both hypoxic and hyperproliferative mechanisms of stabilization of the protein. In wounds, HIF-1α mRNA and target gene expression is highest in basal keratinocytes re-epithelializing the open ulcer which are juxtaposed to capillaries and microvessels (Elson et al. 2000). In cancers, nuclear HIF-1α expression is detected in the viable rim of malignant cells surrounding necrotic core regions consistent with hypoxic stabilization of the protein, and at the periphery of cancers wherein malignant cells are invading and intermingling with the functional organ parenchymal cells consistent with normoxic protein stabilization (Zagzag et al. 2000). Regulation of invasion might be controlled by such HIF-1α target genes as urokinase plasminogen activator receptor, cathepsin-D, or plasminogen activator inhibitor (Fig. 1) (Semenza 2003; Krishnamachary et al. 2003). HIF-1α protein is also expressed in hypoxic endothelial cells (Berra et al. 2000), and HIF-1α up-regulation of proteases may facilitate tissue invasion of nascent angiogenic microvessels in benign conditions, such as wounding, as well as in invasive malignancies.

CONSTITUTIVE GAIN OF HIF-1α AND ANGIOGENESIS

Our initial goal was to model gain of HIF-1α function and to determine the biology of hypoxic up-regulation of target genes in response to one discrete molecular event in a prototypic epithelium, the skin. Given the association of HIF-1α up-regulation with hypoxic angiogenesis, and a potential link to neoplastic proliferation via induction of glucose metabolism and an increase in protease activity, we expected induction of HIF-1α function alone to induce epidermal hyperproliferation and a leaky skin microvasculature. We targeted expression of either a constitutively active mutant lacking the ODD (HIF-1αΔODD) or a wild-type human HIF-1α cDNA to the basal keratinocytes of mouse epidermis using a keratin-14 expression cassette (Elson et al. 2001). Three permanent transgenic mouse lines expressing HIF-1αΔODD and two lines expressing HIF-1α wild-type cDNA were established. Only the K14-HIF-1αΔODD transgenic mice displayed a skin phenotype consisting of a pinker color and prominent vasculature of unfurred ear, paw, and tail skin (Fig. 4), and a slight roughness or irregularity to the coat of furred skin. Notably, this skin phenotype remained stable as the transgenic mice aged, no skin tumors or vascular malformations developed, and the skin remained intact without ulceration, even in areas exposed to grooming or trauma such as ear or chest skin. The lack of progressive vascular phenotype in K14-HIF-1αΔODD transgenic mice contrasted with the skin phenotype of K14-VEGF164 transgenic mice, which developed focal vascular lesions and skin ulcerations as they aged (Thurston et al. 1999).

Despite our expectations, histological examination of skin thickness, proliferation, and differentiation indices were similar in K14-HIF-1αΔODD transgenic mice compared to nontransgenic controls. However, a marked increase in the frequency of dermal blood vessels was apparent in both routine histopathological staining (Elson et al. 2001) and special techniques of vascular visualization. In vivo endothelial cell decoration with the lectin lyso-persicon esculentum reveals blood vessels that are actively perfused, and details vascular morphology (Fig. 4).

Using tissue whole mounts or thick paraffin sections, a three-dimensional picture of blood vessel network organization within the tissue can be visualized and reconstructed, if desired, by confocal microscopy.

Lectin perfusion demonstrated an increase in the number of small-caliber vessels and capillaries immediately beneath the epidermis and surrounding each hair follicle. The latter observation is consistent with keratin-14 expression in the hair outer root sheath cells. Lectin perfusion additionally demonstrated that the architecture in the skin microvessels of K14-HIF-1α∆ODD transgenic mice was normal, but computer analysis of microvascular density revealed a 60% increase in the number of microvessels per unit length area in the transgenic compared to the nontransgenic mice (Elson et al. 2001).

Because the HIF-1α target VEGF-A was an obvious candidate as a mediator of increased vascularity in our model, we determined its expression level quantitatively using real-time RT-PCR (Fig. 5). VEGF-A expression was elevated approximately tenfold in the skin of K14-HIF-1α∆ODD transgenic mice compared to controls. Localization of VEGF-A expression using mRNA in situ hybridization of paraffin-embedded skin tissue sections demonstrated that the epidermal keratinocytes were the source of VEGF-A expression (Elson et al. 2001). VEGF-A is differentially spliced into five isoforms, each of which encodes proteins with distinct affinities for the cell surface heparin sulfates and endothelial cell growth factor receptors (Ferrara 1999). As such, we used isoform-specific PCR real-time RT-PCR analysis to determine the distribution of the five VEGF-A isoforms in the skin of transgenic compared to nontransgenic mice (Fig. 5). This analysis demonstrated an equivalent eightfold up-regulation of each of the five VEGF-A isoforms compared to nontransgenic controls.

VEGF-A has several biological properties distinct from increase in blood vessel number, including increased vascular permeability and up-regulation of tissue inflammation. Indeed, three other transgenic models in which isoforms of VEGF-A were targeted to the epidermis demonstrated increases in microvessel permeability and skin inflammation. Histopathological analysis had already demonstrated a lack of skin inflammation in the K14-HIF-1α∆ODD transgenic mice. To determine the functional characteristics of the HIF-1α-induced microvasculature, we investigated microvessel permeability both in the resting state and in response to a neurogenic inflammatory stimulus. Despite a marked increase in vasculature and in VEGF-A expression, microvascular leakage in the K14-HIF-1α∆ODD transgenic mice was actually slightly less than in nontransgenic controls and contrasted with the marked leakage in similarly treated K14-VEGF164 transgenic mice (Fig. 6).

To explore the mechanisms underlying the increased but quiescent vasculature created by gain of HIF-1α function, we determined the expression level of several angiogenic and growth factors known to regulate microvessel permeability using real-time PCR. We did not detect an increase in either angiopoietin-1 or angiopoietin-2 in the skin of transgenic mice (Elson et al. 2001). The latter finding was surprising, because HIF-1α is known to increase angiopoietin-2 expression in endothelial cells (Oh et al. 1999); here cell context likely underlies lack of induction in epidermal keratinocytes.

At present, the mechanism(s) underlying the normal permeability of the "quiescent" microvasculature associated with HIF-1α gain of function is obscure. One hypothesis is that this phenotype is mediated by an equivalent, "balanced" induction of each VEGF-A isoform. Each VEGF isoform possesses different affinities both for the cell surface and for heparin in the extracellular matrix, and also differential affinities for both the VEGFR2 and the neuropilin-1 and -2 receptors. Thus, a combination of gradients of VEGF-A and orchestrated signaling may underlie the leakage-resistant microvasculature mediated by gain of HIF-1α function. Experimental evidence supporting this notion was recently provided by a report using engineered zinc-finger protein (ZFP) transcription factors specific for the VEGF-A regulatory region (Rebar et al. 2002). Similar to HIF-1α gain of function, cell transfection of VEGF-A ZFPs produced an equivalent induction of all VEGF-A isoforms. Moreover, skin infection with adenoviruses containing VEGF-A ZFP cDNAs produced an increase in dermal microvasculature that was leakage resistant (Rebar et al. 2002). An

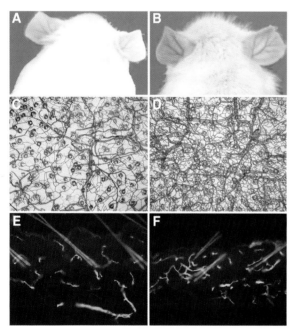

Figure 4. Transgenic mice expressing a constitutively active HIF-1α mutant lacking the ODD display increased vascularity. Expression of the HIF-1α∆ODD cDNA was targeted to the basal keratinocytes of skin using a keratin-14 promoter. Compared to nontransgenic littermate controls (*A*), the transgenic mice evidence a pinker color of unfurred skin, prominent vascularity of the ears, and a slight coat roughness (*B*). Blood vessel delineation by perfusion with either ear whole mounts with biotinylated (*C, D*) or back skin with FITC-labeled lysopersicon esculentum lectin (*E, F*) demonstrates increased microvessel density in the transgenic (*D, F*) compared to the nontransgenic (*C, E*) skin. Cross-sectional skin sections from the back reveal that the microvessels are located just beneath the interfollicular epidermis and surrounding hair follicles (note birefringent diagonal hair shafts in *E* and *F*). (Reprinted, with permission, from Elson et al. 2001.)

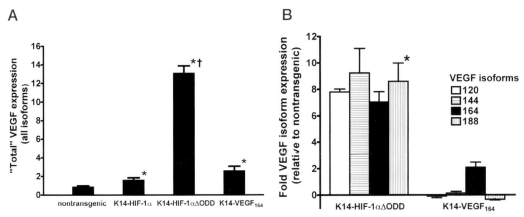

Figure 5. Vascular endothelial growth factor expression in ear skin from transgenic and nontransgenic mice. Real-time RT-PCR analysis using PCR primers and TaqMan probes specific for either total VEGF-A, common exons 3 and 4 shared by all isoforms (*A*), or primer and probe sets specific for four of the five most abundant VEGF-A isoforms (*B*), demonstrate an 8- to 13-fold elevation of total VEGF-A message in the skin of K14-HIF-1α∆ODD transgenic mice. Moreover, each isoform displayed an equivalent 8- to 10-fold elevation in transgenic compared to nontransgenic mice. The levels of total VEGF-A mRNA were greater in the K14-HIF-1α∆ODD compared to K14-VEGF164 transgenic mice. (Reprinted, with permission, from Elson et al. 2001.)

alternative hypothesis in our work is that an additional HIF-1α target is up-regulated in the K14-HIF-1α∆ODD transgenic mice that is independently responsible for mediate maintenance of vascular permeability despite VEGF-A elevation. Induction of an additional target gene could also underlie the permeability maintenance in ZFP-treated mice if a surreptitious target was up-regulated by the engineered transcription factor.

DIFFERENCES BETWEEN TISSUE HYPOXIA AND CURRENT MODEL

It is important to remember that our model encompasses only one facet of hypoxia; that is, gain of HIF-1α function in the epithelial cell. Hypoxic microvasculature has in the past been characterized as leaky. However, *tissue* hypoxia encompasses multiple cell types such as endothelial cells and pericytes of the microvasculature itself, as well as macrophages, mast cells, and fibroblasts (Coussens et al. 1999). There are also multiple physiological alterations such as acid tissue pH, generation of both free radicals and reactive oxygen species, induction of inflammatory cytokines and chemokines, and, in the case of tumors, breaching of the integrity and replacement of portions of the vessel wall by malignant cells (Brown and Giaccia 1998; Carmeliet and Jain 2000; Coussens and Werb 2002). Moreover, additional transcription factors are up-regulated by hypoxia independent of HIF-1α or HIF-2α including Egr-1 (Yan et al. 2000), NFκB (Royds et al. 1998), Jun amino-terminal kinase (Jin et al. 2000), c-Jun (Alfranca et al. 2002), c-fos (Premkumar et al. 2000), p53 (Koumenis et al. 2001), and ATM (Hammond et al. 2002). These disparate elements likely conspire to produce a leaky hypoxic vasculature. Further studies will be needed to determine whether induction of HIF-1α itself or a selected set of HIF-target genes can produce an increase in a functional vasculature to minimize damage and to accelerate function restoration in ischemic tissues.

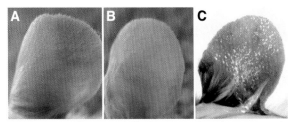

Figure 6. Vascular permeability in transgenic and nontransgenic mice. Nontransgenic (*A*), K14-HIF-1α∆ODD (*B*), and K14-VEGF164 (*C*) transgenic mice were anesthetized, injected with Evans blue dye at time 0, treated twice with mustard oil over 30 minutes, and perfused with paraformaldehyde (Elson et al. 2001). The K14-VEGF164 transgenic mice display a marked leakage of Evans blue dye into interstitial tissue, whereas dye leakage was similar in K14-HIF-1α∆ODD transgenic mice compared to littermate controls. (*C*, Reprinted, with permission, from Thurston et al. 1999 [copyright A.A.A.S.].)

CONCLUSIONS AND PERSPECTIVES

Accumulating evidence suggests that the goal of induction of an increased vasculature without leakage, inflammation, or vascular malformation may be attainable. Manipulation of HIF-1α function is particularly attractive functionally because the molecule possesses the ability to coordinately increase not only the microvasculature, but also metabolic enzymes enabling survival of cells at the limits of oxygen and glucose supply. This approach would be particularly attractive to several different types of ischemic disease ranging from strokes and heart attacks to peripheral vascular arteriosclerosis. A major hurdle is the optimum methodology of induction of HIF-1α function. Gene therapy with short- or long-term expression vectors is one option. Indeed, an adenoviral vector containing an engineered chimeric molecule containing the HIF-1α DNA-binding and the VP16 *trans*-activation domains was shown to increase blood flow in models of rabbit hindlimb ischemia and rat myocardial infarction (Vincent et al. 2000; Shyu et al. 2002). Moreover, the lack of induction of proliferation by HIF-1α gain of function alone minimizes the need to turn off HIF-1α-expressing gene therapy vectors following vascular restoration for fear of oncogenic transformation. Another approach would be to inhibit the function of the prolyl and asparaginyl hydroxylases to affect both HIF-1α protein stability and *trans*-activation function in "watershed" regions adjacent to ischemic zones. Alternatively, an expanded repertoire of ZFP transcription factors regulating a repertoire of downstream targets similar to HIF-1α could be engineered to emulate the latter's ability to increase tissue perfusion and ischemic tolerance. Multimodality therapy with additional effectors of therapeutic angiogenesis such as stem cell precursors could be designed as well. Collectively, these approaches to therapeutic angiogenesis offer the promise of maximal salvage of ischemic tissues with minimal long-term deficits.

ACKNOWLEDGMENTS

The author thanks Ananth Kadambi for comments on the manuscript; David Elson, Gavin Thurston, and Donald McDonald for their collaborative work on the K14-HIF-1αΔODD transgenic mice; and Gregg Semenza for providing data in press. This work was supported by National Institutes of Health grant R01 CA-71398.

REFERENCES

Alfranca A., Gutierrez M.D., Vara A., Aragones J., Vidal F., and Landazuri M.O. 2002. c-Jun and hypoxia-inducible factor 1 functionally cooperate in hypoxia-induced gene transcription. *Mol. Cell. Biol.* **22:** 12.

Benezra R., Rafii S., and Lyden D. 2001. The Id proteins and angiogenesis. *Oncogene* **20:** 8334.

Berra E., Milanini J., Richard D.E., Le Gall M., Vinals F., Gothie E., Roux D., Pages G., and Pouyssegur J. 2000. Signaling angiogenesis via p42/p44 MAP kinase and hypoxia. *Biochem. Pharmacol.* **60:** 1171.

Boucher Y., Leunig M., and Jain R.K. 1996. Tumor angiogenesis and interstitial hypertension. *Cancer Res.* **56:** 4264.

Brown J.M. and Giaccia A.J. 1998. The unique physiology of solid tumors: Opportunities (and problems) for cancer therapy. *Cancer Res.* **58:** 1408.

Bruick R.K. and McKnight S.L. 2001. A conserved family of prolyl-4-hydroxylases that modify HIF. *Science* **294:** 1337.

Carmeliet P. and Jain R. 2000. Angiogenesis in cancer and other diseases. *Nat. Cell Biol.* **407:** 249.

Carmeliet P., Moons L., Luttun A., Vincenti V., Compernolle V., De Mol M., Wu Y., Bono F., Devy L., Beck H., Scholz D., Acker T., DiPalma T., Dewerchin M., Noel A., Stalmans I., Barra A., Blacher S., Vandendriessche T., Ponten A., Eriksson U., Plate K.H., Foidart J.M., Schaper W., Charnock-Jones D.S., Hicklin D.J., Herbert J.M., Collen D., and Persico M.G. 2001. Synergism between vascular endothelial growth factor and placental growth factor contributes to angiogenesis and plasma extravasation in pathological conditions. *Nat. Med.* **7:** 575.

Cockman M., Masson N., Mole D., Jaakkola P., Chang G., Clifford S., Maher E., Pugh C., Ratcliffe P., and Maxwell P. 2000. Hypoxia inducible factor-alpha binding and ubiquitylation by the von Hippel-Lindau tumor suppressor protein. *J. Biol. Chem.* **275:** 25733.

Coussens L.M. and Werb Z. 2002. Inflammation and cancer. *Nature* **420:** 860.

Coussens L.M., Raymond W., Bergers G., Laig-Webster M., Behrendtsen O., Werb Z., Caughey G., and Hanahan D. 1999. Inflammatory mast cells up-regulate angiogenesis during squamous epithelial carcinogenesis. *Genes Dev.* **13:** 1382.

Davis S. and Yancopoulos G.D. 1999. The angiopoietins: Yin and yang in angiogenesis. *Curr. Top. Microbiol. Immunol.* **237:** 173.

Detmar M., Brown L., Schon M., Elicker B., Velasco P., Richard L., Fukumura D., Monsky W., Claffey K., and Jain R. 1998. Increased microvascular density and enhanced leukocyte rolling and adhesion in the skin of VEGF transgenic mice. *J. Investig. Dermatol.* **111:** 1.

Dvorak A.M. and Feng D. 2001. The vesiculo-vacuolar organelle (VVO). A new endothelial cell permeability organelle. *J. Histochem. Cytochem.* **49:** 419.

Dvorak H.F. 2000. VPF/VEGF and the angiogenic response. *Semin. Perinatol.* **24:** 75.

Eliceiri B.P., Paul R., Schwartzberg P.L., Hood J.D., Leng J., and Cheresh D.A. 1999. Selective requirement for Src kinases during VEGF-induced angiogenesis and vascular permeability. *Mol. Cell* **4:** 915.

Eliceiri B.P., Puente X.S., Hood J.D., Stupack D.J., Schlaepfer D.D., Huang X.Z., Sheppard D., and Cheresh D.A. 2002. Src-mediated coupling of focal adhesion kinase to integrin alpha(v)beta5 in vascular endothelial growth factor signaling. *J. Cell Biol.* **157:** 149.

Elkins J.M., Hewitson K.S., McNeill L.A., Seibel J.F., Schlemminger I., Pugh C.W., Ratcliffe P.J., and Schofield C.J. 2003. Structure of factor-inhibiting hypoxia-inducible factor (HIF) reveals mechanism of oxidative modification of HIF-1 alpha. *J. Biol. Chem.* **278:** 1802.

Elson D.A., Ryan H.E., Snow J.W., Johnson R., and Arbeit J.M. 2000. Coordinate up-regulation of hypoxia inducible factor (HIF)-1α and HIF-1 target genes during multi-stage carcinogenesis and wound healing. *Cancer Res.* **60:** 6189.

Elson D.A., Thurston G., Huang L.E., Ginzinger D.G., McDonald D.M., Johnson R.S., and Arbeit J.M. 2001. Induction of hypervascularity without leak or inflammation in transgenic mice overexpressing hypoxia-inducible factor-1α. *Genes Dev.* **15:** 2520.

Epstein S.E., Kornowski R., Fuchs S., and Dvorak H.F. 2001a. Angiogenesis therapy: Amidst the hype, the neglected potential for serious side effects. *Circulation* **104:** 115.

Epstein A.C., Gleadle J.M., McNeill L.A., Hewitson K.S., O'Rourke J., Mole D.R., Mukherji M., Metzen E., Wilson M.I., Dhanda A., Tian Y.M., Masson N., Hamilton D.L., Jaakkola P., Barstead R., Hodgkin J., Maxwell P.H., Pugh C.W., Schofield C.J., and Ratcliffe P.J. 2001b. C. elegans EGL-9 and mammalian homologs define a family of dioxygenases that regulate HIF by prolyl hydroxylation. *Cell* **107:** 43.

Feng D., Nagy J.A., Pyne K., Hammel I., Dvorak H.F., and Dvorak A.M. 1999. Pathways of macromolecular extravasation across microvascular endothelium in response to VPF/VEGF and other vasoactive mediators. *Microcirculation* **6:** 23.

Ferrara N. 1999. Vascular endothelial growth factor: Molecular and biological aspects. *Curr. Top. Microbiol. Immunol.* **237:** 1.

———. 2001. Role of vascular endothelial growth factor in regulation of physiological angiogenesis. *Am. J. Physiol. Cell Physiol.* **280:** C1358.

———. 2002. Role of vascular endothelial growth factor in physiologic and pathologic angiogenesis: Therapeutic implications. *Semin. Oncol.* **29:** 10.

Freedman S.B. and Isner J.M. 2001. Therapeutic angiogenesis for ischemic cardiovascular disease. *J. Mol. Cell. Cardiol.* **33:** 379.

Freedman S.J., Sun Z.Y., Poy F., Kung A.L., Livingston D.M., Wagner G., and Eck M.J. 2002. Structural basis for recruitment of CBP/p300 by hypoxia-inducible factor-1 alpha. *Proc. Natl. Acad. Sci.* **99:** 5367.

Gill M., Dias S., Hattori K., Rivera M.L., Hicklin D., Witte L., Girardi L., Yurt R., Himel H., and Rafii S. 2001. Vascular trauma induces rapid but transient mobilization of VEGFR2(+)AC133(+) endothelial precursor cells. *Circ. Res.* **88:** 167.

Grunstein J., Masbad J., Hickey R., Giordano F., and Johnson R. 2000. Isoforms of vascular endothelial growth factor act in a coordinate fashion to recruit and expand tumor vasculature. *Mol. Cell. Biol.* **20:** 7282.

Hammond E.M., Denko N.C., Dorie M.J., Abraham R.T., and Giaccia A.J. 2002. Hypoxia links ATR and p53 through replication arrest. *Mol. Cell. Biol.* **22:** 1834.

Hara S., Hamada J., Kobayashi C., Kondo Y., and Imura N. 2001. Expression and characterization of hypoxia-inducible factor (HIF) 3alpha in human kidney: Suppression of HIF-mediated gene expression by HIF-3alpha. *Biochem. Biophys. Res. Commun.* **287:** 808.

Hattori K., Heissig B., Wu Y., Dias S., Tejada R., Ferris B., Hicklin D.J., Zhu Z., Bohlen P., Witte L., Hendrikx J., Hackett N.R., Crystal R.G., Moore M.A., Werb Z., Lyden D., and Rafii S. 2002. Placental growth factor reconstitutes hematopoiesis by recruiting VEGFR1(+) stem cells from bone-marrow microenvironment. *Nat. Med.* **8:** 841.

Heissig B., Hattori K., Friedrich M., Rafii S., and Werb Z. 2003. Angiogenesis: Vascular remodeling of the extracellular matrix involves metalloproteinases. *Curr. Opin. Hematol.* **10:** 136.

Heissig B., Hattori K., Dias S., Friedrich M., Ferris B., Hackett N.R., Crystal R.G., Besmer P., Lyden D., Moore M.A., Werb Z., and S. Rafii S. 2002. Recruitment of stem and progenitor cells from the bone marrow niche requires MMP-9 mediated release of kit-ligand. *Cell* **109:** 625.

Huang L.E., Gu J., Schau M., and Bunn H.F. 1998. Regulation of hypoxia-inducible factor 1alpha is mediated by an O2-dependent degradation domain via the ubiquitin-proteasome pathway. *Proc. Natl. Acad. Sci.* **95:** 7987.

Hudson C.C., Liu M., Chiang G.G., Otterness D.M., Loomis D.C., Kaper F., Giaccia A.J., and Abraham R.T. 2002. Regulation of hypoxia-inducible factor 1alpha expression and function by the mammalian target of rapamycin. *Mol. Cell. Biol.* **22:** 7004.

Isaacs J.S., Jung Y.J., Mimnaugh E.G., Martinez A., Cuttitta F., and Neckers L.M. 2002. Hsp90 regulates a von Hippel Lindau-independent hypoxia-inducible factor-1 alpha-degradative pathway. *J. Biol. Chem.* **277:** 29936.

Ivan M., Kondo K., Yang H., Kim W., Valiando J., Ohh M., Salic A., Asara J.M., Lane W.S., and Kaelin W.G., Jr. 2001. HIFalpha targeted for VHL-mediated destruction by proline hydroxylation: Implications for O2 sensing. *Science* **292:** 464.

Iyer N., Kotch L., Agani F., Leung S., Laughner E., Wenger R., Gassmann M., Gearhart J., Lawler A., Yu A., and Semenza G. 1998. Cellular and developmental control of O2 homeostasis by hypoxia-inducible factor 1 alpha. *Genes Dev.* **12:** 149.

Jaakkola P., Mole D.R., Tian Y.M., Wilson M.I., Gielbert J., Gaskell S.J., Kriegsheim A., Hebestreit H.F., Mukherji M., Schofield C.J., Maxwell P.H., Pugh C.W., and Ratcliffe P.J. 2001. Targeting of HIF-alpha to the von Hippel-Lindau ubiquitylation complex by O2-regulated prolyl hydroxylation. *Science* **292:** 468.

Jain R.K. 2001. Normalizing tumor vasculature with anti-angiogenic therapy: A new paradigm for combination therapy. *Nat. Med.* **7:** 987.

Jeong J.W., Bae M.K., Ahn M.Y., Kim S.H., Sohn T.K., Bae M.H., Yoo M.A., Song E.J., Lee K.J., and Kim K.W. 2002. Regulation and destabilization of HIF-1alpha by ARD1-mediated acetylation. *Cell* **111:** 709.

Jiang B.H., Zheng J.Z., Leung S.W., Roe R, and Semenza G.L. 1997. Transactivation and inhibitory domains of hypoxia-inducible factor 1alpha. *J. Biol. Chem.* **272:** 19253.

Jin N., Hatton N., Swartz D.R., Xia X., Harrington M.A., Larsen S.H., and Rhoades R.A. 2000. Hypoxia activates jun-N-terminal kinase, extracellular signal-regulated protein kinase, and p38 kinase in pulmonary arteries. *Am. J. Respir. Cell Mol. Biol.* **23:** 593.

Kaelin W.G., Jr. and Maher E.R. 1998. The VHL tumour-suppressor gene paradigm. *Trends Genet.* **14:** 423.

Karni R., Dor Y., Keshet E., Meyuhas O., and Levitzki A. 2002. Activated pp60c-Src leads to elevated hypoxia-inducible factor (HIF)-1alpha expression under normoxia. *J. Biol. Chem.* **277:** 42919.

Kawamoto A., Gwon H.C., Iwaguro H., Yamaguchi J.I., Uchida S., Masuda H., Silver M., Ma H., Kearney M., Isner J.M., and Asahara T. 2001. Therapeutic potential of ex vivo expanded endothelial progenitor cells for myocardial ischemia. *Circulation* **103:** 634.

Kim W. and Kaelin W.G. 2003. The von Hippel-Lindau tumor suppressor protein: New insights into oxygen sensing and cancer. *Curr. Opin. Genet. Dev.* **13:** 55.

Koh G.Y., Kim I., Kwak H.J., Yun M.J., and Leem J.C. 2002. Biomedical significance of endothelial cell specific growth factor, angiopoietin. *Exp. Mol. Med.* **34:** 1.

Koumenis C., Alarcon R., Hammond E., Sutphin P., Hoffman W., Murphy M., Derr J., Taya Y., Lowe S.W., Kastan M., and Giaccia A. 2001. Regulation of p53 by hypoxia: Dissociation of transcriptional repression and apoptosis from p53-dependent transactivation. *Mol. Cell. Biol.* **21:** 1297.

Krishnamachary B., Berg-Dixon S., Kelly B., Agani F., Feldser D., Ferreira G., Iyer N., LaRusch J., Pak B., Taghavi P., and Semenza G. 2003. Regulation of colon carcinoma cell invasion by hypoxia-inducible factor 1. *Cancer Res.* **63:** 1138.

Lando D., Peet D.J., Gorman J.J., Whelan D.A., Whitelaw M.L., and Bruick R.K. 2002. FIH-1 is an asparaginyl hydroxylase enzyme that regulates the transcriptional activity of hypoxia-inducible factor. *Genes Dev.* **16:** 1466.

Larcher F., Murillas R., Bolontrade M., Conti C.J., and Jorcano J.L. 1998. VEGF/VPF overexpression in skin of transgenic mice induces angiogenesis, vascular hyperpermeability and accelerated tumor development. *Oncogene* **17:** 303.

Lee C., Kim S.J., Jeong D.G., Lee S.M., and Ryu S.E. 2003. Structure of human FIH-1 reveals a unique active site pocket and interaction sites for HIF-1 and von Hippel-Lindau. *J. Biol. Chem.* **278:** 7558.

Lee R., Springer M., Blanco-Bose W., Shaw R., Ursell P., and Blau H. 2000. VEGF gene delivery to myocardium: Deleterious effects of unregulated expression. *Circulation* **102:** 898.

Luttun A., Tjwa M., and Carmeliet P. 2002a. Placental growth factor (PlGF) and its receptor Flt-1 (VEGFR-1): Novel therapeutic targets for angiogenic disorders. *Ann. N.Y. Acad. Sci.* **979:** 80.

Luttun A., Tjwa M., Moons L., Wu Y., Angelillo-Scherrer A., Liao F., Nagy J.A., Hooper A., Priller J., De Klerck B., Compernolle V., Daci E., Bohlen P., Dewerchin M., Herbert J.M., Fava R., Matthys P., Carmeliet G., Collen D., Dvorak H.F., Hicklin D.J., and Carmeliet P. 2002b. Revascularization of ischemic tissues by PlGF treatment, and inhibition of tumor angiogenesis, arthritis and atherosclerosis by anti-Flt1. *Nat. Med.* **8:** 831.

Mahon P.C., Hirota K., and Semenza G.L. 2001. FIH-1: A novel protein that interacts with HIF-1alpha and VHL to mediate repression of HIF-1 transcriptional activity. *Genes Dev.* **15:** 2675.

Maltepe E., Schmidt J.V., Baunoch D., Bradfield C.A., and Simon M.C. 1997. Abnormal angiogenesis and responses to glucose and oxygen deprivation in mice lacking the protein ARNT. *Nature* **386:** 403.

Maxwell P., Wiesener M., Chang G., Clifford S., Vaux E., Cockman M., Wykoff C., Pugh C., Maher E., and Ratcliffe P. 1999. The tumour suppressor protein VHL targets hypoxia-inducible factors for oxygen-dependent proteolysis. *Nature* **399:** 271.

McNeill L.A., Hewitson K.S., Gleadle J.M., Horsfall L.E., Oldham N.J., Maxwell P.H., Pugh C.W., Ratcliffe P.J., and Schofield C.J. 2002. The use of dioxygen by HIF prolyl hydroxylase (PHD1). *Bioorg. Med. Chem. Lett.* **12:** 1547.

Min J.H., Yang H., Ivan M., Gertler F., Kaelin W.G., Jr., and Pavletich N.P. 2002. Structure of an HIF-1alpha-pVHL complex: Hydroxyproline recognition in signaling. *Science* **296:** 1886.

Odorisio T., Schietroma C., Zaccaria M.L., Cianfarani F., Tiveron C., Tatangelo L., Failla C.M., and Zambruno G. 2002. Mice overexpressing placenta growth factor exhibit increased vascularization and vessel permeability. *J. Cell Sci.* **115:** 2559.

Oh H., Takagi H., Suzuma K., Otani A., Matsumura M., and Honda Y. 1999. Hypoxia and vascular endothelial growth factor selectively up-regulate angiopoietin-2 in bovine microvascular endothelial cells. *J. Biol. Chem.* **274:** 15732.

Ohh M., Park C., Ivan M., Hoffman M., Kim T., Huang L., Pavletich N., Chau V., and Kaelin W. 2000. Ubiquitination of hypoxia-inducible factor requires direct binding to the beta-domain of the von Hippel-Lindau protein. *Nat. Cell Biol.* **2:** 423.

Paul R., Zhang Z.G., Eliceiri B.P., Jiang Q., Boccia A.D., Zhang R.L., Chopp M., and Cheresh D.A. 2001. Src deficiency or blockade of Src activity in mice provides cerebral protection following stroke. *Nat. Med.* **7:** 222.

Peichev M., Naiyer A.J., Pereira D., Zhu Z., Lane W.J., Williams M., Oz M.C., Hicklin D.J., Witte L., Moore M.A., and Rafii S. 2000. Expression of VEGFR-2 and AC133 by circulating human CD34(+) cells identifies a population of functional endothelial precursors. *Blood* **95:** 952.

Premkumar D.R., Adhikary G., Overholt J.L., Simonson M.S., Cherniack N.S., and Prabhakar N.R. 2000. Intracellular pathways linking hypoxia to activation of c-fos and AP-1. *Adv. Exp. Med. Biol.* **475:** 101.

Rafii S. 2000. Circulating endothelial precursors: Mystery, reality, and promise. *J. Clin. Invest.* **105:** 17.

Rafii S., Lyden D., Benezra R., Hattori K., and Heissig B. 2002a. Vascular and haematopoietic stem cells: Novel targets for anti-angiogenesis therapy? *Nat. Rev. Cancer* **2:** 826.

Rafii S., Meeus S., Dias S., Hattori K., Heissig B., Shmelkov S., Rafii D., and Lyden D. 2002b. Contribution of marrow-derived progenitors to vascular and cardiac regeneration. *Semin. Cell Dev. Biol.* **13:** 61.

Rebar E.J., Huang Y., Hickey R., Nath A.K., Meoli D., Nath S., Chen B., Xu L., Liang Y., Jamieson A.C., Zhang L., Spratt S.K., Case C.C., Wolffe A., and Giordano F.J. 2002. Induction of angiogenesis in a mouse model using engineered transcription factors. *Nat. Med.* **8:** 1427.

Royds J.A., Dower S.K., Qwarnstrom E.E., and Lewis C.E. 1998. Response of tumour cells to hypoxia: Role of p53 and NFκB. *Mol. Pathol.* **51:** 55.

Ryan H., Lo J., and Johnson R. 1998. HIF-1 alpha is required for solid tumor formation and embryonic vascularization. *EMBO J.* **17:** 3005.

Ryan H., Poloni M., McNulty W., Elson D., Gassmann M., Arbeit J., and Johnson R. 2000. Hypoxia-inducible factor-1alpha is a positive factor in solid tumor growth. *Cancer Res.* **60:** 4010.

Salceda S. and Caro J. 1997. Hypoxia-inducible factor 1alpha (HIF-1alpha) protein is rapidly degraded by the ubiquitin-proteasome system under normoxic conditions. Its stabilization by hypoxia depends on redox-induced changes. *J. Biol. Chem.* **272:** 22642.

Schipani E., Ryan H.E., Didrickson S., Kobayashi T., Knight M., and Johnson R.S. 2001. Hypoxia in cartilage: HIF-1alpha is essential for chondrocyte growth arrest and survival. *Genes Dev.* **15:** 2865.

Semenza, G.L. 2000. Surviving ischemia: Adaptive responses mediated by hypoxia-inducible factor 1. *J Clin. Invest.* **106:** 809.

———. 2002. Signal transduction to hypoxia-inducible factor 1. *Biochem. Pharmacol.* **64:** 993.

———. 2003. Angiogenesis in ischemic and neoplastic disorders. *Annu. Rev. Med.* **54:** 17.

Shyu K.-G., Wang M.-T., Wang B.-W., Chang C.-C., Leu J.-G., Kuan P., and Chang H. 2002. Intramyocardial injection of naked DNA encoding HIF-1α/VP16 hybrid to enhance angiogenesis in an acute myocardial infarction model in the rat. *Cardiovasc. Res.* **54:** 576.

Springer M., Chen A., Kraft P., Bednarski M., and Blau H. 1998. VEGF gene delivery to muscle: Potential role for vasculogenesis in adults. *Mol. Cell* **2:** 549.

Sundberg C., Nagy J.A., Brown L.F., Feng D., Eckelhoefer I.A., Manseau E.J., Dvorak A.M., and Dvorak H.F. 2001. Glomeruloid microvascular proliferation follows adenoviral vascular permeability factor/vascular endothelial growth factor-164 gene delivery. *Am. J. Pathol.* **158:** 1145.

Suri C., McClain J., Thurston G., McDonald D., Zhou H., Oldmixon E., Sato T., and Yancopoulos G. 1998. Increased vascularization in mice overexpressing angiopoietin-1. *Science* **282:** 468.

Talks K., Turley H., Gatter K., Maxwell P., Pugh C., Ratcliffe P., and Harris A. 2000. The expression and distribution of the hypoxia-inducible factors HIF-1alpha and HIF-2alpha in normal human tissues, cancers, and tumor-associated macrophages. *Am. J. Pathol.* **157:** 411.

Thurston G., Suri C., Smith K., McClain J., Sato T., Yancopoulos G.D., and McDonald D. 1999. Leakage-resistant blood vessels in mice transgenically overexpressing angiopoietin-1. *Science* **286:** 2511.

Thurston G., Rudge J.S., Ioffe E., Zhou H., Ross L., Croll S., Glazer N., Holash J., McDonald D., and Yancopoulos G.D. 2000. Angiopoietin-1 protects the adult vasculature against plasma leakage. *Nat. Med.* **6:** 460.

Tian H., Hammer R.E., Matsumoto A.M., Russell D.W., and McKnight S.L. 1998. The hypoxia-responsive transcription factor EPAS1 is essential for catecholamine homeostasis and protection against heart failure during embryonic development. *Genes Dev.* **12:** 3320.

Vincent K., Shyu K., Luo Y., Magner M., Tio R., Jiang C., Gopldberg M., Akita G., Gregory R., and Isner J. 2000. Angiogenesis is induced in a rabbit model of hindlimb ischemia by naked DNA encoding an HIF-1α/VP16 hybrid transcription factor. *Circulation* **102:** 2255.

Yan S.F., Fujita T., Lu J., Okada K., Shan Zou Y., Mackman N., Pinsky D.J., and Stern D.M. 2000. Egr-1, a master switch coordinating upregulation of divergent gene families underlying ischemic stress. *Nat. Med.* **6:** 1355.

Yancopoulos G.D., Davis S., Gale N.W., Rudge J.S., Wiegand S.J., and Holash J. 2000. Vascular-specific growth factors and blood vessel formation. *Nature* **407:** 242.

Yang H. and Kaelin W.G., Jr. 2001. Molecular pathogenesis of the von Hippel-Lindau hereditary cancer syndrome: Implications for oxygen sensing. *Cell Growth Differ.* **12:** 447.

Yurasov S.V., Flasshove M., Rafii S., and Moore M.A. 1996. Density enrichment and characterization of hematopoietic progenitors and stem cells from umbilical cord blood. *Bone Marrow Transplant.* **17:** 517.

Zagzag D., Zhong H., Scalzitti J.M., Laughner E., Simons J.W., and Semenza G.L. 2000. Expression of hypoxia-inducible factor 1alpha in brain tumors: Association with angiogenesis, invasion, and progression. *Cancer* **88:** 2606.

Zundel W., Schindler C., Haas-Kogan D., Koong A., Kaper F., Chen E., Gottschalk A., Ryan H., Johnson R., Jefferson A., Stokoe D., and Giaccia A. 2000. Loss of PTEN facilitates HIF-1-mediated gene expression. *Genes Dev.* **14:** 391.

The Diverse Roles of Integrins and Their Ligands in Angiogenesis

R.O. Hynes, J.C. Lively, J.H. McCarty, D. Taverna, S.E. Francis,*
K. Hodivala-Dilke,† and Q. Xiao

Howard Hughes Medical Institute and Center for Cancer Research, Massachusetts Institute of Technology, Cambridge, Massachusetts 02139

The circulatory system is the first organ established during vertebrate development and is essential for life. During later development and in adult life, the vasculature is a dynamic system, generating additional vessels in response to need. It is generally agreed that the initial vascular network arises by a process termed vasculogenesis in which vascular precursors, angioblasts, generate endothelial cells that migrate and coalesce into the primitive vasculature. Later development of the vascular tree, and essentially all new vessel formation in adult life, arise by branching from this initial network and its derivatives. This process is usually called angiogenesis, although that encompasses diverse processes such as sprouting, branching, subdivision of existing vessels by bridging, and intussusception (Beck and D'Amore 1997; Carmeliet and Collen 1997; Risau 1997). Both vasculogenesis and angiogenesis, in their various forms, require cell proliferation and migration and cell–cell adhesion to form the endothelial tubes of the vasculature. Those endothelial tubes are surrounded by a basement membrane (BM), and adhesion of the endothelial cells to this extracellular matrix (ECM) is also required. Furthermore, additional cells surround the basic endothelial tubes and interact with the BM of the vessels. Depending on the vessel type, these "mural cells" are termed pericytes (small vessels) or smooth muscle cells (larger vessels), although they likely arise from the same cell lineage. These mural cells must also migrate to the developing vessels and form adhesive contacts with the BMs of the vasculature. Therefore, cell migration and cell–cell and cell–matrix adhesion are central to vascular development and remodeling. Much of the cell–matrix adhesion occurs via members of the integrin family, which are also involved in migration of the cells to form the vessels. Other cell adhesion molecules, such as cadherins, and cell interaction receptors, such as the ephrin, Eph, and Notch families, are also involved and may interact with the integrins. In this paper, we focus on integrins and their ligands, predominantly ECM glycoproteins, and consider the diverse roles that they play in vascular development and remodeling.

VASCULAR INTEGRINS AND THEIR LIGANDS

We have previously reviewed the complement of integrins whose expression on vascular cells is well established, along with their ligands (Hynes et al. 1999). The major ones are $\alpha5\beta1$ integrin and its ligand, fibronectin; $\alpha1\beta1$ and $\alpha2\beta1$, which are receptors for laminins and collagens; and the αv integrins, $\alpha v\beta3$ and $\alpha v\beta5$, both of which recognize a variety of ECM proteins containing the recognition sequence Arg-Gly-Asp (RGD). These include vitronectin, fibronectin, thrombospondins, von Willebrand factor (Plow et al. 2000; van der Flier and Sonnenberg 2001), Del1 (Hidai et al. 1998; Penta et al. 1999), and members of the CCN matrix-associated family of proteins (Lau and Lam 1999). $\alpha v\beta3$ is also reported to bind to various fragments of collagens, laminins, and other matrix proteins (see below).

The challenge is to divine which of these integrin–ligand pairs are important regulators or mediators of vascular development. This is of intrinsic interest, but also has relevance to the development of antiangiogenic drugs, since integrins are well suited as targets. They are present on the cell surface and therefore accessible. They bind their ligands with relatively low affinities and commonly recognize short peptide motifs, such as RGD, so that the integrin–ligand interactions can readily be blocked in many cases by peptides or peptidomimetics. These features make integrins accessible drug targets and blocking antibodies and small molecules are already in clinical use as antithrombotics, targeting the major platelet integrin, $\alpha IIb\beta3$ (Scarborough and Gretler 2000), and are in clinical development as anti-inflammatory drugs targeting integrins of the $\beta2$ and $\alpha4$ families on white blood cells. Thus, it is reasonable to contemplate applying similar strategies to vascular integrins. For such approaches to succeed, one needs to know which integrin–ligand pairs do what and which are the most appropriate targets.

$\alpha5\beta1$ AND FIBRONECTIN

We start with a consideration of the integrin–ligand pair that is best established as being essential for vasculogenesis and angiogenesis. Fibronectin is expressed on the earliest-forming vascular networks as a constituent of the BM (Risau and Lemmon 1988; Francis et al. 2002). Genetic ablation of either $\alpha5$ or fibronectin produces early

---

Present addresses: *University of Sheffield, Sheffield, S57AU, United Kingdom; †Cancer Research UK, London, SE1 7EH, United Kingdom.

embryonic lethality in mouse embryos. The phenotypes differ slightly, but a major feature in each case is that the yolk sac and embryonic vascular networks fail to form properly; i.e., vasculogenesis and angiogenesis are defective (George et al. 1993; Yang et al. 1993). In both cases, endothelial cells (defined by expression of PECAM-1/CD31) do develop from their angioblast precursors. In α5-null embryos, endothelial tubes do form, but they are dilated and disrupted and the pattern of vessels is abnormal (Yang et al. 1993; Goh et al. 1997; Francis et al. 2002). The defects seen in fibronectin-null embryos are more severe, ranging from severely disrupted extraembryonic and intraembryonic vessels and heart to a complete absence of a heart, depending on the genetic background (George et al. 1993, 1997; Georges-Labousse et al. 1996). In the most severe cases, the two heart primordia fail to migrate to the midline to form the heart (George et al. 1997). Major deficits in vascular formation are also seen in α5-null embryoid bodies (Taverna and Hynes 2001) and in α5-null teratocarcinomas (Taverna and Hynes 2001; Francis et al. 2002). Fibronectin-null embryoid bodies, like fibronectin-null embryos, show more severe deficits than do their α5-null counterparts (see Table 1) (Francis et al. 2002), but some rescue of vasculogenesis can be achieved by adding back fibronectin to the developing embryoid bodies (Francis et al. 2002). In complete accord with these genetic analyses, antibodies against α5β1 or against fibronectin, or peptides that selectively block their interaction, have been shown to inhibit angiogenesis in several systems, including the chicken chorioallantoic membrane and implanted Matrigel plugs (Kim et al. 2000), and both α5β1 and fibronectin are up-regulated by the angiogenic growth factor FGF-2 (Kim et al. 2000).

It seems clear from these results that α5β1 binding to fibronectin plays an essential role in vasculogenesis and angiogenesis. Given what is known about this receptor–ligand pair, those roles very likely include promotion of cell migration, appropriate ECM (e.g., BM) assembly, and signal transduction into the cell, affecting cell proliferation and survival and other processes. Accordingly, this receptor–ligand pair seems a possible candidate for antiangiogenic drugs. Peptides can be chosen that are highly selective for this pair, even distinguishing them from other integrins that also recognize RGD, as does α5β1 (Koivunen et al. 1993, 1994). The major challenge in targeting α5β1 is likely to be toxicity, since this integrin is relatively widely distributed, as is fibronectin, and side effects are likely.

α1β1 AND α2β1

These two integrins are also expressed by endothelial cells and are up-regulated by vascular endothelial growth factor (VEGF) (Senger et al. 1997). Both these integrins are receptors for a variety of collagens and, in some cells and situations, also for laminins. Antibodies to α1β1 plus α2β1 inhibit angiogenesis in response to VEGF in the chicken chorioallantoic membrane (Senger et al. 1997), and in mouse skin in response to human tumor xenografts (Senger et al. 2002), and block endothelial migration in collagen gels (Senger et al. 2002). Genetic data are in overall agreement; the genes for both α1 and α2 have been knocked out in mice (Gardner et al. 1996; Pozzi et al. 2000; Chen et al. 2002). Neither produces embryonic lethality; the mice are viable and fertile. The double knockout has not yet been studied, which will be necessary because of possible overlapping functions of these two integrins. Tumor growth and angiogenesis are markedly reduced in α1-null mice (Pozzi et al. 2000, 2002), although some of the data suggest that this is an indirect effect involving matrix metalloproteases and release of angiostatin, rather than a simple role in cell–matrix adhesion. We discuss this aspect further below.

Although α1β1 and α2β1 are clearly less essential for angiogenesis than is α5β1, there are several lines of evidence implicating them in various aspects of vascular remodeling or pathological angiogenesis. Further analyses of the mutant mice will be informative, and progress in developing selective low-molecular-weight inhibitors of these two integrins (Knight et al. 1998) will yield further insights into their roles.

αV INTEGRINS

The situation concerning the αv subfamily of integrins is less simply summarized. There are large bodies of data, both using inhibitory antibodies and peptide-based reagents, and from genetic ablation experiments, but, unlike the situation for α5β1 and fibronectin, these two sets of data are not readily reconciled, and it seems clear that the roles of αv integrins in vascular development are more complex. There are five integrins that contain a common αv subunit. Two of these, αvβ3 and αvβ5, are expressed on many angiogenic blood vessels, including tumor vasculature, and have attracted a great deal of attention as potential targets for angiogenesis (Brooks et al. 1994a,b; Eliceiri and Cheresh 1999, 2001). Both αvβ3 and αvβ5 are up-regulated by angiogenic growth factors

Table 1. Effects of Various Mutations on Vascularization of Embryoid Bodies

	Wild type	TSP-null	αv-null	β3-null	α5-null	FN-null	FN-null plus FN
EB diameter μm ± S.E.M.	167 ± 14	170 ± 11	175 ± 11	168 ± 14	188 ± 13	213 ± 18	196 ± 16
% area of EB occupied by CD31+ stain	54.3 ± 2.8	55 ± 4.9	47 ± 3.9	40 ± 4.8	22 ± 2.8	6.0 ± 0.7	26 ± 3.5[a]

[a]$p < 0.05$ relative to FN-null.

ES cells of various genotypes were cultured as embryoid bodies for 11 days with growth factors. The embryoid body diameter and the percent of the area of sections occupied by PECAM/CD31-positive cells was determined. In the case of FN-null ES cells, some were cultured with addition of 100 μg/ml mouse FN.

and promote endothelial adhesion and migration to vitronectin and to other ECM molecules.

The genetic results on this family of integrins, however, show clearly that they are not essential for angiogenesis. All six subunits (αv, β1, β3, β5, β6, and β8) have been knocked out in mice. Null mutants of β3 (Hodivala-Dilke et al. 1999), β5 (Huang et al. 2000), and β6 (Huang et al. 1996) are all viable and fertile with no obvious defects in vascular development. All three double null strains (β3/β5, β3/β6, and β5/β6) are similarly viable, fertile, and, insofar as they have been tested, show no defects in vasculogenesis or angiogenesis (see more below), and embryoid bodies null for αv or β3 show extensive angiogenesis (Table 1) (Francis et al. 2002). The β1-null mutation produces early embryonic lethality, as expected, given that this subunit pairs with a dozen α subunits (Fassler and Meyer 1995; Stephens et al. 1995). That result is not particularly informative vis-à-vis any role for αvβ1, which is a rare integrin, since αv prefers to partner with β3 when it is present. However, β1-null teratocarcinomas fail to develop their own vasculature (Bloch et al. 1997), and this deficit is more serious than that shown by α5-null teratocarcinomas (Taverna and Hynes 2001), thus implicating other β1 integrins in vascular development.

More illuminating are the results of ablation of αv (Bader et al. 1998; McCarty et al. 2002) or β8 (Zhu et al. 2002). These two knockouts have very similar phenotypes; indeed, at this point, there are few clear differences between them. Both are embryonic and postnatal lethals, but the embryos develop normally up to E9.5, including extensive vasculogenesis and angiogenesis in the yolk sac and within the embryo. Shortly thereafter, 60–80% of the null embryos develop placental defects and, progressively, other defects leading to death by E11–E12. The remaining embryos continue to develop to term. Most tissues, including their vasculature, are normal, and the pups are born alive. However, they do not suckle, and they die within a few hours of birth. The major defect seen in both αv-null and β8-null embryos is the progressive development of vascular abnormalities in the brain. As mentioned above, blood vessels develop normally in the rest of the embryo; there must be something special about the function of αvβ8 in the brain, since similar vascular defects are not seen in αvβ3/αvβ5 double knockout embryos (Fig. 1b).

Closer analysis suggests that the defects do indeed lie in the brain parenchyma rather than in the vessels themselves. The vasculature of the brain develops purely by

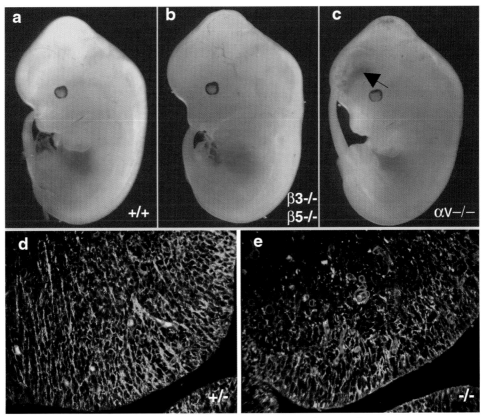

Figure 1. Cerebral hemorrhage in mice lacking αv integrins. (*a–c*) Absence of αvβ3 and αvβ5 does not lead to cerebral hemorrhage. E12.5 embryos, either wild type (*a*), null for both β3 and β5 integrin genes (*b*), or null for all five αv integrins (*c*). Only embryos lacking all αv integrins display grossly visible cerebral hemorrhage (*arrow* in *c*). (*d,e*) E12.5 paraffin-embedded transverse sections of ganglionic eminences from αv-heterozygous control (*d*) or αv-null (*e*) brains were immunolabeled with anti-nestin monoclonal antibody to visualize the neuroepithelial cells (*green*) and with anti-fibronectin antiserum to label the cerebral microvessels (*red*). Elaborate neuroepithelial processes span the ganglionic eminence and make contact points with vessels in the αv$^{+/-}$ control section (*d*). In an αv-null sample (*e*) devoid of significant vessel distension and hemorrhage, note the disorganized neuroepithelial processes and vessels.

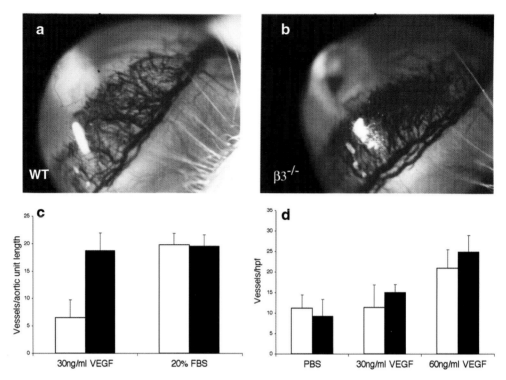

Figure 2. Angiogenesis assays in wild-type and $\beta3^{-/-}$ mice. Wild-type (*a*) and $\beta3^{-/-}$ (*b*) animals showed no quantifiable differences in angiogenesis in response to bFGF pellets in the mouse corneal micropocket angiogenesis assay. (*c, d*) $\beta3^{-/-}$ animals (■) were more responsive than wild-type (□) in response to VEGF stimulation in both aortic ring sprouting assays (*c*, $p<0.001$) and Matrigel plug assays (*d*, $p<0.05$).

angiogenesis from sprouts that develop from the perineural/meningeal vessels in response to VEGF released by the ventricular zone of the developing neural tube (Marin-Padilla 1985). These angiogenic sprouts initially develop normally in the absence of αv or β8, enter the brain parenchyma, lay down basement membrane components, penetrate down into the subventricular zone, branch, and carry blood (Bader et al. 1998; McCarty et al. 2002; Zhu et al. 2002). Thus, endothelial functions initially appear normal. Furthermore, pericytes are recruited to the developing vessels in normal numbers (McCarty et al. 2002; Zhu et al. 2002). However, by E11.5, the vessels in the mutant brains become dilated and progressively convoluted and develop spaces around them and eventually hemorrhage. Electron microscopic analyses show that, whereas the endothelial-pericyte interactions are normal, the surrounding neuroepithelial and glial cells are not closely apposed to the vessels as in normal brains and there is disorganization of the processes of these parenchymal cells (Fig. 1d,e).

The most straightforward interpretation of the results is that αvβ8 expressed in the cells of the neural parenchyma is necessary for proper apposition to, and support of, the developing vessels. This would explain the brain-specific nature of the defect and the absence of similar defects in embryos lacking both the endothelial integrins αvβ3 and αvβ5. Conclusive evidence that αvβ8 is expressed by the neuroepithelial cells and not by the endothelium or pericytes is lacking, although some data do suggest that conclusion (McCarty et al. 2002; Zhu et al. 2002). Whatever the final resolution concerning the detailed basis for the brain-specific αv/β8 defects, it is clear that most developmental angiogenesis proceeds normally in the complete absence of all five αv integrins—they clearly are not essential for developmental angiogenesis.

Until recently, it was possible to consider the possibility that αvβ3 and αvβ5 were necessary for pathological angiogenesis such as in retinopathy of prematurity or in response to tumors. However, that too has now been shown not to be the case (Reynolds et al. 2002; our unpublished data). Studies of mice lacking β3 integrins reveal a normal response to FGF-2 in corneal angiogenesis assays (Fig. 2a,b) and extensive retinal angiogenesis in response to hyperoxia followed by return to normoxia, which produces a burst of VEGF causing proliferative angiogenesis in the retina (the cause of infant blindness in retinopathy of prematurity). Indeed, the β3-null mice show enhanced angiogenesis in Matrigel plugs containing VEGF (Fig. 2c) (Reynolds et al. 2002). They also show enhanced sprouting of vessels from aortic rings in response to VEGF (Fig. 2d) (Reynolds et al. 2002). In fact, these mice, rather than showing a *deficit* in angiogenic responses to VEGF, show an *enhanced response* in all assays tried thus far (see further discussion below).

What then of tumor angiogenesis? β3-null mice, as well as β3/β5 double null mice, consistently support the growth of *larger* tumors (Fig. 3a) with *more* blood vessels (Fig. 3b–e) than do wild-type control mice (Reynolds et al. 2002). Four different transplantable tumors have been investigated, and the results are all in agreement

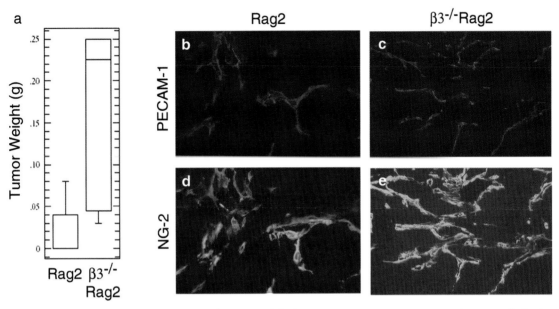

Figure 3. Tumor growth and angiogenesis in Rag2$^{-/-}$ and Rag2$^{-/-}$β3$^{-/-}$ mice. (*a*) A375SM tumors grown subcutaneously for 4 weeks were larger in Rag2β3$^{-/-}$ animals than in Rag2 counterparts ($n = 7$, $p = 0.003$). Tumor sections from Rag2$^{-/-}$ (*b,d*) and Rag2$^{-/-}$β3$^{-/-}$ (*c,e*) mice were double-labeled for vascular markers PECAM-1 (*b,c*) and NG-2 (*d,e*). An increase in vascular staining is shown in tumors from Rag2β3$^{-/-}$ animals.

(Table 2); rather than showing reduced tumor growth and angiogenesis, these mice showed enhancement of both. The results lead to the conclusion that these two integrins are negative regulators of angiogenesis.

That conclusion is not what one might have expected from the observation that these two integrins are up-regulated by angiogenic growth factors and in tumor vasculature (Brooks et al. 1994a,b; Eliceiri and Cheresh 1999, 2001). However, these results are not in conflict—perhaps these integrins are up-regulated precisely in order to regulate the angiogenesis and shut it down at the appropriate point. But there is a body of data using blocking antibodies, peptides, and peptidomimetics based on the RGD recognition motif that is in apparent conflict with the genetic data.

RGD-based low-molecular-weight reagents and certain blocking antibodies can inhibit the in vitro functions of αvβ3 and/or αvβ5 in cell adhesion and migration. These same reagents have been applied to various angiogenic assays, most extensively to angiogenesis in the chorioallantoic membrane in response to growth factors or tumors (Brooks et al. 1994a,b), but also to transplanted tumors (Brooks et al. 1995) and retinal angiogenesis (Friedlander et al. 1996; Hammes et al. 1996). In those studies, the anti-αvβ3 monoclonal antibody, LM609, RGD peptides, and peptidomimetics have been reported to inhibit angiogenesis. Those results, in concert with the expression data mentioned earlier and the in vitro antagonist data, have been interpreted to show that αvβ3 and/or αvβ5 are necessary for angiogenesis and are proangiogenic. That model needs reevaluation in light of the genetic data that clearly show these two integrins are not necessary for angiogenesis and appear to act as suppressors or negative regulators of angiogenesis. We make the alternative proposals that these integrins are indeed negative regulators and that the "blocking reagents" are acting as *agonists* of negative regulatory functions rather than as antagonists of positive functions of these integrins. We develop these models further below, but it is expedient first to discuss some known negative regulators of angiogenesis, which are also integrin ligands.

THROMBOSPONDINS

Among the earliest and best-established endogenous regulators of angiogenesis are thrombospondins (TSPs) 1 and 2. (for reviews, see Lawler 2000; Adams 2001). Fragments and peptides derived from thrombospondins can be shown to inhibit various functions of endothelial cells and in vitro assays of angiogenesis. Whereas the null mutation for TSP-1 does not produce an obvious increase in angiogenesis (Lawler et al. 1998), TSP-2 animals show enhanced vascularity (Kyriakides et al. 1998). TSP-1-null animals support growth of larger tumors and enhanced angiogenesis (Rodriguez-Manzaneque et al. 2001), whereas overexpression of TSP 1 or 2 suppresses tumor angiogenesis (Streit et al. 1999a,b; Rodriguez-Manzaneque et al. 2001). Therefore, the results are in accord that

Table 2. Tumor Growth and Angiogenesis

	WT	β3KO	β3/β5DKO
B16 melanoma	+	++	++
CMT19T lung carcinoma	+	++	++
LS180 adenocarcinoma	+	++	++
A375M melanoma	+	++	++

Tumor cells were inoculated subcutaneously into mice of various genotypes and monitored for tumor growth and for vascularization (vessels per unit area).

TSP-1 and TSP-2 serve as negative regulators of angiogenesis in vivo.

Evidence has been presented in support of several mechanisms for the antiangiogenic effects of thrombospondins. The antiangiogenic fragments and peptides are from the type 1 repeats, which bind to the CD36 cell-surface receptor and trigger apoptosis (Jimenez et al. 2000). The same region of TSP-1 activates TGFβ (Lawler 2000; Murphy-Ullrich and Poczatek 2000; Adams 2001). TGFβ and its receptors clearly play some role in angiogenesis, since null mutations in TGFβ or in TGFβR or a dominant-negative transgene of TGFβR produce vascular defects (Dickson et al. 1995; Oshima et al. 1996; Goumans et al. 1999). TGFβ is well known to stimulate the expression of ECM proteins, including fibronectin and thrombospondin, and of integrins.

It has also been shown that TSP-1 inhibits the activation of MMP-9 (Rodriguez-Manzaneque et al. 2001). Since MMP-9 is known to release VEGF from matrix-bound form, allowing it to bind to its receptor, VEGFR2 (Bergers et al. 2000), TSP-1 acts to suppress the VEGF/VEGFR2 angiogenic stimulus. TSP-1 and 2 both bind and inhibit MMP-2 (Bein and Simons 2000; Yang et al. 2000). As shown below, MMP-2 can have both pro- and antiangiogenic effects. Thus, there are several ways in which thrombospondins could suppress angiogenesis (Fig. 4), and it is not clear which are the most important. Nonetheless, the knockout and transgenic results show clearly that the net result is inhibition of angiogenesis.

FRAGMENTS OF EXTRACELLULAR MATRIX PROTEINS

Stimulated in part by the early results on thrombospondin fragments, Folkman and colleagues have isolated a number of proteolytic fragments that show antiangiogenic activity. The most famous of these are angiostatin, a fragment of plasminogen (O'Reilly et al. 1994), and endostatin, a proteolytic fragment of collagen XVIII (O'Reilly et al. 1997). Their mechanisms of action are not well understood, although both have been reported to bind to integrins (Rehn et al. 2001; Tarui et al. 2001).

Following on this work, several groups have isolated yet additional proteolytic fragments by digesting basement membrane components with matrix metalloproteases. The various isoforms of type IV collagen have proven to be a particularly rich source of antiangiogenic fragments (Colorado et al. 2000; Maeshima et al. 2000; Petitclerc et al. 2000). Like endostatin, which comprises the carboxy-terminal noncollagenous domain of collagen XVIII, analogous carboxy-terminal, noncollagenous fragments of several collagen IV α chains have been isolated and reported to be antiangiogenic. Several of these have been reported to bind to various integrins.

The most extensively studied of these fragments is tumstatin, derived from the α3 (IV) subunit (Maeshima et al. 2000). This fragment, and smaller peptides derived from it, inhibit endothelial proliferation and induce apoptosis. The effects of these reagents are totally dependent on αvβ3, to which they bind (Maeshima et al. 2001, 2002). In contrast, endostatin inhibition of endothelial cell proliferation is independent of αvβ3 (Maeshima et al. 2002). Tumstatin and its derivatives inhibit a signaling pathway of αvβ3 acting through PI3 kinase, Akt, and mTOR to regulate CAP-dependent protein synthesis (Fig. 5). The antiangiogenic segment of tumstatin lacks an RGD sequence and apparently does not compete directly with vitronectin, which binds to αvβ3 to activate the pathway. Tumstatin is an attractive lead for the development of antiangiogenic drugs, although it remains unclear whether it is a true endogenous inhibitor.

MODELS FOR NEGATIVE REGULATION OF ANGIOGENESIS BY αV INTEGRINS

As discussed above, the results of genetic ablation strongly suggest that αvβ3 and αvβ5 integrins act in vivo to suppress angiogenesis. This is in apparent conflict with the inhibition of angiogenesis by antibodies and low-molecular-weight reagents targeting these same integrins.

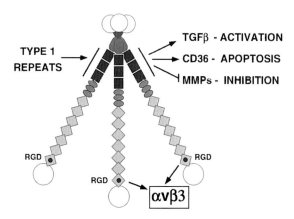

Figure 4. Effects of thrombospondins 1 and 2 on angiogenesis. The type 1 repeats bind TGFβ and, in the case of TSP-1, activate it. They also bind to the CD36 cell-surface receptor on endothelial cells and activate apoptopic pathways. Finally, TSP-1 can inhibit the activation of MMP-9, and both TSP-1 and -2 bind and inhibit MMP-2 (see text for details and references). The last type 3 repeat contains an RGD sequence that can bind to αvβ3; this binding could concentrate thrombospondins at the tips of angiogenic vessels and thus focus their other activities.

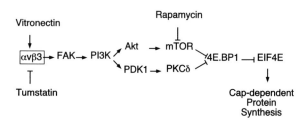

Figure 5. Inhibition of protein synthesis by tumstatin acting through αvβ3 integrin. Tumstatin binds αvβ3 at a site distinct from vitronectin. This leads to inhibition of activation of FAK, PI3 kinase, Akt, and mTOR, relieving inhibition of 4E-BP1 and allowing it to bind and inhibit EIF4E, and thus cap-dependent protein synthesis (Maeshima et al. 2000). Typical integrin ligands such as vitronectin, in contrast, activate the same pathway. Akt is also known to suppress apoptosis, and thus vitronectin promotes cell survival while tumstatin promotes apoptosis.

In this section, we discuss models that may reconcile these two sets of data. Before doing so, it is worth noting that such conflict between genetic and pharmacologic data is not typical. It is not seen for α5β1 and fibronectin, as discussed above, nor for the role of αvβ3 in bone remodeling. Bone resorption is inhibited both by antibodies and RGD-based reagents (Horton 2001) and by ablation of the gene for β3 (McHugh et al. 2000). Therefore, there is something distinct about the functions of αvβ3 and αvβ5 specifically in angiogenesis. We argue the case that their role is as negative regulators and that the so-called antagonists may be acting as agonists of this negative regulatory function.

We consider first the possibility of negative regulation of the VEGF/VEGFR2 signaling pathway by αvβ3. This possibility is suggested by the observation that VEGFR2 (flk-1) is up-regulated in endothelial cells from β3-null mice and can be restored to normal levels by transformation of β3 cDNA (Reynolds et al. 2002). Thus, a negative feedback loop between αvβ3 and VEGF/VEGFR2 signaling appears to exist, and its absence in β3-null mice could contribute to their enhanced responses to VEGF in various assays (Fig. 2) (Reynolds et al. 2002).

A second class of model involves transdominant inhibition of other integrins by αvβ3 and αvβ5 (Fig. 6). It is well established that α5β1 on endothelial cells acts to promote angiogenesis, and reasonably well established that α1β1 and α2β1 are likewise proangiogenic (see earlier sections). Cross talk among integrins is also well documented and can be either positive or negative (Blystone et al. 1999; Schwartz and Ginsberg 2002). In particular, it has been shown that αIIbβ3 can act as a negative regulator of α2β1 and α5β1 within the same cell (Diaz-Gonzalez et al. 1996). Low-molecular-weight "antagonists" of αIIbβ3 inhibit in *trans* the functions of the other two integrins. This is not due to nonspecificity, but to some form of transdominant inhibition, either via signal transduction or by competition for some limiting component. This implies that the antagonist is actually acting as an *agonist* of αIIβ3, triggering intracellular effects. Indeed, there is a lot of evidence that low-molecular-weight reagents such as RGD or RGD mimetics can activate both αIIbβ3 (Du et al. 1993; Peter et al. 2001) and αvβ3 (Legler et al. 2001). It has also been shown that antibodies to αvβ3 can block the functions of α5β1 in phagocytosis (Blystone et al. 1994, 1995, 1999) or in cell migration (Bilato et al. 1997; Simon et al. 1997), and the evidence suggests this is via signaling from the β3 cytoplasmic domain to inhibit CaM kinase II (Bilato et al. 1997; Blystone et al. 1999). Thus, it seems highly plausible that the effects of blocking antibodies or low-molecular-weight ligand mimetics targeting αvβ3 (and by analogy αvβ5) could actually function as agonists for these integrins and thereby inhibit the functions of the proangiogenic β1 integrins (Fig. 6). Removal of the αv integrins by mutation would then be expected to enhance angiogenesis, as observed.

A different sort of negative regulatory model invokes ligands of αvβ3 that are themselves negative regulators. We discussed the compelling evidence that thrombospondins 1 and 2 are negative regulators of angiogene-

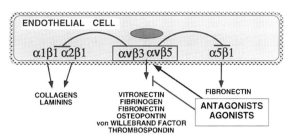

Figure 6. Transdominant inhibition via αvβ3 and αvβ5. On the basis of well-established examples of cross inhibition among integrins, it is proposed that reagents which block ligand-binding by αvβ3 and αvβ5 (i.e., antagonists) can at the same time activate signaling by these same integrins (i.e., act as agonists). See text for further discussion.

sis (Fig. 4). These glycoproteins are also ligands for αvβ3 (Lawler et al. 1988). Therefore, it is possible that αvβ3 on angiogenic endothelial cells could bind and concentrate thrombospondins, thereby focusing their antiangiogenic functions.

Another way in which αvβ3 and αvβ5 could generate and/or focus an antiangiogenic effect is through effects on matrix metalloproteases (MMPs). αvβ3 can bind, localize, and activate MMP2 (Brooks et al. 1996). This could lead to degradation of ECM in the area around the growing tips of angiogenic vessels. Such degradation could enhance angiogenesis, but it could also lead to release of antiangiogenic fragments of ECM, as discussed earlier. These could in turn act via integrins on the endothelial cells. Tumstatin has clearly been shown to bind to αvβ3, inhibit endothelial proliferation, and induce apoptosis (Maeshima et al. 2000, 2001, 2002). Other antiangiogenic ECM fragments have been reported to bind to αvβ3, α5β1, or α2β1. Thus, αvβ3 binding and activation of MMP2 could initiate a negative feedback loop (Fig. 7).

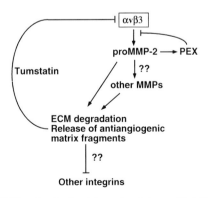

Figure 7. Negative feedback loops involving αvβ3, MMP2, and antiangiogenic matrix fragments. αvβ3 is known to bind and activate MMP2. This matrix metalloprotease can cleave ECM proteins (either directly or indirectly together with other MMPs). Many such fragments are antiangiogenic and are believed to bind integrins. Most particularly, tumstatin (see Fig. 5 and text) is known to bind αvβ3, inhibit protein synthesis, and induce apoptosis in endothelial cells. MMP2 can also cleave itself, releasing a noncatalytic fragment, PEX, that negatively regulates the effects of αvβ3 on MMP2 and is antiangiogenic.

Absence of MMP-2 neither enhances nor suppresses angiogenesis in the RIPTAg model (Bergers et al. 2000), but does cause reduced wound angiogenesis and transplanted tumor growth (Itoh et al. 1998). Those results tend to support a positive rather than a negative role for MMP2 in angiogenesis. That inference is supported by experiments introducing a noncatalytic fragment of MMP-2 (called PEX) into angiogenic systems. Such intervention interferes with binding and activation of MMP-2 by αvβ3 and suppresses angiogenesis and tumor growth (Brooks et al. 1998; Pfeifer et al. 2000). However, since PEX is generated by autocatalysis by MMP-2, it also serves as a negative feedback loop (Fig. 7). It seems likely that the ratio of MMP-2 to PEX and, therefore, the balance between proangiogenic and antiangiogenic effects of this system, may vary over the course of angiogenesis. In fact, this issue of balance between proangiogenic and antiangiogenic stimuli needs to be incorporated into any consideration of the regulation of angiogenesis. It is conceivable that αv integrins could have differing roles at different stages of angiogenesis, depending on the relative concentrations of the various ligands.

The roles of αv integrins in regulating apoptosis of endothelial cells may also encompass both positive and negative effects (see Fig. 8). In the earlier blocking experiments, it was reported that LM609 or RGD peptides induce apoptosis of endothelial cells (Brooks et al. 1994b). This was interpreted in terms of the well-established role of integrins as suppressors of apoptosis (Meredith et al. 1993; Frisch and Screaton 2001). However, it has also been shown that RGD peptides can directly activate caspases without any involvement of integrins (Buckley et al. 1999; Adderley and Fitzgerald 2000). The potential contribution of this mechanism to the apoptosis observed in the angiogenesis inhibition experiments has yet to be evaluated. Finally, it has recently been reported that unligated αvβ3 can activate caspase 8 and induce apoptosis (Stupack et al. 2001) and that this is blocked by ECM ligands of αvβ3. It has been suggested that this might explain the disparity between the pharmacologic and genetic data (Cheresh and Stupack 2002). However, there is no evidence that αvβ3, a highly promiscuous integrin with a plethora of potential ligands, is ever "unligated" during angiogenesis. For this model to explain the pharmacologic data, it would be necessary to claim that αvβ3 blocked by LM609 or RGD peptides is unligated. However, as discussed above, there is good evidence that RGD peptides ligate and activate integrins, including αvβ3 (Du et al. 1993; Legler et al. 2001; Peter et al. 2001). Thus, it remains unclear whether unligation of integrins really plays any role in controlling apoptosis of endothelial cells. It seems more likely that apoptosis is induced by negative regulators such as thrombospondin and tumstatin (see earlier discussion), perhaps acting through αvβ3 in the case of tumstatin or organized by it in the case of thrombospondin.

In conclusion, there exist multiple plausible mechanisms whereby αvβ3 and αvβ5 could be acting as negative regulators of angiogenesis (see Figs. 4–8). Since the genetic data argue strongly for that interpretation, it is reasonable to attempt to incorporate the antibody and small molecule results on the same model. The foregoing discussion shows that that is feasible. In fact, the negative regulation models provide at least as good, if not better, interpretations of the data. That is not to say that all aspects of αv integrin function in angiogenesis are negative. As mentioned earlier, it is conceivable that, at some stages of angiogenesis, these integrins may play a proangiogenic role. For example, it could be that at early stages, these integrins promote endothelial proliferation and migration on substrates such as vitronectin, Del 1, or members of the CCN family of angiogenic regulators. The fact that vitronectin-null mice show decreased wound angiogenesis (Jang et al. 2000) might support this interpretation. However, on balance, it appears that the αv integrins act as suppressors, and most of the pharmacologic data can be incorporated into negative regulation models assuming only that the blocking antibodies, peptides, and peptomimetics act as *agonists* of some functions of these integrins. Although they were originally characterized as antagonists of integrin-mediated adhesion, that does not preclude their being agonists of other functions and, as reviewed above, there is evidence that they can be agonists.

IMPLICATIONS FOR ANTIANGIOGENIC THERAPIES

It remains unclear whether the antibodies and peptides currently under evaluation as antiangiogenic drugs (Dechantsreiter et al. 1999; Gutheil et al. 2000) will prove efficacious—only further tests will tell. However, given the discussion presented here, it seems clear that candidate drugs targeted to αvβ3 and αvβ5 need to be evaluated in more detail than they have been. The likelihood that some of these reagents are acting as agonists of negative regulatory functions rather than as antagonists of positive roles needs to be investigated. In designing the

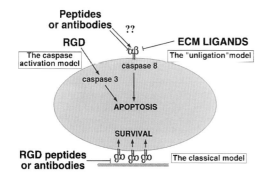

Figure 8. Various models for the effects of αvβ3 integrin on cell survival and apoptosis. In the classical model, integrins bound to their matrix ligands provide signals leading to cell survival (e.g., Akt, see Fig. 5). Interference with this pathway by antagonists of the integrin–ligand interaction removes the survival signals and causes apoptosis. In the caspase activation model, RGD peptides directly activate caspases without involvement of integrins. In the unligation model, unoccupied integrins are proposed to activate caspase 8, and this is blocked by ECM ligands. It is unclear what effect antibodies or RGD peptides have in this case.

best drugs, it will be important to understand how they work. Thus, investigation of signal transduction pathways regulating VEGFR2 expression, transdominant regulation of other integrins, survival, apoptosis, and matrix metalloprotease activities and products (see Figs. 4–8) will all need to be studied to determine, first, how the integrins affect these pathways and, second, how various antagonists/agonists modify them. Only in that way will it be possible to take a rational approach to optimizing the efficacy of the reagents.

ACKNOWLEDGMENTS

We thank Genevieve Hendrey for help in manuscript preparation. The research described here was supported by the Howard Hughes Medical Institute and by grants from the National Cancer Institute and the National Heart, Lung and Blood Institute.

REFERENCES

Adams J.C. 2001. Thrombospondins: Multifunctional regulators of cell interactions. *Annu. Rev. Cell Dev. Biol.* **17:** 25.

Adderley S.R. and Fitzgerald D.J. 2000. Glycoprotein IIb/IIIa antagonists induce apoptosis in rat cardiomyocytes by caspase-3 activation. *J. Biol. Chem.* **275:** 5760.

Bader B.L., Rayburn H., Crowley D., and Hynes R.O. 1998. Extensive vasculogenesis, angiogenesis, and organogenesis precede lethality in mice lacking all αv integrins. *Cell* **95:** 507.

Beck L., Jr., and D'Amore P.A. 1997. Vascular development: Cellular and molecular regulation. *FASEB J.* **11:** 365.

Bein K. and Simons M. 2000. Thrombospondin type 1 repeats interact with matrix metalloproteinase 2. Regulation of metalloproteinase activity. *J. Biol. Chem.* **275:** 32167.

Bergers G., Brekken R., McMahon G., Vu T.H., Itoh T., Tamaki K., Tanzawa K., Thorpe P., Itohara S., Werb Z., and Hanahan D. 2000. Matrix metalloproteinase-9 triggers the angiogenic switch during carcinogenesis. *Nat. Cell Biol.* **2:** 737.

Bilato C., Curto K.A., Monticone R.E., Pauly R.R., White A.J., and Crow M.T. 1997. The inhibition of vascular smooth muscle cell migration by peptide and antibody antagonists of the alphavbeta3 integrin complex is reversed by activated calcium/calmodulin- dependent protein kinase II. *J. Clin. Invest.* **100:** 693.

Bloch W., Forsberg E., Lentini S., Brakebusch C., Martin K., Krell H.W., Weidle U.H., Addicks K., and Fassler R. 1997. Beta 1 integrin is essential for teratoma growth and angiogenesis. *J. Cell Biol.* **139:** 265.

Blystone S.D., Graham I.L., Lindberg F.P., and Brown E.J. 1994. Integrin alpha v beta 3 differentially regulates adhesive and phagocytic functions of the fibronectin receptor alpha 5 beta 1. *J. Cell Biol.* **127:** 1129.

Blystone S.D., Lindberg F.P., LaFlamme S.E., and Brown E.J. 1995. Integrin beta 3 cytoplasmic tail is necessary and sufficient for regulation of alpha 5 beta 1 phagocytosis by alpha v beta 3 and integrin-associated protein. *J. Cell Biol.* **130:** 745.

Blystone S.D., Slater S.E., Williams M.P., Crow M.T., and Brown E.J. 1999. A molecular mechanism of integrin crosstalk: alphavbeta3 suppression of calcium/calmodulin-dependent protein kinase II regulates alpha5beta1 function. *J. Cell Biol.* **145:** 889.

Brooks P.C., Clark R.A., and Cheresh D.A. 1994a. Requirement of vascular integrin alpha v beta 3 for angiogenesis. *Science* **264:** 569.

Brooks P.C., Silletti S., von Schalscha T.L., Friedlander M., and Cheresh D.A. 1998. Disruption of angiogenesis by PEX, a noncatalytic metalloproteinase fragment with integrin binding activity. *Cell* **92:** 391.

Brooks P.C., Stromblad S., Klemke R., Visscher D., Sarkar F.H., and Cheresh D.A. 1995. Antiintegrin alpha v beta 3 blocks human breast cancer growth and angiogenesis in human skin. *J. Clin. Invest.* **96:** 1815.

Brooks P.C., Montgomery A.M., Rosenfeld M., Reisfeld R.A., Hu T., Klier G., and Cheresh D.A. 1994b. Integrin alpha v beta 3 antagonists promote tumor regression by inducing apoptosis of angiogenic blood vessels. *Cell* **79:** 1157.

Brooks P.C., Stromblad S., Sanders L.C., von Schalscha T.L., Aimes R.T., Stetler-Stevenson W.G., Quigley J.P., and Cheresh D.A. 1996. Localization of matrix metalloproteinase MMP-2 to the surface of invasive cells by interaction with integrin alpha v beta 3. *Cell* **85:** 683.

Buckley C.D., Pilling D., Henriquez N.V., Parsonage G., Threlfall K., Scheel-Toellner D., Simmons D.L., Akbar A.N., Lord J.M., and Salmon M. 1999. RGD peptides induce apoptosis by direct caspase-3 activation. *Nature* **397:** 534.

Carmeliet P. and Collen D. 1997. Genetic analysis of blood vessel vormation. *Trends Cardiovasc. Med.* **7:** 271.

Chen J., Diacovo T.G., Grenache D.G., Santoro S.A., and Zutter M.M. 2002. The α2 integrin subunit deficient mouse: A multifaceted phenotype including defects of branching morphogenesis and hemostasis. *Am. J. Physiol.* **161:** 337.

Cheresh D.A. and Stupack D.G. 2002. Integrin-mediated death: An explanation of the integrin-knockout phenotype? *Nat. Med.* **8:** 193.

Colorado P.C., Torre A., Kamphaus G., Maeshima Y., Hopfer H., Takahashi K., Volk R., Zamborsky E.D., Herman S., Sarkar P.K., Ericksen M.B., Dhanabal M., Simons M., Post M., Kufe D.W., Weichselbaum R.R., Sukhatme V.P., and Kalluri R. 2000. Anti-angiogenic cues from vascular basement membrane collagen. *Cancer Res.* **60:** 2520.

Dechantsreiter M.A., Planker E., Matha B., Lohof E., Holzemann G., Jonczyk A., Goodman S.L., and Kessler H. 1999. N-Methylated cyclic RGD peptides as highly active and selective alpha(V)beta(3) integrin antagonists. *J. Med. Chem.* **42:** 3033.

Diaz-Gonzalez F., Forsyth J., Steiner B., and Ginsberg M.H. 1996. Trans-dominant inhibition of integrin function. *Mol. Biol. Cell* **7:** 1939.

Dickson M.C., Martin J.S., Cousins F.M., Kulkarni A.B., Karlsson S., and Akhurst R.J. 1995. Defective haematopoiesis and vasculogenesis in transforming growth factor-beta 1 knock out mice. *Development* **121:** 1845.

Du X., Gu M., Weisel J.W., Nagaswami C., Bennett J.S., Bowditch R., and Ginsberg M.H. 1993. Long range propagation of conformational changes in integrin alpha IIb beta 3. *J. Biol. Chem.* **268:** 23087.

Eliceiri B.P. and Cheresh D.A. 1999. The role of alphav integrins during angiogenesis: Insights into potential mechanisms of action and clinical development. *J. Clin. Invest.* **103:** 1227.

———. 2001. Adhesion events in angiogenesis. *Curr. Opin. Cell Biol.* **13:** 563.

Fassler R. and Meyer M. 1995. Consequences of lack of beta 1 integrin gene expression in mice. *Genes Dev.* **9:** 1896.

Francis S.E., Goh K.L., Hodivala-Dilke K., Bader B.L., Stark M., Davidson D., and Hynes R.O. 2002. Central roles of alpha5beta 1 integrin and fibronectin in vascular development in mouse embryos and embryoid bodies. *Arterioscler. Thromb. Vasc. Biol.* **22:** 927.

Friedlander M., Theesfeld C.L., Sugita M., Fruttiger M., Thomas M.A., Chang S., and Cheresh D.A. 1996. Involvement of integrins alpha v beta 3 and alpha v beta 5 in ocular neovascular diseases. *Proc. Natl. Acad. Sci. U S A* **93:** 9764.

Frisch S.M. and Screaton R.A. 2001. Anoikis mechanisms. *Curr. Opin. Cell Biol.* **13:** 555.

Gardner H., Kreidberg J., Koteliansky V., and Jaenisch R. 1996. Deletion of integrin alpha 1 by homologous recombination permits normal murine development but gives rise to a specific deficit in cell adhesion. *Dev. Biol.* **175:** 301.

George E.L., Baldwin H.S., and Hynes R.O. 1997. Fibronectins are essential for heart and blood vessel morphogenesis but are dispensable for initial specification of precursor cells. *Blood* **90:** 3073.

George E.L., Georges-Labouesse E.N., Patel-King R.S., Rayburn H., and Hynes R.O. 1993. Defects in mesoderm, neural tube and vascular development in mouse embryos lacking fibronectin. *Development* **119:** 1079.

Georges-Labouesse E.N., George E.L., Rayburn H., and Hynes R.O. 1996. Mesodermal development in mouse embryos mutant for fibronectin. *Dev. Dyn.* **207:** 145.

Goh K.L., Yang J.T., and Hynes R.O. 1997. Mesodermal defects and cranial neural crest apoptosis in α5 integrin-null embryos. *Development* **124:** 4309.

Goumans M.J., Zwijsen A., van Rooijen M.A., Huylebroeck D., Roelen B.A., and Mummery C.L. 1999. Transforming growth factor-beta signalling in extraembryonic mesoderm is required for yolk sac vasculogenesis in mice. *Development* **126:** 3473.

Gutheil J.C., Campbell T.N., Pierce P.R., Watkins J.D., Huse W.D., Bodkin D.J., and Cheresh D.A. 2000. Targeted antiangiogenic therapy for cancer using Vitaxin: A humanized monoclonal antibody to the integrin alphavbeta3. *Clin. Cancer Res* **6:** 3056.

Hammes H.P., Brownlee M., Jonczyk A., Sutter A., and Preissner K.T. 1996. Subcutaneous injection of a cyclic peptide antagonist of vitronectin receptor-type integrins inhibits retinal neovascularization. *Nat. Med.* **2:** 529.

Hidai C., Zupancic T., Penta K., Mikhail A., Kawana M., Quertermous E.E., Aoka Y., Fukagawa M., Matsui Y., Platika D., Auerbach R., Hogan B.L., Snodgrass R., and Quertermous T. 1998. Cloning and characterization of developmental endothelial locus-1: An embryonic endothelial cell protein that binds the alphavbeta3 integrin receptor. *Genes Dev.* **12:** 21.

Hodivala-Dilke K.M., McHugh K.P., Tsakiris D.A., Rayburn H., Crowley D., Ullman-Cullere M., Ross F.P., Coller B.S., Teitelbaum S., and Hynes R.O. 1999. Beta3-integrin-deficient mice are a model for Glanzmann thrombasthenia showing placental defects and reduced survival. *J. Clin. Invest.* **103:** 229.

Horton M.A. 2001. Integrin antagonists as inhibitors of bone resorption: Implications for treatment. *Proc. Nutr. Soc.* **60:** 275.

Huang X., Griffiths M., Wu J., Farese R.V., Jr., and Sheppard D. 2000. Normal development, wound healing, and adenovirus susceptibility in beta5-deficient mice. *Mol. Cell. Biol.* **20:** 755.

Huang X.Z., Wu J.F., Cass D., Erle D.J., Corry D., Young S.G., Farese R.V., Jr., and Sheppard D. 1996. Inactivation of the integrin beta 6 subunit gene reveals a role of epithelial integrins in regulating inflammation in the lung and skin. *J. Cell Biol.* **133:** 921.

Hynes R.O., Bader B.L., and Hodivala-Dilke K. 1999. Integrins in vascular development. *Braz. J. Med. Biol. Res.* **32:** 501.

Itoh T., Tanioka M., Yoshida H., Yoshioka T., Nishimoto H., and Itohara S. 1998. Reduced angiogenesis and tumor progression in gelatinase A-deficient mice. *Cancer Res.* **58:** 1048.

Jang Y.C., Tsou R., Gibran N.S., and Isik F.F. 2000. Vitronectin deficiency is associated with increased wound fibrinolysis and decreased microvascular angiogenesis in mice. *Surgery* **127:** 696.

Jimenez B., Volpert O.V., Crawford S.E., Febbraio M., Silverstein R.L., and Bouck N. 2000. Signals leading to apoptosis-dependent inhibition of neovascularization by thrombospondin-1. *Nat. Med.* **6:** 41.

Kim S., Bell K., Mousa S.A., and Varner J.A. 2000. Regulation of angiogenesis in vivo by ligation of integrin alpha5beta1 with the central cell-binding domain of fibronectin. *Am. J. Pathol.* **156:** 1345.

Knight C.G., Morton L.F., Onley D.J., Peachey A.R., Messent A.J., Smethurst P.A., Tuckwell D.S., Farndale R.W., and Barnes M.J. 1998. Identification in collagen type I of an integrin alpha2 beta1-binding site containing an essential GER sequence. *J. Biol. Chem.* **273:** 33287.

Koivunen E., Gay D.A., and Ruoslahti E. 1993. Selection of peptides binding to the alpha 5 beta 1 integrin from phage display library. *J. Biol. Chem.* **268:** 20205.

Koivunen E., Wang B., and Ruoslahti E. 1994. Isolation of a highly specific ligand for the alpha 5 beta 1 integrin from a phage display library. *J. Cell Biol.* **124:** 373.

Kyriakides T.R., Zhu Y.H., Smith L.T., Bain S.D., Yang Z., Lin M.T., Danielson K.G., Iozzo R.V., LaMarca M., McKinney C.E., Ginns E.I., and Bornstein P. 1998. Mice that lack thrombospondin 2 display connective tissue abnormalities that are associated with disordered collagen fibrillogenesis, an increased vascular density, and a bleeding diathesis. *J. Cell Biol.* **140:** 419.

Lau L.F. and Lam S.C. 1999. The CCN family of angiogenic regulators: The integrin connection. *Exp. Cell Res.* **248:** 44.

Lawler J. 2000. The functions of thrombospondin-1 and -2. *Curr. Opin. Cell Biol.* **12:** 634.

Lawler J., Weinstein R., and Hynes R.O. 1988. Cell attachment to thrombospondin: The role of ARG-GLY-ASP, calcium, and integrin receptors. *J. Cell Biol.* **107:** 2351.

Lawler J., Sunday M., Thibert V., Duquette M., George E.L., Rayburn H., and Hynes R.O. 1998. Thrombospondin-1 is required for normal murine pulmonary homeostasis and its absence causes pneumonia. *J. Clin. Invest.* **101:** 982.

Legler D.F., Wiedle G., Ross F.P., and Imhof B.A. 2001. Superactivation of integrin alphavbeta3 by low antagonist concentrations. *J. Cell Sci.* **114:** 1545.

Maeshima Y., Colorado P.C., and Kalluri R. 2000. Two RGD-independent alpha vbeta 3 integrin binding sites on tumstatin regulate distinct anti-tumor properties. *J. Biol. Chem.* **275:** 23745.

Maeshima Y., Sudhakar A., Lively J.C., Ueki K., Kharbanda S., Kahn C.R., Sonenberg N., Hynes R.O., and Kalluri R. 2002. Tumstatin, an endothelial cell-specific inhibitor of protein synthesis. *Science* **295:** 140.

Maeshima Y., Yerramalla U.L., Dhanabal M., Holthaus K.A., Barbashov S., Kharbanda S., Reimer C., Manfredi M., Dickerson W.M., and Kalluri R. 2001. Extracellular matrix-derived peptide binds to alpha(v)beta(3) integrin and inhibits angiogenesis. *J. Biol. Chem.* **276:** 31959.

Marin-Padilla M. 1985. Early vascularization of the embryonic cerebral cortex: Golgi and electron microscopic studies. *J. Comp. Neurol.* **241:** 237.

McCarty J.H., Monahan-Earley, R.A., Dvorak, A.M., Brown, L.F., Keller, M., Gerhardt, H., Rubin, K., Shani, M., Wolburg, H., Bader, B.L., Dvorak, H.F., and Hynes, R.O. 2002. Defective associations between blood vessels and devloping neuronal parenchyma lead to cerebral hemorrhage in mice lacking alpha-v integrins. *Mol. Cell. Biol.* **22:** 7667.

McHugh K.P., Hodivala-Dilke K., Zheng M.H., Namba N., Lam J., Novack D., Feng X., Ross F.P., Hynes R.O., and Teitelbaum S.L. 2000. Mice lacking beta3 integrins are osteosclerotic because of dysfunctional osteoclasts. *J. Clin. Invest.* **105:** 433.

Meredith J.E., Jr., Fazeli B., and Schwartz M.A. 1993. The extracellular matrix as a cell survival factor. *Mol. Biol. Cell* **4:** 953.

Murphy-Ullrich J.E., and Poczatek M. 2000. Activation of latent TGF-beta by thrombospondin-1: Mechanisms and physiology. *Cytokine Growth Factor Rev.* **11:** 59.

O'Reilly M.S., Boehm T., Shing Y., Fukai N., Vasios G., Lane W.S., Flynn E., Birkhead J.R., Olsen B.R., and Folkman J. 1997. Endostatin: An endogenous inhibitor of angiogenesis and tumor growth. *Cell* **88:** 277.

O'Reilly M.S., Holmgren L., Shing Y., Chen C., Rosenthal R.A., Moses M., Lane W.S., Cao Y., Sage E.H., and Folkman J. 1994. Angiostatin: A novel angiogenesis inhibitor that mediates the suppression of metastases by a Lewis lung carcinoma. *Cell* **79:** 315.

Oshima M., Oshima H., and Taketo M.M. 1996. TGF-beta receptor type II deficiency results in defects of yolk sac hematopoiesis and vasculogenesis. *Dev. Biol.* **179:** 297.

Penta K., Varner J.A., Liaw L., Hidai C., Schatzman R., and Quertermous T. 1999. Del1 induces integrin signaling and angiogenesis by ligation of alphaVbeta3. *J. Biol. Chem.* **274:** 11101.

Peter K., Schwarz M., Nordt T., and Bode C. 2001. Intrinsic activating properties of GP IIb/IIIa blockers. *Thromb. Res.* (suppl. 1) **103:** S21.

Petitclerc E., Boutaud A., Prestayko A., Xu J., Sado Y., Ninomiya Y., Sarras M.P., Jr., Hudson B.G., and Brooks P.C. 2000. New functions for non-collagenous domains of human collagen type IV. Novel integrin ligands inhibiting angiogenesis and tumor growth in vivo. *J. Biol. Chem.* **275:** 8051.

Pfeifer A., Kessler T., Silletti S., Cheresh D.A., and Verma I.M. 2000. Suppression of angiogenesis by lentiviral delivery of PEX, a noncatalytic fragment of matrix metalloproteinase 2. *Proc. Natl. Acad. Sci.* **97:** 12227.

Plow E.F., Haas T.A., Zhang L., Loftus J., and Smith J.W. 2000. Ligand binding to integrins. *J. Biol. Chem.* **275:** 21785.

Pozzi A., LeVine W.F., and Gardner H.A. 2002. Low plasma levels of matrix metalloproteinase 9 permit increased tumor angiogenesis. *Oncogene* **21:** 272.

Pozzi A., Moberg P.E., Miles L.A., Wagner S., Soloway P., and Gardner H.A. 2000. Elevated matrix metalloprotease and angiostatin levels in integrin alpha 1 knockout mice cause reduced tumor vascularization. *Proc. Natl. Acad. Sci.* **97:** 2202.

Rehn M., Veikkola T., Kukk-Valdre E., Nakamura H., Ilmonen M., Lombardo C., Pihlajaniemi T., Alitalo K., and Vuori K. 2001. Interaction of endostatin with integrins implicated in angiogenesis. *Proc. Natl. Acad. Sci.* **98:** 1024.

Reynolds L.E., Wyder L., Lively J.C., Taverna D., Robinson S.D., Huang X., Sheppard D., Hynes R.O., and Hodivala-Dilke K.M. 2002. Enhanced pathological angiogenesis in mice lacking beta3 integrin or beta3 and beta5 integrins. *Nat. Med.* **8:** 27.

Risau W. 1997. Mechanisms of angiogenesis. *Nature* **386:** 671.

Risau W. and Lemmon V. 1988. Changes in the vascular extracellular matrix during embryonic vasculogenesis and angiogenesis. *Dev. Biol.* **125:** 441.

Rodriguez-Manzaneque J.C., Lane T.F., Ortega M.A., Hynes R.O., Lawler J., and Iruela-Arispe M.L. 2001. Thrombospondin-1 suppresses spontaneous tumor growth and inhibits activation of matrix metalloproteinase-9 and mobilization of vascular endothelial growth factor. *Proc. Natl. Acad. Sci.* **98:** 12485.

Scarborough R.M. and Gretler D.D. 2000. Platelet glycoprotein IIb-IIIa antagonists as prototypical integrin blockers: Novel parenteral and potential oral antithrombotic agents. *J. Med. Chem.* **43:** 3453.

Schwartz M.A. and Ginsberg M.H. 2002. Networks and cross talk: Integrin signalling spreads. *Nat. Cell Biol.* **4:** E65.

Senger D.R., Claffey K.P., Benes J.E., Perruzzi C.A., Sergiou A.P., and Detmar M. 1997. Angiogenesis promoted by vascular endothelial growth factor: Regulation through alpha1beta1 and alpha2beta1 integrins. *Proc. Natl. Acad. Sci.* **94:** 13612.

Senger D.R., Perruzzi C.A., Streit M., Koteliansky V.E., de Fougerolles A.R., and Detmar M. 2002. The alpha(1)beta(1) and alpha(2)beta(1) integrins provide critical support for vascular endothelial growth factor signaling, endothelial cell migration, and tumor angiogenesis. *Am. J. Pathol.* **160:** 195.

Simon K.O., Nutt E.M., Abraham D.G., Rodan G.A., and Duong L.T. 1997. The alphavbeta3 integrin regulates alpha5beta1-mediated cell migration toward fibronectin. *J. Biol. Chem.* **272:** 29380.

Stephens L.E., Sutherland A.E., Klimanskaya I.V., Andrieux A., Meneses J., Pedersen R.A., and Damsky C.H. 1995. Deletion of beta 1 integrins in mice results in inner cell mass failure and peri-implantation lethality. *Genes Dev.* **9:** 1883.

Streit M., Riccardi L., Velasco P., Brown L.F., Hawighorst T., Bornstein P., and Detmar M. 1999a. Thrombospondin-2: A potent endogenous inhibitor of tumor growth and angiogenesis. *Proc. Natl. Acad. Sci.* **96:** 14888.

Streit M., Velasco P., Brown L.F., Skobe M., Richard L., Riccardi L., Lawler J., and Detmar M. 1999b. Overexpression of thrombospondin-1 decreases angiogenesis and inhibits the growth of human cutaneous squamous cell carcinomas. *Am. J. Pathol.* **155:** 441.

Stupack D.G., Puente X.S., Boutsaboualoy S., Storgard C.M., and Cheresh D.A. 2001. Apoptosis of adherent cells by recruitment of caspase-8 to unligated integrins. *J. Cell Biol.* **155:** 459.

Tarui T., Miles L.A., and Takada Y. 2001. Specific interaction of angiostatin with integrin alpha(v)beta(3) in endothelial cells. *J. Biol. Chem.* **276:** 39562.

Taverna D. and Hynes R.O. 2001. Reduced blood vessel formation and tumor growth in alpha5-integrin-negative teratocarcinomas and embryoid bodies. *Cancer Res.* **61:** 5255.

van der Flier A. and Sonnenberg A. 2001. Function and interactions of integrins. *Cell Tissue Res.* **305:** 285.

Yang J.T., Rayburn H., and Hynes R.O. 1993. Embryonic mesodermal defects in alpha 5 integrin-deficient mice. *Development* **119:** 1093.

Yang Z., Kyriakides T.R., and Bornstein P. 2000. Matricellular proteins as modulators of cell-matrix interactions: Adhesive defect in thrombospondin 2-null fibroblasts is a consequence of increased levels of matrix metalloproteinase-2. *Mol. Biol. Cell* **11:** 3353.

Zhu J., Motejlek K., Wang D., Zang K., Schmidt A., and Reichardt L.F. 2002. beta8 integrins are required for vascular morphogenesis in mouse embryos. *Development* **129:** 2891.

Arteries, Veins, Notch, and VEGF

B.M. WEINSTEIN AND N.D. LAWSON
*Laboratory of Molecular Genetics, NICHD, National Institutes of Health,
Bethesda, Maryland 20892*

Arteries and veins are the two most fundamental blood vessel types: Higher-pressure, oxygenated blood flows outward through arteries, and lower-pressure, deoxygenated blood returns via veins. The existence of these two intertwined yet distinct vascular networks has been appreciated for thousands of years. The difference between arteries and veins was noted by the Greek anatomists Praxagoras and Herophilus in the third century B. C. (Wiltse and Pait 1998). In the 17th century, William Harvey established the functional definition of arteries and veins as vessels that carry blood away from the heart and toward the heart in a continuous circulation (Harvey 1970). Since Harvey's time, there has been a great deal of additional description of arteries and veins and the morphological differences that distinguish them. These morphological differences include the presence of a thicker vascular smooth muscle-containing wall around arterial vessels, and the presence of valves and other specialized structures found primarily within venous vessels (Fig. 1). The origins of the morphological differences between arteries and veins have generally been attributed to physiological factors such as blood flow and hemodynamic pressure (Clark 1918; Gonzales-Crussi 1971; Girard 1973), and it has been assumed that the endothelial cells lining both types of vessels are essentially naive with respect to their arterial or venous identity, at least initially. In recent years, however, information has begun to be forthcoming about the molecular aspects of blood vessel determination challenging prevailing views on the nature and origins of arteries and veins.

ARTERIAL AND VENOUS VASCULAR ENDOTHELIA HAVE DISTINCT MOLECULAR IDENTITIES

Just within the last four years, it has become clear that the vascular endothelial cells that line arteries and veins do in fact have distinct molecular identities, and that this distinction precedes the initiation of blood flow, or even the assembly of a morphologically defined, lumenized vascular tube (Fig. 2). The first such evidence arrived in 1998 with a report describing arterial- and venous-restricted gene expression in mice. During murine embryonic development, the ephrinB2 and EphB4 genes are reciprocally expressed in arteries and veins, respectively (Wang et al. 1998). The functional importance of this dis-

tinction was highlighted by targeted disruption of the ephrinB2 locus, which resulted in improper morphogenesis of both arterial and venous blood vessels and defects in remodeling of the arterial–venous vascular interface (Wang et al. 1998). The symmetrical results of targeted disruption of the EphB4 locus suggested that EphB4 and ephrinB2 act as a defined ligand–receptor pair (Gerety et al. 1999). Introduction of a cytoplasmic deletion allele of ephrinB2 into mice suggested further that signaling between these molecules is bidirectional and that ephrinB2 can act not only as a ligand but also as a receptor (Adams et al. 2001). The idea that ephrin ligands can also have receptor activity has precedent in other studies of ephrin and Eph genes, particularly in relation to nervous system

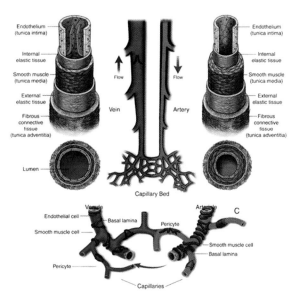

Figure 1. Blood vessels come in two fundamental flavors—arteries and veins. Both types of vessels are composed of an inner endothelium (tunica intima) surrounded by internal elastic tissue, smooth muscle cell layer (tunica media), external elastic tissue, and fibrous connective tissue (tunica adventitia). Larger-caliber arteries have thicker smooth muscle cell layers, and larger veins possess specialized structures such as valves. The two networks of tubes are completely separate at the level of the larger vessels but are linked distally through a system of fine capillaries. (Reprinted, with permission, from Cleaver and Krieg 1999 [copyright Academic Press].)

development (Cowan and Henkemeyer 2002). Vascular expression of other ephrin and Eph genes was also reported, and functional roles for some of these genes in vessel differentiation and patterning were also established (Adams et al. 1999; Helbling et al. 2000). The molecular distinction between arteries and veins represented by arterial-specific expression of ephrinB2 extends to the finest capillaries and includes expression in both the endothelial cells and the vascular smooth muscle cells that surround the arteries (Gale et al. 2001; Shin et al. 2001). Although the original ephrinB2 knockout constructs did not distinguish between vascular endothelial and vascular smooth muscle-specific functions of ephrinB2, a more recent endothelial-specific knockout of ephrinB2 showed that, at least for the early, lethal phenotypes of loss of ephrinB2, it is the expression within vascular endothelial cells that is critical (Gerety and Anderson 2002). This idea is also supported by evidence that ephrinB2 expression in endothelial cells precedes expression in adjacent smooth muscle cells, and suggests that smooth muscle cell expression may depend on signals from endothelium (Gale et al. 2001; Shin et al. 2001).

Although the results above indicate that the arterial–venous (A-V) identity of early blood vessels is genetically programmed and precedes circulatory flow, other recent work suggests that this fate choice is not irreversible and that maintenance of differentiated A-V identity might require components of the vascular wall. Two separate groups recently performed quail–chick grafting experiments to examine the plasticity of A-V endothelial cell fate (Moyon et al. 2001; Othman-Hassan et al. 2001). Portions of embryonic arteries or veins were grafted from quail donors at various stages of development into chick hosts, and the A-V identity of donor cells contributing to different host vessels was assessed using artery- or vein-specific molecular markers. Up until approximately E7, donor cells populate both types of vessels and assume the appropriate molecular identity, but after E7, this plasticity is progressively lost. However, isolated endothelial cells or isolated dissected endothelia were still plastic even in older vessels, suggesting that components of the vascular wall are necessary to maintain, or are sufficient to redirect, the A-V identity of adjacent endothelial cells.

SPECIFICATION OF A-V IDENTITY: THE ROLE OF NOTCH SIGNALING

It is clear from the phenotypes of targeted disruption of ephrin and Eph genes in mice that arterial or venous expression of these genes is critical for the proper formation of arteries and veins (Wang et al. 1998; Adams et al. 1999; Gerety et al. 1999). However, it is equally clear that these genes are not themselves involved in the initial selection of arterial and venous endothelial cell populations. Mice homozygous for an ephrinB2 lacZ knock-in allele still expressed β-galactosidase appropriately in the arterial endothelial compartment, at least initially, although boundaries between arterial and venous endothelial populations were not as well defined and the cell populations did not interact properly (Wang et al. 1998). This indicates that specification of arterial and venous domains must depend on other, upstream factors. What are the factors that lie upstream of ephrinB2 and EphB4?

The zebrafish, a genetically tractable vertebrate with a physically accessible, optically clear embryo, has proven to be a highly useful model for studying vascular development (Roman and Weinstein 2000; Vogel and Weinstein 2000). Zebrafish studies have given us important new insights into the molecular signals regulating A-V identity. In particular, the well-studied Notch signaling pathway (Artavanis-Tsakonas et al. 1999) has been shown to have an important new role in regulating vascular endothelial A-V cell fate determination (Lawson et al. 2001). A variety of Notch signaling genes are expressed in the vasculature. In mouse, Notch1 and 2 are expressed in endothelial cells (Del Amo et al. 1992; Zimrin et al. 1996), and the expression of Notch4 is restricted to endothelial cells (Uyttendaele et al. 1996). Vascular expression of these and other Notch receptors and ligands has been reported to be restricted to arteries in mouse and zebrafish embryos (Shutter et al. 2000; Smithers et al. 2000; Lawson et al. 2001; Villa et al. 2001), suggesting a role for this pathway during arterial differentiation. Mice that lack Jagged1 or Notch1 have abnormal vascular development (Xue et al. 1999; Krebs et al. 2000), whereas expression of an activated form of Notch4 specifically in the vasculature results in defective angiogenic blood vessel growth (Uyttendaele et al. 2001). Although phenotypic studies of targeted disruption of different Notch pathway receptors and ligands in mice have pointed to the importance of the Notch pathway in vascular morphogenesis, these studies have not addressed the specific function of Notch in the vasculature or its role in A-V differentiation.

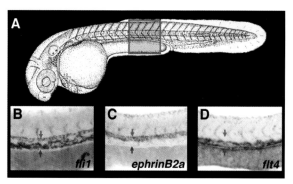

Figure 2. Arterial and venous endothelial cells have molecularly defined identities that are evident prior to circulatory flow or even tubulogenesis. In the zebrafish, the expression of artery markers such as *ephrinB2a* (*C*) and vein markers such as *flt4* (*D*) is evident by in situ hybridization of 25-somite-stage embryos, several hours before circulation begins in the trunk. In fact, expression of *ephrinB2a* within the dorsal aorta begins just as the endothelial cells that have migrated from the lateral mesoderm are aggregating into a cord of cells at the trunk midline. Expression of the pan-endothelial marker *fli1* is shown for comparison (*B*). Box *A* shows approximate location of in situ images, for reference. (*Red arrows*) Dorsal aorta. (*Blue arrows*) Posterior cardinal vein. (*A*, modified, with permission, from Kimmel et al. 1995 [copyright Wiley Interscience].)

Several recent studies in the zebrafish (Lawson et al. 2001, 2002) have shown that Notch signaling promotes arterial differentiation at the expense of venous differentiation during vascular development. As in other vertebrate species, the artery-specific expression of Notch signaling genes such as *notch5* (Kortschak et al. 2001) and *deltaC* (Smithers et al. 2000) in zebrafish blood vessels suggested that Notch might be playing an important role in artery formation. The ability to readily visualize the vasculature and carry out defined functional manipulation of developing embryos permitted a direct test of this idea. Notch signaling was repressed in zebrafish embryos either genetically, using the neurogenic *mindbomb* (*mib*) mutant, or experimentally, by injecting mRNA encoding a dominant-negative DNA-binding mutant of *Xenopus suppressor of hairless* protein (Lawson et al. 2001). In either case, repression of Notch signaling resulted in loss of *ephrinB2a* expression from arteries accompanied by ectopic expansion of normally venous-restricted markers into the arterial domain (Fig. 3). Conversely, activation of Notch signaling suppressed the expression of vein-restricted markers and promoted ectopic expression of *ephrinB2a* and other arterial markers in venous vessels. This activation was accomplished either by heat-shock promoter-driven ubiquitous expression of the Notch1a intracellular domain (Notch1a-ICD) or by Fli1-promoter-driven vascular-specific expression of Notch5-ICD. The latter set of experiments demonstrated the vascular endothelial cell autonomy of Notch-ICD effects, confirming that Notch is in fact acting directly at the level of the vascular endothelial cell itself and not via indirect signals from some other, adjacent Notch-responsive cells or tissues.

What are the functional consequences of a change in A-V molecular identity? In *mib* mutants or DN-Su(H)-injected embryos, reduction in Notch signaling causes the major trunk axial vessels (dorsal aorta and posterior cardinal vein) to display defects in morphogenesis and remodeling, and to form with poorly defined A-V boundaries (Lawson et al. 2001). These phenotypes are similar to those observed in mice with targeted disruption of Notch receptors or ligands (Xue et al. 1999; Krebs et al. 2000). In addition, the ability to visualize circulatory flow patterns in living zebrafish embryos permitted observation of prominent A-V shunts between the dorsal aorta and posterior cardinal vein in embryos with reduced Notch signaling. This reinforces the idea that the formation of distinct, well-demarcated arterial and venous cell populations is perturbed in these embryos. The strong similarity between Notch and ephrin/eph gene knockout phenotypes in mice suggests that ephrinB2 and ephB4 are probably playing a major role in arterial differentiation downstream from Notch signaling.

Potential downstream targets of Notch signaling have also been identified in the vasculature. Zebrafish embryos mutant for the *gridlock* (*grl*) gene possess cranial circulation but lack trunk circulation as a result of defective formation of the trunk lateral and dorsal aortae (Weinstein et al. 1995). Positional cloning of *grl* revealed that it is a basic helix-loop-helix protein of the *hairy/enhancer of split* family of transcriptional repressors (Zhong et al. 2000) similar to the murine HRT2 gene (Nakagawa et al. 1999). Like *notch5* and *ephrinB2*, *grl* is expressed in arterial but not venous blood vessels. HRT2 and *grl* are both targets of Notch signaling, at least in some cellular contexts (Nakagawa et al. 2000; Zhong et al. 2001). Injection of *grl* mRNA into wild-type zebrafish embryos can repress the expression of the vein-restricted marker *flt4* (Zhong et al. 2001), as does Notch activation in vivo (Lawson et al. 2001). Reduction of Grl activity by antisense morpholino oligonucleotides eliminates arterial expression of *ephrinB2a* (Zhong et al. 2001). However, *grl* continues to be expressed in the dorsal aorta in embryos lacking Notch activity despite the fact that arterial differentiation is suppressed in these and they display vascular morphogenesis defects (Lawson et al. 2001). Furthermore, *grl* and *flt4* are expressed simultaneously throughout early vascular development (Thompson et al. 1998; Zhong et al. 2001), and *grl* mutants continue to express *ephrinB2a* in the dorsal aorta (Lawson et al. 2001) despite their vascular morphogenesis defects (Zhong et al. 2000). It should also be noted that *grl* morpholino-injected embryos not only lose dorsal aorta *ephrinB2a* expression, but frequently lose the dorsal aorta entirely (Zhong et al. 2001), unlike *mib* mutants or dominant negative suppressor of hairless–injected embryos (Lawson et al. 2001). Taken together, these results indicate that *grl* plays an important role in formation of the dorsal aorta but suggest that its role as a regulator of A-V differentiation requires further clarification.

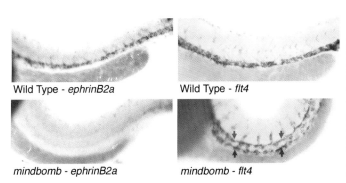

Figure 3. Reduction in Notch signaling perturbs A-V identity. In situ hybridization using artery markers in 30-somite-stage Notch-deficient *mindbomb* (*mib*^{ta52b}) mutant and wild-type sibling zebrafish embryos. (*A*) In wild-type siblings, *ephrinB2a* expression is apparent in the DA. (*B*) In *mib*^{ta52b} mutant embryos, this expression is absent. (*C*) In wild-type siblings, *flt4* expression is restricted to the PCV by the 30-somite stage. (*D*) In *mib*^{ta52b} mutant embryos, expression persists within both the PCV (*blue arrows*) and the DA (*red arrows*). All panels show lateral views of the trunk, dorsal up, anterior to the left. (Modified from Lawson et al. 2001.)

VASCULAR ENDOTHELIAL GROWTH FACTOR ACTS UPSTREAM OF NOTCH

What acts upstream of Notch? Recent studies in zebrafish and mice have revealed a surprising new role for vascular endothelial growth factor (VEGF) signaling upstream of Notch. VEGF-A is a key regulator of vascular development in vertebrates (Ferrara and Gerber 2001). It mediates vascular permeability and functions as an endothelial mitogen and angiogenic inducer. Loss of only a single allele of this gene is lethal in mouse, with severe defects in both vasculogenic and angiogenic blood vessel growth (Carmeliet et al. 1996; Ferrara et al. 1996). A number of studies in mice and zebrafish have provided evidence that another important role of VEGF is to specifically induce the differentiation of arterial blood vessels.

Postnatal retinal angiogenesis was examined in mice selectively expressing individual VEGF isoforms (Stalmans et al. 2002). Most vertebrates have three predominant VEGF-A isoforms (molecular weights 120, 164, and 188 in mice). Mice engineered to express only VEGF164 were healthy and normal, but mice expressing only the VEGF120 or VEGF188 isoforms had reduced viability and exhibited defects in retinal arterial differentiation. The phenotype of VEGF188 mice was particularly interesting, since these mice had relatively normal venular outgrowth but greatly impaired retinal arterial development. In another study, mice were generated in which VEGF164 was overexpressed in cardiac muscle under the control of a myosin heavy chain (MHC) promoter (Visconti et al. 2002). These mice were crossed together with ephrinB2 or EphB4 τ-lacZ knock-in "reporter" mice in order to easily assay arterial or venous endothelial differentiation. MHC-VEGF164 mice had increased numbers of ephrinB2-positive capillaries at the expense of EphB4-positive vessels in the heart, indicating that VEGF expression had selectively promoted the formation of additional arterial vessels.

Another study, examining the role of sensory nerves in guiding the patterning of blood vessels, provided additional evidence for an arteriogenic role for VEGF (Mukouyama et al. 2002). Previous work had documented the fact that nerves and larger vessels co-align in the skin (Martin and Lewis 1989), and Mukouyama et al. examined the molecular basis for this phenomenon. They demonstrated that arteries, but not veins, specifically align with peripheral nerves in embryonic mouse limb skin. Loss of peripheral sensory nerves or Schwann cells leads to defects in arteriogenesis, whereas these same cells can induce arterial marker expression in isolated embryonic endothelial cells when they are co-cultured in vitro. Sensory neurons and glia both express VEGF, and additional in vitro experiments demonstrated that VEGF is necessary and sufficient to mediate the arterial induction effects of these cells. VEGF alone could induce arterial differentiation of isolated embryonic endothelial cells, and at lower doses it did so without inducing proliferation, demonstrating that the increase in *ephrinB2*-positive cells was not due to preferential growth of these cells but reflected direct induction of artery-specific gene programs. Soluble VEGFR2(Flk1)-Fc, a potent VEGF antagonist (Ferrara and Davis-Smyth 1997), blocked the in vitro arterial differentiation effects of peripheral sensory nerves and Schwann cells. Together, these data suggest that peripheral nerves provide a template for the formation of arteries in the skin via local secretion of VEGF.

The murine studies described above are consistent with recent work in the zebrafish demonstrating that VEGF is necessary and sufficient for arterial differentiation (Lawson et al. 2002). The zebrafish data also provide a link to Notch, showing that VEGF acts upstream of Notch signaling in arterial fate determination. Reduction of VEGF activity using antisense morpholino oligonucleotides caused loss of arterial marker expression from the trunk dorsal aorta, ectopic arterial expression of vein markers, and morphologic defects in the aorta and cardinal vein (Fig. 4). All of these phenotypes were similar to those noted for loss of Notch signaling (Lawson et al. 2001). Injection of VEGF mRNA induced ectopic expression of *ephrinB2a* in the posterior cardinal vein. The relationship between VEGF and Notch signaling during arterial differentiation was further tested in several additional experiments (Fig. 5). VEGF mRNA injected into Notch signaling-deficient *mindbomb* mutant embryos did not induce the expression of arterial markers such as *ephrinB2a* in the trunk. This did not reflect a general lack of VEGF responsiveness of the endothelial cells in *mindbomb* mutants, since up-regulation of the VEGF receptor *flk1*, a documented response to VEGF induction (Flamme et al. 1995; Kremer et al. 1997; Liang et al. 2001), was

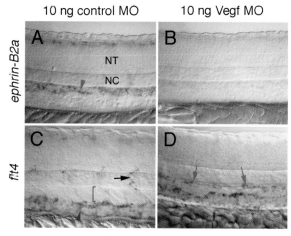

Figure 4. Loss of VEGF function perturbs A-V identity. (*A–D*) Whole-mount in situ hybridization of zebrafish embryos at 26 hpf. Lateral views of the trunk, dorsal up, anterior to the left. (*A*) Normal *ephrinB2a* expression in the dorsal aorta (*red arrowhead*) of an embryo injected with control morpholino; (NC) notochord, (NT) neural tube. (*B*) Loss of DA *ephrinB2a* expression in an embryo injected with 10 ng of a morpholino targeting VEGF. (*C*) Posterior cardinal vein-restricted *flt4* expression (*blue arrowhead*) in control morpholino-injected embryos (*black arrow* indicates normal expression within a segmental vessel). Expression is absent from the region of the dorsal aorta (indicated by *red bracket*). (*D*) Ectopic *flt4* expression in the dorsal aorta in VEGF morpholino-injected embryos. Red arrows indicate expression of *flt4* in cells of the dorsal roof of the dorsal aorta. (Modified from Lawson et al. 2002.)

still observed in VEGF-injected mutants. In contrast, activated Notch induced arterial differentiation just as efficiently in VEGF morpholino-injected embryos as in their control-injected siblings. Taken together, these experiments indicated that VEGF acts upstream of Notch signaling during arterial differentiation.

SPECULATION ON THE ASSEMBLY OF A-V NETWORKS

How can we reconcile this new arterial differentiation role for VEGF with its other well-documented roles as an endothelial mitogen, promigratory factor, and vascular permeability factor (Ferrara and Gerber 2001)? All of these previously documented activities are associated with new vessel growth. Induction of endothelial cell proliferation and migration are necessary to generate the cells that contribute to new vessels and allow them to take up positions in new, sometimes distant locales. Permeabilization of vessels facilitates deposition of new basement membrane, providing a suitable matrix for new vessel growth. Perhaps preferential induction of arterial fate is also an intrinsic feature of new vessel growth in response to VEGF. The results described above suggest a two-phase model for formation of new vascular networks (Fig. 6, panel A). In the first phase, new arteries would be formed in response to VEGF signaling, either by selection from a preexisting vascular plexus or by de novo (angiogenic) growth. In the second phase, venous vessels would emerge to provide return for arterial blood flow and allow the system to function.

This model has a number of attractive features. It is consistent with the observed anatomical colocalization of larger arteries and veins—there are few locales in the body where larger caliber veins and arteries are not found in pairs. It also provides a built-in mechanism for ensuring that venous return routes are provided for blood flowing through arterial vessels and that a functional system is constructed (Fig. 2). This comes as a natural consequence of using a primary, arterial vascular network as the template for a secondary, venous network. Arterial vessels could serve as actual direct templates for the formation of juxtaposed veins via molecular signals transmitted from arterial vessels to venous endothelial progenitors. Alternatively, or in addition, morphogenesis of a venous return system from a surrounding "naive" vascular plexus could be driven by flow dynamics once a defined arterial tree has been generated.

Is there evidence for this two-step model in vivo? In fact, many vessels do form in the predicted manner. More than a century ago, Popoff (1894) published a staged description of the development of the vasculature of the avian yolk sac, and the emergence of the venous system was described in detail by Ishida (1956). In exquisite hand-drawn illustrations, these authors detailed how the vitelline arteries appear first within the vitelline capillary network. The remaining capillaries in the plexus lose their connections to the vitelline artery, and the detached capillaries on either side of the artery then sprout and remodel to form intermedial and then collateral veins above

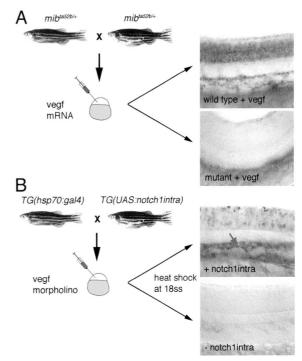

Figure 5. VEGF acts upstream of Notch signaling during arterial differentiation. (*A*) Notch signaling is required downstream of VEGF. Embryos that were derived from mib^{ta52b} heterozygous fish are injected with *vegf* mRNA (*left*). In wild-type embryos, artery marker gene expression is expanded in response to VEGF (*asterisks*), but is absent in mib^{ta52b} mutant embryos, indicating that Notch signaling is required downstream of VEGF for this expansion to occur. (*B*) Activation of the Notch pathway can induce arterial differentiation in the absence of VEGF. Adult fish that carry a heat-inducible *gal4* transgene (TG) (Scheer et al. 2001) are crossed to fish that express an intracellular form of zebrafish *notch1* (*notch1-intra*) (Scheer and Campos-Ortega 1999) that is constitutively active. Notch1-intra is tagged with an epitope to allow subsequent visualization by immunostaining and is driven by a Gal4-responsive promoter (UAS). Embryos from these fish are injected with a morpholino oligonucleotide to reduce VEGF protein levels (Nasevicius et al. 2000) and subsequently heat-shocked at the 18-somite stage to induce expression of the Notch1-intra. All embryos will have reduced VEGF, but a proportion of the embryos will also have activated the Notch pathway because they carry both transgenes. In embryos that have reduced VEGF expression but no exogenous Notch1-intra, *ephrinB2a* is not expressed (*top* panel). Activation of the Notch pathway as a result of exogenous Notch1-intra expression rescues *ephrinB2a* expression (Lawson et al. 2002) (*bottom* panel, *asterisk*). Together, these results indicate that the Notch pathway functions downstream of VEGF to promote arterial differentiation. (Modified from Lawson et al. 2002.)

and alongside the artery (Fig. 6, panels B–D). A more recent study showed that initial vascular innervation of the XY gonad in mice appears to occur almost exclusively via immigration of ephrinB2- and Notch-positive arterial progenitors (Brennan et al. 2002). EphB4-positive venous progenitors are not apparent until later stages. Other recent work has shown that in the zebrafish the primitive intersegmental vessels in the trunk form in two distinct steps, with emergence of arterial and venous sprouts in temporally separate waves (S. Isogai et al., unpubl.). The studies described above in which ephrinB2-τ-lacZ knock-

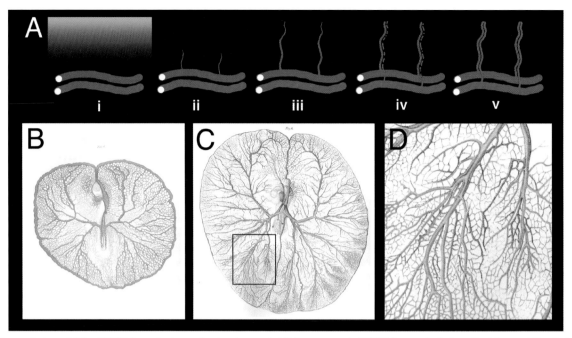

Figure 6. A model for VEGF-dependent vascular development. (*A*) In response to VEGF (*green* shaded gradient in panel *i*), arterial vessels appear (*ii*) and establish an initial defined arterial network (*iii*). A venous vascular network (*blue*) then coalesces adjacent to the arterial vessels (*iv*, *v*) to provide a venous circulatory return system. Evidence that vessels form in this way has existed in the literature for many years. Descriptive studies of the vasculature of the chick yolk sac showed that the yolk sac vasculature develops in a manner strikingly similar to the predictions of this model. (*B*) Initially a complex branched arterial network emerges from the vascular plexus, with venous drainage from much of this system primarily centrifugal, via the marginal vein at the rim of the area vasculosa. (*C*) At later stages, new intermedial and centripetally draining collateral veins emerge. (*D*) Higher magnification of the boxed region in *C* shows how this new venous drainage system takes shape via remodeling of the plexus of vessels surrounding the arteries. (*B–D*, modified from Popoff 1894.)

in mice were used to probe the expression profile of ephrinB2 (Gale et al. 2001; Shin et al. 2001) also revealed that ephrinB2 is selectively expressed "at sites of secondary angiogenesis in the embryo as well as sites of normal and pathological angiogenesis in the adult" (Gale et al. 2001). The vessels in these locales were not exclusively ephrinB2-positive, however; many PECAM-positive but ephrinB2-negative vessels were also seen. The presence of non-arterial vessels likely indicates that the two phases of artery and vein formation occur asynchronously throughout adjacent areas of a tissue. In addition, as noted above for the yolk sac, the emerging arterial system may be surrounded by a plexus of other vessels that lack arterial character, but that are not necessarily part of a morphologically defined venous network. In fact, recent evidence suggests that venous identity may represent a "ground state" in A-V differentiation (Lawson et al. 2001, 2002). When clear temporal separation between artery and vein formation is noted, this might represent situations where the arterial system has an alternative means of drainage initially, or where functioning of the system is not immediately necessary or desirable. The alignment of nerves and arteries in the skin and the role of VEGF in bringing this about (Mukouyama et al. 2002) help to explain how this model can also be used to form a vascular network with a defined and reproducible pattern. VEGF expressed from particular cells, tissues, and organs could direct artery formation in reproducible positions with respect to preexisting anatomical structures. In the skin, peripheral nerves provide VEGF; in the case of other organs and tissues, a VEGF-expressing "template" for artery formation might be provided by other cell types. In the retina, for example, astrocytes expressing VEGF induce vascular outgrowth, with migrating astrocytes leading an outward "front" of vessel assembly (Stone et al. 1995; Zhang et al. 1999).

MANY NEW QUESTIONS

Although we now understand a great deal more about how veins and arteries differ and how these differences come about than we did a few years ago, what we have learned has raised more questions than it has answered. How widely generalizable is the artery-specific role of VEGF, and the two-step model for vessel formation that we have proposed? What are the relative roles of different VEGF isoforms in arterial differentiation and the other functions of VEGF? Are there factors other than VEGF that play a similar arterial differentiation role in different contexts? In this regard, it is interesting to note the recent discovery of a novel endocrine-specific vascular growth factor, EG-VEGF. Does this factor have an arterial inducing activity similar to that of VEGF? If VEGF is primarily driving arterial patterning, what is responsible for the pattern of veins? Are flow dynamics the driving force, or are there novel artery-specific signals for vein assembly? These and many other questions are sure to occupy the minds and experimental activities of vascular biologists in the years to come.

ACKNOWLEDGMENTS

The authors thank Sumio Isogai for a critical reading of this manuscript, and for helpful discussions on avian yolk sac vascular development. This work was supported by National Institutes of Health grant ZO1-HD-01011.

REFERENCES

Adams R.H., Diella F., Hennig S., Helmbacher F., Deutsch U., and Klein R. 2001. The cytoplasmic domain of the ligand ephrinB2 is required for vascular morphogenesis but not cranial neural crest migration. *Cell* **104**: 57.

Adams R.H., Wilkinson G.A., Weiss C., Diella F., Gale N.W., Deutsch U., Risau W., and Klein R. 1999. Roles of ephrinB ligands and EphB receptors in cardiovascular development: Demarcation of arterial/venous domains, vascular morphogenesis, and sprouting angiogenesis. *Genes Dev.* **13**: 295.

Artavanis-Tsakonas S., Rand M.D., and Lake R.J. 1999. Notch signaling: Cell fate control and signal integration in development. *Science* **284**: 770.

Brennan J., Karl J., and Capel B. 2002. Divergent vascular mechanisms downstream of Sry establish the arterial system in the XY gonad. *Dev. Biol.* **244**: 418.

Carmeliet P., Ferreira V., Breier G., Pollefeyt S., Kieckens L., Gertsenstein M., Fahrig M., Vandenhoeck A., Harpal K., Eberhardt C., Declercq C., Pawling J., Moons L., Collen D., Risau W., and Nagy A. 1996. Abnormal blood vessel development and lethality in embryos lacking a single VEGF allele. *Nature* **380**: 435.

Clark E.R. 1918. Studies on the growth of blood vessels in the tail of the frog larvae. *Am. J. Anat.* **23**: 37.

Cleaver O. and Krieg P.A. 1999. Molecular mechanisms of vascular development. In *Heart development* (ed. R.P. Harvey and N. Rosenthal), p. 221. Academic Press, San Diego, California.

Cowan C.A. and Henkemeyer M. 2002. Ephrins in reverse, park and drive. *Trends Cell Biol.* **12**: 339.

Del Amo F.F., Smith D.E., Swiatek P.J., Gendron-Maguire M., Greenspan R.J., McMahon A.P., and Gridley T. 1992. Expression pattern of Motch, a mouse homolog of *Drosophila* Notch, suggests an important role in early postimplantation mouse development. *Development* **115**: 737.

Ferrara N. and Davis-Smyth T. 1997. The biology of vascular endothelial growth factor. *Endocr. Rev.* **18**: 4.

Ferrara N. and Gerber H.P. 2001. The role of vascular endothelial growth factor in angiogenesis. *Acta Haematol.* **106**: 148.

Ferrara N., Carver-Moore K., Chen H., Dowd M., Lu L., O'Shea K.S., Powell-Braxton L., Hillan K.J., and Moore M.W. 1996. Heterozygous embryonic lethality induced by targeted inactivation of the VEGF gene. *Nature* **380**: 439.

Flamme I., von Reutern M., Drexler H.C., Syed-Ali S., and Risau W. 1995. Overexpression of vascular endothelial growth factor in the avian embryo induces hypervascularization and increased vascular permeability without alterations of embryonic pattern formation. *Dev. Biol.* **171**: 399.

Gale N.W., Baluk P., Pan L., Kwan M., Holash J., DeChiara T.M., McDonald D.M., and Yancopoulos G.D. 2001. Ephrin-B2 selectively marks arterial vessels and neovascularization sites in the adult, with expression in both endothelial and smooth- muscle cells. *Dev. Biol.* **230**: 151.

Gerety S.S. and Anderson D.J. 2002. Cardiovascular ephrinB2 function is essential for embryonic angiogenesis. *Development* **129**: 1397.

Gerety S.S., Wang H.U., Chen Z.F., and Anderson D.J. 1999. Symmetrical mutant phenotypes of the receptor EphB4 and its specific transmembrane ligand ephrin-B2 in cardiovascular development. *Mol. Cell* **4**: 403.

Girard H. 1973. Arterial pressure in the chick embryo. *Am. J. Physiol.* **224**: 454.

Gonzales-Crussi F. 1971. Vasculogenesis in the chick embryo. An ultrastructural study. *Am. J. Anat.* **130**: 441.

Harvey W. 1970. *Exercitatio anatomica de motu cordis et sanguinis in animalibus*, 5 edition (transl. by C.D. Leake). Charles C. Thomas, Springfield, Illinois.

Helbling P.M., Saulnier D.M., and Brandli A.W. 2000. The receptor tyrosine kinase EphB4 and ephrin-B ligands restrict angiogenic growth of embryonic veins in *Xenopus laevis*. *Development* **127**: 269.

Ishida Z. 1956. Disvolvigo de kalateralaj vejnoj sur la ovoflavsako de kokojo (in Japanese, Esperanto summary). *Kaibogaku Zasshi* **31**: 334.

Kimmel C.B., Ballard W.W., Kimmel S.R., Ullmann B., and Schilling T.F. 1995. Stages of embryonic development of the zebrafish. *Dev. Dyn.* **203**: 253.

Kortschak R.D., Tamme R., and Lardelli M. 2001. Evolutionary analysis of vertebrate Notch genes. *Dev. Genes Evol.* **211**: 350.

Krebs L.T., Xue Y., Norton C.R., Shutter J.R., Maguire M., Sundberg J.P., Gallahan D., Closson V., Kitajewski J., Callahan R., Smith G.H., Stark K.L., and Gridley T. 2000. Notch signaling is essential for vascular morphogenesis in mice. *Genes Dev.* **14**: 1343.

Kremer C., Breier G., Risau W., and Plate K.H. 1997. Up-regulation of flk-1/vascular endothelial growth factor receptor 2 by its ligand in a cerebral slice culture system. *Cancer Res.* **57**: 852.

Lawson N.D., Vogel A.M., and Weinstein B.M. 2002. Sonic hedgehog and vascular endothelial growth factor act upstream of the Notch signaling pathway during arterial endothelial cell differentiation. *Dev. Cell* **3**: 127.

Lawson N.D., Scheer N., Pham V., Kim C.-H., Chitnis A.B., Campos-Ortega J., and Weinstein B.M. 2001. Notch signaling is required for arterial-venous differentiation during embryonic vascular development. *Development* **128**: 3675.

Liang D., Chang J.R., Chin A.J., Smith A., Kelly C., Weinberg E.S., and Ge R. 2001. The role of vascular endothelial growth factor (VEGF) in vasculogenesis, angiogenesis, and hematopoiesis in zebrafish development. *Mech. Dev.* **108**: 29.

Martin P. and Lewis J. 1989. Origins of the neurovascular bundle: Interactions between developing nerves and blood vessels in embryonic chick skin. *Int. J. Dev. Biol.* **33**: 379.

Moyon D., Pardanaud L., Yuan L., Breant C., and Eichmann A. 2001. Plasticity of endothelial cells during arterial-venous differentiation in the avian embryo. *Development* **128**: 3359.

Mukouyama Y.S., Shin D., Britsch S., Taniguchi M., and Anderson D.J. 2002. Sensory nerves determine the pattern of arterial differentiation and blood vessel branching in the skin. *Cell* **109**: 693.

Nakagawa O., Nakagawa M., Richardson J.A., Olson E.N., and Srivastava D. 1999. HRT1, HRT2, and HRT3: A new subclass of bHLH transcription factors marking specific cardiac, somitic, and pharyngeal arch segments. *Dev. Biol.* **216**: 72.

Nakagawa O., McFadden D.G., Nakagawa M., Yanagisawa H., Hu T., Srivastava D., and Olson E.N. 2000. Members of the HRT family of basic helix-loop-helix proteins act as transcriptional repressors downstream of notch signaling. *Proc. Natl. Acad. Sci.* **97**: 13655.

Nasevicius A., Larson J., and Ekker S.C. 2000. Distinct requirements for zebrafish angiogenesis revealed by a VEGF-A morphant. *Yeast* **17**: 294.

Othman-Hassan K., Patel K., Papoutsi M., Rodriguez-Niedenfuhr M., Christ B., and Wilting J. 2001. Arterial identity of endothelial cells is controlled by local cues. *Dev. Biol.* **237**: 398.

Popoff D. 1894. *Dottersack-gafässe des Huhnes*. Kreidl's Verlag, Wiesbaden, Germany.

Roman B.L. and Weinstein B.M. 2000. Building the vertebrate vasculature: Research is going swimmingly. *Bioessays* **22**: 882.

Scheer N. and Campos-Ortega J.A. 1999. Use of the Gal4-UAS

technique for targeted gene expression in the zebrafish. *Mech. Dev.* **80:** 153.

Scheer N., Groth A., Hans S., and Campos-Ortega J.A. 2001. An instructive function for Notch in promoting gliogenesis in the zebrafish retina. *Development* **128:** 1099.

Shin D., Garcia-Cardena G., Hayashi S., Gerety S., Asahara T., Stavrakis G., Isner J., Folkman J., Gimbrone M.A., Jr., and Anderson D.J. 2001. Expression of ephrinB2 identifies a stable genetic difference between arterial and venous vascular smooth muscle as well as endothelial cells, and marks subsets of microvessels at sites of adult neovascularization. *Dev. Biol.* **230:** 139.

Shutter J.R., Scully S., Fan W., Richards W.G., Kitajewski J., Deblandre G.A., Kintner C.R., and Stark K.L. 2000. Dll4, a novel Notch ligand expressed in arterial endothelium. *Genes Dev.* **14:** 1313.

Smithers L., Haddon C., Jiang Y., and Lewis J. 2000. Sequence and embryonic expression of deltaC in the zebrafish. *Mech. Dev.* **90:** 119.

Stalmans I., Ng Y.S., Rohan R., Fruttiger M., Bouche A., Yuce A., Fujisawa H., Hermans B., Shani M., Jansen S., Hicklin D., Anderson D.J., Gardiner T., Hammes H.P., Moons L., Dewerchin M., Collen D., Carmeliet P., and D'Amore P.A. 2002. Arteriolar and venular patterning in retinas of mice selectively expressing VEGF isoforms. *J. Clin. Invest.* **109:** 327.

Stone J., Itin A., Alon T., Pe'er J., Gnessin H., Chan-Ling T., and Keshet E. 1995. Development of retinal vasculature is mediated by hypoxia-induced vascular endothelial growth factor (VEGF) expression by neuroglia. *J. Neurosci.* **15:** 4738.

Thompson M.A., Ransom D.G., Pratt S.J., MacLennan H., Kieran M.W., Detrich H.W., III, Vail B., Huber T.L., Paw B., Brownlie A.J., Oates A.C., Fritz A., Gates M.A., Amores A., Bahary N., Talbot W.S., Her H., Beier D.R., Postlethwait J.H., and Zon L.I. 1998. The cloche and spadetail genes differentially affect hematopoiesis and vasculogenesis. *Dev. Biol.* **197:** 248.

Uyttendaele H., Ho J., Rossant J., and Kitajewski J. 2001. Vascular patterning defects associated with expression of activated Notch4 in embryonic endothelium. *Proc. Natl. Acad. Sci.* **98:** 5643.

Uyttendaele H., Marazzi G., Wu G., Yan Q., Sassoon D., and Kitajewski J. 1996. Notch4/int-3, a mammary proto-oncogene, is an endothelial cell-specific mammalian Notch gene. *Development* **122:** 2251.

Villa N., Walker L., Lindsell C.E., Gasson J., Iruela-Arispe M.L., and Weinmaster G. 2001. Vascular expression of Notch pathway receptors and ligands is restricted to arterial vessels. *Mech. Dev.* **108:** 161.

Visconti R.P., Richardson C.D., and Sato T.N. 2002. Orchestration of angiogenesis and arteriovenous contribution by angiopoietins and vascular endothelial growth factor (VEGF). *Proc. Natl. Acad. Sci.* **99:** 8219.

Vogel A.M. and Weinstein B.M. 2000. Studying vascular development in the zebrafish. *Trends Cardiovasc. Med.* **10:** 352.

Wang H.U., Chen Z.F., and Anderson D.J. 1998. Molecular distinction and angiogenic interaction between embryonic arteries and veins revealed by ephrin-B2 and its receptor Eph-B4. *Cell* **93:** 741.

Weinstein B.M., Stemple D.L., Driever W., and Fishman M.C. 1995. Gridlock, a localized heritable vascular patterning defect in the zebrafish. *Nat. Med.* **1:** 1143.

Wiltse L.L. and Pait T.G. 1998. Herophilus of Alexandria (325-255 B.C.). The father of anatomy. *Spine* **23:** 1904.

Xue Y., Gao X., Lindsell C.E., Norton C.R., Chang B., Hicks C., Gendron-Maguire M., Rand E.B., Weinmaster G., and Gridley T. 1999. Embryonic lethality and vascular defects in mice lacking the Notch ligand Jagged1. *Hum. Mol. Genet.* **8:** 723.

Zhang Y., Porat R.M., Alon T., Keshet E., and Stone J. 1999. Tissue oxygen levels control astrocyte movement and differentiation in developing retina. *Brain Res. Dev. Brain Res.* **118:** 135.

Zhong T.P., Childs S., Leu J.P., and Fishman M.C. 2001. Gridlock signalling pathway fashions the first embryonic artery. *Nature* **414:** 216.

Zhong T.P., Rosenberg M., Mohideen M.A., Weinstein B., and Fishman M.C. 2000. gridlock, an HLH gene required for assembly of the aorta in zebrafish. *Science* **287:** 1820.

Zimrin A.B., Pepper M.S., McMahon G.A., Nguyen F., Montesano R., and Maciag T. 1996. An antisense oligonucleotide to the notch ligand jagged enhances fibroblast growth factor-induced angiogenesis in vitro. *J. Biol. Chem.* **271:** 32499.

Cell Cycle Signaling and Cardiovascular Disease

E.G. Nabel, M. Boehm, L.M. Akyurek, T. Yoshimoto, M.F. Crook,
M. Olive, H. San, and X. Qu

*Cardiovascular Branch, National Heart, Lung and Blood Institute,
Bethesda, Maryland 20892*

Our laboratory is interested in signaling pathways in vascular smooth muscle cells (vsmcs) as a model system for understanding cell growth and differentiation in normal physiology and disease pathophysiology. We have focused on the mechanisms by which the cyclin-dependent kinases (CDKs) and their inhibitors, the cyclin-dependent kinase inhibitors (CKIs), control mitogen-dependent cell growth within blood vessels and regulate vascular wound repair and remodeling. Here, we discuss the expression and function of the CIP/KIP proteins $p27^{Kip1}$ and $p21^{Cip1}$ in animal models of vascular disease.

CYCLINS, CDKs, AND CKIs

Progression through G_1 phase of the cell cycle and initiation of DNA replication in S phase are regulated by the coordinated activities of the CDKs and their inhibitors, the CKIs (Fig. 1). The CDKs are holoenzyme complexes that contain cyclin regulatory and CDK catalytic subunits. Two families of enzymes, the D- and E-dependent kinases, regulate transit through the G_1/S restriction point. The D-type cyclins (D1, D2, and D3) interact combinatorially with two catalytic partners, CDK4 and CDK6, to yield at least six holoenzymes that are expressed in tissue-specific patterns (Sherr and Roberts 1999). Cyclin E forms a complex with its catalytic partner CDK2. The CKIs, in turn, are classified into two families based on their structure and function. The CIP/KIP proteins, $p21^{Cip1}$, $p27^{Kip1}$, and $p57^{Kip2}$, have common cyclin-CDK binding motifs in their amino-terminal domains. These proteins have broad functions and form inhibitory complexes with cyclin D-, E-, and A-dependent kinases (see el-Diery et al. 1993; Gu et al. 1993; Harper et al. 1993; Polyak et al. 1994; Toyoshima and Hunter 1994; Lee et al. 1995; Matsuoka et al. 1995). In contrast, the INK family of proteins, $p16^{INK4a}$, $p15^{INK4b}$, $p18^{INK4c}$, and $p19^{INK4d}$, contain multiple ankyrin repeats and specifically inhibit the catalytic subunits of CDK4 and CDK6 (see Chan et al. 1995; Hirai et al. 1995).

The cyclins, the CDKs, and the CKIs interact in positive and negative ways to regulate G_1-to-S phase transition (Figs. 2 and 3). The D-cyclins are growth factor sen-

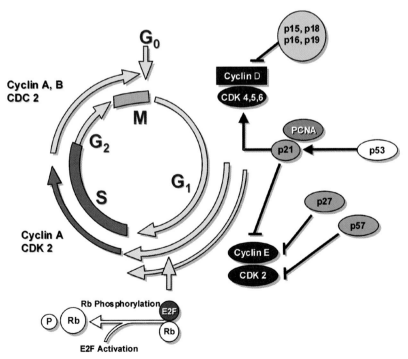

Figure 1. Cell cycle pathways. Mitogen-dependent progression through G_1 phase and initiation of S phase during the mammalian cell cycle are cooperatively regulated by the cyclin-dependent kinases (CDKs) whose activities are controlled by CDK inhibitors (CKIs). CKIs have been assigned to two families on the basis of their structure: the INK4 proteins (p15, p16, p18, p19) and the CIP/KIP proteins (p21, p27, p57). (Reprinted, with permission, from Nabel 2002 [copyright Macmillan].)

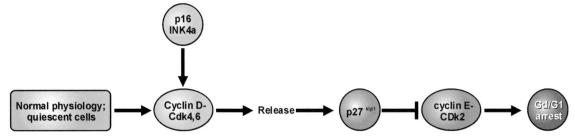

Figure 2. CKI regulation in normal physiology. The CIP/KIP proteins are positive regulators of cyclin-D-dependent kinases. In normal cells, p21^{Cip1} and PCNA are incorporated into higher-order complexes with various CDKs. Likewise, in quiescent cells, the levels of p27^{Kip1} are relatively high, constrain cyclin E–Cdk2, and induce G$_1$ phase arrest. (Reprinted, with permission, from Nabel 2002 [copyright Macmillan].)

sors. When quiescent cells enter the cell cycle, mitogens induce the D-type cyclins (through a Ras-mediated pathway) to assemble with their catalytic partners, CDK4 and CDK6 and a CIP or KIP protein. The cyclin-D-dependent kinase complex sequesters CIP/KIP proteins, thus lowering the threshold of unbound CIP/KIP proteins and facilitating activation of cyclin E–CDK2. Phosphorylation of cyclin-D-dependent kinases in mid G$_1$ and cyclin-E-dependent kinases in late G$_1$ sequentially leads to phosphorylation of Rb, releasing repression of E2F family members and activating genes required for S-phase entry.

The second important function of the cyclin-D-dependent kinases is sequestration of the CIP/KIP proteins during cell proliferation. Most p27^{Kip1} in proliferating cells is sequestered by cyclin-D-dependent kinases. p21^{Cip1} and proliferating cell nuclear antigen (PCNA) are also incorporated into complexes with the CDKs. Assembly of unbound p27^{Kip1} and p21^{Cip1} into complexes with cyclin-D-dependent kinases relieves cyclin E–CDK2 from p27^{Kip1} and p21^{Cip1} inhibition, facilitating cyclin E–CDK2 activation in late G$_1$ phase. Residual p27^{Kip1} and p21^{Cip1} remain bound to cyclin D–CDK complexes throughout subsequent cell cycles. CDK2 phosphorylates p27^{Kip1} on a threonine 187 (T187) residue leading to degradation of p27^{Kip1} by proteosomes in the cytoplasm. This regulatory step also promotes activation of cyclin E–CDK2. In quiescent cells, levels of p27^{Kip1} are high. CIP/KIP proteins are released from cyclin D complexes when mitogens are withdrawn, inhibiting cyclin E–CDK2 activation and inducing G$_1$ arrest.

In summary, a complex set of biochemical pathways regulates the interactions of the cyclins, CDKs, and CKIs in mammalian cells. The inhibitory effects of hypophosphorylated Rb and the CIP/KIP proteins on cell proliferation are overcome by a mitogen-dependent pathway mediated by the D cyclins. During cell proliferation, cyclin D has two major functions: phosphorylation of Rb and sequestration of the CIP/KIP proteins. Once cyclin E–CDK2 is activated, CDK2 phosphorylates p27^{Kip1}, leading to irreversible degradation of the protein and entry into S phase.

CIP/KIP STRUCTURE AND FUNCTION

p27^{Kip1} was purified and cloned from cyclin–CDK holoenzymes (Polyak et al. 1994; Toyoshima and Hunter 1994), and the protein was induced by anti-mitogenic signals and inhibited CDK activity (Koff et al. 1993). The *Kip1* gene in mouse and human comprises two coding exons; there is a third noncoding exon that is 3´ to the translation stop codon (Pietenpol et al. 1995). The p27^{Kip1} promoter has a 170-bp region upstream of the start site that contains two SP1-binding sites. SP1 *trans*-activation confers maximal transcriptional activity (Zhang and Lin 1997). p27^{Kip1} has a CDK-binding domain in the amino-terminal region at position 27-88, a nuclear localization signal in the carboxyl terminus, and three phosphorylation sites (S10, S178, and T187) (Ishida et al. 2000).

p27^{Kip1} is regulated by transcriptional (Servant et al. 2000), translational (Agrawal et al. 1996; Hengst and Reed 1996; Millard et al. 1997; Miskimins et al. 2001), and proteolytic mechanisms. Two proteolytic pathways act in sequence during the cell cycle to control p27^{Kip1} abundance. The first pathway is active in G$_0$/G$_1$ and is triggered by mitogens (Malek et al. 2001). Phosphorylation of p27^{Kip1} on residue S10 by kinase-interacting stathmin (KIS) is activated by mitogens in G$_0$/G$_1$ cells and leads to nuclear export of p27^{Kip1} to the cytoplasm

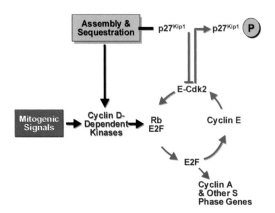

Figure 3. CKI regulation in disease pathophysiology. Mitogenic signals promote the assembly of cyclin D–Cdk4,6 complexes and a CIP/KIP protein. Sequestration of CIP/KIP proteins lowers the inhibitory threshold and facilitates activation of cyclin E–Cdk2 complexes. The cyclin D–CDKs and cyclin E–CDKs contribute sequentially to Rb phosphorylation, canceling Rb repression of E2F family members, and entry into S phase. (Reprinted, with permission, from Nabel 2002 [copyright Macmillan].)

(Boehm et al. 2002). This results in a lower abundance of p27^{Kip1} in the nucleus, activation of cyclin E–CDK2 complexes, and onset of the second pathway for p27^{Kip1} turnover. The second pathway operates in late G$_1$, S, and G$_2$ phases and is dependent on nuclear CDK2 phosphorylation of p27^{Kip1} on residue T187, which creates a binding site for the E3–ubiquitin–protein–ligase complex, SCF (SKP1–cullin–F-box) (Pagano et al. 1995). Ubiquitination of p27^{Kip1} by the SCF complex results in degradation of p27^{Kip1} by the proteasome (Bai et al. 1996; Feldman et al. 1997; Skowyrs et al. 1997). Inactivation of p27^{Kip1} switches cells in mid-G$_1$ from being mitogen dependent to being mitogen independent.

p27^{Kip1} is an inhibitor of cell division. An increase in levels of p27^{Kip1} causes proliferating cells to exit from the cell cycle, and a decrease in p27^{Kip1} is necessary for quiescent cells to resume division. Deletion of *Kip1* in mice results in hyperplasia in several organs, including endocrine tissues (adrenals, gonads, and pituitary gland), the retina, and the thymus (Fero et al. 1996; Kiyokawa et al. 1996; Nakayama et al. 1996). Thymic hyperplasia is associated with increased T-cell proliferation, and the absence of *Kip1* in the spleen selectively enhances proliferation of hematopoietic progenitor cells. These findings suggested at the time that p27^{Kip1} is an essential component of the signaling pathway that connects mitogenic signals to the cell cycle at the G$_1$ restriction point. These findings have subsequently been confirmed in studies of the cardiovascular system as well.

The CKI p21^{Cip1} has been implicated in a broad array of cellular processes. p21^{Cip1}, also known as CIP1 and WAF1, is a dual inhibitor of CDKs and PCNA (Xiong et al. 1993). As a transcriptional target of the tumor-suppressor protein p53, p21^{Cip1} has been shown to be an effector of cell cycle arrest after DNA damage, a pro-apoptotic factor in the cell death program, and a tumor suppressor. *Cip1*-deficient mice undergo normal development, but fibroblasts from these mice display defective G$_1$ arrest in response to DNA damage (Brugarolas et al. 1995; Deng et al. 1995). Following irradiation, the G$_1$ checkpoint in *Cip1* null mice is only partially impaired, indicating a second p53-dependent function that may lead to arrest of G$_1$ cells. In contrast to p53$^{-/-}$ mice, *Cip1*$^{-/-}$ mice do not exhibit early tumorigenesis or failure to induce apoptosis.

The CIP/KIP family is defined by a conserved amino-terminal domain that is sufficient for stable binding to cyclin–CDK complexes and inhibition of CDK protein kinase activity. The CIP/KIP proteins inhibit each of the cyclin–CDK complexes that are essential for G$_1$ progression and S-phase entry. However, these proteins have divergent structures and functions, lending them unique biological properties. p27^{Kip1} and p21^{Cip1} are most divergent in their carboxy-terminal regions. p21^{Cip1} binds PCNA in its carboxyl terminus and inhibits PCNA replication activity (Waga et al. 1994). The CIP/KIP proteins have distinct responses to mitogenic and anti-mitogenic signals. For example, p21^{Cip1} expression is increased in a p53-dependent manner in cells that contain damaged DNA, and in a p53-independent manner in postmitotic, terminally differentiated cells (Halevy et al. 1995; Parker et al. 1995). In contrast, p27^{Kip1} expression increases primarily in response to extracellular postmitogenic signals. For example, normal cells that are grown to high density or are deprived of serum undergo cell cycle arrest and express elevated levels of p27^{Kip1} protein (Coats et al. 1996). Cells exposed to anti-mitogenic factors, such as rapamycin, also express high levels of p27^{Kip1} protein, but the protein is sequestered by cyclin D complexes in a cooperative manner with p16^{INK4a} to cause G$_1$ arrest in cells treated by transforming growth factor-β (TGFβ), as an example. In summary, the structure of the CIP/KIP proteins is characterized by regions of homology and divergence, which dictate distinct functional properties.

EXPRESSION OF THE CKIs IN VASCULAR DISEASE

Several years ago, we made a simple but striking observation regarding the expression of the CIP/KIP proteins in normal and diseased arteries. The CKIs have distinct temporal and spatial patterns of expression in blood vessels (Fig. 4). In normal arteries, p27^{Kip1} is constitutively expressed in medial and intimal vascular smooth muscle cells, but the protein is rapidly down-regulated after vascular injury when mitogen-induced vsmc proliferation proceeds (Tanner et al. 1998). vsmcs synthesize collagen and other extracellular matrix molecules that signal back to vsmcs in autocrine and paracrine feedback loops to cause up-regulation of p27^{Kip1} and down-regulation of cyclin E and CDK2 (Koyama et al. 1996). p27^{Kip1} expression persists in medial and intimal vsmcs after wound repair is complete. In contrast, p21^{Cip1} is not observed in vsmcs of normal arteries, and the protein is upregulated along with p27^{Kip1} in the later phases of arterial wound repair. p16^{INK4a} is not expressed in high levels in normal or injured arteries (Tanner et al. 1998). These patterns of protein expression have been observed in several animal models of vascular injury, including rat and pig. In human coronary arteries, p27^{Kip1} is expressed within medial and intimal vsmcs of normal and atherosclerotic arteries, including vsmcs of plaque angiogenesis (Tanner et al. 1998). p21^{Cip1} is observed only in advanced atherosclerotic lesions, and p16^{INK4a} is not detected by western blot or immunostaining in either normal or diseased coronary arteries. These distinct temporal and spatial patterns of expression suggest that the CIP/KIP CKIs regulate G$_1$- to S-phase transition in vascular cells and promote vascular wound repair and remodeling.

DELETION OF THE *Kip1* LOCI ACCELERATES CELLULAR LESION DEVELOPMENT AND PROGRESSION

These observational studies led us to hypothesize that p27^{Kip1} and p21^{Cip1} are essential for arterial wound repair. To directly test the importance of p27^{Kip1} and p21^{Cip1} pathways in vascular remodeling in vivo, we investigated *Kip1* and *Cip1* and *Kip1/Cip1* null mice subjected to stresses of mechanically induced vascular injury (Boehm et al. 2001). The arterial phenotype of *Kip1* and *Cip1* and *Kip1/Cip1* null mice is normal; however, following vascular injury, a severe phenotype results. We

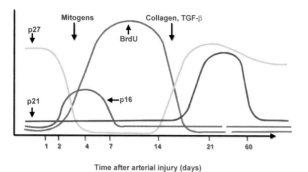

Figure 4. CKI expression in normal and injured blood vessels. p27^{Kip1} levels are high in normal vessels, its levels are quickly down-regulated after injury, and the protein is up-regulated in the later phases of arterial wound repair, leading to successful vessel remodeling. (Adapted, with permission, from Tanner et al. 1998 [copyright Lippincott Williams and Wilkins].)

found a threefold increase in cell number and BrdU incorporation in vsmcs from *Kip1* and *Cip1* null mice compared to wild-type control mice. In vivo vascular injury experiments were performed using a flexible guidewire. Intimal lesion formation was accelerated 2 weeks after injury, with *Kip1/Cip1* double null mice having the most severe phenotype, followed by *Kip1* null and *Cip1* null animals compared to wild-type controls. The intimal lesions were characterized by increased vsmc proliferation and extracellular matrix accumulation (Fig. 5). Interestingly, the *Kip1* null lesions also had an intense inflammatory reaction throughout the vessel wall.

To explore the mechanisms for the inflammation and increased proliferation, two sets of bone marrow transplantation (BMT) experiments were performed. We first investigated the source of the proliferative vsmc—Did they derive from the local artery or from bone marrow progenitors? Transplantation of wild-type marrow into *Kip1* null mice decreased the size of the intimal lesion by about 20%. In contrast, transfer of *Kip1* null marrow into wild-type animals increased lesion size by about 50%, suggesting that the relative contribution of the local artery or bone marrow compartment varied according to the genotype (M. Boehm, pers. comm.). *Kip1* null progenitor cells proliferate at higher indices than wild-type cells; indeed, previous studies have shown that *Kip1* regulates stem cell progenitor proliferation and pool size (Cheng et al. 2000). Likewise, vsmcs in the media of injured arteries cycle at higher rates than vsmcs from wild-type animals.

Next, we are examining the source of inflammation in the *Kip1* null and *Kip1/Cip1* null mice, by crossing *Kip1* null and *Rag-1* null mice. Vascular injury in *Rag-1* null mice led to a significantly reduced vascular lesion, compared to wild-type and *Kip1* null mice, suggesting that inflammation due to T and B cells contributes to lesion formation (M. Boehm, pers. comm.). BMT experiments of *Kip1* null marrow into *Rag-1* null mice, *Rag-1* null marrow into *Kip1* null mice, and the corresponding controls will address the question of the contribution of T- and B-cell proliferation to vascular remodeling in the *Kip1* null animals. In summary, p27^{Kip1} and p21^{Cip1} safeguard against excessive cell proliferation in pathological settings. p27^{Kip1}, more so than p21^{Cip1}, has a major role in the vasculature in protecting tissues from inflammatory injury and promoting arterial wound repair.

In addition, we hypothesized that a deficiency of p27^{Kip1} and p21^{Cip1} would accelerate atherogenesis in apolipoprotein E (apoE)-deficient mice. To test this hypothesis, *Kip1* and *Cip1* null mice were backbred into apoE mice, to create double and triple null animals (L.M. Akyurek et al., pers. comm.). Aortae were extirpated following 6 or 15 weeks of either a normal or high-cholesterol diet. Oil-red-O-stained aortae were analyzed for the percentage of plaque formation. Cross sections were taken at different levels to quantify intima/media area ratios and to detect infiltrating cells into the neointima. Compared to apoE null aortae, both *apoE/Kip1* null and

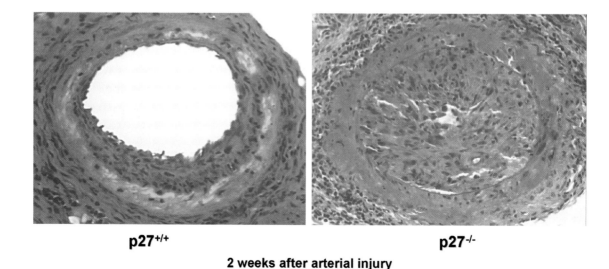

Figure 5. Vascular lesions in *Kip1* null arteries. Femoral arteries from wild-type (*left*) and *Kip1* null mice were injured and analyzed 2 weeks later. The intimal lesion in the *Kip1* null artery is occluded with vsmcs.

apoE/Cip1 null aortae exhibited more atherosclerotic plaque formation following a high-cholesterol regimen. This increase was observed especially in the thoracic and abdominal aortic regions. The lesions in the *apoE/Kip1* null mice were larger and contained more inflammatory cells than the *apoE/Cip1* null animals. Many replicating monocyte-derived macrophages expressing MMP-9 and MMP-12, as well as vsmcs containing lipid particles, were present in the neointima. Hence, a deficiency of $p27^{Kip1}$ and $p21^{Cip1}$, either singly or in combination, accelerates atherogenesis in apoE null mice.

These findings in animal models of vascular disease have direct implications for the design of therapies to treat human vascular disease. Molecular therapies that target the proliferative phase of in-stent restenosis are ideal candidates for drug development combined with device technology. Gene transfer of vectors that encode $p27^{Kip1}$ or $p21^{Cip1}$ into balloon-injured arteries produces a significant reduction in vsmc proliferation and neointimal lesion formation in several animal models of vascular injury (Chang et al. 1995; Yang et al. 1996; Tanner et al. 2000). These studies demonstrated proof of principle that cell cycle inhibition was an effective approach. Advances in bioengineering of stents, combined with cell cycle inhibitors, have provided a new platform for the development of therapies that are poised to seriously address the previously intractable problem of in-stent restenosis. Recent attention has focused on stents coated with drugs that inhibit vsmc proliferation (Fig. 6). Rapamycin- and paclitaxel-coated stents have been approved in Europe and are in the final stages of clinical testing in the United States (Heldman et al. 2001; Morice et al. 2002; for review, see Nabel 2002).

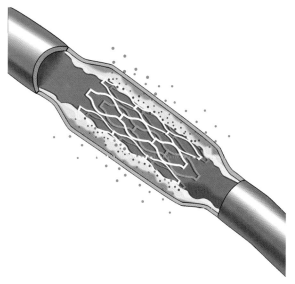

Figure 6. CKIs are molecular targets to treat vascular proliferative diseases. Eluting stents release drugs or plasmid DNA from a polymer coated on the external surface of the stent into endothelial and smooth muscle cells of an artery. These stents are effective in the local treatment of focal vascular lesions. Current prototypes contain rapamycin and paclitaxel, which inhibit cell cyclin progression in vsmcs. (Reprinted, with permission, from Nabel 2002 [copyright Macmillan].)

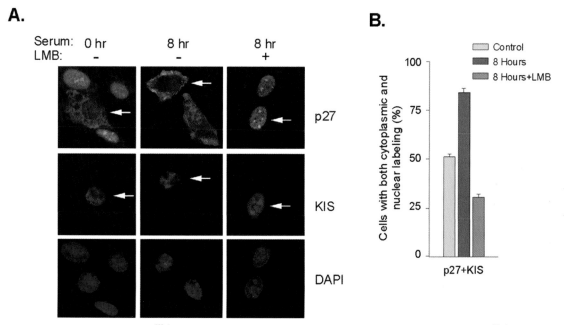

Figure 7. KIS phosphorylation on $p27^{Kip1}$ S10 causes nuclear export. (*A*) Subcellular localization of endogenous $p27^{Kip1}$ in cells transfected with hKIS. Cells were serum-starved, and then serum-stimulated in the absence (–) or presence (+) of leptomycin B. In serum-starved cells, endogenous $p27^{Kip1}$ is nuclear. Serum stimulation leads to redistribution to the cytoplasm. In contrast, in serum-starved cells expressing KIS, endogenous $p27^{Kip1}$ is nuclear and cytoplasmic. After serum stimulation, endogenous 27^{Kip1} is cytoplasmic. The arrows indicate endogenous $p27^{Kip1}$ in cells expressing KIS (*upper panel*) and transfected KIS (*middle panel*). A nuclear DAPI stain is shown in the lower panel. (*B*) Quantitative analysis of the cellular localization of endogenous $p27^{Kip1}$ in cells transfected with hKIS. The results are expressed as the percentage of cells demonstrating both cytoplasmic and nuclear staining. (Adapted, with permission, from Boehm et al. 2002 [copyright Oxford University Press].)

CLONING AND CHARACTERIZATION OF A KINASE THAT PHOSPHORYLATES p27^{Kip1} AND PROMOTES CELL CYCLE PROGRESSION

p27^{Kip1} protein is controlled by its subcellular localization and subsequent degradation. The role of T187 in the phosphorylation of p27^{Kip1} by CDKs is well known (Pagano et al. 1995; Bai et al. 1996; Feldman et al. 1997; Skowyrs et al. 1997). Serine 10 (S10) is another phosphorylation site on p27^{Kip1}, but the kinase that phosphorylates S10 and its effect on cell proliferation had not been defined. Using a yeast two-hybrid approach, we identified the kinase responsible for S10 phosphorylation as the human homolog of KIS or hKIS (human kinase interacting stathmin) (Boehm et al. 2002). hKIS is a nuclear, serine-threonine kinase that binds the carboxy-terminal domain of p27^{Kip1} and phosphorylates it on S10 in vitro and in vivo, promoting its nuclear export to the cytoplasm (Fig. 7). The physiological significance of S10 phosphorylation by hKIS is that it regulates cell cycle progression. Export of p27^{Kip1} to the cytoplasm lowers the abundance of p27^{Kip1} protein in the nucleus and removes cyclin E and CDK2 from nuclear targets, promoting cell cycle progression. We also observed that hKIS is activated by mitogens during G_0/G_1 phase of the cell cycle, and expression of hKIS overcomes growth arrest induced by p27^{Kip1}, suggesting again a role for hKIS in cell cycle progression. Finally, depletion of KIS using siRNA inhibits S10 phosphorylation and enhances growth arrest (Fig. 8). *Kip1* null cells treated with KIS siRNA grow and progress to S/G_2 similar to control treated cells, implicating p27^{Kip1} as the critical target for KIS. In summary, through phosphorylation of p27^{Kip1} on S10, hKIS regulates cell cycle progression in response to mitogens. Additional investigations are analyzing the phenotype of *KIS* null mice (M. Boehm, pers. comm.) and the developmental expression and function of KIS (M. Olive, pers. comm.).

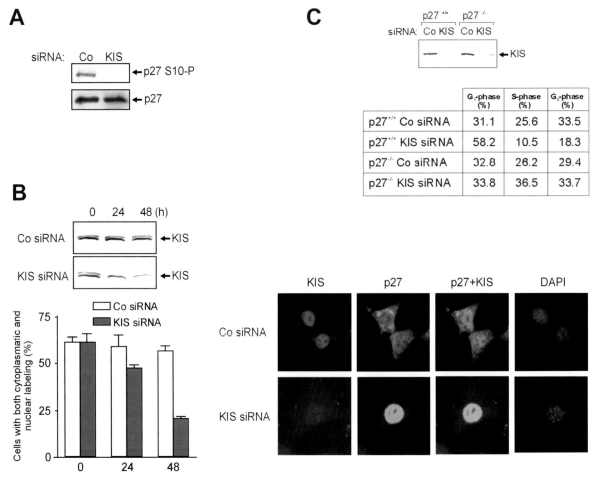

Figure 8. KIS is required for S10 phosphorylation. (*A*) Depletion of hKIS causes an absence of S10 phosphorylation. (*B*) KIS is required for nuclear export of p27^{Kip1}. Cells were transfected with KIS siRNA or Co siRNA and immunoblotting of cell lysates was performed with antibodies to KIS (*upper left panel*). Immunofluorescence for p27^{Kip1} was also performed, and the number of cells expressing cytoplasmic and nuclear p27^{Kip1} at 0, 24, and 48 hours was counted (*lower left panel*). Cells treated with Co or KIS siRNA and Co siRNA were immunostained for KIS and p27^{Kip1} and KIS 48 hours after transfection and examined by confocal microscopy (*right panel*). (Co) Control. (*C*) p27^{Kip1} is a critical target for KIS. p27$^{+/+}$ and p27$^{-/-}$ fibroblasts were treated with Co or KIS siRNA and harvested 48 hours later. Cell lysates were immunoblotted with KIS antibodies (*upper panel*), and FACS analysis was performed (*lower panel*). (Adapted, with permission, from Boehm et al. 2002 [copyright Oxford University Press].)

CONCLUSIONS

The CIP/KIP CDK inhibitors, $p27^{Kip1}$ and $p21^{Cip1}$, are essential regulators of cellular proliferation in the vasculature. A deficiency of these proteins leads to a loss of growth control and apoptosis and a failure to repair arterial wounds. Understanding the signaling pathways and molecular controls of arterial wound repair is critical to dissection of the pathogenesis of vascular diseases. Furthermore, these CKIs represent a class of therapeutic agents that have a broad range of applications in the cardiovascular system. Gain-of-function approaches might prove to be effective therapies for vascular proliferative and inflammatory disorders. We are pursuing the molecular and cellular biology of these CKIs as a model system for understanding the regulation of growth control and differentiation in the cardiovascular system.

REFERENCES

Agrawal D., Hauser P., McPherson F., Dong F., Garcia A., and Pledger W.J. 1996. Repression of $p27^{Kip1}$ synthesis by platelet-derived growth factor in BALB/c 3T3 cells. *Mol. Cell Biol.* **16:** 4327.

Bai C., Sen P., Hofmann K., Goebl M., Harper J.W., and Elledge S.J. 1996. SKP1 connects cell cycle regulation to the ubiquitin proteolysis machinery through a novel motif, the F-box. *Cell* **86:** 263.

Boehm M., Yoshimoto T., Crook M.F., Nallamshetty S., True A., Nabel G.J., and Nabel E.G. 2002. A growth factor-dependent nuclear kinase phosphorylates $p27^{Kip1}$ and regulates cell cycle progression. *EMBO J.* **21:** 3390.

Boehm M., True A.L., San H., Akyurek L.M., Tashiro J., Crook M.J., Nabel G.J., and Nabel E.G. 2001. Deletion of the $p27^{Kip1}$ and $p21^{Cip1}$ loci accelerates cellular proliferation and impairs arterial wound repair. *Circulation* **104:** 1553. (Abstr.)

Brugarolas J., Chandrasekaran C., Gordon J., Beach D., Jacks T., and Hannon G.J. 1995. Radiation-induced cell cycle arrest compromised by p21 deficiency. *Nature* **377:** 552.

Chan F.K., Zhang J., Chen L., Shapiro D.N., and Winoto A. 1995. Identification of human/mouse p19, a novel CDK4/CDK6 inhibitor with homology to $p16^{ink4}$. *Mol. Cell. Biol.* **15:** 2682.

Chang M.W., Barr E., Lu M.M., Barton K., and Leiden J.M. 1995. Adenovirus-mediated overexpression of the cyclin/cyclin-dependent kinase inhibitor, p21 inhibits vascular smooth muscle cell proliferation and neointima formation in the rat carotid artery model of balloon angioplasty. *J. Clin. Invest.* **96:** 2260.

Cheng T., Rodrigues N., Dombkowski D., Stier S., and Scadden D.T. 2000. Stem cell repopulation efficiency but not pool size is governed by $p27^{kip1}$. *Nat. Med.* **6:** 1235.

Coats S., Flannagan W.M., Nourse J., and Roberts J.M. 1996. Requirement of $p27^{Kip1}$ for restriction point control of the fibroblast cell cycle. *Science* **272:** 877.

Deng C., Zhang P., Harper J.W., Elledge S.J., and Leder P. 1995. Mice lacking $p21^{CIP1/WAF1}$ undergo normal development but are defective in G1 checkpoint control. *Cell* **82:** 675.

el-Diery W.S., Tokino T., Velculescu V.E., Levy D.B., Parsons R., Trent J.M., Lin D., Mercer W.E., Kinzler K.W., and Vogelstein B. 1993. WAF1, a potential mediator of p53 tumor suppression. *Cell* **75:** 817.

Feldman R.M., Correll C.C., Kaplan K.B., and Deshaies R.J. 1997. A complex of Cdc4p, Skp1p, and Cdc53p/cullin catalyzes ubiquitination of the phosphorylated CDK inhibitor Sic1p. *Cell* **91:** 221.

Fero M.L., Rivkin M., Tasch M., Porter P., Carow C.E., Firpo E., Polyak K., Tsai L.-H., Vroudy V., Perlmutter R.M., Kaushansky K., and Roberts J.M. 1996. A syndrome of multiorgan hyperplasia with features of gigantism, tumorigenesis, and female sterility in $p27^{Kip1}$-deficient mice. *Cell* **85:** 733.

Gu Y., Turek C.W., and Morgan D.O. 1993. Inhibition of CDK2 activity in vivo by an associated 20K regulatory subunit. *Nature* **366:** 707.

Halevy O., Novitch B., Spicer D.B., Skapek S.X., Rhee J., Hannon G.J., Beach D., and Lassar A.B. 1995. Correlation of terminal cell cycle arrest of skeletal muscle with induction of p21 by MyoD. *Science* **267:** 1018.

Harper J.W., Adami G.R., Wei N., Keyomarsi K., and Elledge S.J. 1993. The p21 cdk-interacting protein Cip1 is a potent inhibitor of G1 cyclin-dependent kinases. *Cell* **75:** 805.

Heldman A.W., Cheng L., Jenkins G.M., Heller P.F., Kim D.W., Ware M., Jr., Nater C., Hurban R.H., Rezai B., Abella B.S., Bunge K.E., Kinsella J.L., Sollott S.J., Lakatta E.G., Brinker J.A., Hunter W.L., and Froehlich J.P. 2001. Paclitaxel stent coating inhibits neointimal hyperplasia at 4 weeks in a porcine model of coronary restenosis. *Circulation* **103:** 2289.

Hengst L. and Reed S.I. 1996. Inhibitors of the Cip/Kip family. *Curr. Top. Microbiol. Immunol.* **227:** 25.

Hirai H., Roussel M.F., Kato J.Y., Ashmun R.A., and Sherr C.J. 1995. Novel INK4 proteins, p19 and p18, are specific inhibitors of the cyclin D-dependent kinases CDK4 and CDK6. *Mol. Cell. Biol.* **15:** 2672.

Ishida N., Kitagawa M., Hatakeyama S., and Nakayama K.-I. 2000. Phosphorylation at serine 10, a major phosphorylation site of $p27^{Kip1}$, increases its protein stability. *J. Biol. Chem.* **275:** 25146.

Kiyokawa H., Kineman R.D., Manova-Todorova K.O., Soares V.C., Hoffman E.S., Ono M., Khanam D., Hayday A.C., Frohman L.A., and Koff A. 1996. Enhanced growth of mice lacking the cyclin-dependent kinase inhibitor function of $p27^{Kip1}$. *Cell* **85:** 721.

Koff A., Ohtsuki M., Polyak K., Roberts J.M., and Massagué J. 1993. Negative regulation of G1 in mammalian cells: Inhibition of cyclin E-dependent kinase by TGF-β. *Science* **260:** 536.

Koyama H., Raines E.W., Bornfeldt K.E., Roberts J.M., and Ross R. 1996. Fibrillar collagen inhibits arterial smooth muscle proliferation through regulation of Cdk2 inhibitors. *Cell* **87:** 1069.

Lee M.H., Reynisdóttir I., and Massagué J. 1995. Cloning of $p57^{Kip2}$, a cyclin-dependent kinase inhibitor with unique domain structure and tissue distribution. *Genes Dev.* **9:** 639.

Malek N.P., Sundberg H., McGrew S., Nakayama K., Kyriakidis T.R., and Roberts J.M. 2001. A mouse knock-in model exposes sequential proteolytic pathways that regulate $p27^{Kip1}$ in G1 and S phase. *Nature* **413:** 323.

Matsuoka S., Edwards M., Bai C., Parker S., Zhang P., Baldini A., Harper J.W., and Elledge S.J. 1995. $p57^{KIP2}$, a structurally distinct member of the $p21^{CIP1}$ cdk inhibitor family, is a candidate tumor suppressor gene. *Genes Dev.* **9:** 650.

Millard S.S., Yan J.S., Nguyen H., Pagano M., Kiyokawa H., and Koff A. 1997. Enhanced ribosomal association of $p27^{Kip1}$ mRNA is a mechanism contributing to accumulation during growth arrest. *J. Biol. Chem.* **272:** 7093.

Miskimins W.K., Wang G., Hawkinson M., and Miskimins R. 2001. Control of cyclin-dependent kinase inhibitor p27 expression by cap-independent translation. *Mol. Cell. Biol.* **21:** 4960.

Morice M.C., Serruys P.W., Sousa J.E., Fajadet J., Ban Hayashi E., Perin M., Colombo A., Schuler G., Barragan P., Guagliumi G., Molnar F., Falotico R., and RAVEL Study Group. 2002. A randomized comparison of a sirolimus-eluting stent with a standard stent for coronary revascularization. *N. Engl. J. Med.* **23:** 1773.

Nabel E.G. 2002. CDKs and CKIs: Molecular targets for tissue remodeling. *Nat. Rev. Drug Discov.* **1:** 587.

Nakayama K., Ishida N., Shirane M., Inomata A., Inoue T., Shishido N., Horii I., Loh D.Y., and Nakayama K.-I. 1996. Mice lacking $p27^{Kip1}$ display increased body size, multiple organ hyperplasia, retinal dysplasia, and pituitary tumors. *Cell* **85:** 707.

Pagano M., Tam S.W., Theodoras A.M., Beer-Romero P., Del

Sal G., Chau V., Yew P.R., Draetta F.G., and Rolfe M. 1995. Role of the ubiquitin-proteasome pathway in regulating abundance of the cyclin-dependent kinase inhibitor p27. *Science* **269**: 682.

Parker S.B., Eichele G., Zhang P., Rawls A., Sands A.T., Bradley A., Olson E.N., Harper J.W., and Elledge S.J. 1995. p53-independent expression of p21Cip1 in muscle and other terminally differentiating cells. *Science* **267**: 1024.

Pietenpol J., Bohlander S., Sato Y., Papadopoulos N., Liu B., Friedman C., Trask B.J., Roberts J.M., Kinzler K.W., and Rowley J.D. 1995. Assignment of the human p27^{Kip1} gene to 12p13 and its analysis in leukemias. *Cancer Res.* **55**: 1206.

Polyak K., Lee M.-H., Erdjument-Bromage H., Koff A., Roberts J.M., Tempst P., and Massague J. 1994. Cloning of p27^{Kip1}, a cyclin-dependent kinase inhibitor and a potential mediator of extracellular antimitogenic signals. *Cell* **78**: 59.

Servant M.J., Coulombe P., Turgeon B., and Meloche S. 2000. Differential regulation of p27^{Kip1} expression by mitogenic and hypertrophic factors: Involvement of transcriptional and post-transcriptional mechanisms. *J. Cell Biol.* **148**: 543.

Sherr C.J. and Roberts J.M. 1999. CDK inhibitors: Positive and negative regulators of G$_1$-phase progression. *Genes Dev.* **13**:1501.

Skowyrs D., Craig K.L., Tylers M., Elledge S.J. and Harper J. W. 1997. F-box proteins are receptors that recruit phosphorylated substrates to the SCF ubiquitin-ligase complex. *Cell* **91**: 209.

Tanner F.C., Yang Z.Y., Duckers E., Gordon D., Nabel G.J., and Nabel E.G. 1998. Expression of cyclin-dependent kinase inhibitors in vascular disease. *Circ. Res.* **82**: 396.

Tanner F.C., Boehm M., Akyurek L.M., San H., Yang Z.Y., Tashiro J., Nabel G.J., and Nabel E.G. 2000. Differential effects of the cyclin-dependent kinase inhibitors p27^{Kip1}, p21^{Cip1}, and p16^{Ink4} on vascular smooth muscle cell proliferation. *Circulation* **101**: 2022.

Toyoshima H. and Hunter T. 1994. p27, a novel inhibitor of G1 cyclin/cdk protein kinase activity, is related to p21. *Cell* **78**: 67.

Waga S., Hannon G., Beach D., and Stillman B. 1994. The p21 inhibitor of cyclin-dependent kinases controls DNA replication by interaction with PCNA. *Nature* **369**: 574.

Xiong Y., Hannon G.J., Zhang D., Casso D., Kobayashi R., and Beach D. 1993. p21 is a universal inhibitor of cyclin kinases. *Nature* **366**: 701.

Yang Z.Y., Simari R., Perkins N., San H., Gordon D., Nabel G.J., and Nabel E.G. 1996. Role of the p21 cyclin-dependent kinase inhibitor in limiting intimal cell proliferation in response to arterial injury. *Proc. Natl. Acad. Sci.* **93**: 7905.

Zhang Y. and Lin S.C. 1997. Molecular characterization of the cyclin-dependent kinase inhibitor p27 promoter. *Biochim. Biophys. Acta* **1353**: 307.

Selective Functions of Angiopoietins and Vascular Endothelial Growth Factor on Blood Vessels: The Concept of "Vascular Domain"

T.N. Sato, S. Loughna, E.C. Davis*, R.P. Visconti, and C.D. Richardson

*The Sato Laboratory and *Department of Cell Biology, The University of Texas Southwestern Medical Center, Dallas, Texas 75390*

Blood vessel growth and remodeling are critical processes for both growth and establishment of functional organs. Furthermore, reestablishment of an intricate network of blood vessels is a requisite process for regaining the functionality of many diseased organs (Folkman 1995; Isner 2002). In past years, many factors that induce blood vessel formation have been discovered and are collectively known as angiogenic factors.

Angiopoietins and vascular endothelial growth factor (VEGF) represent two major classes of angiogenic factors (Ferrara et al. 1995; Yancopoulos et al. 2000). They exhibit both distinct and collaborative activities during the formation of blood vessels (Suri et al. 1998; Thurston et al. 1999; Visconti et al. 2002). These distinctive and collaborative functions of the angiopoietins and VEGF were conventionally considered to be universal for all types of blood vessels. However, we have recently found experimental evidence that challenges the universality of some of these angiogenic factors (Visconti et al. 2002). In this paper, we summarize the results that led to this new concept and discuss potential biological and clinical implications.

HANDEDNESS AND POLARIZED FUNCTION OF ANGIOPOIETINS AND THEIR RECEPTORS

Angiopoietins form a family of secreted glycoproteins that exhibit a unique domain structure: coiled-coil amino-terminal and fibrinogen-like carboxy-terminal domains (Davis et al. 1996). There are four distinct but closely related angiopoietins: angiopoietin-1 (Ang-1), angiopoietin-2 (Ang-2), angiopoietin-3 (Ang-3), and angiopoietin-4 (Ang-4) (Davis et al. 1996; Maisonpierre et al. 1997; Valenzuela et al. 1999). They are synthesized and secreted by perivascular cells such as smooth muscle cells, pericytes, and other cells that reside in the vessel wall (Davis et al. 1996; Suri et al. 1996; Maisonpierre et al. 1997). The secreted angiopoietins bind to endothelial-specific receptor-type tyrosine kinase, Tie2, and control the behavior of endothelial cells during vascular development and blood vessel formation and remodeling in pathological conditions (Suri et al. 1996; Maisonpierre et al. 1997). On the basis of this initial description, it has generally been believed that the angiopoietins are paracrine factors that work on blood vessels of all types.

However, we have recently challenged this generally accepted model.

Asymmetrical Expression of Angiopoietin-1 during Development

The spatial expression pattern of Ang-1 during vascular development was studied in a more careful manner in a search for any type of asymmetrical or polarized expression patterns that may indicate a unique and selective function of Ang-1 (Loughna and Sato 2001). Ang-1 was most abundantly expressed in the developing sinus venosus from which the major venous vascular network originates at the earliest stage of cardiovascular development (Fig. 1a).

Ang-1 expression was detected in a symmetrical manner at E8.5 embryonic heart (Fig. 1b,c). However, the expression pattern becomes asymmetrical later in development (Fig. 1d,e).

Whether this handedness of the Ang-1 expression pattern is related to left- or right-hand sides of the whole body plan was also tested. This was investigated by examining the expression of Ang-1 in mutant mouse embryos that exhibit inversion of the left and right sides of the body. In both *inv* and GDF-1-deficient mice, the Ang-1 negative side was found to be on the right-hand side, therefore inverted as compared to normal mice (Fig. 1f–i).

Polarized Expression of Angiopoietin-1 in the Vessel Wall

The possibility that Ang-1 expression may be polarized even in the peripheral blood vessels was also studied. It was found that Ang-1 expression seems more abundant around the developing venous system as compared to the arterial system (Fig. 2). Furthermore, Ang-1 expression was primarily detected at the ventral side of the vessel wall (Fig. 2). This, and the above-described results, clearly showed that Ang-1 expression is polarized in both central and peripheral vasculature.

Functional Significance of the Asymmetrical and Polarized Expression Patterns of Angiopoietins

Putative functional significance of the asymmetrical and polarized expression pattern of Ang-1 was studied (Loughna and Sato 2001). Because the Ang-1-deficient mouse embryos do not show any asymmetrical vascular phenotypes, we hypothesized that there may be other angiopoietin or angiopoietin-like factors that work together with Ang-1 to act selectively on the vascular network on one side of the body but not on the other.

To this end, Ang-1 and Tie1 double-knockout mice were found to show a specific asymmetrical vascular defect (Fig. 3) (Loughna and Sato 2001). Tie1 is another endothelial-specific receptor tyrosine kinase that is structurally related to Tie2 with no known specific ligands (Fig. 3a) (Partanen et al. 1992; Sato et al. 1993).

In Ang-1$^{-/-}$Tie1$^{-/-}$ mice, the cardinal vein on the right-hand side was completely missing (Fig. 3b–d). The cardinal vein on the left-hand-side was intact (Fig. 3b–d). This phenotype is consistent with the asymmetrical expression pattern of Ang1 (Fig. 1). This finding is also consistent with the hypothesis that the right-hand side-specific expression of Ang-1, in combination with a putative ligand for Tie1, controls the selective establishment of the right-hand-side venous system.

The structural nature of the defective right-hand-side cardinal vein was studied by electron microscopy (EM) (Fig. 4). The intact left-hand-side cardinal vein consisted of one or two endothelial cells that form a single large lumen (Fig. 4a). However, the defective cardinal vein on the right-hand side exhibited small multiple lumens (Fig. 4b–d). Some of the lumens were composed of an abnormally large number of endothelial cells (Fig. 4c,d). Sometimes, a single endothelial cell was involved in the walls of two adjacent lumens (Fig. 4d). This EM analysis suggests that the defect may be due to abnormal endothelial cell–cell recognition during the establishment of the right-hand-side cardinal vein in the Ang-1$^{-/-}$ Tie1$^{-/-}$ embryo.

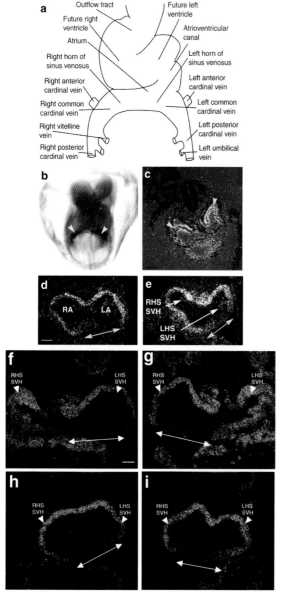

Figure 1. Polarized expression of Ang-1 in the embryonic atrium and sinus venosus. (*a*) Schematic diagram of the heart and connecting major vessels at E9.5 (modified from Edwards 1998). (*b, c*) In situ hybridization whole-mount (*b*) and section (*c*) of an E8.5 embryo. Ang-1 is expressed symmetrically at the sinus venosus of the heart (*yellow arrowheads*). (*d, e*) In situ hybridization on tissue sections of an E9.5 embryo. Ang-1 is expressed in the atrium (*d*) and sinus venosus (*e*), except to their ventral left walls (*yellow arrows*). (*f*) In situ hybridization on a tissue section from an E9.5 wild-type embryo. Ang-1 is expressed in the sinus venosus, except to the ventral left wall (*white arrow*). (*g*) In situ hybridization on a tissue section from inv mice at E9.5. The heart looping in this mutant embryo is inverted. Ang-1 expression pattern is also inverted. (*h*) In situ hybridization on a tissue section from a GDF1 null mutant E9.5 embryo. The heart of this embryo looped correctly to the right and Ang-1 is expressed as normal, absent to the left ventral wall of the left-hand-side horn (*white arrow*). (*i*) In situ hybridization on a tissue section from another GDF1 null mutant E9.5 embryo. The heart of this embryo did not undergo correct looping to the right, but underwent looping to the left. Ang-1 expression was found to correlate with the inverted looping, as it was expressed throughout the heart except to the right ventral myocardial wall of the sinus venosus (*white arrow*). (LHS SVH) Left-hand-side sinus venosus horn; (RHS SVH) Right-hand-side sinus venosus horn. Bar, 100 µm. (Adapted, with permission, from Loughna and Sato 2001 [copyright Elsevier Science].)

ANGIOPOIETIN-1 AS A VENOUS-SPECIFIC REGULATOR?

Another interesting point of the above study is the relatively specific sensitivity of the venous system to the lack of Ang-1 and Tie1 as compared to the arterial system (Fig. 4). This finding is consistent with the fact that Ang-1 is preferentially expressed around veins (Fig. 2) and thus suggests that Ang-1 is primarily involved in the establishment of the venous system.

ARTERIAL AND VENOUS SELECTIVITY OF ANGIOPOIETINS AND VEGF

On the basis of the above evidence for Ang-1 as a venous-specific regulator, we have tested whether the angiopoietin or another class of angiogenic factor, VEGF, can induce a specific type (i.e., venous vs. arterial) of blood vessel formation when ectopically expressed in vivo (Visconti et al. 2002).

Each angiopoietin or VEGF was individually and specifically expressed in cardiac myocytes in transgenic mice. Since angiopoietins and VEGF show collaborative effects on angiogensis, combinations of these classes of factors were also expressed in the transgenic heart.

Collaborative Functions of Angiopoietins and VEGFs

Neither Ang-1 nor Ang-2 alone was effective in inducing angiogenesis in heart (Fig. 5a, b). VEGF was effective and Ang-2 enhanced the effect of VEGF-mediated angiogenesis (Fig. 5a,b). In contrast, Ang-1 blocked both VEGF- and VEGF/Ang-2-mediated angiogenesis. This is in contrast to what was previously shown in skin, where Ang-1 enhanced VEGF-mediated angiogenesis. These results suggest that the effect of angiopoietins may be dependent on the organ microenvironment.

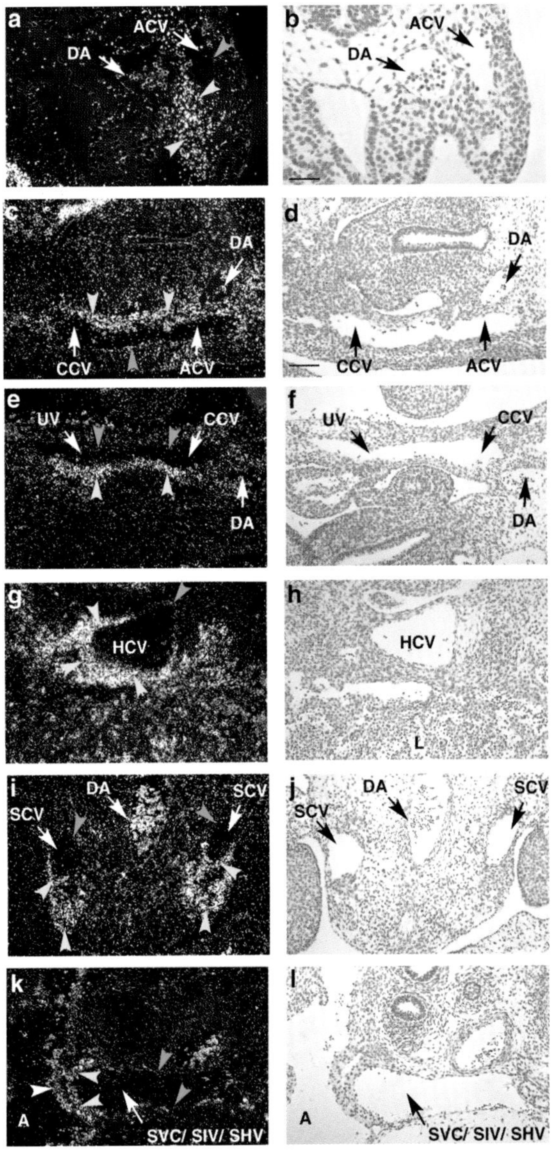

Figure 2. Highly polarized expression pattern of Ang-1 in developing peripheral vessels. Dark field (*a, c, e, g, i, k*) and bright field (*b, d, f, h, j, l*) are shown. (*a, b*) At E9.5, Ang-1 expression is restricted to the ventral mesenchymal/smooth muscle wall of the anterior cardinal vein (ACV) (*yellow arrowheads*) prior to communication with the common cardinal vein. The rest of the anterior cardinal vein wall does not express Ang-1 (*blue arrowhead*). (*c, d*) Upon communication of the common cardinal vein (CCV) and anterior cardinal vein at E10.5, Ang-1 expression is detected in the medial wall of these vessels (*yellow arrowheads*). The lateral wall is negative for Ang-1 expression (*blue arrowhead*). (*e, f*) As the umbilical vein (UV) and common cardinal vein join at E10.5, Ang-1 expression is seen in the medial wall (*yellow arrowheads*), whereas the rest of the vessel wall is negative (*blue arrowheads*). (*g, h*) Expression is seen in the wall and surrounding mesenchyme of the heptocardiac vein (HCV) (*yellow arrowheads*) at E11.5, except to the dorsal wall (*blue arrowhead*). (*i, j*) At E11.5, Ang-1 expression is restricted to the ventral mesenchymal wall of the subcardinal veins (SCV) (*yellow arrowheads*). The remainder of the subcardinal vein walls do not express Ang-1 (*blue arrowheads*). (*k, l*) As the left-side superior vena cava (SVC), superior intercostal vein (SIV), and superior hemizygous vein (SHV) communicate at E13.5, expression is seen only in the ventral wall (*yellow arrowheads*). White arrowhead denotes expression in the wall of the adjacent atrium (A). Ang-1 expression is not detected in the rest of the vessel wall (*blue arrowheads*). (DA) dorsal aorta; (L) liver. Bars, 50 μm (*a, b*) and 100 μm (*c–l*).

In addition to regulation of capillary numbers, certain features of capillary morphology were also controlled by VEGF and Ang-2. The capillaries formed by VEGF exhibited relatively larger lumens (Fig. 5a). Furthermore, the VEGF/Ang2 combination induced abnormally clustered capillaries (Fig. 5a,c). The endothelial nature of these abnormal capillaries was confirmed by their expression of both CD31 (PECAM-1) and Flk1 (Fig. 5a,c).

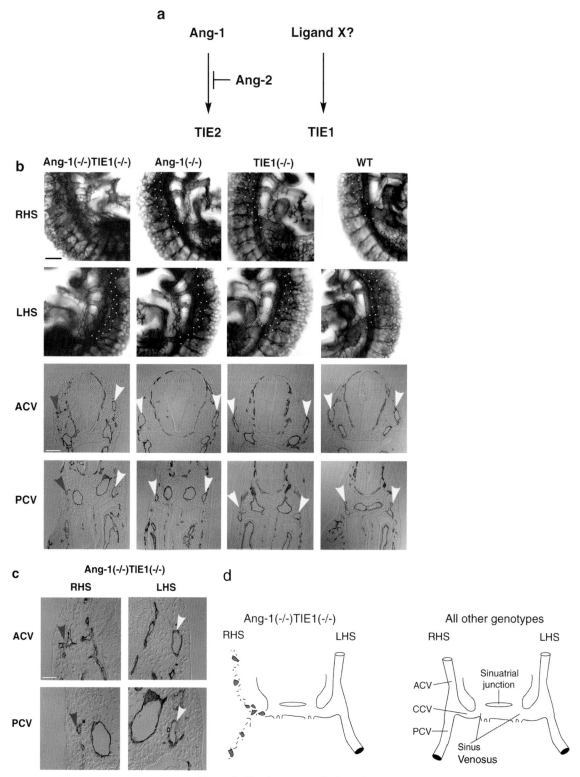

Figure 3. (*See facing page for legend.*)

Arteriovenous Selectivity of Angiopoietins and VEGFs

It was also found that VEGF-induced angiogenesis is primarily due to an increased number of arterial-type capillaries (Fig. 6a), which was accompanied by a decrease in venous-type capillaries (Fig. 6b). Although Ang-2 enhanced VEGF-mediated angiogenesis, it inhibited the increase in arterial-type capillaries (Fig. 6a). No significant increase in venous-type capillaries accompanied the Ang-2-mediated decrease in arterial-type capillaries (Fig. 6b). As a result, the majority of the abnormally clustered capillaries were found to be neither arterial nor venous types (Fig. 6a,b).

In summary, VEGF primarily induces the establishment of new arterial-type capillaries (i.e., arteriogenesis) (Fig.7). These capillaries exhibit slightly larger lumen as compared to normal capillaries (Fig. 7). Ang-2 can further increase capillary numbers, and these capillaries are abnormally clustered together (Fig. 7). However, Ang-2 inhibits VEGF-mediated establishment of new arterial-type capillaries without inducing the formation of new venous-type capillaries (Fig. 7). In contrast, Ang-1 can neutralize both VEGF and VEGF/Ang-2-mediated angiogenesis and arteriogenesis (Fig. 7).

BIOLOGICAL IMPLICATIONS OF THE SELECTIVITY OF ANGIOGENIC FACTORS

Herein, we have discussed the experimental results that support the idea of the existence of vascular-type selectivity by two major classes of angiogenic factors, angiopoietins and VEGF (Fig. 8). The biological significance of each type of vascular selectivity is discussed in the following sections. Furthermore, potential existence of other types of selectivity that may critically contribute to new blood vessel formation, regression, and remodeling is discussed.

Handedness of Vascular Development

The vascular network initially arises as a symmetrical form during embryogenesis (Langman 1990). However, this symmetrical network becomes remodeled during development, and a mature asymmetrical network is eventually established (Langman 1990). This remodeling process involves both the growth and regression of specific vascular branches. Furthermore, the directionality of the growth of branches also influences the final form of the vascular network.

No specific molecular mechanisms have been proposed that explain how the asymmetric pattern of the vascular network is established. Our study discussed above may implicate the involvement of angiopoietin and Tie receptor pathways in the establishment of this network.

How do Ang-1 and Tie1 regulate the establishment of an asymmetrical venous network pattern? Although Ang-1-expressing cells in the sinus venosus are asymmetrically distributed, Ang-1$^{-/-}$ embryos do not exhibit any asymmetric vascular phenotype. This requires the lack of Tie1. Therefore, it is possible that a putative ligand for Tie1 may be involved in establishing the right-hand-side venous system, in combination with Ang-1 function.

According to our EM studies, the Ang-1/Tie1 pathway seems to control proper endothelial cell–cell recognition. Therefore, it is possible that this pathway controls the expression of a certain class of endothelial cell adhesion molecules. Alternatively, it is also possible that this pathway controls the proper embedding of the endothelial cells in an extracellular matrix that is permissive for the cells to recognize each other.

Another interesting point is that the vessels sensitive to the Ang-1/Tie1 pathway are restricted to the venous system. However, arterial vessels are also known to undergo asymmetrical development. Thus, there must be other molecular pathways that effect a selectivity for the vascular network on one side of the body.

Until now, no other angiogenic factors have been investigated for their potential asymmetric distribution in the developing vascular system. Once such studies are conducted in an extensive manner, other classes of angiogenic factors may also be identified as being involved in the establishment of an asymmetric vascular network pattern.

Another question is how the establishment of this asymmetric vascular network is related to the determination of left- and right-hand sides of the whole body. There have been extensive genetic and molecular studies on the

Figure 3. Requirement of Ang-1 and TIE1 specifically for normal right-side venous formation at E9.5 (*a*) Ang-1 is an activating ligand for the TIE2 receptor, whereas Ang-2 is an antagonist for this pathway. Tie1 is an orphan receptor. (*b*) Whole embryos immunostained with a pan-endothelial marker CD31. A distinctive normal RHS anterior cardinal vein can be seen in all genotypes (outlined with *yellow dotted line*) except Ang1(–/–)TIE1(–/–), where no continuous vessel lumen could be detected. Instead, a network of small disorganized vessels is detected (*red arrows*). On the LHS all genotypes have a normal anterior cardinal vein (*yellow dotted line*). This was confirmed on sectioning, with analysis using differential interference contrast (DIC) optics. A normal anterior cardinal vein (ACV) and posterior cardinal vein (PCV) can be seen for all genotypes on the left side and right side (*yellow arrows*) except for Ang1(–/–)TIE1(–/–), where both of these vessels are abnormal specifically on the RHS (*red arrows*). Three to ten embryos of each genotype were analyzed, each of 20–25 somites. Bar, 100 µm. (*c*) High magnification of CD31 immunostained Ang1(–/–)TIE1(–/–) sections using DIC optics. Red arrows point to abnormal anterior and posterior cardinal veins on the RHS. Note the cluster of CD31-positive cells and the very small discontinuous lumen. On the LHS, both vessels exhibit normal large and distinctive lumens (*yellow arrows*). Bar, 25 µm. (*d*) Schematic diagram showing the failure to form the RHS anterior cardinal and posterior cardinal veins in the Ang1(–/–) TIE1(–/–) knockout. Scattered endothelial cells (*black dots*) were seen and only the small discontinuous vessel lumens (*red*) were observed. All other genotypes had normal anterior and posterior cardinal veins on both the RHS and LHS. (ACV) Anterior cardinal vein; (CCV) common cardinal vein; (PCV) posterior cardinal vein. In addition to the presented genotypes, Ang1(–/–)TIE2(–/–), Ang2(–/–)TIE2(–/–), and Ang2(–/–)TIE1(–/–) embryos and all the double heterozygous embryos were also studied, and they exhibit no difference in the RHS and LHS vascular system at E9.5 (data not shown). (Adapted, with permission, from Loughna and Sato 2001 [copyright Elsevier Science].)

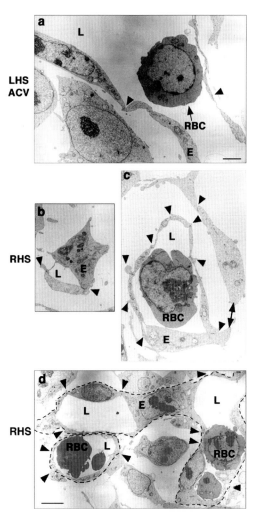

Figure 4. Electron micrographs of the anterior cardinal veins from an Ang-1/TIE1 null mutant. (*a*) A segment of the left anterior cardinal vein shows normal endothelial cell morphology and number of junctions. (*b*) Two small endothelial cells, one with aberrant square morphology, join to form a lumen too small for the passage of a red blood cell. (*c*) A red blood cell can be seen in the lumen of an abnormal vessel, which is composed of eight endothelial cells. The double-headed arrow denotes a region that was in continuity in a previous serial section. (*d*) An abnormally square-shaped endothelial cell can be seen contributing to two separate lumens, which are outlined by a dashed line. A third lumen, also outlined, is in proximity. (E) Endothelial cell; (L) lumen; (LHS ACV) left-hand-side anterior cardinal vein; (RBC) red blood cell; (RHS) right-hand-side. Arrowheads denote cell–cell junctions. Bars, 100 μm (*a–c*) and 300 μm (*d*).

establishment of left- and right-handedness, and genetic and signaling pathways involved in this type of body plan have become more clear (Yost 1999, 2001; Patterson 2001). One of our future goals would be to tie together the regulatory pathways underlying the establishment of the asymmetrical vascular network and the synthesis of the whole body plan.

Polarity of the Vascular Network

Another selectivity is exemplified by the ventral-side-specific expression of Ang-1 in the vessel wall during vascular development. This is potentially interesting since smooth muscle (SM) precursor cells are known to be localized at the ventral side of the vessel wall during vascular development (Hungerford et al. 1997; Liu et al. 2000; Sato 2000). These SM precursor cells migrate upward and eventually invest the entire vessel wall as the vessels mature (Liu et al. 2000; Sato 2000). An interesting possibility is that the Ang-1-positive cells at the dorsal side of the vessel wall represent these putative SM precursor cells. Therefore, this positioning of SM precursor cells and/or Ang-1-expressing cells along the vessel wall determines a certain aspect of vascular polarity.

What is the signal that permits the specific positioning of SM precursor cells or Ang-1-positive cells at the dorsal aspect of the vessel wall? Although this putative signal is not known, some insight was recently obtained for the signal that controls the migration of the dorsally localized SM precursor cells upward along the vessel wall (Liu et al. 2000; Sato 2000).

The vascular network is a highly polarized structure. This structural polarity is primarily determined by the position of each vascular branching point along the vasculature and the directionality of each branch (Loughna and Sato 2001). What determines the position of a vascular branch? What determines the direction that each branch takes? Is the structural polarity of the vascular network related to the polarized SM precursor localization discussed above? The answers to these questions are expected to contribute significantly to our understanding of the fundamental principles underlying the establishment of the vascular system.

Arterial Versus Venous Network

Our studies discussed above clearly show that specific and distinct signals dictate the establishment of arteries and veins (Fig. 7). Our study shows that VEGF specifically directs the formation of arteries. Ang-2 seems to suppress the establishment of arterial identity of the vessels induced by VEGF. However, the vessels that failed to become arteries due to the presence of Ang-2 do not automatically assume a venous identity. These vessels remain uncommitted to either vascular type. Ang-1 also inhibits VEGF-induced arterial vessel formation. However, unlike Ang-2, the vessels that have lost arterial identity due to the presence of Ang-1 resume a venous identity.

The interplay between angiopoietins and VEGF discussed above suggests a possible mechanism underlying the establishment of arteries and veins during development. It is proposed that the default state of the most primitive vessels is the venous type (Lawson et al. 2001; Zhong et al. 2001). Therefore, our study suggests that VEGF is either a direct or permissive inducer of the establishment of arteries. This could be accomplished by directly inducing the cascades of gene expression required for the establishment of arterial identity of the endothelial cells. Alternatively, VEGF could act as a permissive inducer by allowing endothelial cells to respond to putative artery-inducing signals. It is also possible that VEGF works as a survival factor preferentially for arterial-type endothelial cells. Recently, Notch signaling has

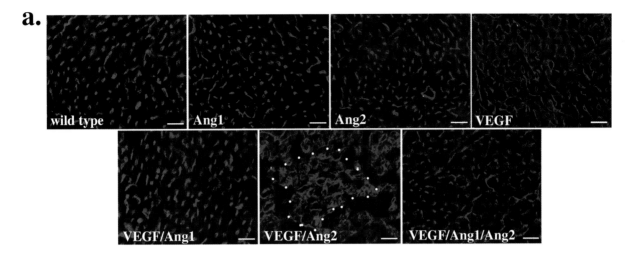

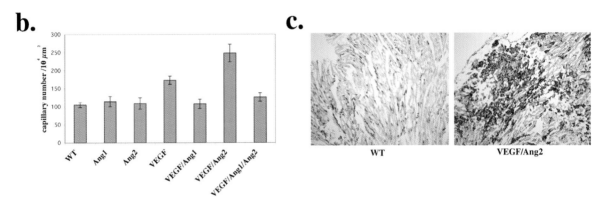

Figure 5. Combinatorial regulation of angiogenesis by VEGF and angiopoietins. (*a*) Anti-CD31/PECAM-1 antibody staining of the adult heart from each transgenic line and wild-type control. Genotype of each line is indicated at the bottom left (Ang1: αMHC::Ang1, Ang2: αMHC::Ang2, VEGF: αMHC::VEGF, VEGF/Ang1: αMHC::VEGF;αMHC::Ang1 double transgenic line, VEGF/Ang2: αMHC::VEGF;αMHC::Ang2 double transgenic line, VEGF/Ang1/Ang2: αMHC::VEGF;αMHC::Ang1;αMHC::Ang2 triple transgenic line). An example of the aberrant clusters of capillaries in the VEGF/Ang2 double transgenic heart is outlined by the dotted line. Bar, 25 μm. (*b*) Quantitation of capillary density. Capillary lumens in each transgenic line and wild-type control hearts were counted (per 10^4 μm^2 field) from 6–13 randomly selected fields. (*c*) Flk1::lacZ staining of the wild-type control and VEGF/Ang2 transgenic lines.

been implicated as an inductive signal for the establishment of arterial-type vessels (Lawson et al. 2001; Zhong et al. 2001). It is possible, therefore, that VEGF may work via a Notch pathway to induce arterial-type vessels.

Ang-2 suppresses the VEGF-mediated establishment of arteries. At the same time, Ang-2 suppresses the putative default pathway that maintains the venous identity of vessels. However, Ang-2 does not suppress VEGF-induced angiogenesis. In fact, Ang-2 enhances VEGF-mediated angiogenesis. Therefore, it seems that the VEGF-mediated angiogenic pathway is, in part, distinct from the pathway mediating induction of the establishment of arteries. Alternatively, there are two distinct angiogenic pathways: one mediating both angiogenesis and arteriogenesis that is primarily controlled by VEGF and the other, mediating only angiogenesis but in compensation for the loss of both arterial and venous identities that is controlled by both VEGF and Ang-2 in a combinatorial manner.

Ang-1 inhibits both VEGF-mediated arteriogenesis and angiogenesis. The Ang-1-mediated inhibition of arteriogenesis results in the default state of primitive vessels (i.e., venous type). Therefore, it is possible that Ang-1 is required to maintain the venous identity of the vessels. This notion is consistent with our finding that endogenous Ang-1 is primarily localized around veins during vascular development.

More Selectivity

In addition to the types of selectivity discussed above, there are other types of selectivity that could potentially exist and contribute significantly to the development of the vascular system. A few examples are discussed in the following sections.

Organ selectivity. One of the primary physiological functions of the vascular system is to control growth and maintenance of organs by providing sufficient oxygen and nutrients. The vascular system also functions as a "gate-keeper" for organs by selectively transporting

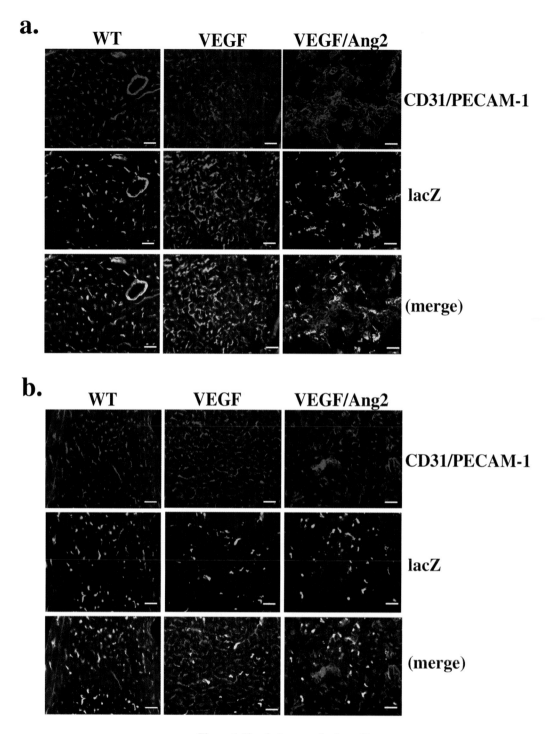

Figure 6. (*See facing page for legend.*)

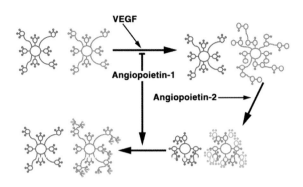

Figure 7. Orchestration of angiogenesis and arteriovenous contribution by angiopoietins and VEGF. Veins (*blue*), arteries (*red*), and uncommitted vessels (*green*) are color-coded.

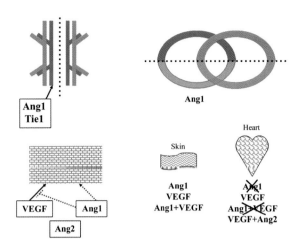

Figure 8. Selectivity of angiopoietins and VEGF. Ang-1 and Tie1 are selective for the left-hand-side venous system. Ang-1 is in general selective for the ventral side of the lumen of the veins. VEGF functions as an arteriogenesis inducer, and the angiopoietins modulate this activity. Whether or not each growth factor functions angiogenically is dependent on the organ microenvironment (e.g., skin vs. heart).

molecules to the organ environment from the circulation. The vascularization within each organ is controlled both temporally and spatially to allow the development of a vascular network suited for optimal growth and acquisition of the physiological functions of the organ. Therefore, the vascular system in each organ may be highly specialized in both its structure and function.

The specialized structure and function of each vascular system may require specific combinations of angiogenic growth factors in the organ microenvironment. It is also possible that vascular cells such as endothelial and smooth muscle cells that form the vessels in each organ may be phenotypically and molecularly distinct. These distinct vascular cells may respond to each growth factor in a highly selective manner.

For example, we have previously shown that endothelial cells in the skin respond to Ang-1 and contribute to the formation of new blood vessels. However, as shown above, endothelial cells in the heart do not respond to Ang-1. Endothelial cells in the heart may thus be phenotypically distinct from those in the skin. It is also possible that the microenvironment in the skin is permissive for Ang-1 to function as an angiogenic growth factor, but the cardiac microenvironment may be nonpermissive.

One of our future goals would be to systematically dissect the molecular phenotypes of the endothelial and other vascular cells in each organ. It would also be important to molecularly define the organ microenvironment in relation to angiogenesis. These future studies are expected to furnish a better understanding of how the vascular system and microenvironment of each organ communicate to harmonize each other's growth and function.

Capillary versus larger vessels. Capillaries are formed by a single layer of endothelial cells. In contrast, larger vessels are normally invested by thick layers of smooth muscle cells together with many extracellular matrix components. Capillaries, therefore, are more likely to be more responsive to angiogenic signals. Capillaries are also more capable of transporting circulatory elements to the organ environment, as they have direct contacts. Therefore, it is possible that capillary endothelial cells have developed the specific molecular machinery that allows them to perform the tasks discussed above. Alternatively, both capillary and large vessel endothelial cells may have the same molecular machinery, but the physical blockage surrounding the larger vessels may not permit those endothelial cells to perform such tasks. This is another area where very few molecular studies have been invested.

The concept of "vascular domains." As discussed above, the vascular system is highly selective for its response to the microenvironment. It is also clear that the vasculature is no longer considered a uniform network of vessels. There are many types of vessels that are molecularly distinct: arteries vs. veins, capillaries vs. larger vessels, vessels in one organ vs. vessels in another organ, right-hand-side vessels vs. left-hand-side vessels. Fur-

Figure 6. Regulation of the dominance of vascular types by VEGF and VEGF/Ang2. (*a*) CD31/PECAM-1 (*top*) and β-gal (*middle*) immunolabeling of each transgenic line, αMHC::VEGF (VEGF), αMHC::VEGF;αMHC::Ang2 (VEGF/Ang2), and wild-type control (WT) crossed to the arterial endothelial reporter line, ephrinB2$^{taulacZ/+}$. The merged images are shown at the bottom. (*b*) CD31/PECAM-1 (*top*) and β-gal (*bottom*) immunolabeling of each transgenic line, αMHC::VEGF (VEGF), αMHC::VEGF;aMHC::Ang2 (VEGF/Ang2) and wild-type (WT) crossed to the venous endothelial reporter line, ephB4$^{taulacZ/+}$. The merged images are shown at the bottom. The percentages of either ephrinB2$^+$ or ephB4$^+$ capillaries were calculated by comparing the number of CD31/PECAM-1-positive and the β-gal-immunopositive capillary lumens in the same section. Percentages of ephrinB2$^+$ (i.e., arterial-type) capillaries: WT (50±3.1%), VEGF (96±6.4%), VEGF/Ang2 (55±17). Percentages of ephB4$^+$ (i.e., venous-type) capillaries: WT (36±6.4%), VEGF (10±2.0%), VEGF/Ang2 (12±4%). Bar, 25 μm. (Adapted, with permission, from Visconti et al. 2002 [copyright PNAS].)

thermore, there are clearly distinct subregions within one type of vessels: a region that is permissive for growth, a region where new smooth muscle cells originate, a region that operates to exchange molecules between the circulation and the organ microenvironment, and so on. Therefore, the vascular system consists of numerous specialized domains. More extensive molecular studies characterizing the specialized structure and function of each of these "vascular domains" are required for a more complete understanding of the development and function of the vascular system and its contribution to the growth and physiology of the entire body.

CLINICAL IMPLICATIONS

Many human diseases involve pathologies of the vascular system: cancer, heart disease, atherosclerosis, stroke, Alzheimer disease, AIDS, diabetes, and aging (Folkman 1995). Certain pathological features of these diseases could be corrected by manipulating the vascular system. In cancer, it would be beneficial to suppress the growth of the tumor blood vessels to limit the growth of cancer cells by starvation. In heart failure or stroke, it would be beneficial to directly grow new blood vessels or to create an organ environment permissive for rapid growth of new vessels. The growth of new functional vessels may facilitate the recovery of a failed heart or brain.

However, we need to understand what types of vessels are most suited to facilitate the recovery of such failed organs. We also need to know which signals are most efficient in growing vessels in each specific organ microenvironment. We need to know the sensitivity of tumor vessels to each angiogenic pathway so that we can block the most sensitive and critical one. It would be most beneficial if we could target the specific vascular domain that is most critical for its growth, degeneration, or physiological functionality. Therefore, our better understanding of the molecular nature of the selectivity of vessels, and the vascular domains within individual vessels, will significantly contribute to our devising the most effective therapies.

ACKNOWLEDGMENTS

The work presented here has been funded by the National Institutes of Health.

REFERENCES

Davis S., Aldrich T.H., Jones P.F., Acheson A., Bruno J., Radjiewski C., Compton D., Maisonpierre P.C., and Yancopoulos G.D. 1996. Isolation of angiopoietin-1, a ligand for the TIE2 receptor, by secretion-trap expression cloning. *Cell* **87**: 1161.
Edwards W.D. 1998. Applied anatomy of the heart. In *Mayo Clinic practice of cardiology* (ed. E.R. Giuliani et al.), p. 422. Mosby, St. Louis, Missouri.
Ferrara N., Heinsohn H., Walter C.E., Buting S., and Thomas G.R. 1995. The regulation of blood vessel growth by vascular endothelial growth factor. *Ann. N.Y. Acad. Sci.* **752**: 246.
Folkman J. 1995. Angiogenesis in cancer, vascular, rheumatoid and other disease. *Nat. Med.* **1**: 27.
Hungerford J.E., Hoeffler J.P., Bowers C.W., Dahm L.M., Falchetto R., Shabanowtz J., Hunt D.F., and Little C.D. 1997. Identification of a novel marker for primordial smooth muscle and its differential expression pattern in contractile vs noncontractile cells. *J. Cell Biol.* **137**: 925.
Isner J.M. 2002. Myocardial gene therapy. *Nature* **415**: 234.
Langman J. 1990. Langman's medical embryology, 6th edition (ed. T.W. Sadler). Williams and Wilkins, Baltimore, Maryland.
Lawson N.D., Scheer N., Pham V.N., Kim C.H., Chitnis A.B., Campos-Ortega J.A., and Weinstein B.M. 2001. Notch signaling is required for arterial-venous differentiation during embryonic vascular development. *Development* **128**: 3675.
Liu Y., Wada R., Yamashita T., Mi Y., Deng C.X., Hobson J.P., Rosenfeldt H.M., Nava V.E., Chae S.S., Lee M.J., Liu C.H., Hla T., Spiegel S., and Proia R.L. 2000. Edg-1, the G protein-coupled receptor for sphingosine-1-phosphate, is essential for vascular maturation. *J. Clin. Invest.* **106**: 951.
Loughna S. and Sato T.N. 2001. A combinatorial role of angiopoietin-1 and orphan receptor TIE1 pathways in establishing vascular polarity during angiogenesis. *Mol. Cell* **7**: 233.
Maisonpierre P.C., Suri C., Jones P.F., Bartunkova S., Wiegand S.J., Radziejewski C., Compton D., McClain J., Aldrich T.H., Papadapoulos N., Daly T.J., Davis S., Sato T.N., and Yancopoulos G.D. 1997. Angiopoietin-2, a natural antagonist for Tie2 that disrupts in vivo angiogenesis. *Science* **277**: 55.
Partanen J., Armstrong E., Makela T.P., Korhonen J., Sandberg M., Renkonen R., Knuutila S., Huebner K., and Alitalo K. 1992. A novel endothelial cell surface receptor tyrosine kinase with extracellular epidermal growth factor homology domains. *Mol. Cell. Biol.* **12**: 1698.
Patterson M. 2001. Two-step regulation of asymmetry. *Nat. Rev. Genet.* **2**: 157.
Sato T.N. 2000. A new role of lipid receptors in vascular and cardiac morphogenesis. *J. Clin. Invest.* **106**: 939.
Sato T.N., Qin Y., Kozak C.A., and Audus K.L. 1993. tie-1 and tie-2 define another class of putative receptor tyrosine kinase genes expressed in early embryonic vascular system. *Proc. Natl. Acad. Sci.* **90**: 9355.
Suri C., Jones P.F., Patan S., Bartunkova S., Maisonpierre P.C., Davis S., Sato T.N., and Yancopoulos G.D. 1996. Requisite role of angiopoietin-1, a ligand for the TIE2 receptor, during embryonic angiogenesis. *Cell* **87**: 1171.
Suri C., McClain J., Thurston G., McDonald D.M., Zhou H., Oldmixon E.H., Sato T.N., and Yancopoulos G.D. 1998. Increased vascularization in mice overexpressing angiopoietin-1. *Science* **282**: 468.
Thurston G., Suri C., Smith K., McClain J., Sato T.N., Yancopoulos G.D., and McDonald D.M. 1999. Leakage-resistant blood vessels in mice transgenically overexpressing angiopoietin-1. *Science* **286**: 2511.
Valenzuela D.M., Griffiths J.A., Rojas J., Aldrich T.H., Jones P.F., Zhou H., McClain J., Copeland N.G., Gilbert D.J., Jenkins N.A., Huang T., Papadopoulos N., Maisonpierre P.C., Davis S., and Yancopoulos G.D. 1999. Angiopoietins 3 and 4: Diverging gene counterparts in mice and humans. *Proc. Natl. Acad. Sci.* **96**: 1904.
Visconti R.P., Richardson C.D., and Sato T.N. 2002. Orchestration of angiogenesis and arteriovenous contribution by angiopoietins and vascular endothelial growth factor (VEGF). *Proc. Natl. Acad. Sci.* **99**: 8219.
Yancopoulos G.D., Davis S., Gale N.W., Rudge J.S., Wiegand S.J., and Holash J. 2000. Vascular-specific growth factors and blood vessel formation. *Nature* **407**: 242.
Yost H.J. 1999. Diverse initiation in a conserved left-right pathway? *Curr. Opin. Genet. Dev.* **9**: 422.
———. 2001. Establishment of left-right asymmetry. *Int. Rev. Cytol.* **203**: 357.
Zhong T.P., Childs S., Leu J.P., and Fishman M.C. 2001. Gridlock signalling pathway fashions the first embryonic artery. *Nature* **414**: 216.

Survival Mechanisms of VEGF and PlGF during Microvascular Remodeling

J.N. Upalakalin,* I. Hemo,† C. Dehio,‡ E. Keshet,† and L.E. Benjamin*

*Department of Pathology, Beth Israel Deaconess Medical Center, Harvard Medical School, Boston, Massachusetts 02215; †Department of Molecular Biology, The Hebrew University-Hadassah Medical School, Jerusalem, Israel 91120; ‡Division of Molecular Microbiology, Biozentrum of the University of Basel, CH-4056 Basel, Switzerland

Angiogenesis, the formation of new blood vessels from preexisting vessels, is a complex and tightly regulated process involving endothelial cell migration, proliferation, and survival. The roles of endothelial cell migration and proliferation have been well established; however, the role of survival has only recently been recognized. Many mediators of vascular development have been identified and include growth factors, cytokines, membrane-bound molecules, and hemodynamic forces. They all play important roles during development, eliciting unique changes in a spatial and temporal manner. Several of them have also been shown to influence endothelial cell survival, and the signals they induce tend to overlap, thus assuring the survival of endothelial cells under a variety of adverse conditions. The importance of having a coordinated regulation between survival and apoptosis cannot be overemphasized. Survival selects desirable cells and maintains cell number, while apoptosis eliminates abnormal, misplaced, or nonfunctional cells; decreases cell number; and removes vestigial structures (Jacobson et al. 1997). Coordination of these processes contributes to the structural remodeling that occurs in every organ. It is particularly important for vascular development and constitutes a major aspect of angiogenesis. Disruption in either the survival signaling pathway, such as loss of the vascular endothelial growth factor gene (Carmeliet et al. 1996; Ferrara et al. 1996), or the programmed cell death pathway, such as loss of the caspase-3 gene (Kuida et al. 1996), is associated with increased morbidity and mortality. Moreover, unchecked vascular growth leads to cardiovascular defects and vascular neoplasms, whereas increased apoptosis causes embryonal death, hemorrhage, and vascular defects (Pietsch et al. 1997; Gerber et al. 1999; Fisher et al. 2000).

SURVIVAL SIGNALING DURING SPROUTING AND REMODELING

Normal blood vessel development begins with a sprouting and vessel formation stage, followed by remodeling and vessel maturation. Figure 1 illustrates some of the changes that occur during remodeling. Figure 1a typifies the early sprouting blood vessels of the embryo, neonatal retina. The larger blood vessels that represent arterioles (green) and venules (red) are connected by an unorganized microvasculature that has not yet clearly established an organized hierarchy of arterial to venous circulation (shown in blue). Active sprouting is illustrated in Figure 1a at the leading edge and is the principal method of initial plexus formation. After remodeling, this same vasculature resembles that shown in Figure 1b. The arterial and venous organization is completed so that oxygenated blood flows efficiently from arteriole (green) to venule (red) and the microvasculature has been trimmed, leaving behind only the minimal number of blood vessels needed to supply the surrounding tissue with an efficient blood flow and sufficient nutrients.

During sprouting, the endothelial cell must leave the vessel wall and invade the surrounding tissue. Many studies have shown that for successful sprouting to occur, there must be coordination of cytoskeletal changes for motility as well as secretion of proteolytic enzymes for matrix degradation (Folkman and D'Amore 1996; Risau 1997; Patan 2000). In this tenuous phase, specific integrins such as $\alpha v \beta 3$ play an important role in contacting the provisional matrix and providing survival signals (Varner and Cheresh 1996). In addition, we have found that drugs which interfere with small GTPases induce selective apoptosis in sprouting endothelia. This suggests that there may be stage-specific controls on GTPase functions. Once tube formation has transformed the sprout into a more functional blood vessel, a more substantial but still provisional basement membrane has been laid and a more elaborate set of integrin and growth factor signals maintain survival.

Not all of the initial blood vessels will survive to be included in the final vasculature. During the remodeling of

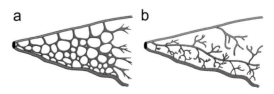

Figure 1. Schematic images of pre- and post-remodeling microvasculature. (*a*) The large blood vessels are shown with arterioles in green and venules in red. The blue unremodeled vascular web is transformed into the differentiated structures shown in *b*.

immature vessels into a mature vascular plexus, some endothelial cells survive, while others undergo programmed cell death or apoptosis. Similarly, in the developing microvasculature, survival signals determine whether a new sprout or a new vessel will become a permanent part of the vasculature. During this period, many studies have shown that soluble survival factors are critical, and, in particular, vascular endothelial growth factor A (VEGF-A) is critical during this stage of normal development (see below). This dependence is relieved by association with mural cells and the stabilization of the vessel wall (Darland and D'Amore 1999).

SURVIVAL MECHANISMS IN THE DEVELOPING MICROVASCULATURE

As a part of the remodeling program, the newly formed microvasculature is adjusted by supply and demand. The final size and density of the microvasculature are fine-tuned to appropriately provide nutrients based on the physiological needs of the surrounding tissue. One way the forming blood vessels are adjusted to appropriately match the needs of the surrounding tissue is through the use of a bifunctional protein whose expression in response to tissue ischemia stimulates proliferations and survival of new blood vessels (Neufeld et al. 1999). Originally identified as vascular permeability factor, VEGF-A is now recognized as one of the most potent regulators of angiogenesis. It is highly expressed in many cell types during development but is barely detectable in the normally quiescent tissues of the adult (Berse et al. 1992). In addition to stimulating endothelial cell proliferation and migration, it exerts a vascular protective effect through the promotion and maintenance of endothelial cell survival (Zachary 2001). When there is not enough oxygen or other nutrients to supply the surrounding tissue, the endothelial cell growth factor VEGF-A is up-regulated and results in endothelial cell proliferation and vessel assembly. When the vessels become patent and the hypoxic stimulus relieved, VEGF-A is down-regulated accordingly so that supply matches demand (Stone et al. 1995). The oxygen-sensing mechanisms triggered by tissue hypoxia stimulate VEGF-A mRNA stability, transcription, and translation, making this process exceedingly sensitive and efficient (Shweiki et al. 1992; Ikeda et al. 1995; Levy et al. 1995; Stein et al. 1995, 1998; Forsythe et al. 1996). In contrast, during the remodeling phase, a relatively plastic period in angiogenesis, an abundance of oxygen indicates an overabundant vasculature, and VEGF-A is concomitantly down-regulated. This causes vessel regression of entire vessel segments as a result of loss of VEGF-A survival factor function.

The critical nature of VEGF-A in this vascular regression has been extensively documented in experiments that model the disease retinopathy of prematurity (ROP), a blindness-inducing disease that begins with an excessive regression during an otherwise normal developmental remodeling (McLeod et al. 1996). Animal models for this disease utilize oxygen to induce this regression, much like the oxygen given to the premature infants afflicted (Smith et al. 1994; Alon et al. 1995; Simpson et al. 1999). The oxygen reduces the endogenous levels of VEGF-A, and loss of the survival function of this factor is enough to cause death of the existing blood vessels above and beyond the loss of its mitogenic function that arrests endothelial cell proliferation (Alon et al. 1995; Pierce et al. 1995; Robbins et al. 1998). Exposure of newborn rats to brief periods of hyperoxia decreases endogenous production of VEGF-A. Without adequate VEGF-A, the viability of endothelial cells can no longer be maintained. This results in endothelial cell apoptosis and vessel regression. Upon returning to normal oxygen concentrations, the undervascularized retina becomes hypoxic and stimulates up-regulation of VEGF-A and neovascularization. Studies using antibody and tyrosine kinase inhibitors to VEGF receptors have clearly shown that this neovascularization can be prevented by blocking VEGF-A signaling (Seo et al. 1999; Sone et al. 1999; Ozaki et al. 2000). Although this burst of VEGF-A increases endothelial cell proliferation and vessel assembly, the vessels that form are leaky, hemorrhagic, and disorganized. However, intraocular injection of VEGF-A prior to the onset of hyperoxia can prevent this injury (Alon et al. 1995). The exogenous VEGF-A that is administered compensates for the loss of endogenous VEGF-A and rescues the retinal vasculature. This demonstrates the critical role of VEGF-A in the survival of newly formed blood vessels.

We have extended this analysis to the VEGF-related factor placenta growth factor (PlGF). PlGF is found predominantly in the placenta and has been shown to increase the survival of trophoblasts in vitro. It activates the SAPK signaling pathway to protect trophoblasts from growth-factor-withdrawal-induced apoptosis. This suggests it may play an important role in the normal development and aging of placenta (Desai et al. 1999). It was found to be up-regulated in fetal growth retardation (Khaliq et al. 1996). Two isoforms are found in humans, PlGF-1 and PlGF-2. The single isoform found in mouse is homologous to PlGF-2. Both isoforms bind VEGF receptor 1 (VEGFR-1), but PlGF-2 also binds neuropilin-1 (Nrp-1). Gene-targeting studies have shown that both of these receptors, in addition to VEGF receptor 2 (VEGFR-2), are essential for vascular development (Carmeliet et al. 1996; Ferrara et al. 1996; Soker et al. 1998; Takashima et al. 2002). Recent studies have shown that PlGF-1 is a natural antagonist to VEGF-A (Eriksson et al. 2002). However, we and other workers have found that PlGF-2 can mimic or augment VEGF-A functions (Carmeliet et al. 2001; Adini et al. 2002). PlGF and its receptor VEGFR-1 are required for VEGF-A-driven pathological angiogenesis (Carmeliet et al. 2001; Luttun et al. 2002). Perhaps this is due in part to the ability of PlGF to mobilize hematopoietic stem cells (Hattori et al. 2002). In addition to forming homodimers, PlGF can also heterodimerize with VEGF-A. In the retinal survival assay shown in Figure 2, we observed that VEGF-A and PlGF-2 homodimers, and the VEGF/PlGF-2 heterodimer, can protect retinal blood vessels from regression in hyperoxia. In these studies, we found PlGF-2 homodimers to

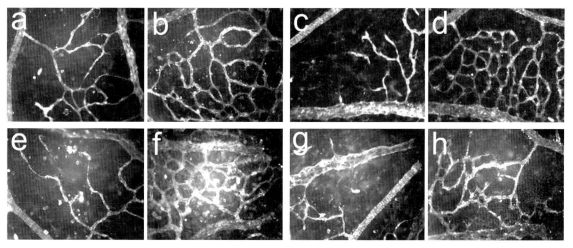

Figure 2. Survival assay in retinal blood vessels following hyperoxia with and without VEGF and PlGF growth factors. Panels *a*, *c*, *e*, and *g* show the uninjected eyes from the same animals as *b*, *d*, *f*, and *h*, respectively. Panel *b* shows rescue from oxygen-induced regression by VEGF/VEGF, (*d*) VEGF/PlGF, (*f*) PlGF/PlGF, and (*h*) VEGF-E. Note that although VEGF/PlGF rescues the vessels effectively, even in this assay capillaries appear enlarged. In contrast, PlGF/PlGF gives a vasculature that appears completely unaltered, although the oxygen-induced damage in the contralateral eye (*e*) is severe.

have optimal activity. In contrast, the VEGF-related protein, VEGF-E, which only binds VEGFR-2, is less effective at providing survival in this assay (Fig. 2 g,h).

Because of the receptor-binding profile of PlGF-2, and the null phenotype of the Nrp-deficient embryos, we hypothesize that NRP1 may be involved in survival signaling during remodeling. Support for this hypothesis can be found in the tumor literature, where VEGF-A has been shown to be an autocrine survival factor for breast carcinoma cells lacking VEGFR-1 and VEGFR-2, but expressing Nrp-1 (Bachelder et al. 2001). It appears to activate the phosphatidylinositide 3-OH kinase (PI3K) pathway, but whether it is working alone or in concert with other receptors to mediate VEGF-A signaling remains unknown.

PERICYTES RELIEVE THE IMMATURE VASCULATURE OF THE NEED FOR SOLUBLE VEGF

As remodeling proceeds, immature vessels acquire a pericyte/smooth muscle cell coverage. The endothelial cell–pericyte association stabilizes the vessels and makes them less sensitive to fluctuations in oxygen tension and VEGF levels. In addition, endothelial cells release platelet-derived growth factor-BB (PDGF-BB), which mediates pericyte function by stimulating pericyte coverage of the vessel (Hellstrom et al. 1999; Hirschi et al. 1999). This, in turn, enhances endothelial cell survival and decreases endothelial cell proliferation (Orlidge and D'Amore 1987; Benjamin et al. 1999; Reinmuth et al. 2001). Mice deficient in both pdgfB and its receptor pdgfβ receptor have defects in pericyte coverage that lead to hemorrhage and vascular dysfunction (Leveen et al. 1994; Lindahl et al. 1997; Hellstrom et al. 1999). Developmental studies of pericyte function in the retina revealed that ectopic PDGF-BB disrupted both pericyte–endothelial cell interactions and remodeling. The hypothesis that emerged from these studies is that proper pericyte–endothelial cell interactions are required for remodeling. Further studies of the remodeling process demonstrated that VEGF-A survival function is only required until the vessels have come in contact with mature pericytes (those expressing a smooth muscle actin, SMA) (Benjamin et al. 1999). Thus, association of pericytes marks the end of the plasticity window and is one hallmark of vessel maturation (Benjamin et al. 1998, 1999; Darland and D'Amore 1999). It is clear that pericyte/smooth muscle-covered blood vessels still retain the ability to enlarge, a process termed arteriogenesis, but these vessels are less sensitive to fluctuations in VEGF-A levels and ischemia in general for both proliferation and regression.

SURVIVAL MECHANISMS IN TUMOR ANGIOGENESIS

Tumor vessels resemble the immature blood vessels of development. They lack pericytes and are susceptible to survival signaling through angiogenic factors. VEGF-A has been implicated in tumor angiogenesis and is likely an important survival factor for tumor vessels as it is for normal vessels. Because of this dependence on angiogenesis, an important component of antitumor therapy is the use of agents that antagonize positive regulators of angiogenesis, such as VEGF, or promote negative regulators of angiogenesis, such as angiostatin, endostatin, and thrombospondin-1 (Guo et al. 1997; O'Reilly et al. 1997; Sim et al. 1997; Isner and Asahara 1999).

Many tumor cell lines secrete VEGF-A in vitro (Senger et al. 1986). In situ hybridization studies show VEGF-A mRNA is up-regulated in most human tumors, including lung, breast, gastrointestinal, kidney, bladder, ovarian, and endometrial carcinomas (Ferrara 1999). Expression of

VEGF-A is probably regulated by oxygen tension, similar to physiological conditions. Central regions of solid tumors are often hypoxic and may become necrotic due to inadequate oxygen (Papetti and Herman 2002). This hypoxia stimulates the cells surrounding these regions to increase VEGF-A expression, thus inducing the survival of endothelial cells as well as tumor angiogenesis. Most of these cells secreting VEGF-A are tumor cells, but stromal cells also appear to be important contributors (Fukumura et al. 1998). Loss of VEGF-A activity or VEGF-A receptor signaling has been shown to reduce tumor angiogenesis and suppress tumor growth in vivo (Kim et al. 1993; Millauer et al. 1996; Cao et al. 1998). Similarly, PlGF expression has been reported in hypervascular primary brain tumors and hypervascular renal cell carcinomas (Nomura et al. 1998; Donnini et al. 1999).

Experiments to test the hypothesis that VEGF-A and PlGF also provide survival functions to tumor blood vessels utilized tetracycline-regulated expression vectors to control VEGF-A and PlGF levels. After the tumor was well-established with high levels of either VEGF-A or PlGF factors, tetracycline administration to the host animal caused abrupt withdrawal of these factors and led to endothelial cell apoptosis and blood vessel regression (Benjamin and Keshet 1997; Adini et al. 2002). PlGF also provided survival to tumor macrophages in these studies. An example of these data is shown in Figure 3: Following PlGF withdrawal, TUNEL staining for apoptotic cells in combination with lectin staining for blood vessels and macrophages demonstrated that these cells were the ones dying after 36 hours. In the treatment of human prostate cancer, androgen-ablation therapy down-regulates VEGF-A production in normal, hyperplastic, and cancerous prostate glands. Similar to the forced regression in the xenograft models of tetracycline-controlled VEGF-A, this loss increases endothelial cell apoptosis. A clear correlation was made between the blood vessels that had apoptotic endothelium and those lacking pericytes (Benjamin et al. 1999). In tumor studies exploring the function of angiopoietin-2 (Ang2), it was reported that in the face of high Ang2 expression, which leads to pericyte detachment, tumor blood vessel survival depended on the level of exogenous VEGF-A produced in the surrounding tissue (Holash et al. 1999a, b).

VEGF-A AND PlGF SIGNALING MECHANISMS

The molecular nature of VEGF-A survival function and the signaling involved have been extensively investigated in tissue culture models. VEGF-A activates the phosphorylation of the survival kinase Akt1 (Mazure et al. 1997; Fujio and Walsh 1999; Jiang et al. 2000; Morales-Ruiz et al. 2000; Wu et al. 2000). This activation can be dependent on functional VE-cadherin (Carmeliet et al. 1999). Blocking PI3K or Akt signaling enhances apoptosis; on the other hand, constitutively activating Akt promotes survival (Gratton et al. 2001). VEGF-A up-regulates PI3K signaling to Akt1, which further up-regulates ICAM (Radisavljevic et al. 2000) and eNOS (Dimmeler et al. 1999; Fulton et al. 1999; Michell et al. 1999), alters cell shape and motility (Morales-Ruiz et al. 2000), and promotes survival in vitro. In a three-dimensional tube formation in vitro assay, Ilan et al. (1998) demonstrated that PMA-induced tube formation required PI3K signaling to Akt and MAPK activation, and that VEGF-A was able to protect the tubes when PMA was removed via the inhibition of apoptosis. VEGF-A was unable to activate tube formation or activate Akt prior to tube formation in 3D collagen, yet after tube formation, VEGF-A maintained survival and coordinately activated MAPK (Ilan et al. 1998). VEGF-A up-regulates expression of the caspase inhibitor survivin and the anti-apoptotic protein Bcl-2, and inhibits the pro-apoptotic protein Bad (Gerber et al. 1998; Nor et al. 1999; O'Connor et al. 2000). PlGF survival signaling has not been extensively explored; however, there is evidence that PlGF induces some of the same molecular responses as VEGF-A. For example, survivin gene expression in primary endothelial cell culture is up-regulated by PlGF (Benjamin and Keshet 1997; Adini et al. 2002). Since PlGF is a survival factor for both ECs and macrophages, it will be interesting to investigate the pathways involved in macrophage survival also.

CONCLUSIONS

The formation of the vascular system, like any organ, is a complex process requiring the spatial and temporal regulation of gene expression and cell–cell interactions. The controls of survival and apoptosis are central for the

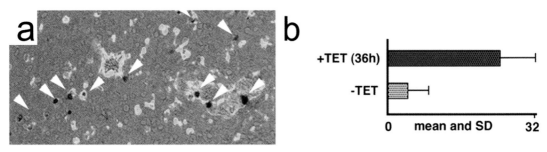

Figure 3. Double staining for TUNEL and blood vessels is shown. Analysis of apoptosis in tumors expressing high levels of *mPlGF* 36 hours after *mPlGF* withdrawal. (*a*) An overlay of fluorescent lectin staining and POD TUNEL staining of the same field shows the apoptotic cells in dark red and the vascular cells in yellow. (*b*) Quantitation of apoptosis is shown for tumors before (control) and after *mPlGF* withdrawal for 36 hours (means = 4.37 in –TET control, and 24.19 in the 36h *mPlGF* withdrawal) per high power field. *p* <0.0001. (Reprinted, with permission, from Adini et al. 2002 [copyright American Association for Cancer Research].).

proper morphogenesis and stability of this organ. Survival and apoptotic signaling pathways are highly redundant, reflecting their central importance for the growth and development of complex organisms (Dragovich et al. 1998). However, the molecular and cellular cues that stimulate these pathways in blood vessels are not only unique to this organ, but are also unique from stage to stage of vessel formation. This review has focused on the controls of survival during microvascular remodeling in normal and pathological settings. At this stage, soluble factors such as VEGF and PlGF, and cellular interactions between endothelial cells and pericytes, have been found to be critical. No doubt there are equally critical stimuli for earlier stages of blood vessel development, and mature blood vessels that have yet to be explored in detail.

REFERENCES

Adini A., Kornaga T., Firoozbakht F., and Benjamin L.E. 2002. Placental growth factor is a survival factor for tumor endothelial cells and macrophages. *Cancer Res.* **62:** 2749.

Alon T., Hemo I., Itin A., Pe'er J., Stone J., and Keshet E. 1995. Vascular endothelial growth factor acts as a survival factor for newly formed retinal vessels and has implications for retinopathy of prematurity. *Nat. Med.* **1:** 1024.

Bachelder R.E., Crago A., Chung J., Wendt M.A., Shaw L.M., Robinson G., and Mercurio A.M. 2001. Vascular endothelial growth factor is an autocrine survival factor for neuropilin-expressing breast carcinoma cells. *Cancer Res.* **61:** 5736.

Benjamin L.E., and Keshet E. 1997. Conditional switching of vascular endothelial growth factor (VEGF) expression in tumors: Induction of endothelial cell shedding and regression of hemangioblastoma-like vessels by VEGF withdrawal. *Proc. Natl. Acad. Sci.* **94:** 8761.

Benjamin L.E., Hemo I., and Keshet E. 1998. A plasticity window for blood vessel remodelling is defined by pericyte coverage of the preformed endothelial network and is regulated by PDGF- B and VEGF. *Development* **125:** 1591.

Benjamin L.E., Golijanin D., Itin A., Pode D., and Keshet E. 1999. Selective ablation of immature blood vessels in established human tumors follows vascular endothelial growth factor withdrawal (comments). *J. Clin. Invest.* **103:** 159.

Berse B., Brown L.F., Van De Water L., Dvorak H.F., and Senger D.R. 1992. Vascular permeability factor (vascular endothelial growth factor) gene is expressed differentially in normal tissues, macrophages, and tumors. *Mol. Biol. Cell* **3:** 211.

Cao Y., Linden P., Farnebo J., Cao R., Eriksson A., Kumar V., Qi J.H., Claesson-Welsh L., and Alitalo K. 1998. Vascular endothelial growth factor C induces angiogenesis in vivo. *Proc. Natl. Acad. Sci.* **95:** 14389.

Carmeliet P., Ferreira V., Breier G., Pollefeyt S., Kieckens L., Gertsenstein M., Fahrig M., Vandenhoeck A., Harpal K., Eberhardt C., Declercq C., Pawling J., Moons L., Collen D., Risau W., and Nagy A. 1996. Abnormal blood vessel development and lethality in embryos lacking a single VEGF allele. *Nature* **380:** 435.

Carmeliet P., Lampugnani M.G., Moons L., Breviario F., Compernolle V., Bono F., Balconi G., Spagnuolo R., Oostuyse B., Dewerchin M., Zanetti A., Angellilo A., Mattot V., Nuyens D., Lutgens E., Clotman F., de Ruiter M.C., Gittenberger-de Groot A., Poelmann R., Lupu F., Herbert J.M., Collen D., and Dejana E. 1999. Targeted deficiency or cytosolic truncation of the VE-cadherin gene in mice impairs VEGF-mediated endothelial survival and angiogenesis. *Cell* **98:** 147.

Carmeliet P., Moons L., Luttun A., Vincenti V., Compernolle V., De Mol M., Wu Y., Bono F., Devy L., Beck H., Scholz D., Acker T., DiPalma T., Dewerchin M., Noel A., Stalmans I., Barra A., Blacher S., Vandendriessche T., Ponten A., Eriksson U., Plate K.H., Foidart J.M., Schaper W., Charnock-Jones D.S., Hicklin D.J., Herbert J.M., Collen D., and Persico M.G. 2001. Synergism between vascular endothelial growth factor and placental growth factor contributes to angiogenesis and plasma extravasation in pathological conditions. *Nat. Med.* **7:** 575.

Darland D.C., and D'Amore P.A. 1999. Blood vessel maturation: Vascular development comes of age (comment). *J. Clin. Invest.* **103:** 157.

Desai J., Holt-Shore V., Torry R.J., Caudle M.R., and Torry D.S. 1999. Signal transduction and biological function of placenta growth factor in primary human trophoblast. *Biol. Reprod.* **60:** 887.

Dimmeler S., Fleming I., Fisslthaler B., Hermann C., Busse R., and Zeiher A.M. 1999. Activation of nitric oxide synthase in endothelial cells by Akt-dependent phosphorylation. *Nature* **399:** 601.

Donnini S., Machein M.R., Plate K.H., and Weich H.A. 1999. Expression and localization of placenta growth factor and PlGF receptors in human meningiomas. *J. Pathol.* **189:** 66.

Dragovich T., Rudin C.M., and Thompson C.B. 1998. Signal transduction pathways that regulate cell survival and cell death. *Oncogene* **17:** 3207.

Eriksson A., Cao R., Pawliuk R., Berg S.-M., Tsang M.L., Zhou D., Fleet C., Tritsaris K., SDissing S., Leboulch P., and Cao Y. 2002. Placenta growth factor-1 antagonizes VEGF-induced angiogenesis and tumor growth by the formation of functionally inactive PlGF-1/VEGF heterodimers. *Cancer Cell* **1:** 99.

Ferrara N. 1999. Molecular and biological properties of vascular endothelial growth factor. *J. Mol. Med.* **77:** 527.

Ferrara N., Carver-Moore K., Chen H., Dowd M., Lu L., O'Shea K.S., Powell-Braxton L., Hillan K.J., and Moore M.W. 1996. Heterozygous embryonic lethality induced by targeted inactivation of the VEGF gene. *Nature* **380:** 439.

Fisher S.A., Langille B.L., and Srivastava D. 2000. Apoptosis during cardiovascular development. *Circ. Res.* **87:** 856.

Folkman J. and D'Amore P.A. 1996. Blood vessel formation: What is its molecular basis? (comment). *Cell* **87:** 1153.

Forsythe J.A., Jiang B.H., Iyer N.V., Agani F., Leung S.W., Koos R.D., and Semenza G.L. 1996. Activation of vascular endothelial growth factor gene transcription by hypoxia-inducible factor 1. *Mol. Cell. Biol.* **16:** 4604.

Fujio Y. and Walsh K. 1999. Akt mediates cytoprotection of endothelial cells by vascular endothelial growth factor in an anchorage-dependent manner. *J. Biol. Chem.* **274:** 16349.

Fukumura D., Xavier R., Sugiura T., Chen Y., Park E.C., Lu N., Selig M., Nielsen G., Taksir T., Jain R.K., and Seed B. 1998. Tumor induction of VEGF promoter activity in stromal cells. *Cell* **94:** 715.

Fulton D., Gratton J.P., McCabe T.J., Fontana J., Fujio Y., Walsh K., Franke T.F., Papapetropoulos A., and Sessa W.C. 1999. Regulation of endothelium-derived nitric oxide production by the protein kinase Akt (erratum in *Nature* [1999] **400:** 792). *Nature* **399:** 597.

Gerber H.P., Dixit V., and Ferrara N. 1998. Vascular endothelial growth factor induces expression of the antiapoptotic proteins Bcl-2 and A1 in vascular endothelial cells. *J. Biol. Chem.* **273:** 13313.

Gerber H.P., Hillan K.J., Ryan A.M., Kowalski J., Keller G.A., Rangell L., Wright B.D., Radtke F., Aguet M., and Ferrara N. 1999. VEGF is required for growth and survival in neonatal mice. *Development* **126:** 1149.

Gratton J.P., Morales-Ruiz M., Kureishi Y., Fulton D., Walsh K., and Sessa W.C. 2001. Akt down-regulation of p38 signaling provides a novel mechanism of vascular endothelial growth factor-mediated cytoprotection in endothelial cells. *J. Biol. Chem.* **276:** 30359.

Guo N., Krutzsch H.C., Inman J.K., and Roberts D.D. 1997. Thrombospondin 1 and type I repeat peptides of thrombospondin 1 specifically induce apoptosis of endothelial cells. *Cancer Res.* **57:** 1735.

Hattori K., Heissig B., Wu Y., Dias S., Tejada R., Ferris B., Hicklin D.J., Zhu Z., Bohlen P., Witte L., Hendrikx J., Hackett N.R., Crystal R.G., Moore M.A., Werb Z., Lyden D., and

Rafii S. 2002. Placental growth factor reconstitutes hematopoiesis by recruiting VEGFR1(+) stem cells from bone-marrow microenvironment. *Nat. Med.* **1:** 1.

Hellstrom M., Kal n M., Lindahl P., Abramsson A., and Betsholtz C. 1999. Role of PDGF-B and PDGFR-beta in recruitment of vascular smooth muscle cells and pericytes during embryonic blood vessel formation in the mouse. *Development* **126:** 3047.

Hirschi K.K., Rohovsky S.A., Beck L.H., Smith S.R., and D'Amore P.A. 1999. Endothelial cells modulate the proliferation of mural cell precursors via platelet-derived growth factor-BB and heterotypic cell contact. *Circ. Res.* **84:** 298.

Holash J., Wiegand S.J., and Yancopoulos G.D. 1999a. New model of tumor angiogenesis: Dynamic balance between vessel regression and growth mediated by angiopoietins and VEGF. *Oncogene* **18:** 5356.

Holash J., Maisonpierre P.C., Compton D., Boland P., Alexander C.R., Zagzag D., Yancopoulos G.D., and Wiegand S.J. 1999b. Vessel cooption, regression, and growth in tumors mediated by angiopoietins and VEGF. *Science* **284:** 1994.

Ikeda E., Achen M.G., Breier G., and Risau W. 1995. Hypoxia-induced transcriptional activation and increased mRNA stability of vascular endothelial growth factor in C6 glioma cells. *J. Biol. Chem.* **270:** 19761.

Ilan N., Mahooti S., and Madri J.A. 1998. Distinct signal transduction pathways are utilized during the tube formation and survival phases of in vitro angiogenesis. *J. Cell. Sci.* **111:** 3621.

Isner J.M. and Asahara T. 1999. Angiogenesis and vasculogenesis as therapeutic strategies for postnatal neovascularization. *J. Clin. Invest.* **103:** 1231.

Jacobson M.D., Weil M., and Raff M.C. 1997. Programmed cell death in animal development. *Cell* **88:** 347.

Jiang B.H., Zheng J.Z., Aoki M., and Vogt P.K. 2000. Phosphatidylinositol 3-kinase signaling mediates angiogenesis and expression of vascular endothelial growth factor in endothelial cells. *Proc. Natl. Acad. Sci.* **97:** 1749.

Khaliq A., Li X.F., Shams M., Sisi P., Acevedo C.A., Whittle M.J., Weich H., and Ahmed A. 1996. Localisation of placenta growth factor (PlGF) in human term placenta. *Growth Factors* **13:** 243.

Kim K.J., Li B., Winer J., Armanini M., Gillett N., Phillips H.S., and Ferrara N. 1993. Inhibition of vascular endothelial growth factor-induced angiogenesis suppresses tumour growth in vivo. *Nature* **362:** 841.

Kuida K., Zheng T.S., Na S., Kuan C., Yang D., Karasuyama H., Rakic P., and Flavell R.A. 1996. Decreased apoptosis in the brain and premature lethality in CPP32-deficient mice. *Nature* **384:** 368.

Leveen P., Pekny M., Gebre Medhin S., Swolin B., Larsson E., and Betsholtz C. 1994. Mice deficient for PDGF B show renal, cardiovascular, and hematological abnormalities. *Genes Dev.* **8:** 1875.

Levy A.P., Levy N.S., Wegner S., and Goldberg M.A. 1995. Transcriptional regulation of the rat vascular endothelial growth factor gene by hypoxia. *J. Biol. Chem.* **270:** 13333.

Lindahl P., Johansson B.R., Leveen P., and Betsholtz C. 1997. Pericyte loss and microaneurysm formation in PDGF-B-deficient mice. *Science* **277:** 242.

Luttun A., Tjwa M., Moons L., Wu Y., Angelillo-Scherrer A., Liao F., Nagy J.A., Hooper A., Priller J., De Klerck B., Compernolle V., Daci E., Bohlen P., Dewerchin M., Herbert J.M., Fava R., Matthys P., Carmeliet G., Collen D., Dvorak H.F., Hicklin D.J., and Carmeliet P. 2002. Revascularization of ischemic tissues by PlGF treatment, and inhibition of tumor angiogenesis, arthritis and atherosclerosis by anti-Flt1. *Nat. Med.* **8:** 831.

Mazure N.M., Chen E.Y., Laderoute K.R., and Giaccia A.J. 1997. Induction of vascular endothelial growth factor by hypoxia is modulated by a phosphatidylinositol 3-kinase/Akt signaling pathway in Ha-ras-transformed cells through a hypoxia inducible factor-1 transcriptional element. *Blood* **90:** 3322.

McLeod D.S., Brownstein R., and Lutty G.A. 1996. Vaso-obliteration in the canine model of oxygen-induced retinopathy. *Invest. Ophthalmol. Vis. Sci.* **37:** 300.

Michell B.J., Griffiths J.E., Mitchelhill K.I., Rodriguez-Crespo I., Tiganis T., Bozinovski S., de Montellano P.R., Kemp B.E., and Pearson R.B. 1999. The Akt kinase signals directly to endothelial nitric oxide synthase. *Curr. Biol.* **9:** 845.

Millauer B., Longhi M.P., Plate K.H., Shawver L.K., Risau W., Ullrich A., and Strawn L.M. 1996. Dominant-negative inhibition of Flk-1 suppresses the growth of many tumor types in vivo. *Cancer Res.* **56:** 1615.

Morales-Ruiz M., Fulton D., Sowa G., Languino L.R., Fujio Y., Walsh K., and Sessa W.C. 2000. Vascular endothelial growth factor-stimulated actin reorganization and migration of endothelial cells is regulated via the serine/threonine kinase Akt. *Circ. Res.* **86:** 892.

Neufeld G., Cohen T., Gengrinovitch S., and Poltorak Z. 1999. Vascular endothelial growth factor (VEGF) and its receptors. *FASEB J.* **13:** 9.

Nomura M., Yamagishi S., Harada S., Yamashima T., Yamashita J., and Yamamoto H. 1998. Placenta growth factor (PlGF) mRNA expression in brain tumors. *J. Neurooncol.* **40:** 123.

Nor J.E., Christensen J., Mooney D.J., and Polverini P.J. 1999. Vascular endothelial growth factor (VEGF)-mediated angiogenesis is associated with enhanced endothelial cell survival and induction of Bcl-2 expression. *Am. J. Pathol.* **154:** 375.

O'Connor D.S., Schechner J.S., Adida C., Mesri M., Rothermel A.L., Li F., Nath A.K., Pober J.S., and Altieri D.C. 2000. Control of apoptosis during angiogenesis by survivin expression in endothelial cells. *Am. J. Pathol.* **156:** 393.

O'Reilly M.S., Boehm T., Shing Y., Fukai N., Vasios G., Lane W.S., Flynn E., Birkhead J.R., Olsen B.R., and Folkman J. 1997. Endostatin: An endogenous inhibitor of angiogenesis and tumor growth. *Cell* **88:** 277.

Orlidge A. and D'Amore P.A. 1987. Inhibition of capillary endothelial cell growth by pericytes and smooth muscle cells. *J. Cell Biol.* **105:** 1455.

Ozaki H., Seo M.S., Ozaki K., Yamada H., Yamada E., Okamoto N., Hofmann F., Wood J.M., and Campochiaro P.A. 2000. Blockade of vascular endothelial cell growth factor receptor signaling is sufficient to completely prevent retinal neovascularization. *Am. J. Pathol.* **156:** 697.

Papetti M. and Herman I.M. 2002. Mechanisms of normal and tumor-derived angiogenesis. *Am. J. Physiol. Cell Physiol.* **282:** C947.

Patan S. 2000. Vasculogenesis and angiogenesis as mechanisms of vascular network formation, growth and remodeling. *J. Neurooncol.* **50:** 1.

Pierce E.A., Avery R.L., Foley E.D., Aiello L.P., and Smith L.E. 1995. Vascular endothelial growth factor/vascular permeability factor expression in a mouse model of retinal neovascularization. *Proc. Natl. Acad. Sci.* **92:** 905.

Pietsch T., Valter M.M., Wolf H.K., von Deimling A., Huang H.J., Cavenee W.K., and Wiestler O.D. 1997. Expression and distribution of vascular endothelial growth factor protein in human brain tumors. *Acta. Neuropathol.* **93:** 109.

Radisavljevic Z., Avraham H., and Avraham S. 2000. Vascular endothelial growth factor regulates ICAM-1 expression via phosophatidylinositol 3 OH kinase-/AKT/Nitric oxide pathway and modulates migration of brain microvascular endothelial cells. *J. Biol. Chem.* **275:** 20770.

Reinmuth N., Liu W., Jung Y.D., Ahmad S.A., Shaheen R.M., Fan F., Bucana C.D., McMahon G., Gallick G.E., and Ellis L.M. 2001. Induction of VEGF in perivascular cells defines a potential paracrine mechanism for endothelial cell survival. *FASEB J.* **15:** 1239.

Risau W. 1997. Mechanisms of angiogenesis. *Nature* **386:** 671.

Robbins S.G., Rajaratnam V.S., and Penn J.S. 1998. Evidence for upregulation and redistribution of vascular endothelial growth factor (VEGF) receptors flt-1 and flk-1 in the oxygen-injured rat retina. *Growth Factors* **16:** 1.

Senger D.R., Perruzzi C.A., Feder J., and Dvorak H.F. 1986. A highly conserved vascular permeability factor secreted by a variety of human and rodent tumor cell lines. *Cancer Res.* **46:** 5629.

Seo M.S., Kwak N., Ozaki H., Yamada H., Okamoto N., Yamada E., Fabbro D., Hofmann F., Wood J.M., and Campochiaro P.A. 1999. Dramatic inhibition of retinal and choroidal neovascularization by oral administration of a kinase inhibitor. *Am. J. Pathol.* **154:** 1743.

Shweiki D., Itin A., Soffer D., and Keshet E. 1992. Vascular endothelial growth factor induced by hypoxia may mediate hypoxia-initiated angiogenesis. *Nature* **359:** 843.

Sim B.K., O'Reilly M.S., Liang H., Fortier A.H., He W., Madsen J.W., Lapcevich R., and Nacy C.A. 1997. A recombinant human angiostatin protein inhibits experimental primary and metastatic cancer. *Cancer Res.* **57:** 1329.

Simpson D.A., Murphy G.M., Bhaduri T., Gardiner T.A., Archer D.B., and Stitt A.W. 1999. Expression of the VEGF gene family during retinal vaso-obliteration and hypoxia. *Biochem. Biophys. Res. Commun.* **262:** 333.

Smith L.E., Wesolowski E., McLellan A., Kostyk S.K., D'Amato R., Sullivan R., and D'Amore P.A. 1994. Oxygen-induced retinopathy in the mouse. *Invest. Ophthalmol. Vis. Sci.* **35:** 101.

Soker S., Takashima S., Miao H.Q., Neufeld G., and Klagsbrun M. 1998. Neuropilin-1 is expressed by endothelial and tumor cells as an isoform-specific receptor for vascular endothelial growth factor. *Cell* **92:** 735.

Sone H., Kawakami Y., Segawa T., Okuda Y., Sekine Y., Honmura S., Suzuki H., Yamashita K., and Yamada N. 1999. Effects of intraocular or systemic administration of neutralizing antibody against vascular endothelial growth factor on the murine experimental model of retinopathy. *Life Sci.* **65:** 2573.

Stein I., Neeman M., Shweiki D., Itin A., and Keshet E. 1995. Stabilization of vascular endothelial growth factor mRNA by hypoxia and hypoglycemia and coregulation with other ischemia-induced genes. *Mol. Cell. Biol.* **15:** 5363.

Stein I., Itin A., Einat P., Skaliter R., Grossman Z., and Keshet E. 1998. Translation of vascular endothelial growth factor mRNA by internal ribosome entry: Implications for translation under hypoxia. *Mol. Cell. Biol.* **18:** 3112.

Stone J., Itin A., Alon T., Pe'er J., Gnessin H., Chan-Ling T., and Keshet E. 1995. Development of retinal vasculature is mediated by hypoxia-induced vascular endothelial growth factor (VEGF) expression by neuroglia. *J. Neurosci.* **15:** 4738.

Takashima S., Kitakaze M., Asakura M., Asanuma H., Sanada S., Tashiro F., Niwa H., Miyazaki Ji J., Hirota S., Kitamura Y., Kitsukawa T., Fujisawa H., Klagsbrun M., and Hori M. 2002. Targeting of both mouse neuropilin-1 and neuropilin-2 genes severely impairs developmental yolk sac and embryonic angiogenesis. *Proc. Natl. Acad. Sci.* **99:** 3657.

Varner J.A. and Cheresh D.A. 1996. Tumor angiogenesis and the role of vascular cell integrin alphavbeta3. *Important Adv. Oncol.* **1996:** 69.

Wu L.W., Mayo L.D., Dunbar J.D., Kessler K.M., Baerwald M.R., Jaffe E.A., Wang D., Warren R.S., and Donner D.B. 2000. Utilization of distinct signaling pathways by receptors for vascular endothelial cell growth factor and other mitogens in the induction of endothelial cell proliferation. *J. Biol. Chem.* **275:** 5096.

Zachary I. 2001. Signaling mechanisms mediating vascular protective actions of vascular endothelial growth factor. *Am. J. Physiol. Cell. Physiol.* **280:** C1375.

# Molecular Mechanisms of Lymphangiogenesis

T. MÄKINEN AND K. ALITALO

Molecular/Cancer Biology Laboratory and Ludwig Institute for Cancer Research, Haartman Institute and Helsinki University Hospital, University of Helsinki, 00014 Helsinki, Finland

Blood and lymphatic vessels together form a circulatory system, which allows the transportation of metabolic substances, cells, and proteins in the body. A major role of the lymphatic vasculature is to return an excess of the protein-rich interstitial fluid to the blood circulation. In addition, the lymphatic vasculature is an important part of the immune system, as it filters lymph and its antigens through the lymph nodes. Lymphatic vessels also serve as one of the major routes for absorption of lipids from the gut (Witte et al. 2001). Until recently, studies of the lymphatic vessels were hampered due to the lack of specific markers, but recently several such markers have been identified. In addition, recent studies have indicated an important role for the lymphatic vessels in certain developmental disorders, such as lymphedema, and as a route for the metastasis of malignant tumors. These findings have brought lymphatic vascular biology to the forefront of cardiovascular research.

Two members of the vascular endothelial growth factor (VEGF) family of growth factors, VEGF-C and VEGF-D, have been shown to stimulate the growth of lymphatic vessels via the lymphatic endothelial cell receptor, VEGFR-3 (Jeltsch et al. 1997; Veikkola et al. 2001). Although VEGFR-3 is required for the development of the blood vascular system during early embryonic development, later on its expression becomes restricted to lymphatic endothelial cells and it thus serves as a marker of these cells in adult tissues (Kaipainen et al. 1995; Dumont et al. 1998). Recent studies demonstrate that VEGFR-3 and its ligands are involved in the development of lymphedema and lymphatic metastasis (for review, see Karkkainen et al. 2001a; Karpanen and Alitalo 2001).

Studies in our laboratory have been targeted to the molecular mechanisms regulating the development and growth of the lymphatic vessels, concentrating on the role of the VEGFR-3 signal transduction pathway in these processes. We have shown that VEGFR-3 mediates signals for the proliferation, survival, and migration of lymphatic endothelial cells and demonstrated that VEGFR-3 function is required for the normal development of the lymphatic vasculature (Mäkinen et al. 2001a,b). Blocking of VEGFR-3 signaling inhibited specifically fetal lymphangiogenesis without affecting blood vessel development, indicating that distinct molecular mechanisms control the development of the two vascular systems. To find the basis for such a specificity in signaling, we have now extended our studies to the characterization of the genetic programs that determine the identities of blood vascular and lymphatic endothelial cells. Our studies give new insights into the phenotypic diversity of endothelial cells and reveal new potential vascular markers, some of which could provide important targets for the treatment of diseases characterized by abnormal angiogenesis or lymphangiogenesis.

DEVELOPMENT OF THE LYMPHATIC VASCULATURE

During embryogenesis, the cardiovascular system is the first functional organ system, and its development and growth are essential for the growth of the embryo. The endothelial cells lining the blood vessels differentiate from mesodermal precursor cells, angioblasts, to form a primitive vascular network in the process called vasculogenesis (Risau and Flamme 1995; Risau 1997). The primitive plexus of homogeneously sized endothelial tubes is then extensively remodeled into a more mature vessel network consisting of a hierarchy of larger and smaller blood vessels (Fig. 1). This remodeling occurs both by regression of some vessels and by angiogenesis, which consists of the sprouting, splitting, or fusion of preexisting vessels to adjust the vascular density according to the metabolic requirements of the respective tissue or organ (Risau 1997; Carmeliet 2000). Already before the onset of the blood circulation, the endothelial tubes are specified as two parallel but distinct vascular networks, the arteries and the veins. The normal sprouting and branching of the maturing vasculature is accompanied by repulsive signals that prevent arteries and veins from fusing together. Soon after the formation of the vessels, endothelial tubes become surrounded by pericytes or vascular smooth muscle cells (Fig. 1). Pericytes and endothelial cells share a common basement membrane (BM) containing fibronectin, laminin, collagens, and heparan sulfate proteoglycans. The composition of the vascular BM varies depending on the type of the vessel and on the tissue environment. Failure to form a proper BM can lead to decreased pericyte adhesion and migration and to defects in vessel stability, indicating an important role for the BM in normal vascular development (Beltramo et al. 2002; Thyboll et al. 2002).

The lymphatic vessels arise after the establishment of the blood circulatory system. In the early 20th century, Florence Sabin described the growth of lymphatic vessels in the pig fetus as centrifugal sprouting from the lymph sacs which are formed in the vicinity of certain fetal veins (Sabin 1902,1909). Recent molecular biological data support Sabin's theory of the venous origin of lymphatic ves-

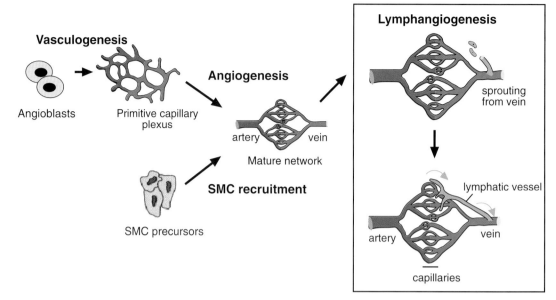

Figure 1. Development of the embryonic blood and lymphatic vessels. During vasculogenesis the endothelial precursor cells, angioblasts, form a primitive capillary plexus which is then further remodeled via angiogenesis. Vessel stabilization requires the recruitment of smooth muscle cells (SMC). Lymphatic vessels arise by sprouting from the veins, or they differentiate from mesodermal precursor cells. (Adapted, with permission, from Alitalo and Carmeliet 2002 [copyright Elsevier Science].)

sels (Dumont et al. 1998; Wigle and Oliver 1999; Wigle et al. 2002). A homeobox transcription factor Prox-1 has been found to have an important role in the formation of the lymph sacs by endothelial cell sprouting from the veins. At the stage when lymphatic development starts, Prox-1 is expressed in a subpopulation of endothelial cells of the anterior cardinal vein. These cells then migrate out and give rise to the lymphatic endothelial cells (Wigle and Oliver 1999). Targeted inactivation of Prox-1 leads to the arrest of the lymphatic endothelial budding process, whereas blood vessel development is not affected (Wigle and Oliver 1999).

Another old theory postulates that the lymphatic endothelial cells are derived from mesenchymal precursor cells (Huntington and McClure 1908; Kampmeier 1912). This theory is supported by the recent finding of these precursors, lymphangioblasts, in avian embryonic mesenchyme (Schneider et al. 1999; Papoutsi et al. 2001). Therefore, at least during embryogenesis, both the sprouting lymphangiogenesis and the mesenchymal cell differentiation into lymphatic endothelial cells may contribute to the formation of the lymphatic vessels (Fig. 1).

MOLECULAR REGULATION OF LYMPHANGIOGENESIS

VEGF, acting mainly via the VEGFR-1 and VEGFR-2 receptor tyrosine kinase receptors, is one of the most important regulators of both physiological and pathological angiogenesis (for review, see Ferrara 1999). In contrast, two other members of the VEGF family, VEGF-C and VEGF-D, have been shown to induce the growth of lymphatic vessels via the lymphatic endothelial cell receptor VEGFR-3 (Jeltsch et al. 1997; Veikkola et al. 2001). VEGF-C and VEGF-D are expressed as prepro-proteins which are proteolytically processed into polypeptides with increasing affinity toward VEGFR-3, and only their fully processed forms can bind to and activate VEGFR-2 (Joukov et al. 1997; Stacker et al. 1999). During embryonic development, VEGF-C is expressed in the mesenchyme surrounding the developing lymphatic vessels (Kukk et al. 1996). This suggested its possible role in the regulation of lymphangiogenesis. Subsequently, both VEGF-C and VEGF-D were shown to induce lymphatic growth in vivo (Jeltsch et al. 1997; Oh et al. 1997; Veikkola et al. 2001) and to induce the proliferation and migration of (lymphatic) endothelial cells in vitro (Marconcini et al. 1999; Mäkinen et al. 2001b). At higher concentrations and under specific conditions, these factors can also stimulate the proliferation and migration of blood vascular endothelial cells and increase vascular permeability and angiogenesis (Joukov et al. 1997; Oh et al. 1997; Witzenbichler et al. 1998; Mäkinen et al. 2001b). The VEGF-C/D-induced effects on blood vessel endothelium are presumably mediated via VEGFR-2 activation, suggesting an important role for specific proteolytic enzymes in regulating the lymphangiogenic versus angiogenic potentials of VEGF-C and VEGF-D.

VEGFR-3 is first expressed in developing blood vessels and is required for normal vascular development before midgestation. In the absence of VEGFR-3, the embryos undergo normal vasculogenesis, but the remodeling and maturation of primary vascular networks is defective, which leads to death in utero around E10 (Dumont et al. 1998). However, after the formation of the lymphatic vessels, VEGFR-3 is detected almost exclu-

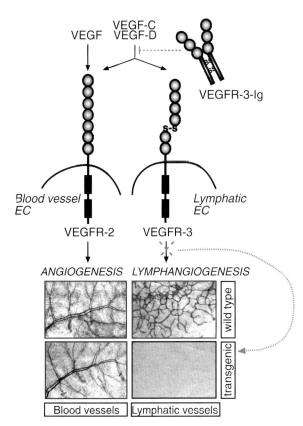

Figure 2. Inhibition of VEGFR-3 signaling by the transgenic expression of a soluble VEGFR-3-immunoglobulin (Ig)-fusion protein. Visualization of dermal blood and lymphatic vessels in wild-type and transgenic mice. Inhibition of VEGFR-3 signaling leads to a specific inhibition of lymphangiogenesis.

sively in the lymphatic endothelium (Kaipainen et al. 1995; Dumont et al. 1998). Because the $VEGFR-3^{-/-}$ embryos die before the emergence of the lymphatic vessels, these mice did not provide any information on the role of VEGFR-3 in the development of lymphatic vessels (Dumont et al. 1998). To address this question, we used a soluble VEGFR-3 as a competitive inhibitor of the VEGF-C/D - VEGFR-3 signaling (Fig. 2). Mice expressing a soluble VEGFR-3 immunoglobulin-fusion protein in the skin keratinocytes were found to be devoid of the dermal lymphatic vessels whereas the blood vasculature remained normal, showing a strikingly specific effect on the lymphatic endothelium (Mäkinen et al. 2001a). The early development of the lymphatic vasculature was not disturbed in the transgenic mice. However, in parallel with a significant increase in transgene-encoded protein expression, the dermal lymphatic vessels regressed in the transgenic embryos via apoptosis of lymphatic endothelial cells. Therefore, our results suggest that continuous VEGFR-3 signaling is required not only for the proliferation, but also for the survival, of the lymphatic endothelial cells and thus for both the development and maintenance of the lymphatic vasculature during embryonic development.

Neutralizing levels of circulating soluble VEGFR-3-Ig protein were also associated with a transient loss of lymphatic vessels in several internal organs of the neonatal transgenic mice. However, after three weeks of age, partial regeneration of the lymphatic vessels was observed, suggesting that alternative pathways may be involved in mediating the growth of the lymphatic vessels. For example, the maturing extracellular matrix may provide signals for the survival and proliferation of lymphatic endothelial cells, and new lymphatic vessels may grow alongside the existing blood vessels while receiving growth signals from them. Infection of newborn mouse pups with a VEGFR-3-Ig-encoding adenovirus resulted also in the regression of the lymphatic vessels, but the vessels regenerated in a few weeks despite the high serum levels of the VEGFR-3-Ig protein (T. Karpanen et al., unpubl.). However, in adult mice AdVEGFR-3-Ig failed to induce the regression of normal lymphatic vessels (Karpanen et al. 2001 and unpubl.; He et al. 2002). These results suggest that the lymphatic vessels undergo some type of maturation process, after which their growth and survival may no longer be dependent on VEGF-C/D. Pericyte recruitment is known to act as a signal for blood vessel maturation, after which the endothelial cells can survive in the absence of VEGF (Benjamin et al. 1998).

Recent studies indicate a significant interplay between the angiopoietins and VEGFs in the formation of the blood and lymphatic vessels. Low VEGF levels combined with high expression of Ang-2 lead to vascular regression, whereas Ang-1 and VEGF seem to cooperate in promoting angiogenesis. Interestingly, Ang-1 protects the vasculature against VEGF-induced plasma leakage (Thurston et al. 1999, 2000). On the other hand, Ang-2 may be involved in the regulation of lymphangiogenesis, since inactivation of the *Ang2* gene leads to nonfunctional and disconnected lymphatic vessels in mouse embryos (Gale et al. 2002). Cross-regulation also exists, as both VEGF and VEGF-C increase Ang-2 mRNA in endothelial cells (Mandriota and Pepper 1998; Oh et al. 1999; T. Veikkola et al., in prep.).

LYMPHATIC VESSELS IN TUMORS AND LYMPHATIC METASTASIS

Tumor metastasis by direct invasion or via blood or lymphatic vessels is one of the leading causes of death in cancer patients (Plate 2001). Although hematogenous metastasis via blood vessels is well recognized, for many tumors the most common pathway for metastatic spread is via the lymphatic vessels. In addition, the lymphatic spread of tumor cells to regional lymph nodes is an indicator of tumor aggressiveness. In general, it is thought that a high interstitial pressure of tumors prevents the growth of lymphatic vessels into the tumor cell mass (Carmeliet and Jain 2000). However, the studies of the involvement of the lymphatic vessels in tumor growth were hampered for a long time due to absence of specific markers of the lymphatic endothelium.

Several experimental studies have recently demonstrated an important role for lymphatic vessels as a route for tumor metastasis. Overexpression of the VEGFR-3 ligands, VEGF-C or VEGF-D, in tumor cells induced lymphangiogenesis in the tumor periphery, and in some cases also intralymphatic tumor growth and metastasis were obtained. These studies were done using transfected cells (Karpanen et al. 2001; Skobe et al. 2001; Stacker et al. 2001) or transgenic mice (Mandriota et al. 2001), but correlation of VEGF-C expression with increased lymphatic metastasis has also been reported in several human cancers (Ohta et al. 1999; Tsurusaki et al. 1999; Yonemura et al. 1999). In addition, when highly metastatic human tumor cells were selected in mice by serial selection via the lymph nodes, the selected cells expressed enhanced levels of VEGF-C (He et al. 2002). These studies suggest that VEGF-C- or VEGF-D-secreting tumors can activate the lymphatic vessels, which may then facilitate metastasis of tumor cells. Interestingly, tumor lymphangiogenesis and metastasis were inhibited by blocking VEGF-C or VEGF-D signaling by neutralizing antibodies against VEGFR-3 or by soluble VEGFR-3 protein (Karpanen et al. 2001; Stacker et al. 2001; He et al. 2002). This suggests that inhibition of VEGFR-3 signaling may provide a novel strategy to inhibit metastasis via lymphatic vessels.

LYMPHEDEMA

Lymphedema is a disorder characterized by insufficiency of the lymphatic system, which leads to accumulation of a protein-rich fluid in the tissues and to a disfiguring and disabling swelling of the extremities. In the chronic condition the patients also suffer from tissue fibrosis, adipose degeneration, impaired wound healing, and susceptibility to infections (Witte et al. 2001). Noninherited secondary or aquired lymphedema develops when the lymphatic vessels are damaged, for example, by surgery or radiation therapy, or obstructed by filarial infection. Inherited primary lymphedema is usually due to hypoplasia or aplasia of the superficial or subcutaneous lympatic vessels.

In some families of hereditary lymphedema, heterozygous missense mutations were found in the *VEGFR-3* gene. All the lymphedema-associated mutations inactivated the tyrosine kinase domain and therefore led to insufficient VEGFR-3 signaling (Irrthum et al. 2000; Karkkainen et al. 2000). However, primary lymphedemas are a heterogeneous group of disorders that are sometimes associated with additional malformations in other organs. Linkage to other genes has been found, although in some cases none has yet been identified. Genetically modified mouse mutants have provided important tools for the studies of molecular mechanisms involved in lymphatic growth and in the development of lymphedema (Table 1). A mouse model for human primary lymphedema was obtained using N-ethyl-N-nitrosourea-induced chemical mutagenesis. As in human lymphedema, heterozygous mice carrying an inactivating missense mutation in the *VEGFR-3* gene had hypoplastic dermal lymphatic vessels (Karkkainen et al. 2001b). In addition, the mice had leaky intestinal lymphatics, which resulted in accumulation of intraperitoneal chylous fluid after suckling. Since the mice carrying only one functional *VEGFR-3* allele appeared normal (Dumont et al. 1998), whereas the tyrosine kinase inactivating mutation was responsible for lymphedema, the phenotype was considered to arise due to a "dominant-negative" effect by the kinase inactive receptor. Despite this, gene therapy using adenoviral VEGF-C induced lymphangiogenesis in the lymphedema mice, suggesting its therapeutic application also for the human disease (Karkkainen et al. 2001b). The absence of dermal lymphatic vessels in the K14-VEGFR-3-Ig mice was associated with fibrosis, increased fluid accumulation in the skin and subcutaneous tissue, and swelling of the feet (Mäkinen et al. 2001a). This phenotype thus resembles human lymphedema, and the data support the view that defective VEGFR-3 signaling can cause the disease.

We have evidence that *VEGFR-3* haploinsufficiency can cause a lymphedema-like phenotype in mice (T. Mäkinen and K. Alitalo, unpubl.). The dependence of the phenotype on the genetic background resembles the observed variability in the penetrance of human lymphedema and suggests that genes modulating, e.g., VEGFR-3, signaling pathways may significantly contribute to the development of the disease. Similar genetic background-dependent lymphatic phenotype, determined by accumulation of chylous ascites, has been previously observed in mice deficient of transcription factor Sox18 (Pennisi et al. 2000). In endothelial cells, Sox18 interacts with a MADS-family transcription factor MEF2C, which potentiates its transcriptional activity (Hosking et al. 2001). Interestingly, targeted mutagenesis of MEF2C leads to a similar phenotype as observed in the VEGFR-3-deficient mice. MEF2C-deficient embryos die at E9.5–10 due to defects in the remodeling of the vasculature and abnormal endocardiogenesis (Lin et al. 1998; Bi et al. 1999). Although the endocardial defects may be caused by a significant decrease in Ang-1 and VEGF expression in the myocardium (Bi et al. 1999), decreased VEGFR-3 expression may contribute to the failure in vascular remodeling. In line with this hypothesis, the VEGFR-3 promoter region contains a conserved binding motif for MEF2 (Iljin et al. 2001). Therefore, both Sox18 and MEF2C may participate in a common transcriptional program involved in the development of the lymphatic vasculature.

MOLECULAR IDENTITY OF LYMPHATIC VERSUS BLOOD VASCULAR ENDOTHELIAL CELLS

In addition to neovascularization, the endothelium has a variety of other important functions in, e.g., regulation of the vascular tone, maintenance of the blood-brain barrier, and inflammatory responses. For example, endothelial cells recruit leukocytes to inflammatory foci, and specialized endothelial cells are responsible for the homing of lymphocytes to the secondary lymphoid organs (Biedermann 2001). Reflecting such functional diversity, the

Table 1. Mouse Mutants Having a Lymphatic Phenotype

Gene	Mouse model[a]	Lymphatic phenotype	Lethality[b]	Reference
Prox-1	KO	defective lymphatic sprouting and differentiation (–/–)	E14.5	Wigle and Oliver (1999)
		accumulation of chylous ascites (+/–)	P2-3	
Integrin α9	KO	chylothorax (–/–)	P6-12	Huang et al. (2000)
Ang-2	KO	disconnected lymphatic vessels, accumulation of chylous ascites		N. Gale et al. (pers. comm.)
VEGFR-3	Mut (kinase dead)	accumulation of chylous ascites, hypoplasia of dermal lymphatic vessels (+/mut)	viable	Karkkainen et al. (2001b)
Net transcription factor	Mut (loss of the Ets DNA-binding domain)	chylothorax (mut/mut)	P2	Ayadi et al. (2001)
Sox18 transcription factor	Mut (truncated trans-activation domain)	accumulation of chylous ascites (mut/mut)	before weaning	Pennisi et al. (2000)
VEGF-C	TG (K14 promoter)	hyperplasia of dermal lymphatic vessels	viable	Jeltsch et al. (1997)
	(insulin promoter)	hyperplasia of lymphatic vessels around the islets of Langerhans	viable	Mandriota et al. (2001)
VEGF-D	TG (K14 promoter)	hyperplasia of dermal lymphatic vessels	viable	Veikkola et al. (2001)
VEGFR-3-Ig (VEGF-C/D "trap")	TG (K14 promoter)	lack of dermal lymphatic vessels	viable	Mäkinen et al. (2001a)

[a]KO = knock-out (gene inactivation), Mut = mutagenesis (chemically induced, VEGFR-3) / gene-targeted (Net) / spontaneous (Sox18), TG = transgenic (gene overexpression).
[b]E = embryonic day, P = postnatal day.

antigenic profile of the vascular endothelium varies in different types of vessels and in different organs.

Lymphatic marker identification has enabled detailed studies of the lymphatic vessels during the last few years. These lymphatic endothelium-specific molecules include VEGFR-3 receptor tyrosine kinase, the Prox-1 transcription factor, the hyaluronen receptor LYVE-1, and the membrane mucoprotein podoplanin (Table 2) (Jussila et al. 1998; Banerji et al. 1999; Breiteneder-Geleff et al. 1999; Wigle and Oliver 1999). None of these molecules is specific for endothelial cells and, on the other hand, they may be expressed only in a subset of the lymphatic vessels. In addition, the function of many of these molecules in lymphatic endothelial cell biology remains to be characterized.

Identification of differentially expressed genes between blood vascular and lymphatic endothelial cells may provide new potential vascular markers and also give a better understanding of the structural and functional properties of these cells. In our studies, we have found that the microvascular endothelial cell cultures consist of distinct lineages of blood vascular and lymphatic endothelial cells (Fig. 3). The lymphatic endothelial cells (LECs) grew in distinct islands surrounded by blood vascular endothelial cells (BECs). We used antibodies against VEGFR-3 or podoplanin and magnetic microbeads for the separation of BECs and LECs, which were then grown as separate cell cultures (Mäkinen et al. 2001b). No interconversion between the cell types was detected in these cultures. In an independent study by Kriehuber et al. (2001), the BECs and LECs were separated from dermal cell suspensions by flow cytometry using antibodies against podoplanin and CD34 (Kriehuber et al. 2001). In agreement with our results, their studies demonstrated that the LECs and BECs constitute stable and specialized EC lineages.

Table 2. Examples of Lymphatic Endothelial Cell Markers

Protein	Function	Reference
VEGFR-3	receptor tyrosine kinase; development of blood vasculature, proliferation, survival and migration of LECs	Jussila et al. (1998); Mäkinen et al. (2001a,b)
LYVE-1	lymphatic hyaluronan receptor	Banerji et al. (1999)
Prox-1	homeobox transcription factor; differentiation of LECs	Wigle and Oliver (1999)
Desmoplakin I and II	intercellular adherens junction molecule	Ebata et al. (2001)
Podoplanin	integral membrane protein	Breiteneder-Geleff et al. (1999)
D6	β-chemokine receptor	Nibbs et al. (2001)
Macrophage mannose receptor I	endocytotic/phagocytotic receptor in macrophages	Irjala et al. (2001)

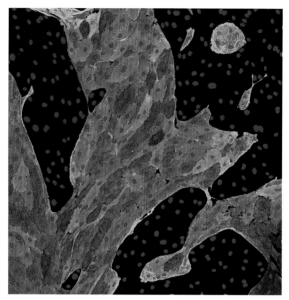

Figure 3. Human dermal microvascular endothelial cells in culture. The lymphatic endothelial cells (stained in green using antibodies against podoplanin) grow in distinct islands surrounded by blood vascular endothelial cells. The nuclei of all cells have been stained blue.

After establishing the isolation and culture of BECs and LECs, we have addressed the question of the genetic program controlling the identity of these two cell lineages using a gene profiling approach. We have identified so far about 300 genes that are differentially expressed between the cultured BECs and LECs (Petrova et al. 2002). The most striking differences were detected in genes that are involved in cytoskeletal and cell–cell or cell–matrix interactions, as well as in the expression of proinflammatory cytokines and chemokines.

Prox-1 AS A FATE-DETERMINING FACTOR FOR LYMPHATIC ENDOTHELIAL CELLS

Homeodomain transcription factors are important regulators of cellular fates and development. Gene-targeting studies have demonstrated that the Prox-1 homeobox transcription factor has a crucial role in the development of the lymphatic vasculature (Wigle and Oliver 1999). Prox-1 was found to be expressed specifically in the cultured LECs. However, when introduced into the BECs using adenoviral gene transfer, Prox-1 induced the expression of certain lymphatic-specific genes, including VEGFR-3 (Petrova et al. 2002). Interestingly, Prox-1 overexpression also resulted in the suppression of ~40% of the genes characteristic of the BECs. These results show that Prox-1 is able to reprogram gene expression in BECs toward the identity of LECs and suggest a role for Prox-1 as one of the fate-determining factors for the LECs.

Our findings and those reported by Wigle et al. (2002) suggest that in the absence of Prox-1 the default endothelial cell differentiation state corresponds to the blood endothelial phenotype and that Prox-1 is required for the establishment of lymphatic endothelial cell identity. These results help to explain the defective lymphatic differentiation and sprouting observed in the Prox-1 null mice (Wigle and Oliver 1999). The first lymphatic sprouts normally bud from the anterior cardinal vein at E10.5, but this budding is arrested in $Prox-1^{-/-}$ embryos at E11.5. In contrast to the wild-type embryos, in $Prox-1^{-/-}$ embryos the endothelial cells that bud off from the cardinal vein at E11.5–E12 do not express the lymphatic endothelial cell markers VEGFR-3, LYVE-1, or SLC. Instead, the mutant cells expressed blood vascular markers laminin and CD34, suggesting that these cells were not committed to the lymphatic endothelial cell lineage in the absence of Prox-1 (Wigle et al. 2002).

Gene targeting also revealed that Prox-1 haploinsufficiency is associated with the accumulation of lymphatic chylous fluid and neonatal death, suggesting that the level of Prox-1 is critical for proper lymphatic development (Wigle and Oliver 1999). We also found that Prox-1 upregulated the expression of genes associated with S-phase progression (Petrova et al. 2002), suggesting that after the differentiation of lymphatic endothelial cells, Prox-1 may also be essential for the maintenance of their proliferation during development.

CONCLUSIONS

Lymphatic vessels are essential for the maintenance of normal fluid balance and immune surveillance, but they also provide a pathway for metastasis in many types of cancers. The VEGFR-3 receptor tyrosine kinase is expressed in the lymphatic endothelium and its ligands VEGF-C and VEGF-D induce lymphangiogenesis. VEGFR-3 mediates signals for the survival, migration, and proliferation of lymphatic endothelial cells, and these signals are essential for the normal development of the lymphatic vessels during embryonic development. Inhibition of VEGFR-3 signaling by using a soluble VEGFR-3 protein resulted in a specific and complete inhibition of lymphangiogenesis in a mouse model. Additionally, in tumor models the soluble VEGFR-3 specifically inhibited lymphangiogenesis and tumor spread via the lymphatic vessels. In cancers that show lymphogenous metastasis, the blocking of the VEGFR-3 signaling pathway may thus provide a useful strategy to inhibit metastasis.

Until now, only a few lymphatic endothelial cell lines have been available for molecular biological studies, and these cells were mainly derived from lymphatic tumors. The ability to isolate and maintain primary cultures of dermal blood vascular and lymphatic endothelial cells (BECs and LECs, respectively) now allows the characterization of the molecular properties of these cells. Using gene expression profiling, we have identified genes that distinguish these two cell lineages. We have also shown that the homeobox transcription factor Prox-1 is involved in the establishment of the molecular identity of the LECs. Molecular discrimination of the blood vascular and lymphatic vessels is essential in studies of diseases involving these vessels and in the targeted treatment of such diseases. Thus the BEC- and LEC-specific molecules may provide new targets for the treatment of such diseases.

Further challenges in the lymphatic research include the identification of new molecular regulators of lymphangiogenesis. The ability to isolate and to culture LECs facilitates the cloning of novel LEC-specific genes, the characterization of which should give us a better understanding of the mechanisms of lymphangiogenesis. The functional analysis of the new lymphangiogenic regulators may also give us novel insights into the pathogenesis of diseases of the lymphatic system.

ACKNOWLEDGMENTS

The studies in our laboratory have been supported by grants from the Helsinki University Central Hospital, the Finnish Technology Development Center, the Academy of Finland, The Novo Nordisk Foundation, Finnish Cancer Organizations, and the Ludwig Institute for Cancer Research.

REFERENCES

Alitalo K. and Carmeliet P. 2002. Molecular mechanisms of lymphangiogenesis in health and disease. *Cancer Cell* **1:** 219.

Ayadi A., Zheng H., Sobieszczuk P., Buchwalter G., Moerman P.,m alitalo K., and Wasylyk B. 2001 Net-targeted mutant mice develop a vascular phenotype and up-regulate egr-1. *EMBO J.* **20:** 5139.

Banerji S., Ni J., Wang S.X., Clasper S., Su J., Tammi R., Jones M., and Jackson D.G. 1999. LYVE-1, a new homologue of the CD44 glycoprotein, is a lymph-specific receptor for hyaluronan. *J. Cell Biol.* **144:** 789.

Beltramo E., Pomero F., Allione A., D'Alu F., Ponte E., and Porta M. 2002. Pericyte adhesion is impaired on extracellular matrix produced by endothelial cells in high hexose concentrations. *Diabetologia* **45:** 416.

Benjamin L.E., Hemo I., and Keshet E. 1998. A plasticity window for blood vessel remodelling is defined by pericyte coverage of the preformed endothelial network and is regulated by PDGF-B and VEGF. *Development* **125:** 1591.

Bi W., Drake C.J., and Schwarz J.J. 1999. The transcription factor MEF2C-null mouse exhibits complex vascular malformations and reduced cardiac expression of angiopoietin 1 and VEGF. *Dev. Biol.* **211:** 255.

Biedermann B.C. 2001. Vascular endothelium: Checkpoint for inflammation and immunity. *News Physiol. Sci.* **16:** 84.

Breiteneder-Geleff S., Soleiman A., Kowalski H., Horvat R., Amann G., Kriehuber E., Diem K., Weninger W., Tschachler E., Alitalo K., and Kerjaschki D. 1999. Angiosarcomas express mixed endothelial phenotypes of blood and lymphatic capillaries: Podoplanin as a specific marker for lymphatic endothelium. *Am. J. Pathol.* **154:** 385.

Carmeliet P. 2000. Mechanisms of angiogenesis and arteriogenesis. *Nat. Med.* **6:** 389.

Carmeliet P. and Jain R.K. 2000. Angiogenesis in cancer and other diseases. *Nature* **407:** 249.

Dumont D.J., Jussila L., Taipale J., Lymboussaki A., Mustonen T., Pajusola K., Breitman M., and Alitalo K. 1998. Cardiovascular failure in mouse embryos deficient in VEGF receptor-3. *Science* **282:** 946.

Ebata N., Nodasaka Y., Sawa Y., Yamaoka Y., Makino S., Totsuka Y., and Yoshida S. 2001. Desmoplakin as a specific marker of lymphatic vessels. *Microvasc. Res.* **61:** 40-48.

Ferrara N. 1999. Molecular and biological properties of vascular endothelial growth factor. *J. Mol. Med.* **77:** 527.

Gale N.W., Thurston G., Hackett S.F., Renard R., Wang Q., McClain J., Martin C., Witte C., Witte M.H., Jackson D., Suri C., Campochiaro P.A., Wiegand S.J., and Yancopoulos G.D. 2002. Angiopoietin-2 is required for postnatal angiogenesis and lymphatic patterning, and only the latter role is rescued by Angiopoietin-1. *Dev. Cell* **3:** 411.

He Y., Kozaki K., Karpanen T., Koshikawa K., Yla-Herttuala S., Takahashi T., and Alitalo K. 2002. Suppression of tumor lymphangiogenesis and lymph node metastasis by blocking vascular endothelial growth factor receptor 3 signaling. *J. Natl. Cancer Inst.* **5:** 819.

Hosking B.M., Wang S.C., Chen S.L., Penning S., Koopman P., and Muscat G.E. 2001. SOX18 directly interacts with MEF2C in endothelial cells. *Biochem. Biophys. Res. Commun.* **287:** 493.

Huang X.Z., Wu J.F., Ferrando R., Lee J.H., Wang Y.L, Farese R.V., Jr., and Sheppard D. 2000. Fatal bilateral chylothorax in mice lacking the integrin $\alpha 9\beta 1$. *Mol. Cell. Biol.* **20:** 5208.

Huntington G.S. and McClure C.F.W. 1908. The anatomy and development of the jugular lymph sac in the domestic cat (*Felis domestica*). *Anat. Rec.* **2:** 1.

Iljin K., Karkkainen M.J., Lawrence E.C., Kimak M.A., Uutela M., Taipale J., Pajusola K., Alhonen L., Halmekyto M., Finegold D.N., Ferrell R.E., and Alitalo K. 2001. VEGFR3 gene structure, regulatory region, and sequence polymorphisms. *FASEB J.* **15:** 1028.

Irjala H., Johansson E.L., Grenman R., Alanen K., Salmi M., and Jalkanen S. 2001. Mannose receptor is a novel ligand for L-selectin and mediates lymphocyte binding to lymphatic endothelium. *J. Exp. Med.* **194:** 1033.

Irrthum A., Karkkainen M.J., Devriendt K., Alitalo K., and Vikkula M. 2000. Congenital hereditary lymphedema caused by a mutation that inactivates VEGFR3 tyrosine kinase. *Am. J. Hum. Genet.* **67:** 295.

Jeltsch M., Kaipainen A., Joukov V., Meng X., Lakso M., Rauvala H., Swartz M., Fukumura D., Jain R.K., and Alitalo K. 1997. Hyperplasia of lymphatic vessels in VEGF-C transgenic mice. *Science* **276:** 1423.

Joukov V., Sorsa T., Kumar V., Jeltsch M., Claesson-Welsh L., Cao Y., Saksela O., Kalkkinen N., and Alitalo K. 1997. Proteolytic processing regulates receptor specificity and activity of VEGF-C. *EMBO J.* **16:** 3898.

Jussila L., Valtola R., Partanen T.A., Salven P., Heikkila P., Matikainen M.T., Renkonen R., Kaipainen A., Detmar M., Tschachler E., Alitalo R., and Alitalo K. 1998. Lymphatic endothelium and Kaposi's sarcoma spindle cells detected by antibodies against the vascular endothelial growth factor receptor-3. *Cancer Res.* **58:** 1599.

Kaipainen A., Korhonen J., Mustonen T., van Hinsbergh V.M., Fang G.-H., Dumont D., Breitman M. and Alitalo K. 1995. Expression of the *fms*-like tyrosine kinase FLT4 gene becomes restricted to lymphatic endothelium during development. *Proc. Natl. Acad. Sci.* **92:** 3566.

Kampmeier O.F. 1912. The value of the injection method in the study of lymphatic development. *Anat. Rec.* **6:** 223.

Karkkainen M.J., Jussila L., Ferrell R.E., Finegold D.N., and Alitalo K. 2001a. Molecular regulation of lymphangiogenesis and targets for tissue oedema. *Trends Mol. Med.* **7:** 18.

Karkkainen M.J., Ferrell R.E., Lawrence E.C., Kimak M.A., Levinson K.L., Alitalo K., and Finegold D.N. 2000. Missense mutations interfere with VEGFR-3 signaling in primary lymphoedema. *Nat. Genet.* **25:** 153.

Karkkainen M.J., Saaristo A., Jussila L., Karila K.A., Lawrence E.C., Pajusola K., Bueler H., Eichmann A., Kauppinen R., Kettunen M.I., Yla-Herttuala S., Finegold D.N., Ferrell R.E., and Alitalo K. 2001b. A model for gene therapy of human hereditary lymphedema. *Proc. Natl. Acad. Sci.* **98:** 12677.

Karpanen T. and Alitalo K. 2001. Lymphatic vessels as targets of tumor therapy? *J. Exp. Med.* **194:** F37.

Karpanen T., Egeblad M., Karkkainen M.J., Kubo H., Ylä-Herttuala S., Jäättelä M., and Alitalo K. 2001. Vacular endothelial growth factor C promotes tumor lymphangiogenesis and intralymphatic tumor growth. *Cancer Res.* **61:** 1786.

Kriehuber E., Breiteneder-Geleff S., Groeger M., Soleiman A., Schoppmann S.F., Stingl G., Kerjaschki D., Maurer D. 2001.

Isolation and characterization of dermal lymphatic and blood endothelial cells reveal stable and functionally specialized cell lineages. *J. Exp. Med.* **194:** 797.

Kukk E., Lymboussaki A., Taira S., Kaipainen A., Jeltsch M., Joukov V., and Alitalo K. 1996. VEGF-C receptor binding and pattern of expression with VEGFR-3 suggest a role in lymphatic vascular development. *Development* **122:** 3829.

Lin Q., Lu J., Yanagisawa H., Webb R., Lyons G.E., Richardson J.A., and Olson E.N. 1998. Requirement of the MADS-box transcription factor MEF2C for vascular development. *Development* **125:** 4565.

Mäkinen T., Jussila L., Veikkola T., Karpanen T., Kettunen M.I., Pulkkanen K.J., Kauppinen R., Jackson D.G., Kubo H., Nishikawa S.-I., Ylä-Herttuala S., and Alitalo K. 2001a. Inhibition of lymphangiogenesis with resulting lymphedema in transgenic mice expressing soluble VEGF receptor-3. *Nat. Med.* **7:** 199.

Mäkinen T., Veikkola T., Mustjoki S., Karpanen T., Wise L., Mercer A., Catimel B., Nice E.C., Kowalski H., Kerjaschki D., Stacker S.A., Achen M.G., and Alitalo K. 2001b. Isolated lymphatic endothelial cells transduce growth, survival and migratory signals via the VEGF-C/D receptor VEGFR-3. *EMBO J.* **20:** 4762.

Mandriota S.J. and Pepper M.S. 1998. Regulation of angiopoietin-2 mRNA levels in bovine microvascular endothelial cells by cytokines and hypoxia. *Circ. Res.* **83:** 852.

Mandriota S.J., Jussila L., Jeltsch M., Compagni A., Baetens D., Prevo R., Banerji S., Huarte J., Montesano R., Jackson D.G., Orci L., Alitalo K., Christofori G., and Pepper M.S. 2001. Vascular endothelial growth factor-C-mediated lymphangiogenesis promotes tumour metastasis. *EMBO J.* **20:** 672.

Marconcini L., Marchio S., Morbidelli L., Cartocci E., Albini A., Ziche M., Bussolino F., and Oliviero S. 1999. c-fos-induced growth factor/vascular endothelial growth factor D induces angiogenesis in vivo and in vitro. *Proc. Natl. Acad. Sci.* **96:** 9671.

Nibbs R.J., Kriehuber E., Ponath P.D., Parent D., Qin S., Campbell J.D., Henderson A., Kerjaschki D., Maurer D., Graham G.J., and Rot A. 2001. The beta-chemokine receptor D6 is expressed by lymphatic endothelium and a subset of vascular tumors. *Am. J. Pathol.* **158:** 867.

Oh H., Takagi H., Suzuma K., Otani A., Matsumura M., and Honda Y. 1999. Hypoxia and vascular endothelial growth factor selectively up-regulate angiopoietin-2 in bovine microvascular endothelial cells. *J. Biol. Chem.* **274:** 15732.

Oh S.J., Jeltsch M.M., Birkenhager R., McCarthy J.E., Weich H.A., Christ B., Alitalo K., and Wilting J. 1997. VEGF and VEGF-C: Specific induction of angiogenesis and lymphangiogenesis in the differentiated avian chorioallantoic membrane. *Dev. Biol.* **188:** 96.

Ohta Y., Shridhar V., Bright R.K., Kalemkerian G.P., Du W., Carbone M., Watanabe Y., and Pass H.I. 1999. VEGF and VEGF type C play an important role in angiogenesis and lymphangiogenesis in human malignant mesothelioma tumours. *Br. J. Cancer* **81:** 54.

Papoutsi M., Tomarev S.I., Eichmann A., Prols F., Christ B., and Wilting J. 2001. Endogenous origin of the lymphatics in the avian chorioallantoic membrane. *Dev. Dyn.* **222:** 238.

Pennisi D., Gardner J., Chambers D., Hosking B., Peters J., Muscat G., Abbott C., and Koopman P. 2000. Mutations in Sox18 underlie cardiovascular and hair follicle defects in ragged mice. *Nat. Genet.* **24:** 434.

Petrova T.V., Mäkinen T., Mäkelä T.P., Saarela J., Virtanen I., Ferrell R.E., Finegold D.N., Kerjaschki D., Ylä-Herttuala S., and Alitalo K. 2002. Lymphatic endothelial reprogramming of vascular endothelial cells by the Prox-1 home box transcription factor. *EMBO J.* **21:** 4593.

Plate K.H. 2001. From angiogenesis to lymphangiogenesis. *Nat. Med.* **7:** 151.

Risau W. 1997. Mechanisms of angiogenesis. *Nature* **386:** 671.

Risau W. and Flamme I. 1995. Vasculogenesis. *Annu. Rev. Cell Dev. Biol.* **11:** 73.

Sabin F.R. 1902. On the origin of the lymphatic system from the veins and the development of the lymph hearts and thoracic duct in the pig. *Am. J. Anat.* **1:** 367.

———. 1909. The lymphatic system in human embryos, with a consideration of the morphology of the system as a whole. *Am. J. Anat.* **9:** 43.

Schneider M., Othman-Hassan K., Christ B., and Wilting J. 1999. Lymphangioblasts in the avian wing bud. *Dev. Dyn.* **216:** 311.

Skobe M., Hawighoest T., Jackson D.G., Prevo R., Janes L., Velasco P., Riccardi L., Alitalo K., Claffey K., and Detmar M. 2001. Induction of tumor lymphangiogenesis by VEGF-C promotes breast cancer metastasis. *Nat. Med.* **7:** 192.

Stacker S.A., Caesar C., Baldwin M.E., Thornton G.E., Williams R.A., Prevo R., Jackson D.G., Nishikawa S.-I., Kubo H., and Achen M.G. 2001. Vascular endothelial growth factor-D promotes the metastatic spread of cancer via the lymphatics. *Nat. Med.* **7:** 186.

Stacker S.A., Stenvers K., Caesar C., Vitali A., Domagala T., Nice E., Roufail S., Simpson R.J., Moritz R., Karpanen T., Alitalo K., and Achen M.G. 1999. Biosynthesis of vascular endothelial growth factor-D involves proteolytic processing which generates non-covalent homodimers. *J. Biol. Chem.* **274:** 32127.

Thurston G., Suri C., Smith K., McClain J., Sato T.N., Yancopoulos G.D., and McDonald D.M. 1999. Leakage-resistant blood vessels in mice transgenically overexpressing angiopoietin-1. *Science* **286:** 2511.

Thurston G., Rudge J.S., Ioffe E., Zhou H., Ross L., Croll S.D., Glazer N., Holash J., McDonald K., and Yancopoulos G.D. 2000. Angiopoietin-1 protects the adult vasculature against plasma leakage. *Nat. Med.* **6:** 460.

Thyboll J., Kortesmaa J., Cao R., Soininen R., Wang L., Iivanainen A., Sorokin L., Risling M., Cao Y., and Tryggvason K. 2002. Deletion of the laminin alpha4 chain leads to impaired microvessel maturation. *Mol. Cell. Biol.* **22:** 1194.

Tsurusaki T., Kanda S., Sakai H., Kanetake H., Saito Y., Alitalo K., and Koji T. 1999. Vascular endothelial growth factor-C expression in human prostatic carcinoma and its relationship to lymph node metastasis. *Br. J. Cancer* **80:** 309.

Veikkola T., Jussila L., Makinen T., Karpanen T., Jeltsch M., Petrova T.V., Kubo H., Thurston G., McDonald D.M., Achen M.G., Stacker S.A., and Alitalo K. 2001. Signalling via vascular endothelial growth factor receptor-3 is sufficient for lymphangiogenesis in transgenic mice. *EMBO J.* **20:** 1223.

Wigle J.T. and Oliver G. 1999. Prox1 function is required for the development of the murine lymphatic system. *Cell* **98:** 769.

Wigle J.T., Harvey N., Detmar M., Lagutina I., Grosveld G., Gunn M.D., Jackson D.G., and Oliver G. 2002. An essential role for Prox1 in the induction of the lymphatic endothelial cell phenotype. *EMBO J.* **21:** 1505.

Witte M.H., Bernas M.J., Martin C.P., and Witte C.L. 2001. Lymphangiogenesis and lymphangiodysplasia: From molecular to clinical lymphology. *Microsc. Res. Tech.* **55:** 122.

Witzenbichler B., Asahara T., Murohara T., Silver M., Spyridopoulos I., Magner M., Principe N., Kearney M., Hu J.S., and Isner J.M. 1998. Vascular endothelial growth factor-C (VEGF-C/VEGF-2) promotes angiogenesis in the setting of tissue ischemia. *Am. J. Pathol.* **153:** 381.

Yonemura Y., Endo Y., Fujita H., Fushida S., Ninomiya I., Bandou E., Taniguchi K., Miwa K., Ohoyama S., Sugiyama K., and Sasaki T. 1999. Role of vascular endothelial growth factor C expression in the development of lymph node metastasis in gastric cancer. *Clin. Cancer Res.* **5:** 1823.

Protease-activated Receptors in the Cardiovascular System

S.R. COUGHLIN

Cardiovascular Research Institute, Department of Medicine and Department of Cellular and Molecular Pharmacology, University of California, San Francisco, California 94143

The coagulation protease thrombin regulates platelet aggregation and a host of responses in endothelial cells and other cell types. How does thrombin, a protease, elicit cellular responses like a traditional hormone? Protease-activated receptors (PARs) provide an answer. PARs are G-protein-coupled receptors that have evolved a mechanism to convert an extracellular proteolytic event into a transmembrane signal. In essence, PARs are peptide receptors that carry their own ligands, which remain silent until activated by receptor cleavage. Overall, it appears that PARs, together with the coagulation cascade, serve to link tissue injury to appropriate cellular responses, but PARs also appear to serve important functions in other contexts.

This paper provides background on the molecular and cellular biology of coagulation protease generation and PAR activation, then focuses on recent studies that begin to define the role of PARs in specific cell types and processes in vivo. Two areas are emphasized.

First, recent studies that provide a working model for the role of PARs in mediating activation of platelets and endothelial cells by coagulation proteases are discussed. These support the general notion that PARs help to orchestrate a coordinated hemostatic and inflammatory response to tissue injury. Thrombin, platelets, and endothelial cells play central roles in thrombosis, which causes myocardial infarction and other pathological events. These same actors may also contribute to excessive inflammation and tissue damage in sepsis and other processes. Thus, understanding the role of PARs in vivo may suggest new strategies for the development of drugs aimed at the prevention or treatment of thrombosis and perhaps inflammation.

Second, recent studies demonstrate that PARs contribute to embryonic development by playing an as yet poorly understood role in the proper formation of blood vessels. At face value, these and other studies suggest a novel role for the "coagulation" cascade itself in embryonic development.

THROMBIN GENERATION

Because this review focuses on cellular responses to coagulation proteases, it is useful to review how, when, and where these proteases are generated. Thrombin is the main effector protease of the coagulation cascade, a series of zymogen conversions that is triggered when circulating coagulation factors contact tissue factor. Tissue factor, a type-I integral membrane protein, binds the protease factor VIIa; the tissue factor VIIa complex then binds the zymogen factor X and cleaves it to form the active protease factor Xa. Factor Xa (with the assistance of the cofactor factor Va) then converts the zymogen prothrombin to active thrombin. Other zymogen conversions provide positive feedback loops that accelerate and amplify thrombin production. Importantly, a variety of inhibitors ensure that active thrombin is short-lived in the circulation and, in the context of a normal endothelium, thrombin activates the protein C pathway, a negative feedback loop that terminates thrombin formation. Thus, thrombin is thought to act near the site of its production (Colman et al. 1994; Esmon et al. 1999).

Epithelial cells, macrophages, and other cell types that are normally not in contact with blood express tissue factor constitutively. Classically, disruption of vascular integrity allows plasma coagulation factors to contact extravascular tissue factor, thereby triggering thrombin generation. Circulating monocytes and vascular endothelial cells—cells that are normally bathed in coagulation factors—can express low levels of tissue factor when activated by endotoxin, cytokines, and other stimuli (Bevilacqua and Gimbrone 1987). Tissue factor is also found in association with cell-derived microparticles; the source of this tissue factor and these microparticles is as yet incompletely characterized. Activated platelets and endothelial cells can bind and localize tissue-factor-bearing monocytes and microparticles to sites of vascular injury, thereby concentrating tissue factor and activating the coagulation cascade (Osterud 1998; Giesen et al. 1999). Like the classical paradigm above, this scenario links tissue injury or inflammation to activation of coagulation proteases. Taken together, these considerations cast thrombin as a local mediator of responses to vascular injury or inflammation and suggest that PAR activation by thrombin might function in this context.

THROMBIN'S ACTIONS ON CELLS

Thrombin has a host of direct actions on cells (Fig. 1) (Coughlin 2000). Thrombin is perhaps the most effective activator of platelets ex vivo (Davey and Lüscher 1967), and platelet responses to thrombin appear designed to pro-

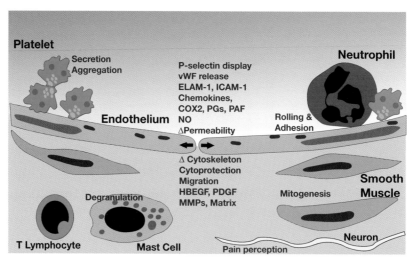

Figure. 1. Thrombin's cellular actions in the context of a blood vessel. Thrombin is a multifunctional serine protease generated at sites of vascular injury. It triggers platelet secretion and aggregation and elicits a host of responses in the vascular endothelium. These include display of adhesion molecules for platelets and leukocytes, production of chemokines and other leukocyte activators, as well as cell shape and permeability changes. Other endothelial responses include migration and production of growth factors, as well as matrix metalloproteinases and matrix proteins themselves. Thrombin triggers a similar set of responses in smooth muscle cells and is also mitogenic for these and other mesenchymal cells. Thrombin also affects T-cell signaling. These cellular actions of thrombin suggest that it helps to orchestrate hemostatic, inflammatory, and perhaps proliferative responses to tissue injury. Evidence that thrombin signaling in endothelial cells is important for the development of blood vessels in the embryo is also emerging, and a role for a subset of endothelial responses to thrombin in vascular remodeling can be imagined. Many of thrombin's actions on cells appear to be mediated by protease-activated receptors (PARs).

mote not only hemostasis but also inflammation and cell proliferation. These responses include (1) shape change from smooth discs to spheres with numerous filopodia, (2) release of stored granules' contents that further promote platelet activation and aggregation (ADP, serotonin, coagulation factor V, vWF, and fibrinogen), (3) release of other granule constituents such as chemokines and growth factors, (4) synthesis and release of the platelet activator thromboxane A2, (5) mobilization of P-selectin and CD40L to their surface (P-selectin is an adhesive receptor for platelets and leukocytes, and CD40L is a TNF-like agonist), and other changes on the platelet surface that promote the assembly of coagulation factor complexes to amplify local thrombin generation, and (6) activation of the key integrin α_{IIb}/β_3 (Hamberg et al. 1975; Stenberg et al. 1985; Sims et al. 1989; Brass et al. 1997; Henn et al. 1998; Hughes and Pfaff 1998). The latter binds fibrinogen and vWF to mediate platelet–platelet interaction and hence formation of the aggregates that plug damaged vessels (Colman et al. 1994).

In cultured endothelial cells, thrombin causes release of vWF (Hattori et al. 1989), display of P-selectin on the plasma membrane (Hattori et al. 1989), and production of chemokines—actions thought to trigger binding of platelets and leukocytes to the endothelial surface in vivo (Frenette et al. 1996; Subramaniam et al. 1996). Endothelial cells change shape and endothelial monolayers show increased permeability in response to thrombin (Lum and Malik 1994) —actions predicted to promote local transudation of plasma proteins and edema (Cirino et al. 1996). Thrombin can also regulate blood vessel diameter by endothelial-dependent vasodilation; in the absence of endothelium, thrombin's actions on smooth muscle cells evoke vasoconstriction. In fibroblast and vascular smooth muscle cell cultures, thrombin regulates cytokine production and is mitogenic, and thrombin triggers calcium signaling and other responses in T lymphocytes.

Many of the cellular responses to thrombin described above are consistent with the view that thrombin signaling connects tissue damage to appropriate hemostatic and inflammatory responses, and this notion fits well with thrombin's being generated at sites of vascular injury. Several of thrombin's actions on endothelial cells and other cell types suggest that thrombin signaling might also play a role in contexts other than tissue injury; for example, in leukocyte transmigration, vascular remodeling, and/or angiogenesis. Identification of the receptors that mediate cellular responses to thrombin provides an opportunity to test these ideas.

HOW DOES THROMBIN TRIGGER CELLULAR RESPONSES?

Thrombin signaling is mediated at least in part by G-protein-coupled protease-activated receptors (PARs) (Rasmussen et al. 1991; Vu et al. 1991a; Coughlin 1999). PAR1, the prototype for this family, is activated when thrombin binds to and cleaves its amino-terminal exodomain at a specific site. This cleavage event unmasks a new amino terminus that then serves as a tethered ligand, binding intramolecularly to the body of the receptor to effect transmembrane signaling (Fig. 2) (Vu et al. 1991a). Intermolecular ligation of PARs can occur but, not surprisingly, appears to be less efficient than intramolecular ligation (Chen et al. 1994; O'Brien et al. 2000). In support of this model, PAR1–thrombin interactions are accounted for by the sequences surrounding the thrombin cleavage

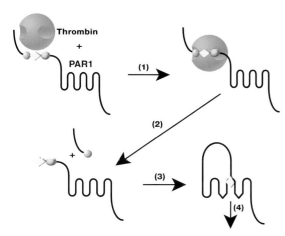

Figure 2. Proteolytic mechanism of PAR1 activation. (1) Thrombin (*large green sphere*) recognizes the amino-terminal exodomain of the G-protein-coupled thrombin receptor PAR1. This interaction utilizes two sites, one amino-(P1-P4; *small blue sphere*) and the other carboxyl-terminal (P9'-P14'; *small pink oval*) to the thrombin cleavage site. The latter sequence resembles the carboxyl tail of the thrombin inhibitor hirudin and binds to thrombin in an analogous manner (Liu et al. 1991; Vu et al. 1991a; Mathews et al. 1994). (2) Thrombin cleaves the peptide bond between receptor residues Arg-41 and Ser-42. (3) This serves to unmask a new amino terminus beginning with the sequence SFLLRN (*yellow diamond*) that functions as a tethered ligand, docking intramolecularly with the body of the receptor to effect transmembrane signaling (4). Synthetic SFLLRN peptide, which mimics the tethered ligand sequence, will function as an agonist independent of receptor cleavage. Thus, PAR1 is, in essence, a peptide receptor that carries its own ligand, the latter being active only after receptor cleavage (Vu et al. 1991).

site in PAR1's amino-terminal exodomain; cleavage at that site is both necessary and sufficient for PAR1 activation, and synthetic peptides that mimic the PAR1 tethered ligand can activate the receptor independent of protease and receptor cleavage (Vu et al. 1991a). Thus, PAR1 can be viewed as a peptide receptor that carries its own ligand within the PAR1 amino-terminal exodomain; the ligand remains dormant until activated by receptor cleavage.

PAR1 can couple to members of the G12/13, Gq, and Gi/z families and hence to a host of intracellular signaling pathways (Fig. 3). Such pluripotent signaling fits well with the known effects of thrombin on platelets, endothelia, and other cells. For example, in mouse platelets, Gq is necessary for platelet secretion and aggregation in response to thrombin but is not necessary for thrombin-triggered shape change (Offermanns et al. 1997b), which may be mediated by G12/13 (Klages et al. 1999). Gz, a Gi family member, is critical for epinephrine's ability to both inhibit cAMP formation and increase the sensitivity of mouse platelets to activation by other agonists (Yang et al. 2000). Thus, PAR1 coupling to Gi/z may play an epinephrine-like role in human platelets.

HOW DO CELLS ACCOMMODATE THE IRREVERSIBILITY OF PAR1 ACTIVATION?

Cleavage of PAR1 is irreversible, and the "peptide agonist" unmasked by cleavage remains tethered to the receptor. How is PAR1 signaling terminated? In addition, thrombin is an enzyme, implying that one thrombin molecule might cleave and activate several molecules of PAR1. How then does PAR1 mediate responses that are dependent on thrombin concentration?

Like other G-protein-coupled receptors, activated PAR1 is rapidly uncoupled from signaling and internalized by phosphorylation-dependent mechanisms (Hoxie et al. 1993; Ishii et al. 1994; Lefkowitz et al. 1998). Instead of recycling, internalized PAR1 is sorted to lysosomes with remarkable efficiency and degraded (Hoxie et al. 1993; Hein et al. 1994; Trejo et al. 1998; Trejo and Coughlin 1999). The rapid shutoff of activated PAR1 may explain how PAR1 can mediate responses that are proportional to thrombin concentration (Ishii et al. 1993). Each cleaved receptor is active for only a finite time and therefore triggers production of some average quantity of second messenger (e.g., inositol trisphosphate) before shutting off. Because the second messenger is itself cleared, the level of second messenger achieved should be proportional to the rate at which receptors are activated and hence to thrombin concentration.

If PAR1 is used once and then discarded, how is responsiveness to thrombin maintained? In fibroblasts and endothelial cells, naive PAR1 molecules are delivered to the cell surface from a preformed intracellular pool (Hein et al. 1994). In contrast, in megakaryocyte-like cell lines, recovery of PAR1 signaling requires new protein synthesis (Hoxie et al. 1993; Hein et al. 1994). Perhaps there is no need for a special resensitization mechanism in platelets because, once incorporated into a clot, they are presumably not reused.

A FAMILY OF PARS

Four PARs are known. Human PAR1 (Rasmussen et al. 1991; Vu et al. 1991a), PAR3 (Ishihara et al. 1997), and PAR4 (Kahn et al. 1998; Xu et al. 1998) can be activated by thrombin. PAR2 can be activated by trypsin (Nystedt et al. 1994) and tryptase (Molino et al. 1997) as well as by coagulation factors VIIa and Xa (Camerer et al. 1996, 2000) but not by thrombin. The latter finding is notable because it suggests that PAR2, although not responsive to thrombin, may serve to sense activation of the coagulation cascade in some settings.

Each of the four known PARs has a basic residue at P1 in the activating cleavage site, and in principle, any trypsin-like protease might trigger PAR signaling. Moreover, cofactors that localize proteases to the cell surface can help orchestrate PAR activation (Camerer et al. 2000; Nakanishi-Matsui et al. 2000). Thus, the full repertoire of proteases that signal through PARs remains to be defined. It is worth stating that simply demonstrating that a protease can activate a PAR does not establish physiological relevance. However, the amino-terminal exodomains of PAR1 and PAR3 have thrombin-interacting sequences both amino- and carboxy-terminal to the thrombin cleavage site (Fig. 2). The carboxy-terminal sequence resembles the carboxyl tail of the leech anticoagulant hirudin and, like the latter, binds to thrombin's fibrinogen-binding exosite; this interaction is important for receptor cleavage at low concentrations of thrombin (Coughlin

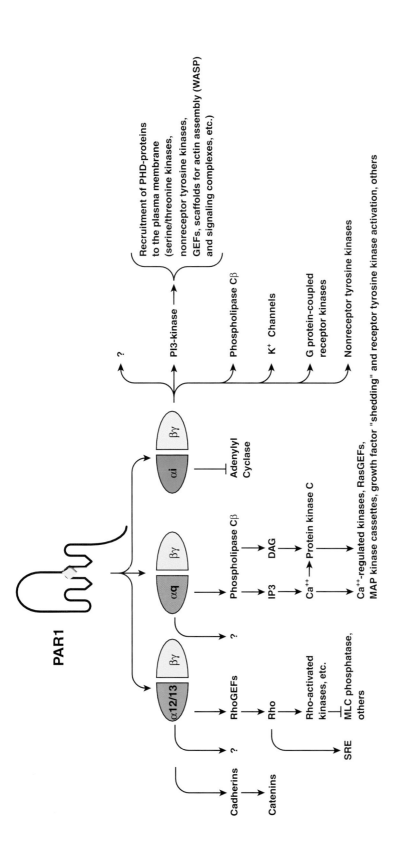

Figure 3. Intracellular signaling pathways activated by PAR1. PAR1 can couple to members of the G12/13, Gq, and Gi/z families (Hung et al. 1992a; Offermanns et al. 1994; Barr et al. 1997) to effect a substantial network of signaling pathways. The α subunits of G12 and 13 bind RhoGEFs (guanine nucleotide exchange factors, which activate small G proteins) (Hart et al. 1998; Fukuhara et al. 1999), providing a path to Rho-dependent cytoskeletal responses that are likely involved in shape change in platelets (Klages et al. 1999) and in permeability and migration in endothelial cells (Offermanns et al. 1997a; Vouret-Craviari et al. 1998). They may also interact with cadherins to release β-catenin, thereby regulating transcription and cell–cell interactions (Kaplan et al. 2001; Meigs et al. 2001, 2002). Gαq activates phospholipase Cβ, thereby triggering phosphoinositide hydrolysis, calcium mobilization, and protein kinase C activation (Taylor et al. 1991). This provides a path to calcium-regulated kinases and phosphatases, GEFs, and MAP kinase cassettes and other proteins that mediate cellular responses ranging from granule secretion, integrin activation, and aggregation in platelets (Offermanns et al. 1997) to transcriptional responses in endothelial and mesenchymal cells. Gαi inhibits adenylyl cyclase, an action known to promote platelet responses. Gβγ subunits can activate phosphoinositide-3 kinase (Stoyanov et al. 1995) and other lipid-modifying enzymes, protein kinases, and channels (Clapham and Neer 1997). Phosphoinositide-3 kinase modifies the inner leaflet of the plasma membrane to provide attachment sites for a host of signaling proteins (Leevers et al. 1999). PAR1 activation can also activate cell surface "sheddases" that liberate ligands for receptor tyrosine kinases, providing a link between thrombin and receptors involved in cell growth and differentiation (Prenzel et al. 1999). The pleiotrophic effect of PAR1 activation on cellular signaling is consistent with thrombin's diverse effects on cells.

1998). The presence of such extended thrombin-interacting sequences in PAR1 and PAR3 suggests that these receptors evolved to mediate responses to thrombin versus other proteases. A hirudin-like sequence is not evident in PAR4, and PAR4 indeed requires higher thrombin concentrations for activation than do the other receptors (Kahn et al. 1998; Xu et al. 1998). The lack of a thrombin-binding, hirudin-like sequence in PAR4 raises the question of whether PAR4 evolved to mediate responses to proteases other than thrombin, but cofactors that localize thrombin to the platelet surface can obviate the need for a hirudin-like domain in PAR4 itself (see below).

PARS AND PLATELET ACTIVATION

Recent studies provide a working model of thrombin signaling in human and mouse platelets and reveal both curious species differences and a variation on the paradigm for PAR activation (Fig. 4).

Thrombin Receptors in Human Platelets

Human platelets express PAR1 and PAR4, and activation of either is sufficient to trigger platelet secretion and aggregation (Vu et al. 1991a; Xu et al. 1998; Kahn et al. 1999). Antibodies to the thrombin interaction site in PAR1 blocked receptor cleavage and platelet activation at low but not high concentrations of thrombin (Brass et al. 1992; Hung et al. 1992b; Kahn et al. 1999). Similar results were obtained with a PAR1 antagonist (Bernatowicz et al. 1996; Kahn et al. 1999). In contrast, PAR4-blocking antibodies by themselves had no effect on platelet activation by thrombin, but when these were combined with PAR1 blockade, platelet activation was markedly inhibited even at high concentrations of thrombin (Kahn et al. 1999). Thus, PAR1 appears to be the major thrombin receptor on human platelets and mediates activation at low thrombin concentrations. In the absence of PAR1 function, PAR4 can mediate platelet activation, but only at high thrombin concentrations.

If PAR1 is sufficient to mediate platelet activation at low concentrations of thrombin, what does PAR4 contribute? It is possible that PAR4 simply provides some redundancy in an important system or that PAR1 and PAR4 interact. It is equally possible that PAR4 makes unique contributions to platelet function. PAR4 may allow platelets to respond to proteases other than thrombin. In this regard, PAR4 can mediate platelet activation by cathepsin G (Selak et al. 1988; Sambrano et al. 2000), a granzyme released by activated neutrophils; whether this is physiologically important is not known. Differences in G protein coupling also exist (Faruqi et al. 2000), and PAR4 is activated and shut off more slowly than PAR1 (Shapiro et al. 2000). The tempo of calcium signaling to thrombin in human platelets appears to represent the sum of contributions of both receptors (Covic et al. 2000; Shapiro et al. 2000).

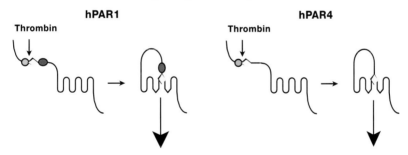

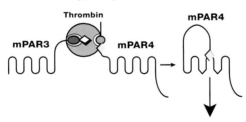

Figure 4. Thrombin receptors in human and mouse platelets. (*Top*) Human platelets express PAR1 and PAR4, and available data suggest that these receptors can independently mediate thrombin signaling. PAR1 can mediate activation of human platelets by low thrombin concentrations. In the absence of PAR1 function, PAR4 can mediate activation of human platelets but requires high thrombin concentrations. (*Bottom*) Mouse platelets express PAR3 and PAR4, and, surprisingly, it appears that mPAR3, rather than itself mediating transmembrane signaling, functions as a cofactor that supports cleavage and activation of mPAR4 at low thrombin concentrations. The cartoon shows mPAR3 presenting thrombin (*green sphere*) to mPAR4. In this model, thrombin has remained bound to a cleaved mPAR3 via interaction with PAR3's hirudin-like sequence (*pink oval*). Whether a ternary complex between mPAR3, thrombin, and mPAR4 actually forms is unknown. It is possible that PAR3 simply increases the effective concentration of thrombin at the platelet surface. These species differences are fortunate in that they provide an opportunity to address key questions regarding the PAR signaling in a convenient way in mouse models (see text).

In addition to interacting with PARs, thrombin also binds to glycoprotein Ibα (GPIbα) on the surface of human platelets (Okamura et al. 1978). GPIbα is part of a multifunctional protein complex that also binds vWF and P-selectin (Andrews et al. 1999). Platelets from patients with Bernard-Soulier syndrome lack surface GPIbα and show decreased thrombin responsiveness, but such platelets are also structurally abnormal (De Marco et al. 1990; Ware et al. 1990; De Marco et al. 1991; Ruggeri 1991). However, antibodies that block thrombin binding to GPIbα attenuate platelet activation by thrombin and decrease PAR1 cleavage (De Marco et al. 1991, 1994; De Candia et al. 2000). These and other observations suggest a model in which GPIbα may serve as a cofactor for thrombin cleavage of PARs on human platelets, analogous to PAR3's role as a cofactor for PAR4 in mouse platelets (Nakanishi-Matsui et al. 2000). Attempts to demonstrate cofactor activity of the GPIb complex by coexpressing it together with PARs have been unsuccessful (Nakanishi-Matsui et al. 2000), but the significance of such negative results is unclear. Clearly, more work is needed on this question.

Thrombin Receptors in Mouse Platelets

Unlike human platelets, mouse platelets do not express PAR1. Instead, mouse platelets utilize PAR3 and PAR4 for thrombin signaling (Kahn et al. 1998). Platelets from PAR3-deficient mice fail to respond to low concentrations of thrombin and have delayed responses at high concentrations of thrombin (Kahn et al. 1998). Thus, PAR3 is important for thrombin responsiveness in mouse platelets, but, intriguingly, mPAR3 appears unable to mediate transmembrane signaling directly (Nakanishi-Matsui et al. 2000). Instead, mPAR3 functions as a cofactor for cleavage and activation of mPAR4 at low thrombin concentrations (Nakanishi-Matsui et al. 2000). This model predicts that, absent other thrombin receptors, all thrombin signaling in mouse platelets should be PAR4-dependent. Indeed, platelets from PAR4-deficient mice were unresponsive to even very high concentrations of thrombin (Sambrano et al. 2001). The relationship between mPAR3 and mPAR4 represents an interesting type of interaction between G-protein-coupled receptors in which one receptor acts as an accessory protein that aids "ligation" of another (Fig. 4). It also emphasizes the potential importance of cofactors that localize proteases to the cell surface in orchestrating activation of PARs.

Potential Role of Platelet PARs in Thrombosis: Utility of Mouse Models

Thrombosis of the arteries that supply the heart, brain, and other vital organs is a major cause of morbidity and mortality. Both thrombin and platelets are clearly important in arterial thrombosis, but a host of mediators and processes contribute. Thrombin itself has many actions, and platelets respond to multiple agonists. Thus, the relative importance of thrombin signaling in platelets per se in thrombosis is unknown. Mouse models provide an interesting tool for probing this issue. Platelets from PAR4-deficient mice show no thrombin signaling, but thrombin signaling in most other cell types in these mice is intact because of PAR1 expression. Although they are relatively healthy and show no spontaneous bleeding, PAR4-deficient mice have remarkably long tail-bleeding times. They are also protected against thrombosis of mesenteric arterioles induced by ferric chloride injury and against thromboplastin-induced pulmonary embolism (Sambrano et al. 2001; Weiss et al. 2002). Thus, thrombin signaling in platelets is necessary for normal hemostasis and plays an important role in several models of thrombosis despite the multiple potentially redundant pathways involved in these processes.

Is inhibition of thrombin signaling in platelets a rational strategy for the development of antithrombotic drugs? Complete ablation of thrombin signaling in platelets like that achieved in PAR4-deficient mice would be difficult to achieve pharmacologically in man because human platelets have two receptors, PAR1 and PAR4, that can both independently mediate thrombin signaling. However, it is possible that complete inhibition of thrombin signaling is not required for an antithrombotic effect. PAR3-deficient mouse platelets show little or no response to low concentrations of thrombin and delayed responses to high concentrations of thrombin, much like human platelets in which PAR1 has been inhibited (Brass et al. 1992; Hung et al. 1992b; Kahn et al. 1998, 1999). Perhaps surprisingly, PAR3-deficient mice showed levels of protection in the mesenteric thrombosis and pulmonary embolism models mentioned above that were similar to those seen with PAR4 deficiency (Weiss et al. 2002). Thus, at least in mouse models, partial inhibition of thrombin signaling in platelets appears to be sufficient for an antithrombotic effect. In addition, administration of a PAR1 antiserum to monkeys was protective in a model of carotid artery thrombosis (Cook et al. 1995). Taken together, these observations may help stimulate the development of better reagents for probing the utility of PAR1 inhibition in relevant models.

PARS IN ENDOTHELIAL ACTIVATION

PAR1 appears to be the major mediator of thrombin signaling in vascular endothelial cells in both mouse and man, and most of the actions of thrombin on endothelial cells cited above have been reproduced using PAR1 agonist peptide. Endothelial cells also express PAR2, which may serve to sense tryptase released from mast cells (Molino et al. 1997) or coagulation proteases VIIa and Xa (Camerer et al. 2000).

What functions do endothelial PARs serve? One possible scenario based on the known actions of thrombin on endothelial cells follows. Tissue injury triggers local generation of coagulation proteases and/or release of mast cell tryptase, which, via PARs, activate endothelial cells (Figs. 1 and 5). The activated endothelium displays adhesion molecules and produces platelet-activating factor, a potent platelet and leukocyte activator (Zimmerman et al. 1996), and chemokines such as IL-8; this serves to recruit platelets and leukocytes to the site. The activated endothelium also becomes more permeable, which may allow increased access of plasma proteins such as complement and immunoglobulins to the extravascular space (Johnson et al. 1998). Thus, PARs may link tissue injury to endothelial responses that bring effector cells and proteins to sites of damage or infection.

Links between coagulation and inflammation are well known, and positive feedback loops in which thrombin triggers endothelial responses that beget additional local thrombin generation and endothelial activation can be envisioned (Fig. 5). On a microscopic scale, such mechanisms may be beneficial for isolating bacteria that inoculate a wound. However, disseminated intravascular coagulation with microvascular thrombosis and tissue damage can occur in the setting of a strong systemic inflammatory stimulus (e.g., sepsis) and/or deficiencies that disinhibit thrombin production (e.g., protein C deficiency, protein S deficiency, and factor V Leiden).

PAR-deficient mice provide an opportunity to explore the role of endothelial PARs in inflammatory responses. PAR1 appears to be the major thrombin receptor in endothelial cells in both mouse and humans, and because PAR1 is not expressed in mouse platelets, PAR1-deficient mice offer an opportunity to perturb thrombin signaling in endothelial cells without perturbing platelet signaling, thereby "isolating" the contribution of endothelial activation by thrombin in models of thrombosis and inflammation. PAR1 deficiency was associated with diminished leukocyte infiltration and renal damage in a mouse model of antibody-mediated glomerulonephritis (Cunningham et al. 2000). Similarly, PAR2 appears to be the major mediator of Xa signaling in endothelial cells (Camerer et al. 2002), and PAR2 deficiency decreased early leukocyte rolling in exteriorized cremasteric preparations (Lindner et al. 2000). Further studies to define the relative importance of PARs in inflammatory responses are warranted.

THROMBIN SIGNALING IN EMBRYONIC DEVELOPMENT

Approximately half of PAR1-deficient embryos die at midgestation (Connolly et al. 1996; Griffin et al. 2001) with pericardial edema and bleeding. Vasculogenesis appeared normal in these embryos, but varying degrees of delayed maturation of the yolk sac vasculature were noted; whether such changes are primary or secondary is unclear. It is difficult to ascribe bleeding and death of PAR1-deficient embryos to a platelet function defect.

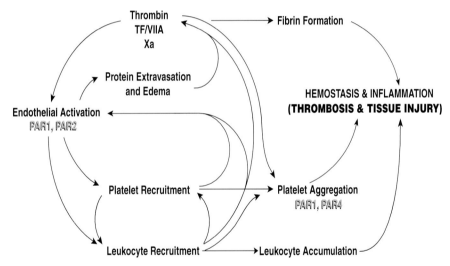

Figure 5. Links between coagulation and inflammation: possible role of PARs. Thrombin activates endothelial PAR1. Factor Xa and/or tissue factor/factor VIIa complex (TF/VIIa) may activate endothelial PAR2. PAR signaling up-regulates adhesion molecules on the endothelial surface and triggers production of lipid mediators and chemokines (Colotta et al. 1994; Ali et al. 1996; Ueno et al. 1996; Zimmerman et al. 1996). This leads to adhesion, rolling, and eventually attachment and activation of platelets and leukocytes at the endothelial surface. Local concentration of tissue-factor-bearing leukocytes and microparticles (Osterud 1998; Giesen et al. 1999) and platelet procoagulant activity (Sims et al. 1989) may promote further thrombin generation (Palabrica et al. 1992). Thrombin also increases endothelial permeability, and PAR activation triggers edema formation at least in part by directly or indirectly triggering mast cell degranulation (Cirino et al. 1996). Increased permeability may also promote extravascular thrombin generation as plasma coagulation factors contact extravascular tissue factor. Platelets and leukocytes can directly activate endothelial cells by presenting CD40L and other mediators, thereby up-regulating not only adhesion molecules and cytokines (Henn et al. 1998) but also tissue factor and PAR2 (Bevilacqua and Gimbrone 1987; Nystedt et al. 1996). Leukocytes and platelets can themselves interact via P-selectin (Palabrica et al. 1992), and neutrophils can activate platelets by release of the granzyme cathepsin G. Localized, such a system may be useful both for effecting hemostasis and for isolating and destroying pathogens that inoculate a wound. Undamped, such a system may lead to tissue damage from leukocyte products or from ischemia due to microvascular thrombosis. Undamped positive feedback between coagulation and inflammation is made more likely by cellular responses to endotoxin (Esmon 2000) and by genetic deficiencies in natural anticoagulant pathways (Inbal et al. 1997) and is associated with hemorrhagic infarction of tissues. Clearly, a host of cell types and signaling systems orchestrate coagulation and inflammatory responses, and the relative importance of PARs in connecting these responses remains to be tested. The recent demonstration that administration of the natural anticoagulant activated protein C decreased the probability of death in sepsis will likely stimulate work in this area.

PAR1 does not function in mouse platelets (Connolly et al. 1994; Derian et al. 1995; Connolly et al. 1996), and platelets are not required for normal development through midgestation (Shivdasani et al. 1995). At midgestation, PAR1 is expressed most abundantly in endocardial and endothelial cells, and a transgene in which the endothelial-specific Tie2 promoter drove PAR1 expression prevented death of PAR1-deficient embryos (Griffin et al. 2001). These data suggest that death of PAR1-deficient embryos is due to a lack of PAR1 signaling in endothelial cells. Several questions arise.

What does PAR1 sense biochemically—thrombin or something novel? One logical approach to this question is to compare the phenotype of PAR1 deficiency to those of coagulation-factor knockouts (Bugge et al. 1996; Carmeliet et al. 1996; Cui et al. 1996; Sun et al. 1998; Xue et al. 1998). Approximately half of factor V- and prothrombin-deficient embryos die at midgestation with phenotypes similar to that of PAR1-deficient embryos (Cui et al. 1996; Sun et al. 1998; Xue et al. 1998). Thus, at first glance, one might envision a simple pathway in which the coagulation cascade functions to activate PAR1. However, combined deficiency of factor V and PAR1 resulted in synthetic lethality. This genetic interaction clearly demonstrates that actions of factor V that are independent of PAR1 are important for embryonic development, at least in PAR1's absence. Of note, knockout of tissue factor, the trigger for coagulation, results in death of 85–100% of embryos at midgestation (Bugge et al. 1996; Carmeliet et al. 1996; Toomey et al. 1996). Taken together, available data are consistent with a working model in which the coagulation cascade activates both PAR1 and other targets during embryonic development (Fig. 6). In this model, the less severe phenotypes associated with deficiencies of the soluble coagulation factors versus tissue factor are explained by transfer of enough maternal coagulation factors to effect partial rescue of the embryo. This model can be probed by determining whether combined deficiencies of PAR1 and other candidate targets recapitulate the phenotype of the tissue-factor knockout.

What does PAR1 sense physiologically during development? If it is indeed activation of the coagulation cascade, is this system about sensing occasional breaks that occur during remodeling of the embryonic vasculature, analogous to hemostasis in the adult? Or does the system have a more sustained function? One might imagine a role in monitoring the leakiness of developing vessels so as to regulate the formation of endothelial junctions, matrix production, or pericyte recruitment. A role in simply allowing endothelial cells to sense the presence of plasma might also be imagined. The transition from an endothe-

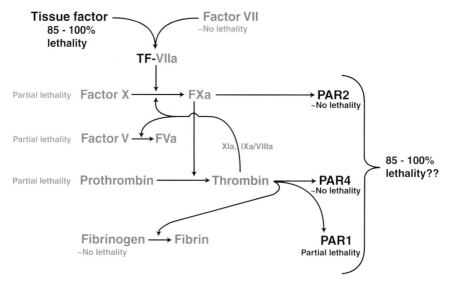

Figure 6. Coagulation factors and their targets: knockout phenotypes at midgestation. An outline of the coagulation cascade and some of its effectors, along with the extent of embryonic lethality at midgestation for individual knockouts, is shown. PAR1 deficiency is associated with death of approximately half of embryos at midgestation. Death is associated with pericardial edema, bleeding from multiple sites, and variable levels of abnormal maturation of the yolk sac vasculature. Transgene rescue studies suggest that death is due to loss of PAR1 function in endothelial cells. Does this imply a new role for coagulation proteases? At first glance, a comparison of the PAR1 phenotype with those of coagulation factor knockouts is confusing. Lack of tissue factor, the integral membrane protein that triggers coagulation, causes death of embryos at midgestation that is associated with bleeding and an abnormal yolk sac vasculature, but the phenotype is more penetrant than that seen with PAR1 (85–100% vs. ~50%). In contrast, phenotypes associated with lack of downstream soluble coagulation factors (shown in *blue*) vary widely. This paradox may well be explained by variable rescue of coagulation null embryos by maternal coagulation factors that enter the embryonic circulation (studies of the development of coagulation-factor-deficient embryos have been done by necessity in heterozygous dams). This does not, of course, explain why the tissue-factor phenotype is more severe than that of PAR1. The recent demonstration that combined Factor-V-deficiency and PAR1-deficiency results in synthetic lethality shows that coagulation cascade targets other than PAR1 are important for development (at least in PAR1's absence). Other PARs are prime candidates. This hypothesis predicts that combined deficiencies of the various PARs should recapitulate the tissue-factor phenotype and is now testable. These findings raise a host of questions (see text). Investigation of the roles of PARs and coagulation proteases in the embryo promises to reveal novel regulatory mechanisms that govern blood vessel development as well as unexpected roles for protease signaling.

lial cord to a tube might be confirmed by sensing plasma flowing in the new lumen in this way.

Last, which of the many endothelial responses to PAR activation are important for embryonic development? There is no dearth of candidates. It is interesting to note that $G\alpha_{13}$-deficient embryos die at midgestation with defective vascular development (Offermanns et al. 1997a) and PAR1 does activate G12/13 signaling among other pathways. Whether defective signaling in endothelial cells is indeed responsible for the death of $G\alpha_{13}$-deficient embryos and whether decreased $G\alpha_{13}$ signaling in endothelial cells contributes to the phenotype of PAR1-deficient embryos remains to be explored, along with many other hypotheses. Regardless, it is clear that PAR signaling contributes to embryonic development in a way that is unrelated to hemostasis in any usual sense. This is exciting because it appears to point to a novel role for the "coagulation" cascade in a context other than mediating responses to tissue injury.

FUTURE DIRECTIONS

The studies cited above raise a host of questions regarding both the molecular mechanisms of PAR activation and protease signaling and the roles of such signaling in vivo.

How general are PAR–PAR interactions, and is oligomerization involved? To what extent do cofactors increase the diversity of proteases to which cells can respond via PARs? What are the physiological activators of PARs in different tissues and settings? It is worth noting that many trypsin-like proteases can activate PARs in vitro; establishing physiological significance for such phenomena is often difficult.

Do PARs account for signaling by thrombin and other coagulation proteases, or will other paradigms be identified? Because PAR1 and PAR4 are the only PARs known to mediate transmembrane signaling in response to thrombin in the mouse, the presence or absence of residual thrombin signaling in cells from mice or mouse embryos deficient in both PAR1 and PAR4 will be telling.

Will PAR1 antagonists prove to be useful in the prevention or treatment of thrombosis? At a practical level, a more robust exploration of this question in appropriate models will require the development of potent blocking antibodies or antagonists. Although blocking the binding of the tethered ligand with a competitive antagonist may be difficult, some success has been achieved (Bernatowicz et al. 1996; Andrade-Gordon et al. 1999). Noncompetitive antagonists represent a possible alternative; clopidogrel and ticlopidine, noncompetitive antagonists of a platelet ADP receptor, provide precedent that is encouraging at several levels.

Will PARs prove significant among the panoply of signaling systems and cell types that orchestrate inflammatory responses? Efforts to define the relative contribution of PARs in endothelial cells and leukocytes are just beginning.

PARs link tissue injury to cellular responses that might be viewed as part of an innate immune response for walling off and killing invading bacteria. One might also want to use this capacity to link tissue injury to cellular signaling for a variety of other purposes. For example, might such a system contribute to the link between tissue injury and perception of pain? Recent studies suggest this may be the case (Vergnolle et al. 2001a,b). Might PAR antagonists have a role as analgesics? Similarly, might one use this system that connects tissue injury to cellular responses to regulate the decision to mount an adaptive immune response? Does PAR signaling influence antigen presentation? Imagination rapidly outpaces available data.

Sorting out the role of the coagulation cascade and PARs in embryonic development remains an interesting challenge. Several of the specific questions are discussed above. It will be important to determine to what extent multiple PARs serve redundant functions in individual cell types. From what is known of PAR expression and signaling, some redundancy is virtually certain. We hope and expect that mice with combined deficiencies of several PARs will reveal as yet unexpected roles for protease signaling.

ACKNOWLEDGMENTS

Thanks to many colleagues and to the members of my laboratory for helpful discussions.

REFERENCES

Ali H., Tomhave E.D., Richardson R.M., Haribabu B., and Snyderman R. 1996. Thrombin primes responsiveness of selective chemoattractant receptors at a site distal to G protein activation. *J. Biol. Chem.* **271:** 3200.

Andrade-Gordon P., Maryanoff B.E., Derian C.K., Zhang H.C., Addo M.F., Darrow A.L., Eckardt A.J., Hoekstra W.J., McComsey D.F., Oksenberg D., Reynolds E.E., Santulli R.J., Scarborough R.M., Smith C.E., and White K.B. 1999. Design, synthesis, and biological characterization of a peptide-mimetic antagonist for a tethered-ligand receptor. *Proc. Natl. Acad. Sci.* **96:** 12257.

Andrews R.K., Shen Y., Gardiner E.E., Dong J.F., López J.A., and Berndt M.C. 1999. The glycoprotein Ib-IX-V complex in platelet adhesion and signaling. *Thromb. Haemostasis* **82:** 357.

Barr A.J., Brass L.F., and Manning D.R. 1997. Reconstitution of receptors and GTP-binding regulatory proteins (G proteins) in Sf9 cells. A direct evaluation of selectivity in receptor·G protein coupling. *J. Biol. Chem.* **272:** 2223.

Bernatowicz M.S., Klimas C.E., Hartl K.S., Peluso M., Allegretto N.J., and Seiler S.M. 1996. Development of potent thrombin receptor antagonist peptides. *J. Med. Chem.* **39:** 4879.

Bevilacqua M.P. and Gimbrone M.A., Jr. 1987. Inducible endothelial functions in inflammation and coagulation. *Semin. Thromb. Hemost.* **13:** 425.

Brass L.F., Manning D.R., Cichowski K., and Abrams C.S. 1997. Signaling through G proteins in platelets: To the integrins and beyond. *Thromb. Haemostasis* **78:** 581.

Brass L.F., Vassallo R.J., Belmonte E., Ahuja M., Cichowski K., and Hoxie J.A. 1992. Structure and function of the human platelet thrombin receptor. Studies using monoclonal antibodies directed against a defined domain within the receptor N terminus. *J. Biol. Chem.* **267:** 13795.

Bugge T.H., Xiao Q., Kombrinck K.W., Flick M.J., Holmbäck K., Danton M.J., Colbert M.C., Witte D.P., Fujikawa K.,

Davie E.W., and Degen J.L. 1996. Fatal embryonic bleeding events in mice lacking tissue factor, the cell-associated initiator of blood coagulation. *Proc. Natl. Acad. Sci.* **93:** 6258.

Camerer E., Huang W., and Coughlin S.R. 2000. Tissue factor- and factor X-dependent activation of protease-activated receptor 2 by factor VIIa. *Proc. Natl. Acad. Sci.* **97:** 5255.

Camerer E., Røttingen J.A., Iversen J.G., and Prydz H. 1996. Coagulation factors VII and X induce Ca^{++} oscillations in Madin-Darby canine kidney cells only when proteolytically active. *J. Biol. Chem.* **271:** 29034.

Camerer E., Kataoka H., Kahn M., Lease K., and Coughlin S.R. 2002. Genetic evidence that protease-activated receptors mediate Factor Xa signaling in endothelial cells. *J. Biol. Chem.* **277:** 16081.

Carmeliet P., Mackman N., Moons L., Luther T., Gressens P., Van V.I., Demunck H., Kasper M., Breier G., Evrard P., Muller M., Risau W., Edgington T., and Collen D. 1996. Role of tissue factor in embryonic blood vessel development. *Nature* **383:** 73.

Chen J., Ishii M., Wang L., Ishii K., and Coughlin S.R. 1994. Thrombin receptor activation: Confirmation of the intramolecular tethered liganding hypothesis and discovery of an alternative intermolecular liganding mode. *J. Biol. Chem.* **269:** 16041.

Cirino G., Cicala C., Bucci M.R., Sorrentino L., Maraganore J.M., and Stone S.R. 1996. Thrombin functions as an inflammatory mediator through activation of its receptor. *J. Exp. Med.* **183:** 821.

Clapham D.E. and Neer E.J. 1997. G protein beta gamma subunits. *Annu. Rev. Pharmacol. Toxicol.* **37:** 167.

Colman R.W., Marder V.J., Salzman E.W., and Hirsh J. 1994. Overview of hemostasis. In *Hemostasis and thrombosis* (ed. R.W. Colman et al.), p. 3. Lippincott, Philadelphia, Pennsylvania.

Colotta F., Sciacca F.L., Sironi M., Luini W., Rabiet M.J., and Mantovani A. 1994. Expression of monocyte chemotactic protein-1 by monocytes and endothelial cells exposed to thrombin. *Am. J. Pathol.* **144:** 975.

Connolly A.J., Ishihara H., Kahn M.L., Farese R.V., Jr., and Coughlin S.R. 1996. Role of the thrombin receptor in development and evidence for a second receptor. *Nature* **381:** 516.

Connolly T.M., Condra C., Feng D.M., Cook J.J., Stranieri M.T., Reilly C.F., Nutt R.F., and Gould R.J. 1994. Species variability in platelet and other cellular responsiveness to thrombin receptor-derived peptides. *Thromb. Haemostasis* **72:** 627.

Cook J.J., Sitko G.R., Bednar B., Condra C., Mellott M.J., Feng D.M., Nutt R.F., Shafer J.A., Gould R.J., and Connolly T.M. 1995. An antibody against the exosite of the cloned thrombin receptor inhibits experimental arterial thrombosis in the African green monkey. *Circulation* **91:** 2961.

Coughlin S.R. 1998. How thrombin 'talks' to cells: Molecular mechanisms and roles in vivo (Sol Sherry lecture in thrombosis). *Arterioscler. Thromb. Vasc. Biol.* **18:** 514.

———. 1999. How the protease thrombin talks to cells. *Proc. Natl. Acad. Sci.* **96:** 11023.

———. 2000. Thrombin signalling and protease-activated receptors. *Nature* **407:** 258.

Covic L., Gresser A.L., and Kuliopulos A. 2000. Biphasic kinetics of activation and signaling for PAR1 and PAR4 thrombin receptors in platelets. *Biochemistry* **39:** 5458.

Cui J., O'Shea K.S., Purkayastha A., Saunders T.L., and Ginsburg D. 1996. Fatal haemorrhage and incomplete block to embryogenesis in mice lacking coagulation factor V. *Nature* **384:** 66.

Cunningham M.A., Rondeau E., Chen X., Coughlin S.R., Holdsworth S.R., and Tipping P.G. 2000. Protease-activated receptor 1 mediates thrombin-dependent, cell-mediated renal inflammation in crescentic glomerulonephritis. *J. Exp. Med.* **191:** 455.

Davey M.G. and Lüscher E.F. 1967. Actions of thrombin and other coagulant and proteolytic enzymes on blood platelets. *Nature* **216:** 857.

De Candia E., Hall S.W., Rutella S., Landolfi R., Andrews R.K., and De Cristofaro R. 2001. Binding of thrombin to the glycoprotein Ib accelerates the hydrolysis of Par-1 on intact platelets. *J. Biol. Chem.* **276:** 4692.

De Marco L., Mazzucato M., Masotti A., and Ruggeri Z.M. 1994. Localization and characterization of an alpha-thrombin-binding site on platelet glycoprotein Ib alpha. *J. Biol. Chem.* **269:** 6478.

De Marco L., Mazzucato M., Masotti A., Fenton J.W., II, and Ruggeri Z.M. 1991. Function of glycoprotein Ib alpha in platelet activation induced by alpha-thrombin. *J. Biol. Chem.* **266:** 23776.

De Marco L., Mazzucato M., Fabris F., De Roia D., Coser P., Girolami A., Vicente V., and Ruggeri Z.M. 1990. Variant Bernard-Soulier syndrome type bolzano. A congenital bleeding disorder due to a structural and functional abnormality of the platelet glycoprotein Ib-IX complex. *J. Clin. Invest.* **86:** 25.

Derian C.K., Santulli R.J., Tomko K.A., Haertlein B.J., and Andrade-Gordon P. 1995. Species differences in platelet responses to thrombin and SFLLRN. Receptor-mediated calcium mobilization and aggregation and regulation by protein kinases. *Thromb. Res.* **6:** 505.

Esmon C.T. 2000. Introduction: Are natural anticoagulants candidates for modulating the inflammatory response to endotoxin? *Blood* **95:** 1113.

Esmon C.T., Fukudome K., Mather T., Bode W., Regan L.M., Stearns-Kurosawa D.J., and Kurosawa S. 1999. Inflammation, sepsis, and coagulation. *Haematologica* **84:** 254.

Faruqi T.R., Weiss E.J., Shapiro M.J., Hung W., and Coughlin S.R. 2000. Structure-function analysis of protease-activated receptor-4 tethered tigand peptides: Determinants of specificity and utility in assays of receptor function. *J. Biol. Chem.* **275:** 19728.

Frenette P.S., Mayadas T.N., Rayburn H., Hynes R.O., and Wagner D.D. 1996. Susceptibility to infection and altered hematopoiesis in mice deficient in both P- and E-selectins. *Cell* **84:** 563.

Fukuhara S., Murga C., Zohar M., Igishi T., and Gutkind J.S. 1999. A novel PDZ domain containing guanine nucleotide exchange factor links heterotrimeric G proteins to Rho. *J. Biol. Chem.* **274:** 5868.

Giesen P.L., Rauch U., Bohrmann B., Kling D., Roqué M., Fallon J.T., Badimon J.J., Himber J., Riederer M.A., and Nemerson Y. 1999. Blood-borne tissue factor: Another view of thrombosis. *Proc. Natl. Acad. Sci.* **96:** 2311.

Griffin C.T., Srinivasan Y., Zheng Y.W., Huang W., and Coughlin S.R. 2001. A role for thrombin receptor signaling in endothelial cells during embryonic development. *Science* **293:** 1666.

Hamberg M., Svensson J., and Samuelsson B. 1975. Thromboxanes: A new group of biologically active compounds derived from prostaglandin endoperoxides. *Proc. Natl. Acad. Sci.* **72:** 2994.

Hart M.J., Jiang X., Kozasa T., Roscoe W., Singer W.D., Gilman A.G., Sternweis P.C., and Bollag G. 1998. Direct stimulation of the guanine nucleotide exchange activity of p115 RhoGEF by Galpha13 (comments). *Science* **280:** 2112.

Hattori R., Hamilton K.K., Fugate R.D., McEver R.P., and Sims P.J. 1989. Stimulated secretion of endothelial von Willebrand factor is accompanied by rapid redistribution to the cell surface of the intracellular granule membrane protein GMP-140. *J. Biol. Chem.* **264:** 7768.

Hein L., Ishii K., Coughlin S.R., and Kobilka B.K. 1994. Intracellular targeting and trafficking of thrombin receptors: A novel mechanism for resensitization of a G protein-coupled receptor. *J. Biol. Chem.* **269:** 27719.

Henn V., Slupsky J.R., Gräfe M., Anagnostopoulos I., Förster R., Müller-Berghaus G., and Kroczek R.A. 1998. CD40 ligand on activated platelets triggers an inflammatory reaction of endothelial cells. *Nature* **391:** 591.

Hoxie J.A., Ahuja M., Belmonte E., Pizarro S., Parton R., and Brass L.F. 1993. Internalization and recycling of activated thrombin receptors. *J. Biol. Chem.* **268:** 13756.

Hughes P.E. and Pfaff M. 1998. Integrin affinity modulation.

Trends Cell Biol. **8:** 359.

Hung D.T., Wong Y.H., Vu T.-K.H., and Coughlin S.R. 1992a. The cloned platelet thrombin receptor couples to at least two distinct effectors to stimulate both phosphoinositide hydrolysis and inhibit adenylyl cyclase. *J. Biol. Chem.* **353:** 20831.

Hung D.T., Vu T.K., Wheaton V.I., Ishii K., and Coughlin S.R. 1992b. Cloned platelet thrombin receptor is necessary for thrombin-induced platelet activation. *J. Clin. Invest.* **89:** 1350.

Inbal A., Kenet G., Zivelin A., Yermiyahu T., Bronstein T., Sheinfeld T., Tamari H., Gitel S., Eshel G., Duchemin J., Aiach M., and Seligsohn U. 1997. Purpura fulminans induced by disseminated intravascular coagulation following infection in 2 unrelated children with double heterozygosity for factor V Leiden and protein S deficiency. *Thromb. Haemostasis* **77:** 1086.

Ishihara H., Connolly A.J., Zeng D., Kahn M.L., Zheng Y.W., Timmons C., Tram T., and Coughlin S.R. 1997. Protease-activated receptor 3 is a second thrombin receptor in humans. *Nature* **386:** 502.

Ishii K., Hein L., Kobilka B., and Coughlin S.R. 1993. Kinetics of thrombin receptor cleavage on intact cells. Relation to signaling. *J. Biol. Chem.* **268:** 9780.

Ishii K., Chen J., Ishii M., Koch W.J., Freedman N.J., Lefkowitz R.J., and Coughlin S.R. 1994. Inhibition of thrombin receptor signaling by a G protein-coupled receptor kinase. Functional specificity among G protein-coupled receptor kinases. *J. Biol. Chem.* **269:** 1125.

Johnson K., Choi Y., DeGroot E., Samuels I., Creasey A., and Aarden L. 1998. Potential mechanisms for a proinflammatory vascular cytokine response to coagulation activation. *J. Immunol.* **160:** 5130.

Kahn M.L., Nakanishi-Matsui M., Shapiro M.J., Ishihara H., and Coughlin S.R. 1999. Protease-activated receptors 1 and 4 mediate activation of human platelets by thrombin. *J. Clin. Invest.* **103:** 879.

Kahn M.L., Zheng Y.W., Huang W., Bigornia V., Zeng D., Moff S., Farese R.V., Jr., Tam C., and Coughlin S.R. 1998. A dual thrombin receptor system for platelet activation. *Nature* **394:** 690.

Kaplan D.D., Meigs T.E., and Casey P.J. 2001. Distinct regions of the cadherin cytoplasmic domain are essential for functional interaction with Galpha 12 and beta-catenin. *J. Biol. Chem.* **276:** 44037.

Klages B., Brandt U., Simon M.I., Schultz G., and Offermanns S. 1999. Activation of G12/G13 results in shape change and rho/rho kinase-mediated myosin light chain phosphorylation in mouse platelets. *J. Cell Biol.* **144:** 745.

Kozasa T., Jiang X., Hart M.J., Sternweis P.M., Singer W.D., Gilman A.G., Bollag G., and Sternweis P.C. 1998. p115 RhoGEF, a GTPase activating protein for Galpha12 and Galpha13 (comments). *Science* **280:** 2109.

Leevers S.J., Vanhaesebroeck B., and Waterfield M.D. 1999. Signalling through phosphoinositide 3-kinases: The lipids take centre stage. *Curr. Opin. Cell Biol.* **11:** 219.

Lefkowitz R.J., Pitcher J., Krueger K., and Daaka Y. 1998. Mechanisms of beta-adrenergic receptor desensitization and resensitization. *Adv. Pharmacol.* **42:** 416.

Lindner J.R., Kahn M.L., Coughlin S.R., Sambrano G.R., Schauble E., Bernstein D., Foy D., Hafezi-Moghadam A., and Ley K. 2000. Delayed onset of inflammation in protease-activated receptor-2- deficient mice. *J. Immunol.* **165:** 6504.

Liu L., Vu T.-K.H., Esmon C.T., and Coughlin S.R. 1991. The region of the thrombin receptor resembling hirudin binds to thrombin and alters enzyme specificity. *J. Biol. Chem.* **266:** 16977.

Lum H. and Malik A.B. 1994. Regulation of vascular endothelial barrier function. *Am. J. Physiol.* **267:** L223.

Mathews I.I., Padmanabhan K.P., Ganesh V., Tulinsky A., Ishii M., Chen J., Turck C.W., Coughlin R., and Fenton J.W., II. 1994. Crystallographic structures of thrombin complexed with thrombin receptor peptides: Existence of expected and novel binding modes. *Biochemistry* **33:** 3266.

Meigs T.E., Fields T.A., McKee D.D., and Casey P.J. 2001. Interaction of Galpha 12 and Galpha 13 with the cytoplasmic domain of cadherin provides a mechanism for beta-catenin release. *Proc. Natl. Acad. Sci.* **98:** 519.

Meigs T.E., Fedor-Chaiken M., Kaplan D.D., Brackenbury R., and Casey P.J. 2002. Galpha 12 and Galpha 13 negatively regulate the adhesive functions of cadherin. *J. Biol. Chem.* **25:** 25.

Molino M., Barnathan E.S., Numerof R., Clark J., Dreyer M., Cumashi A., Hoxie J.A., Schechter N., Woolkalis M., and Brass L.F. 1997. Interactions of mast cell tryptase with thrombin receptors and PAR-2. *J. Biol. Chem.* **272:** 4043.

Nakanishi-Matsui M., Zheng Y.W., Sulciner D.J., Weiss E.J., Ludeman M.L., and Coughlin S.R. 2000. PAR3 is a cofactor for PAR4 activation by thrombin. *Nature* **404:** 609.

Nystedt S., Ramakrishnan V., and Sundelin J. 1996. The proteinase-activated receptor 2 is induced by inflammatory mediators in human endothelial cells. Comparison with the thrombin receptor. *J. Biol. Chem.* **271:** 14910.

Nystedt S., Emilsson K., Wahlestedt C., and Sundelin J. 1994. Molecular cloning of a potential proteinase activated receptor (comments). *Proc. Natl. Acad. Sci.* **91:** 9208.

O'Brien P.J., Prevost N., Molino M., Hollinger M.K., Woolkalis M.J., Woulfe D.S., and Brass L.F. 2000. Thrombin responses in human endothelial cells. Contributions from receptors other than PAR1 include the transactivation of PAR2 by thrombin-cleaved PAR1. *J. Biol. Chem.* **275:** 13502.

Offermanns S., Laugwitz K.-L., Spicher K., and Schultz G. 1994. G proteins of the G12 family are activated via thromboxane A2 and thrombin receptors in human platelets. *Proc. Natl. Acad. Sci.* **91:** 504.

Offermanns S., Mancino V., Revel J.-P., and Simon M.I. 1997a. Vascular system defects and impaired cell chemokinesis as a result of Gα13 deficiency. *Science* **275:** 533.

Offermanns S., Toombs C.F., Hu Y.H., and Simon M.I. 1997b. Defective platelet activation in G alpha(q)-deficient mice. *Nature* **389:** 183.

Okamura T., Hasitz M., and Jamieson G.A. 1978. Platelet glycocalicin: Interaction with thrombin and role as thrombin receptor on the platelet surface. *J. Biol. Chem.* **253:** 3435.

Osterud B. 1998. Tissue factor expression by monocytes: Regulation and pathophysiological roles. *Blood Coagul. Fibrinolysis* (suppl. 1) **9:** S9.

Palabrica T., Lobb R., Furie B.C., Aronovitz M., Benjamin C., Hsu Y.M., Sajer S.A., and Furie B. 1992. Leukocyte accumulation promoting fibrin deposition is mediated in vivo by P-selectin on adherent platelets. *Nature* **359:** 848.

Prenzel N., Zwick E., Daub H., Leserer M., Abraham R., Wallasch C., and Ullrich A. 1999. EGF receptor transactivation by G-protein-coupled receptors requires metalloproteinase cleavage of proHB-EGF. *Nature* **402:** 884.

Rasmussen U.B., Vouret-Craviari V., Jallat S., Schlesinger Y., Pagers G., Pavirani A., Lecocq J.P., Pouyssegur J., and Van Obberghen-Schilling E. 1991. cDNA cloning and expression of a hamster alpha-thrombin receptor coupled to Ca^{++} mobilization. *FEBS Lett.* **288:** 123.

Ruggeri Z.M. 1991. The platelet glycoprotein Ib-IX complex. *Prog. Hemostasis Thromb.* **10:** 35.

Sambrano G.R., Weiss E.J., Zheng Y.W., Huang W., and Coughlin S.R. 2001. Role of thrombin signalling in platelets in haemostasis and thrombosis. *Nature* **413:** 74.

Sambrano G.R., Huang W., Faruqi T., Mahrus S., Craik C., and Coughlin S.R. 2000. Cathepsin G activates protease-activated receptor-4 in human platelets. *J. Biol. Chem.* **275:** 6819.

Selak M.A., Chignard M., and Smith J.B. 1988. Cathepsin G is a strong platelet agonist released by neutrophils. *Biochem. J.* **251:** 293.

Shapiro M.J., Weiss E.J., Faruqi T.R., and Coughlin S.R. 2000. Protease-activated receptors 1 and 4 are shut off with distinct kinetics after activation by thrombin. *J. Biol. Chem.* **275:** 25216.

Shivdasani R.A., Rosenblatt M.F., Zucker F.D., Jackson C.W., Hunt P., Saris C.J., and Orkin S.H. 1995. Transcription factor NF-E2 is required for platelet formation independent of the actions of thrombopoietin/MGDF in megakaryocyte development. *Cell* **81:** 695.

Sims P.J., Wiedmer T., Esmon C.T., Weiss H.J., and Shattil S.J. 1989. Assembly of the platelet prothrombinase complex is linked to vesiculation of the platelet plasma membrane. *J. Biol. Chem.* **264:** 17049.

Stenberg P.E., McEver R.P., Shuman M.A., Jacques Y.V., and Bainton D.F. 1985. A platelet alpha-granule membrane protein (GMP-140) is expressed on the plasma membrane after activation. *J. Cell Biol.* **101:** 880.

Stoyanov B., Volinia S., Hanck T., Rubio I., Loubtchenkov M., Malek D., Stoyanova S., Vanhaesebroeck B., Dhand R., Nürnberg B. et al. 1995. Cloning and characterization of a G protein-activated human phosphoinositide-3 kinase. *Science* **269:** 690.

Subramaniam M., Frenette P.S., Saffaripour S., Johnson R.C., Hynes R.O., and Wagner D.D. 1996. Defects in hemostasis in P-selectin-deficient mice. *Blood* **87:** 1238.

Sun W.Y., Witte D.P., Degen J.L., Colbert M.C., Burkart M.C., Holmbäck K., Xiao Q., Bugge T.H., and Degen S.J. 1998. Prothrombin deficiency results in embryonic and neonatal lethality in mice. *Proc. Natl. Acad. Sci.* **95:** 7597.

Taylor S., Chae H., Rhee S.-G., and Exton J. 1991. Activation of the B1 isozyme of phospholipase C by a subunits of the Gq class of G proteins. *Nature* **350:** 516.

Toomey J.R., Kratzer K.E., Lasky N.M., Stanton J.J., and Broze G.J., Jr. 1996. Targeted disruption of the murine tissue factor gene results in embryonic lethality. *Blood* **88:** 1583.

Trejo J. and Coughlin S.R. 1999. The cytoplasmic tails of protease-activated receptor-1 and substance P receptor specify sorting to lysosomes versus recycling. *J. Biol. Chem.* **274:** 2216.

Trejo J., Hammes S.R., and Coughlin S.R. 1998. Termination of signaling by protease-activated receptor-1 is linked to lysosomal sorting. *Proc. Natl. Acad. Sci.* **95:** 13698.

Ueno A., Murakami K., Yamanouchi K., Watanabe M., and Kondo T. 1996. Thrombin stimulates production of interleukin-8 in human umbilical vein endothelial cells. *Immunology* **88:** 76.

Vergnolle N., Wallace J.L., Bunnett N.W., and Hollenberg M.D. 2001a. Protease activated receptors in inflammation, neuronal signaling and pain. *Trends Pharmacol. Sci.* **22:** 146.

Vergnolle N., Bunnett N.W., Sharkey K.A., Brussee V., Compton S.J., Grady E.F., Cirino G., Gerard N., Basbaum A.I., Andrade-Gordon P., Hollenberg M.D., and Wallace J.L. 2001b. Proteinase-activated receptor-2 and hyperalgesia: A novel pain pathway. *Nat. Med.* **7:** 821.

Vouret-Craviari V., Boquet P., Pouysségur J., and Van Obberghen-Schilling E. 1998. Regulation of the actin cytoskeleton by thrombin in human endothelial cells: Role of Rho proteins in endothelial barrier function. *Mol. Biol. Cell* **9:** 2639.

Vu T.-K.H., Hung D.T., Wheaton V.I., and Coughlin S.R. 1991a. Molecular cloning of a functional thrombin receptor reveals a novel proteolytic mechanism of receptor activation. *Cell* **64:** 1057.

Vu T.-K.H., Wheaton V.I., Hung D.T., and Coughlin S.R. 1991b. Domains specifying thrombin-receptor interaction. *Nature* **353:** 674.

Ware J., Russell S.R., Vicente V., Scharf R.E., Tomer A., McMillan R., and Ruggeri Z.M. 1990. Nonsense mutation in the glycoprotein Ib alpha coding sequence associated with Bernard-Soulier syndrome. *Proc. Natl. Acad. Sci.* **87:** 2026.

Weiss E.J., Hamilton J.R., Lease K.E., and Coughlin S.R. 2002. Protection against thrombosis in mice lacking PAR3. *Blood* **100:** 3240.

Xu W.F., Andersen H., Whitmore T.E., Presnell S.R., Yee D.P., Ching A., Gilbert T., Davie E.W., and Foster D.C. 1998. Cloning and characterization of human protease-activated receptor 4. *Proc. Natl. Acad. Sci.* **95:** 6642.

Xue J., Wu Q., Westfield L.A., Tuley E.A., Lu D., Zhang Q., Shim K., Zheng X., and Sadler J.E. 1998. Incomplete embryonic lethality and fatal neonatal hemorrhage caused by prothrombin deficiency in mice. *Proc. Natl. Acad. Sci.* **95:** 7603.

Yang J., Wu J., Kowalska M.A., Dalvi A., Prevost N., O'Brien P.J., Manning D., Poncz M., Lucki I., Blendy J.A., and Brass L.F. 2000. Loss of signaling through the G protein, Gz, results in abnormal platelet activation and altered responses to psychoactive drugs. *Proc. Natl. Acad. Sci.* **97:** 9984.

Zimmerman G.A., Elstad M.R., Lorant D.E., McLntyre T.M., Prescott S.M., Topham M.K., Weyrich A.S., and Whatley R.E. 1996. Platelet-activating factor (PAF): Signalling and adhesion in cell-cell interactions. *Adv. Exp. Med. Biol.* **416:** 297.

Migration of Monocytes/Macrophages In Vitro and In Vivo Is Accompanied by MMP12-dependent Tunnel Formation and by Neovascularization

M. ANGHELINA, A. SCHMEISSER,* P. KRISHNAN, L. MOLDOVAN, R.H. STRASSER,*
AND N.I. MOLDOVAN

*Dorothy M. Davis Heart and Lung Research Institute, The Ohio State University Medical Center, Columbus, Ohio 43210; *Heart Center, Medical Clinic II, Technical University of Dresden, Dresden, Germany, D-01307*

Migration of cells within tissues often requires the proteolytic destruction of the extracellular matrix (ECM). There is a large body of evidence that shows the involvement of matrix degrading enzymes, in particular the metalloproteases (MMP) and their inhibitors (TIMPS), in many physiological and pathological processes related to cell movement in general (Amorino et al. 1998; Hiraoka et al. 1998; Murphy and Gavrilovic 1999; Shapiro 1999; Stetler-Stevenson 1999; Levine et al. 2000) and to angiogenesis in particular (Werb et al. 1999; Moldovan et al. 2000; Libby and Schonbeck 2001). However, the fate of the tracks created within the matrix by migrating cells (Fig. 1a), which we called tunnels (Moldovan et al. 2000; Moldovan 2002), remain largely unexplored. Do they really exist? How long do they survive after a cell moves away? Do they collapse, are they passively filled by other proteins, or is there an active mechanism of tissue repair behind the migrating cells?

By their very definition, when the tunnels are well preserved morphologically, they should look in cross-section like empty spaces of about a cell's diameter, or like proteolysis rims surrounding the cells (Fig. 1b). Alternatively, the tunnels may become paved by cells attached to their walls (Fig. 1c), and then they may present themselves in histological sections as rings staining positive for cellular antigens, other than endothelial ones (Fig. 1d).

The chances to detect clear-cut, well-preserved tunnels depend on the density of the matrix. Additionally, the speed of cell movement will be reflected in the efficiency of their proteolytic machinery, and therefore in the likelihood of seeing the proteolytic rims around them. Slowly moving cells may simply push apart the fibers of the ECM, squeezing between them rather than lysing them.

Therefore, the place to look for tunneling and for its consequences is in denser tissues. A fibrin clot penetrated by inflammatory cells, or a collagenous matrix as found in a healing wound, is more likely to display well-preserved tunnels than the vitreous, for example. Additionally, holes in the thin ECM layers, as in the basement membranes or elastic laminae of vessels, could sometimes be, in fact, tracks of migrating cells. Nevertheless, a relative homogeneity of the tissue, enough tridimensional extension, and a good morphological preservation of the specimens are also conditions for the reliable identification of these structural modifications of the ECM associated with cell movement, i.e., of the tunnels.

These tunnels have been noted in the past, but little attention has been paid to them. Most probably, they were considered artifacts of tissue processing. However, tunnel-like structures were clearly observed in a study describing the penetration of fibrin gels by macrophages (Castellucci et al. 1988), and in an analysis of migration of endothelial cells in an angiogenic setting (Nicosia et al. 1984).

MORPHOLOGY OF TUNNELS

The best evidence we have obtained regarding the ability of inflammatory cells to drill tunnels in situ was de-

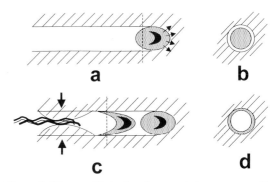

Figure 1. Diagram illustrating the expected appearance of tunneling in histological sections. (*a*) The "empty" (or of reduced density) space created behind a cell migrating in a dense extracellular matrix, as an effect of localized secretion of proteolytic enzymes, and viewed in longitudinal section. The dashed line indicates the plane of section discussed in *b*. (*b*) Cross-section through a tunnel. The proteolytic rim of about a cell's diameter surrounds the cell fragment retained in the section. (*c*) Likely fates of a tunnel: colonized by other cell(s) of same or different phenotype; collapsed under the effect of lateral pressure; filled with other (ECM) proteins, either passively or actively. The dashed line indicates the sectioning plane discussed in *d*. (*d*) Cross-section through a tunnel colonized by a non-endothelial cell. The lumen might look like the one of a capillary, but the cell would bear non-endothelial markers. The presence of a plasmalemma on the luminal side, defined by specific markers, would be indicative of the fact that this is not an intracellular vacuole.

rived from the use of a modified plastic chamber (Dvorak et al. 1987) filled with Matrigel supplemented with various chemoattractants and implanted subcutaneously in mice. The advantages of Matrigel are: (1) It is a homogeneous and compact medium, (2) it has a composition similar to the basement membrane of endothelial cells, and (3) it was already used as support for in vivo angiogenesis assays (Passaniti et al. 1992). The placement of the chamber in a wounding setting in vivo allows the recruitment of those inflammatory cells naturally involved in ECM penetration.

After 5 days, when the chambers were retrieved, dismantled, and analyzed by regular histological techniques, we found various levels of infiltration. Two populations of cells were found in the chambers. One did not penetrate the gel, and instead remained in the space between gel and filter. The other actively penetrated the gel and morphologically presented a monocyte/macrophage (MC/Mph) phenotype, with kidney-like nucleus and a larger size. Polymorphonuclear leukocytes were also noted among the MC/Mph. The exact nature of all these cells, and their proportions, are currently being analyzed in our laboratory.

Tunnels were also present in the gel. They were tubular empty spaces of various apparent lengths and about a cell's diameter. These structures were associated with mononuclear cells and were located behind their advancing front, giving to the sections a Swiss-cheese-like aspect (data not shown). The tunnels were straight or bent, often displaying branching.

In some instances, the cells at the leading end of the tunnel could be identified, and they showed a MC/Mph phenotype (Fig. 2a). Usually, a given tunnel seemed to be populated by multiple cells, and we observed a tendency for the recurrent use of the same tunnel by many cells. This characteristic of the tunneling process, observed in other in vivo and in vitro instances (see below), suggests a possible advantage for cohort penetration of ECM, as opposed to random migration, in terms of efficiency of the migration (speed and energy expenditure).

A surprising finding was the presence of free erythrocytes, both beneath the filter and within the tunnels, sometimes in a striking linear pattern, without microscopic evidence of capillaries present in these chambers. This observation raises the possibility that the erythrocytes may squeeze in the tunnels as an effect of the pressure of the extracellular fluid, possibly combined with a "suction" (depression) produced by the advancing MC/Mph.

TUNNELING IN VITRO

We were able to reproduce the tunneling effect in vitro. To this end, we used the monocyte cell line THP-1. We incubated these cells in wells 3 mm in diameter, made in a 1-mm Matrigel layer, near other wells filled with a 100 ng/ml MCP-1 solution. Within about a week, the cells started to migrate and to develop "spikes" composed of columns of advancing cells (Fig. 2). These were oriented more toward the MCP-1 source (Fig. 2b), as compared to the opposite direction (Fig. 2a). These columns contained monocytes either in contact with each other or separated. Those cells placed at the tips of the spikes usually displayed an elongated, migratory phenotype.

Regarding the mechanism of penetration, other investigators have shown that migration of macrophages in Matrigel is dependent on the large-spectrum matrix metalloprotease MMP12 (Shipley et al. 1996).

Based on observation done on implanted chambers, we tested the hypothesis that the erythrocytes can penetrate a matrix without the contribution of capillaries, solely as an effect of the guiding by MC/Mph. On top of Matrigel layers preincubated for 5 days with THP-1 cells in the presence of MCP-1, and then washed free of monocytes, we added diluted human volunteer blood. After one week, we identified both an increased tunneling activity of the

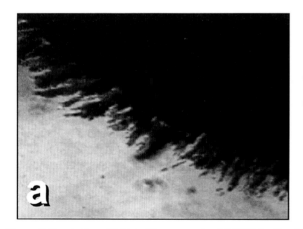
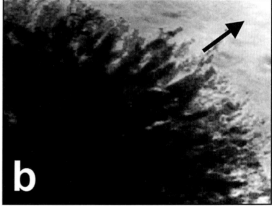

Figure 2. Penetration of Matrigel by monocytes. (*a*) THP-1 cells added to wells cut in Matrigel (*black* areas). Under the combined effect of cell proliferation and migration, the cells penetrate the gel radially in cohorts of cells mostly following one after another, presumably as a consequence of facilitated migration produced by the cells at the front of the advancing column. (*b*) In the neighborhood of the well containing the cells, there was another one containing an MCP-1 solution. In this case, the length of columns was increased, while their density decreased, suggesting active migration along the tunnels. The arrow indicates the direction of the MCP-1 chemotactic gradient. Magnification, 40×.

gel by the remaining monocytes and columns of erythrocytes penetrating the gel (N.I. Moldovan et al., in prep.). Combined with the in vivo data, this observation provides an alternative interpretation for the known presence of free erythrocytes associated with angiogenic fields, such as those found in the granulation tissue or in tumors (Zamir and Silver 1992; McDonald and Foss 2000). So far, it was generally assumed that the erythrocyte extravasation is the effect of "capillary immaturity" or leakiness (Hashizume et al. 2000). Instead, their presence and distribution away from the capillary of origin may be indicative of facilitated relocation within the tunneling network created by the concurrent inflammatory cells.

CO-PENETRATION OF MATRIGEL PLUGS BY MC/MPH AND CPEC

Macrophages in general (Sunderkotter et al. 1994; Dahlqvist et al. 1999; Ono et al. 1999; Torisu et al. 2000; Lingen 2001; Nesbit et al. 2001), and the macrophage-specific matrix metalloprotease 12 (MMP12, metalloelastase) expression in particular (Madlener et al. 1998; Moldovan et al. 2000), were often associated with angiogenesis. Advancement of capillary sprouts may benefit from the assistance of specialized phagocytes, because an intense phagocytic activity is needed in that setting. This activity was shown even in pure endothelial cultures (Meyer et al. 1997). To address the cooperation between inflammatory and endothelial cells in angiogenesis, we examined Matrigel plugs injected subcutaneously in transgenic mice expressing β-galactosidase under the endothelial-specific Tie2 promoter, an in vivo experimental model for induction of inflammatory angiogenesis (Asahara et al. 1999).

We found at the periphery of plugs retrieved after 10 days Tie2-positive microvessels, accompanied by a massive perivascular infiltrate. Deeper in the gel, however, the mononuclear cells were no more associated with microvessels, but they still displayed a linear alignment. We examined the distribution of the mouse macrophage specific antigen F4/80 (Austyn and Gordon 1981), and we found that many of the cells within cords stained positive for this marker. Concurrently, isolated Tie2-positive mononuclear cells, possibly circulating precursor endothelial cells (CPEC), were intimately associated with the penetrating F4/80-positive MC/Mph, suggesting a cooperative migration and colonization of tunnels (N.I. Moldovan et al., in prep.).

TUNNELING IN VIVO

We assessed the pattern and mechanism of MC/Mph penetration in a hard tissue by analyzing the hearts of transgenic mice expressing in cardiomyocytes the monocyte chemoattractant protein 1 (MCP-1), under the myosin heavy-chain promoter. These mice develop an inflammatory cardiomyopathy and die of heart failure (Kolattukudy et al. 1998), as a result of the destructive activity of a massive infiltration by mononuclear cells, in particular MC/Mph (Moldovan et al. 2000). We immunostained the hearts of these mice for MMP12 and found that the enzyme is strongly expressed, particularly in the subendocardium. Interestingly, the pattern of distribution of MMP12 was not only cell-associated (Fig. 3a, inset), as expected, but also followed extracellular linear and elliptical tracks. This observation may be related to the known ability of secreted, immunologically intact MMP12 to bind components of ECM (Curci et al. 1998). These structures closely resembled the tunnels described in vitro. We could detect MMP12-positive macrophages present in the tunnels, and other mononuclear cells colonizing them as well. The tunnels stained positive for an antigen associated with, or derived from, this enzyme even in the absence of the secreting cells. The tunnels were different from capillaries, and this was most obvious where the two networks appeared together but were spatially distinct (Moldovan et al. 2000).

Immunostaining for the pan-leukocyte antigen CD18 (present in the Mac-1 complex) revealed a number of unexpected findings. A tunnel crossing a cardiomyocyte in parallel with a capillary (Fig. 3a) stained positive for CD18 on its entire length, indicating that it can be covered on long portions by the membrane(s) and cytoplasm of leukocytes. This possibility was supported by observations on oblical or transversal sections (Fig. 3b), which displayed cells in endothelial positions, but staining positive for this leukocyte antigen. Albeit less frequently, we noticed the same pattern in sections stained for Mac-3, an adult macrophage antigen (Fig. 3b, inset). At the other end of the differentiation spectrum, we found tunnels staining positive for CD34, in the neighborhood of non-stained capillaries (Fig. 3d). Moreover, the presence of CD34 antigen on isolated interstitial cells, as shown in the micrograph, is an argument for their hematopoietic origin, rather than an "activated" endothelial phenotype.

We then asked whether these leukocyte-covered tunnels may have a functional significance for the microcirculation. We did find CD18-positive tunnels containing erythrocytes (Fig. 3c). This would argue for the connection of tunnels to the microcirculatory system, in a way yet to be explained. Although the possibility remains that in Figure 3c and d, the damaged endothelium has been replaced by leukocytes, observations such as in Figure 3a incline the interpretation toward de novo formation of tunnels by mononuclear leukocytes, and to their covering by the same cells. The presence of a CD34 marker in sparse cells and in the intima of some, but not all, microvessels in the MCP-1 mouse hearts prompted us to look for other markers, also considered to be signs of neovascularization. An example is Thy-1, a hematopoietic molecule described on the endothelium during angiogenesis (Lee et al. 1998). We found Thy-1 positivity on capillaries and on isolated interstitial cells, or on those cells apparently in the process of insertion into the intima of microvessels (data not shown).

Combined with the presence of Tie2-positive mononuclear cells found in tunnels, as described before, these data open the likely possibility that the tunneling mechanism is instrumental for the tissue engraftment of CPEC. At present, there are two different views of the notion of

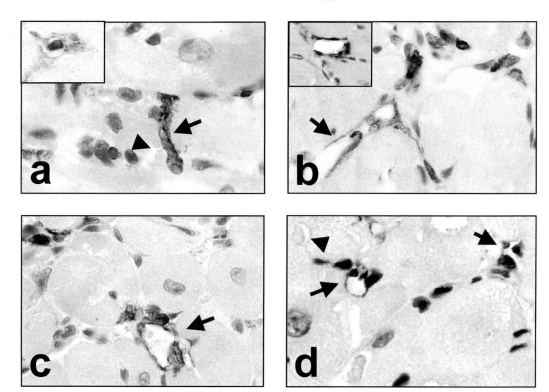

Figure 3. In vivo evidence for tunnel formation and covering. The hearts of MCP-1 mice were analyzed by immunocytochemistry (*brown*) with hematoxylin counterstaining (*blue*). (*a*) CD18 (pan leukocyte antigen) positive structure (*arrow*) running in parallel with a capillary (erythrocytes, *arrowhead*). Note the nucleus of the penetrating cells (*blue*). The inset depicts a MC/MPh, staining positive for MMP12 and presenting the surrounding proteolysis rim. (*b*) Tubular empty spaces in the myocardium, edged by CD18-positive cells (*arrow*). (*Inset*) Lumen covered by Mac-3-positive macrophages. (*c*) Presence of erythrocytes within a CD18-positive lumen (*arrow*). (*d*) Tunnels populated by CD34-positive cells (*arrows*) in a region where the bona fide capillaries identified by erythrocyte content (*arrowhead*) are CD34-negative. Magnification, 120×.

CPEC (Graf 2002). On one hand, these cells are considered bone-marrow-derived, poorly differentiated but committed endothelial precursors of angioblast type (Asahara et al. 1999; Takahashi et al. 1999; Rafii 2000; Schatteman et al. 2000). On the other hand, the nature of CPEC is considered more similar to that of adult monocytes, apparently able to transdifferentiate, given the appropriate conditions, in endothelial-like cells (Fernandez et al. 2000; Harraz et al. 2001; Schmeisser et al. 2001).

In any case, the classical sprouting mechanism of angiogenesis can hardly accommodate these data, unless one admits that, for some unspecified reason, the new microvessels prefer to acquire their endothelium from circulation, rather than from the proliferation of their own preexistent intimal cells, as supposed so far.

The fact is that new angiogenic foci can develop distantly and disconnected from a nearby capillary, in a process of "seeding" of the tissue by angioblasts. A question raised by such a possibility is how these CPEC, or angioblasts, travel to the appropriate position for nesting. Another unsolved problem is whether CPEC have the ability to penetrate by themselves; i.e., whether they have the mature enzymatic equipment for a deep tissue engraftment. We consider it likely that CPEC belong to an inflammatory type of cellular infiltrate, often associated with angiogenesis. Therefore, their proteolytic needs may be supplied by the co-migrating cells, which would penetrate the ECM first, followed or accompanied by CPEC. If this is the case, this might be in fact one of the major physiologic functions of tunneling (Moldovan 2002). Moreover, the tunnels may become incidentally covered by MC/Mph, which could further acquire by trans-differentiation an endothelial function, or even phenotype, especially if the tunnel becomes connected to a blood supply.

TUNNELING IN A CLINICAL SETTING

From both fundamental and clinical perspectives, it is important to determine whether the concepts developed so far (formation of tunnels, their covering by cells, co-migration of cells with different phenotypes, facilitation of angiogenesis by the tunneling mechanism) are relevant to human biology and pathology. MC/Mph recruitment is a prerequisite of atherosclerotic plaque formation. Moreover, the hematopoietic stem-like cells have been shown to contribute to the smooth muscle cell population of the plaque (Sata et al. 2002). Because the development of atherosclerotic plaques is accompanied by neovessel formation (Cliff and Schoefl 1983; Kamat et al. 1987; O'Brien et al. 1994; Kumamoto et al. 1995; Moulton et al. 1999), it is logical to expect that CPEC may also be recruited in the plaque and may contribute to angiogenesis.

Therefore, an atherosclerotic plaque is an appropriate place to look for a correlation between MC/Mph migration, tunnel formation, angiogenesis, and other processes taking place in this setting.

To this end, we analyzed by immunocytochemistry human carotid plaques, collected by trans-esophageal atherectomy. We detected a panel of molecular markers in the neointimal microvessels of these plaques. They ranged from the endothelial phenotype (vWF, Tie2) to hematopoietic cells (CD34, CD45), pan-leukocyte markers (CD80), monocytes (CD14), and macrophages (CD68, MIF-1α, MMP12). These markers have been found to cover the lumens of various microvessels (in an intimal position) in different parts of the plaques, sometimes overlapping (A. Schmeisser et al., in prep.). These data suggest that the plaque's microvessels display a remarkable heterogeneity regarding the composition of their endothelial lining, and give support to other recent findings, which consider that a significant contribution to neovascularization is provided by CPEC-related leukocytes, as discussed above.

Next, we investigated the reciprocal distribution of macrophage-specific MMP12 and its association with angiogenic markers in the chronic inflammatory infiltrate of a selected atherosclerotic carotid specimen (Fig. 4). In Figure 4a we present the close association found between MMP12-producing macrophages, in a column-like disposition, and the tip of a microvessel placed very close to the luminal side of the plaque. This association suggests a facilitator role of metalloelastase (MMP12) in the advancement of capillary tips, as described before in an experimental animal setting (Madlener et al. 1998).

The anti-MMP12 antibody stained heavily two regions of the section. One displayed an extensive destruction of the elastic laminae and intraplaque hemorrhage (Fig. 4c). The consistent "background" staining, associated with layered structures and otherwise absent in controls where the first antibody was omitted (Fig. 4b), may be due to the presence of extracellularly secreted elastase bound to ECM material (Curci et al. 1998). The other instance where we found a high level of immunostaining for MMP12 was a region rich in macrophages, also displaying empty spaces, evocative of tunnels (Fig. 4c, inset). In this particular case, the "holes" in the section may also well be represented by intracellular vacuoles or extracted lipid droplets of macrophages, which in oblique section may look like tunnels covered by a MMP12-positive cytoplasm. Obviously, this would be a confounding factor, which complicates, in this particular condition, the detection of tunnels. Nevertheless, what makes this possibility unlikely is the very extended longitudinal axis (as compared with the transverse axis) of this "vacuole," combined with the presence of multiple (four) elongated nuclei facing its lumen. Another strong argument that what we are describing is not a vacuole is the presence of an erythrocyte in the right side of this lumen, visible at higher magnifications. Yet, the cells on the left side clearly stain for MMP12, suggesting the presence of macrophages both in the process of matrix penetration, and in that of covering the microvessel-like structures.

Paired with Figure 4c is a parallel section, stained for CD34. Free-floating erythrocytes are found accompanied by isolated CD34$^+$ cells (Fig. 4d). In other instances, these cells which appear to be of bone-marrow origin displayed a lumen, but occurred assembled in a tube-like structure (Fig. 4d, inset).

CONCLUSIONS AND PERSPECTIVES

The tunneling mechanism brings a new example of intercellular cooperation in the performance of a complex biological process, such as angiogenesis. Both paradigms that currently dominate the angiogenesis research (sprouting, and the one based on contribution of CPEC) would benefit from the tunneling model.

The tunneling may answer a critical issue in angiogenesis which takes place with the contribution of CPEC; (i.e., adult vasculogenesis): What is the mechanism of tissue engraftment of these angioblast-like cells? Are they able to penetrate ECM? Obviously they do so, leaving the nearby intact capillaries and going deeper in the tissue to create new angiogenic foci. Apparently, they travel as part of an inflammatory infiltrate containing CD34$^+$ and CD34$^-$ cells (Harraz et al. 2001). A facilitated migration in cohorts formed by tunneling, as described here, is a likely candidate mechanism. Consequently, in order to control the efficiency and localization of tissue engraftment of CPEC, one may target the "driving force" of their translocation, i.e., the tunneling cells.

Equally important, the penetration of capillary sprouts in the ECM was considered to rely exclusively on the proteases synthesized by the endothelial cells themselves. However, in instances such as the extension of the vasa vasorum through the medial elastic layers of the arteries, the proteolytic spectrum of EC may not be sufficient. In this instance, the presence of metalloelastase may be required, and the cells specifically able to produce it are the MC/Mph.

The presence of MMP12-producing cells is also important from the standpoint of plaque stability. If the tunnels start at the luminal side of the artery, as seems to be the case, they may become filled with more and more blood. This may happen particularly if/when the patient has increased blood pressure, and by enlargement they may induce intraplaque hemorrhage. Alternatively, the tunnels preexisting in the plaque as a network of empty spaces may become filled with blood when the outermost ones become connected to the lumen. For the above reasons, the new knowledge gained with respect to the tunneling process may have beneficial basic and clinical consequences.

ACKNOWLEDGMENTS

The authors are grateful to Dr. P. Goldschmidt-Clermont for advice in the early stages of the project, and to Dr. P. E. Kolattukudy for providing the MCP-1 transgenic mice. This work was supported by National Institutes of Health grant R01-HL-65983.

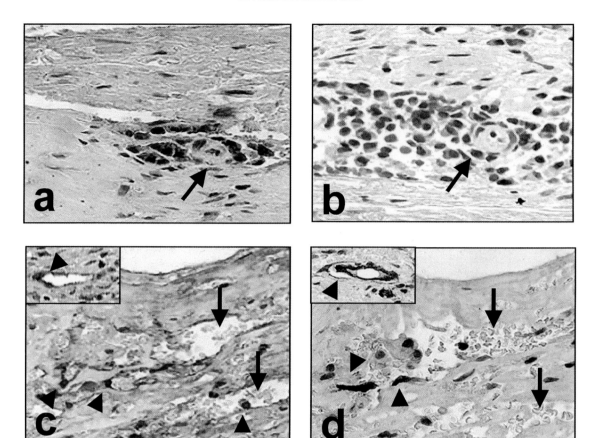

Figure 4. Expression of MMP12 and correlated effects in human carotid atherosclerotic plaque specimens. (*a*) Association of MMP12 expression in macrophages with the advancing tips of a microvessel in the neointima of the plaque. (*b*) Negative control for the immunostaining (omission of the first antibody). (*c*) MMP12 immunoreactivity in another region of the same lesioned carotid that displayed infiltrating MMP12-positive macrophages (*arrowheads*) and erythrocytes-filled MMP12-positive cavities (*arrows*). (*Inset*) The lumen of a microvessel bordered by MMP12-positive cells (*arrowhead*). (*d*) Parallel section, displaying the same free erythrocyte pools as in *c* (*arrows*) associated with isolated CD34-positive mononuclear cells (*arrowheads*). (*Inset*) A CD34$^+$ lumen inserted in a larger cellular sheath, suggestive for tunneling. Magnification, 120×.

REFERENCES

Amorino G.P. and Hoover R.L. 1998. Interactions of monocytic cells with human endothelial cells stimulate monocytic metalloproteinase production. *Am. J. Pathol.* **152:** 199.

Asahara T., Takahashi T., Masuda H., Kalka C., Chen D., Iwaguro H., Inai Y., Silver M., and Isner J.M. 1999. VEGF contributes to postnatal neovascularization by mobilizing bone marrow-derived endothelial progenitor cells. *EMBO J.* **18:** 3964.

Austyn J.M. and Gordon S. 1981. F4/80, a monoclonal antibody directed specifically against the mouse macrophage. *Eur. J. Immunol.* **11:** 805.

Castellucci M. and Montesano R. 1988. Phorbol ester stimulates macrophage invasion of fibrin matrices. *Anat. Rec.* **220:** 1.

Cliff W.J. and Schoefl G.I. 1983. Pathological vascularization of the coronary intima. *Ciba Found. Symp.* **100:** 207.

Curci J.A., Liao S., Huffman M.D., Shapiro S.D., and Thompson R.W. 1998. Expression and localization of macrophage elastase (matrix metalloproteinase-12) in abdominal aortic aneurysms. *J. Clin. Invest.* **102:** 1900.

Dahlqvist K., Umemoto E.Y., Brokaw J.J., Dupuis M., and McDonald D.M. 1999. Tissue macrophages associated with angiogenesis in chronic airway inflammation in rats. *Am. J. Respir. Cell Mol. Biol.* **20:** 237.

Dvorak H.F., Harvey V.S., Estrella P., Brown L.F., McDonagh J., and Dvorak A.M. 1987. Fibrin containing gels induce angiogenesis. Implications for tumor stroma generation and wound healing. *Lab. Invest.* **57:** 673.

Fernandez P.B., Lucibello F.C., Gehling U.M., Lindemann K., Weidner N., Zuzarte M.L., Adamkiewicz J., Elsasser H.P., Muller R., and Havemann K. 2000. Endothelial-like cells derived from human CD14 positive monocytes. *Differentiation* **65:** 287.

Galis Z.S. and Khatri J.J. 2002. Matrix metalloproteinases in vascular remodeling and atherogenesis: The good, the bad, and the ugly. *Circ. Res.* **90:** 251.

Graf T. 2002. Differentiation plasticity of hematopoietic cells. *Blood* **99:** 3089.

Harraz M., Jiao C., Hanlon H.D., Hartley R.S., and Schatteman G.C. 2001. CD34$^-$ blood-derived human endothelial cell progenitors. *Stem Cells* **19:** 304.

Hashizume H., Baluk P., Morikawa S., McLean J.W., Thurston G., Roberge S., Jain R.K., and McDonald D.M. 2000. Openings between defective endothelial cells explain tumor vessel leakiness. *Am. J. Pathol.* **156:** 1363.

Hiraoka N., Allen E., Apel I.J., Gyetko M.R., and Weiss S.J. 1998. Matrix metalloproteinases regulate neovascularization by acting as pericellular fibrinolysins. *Cell* **95:** 365.

Kamat B.R., Galli S.J., Barger A.C., Lainey L.L., and Silverman K.J. 1987. Neovascularization and coronary atherosclerotic plaque: Cinematographic localization and quantitative histologic analysis. *Hum. Pathol.* **18:** 1036.

Kolattukudy P.E., Quach T., Bergese S., Breckenridge S., Hensley J., Altschuld R., Gordillo G., Klenotic S., Orosz C., and Parker-Thornburg J. 1998. Myocarditis induced by targeted expression of the MCP-1 gene in murine cardiac muscle. *Am. J. Pathol.* **152:** 101.

Kumamoto M., Nakashima Y., and Sueishi K. 1995. Intimal neovascularization in human coronary atherosclerosis: Its origin and pathophysiological significance. *Hum. Pathol.* **26:** 450.

Lee W.S., Jain M.K., Arkonac B.M., Zhang D., Shaw S.Y., Kashiki S., Maemura K., Lee S.L., Hollenberg N.K., Lee M.E., and Haber E. 1998. Thy-1, a novel marker for angiogenesis upregulated by inflammatory cytokines. *Circ. Res.* **82:** 845.

Levine H.A., Sleeman B.D., and Nilsen-Hamilton M. 2000. A mathematical model for the roles of pericytes and macrophages in the initiation of angiogenesis. I. The role of protease inhibitors in preventing angiogenesis. *Math. Biosci.* **168:** 77.

Libby P. and Schonbeck U. 2001. Drilling for oxygen: Angiogenesis involves proteolysis of the extracellular matrix. *Circ. Res.* **89:** 195.

Lingen M.W. 2001. Role of leukocytes and endothelial cells in the development of angiogenesis in inflammation and wound healing. *Arch. Pathol. Lab. Med.* **125:** 67.

Madlener M., Parks W.C., and Werner S. 1998. Matrix metalloproteinases (MMPs) and their physiological inhibitors (TIMPs) are differentially expressed during excisional skin wound repair. *Exp. Cell Res.* **242:** 201.

McDonald D.M. and Foss A.J. 2000. Endothelial cells of tumor vessels: Abnormal but not absent. *Cancer Metastasis Rev.* **19:** 109.

Meyer G.T., Matthias L.J., Noack L., Vadas M.A., and Gamble J.R. 1997. Lumen formation during angiogenesis in vitro involves phagocytic activity, formation and secretion of vacuoles, cell death, and capillary tube remodelling by different populations of endothelial cells. *Anat. Rec.* **249:** 327.

Moldovan N.I. 2002. Role of monocytes and macrophages in adult angiogenesis: A light at the tunnel's end. *J. Hematother. Stem Cell Res.* **11:** 179.

Moldovan N.I., Goldschmidt-Clermont P.J., Parker-Thornburg J., Shapiro S.D., and Kolattukudy P.E. 2000. Contribution of monocytes/macrophages to compensatory neovascularization: The drilling of metalloelastase-positive tunnels in ischemic myocardium. *Circ. Res.* **87:** 378.

Moulton K.S., Heller E., Konerding M.A., Flynn E., Palinski W., and Folkman J. 1999. Angiogenesis inhibitors endostatin or TNP-470 reduce intimal neovascularization and plaque growth in apolipoprotein E-deficient mice. *Circulation* **99:** 1726.

Murphy G. and Gavrilovic J. 1999. Proteolysis and cell migration: Creating a path? *Curr. Opin. Cell Biol.* **11:** 614.

Nesbit M., Schaider H., Miller T.H., and Herlyn M. 2001. Low-level monocyte chemoattractant protein-1 stimulation of monocytes leads to tumor formation in nontumorigenic melanoma cells. *J. Immunol.* **166:** 6483.

Nicosia R.F., McCormick J.F., and Bielunas J. 1984. The formation of endothelial webs and channels in plasma clot culture. *Scanning Electron Microsc.* **Pt 2:** 793.

O'Brien E.R., Garvin M.R., Dev R., Stewart D.K., Hinohara T., Simpson J.B., and Schwartz S.M. 1994. Angiogenesis in human coronary atherosclerotic plaques. *Am. J. Pathol.* **145:** 883.

Ono M., Torisu H., Fukushi J., Nishie A., and Kuwano M. 1999. Biological implications of macrophage infiltration in human tumor angiogenesis. *Cancer Chemother. Pharmacol.* (suppl.) **43:** S69.

Passaniti A., Taylor R.M., Pili R., Guo Y., Long P.V., Haney J.A., Pauly R.R., Grant D.S., and Martin G.R. 1992. A simple, quantitative method for assessing angiogenesis and antiangiogenic agents using reconstituted basement membrane, heparin, and fibroblast growth factor. *Lab. Invest.* **67:** 519.

Rafii S. 2000. Circulating endothelial precursors: Mystery, reality, and promise. *J. Clin. Invest.* **105:** 17.

Sata M., Saiura A., Kunisato A., Tojo A., Okada S., Tokuhisa T., Hirai H., Makuuchi M., Hirata Y., and Nagai R. 2002. Hematopoietic stem cells differentiate into vascular cells that participate in the pathogenesis of atherosclerosis. *Nat. Med.* **8:** 403.

Schatteman G.C., Hanlon H.D., Jiao C., Dodds S.G., and Christy B.A. 2000. Blood-derived angioblasts accelerate blood-flow restoration in diabetic mice. *J. Clin. Invest.* **106:** 571.

Schmeisser A., Garlichs C.D., Zhang H., Eskafi S., Graffy C., Ludwig J., Strasser R.H., and Daniel W.G. 2001. Monocytes coexpress endothelial and macrophagocytic lineage markers and form cord-like structures in Matrigel under angiogenic conditions. *Cardiovasc. Res.* **49:** 671.

Shapiro S.D. 1999. Diverse roles of macrophage matrix metalloproteinases in tissue destruction and tumor growth. *Thromb. Haemostasis* **82:** 846.

Shipley J.M., Wesselschmidt R.L., Kobayashi D.K., Ley T.J., and Shapiro S.D. 1996. Metalloelastase is required for macrophage-mediated proteolysis and matrix invasion in mice. *Proc. Natl. Acad. Sci.* **93:** 3942.

Stetler-Stevenson W.G. 1999. Matrix metalloproteinases in angiogenesis: A moving target for therapeutic intervention. *J. Clin. Invest.* **103:** 1237.

Sunderkotter C., Steinbrink K., Goebeler M., Bhardwaj R., and Sorg C. 1994. Macrophages and angiogenesis. *J. Leukoc. Biol.* **55:** 410.

Takahashi T., Kalka C., Masuda H., Chen D., Silver M., Kearney M., Magner M., Isner J.M., and Asahara T. 1999. Ischemia- and cytokine-induced mobilization of bone marrow-derived endothelial progenitor cells for neovascularization. *Nat. Med.* **5:** 434.

Torisu H., Ono M., Kiryu H., Furue M., Ohmoto Y., Nakayama J., Nishioka Y., Sone S., and Kuwano M. 2000. Macrophage infiltration correlates with tumor stage and angiogenesis in human malignant melanoma: Possible involvement of TNFalpha and IL-1alpha. *Int. J. Cancer* **85:** 182.

Werb Z., Vu T.H., Rinkenberger J.L., and Coussens L.M. 1999. Matrix-degrading proteases and angiogenesis during development and tumor formation. *APMIS* **107:** 11.

Zamir M. and Silver M.D. 1992. 'Hemorrhagic' and microvascular phenomena within the arterial wall. *Can. J. Cardiol.* **8:** 981.

The Role of EG-VEGF in the Regulation of Angiogenesis in Endocrine Glands

J. LeCouter, R. Lin, and N. Ferrara

Department of Molecular Oncology, Genentech, Inc., South San Francisco, California 94080

The cardiovascular system is the first organ system to develop and reach a functional state in an embryo (Hamilton et al. 1962). The initial steps in blood vessel development consist of vasculogenesis, the differentiation of endothelial cell precursors, the angioblasts, from the hemangioblasts (Risau and Flamme 1995). The juvenile vascular system evolves from the primary capillary plexus by subsequent pruning and reorganization of endothelial cells in a process called angiogenesis (Risau 1997). More recent evidence suggests that incorporation of bone-marrow-derived endothelial precursor cells contributes to the growing vessels, complementing the sprouting of resident endothelial cells (Asahara et al. 1997; Rafii et al. 2002). The development of a vascular supply is also essential in the adult for wound healing and reproductive functions (Folkman 1995). In addition, angiogenesis is implicated in the pathogenesis of a variety of proliferative disorders: age-related macular degeneration, proliferative retinopathies, tumors, rheumatoid arthritis, and psoriasis (Garner 1994; Folkman 1995). The process of angiogenesis has therefore become the focus of a variety of strategies to increase or create new vascular supply, reduce vascular damage, or compromise in order to preserve functional tissue in ischemic disease, but also, conversely, to develop means of inhibiting angiogenesis in other pathological conditions associated with inappropriate growth, including cancer.

Several potential regulators of angiogenesis have been identified, including aFGF, bFGF, TGFα, TGFβ, HGF/SF, TNFα, angiogenin, and IL-8 (Folkman and Shing 1992; Risau 1997), and the angiopoietins (Suri et al. 1996; Maisonpierre et al. 1997). Recent studies suggest a coordinated role for the vascular endothelial growth factor (VEGF) receptors, Tie1, Tie2, and the PDGFβ receptor in the correct assembly and morphogenesis of the vessel wall (Yancopoulos et al. 2000).

VEGF, the first endothelial cell-specific mitogen to be identified, is a potent endothelial chemoattractant and angiogenesis inducer in vivo (Leung et al. 1989). The specificity of VEGF is attributed to the rather cell-type-restricted expression of its receptors, VEGFR-1 (Flt-1) and VEGFR-2 (KDR/Flk) (Ferrara and Davis-Smyth 1997). VEGF and its receptors have been extensively investigated over the last decade (Ferrara and Alitalo 1999). The key role of VEGF in the development of the vascular system is underscored by embryonic lethality following inactivation of a single VEGF allele (Ferrara et al. 1996). VEGF is also critical for corpus luteum development (Ferrara et al. 1998), postnatal organ growth (Gerber et al. 1999b), and endochondral bone formation (Gerber et al. 1999a). Currently, inhibition of VEGF activity is being pursued as a strategy to treat tumors (Kim et al. 1993; Presta et al. 1997; Wood et al. 2000) and intraocular neovascular syndromes (Aiello et al. 1994).

Endothelial cell phenotype, functions, and growth characteristics are different among various organs, whereas the expression of VEGF, as well as that of the Tie2 ligands, is almost ubiquitous. This paradox led to the suggestion that, although VEGF signaling may be a rate-limiting step in a variety of circumstances, local regulatory pathways modulate endothelial cell growth and function in a tissue- or organ-specific manner (LeCouter et al. 2001). An understanding of the network of tissue-specific signals that control or coordinate angiogenesis, vasculogenesis, and vascular regression will be key for harnessing their therapeutic value.

POTENTIAL TISSUE- OR ORGAN-SPECIFIC REGULATION OF ENDOTHELIAL FUNCTION

The endothelium of the various vascular beds is diverse and distinct with respect to the specific tissue compartment served by the vessels (Simionescu and Simionescu 1988). The vascular-bed-specific expression of a variety of molecules, including cell surface antigens, has been evaluated by several investigators (Aird et al. 1997). The most recent work has assessed in vivo selective binding of peptides displayed on phage, or assignment of so-called vascular address (Ruoslahti and Rajotte 2000). The morphology and architecture of endothelial cells also differs among the capillary beds. The extent of the plasmalemmal vesicles within the endothelial cells, for example, varies in different organs, as does the presence of discontinuities in the plasma membrane, referred to as fenestrae. Fenestrae are associated with highly permeable vessels such as those serving the endocrine tissues (Palade et al. 1979). The biochemical and structural properties of the endothelial cells and vessels obviously have important implications for tissue function. It is therefore expected that the tissue parenchyma could affect the phenotype of the vessels that are established during organ development.

The contribution of the tissue "microenvironment" to endothelial cell phenotype has been clearly demonstrated by assessing the physical and functional features of vessels that develop within distinct grafted tissues (Stewart and Wiley 1981) as well as by using tumor xenografts established at various anatomic sites (Dellian et al. 1996; Roberts et al. 1998). Elegant studies, for example, have demonstrated that endothelial cells grafted into nervous tissue acquire blood–brain barrier characteristics (Stewart and Wiley 1981). Thus, the phenotypic and growth properties of endothelial cells are shaped by signals in the microenvironment, although the molecular nature of these cues remains undefined. A network of instructive signals incorporating cell–cell contacts, extracellular matrix (ECM) components, and secreted proteins in the tissue environment is likely to be responsible for shaping the growth, structure, and function of the endothelium.

EG-VEGF AS THE FIRST EXAMPLE OF TISSUE-SPECIFIC ANGIOGENIC FACTOR

We recently reported the identification of a novel endothelial cell growth factor with a unique selective activity and very distinct expression pattern. Human EG-VEGF (endocrine-gland-derived vascular endothelial growth factor), also known as prokineticin-1 (Li et al. 2001), was identified in a functional screen for novel endothelial cell mitogens from a library of human secreted proteins (LeCouter et al. 2001). Mature human EG-VEGF is an 86-amino-acid peptide, including 10 cysteines, that displays a remarkably high degree of homology (80%) with a nontoxic protein purified from the black mamba snake known as protein A (VPRA) (Joubert and Strydom 1980). Notably, the number and span of cysteines of EG-VEGF are completely conserved. Therefore, the molecule is predicted to adopt a colipase fold motif, as reported for the snake VPRA structure (Boisbouvier et al. 1998). Angiogenic factors, including bFGF and VEGF, interact with ECM components such as heparin sulfate proteoglycans, and this interaction is thought to regulate the bioavailability and activity of these molecules (Klagsbrun 1992). EG-VEGF is a basic molecule, pI 8.6, that binds to heparin and thus it may be sequestered in the extracellular compartment in vivo. Although EG-VEGF is not structurally related to VEGF, the functional activity of this molecule is indistinguishable from VEGF, but only in select cell and tissue contexts, as discussed below. Consistent with the bioactivity data, EG-VEGF mediates the activation of the same signal transduction pathways as VEGF in endothelial cells, although, as noted below, the EG-VEGF receptors are not tyrosine kinases but rather G-protein-coupled receptors.

EG-VEGF was selectively mitogenic, motogenic, and morphogenic for the primary adrenal cortical capillary endothelial (ACE) cells that served as the target for our bioassay (LeCouter et al. 2001). The mitogenic response in ACE cultures was comparable to that elicited by VEGF. However, in contrast to EG-VEGF, VEGF was also a potent mitogen for all of the endothelial cell types tested, including those from aorta, umbilical vein, and dermal microvasculature. EG-VEGF also induced strong and reproducible cell migration responses, again restricted to adrenal cortical endothelial cells. Thus far, the target cell selectivity demonstrated by EG-VEGF is unique among the known endothelial cell mitogens.

Endothelial cells resident in endocrine glands, including the adrenal cortex, often display a fenestrated phenotype that is found also in the choroid plexus, the gastrointestinal tract, and several tumors (Simionescu and Simionescu 1988). The fenestrae are highly permeable to fluid and small solutes and are thought to facilitate the large exchange of materials between interstitial fluid and plasma, such as that occurring in steroid-producing as well as other endocrine glands like pancreatic islets (Palade et al. 1979). VEGF has been reported to induce fenestration, in vivo and in vitro (Roberts and Palade 1997; Esser et al. 1998). Interestingly, EG-VEGF could induce fenestrae in endothelial cells alone or in combination with VEGF. Treating ACE cultures with both factors resulted in an additive response, indicating that these molecules may also cooperate in vivo to induce the fenestrated phenotype. This is intriguing because, particularly in the primate ovary, VEGF and EG-VEGF expression appear to be, at least in part, overlapping (LeCouter et al. 2001).

In human and nonhuman primates, EG-VEGF expression is largely restricted to the steroidogenic glands, ovary, testis, adrenal cortex, and placenta. In the testis, EG-VEGF expression is localized to the testosterone-producing Leydig cells and overlaps with VEGF expression. In the ovary, which exhibits cyclic growth and angiogenesis, the EG-VEGF expression pattern is particularly striking: Expression is found throughout the ovarian stroma and also within the theca interna layer of the follicle. VEGF is also expressed in the theca but very little in the stroma. The developing follicle initially acquires a capillary network within this layer, coincident with the induction of steroidogenic property (Plendl 2000). EG-VEGF may therefore cooperate with VEGF in the induction of follicular angiogenesis. Unlike VEGF, EG-VEGF shows little expression in the corpus luteum. Similar to VEGF (Shweiki et al. 1992; Shima et al. 1995), expression of EG-VEGF is controlled by low oxygen tension via a putative HIF-1α-dependent mechanism, and it will be interesting to assess whether it is hormonally regulated (Shifren et al. 1996; Neulen et al. 1998) as well.

Our findings from cell culture systems relating to the selectivity of EG-VEGF were predictive of its in vivo activities. Although VEGF elicited the expected, strong angiogenic response in the cornea and in the skeletal muscle, the vessels in these tissues did not respond to EG-VEGF. However, injection of adenovirus encoding EG-VEGF into the ovary resulted in a robust angiogenic response, with new and leaky vessels appearing within one week following injection. The histological picture of the EG-VEGF response was indistinguishable from that observed in VEGF-treated ovaries. Therefore, EG-VEGF is a vascular-bed-specific angiogenic molecule. The selective action of a given ligand may be achieved by several means. The expression of the ligand may be restricted, whereas the receptor is widely expressed. Conversely, the ligand may be produced in any tissues, but the receptor expression is restricted. Alternatively, expression of both ligand and its receptor may be restricted, thereby achieving selective activity. We have clearly demonstrated that human EG-VEGF mRNA expression is restricted to steroidogenic tissue, and our data in cell culture systems and in vivo are consistent with the restricted expression of the EG-VEGF receptor within these tissues.

EG-VEGF SIGNALING IN ENDOCRINE ENDOTHELIAL CELLS

Several signaling pathways are activated upon EG-VEGF stimulation in bovine adrenal cortex capillary endothelial cells (Lin et al. 2002). EG-VEGF led to a time- and dose-dependent phosphorylation of p44/42 MAPK. Pretreatment of ACE cells with PD98059 inhibited such phosphorylation and resulted in the suppression of EG-VEGF-induced proliferation and migration in ACE cells, suggesting a crucial role for MAPK as a downstream activator of EG-VEGF signaling. EG-VEGF also increased the phosphorylation status of Akt in a phosphatidylinositol-3 kinase (PI3K)-dependent manner. Consistent with this effect, EG-VEGF was a potent survival factor for ACE cells. Interestingly, a homolog of EG-VEGF, referred to as the mammalian Bv8 peptide, was found to act as a neuronal survival factor through the activation of MAPK and PI3K pathways in the central nervous system (Melchiorri et al. 2001). In addition to these downstream effectors, we identified endothelial nitric-oxide synthase (eNOS) as one of the targets of Akt activation in ACE cells (Lin et al. 2002). VEGF has been previously shown to induce the release of NO from endothelial cells (Yang et al. 1998), and several lines of evidence indicate that eNOS is an important modulator of angiogenesis and vascular tone (Murohara et al. 1998). Thus, EG-VEGF stimulation led to proliferation, migration, and survival of responsive endothelial cells through the activation of the MAPK p44/42 and PI3K signaling pathways.

Notwithstanding the fact that the effects of EG-VEGF on ACE cells are very similar to those elicited by VEGF, EG-VEGF stimulation was not associated with any detectable induction of receptor tyrosine phosphorylation. This indicated that, in contrast to VEGFR-1 and VEGFR-2, the EG-VEGF receptor is not a tyrosine kinase type, and indeed, there is now clear evidence that the effects of EG-VEGF are mediated by a G-protein-coupled receptor. EG-VEGF-induced MAPK phosphorylation was blocked by pretreatment with pertussis toxin, indicating that heterotrimeric G protein, $G\alpha_i$, plays an important role in mediating this activation (Lin et al. 2002). Two highly homologous proteins with seven-transmembrane spans have been recently identified as the receptors for EG-VEGF (Masuda et al. 2002). These two putative EG-VEGF receptors belong to the neuropeptide Y subfamily, and display 80–90% identity with each other. Treatment of cells expressing these receptors with EG-VEGF induced mobilization of calcium, stimulation of phosphoinositide turnover, and activation of the MAPK signaling pathway. Thus, EG-VEGF binding to these putative receptors results in the mobilization and activation of multiple key effectors, including calcium, phosphoinositol, MAPK, and PI3K.

POSSIBLE ROLE OF EG-VEGF IN PATHOLOGICAL CONDITIONS

EG-VEGF may have considerable pathophysiological and therapeutic significance for several endocrine disorders characterized by excessive angiogenesis and for certain endocrine gland tumors. We have recently evaluated the expression of EG-VEGF in a panel of human ovary specimens isolated from patients diagnosed with polycystic ovary syndrome (PCOS) (N. Ferrara et al., in prep.). This condition is a major cause of infertility, affecting 5–10% of women of reproductive age (Goldziher and Green 1962; Dunaif and Thomas 2001). It is known that the increase in ovary mass in PCOS is accompanied by extensive angiogenesis in the stroma and theca (Yen 1999). Our previous studies have demonstrated that adenovirus-mediated delivery of EG-VEGF in the ovary elicits an angiogenic response and cyst formation, similar to that induced by VEGF (LeCouter et al. 2001). Although both EG-VEGF and VEGF are expressed in the PCOS ovaries, their expression patterns are very distinct, essentially nonoverlapping. VEGF mRNA expression is associated with the granulosa cell layer. EG-VEGF expression is strongly correlated with the hyperplastic stroma and thecal cells, indicating that this molecule may participate in the hyperplastic/angiogenic changes that occur in PCOS. We propose that, in addition to the potentially coordinating activities that promote angiogenesis in the normal human ovary, VEGF and EG-VEGF may also jointly contribute to the development of PCOS.

We are currently assessing the expression of EG-VEGF and VEGF in a series of adrenal gland tumor specimens (our unpublished observations). Again, the expression pattern of EG-VEGF is rather distinct from that exhibited by VEGF. In adrenal-medulla-derived tumors (pheochromocytomas), VEGF is highly expressed throughout, whereas EG-VEGF is undetectable. In contrast, carcinomas of cortical origin express both VEGF and EG-VEGF. These data exemplify the concept that therapies to affect therapeutic angiogenesis or antiangiogenesis may require knowledge of the tissue-specific molecules that regulate and maintain the distinct vascular beds. Currently, we are examining the expression and potential role of EG-VEGF in other endocrine disease states, including pathologies of the placenta. The therapeutic values of EG-VEGF neutralizing reagents, and of EG-VEGF receptor-specific agonists or antagonists, remain to be demonstrated.

CONCLUSIONS

The identification of EG-VEGF permits a novel concept for the tissue-specific regulation of angiogenesis. Steroidogenic glands appear to have developed highly specific, local mechanisms, possibly to fine-tune and complement the action of the ubiquitous VEGF/VEGF receptor system. That other tissue-specific angiogenic molecules exist is an appealing concept, but remains to be demonstrated. Attempts to identify other tissue-specific angiogenic molecules may rely on sequence similarities or tissue-specific expression. However, these molecules may be unrelated in primary amino acid structure, may be expressed at low levels in the tissue, especially if they have a maintenance role, and may be regulated by sequestration in the ECM. Other functional screens, analogous to the assay that identified EG-VEGF, may reveal novel selective angiogenic factors.

The tissues that express EG-VEGF are characterized by highly permeable, often fenestrated vessels, high metabolic demand, and a relatively high endothelial cell turnover. In the testis, 3% of endothelial cells are labeled with BrdU (Collin and Bergh 1996) in comparison with other tissues that turn over at an extremely low rate. The high metabolic demand of steroid hormone synthesis and spermatogenesis may explain this capacity for endothelial cell proliferation exhibited within the testis.

Recent studies suggest the existence of VEGF-independent pathways that remain to be identified. Early results of clinical trials with VEGF inhibitors have provided initial evidence of clinical efficacy. However, progression eventually occurs in some patients (Ferrara 2000). It is tempting to speculate that the ability of certain tumors to escape anti-VEGF or other antiangiogenic strategies may be, at least in part, related to the expression of organ- or tumor-specific angiogenic factors.

VEGF is a potent angiogenic molecule that may dominate the initiation of neovascularization. The existence of EG-VEGF and possibly other tissue-specific angiogenic mitogens permits an additional layer of signaling refinement to establish and maintain the differentiated endothelial structure and function. The concept of tissue-specific angiogenic factors, along with the phenotypic plasticity of vascular bed endothelial cells, potentially offers a new avenue to regulate tissue growth and function. These also suggest novel approaches to salvage or regenerate tissue. For example, mitogens specific for the endothelium of cardiac or skeletal muscle would be potentially of great therapeutic significance. A principal benefit of tissue-specific angiogenic therapeutics could be the elimination of systemic, undesired effects associated with broad-spectrum angiogenic molecules.

ACKNOWLEDGMENT

We thank N. van Bruggen for critically reading the manuscript.

REFERENCES

Aiello L.P., Avery R.L., Arrigg P.G., Keyt B.A., Jampel H.D., Shah S.T., Pasquale L.R., Thieme H., Iwamoto M.A., Park J.E., Nguyen H., Aiello L.M., Ferrara N., and King G.L. 1994. Vascular endothelial growth factor in ocular fluid of patients with diabetic retinopathy and other retinal disorders. *N. Engl. J. Med.* **331:** 1480.

Aird W.C., Edelberg J.M., Weiler-Guettler H., Simmons W.W., Smith T.W., and Rosenberg R.D. 1997. Vascular bed-specific expression of an endothelial cell gene is programmed by the tissue microenvironment. *J. Cell Biol.* **138:** 1117.

Asahara T., Murohara T., Sullivan A., Silver M., van der Zee R., Li T., Witzenbichler B., Schatteman G., and Isner J.M. 1997. Isolation of putative progenitor endothelial cells for angiogenesis. *Science* **275:** 964.

Boisbouvier J., Albrand J.P., Blackledge M., Jaquinod M., Schweitz H., Lazdunski M., and Marion D. 1998. A structural homologue of colipase in black mamba venom revealed by NMR floating disulphide bridge analysis. *J. Mol. Biol.* **283:** 205.

Collin O. and Bergh A. 1996. Leydig cells secrete factors which increase vascular permeability and endothelial cell proliferation. *Int. J. Androl.* **19:** 221.

Dellian M., Witwer B.P., Salehi H.A., Yuan F., and Jain R.K. 1996. Quantitation and physiological characterization of angiogenic vessels in mice: Effect of basic fibroblast growth factor, vascular endothelial growth factor/vascular permeability factor, and host microenvironment. *Am. J. Pathol.* **149:** 59.

Dunaif A. and Thomas A. 2001. Current concepts in the polycystic ovary syndrome. *Annu. Rev. Med.* **52:** 401.

Esser S., Wolburg K., Wolburg H., Breier G., Kurzchalia T., and Risau W. 1998. Vascular endothelial growth factor induces endothelial fenestrations in vitro. *J. Cell Biol.* **140:** 947.

Ferrara N. 2000. VEGF: An update on biological and therapeutic aspects. *Curr. Opin. Biotechnol.* **11:** 617.

Ferrara N. and Alitalo K. 1999. Clinical applications of angiogenic growth factors and their inhibitors. *Nat. Med.* **5:** 1359.

Ferrara N. and Davis-Smyth T. 1997. The biology of vascular endothelial growth factor. *Endocr. Rev.* **18:** 4.

Ferrara N., Carver-Moore K., Chen H., Dowd M., Lu L., O'Shea K.S., Powell-Braxton L., Hillan K.J., and Moore M.W. 1996. Heterozygous embryonic lethality induced by targeted inactivation of the VEGF gene. *Nature* **380:** 439.

Ferrara N., Chen H., Davis-Smyth T., Gerber H.-P., Nguyen T.-N., Peers D., Chisholm V., Hillan K.J., and Schwall R.H. 1998. Vascular endothelial growth factor is essential for corpus luteum angiogenesis. *Nat. Med.* **4:** 336.

Folkman J. 1995. Angiogenesis in cancer, vascular, rheumatoid and other disease. *Nat. Med.* **1:** 27.

Folkman J. and Shing Y. 1992. Angiogenesis. *J. Biol. Chem.* **267:** 10931.

Garner A. 1994. Vascular diseases. In *Pathobiology of ocular disease* (ed. A. Garner and G.K. Klintworth), p. 1625. Marcel Dekker, New York.

Gerber H.P., Vu T.H., Ryan A.M., Kowalski J., Werb Z., and Ferrara N. 1999a. VEGF couples hypertrophic cartilage remodeling, ossification and angiogenesis during endochondral bone formation. *Nat. Med.* **5:** 623.

Gerber H.P., Hillan K.J., Ryan A.M., Kowalski J., Keller G.-A., Rangell L., Wright B.D., Radtke F., Aguet M., and Ferrara N. 1999b. VEGF is required for growth and survival in neonatal mice. *Development* **126:** 1149.

Goldziher J.W. and Green J.A. 1962. The polycistic ovary. I. Clinical and histologic features. *J. Clin. Endocrinol. Metab.* **22:** 325.

Hamilton W.J., Boyd J.D., and Mossmann H.W. 1962. *Human embryology*. Williams and Wilkins, Baltimore, Maryland.

Joubert F.J. and Strydom D.J. 1980. Snake venom. The amino acid sequence of protein A from *Dendroaspis polylepis polylepis* (black mamba) venom. *Hoppe-Seyler's Z. Physiol. Chem.* **361:** 1787.

Kim K.J., Li B., Winer J., Armanini M., Gillett N., Phillips H.S., and Ferrara N. 1993. Inhibition of vascular endothelial growth factor-induced angiogenesis suppresses tumor growth in vivo. *Nature* **362:** 841.

Klagsbrun M. 1992. Mediators of angiogenesis: The biological significance of basic fibroblast growth factor (bFGF)-heparin and heparan sulfate interactions. *Semin. Cancer Biol.* **3:** 81.

LeCouter J., Kowalski J., Foster J., Hass P., Zhang Z., Dillard-Telm L., Frantz G., Rangell L., DeGuzman L., Keller G.-A., Peale F., Gurney A., Hillan K.J., and Ferrara N. 2001. Identification of an angiogenic mitogen selective for endocrine gland endothelium. *Nature* **412:** 877.

Leung D.W., Cachianes G., Kuang W.J., Goeddel D.V., and Ferrara N. 1989. Vascular endothelial growth factor is a secreted angiogenic mitogen. *Science* **246:** 1306.

Li M., Bullock C.M., Knauer D.J., Ehlert F.J., and Zhou Q.Y. 2001. Identification of two prokineticin cDNAs: Recombinant proteins potently contract gestrointestinal smooth muscle. *Mol. Pharmacol.* **59**: 692.

Lin R., LeCouter J., Kowalski J., and Ferrara N. 2002. Characterization of EG-VEGF signaling in adrenal cortex capillary endothelial cells. *J. Biol. Chem.* **277**: 8724.

Maisonpierre P.C., Suri C., Jones P.F., Bartunkova S., Wiegend S.J., Radziejewski C., Compton D., McClain J., Aldrich T.H., Papadopulos N., Daly T.J., Davis S., Sato T.N., and Yancopoulos G.D. 1997. Angiopoietin-2, a natural antagonist for Tie-2 that disrupts in vivo angiogenesis. *Science* **277**: 55.

Masuda Y., Takatsu Y., Terao Y., Kumano S., Ishibashi Y., Suenaga M., Abe M., Fukusumi S., Watanabe T., Shintani Y., Yamada T., Hinuma S., Inatomi N., Ohtaki T., Onda H., and Fujino M. 2002. Isolation and identification of EG-VEGF/prokineticins as cognate ligands for two orphan G-protein-coupled receptors. *Biochem. Biophys. Res. Commun.* **293**: 396.

Melchiorri D., Bruno V., Besong G., Ngomba R. T., Cuomo L., De Blasi A., Copani A., Moschella C., Storto M., Nicoletti F., Lepperdinger G., and Passarelli F. 2001. The mammalian homologue of the novel peptide Bv8 is expressed in the central nervous system and supports neuronal survival by activating the MAP kinase/PI-3-kinase pathways. *Eur. J. Neurosci.* **13**: 1694.

Murohara T., Asahara T., Silver M., Bauters C., Masuda H., Kalka C., Kearney M., Chen D., Symes J.F., Fishman M.C., Huang P.L., and Isner J.M. 1998. Nitric oxide synthase modulates angiogenesis in response to tissue ischemia. *J. Clin. Invest.* **101**: 2567.

Neulen J., Raczek S., Pogorzelski M., Grunwald K., Yeo T.K., Dvorak H.F., Weich H.A., and Breckwoldt M. 1998. Secretion of vascular endothelial growth factor/vascular permeability factor from human luteinized granulosa cells is human chorionic gonadotrophin dependent. *Mol. Hum. Reprod.* **4**: 203.

Palade G.E., Simionescu M., and Simionescu N. 1979. Structural aspects of the permeability of the microvascular endothelium. *Acta Physiol. Scand. Suppl.* **463**: 11.

Plendl J. 2000. Angiogenesis and vascular regression in the ovary. *Anat. Histol. Embryol.* **29**: 257.

Presta L.G., Chen H., O'Connor S.J., Chisholm V., Meng Y.G., Krummen L., Winkler M., and Ferrara N. 1997. Humanization of an anti-VEGF monoclonal antibody for the therapy of solid tumors and other disorders. *Cancer Res.* **57**: 4593.

Rafii S., Meeus S., Dias S., Hattori K., Heissig B., Shmelkov S., Rafii D., and Lyden D. 2002. Contribution of marrow-derived progenitors to vascular and cardiac regeneration. *Semin. Cell Dev. Biol.* **13**: 61.

Risau W. 1997. Mechanisms of angiogenesis. *Nature* **386**: 671.

Risau W. and Flamme I. 1995. Vasculogenesis. *Annu. Rev. Cell Dev. Biol.* **11**: 73.

Roberts W.G. and Palade G.E. 1997. Neovasculature induced by vascular endothelial growth factor is fenestrated. *Cancer Res.* **57**: 765.

Roberts W.G., Delaat J., Nagane M., Huang S., Cavenee W.K., and Palade G.E. 1998. Host microvasculature influence on tumor vascular morphology and endothelial gene expression. *Am. J. Pathol.* **153**: 1239.

Ruoslahti E. and Rajotte D. 2000. An address system in the vasculature of normal tissues and tumors. *Annu. Rev. Immunol.* **18**: 813.

Shifren J.L., Tseng J.F., Zaloudek C.J., Ryan I.P., Meng Y.G., Ferrara N., Jaffe R.B., and Taylor R.N. 1996. Ovarian steroid regulation of vascular endothelial growth factor in the human endometrium: Implications for angiogenesis during the menstrual cycle and in the pathogenesis of endometriosis. *J. Clin. Endocrinol. Metab.* **81**: 3112.

Shima D.T., Adamis A.P., Ferrara N., Yeo K.T., Yeo T.K., Allende R., Folkman J., and D'Amore P.A. 1995. Hypoxic induction of endothelial cell growth factors in retinal cells: Identification and characterization of vascular endothelial growth factor (VEGF) as the mitogen. *Mol. Med.* **1**: 182.

Shweiki D., Itin A., Soffer D., and Keshet E. 1992. Vascular endothelial growth factor induced by hypoxia may mediate hypoxia-initiated angiogenesis. *Nature* **359**: 843.

Simionescu N. and Simionescu M. 1988. The cardiovascular system. In *Cell and tissue biology* (ed. L. Weiss), p. 355. Urban and Schwarzenberg, Baltimore, Maryland.

Stewart P.A. and Wiley M.J. 1981. Developing nervous tissue induces formation of blood-brain barrier characteristics in invading endothelial cells: A study using quail–chick transplantation chimeras. *Dev. Biol.* **84**: 183.

Suri C., Jones P.F., Patan S., Bartunkova S., Maisonpierre P.C., Davis S., Sato T.N., and Yancopoulos G.D. 1996. Requisite role of angiopoietin-1, a ligand for the TIE2 receptor, during embryonic angiogenesis. *Cell* **87**: 1171.

Wood J.M., Bold G., Buchdunger E., Cozens R., Ferrari S., Frei J., Hofmann F., Mestan J., Mett H., O'Reilly T., Persohn E., Rosel J., Schnell C., Stover D., Theuer A., Towbin H., Wenger F., Woods-Cook K., Menrad A., Siemeister G., Schirner M., Thierauch K.H., Schneider M.R., Drevs J., Martiny-Baron G., and Totzke F. 2000. PTK787/ZK 222584, a novel and potent inhibitor of vascular endothelial growth factor receptor tyrosine kinases, impairs vascular endothelial growth factor-induced responses and tumor growth after oral administration. *Cancer Res.* **60**: 2178.

Yancopoulos G.D., Davis S., Gale N.W., Rudge J.S., Wiegand S.J., and Holash J. 2000. Vascular-specific growth factors and blood vessel formation. *Nature* **407**: 242.

Yang R., Bunting S., Ko A., Keyt B.A., Modi N.B., Zioncheck T.F., Ferrara N., and Jin H. 1998. Substantially attenuated hemodynamic responses to *Escherichia coli*-derived vascular endothelial growth factor given by intravenous infusion compared with bolus injection. *J. Pharmacol. Exp. Ther.* **284**: 103.

Yen S.S.C. 1999. Polycystic ovary syndrome (hyperandrogenic chronic anovulation). In *Reproductive endocrinology* (ed. S.S.C. Yen et al.), p. 436. W.B. Saunders, Philadelphia, Pennsylvania.

Profiling the Molecular Diversity of Blood Vessels

R. Pasqualini and W. Arap

The University of Texas, M. D. Anderson Cancer Center, Houston, Texas 77030

Over the past few years, work by our laboratory and others has uncovered a vascular address system based on the ability to fingerprint the molecular heterogeneity of the vasculature in vivo (Kolonin et al. 2001; Pasqualini and Arap 2002). By using a random peptide library system (Smith and Scott 1985, 1993), we were able to select probes based on their ability to target normal blood vessels or blood vessels of tumors. Sophisticated imaging technology for determining the distribution of such targeted probes in vivo, their organ specificity, and cellular receptors has contributed to the definition of differences in the normal vasculature and alterations observed due to multiple abnormalities in tumor blood vessels (Hashizume et al. 2000). The spatial and cellular diversity of the endothelium in the context of cancer is likely to suggest novel therapeutic strategies. It has become clear that tumor blood vessel cells express angiogenic markers that are not detectable in normal blood vessels (Brooks et al. 1994; Arap et al. 1998a). Moreover, many vascular receptors are present in restricted—but highly specific and accessible—locations within different tissues (Pasqualini and Arap 2002). Finally, the genetic progression of malignant cells is paralleled by angiogenesis-related epigenetic changes in the cells that form the endothelium; thus, cancer-specific abnormalities may develop in the nonmalignant cells forming tumor blood vessels. Molecular targets relevant for disease progression may often be overlooked in high-throughput sequencing or gene array approaches that do not take into account the molecular diversity and heterogeneity of the tumor vasculature. Taken together, our previous work—among others—has shown (1) that the vascular endothelium of organs is modified in a manner that is intrinsic to the tissue microenvironment and (2) that the development of most diseases is accompanied by specific abnormalities in the cells forming tumor blood vessels.

PHAGE DISPLAY LIBRARIES AND TARGETING TECHNOLOGY

We have perfected a sophisticated system to identify peptides and probes that can bind to molecular targets of interest. Mechanistic context and the expression pattern of such markers in human tissue have revealed unexpected properties of blood vessels. Interestingly, certain cell surface receptors expressed in tumor blood vessels had been originally identified as tumor-associated antigens. However, although tumor and angiogenic endothelial cells are often antigenically distinguishable from their counterparts in normal tissues, only a limited number of receptors have thus far been identified in a context that is functionally relevant for targeted delivery.

The in vivo screening method we developed relies on the identification of peptides that home to specific organs or tissues. These ligands are functionally selected after intravenous administration of a phage display random peptide library (Kolonin et al. 2001). This work has uncovered multiple vascular addresses (Rajotte and Ruoslahti 1999; Pasqualini et al. 2000). Peptides that home to tumor vasculature have been used as carriers to deliver cytotoxic drugs (Arap et al. 1998b), proapoptotic peptides (Ellerby et al. 1999), cytokines (Curnis et al. 2000), protease inhibitors (Koivunen et al. 1999), and gene therapy vectors (Wickham et al. 1997; Trepel et al. 2000). The coupling of homing peptides can yield targeted agents that, when compared to the parental compounds, show improved therapeutic index. In vivo phage screenings in humans will develop and use probes for targeting therapies and to identify diagnostic and prognostic factors. Moreover, the diversity of the vasculature may have implications in responses observed after therapy (in particular when antiangiogenic drugs are used).

To map vascular receptors, we had relied for some time on isolating ligands and receptors in mice following attempts to identify putative human homologs. However, when such homologs are found, the question remains as to whether targeting will be observed in the context of human vasculature. In humans, cell surface receptors may have a very different distribution and function than in mice. The existence of differences among species means that any data derived from mouse studies must be carefully validated before being translated to human studies.

MAPPING HUMAN VASCULATURE DIRECTLY IN PATIENTS

Recently, we have reported the first in vivo screening of a peptide library in a patient. We surveyed 47,160 motifs that localized to different organs. This large-scale screening shows that the tissue distribution of circulating peptides is nonrandom (Arap et al. 2002). High-throughput analysis of the motifs revealed similarities to ligands for differentially expressed cell surface proteins. We isolated from the prostate a potential mimetope of interleukin 11 (IL-11). To test the tissue specificity of the IL-11 peptide mimic, we developed a phage overlay assay to evaluate receptor–ligand interactions in tissue sections. We chose the motif RRAGGS for our studies (Arap et al. 2002). We showed by phage overlay on human tissue sections that the prostate-homing phage displaying an IL-11 peptide mimic specifically bound to the endothelium and to the epithelium of normal prostate, but not to control organs, such as skin. In contrast, a phage selected from the skin (displaying the motif HGGVG) did not bind to prostate tissue; however, this phage specifically recognized blood vessels in the skin. Moreover, the immunostaining pattern obtained with an antibody against human IL-11Rα on normal prostate tissue is undistinguishable from that of the CGRRAGGSC-displaying phage overlay; a control antibody showed no staining in prostate tissue. These findings were recapitulated in multiple tissue sections obtained from several different patients. Finally, using a ligand–receptor-binding assay in vitro, we demonstrate the interaction of the CGRRAGGSC-displaying phage with immobilized IL-11Rα at the protein–protein level. Such binding is specific, since it was inhibited by the native IL-11 ligand in a concentration-dependent manner (Arap et al. 2002). The potential of phage display to identify targeting sequences and receptors has hardly been fully explored. We are now expanding our studies and hope to validate multiple receptor–ligand pairs based on this strategy.

TARGETING ENDOTHELIUM-DERIVED HUMAN CELLS

Many cell surface receptors require the cell membrane microenvironment to function so that protein–protein interaction may occur. Combinatorial approaches may allow the selection of cell membrane ligands in a functional assay and without any bias about the cellular surface receptor. Therefore, even as-yet-unidentified receptors may be targeted in this manner.

We have developed a novel approach for the screening of cell-surface-binding peptides from phage libraries. Biopanning and rapid analysis of selective interactive ligands (termed BRASIL) is based on differential centrifugation in which a cell suspension incubated with phage in an aqueous upper phase is centrifuged through a nonmiscible organic lower phase. This single-step organic phase separation is faster, more sensitive, and more specific than current methods that rely on washing steps or limiting dilution. As a proof of concept, we screened human endothelial cells stimulated with vascular endothelial growth factor (VEGF), constructed a peptide-based ligand–receptor map of that VEGFR family, and validated a new targeting ligand (Giordano et al. 2001). Mapping ligand–receptor interactions by BRASIL may allow an understanding of binding requirements for other endothelial cell surface-receptor families and enable isolation of a panel of peptides for endothelium-derived cell targeting applications.

We have also performed in vivo screenings to characterize the vasculature of highly specialized tissues such as the placenta (Kolonin et al. 2002). In mammals, the interconnection between maternal and fetal molecules takes place through multiple distinct cell layers within the placenta (Rugh 1990). Teratogens that affect the placenta seem to function by inhibiting selective transporters which are necessary for the transfer of nutrients to the embryo (Maranghi et al. 1998). The functional role of the molecules that mediate maternal–fetal exchange in development is yet to be established. We have adapted in vivo phage display for the study of targets relevant in placental transport. We have selected teratogenic ligands that bind nonendothelial receptors expressed on the placental epithelium, the site where the molecular exchange occurs between the mother and the fetus. We used complementary approaches to demonstrate that a peptide motif, TPKTSVT, targets the FcRn/β_2m receptor complex that mediates maternal–fetal IgG transport (Ghetie and Ward 2000). Administration of this peptide delayed embryogenesis and led to pregnancy interruption. Such teratogenic effects are likely to be secondary to the evident placental thrombosis and ischemia that are triggered by the peptide treatment. This study represents the first in vivo screening of a phage display library on nonendothelial tissues. Future implications of these findings include the basis for high-throughput identification of placental receptors prone to be affected by teratogenic compounds and also the design of safer drugs that could be used during pregnancy.

Recognition of molecular diversity in human disease is required for the development of targeted therapies. Validated ligands may be used as carriers for imaging or therapeutic agents. Moreover, the ligands themselves may be used as either drug discovery leads or for therapeutic modulation of their corresponding receptor(s). Finally, another application of the selected targeted ligands is identification of their vascular receptors.

CONCLUSIONS

Proteomics can be defined as the systematic analysis of the proteins in biological samples that aims to document the overall distribution of proteins in cells, identify and characterize individual proteins of interest, and ultimately to elucidate their relationships and functional roles. Vascular proteomics is the molecular phenotyping of cells forming blood vessels at the protein–protein interaction level. Exploiting the molecular diversity of cell surface receptors expressed in the human endothelium may lead to a ligand–receptor targeting map of the vasculature. We are developing integrated, combinatorial library platform technologies whose goal is to enable the identification, validation, and prioritization of molecular targets in human blood vessels. Targeting the differential protein expression in the vasculature associated with normal tissues or diseases with an angiogenesis component (these include cancer, arthritis, diabetes, and cardiovascular diseases) may suggest novel therapeutic strategies. Our long-term goal is to translate a functional map of molecular targets and biomarkers into clinical applications.

ACKNOWLEDGMENTS

This work was supported by the National Institutes of Health (CA-90270 and CA-82976 to R.P.; CA-90270 and CA-90810 to W.A.) and by a Gillson-Longenbaugh Foundation Award (to R.P. and W.A.).

REFERENCES

Arap W., Pasqualini R., and Ruoslahti E. 1998a. Chemotherapy targeted to tumor vasculature. *Curr. Opin. Oncol.* **10**: 560.

———. 1998b. Cancer treatment by targeted drug delivery to tumor vasculature. *Science* **279**: 377.

Arap W., Kolonin M.G., Trepel M., Lahdenranta J., Cardo-Vila M., Giordano R.J., Mintz P.J., Ardelt P.U., Yao V.J., Vidal C.I., Chen L., Flamm A., Valtanen H., Weavind L.M., Hicks M.E., Pollock R.E., Botz G.H., Bucana C.D., Koivunen E., Cahill D., Troncoso P., Baggerly K.A., Pentz R.D., Do K.A., Logothetis C.J., and Pasqualini R. 2002. Steps toward mapping human vasculature by in vivo phage display. *Nat. Med.* **8**: 121.

Brooks P.C., Clark R.A., and Cheresh D.A. 1994. Requirement of vascular integrin $\alpha v \beta 3$ for angiogenesis. *Science* **264**: 569.

Curnis F., Sacchi A., Borgna L., Magni F., Gasparri A., and Corti A. 2000. Enhancement of tumor necrosis factor α antitumor immunotherapeutic properties by targeted delivery to aminopeptidase. *Nat. Biotechnol.* **18**: 1185.

Ellerby M., Arap W., Andrusiak R., Del Rio G., Kain R., Ruoslahti E., Bredesen D., and Pasqualini R. 1999. Targeted proapoptotic peptides for cancer therapy. *Nat. Med.* **9**: 1032.

Ghetie V. and Ward E.S. 2000. Multiple roles for the major histocompatibility complex class I-related receptor FcRn. *Annu. Rev. Immunol.* **18**: 739.

Giordano R., Cardo-Vila M., Lahdenranta J., Pasqualini R., and Arap W. 2001. Biopanning and rapid analysis of selective and interactive ligands. *Nat. Med.* **11**: 1249.

Hashizume H., Baluk P., Morikawa S., McLean J.W., Thurston G., Roberge S., Jain R.K., and McDonald D.M. 2000. Openings between defective endothelial cells explain tumor vessel leakiness. *Am. J. Pathol.* **156**: 1363.

Kolonin M.G., Pasqualini R., and Arap W. 2001. Molecular addresses in blood vessels as targets for therapy. *Curr. Opin. Chem. Biol.* **5**: 308.

———. 2002. Teratogenicity induced by targeting a placental immunoglobulin transporter. *Proc. Natl. Acad. Sci.* **99**: 13055.

Koivunen E., Arap W., Valtanen H., Rainisalo A., Medina O.P., Heikkila P., Kantor C., Gahmberg C.G., Salo T., Konttinen Y.T., Sorsa T., Ruoslahti E., and Pasqualini R. 1999. Tumor targeting with a selective gelatinase inhibitor. *Nat. Biotechnol.* **8**: 768.

Maranghi F., Macri C., Ricciardi C., Stazi A.V., and Mantovani A. 1998. Evaluation of the placenta: Suggestions for a greater role in developmental toxicology. *Adv. Exp. Med. Biol.* **444**: 129.

Pasqualini R. and Arap W. 2002. Vascular targeting. In *Encyclopedia of cancer* (ed. J.R. Berino), p. 501. Elsevier Science, New York.

Pasqualini R., Koivunen E., Kain R., Lahdenranta J., Shapiro L., Sakamoto M., Stryn A., Arap W., and Ruoslahti E. 2000. Aminopeptidase N is a receptor for tumor-homing peptides and a target for inhibiting angiogenesis. *Cancer Res.* **60**: 722.

Rajotte D. and Ruoslahti E. 1999. Membrane dipeptidase is the receptor for a lung-targeting peptide identified by in vivo phage display. *J. Biol. Chem.* **274**: 11593.

Rugh R. 1990. *The mouse: Its reproduction and development.* Oxford Science Publications, Oxford, United Kingdom.

Smith G.P. and Scott J.K. 1985. Searching for peptide ligands with an epitope library. *Science* **228**: 1315.

———. 1993. Libraries of peptides and proteins displayed in filamentous phage. *Methods Enzymol.* **21**: 228.

Trepel M., Grifman M., Weitzman M.D., and Pasqualini R. 2000. Molecular adaptors for vascular targeted adenoviral gene delivery. *Hum. Gene Ther.* **11**: 1971.

Wickham T.J., Haskard D., Segal D., and Kovesdi I. 1997. Targeting endothelium for gene therapy via receptors up-regulated during angiogenesis and inflammation. *Cancer Immunol. Immunother.* **45**: 149.

VEGF-A Induces Angiogenesis, Arteriogenesis, Lymphangiogenesis, and Vascular Malformations

J.A. NAGY, E. VASILE, D. FENG, C. SUNDBERG,* L.F. BROWN,
E.J. MANSEAU, A.M. DVORAK, AND H.F. DVORAK

Departments of Pathology, Beth Israel Deaconess Medical Center and Harvard Medical School, Boston, Massachusetts 02215

The mature vasculature develops initially from the twin processes of vasculogenesis (de novo formation of blood vessels from primitive precursors) and angiogenesis (formation of new blood vessels from preexisting vessels) (Folkman and D'Amore 1996; Beck and D'Amore 1997; Folkman 1997; Hanahan 1997; Risau 1997; Gale and Yancopoulos 1999). The resulting blood vessels that come to supply normal adult tissues are distributed at regular and closely spaced intervals and are organized into a hierarchy of elastic and muscular arteries, arterioles, capillaries, post-capillary venules, and small and large veins. Each of these vessel types has a characteristic size and structure that allow it to perform specialized functions. Thus, muscular arteries and arterioles regulate pressure; capillaries are the major site of molecular exchange; and venules and small veins are the primary vascular segments responsive to inflammation, whether *humoral* (plasma extravasation initiated by histamine, VEGF-A, etc.) or *cellular* (inflammatory cell extravasation).

The blood vessels that supply tumors differ markedly from those of normal tissues (Warren 1979; Dvorak et al. 1991, 2000). Tumor vessels are nonuniformly distributed, branch irregularly, form A-V shunts, are structurally heterogeneous, and do not conform to a clear hierarchical pattern. They overexpress the two high-affinity VEGF-A receptors (Brown et al. 1997; Dvorak et al. 1999), are lined by actively dividing endothelial cells, and are often deficient in pericytes and smooth muscle cells (Benjamin et al. 1998, 1999). Tumor blood vessels function ineffectively, providing only marginal gas exchange, nutrition, and waste removal. They are also hyperpermeable to circulating macromolecules with resulting plasma extravasation, edema, and extravascular fibrin deposition (Dvorak et al. 1979, 1984, 1988; Jain 1990).

The differences in organization, structure, and function that distinguish normal from tumor vessels imply substantial differences in their generation. The normal vasculature results from the concerted and coordinated activities of numerous cytokines, including at least VEGF-A, FGF-2, PDGF B, Ang-1, Ang-2, and their receptors (Folkman 1997; Hanahan 1997; Yancopoulos et al. 1998). Other members of the VEGF family, Ephrins,

TGFβ, Id proteins, etc., also have important though as yet incompletely understood roles, and additional cytokines and receptors likely await discovery. Tumor cells, of course, express many of these same cytokines. Therefore, to account for the observed differences between normal and tumor vessels, we have proposed the following general hypothesis:

> Normal blood vessels arise from the balanced expression of numerous angiogenic cytokines and inhibitors in appropriate amounts and in ordered sequence. Tumor angiogenesis represents a deviation from this normal pathway and is characterized by unbalanced expression of one or a few individual angiogenic cytokines, but without the checks and balances of modifying cytokines and inhibitors that characterize normal vascular development.

If this hypothesis is valid, it should be possible to mimic tumor angiogenesis by overexpressing in tissues one or another angiogenic cytokine that tumor cells overexpress. In this way it should be possible to generate surrogate tumor blood vessels; i.e., vessels that resemble tumor vessels in every respect but that are generated independent of tumor cells. Such experiments should achieve a number of important goals: (1) identify the cytokine(s) that are necessary to generate tumor vessels, (2) elucidate the steps and mechanisms of tumor angiogenesis, (3) provide a rich supply of tumor vessel surrogates for molecular characterization, allowing identification of useful targets for tumor imaging and antiangiogenesis or antivascular therapy, (4) determine the effects of cytokine deprivation on newly induced tumor vessel surrogates following cessation of cytokine expression, (5) compare the steps and mechanisms of tumor angiogenesis with those of other examples of pathological angiogenesis (e.g., wound healing, inflammation) and of normal developmental angiogenesis.

ADENOVIRAL VECTORS AS TOOLS FOR STABLE EXPRESSION OF ANGIOGENIC CYTOKINES IN VIVO

Our first goal was to establish a robust, low-background system for expressing individual cytokines in vivo at steady-state levels for a sufficient time to induce and sustain angiogenesis. Several different assays have been used for this purpose, including the chick chorioallantoic membrane (CAM), the cornea, and Matrigel plug

---

*Present address: Department of Medical Biochemistry and Microbiology, Uppsala University, BMC, Box 575, SE-751 23, Uppsala, Sweden.

implants. These systems do permit study of individual angiogenic factors but suffer from significant shortcomings. Because cytokines are soluble in body fluids and diffuse rapidly from application sites, test substances must be incorporated in slow-release polymers which are themselves foreign bodies that require surgical implantation. The CAM assay is performed in birds and involves a developing, nonembryonic tissue with an already high background of growing blood vessels. The cornea assay is performed in mammals but in a tissue that is not normally vascularized; therefore, powerful but as yet poorly understood forces that normally prevent angiogenesis must be overcome for new blood vessels to develop.

For these reasons, we turned, with the help of Dr. Richard Mulligan and the Harvard Gene Therapy Initiative, to nonreplicating adenoviral vectors as tools for studying the vascular response induced by different cytokines. Adenoviral vectors are readily engineered and allow local introduction and expression of inserted cytokines in any accessible tissue of adult or embryonic mice. They offer a number of additional advantages: They infect nondividing host cells; expression is maintained at steady levels for ~ 2 weeks; cytokine levels are easily adjusted by varying the dose of injected virus or choice of promoter; control empty cassettes or vectors that express noninflammatory cytokines give extremely low backgrounds; foreign matrix need not be implanted; and combinations of cytokines can be studied, together or in sequence. Immunodeficient animals are generally used to avoid an immune response against the viral vector.

THE VASCULAR RESPONSE TO VEGF-A

VEGF-A is arguably the single most important angiogenic cytokine. Both copies of this gene are necessary for normal vascular development (Carmeliet et al. 1996; Ferrara et al. 1996; Carmeliet and Collen 1999). Also, the VEGF-A^{164} isoform (the murine analog of human VEGF-A^{165}) and its receptors are overexpressed in the great majority of malignant human and animal tumors, in healing wounds, and in chronic inflammatory diseases such as psoriasis and rheumatoid arthritis (Dvorak et al. 1995, 1999; Brown et al. 1997). Finally, interference with either VEGF-A^{164} or its receptors inhibits tumor angiogenesis and often provokes extensive ischemic tumor necrosis (for review, see Dvorak et al. 1995, 1999; Brown et al. 1997). For all of these reasons, we began our studies with an adenoviral vector engineered to express VEGF-A^{164}.

ANGIOGENESIS

The Initial Response to Ad-VEGF-A^{164}

Within hours of injection into any of several different tissues of nude mice or rats, infected host cells began to express VEGF-A mRNA and continued to do so at steady levels for 10–14 days (Pettersson et al. 2000); thereafter, expression declined gradually and, by ~6 weeks, was no longer detectable by in situ hybridization.

The initial response (1–3 days) was similar in all tissues studied, but, for convenient direct visual monitoring, we performed most of our studies in ear skin (Fig. 1). Consistent with VEGF-A's potency as a vascular permeabilizing agent, preexisting microvessels were rendered hyperpermeable to plasma and plasma proteins, resulting in tissue edema, clotting of extravasated plasma fibrinogen, and deposition of an extravascular pro-angiogenic fibrin gel provisional matrix. Control adenoviral vectors that expressed Lac Z or green fluorescent protein induced no detectable gross or microscopic response.

To investigate the anatomic pathway(s) by which macromolecules extravasated from VEGF-A permeabilized vessels, we injected anionic ferritin intravenously into nude mice at various times after injecting Ad-VEGF-A^{164} into ear skin and followed its passage across the vascular wall by transmission electron microscopy. Ferritin, itself a plasma protein, can be visualized directly by electron microscopy. As anticipated from our earlier studies involving local injection of VEGF-A protein (Feng et al. 1999), ferritin tracer extravasated from venules by a trans-endothelial cell pathway, that provided by the vesiculo-vacuolar organelle (VVO). VVOs are grape-like clusters of uncoated, largely para-junctional, cytoplasmic vesicles and vacuoles that traverse endothelial cytoplasm

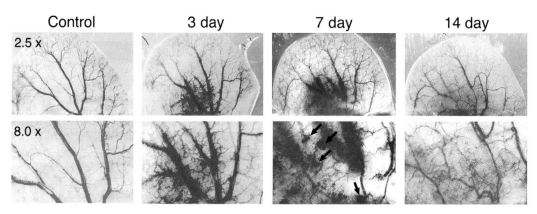

Figure 1. Angiogenic response induced in nude mouse ears at indicated intervals and magnifications after injection of 2.5 × 10^8 pfu Ad-VEGF-A^{164}. Several glomeruloid bodies are indicated by arrows.

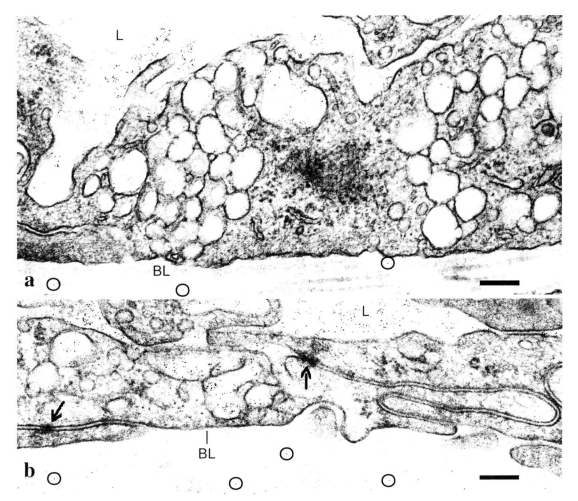

Figure 2. Electron micrographs 3 days after local Ad-VEGF-A^{164} injection and 20 minutes after i.v. ferritin tracer. (*a.*) Hyperpermeable venule lined by endothelial cells of normal height, illustrating two prominent VVOs (collections of vesicles/vacuoles) that span endothelial cell cytoplasm from lumen (L) to abluminal basal lamina (BL). Note ferritin particles in vascular lumen, VVO vesicles and vacuoles, and BL (*black dots*, several encircled). (*b*). Mother vessel with greatly thinned endothelium spanned by four or fewer VVO vesicles/vacuoles. Note ferritin in these vesicles and vacuoles but tight junctions (*arrows*) remain closed and do not admit tracer. Bars, 200 nm. (Reprinted, with permission, from Pettersson et al. 2000 [copyright Lippincott Williams and Wilkins].).

from lumen to ablumen (Fig. 2) (Dvorak et al. 1996; Feng et al. 1996, 1999, 2000a). They were originally described in endothelial cells lining tumor vessels and normal venules (Kohn et al. 1992). The individual vesicles and vacuoles that comprise VVOs are linked to each other, and to the luminal and abluminal plasma membranes, by stomata that are normally closed by thin diaphragms. By some as yet unknown mechanism, VEGF-A^{164} (and other vasoactive mediators) causes these stomata to open, providing a trans-endothelial cell pathway through which plasma proteins extravasate. In contrast, inter-endothelial cell junctions remained tightly closed and did not admit tracer ferritin, even in the presence of abundant VEGF-A.

"Mother" Vessels

By 18 hours, enlarged, thin-walled, serpentine, pericyte-poor, hyperpermeable, and strongly VEGFR-2-positive sinusoids appeared, structures to which we have given the name mother vessels (Fig. 3). Vessels of this description are commonly found in human and animal tumors (Feng et al. 2000a) and are also transiently present in healing wounds (Brown et al. 1992).

Mother vessels arose from the enlargement of preexisting microvessels, primarily venules. By 18–24 hours, mean cross-sectional microvessel area had increased by about fourfold, mean vessel perimeter by about twofold, and the percentage of dermis occupied by microvessels by four- to sevenfold (Pettersson et al. 2000). Mother vessels formed by a multistage process that involved proteolytic digestion of vascular basement membranes, pericyte detachment, and spreading and thinning of endothelial cells to cover the resulting, greatly expanded surface; for at least the first 24 hours, mother vessels formed without endothelial cell or pericyte division.

The vascular basement membrane is an especially important impediment to microvessel enlargement. Intact basement membranes are noncompliant (inelastic) structures that limit vascular enlargement to an ~ 30% increase

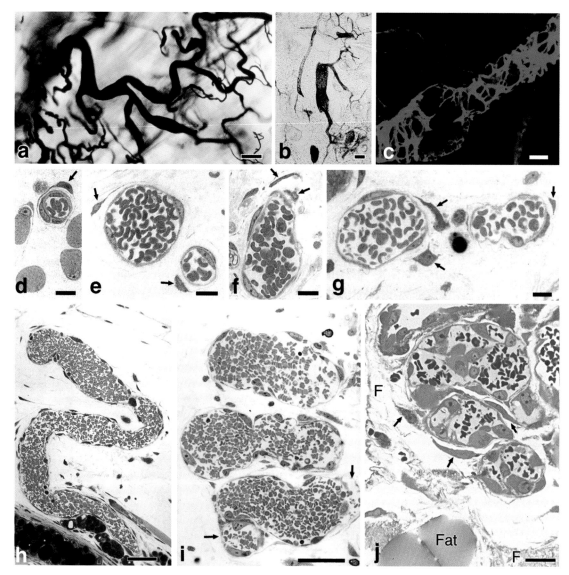

Figure 3. Mother vessels induced by Ad-VEGF-A^{164} in nude mouse ears. (*a, b*) Whole mounts illustrate colloidal carbon-perfused vascular beds. Mother vessels appear as enlarged segments of much smaller, normal venules. (*c*) Confocal microscopic image of mother vessel stained for pericytes with an antibody to α smooth muscle actin. Note incomplete pericyte covering, especially over segments of greatest vessel enlargement. (*d–g*) Mother vessel formation at 18 hours after Ad-VEGF-A^{164}. A normal small venule remains in panel *d* (*arrow*), whereas larger, developing mother vessels with detaching pericytes (*arrows*) are evident in *e–g*. (*h*) Three-day mother vessel illustrating typical serpentine pattern. (*i*) Similar mother vessel at 4 days cut in cross section at three levels. Note highly irregular luminal surface and endothelial cell bridging to form secondary lumens (*arrows*). (*j*) Mother vessels at 5 days after Ad-VEGF-A^{164} injection. Activated endothelial cells extend processes, forming trans-luminal bridges that create multiple smaller lumens. Note partially detached pericytes (*arrows*). Bars: *a, b*, 100 μm; *c, h, i*, 50 μm; *d–g*, 10 μm; *j*, 15 μm. (*d–j*, Reprinted, with permission, from Pettersson et al. 2000 [copyright Lippincott Williams & Wilkins].)

in cross-sectional area (Swayne and Smaje 1989; Swayne et al. 1989). Therefore, structural changes in the vascular basement membrane were necessary to accommodate the dramatic increases in microvessel area that followed within 18 hours of Ad-VEGF-A^{164} injection. Immunohistochemistry with antibodies to laminin demonstrated reduced or absent staining for basement membrane components in forming mother vessels, indicative of basement membrane degradation (Pettersson et al. 2000). VEGF-A is known to up-regulate the expression of endothelial cell proteases that are capable of degrading vascular basement membranes (Unemori et al. 1992). Basement membrane degradation has been thought to be essential for sprout formation (Grant and Kleinman 1997), and our findings here add an additional consequence, that of permitting the vascular expansion necessary to form mother vessels.

Role of VVOs in Mother Vessel Formation

As noted above, VVOs provided the primary anatomic pathway by which plasma proteins extravasated from venules in response to Ad-VEGF-A^{164} (Kohn et al. 1992; Dvorak et al. 1996; Feng et al. 1996, 2000a,b; Folkman

Table 1. Loss of VVOs and VVO Vesicles and Vacuoles Accompanies Mother Vessel Formation Induced by Injection of 2.5×10^8 pfu Ad-VEGF-A^{164} into Nude Mouse Ear Skin

Days	No. of vessel cross-sections studied	No. of VVOs per μm vessel perimeter	No. vesicles or vacuoles per VVO	No. of VVO vesicles per μm vessel perimeter
0	22	0.74 ± 0.05	11.38 + 0.46	8.2 ± 0.53
1	18	0.46 ± 0.03	7.36 ± 0.30*	3.4 ± 0.29**
3	20	0.38 ± 0.04***	5.85 ± 0.15***	2.2 ± 0.26***

Statistical differences from day-zero values were calculated with Dunn's multiple comparisons test and are expressed as follows: *$p \leq 0.05$; **$p \leq 0.01$; ***$p \leq 0.001$.

and D'Amore 1996.). However, VVOs were also found to have a second critical function, that of providing an intracellular store of membrane that could be mobilized to permit rapid vessel enlargement without new membrane synthesis. Morphometry revealed that normal venular endothelium stored an amount of membrane in VVOs that was equivalent to 2.25 times that of plasma membrane. Therefore, transfer of VVO membranes to the cell surface has the potential to provide a more than threefold increase in endothelial cell perimeter without new membrane synthesis. Consistent with this possibility, mother vessel formation was accompanied by progressive reduction in the number of VVOs, in the number of VVO vesicles and vacuoles per VVO, and in the total number of VVO vesicles and vacuoles (Table 1). Our findings, therefore, indicate that the endothelial cell spreading and thinning that accompany mother vessel formation can be accommodated by transfer of intracellular VVO membrane to the cell surface.

Although the endothelial cells of mother vessels have fewer and smaller VVOs than normal venular endothelium, they remained hyperpermeable to macromolecular tracers. This finding is readily explained by the fact that, because of considerable endothelial cell thinning, extravasating macromolecules need pass through many fewer interconnected vesicles or vacuoles than in normal venular endothelial cells (Fig. 2); i.e., the trans-endothelial cell path is considerably shortened in mother vessels. Mother vessels continued to form for 3–5 days, and, after 48 hours, mother vessel enlargement was accompanied by extensive division of both endothelial cells and pericytes.

Mother Vessel Progeny: Daughter Capillaries and Glomeruloid Bodies

Mother vessels are transitional forms that evolved along several different pathways (Pettersson et al. 2000). Many developed into daughter capillaries by a process of "bridging" (Fig. 3i,j; Fig. 4). Lining endothelial cells projected cytoplasmic processes into and across mother vessel lumens; these transluminal projections subsequently formed "bridges" that divided blood flow into smaller-sized channels. Over the course of several days, some of these separated from each other to form individual, smaller-caliber "daughter" vessels that were indistinguishable from normal capillaries.

Glomeruloid bodies (GB) are a second mother vessel derivative (Fig. 5) (Pettersson et al. 2000; Sundberg et al. 2001). They are poorly organized vascular structures that resemble renal glomeruli (hence the name) and are a common feature of several human tumors, particularly glioblastoma multiforme (Lantos et al. 1997), and certain vascular malformations (Fig. 5) (McKee 1996; Lantos et al. 1997). GB precursors were first recognized at ~3 days as focal collections of large, primitive cells in the endothelial cell lining of mother vessels (Fig. 5c–f). It remains to be determined whether these cells arose locally from vascular endothelium or were instead derived from circulating bone marrow precursors (Asahara et al. 1997; Rafii 2000; Rafii et al. 2002). Whatever their source, they proliferated rapidly (as determined by [^{3}H]thymidine incorporation) and extended inward into mother vessel lumens as well as outward into the extravascular connective tissue, forming nodules that were often centered around VEGF-A^{164}-expressing cells. Initially, all of the cells comprising GB bore endothelial cell markers, but, by 7–10 days, pericytes were also prominent and organized themselves around small clusters of endothelial cells to form primitive microvessels (Sundberg et al. 2001). As they expanded, GB severely compromised the mother vessels in which they had arisen, reducing and dividing originally large single vascular lumens into multiple, smaller channels (Fig. 5g–j). Blood flow through mature GB was greatly compromised.

After about 14 days, as VEGF-A^{164} expression began to wane, GB progressively devolved by a process that involved both apoptosis and reorganization of endothelial cells and pericytes to form normal appearing capillaries (Fig. 5k,l). Taken together, these experiments provided the first animal model for generating GB, offered important insights into the mechanisms of GB formation and devolution, and demonstrated that VEGF-A^{164} is both sufficient for GB generation and necessary for their maintenance.

By ~ 8 weeks, VEGF-A^{164} expression was no longer detectable, and the angiogenic response had largely resolved with a return of microvascular density to near-normal levels.

ARTERIO-VENOUS MALFORMATIONS

Instead of evolving into daughter capillaries or GB, some mother vessels retained their large size, becoming stabilized by acquiring a coat of smooth muscle cells and/or collagen-expressing fibroblasts (Pettersson et al. 2000). However, such vessels were readily distinguished from normal arteries and veins by their inappropriate large size (occasionally approaching the 1-mm diameter of the renal artery), by their thinner and often asymmetric muscular coat, and in other cases by the presence of

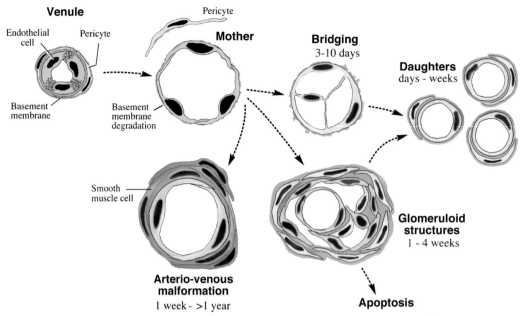

Figure 4. Schematic diagram of angiogenic response induced by Ad-VEGF-A^{164}.

perivascular fibrosis (Fig. 6). Vessels of this description closely resemble those found in arterio-venous malformations and in some benign vascular tumors (McKee 1996). Once formed, arterio-venous malformations persisted indefinitely (for more than a year), long after VEGF-A^{164} expression had ceased. Thus, in contrast to the angiogenic response induced by VEGF-A^{164}, these malformations were VEGF-A^{164}-independent.

ARTERIOGENESIS

The most prominent early blood vascular response to Ad-VEGF-A^{164} was that of angiogenesis. In parallel, however, small arteries and arterioles also responded with endothelial cell and smooth muscle cell replication (Fig. 7). This response led to an increase in the size of well-differentiated arteries. Like the arterio-venous malformations, these remodeled arteries also persisted indefinitely and therefore, once formed, did not require VEGF-A for maintenance.

LYMPHANGIOGENESIS

Unexpectedly, VEGF-A also stimulated the proliferation of lymphatic endothelium, leading to the formation of greatly enlarged lymphatic vessels. We have named these structures giant lymphatics. Lymphangiogenesis developed in parallel with angiogenesis but at a slower pace (Nagy et al. 2001, 2002b). Individual enlarged lymphatics were first seen at ~ 3 days after injection of Ad-VEGF-A^{164} and increased in size and number over the course of several weeks. This resulted in a dense meshwork of tortuous, irregularly shaped lymphatic channels that expanded to occupy large portions of the dermis (Fig. 8).

Unlike venules, normal lymphatics are thin-walled structures that lack a well-developed basement membrane (Fig. 5g), have few pericytes, and are capable of considerable enlargement in response to accumulations of edema (Leak et al. 1978; Feng et al. 2002). Therefore, lymphatics were able to enlarge without the basement membrane degradation, pericyte detachment, and endothelial cell thinning that were necessary for mother blood vessel formation. Formation of mother lymphatics was largely attributable to replication of lymphatic endothelial cells. Like the endothelium that lines blood vessels, lymphatic endothelial cells normally do not replicate. However, in response to Ad-VEGF-A^{164}, lymphatic endothelium began to incorporate [^{3}H]thymidine at 2 days and, at the height of the lymphangiogenic response, ~15% of lymphatic endothelial cells were labeled (Nagy et al. 2002b).

The function of the enlarged lymphatics induced by Ad-VEGF-A^{164} was evaluated by following their ability to clear colloidal carbon. In normal mice, or in mice injected with a control adenoviral vector such as Ad-LacZ, injected carbon was largely cleared from the ear within a few minutes. In contrast, in the Ad-VEGF-A^{164}-induced lymphatic network, tracer outflow was greatly retarded and substantial amounts of colloidal carbon remained within giant lymphatics for more than a day. Giant lymphatics of this description were evident by 5–7 days and persisted indefinitely, in some instances for > 1 year, independent of exogenous VEGF-A^{164}.

CONCLUSIONS

The data summarized here call attention to the remarkable heterogeneity of the vascular response induced by a single cytokine, VEGF-A^{164}. Using adenoviral vectors as a powerful, low-background delivery system and the nude mouse ear as a convenient site for monitoring the rise and subsequent regression of the angiogenic re-

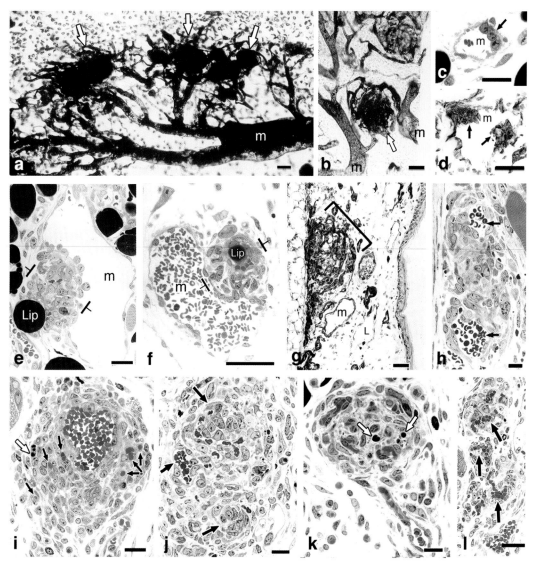

Figure 5. Glomeruloid body (GB) formation and subsequent devolution. (*a, b*) Whole mounts of ears from mice perfused with colloidal carbon. Note mature GB (*white arrows*) supplied by afferent and efferent vessels connected to mother vessels (m). *(c)* 1-μm Epon section illustrating focal accumulation of large primitive cells (*arrow*) in a developing mother vessel (m). (*d*) Immunohistochemistry demonstrates that such cells, here projecting into the mother vessel lumen, stain with an endothelial cell marker, CD 31. (*e, f*) Primitive GB develop as focal nodules (*between brackets*) as the result of cell proliferation in the wall of a mother vessel (m) and extend both into the lumen and out into the extravascular connective tissue. (Lip) Adipocyte. (*g*) Immunohistochemical staining for entactin demonstrates abundant basement membrane in a large GB (*bracket*) as well as in a mother vessel (m) and, conversely, very little in a lymphatic (L). (*h, i*) Maturing GB encroach on mother vessels, dividing them into multiple, much smaller lumens (*arrows*). Early apoptosis indicated by white arrow. (*j–k*) Devolving GB reorganizing into more normal-appearing microvessels (*black arrows*). White arrows indicate apoptotic bodies in *k*. (*l*) End-stage GB devolving into normal microvessels (*arrows*). Bars: *a, b,* 50 μm; *c–f* and *h–l,* 25 μm; *g,* 100 μm. (Reprinted, with permission, from Sundberg et al. 2001 [copyright American Society for Investigation Pathology].)

sponse, we have demonstrated that VEGF-A^{164} induces angiogenesis according to a highly reproducible series of steps: tissue edema, extravascular fibrin deposition, degradation of venular basement membranes, detachment of pericytes, and thinning and spreading of endothelial cells to form mother vessels. Mother vessels, in turn, divided into smaller daughter capillaries by transluminal bridging or evolved into glomeruloid bodies. As VEGF-A^{164} expression declined, GB devolved into normal-appearing capillaries and, by 2 months, microvascular density had returned to near-normal levels as the result of vascular apoptosis. Together, these findings have elucidated the steps and some of the mechanisms by which VEGF-A^{164} induces angiogenesis and have provided a temporal framework for dissecting apart the molecular events that underlie the formation of each type of vessel. In addition to angiogenesis, VEGF-A^{164} induced arteriogenesis, lymphangiogenesis, and arterio-venous malformations. Moreover, whereas angiogenesis declined in parallel with declining VEGF-A^{164} expression, new arteries, lymphatics, and vascular malformations, once formed, became VEGF-A^{164}-independent and persisted

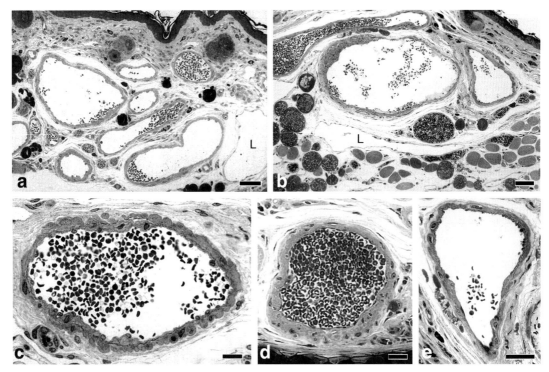

Figure 6. Arterio-venous malformations at 35 days (*a, c*) and 131 days (*b, d, e*) after injection of Ad-VEGF-A[164] into ear skin. Note greatly enlarged, irregularly shaped vessels with asymmetric distribution of smooth muscle cells (*b, d*) and perivascular fibrosis (*d, e*). (L) Lymphatic. Bars: *a, b, e*, 50 μm; *c, d*, 20 μm. (Panels *a* and *c* reprinted, with permission, from Pettersson et al. 2000 [copyright Lippincott Williams and Wilkins].)

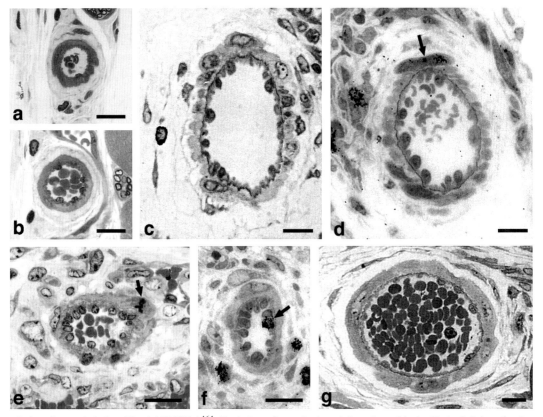

Figure 7. Arteriogenesis induced by Ad-VEGF-A[164]. Typical small arteries in uninjected ear skin (*a, b*). At 6–7 days (*c–g*), arteries are enlarged, and their endothelium and smooth muscle cells display mitotic figures (*e, arrow*) or incorporate [³H]thymidine (*d, f, arrows*). Bar, 15 μm.

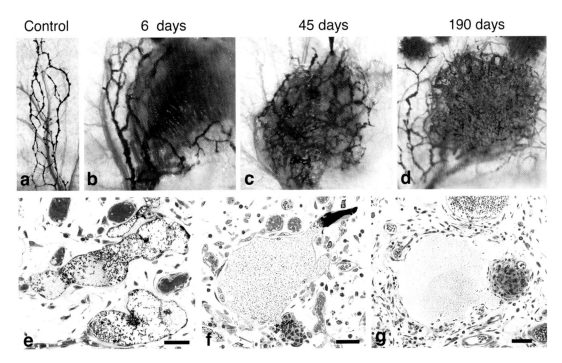

Figure 8. (a–d) Greatly enlarged "mother" lymphatics generated in ear skin at indicated times following local injection of Ad-VEGF-A^{164}. Lymphatics were perfused with colloidal carbon and viewed macroscopically. (e–g) One-micron Epon sections of mother lymphatics at 6, 7, and 14 days after Ad-VEGF-A^{164}. Lymphatics were perfused with colloidal carbon 20 minutes before harvest and, unlike normal lymphatics, still retain abundant tracer. Bars, 50 μm.

indefinitely in the absence of exogenous VEGF-A^{164}. Together, these findings contribute to an understanding of tumor angiogenesis and to the generation of hemangiomas and arterio-vascular malformations as they occur in patients.

Several of these points require amplification. First, as we had hypothesized, VEGF-A^{164} generated at least two types of surrogate tumor blood vessels, mother vessels and glomeruloid bodies. These developed with a consistent and highly reproducible chronology such that, at appropriate intervals after administering Ad-VEGF-A^{164}, each was present in sufficient numbers for purification without the need to separate them from overwhelming numbers of tumor cells as would be the case if they had to be isolated from tumors. Our goal is to characterize these surrogate vessels at the molecular level in order to discover unique markers that may distinguish them, and therefore tumor vessels, from the normal vasculature. We hope that such markers will serve as targets for diagnosis and antiangiogenesis or antivascular therapy.

The VEGF-A^{164} dependency of the angiogenic response is of particular interest. Benjamin and Keshet have shown that VEGF-A^{164} is a survival factor for newly formed tumor blood vessels and that withdrawal of VEGF-A^{164} leads to the selective apoptosis of microvessels that lack a pericyte coating (Benjamin et al. 1999). We have now extended these findings to demonstrate that maintenance of the entire VEGF-A^{164}-induced angiogenic response requires continuing VEGF-A^{164} stimulation. Mother vessels, with relatively reduced numbers of pericytes, but also pericyte-rich glomeruloid bodies and daughter capillaries apparently coated with a full complement of pericytes, all regressed in response to declining VEGF-A^{164} expression. Additional factors besides pericytes are therefore likely to be involved in VEGF-A^{164}-independent vascular survival.

In addition to stimulating the formation of microvessels, VEGF-A^{164} also induced the formation of arteriovenous malformations as an alternative pathway of mother vessel evolution. Thin-walled mother vessels lacking pericyte and basement membrane support are subject to thrombosis or collapse in response to elevated interstitial tissue pressure. Stabilization was accomplished by acquisition of a coat of supporting cells (pericytes, smooth muscle cells) or collagen. Thus, our experiments provide an animal model for generating vascular malformations and suggest that the malformations observed in patients are induced by local overexpression of VEGF-A^{164}.

Also of interest is the finding that VEGF-A^{164} induced arteriogenesis, i.e., the generation of larger but structurally normal arteries from preexisting arteries and arterioles. Unlike angiogenesis, arteriogenesis has been thought to result from mechanical forces rather than from the actions of cytokines (Buschmann and Schaper 1999). Thus, our finding that VEGF-A^{164} induces arteriogenesis may be important clinically in that arteriogenesis, rather than angiogenesis, is the desired therapeutic response in the treatment of coronary heart disease or limb ischemia. Moreover, in contrast to the transient angiogenic response, VEGF-A^{164}-induced arteriogenesis persisted for a considerable period of time. It is possible, therefore, that the clinical benefits that have been reported for gene therapy with angiogenic cytokines may have resulted not from angiogenesis but from unrecognized arteriogenesis (Baumgartner et al. 1998; Losordo et al. 1998).

The finding that VEGF-A^{164} induced lymphangiogenesis was also unanticipated. Another member of the VEGF-A^{164} family, VEGF-C, is essential for the normal development of lymphatics (Joukov et al. 1997; Oh et al. 1997), and we therefore considered the possibility that VEGF-A^{164}-induced lymphangiogenesis resulted from a secondary up-regulation of VEGF-C expression. However, preliminary experiments provide no support for this possibility, and it seems more likely that VEGF-A^{164} acts directly on lymphatic endothelial cells, perhaps by activating the VEGFR-2 receptor that we found these cells to express (Feng et al. 2000b). In contrast to the normal lymphatics that VEGF-C induces, those induced by VEGF-A^{164} are highly abnormal in both structure and function. Further experiments will be required to work out the mechanisms involved. Whatever these may be, our findings predict that malignant tumors, because they express large amounts of VEGF-A^{164}, will also induce the formation of lymphatics. This prediction is counterintuitive in that traditional teaching holds that tumors are not lymphangiogenic. However, this issue must now be reinvestigated in light of the fact that any lymphatics induced by VEGF-A^{164} are likely to be abnormal in structure and may function poorly, if at all, in the high interstitial tissue pressure environment that is characteristic of tumors (Jain 1990).

Taken together, our work demonstrates that a single cytokine, VEGF-A^{164}, induces a remarkably diverse series of vascular responses. Perhaps the simplest explanation of this diversity is that VEGF-A^{164} acts on the endothelial cells of several types of vessels (venules, small arteries, lymphatics) and that this initial response must be shaped and refined by secondary factors to induce different outcomes. Curiously, all of these outcomes occur, side by side, in single tissues such as the mouse ear. Discovering the responsible secondary factors and elucidating their mode of action will keep us and other workers busy for the foreseeable future. However, the effort expended is likely to yield a rich reward in terms of understanding normal vascular development and for developing both pro- and antiangiogenic therapies that will have use in the clinic.

ACKNOWLEDGMENTS

This work was supported in part by U.S. Public Health Service grants CA-50453, HL-59316, CA-92644, AI-33372, and AI-44066; by a contract from the National Foundation for Cancer Research; and by grants from the Swedish Cancer Society, Swedish Medical Research Council, Konung Gustaf V:s 80-årsfond (to C.S.).

REFERENCES

Asahara T., Murohara T., Sullivan A., Silver M., van der Zee R., Li T., Witzenbichler B., Schatteman G., and Isner J.M. 1997. Isolation of putative progenitor endothelial cells for angiogenesis. *Science* **275:** 964.

Baumgartner I., Pieczek A., Manor O., Blair R., Kearney M., Walsh K., and Isner J.M. 1998. Constitutive expression of phVEGF165 after intramuscular gene transfer promotes collateral vessel development in patients with critical limb ischemia (comments). *Circulation* **97:** 1114.

Beck L., Jr. and D'Amore P.A. 1997. Vascular development: Cellular and molecular regulation. *FASEB J.* **11:** 365.

Benjamin L.E., Hemo I., and Keshet E. 1998. A plasticity window for blood vessel remodelling is defined by pericyte coverage of the preformed endothelial network and is regulated by PDGF- B and VEGF. *Development* **125:** 1591.

Benjamin L.E., Golijanin D., Itin A., Pode D., and Keshet E. 1999. Selective ablation of immature blood vessels in established human tumors follows vascular endothelial growth factor withdrawal (comments). *J. Clin. Invest.* **103:** 159.

Brown L.F., Detmar M., Claffey K., Nagy J.A., Feng D., Dvorak A.M., and Dvorak H.F. 1997. Vascular permeability factor/vascular endothelial growth factor: A multifunctional angiogenic cytokine. *EXS* **79:** 233.

Brown L.F., Yeo K.T., Berse B., Yeo T.K., Senger D.R., Dvorak H.F., and van de Water L. 1992. Expression of vascular permeability factor (vascular endothelial growth factor) by epidermal keratinocytes during wound healing. *J. Exp. Med.* **176:** 1375.

Buschmann I. and Schaper W. 1999. Arteriogenesis versus angiogeneseis: Two mechanisms of vessel growth. *News Physiol. Sci.* **14:** 121.

Carmeliet P. and Collen D. 1999. Role of vascular endothelial growth factor and vascular endothelial growth factor receptors in vascular development. *Curr. Top. Microbiol. Immunol.* **237:** 133.

Carmeliet P., Ferreira V., Breier G., Pollefeyt S., Kieckens L., Gertsenstein M., Fahrig M., Vandenhoeck A., Harpal K., Eberhardt C., Declercq C., Pawling J., Moons L., Collen D., Risau W., and Nagy A. 1996. Abnormal blood vessel development and lethality in embryos lacking a single VEGF allele. *Nature* **380:** 435.

Dvorak A.M., Kohn S., Morgan E.S., Fox P., Nagy J.A., and Dvorak H.F. 1996. The vesiculo-vacuolar organelle (VVO): A distinct endothelial cell structure that provides a transcellular pathway for macromolecular extravasation. *J. Leukoc. Biol.* **59:** 100.

Dvorak H.F., Harvey V.S., and McDonagh J. 1984. Quantitation of fibrinogen influx and fibrin deposition and turnover in line 1 and line 10 guinea pig carcinomas. *Cancer Res.* **44:** 3348.

Dvorak H.F., Nagy J.A., and Dvorak A.M. 1991. Structure of solid tumors and their vasculature: Implications for therapy with monoclonal antibodies. *Cancer Cells* **3:** 77.

Dvorak H.F., Brown L.F., Detmar M., and Dvorak A.M. 1995. Vascular permeability factor/vascular endothelial growth factor, microvascular hyperpermeability, and angiogenesis. *Am. J. Pathol.* **146:** 1029.

Dvorak H.F., Nagy J.A., Dvorak J.T., and Dvorak A.M. 1988. Identification and characterization of the blood vessels of solid tumors that are leaky to circulating macromolecules. *Am. J. Pathol.* **133:** 95.

Dvorak H.F., Nagy J., Feng D., and Dvorak A. 2000. Tumor architecture and targeted delivery. In *Radioimmunotherapy of cancer* (ed. P. Abrams and A. Fritzberg), p. 107. Marcel Dekker, New York.

Dvorak H.F., Nagy J.A., Feng D., Brown L.F., and Dvorak A.M. 1999. Vascular permeability factor/vascular endothelial growth factor and the significance of microvascular hyperpermeability in angiogenesis. *Curr. Top. Microbiol. Immunol.* **237:** 97.

Dvorak H.F., Orenstein N.S., Carvalho A.C., Churchill W.H., Dvorak A.M., Galli S.J., Feder J., Bitzer A.M., Rypysc J., and Giovinco P. 1979. Induction of a fibrin-gel investment: An early event in line 10 hepatocarcinoma growth mediated by tumor-secreted products. *J. Immunol.* **122:** 166.

Feng D., Nagy J.A., Dvorak A.M., and Dvorak H.F. 2000a. Different pathways of macromolecule extravasation from hyper-

permeable tumor vessels. *Microvasc. Res.* **59:** 24.
Feng D., Nagy J., Dvorak H., and Dvorak A. 2002. Ultrastructural studies define soluble macromolecular, particulate, and cellular transendothelial cell pathways in venules, lymphatic vessels, and tumor-associated microvessels in man and animals. *Microsc. Res. Tech.* **57:** 289.
Feng D., Nagy J.A., Hipp J., Dvorak H.F., and Dvorak A.M. 1996. Vesiculo-vacuolar organelles and the regulation of venule permeability to macromolecules by vascular permeability factor, histamine, and serotonin. *J. Exp. Med.* **183:** 1981.
Feng D., Nagy J.A., Pyne K., Hammel I., Dvorak H.F., and Dvorak A.M. 1999. Pathways of macromolecular extravasation across microvascular endothelium in response to VPF/VEGF and other vasoactive mediators. *Microcirculation* **6:** 23.
Feng D., Nagy J.A., Brekken R.A., Pettersson A., Manseau E.J., Pyne K., Mulligan R., Thorpe P.E., Dvorak H.F., and Dvorak A.M. 2000b. Ultrastructural localization of the vascular permeability factor/vascular endothelial growth factor (VPF/VEGF) receptor-2 (FLK-1, KDR) in normal mouse kidney and in the hyperpermeable vessels induced by VPF/VEGF-expressing tumors and adenoviral vectors. *J. Histochem. Cytochem.* **48:** 545.
Ferrara N., Carver Moore K., Chen H., Dowd M., Lu L., O'Shea K.S., Powell Braxton L., Hillan K.J., and Moore M.W. 1996. Heterozygous embryonic lethality induced by targeted inactivation of the VEGF gene. *Nature* **380:** 439.
Folkman J. 1997. Angiogenesis and angiogenesis inhibition: An overview. *EXS* **79:** 1.
Folkman J. and D'Amore P.A. 1996. Blood vessel formation: What is its molecular basis (comment)? *Cell* **87:** 1153.
Gale N.W. and Yancopoulos G.D. 1999. Growth factors acting via endothelial cell-specific receptor tyrosine kinases: VEGFs, angiopoietins, and ephrins in vascular development. *Genes Dev.* **13:** 1055.
Grant D.S. and Kleinman H.K. 1997. Regulation of capillary formation by laminin and other components of the extracellular matrix. *EXS* **79:** 317.
Hanahan D. 1997. Signaling vascular morphogenesis and maintenance. *Science* **277:** 48.
Jain R.K. 1990. Physiological barriers to delivery of monoclonal antibodies and other macromolecules in tumors. *Cancer Res.* **50:** 814s.
Joukov V., Kaipainen A., Jeltsch M., Pajusola K., Olofsson B., Kumar V., Eriksson U., and Alitalo K. 1997. Vascular endothelial growth factors VEGF-B and VEGF-C. *J. Cell. Physiol.* **173:** 211.
Kohn S., Nagy J.A., Dvorak H.F., and Dvorak A.M. 1992. Pathways of macromolecular tracer transport across venules and small veins. Structural basis for the hyperpermeability of tumor blood vessels. *Lab. Invest.* **67:** 596.
Lantos P., VandenBerg S., and Kleihues P. 1997. Tumors of the nervous system. In *Greenfield's neuropathology* (ed. D. Graham and P. Lantos), p. 583. Arnold, London.
Leak L.V., Schannahan A., Scully H., and Daggett W.M. 1978. Lymphatic vessels of the mammalian heart. *Anat. Rec.* **191:** 183.
Losordo D.W., Vale P.R., Symes J.F., Dunnington C.H., Esakof D.D., Maysky M., Ashare A.B., Lathi K., and Isner J.M. 1998. Gene therapy for myocardial angiogenesis: Initial clinical results with direct myocardial injection of phVEGF165 as sole therapy for myocardial ischemia. *Circulation* **98:** 2800.

McKee P. 1996. *Pathology of the skin with clinical correlations*, p. 16.57. Mosby International, London.
Nagy J.A., Vasile E., Brown L.F., Manseau E.J., Eckelhoefer I.A., Bliss S.H., Dvorak A.M., and Dvorak H.F. 2002a. Vascular permeability factor/vascular endothelial growth factor-A (VPF/VEGF, VEGF-A) induces lymphangiogenesis as well as angiogenesis. *FASEB J.* **16:** A367.
Nagy J.A., Vasile E., Feng D., Sundberg C., Brown L.F., Detmar M.J., Lawitts J.A., Benjamin L., Tan X., Manseau E.J., Dvorak A.M., and Dvorak H.F. 2002b. Vascular permeability factor/vascular endothelial growth factor induces lymphangiogenesis as well as angiogenesis. *J. Exp. Med.* **196:** 1497.
Nagy J.A., Vasile E., Brown L.F., Feng D., Sundberg C., Manseau E.J., Eckelhoefer I.A., Bliss S.H., Lawitts J., Dvorak A.M., and Dvorak H.F. 2001. Vascular permeability factor/vascular endothelial growth factor-A (VPF/VEGF, VEGF-A) induces lymphangiogenesis as well as angiogenesis. In *Abstracts from the Annual Meeting of the American Society for Cell Biology*, L69.
Oh S.J., Jeltsch M.M., Birkenhager R., McCarthy J.E., Weich H.A., Christ B., Alitalo K., and Wilting J. 1997. VEGF and VEGF-C: Specific induction of angiogenesis and lymphangiogenesis in the differentiated avian chorioallantoic membrane. *Dev. Biol.* **188:** 96.
Pettersson A., Nagy J.A., Brown L.F., Sundberg C., Morgan E., Jungles S., Carter R., Krieger J.E., Manseau E.J., Harvey V.S., Eckelhoefer I.A., Feng D., Dvorak A.M., Mulligan R.C., and Dvorak H.F. 2000. Heterogeneity of the angiogenic response induced in different normal adult tissues by vascular permeability factor/vascular endothelial growth factor. *Lab. Invest.* **80:** 99.
Rafii S. 2000. Circulating endothelial precursors: Mystery, reality, and promise. *J. Clin. Invest.* **105:** 17.
Rafii S., Meeus S., Dias S., Hattori K., Heissig B., Shmelkov S., Rafii D., and Lyden D. 2002. Contribution of marrow-derived progenitors to vascular and cardiac regeneration. *Semin. Cell Dev. Biol.* **13:** 61.
Risau W. 1997. Mechanisms of angiogenesis. *Nature* **386:** 671.
Sundberg C., Nagy J.A., Brown L.F., Feng D., Eckelhoefer I.A., Manseau E.J., Dvorak A.M., and Dvorak H.F. 2001. Glomeruloid microvascular proliferation follows adenoviral vascular permeability factor/vascular endothelial growth factor-164 gene delivery. *Am. J. Pathol.* **158:** 1145.
Swayne G.T. and Smaje L.H. 1989. Dynamic compliance of single perfused frog mesenteric capillaries and rat venules: A filtration coefficient correction. *Int. J. Microcirc. Clin. Exp.* **8:** 43.
Swayne G.T., Smaje L.H., and Bergel D.H. 1989. Distensibility of single capillaries and venules in the rat and frog mesentery. *Int. J. Microcirc. Clin. Exp.* **8:** 25.
Unemori E.N., Ferrara N., Bauer E.A., and Amento E.P. 1992. Vascular endothelial growth factor induces interstitial collagenase expression in human endothelial cells. *J. Cell. Physiol.* **153:** 557.
Warren B.A. 1979. The vascular morphology of tumors. In *Tumor blood circulation: Angiogenesis, vascular morphology and blood flow of experimental and human tumors* (ed. H.-I. Peterson), p. 1. CRC Press, Boca Raton, Florida.
Yancopoulos G.D., Klagsbrun M., and Folkman J. 1998. Vasculogenesis, angiogenesis, and growth factors: Ephrins enter the fray at the border (comment). *Cell* **93:** 661.

Angiogenesis and Lymphangiogenesis in Tumors: Insights from Intravital Microscopy

R.K. JAIN

Edwin L. Steele Laboratory, Department of Radiation Oncology, Massachusetts General Hospital and Harvard Medical School, Boston, Massachusetts 02114

A solid tumor is an organ-like structure composed of cancer cells and host stromal cells, embedded in an extracellular matrix and bathed in interstitial fluid (Jain et al. 2002). These cells are nourished by blood vessels made up of vascular endothelial cells. In addition to supplying nutrients and removing waste products, these blood vessels serve as a conduit for the spread of cancer cells to distal sites (hematogenic metastases) (Carmeliet and Jain 2000). A second vascular system associated with solid tumors is made up of lymphatic endothelial cells. In addition to absorbing excess interstitial fluid and macromolecules to maintain homeostasis, the lymphatic system serves as a conduit for host immune cells and cancer cells to reach the draining lymph nodes (Alitalo and Carmeliet 2002; Jain and Fenton 2002). Thus, blocking the formation and function of these two vascular systems has the potential to arrest tumor growth and metastasis (Folkman 2000).

Here, I briefly discuss insights revealed by intravital microscopy (IVM) into the molecular, cellular, anatomical, and physiological workings of these two systems and how these insights have facilitated the development of improved strategies for cancer detection and treatment (Jain et al. 2002). This account is by no means an exhaustive review of work in this field. That is beyond the scope of this paper. Here, I focus largely on the work done in my laboratory on the role of host–tumor interactions in the integrative pathophysiology of tumors.

INTRAVITAL MICROSCOPY

IVM is a powerful optical imaging technique that allows continuous, noninvasive monitoring of molecular, cellular, and physiological events in an intact living tissue with a resolution of 1 to 10 microns. Such resolution is currently not possible with non-optical imaging techniques such as MRI, PET, or CT. Other techniques, such as in situ hybridization, immunohistochemistry, autoradiography, and electron microscopy, provide high spatial resolution but require tissue excision and, hence, miss the temporal dynamics of in vivo tumor growth.

IVM requires an appropriate animal model, a molecular probe that can be optically detected, a microscope equipped with a detection system, and a computer to process and analyze the data to yield parameters of interest (Fig. 1).

Animal models: Our laboratory has employed three types of animal models for tissue preparation: chronic windows, such as those placed in the dorsal skin (Leunig et al. 1992) or the cranium (Yuan et al. 1994) of the mice; acute preparations such as the liver (Fukumura et al. 1997) and the mammary gland (Monsky et al. 2002); and in situ preparations such as the tail (Leu et al. 1994) and the ear (Fukumura et al. 1998) (Fig. 1a).

Molecular probes and optical detection: The molecular probe can be an injected fluorescent molecule (e.g., FITC-dextran) (Nugent and Jain 1984) or an endogenous fluorescent reporter (e.g., GFP) (Fukumura et al. 1998). The probe may change its optical property in response to the local microenvironment (e.g., pH, pO_2) (Helmlinger et al. 1997a) or enzyme activity (Weissleder 2002) (Fig. 1b). The optical signal can be detected using a microscope equipped with a photomultiplier tube or a sensitive camera (Fig. 1c). The standard epifluorescence or confocal microscope can be used to image up to 100–150 microns deep in a tissue in vivo. Recently, our laboratory has adapted multiphoton laser scanning microscopy for intravital imaging up to 700 microns deep in tissue (Brown et al. 2001). As discussed below, this exciting development has provided new insights into the host–tumor interactions and the function of the blood and lymphatic vascular systems deep inside a tumor.

Data analysis: Computer-assisted image analysis has allowed us to extract molecular, cellular, anatomical, and functional parameters of interest. The molecular parameters that we have measured include micropharmacokinetics (Nugent and Jain 1984), pH and pO_2 (Helmlinger et al. 1997a), as well as promoter and enzyme activity (Fukumura et al. 1998). Cellular parameters include measures of adhesion and migration rates of cancer cells (Shioda et al. 1997), immune cells (Melder et al. 1996), and host stromal cells (Brown et al. 2001). Anatomical parameters include tumor size; diameter, length, and branching patterns of blood (Leunig et al. 1992) and lymphatic vessels (Padera et al. 2002); as well as the size of the "pores" in the blood vessel wall (Hobbs et al. 1998) and the extracellular matrix (Pluen et al. 2001). Functional parameters include blood (Leunig et al. 1992) and lymph (Berk et al. 1996; Swartz et al. 1996) flow rates; vascular permeability (Yuan et al. 1996); and interstitial diffusion, convection, and binding (Chary and Jain 1989; Berk et al. 1997; Pluen et al. 2001).

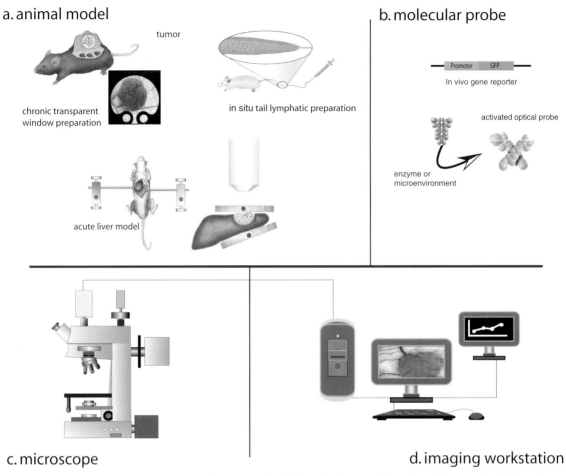

Figure 1. Intravital microscopy (IVM) has provided powerful insight into the inner workings of a solid tumor, including angiogenesis and lymphangiogenesis. IVM requires four components: (*a*) a tissue preparation that allows imaging of molecular, cellular, anatomical, and functional characteristics; (*b*) a molecular probe that can be detected; (*c*) a microscope equipped with an excitation and detection system; and (*d*) a computer-assisted image analysis system to calculate parameters of interest. (Adapted from Jain et al. 2002.)

ANGIOGENESIS AND VASCULAR FUNCTION: ROLE OF HOST–TUMOR INTERACTIONS

A solid tumor can be nourished by the host vessels co-opted by the growing neoplastic cells and/or by the newly formed vessels. The formation of new vessels occurs by sprouting, intussusception, and/or recruitment of circulating endothelial precursor cells mobilized from the bone marrow or peripheral blood (Fig. 2) (Carmeliet and Jain 2000).

Various positive and negative regulators orchestrate different steps in vessel formation and their function (Carmeliet and Jain 2000). Among these, vascular endothelial growth factor (VEGF) is perhaps the most critical molecule. VEGF increases vascular permeability, promotes migration and proliferation of endothelial cells (EC), serves as an EC survival factor, and is known to up-regulate leukocyte adhesion molecules on ECs (Melder et al. 1996; Carmeliet and Jain 2000; Dvorak et al. 2000; Ferrara 2001).

The production and/or release of VEGF and other angiogenic regulators can be triggered by injury, metabolic stress (e.g., low pO_2, low pH), mechanical stress (e.g., shear stress), soluble mediators (e.g., hormones, growth factors, cytokines), immune/inflammatory response (e.g., immune/inflammatory cells infiltrating the tumor), and genetic mutations (Bouck et al. 1996; Hanahan and Weinberg 2000; Xu et al. 2002). These molecules can originate from cancer cells, endothelial cells, stromal cells, blood, or the extracellular matrix (Fukumura et al. 1998; Helmlinger et al. 2000; Brown et al. 2001).

To discern the role of host cells in VEGF production, we engineered transgenic mice that express GFP under the control of the VEGF promoter (Fukumura et al. 1998). To test the validity of this model in reporting VEGF promoter activity, we created a wound in the ears of these mice that was 2 mm in diameter. Since we know that wound healing is accompanied by VEGF-driven angiogenesis, we expected the wound to become fluorescent during the healing process. Indeed, approximately one week after wound creation, the periphery of the wound showed fluorescence. After two weeks, the majority of the wound became fluorescent. Examination of the fluorescent regions showed that the VEGF promoter

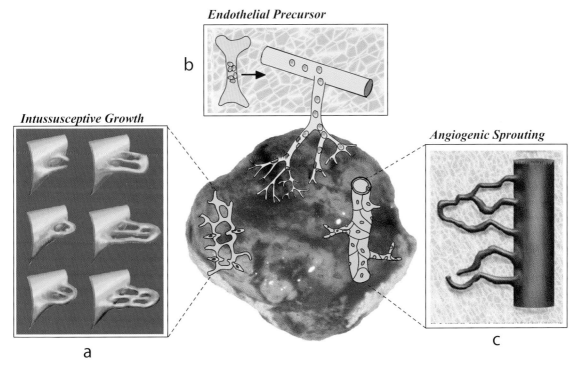

Figure 2. Cellular mechanisms of vascularization in tumors. (*a*) Intussusception: tumor vessels enlarge and an interstitial tissue column grows in the enlarged lumen, expanding the network; (*b*) vasculogenesis: endothelial precursor cells mobilized from the bone marrow or peripheral blood contribute to the endothelial lining of blood vessels; (*c*) "sprouting" angiogenesis: the existing vascular network expands by forming sprouts; and (*d*) co-option (not shown): tumor cells grow around existing vessels to form "perivascular cuffs." (Adapted from Carmeliet and Jain 2000.)

activity coincided with the formation of new vessels in the wound. Thus, this model was ideal for our purpose of monitoring changes associated with VEGF production.

To determine whether the tumor itself can trigger VEGF promoter activity in the host stromal cells, we implanted wild-type cancer cells (tumors that do not harbor the gene for GFP) in the dorsal windows in these transgenic mice. This allowed us to observe fluorescence that originated only from the host stromal cells of the VEGF-GFP mice. As shown in Figure 3, the periphery of the tumor became highly fluorescent in one week (Fig. 3A), and the fluorescent region continued to enlarge to fill the whole tumor in about two weeks (Fig. 3B). These data showed for the first time that the tumor can activate VEGF promoter in the host stromal cells.

This exciting finding led to a number of questions: What are these stromal cells? How do they become activated? How far do they migrate into the tumor? And at what rate? What can we learn from the dynamics of their interaction with the angiogenic vessels? Finally, how much VEGF do they make?

We originally learned that these fluorescent cells have a fibroblast-like morphology (Fukumura et al. 1998). Multiphoton microscopy allowed us to track these cells deep inside the tumor. We found that there is an abundance of these activated cells at the host–tumor interface (Fig. 3C) and then there is a precipitous drop in the number of cells as one moves away from the interface toward the center of the tumor (Fig. 3D) (Brown et al. 2001). Intriguingly, these activated cells closely surround the blood vessels, giving the impression that they are acting as pericytes (Fig. 3D). The most exciting finding was that these cells, in some cases, may lead the endothelial cell instead of following it (Fig. 3E). These preliminary findings suggest that, contrary to the widely accepted hypothesis that endothelial cells are the principal players in angiogenesis, activated stromal cells may be equally or even more important. These findings also showed, for the first time, that the host stromal cells are not passive bystanders in the process of angiogenesis, but rather are active participants in the process.

To determine the relative contribution of the host stromal cells to VEGF production, we next implanted VEGF$^{-/-}$ and wild-type embryonic stem (ES) cells in the dorsal windows in mice (Tsuzuki et al. 2000). We reasoned that the VEGF level in the VEGF$^{-/-}$ ES cell-derived tumors would emanate strictly from the host stromal cells. We found that the VEGF protein level in the VEGF$^{-/-}$ ES cell-derived teratomas (that is the VEGF attributable to host stromal cells) was ~50% compared to their wild-type counterpart. This showed, again for the first time, that the host cells could produce a significant fraction of VEGF in tumors.

We then asked whether the host cells could also contribute to the production of antiangiogenic molecules. To answer this question, we implanted a human gall bladder tumor in two different sites, reflecting two different populations of host cells: subcutaneously (an ectopic site)

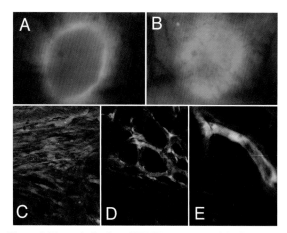

Figure 3. Tumor induction of VEGF promoter activity in stromal cells. The expression of VEGF in host cells was examined using transgenic mice expressing GFP under the control of VEGF promoter. (*A*) A murine mammary carcinoma xenograft shows host cell VEGF expression mainly at the periphery of the tumor after 1 week. (*B*) After 2 weeks, the VEGF-expressing host cells have infiltrated the tumor. (Reproduced from Fukumura et al. 1998.) (*C*) A GFP-expressing layer of host cells can be seen at the host–tumor interface. (*D,E*) The VEGF-expressing host cells surround and may even lead the endothelial cells. (Reprinted from Brown et al. 2001.) (*Green*) GFP; (*orange*) blood flow marker.

and in the gall bladder (the orthotopic site) (Gohongi et al. 1999). To determine the effects of endogenous antiangiogenic molecules produced by each of these tumors, we examined angiogenesis and tumor growth rates in secondary tumors growing in the brain. If antiangiogenic molecules were indeed produced, we would expect that the secondary tumor site would show suppression of angiogenesis and retardation of tumor growth. Indeed, compared to the ectopically implanted gall bladder tumors, the orthotopic gall bladder tumors showed increased suppression of angiogenesis and tumor growth in a secondary tumor observed through a cranial window. We also discovered that the antiangiogenic molecule responsible for this suppression was TGFβ. TGFβ is present in abundant quantities in human gall bladder tumors (Gohongi et al. 1999). Thus, these data showed, for the first time, that the host–tumor interaction regulates the production of angiostatic molecules.

Another striking example of the role of the host cells in the production of both pro- and antiangiogenic molecules came from our recent studies on the antiangiogenic effects of Herceptin treatment of breast cancer (Izumi et al. 2002). Herceptin is known to down-regulate VEGF production in her2/neu overexpressing cells in vitro. Surprisingly, Herceptin did not change the level of VEGF significantly in the tumors derived from these cells in vivo. Further examination showed this unexpected finding to be due to the increased production of compensatory VEGF by the host stromal cells during Herceptin treatment. Moreover, Herceptin altered the level of four additional angiogenic regulators: angiopoetin1, TGFα, PAI-1, and thrombospondin-1. Presumably, this occurs by the involvement of host stromal cells.

These three examples highlight the importance of the host cells in tumor angiogenesis, growth, and response to therapy. Because the host stromal cells differ from one organ to the next, we would expect the tumor biology to be dependent on the host–tumor interactions. Indeed, we have shown that gene expression, angiogenesis, vascular permeability, leukocyte adhesion, blood velocity, and interstitial diffusion change when the same neoplastic cells are implanted in different sites (Yuan et al. 1994; Dellian et al. 1996; Fukumura et al. 1997; Hobbs et al. 1998; Gohongi et al. 1999; Pluen et al. 2001; Monsky et al. 2002). For example, a human glioma cell line yields a leaky tumor when grown subcutaneously but not when grown in the brain (Fig. 4A,B). In the brain, the same cell line leads to a tumor with the blood–brain-barrier-like properties (Yuan et al. 1994).

As another example, a human melanoma cell line yields a collagen-rich tumor when grown subcutaneously and a collagen-poor tumor in the cranial window (Fig. 4C,D) (Pluen et al. 2001). In concert with our previous finding that collagen can hinder the diffusion of macromolecules in tumors (Netti et al. 2000), we found that the diffusion of macromolecules in this melanoma was lower in the subcutaneous implant compared to the cranial implant (Pluen et al. 2001).

Since the delivery of drugs in a tumor depends on vascular permeability and interstitial diffusion, the accretion of drugs and, hence, tumor response is likely to be different in the tumors growing in different sites. Thus, the role of the host–tumor interaction must be taken into account in extrapolating preclinical findings to the clinical setting (Fidler 2001).

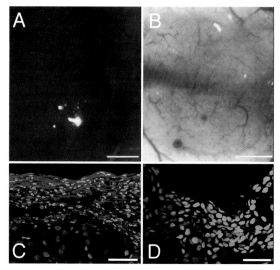

Figure 4. Effect of the host–tumor interaction on vascular permeability (*A,B*) and extracellular matrix production (*C,D*). A human glioma is leaky when grown in the subcutaneous space (*A*) and not leaky when grown in the cranium (*B*). (Adapted from Yuan et al. 1994.) A human melanoma xenograft has an abundance of collagen when grown in the sc space (*C*) and a paucity of collagen when grown in the cranium (*D*). (Reprinted from Pluen et al. 2001.) (*Green*) Lissamine green dye; (*red*) collagen; (*blue*) cell nucleus.

LYMPHANGIOGENESIS AND LYMPHATIC FUNCTION: ROLE IN METASTASIS

In most normal tissues, extravasated plasma, macromolecules, and cells are taken up by the lymphatic system and brought back to the central circulation. It is widely accepted that lymphatic vessels are present in the tumor margin and the peri-tumor tissue (Fig. 5). Indeed, invasion of peri-tumor lymphatics is considered to be a poor prognostic factor for a number of neoplasms (e.g., breast, colorectal, and endometrial cancers), and lymphatic metastasis is a major cause of morbidity and mortality for others (e.g., melanoma, head, neck, lung, and cervical cancer) (www.uptodate.com). The hotly debated issue for nearly a century has been whether anatomically defined lymphatic vessels are present within solid tumors (Gullino 1975), and if so, whether they function (Fig. 5) (Leu et al. 2000). Currently available immunohistochemical markers stain structures in some tumors that resemble lymphatic vessels. However, many of these markers lack specificity (Mouta Carreira et al. 2001; Jain and Fenton 2002). Furthermore, it is not clear whether they stain functional lymphatic vessels, endothelial cells from remnant lymphatic vessels, or some other structures; e.g., preferential fluid channels (Boucher et al. 1998). From our previous work on transplanted tumors in animals and spontaneous tumors in patients, we know that all tumors examined to date exhibit interstitial hypertension (Jain 1994), a clear indicator of the impaired lymphatic vessels. On the basis of these diverse sets of data, we hypothesized that a tumor has the capacity to co-opt or form new lymphatic vessels, but the mechanical stress generated by proliferating cancer cells and the remodeling of the extracellular matrix may compress the intra-tumor lymphatics, rendering them impaired (Helmlinger et al. 1997b). The lymphatic vessels in the tumor margin, however, remain functional and available to facilitate lymphatic dissemination (Jain and Fenton 2002).

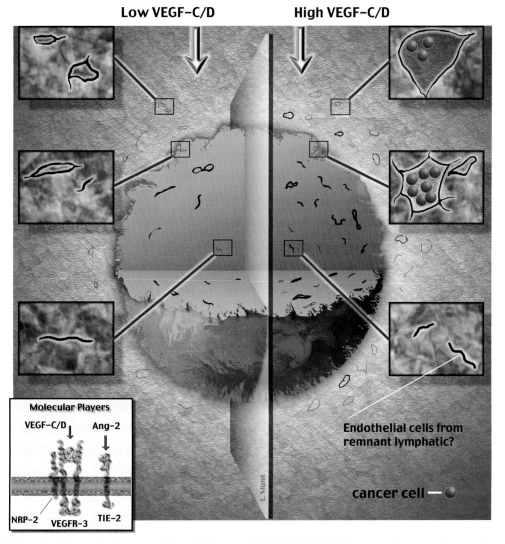

Figure 5. Schematic of lymphatics in tumors in low (*left*) versus high (*right*) VEGF-C/-D expressing tumors. Increase in the surface area of the functional lymphatics in the periphery contributes to lymphatic metastasis. Lower left-hand inset shows some of the molecular players in lymphangiogenesis. (Reprinted from Jain and Fenton 2002.)

A prerequisite to test this hypothesis is the availability of an orthotopic tumor model that reliably produces lymphatic metastasis and is amenable to immunohistochemical and functional evaluation of lymphatics. To this end, we overexpressed VEGF-C, a known molecule involved in angiogenesis and lymphangiogenesis in two different tumor lines (T241 and B16F10), and implanted them orthotopically in mice (Padera et al. 2002). We found an approximately threefold increase in the lymph node metastases from these VEGF-C-overexpressing lines, compared to their mock-transfected counterparts. We then stained these tumors with a pan-endothelial marker (MECA32) and putative lymphatic markers (LYVE-1, Prox 1) and examined their lymphatic function by lymphangiography and by measuring their interstitial pressure. Lymphangiography involves injecting a macromolecular tracer in the interstitial compartment of a tissue. This macromolecular tracer is taken up by the nearby lymphatics and can be visualized by IVM, if optically accessible, or by immunostaining.

In support of our hypothesis, we found that in both mock-transfected and VEGF-C-overexpressing tumors, functional lymphatics were present only in the tumor margin and not intratumorally (Padera et al. 2002). Overexpression of VEGF-C did not lower the intra-tumor pressure either, again supporting the notion that intra-tumor lymphatics induced by VEGF-C, if any, are not functional; otherwise they would have drained the tumor interstitial fluid and lowered the pressure. We also found that the diameter of the lymphatics in the tumor margin was larger in the mock-transfected tumors compared to that in the normal host tissue (Leu et al. 2000), and the diameter increased even further in the VEGF-C-overexpressing tumors (Padera et al. 2002). This lymphatic hyperplasia was reminiscent of the hyperplasia in the skin lymphatics we had observed earlier in the transgenic mice overexpressing VEGF-C under the control of the K14 promoter (Jeltsch et al. 1997). We therefore conclude that the increased surface area of lymphatics in the tumor margin provides adequate opportunity for the neoplastic cells to disseminate to the nearby lymph nodes.

Is this also true for spontaneous tumors? We found that the spontaneously arising tumors in aging mice also have the functional lymphatics confined to their margin. Furthermore, primary and secondary liver tumors as well as lung tumors in patients exhibit LYVE-1/Prox 1 double-stained vessels (indicating the presence of lymphatics) only in the tumor margin or the peri-tumor tissue. In the case of lung adenocarcinomas, all tumors exhibited interstitial hypertension, supporting the lack of lymphatic function within these tumors. Moreover, the level of hypertension or the intra-tumor LYVE-1/Prox 1 staining did not correlate with the incidence of lymph node metastasis or the patient survival. These data collectively suggest that we need to aggressively treat the lymphatics specifically in the tumor margin to stop further dissemination of cancer cells through the lymphatic system (Padera et al. 2002).

The challenge is how to target lymphatics associated with tumors without compromising normal lymphatic function (Gershenwald and Fidler 2002). The most obvious candidates for this are VEGF-C and -D, and their common receptor VEGFR3, especially because of their association with lymphogenic metastasis in a variety of human tumors (Jain and Fenton 2002). Although these three molecular players are also involved in angiogenesis, reagents that block VEGF-C/-D signaling (e.g., antibodies, soluble VEGFR3) have shown promise in preventing lymph node metastases in transplanted tumors when treated at the time of tumor implantation (Jain and Padera 2002). Unfortunately, in the clinical setting, lymph node micro-metastases may already be present by the time the primary tumor is currently detectable (about 1 cm in diameter). Thus, strategies that target both the formation of new lymphatic vessels (anti-lymphangiogenesis) and the already established lymphatics (anti-lymphatic vessels) associated with the primary and metastatic tumors are needed. The in vivo phage display technology has shown promise in identifying the "zip codes" on the tumor-associated lymphatic vessels (Ruoslahti 2002). Further progress will result from a better understanding of the molecular and cellular players involved in lymphatic vessel formation and function (Alitalo and Carmeliet 2002).

As in vascular angiogenesis, a constellation of positive and negative regulators, such as angiopoietins (Yancopoulos et al. 2000), and other receptors, such as chemokine receptors and neuropilins, may be involved in lymphangiogenesis, and cellular mechanisms analogous to co-option, intussusception, sprouting, and vasculogenesis may operate in lymphatic growth. Similar to the recently discovered organ-specific angiogenic molecule (EG-VEGF) (LeCouter et al. 2001) and endothelial precursor cells (Lyden et al. 2001; Isner 2002), there may be organ-specific lymphangiogenic molecules and lymphatic endothelial precursor cells that contribute to tumor-associated lymphangiogenesis (Jain and Padera 2003). Moreover, the proteolytic processing of lymphangiogenic molecules, as well as the phenotype and function of the resulting lymphatics, may depend not only on the tumor type, but also on the host organ in which the tumor is growing (Kadambi et al. 2001; Alitalo and Carmeliet 2002). Perhaps the area that needs the most attention is the identification of the molecular and/or mechanical signals that trigger the lymphangiogenic switch. Because lymphatic vessels help maintain the balance of fluid in tissues, hydrostatic pressure is a likely trigger. Whether the lymphatic hyperplasia seen in the tumor margins is a response to the elevated hydrostatic pressure in tumors and whether the newly formed lymphatics are able to remain open and carry cancer cells are questions awaiting answers.

NORMALIZING TUMOR VASCULATURE USING ANTIANGIOGENIC THERAPY: A NEW PARADIGM FOR COMBINATION THERAPY

Two major problems currently plague the nonsurgical treatment of malignant solid tumors. First, abnormal tumor vasculature and the extracellular matrix create barriers to the delivery of therapeutics and oxygen (a key radiation sensitizer) at effective concentrations to all cancer cells (Jain 1998). Second, inherent or acquired resistance resulting from genetic and epigenetic mechanisms reduces the effectiveness of conventional as well as novel therapies (McCormick 2001). The antiangiogenic and antivascular approaches have the potential to circumvent, or at least reduce the impact of, these problems. The vascular endothelial cells are easily accessible to blood-borne drugs, and these cells are presumably genetically stable. In addition, each EC provides nourishment to multiple cancer cells, thus creating "therapeutic amplification." However, the phenotypic heterogeneity of the EC population can compromise the effectiveness of antivascular therapy (Fidler 2001). Similarly, the dependence of ECs on multiple angiogenic molecules can limit the effectiveness of various antiangiogenic therapies when used alone (Carmeliet and Jain 2000; Jain and Carmeliet 2001; Izumi et al. 2002). Thus, it is now widely accepted that these therapies will be used in combination with conventional or molecularly targeted approaches.

The challenge now is how to combine these therapies optimally in patients. Delivery of drugs or oxygen can be compromised if the antiangiogenic/vascular therapy completely destroys the tumor vasculature. On the other hand, if we could selectively prune only the inefficient and immature component of the abnormal tumor vasculature by restoring the balance of pro- and antiangiogenic molecules, the remaining vasculature would become more like normal vasculature and more efficient for drug delivery (Fig. 6). We propose that judiciously applied antiangiogenic therapy has the potential for such a "normalization" of the tumor vasculature (Jain 2001).

We had the first glimpse of the normalization window when we imaged the vasculature of an androgen-dependent tumor in the mouse dorsal window using IVM following androgen withdrawal (Fig. 7) (Jain et al. 1998). The vasculature of this tumor is highly leaky, and the av-

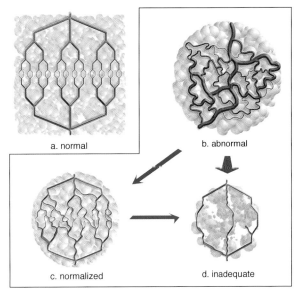

Figure 6. Normalization of the tumor vasculature. (*a*) Normal vessels are well organized with hierarchical branching patterns and diameters. (*b*) In contrast, tumor vessels are tortuous with increased vessel diameter and permeability, and chaotic branching patterns. (*c*) Antiangiogenic therapy can normalize the tumor vascular network by pruning the immature vessels. (*d*) Continued therapy can reduce the vasculature to the point that it cannot support tumor growth. (Reprinted from Jain 2001.)

erage diameter of these leaky and tortuous vessels increases with tumor growth. Within one day of castration, the tumor endothelial cells begin to undergo apoptosis, and the caliber, tortuousity, permeability, and adhesiveness of the remaining vasculature begin to decrease. Within 2–3 days, both ECs and cancer cells exhibit apoptosis, and the tumor begins to regress. However, similar to the hormone-dependent human tumors, in a couple of weeks after the hormone-refractory cancer cells begin to

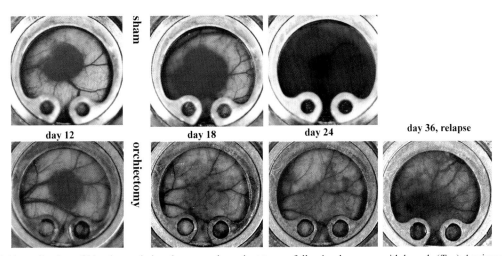

Figure 7. Normalization of blood vessels in a hormone-dependent tumor following hormone withdrawal. (*Top*) Angiogenesis and growth of an androgen-dependent tumor in the dorsal window. (*Bottom*) Regression and relapse of the tumor following castration on day 15. Note the decrease in angiogenesis after castration and a second wave of angiogenesis during relapse. The vessels develop a more normal phenotype during regression. (Reprinted from Jain et al. 1998.)

proliferate, a second wave of angiogenesis ensues and the tumor relapses (Fig. 7). The vessels in the relapsed tumor are as abnormal as in the tumor prior to hormone withdrawal. Since hormone withdrawal lowers the expression of VEGF in cancer cells, and VEGF is a survival factor for the immature vessels, we attributed the pruning of the tumor vasculature to this decrease in VEGF. Furthermore, the compensatory production of VEGF by the host stromal cells and the hormone-independent cancer cells presumably contributes to the renewed level of VEGF in the relapsing tumors. These data suggest that an indirect antiangiogenic (anti-VEGF) therapy can normalize the tumor vasculature.

We observed a similar normalization during Herceptin treatment in a *her2/neu* overexpressing human breast tumor xenograft (Izumi et al. 2002). In tumors treated with a control antibody, the diameter and permeability of vessels increase during tumor growth. On the other hand, Herceptin treatment arrested the increase in vessel diameter and caused the vascular permeability to decrease. Using gene-chip analysis, we discovered that Herceptin acts as a cocktail of five antiangiogenic molecules. Thus, we attributed this normalization effect to Herceptin's indirect antiangiogenic activity.

Finally, we tested our normalization hypothesis using a direct antiangiogenic agent—an antibody against VEGFR2, the key signaling receptor for VEGF on ECs (Kadambi et al. 2001). As shown in Figure 8, within one day after the antibody treatment, the less efficient vessels are pruned (Jain and Carmeliet 2001). As expected, long-term treatment leads to an inadequate vasculature unable to sustain tumor growth.

One major consequence of the abnormal tumor vasculature is impaired blood flow, interstitial hypertension, and hypoxia in tumors. We therefore reasoned that this normalization may decrease hypertension and hypoxia in tumors. Indeed, we have found this to be the case for some tumors undergoing anti-VEGF treatment (Lee et al. 2000). It is likely that normalization induced by treatment with an anti-VEGFR2 antibody led to the long-term cures we recently reported following radiation treatment in human tumor xenografts in mice (Kozin et al. 2001).

On the basis of the normalization concept, we would predict that lower doses of anti-VEGF antibodies (that do not destroy the tumor vasculature completely but render it more efficient) would be more effective than very high doses when combined with chemotherapy. On the other hand, we would recommend the higher dose for monotherapy. Two recent clinical trials support our predictions. The Genentech anti-VEGF antibody, Avastin, trial shows that a 5-mg dose is more effective than a 10-mg dose for combination therapy of colon cancers. However, the National Cancer Institute Avastin trial shows that a 10-mg dose is more effective against the renal cell carcinoma when used as a monotherapy (W. Novatny, pers. comm.).

CONCLUDING COMMENTS

We are in the golden age of biology and medicine. Our nation's investment in cancer research has provided phenomenal insights that have led to the identification of many genes associated with oncogenesis and angiogenesis. The challenge now is to decipher the function of these genes not only in vitro or in silico, but ultimately in an intact host. A second challenge is to characterize and overcome the physiological barriers to delivery and to improve the efficacy of conventional and novel therapies. Over the past 20 years, my colleagues and I have tried to meet these challenges using intravital microscopy. This "physiomic" approach, coupled with genomic and proteomic approaches, has provided powerful insights into the complex pathophysiology of tumors, revealed the crucial role of host–tumor interactions in the biology of tumors and their responses to therapy, and suggested a counterintuitive approach to normalize the abnormal tumor vasculature using antiangiogenic therapies. With the breathtaking pace of discoveries in vascular and lymphatic biology being made, this is an exciting time for the field.

ACKNOWLEDGMENTS

I am grateful to the National Cancer Institute for 23 years of continuous support. I thank E. Brown, E. di Tomaso, D. Fukumura, L.L. Munn, T. Padera, and J. Samson for their help in the preparation of this manuscript. The work described here would not have been possible without the dedicated efforts of more than 100 former and current coworkers who have shared this exciting journey through the world of solid tumors with me.

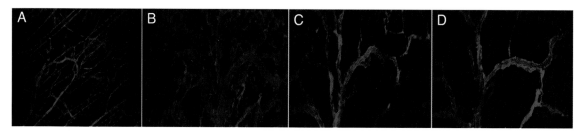

Figure 8. Normalization of tumor vessels following treatment with an anti-VEGFR2 antibody. (*A*) Normal vasculature. (*B*) Tumor vasculature prior to treatment. (*C*) Normalized vasculature within 1 day of treatment. (*D*) Inadequate vasculature after 3 days of treatment. (Reprinted from Jain and Carmeliet 2001.)

REFERENCES

Alitalo K. and Carmeliet P. 2002. Molecular mechanisms of lymphangiogenesis in health and disease. *Cancer Cell* **1:** 219.

Berk D.A., Swartz M.A., Leu A.J., and Jain R.K. 1996. Transport in lymphatic capillaries. II. Microscopic velocity measurement with fluorescence photobleaching. *Am. J. Physiol.* **270:** H330.

Berk D.A., Yuan F., Leunig M., and Jain R.K. 1997. Direct in vivo measurement of targeted binding in a human tumor xenograft. *Proc. Natl. Acad. Sci.* **94:** 1785.

Boucher Y., Brekken C., Netti P.A., Baxter L.T., and Jain R.K. 1998. Intratumoral infusion of fluid: Estimation of hydraulic conductivity and implications for the delivery of therapeutic agents. *Br. J. Cancer* **78:** 1442.

Bouck N., Stellmach V., and Hsu S.C. 1996. How tumors become angiogenic. *Adv. Cancer Res.* **69:** 135.

Brown E.B., Campbell R.B., Tsuzuki Y., Xu L., Carmeliet P., Fukumura D., and Jain R.K. 2001. In vivo measurement of gene expression, angiogenesis and physiological function in tumors using multiphoton laser scanning microscopy. *Nat. Med.* **7:** 864.

Carmeliet P. and Jain R.K. 2000. Angiogenesis in cancer and other diseases. *Nature* **407:** 249.

Chary S.R. and Jain R.K. 1989. Direct measurement of interstitial convection and diffusion of albumin in normal and neoplastic tissues by fluorescence photobleaching. *Proc. Natl. Acad. Sci.* **86:** 5385.

Dellian M., Witwer B.P., Salehi H.A., Yuan F., and Jain R.K. 1996. Quantitation and physiological characterization of bFGF and VEGF/VPF induced vessels in mice: Effect of microenvironment on angiogenesis. *Am. J. Pathol.* **149:** 59.

Dvorak H.F., Nagy J.A., Feng D., and Dvorak A.M. 2000. Tumor architecture and targeted delivery. In *Radioimmunotherapy of cancer* (ed. P.G. Abrams and A.R. Fritzberg), p. 107. Marcel Dekker, New York.

Ferrara N. 2001. Role of vascular endothelial growth factor in regulation of physiological angiogenesis. *Am. J. Physiol.* **280:** C1358.

Fidler I.J. 2001. Angiogenic heterogeneity: Regulation of neoplastic angiogenesis by the organ microenvironment. *J. Natl. Cancer Inst.* **93:** 1040.

Folkman J. 2000. Tumor angiogenesis. In *Cancer medicine* (ed. J.F. Holland et al.), p. 132. B.C. Decker, Ontario, Canada.

Fukumura D., Yuan F., Monsky W.L., Chen Y., and Jain R.K. 1997. Effect of host microenvironment on the microcirculation of human colon adenocarcinoma. *Am. J. Pathol.* **151:** 679.

Fukumura D., Xavier R., Sugiura T., Chen Y., Parks E., Lu N., Selig M., Nielsen G., Taksir T., Jain R., and Seed B. 1998. Tumor induction of VEGF promoter activity in stromal cells. *Cell* **94:** 715.

Gershenwald J.E. and Fidler I.J. 2002. Targeting lymphatic metastasis. *Science* **296:** 1811.

Gohongi T., Fukumura D., Boucher Y., Yun C., Soff G.A., Compton C., Todoroki T., and Jain R.K. 1999. Tumor-host interactions in the gall bladder suppress distal angiogenesis and tumor growth: Involvement of transforming growth factor beta 1. *Nat. Med.* **5:** 1203.

Gullino P. 1975. Extracellular compartments of solid tumors. In *Cancer* (ed. F. Becker), p. 327. Plenum Press, New York.

Hanahan D. and Weinberg R.A. 2000. The hallmarks of cancer. *Cell* **100:** 57.

Helmlinger G., Yuan F., Dellian M., and Jain R.K. 1997a. Interstitial pH and pO_2 gradients in solid tumors *in vivo*: High-resolution measurements reveal a lack of correlation. *Nat. Med.* **3:** 177.

Helmlinger G., Endo M., Ferrara N., Hlatky L., and Jain R.K. 2000. Formation of endothelial cell networks. *Nature* **405:** 139.

Helmlinger G., Netti P.A., Lichtenbeld H.C., Melder R.J., and Jain R.K. 1997b. Solid stress inhibits the growth of multicellular tumor spheroids. *Nat. Biotechnol.* **15:** 778.

Hobbs S.K., Monsky W.L., Yuan F., Roberts W.G., Griffith L., Torchilin V.P., and Jain R.K. 1998. Regulation of transport pathways in tumor vessels: Role of tumor type and microenvironment. *Proc. Natl. Acad. Sci.* **95:** 4607.

Isner J.M. 2002. Myocardial gene therapy. *Nature* **415:** 234.

Izumi Y., Xu L., di Tomaso E., Fukumura D., and Jain R.K. 2002. Herceptin acts as an anti-angiogenic cocktail. *Nature* **416:** 279.

Jain R.K. 1994. Barriers to drug delivery in solid tumors. *Sci. Am.* **271:** 58.

———. 1998. The next frontier of molecular medicine: Delivery of therapeutics. *Nat. Med.* **4:** 655.

———. 2001. Normalizing tumor vasculature with anti-angiogenic therapy: A new paradigm for combination therapy. *Nat. Med.* **7:** 987.

Jain R.K. and Carmeliet P. 2001. Vessels of death or life. *Sci. Am.* **285:** 38.

Jain R.K. and Fenton B.T. 2002. Intra-tumor lymphatic vessels: A case of mistaken identity or malfunction? *J. Natl. Cancer Inst.* **94:** 417.

Jain R.K. and Padera T.P. 2002. Prevention of lymphatic metastasis by anti-lymphangiogenic therapy. *J. Natl. Cancer Inst.* **94:** 785.

———. 2003. Lymphatics make the break. *Science* **299:** 209.

Jain R.K., Munn L., and Fukumura D. 2002. Dissecting tumors physiology using intravital microscopy. *Nat. Rev. Cancer* **2:** 266.

Jain R.K., Safabakhsh N., Sckell A., Chen Y., Benjamin L.A., Yuan F., Keshet E., and Jiang P. 1998. Endothelial cell death, angiogenesis, and microvascular function after castration in an androgen-dependent tumor: Role of vascular endothelial growth factor. *Proc. Natl. Acad. Sci.* **95:** 10820.

Jeltsch M., Kaipainen A., Joukov V., Meng X., Lakso M., Rauvala H., Swartz M., Fukumura D., Jain R.K., and Alitalo K. 1997. Hyperplasia of lymphatic vessels in VEGF-C transgenic mice. *Science* **276:** 1423.

Kadambi A., Mouta Carreira C., Yun C.O., Padera T.P., Dolmans D.E., Carmeliet P., Fukumura D., and Jain R.K. 2001. Vascular endothelial growth factor (VEGF)-C differentially affects tumor vascular function and leukocyte recruitment: Role of VEGF-receptor 2 and host VEGF-A. *Cancer Res.* **61:** 2404.

Kozin S.V., Boucher Y., Hicklin D.J., Bohlen P., Jain R.K., and Suit H.D. 2001. Vascular endothelial growth factor receptor 2-blocking antibody potentiates radiation-induced long-term control of human tumor xenografts. *Cancer Res.* **61:** 39.

LeCouter J., Kowalski J., Foster J., Hass P., Zhang Z., Dillard-Telm L., Frantz G., Rangell L., DeGuzman L., Keller G.A., Peale F., Gurney A., Hillan K.J., and Ferrara N. 2001. Identification of an angiogenic mitogen selective for endocrine gland endothelium. *Nature* **412:** 877.

Lee C.G., Heijn M., di Tomaso E., Griffon-Etienne G., Ancukiewicz C., Koike C., Park K.R., Ferrara N., Jain R.K., Suit H.D., and Boucher Y. 2000. Anti-vascular endothelial growth factor treatment augments tumor radiation response under normoxic or hypoxic conditions. *Cancer Res.* **60:** 5565.

Leu A.J., Berk D.A., Yuan F., and Jain R.K. 1994. Flow velocity in the superficial lymphatic network of the mouse tail. *Am. J. Physiol.* **267:** H1507.

Leu A.J., Berk D.A., Lymboussaki A., Alitalo K., and Jain R.K. 2000. Absence of functional lymphatics within a murine sarcoma: A molecular and functional evaluation. *Cancer Res.* **60:** 4324.

Leunig M., Yuan F., Menger M.D., Boucher Y., Goetz A.E., Messmer K., and Jain R.K. 1992. Angiogenesis, microvascular architecture, microhemodynamics, and interstitial fluid pressure during early growth of human adenocarcinoma LS174T in SCID mice. *Cancer Res.* **52:** 6553.

Lyden D., Hattori K., Dias S., Costa C., Blaikie P., Butros L., Chadburn A., Heissig B., Marks W., Witte L., Wu Y., Hicklin D., Zhu Z., Hackett N.R., Crystal R.G., Moore M.A., Hajjar K.A., Manova K., Benezra R., and Rafii S. 2001. Impaired recruitment of bone-marrow-derived endothelial and hematopoietic precursor cells blocks tumor angiogenesis and growth. *Nat. Med.* **7:** 1194.

McCormick F. 2001. New-age drug meets resistance. *Nature* **412:** 281.

Melder R.J., Koenig G.C., Witwer B.P., Safabakhsh N., Munn L.L., and Jain R.K. 1996. During angiogenesis, vascular endothelial growth factor and basic fibroblast growth factor regulate natural killer cell adhesion to tumor endothelium. *Nat. Med.* **2:** 992.

Monsky W.L., Carreira C.M., Tsuzuki Y., Gohongi T., Fukumura D., and Jain R.K. 2002. Role of host microenvironment in angiogenesis and microvascular functions in human breast cancer xenografts: Mammary fat pad vs. cranial tumors. *Clin. Cancer Res.* **61:** 1008.

Mouta Carreira C., Nasser S.M., di Tomaso E., Padera T.P., Boucher Y., Tomarev S.I., and Jain R.K. 2001. LYVE-1 is not restricted to the lymph vessels: Expression in normal liver blood sinusoids and down-regulation in human liver cancer and cirrhosis. *Cancer Res.* **61:** 8079.

Netti P.A., Berk D.A., Swartz M.A., Grodzinsky A.J., and Jain R.K. 2000. Role of extracellular matrix assembly in interstitial transport in solid tumors. *Cancer Res.* **60:** 2497.

Nugent L.J. and Jain R.K. 1984. Extravascular diffusion in normal and neoplastic tissues. *Cancer Res.* **44:** 238.

Padera T.P., Kadambi A., di Tomaso E., Mouta-Carreira C., Brown E.B., Boucher Y., Choi N.C., Mathisen D., Wain J., Mark E.J., Munn L.L., and Jain R.K. 2002. Lymphatic metastasis in the absence of functional intratumor lymphatics. *Science* **296:** 1883.

Pluen A., Boucher Y., Ramanujan S., McKee T.D., Gohongi T., di Tomaso E., Brown E.B., Izumi Y., Campbell R.B., Berk D.A., and Jain R.K. 2001. Role of tumor-host interactions in interstitial diffusion of macromolecules: Cranial vs. subcutaneous tumors. *Proc. Natl. Acad. Sci.* **98:** 4628.

Ruoslahti E. 2002. Specialization of tumor vasculature. *Nat. Rev. Cancer* **2:** 83.

Shioda T., Munn L., Fenner M., Jain R., and Isselbacher K. 1997. Early events of metastasis in the microcirculation involve changes in gene expression of cancer cells: Tracking mRNA levels of metastasizing cancer cells in the chick embryo chorioallantoic membrane. *Am. J. Pathol.* **150:** 2099.

Swartz M.A., Berk D.A., and Jain R.K. 1996. Transport in lymphatic capillaries. I. Macroscopic measurements using residence time distribution theory. *Am. J. Physiol.* **270:** H324.

Tsuzuki Y., Fukumura D., Oosthuyse B., Koike C., Carmeliet P., and Jain R.K. 2000. Vascular endothelial growth factor (VEGF) modulation by targeting HIF-1α→HRE→VEGF cascade differentially regulates vascular response and growth rate in tumors. *Cancer Res.* **60:** 6248.

Weissleder R. 2002. Scaling down imaging: Molecular mapping of cancer in mice. *Nat. Rev. Cancer* **2:** 11.

Xu L., Fukumura D., and Jain R.K. 2002. Acidic extracellular pH induces VEGF in human glioblastoma cells via AP-1 and requires ERK1/2 MAPK signaling pathway. *J. Biol. Chem.* **277:** 11368.

Yancopoulos G.D., Davis S., and Gale N.W. 2000. Vascular-specific growth factors and blood vessel formation. *Nature* **407:** 242.

Yuan F., Chen Y., Dellian M., Safabakhsh N., Ferrara N., and Jain R.K. 1996. Time-dependent vascular regression and permeability changes in established human tumor xenografts induced by an anti-vascular endothelial growth factor/vascular permeability factor antibody. *Proc. Natl. Acad. Sci.* **93:** 14765.

Yuan F., Salehi H.A., Boucher Y., Vasthare U.S., Tuma R.F., and Jain R.K. 1994. Vascular permeability and microcirculation of gliomas and mammary carcinomas transplanted in rat and mouse cranial windows. *Cancer Res.* **54:** 4564.

A Genetic Approach to Understanding Tumor Angiogenesis

R. Benezra,* P. de Candia,* H. Li,* E. Romero,* D. Lyden,† S. Rafii,‡ and M. Ruzinova*

*Department of Cell Biology and †Department of Pediatrics, Memorial Sloan-Kettering Cancer Center, New York, New York 10021; ‡Department of Medicine-Hematology/Oncology, New York Medical Center, New York, New York 10021

There is a widely held belief in the field of cancer biology, proposed originally by Dr. Judah Folkman, that tumors cannot grow beyond a very limited size if the microvessel density within the tumor is sufficiently low. Therefore, inhibition of the formation of blood vessels into the tumor bed is likely to be an effective anticancer therapy. With a large number of antiangiogenic drugs already in the clinic, the results beginning to trickle in suggest that this first generation of therapeutics might not be all that was hoped for. Although it is too early to draw firm conclusions, we can at least begin to think of ways to improve the effectiveness of these therapies.

We have taken a genetic approach to the dissection of the effects of antiangiogenesis on tumor growth in mouse models. We have done this by the disruption of two genes (Id1 and Id3) shown to be essential for tumor angiogenesis but dispensable for normal angiogenic processes (Lyden et al. 1999). We have characterized the effects of loss of these gene products on the growth and metastasis of tumors initiated by subcutaneous inoculation, by overexpression of oncogenes, and by the loss of tumor suppressor genes. These systems have demonstrated quite clearly that there is a broad distribution of effects on the growth of a tumor that results from antiangiogenic interventions. Some tumors respond as predicted by showing lower growth after hypoxic stress; others, however, grow to rather large sizes despite extensive hemorrhage and necrosis internal to the tumor mass by what appears to be an effective utilization of peripheral vasculature. A viable rim of cells that surround a necrotic core is an attractive target for chemotherapeutic intervention and points to the likelihood that antiangiogenic and anti-cell-based therapies will synergize to inhibit the growth of tumors. Finally, the Id knockout model, by providing an animal in which tumor vasculature is normally blocked, has allowed unequivocal demonstration that bone-marrow-derived precursor cells are critical components in the establishment of a functional tumor vasculature (Lyden et al. 2001). These results, overall, have significant implications both in the basic biology of tumor vascularization and in the application to antiangiogenic intervention.

BASIC BIOLOGY OF THE Id PROTEINS

How was it determined that the Id proteins are critical components of tumor angiogenesis? The history of this discovery provides an excellent example of how the path from basic science to clinical implication can be unexpectedly direct. The Id proteins were originally defined as naturally occurring dominant negative antagonists of the basic-helix-loop-helix (bHLH) family of transcription factors (for review, see Benezra 2001). The bHLH proteins contain a cluster of amino acids rich in basic residues adjacent to an HLH dimerization motif that mediates DNA binding of homodimeric and heterodimeric bHLH complexes. Because Id lacks the basic DNA-binding domain, heterodimers between Id and bHLH proteins cannot bind DNA (see Fig. 1). This dominant negative mode of inhibition of DNA-binding activity is widely used in the cell, as it is also employed by members of the leucine zipper and homeodomain protein families (Treacy et al. 1991; Ron and Habener 1992).

Biochemical and genetic data have established that the primary method whereby Id exerts its dominant negative effect is to sequester the ubiquitously expressed bHLH proteins referred to as E proteins and to prevent them from binding DNA either alone or as heterodimers with tissue-restricted bHLH proteins. Specifically, endogenous Id1 protein present in an undifferentiated myoblast quantitatively sequesters the available pool of E2A proteins, forces dimers between MyoD and E47 (a splice product of the E2A gene) to resist Id1 inhibition, and causes the transcriptional inhibitory effects of Id1 and Id3 to be overcome by the overexpression of E proteins. In addition, the postnatal lethality observed in mice lacking products of the E2A gene is partially suppressed by disruption of the Id1 gene. Thus, by sequestering E proteins that are required to enhance transcriptional activity of bHLH proteins in multiple cell types, the Id proteins can control the activity of bHLH proteins in diverse lineages.

Id1 and Id3 are co-expressed temporally and spatially during murine neurogenesis and angiogenesis, two processes that are affected in the Id1/Id3 double knockout animals (Jen et al. 1997; Lyden et al. 1999). In general terms, Id1, 2, and 3 are expressed in dividing neuroblasts in the central nervous system (CNS) up to about E12.5, after which time Id2 expression appears in the presumptive neurons that were undergoing maturation in both the future cerebellum and the cerebral cortex. Id4, which, unlike the other Id genes is localized exclusively in the CNS and peripheral nervous system, has a pattern of expression that is often complementary to Id1 and Id3 and is

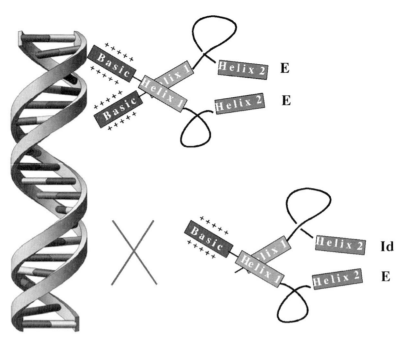

Figure 1. Mechanism of Id activity. The primary target of the Id proteins in the cell are the E proteins (E) which homodimerize in lymphoid cells via their helix-loop-helix (HLH) domains and bind negatively charged DNA through contacts with the two basic clusters of amino acids that carry positive charges under physiological conditions. E proteins also heterodimerize with other bHLH proteins in many nonlymphoid tissues and bind DNA similarly. Id can associate with E proteins via its HLH domain, but because it lacks a basic region, the E–Id heterodimer is incapable of binding DNA. Geometries depicted are used to show the essential features of the model. (Reprinted, with permission, from Benezra 2001 [copyright Elsevier Science].)

found in regions undergoing neuronal maturation. In adult animals, Id1 and Id3 are no longer expressed in the brain, but Id2 expression persists in the Purkinje cells of the cerebellum and to a lesser extent in the mitral cells of the olfactory bulb in layer five of the neocortex (Neuman et al. 1993; Jen et al. 1997). Given the similarities and biochemical activity of Id1, 2, and 3 and the known role of bHLH proteins in neural development, it was not surprising to observe premature neural differentiation in Id1,3 knockout animals when Id2 expression was lost in dividing neuroblasts (~E12.5) (Lyden et al. 1999).

A role for Id in angiogenesis was unexpected, however. Angiogenesis, the branching and spreading of capillaries from the primary vascular plexus, occurs both in the yolk sac and in the embryo proper, particularly in the brain. Signaling from both the VEGF and Tie-2 receptors has been implicated in the process as well as during tumor angiogenesis (for review, see Yancopoulos et al. 2000). Little is known of the involvement of bHLH proteins, however, during these processes. bHLH proteins that contain a second dimerization motif (referred to as a PAS domain) are expressed in endothelial cells (HIF-1α, EPAS, HRF) but are unlikely to be regulated by Id, as the specificity of these proteins is dictated by the PAS motif and no HLH/bHLH-PAS interactions have been documented. More importantly, HIF-1α and EPAS have been shown to up-regulate VEGF expression (for review, see Levy et al. 1997), and loss of Id function leads to a decrease in VEGF expression, implying, counter to its normal mode of action, that Id would enhance the activity of these transcription factors directly or indirectly. bHLH-EC2 is expressed in endothelial cells and lacks a PAS domain (Quertermous et al. 1994), making it a possible candidate for Id regulation. At least one copy of the Id1 or Id3 genes is required in mice to prevent embryonic lethality associated with premature neural differentiation and angiogenic defects in the brain. Interestingly, Id1 and Id3 but not Id2 are expressed in the endothelial cells in the brain whereas Id1, 2, and 3 are all expressed in endothelial cells throughout the rest of the embryo during development, perhaps accounting for the brain specificity of the phenotype (Lyden et al. 1999). Lethality may be due to intraventricular hemorrhage, which contributes to death in humans and rodents.

ROLE OF Id IN ANGIOGENESIS WITHIN TUMORS

Id1 and 3 expression has been detected in the endothelial cells of vessels invading tumors but not in resting vessels within normal tissues (Lyden et al. 1999). This observation, coupled with the genetics described above, led us to examine the Id knockout mice for their ability to support the growth of tumor xenografts. Strikingly, three different tumors fail to grow and/or metastasize in $Id1^{+/-}Id3^{-/-}$ animals (see Fig. 2), and any tumor growth present showed poor vascularization and extensive necrosis. Blood vessels in Id knockout animals lack the ability to branch and sprout and form a normal-caliber lumen during tumor progression. These processes depend on the

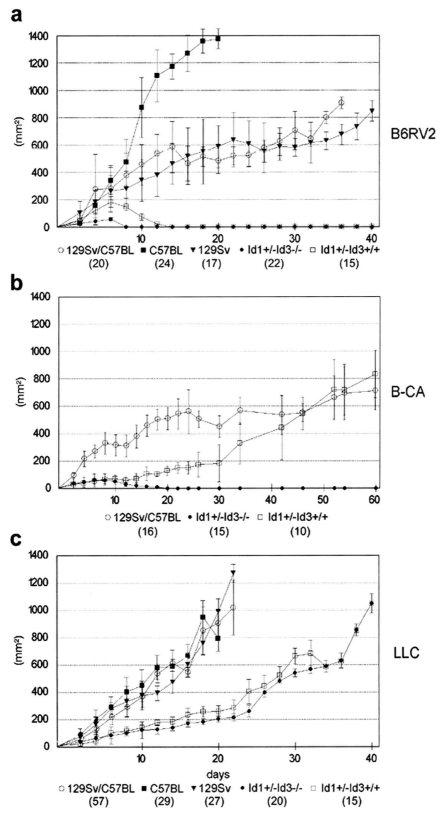

Figure 2. Tumor growth is inhibited in Id knockout mice. (*a–c*) Wild-type and Id knockout mice were injected intradermally with tumor cell lines as indicated. Tumor surface area and the mean and standard deviations are shown. Number of animals is given in parentheses. (Reprinted, with permission, from Lyden et al. 1999 [copyright Macmillan].)

proteolysis and remodeling of the extracellular matrix, and indeed, pronounced thickening of the extracellular matrix surrounding endothelial cells in Id knockout animals was observed. Moreover, little if any $\alpha v\beta 3$ integrin or its associated metalloproteinase MMP2 was detected in the stunted vessels present in the residual tumor mass of $Id1^{+/-}Id3^{-/-}$ animals. Thus, Id expression may directly or indirectly control the expression of $\alpha v\beta 3$ integrin during tumor-induced angiogenesis and the recruitment or expression of MMP2 metalloproteinase. Although $\alpha v\beta 3$ integrins may have both positive and negative roles in tumor angiogenesis (Brooks et al. 1994; Reynolds et al. 2002), the misregulation of other genes besides $\alpha v\beta 3$ is probably contributing to the phenotype. Although certain tumors such as Lewis lung carcinoma continue to grow (albeit more slowly in Id knockout animals), there was nonetheless a profound suppression of metastasis in these animals even when the tumors were of equal volumes in both genotypes. This could be due to poor vascularization of the primary tumor, which would prevent egress of the tumor cell. However, when primary tumor cells were injected in the tail veins of $Id1^{-/-}Id3^{+/-}$ animals and wild-type controls, thereby providing a good vasculature for metastatic spread of both genotypes, only the wild-type animals demonstrated metastatic lesions. This suggests that the loss of Id function has, in some circumstances, a more profound effect on the ability of a tumor to grow at sites of metastasis, perhaps due to a more stringent requirement for an intact vasculature at those locations. In any event, these results point to the possible importance of antiangiogenic intervention in metastatic spread of tumors even when primary tumors are only marginally affected.

More physiologically relevant tumor models than xenografts are now at our disposal, and we have begun to test the effects of loss of Id on these systems as well. Animals overexpressing oncogenes, or that have disrupted tumor suppressor genes, have been examined. In preliminary experiments, it appears that there is a consistent upregulation of Id1 and Id3 in the endothelial cell of high-grade tumors and that loss of Id leads to extensive hemorrhage and necrosis within these tumors. The response of the tumor, however, varies considerably, with some lesions regressing completely and others growing extensively with a viable rim of cells surrounding a centrally necrotic core. Although this is similar to the effects of loss of Id in xenografts, the disrupted vasculature looks very different in these two settings. The xenografts $Id1^{-/-}Id3^{+/-}$ are mostly avascular, with some stunted and occluded vessels observed. In the vasculature of spontaneous tumors, the lesions look more like the embryonic state with gross dilation and aggregation of the endothelial cells. Array analyses are now being employed to determine which genes are affected by loss of Id, but at a minimum, it points to the differences in vasculature of xenografts versus spontaneous tumors under antiangiogenic stress. Such observations suggest that the clinical effects of antiangiogenic therapies in spontaneous human tumors might differ from the responses observed in xenograft models.

Id IS ESSENTIAL FOR THE MOBILIZATION OF BONE-MARROW-DERIVED ENDOTHELIAL PRECURSORS

The Id knockout model has been a valuable tool in assessing the role of bone-marrow-derived precursor cells in tumor angiogenesis. Since the $Id1^{-/-}Id3^{+/-}$ animals fail to develop a vasculature capable of supporting tumor growth and metastasis, transplantation with β-galactosidase-expressing bone marrow from wild-type animals allows a rapid assessment of the recovery of tumor growth and the derivation of any endothelial cells that make it to the tumor bed (see Fig. 3). As reported previously, we have observed after lethal irradiation of $Id1^{-/-}Id3^{+/-}$ mice that wild-type bone marrow can completely rescue the growth kinetics of subcutaneous tumors in these animals and that the majority of blood vessels that make it to the tumor bed express β-galactosidase and are therefore bone-marrow derived (Lyden et al. 2001). It could be argued that this result was merely a consequence of having "frozen" the peripheral vasculature in the Id knockouts, and bone-marrow-derived cells not normally used were now the only source of tumor endothelium. This possibility was ruled out by the observation that in a similar transplant protocol, in which wild-type bone marrow was used to reconstitute a lethally irradiated wild animal, a very high percentage of donor-derived bone marrow endothelium was found in the tumor mass. Whereas these results suggest that bone-marrow-derived precursor cells are sufficient for the vascularization of the tumor mass, the necessity for these cells was established by doing the converse experiment in which $Id1^{-/-}Id3^{+/-}$ marrow was used to reconstitute a lethally irradiated wild-type mouse. In these animals, a long delay in the onset of growth of a subcutaneous tumor graft was observed. Subsequent tumor growth was likely due to the recovery of residual

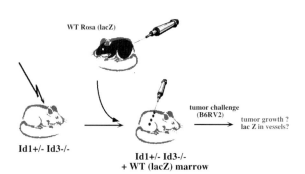

Figure 3. Bone marrow transplantation scheme. Since $Id1^{-/-}Id3^{+/-}$ animals fail to support the vascularization of subcutaneous tumors, this model allowed a functional test of the hypothesis that bone-marrow-derived precursor cells are critical components of tumor neovascularization. After lethal irradiation, the $Id1^{-/-}Id3^{+/-}$ animals were rescued with bone-marrow derived from ROSA-26 animals that is marked with β-galactosidase. After 2–3 weeks' recovery, the animals are challenged with a subcutaneous B6RV2 lymphoma inoculum. Tumor progression and staining of the vessels within the tumors for β-galactosidase are then performed. As described in the text, tumor growth is restored in the $Id1^{-/-}Id3^{+/-}$ transplanted animals, and the vast majority of the vessels are positive for β-galactosidase.

wild-type bone marrow, but at a minimum, it is clear that bone-marrow-derived precursor cells are necessary in the early phases of tumor growth.

What is the function of Id in the bone-marrow-derived precursor cells when vascular endothelial growth factor (VEGF) levels are increased in the circulation? Id expression appears to be required for the mobilization of VEGF-R2-positive cells into the periphery, since $Id1^{-/-}$ $Id3^{+/-}$ animals fail to generate a significant number of these cells in the peripheral blood mononuclear fraction after elevation of VEGF levels either by adeno-VEGF infection or by tumor development (Lyden et al. 2001). These cells are likely to contain progenitors (and not simply reflect mature endothelium), since they have what is referred to as to late outgrowth potential or the ability to commit to the endothelial cell lineage only after multiple rounds of division in cell culture assays. In addition to the mobilization of the VEGF-R2 fraction, Id appears to be required for the expansion (at least in vitro) of VEGF-R1-positive bone-marrow-derived cells that seem to decorate the newly forming vessels within the tumor xenografts. These VEGF-R1-positive cells express monomyelocytic markers and therefore may represent a hematopoietic population that is required for normal blood vessel development. Such tunneling by myeloid cells has been reported in other systems (Moldovan et al. 2000). That both the VEGF-R1 and VEGF-R2 populations are important for the formation of an intact vasculature is supported by antibody neutralization experiments in which synergistic effects on tumor growth and vascular integrity are observed when monoclonal antibodies against VEGF-R1 and VEGF-R2 are used in combination (Lyden et al. 2001). Thus, by short-circuiting both the VEGF-R1 and VEGF-R2 pathways, loss of Id can lead to a profound inhibition of the ability of the host to utilize bone-marrow-derived precursor cells (VEGF-R2 positive) and supporting hematopoietic cells (VEGF-R1 positive) for the formation of a tumor vasculature (see Fig. 4). A central question now raised by the studies is, What are the downstream effects of loss of Id in the bone marrow that leads to such defects? Array analyses on wild-type and Id-deficient endothelium are now being performed and will likely shed new light on the mechanisms of tumor-induced endothelial cell mobilization and maturation.

The observation that bone-marrow-derived endothelial cell precursors are critical components of tumor vascularization may have important prognostic and therapeutic implications in cancer treatment. However, before such application should be undertaken, it will first be critical to determine whether these bone-marrow-derived cells are also utilized in tumors that arise in non-subcutaneous models. Indeed, the use of "peripheral vasculature" to feed a tumor may be exquisitely sensitive to the location of the tumor growth, and the utilization of this vasculature in the subcutaneous model and tumors that initiate by the overexpression of an oncogene in the tissue may be dramatically different. The grade of a tumor, reflecting its differentiation status and proliferation rate, may also contribute to the source of endothelial cells used to support its growth. The determination of whether bone-marrow-derived precursor cells are used in tumors that develop as a result of genetic predisposition, i.e., overexpression of an oncogene or loss of a tumor suppressor gene, is currently being performed using the bone marrow transplantation assay as described above.

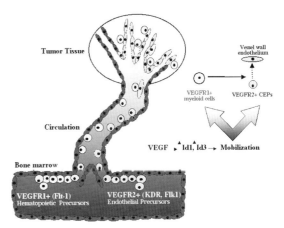

Figure 4. Model for bone marrow supporting early tumor angiogenesis. Co-mobilization and recruitment of VEGFR1 + hematopoietic precursor cells and VEGFR2 + endothelial precursor cells contribute to tumor angiogenesis. VEGF-induced mobilization and activation of VEGFR1 + myeloid precursor cells result in the release of growth factors such as VEGF and activated MMP-9. This facilitates integration of recruited CEPs into neo-vessels. In turn, VEGF induces release of cytokines by endothelial cells, including GM-CSF, G-CSF, and IL6 that maintains and supports the proliferation of myeloid progenitor cells. (Reprinted, with permission, from Benezra et al. 2001 [copyright Nature Publishing Group].)

Assuming some level of use of the bone marrow is observed in these models, the confirmation in human systems will suggest a range of therapeutic applications. Confirmation could be achieved either by carefully examining the VEGF-R2-positive cell population in the peripheral blood of cancer versus normal subjects. Alternatively, gender-mismatched bone-marrow transplant recipients who develop secondary tumors should be examined for the presence of endothelial cells with the sex chromosome complement of the donor. If confirmed that bone-marrow-derived cells are contributing to the vasculature of human tumors, then the levels of circulating precursor cells may be important surrogate markers for the effectiveness of antiangiogenic treatments. Ex vivo treatments of bone marrow with endothelial cell-toxic agents (perhaps too toxic for systemic therapies) or depletion of endothelial cell precursors from the marrow may, upon reintroducing the marrow into a patient, lead to a decrease in the vascularization of a tumor and the inhibition of its growth. Ex vivo genetic manipulation of endothelial cell precursors to allow them to acquire certain antitumor activities can also be imagined in the future.

OUTLOOK FOR Id AS TARGET

What are the prospects for targeting Id as a means of inhibiting tumor angiogenesis? Since Id works by associating with other proteins, one is faced with the difficult task of inhibiting a protein–protein interaction with a drug. It could be argued that if a target is sufficiently attractive, extensive, even high-risk, efforts on hitting this target are justified. Why is Id attractive? First, as shown by immunohistochemistry and in situ hybridization studies, Id1 and Id3 are expressed specifically in tumor endothelium and are not present (and therefore are not essential components) in resting vasculature and normal tissues. In addition, Id is rarely expressly in normal adult tissue but often expressed in tumors derived from those tissues, making it possible that an anti-Id therapy will inhibit both tumor vasculature and the tumor itself. Complete loss of Id function is not required to induce dramatic inhibition of tumor angiogenesis as evidenced by the phenotypes observed in animals missing even a single copy of the four Id1,3 gene copies. The primary mechanism of action of the Id family is known to be its association with E protein, so screens designed to target this interaction can be easily devised. Finally, the validation of Id as an important target for antiangiogenic drug design has been established not only in subcutaneous tumor models, but also in tumors that develop as a result of oncogene overexpression and loss of tumor suppressor genes, much more physiologically relevant models of human disease.

How can Id be efficiently inhibited in an adult? Here, of course, we are forced to speculate, but hopefully with an eye toward novel approaches that can be developed in the not-too-distant future. The first task will be to identify a lead compound that can interact with Id at high efficiency. This could be achieved by screening conventional small molecule libraries, or perhaps libraries of peptidomimics used recently to target the myc-max interaction. Improved affinities could be achieved by standard modification regimes or, if structural information is obtained, by rational drug design. Larger molecule strategies, i.e., anti-sense or siRNA, can also be envisioned, but here the ability to convert these compounds into drugs may still pose a significant hurdle. Delivery of any identifiable molecules to the critical tissue (i.e., tumor endothelium) may have been made simpler by the identification of tumor endothelial cell "zip codes." These zip codes are receptors for short peptides identified by phage display (Arap et al. 1998). These peptides bind, are internalized, and are concentrated in the nucleus of the tumor endothelium, thus far in xenograft models. Therefore, one can envision chemically modifying these peptides to deliver an anti-Id small molecule or perhaps large molecule and thereby direct these compounds with high efficiencies to the critical tissue type.

If Id itself turns out not to be "drugable," then other targets downstream from Id may be useful. Since Id itself regulates the activity of transcription factors, it can be anticipated that loss of Id will lead to changes in gene expression in endothelial cells. Genes that are down-regulated in endothelial cells as a result of loss of Id may themselves be required for normal endothelial cell development in tumors and may therefore be good targets for antiangiogenic drug designs. The importance of these markers will need to be established by loss-of-function experiments (alone and in combination perhaps), that attempt to phenocopy the Id knockouts. Particular attention should be paid to molecules that are easily targeted such as surface markers or enzymes. In this way, a mimic of the Id knockout may be achieved and provide a useful therapeutic approach to the inhibition of blood vessel formation within human tumors.

ACKNOWLEDGMENTS

This work was supported by grants from the Breast Cancer Research Foundation and the National Institutes of Health.

REFERENCES

Arap W., Pasqualini R., and Ruoslahti E. 1998. Chemotherapy targeted to tumor vasculature. *Curr. Opin. Oncol.* **10**: 560.

Benezra R. 2001. Role of Id proteins in embryonic and tumor angiogenesis. *Trends Cardiovasc. Res.* **11**: 237.

Benezra R., Rafii S., and Lyden D. 2001. The Id proteins and angiogenesis. *Oncogene* **20**: 8334.

Brooks P.C., Clark R.A., and Cheresh D.A. 1994. Requirement of vascular integrin alpha v beta 3 for angiogenesis. *Science* **264**: 569.

Jen Y., Manova K., and Benezra R. 1997. Each member of the Id gene family exhibits a unique expression pattern in mouse gastrulation and neurogenesis. *Dev. Dyn.* **208**: 92.

Levy A.P., Levy N.S., Iliopoulos O., Jiang C., Kaplin W.G., Jr., and Goldberg M.A. 1997. Regulation of vascular endothelial growth factor by hypoxia and its modulation by the von Hippel-Lindau tumor suppressor gene. *Kidney Int.* **51**: 575.

Lyden D., Young A.Z., Zagzag D., Yan W., Gerald W., O'Reilly R., Bader B.L., Hynes R.O., Zhuang Y., Manova K., and Benezra R. 1999. Id1 and Id3 are required for neurogenesis, angiogenesis and vascularization of tumour xenografts (comments). *Nature* **401**: 670.

Lyden D., Hattori K., Dias S., Costa C., Blaikie P., Butros L., Chadburn A., Heissig B., Marks W., Witte L., Wu Y., Hicklin D., Zhu Z., Hackett N.R., Crystal R.G., Moore M.A., Hajjar K.A., Manova K., Benezra R., and Rafii S. 2001. Impaired recruitment of bone-marrow-derived endothelial and hematopoietic precursor cells blocks tumor angiogenesis and growth. *Nat. Med.* **7**: 1194.

Moldovan N.I., Goldschmidt-Clermont P.J., Parker-Thornburg J., Shapiro S.D., and Kolattukudy P.E. 2000. Contribution of monocytes/macrophages to compensatory neovascularization: The drilling of metalloelastase-positive tunnels in ischemic myocardium. *Circ. Res.* **87**: 378.

Neuman T., Keen A., Zuber M.X., Kristjansson G.I., Gruss P., and Nornes H.O. 1993. Neuronal expression of regulatory helix-loop-helix factor Id2 gene in mouse. *Dev. Biol.* **160**: 186.

Quertermous E.E., Hidai H., Blanar M.A., and Quertermous T. 1994. Cloning and characterization of a basic helix-loop-helix protein expressed in early mesoderm and the developing somites. *Proc. Natl. Acad. Sci.* **91**: 7066.

Reynolds L.E., Wyder L., Lively J.C., Taverna D., Robinson S.D., Huang X., Sheppard D., Hynes R.O., and Hodivala-Dilke K.M. 2002. Enhanced pathological angiogenesis in mice lacking beta3 integrin or beta3 and beta5 integrins. *Nat. Med.* **8**: 27.

Ron D. and Habener J.F. 1992. CHOP, a novel developmentally regulated nuclear protein that dimerizes with transcription factors C/EBP and LAP and functions as dominant-negative inhibitor of gene transcription. *Genes Dev.* **6**: 439.

Treacy M.N., He X., and Rosenfeld M.G. 1991. I-POU: A POU-domain protein that inhibits neuron-specific gene activation. *Nature* **350**: 577.

Yancopoulos G.D., Davis S., Gale N.W., Rudge J.S., Wiegand S.J., and Holash J. 2000. Vascular-specific growth factors and blood vessel formation. *Nature* **407**: 242.

Discovery of Type IV Collagen Non-collagenous Domains as Novel Integrin Ligands and Endogenous Inhibitors of Angiogenesis

R. KALLURI

Program in Matrix Biology, Department of Medicine and the Cancer Center, Beth Israel Deaconess Medical Center and Harvard Medical School, Boston, Massachusetts 02215

Type IV collagen is a major constituent of all mammalian basement membranes, including basement membranes of the vasculature. Type IV collagen is an ancient matrix scaffold, and it is one of the two collagens present in *Caenorhabditis elegans* and thus is ~600 million years old. Type IV collagen forms a spider web-like suprastructure that interacts with other components of molecules such as laminins, perlecans, and nidogen (entactin) to form a highly cross-linked and insoluble basement membrane usually present on the basolateral aspect of various cells. Type IV collagen interacts with cells via adhesion molecules such as integrins, and these interactions are considered important for establishing a productive cell–matrix interaction and subsequent intracellular signaling. The significance of type IV collagen in the regulation of angiogenesis has been studied for more than 15 years now, and whereas the original work focused on positive regulation, in recent years, type IV collagen has been identified as an important negative regulator of angiogenesis, potentially mediated by integrins. Degradation of type IV collagen can occur during various physiological and pathological processes that generate smaller collagen-derived fragments. These degradation fragments are likely involved in bioactivities differing from their original intact type IV collagen molecule. This review highlights the properties of type IV collagen degradation products and, in particular, describes the discoveries of arresten, canstatin, and tumstatin as novel integrin ligands and inhibitors of angiogenesis.

ANGIOGENESIS AND BASEMENT MEMBRANES

Development of the vertebrate cardiovascular system is associated with angiogenesis (Folkman 1995, 2001). Angiogenesis, the formation of new capillaries from preexisting blood vessels, is an event critical to many normal physiological processes (Folkman 1995, 2001; Hanahan and Folkman 1996). Angiogenesis is also essential for the progression of various pathological disorders including diabetic retinopathy and rheumatoid arthritis, as well as tumor growth and metastasis (Carmeliet and Jain 2000; Folkman 2001; Jain and Carmeliet 2001). The switch to an angiogenic phenotype requires both up-regulation of angiogenic stimulators and down-regulation of angiogenesis inhibitors (Hanahan and Folkman 1996; Eliceiri and Cheresh 2001). The induction and resolution phases of angiogenesis are self-contained processes in the human body. The induction of angiogenesis in the human body is presumably prevented due to the existence of an angiogenic barrier, and when new blood capillaries are required, this barrier is overcome and the induction of angiogenesis ensues (Bouck 1992, Hanahan and Folkman 1996). Several investigators have proposed various endogenous factors that help the human body maintain an angiogenically quiescent state (Bouck 1990, 1992; Cheresh 1998; Eliceiri and Cheresh 2001). A balance between the induction and resolution phases of angiogenesis is pivotal for organ homeostasis. What constitutes the angiogenic barrier in the human body is an active area of investigation due to the promise of the discovery of endogenous inhibitors of angiogenesis (Folkman and Shing 1992; Folkman 1995, 2001).

Vascular endothelial growth factor (VEGF) and basic fibroblast growth factor (bFGF) are among the major inducers of angiogenesis (Folkman et al. 1988; Folkman 1995, 2001; Ferrara 2000). To date, a number of angiogenesis inhibitors have been identified, and certain factors such as thrombospondin (Dameron et al. 1994; Crawford et al. 1998; Lawler 2000; Doll et al. 2001; Lawler et al. 2001; Rodriguez-Manzaneque et al. 2001), angiostatin (O'Reilly et al. 1994), endostatin (O'Reilly et al. 1997), canstatin (Kamphaus et al. 2000), arresten (Colorado et al. 2000), and tumstatin (Maeshima et al. 2000b) are found endogenously and generated in vivo. In cancer and several other diseases, uncontrolled capillary growth can have dramatic pathological consequences. At the start, a tumor is just a mass of transformed cells, and in order for the cells to proliferate beyond a few mm^3, they require a constant blood supply (Folkman 2001). Without the blood supply, tumor cells will not receive the necessary nutrients, and thus potentially will die. Therefore, cancer-oriented angiogenesis research in many aspects is directed toward identifying inhibitors that stop the flow of blood to the tumor, thus killing the tumor (Folkman 1972, 2001; Bouck 1992; Hanahan and Folkman 1996; Cheresh 1998; Carmeliet and Jain 2000; Hanahan and Weinberg 2000; Jain and Carmeliet 2001).

Three key components of a blood vessel/capillary are endothelial cells, vascular basement membrane (VBM), and vascular smooth muscle cells/pericytes. Basement membranes (BM) are a specialized meshwork of extracellular matrix (ECM) molecules, which are crucial for the development of metazoans (Paulsson 1992; Yurchenco and O'Rear 1994; Timpl 1996; Li et al. 2002). Along with providing structural support, VBM is also speculated to modulate cell behavior. Its remodeling, as defined by VBM synthesis and degradation, generates disparate signals to the cells (Bergers et al. 2000; Colorado et al. 2000; Egeblad and Werb 2002).

In a general sense, BM are thin layers of a specialized ECM that provide the supporting structure on which epithelial and endothelial cells grow and reside. In addition, they are closely associated with muscle fibers, fat mass, and peripheral nerves. BM are composed of about 50 different macromolecules; the most abundant molecules are type IV collagen, laminin, heparan sulfate proteoglycans (perlecans), fibronectin, and nidogen (Fig. 1A). BM molecules are quite large in molecular size and have the capacity to self-aggregate to form higher-order structures (Yurchenco and O'Rear 1994; Li et al. 2002; Yurchenco et al. 2002). Type IV collagen protomers self-associate via their ends and their middle triple helical regions to form a spider web scaffold-like structure that interacts with the laminin network, which by itself also has the ca-

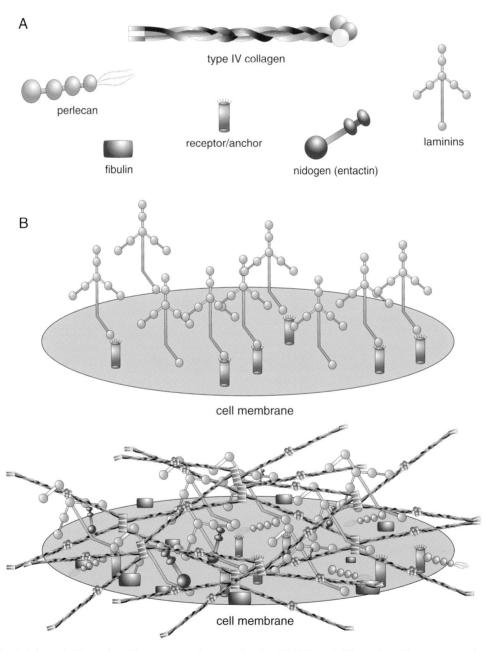

Figure 1. (*A*) Schematic illustration of basement membrane molecules. (*B*) Schematic illustration of basement membrane assembly, beginning with laminin assembly on the cell surface with potential help from receptor anchors, such as β1 integrins. (Illustration by Lori Siniski.)

pacity to self-assemble (Fig. 1B). Nidogen, also known as entactin, bridges the interactions between the type IV collagen and laminin network (Fig. 1B) (Yurchenco and O'Rear 1994; Li et al. 2002; Yurchenco et al. 2002). Perlecan and many other BM components interact with the type IV collagen-nidogen-laminin network to provide a highly cross-linked, amorphous-looking, insoluble organized structure usually on the basolateral side of various cell types (Fig. 1B). This structural property of BM provides tensile strength and framework to a given tissue.

GENE ORGANIZATION, STRUCTURE, AND FUNCTION OF TYPE IV COLLAGEN

Type IV collagen is composed of six distinct gene products, namely, α1–α6 (Fig. 2A). This collagen is present in almost all organisms except for sponges, and thus is evolutionarily constant. The α1 and α2 isoforms are ubiquitously present in human BM, whereas the other four isoforms exhibit restricted distributions (Hudson et al. 1993; Kalluri et al. 1997). Type IV collagen promotes

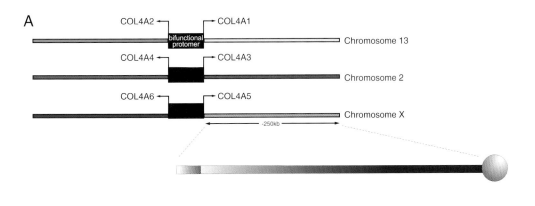

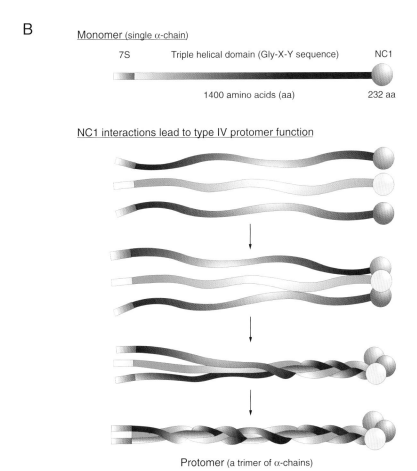

Figure 2. (*A*) Schematic illustration of type IV collagen gene organization. (*B*) α-chain structure and protomeric assembly. (Illustration by Lori Siniski.)

cell adhesion, migration, differentiation, and growth, while playing a crucial role in angiogenesis (Form et al. 1986; Madri and Pratt 1986; Ingber and Folkman 1988; Madri 1997; Haas and Madri 1999). The non-collagenous domains of type IV collagen have been implicated as important for the assembly of type IV collagen and also other functions that regulate cell behavior directly (Furcht 1984, 1986; Herbst et al. 1988; Tsilibary et al. 1988, 1990; Chelberg et al. 1990; Cameron et al. 1991; Miles et al. 1995; Zeisberg et al. 2001).

The full complement of type IV collagen in mammals is derived from six genetically distinct α-chain polypeptides (Hudson et al. 1993; Kalluri and Cosgrove 2000). All α-chains exhibit similar domain structures and share between 50% and 70% homology at the amino acid level (Hudson et al. 1993). The α-chains are composed of three domains: the amino-terminal 7S domain, the middle triple-helical domain, and the carboxy-terminal globular non-collagenous domain (NC1) (Fig. 2A). The 7S domain can be 13–143 amino acids in length, depending on the α-chain. The triple helical domain is long and between 1200 and 1400 amino acids, with about 22 interruptions in the classic Gly-X-Y sequence. The NC1 domain is about 230 amino acids in length. The NC1 domain is considered important for the assembly of type IV collagen heterotrimer structure. The assembly of a particular heterotrimer starts when the three NC1 domains initiate a yet unknown molecular interaction between the α-chains in the carboxy-terminal region. This process initiates the trimerization process, which proceeds like a zipper from the carboxy-terminal end, and this results in a fully assembled heterotrimer (Fig. 2B). The assembled trimer is flexible and can bend at many triple helical interruption points in the molecule. The next step in the assembly of type IV collagen is the type IV collagen dimer formation. Two type IV collagen protomers associate via their carboxy-terminal NC1 trimers to form an NC1 hexamer (Fig. 3). Next, four heterotrimers interact at the glycosylated amino-terminal 7S region to form 7S tetramers (Fig. 3). These interactions form the nucleus for a type IV collagen scaffold. The scaffold evolves into a type IV collagen suprastructure with the help of end-to-end associations and also lateral associations between type IV collagen protomer (Yurchenco and O'Rear 1994). This network of type IV collagen interacts with a network of laminin mediated by nidogen/entactin to initiate the formation of a BM-like structure (Fig. 1). Perlecan, fibulins, and other collagen types such as type VI, type VIII, type XV, and type XVIII, interact with the type IV collagen and laminin networks to build the final structure of a BM (Fig. 1). The complexity and large size associated with all BM molecules result in a structure that is highly cross-linked and insoluble in physiological solutions.

With six different type IV collagen α-chains known today, there are 56 combinations of trimers possible. Which three α-chains form a protomer is still unclear, although biochemical studies suggest that α1 and α2 can co-assemble to give a protomer with the α-composition of a 2:1 ratio of α1 and α2 (Timpl 1996). This was recently validated by NC1 hexamer crystal structure. The specificity of protomer formation involving other four α-chains is still largely speculative and derived from NC1 domain interactions (Gunwar et al. 1998; Boutaud et al. 2000; Borza et al. 2001; Sundaramoorthy et al. 2002). The α1 and α2 isoforms are ubiquitously present in human BM (Hudson et al. 1993). The other four isoforms exhibit more tissue- and organ-specific distributions (Hudson et al. 1993; Kalluri et al. 1997).

EARLY STUDIES WITH BASEMENT MEMBRANES, TYPE IV COLLAGEN, AND ANGIOGENESIS

The possible role for ECM and BM-like material in angiogenesis was described by Clark et al. in 1934 (Clark and Clark 1934). They suggest in a camera lucida study of physiological neovascularization that growing sprouts become functional capillary tubes when the perivascular matrix changes to a tissue substance resembling a "soft gel." Endothelial cells are surrounded and rest on capillary BM composed of several matrix molecules (Form et al. 1986; Madri and Pratt 1986; Madri 1997).

Ingber and Folkman further suggest that FGF-stimulated endothelial cells may be "switched" between growth, differentiation, and involution modes during angiogenesis by altering the adhesivity or mechanical integrity of their ECM (Ingber and Folkman 1989). The endothelial cells are normally quiescent while resting on the capillary BM composed of type IV collagen, laminin, heparan sulfate proteoglycans, and nidogen. This suggests that, primarily, the signals originating from capillary BM at this point induce growth inhibition and promote an environment which facilitates cell–cell adhesion. Once the angiogenic response is induced, presumably by growth factors such as VEGF and bFGF, capillary BM is degraded by several matrix-degrading enzymes. Such degradation serves multiple purposes, including liberation of endothelial cells to migrate and proliferate, liberation of sequestered growth factors such as VEGF, and detachment of pericytes (Fig. 4). The detachment of endothelial cells from the BM dramatically changes the microenvironment of endothelial cells. The cells now are in direct contact with interstitial provisional matrix such as vitronectin, fibronectin, type I collagen, and thrombin (Form et al. 1986; Hynes 1996; Madri 1997). The endothelial cells at the same time are producing many of these matrix molecules. These observations strongly suggest that provisional matrix provides proliferative cues to the endothelial cells while the BM matrix provides growth inhibitory influence (Fig. 4). Further studies are required to understand this mechanism. Nevertheless, it is now well established that the inductive phase and the resolution phase of angiogenesis are under the influence of temporally ordered matrix synthesis. Such endothelial cell–matrix interactions provide dynamic regulation of endothelial cells during the process of angiogenesis.

A strong indication for the importance of type IV collagen came from studies with inhibitors of BM collagen synthesis. In these studies, inhibitors such as GPA 1734,

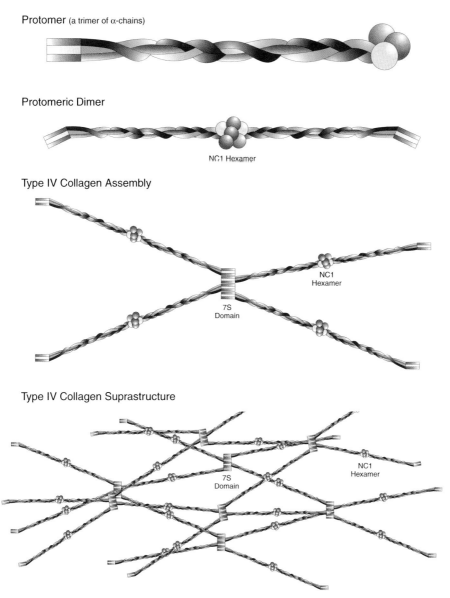

Figure 3. Schematic illustration of type IV collagen assembly. (Illustration by Lori Siniski.)

D609, cis-hydroxyproline, and β-aminopropionitrile were used to inhibit collagen biosynthesis and prevent endothelial tube formation, angiogenesis, and tumor growth (Oberbaumer et al. 1982; Ingber and Folkman 1988; Maragoudakis et al. 1993; Grant et al. 1994). Such studies further supported the notion that disruption of BM collagen assembly can result in the inhibition of angiogenesis.

In this regard, studies with the NC1 domains of type IV collagen require special attention. As indicated earlier in this review, type IV collagen protomer assembly is speculated to initiate from the carboxy-terminal globular NC1 domain of the molecule (Fig. 2B). The NC1 domain hexamer formation during the process of protomeric dimer formation is also mediated by individual NC1 domains (Fig. 3). The crystal structure studies of type IV collagen NC1 domains isolated from the human placenta or bovine lens capsule show that NC1 domains can fold into a novel tertiary structure utilizing β-strands between two homologous subdomains. The trimers and subsequent hexamers are assembled through a domain swap between the subdomains (Sundaramoorthy et al. 2002; Than et al. 2002). These studies provide structural validation that NC1 domains are critical for the assembly of type IV collagen network suprastructure. This network interacts with a laminin network with help from nidogen, which subsequently helps initiate the complex formation of the sheet-like BM structure, which includes many other macromolecules such as fibulins and perlecans. Therefore, NC1 domain interactions can be considered as pivotal for the assembly of BM (Figs. 1 and 2).

In this regard, culture experiments using endothelial cells and matrix molecules such as laminin and type IV collagen suggest that increasing the ratio of type IV collagen to that of laminin results in a progressive decrease

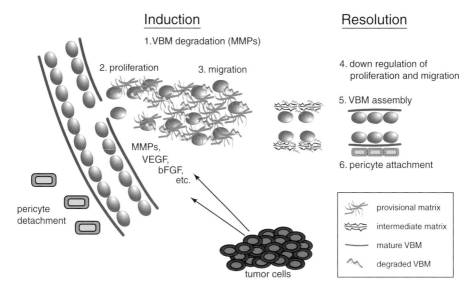

Figure 4. Vascular basement membrane (VBM) is an important structural component of blood vessels and modulates the behavior of vascular endothelial cells during angiogenesis. In response to growth factors such as vascular endothelial growth factor (VEGF) and basic fibroblast growth factor (bFGF), and the release of matrix-degrading enzymes such as matrix metalloproteinases (MMPs), VBM undergoes many degradative and structural changes. This transition from mature VBM to provisional matrix that interacts with the proliferating and migrating endothelial cells is also associated with liberation of many fragments of matrix proteins; some of these have recently been implicated as endogenous regulators of angiogenesis, such as arresten, canstatin, tumstatin, metastatin, and endostatin.

in the proliferative rate of microvascular endothelial cells (Form et al. 1986). Conceivably, the NC1 domain exposed within the type IV collagen could induce an antiproliferative effect on microvascular endothelial cells (see below). In a different study, Furcht et al. demonstrated that triple helical fragments of type IV collagen were nearly as active as intact type IV collagen in stimulating endothelial cell adhesion and migration, whereas the carboxyl globular domain (NC1 domain) was less active in stimulating endothelial cell migration (Furcht 1986; Herbst et al. 1988; Tsilibary et al. 1988). Furcht et al. go on to suggest that such differences may play an important role in angiogenesis and large blood vessels (Herbst et al. 1988). Tsilibary and Furcht groups show that heparin binds type IV collagen NC1 domain ($\alpha 1$ chain). The heparin-binding site coincides with a peptide sequence that is important for the "assembly" of type IV collagen (Tsilibary et al. 1988). The Col $\alpha 1$ NC1 domain, as well as a 12-amino-acid peptide within the NC1 domain, disrupts the assembly of BM type IV collagen (Tsilibary et al. 1988). This study was an important demonstration that the NC1 domain was capable of disrupting the assembly of type IV collagen, further supporting the original studies first published by Kuhn and Timpl implicating the NC1 domain in the protomer assembly of type IV collagen (Timpl et al. 1981; Weber et al. 1984; Engel et al. 1985). Several years later, in 1994, Sarras's group published similar results using a hydra system, where they show that type IV collagen NC1 monomer and also 7S domain were effective in blocking hydra aggregate development and mesoglea (primitive BM material) formation (Zhang et al. 1994). Interestingly, although these studies were similar to the studies published by the Tsilibary and Furcht group a few years earlier, the hydra manuscript does not cite any of these earlier references (Zhang et al. 1994).

DISCOVERY OF TYPE IV COLLAGEN NC1 DOMAINS AS ENDOGENOUS INHIBITORS OF ANGIOGENESIS

In 1996, three independent lines of evidence led us to launch a new investigation into the role of VBM in the regulation of angiogenesis: (1) the observation that VBM proteins are associated with resting and growth-arrested capillary endothelial cells, while proliferating and migrating endothelial cells are under the influence of provisional matrix molecules such as vitronectin and fibronectin (Fig. 4 and Table 1); (2) studies by the Madri group and the Furcht group which suggest that subdomains of type IV collagen exhibit different properties as compared to the intact full-length molecule; and (3) the demonstration by the Furcht group that NC1 domain can disrupt the assembly of type IV collagen.

Our hypothesis centered on the idea that during the inductive phase of angiogenesis involving the degradation of VBM, proteolytic degradation fragments of BM are liberated with the potential for positive and negative regulation of capillary endothelial proliferation and migration. Such a potential role for matrix fragments could provide the regulatory balance for an organized formation of new capillaries. To test this hypothesis, we isolated BM preparations from human placenta, amnion, and testis and subjected them to various proteases identified as present in significantly higher amounts in the tumor microenvironment. Such enzymes include MMPs, elastase, cathes-

Table 1. Extracellular Matrix Transition during Angiogenesis

Vascular basement membrane	Provisional matrix
Type IV collagen	vitronectin
Laminin	fibronectin
Heparan sulfate proteoglycan (perlecan)	fibrin
Nidogen/entactin	osteopontin
Angrin	thrombin
Fibronectin	type 1 collagen
Fibulins	
Type XV collagen	
Type XVIII collagen	

pins, and serine proteinases (Ray and Stetler-Stevenson 1994; Gershtein et al. 2001; Krepela 2001; Tang et al. 2001; Berchem et al. 2002; Coussens et al. 2002; Egeblad and Werb 2002; Ghosh et al. 2002; Kaufmann et al. 2002; Levicar et al. 2002; Staack et al. 2002). In our initial studies, the BM preparation was subjected to the tumor microenvironment-associated MMPs such as MMP-2, MMP-7, MMP-3, MMP-9, and MMP-13, and the degradation fragments in the supernatant were separated using anion exchange chromatography, gel filtration, and HPLC (Fig. 5). Several fragments were recovered, using gel filtration, anion exchange chromatography, amino-terminal amino acid sequence analysis, and western blotting using antibodies to all six NC1 domains of type IV collagen (Fig. 5). These studies confirmed earlier reports which suggest that certain MMPs can cleave type IV collagen and liberate NC1 domain fragments (Egeblad and Werb 2002). All six α-chain-derived NC1 domains were identified as degradation products resulting from the digestion of BM preparation by the cocktail of MMPs (see above) (Fig. 5). Further purification of these NC1 domains from each other was accomplished with C-18 hy-

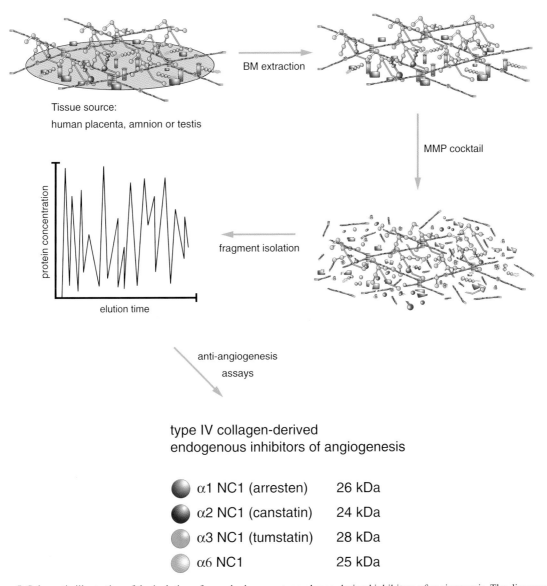

Figure 5. Schematic illustration of the isolation of vascular basement membrane-derived inhibitors of angiogenesis. The discovery of α6NC1 was published by the Brooks group (Peticlerc et al. 2000). Our research on the α6NC1 domain is unpublished. (Illustration by Lori Siniski.)

drophobic HPLC separation using acetonitrile gradient in conjunction with affinity purification using NC1 domain-specific antibodies to each of the α-chains (Colorado et al. 2000; Kamphaus et al. 2000; Maeshima et al. 2000b). All six NC1 domains were isolated to homogeneity and tested for their capacity to regulate endothelial tube formation. The α1, α2, and α3 NC1 domains exhibited significant inhibition of endothelial tube formation, whereas the α4 and α5 NC1 domains did not demonstrate this property. The α6 NC1 domain revealed a weaker, yet inhibitory, activity in this assay. Several other fragments derived from the triple helix and the 7S domain of type IV collagen did not demonstrate inhibition of tube formation. A few fragments revealed a weak stimulation of endothelial cell proliferation. The rest of this review highlights the antiangiogenic property of α1, α2, and α3 NC1 domains of type IV collagen (Fig. 5). There are at least 25 different collagens known today, which are made of one or more different α-chains, providing further complexity to the collagen family of proteins. The NC1 domain fragment of α1 chain of type XVIII collagen is known as endostatin. We named the α1(IV)NC1 as arresten, α2(IV)NC1 as canstatin, and α3(IV)NC1 as tumstatin.

ARRESTEN

In Vitro Antiangiogenic Activity

The native human arresten (26 kD) was isolated from human placenta and amnion and used in the endothelial tube formation assay, proliferation assay, and migration assays. These experiments demonstrate the antiangiogenic capacity of human arresten. Human arresten was produced both in *Escherichia coli* as a fusion protein with a carboxy-terminal six-histidine tag, and also as a secreted soluble protein in 293 embryonic kidney cells. Recombinant arresten selectively inhibits tube formation, proliferation, and migration of mouse and human endothelial cells, with no effect on tumor cells, primary epithelial cell lines, and primary fibroblasts (Colorado et al. 2000).

In Vivo Antiangiogenic Activity

Recombinant arresten inhibits the growth of small and large renal cell carcinoma xenograft (786-0) tumors, and PC-xenograft tumors on nude mice at doses as low as 4 mg/kg/day/i.p. This decrease in tumor size was consistent with a decrease in CD31-positive vasculature (Colorado et al. 2000).

Mechanism of Action

Arresten binds to heparin, and this property is speculated to be important for the antiangiogenic activity of arresten. Arresten also binds to heparan sulfate proteoglycan, similar to endostatin (Colorado et al. 2000). Endothelial cell-binding studies suggest that arresten binds to α1β1 integrin on proliferating endothelial cells (Colorado et al. 2000). These studies suggest that the negative regulation of angiogenesis by arresten is mediated by α1β1 integrin. Two independent studies support this contention. Senger et al. (1997) demonstrate that VEGF induces the expression of α1β1 and α2β1 integrin on endothelial cells. In the same study, this group demonstrates that α1 integrin-neutralizing antibodies can inhibit VEGF-driven angiogenesis, without detectable effect on preexisting vasculature (Senger et al. 1997). Pozzi et al. (2000, 2002) demonstrate that tumors grow slower in α1 integrin null mice, when compared to the wild-type mice. Collectively, all of these studies suggest that arresten can potentially act as an endogenous inhibitor of α1β1-induced angiogenesis.

CANSTATIN

In Vitro Antiangiogenic Activity

The native human canstatin (24 kD) was isolated from human placenta and amnion and used in the endothelial tube formation assay, proliferation assays, and migration assays. These experiments demonstrate the antiangiogenic capacity of human canstatin. Human canstatin was produced both in *E. coli* as a fusion protein with a carboxy-terminal six-histidine tag and also as a secreted soluble protein in 293 embryonic kidney cells. Recombinant canstatin selectively inhibits tube formation, proliferation, and migration of mouse and human endothelial cells, with no effect on tumor cells, primary epithelial cell lines (Kamphaus et al. 2000), and also primary fibroblast cultures. The antiangiogenic activity of canstatin was also independently validated in a subsequent publication by Petitclerc et al. (2000). In this publication, the group shows that recombinant canstatin, tumstatin, and the α6(IV)NC1 domain can inhibit angiogenesis in a CAM assay. These investigators also show that canstatin can bind to αVβ3 integrin on endothelial cells (Petitclerc et al. 2000).

In Vivo Antiangiogenic Activity

Recombinant canstatin inhibits the growth of small and large renal cell carcinoma (786-0) tumors by four- and threefold with respect to placebo-treated mice. Established human prostate (PC-3) tumors in severe combined immunodeficient (SCID) mice or athymic (nu/nu) mice exhibited fractional tumor volumes of about threefold less than placebo-treated mice when treated with 3 mg/kg canstatin. This decrease in tumor size was consistent with a decrease in CD31-positive vasculature (Kamphaus et al. 2000). Recent studies using syngenic mouse Lewis lung tumors (LLC) in C57BL/6 mice and orthotopic MDA-MB 435 breast tumors in nude mice also reveal significant tumor inhibition with recombinant canstatin.

Mechanism of Action

The mechanism by which canstatin inhibits proliferation and migration of endothelial cells was further investigated. Kamphaus et al. assessed the effect of canstatin on ERK activation induced by 20% fetal bovine serum and endothelial mitogens. These results show that canstatin does not primarily work by inhibiting VEGF or bFGF activation of ERK (Kamphaus et al. 2000). Alternatively, canstatin specifically induced apoptosis of endothelial cells with insignificant effect on non-endothelial cells. Apoptosis of endothelial cells by canstatin is mediated by induction of a steady decrease in FLIP protein (an anti-apoptotic protein associated with the FAS-mediated apoptosis pathway) levels in the presence of 20% fetal calf serum, bFGF, and VEGF. Interestingly, canstatin did not affect FLIP levels in the absence of growth factors. Canstatin's lack of effect on endothelial cells in the absence of growth factors may indicate that only proliferative endothelium is targeted (Kamphaus et al. 2000). Recent studies suggest that canstatin binds to αVβ3 integrin and also α3β1 integrin on proliferating endothelial cells. Whether this binding is important for the antiangiogenic activity of canstatin is still unknown.

TUMSTATIN

Anti-Melanoma Activity

The research group lead by Kefalides identified a synthetic peptide encompassing residues 183–205 of the α3 chain NC1 domain which specifically inhibits activation of polymorphonuclear leukocytes (Monboisse et al. 1994). Subsequently, they show that an -SNS- triplet unique to this peptide provides the specificity for this activity. Other publications have demonstrated that this peptide binds to a CD47/αVβ3 integrin complex, promotes adhesion and chemotaxis, and inhibits proliferation of various melanoma and other epithelial tumor cell lines (Han et al. 1997; Shahan et al. 1999a,b; 2000). This interaction is suggested to stimulate focal adhesion kinase (FAK), and phosphatidylinositol 3-kinase (PI3K) phosphorylation and to induce the inhibition of melanoma and fibrosarcoma cell proliferation (Pasco et al. 2000a,b). In addition, this peptide mediates a decrease in the expression of MT1-MMP and also a decrease in the activated membrane-bound MMP-2. Such a decrease in MMP expression is speculated to mediate the decrease in the invasion of tumor cells by this peptide. The 185–205 peptide or a recombinant fragment containing this peptide does not exhibit any effect on endothelial cells, but rather demonstrates an anti-melanoma cell activity (Maeshima et al. 2000b). The full-length tumstatin, which contains the 185–203 peptide sequence, does not have any effect on tumor cells (Maeshima et al. 2000b), therefore, these results suggest that the antitumor cell activity within the α3NC1 domain (tumstatin) is cryptic and requires further degradation to release this fragment or further structural alteration to expose this site (Maeshima et al. 2000b).

In Vitro Antiangiogenic Activity

The antiangiogenic activity of human tumstatin was evaluated using recombinant protein produced in *E coli*, *Pichia* yeast, and 293 embryonic kidney cells, and also native human protein isolated from human testis BM (Maeshima et al. 2000b). Human tumstatin inhibits proliferation of human, bovine, and mouse endothelial cells and induces apoptosis (Maeshima et al. 2000b). Tumstatin inhibits proliferation via G_1 arrest of VEGF- and FGF-stimulated proliferating endothelial cells (Maeshima et al. 2000a,b; 2001a,b). Tumstatin-mediated apoptosis of proliferating endothelial cells is associated with up-regulation of caspase 3 (Maeshima et al. 2000a,b; 2001a,b). Biochemical studies demonstrate that the antiangiogenic activity of tumstatin is not dependent on disulfide linkage between the 12 cysteines in this molecule. Reduction and alkylation experiments show that the loss of the disulfide bond did not alter the activity of tumstatin (Maeshima et al. 2000a,b; 2001a,b). Site-directed mutagenesis experiments demonstrate that the entire antiangiogenic activity of tumstatin is located between 54–132 amino acids in the fragment termed tum-5 (Maeshima et al. 2000a,b; 2001a,b). Further analysis led to the identification of a 25-amino-acid peptide between 74–98 amino acids which contained the entire antiangiogenic activity (Maeshima et al. 2000a,b; 2001a,b). At equal molar concentrations, the full-length tumstatin exhibits similar activity as the 25-amino-acid peptide termed as T7 (Maeshima et al. 2000a,b; 2001a,b). All of the antiangiogenic activity associated with full-length tumstatin is reproducible with the T7 peptide (Maeshima et al. 2000a,b; 2001a,b).

In Vivo Antiangiogenic Activity

The full-length tumstatin (28 kD), the tum-5 fragment, and the T7 peptide show significant inhibition angiogenesis in Matrigel plug assays (Maeshima et al. 2000a,b; 2001a,b). Tumstatin and its active fragments (tum-5, T7, and T8) show significant inhibition of human tumor xenografts in nude mice, syngenic LLC tumors in C57BL/6 mice, and orthotopic MDA-MB 435 breast tumor in nude mice at doses as low as 1 mg/kg/day/i.p or i.v. The tumor growth inhibition is associated with a decrease in the CD-31-positive vasculature.

Mechanism of Action

Tumstatin (α3[IV]NC1) binds to endothelial cells via αVβ3 integrin and α6β1 integrin (Maeshima et al. 2000a,b; 2001a,b). The αVβ3-binding activity for tumstatin was also demonstrated by Petitclerc et al. (2000). We show that the binding to αVβ3 is pivotal for the antiangiogenic activity associated with tumstatin, and that the activity is restricted to amino acids 54–132 (tum-5) within the 244-amino-acid tumstatin using deletion mutagenesis (Maeshima et al. 2000a; 2001a,b). Additionally, αVβ3 binding to tumstatin is mediated via a mechanism independent of the RGD-containing amino acid sequence and vitronectin/fibronectin binding of proliferating endothelial cells. The antiangiogenic activity of tumstatin can be further localized to a 25-amino-acid region within tum-5, and a synthetic peptide representing this region preserves "full" antiangiogenic activity and binds to αVβ3 integrin (Maeshima et al. 2000a, 2001a,b). Cell biological experiments demonstrate that the antiangiogenic activity of tumstatin is dependent on αVβ3 binding on proliferating endothelial cells (Maeshima et al. 2000a; 2001a,b). These experiments support the notion that through the action of endogenous inhibitors such as tumstatin, αVβ3 integrin could also function as a negative regulator of angiogenesis (Maeshima et al. 2001a; Stupack et al. 2001; Cheresh and Stupack 2002; Hynes 2002; Stupack and Cheresh 2002).

Tumstatin induces apoptosis of proliferating endothelial cells (Maeshima et al. 2002). Maeshima et al. demonstrate that tumstatin is an αVβ3 integrin-dependent inhibitor of cap-dependent translation mediated through negative regulation of mTOR signaling. In essence, tumstatin is an endothelial-cell-specific inhibitor of cap-dependent translation. The mTOR inhibitory property of tumstatin mimics that of rapamycin, except unlike rapamycin, which is pan-specific, tumstatin is specific for only endothelial cells (Maeshima et al. 2002). The action on mTOR is mediated by the FAK-PI3K-Akt pathway downstream from αVβ3 integrin (Maeshima et al. 2002).

CONCLUSION

BM are a complex network of large proteins that provide both structural and functional support to various cells, including capillary endothelial cells. While in the assembled form these proteins exhibit one set of properties, but it is clear now that degradation of these precursor proteins releases and exposes novel cryptic domains which possess, in some cases, an entirely opposite activity, or novel properties not present in the assembled form. Arresten, canstatin, and tumstatin are examples of such protein domains. These discoveries add to the growing list of inhibitors or activators derived from precursor proteins, such as endostatin and angiostatin. These discoveries constitute a new paradigm highlighting the disparate regulation of cell behavior by a signal protein, a concept that goes beyond just angiogenesis research.

ACKNOWLEDGMENTS

The work presented here was supported by National Institutes of Health grants 55001 and 51711, and by funds from the Program in Matrix Biology at Beth Israel Deaconess Medical Center. Arresten, canstatin, and tumstatin are licensed to Ilex Oncology for clinical development. The author and the institution both hold equity in Ilex Oncology. I wish to thank Lori Siniski for help in designing the illustrations and preparing this manuscript.

REFERENCES

Berchem G., Glondu M., Gleizes M., Brouillet J.P., Vignon F., Garcia M., and Liaudet-Coopman E. 2002. Cathepsin-D affects multiple tumor progression steps in vivo: Proliferation, angiogenesis and apoptosis. *Oncogene* **21:** 5951.
Bergers G., Brekken R., McMahon G., Vu T.H., Itoh T., Tamaki K., Tanzawa K., Thorpe P., Itohara S., Werb Z., and Hanahan D. 2000. Matrix metalloproteinase-9 triggers the angiogenic switch during carcinogenesis. *Nat. Cell Biol.* **2:** 737.
Borza D.B., Bondar O., Ninomiya Y., Sado Y., Naito I., Todd P., and Hudson B.G. 2001. The NC1 domain of collagen IV encodes a novel network composed of the alpha 1, alpha 2, alpha 5, and alpha 6 chains in smooth muscle basement membranes. *J. Biol. Chem.* **276:** 28532.
Bouck N. 1990. Tumor angiogenesis: The role of oncogenes and tumor suppressor genes. *Cancer Cells* **2:** 179.
———. 1992. Angiogenesis: A mechanism by which oncogenes and tumor suppressor genes regulate tumorigenesis. *Cancer Treat. Res.* **63:** 359.
Boutaud A., Borza D.B., Bondar O., Gunwar S., Netzer K.O., Singh N., Ninomiya Y., Sado Y., Noelken M.E., and Hudson B.G. 2000. Type IV collagen of the glomerular basement membrane: Evidence that the chain specificity of network assembly is encoded by the noncollagenous NC1 domains. *J. Biol. Chem.* **275:** 30716.
Cameron J.D., Skubitz A.P., and Furcht L.T. 1991. Type IV collagen and corneal epithelial adhesion and migration. Effects of type IV collagen fragments and synthetic peptides on rabbit corneal epithelial cell adhesion and migration in vitro. *Invest. Ophthalmol. Vis. Sci.* **32:** 2766.
Carmeliet P. and Jain R.K. 2000. Angiogenesis in cancer and other diseases. *Nature* **407:** 249.
Chelberg M.K., McCarthy J.B., Skubitz A.P., Furcht L.T., and Tsilibary E.C. 1990. Characterization of a synthetic peptide from type IV collagen that promotes melanoma cell adhesion, spreading, and motility. *J. Cell Biol.* **111:** 261.
Cheresh D.A. 1998. Death to a blood vessel, death to a tumor. *Nat. Med.* **4:** 395.
Cheresh D.A. and Stupack D.G. 2002. Integrin-mediated death: An explanation of the integrin-knockout phenotype? *Nat. Med.* **8:** 193.
Clark E.R. and Clark E.L. 1934. Microscopic observation on the growth of blood capillaries in the living organisms. *Am. J. Anat.* **64:** 251.
Colorado P.C., Torre A., Kamphaus G., Maeshima Y., Hopfer H., Takahashi K., Volk R., Zamborsky E.D., Herman S., Ericksen M.B., Dhanabal M., Simons M., Post M., Kufe D.W., Weichselbaum R.R., Sukhatme V.P., and Kalluri R. 2000. Anti-angiogenic cues from vascular basement membrane collagen. *Cancer Res.* **60:** 2520.
Coussens L.M., Fingleton B., and Matrisian L.M. 2002. Matrix metalloproteinase inhibitors and cancer: Trials and tribulations. *Science* **295:** 2387.
Crawford S.E., Stellmach V., Murphy-Ullrich J.E., Ribeiro S.M., Lawler J., Hynes R.O., Boivin G.P., and Bouck N. 1998. Thrombospondin-1 is a major activator of TGF-beta1 in vivo. *Cell* **93:** 1159.
Dameron K.M., Volpert O.V., Tainsky M.A., and Bouck N. 1994. Control of angiogenesis in fibroblasts by p53 regulation of thrombospondin-1. *Science* **265:** 1582.
Doll J.A., Reiher F.K., Crawford S.E., Pins M.R., Campbell S.C., and Bouck N.P. 2001. Thrombospondin-1, vascular endothelial growth factor and fibroblast growth factor-2 are key functional regulators of angiogenesis in the prostate. *Prostate* **49:** 293.
Egeblad M. and Werb Z. 2002. New functions for the matrix metalloproteinases in cancer progression. *Nat. Rev. Cancer* **2:** 161.
Eliceiri B.P. and Cheresh D.A. 2001. Adhesion events in angiogenesis. *Curr. Opin. Cell Biol.* **13:** 563.
Engel J., Furthmayr H., Odermatt E., von der Mark H., Aumailley M., Fleischmajer R., and Timpl R. 1985. Structure and macromolecular organization of type VI collagen. *Ann. N.Y. Acad. Sci.* **460:** 25.
Ferrara N. 2000. Vascular endothelial growth factor and the regulation of angiogenesis. *Recent Prog. Horm. Res.* **55:** 15.
Folkman J. 1972. Anti-angiogenesis: New concept for therapy of solid tumors. *Ann. Surg.* **175:** 409.
———. 1995. Angiogenesis in cancer, vascular, rheumatoid and other disease. *Nat. Med.* **1:** 27.
———. 2001. Angiogenesis-dependent diseases. *Semin. Oncol.* **28:** 536.

Folkman J. and Shing Y. 1992. Angiogenesis. *J. Biol. Chem.* **267:** 10931.

Folkman, J., Klagsbrun M., Sasse J., Wadzinski M., Ingber D., and Vlodavsky I. 1988. A heparin-binding angiogenic protein—basic fibroblast growth factor—is stored within basement membrane. *Am. J. Pathol.* **130:** 393.

Form D.M., Pratt B.M., and Madri J.A. 1986. Endothelial cell proliferation during angiogenesis. In vitro modulation by basement membrane components. *Lab. Invest.* **55:** 521.

Furcht L.T. 1984. Role of cell adhesion molecules in promoting migration of normal and malignant cells. *Prog. Clin. Biol. Res.* **149:** 15.

———. 1986. Critical factors controlling angiogenesis: Cell products, cell matrix, and growth factors. *Lab. Invest.* **55:** 505.

Gershtein E.S., Medvedeva S.V., Babkina I.V., Kushlinskii N.E., and Trapeznikov N.N. 2001. Tissue- and urokinase-type plasminogen activators and type 1 plasminogen activator inhibitor in melanomas and benign skin pigment neoplasms. *Bull. Exp. Biol. Med.* **132:** 670.

Ghosh S., Wu Y., and Stack M.S. 2002. Ovarian cancer-associated proteinases. *Cancer Treat. Res.* **107:** 331.

Grant D.S., Kibbey M.C., Kinsella J.L., Cid M.C., and Kleinman H.K. 1994. The role of basement membrane in angiogenesis and tumor growth. *Pathol. Res. Pract.* **190:** 854.

Gunwar S., Ballester F., Noelken M.E., Sado Y., Ninomiya Y., and Hudson B.G. 1998. Glomerular basement membrane. Identification of a novel disulfide-cross- linked network of alpha3, alpha4, and alpha5 chains of type IV collagen and its implications for the pathogenesis of Alport syndrome. *J. Biol. Chem.* **273:** 8767.

Haas T.L. and Madri J.A. 1999. Extracellular matrix-driven matrix metalloproteinase production in endothelial cells: Implications for angiogenesis. *Trends Cardiovasc. Med.* **9:** 70.

Han J., Ohno N., Pasco S., Monboisse J.C., Borel J.P., and Kefalides N.A. 1997. A cell binding domain from the alpha3 chain of type IV collagen inhibits proliferation of melanoma cells. *J. Biol. Chem.* **272:** 20395.

Hanahan D. and Folkman J. 1996. Patterns and emerging mechanisms of the angiogenic switch during tumorigenesis. *Cell* **86:** 353.

Hanahan D. and Weinberg R.A. 2000. The hallmarks of cancer. *Cell* **100:** 57.

Herbst T.J., McCarthy J.B., Tsilibary E.C., and Furcht L.T. 1988. Differential effects of laminin, intact type IV collagen, and specific domains of type IV collagen on endothelial cell adhesion and migration. *J. Cell Biol.* **106:** 1365.

Hudson B.G., Reeders S.T., and Tryggvason K. 1993. Type IV collagen: Structure, gene organization, and role in human diseases. Molecular basis of Goodpasture and Alport syndromes and diffuse leiomyomatosis. *J. Biol. Chem.* **268:** 26033.

Hynes R.O. 1996. Targeted mutations in cell adhesion genes: What have we learned from them? *Dev. Biol.* **180:** 402.

———. 2002. A reevaluation of integrins as regulators of angiogenesis. *Nat. Med.* **8:** 918.

Ingber D.E. and Folkman J. 1988. Inhibition of angiogenesis through modulation of collagen metabolism. *Lab. Invest.* **59:** 44.

———. 1989. Mechanochemical switching between growth and differentiation during fibroblast growth factor-stimulated angiogenesis in vitro: Role of extracellular matrix. *J. Cell Biol.* **109:** 317.

Jain R.K. and Carmeliet P.F. 2001. Vessels of death or life. *Sci. Am.* **285:** 38.

Kalluri R. and Cosgrove D. 2000. Assembly of type IV collagen. Insights from alpha3(IV) collagen-deficient mice. *J. Biol. Chem.* **275:** 12719.

Kalluri R., Shield C.F., Todd P., Hudson B.G., and Neilson E.G. 1997. Isoform switching of type IV collagen is developmentally arrested in X- linked Alport syndrome leading to increased susceptibility of renal basement membranes to endoproteolysis. *J. Clin. Invest.* **99:** 2470.

Kamphaus G.D., Colorado P.C., Panka D.J., Hopfer H., Ramchandran R., Torre A., Maeshima Y., Mier J.W., Sukhatme V.P., and Kalluri R. 2000. Canstatin, a novel matrix-derived inhibitor of angiogenesis and tumor growth. *J. Biol. Chem.* **275:** 1209.

Kaufmann R., Junker U., Junker K., Nuske K., Ranke C., Zieger M., and Scheele J. 2002. The serine proteinase thrombin promotes migration of human renal carcinoma cells by a PKA-dependent mechanism. *Cancer Lett.* **180:** 183.

Krepela E. 2001. Cysteine proteinases in tumor cell growth and apoptosis. *Neoplasma* **48:** 332.

Lawler J. 2000. The functions of thrombospondin-1 and-2. *Curr. Opin. Cell Biol.* **12:** 634.

Lawler J., Miao W.M., Duquette M., Bouck N., Bronson R.T., and Hynes R.O. 2001. Thrombospondin-1 gene expression affects survival and tumor spectrum of p53-deficient mice. *Am. J. Pathol.* **159:** 1949.

Levicar N., Kos J., Blejec A., Golouh R., Vrhovec I., Frkovic-Grazio S., and Lah T.T. 2002. Comparison of potential biological markers cathepsin B, cathepsin L, stefin A and stefin B with urokinase and plasminogen activator inhibitor-1 and clinicopathological data of breast carcinoma patients. *Cancer Detect. Prev.* **26:** 42.

Li S., Harrison D., Carbonetto S., Fassler R., Smyth N., Edgar D., and Yurchenco P.D. 2002. Matrix assembly, regulation, and survival functions of laminin and its receptors in embryonic stem cell differentiation. *J. Cell Biol.* **157:** 1279.

Madri J.A. 1997. Extracellular matrix modulation of vascular cell behaviour. *Transpl. Immunol.* **5:** 179.

Madri J.A. and Pratt B.M. 1986. Endothelial cell-matrix interactions: In vitro models of angiogenesis. *J. Histochem. Cytochem.* **34:** 85.

Maeshima, Y., Colorado P.C., and Kalluri R. 2000a. Two RGD-independent alpha v beta 3 integrin binding sites on tumstatin regulate distinct anti-tumor properties. *J. Biol. Chem.* **275:** 23745.

Maeshima Y., Manfredi M., Reimer C., Holthaus K.A., Hopfer H., Chandamuri B.R., Kharbanda S., and Kalluri R. 2001a. Identification of the anti-angiogenic site within vascular basement membrane derived tumstatin. *J. Biol. Chem.* **276:** 15240.

Maeshima Y., Sudhakar A., Lively J.C., Ueki K., Kharbanda S., Kahn C.R., Sonenberg N., Hynes R.O., and Kalluri R. 2002. Tumstatin, an endothelial cell-specific inhibitor of protein synthesis. *Science* **295:** 140.

Maeshima Y., Colorado P.C., Torre A., Holthaus K.A., Grunkemeyer J.A., Ericksen M.D., Hopfer H., Xiao Y., Stillman I.E., and Kalluri R. 2000b. Distinct anti-tumor properties of a type IV collagen domain derived from basement membrane. *J. Biol. Chem.* **275:** 21340.

Maeshima Y., Yerramalla U.L., Dhanabal M., Holthaus K.A., Barbashov S., Kharbanda S., Reimer C., Manfredi M., Dickerson W.M., and Kalluri R. 2001b. Extracellular matrix-derived peptide binds to alpha(v)beta(3) integrin and inhibits angiogenesis. *J. Biol. Chem.* **276:** 31959.

Maragoudakis M.E., Missirlis E., Karakiulakis G.D., Sarmonica M., Bastakis M., and Tsopanoglou N. 1993. Basement membrane biosynthesis as a target for developing inhibitors of angiogenesis with anti-tumor properties. *Kidney Int.* **43:** 147.

Miles A.J., Knutson J.R., Skubitz A.P., Furcht L.T., McCarthy J.B., and Fields G.B. 1995. A peptide model of basement membrane collagen alpha 1 (IV) 531-543 binds the alpha 3 beta 1 integrin. *J. Biol. Chem.* **270:** 29047.

Monboisse J.C., Garnotel R., Bellon G., Ohno N., Perreau C., Borel J.P., and Kefalides N.A. 1994. The alpha 3 chain of type IV collagen prevents activation of human polymorphonuclear leukocytes. *J. Biol. Chem.* **269:** 25475.

Oberbaumer I., Wiedemann H., Timpl R., and Kuhn K. 1982. Shape and assembly of type IV procollagen obtained from cell culture. *EMBO J.* **1:** 805.

O'Reilly M.S., Boehm T., Shing Y., Fukai N., Vasios G., Lane W.S., Flynn E., Birkhead J.R., Olsen B.R., and Folkman J. 1997. Endostatin: an endogenous inhibitor of angiogenesis and tumor growth. *Cell* **88:** 277.

O'Reilly M.S., Holmgren L., Shing Y., Chen C., Rosenthal R.A., Moses M., Lane W.S., Cao Y., Sage E.H., and Folkman J. 1994. Angiostatin: A novel angiogenesis inhibitor that mediates the suppression of metastases by a Lewis lung carcinoma (comments). *Cell* **79:** 315.

Pasco S., Monboisse J.C., and Kieffer N. 2000a. The alpha 3(IV)185-206 peptide from noncollagenous domain 1 of type IV collagen interacts with a novel binding site on the beta 3 subunit of integrin alpha vbeta 3 and stimulates focal adhesion kinase and phosphatidylinositol 3-kinase phosphorylation. *J. Biol. Chem.* **275:** 32999.

Pasco S., Han J., Gillery P., Bellon G., Maquart F.X., Borel J.P., Kefalides N.A., and Monboisse J.C. 2000b. A specific sequence of the noncollagenous domain of the alpha3(IV) chain of type IV collagen inhibits expression and activation of matrix metalloproteinases by tumor cells. *Cancer Res.* **60:** 467.

Paulsson M. 1992. Basement membrane proteins: Structure, assembly, and cellular interactions. *Crit. Rev. Biochem. Mol. Biol.* **27:** 93.

Petitclerc E., Boutaud A., Prestayko A., Xu J., Sado Y., Ninomiya Y., Sarras M.P., Jr., Hudson B.G., and Brooks P.C. 2000. New functions for non-collagenous domains of human collagen type IV. Novel integrin ligands inhibiting angiogenesis and tumor growth in vivo. *J. Biol. Chem.* **275:** 8051.

Pozzi A., LeVine W.F., and Gardner H.A. 2002. Low plasma levels of matrix metalloproteinase 9 permit increased tumor angiogenesis. *Oncogene* **21:** 272.

Pozzi A., Moberg P.E., Miles L.A., Wagner S., Soloway P., and Gardner H.A. 2000. Elevated matrix metalloprotease and angiostatin levels in integrin alpha 1 knockout mice cause reduced tumor vascularization. *Proc. Natl. Acad. Sci.* **97:** 2202.

Ray J.M. and Stetler-Stevenson W.G. 1994. The role of matrix metalloproteases and their inhibitors in tumour invasion, metastasis and angiogenesis. *Eur. Respir. J.* **7:** 2062.

Rodriguez-Manzaneque J.C., Lane T.F., Ortega M.A., Hynes R.O., Lawler J., and Iruela-Arispe M.L. 2001. Thrombospondin-1 suppresses spontaneous tumor growth and inhibits activation of matrix metalloproteinase-9 and mobilization of vascular endothelial growth factor. *Proc. Natl. Acad. Sci.* **98:** 12485.

Senger D.R., Claffey K.P., Benes J.E., Perruzzi C.A., Sergiou A.P., and Detmar M. 1997. Angiogenesis promoted by vascular endothelial growth factor: Regulation through alpha1beta1 and alpha2beta1 integrins. *Proc. Natl. Acad. Sci.* **94:** 13612.

Shahan T.A., Fawzi A., Bellon G., Monboisse J.C., and Kefalides N.A. 2000. Regulation of tumor cell chemotaxis by type IV collagen is mediated by a Ca(2+)-dependent mechanism requiring CD47 and the integrin alpha(V)beta(3). *J. Biol. Chem.* **275:** 4796.

Shahan T.A., Ohno N., Pasco S., Borel J.P., Monboisse J.C., and Kefalides N.A. 1999a. Inhibition of tumor cell proliferation by type IV collagen requires increased levels of cAMP. *Connect. Tissue Res.* **40:** 221.

Shahan T.A., Ziaie Z., Pasco S., Fawzi A., Bellon G., Monboisse J.C., and Kefalides N.A. 1999b. Identification of CD47/integrin-associated protein and alpha(v)beta3 as two receptors for the alpha3(IV) chain of type IV collagen on tumor cells. *Cancer Res.* **59:** 4584.

Staack A., Koenig F., Daniltchenko D., Hauptmann S., Loening S.A., Schnorr D., and Jung K. 2002. Cathepsins B, H, and L activities in urine of patients with transitional cell carcinoma of the bladder. *Urology* **59:** 308.

Stupack D.G. and Cheresh D.A. 2002. ECM remodeling regulates angiogenesis: Endothelial integrins look for new ligands. *Sci. STKE* **2002:** E7.

Stupack D.G., Puente X.S., Boutsaboualoy S., Storgard C.M., and Cheresh D.A. 2001. Apoptosis of adherent cells by recruitment of caspase-8 to unligated integrins. *J. Cell Biol.* **155:** 459.

Sundaramoorthy M., Meiyappan M., Todd P., and Hudson B.G. 2002. Crystal structure of NC1 domains. structural basis for type iv collagen assembly in basement membranes. *J. Biol. Chem.* **277:** 31142.

Tang W.H., Friess H., Kekis P.B., Martignoni M.E., Fukuda A., Roggo A., Zimmerman A., and Buchler M.W. 2001. Serine proteinase activation in esophageal cancer. *Anticancer Res.* **21:** 2249.

Than M.E., Henrich S., Huber R., Ries A., Mann K., Kuhn K., Timpl R., Bourenkov G.P., Bartunik H.D., and Bode W. 2002. The 1.9-Å crystal structure of the noncollagenous (NC1) domain of human placenta collagen IV shows stabilization via a novel type of covalent Met-Lys cross-link. *Proc. Natl. Acad. Sci.* **99:** 6607.

Timpl R. 1996. Macromolecular organization of basement membranes. *Curr. Opin. Cell Biol.* **8:** 618.

Timpl R., Wiedemann H., van Delden V., Furthmayr H., and Kuhn K. 1981. A network model for the organization of type IV collagen molecules in basement membranes. *Eur. J. Biochem.* **120:** 203.

Tsilibary E.C., Koliakos G.G., Charonis A.S., Vogel A.M., Reger L.A., and Furcht L.T. 1988. Heparin type IV collagen interactions: Equilibrium binding and inhibition of type IV collagen self-assembly. *J. Biol. Chem.* **263:** 19112.

Tsilibary, E.C., Reger, L.A., Vogel, A.M., Koliakos, G.G., Anderson, S.S., Charonis, A.S., Alegre, J.N., and Furcht L.T. 1990. Identification of a multifunctional, cell-binding peptide sequence from the a1(NC1) of type IV collagen. *J. Cell Biol.* **111:** 1583.

Weber S., Engel J., Wiedemann H., Glanville R.W., and Timpl R. 1984. Subunit structure and assembly of the globular domain of basement- membrane collagen type IV. *Eur. J. Biochem.* **139:** 401.

Yurchenco P.D. and O'Rear J.J. 1994. Basal lamina assembly. *Curr. Opin. Cell Biol.* **6:** 674.

Yurchenco P.D., Smirnov S., and Mathus T. 2002. Analysis of basement membrane self-assembly and cellular interactions with native and recombinant glycoproteins. *Methods Cell Biol.* **69:** 111.

Zeisberg M., Bonner G., Maeshima Y., Colorado P., Muller G.A., Strutz F., and Kalluri R. 2001. Renal fibrosis: Collagen composition and assembly regulates epithelial-mesenchymal transdifferentiation. *Am. J. Pathol.* **159:** 1313.

Zhang X., Hudson B.G., and Sarras M.P., Jr. 1994. Hydra cell aggregate development is blocked by selective fragments of fibronectin and type IV collagen. *Dev. Biol.* **164:** 10.

Complementary and Coordinated Roles of the VEGFs and Angiopoietins during Normal and Pathologic Vascular Formation

N.W. GALE, G. THURSTON, S. DAVIS, S.J. WIEGAND, J. HOLASH,
J.S. RUDGE, AND G.D. YANCOPOULOS

Regeneron Pharmaceuticals, Inc., Tarrytown, New York 10591

Mammalian organisms depend on their vasculature to deliver nutrients and oxygen to all of their tissues, to transport products (such as hormones and antibodies) from certain cells to distant parts of the body, and to carry away waste products. The development of a functioning vasculature, as well as its proper integration into the tissues it serves, depends on myriad interactions and communications between the many cell types involved. Although a large number of signals are involved in mediating these intercellular communications, a great deal of focus has been directed to growth factors that are members of either the vascular endothelial growth factor (VEGF) family or the angiopoietin family. Why the focus on these two families of growth factors? First of all, these two families of growth factors are unique in that they act via receptors that are largely restricted to the vasculature endothelium—this very restricted distribution of their receptors indicates that these two families of growth factors evolved to play very particular roles specifically involving the vasculature. Moreover, genetic approaches—involving gene knockouts and transgenic overexpression in mice—have spectacularly confirmed the very critical and very specific roles played by members of these two growth factor families during vascular development. Thus, the focus on the VEGFs and angiopoietins seems well-placed based on their action via vascular-specific receptors and the confirmation of their critical and specific vascular roles based on genetic studies in mice. Since the VEGFs have been extensively dealt with in a number of excellent reviews (Eriksson and Alitalo 1999; Ferrara 1999; Yancopoulos et al. 2000; Carmeliet et al. 2001), this review highlights work from our laboratory regarding the angiopoietins, although much of this work is presented in the context of the complementary and reciprocal actions of the angiopoietins as compared to the VEGFs.

MOLECULAR CLONING OF THE Tie RECEPTORS AND THEIR ANGIOPOIETIN LIGANDS

Although the VEGFs utilize a number of accessory receptor components such as the neuropilins, the primary actions of the VEGFs appear to be mediated via three closely related receptor tyrosine kinases, now referred to as VEGF receptor 1 (VEGFR-1, previously known as Flt-1), VEGF receptor 2 (VEGFR-2, previously known as KDR or Flk-1), and VEGF receptor 3 (VEGFR-3, previously known as Flt-3) (Eriksson and Alitalo 1999; Ferrara 1999; Yancopoulos et al. 2000; Carmeliet et al. 2001). The various VEGFs have an overlapping set of specificities for the three VEGF receptors (Fig. 1A). These VEGF receptors are conventional members of the receptor tyrosine kinase superfamily, which also includes as members the receptors for the epidermal growth factors (EGFs), the fibroblast growth factors (FGFs), the platelet-derived growth factors (PDGFs), and many other key growth factors. The critical distinguishing feature of the VEGF receptors is their cellular distribution. That is, the VEGF receptors are unlike the other aforementioned growth factor

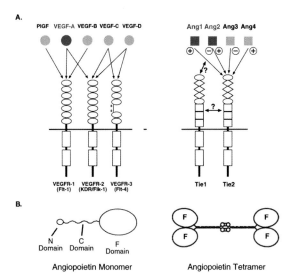

Figure 1. (*A*) Schematic summary of interactions of VEGFs with their receptors, and of angiopoietins with their Tie receptors. (*B*) On left, schematic view of angiopoietin monomer indicating amino-terminal domain (N domain), coil-coil domain (C domain), and fibrinogen-like domain (F domain). The F domain is the receptor-binding portion of this complex ligand, whereas the C domain serves to dimerize two F domains, and the N domain acts to further multimerize these dimers into tetramers, as shown on the right, or even higher-order structures (not shown).

receptors in that they are largely restricted to the vascular endothelium, both during development and in the adult. This very restricted distribution of their receptors indicates that the VEGFs evolved to play unique roles that very specifically involve the vasculature.

Because of an appreciation of how specifically important the VEGFs are for vascular development, and the realization that this specificity directly resulted from the restricted distribution of their receptors to the vasculature, we and other workers searched for additional families of receptor tyrosine kinases that might, like the VEGF receptors, be largely restricted to the vasculature. These efforts led to the discovery of a novel two-member family of receptor tyrosine kinases, now known as the Ties, that were indeed as restricted to the vasculature as were the VEGF receptors (Korhonen et al. 1992; Dumont et al. 1993; Iwama et al. 1993; Maisonpierre et al. 1993; Sato et al. 1993). At the time of their discovery, the Tie1 and Tie2 receptors were referred to as "orphans," since their binding partners had not yet been identified. However, it was presumed that the unidentified binding partners for these receptors would be growth factors specific for the vasculature.

To identify their presumably very interesting ligands, we converted the Tie receptor ectodomains into detection reagents that we used to identify sources of potential ligands, and then we used a novel expression cloning strategy (termed Secretory Trap expression cloning) in order to molecularly clone the first ligand for the Ties, which we termed angiopoietin-1 (Davis et al. 1996). We then cloned additional angiopoietins based on their homology with angiopoietin-1 (Maisonpierre et al. 1997; Valenzuela et al. 1999). All of the known angiopoietins bind primarily to the same Tie receptor, Tie2, and it is unclear whether there are independent ligands for the second Tie receptor, Tie1, or (as currently seems more likely) whether the known angiopoietins can in some way or under some circumstances also engage Tie1, perhaps as a second component in a heteromerized complex (Fig. 1A). Interestingly, whereas angiopoietin-1 is an obligate activator of its Tie2 receptor, angiopoietin-2 seems to be a more complex regulator of this receptor; that is, under some conditions it seems to activate Tie2, whereas under other conditions it may act as a blocker of this receptor (Maisonpierre et al. 1997).

UNIQUE MODULAR STRUCTURE OF THE ANGIOPOIETINS: BINDING AND MULTIMERIZATION MOTIFS

The angiopoietins have a modular structure unlike that of any previously characterized growth factor (Davis et al. 2003). This modular structure consists of a receptor-binding domain, a dimerization motif, and a superclustering motif that forms variable-sized multimers (Fig. 1B). Genetic engineering of precise multimers of the receptor-binding domain of angiopoietin-1, using surrogate multimerization motifs, reveals that tetramers (Fig. 1B) are minimally required for activating endothelial Tie2 receptors, whereas engineered dimers can antagonize endothelial Tie2 receptors (Davis et al. 2003). Surprisingly, angiopoietin-2 has a modular structure and multimerization state similar to that of angiopoietin-1, and its dual agonist/antagonist activities appear to be encoded in its receptor-binding domain (Davis et al. 2003).

INSIGHTS FROM KNOCKOUTS AND TRANSGENICS OF ANGIOPOIETIN-1: ROLES IN VESSEL MATURATION, STABILIZING THE VESSEL WALL, AND REGULATING VESSEL SIZE

The most important insights into the normal roles of angiopoietin-1 and its Tie2 receptor came from the analysis of mice engineered to lack these gene products (Dumont et al. 1994; Sato et al. 1995; Suri et al. 1996). Unlike mouse embryos lacking VEGF or VEGFR-2, embryos lacking angiopoietin-1 or Tie2 develop a rather normal primary vasculature. However, this vasculature fails to undergo normal further remodeling. The most prominent defects are in the heart, with problems in the associations between the endocardium and underlying myocardium, as well as in trabecula formation, and also in the remodeling of many vascular beds into large and small vessels. In these vascular beds, as in the heart, ultrastructural analysis suggests that endothelial cells fail to associate appropriately with underlying support cells, which are the cells that provide the angiopoietin-1 protein that acts on endothelial Tie2 receptors (Suri et al. 1996). This finding has led to the suggestion that angiopoietin-1 via Tie2 does not supply an instructive signal that actually directs specific vascular remodeling events, but rather plays more of a permissive role by optimizing the manner in which endothelial cells integrate with supporting cells, thus allowing the cells to receive other critical signals from their environment (Suri et al. 1996). Altogether, insights from the analysis of mice lacking angiopoietin-1 led to the suggestion that it played a key role, complementary to that of VEGF, to allow vessel maturation and vessel wall stabilization (Fig. 2).

Transgenic overexpression of angiopoietin-1 in the skin resulted in a dramatic hypervascularization phenotype (Suri et al. 1998; Thurston et al. 1999). Although there are modest increases in vessel number, the most dramatic increase is in vessel size. In contrast, VEGF in similar models primarily increases vessel number. These findings suggest that angiopoietin-1 may promote circumferential vessel growth as opposed to sproutive growth. Combining transgenic overexpression of angiopoietin-1 and VEGF leads to unprecedented increases in vascularity, which result from a combination of increases in both vessel size and number (Thurston et al. 1999). The vascular patterns induced by the combination are still obviously abnormal in morphology, suggesting that much must be learned about exploiting even this growth factor combination in therapeutic settings so as to grow normal vessels.

Recent insights based on delivery of angiopoietin-1 protein to newborn and adult mice confirm that it has very different vascular growth effects than does VEGF (G. Thurston et al., in prep.). Whereas VEGF primarily promotes angiogenic sprouting, angiopoietin-1 primarily

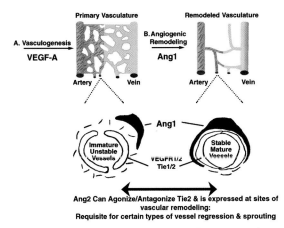

Figure 2. Schematic summary of the complementary and coordinated roles of the VEGFs and angiopoietins during vascular development. VEGF-A acts initially to form the primitive primary vasculature, angiopoietin-1 acts to mature and stabilize this primitive vasculature, in part by optimizing interactions between endothelial cells and their surrounding support cells. Angiopoietin-2 is expressed, and acts, at sites of subsequent remodeling of a previously stable vasculature and is requisite for certain types of vascular regressions and sprouting.

promotes circumferential vessel enlargement in the complete absence of angiogenic sprouting. Interestingly, these effects of angiopoietin-1 seem to be regulated in a stage- and segment-specific manner. That is, angiopoietin-1 can primarily promote circumferential enlargement of "plastic" and "immature" vessels, as opposed to mature vessels. Moreover, it promotes enlargement of venous vessels but not arterial vessels.

ANGIOPOIETIN-1 CAN ALSO OPPOSE VASCULAR PERMEABILITY ACTIONS OF VEGF

In addition to their disparate effects on vascular morphology, transgenic overexpression of angiopoietin-1 and VEGF also leads to dramatically distinct effects on vascular function and integrity. Transgenic VEGF produces immature, leaky, inflamed, and hemorrhagic vessels (Detmar et al. 1998; Larcher et al. 1998; Thurston et al. 1999). On the other hand, transgenic angiopoietin-1 results in vessels that are actually resistant to leak, whether the leak is induced by VEGF or by inflammatory agents (Thurston et al. 1999). This resistance appears related to the ability of angiopoietin-1 to maximize interactions between endothelial cells and their surrounding support cells and matrix, as the angiopoietin-1 vessels are resistant to treatments that normally create holes in the endothelial cell barrier (Thurston et al. 1999). These findings suggested that angiopoietin-1 might act counter to VEGF as an anti-permeability factor, and raised an assortment of therapeutic possibilities (Thurston et al. 1999). There are numerous disease processes—ranging from diabetic retinopathy to inflammation to brain edema following ischemic stroke—in which vessels become damaged and leaky, and an agent that was able to repair the damage and prevent the leak could have enormous therapeutic benefit. To be considered for such applications, angiopoietin-1 would have to exhibit its anti-leak actions not only when applied transgenically during vessel development, but also when acutely administered to the adult animal. Furthermore, angiopoietin-1 would not only have to be able to protect against leak acutely, but also to accomplish this without causing acute changes in vascular morphology. Supporting the clinical potential of angiopoietin-1 in settings of vascular leak, acute adenoviral administration of angiopoietin-1 to adult animals demonstrated that angiopoietin-1 could indeed acutely protect the adult vasculature from vascular leak, without inducing immediate changes in vascular morphology (Thurston et al. 2000).

ANGIOPOIETIN-2: COMPLEX REGULATOR OF Tie2 WITH DIVERSE ROLES IN POSTNATAL VASCULAR REMODELING

Angiopoietin-2 was cloned on the basis of its homology with angiopoietin-1, and it displayed similarly high affinity for the Tie2 receptor (Maisonpierre et al. 1997). Angiopoietin-2 differed from angiopoietin-1 in that, depending on the cell examined, angiopoietin-2 could either activate or antagonize the Tie2 receptor on cultured cells (Maisonpierre et al. 1997). Initial insights into the function of angiopoietin-2 came from the realization that it was dramatically induced in the endothelium of vessels undergoing active remodeling, such as sprouting or regressing vessels in the ovary (Maisonpierre et al. 1997; Goede et al. 1998), or in tumors (Stratmann et al. 1998; Holash et al. 1999a,b; Zagzag et al. 1999). These findings, together with the possibility that angiopoietin-2 could act as a dual Tie2 agonist/antagonist, led to the hypothesis (Fig. 2) that angiopoietin-2 might play a key (but complex) role at sites of angiogenic remodeling (Maisonpierre et al. 1997; Holash et al. 1999a,b; Zagzag et al. 1999). This possibility was recently confirmed by analysis of mice knocked out for angiopoietin-2 (Gale et al. 2002). This analysis revealed that angiopoietin-2, unlike VEGF and angiopoietin-1, is not requisite during embryonic vascular development, but instead is necessary during subsequent postnatal vascular remodeling. Specifically, postnatal vascular remodeling was explored in the neonatal eye, one of the most thoroughly studied sites of postnatal vascular remodeling (Alon et al. 1995; Stone et al. 1995, 1996; Benjamin et al. 1998; Ito and Yoshioka 1999; Hackett et al. 2000). During eye development, an initial vasculature, known as the hyaloid vasculature, is formed which nourishes the lens, while the retina is initially avascular. In the rodent, in the first few weeks after birth, the initial hyaloid vasculature regresses, while sprouts from the central artery of the eye produce a retinal vasculature; these simultaneous vascular regression and sprouting phenomena are thought to be coupled in some manner. In mice lacking angiopoietin-2, the initial eye vasculature at birth appears normal (i.e., the hyaloid vasculature), indicating that angiopoietin-2 is dispensable for formation of the initial vasculature (Gale et al. 2002). However, the mice fail to undergo normal remodeling of the eye vasculature in the first two postnatal weeks; that is, their hyaloid vasculature does not regress, nor does sprouting from the central artery occur to form the retinal vasculature (Gale et al. 2002). Moreover, in these mice, the angiopoietin-2 gene was replaced by a reporter gene (β-galactosidase), allowing high-resolution detection of sites of angiopoietin-2 expression (Gale et al. 2002). This reporter gene approach revealed that angiopoietin-2 was indeed highly expressed precisely at the sites of vascular remodeling—within the hyaloid vessels during regression, as well as at the site of sprouting from the central artery.

Altogether, these data confirm that angiopoietin-2 plays a critical and requisite role at sites of vascular remodeling, both during vessel regression and during vessel sprouting. Thus, our studies reveal that angiopoietin-2 is the first angiogenic factor genetically confirmed to be dispensable for embryonic angiogenesis but specifically required for normal postnatal vascular remodeling. It remains unclear as to whether angiopoietin-2 is acting as an agonist or antagonist of the Tie2 receptor during these processes. In addition, subsequent studies have revealed that certain types of pathological angiogenesis can proceed in the angiopoietin-2 knockouts, indicating that although it is required for some normal forms of vascular remodeling and angiogenesis, it may not be required for all.

ANGIOPOIETIN-2: UNEXPECTED REQUISITE ROLE DURING LYMPHATIC VESSEL DEVELOPMENT

As noted above, prior observations that angiopoietin-2 was expressed at sites of vascular remodeling presaged the finding in the mouse knockouts that this factor was required for certain types of postnatal vascular remodeling. There were no such prior clues to suggest that angiopoietin-2 might also play a crucial role during lymphatic vessel development. This was a completely unexpected finding that resulted from an obvious abnormality noted in the mice lacking angiopoietin-2 (Gale et al. 2002). These knockout mice, shortly after feeding, developed engorged abdomens filled with milky fluid. Subsequent analyses revealed that this corresponded to profound chylous ascites due to malfunctioning lymphatic lacteals in the intestines. Moreover, the mice developed widespread lymphatic dysfunction, characterized by widespread tissue edema and correlating with morphologically abnormal lymphatics at every site examined. Thus, genetic deletion of angiopoietin-2 results in profound and widespread defects in the patterning and function of the lymphatic vasculature (Gale et al. 2002).

To learn more about the mechanism of action of angiopoietin-2, we generated mice in which the angiopoietin-2 gene was replaced with cDNA encoding angiopoietin-1 (Gale et al. 2002). Surprisingly, angiopoietin-1 completely rescued the lymphatic defects in mice lacking angiopoietin-2, indicating that angiopoietin-2 acts as a Tie2 agonist in the lymphatic vasculature.

Because some of the vascular defects seen in mice lacking angiopoietin-1 have been attributed to disrupted interactions between the vascular endothelium and supporting smooth muscle cells, we examined the lymphatics in the angiopoietin-2 knockout mice for their smooth muscle investiture. Indeed, whereas the well-defined lymphatic channels seen in control pups were closely enveloped by smooth muscle cells, the disorganized lymphatic networks found in angiopoietin-2 knockout mice were often surrounded by poorly associated clusters of smooth muscle cells. These findings are consistent with a model in which local angiopoietin-2 expression, provided by the lymphatics themselves and/or by adjacent large blood vessels, acts on Tie2 receptors within the lymphatics in a manner that is necessary for proper lymphatic development. On the basis of the gene rescue studies in which the angiopoietin-2 coding region is replaced with that of angiopoietin-1, it appears that angiopoietin-2 is acting as an agonist of Tie2 during lymphatic development, perhaps by promoting interactions between lymphatic endothelium and smooth muscle, just as angiopoietin-1 seems to do for the blood vessel development.

Previous studies demonstrated that members of the VEGF family, most likely both VEGF-C and VEGF-D working via VEGFR-3, play critical roles in the development of the lymphatic vasculature (Kukk et al. 1996; Jeltsch et al. 1997; Karkkainen et al. 2000; Makinen et al. 2001; Veikkola et al. 2001). Just as earlier work revealed that members of the VEGF and angiopoietin families (i.e., VEGF-A and angiopoietin-1) are obligate partners during the development of blood vasculature (Dumont et al. 1994; Sato et al. 1995; Suri et al. 1996), our current findings suggest that VEGF-C and VEGF-D also obligately require angiopoietin-2 in order to form functional lymphatics (Fig. 3). As appears to be the case for the blood vasculature, the angiopoietins are not required for the initiation of lymphatic vascular development, unlike other key lymphatic regulators such as the transcription factor Prox-1 (Wigle and Oliver 1999) or the VEGF-C/VEGFR-3 pathway (Kukk et al. 1996; Jeltsch et al. 1997; Karkkainen et al. 2000; Makinen et al. 2001; Veikkola et al. 2001). Rather, angiopoietin-2 seems to play a key role in subsequent remodeling and maturation of the lymphatics, in a manner that is absolutely required for their normal function, as suggested for angiopoietin-1 within the blood vasculature (Fig. 3). Because angiopoietin-1 is able to rescue the lymphatic defect, angiopoietin-2 appears to be acting as an activating agonist in this situation. As has also been proposed for angiopoietin-1 and the blood vasculature, the role of angiopoietin-2 in

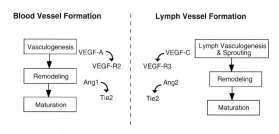

Figure 3. Schematic indicating that just as VEGF-A and angiopoietin-1 collaborate during formation of the blood vasculature, VEGF-C/D and angiopoietin-2 act together in analogous ways during formation of the lymphatic vasculature.

the lymphatic vasculature may well involve the optimizing of interactions between endothelial cells and surrounding smooth muscle cells. Thus, our studies demonstrate that members of the VEGF and angiopoietin families work together not only during development of the blood vasculature, but also during development of the lymphatic vasculature.

ROLES OF ANGIOPOIETINS AND VEGFs IN TUMOR ANGIOGENESIS; EFFICACY OF VEGF TRAP IN TUMOR MODELS

Previous studies had found that angiopoietin-2, as well as the Tie receptors, is dramatically induced within the endothelium during tumor angiogenesis (Stratmann et al. 1998; Holash et al. 1999a,b; Zagzag et al. 1999). This led to the proposal that the angiopoietin/Tie system might be as useful a target for tumor angiogenesis as the VEGF system. More recent studies have not validated the utility of this system for tumor angiogenesis, and we have not been able to demonstrate dramatic effects of promoting or blocking Tie receptor function during tumor angiogenesis. Similarly, tumors can grow in mice lacking angiopoietin-2, although there may be subtle alterations in the resulting tumor vessels. All this has left us with the conclusion that the VEGF pathway remains the best validated target pathway for approaches aimed at controlling tumor growth by blocking required tumor angiogenesis. To take advantage of this realization, we developed a very potent VEGF blocker termed the VEGF trap (Holash et al. 2002). Recent application of this agent indicates that it may be very useful in the treatment of angiogenesis-dependent tumors (Holash et al. 2002; Kim et al. 2002), as well as in other settings in which VEGF and associated vascular leak or angiogenesis may be causing clinical problems, such as in diabetic retinopathy, age-related macular edema, endometriosis, and tumor-associated ascites and effusions.

CONCLUSIONS

Our findings reveal that the angiopoietins are important modulators of blood and lymphatic vessel formation and function, working in a collaborative and cooperative manner with members of the VEGF family. For the blood vasculature, angiopoietin-1 seems to work subsequent to VEGF-A so as to promote vessel maturation and vessel wall function. Interestingly, angiopoietin-1 seems to act reciprocally as compared to VEGF-A with regard to vascular leak—with VEGF-A promoting leak and permeability, and angiopoietin-1 seemingly opposing these actions. Similarly, in terms of the regulation of vascular growth, VEGF-A and angiopoietin-1 also appear to act quite differently, with VEGF-A promoting angiogenic sprouting and angiopoietin-1 instead regulating circumferential vessel size.

In terms of the blood vasculature, angiopoietin-2 is unlike VEGF-A and angiopoietin-1 in that it is dispensable for normal embryonic vascular development, since mice lacking angiopoietin-2 are born and appear quite normal initially. However, as predicted by observations that angiopoietin-2 is highly induced at sites of vascular remodeling, this factor is indeed required for certain types of postnatal vascular remodelings, such as the normal regressions and sprouting seen in the eye after birth. Surprisingly, however, angiopoietin-2 may not be absolutely required for all types of postnatal angiogenesis, consistent with observations that the angiopoietin/Tie system may not be as critical for tumor angiogenesis as the VEGF system.

Just as VEGF-A and angiopoietin-1 seem to collaborate during initial formation of the blood vasculature, it appears as if VEGF-C/D and angiopoietin-2 collaborate in similar manner during formation of the lymphatic vasculature. Many of the features of this collaboration in the lymphatic vasculature are reminiscent of those that characterize the collaboration in the blood vasculature. That is, the VEGF family members seemingly play key initiating roles in both blood and lymphatic vessel formation, whereas the angiopoietin family members seem to play subsequent maturation roles, potentially involving the optimization of endothelial–smooth muscle cell interactions. It is perhaps not surprising that nature seems to have duplicated the players and roles it uses to produce these two different types of vasculatures.

On the basis of our work with the VEGFs and the angiopoietins, it becomes clear that there are a large number of critical growth factors involved in the physiological regulation of blood vessel formation, and the actions of these molecular players must be very carefully orchestrated in terms of time, space, and dose so as to form a functioning vascular network. The complexity of the process makes ongoing therapeutic efforts aimed at growing new vascular networks to treat ischemic disease appear quite challenging. For example, delivery of just VEGF, or even delivery of an imbalance of VEGF compared to angiopoietin-1, has the potential to cause more harm (by forming malfunctioning vessels prone to leak and hemorrhage) than good. The same sorts of complexities must now be considered in attempts to therapeutically promote lymphatic vessel growth so as to treat certain edematous conditions. It appears as if there is more to learn about how various combinations of factors interact during vessel formation in order to exploit these factors for the therapeutic growth of vessels, whether it be blood vessels in settings of ischemia, or lymphatic vessels in settings of lymphedema.

Although the complexities of vascular formation create major challenges for those trying to therapeutically grow vessels, these same complexities may work in favor of therapeutic approaches aimed at blocking vessel growth (so as to benefit diseases ranging from cancer to endometriosis, or neovascularization conditions of the eye such as occur in settings of diabetic retinopathy or age-related macular degeneration). That is, blockade of many different molecular players may all result in the blunting of vessel formation. There is no doubt that VEGF is the best-validated target for antiangiogenesis therapies, based on overwhelming genetic, mechanistic, and animal efficacy data.

Recent efforts also suggest heretofore unimagined applications for vascular growth factors. For example, the possibility that angiopoietin-1 may help prevent or repair damaged and leaky vessels offers therapeutic hope for an assortment of unmet clinical needs, such as in the vascular leak which creates major problems in diabetic retinopathy, acute macular degeneration, ischemia/reperfusion injury as occurs following strokes and ARDS, or in inflammatory settings. The continued discovery and characterization of the molecular players regulating vessel formation are sure to lead to additional unexpected therapeutic opportunities, as well as to the refinement of current therapeutic approaches aimed at growing or blocking vessel formation.

REFERENCES

Alon T., Hemo I., Itin A., Pe'er J., Stone J., and Keshet E. 1995. Vascular endothelial growth factor acts as a survival factor for newly formed retinal vessels and has implications for retinopathy of prematurity. *Nat. Med.* **1:** 1024.

Benjamin L.E., Hemo I., and Keshet E. 1998. A plasticity window for blood vessel remodelling is defined by pericyte coverage of the preformed endothelial network and is regulated by PDGF-B and VEGF. *Development* **125:** 1591.

Carmeliet P., Moons L., Luttun A., Vincenti V., Compernolle V., De Mol M., Wu Y., Bono F., Devy L., Beck H., Scholz D., Acker T., DiPalma T., Dewerchin M., Noel A., Stalmans I., Barra A., Blacher S., Vandendriessche T., Ponten A., Eriksson U., Plate K.H., Foidart J.M., Schaper W., Charnock-Jones D.S., Hicklin D.J., Herbert J.M., Collen D., and Persico M.G. 2001. Synergism between vascular endothelial growth factor and placental growth factor contributes to angiogenesis and plasma extravasation in pathological conditions. *Nat. Med.* **7:** 575.

Davis S., Aldrich T.H., Jones P.F., Acheson A., Compton D., Vivek J., Ryan T., Bruno J., Radjiewski C., Maisonpierre P.C., and Yancopoulos G.D. 1996. Isolation of angiopoietin-1, a ligand for the TIE2 receptor, by secretion-trap expression cloning. *Cell* **87:** 1161.

Davis S., Papadopoulos N., Aldrich T.H., Maisonpierre P.C., Huang T., Kovac L., Xu A., Leidich R., Radziejewska E., Goldberg J., Jain V., Bailey K., Karow M., Fandl J., Sameulsson S.J., Ioffe E., Rudge J.S., Daly T.J., Radziejewski C., and Yancopoulos G.D. 2003. Angiopoietins as modular growth factors with distinct functionally essential domains for receptor binding, dimerization, and superclustering. *Nat. Struct. Biol.* **10:** 38.

Detmar M., Brown L.F., Schon M.P., Elicker B.M., Velasco P., Richard L., Fukumura D., Monsky W., Claffey K.P., and Jain R.K. 1998. Increased microvascular density and enhanced leukocyte rolling and adhesion in the skin of VEGF transgenic mice. *J. Invest. Dermatol.* **111:** 1.

Dumont D.J., Gradwohl G.J., Fong G.-H., Auerbach R., and Breitman M.L. 1993. The endothelial-specific receptor tyrosine kinase, *tek*, is a member of a new subfamily of receptors. *Oncogene* **8:** 1293.

Dumont D.J., Gradwohl G., Fong G.-H., Puri M.C., Gerstenstein M., Auerbach A., and Breitman M.L. 1994. Dominant-negative and targeted null mutations in the endothelial receptor tyrosine kinase, tek, reveal a critical role in vasculogenesis of the embryo. *Genes Dev.* **8:** 1897.

Eriksson U. and Alitalo K. 1999. Structure, expression and receptor-binding properties of novel vascular endothelial growth factors. *Curr. Top. Microbiol. Immunol.* **237:** 41.

Ferrara N. 1999. Vascular endothelial growth factor: Molecular and biological aspects. *Curr. Top. Microbiol. Immunol.* **237:** 1.

Gale N.W., Thurston G., Hackett S.F., Renard R., Wang Q., McClain J., Martin C., Witte C., Witte M.H., Jackson D., Suri C., Campochiaro P.A., Wiegand S.J., and Yancopoulos G.D. 2002. Angiopoietin-2 is required for postnatal angiogenesis and lymphatic patterning, and only the latter role is rescued by angiopoietin-1. *Dev. Cell* **3:** 411.

Goede V., Schmidt T., Kimmina S., Kozian D., and Augustin H.G. 1998. Analysis of blood vessel maturation processes during cyclic ovarian angiogenesis. *Lab. Invest.* **78:** 1385.

Hackett S.F., Ozaki H., Strauss R.W., Wahlin K., Suri C., Maisonpierre P., Yancopoulos G., and Campochiaro P.A. 2000. Angiopoietin 2 expression in the retina: Upregulation during physiologic and pathologic neovascularization. *J. Cell. Physiol.* **184:** 275.

Holash J., Wiegand S.J., and Yancopoulos G.D. 1999a. New model of tumor angiogenesis: Dynamic balance between vessel regression and growth mediated by angiopoietins and VEGF. *Oncogene* **18:** 5356.

Holash J., Maisonpierre P.C., Compton D., Boland P., Alexander C.R., Zagzag D., Yancopoulos G.D., and Wiegand S.J. 1999b. Vessel cooption, regression, and growth in tumors mediated by angiopoietins and VEGF. *Science* **284:** 1994.

Holash J., Davis S., Papadopoulos N., Croll S.D., Ho L., Russell M., Boland P., Leidich R., Hylton D., Burova E., Ioffe E., Huang T., Radziejewski C., Bailey K., Fandl J.P., Daly T., Wiegand S.J., Yancopoulos G.D., and Rudge J.S. 2002. VEGF-Trap: A VEGF blocker with potent antitumor effects. *Proc. Natl. Acad. Sci.* **99:** 11393.

Ito M. and Yoshioka M. 1999. Regression of the hyaloid vessels and pupillary membrane of the mouse. *Anat. Embryol.* **200:** 403.

Iwama A., Hamaguchi I., Hashiyama M., Murayama Y., Yasunaga K., and Suda T. 1993. Molecular cloning and characterization of mouse *Tie* and *Tek* receptor tyrosine kinase genes and their expression in hematopoietic stem cells. *Biochem. Biophys. Res. Commun.* **195:** 301.

Jeltsch M., Kaipainen A., Joukov V., Meng X., Lakso M., Rauvala H., Swartz M., Fukumura D., Jain R.K., and Alitalo K. 1997. Hyperplasia of lymphatic vessels in VEGF-C transgenic mice. *Science* **276:** 1423.

Karkkainen M.J., Ferrell R.E., Lawrence E.C., Kimak M.A., Levinson K.L., McTigue M.A., Alitalo K., and Finegold D.N. 2000. Missense mutations interfere with VEGFR-3 signalling in primary lymphoedema. *Nat. Genet.* **25:** 153.

Kim E.S., Serur A., Huang J., Manley C.A., McCrudden K.W., Frischer J.S., Soffer S.Z., Ring L., New T., Zabisky S., Rudge J.S., Holash J., Yancopoulos G.D., and Kandel J.J., and Yamashiro D.J. 2002. Potent VEGF blockade causes regression of coopted vessels in a model of neuroblastoma. *Proc. Natl. Acad. Sci.* **99:** 11399.

Korhonen J., Partanen J., Armstrong E., Vaahtokari A., Elenius K., Jalkanen M., and Alitalo K. 1992. Enhanced expression of the *tie* receptor tyrosine kinase in endothelial cells during neovascularization. *Blood* **80:** 2548.

Kukk E., Lymboussaki A., Taira S., Kaipainen A., Jeltsch M., Joukov V., and Alitalo K. 1996. VEGF-C receptor binding and pattern of expression with VEGFR-3 suggests a role in lymphatic vascular development. *Development* **122:** 3829.

Larcher F., Murillas R., Bolontrade M., Conti C.J., and Jorcano J.L. 1998. VEGF/VPF overexpression in skin of transgenic mice induces angiogenesis, vascular hyperpermeability and accelerated tumor development. *Oncogene* **17:** 303.

Maisonpierre P.C., Goldfarb M., Yancopoulos G.D., and Gao G. 1993. Distinct rat genes with related profiles of expression define a TIE receptor tyrosine kinase family. *Oncogene* **8:** 1631.

Maisonpierre P.C., Suri C., Jones P.F., Bartunkova S., Wiegand S.J., Radziejewski C., Compton D., McClain J., Aldrich T.H., Papadopoulos N., Daly T.J., Davis S., Sato T.N., and Yancopoulos G.D. 1997. Angiopoietin-2, a natural antagonist for Tie2 that disrupts in vivo angiogenesis (comments). *Science* **277:** 55.

Makinen T., Jussila L., Veikkola T., Karpanen T., Kettunen M.I., Pulkkanen K.J., Kauppinen R., Jackson D.G., Kubo H., Nishikawa S., Yla-Herttuala S., and Alitalo K. 2001. Inhibition of lymphangiogenesis with resulting lymphedema in transgenic mice expressing soluble VEGF receptor-3. *Nat. Med.* **7:** 199.

Sato T.N., Qin Y., Kozak C.A., and Audus K.L. 1993. tie-1 and *tie-2* define another class of putative receptor tyrosine kinase genes expressed in early embryonic vascular system. *Proc. Natl. Acad. Sci.* **90:** 9355.

Sato T.N., Tozawa Y., Deutsch U., Wolburg-Buchholz K., Fujiwara Y., Gendron-Maguire M., Gridley T., Wolburg H., Risau W., and Qin Y. 1995. Distinct roles of the receptor tyrosine kinases Tie-1 and Tie-2 in blood vessel formation. *Nature* **376:** 70.

Stone J., Chan-Ling T., Pe'er J., Itin A., Gnessin H., and Keshet E. 1996. Roles of vascular endothelial growth factor and astrocyte degeneration in the genesis of retinopathy of prematurity. *Invest. Ophthalmol. Vis. Sci.* **37:** 290.

Stone J., Itin A., Alon T., Pe'er J., Gnessin H., Chan Ling T., and Keshet E. 1995. Development of retinal vasculature is mediated by hypoxia-induced vascular endothelial growth factor (VEGF) expression by neuroglia. *J. Neurosci.* **15:** 4738.

Stratmann A., Risau W., and Plate K.H. 1998. Cell type-specific expression of angiopoietin-1 and angiopoietin-2 suggests a role in glioblastoma angiogenesis. *Am. J. Pathol.* **153:** 1459.

Suri C., McLain J., Thurston G., McDonald D.M., Oldmixon E.H., Sato T.N., and Yancopoulos G.D. 1998. Angiopoietin-1 promotes increased vascularization in vivo. *Science* **282:** 468.

Suri C., Jones P.F., Patan S., Bartunkova S., Maisonpierre P.C., Davis S., Sato T.N., and Yancopoulos G.D. 1996. Requisite role of angiopoietin-1, a ligand for the Tie2 receptor, during embryonic angiogenesis. *Cell* **87:** 1171.

Thurston G., Suri C., Smith K., McClain J., Sato T.N., Yancopoulos G.D., and McDonald D.M. 1999. Leakage-resistant blood vessels in mice transgenically overexpressing angiopoietin-1. *Science* **286:** 2511.

Thurston G., Rudge J.S., Ioffe E., Zhou H., Ross L., Croll S.D., Glazer N., Holash J., McDonald D.M., and Yancopoulos G.D. 2000. Angiopoietin-1 protects the adult vasculature against plasma leakage. *Nat. Med.* **6:** 460.

Valenzuela D., Griffiths J., Rojas J., Aldrich T.H., Jones P.F., Zhou H., McClain J., Copeland N.G., Gilbert D.J., Jenkins N.A., Huanh T., Papadopoulos N., Maisonpierre P.C., Davis S., and Yancopoulos G.D. 1999. Angiopoietins 3 and 4: Diverging gene counterparts in mouse and man. *Proc. Natl. Acad. Sci.* **96:** 1904.

Veikkola T., Jussila L., Makinen T., Karpanen T., Jeltsch M., Petrova T.V., Rudge H., Thurston G., McDonald D.M., Achen M.G., Stacker S.A., and Alitalo K. 2001. Signalling via vascular endothelial growth factor receptor-3 is sufficient for lymphangiogenesis in transgenic mice. *EMBO J.* **20:** 1223.

Wigle J.T. and Oliver G. 1999. Prox1 function is required for the development of the murine lymphatic system. *Cell* **98:** 769.

Yancopoulos G.D., Davis S., Gale N.W., Rudge J.S., Wiegand S.W., and Holash J. 2000. Vascular-specific growth factors and blood vessel formation. *Nature* **407:** 242.

Zagzag D., Hooper A., Friedlander D.R., Chan W., Holash J., Wiegand S.J., Yancopoulos G.D., and Grumet M. 1999. In situ expression of angiopoietins in astrocytomas identifies angiopoietin-2 as an early marker of tumor angiogenesis. *Exp. Neurol.* **159:** 391.

Involvement of G Proteins in Vascular Permeability Factor/ Vascular Endothelial Growth Factor Signaling

D. MUKHOPADHYAY AND H. ZENG

Department of Pathology, Beth Israel Deaconess Medical Center and Harvard Medical School, Boston, Massachusetts 02215

In order to grow beyond minimal size, tumors must generate a new vascular supply for purposes of gas exchange, cell nutrition, and waste disposal (Folkman 1971, 1996; Folkman and Klagsbrun 1987; Folkman et al. 1989). They do so by secreting angiogenic cytokines that induce the formation of new blood vessels (Folkman and Klagsbrun 1987; Folkman et al. 1989; Risau 1997; Dvorak et al. 1999). Tumor-secreted angiogenic cytokines include fibroblast growth factor, platelet-derived growth factor B, and vascular permeability factor/vascular endothelial growth factor (VPF/VEGF) (Senger et al. 1986; Vlodavsky et al. 1991; Benezra et al. 1993; Risau 1997). VPF/VEGF is likely the most important of these because it is expressed abundantly by a wide variety of human and animal tumors and because of its potency, selectivity for endothelial cells, and ability to regulate most and perhaps all of the steps in the angiogenic cascade (Dvorak et al. 1979,1984,1999; Dvorak 1990; Risau 1997; Ferrara 1999). Moreover, a number of other angiogenic cytokines act, at least in part, by up-regulating VPF/VEGF expression (Seghezzi et al. 1998; Dvorak et al. 1999). VPF/VEGF extensively reprograms endothelial cell expression of proteases, integrins, and glucose transporters; stimulates endothelial cell migration and division; protects endothelial cells from apoptosis and senescence; and induces angiogenesis in both in vitro and in vivo models (for review, see Leung et al. 1989; Risau 1997; Dvorak et al. 1999; Ferrara 1999). In addition, VPF/VEGF is the only angiogenic cytokine identified thus far that renders microvessels hyperpermeable to circulating macromolecules, a characteristic feature of angiogenic blood vessels (Dvorak et al. 1979, 1984; Senger et al. 1983, 1986; Dvorak 1990).

Most of VPF/VEGF's biological activities are thought to be mediated by its interaction with two high-affinity receptor tyrosine kinases, Flt-1 (VEGFR-1) and KDR (VEGFR-2, flk-1 in mice) (Terman et al. 1992; Millauer et al. 1993; Quinn et al. 1993; Fong et al. 1995; Shalaby et al. 1997). A third receptor, neuropilin, has been recognized, but little is as yet known about its capacity to initiate endothelial cell signaling (Soker et al. 1998; Gagnon et al. 2000). Both Flt-1 and KDR are selectively expressed on vascular endothelium but bind VPF/VEGF with different affinities; thus, Flt-1 binds VPF/VEGF with a K_d of ~10 pM whereas the K_d for KDR binding is 400–900 pM (Waltenberger et al. 1994; Joukov et al. 1997). Both receptors possess tyrosine-kinase domains, potential ATP-binding sites, and long kinase-insert regions that contain phosphorylation sites with binding capacity for different signaling molecules. Flt-1 and KDR also have different ligand specificities. Thus, Flt-1 interacts with VPF/VEGF (also known as VEGF-A) and with two other members of the VPF/VEGF family, PlGF and VEGF-B. KDR, on the other hand, interacts with VEGF-C and VEGF-D, in addition to VPF/VEGF (Petrova et al. 1999). Both Flt-1 and KDR are essential for normal vascular development (Fong et al. 1995; Shalaby et al. 1997).

At present, the signaling cascades following VPF/VEGF interaction with cultured endothelial cells (EC) are only partially understood but are known to involve a series of protein phosphorylations, beginning with receptor phosphorylation, subsequently with tyrosine phosphorylation of phospholipase C-γ (PLC-γ) and phosphatidylinositol 3-kinase (PI3K) (for review, see English et al. 1999; Petrova et al. 1999). Like other endothelial cell agonists such as thrombin and histamine, VPF/VEGF activates protein kinase C (PKC), increases $[Ca^{++}]_i$, and stimulates inositol-1,4,5-triphosphate (IP3) accumulation (Brock et al. 1991).

DISTINCT SIGNALING PATHWAYS OF KDR AND Flt-1 IN VASCULAR ENDOTHELIUM

To dissect the respective signaling pathways and biological functions mediated by these receptors in primary endothelial cells with these two receptors intact, we developed a chimeric-receptor system in which the amino terminus of the epidermal growth factor receptor (EGFR) was fused to the transmembrane domain and intracellular domain of KDR and Flt-1 (Fig. 1). We observed that KDR, but not Flt-1, was responsible for VPF/VEGF-induced human umbilical vein endothelial cell (HUVEC) proliferation and migration. Moreover, Flt-1 showed an inhibitory effect on KDR-mediated proliferation, but not migration (Zeng et al. 2001a,b). We also demonstrated that the inhibitory function of Flt-1 was mediated through the PI3K-dependent pathway, since inhibitors of PI3K as well as dominant negative mutant of p85 (PI3K subunit) reversed the inhibition, whereas a constitutive active mutant of p110 introduced the inhibition to HUVEC-EGDR. We also observed that, in VPF/VEGF-stimulated HUVEC, the Flt-1/EGLT-mediated down-modulation of

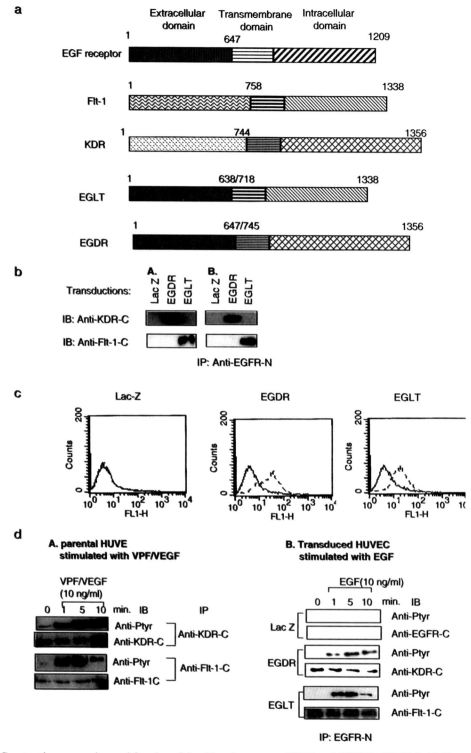

Figure 1. Construction, expression, and function of the chimeric receptors EGDR and EGLT in HUVEC. (*a*) The extracellular domain of the epidermal growth factor receptor (EGFR) was fused to the transmembrane and intracellular domains of KDR or Flt-1 to create the chimeric fusion receptors EGDR and EGLT, respectively. (*b*) Extracts of HUVEC that had been transduced with LacZ, EGDR, or EGLT were immunoblotted with antibodies against the carboxyl terminus of KDR (Anti-KDR-C) or Flt-1 (Anti-Flt-1-C), with (panel *B*) or without (panel *A*) prior immunoprecipitation (IP) of the extracts with an antibody against the EGFR amino terminus (EGFR-N). (*c*) FACS analysis of HUVEC transduced with LacZ, EGDR, or EGLT with fluorescent antibodies against EGFR-N (*dashed line*) or mouse IgG (*solid line*). (*d*) Parental HUVEC (2×10^6 and 6×10^6 cells for KDR and Flt-1, respectively) were stimulated with 10 ng/ml VPF/VEGF (panel *A*), or HUVEC transduced with LacZ, EGDR, or EGLT stimulated with 10 ng/ml EGF (panel *B*). Cell extracts were immunoprecipitated (IP) with antibodies against KDR-C, Flt-1-C, or EGFR-N termini, as indicated, and were immunoblotted (IB) with antibody to phosphotyrosine. Blots were then stripped and reprobed with antibodies against KDR-C, Flt-1-C, or EGFR-C, as indicated. Due to the low expression level of Flt-1, 1.5 mg, instead of 0.5 mg, of protein was used to detect Flt-1 phosphorylation.

KDR/EGDR signaling was at the same time as or before intracellular Ca^{++} mobilization, but after KDR/EGDR phosphorylation. By mutational analysis, we further identified that the Tyr-794 residue of Flt-1 was essential for its antiproliferative effect. Taken together, these studies contribute significantly to our understanding of the signaling pathways and biological functions triggered by KDR and Flt-1 and describe a unique mechanism in which PI3K acts as a mediator of antiproliferation in primary vascular endothelium (Zeng et al. 2001a).

INVOLVEMENT OF PLC IN VPF/VEGF SIGNALING

It was shown previously that both BAPTA/AM, an intracellular Ca^{++} chelator, and U73122, a specific inhibitor of the PLC family, can inhibit VPF/VEGF-stimulated HUVEC proliferation and migration (Wu et al. 2000; Zeng et al. 2001a). U73122 was also shown to inhibit VPF/VEGF-mediated increasing intracellular $[Ca^{++}]$ (Zeng et al. 2001a). However, increasing $[Ca^{++}]_i$ starts after ~40 seconds, whereas phosphorylation of PLC-γ occurs after 5 minutes of VPF/VEGF stimulation. Therefore, increasing $[Ca^{++}]_i$ prior to phosphorylation of PLC-γ (Guo et al. 1995; Zeng et al. 2001a) implies that the phosphorylation of PLC-γ modulated by the receptor tyrosine kinase might not be responsible for VPF/VEGF-stimulated intracellular Ca^{++} release. These results tempted us to speculate that the increasing $[Ca^{++}]_i$ might be mediated by PLC-β family members, which can be turned on within seconds. To test this hypothesis, we measured the catalytic activities of PLC-β1, β2, β3, and also PLC-γ in cellular extracts prepared from HUVEC suspension stimulated with VPF/VEGF for 20 seconds, 40 seconds, 1 minute, 2 minutes, and 5 minutes. Surprisingly, the catalytic activity of PLC-β3 started to increase at 20 seconds, reached a maximal level at 40 seconds, and dropped to the baseline within 1 minute after stimulation (Fig. 2). On the other hand, there was no significant change in the catalytic activity of PLC-β1, β2, or PLC-γ (Fig 2). Similar results were obtained with HUVEC in adhesive cultures stimulated with VPF/VEGF (data not shown). These results clearly indicate that it is the PLC-β3, not PLC-γ, that is activated by VPF/VEGF and can be a mediator for the signaling of increasing $[Ca^{++}]_i$.

HETEROTRIMERIC G PROTEINS ARE INVOLVED IN VPF/VEGF-INDUCED PROLIFERATION AND MIGRATION SIGNALING

It is known that PLC-β can only be activated by heterotrimeric G proteins (Rhee and Bae 1997). Therefore, we investigated whether heterotrimeric G proteins are involved in this pathway. Pertussis toxin, which inhibits Gi and Go families, did not affect the VPF/VEGF-induced $[Ca^{++}]_i$ increase (data not shown). To test the involvement of Gq in the VPF/VEGF-induced signaling pathway, we synthesized a fusion peptide TAT-FLAG-Gp (TATFGp), which consists of Gp Antagonist-2A, a Gq-specific peptide antagonist (Mukai et al. 1992; Hunt et al. 1999); TAT, the amino-terminal 11-amino-acid peptide from the human immunodeficiency virus (HIV) protein Tat; and an epitope tagged-FLAG sequence (Fig. 3a). The TAT peptide was previously shown to efficiently deliver proteins or peptides into mammalian cells or tissues (Schwarze et al. 1999; Datta et al. 2001). An unrelated control peptide, TAT-FLAG-VHL150 (TATFVHL150), in which TAT peptide was fused with VHL152-171, was also synthesized. Pretreatment with the TATFGp peptide (2 μM), but not TAT-FLAG-VHL150 (5 μM), blocked $[Ca^{++}]_i$ increase in HUVEC stimulated with VPF/VEGF (Fig. 3b). To confirm that this peptide is specific for Gq family proteins, HUVEC were pretreated with TATFGp peptide and then stimulated with thrombin, which is known to induce increasing $[Ca^{++}]_i$ through the Gi protein (Gennity and Siess 1991). As shown in Figure 3c, the peptide did not inhibit increasing $[Ca^{++}]_i$ in HUVEC stimulated by thrombin, indicating the specificity of TATFGp peptide for Gq family proteins. We also tested the effect of this peptide on PLC-γ-mediated Ca^{++} release in wild-type chicken B-cell line DT40 cells, which are known to be activated by PLC-γ after B-cell receptor stimulation by IgM (Takata and Kurosaki 1996). As indicated in Figure 3d, TATFGp at 5 μM had no effect on the $[Ca^{++}]_i$ increase. These data clearly indicate that, in HUVEC stimulated with VPF/VEGF, $[Ca^{++}]_i$ increase is mediated mostly by the Gq-PLC-β3 pathway. To test whether this pathway is involved in HUVEC proliferation and migration stimulated by VPF/VEGF, the cells were pretreated with different concentrations of TATFGp and TATFVHL150 before VPF/VEGF stimulation. As shown in Figure 4, a and b, the TATFGp peptide caused an ~50% inhibition of proliferation and migration at 2 μM and 5 μM, and a complete inhibition at 10 μM. However, the control peptide, TATFVHL150, had no effect.

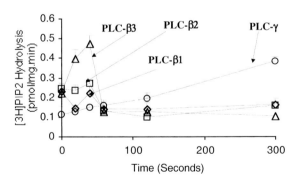

Figure 2. Activation of PLC isoforms in HUVEC stimulated with VPF/VEGF. Phospholipase activities were measured after immunoprecipitation of each PLC isoform after stimulation of VPF/VEGF. 80% confluent HUVECs were serum-starved in EBM/0.1% FBS overnight and detached from the plate by incubating in collagenase solution as described elsewhere. Cells were washed with EBM/0.1% FBS and stimulated with 10 ng/ml of VPF/VEGF for 20 seconds, 40 seconds, 1, 2, and 5 minutes (for the adherent state, no collagenase treatment). Cells were lysed and the PLC catalytic assays were carried out following immunoprecipitations using antibodies against PLC-β1, β2, β3, and PLC-γ. The data shown represent the means with S.D. of quadruplicate samples in at least three independent experiments.

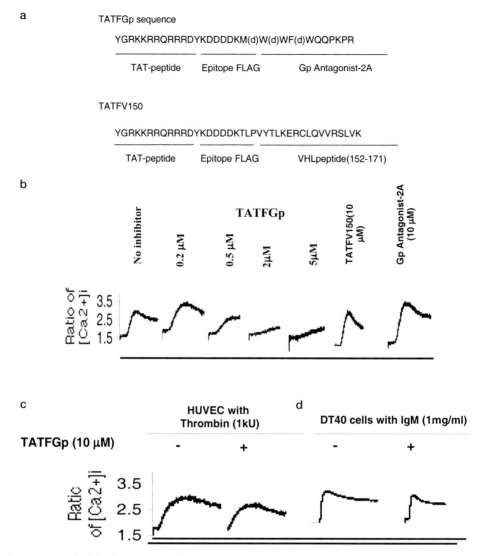

Figure 3. Gq protein required for increasing [Ca^{++}]$_i$ in HUVEC stimulated with VPF/VEGF. (*a*) Sequence of TATFGp and TAT-FVHL150. (d)W represents D-amino acid. (*b*) Dose response effect of TATFGp on increasing [Ca^{++}]$_i$ in HUVEC stimulated with 10 ng/ml VPF/VEGF. (*c*) Increasing [Ca^{++}]$_i$ in HUVEC stimulated with 1 unit of thrombin. (*d*) Increasing [Ca^{++}]$_i$ in wild-type DT40 cells stimulated with 1 µg/ml IgM. Increasing [Ca^{++}]$_i$ was assayed as described elsewhere (Brock et al. 1991). Peptide was added 2 minutes prior to the addition of ligand.

BLOCKING G-PROTEIN ACTIVATION CAN IMPAIR PLC-β3 ACTIVATION BY VPF/VEGF

To further confirm that heterotrimeric Gq protein is involved in the activation of PLC-β3, HUVEC were pretreated with different concentrations of TATFGp, followed by VPF/VEGF stimulation for 40 seconds, and then the cellular extracts were subjected to immunoprecipitation with antibody against PLC-β3 and assayed for catalytic activity. Interestingly, TATFGp completely inhibited the PIP$_2$ hydrolysis mediated by PLC-β3 at 2 µM (data not shown). The higher concentration of peptide required for inhibiting VPF/VEGF-induced proliferation and migration than that for [Ca^{++}]$_i$ increase and PLC-β3 activation may be due to the fact that the proliferation and migration assays were carried out in adherent cells, whereas other assays were performed in suspended cells.

Like most of the protein tyrosine kinase receptors, both KDR and Flt-1 can be tyrosine-phosphorylated after VPF/VEGF treatment (Millauer et al. 1993; Fong et al. 1995). Interestingly, phosphorylation of KDR was also completely inhibited by TATFGp at 5 µM in HUVEC stimulated with VPF/VEGF (Fig. 4c). Treatment with the control peptide did not affect phosphorylation of KDR. As expected, phosphorylation of MAPK which is required for proliferation, but not migration of HUVEC stimulated with VPF/VEGF (Takahashi et al. 1999; Zeng et al. 2001a,b), was completely inhibited by TATFGp at 10 µM (data not shown). Taken together, these findings clearly indicate that the heterotrimeric Gq family proteins and PLC-β3 are required for VPF/VEGF-mediated HUVEC proliferation and migration.

Usually, G protein is activated by its α-subunit interaction with ligand-bound receptor. The dissociated α and βγ subunits trigger the downstream pathways. It is known

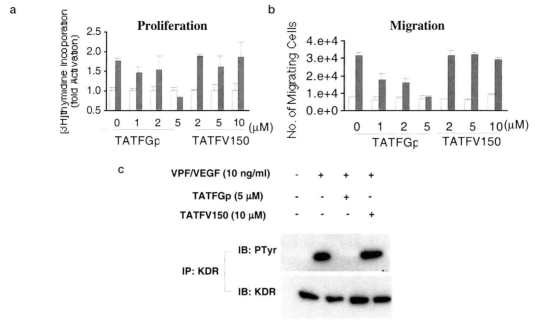

Figure 4. Effect of TATFGp on VPF/VEGF-stimulated signal pathways in HUVEC. (*a*) Proliferation assay. (*b*) Migration assay. (*c*) Immunoprecipitation followed by western blot of phosphorylation of KDR. Proliferation and migration assays were described previously (Zeng et al. 2001a).

that the α subunit and the βγ subunit of the Gq subfamily activate PLC-β isozymes differently (Rhee and Bae 1997). Gβγ subunit activates PLC-β isoforms in an order of PLC-β3 > PLC-β2 > PLC-β1; whereas Gqα stimulates PLC-β in an order of PLC-β1 > PLC-β3 > PLC-β2 (Rhee and Bae 1997). To investigate which G-protein subunit is required for the VPF/VEGF signaling pathway, we used the Gqα- and Gβγ-specific minigene, hGqα359, and hβARK1(495), which encode the carboxy-terminal portions of human Gqα and G-protein-coupled receptor kinase 2 (GRK2 or βARK1), respectively (Koch et al. 1994; Akhter et al. 1998). Gqα395 contains the carboxy-terminal 59 amino acids of the human Gqα subunit that interacts with the intracellular domain of agonist-occupied receptors, and its expression specifically inhibits the interaction of Gq to Gq-coupled receptors and prevents the activation of Gq protein (Akhter et al. 1998). hβARK1(495) contains the carboxy-terminal domain of human βARK1 that physically interacts with free Gβγ, and therefore acts as a specific intracellular Gβγ antagonist and inhibits Gβγ-mediated downstream events (Koch et al. 1994). These minigenes were subcloned into a retroviral expression vector and overexpressed in HUVEC (data not shown). We utilized these minigenes to examine the role of G-protein subunits in VPF/VEGF-induced cell signaling. Figure 5, a and b, clearly indicates that overexpression of human βARK1(495) completely inhibits the proliferation and migration of HUVEC stimulated with VPF/VEGF, but not bFGF. On the other hand, hGqα395 minigene has no effect in either assay (data not shown). As expected, overexpression of hβARK1(495) completely inhibited VPF/VEGF-stimulated phosphorylation of MAPK in HUVEC (Fig. 5c). Furthermore, overexpression of hβARK1(495) also blocked intracellular Ca^{++} re-

lease and the PLC-β3 activation mediated by VPF/VEGF; again the hGqα395 minigene failed to do so (Fig. 5d, data not shown). However, HUVEC overexpressed with either hGqα395 or hβARK(495) shows no inhibition of phosphorylation of KDR when stimulated with VPF/VEGF (data not shown). These data clearly suggest that the involvement of the Gβγ subunit in VPF/VEGF mediated signaling pathways that lead to HUVEC proliferation and migration. At this point, it is not clear whether Gqα is required for VPF/VEGF-induced cell signaling, since the Gqα minigene we have utilized only inhibits the interaction of Gα with G-protein coupling receptor. Therefore, at least, we can suggest that a Gq coupling receptor is not required for VPF/VEGF-mediated activation of Gβγ.

CONCLUSIONS

VPF/VEGF is an important, multifunctional angiogenic cytokine that exerts a variety of biological activities on vascular endothelium. These include induction of microvascular hyperpermeability; stimulation of proliferation and migration; significant reprogramming of gene expression; endothelial cell survival; and prevention of senescence (for review, see Risau 1997; Dvorak et al. 1999; Ferrara 1999; Petrova et al. 1999). All of these functions are thought to be mediated by two receptor tyrosine kinases, KDR and Flt-1, that are selectively expressed on vascular endothelium and that are up-regulated at sites of VPF/VEGF overexpression as in tumors, healing wounds, chronic inflammation, etc. (for review, see Risau 1997). Because both receptors are expressed on vascular endothelium, it has been difficult to define the

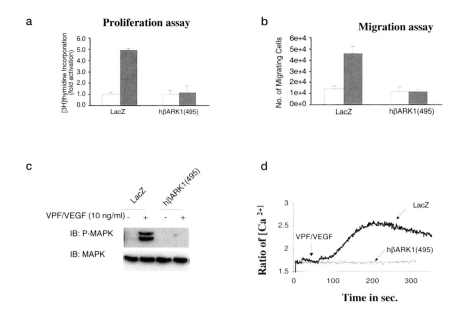

Figure 5. Gβγ subunit is required for VPF/VEGF-stimulated signaling pathways in HUVEC. (*a*) Proliferation assay. (*b*) Migration assay. (*c*) Western blot of phosphorylation of MAPK. (*d*) Increasing [Ca^{++}]$_i$ in HUVEC infected with retrovirus containing LacZ, hGqα395, or hβARK1(495), respectively. The human Gqα395 and βARK1(495) minigenes were obtained by RT-PCR from HUVEC mRNA with oligonucleotides starting at amino acid 395 of Gqα and 495 of βARK1, respectively. The PCR products were subcloned in retrovirus vector (pMMP) to overexpress proteins in HUVEC and sequences were confirmed by DNA sequencing. To prepare the retrovirus, 293T cells were seeded at 6 × 10^6 per T100 plate 24 hours before transfection. 20 μg of pMMP-Gqα395, pMMP-βARK1(495), or pMMP-LacZ; 15 μg of pMD.MLV gag.pol, and 5 μg of pMD.G DNA, encoding the cDNAs of the proteins which are required for virus packaging (kindly provided by Dr. Mulligan), were mixed in 500 μl of water with 62 μl of 2 M CaCl$_2$, 500 μl of 2× HBS (280 mM NaCl, 10 mM KCl, 1.5 mM Na$_2$HPO$_4$, 12 mM dextrose, and 50 mM HEPES, pH 7.05) were added to the DNA mixture and incubated at room temperature for 20 minutes. The DNA mixture was added dropwise to 293T cells. Medium was changed after 16 hours. Retrovirus was isolated and used immediately for infection.

respective roles of each in mediating the various signaling events and biological activities induced in endothelium by VPF/VEGF. Current information, therefore, has been gleaned largely from studies with a cell line, PAE, that does not express either receptor unless engineered to do so (Waltenberger et al. 1994; Joukov et al. 1997); from studies with PlGF, a ligand that binds Flt-1 but not KDR; from the use of a Flt-1-specific antibody (Kanno et al. 2000); from studies with antisense oligonucleotides that block Flt-1 expression (Bernatchez et al. 1999); and from VPF/VEGF mutants that specifically bind to Flt-1 (Gille et al. 2000a,b).

We sought to delineate the respective roles of KDR and Flt-1 in early-passage endothelial cells in which both receptors remained intact and functional. To that end, we engineered chimeric receptors, fusing the extracellular domain of the EGF receptor with the transmembrane and intracellular domains of either KDR or Flt-1. We transduced these constructs (or, as a control, LacZ) into HUVEC with a retroviral vector. This approach was feasible because, under the conditions of our experiments (≤ 80% confluence), HUVEC did not express the EGFR and did not respond to EGF. This strategy was also attractive because endogenous KDR and Flt-1 persisted in HUVEC transduced with either or both chimeric receptors; therefore, the signaling and biological responses induced by VPF/VEGF or EGF could be compared in the same cells. Available data suggest that KDR and Flt-1 have different, perhaps complementary, roles in vasculogenesis and angiogenesis. However, many of the data have been obtained from immortalized cell lines that may differ significantly in behavior from that of early-passage cells derived from primary endothelial cultures. Therefore, in order to elucidate the respective roles of KDR and Flt-1 in early-passage HUVEC expressing both KDR and Flt-1, we engineered chimeric constructs of both receptors, replacing the extracellular domain of each with the extracellular domain of EGFR. We used retroviral vectors to express these chimeric receptors in HUVEC that expressed both KDR and Flt-1 but not EGFR. Using this system, we demonstrated that HUVEC proliferation and migration were mediated exclusively through the KDR signaling pathway, a conclusion consistent with that of other workers (Waltenberger et al. 1994; Bernatchez et al. 1999; Gille et al. 2000a; Kanno et al. 2000). Interestingly, however, Flt-1 activation mediated a distinctive inhibitory signaling pathway through PI3K that down-regulated the cell proliferation pathway triggered by KDR. Moreover, Flt-1 was found to interact with the p85 subunit of PI3K, whereas no association was observed between KDR and p85 (Zeng et al. 2001a). These results corroborated the findings of Cunningham et al. (1995), which suggested an association between Flt-1 and p85 using the yeast two-hybrid system. Furthermore, PI3K (p110α) knockout mice (Pik3ca$^{del/del}$) demonstrated extravasation blood, suggestive of defective angiogenesis. These embryos are developmentally retarded and die between E9.5 and E10.5 (Bi et al. 1999). The similarities of the Pik3ca$^{del/del}$ phenotype to that of Flt-1 knockout mice are consistent with our results that PI3K is involved in the Flt-1 signaling pathway.

The function of Flt-1 has been much less clear. One proposal has been that Flt-1 is a decoy receptor rather than a signal transducer because Flt-1 kinase domain null mice develop normally (Hiratsuka et al. 1998), in contrast to Flt-1 knockout mice, which are embryonic lethal (Petrova et al. 1999). It was also reported that VPF/VEGF failed to stimulate Flt-1 tyrosine phosphorylation in PAE engineered to overexpress Flt-1 and did so poorly in HUVEC (de Vries et al. 1992; Waltenberger et al. 1994). However, the expression level of Flt-1 in HUVEC is only about one-tenth that of KDR (Waltenberger et al. 1994; Joukov et al. 1997), and therefore measurement of Flt-1 phosphorylation is difficult and may have been underestimated. To take account of the large differences in receptor expression, we used three times as many HUVEC for measurement of phosphorylation of Flt-1 as for KDR and had no difficulty in demonstrating phosphorylation of both receptors (Fig. 1d, panel A).

So far we described that we have engineered chimeric receptors to distinguish the signaling events triggered by KDR and Flt-1 in early-passage HUVEC. These tools have allowed us to identify a number of downstream signaling pathways that are stimulated by activation of KDR and Flt-1. KDR-mediated proliferation and migration involve activation of PLC, whereas Flt-1 was found to exert an inhibitory effect on HUVEC proliferation, but not on migration, through the PI3K pathway. The Flt-1-mediated antiproliferative pathway acts after KDR phosphorylation but at or before KDR-mediated intracellular Ca^{++} mobilization. This study thus represents the first direct analysis of Flt-1 and KDR function in early-passage EC and has demonstrated "cross-talk" between the pathways mediated by these two receptors.

To examine the involvement of CDC42, which belongs to the Ras superfamily of small GTPases, in VPF/VEGF-mediated EC function, very recently we showed that the dominant negative mutant of CDC42, CDC42-17N, increases proliferation but has no effect on migration in HUVEC stimulated with VPF/VEGF (Zeng et al. 2002). CDC42-17N also increases the kinetic slop and the maximum response of VPF/VEGF-stimulated intracellular Ca^{++} mobilization, but has no effect on KDR phosphorylation and MAPK phosphorylation. Furthermore, we also showed that CDC42 can be activated by VPF/VEGF. Utilizing the chimeric receptors (EGDR and EGLT), we also demonstrated that CDC42 is activated by EGLT, but not by EGDR (Zeng et al. 2002). Surprisingly, CDC42 activation by EGLT can be inhibited by pertussis toxin, the free Gβγ-specific sequestering minigene (hβARK[495]), PI3K inhibitors (wortmannin and LY294002), and p85 dominant negative mutant (p85DN), but neither by a PLC inhibitor (U73122), nor by an intracellular Ca^{++} chelator BAPTA/AM. Further studies from our laboratory indicated that pertussis toxin also increases the kinetic slop and the maximum response of intracellular Ca^{++} mobilization and proliferation in HUVEC stimulated with VPF/VEGF. Taken together, our study suggested that pertussis toxin-sensitive G proteins, Gβγ subunit, and CDC42 are involved in the signaling pathway of Flt-1 that down-regulates HUVEC proliferation (Zeng et al. 2002).

The role of G proteins in VPF/VEGF signaling is rather novel, considering that the receptors which lead to signaling are mostly receptor tyrosine kinases. The ability to block VPF/VEGF-stimulated HUVEC proliferation and migration with a Gq-specific antagonist fusion peptide and βARK1 minigene implies the existence of a novel signaling pathway. In this pathway, VPF/VEGF binds to its receptor and activates Gq protein by an unknown mechanism, in which a Gq coupling receptor is not required. Interestingly, we also described that the inhibition of Gq activation can block KDR phosphorylation, which suggests a unique pathway by which Gq modulates a receptor tyrosine kinase activation. The free Gβγ subunit activates PLC-β3 followed by increasing $[Ca^{++}]_i$ that results in proliferation and migration of endothelial cells and thus angiogenesis. Our results are somehow consistent with the observation that Gi might be involved in VPF/VEGF-mediated macrophage migration which was mediated by Flt-1 (Barleon et al. 1996), since this migration was inhibited by pertussis toxin (Barleon et al. 1996). However, as reported recently, it is KDR, not Flt-1, that mediates HUVEC proliferation and migration through the PLC-Ca^{++} pathway (Zeng et al. 2001a,b). These results thus suggest that Gq-Gβγ-PLCβ3 are required for KDR function and describe for the first time that G protein can be activated by VPF/VEGF. Our data also indicate a new pathway for G-protein activation, in which a tyrosine kinase receptor, not a G-protein-coupled receptor, is involved. Furthermore, KDR had been shown to be specific and crucially important for solid tumor progression (Millauer et al. 1994), and the control of VPF/VEGF function in angiogenesis is a major goal of present tumor research. These studies show that targeting the Gβγ subunit, as here we have utilized either the peptide antagonist (TATFGp) or the βARK1 minigene, might be suitable for developing specific antiangiogenic drugs by blocking the VPF/VEGF signaling pathway. However, neither the KDR nor Flt-1 pathways are as yet fully defined, and further work is needed to demonstrate the complete complement of signaling steps and regulatory pathways that govern such other VPF/VEGF-mediated effects on EC as increased microvascular permeability, gene expression reprogramming, survival, and prevention of senescence.

ACKNOWLEDGMENTS

This work was partly supported by National Institutes of Health grants HL-70567 and CA-78383 and grants from the Department of Defense Breast Cancer Program and the Massachusetts Department of Public Health to D.M. H.Z. is a National Research Service Award fellow. D.M. is a Eugene P. Schonfeld National Kidney Cancer Association Medical Research Awardee.

REFERENCES

Akhter S.A., Luttrell L.M., Rockman H.A., Iaccarino G., Lefkowitz R.J., and Koch W.J. 1998. Targeting the receptor-Gq interface to inhibit in vivo pressure overload myocardial hypertrophy. *Science* **280:** 574.

Barleon B., Sozzani S., Zhou D., Weich H.A., Mantovani A., and Marme D. 1996. Migration of human monocytes in response to vascular endothelial growth factor (VEGF) is mediated via the VEGF receptor flt-1. *Blood* **87:** 3336.

Benezra M., Vlodasky I., Ishai-Michaeli R., Neufeld G., and Bar-Shavit R. 1993. Thrombin-induced release of active basic fibroblast growth factor-heparan sulfate complexes from subendothelial extracellular matrix. *Blood* **81:** 3324.

Bernatchez P.N., Soker S., and Sirois M.G. 1999. Vascular endothelial growth factor effect on endothelial cell proliferation, migration and platelet-activating factor synthesis is Flk-1-dependent. *J. Biol. Chem.* **274:** 31047.

Bi L., Okabe I., Bernard D.J., Wynshaw-Boris A., and Nussbaum R.L. 1999. Proliferative defect and embryonic lethality in mice homozygous for a deletion in the p110alpha subunit of phosphoinositide 3-kinase. *J. Biol. Chem.* **274:** 10963.

Brock T.A., Dvorak H.F., and Senger D.R. 1991. Tumor-secreted vascular permeability factor increases cytosolic Ca^{++} and von Willebrand factor release in human endothelial cells. *Am. J. Pathol.* **138:** 213.

Cunningham S.A., Waxham M.N., Arrate P.M., and Brock T.A. 1995. Interaction of the Flt-1 tyrosine kinase receptor with the p85 subunit of phosphatidylinositol 3-kinase. *J. Biol. Chem.* **270:** 20254.

Datta K., Sundberg C., Karumanchi S.A., and Mukhopadhyay D. 2001. The 104-123 amino acid sequence of the beta-domain of von Hippel-Lindau gene product is sufficient to inhibit renal tumor growth and invasion. *Cancer Res.* **61:** 1768.

de Vries C., Escobedo J.A., Ueno H., Houck K., Ferrara N., and Williams L.T. 1992. The fms-like tyrosine kinase, a receptor for vascular endothelial growth factor. *Science* **255:** 989.

Dvorak H.F. 1990. Leaky tumor vessels: Consequences for tumor stroma generation and for solid tumor therapy. *Prog. Clin. Biol. Res.* **354A:** 317.

Dvorak, H.F., Senger D.R., and Dvorak A.M. 1984. Fibrin formation: Implications for tumor growth and metastasis. *Dev. Oncol.* **22:** 96.

Dvorak H.F., Nagy J.A., Feng D., Brown L.F., and Dvorak A.M. 1999. Vascular permeability factor/vascular endothelial growth factor and the significance of microvascular hyperpermeability in angiogenesis. *Curr. Top. Microbiol. Immunol.* **237:** 97.

Dvorak H.F., Orenstein N.S., Carvalho A.C., Churchill W.H., Dvorak A.M., Galli S.J., Feder J., Bitzer A.M., Rypysc J., and Giovinco P. 1979. Induction of a fibrin-gel investment: An early event in line 10 hepatocarcinoma growth mediated by tumor-secreted products. *J. Immunol.* **122:** 166.

English J., Pearson G., Wilsbacher J., Swantek J., Karandikar M., Xu S., and Cobb M.H. 1999. New insights into the control of MAP kinase pathways. *Exp. Cell Res.* **253:** 255.

Ferrara N. 1999. Vascular endothelial growth factor: Molecular and biological aspects. *Curr. Top. Microbiol. Immunol.* **237:** 1.

Folkman J. 1971. Tumor angiogenesis: Therapeutic implications. *N. Engl. J. Med.* **285:** 1182.

———. 1996. Fighting cancer by attacking its blood supply. *Sci. Am.* **275:** 150.

Folkman J. and Klagsbrun M. 1987. Angiogenic factors. *Science* **235:** 442.

Folkman J., Watson K., Ingber D., and Hanahan D. 1989. Induction of angiogenesis during the transition from hyperplasia to neoplasia. *Nature* **339:** 58.

Fong G.H., Rossant J., Gertsenstein M., and Breitman M.L. 1995. Role of the Flt-1 receptor tyrosine kinase in regulating the assembly of vascular endothelium. *Nature* **376:** 66.

Gagnon M.L., Bielenberg D.R., Gechtman Z., Miao H.Q., Takashima S., Soker S., and Klagsbrun M. 2000. Identification of a natural soluble neuropilin-1 that binds vascular endothelial growth factor: In vivo expression and antitumor activity. *Proc. Natl. Acad. Sci.* **9:** 2573.

Gennity J.M. and Siess W. 1991. Thrombin inhibits the pertussis-toxin-dependent ADP-ribosylation of a novel soluble Gi-protein in human platelets. *Biochem. J.* **279:** 643.

Gille H., Kowalski J., Yu L., Chen H., Pisabarro M.T., Davis-Smyth T., and Ferrara N. 2000a. A repressor sequence in the juxtamembrane domain of Flt-1 (VEGFR-1) constitutively inhibits vascular endothelial growth factor-dependent phosphatidylinositol 3´-kinase activation and endothelial cell migration. *EMBO J.* **19:** 4064.

Gille H., Kowalski J., Li B., LeCouter J., Moffat B., Zioncheck T.F., Pelletier N., and Ferrara N. 2000b. Analysis of biological effects and signaling properties of Flt-1 and KDR: A reassessment using novel highly receptor-specific VEGF mutants. *J. Biol. Chem.* **31:** 31.

Guo D., Jia Q., Song H., Warren R., and Donner D. 1995. Vascular endothelial cell growth factor promotes tyrosine phosphorylation of mediators of signal transduction that contain SH2 domains. Association with endothelial cell proliferation. *J. Biol. Chem.* **270:** 6729.

Hiratsuka S., Minowa O., Kuno J., Noda T., and Shibuya M. 1998. Flt-1 lacking the tyrosine kinase domain is sufficient for normal development and angiogenesis in mice. *Proc. Natl. Acad. Sci.* **95:** 9349.

Hunt R.A., Bhat G.J., and Baker K.M. 1999. Angiotensin II-stimulated induction of sis-inducing factor is mediated by pertussis toxin-insensitive G(q) proteins in cardiac myocytes. *Hypertension* **34:** 603.

Joukov V., Sorsa T., Kumar V., Jeltsch M., Claesson-Welsh L., Cao Y., Saksela O., Kalkkinen N., and Alitalo K. 1997. Proteolytic processing regulates receptor specificity and activity of VEGF-C. *EMBO J.* **16:** 3898.

Kanno S., Oda N., Abe M., Terai Y., Ito M., Shitara K., Tabayashi K., Shibuya M., and Sato Y. 2000. Roles of two VEGF receptors, Flt-1 and KDR, in the signal transduction of VEGF effects in human vascular endothelial cells. *Oncogene* **19:** 2138.

Koch W.J., Hawes B.E., Inglese J., Luttrel L.M., and Lefkowitz R.J. 1994. Cellular expression of the carboxyl terminus of a G protein-coupled receptor kinase attenuates G beta gamma-mediated signaling. *J. Biol. Chem.* **269:** 6193.

Leung D.W., Cachianes G., Kuang W.J., Goeddel D.V., and Ferrara N. 1989. Vascular endothelial growth factor is a secreted angiogenic mitogen. *Science* **246:** 1306.

Millauer B., Wizigmann-Voos S., Schnurch H., Martinez R., Meller N.P.H., Risau W., and Ullrich A. 1993. High affinity VEGF binding and developmental expression suggest Flk-1 as a major regulator of vasculogenesis and angiogenesis. *Cell* **72:** 835.

Mukai H., Munekata E., and Higashijima T. 1992. G protein antagonists. A novel hydrophobic peptide competes with receptor for G protein binding. *J. Biol. Chem.* **267:** 16237.

Petrova T.V., Makinen T., and Alitalo K. 1999. Signaling via vascular endothelial growth factor receptors. *Exp. Cell Res.* **253:** 117.

Quinn T.P., Peters K.G., De Vries C., Ferrara N., and Williams L.T. 1993. Fetal liver kinase 1 is a receptor for vascular endothelial growth factor and is selectively expressed in vascular endothelium. *Proc. Natl. Acad. Sci.* **90:** 7533.

Rhee S.G. and Bae Y.S. 1997. Regulation of phosphoinositide-specific phospholipase C isozymes. *J. Biol. Chem.* **272:** 15045.

Risau W. 1997. Mechanisms of angiogenesis. *Nature* **386:** 671.

Schwarze S.R., Ho A., Vocero-Akbani A., and Dowdy S.F. 1999. In vivo protein transduction: Delivery of a biologically active protein into the mouse. *Science* **285:** 1569.

Seghezzi G., Patel S., Ren C.J., Gualandris A., Pintucci G., Robbins E.S., Shapiro R.L., Galloway A.C., Rifkin D.B., and Mignatti P. 1998. Fibroblast growth factor-2 (FGF-2) induces vascular endothelial growth factor (VEGF) expression in the endothelial cells of forming capillaries: An autocrine mechanism contributing to angiogenesis. *J. Cell Biol.* **141:** 1659.

Senger D.R., Perruzzi C.A., Feder J., and Dvorak H.F. 1986. A highly conserved vascular permeability factor secreted by a variety of human and rodent tumor cell lines. *Cancer Res.* **46:** 5629.

Senger D.R., Galli S.J., Dvorak A.M., Perruzzi C.A., Harvey V.S., and Dvorak H.F. 1983. Tumor cells secrete a vascular permeability factor that promotes accumulation of ascites

fluid. *Science* **219**: 983.
Shalaby F., Ho J., Stanford W.L., Fischer K.D., Schuh A.C., Schwartz L., Bernstein A., and Rossant J. 1997. A requirement for Flk1 in primitive and definitive hematopoiesis and vasculogenesis. *Cell* **89**: 981.
Soker S., Takashima S., Miao H.Q., Neufeld G., and Klagsbrun M. 1998. Neuropilin-1 is expressed by endothelial and tumor cells as an isoform-specific receptor for vascular endothelial growth factor. *Cell* **92**: 735.
Takahashi T., Ueno H., and Shibuya M. 1999. VEGF activates protein kinase C-dependent, but Ras-independent Raf-MEK-MAP kinase pathway for DNA synthesis in primary endothelial cells. *Oncogene* **18**: 2221.
Takata M. and Kurosaki T. 1996. A role for Bruton's tyrosine kinase in B cell antigen receptor-mediated activation of phospholipase C-gamma 2. *J. Exp. Med.* **184**: 31.
Terman B., Dougher-Vermazen M., Carrion M., Dimitrov D., Armellino D., Gospodarowicz D., and Bohlen P. 1992. Identification of the KDR tyrosine kinase as a receptor for vascular endothelial cell growth factor. *Biochem. Biophys. Res. Commun.* **187**: 1579.
Vlodavsky I., Fuks Z., Ishai-Michaeli R., Bashkin P., Levi E., Korner G., Bar-Shavit R., and Klagsbrun M. 1991. Extracellular matrix-resident basic fibroblast growth factor: Implication for the control of angiogenesis. *J. Cell. Biochem.* **45**: 167.
Waltenberger J., Claesson-Welsh L., Siegbahn A., Shibuya M., and Heldin C.H. 1994. Different signal transduction properties of KDR and Flt-1, two receptors for vascular endothelial growth factor. *J. Biol. Chem.* **269**: 26988.
Wu L.W., Mayo L.D., Dunbar J.D., Kessler K.M., Ozes O.N., Warren R.S., and Donner D.B. 2000. VRAP is an adaptor protein that binds KDR, a receptor for vascular endothelial cell growth factor. *J. Biol. Chem.* **275**: 6059.
Zeng H., Dvorak H.F., and Mukhopadhyay D. 2001a. VPF/VEGF receptor-1 down modulates VPF/VEGF receptor-2 mediated endothelial cell proliferation, but not migration, through phosphatidylinositol 3-kinase dependent pathways. *J. Biol. Chem.* **276**: 26969.
Zeng H., Sanyal S., and Mukhopadhyay D. 2001b. Tyrosine residues 951 and 1059 of vascular endothelial growth factor receptor-2 (kdr) are essential for vascular permeability factor/vascular endothelial growth factor-induced endothelium migration and proliferation, respectively. *J. Biol. Chem.* **276**: 32714.
Zeng H., Zhao D., and Mukhopadhyay D. 2002. Flt-1-mediated down-regulation of endothelial cell proliferation through pertussis toxin-sensitive G proteins, beta gamma subunits, small GTPase CDC42, and partly by Rac-1. *J. Biol. Chem.* **277**: 4003.

Targeted Delivery of Mutant Raf Kinase to Neovessels Causes Tumor Regression

J.D. HOOD AND D.A. CHERESH
The Scripps Research Institute, Departments of Immunology, La Jolla, California 92037

Endothelial cells are the key regulators of vascular tone, permeability, and neovascular growth. Because of the enormous biological impact of these processes, the regulation of endothelial cell function represents a key therapeutic target in a wide array of diseases ranging from overt vascular diseases, such as hypertension and stroke, to tumor growth and metastases. As such, the focus of many laboratories is the elucidation of the signal transduction pathways within endothelial cells which regulate their behavior. This has contributed to an enhanced understanding of endothelial cell biology and to the discovery of novel therapeutics for diseases. Recently, particular attention has been paid to the processes regulating angiogenesis, the sprouting of endothelial cells from a preexisting blood vessel in order to form a new vessel. This strategy has led to recent clinical successes with compounds such as Avastin (Langmuir et al. 2002), a monoclonal antibody that blocks vascular endothelial growth factor (VEGF) signaling. Here we examine the results of recent studies from our laboratory and others revealing a role for integrins and growth factors in regulating angiogenesis and how modulation of these molecules and their downstream signaling might be used therapeutically.

REGULATION OF ANGIOGENESIS BY INTEGRINS

During angiogenesis, endothelial cells switch from a stationary, nonproliferative phenotype to a rapidly proliferating, highly migratory one (Folkman 2001). This is a necessary and potentially rate-limiting step in tumor formation, inflammatory diseases, and tissue regeneration (Folkman 2001). Angiogenesis is initiated by local tissue availability of growth factors, such as VEGF and basic fibroblast growth factor (bFGF), that bind specific receptors on endothelial cells initiating the activation of intracellular signaling cascades (Yancopoulos et al. 1998). In addition, the concomitant interaction between endothelial cells (ECs) and extracellular matrix (ECM) molecules stimulates a convergent signal transduction cascade that integrates with growth factor receptor signaling to induce adhesion-dependent migration, survival, and angiogenesis (Eliceiri 2001).

Cell adhesion to the ECM is mediated by integrins, a family of heterodimeric transmembrane glycoproteins composed of α and β subunits (Giancotti and Ruoslahti 1999). Different combinations of single α and β subunits dimerize to form at least 25 different integrin receptors in mammals (Giancotti and Ruoslahti 1999). The biological significance of the range of ECM-integrin specificities during cell adhesion is not clear. Although integrins support distinct cell–ECM interactions for endothelial cell adhesion, the identification of mechanisms by which specific subsets of integrins mediate development and angiogenesis remains elusive (Eliceiri and Cheresh 2001).

In addition to regulating cell adhesion to the ECM, integrins relay molecular cues regarding the cellular environment to the intracellular signaling machinery that influence cell shape, survival, proliferation, gene transcription, and migration (Hood and Cheresh 2002). For example, integrin ligation induces a wide range of signaling events, including the activation of Ras, MAP kinase, focal adhesion kinase (FAK), Src, Rac/Rho/cdc42 GTPases, protein kinase C, and phosphatidylinositol 3-kinase (Aplin et al. 1998). In addition, integrin ligation increases intracellular pH, calcium levels, inositol lipid synthesis, cyclin synthesis, and the expression of immediate-early genes, and promotes cell survival (Ridley 2000; Schwartz and Ginsberg 2002). Furthermore, many of the signaling pathways and effectors activated by integrin ligation are also activated following growth factor stimulation. This points to the possibility that integrin and growth factor-mediated cellular responses synergize and coordinate biochemical responses.

The physiological importance of integrins during angiogenesis has been extensively studied in the case of αv integrins and certain $\beta 1$ integrins. Antagonists of integrins $\alpha v \beta 3$ and $\alpha v \beta 5$ block growth factor- and tumor-induced angiogenesis in multiple animal models (Brooks et al. 1994; Kumar et al. 2001). Furthermore, recent data from clinical trials suggest that antagonists of $\alpha v \beta 3$ and/or $\alpha v \beta 5$ may have a clinical benefit in humans with solid tumors (Guthiel et al. 2000; Kumar et al. 2001). In addition, antagonists of $\alpha 5 \beta 1$ block angiogenesis and the resulting tumor growth in various animal models (Kim et al. 2000). Moreover, endostatin, a naturally occurring endogenous inhibitor of angiogenesis (O'Reilly et al. 1997), binds to $\alpha 5 \beta 1$ on ECs (Rehn et al. 2001).

Intriguingly, although both VEGF and bFGF require ligation of αv integrins for angiogenesis to progress, they are dependent on ligation of distinct αv-integrins (Friedlander et al. 1995). This suggests that although both growth factors induce the signals necessary for new blood vessel formation, the signal pathways they activate are distinct. For example, VEGF- and bFGF-induced angiogenesis are each inhibited by antagonists of the separate yet functionally related αv integrins, αvβ3 and αvβ5, respectively (Friedlander et al. 1995). In vivo studies, using both the chick chorioallantoic membrane and rabbit corneal eye pocket angiogenesis assay, reveal that an anti-αvβ3 monoclonal antibody blocks bFGF-induced angiogenesis, whereas an anti-αvβ5 antibody blocks VEGF-induced angiogenesis (Friedlander et al. 1995). Furthermore, inhibition of protein kinase C (Friedlander et al. 1995) or Src kinase (Eliceiri et al. 1999) disrupts VEGF-induced signaling specifically, but does not affect bFGF-induced angiogenesis. Although αvβ5 antagonists do not affect bFGF-induced neovascularization, αvβ3 antagonists can inhibit up to 50% of VEGF-induced angiogenesis (Friedlander et al. 1995). This observation is consistent with the finding that αvβ3 is up-regulated in angiogenic vessels irrespective of the growth factor stimuli and that VEGF promotes αvβ3-mediated endothelial cell adhesion and migration in vitro (Senger et al. 1996).

In addition to growth factor-mediated migration and proliferation, ECM composition likely influences the fate and integrity of newly sprouting blood vessels based on the expression of specific integrins on the EC surface (Alon et al. 1995; Stupack et al. 2001). In fact, recent reports revealed that certain unligated integrins could induce "integrin-mediated death" (Stupack et al. 2001), which may regulate angiogenesis by promoting EC apoptosis, thereby only permitting the survival of appropriately positioned ECs during vascular remodeling. These results may be of particular relevance during pathological angiogenesis when multiple growth factors and ECM components contribute to the growth and morphology of angiogenic blood vessels to support tumor proliferation or maintain the inflammatory state.

Raf AS AN ANTIANGIOGENIC TARGET

Although distinct αv-integrins and kinases such as Src are differentially required by VEGF and bFGF to regulate angiogenesis, recent studies indicate that Raf kinase is a key signaling intermediary for either pathway downstream of both growth factor receptors and integrins (Fig. 1) (Hood et al. 2002). Because Raf is one of a handful of proto-oncogenes, it is not surprising that its activity is critically related with growth regulation in events such as angiogenesis; be it proliferation, differentiation, or survival (Kolch 2000).

The Ras-Raf-MEK-ERK cascade was the first signaling pathway to be entirely mapped from the cell membrane to the cell nucleus (Kolch 2000), and a key mechanism by which adhesion events influence cellular behavior is based on integrin-mediated activation of the ERK cascade (Renshaw et al. 1997). A number of studies have characterized the intracellular pathways leading to ERK activation. Since the biological outcome is determined by the strength and duration of the activation of this pathway (Marshall 1995), it is tightly regulated at all

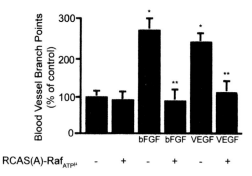

Figure 1. ATP$^\mu$-Raf blocks FGF and VEGF-mediated angiogenesis. Chick CAMs (10 days) were exposed to filter paper disks saturated with RCAS-ATP$^\mu$-Raf followed by stimulation with either bFGF or VEGF (2 μg/ml) for 72 hours. Blood vessels were enumerated by counting vessel branch points in a double-blinded manner. Each bar represents the mean ± S.E.M. of 24 replicates.

levels, with the most intricate controls operating at the level of Raf (Kolch 2000). First, growth factor receptors or integrins recruit adapter proteins that convert Ras to its GTP-bound form, which then recruits Raf-1 to the cytoplasmic membrane where it can be activated. Although the mechanisms of Raf regulation are incompletely understood, they are thought to involve phosphorylation by protein kinases such as p21-activated kinase (PAK) and Src, which are activated in convergent cascades by growth factor receptors together with integrins (Fabian et al. 1993; King et al. 1998; del Pozo et al. 2000; Schlessinger 2000). Raf-1 phosphorylates and activates MEK, which directly activates MAP kinases such as ERK1 and ERK2, leading to phosphorylation of cytoplasmic and nuclear targets (Pearson et al. 2001).

Raf activation in response to growth factors depends on phosphorylation at a number of key sites. One such site is serines 338/339, which can be phosphorylated by PAK (King et al. 1998). Integrin-mediated adhesion can activate PAK through its effector Rac (del Pozo et al. 2000), which enables PAK to activate Raf-1 by phosphorylation of serines 338/339 (King et al. 1998; del Pozo et al. 2000). Notably, phosphorylation of these sites together with Ras activity may be sufficient for Raf-1 activity induced by integrin ligation (Chaudhary et al. 2000). A recent study indicated that inhibition of PAK prevents bFGF-mediated angiogenesis in the chick CAM (Kiosses et al. 2002), perhaps through inhibiting Raf activity.

Recent reports indicate that Raf may be able to activate a wide range of important substrates other than MEK (Murakami and Morrison 2001). For example, Raf-1 activation can lead to ERK-independent phosphorylation of Bad and inhibition of stress-induced apoptosis (Wang et al. 1996). Consistent with this finding, mice homozygous null in either Raf-1 or B-Raf die by embryonic day 12.5 due to vascular defects and high levels of cellular apoptosis, despite showing no apparent impairment in cellular ERK activity (Wojnowski et al. 1997; Huser et al. 2001). Similarly, recent reports (Hood et al. 2002) indicate that Raf kinase inhibition during tumor formation results in neovascular apoptosis leading to disease reversal.

TARGETED DELIVERY OF GENES ENCODING MUTANT Raf TO ANGIOGENIC VESSELS RESULTS IN NEOVASCULAR APOPTOSIS AND TUMOR REGRESSION

Because Raf is a key intermediate both in driving proliferation and migration and in promoting survival during angiogenesis, it is an attractive antiangiogenic therapeutic target. To inhibit Raf kinase selectively within angiogenic blood vessels, a recent study exploited a novel nonviral vector system targeted to the integrin $\alpha v\beta 3$ (Hood et al. 2002). Vascular targeting offers several advantages over general tissue targeting or tumor targeting per se. These include accessibility, common targets for numerous disease states, and a genetically stable target cell. Intravascular delivery of viral or nonviral DNA complexes results mostly in gene expression in vascularly accessible cells, such as ECs (Liu et al. 1997). Therefore, vectors directed at vascular epitopes have advantageous access to their targets. Furthermore, whereas extravascular targets are likely to show great diversity dependent on tumor or disease type, neovessels selectively up-regulate common epitopes independent of disease stimuli (Arap et al. 1998). Therefore, the expectation is that one targeting vector might be used for any of a number of angiogenesis-dependent diseases. Finally, whereas genetically unstable tumor cells might rapidly develop resistance to any genetic intervention applied, genetically stable vascular cells would be expected to respond to repeated dosing (Folkman 2001). These factors have given vascular targeting therapeutic promise for the delivery of drugs (Arap et al. 1998) and radionuclides (Sipkins et al. 1998). Similarly, viral vectors (Blezinger et al. 1999), liposomes (Wang and Becker 1997), and naked DNA (Losordo et al. 1998) have been used for delivery of therapeutic genes to vascular tissue, but none of these approaches is specific for ECs.

Endothelial cells in different organs are not identical (Arap et al. 1998). Inflammatory sites as well as organs and tissues express distinct surface markers and during vascular remodeling and angiogenesis, ECs show increased expression of several cell-surface molecules that potentiate cell invasion and proliferation (Eliceiri and Cheresh 2001). One such molecule is the vitronectin receptor, integrin $\alpha v\beta 3$ (Brooks et al. 1994). In addition to its role in cell matrix recognition, $\alpha v\beta 3$ may be of particular use in gene delivery strategies, since this receptor potentiates the internalization of adenovirus (Wickham et al. 1993), foot-and-mouth disease virus (Berinstein et al. 1995), and rotavirus (Guerrero et al. 2000), thereby facilitating gene transfer. Its preferential expression on angiogenic endothelium and its key contribution to viral internalization make the integrin $\alpha v\beta 3$ an attractive target for nonviral gene delivery. To this end, a recent study reported the synthesis of a cationic polymerized lipid-based nanoparticle (NP) that was covalently coupled to a small organic $\alpha v\beta 3$ ligand ($\alpha v\beta 3$-NP) (Fig. 2) (Hood et al. 2002). Importantly, the $\alpha v\beta 3$-binding ligand was chosen on the basis of its selectivity for $\alpha v\beta 3$ in both receptor-binding studies and cell adhesion experiments, with an IC_{50} of 0.04 μM for purified $\alpha v\beta 3$ compared to IC_{50} of 5.5 μM for $\alpha v\beta 5$ and 2.1 μM for $\alpha IIb\beta 3$. In cell adhesion experiments, this compound was 100-fold more potent at disrupting $\alpha v\beta 3$-mediated than $\alpha v\beta 5$-mediated cell attachment to vitronectin (0.33 μM vs. 30 μM, respectively). Notably, the increase in avidity due to the presentation of the ligand in the multivalent format of the NP resulted in an 80-fold reduction in the IC_{50} for the particle in cell adhesion experiments.

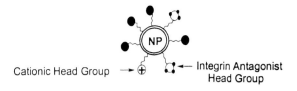

Figure 2. Schematic representation of nanoparticle.

To determine whether the $\alpha v\beta 3$-NP could deliver genes to angiogenic tumor-associated blood vessels, either $\alpha v\beta 3$-NP or a nontargeted nanoparticle (nt-NP) was injected complexed to the gene encoding firefly luciferase into the tail vein of mice bearing $\alpha v\beta 3$-negative M21-L melanomas. After 24 hours, maximal luciferase activity was detected in tumors after injection of NP coupled to 25 μg of luciferase (Fig. 3, inset). At this dose, minimal luciferase was detected in the lung and heart (Fig. 4), and no detectable expression was found in the liver, brain, kidney, skeletal muscle, spleen, and bladder (not shown). Tumor-specific luciferase expression was completely blocked when mice were co-injected with a 20-fold molar excess of the soluble $\alpha v\beta 3$-targeting ligand as a competitive inhibitor of integrin binding, indicating that the interaction between the NP and the integrin mediated gene uptake into the cell (Fig. 4).

Because Raf appears to play a central role in neovascularization (Fig. 1), a mutant form of Raf-1 that fails to bind ATP (ATPμ-Raf) (Heidecker et al. 1990) and that blocks EC Raf activity in cultured ECs (Hood et al. 2002), was evaluated for antitumor effects. The gene encoding FLAG-tagged ATPμ-Raf was complexed to the $\alpha v\beta 3$-NP ($\alpha v\beta 3$-NP/Raf[–]) and injected into mice bearing large subcutaneous $\alpha v\beta 3$-null M21-L melanomas. After 24 or 72 hours, the tumors were removed and costained with anti-VE-cadherin antibody as a blood vessel marker and anti-FLAG to detect gene expression. The tumors were also evaluated for TUNEL staining to mark apoptotic cells, since suppression of Raf activity was reported to promote apoptosis (Wang et al. 1996). Twenty-four hours after injection of $\alpha v\beta 3$-NP/Raf(–), TUNEL-positive cells were only detected among the vessels that had been transduced (FLAG) (Fig. 4A). To assess the impact of ATPμ-Raf on tumor cell viability, cryosections taken from tumors 72 hours after treatment were stained and examined at lower magnification (100×) to evaluate both blood vessels and surrounding tumor parenchyma. In addition to the apoptosis among the blood vessels (VE-cadherin positive), there were concentric rings of apopto-

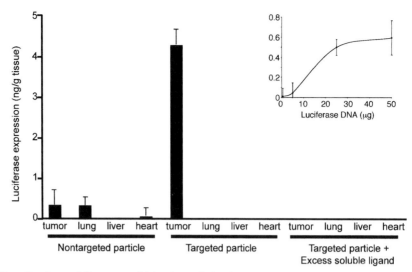

Figure 3. αvβ3-NP-mediated gene delivery to αvβ3-bearing cells in vivo. Athymic WEHI mice were subcutaneously injected with M21-L cells (5 × 10⁶) and tumors were allowed to grow to ~100 mm³. Mice were then injected i.v. with 450 nanomoles of NP electrostatically coupled to 25 μg (Hood et al. 2002) of plasmid expressing firefly luciferase, and one group received a co-injection of 20-fold molar excess of the soluble αvβ3-targeting ligand. After 24 hours, mice were sacrificed, tissues surgically removed, and luciferase activity quantified. Inset shows luciferase expression as a function of DNA dose injected. Each bar represents the mean ± S.D. of five replicates. (Reprinted, with permission, from Hood et. al. 2002 [copyright American Association for the Advancement of Science].)

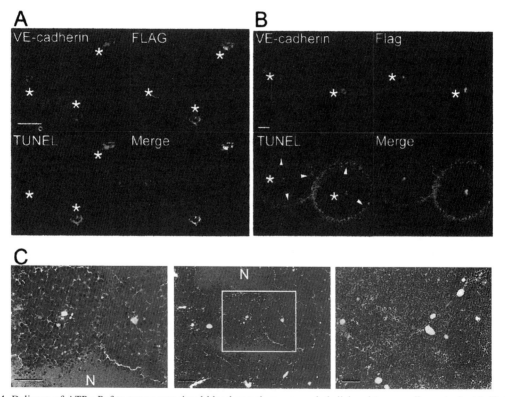

Figure 4. Delivery of ATPμ-Raf to tumor-associated blood vessels causes endothelial and tumor cell apoptosis. (*A–C*) Athymic WEHI mice were subcutaneously implanted with M21-L melanoma, and tumors were allowed to grow to ~400 mm³. Mice were then given a single i.v. injection of αvβ3-NP-Raf(–). Control animals were injected with the αvβ3-NP coupled to a shuttle vector. After 24 or 72 hours, mice were sacrificed, their tumors resected, fixed, sectioned, and stained (Hood et al. 2002). (*A*) Tumors harvested 24 hours after treatment were immunostained for VE-cadherin (endothelial cells), FLAG (gene expression), and TUNEL (apoptosis). Bar, 50 μM. Asterisks denote blood vessels. (*B*) Tumors harvested 72 hours after treatment were stained as above. Bar, 50 μM. Arrowheads denote ring of tumor cells undergoing apoptosis. (*C*) Tumors harvested 72 hours after treatment with αvβ3-NP-Raf(–) (*left* and *center* panels) or controls (*right* panel) were stained with hematoxylin and eosin. Necrotic tissues are denoted by N. Bar, 50 μM, *left* panel; 100 μM, *center* and *right* panels.

sis (TUNEL) among the tumor cells proximal to each apoptotic vessel (Fig. 4B). Accordingly, hematoxylin and eosin revealed extensive tumor necrosis (Fig. 4C), which was found in concentric rings 8–10 cells in diameter, which roughly corresponds to the diffusion distance of a single vessel (Folkman 2001).

To further test the therapeutic efficacy of this construct, this approach was evaluated for efficacy against established syngeneic pulmonary metastases of a colon carcinoma. To this end, murine CT-26 carcinoma cells were injected intravenously into Balb/c mice. This procedure typically results in the formation of experimental lung metastases within 4 days (Xiang et al. 1997) and animal death by 24 days due to extensive pulmonary metastases. This tumor model is typically used to evaluate the anti-metastatic capacity of therapies injected soon after initial tumor injection. However, in this study (Hood and Cheresh 2002), the pulmonary metastases were allowed to establish for 10 days prior to treatment with the NP/gene complexes to ensure that all animals contained actively growing lung tumors. Control mice treated with PBS, αvβ3-NP complexed to a control vector, or a nt-NP/Raf(–) showed extensive tumor burden in the lungs (Fig. 5). In contrast, mice treated with αvβ3-NP/Raf(–) displayed little or no visible tumor metastases (Fig. 5), as demonstrated by a significant reduction in wet lung weight (Fig. 5). Mice injected with αvβ3-NP/Raf(–) along with a 20-fold molar excess of competing soluble targeting ligand had a tumor burden similar to that in control mice, demonstrating that this response is αvβ3-specific (Fig. 5). In a parallel study in which mice were euthanized and tumor volume established during the course of the experiment, αvβ3-NP-Raf(–) was shown to cause regression of pulmonary metastases (Fig. 5).

In summary, we have shown that pronounced tumor regressions can be achieved in mice by systemic delivery of an antiangiogenic gene (Raf-1) that is targeted to the tumor vasculature. Several components of this strategy likely contribute to its antitumor activity, and these may be useful for similar treatments in humans. First, the NP used in this study has multivalent targeting of integrin αvβ3 that selectively delivers genes to angiogenic blood vessels. A similar particle containing gadolinium and the anti-αvβ3 targeting antibody, LM609, has been successfully used to image angiogenic blood vessels in tumor-bearing rabbits (Sipkins et al. 1998). Second, the mutant Raf-1 gene delivered to these tissues influences the signaling cascades of two prominent angiogenic growth factors, bFGF and VEGF. The robust pro-apoptotic activity of this gene is consistent with previous studies revealing a role for Raf-1 in promoting cell survival (Huser et al. 2001). Finally, because nanoparticles are less immunogenic than viral vectors, it may be feasible to deliver therapeutic genes repeatedly to angiogenic blood vessels for sustained treatment of diseases that depend on angiogenesis and vascular remodeling.

CONCLUSIONS

Recent efforts have focused on finding signal transduction pathways regulating angiogenesis and applying these findings to the treatment of disease states. One such target is Raf kinase. Recent reports indicate that Raf plays a central role for both proliferative and survival signals downstream from either VEGF or bFGF. These findings were extended therapeutically using a vascular targeting strategy which selectively delivers to angiogenic vessels. Using this vector, delivery of mutant null forms of Raf ki-

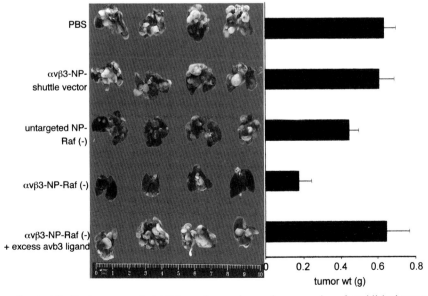

Figure 5. Delivery of mutant Raf to tumor vessels inhibits angiogenesis, causing regression of established tumors. (*A*) Pulmonary metastases of CT-26 colon carcinoma cells were formed in Balb/c mice by i.v. injection (Xiang et al. 1997). Metastatic tumors were allowed to grow for 10 days before mice were injected i.v. on days 10 and 17. Lungs were harvested on day 24 (*B*) or at indicated time points (*C*), weighed (*B–C*) (Xiang et al. 1997), and photographed (*D*). Each bar represents the mean ± S.D. of 6–8 mice. $p<0.05$.

nase resulted in apoptosis of angiogenic endothelial and subsequent tumor apoptosis, leading to regression of established metastatic tumors. Future work in this area will undoubtedly include the evaluation of other genes using this targeted gene delivery strategy and extending the Raf findings to other angiogenesis-dependent disease states such as rheumatoid arthritis and blinding eye diseases.

REFERENCES

Alon T., Hemo I., Itin A., Pe'er J., Stone J., and Keshet E. 1995. Vascular endothelial growth factor acts as a survival factor for newly formed retinal vessels and has implications for retinopathy of prematurity. *Nat. Med.* **1**: 1024.

Aplin A.E., Howe A., Alahari S.K., and Juliano R.L. 1998. Signal transduction and signal modulation by cell adhesion receptors: The role of integrins, cadherins, immunoglobulin-cell adhesion molecules, and selectins. *Pharmacol. Rev.* **50**: 197.

Arap W., Pasqualini R., and Ruoslahti E. 1998. Chemotherapy targeted to tumor vasculature. *Curr. Opin. Oncol.* **10**: 560.

Berinstein A., Roivainen M., Hovi T., Mason P.W., and Baxt B. 1995. Antibodies to the vitronectin receptor (integrin alpha V beta 3) inhibit binding and infection of foot-and-mouth disease virus to cultured cells. *J. Virol.* **69**: 2664.

Blezinger P., Wang J., Gondo M., Quezada A., Mehrens D., French M., Singhal A., Sullivan S., Rolland A., Ralston R., and Min W. 1999. Systemic inhibition of tumor growth and tumor metastases by intramuscular administration of the endostatin gene. *Nat. Biotechnol.* **17**: 343.

Brooks P.C., Montgomery A.M., Rosenfeld M., Reisfeld R.A., Hu T., Klier G., and Cheresh D.A. 1994. Integrin alpha v beta 3 antagonists promote tumor regression by inducing apoptosis of angiogenic blood vessels. *Cell* **79**: 1157.

Chaudhary A., King W.G., Mattaliano M.D., Frost J.A., Diaz B., Morrison D.K., Cobb M.H., Marshall M.S., and Brugge J.S. 2000. Phosphatidylinositol 3-kinase regulates Raf1 through Pak phosphorylation of serine 338. *Curr. Biol.* **10**: 551.

del Pozo M.A., Price L.S., Alderson N.B., Ren X.D., and Schwartz M.A. 2000. Adhesion to the extracellular matrix regulates the coupling of the small GTPase Rac to its effector PAK. *EMBO J.* **19**: 2008.

Eliceiri B.P. 2001. Integrin and growth factor receptor crosstalk. *Circ. Res.* **89**: 1104.

Eliceiri B.P. and Cheresh D.A. 2001. Adhesion events in angiogenesis. *Curr. Opin. Cell Biol.* **13**: 563.

Eliceiri B.P., Paul R., Schwartzberg P.L., Hood J.D., Leng J., and Cheresh D.A. 1999. Selective requirement for Src kinases during VEGF-induced angiogenesis and vascular permeability. *Mol. Cell* **4**: 915.

Fabian J.R., Daar I.O., and Morrison D.K. 1993. Critical tyrosine residues regulate the enzymatic and biological activity of Raf-1 kinase. *Mol. Cell. Biol.* **13**: 7170.

Folkman J. 2001. Angiogenesis-dependent diseases. *Semin. Oncol.* **28**: 536.

Friedlander M., Brooks P.C., Shaffer R.W., Kincaid C.M., Varner J.A., and Cheresh D.A. 1995. Definition of two angiogenic pathways by distinct alpha v integrins. *Science* **270**: 1500.

Giancotti F.G. and Ruoslahti E. 1999. Integrin signaling. *Science* **285**: 1028.

Guerrero C.A., Mendez E., Zarate S., Isa P., Lopez S., and Arias C.F. 2000. Integrin alpha(v)beta(3) mediates rotavirus cell entry. *Proc. Natl. Acad. Sci.* **97**: 14644.

Gutheil J.C., Campbell T.N., Pierce P.R., Watkin J.D., Huse W.D., Bodkin D.J., and Cheresh D.A. 2000. Targeted antiangiogenic therapy for cancer using Vitaxin: A humanized monoclonal antibody to the integrin αvβ3. *Clin. Cancer Res.* **6**: 3056.

Heidecker G., Huleihel M., Cleveland J.L., Kolch W., Beck T.W., Lloyd P., Pawson T., and Rapp U.R. 1990. Mutational activation of c-raf-1 and definition of the minimal transforming sequence. *Mol. Cell. Biol.* **10**: 2503.

Hood J.D. and Cheresh D.A. 2002. Role of integrins in cell invasion and migration. *Nat. Rev. Cancer* **2**: 91.

Hood J., Bednarski M.D., Frausto R., Guccione S., Reisfeld R.A., Xiang R., and Cheresh D.A. 2002. Tumor regression by targeted gene delivery to the neovasculature. *Science* **296**: 2404.

Huser M., Luckett J., Chiloeches A., Mercer K., Iwobi M., Giblett S., Sun X.M., Brown J., Marais R., and Pritchard C. 2001. MEK kinase activity is not necessary for Raf-1 function. *EMBO J.* **20**: 1940.

Kim S., Bell K., Mousa S.A., and Varner J.A. 2000. Regulation of angiogenesis in vivo by ligation of integrin alpha5beta1 with the central cell-binding domain of fibronectin. *Am. J. Pathol.* **156**: 1345.

King A.J., Sun H., Diaz B., Barnard D., Miao W., Bagrodia S., and Marshall M.S. 1998. The protein kinase Pak3 positively regulates Raf-1 activity through phosphorylation of serine 338. *Nature* **396**: 180.

Kiosses W.B., Hood J., Yang S., Gerritsen M.E., Cheresh D.A., Alderson N., and Schwartz M.A. 2002. A dominant-negative p65 PAK peptide inhibits angiogenesis. *Circ. Res.* **90**: 697.

Kolch W. 2000. Meaningful relationships: The regulation of the Ras/Raf/MEK/ERK pathway by protein interactions. *Biochem. J.* **351**: 289.

Kumar C.C., Malkowski M., Yin Z., Tanghetti E., Yaremko B., Nechuta T., Varner J., Liu M., Smith E.M., Neustadt B., Presta M., and Armstrong L. 2001. Inhibition of angiogenesis and tumor growth by SCH221153, a dual alpha(v)beta3 and alpha(v)beta5 integrin receptor antagonist. *Cancer Res.* **61**: 2232.

Langmuir V.K., Cobleigh M.A., Herbst R.S., Holmgren E., Hurwitz H., Kabbinava F., Miller K., and Novotny W. 2002. Successful long-term therapy with bevacizumab (Avastin™) in solid tumors. In *Abstracts from the 38th Annual Meeting of the American Society of Clinical Oncology* (Orlando, Florida), p. 32.

Liu Y., Mounkes L.C., Liggitt H.D., Brown C.S., Solodin I., Heath T.D., and Debs R.J. 1997. Factors influencing the efficiency of cationic liposome-mediated intravenous gene delivery. *Nat. Biotechnol.* **15**: 167.

Losordo D.W., Vale P.R., Symes J.F., Dunnington C.H., Esakof D.D., Maysky M., Ashare A.B., Lathi K., and Isner J.M. 1998. Gene therapy for myocardial angiogenesis: Initial clinical results with direct myocardial injection of phVEGF165 as sole therapy for myocardial ischemia. *Circulation* **98**: 2800.

Marshall C.J. 1995. Specificity of receptor tyrosine kinase signaling: Transient versus sustained extracellular signal-regulated kinase activation. *Cell* **80**: 179.

Murakami M.S. and Morrison D.K. 2001. Raf-1 without MEK? *Sci. STKE* **2001**: PE30.

O'Reilly M.S., Boehm T., Shing Y., Fukai N., Vasios G., Lane W.S., Flynn E., Birkhead J.R., Olsen B.R., and Folkman J. 1997. Endostatin: An endogenous inhibitor of angiogenesis and tumor growth. *Cell* **88**: 277.

Pearson G., Robinson F., Beers Gibson T., Xu B.E., Karandikar M., Berman K., and Cobb M.H. 2001. Mitogen-activated protein (MAP) kinase pathways: Regulation and physiological functions. *Endocr. Rev.* **22**: 153.

Rehn M., Veikkola T., Kukk-Valdre E., Nakamura H., Ilmonen M., Lombardo C., Pihlajaniemi T., Alitalo K., and Vuori K. 2001. Interaction of endostatin with integrins implicated in angiogenesis. *Proc. Natl. Acad. Sci.* **98**: 1024.

Renshaw M.W., Ren X.D., and Schwartz M.A. 1997. Growth factor activation of MAP kinase requires cell adhesion. *EMBO J.* **16**: 5592.

Ridley A. 2000. Rho GTPases. Integrating integrin signaling. *J. Cell Biol.* **150**: F107.

Schlessinger J. 2000. New roles for Src kinases in control of cell survival and angiogenesis. *Cell* **100**: 293.

Schwartz M.A. and Ginsberg M.H. 2002. Networks and crosstalk: Integrin signalling spreads. *Nat. Cell Biol.* **4**: E65.

Senger D.R., Ledbetter S.R., Claffey K.P., Papadopoulos-Sergiou A., Peruzzi C.A., and Detmar M. 1996. Stimulation of endothelial cell migration by vascular permeability factor/vascular endothelial growth factor through cooperative mechanisms involving the alphavbeta3 integrin, osteopontin, and thrombin. *Am. J. Pathol.* **149:** 293.

Sipkins D.A., Cheresh D.A., Kazemi M.R., Nevin L.M., Bednarski M.D., and Li K.C. 1998. Detection of tumor angiogenesis in vivo by alphaVbeta3-targeted magnetic resonance imaging. *Nat. Med.* **4:** 623.

Stupack D.G., Puente X.S., Boutsaboualoy S., Storgard C.M., and Cheresh D.A. 2001. Apoptosis of adherent cells by recruitment of caspase-8 to unligated integrins. *J. Cell Biol.* **155:** 459.

Wang H.G., Rapp U.R., and Reed J.C. 1996. Bcl-2 targets the protein kinase Raf-1 to mitochondria. *Cell* **87:** 629.

Wang Y. and Becker D. 1997. Antisense targeting of basic fibroblast growth factor and fibroblast growth factor receptor-1 in human melanomas blocks intratumoral angiogenesis and tumor growth. *Nat. Med.* **3:** 887.

Wickham T.J., Mathias P., Cheresh D.A., and Nemerow G.R. 1993. Integrins alpha v beta 3 and alpha v beta 5 promote adenovirus internalization but not virus attachment. *Cell* **73:** 309.

Wojnowski L., Zimmer A.M., Beck T.W., Hahn H., Bernal R., Rapp U.R., and Zimmer A. 1997. Endothelial apoptosis in Braf-deficient mice. *Nat. Genet.* **16:** 293.

Xiang R., Lode H.N., Dolman C.S., Dreier T., Varki N.M., Qian X., Lo K.M., Lan Y., Super M., Gillies S.D., and Reisfeld R.A. 1997. Elimination of established murine colon carcinoma metastases by antibody-interleukin 2 fusion protein therapy. *Cancer Res.* **57:** 4948.

Yancopoulos G.D., Klagsbrun M., and Folkman J. 1998. Vasculogenesis, angiogenesis, and growth factors: Ephrins enter the fray at the border. *Cell* **93:** 661.

Combining Antiangiogenic Agents with Metronomic Chemotherapy Enhances Efficacy against Late-stage Pancreatic Islet Carcinomas in Mice

G. Bergers*¶ and D. Hanahan†‡¶

*Department of Neurological Surgery and †Department of Biochemistry & Biophysics, Comprehensive ‡Diabetes and ¶Cancer Centers, University of California, San Francisco, California 94143

Angiogenesis, the formation of new blood vessels, is essential for most tumors to expand (Hanahan and Folkman 1996). Conversely, inhibition of this complex process can demonstrably restrict tumor growth and even elicit tumor shrinkage or regression in animal models (Folkman 2000). Thus, it is not surprising that widespread efforts in industry and academia have focused on the generation of antiangiogenic drugs that target the endothelial cells, the structural constituents of the vascular system. Antiangiogenic therapy takes advantage of the fact that tumor endothelial cells are chronically activated to proliferate and migrate in the course of assembling a neovasculature, in contrast to the largely quiescent endothelium found in the vessels of normal organs. Most of the current angiogenesis inhibitors were identified by their ability to block proliferation or migration of cultured tumor endothelial cells, typically resulting in endothelial cell apoptosis.

The concept of targeting the tumor vasculature as a means to disrupt tumors contrasts with conventional treatment strategies, which typically seek to kill as many proliferating tumor cells as possible by treating tumor-bearing individuals with maximal tolerated doses (MTD) of ionizing radiation or cytotoxic agents that damage DNA or disrupt microtubule functions during mitosis (Kerbel et al. 2002). Such harsh treatment protocols must be alternated with treatment-free periods to permit recovery of normal cells such as bone marrow progenitors and gut mucosal cells. Although often initially efficacious in the form of clinical remissions, such MTD therapies typically elicit recurrent tumors that are insensitive to that therapeutic.

Recently, a new twist on chemotherapeutic trial design has been developed by two groups (Browder et al. 2000; Klement et al. 2000; Kerbel et al. 2002), who discovered that chemotherapeutics, when given in much different dosing schedules, can act as antiangiogenic agents, even in tumors that are resistant to the particular drug. Both the Folkman and Kerbel laboratories showed that by altering the dosing regimen to one of regular inoculations without rest periods at lower (1/3–1/10 of MTD) concentrations, so-called 'metronomic' or antiangiogenic chemotherapy dosing, traditional cytotoxic drugs could produce antiangiogenic effects in xenotransplant models, even against drug-resistant tumors (Browder et al. 2000; Hanahan et al. 2000; Klement et al. 2000). How is such an effect possible? What had been historically ignored with chemotherapeutic strategies was the possibility that proliferating tumor endothelial cells could be killed by cytotoxic drugs. Cytotoxic killing was not evident because their slower doubling times and intact DNA damage response system rendered angiogenic endothelial cells relatively insensitive to episodic high-dose treatment, since they could arrest growth and repair the damage. By increasing the frequency and demanding regularity of treatment, the slowly proliferating endothelial cells could not so readily repair themselves, as they apparently do during the traditional rest periods, resulting in endothelial apoptosis and impaired angiogenesis. Elimination of the rest periods obligated reduction in the doses to avoid myelosuppression, which had the added effect of reducing (in some regimens) other toxic side effects. Kerbel and colleagues further demonstrated that the efficacy of a metronomic chemotherapy could be enhanced by combining it with vascular endothelial growth factor (VEGF) receptor-2-blocking antibodies (VEGFRI). The rationale was that the effects of metronomic chemotherapy on activated endothelial cells might be selectively amplified in the presence of drugs that target survival mechanisms of endothelial cells (Klement et al. 2000). VEGF is known both to stimulate angiogenesis and to prevent endothelial cell apoptosis (Ferrara and Alitalo 1999). Browder and colleagues showed that metronomic chemotherapy could also be beneficially combined with another antiangiogenic agent, TNP-470. Both groups argued that the endothelial cells, being genetically stable, would not prove so susceptible to acquired drug resistance, thus providing a means to finesse the otherwise inevitable drug resistance and relapse that result from chemotherapy.

Motivated by these results, we have asked how metronomic chemotherapy performs in combination with targeted antiangiogenic agents in a genetically engineered mouse model (GEM) of multistep pancreatic islet carcinogenesis, the Rip1Tag2 mouse model. We present here trials of a cytotoxic drug (vinblastine or cyclophosphamide) in a low-dose metronomic regimen, both alone and in combination with a VEGF receptor inhibitor, SU5416, or a broad-spectrum matrix metalloprotease (MMP) inhibitor, BB-94. These combinatorial trials were motivated by both previously published and new data indicating that each class of agent (metronomic chemotherapeutic, VEGF-R inhibitor, and MMP inhibitor) was efficacious when used to treat early-stage disease, but poorly effective in treating well-established solid tumors in this model. Our postulate was that combinatorial therapy might provide an inroad into such otherwise refractory large tumors, and the results support the potential of this strategy for treating end-stage disease.

THE RIP1TAG2 MOUSE MODEL OF MULTISTAGE CARCINOGENESIS ALLOWS DRUG EVALUATION AT DISCRETE STAGES OF TUMOR PROGRESSION

Translation of preclinical results using animal models to clinical trials have typically proved challenging, both in extrapolating experimental design and in interpreting the predictive value of the animal data. In particular, traditional xenograft mouse models for experimental therapeutic trials of anticancer agents have a spotty history in their predictive value. Much has been attributed to the differences in pharmacokinetics consequent to distinctive mouse and human metabolism. Another profound difference, often unmentioned, is that traditional xenotransplant models involve cultured tumor cells that are inoculated into different sites, most frequently subcutaneous, where tumor cells assemble into nodules and grow. In contrast, GEM carrying oncogenes or disruptions in tumor suppressors can develop tumors de novo, much like human cancers, originating out of once-normal cells in their natural tissue microenvironments, typically in multistep pathways. These models are providing new insight into mechanisms of carcinogenesis, and some are already demonstrating potential value in evaluating targeted therapies for different types of cancer. Among such GEM, the Rip1Tag2 model (Hanahan 1985) has proved to be particularly amenable toward investigation of the efficacy of candidate drugs and treatment regimens at distinct stages of cancer progression.

Rip1Tag2 mice express the SV40 T-antigen oncogenes under the control of the rat insulin promoter in the β cells populating approximately 400 islets of Langerhans in the pancreas; the mice consequently develop islet cell carcinomas in a multistep pathway. Although oncogene expression ensues in the embryo, morphologically normal islets are formed. Aberrant hyperproliferation starts at 3–4 weeks of age, producing hyperplastic and dysplastic islets, initially with a quiescent vasculature. Then, at 5–7 weeks of age, angiogenesis is switched on in a subset of these dysplastic islets from which solid tumors, encapsulated or invasive, arise as soon as the mice reach an age of about 9–10 weeks (Fig. 1). Most of the transgenic mice die between 13 and 14 weeks of age as a combined consequence of tumor burden and hyperinsulinemia (Fig. 1).

ANTIANGIOGENIC AGENTS SHOW STAGE-SPECIFIC EFFICACY DEPENDENT ON DISEASE PROGRESSION

We have demonstrated the utility of the Rip1Tag2 model as a platform for experimental therapeutic trials (Parangi et al. 1996; Bergers et al. 1999, 2000), taking advantage of the multistage pathway of tumorigenesis to establish three distinctive trial protocols. These trials are designed to investigate whether specific inhibitors can (*1*) prevent the angiogenic switch in premalignant lesions (prevention trial, 'PT'), (*2*) intervene in the rapid expansion of small tumors (intervention trial, 'IT'), or (*3*) regress or stabilize bulky, end-stage tumors (regression trial, 'RT') (Fig. 1). In the PT, treatment commences at 5 weeks of age, when angiogenic switching in hyperplastic islets begins, and continues to 10.5 weeks. We then physically isolate and count angiogenic islets and nascent tumors, which can be distinguished from nonangiogenic islets by their hemorrhagic nature, which produces red-colored nodules. In the IT, treatment is started later, at 10 weeks, when solid tumors are just forming, and continues until 13.5 weeks of age. At this time mice are moribund with substantial tumor burden. Tumors are dissected and tumor number and cumulative volumes (burden) are determined. The same procedure is undertaken in the RT,

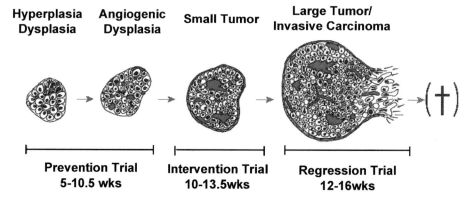

Figure 1. Overview of tumor progression and experimental therapeutic trials in the Rip1Tag2 model of pancreatic islet carcinogenesis. Expression of the SV40-T antigens in pancreatic β cells of the islet of Langerhans in Rip1Tag2 mice elicits a multistage pathway beginning as normal islets that progress to hyperplasic islets, dysplasic islets, angiogenic (dysplastic) islets, and then to islet tumors. Based on the synchronicity of progression as a function of age, three therapeutic trials were designed to target distinct stages of disease progression. The three trials aimed to prevent angiogenesis in early lesions (prevention), to reduce growth of small tumors (intervention) or to regress bulky end-stage tumors (regression).

only the trial starts at 12 weeks of age, when mice carry a substantial tumor burden and continues until the mice die at 13.5–14 weeks, or until a defined endpoint of 16 weeks, if the treatment is able to extend life span.

In the course of evaluating a series of candidate angiogenesis inhibitors using these three trial designs, a provocative concept emerged, which is that certain antiangiogenic inhibitors were found to be most effective in treating early-to-mid-stage disease, whereas others were more effective on late-stage disease (Bergers et al. 1999). Among the inhibitors of the first group, two classes of synthetic antiangiogenic agents are shown in Figure 2: (1) SU5416, an inhibitor of the two receptor tyrosine kinases, VEGF-R1 and -R2, that signal in response to the angiogenic growth factor VEGF-A and (2) two broad-spectrum metalloprotease inhibitors (MMPI), BB-94 (batimastat, British Biotech.) and BAY 12-9566 (Bayer Corporation), which inhibit a variety of matrix metalloproteinases, including MMP-2, MMP-9, MMP-3, and MMP-8 (Sternlicht and Bergers 2000). We previously reported on BB-94 in all three trials (Bergers et al. 1999) and on SU-5416 in prevention and intervention trials (Bergers et al. 2000). The regression trials of SU5416 are presented here for the first time, as are all three trials of BAY 12-9566. Regimen and dose schedules are summarized in Table 1.

The trials of SU5416 demonstrated that this agent was highly effective in blocking the initial angiogenic switch induced in the previously quiescent islet capillaries (Fig. 2A, PT) and at impairing the explosive growth of nascent tumors (Fig. 2A, IT) indicative of a crucial role for VEGF signaling in islet tumorigenesis (Fig. 2A, PT). Consistent with this conclusion, we recently showed that targeted deletion of the VEGF-A ligand gene in the pancreatic β cells resulted in an almost complete block of angiogenic switching in progenitor hyperplastic/dysplastic islets (Inoue et al. 2002). Furthermore, tumor growth was severely affected in mice whose oncogene-expressing islets lacked VEGF. Most tumors were small with an impaired angiogenic phenotype, containing avascular, necrotic cores, with viable tumor cells present only in a thin layer at the periphery. These results support the notion that tumor growth indeed is angiogenesis-dependent. One question that could not be answered by the cell-type-specific gene-knockout approach is whether VEGF signaling remains a critical factor at late stages of disease progression. Pharmacological intervention, in contrast, has allowed this question to be addressed. When SU5416 therapy was initiated against early-stage tumors in an IT, it was effective, reducing tumor burden by ~70%. In marked contrast, this VEGF receptor inhibitor had little effect on end-stage tumors in a RT (Fig. 2A, RT). Although SU5416 was able to achieve a modest extension of life span (from 14 to16 weeks), it was not able to elicit regression of tumor mass, or even to produce 'stable disease'.

We similarly tested the two MMPIs, BB-94 and BAY12-9566, at distinct stages of tumorigenesis, and found that each produced an efficacy profile similar to SU5416. Both were effective at reducing the incidence of angiogenic switching by ~50% (Fig. 2B, PT), and each restrained growth of small solid tumors by ~70–80% (Fig. 2B, IT) As with SU5416, neither MMPI was able to regress bulky end-stage tumors (Fig. 2B, RT), a result that is likely relevant to the experiences of MMP-I in clinical trials. Three MMPI, including a relative of BB-94 (Marimastat) and BAY 12-9566 were withdrawn from clinical trials due to disappointing responses; (Sternlicht and Bergers 2000; Coussens et al. 2002).

Notably, the similar stage-specific efficacy patterns seen with a VEGFR inhibitor and two MMPI is likely not coincidental. MMP-9, a target protease of these MMPI, is now appreciated to be a regulator of VEGF-A bioavailability in the pancreatic islets. The data support a model whereby MMP-9 produced by leucocytes infiltrating dysplastic islets is released and activated concomitant with the angiogenic switch, thereby mobilizing constitutively expressed but physically sequestered VEGFA, which in turn triggers an angiogenic response in the islet capillaries (Bergers et al. 2000). MMP-9 deficient-Rip1-Tag2 mice, as well as mice that are treated early with BB-94 show a similarly reduced frequency of angiogenic switching and comparable reductions in tumor growth. Interestingly, although MMPs were originally identified as molecules that affect invasion, neither genetic nor functional ablation of MMPs prevented the eventual histologic progression to invasive carcinomas.

In summary, both a VEGF-RI and two MMPIs were efficacious in treating premalignant disease and small tumors, but none was able to regress bulky end-stage tumors. This later stage is the situation, however, in which clinical trials are typically performed, and the state where the need for better cancer therapies is particularly acute. Most patients who enter clinical trials have bulky disease, having failed multiple conventional therapies. Our results predict that MMPI as well as both VEGF and VEGF-R inhibitors will not prove generally efficacious against well-established solid tumors, if one accepts the proposition that the Rip1Tag2 model can serve as a general prototype for therapeutic strategies that target the tumor vasculature. Certainly the results with this model are consistent with the poor responses of MMPI in the clinic and predict that VEGF/VEGFR inhibitors will similarly prove most effective against early-stage disease in human trials. Notably, the VEGFR inhibitor we tested here, SU5416, has recently been withdrawn from clinical trials (http://www.pharmaciaoncology.com/popup.asp?ptype=ct_SU5416&link=CT) based on a number of unelaborated considerations, most likely including poor responses against late-stage disease.

Given the imperative to develop drugs that are effective against late-stage tumors, an increasingly common strategy is to combine experimental agents such as these with conventional cytotoxic drugs delivered in MTD regimens. Indeed, such trials were ongoing both with BAY 12-9655 and SU5416, apparently without great benefit. We became interested in further evaluating the prospects of metronomic chemotherapy, based on the results and considerations discussed above, and describe below trials of metronomic chemotherapy, involving two cytotoxic drugs delivered alone, and in combination with the antiangiogenesis inhibitors just discussed.

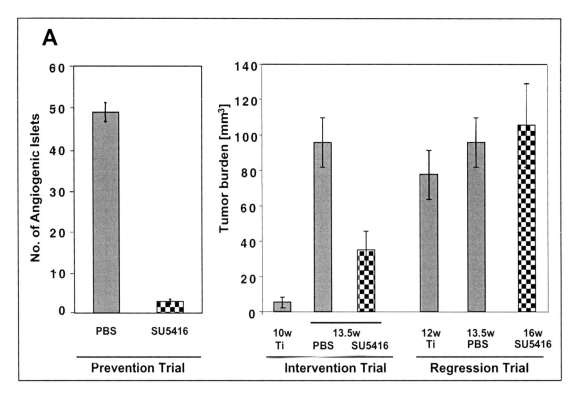

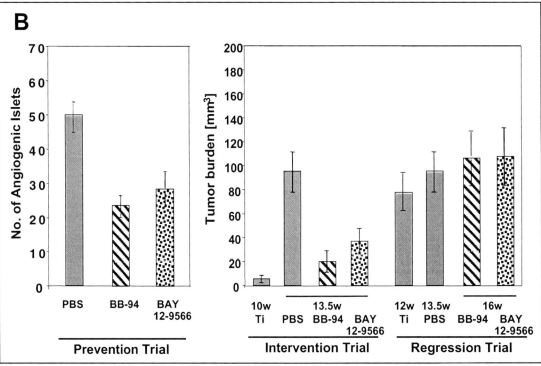

Figure 2. The VEGF-R inhibitor SU5416 and two MMP inhibitors, BB-94 and BAY12-9566, show stage-specific efficacy in the RIP1Tag2 model of carcinogenesis. Rip1Tag2 mice were treated at distinct stages of disease progression (Prevention, Intervention, Regression Trials). The average numbers ± S.E.M. of angiogenic islets of control (PBS) and treated mice are shown in the prevention trial, whereas average tumor burden ± S.E.M. of control and treated animals is shown in the intervention and regression trial. Cohorts of 5–21 animals were used. Statistical analysis was performed with a two-tailed, unpaired Mann-Whitney test, comparing experimental groups to PBS-injected control mice. Tumor burdens of experimental groups in the regression trial were compared to the tumor burden of 12-week-old Rip1Tag2 mice (Ti). p values less than 0.1 are considered statistically significant. (*A*) SU5416 was very efficacious in the prevention and intervention trial but had only modest effects on tumor reduction in the regression trial (p values of SU5416 PT= 0.0001; SU5416 IT= 0.0124; SU5416 RT= 0.1807). (*B*) The two MMPIs have a comparable efficacy profile to SU5416, being efficacious in early but not late stages (p-values of BB-94 PT = 0.0007, BAY12-9566=0.021; BB-94 IT= 0.0002, BAY12-9566 IT= 0.0003; BB-94 RT= 0.2029; BAY 12-9566 RT= 0.34).

Table 1: Cytotoxic and Antiangiogenic Dosing Regimens

Drug	Dosing and Administration Route			Reference
Vinblastine MD	0.75 mg/m^2	i.p.	bolus	Klement et al. (2000)
	1.5 mg/m^2	i.p.	3rd day	
Cyclophosphamide (CTX) MD	10 mg/kg	drinking water	Q.D.	Man et. al. (2002)
SU5416	50–100 mg/kg	s.c.	Q.O.D.	Bergers et al. (2000)
BAY-12-9566	100 mg/kg	p.o.	Q.D.	Bayer Corp., (pers. comm.)
BB-94	30 mg/kg	i.p.	Q.D.	Bergers et al. (2000)

RipTag2 mice were treated with one of two cytotoxic drugs and/or one of two angiogenesis inhibitors as indicated above. Treated and control mice were sacrificed at the end of each trial and pancreases collected for gross and/or histopathology, in particular, quantitation of the angiogenic switching frequency or tumor burden. (MD, metronomic dosing; i.p., intraperitoneal; s.c., subcutaneous; p.o., oral; Q.D., daily). To evaluate a prevention trial (5–10.5-wk treatment), islets were isolated by retrograde perfusion and collagenase P digestion of the pancreas, and sorted under darkfield illumination in a dissecting microscope; angiogenic islets were scored by their red flecked color, and counted. For intervention trials (10–13.5-wk treatment) and regression trials (12–16-wk treatment), tumors were physically excised using fine surgery tools and a dissecting microscope. Tumor burden of a mouse was calculated by summing the tumor volume (volume [mm^3] = 0.52 × [width]^2 × [length]) of every pancreatic tumor >1 mm in diameter. The details of the methods can be found in Bergers et al. (1999, 2000).

ASSESSING THE PROSPECTS OF METRONOMIC CHEMOTHERAPY AGAINST END-STAGE DISEASE

We sought to improve the efficacy of the MMP-I BB-94 and the VEGFR-I SU5416 in late-stage pancreatic islet tumors by combining them with the cytotoxic agents vinblastine and cyclophosphamide (CTX) delivered in metronomic chemotherapy regimens. We asked whether a combinatorial therapy was able to reduce tumor burden and extend life span in a regression trial. Thus, we treated 12-week old mice that bear substantial initial tumor burden (T_i, Fig. 3A,B) and will live no longer than another 2 weeks, and assessed survival and tumor burden. The trials were arbitrarily ended after 4 weeks of treatment (at 16 weeks) unless the mice died sooner, as is the case for untreated controls. The cytotoxic drugs were supplied in a metronomic regime alone, or in combination with SU5416 or BB-94. The trial results are summarized in Figure 3 (A and B); Table 1 indicates regimen and dosing schedules. The single-agent trials indicate little or no efficacy for vinblastine, whereas CTX had demonstrable efficacy when delivered continuously in the drinking water. Development of the oral dosing protocol for mice by the Kerbel group was motivated by the routine oral delivery of CTX in humans (Kerbel et al. 2002); this oral dosing protocol in mice has demonstrated efficacy both in traditional xenotransplant models and in the Rip1-Tag2 model (Man et al. 2002).

As shown and discussed above, neither BB-94 nor SU5416 was able to stop continuing tumor growth in a regression trial as a single agent. In contrast, the combination trials with vinblastine showed benefit, producing an apparent condition of stable disease, where continuing growth was restricted, but the tumors did not regress. Remarkably, the combination of oral CTX with both drugs reduced tumor burden by ~50–60% compared to the initial tumor burden at 12 weeks, indicative of tumor regression (Figs. 4 and 5). CTX alone had potential to stabilize disease in some individuals, but 50% of the mice were dead at the end of the regression trial. Thus, the metronomic regimen of oral CTX produced clear and convincing combinatorial efficacy, given that none of the three agents (CTX, BB-94, or SU-5416) was able to significantly regress late-stage tumors alone.

We also assessed survival in each cohort of treated or control mice. Metronomic regimens involving CTX or vinblastine alone resulted in about 50% survival at 16 weeks, whereas all untreated controls were already dead by this time (Figs. 4 and 5). BB-94 treatment, either alone or in combination with vinblastine, also resulted in a median survival rate of 50% at 16 weeks (Fig. 5A). In contrast, combined administration of CTX and BB-94 significantly improved survival and reduced tumor growth in comparison to single treatments, and therefore proved to be the best combination of all variations tested here (Fig. 5B). Interestingly, tumor burden and survival did not always correlate. Although SU5416 + CTX combination therapy reduced tumor burden, it did not improve survival compared to SU5416 treatment as a single agent (Fig. 4B). The other combination, of SU5416 with vinblastine, caused substantial toxicity, leaving only 30% of the mice alive at 16 weeks (Fig. 4A), and thus was worse than either single agent alone. We and other investigators have noticed that SU5416 causes hemorrhage formation in the lungs due to uncharacterized side effects that were seemingly exacerbated in the presence of vinblastine. These results suggest that the benefits of combinatorial therapies involving cytotoxic drugs delivered in metronomic regimes may be combination- and regimen-specific, because we have been able to improve efficacy of an MMPI but not of SU-5416, when considering survival as the factor.

CONCLUSIONS

We have shown using a prototypical model of multi-stage tumorigenesis that combinatorial therapies involving antiangiogenic agents and metronomic chemotherapy (itself demonstrably antiangiogenic) can produce tumor regression and increased survival in mice with late-stage pancreatic islet carcinomas. These results are encouraging, given our sobering observation that none of the single agents (metronomic chemotherapeutic, VEGF-RI, or MMPI) was effective alone against late-stage disease,

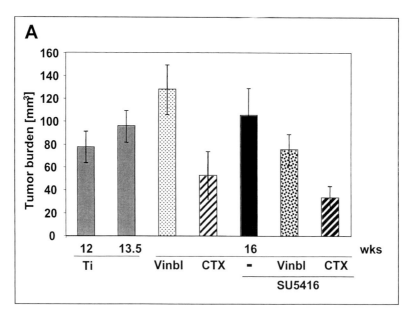

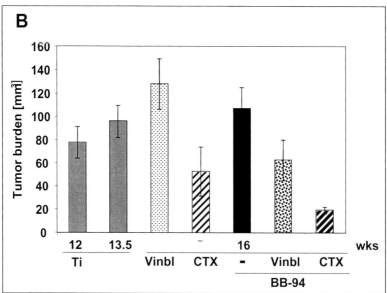

Figure 3. Metronomic chemotherapy enhances the efficacy of VEGF-RI and MMPI in a late-stage regression trial in Rip1Tag2 mice. Starting at 12 weeks, mice with substantial tumor burden were treated for 4 weeks with SU5416 (*A*) or BB-94 (*B*), alone, or in combination with metronomic, low-dose cyclophosphamide (CTX) or Vinblastine (Vinbl). Average tumor burden ± s.e.m. of 12-week-old control mice (Ti) and 16 wk old treated animals are shown in the regression trial. Cohorts of 5–21 animals were used. Statistical analysis was performed with a two-tailed, unpaired Mann-Whitney test comparing experimental groups to the initial tumor burden of 12-week-old Rip1Tag2 mice (Ti). p values less than 0.1 are considered statistically significant. SU5416 (*A*) or BB-94 (*B*) alone did not stop tumor growth, whereas SU5416 or BB-94 + vinblastine produced "stable disease," and SU5416 or BB94 + CTX elicited tumor regression (p-values of SU5416= 0.999; SU5416/CTX= 0.048; SU5416/Vinbl.= 0.7573; BB-94= 0.2029; BB-94/CTX= 0.0026, BB-94/Vinbl.= 0.5113).

congruent with disappointing clinical trials with the latter two classes of agents. Thus, we are hopeful that similar benefits could be realized in the clinic, in particular by combining MMPI or VEGF-RI with low-dose metronomic chemotherapy rendering effective drugs otherwise destined for failure. Although we have not yet comprehensively evaluated MTD regimens in similar combination with MMPI or VEGF-RI, we suspect that metronomic low-dose regimens will have demonstrable benefit over traditional regimes, particularly when combined with other agents intended to target endothelial cells and thereby inhibit angiogenesis. There is a further rationale in favor of metronomic dosing designs in combination with MMPIs. It has recently been shown that pharmacological inhibition (or genetic ablation) of MMP-9 impairs recovery from high-dose chemotherapy, by restricting mobilization of bone marrow progenitors (Heissig et al. 2002); repopulation of the hematopoeitic system from the bone marrow is crucial for resolving myleosuppression and obviating susceptibility to acute infection consequent to high-dose chemotherapy. Interestingly, the combination of metronomic low-dose CTX with the MMPI BB-94

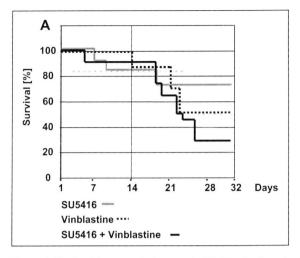
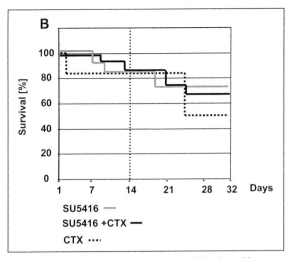

Figure 4. Kaplan-Meyer survival curves in Rip1Tag2 mice submitted to the VEGFRI SU5416 alone or in combination with metronomic chemotherapy. Survival curves of Rip1Tag2 mice treated with SU5416 alone or in combination with vinblastine (*A*) or CTX (*B*) are shown over a period of 32 days in the regression trial. All untreated mice were dead 2 weeks after the trial started. Cohorts of 6–18 mice were used. There was no survival benefit of SU 5416 combined with either Vinbl. (*A*) or CTX (*B*) in the 4-week treatment period.

had the best survival benefit in our study, suggesting that this regimen was not impairing homeostasis of the hematopoeitic system.

The results presented here give further encouragement to combinatorial trial designs if one recalls that we demonstrated enhanced efficacy in a relatively short treatment period of 4 weeks, in a very stringent multifocal model, where physiological stresses of hyperinsulinemia exacerbate local effects of tumor burden. Rip1Tag2 mice become hyperinsulinemic at endstage, and if the tumors do not stabilize and regress fast enough, the mice can develop acute hypoglycemic shock and die. This may explain why a subset of Rip1Tag2 micc in every treatment cohort typically drop out near the beginning of every regression trial. Taking the unusual characteristic of this model into consideration, we expect that metronomic low-dose CTX combined with VEGF/VEGFR inhibitors will prove to confer better survival if tumor-bearing mice are treated for longer periods of time, particularly for other types of cancer where the physiological burden of hyperinsulinemia is not a factor, or perhaps in islet carcinomas when hyperinsulinemia is managed pharmacologically. This argument is supported by the findings of the Kerbel group, who initially observed continuing tumor growth in trials involving oral CTX plus an anti-VEGFR antibody in xenotransplant mice, before disease stabilization ensued after ~60 days of treatment (Man et al. 2002). Furthermore, other VEGFR-I may not show some of the

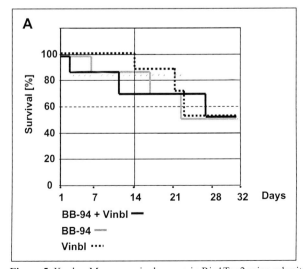
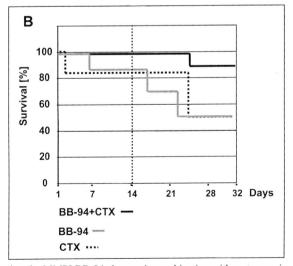

Figure 5. Kaplan-Meyer survival curves in Rip1Tag2 mice submitted to the MMPI BB-94 alone or in combination with metronomic chemotherapy. Survival curves of Rip1Tag2 mice treated with BB-94 alone or in combination with Vinblastine (*A*) or CTX (*B*) are shown over a period of 32 days in the regression trial. All untreated mice were dead 2 weeks after the trial started. Cohorts of 6–12 mice were used. BB-94 combined with Vinblastine (*A*) had no survival benefit, but when combined with CTX (*B*), survival was significantly improved.

toxicities associated with SU5416, some of which are suspected to reflect its particular chemistry rather than its actions against the VEGF receptors. In view of these considerations, we conclude there is good reason to continue investigating combinations of metronomic low-dose chemotherapy with VEGF/VEGFR inhibitors.

In summary, the results presented herein support the notion that low-dose metronomic chemotherapy in appropriate combinations can substantially improve current treatment modalities and thereby render otherwise poorly effective angiogenesis inhibitors demonstrably beneficial. We demonstrate for the first time that MMPIs can be efficacious in late-stage disease when used in conjunction with chemotherapy if the metronomic dosing schedule is applied. Thus, our data beg the proposition that MMPIs could be resurrected clinically as cancer therapeutics if they were tested in combination with metronomic chemotherapy; in contrast, we have reservations about MTD chemotherapy and suggest it will not demonstrate such combinatorial efficacy and low toxicity against well-established human tumors. We hope this proposition will be tested as well. Beyond the potential for improving efficacy of MMPIs, the work reported here, as well as the previous reports from the Folkman and Kerbel labs (Browder et al. 2000; Klement et al. 2000, 2002; Man et al. 2002), indicates that other classes of angiogenesis inhibitors will benefit from combination with metronomic, low-dose cytotoic therapy. Perhaps such combination therapies will prove able to destabilize and slowly regress well-established human tumors much as they can in the de novo islet carcinomas of the pancreas in Rip1Tag2 mice, producing a chronic, manageable disease in some cases, and potentially resolving tumors altogether in others. More generally, one might ask whether other distinct classes of targeted therapies (e.g. the receptor tyrosine kinase inhibitors Herceptin and Aressa) that are currently used with conventional MTD chemotherapy might also be beneficially combined with metronomic chemotherapy, thereby targeting both tumor cells and their supporting vasculature. Indeed, Herceptin has recently been reported to be antiangiogenic (Izumi et al. 2002). Such possibilities deserve further consideration, as modified trial designs involving the combination of new target-selective drugs with metronomic low-dose cytotoxic therapy might provide inroads into otherwise intractable disease states.

ACKNOWLEDGMENTS

We thank SUGEN Inc., South San Francisco, California, for SU5416; British Biotech Pharmaceuticals, Oxford, United Kingdom, for BB-94 (batimastat); and Bayer Corp. West Haven, Connecticut, for BAY-12-9655. We also thank E. Soliven and J. Imperio for excellent technical assistance, and A. McMillan for help with statistical analysis. This work was funded by grants from the National Cancer Institute and by Bill Bowes, whom we thank for his visionary support and encouragement. Correspondence may be addressed to either author.

REFERENCES

Bergers G., Javaherian K., Lo K.-M., Folkman J., and Hanahan D. 1999. Effects of angiogenesis inhibitors on multistage carcinogenesis in mice. *Science* **284**: 808.

Bergers G., Brekken R., McMahon J., Vu T., Itoh T., Tamaki K., Tanzawa K., Thorpe P., Itohara S., Werb Z., and Hanahan D. 2000. Matrix Metalloproteinase-9 triggers the angiogenic switch during carcinogenesis. *Nat. Cell Biol.* **2**: 737.

Browder T., Butterfield C.E., Kräling B.M., Shi B., Marshall B., O'Reilly M.S., and Folkman J. 2000. Antiangiogenic scheduling of chemotherapy improves efficacy against experimental drug-resistant cancer. *Cancer Res.* **60**: 1878.

Coussens L.M., Fingleton B., and Matrisian M. 2002. Matrix metalloproteinase inhibitors and cancer: Trials and tribulations. *Science* **295**: 2387.

Ferrara N. and Alitalo K. 1999. Clinical applications of angiogenic growth factors and their inhibitors. *Nat. Med.* **5**: 1359.

Folkman J. 2000. Tumor angiogenesis. In *Cancer medicine*, p. 2546. 5th edition (ed. J.F. Holland et al.), B.C. Decker, Hamilton, Ontario.

Folkman J. 2000. 'Tumor angiogenesis', 5th ed., in H. e. al., (ed.), *Cancer Medicine*, p. 2546. B.C. Decker, Hamilton, Ontario.

Hanahan D. 1985. Heritable formation of pancreatic β-cell tumors in transgenic mice harboring recombinant insulin/simian virus 40 oncogenes. *Nature* **315**: 115.

Hanahan D., Bergers G., and Bergsland E. 2000. Less is more, regularly: Metronomic dosing of cytotoxic drugs can target tumor angiogenesis in mice. *J. Clin. Invest.* **105**: 1045.

Hanahan D. and Folkman J. 1996. Patterns and emerging mechanisms of the angiogenic switch during tumorigenesis. *Cell* **86**: 353.

Heissig B., Hattori K., Dias S., Friedrich M., Ferris B., Hackett N., Crystal R., Besmer P., Lyden D., Moore M., Werb Z., and Rafii S. 2002. Recruitment of stem and progenitor cells from the bone marrow niche requires mmp-9 mediated release of kit-ligand. *Cell* **109**: 625.

Inoue M., Hager J., Ferrara N., Miller D., Basilico C., Gerber H.-P., and Hanahan D. 2002. VEGF- has a critical non-redundant role in angiogenic switching and pancreatic β-cell carcinogenesis. *Cancer Cell* **1**: 193.

Izumi Y., Xu L., di Tomaso E., Fukumura D., and Jain R. 2002. Tumour biology: Herceptin acts as an antiangiogenic cocktail. *Nature* **416**: 279.

Kerbel R.S., Klement G., Pritchard K.I., and Kamen B. 2002. Continuous low-dose anti-angiogenic/metronomic chemotherapy: From the research laboratory into oncologic clinic. *Ann. Oncol.* **13**: 12.

Klement G., Baruchel S., Rak J., Man S., Clark K., Hicklin D.J., Bohlen P., and Kerbel R.S. 2000. Chronically sustained regressions of human tumor xenografts in the absence of overt toxicity by continuous low-dose vinblastine and anti-VEGF receptor-2 antibody therapy. *J. Clinical Invest.* **105**: R15,

Klement G., Huang P., Mayer B., Green S.K., Man S., Bohlen P., Hicklion D., and Kerbel R.S. 2002. Differences in therapeutic indexes of combination metronomic chemotherapy and an anti-VEGFR-2 antibody in multidrug-resistant human breast cancer xenografts. *Clin. Cancer Res.* **8**: 221.

Man S., Bocci G., Francia G., Green S.K., Jothy S., Bergers G., Hanahan D., Bohlien P., Hicklin D., and Kerbel R.S. 2002. Anti-tumor effects in mice low-dose (metronomic) cyclophosphamide administered continously through the drinking water. *Cancer Res.* **62**: 2731.

Parangi S., O'Reilly M., Christofori G., Holmgren L., Grosfeld J., Folkman J., and Hanahan D. 1996. Antiangiogenic therapy of transgenic mice impairs de novo tumor growth. *Proc. Natl. Acad. Sci.* **93**: 2002.

Sternlicht M. and Bergers G. 2000. Matrix metalloproteinases as emerging targets in cancer: Status and prospects. *Emerg. Ther. Targets* **4**: 609.

Zebrafish: The Complete Cardiovascular Compendium

C.A. MACRAE AND M.C. FISHMAN

Cardiovascular Research Center, Massachusetts General Hospital and Harvard Medical School, Charlestown, Massachusetts 02129

The genetic screen provides a powerful pyramid of biological insight. As shown in Figure 1, at the pinnacle is the opportunity to understand the essential logic of a developmental system, obtained during the initial phenotypic analysis. What are the essential and interpretable decisions? There had been a sense, prior to the work in the zebrafish, that organ development occurred too late to be clearly dissected through a screen; that the mutations that affected heart development would be too pleiotropic, and hence uninformative. Fortunately, it turned out that the screens have been as successful in revealing unexpected steps to organogenesis as about any developmental processes (Fishman and Olson 1997). The next layer, as shown in Figure 1, the cloning of mutated genes, required the establishment of a genomic infrastructure for zebrafish. Now generated, this too has provided novel insights to new genes or families of genes (Fishman 2001). In some cases, these genes already have been used as entrance points to pathways such as *gridlock*, which, embedded in a Notch pathway, culminates in generation of the arterial cell fate (Zhong et al. 2000, 2001). One of the more surprising attributes of the zebrafish embryo is how well some of the mutations model human disease (Shin and Fishman 2002), thereby providing mechanisms and candidate genes for common human disorders. We address some of these insights here.

ASSEMBLY AND GROWTH OF THE CARDIOVASCULAR SYSTEM IN ZEBRAFISH

During evolution, the vertebrate cardiovascular system most likely arose as an integrated entity, including an endothelial-lined circulation and a multichambered heart, driving a unidirectional flow of blood under physiological controls (Fishman and Olson 1997). Ancestral chordates, presumably resembling contemporary amphioxus or tunicates, would have had a single-chambered, thin-walled heart, and pacemaker function alternating between ends, so that blood would flow first in one direction and then the other. Although the circulation in these organisms, in contrast to *Drosophila*, may have been within a closed vascular system, it is unlikely to have been lined by endothelium, the interface between blood constituents and vascular wall (Warren et al. 2000). Several elements came into play to guarantee the unidirectional flow of blood, which in turn, permitted localized, efficient oxygenation at respiratory organs (Fishman and Chien 1997; Fishman and Olson 1997). One is the development of a more muscular ventricle, designed to generate high pressures, as opposed to the atrium, which acts as a capacitance vessel collecting venous return. A single localized pacemaker, along with nodes and specialized conduction system within the heart, ensures that chambers contract in a paced and sequential manner, and couples the circulation to the changing activities of the whole organism. The functional specialization of the vessels and their endothelial linings generates a repertoire of arteries, capillaries, and veins, with features particular to each organ or vascular bed, and provides additional mechanisms for global and regional regulation (Arap et al. 2002). These are associated with increased intravascular pressures which, through ultrafiltration and selective reabsorption in the kidney, enable the homeostasis of extracellular fluid volume and composition. Genetic screens in zebrafish have begun to reveal both the unitary components of the cardiovascular system and the logic for many of these elements (Stainier et al. 1996; Chen and Fishman 2000).

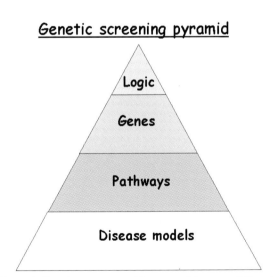

Figure 1. Genetic screens provide a powerful means of dissecting biological processes. The phenotypes themselves often reveal much about the underlying logic of the particular phenomenon. With the identification of the responsible gene defect, critical functions are attributed to specific loci; pathways can then be entered and testable models created.

Certain general principles have emerged which underpin the assembly of this complex system. Prominent among these is the interplay of use-dependent and use-independent mechanisms, in some ways akin to the establishment and subsequent remodeling of the nervous system. Blood pressure certainly regulates the final structure of the adult cardiovascular system. It is apparent that some of the nascent structures also depend on use (Serluca et al. 2002). Other unitary elements of the heart and vascular tree seem to be independent of electrical or hemodynamic forces. All of these would have been difficult to discern in amniotes because of the dependence of these organisms on a functional circulation for survival. In the zebrafish embryo, which lives without a circulation for days, subsisting by diffusion, mutations primarily affecting heart or vessels can be directly observed and evaluated without confusion of the phenotypes due to secondary degeneration (Driever and Fishman 1996).

CHAMBER FORM AND PATTERN

We focus here on the later organotypic pathways that fashion the global structure of the heart, rather than on cell fate decisions, which designate cardiocytes and subdivide them into atrial and ventricular precursors.

We have found that the heart as an organ is designed along two axes (Fig. 2), antero-posterior and concentric (Fishman and Chien 1997). In the antero-posterior direction, the ventricle's placement downstream from the atrium requires heart and soul (*has*). Mutation of *has* causes the remarkable phenotype of ventricle forming within the atrium (Fig. 3). Thus, atrial and ventricular fates are achieved, as is the assembly of cardiac chambers, but the placement of these chambers is incorrect. The atrium and ventricle beat independently and the heart is small (Stainier et al. 1996). The defect first is manifest just as the heart tube is assembling at the midline from bilateral precursors. This occurs normally by a process of fusion, which progresses from posterior to anterior, to generate a single midline cone based on the yolk. Normally posterior fusion anticipates anterior, but in *has* mutant embryos, posterior fusion is delayed (Stainier et al. 1993; Peterson et al. 2001). *has* mutant embryos also have a defect in formation of the central nervous system with disruption of the layering of epithelia in the neural plate and in the retina. We sought to dissect the cardiac from the central nervous system phenotype using a small-molecule screen. We identified concentramide as a molecule that causes the cardiac *has* phenotype but without nervous system degeneration. The mechanism of action also appears to affect antero-posterior polarity at the

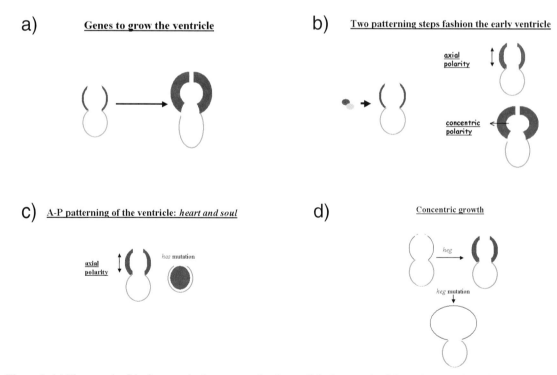

Figure 2. (*a*) The growth of the heart, and other organs, closely parallels the growth of the entire organism, although the regulation of this coupling is unknown. (*b*) The early growth of the heart can be demonstrated to occur along at least two discrete axes. These patterning steps, antero-posterior or axial growth and concentric growth, have distinctive mechanisms and regulation. (*c*) Antero-posterior growth patterning is disrupted in the *heart and soul* (*has*) mutant, such that the ventricle forms within the atrium. The specification of atrial and ventricular cell fates occurs normally, but as a result of abnormalities in the fusion of the cardiac primordia, the relative placement of these chambers is dramatically perturbed. (*d*) Concentric growth is selectively disrupted in a series of mutants of which *heart of glass* (*heg*) is the archetype. Inactivation of the membrane-bound form of this novel protein results in the accretion of myocytes in a single layer, resulting in a thin-walled and enormously dilated heart.

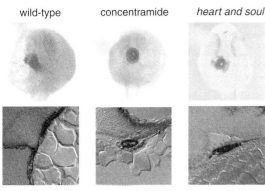

Figure 3. The ventricle appears to develop normally but is encapsulated within the atrium in the mutant *heart and soul*. This cardiac phenotype is recapitulated by the small molecule concentramide. The upper panel in this figure demonstrates whole-mount in situ using *myosin light chain 2* riboprobe in wild-type, concentramide-treated, and mutant larvae. The atrium and ventricle are oriented sequentially in the wild-type larvae, but in both the concentramide-treated and the *has* mutant larvae the ventricle can be seen to lie within the atrium. This is confirmed in the sections in the lower panel where the ventricle is highlighted using *ventricular myosin heavy chain* riboprobe and the atrium is stained with the S46 antibody.

time of generation of the primitive heart tube, in that the heart field is displaced anteriorly with respect to the anterior end of the notochord, and manifest by delay in fusion of the posterior end of the cardiac cone (Peterson et al. 2001). We cloned *has*, which is the PKCλ gene (Horne-Badovinac et al. 2001; Peterson et al. 2001). Interestingly, the orthologous gene in nematode (PKC-3) is essential for antero-posterior patterning of the early embryo. Concentramide does not act via PKCλ, but the chemical and genetic pathways appear to converge upon a critical antero-posterior patterning step (Peterson et al. 2001).

A second axis for ventricle growth patterning is concentric, resulting in thickening of the myocardium between endocardial and epicardial surfaces (Fig. 2). Trabeculation and thickening of the mammalian myocardium is controlled, in part, by the signaling molecule *neuregulin*, which binds to *erb b3* and *erb b4* receptors expressed on the myocardial cells. Mutation of these genes in mice causes the heart to remain thin-walled and small (Gassmann et al. 1995; Lee et al. 1995). Several mutations in zebrafish seem not to block the addition of cells per se, but rather the growth pattern. These three mutants, *heart of glass (heg), santa*, and *valentine*, all manifest a similar phenotype, in which the ventricle forms and contracts but becomes enormously dilated as all new cells are added in a single layer, along the longitudinal axis of the heart rather than perpendicular to it (Stainier et al. 1996; Chen and Fishman 2000). We have recently cloned *heg*, which is a novel gene, and encodes, through alternative splicing, both a membrane-associated form and an apparently secreted form of the final protein. By selective interference with each splice variant, using morpholino antisense oligonucleotides, it appears that it is the membrane-anchored isoform that is critical for heart development (J. Mably and M.C. Fishman, unpubl.). The *heg* gene is expressed in the endocardial layer, and there is evidence that known signals to the myocardial layer are disrupted in *heg* mutants. These preliminary data suggest that endocardial signals control the growth pattern of the ventricular wall, as well as its trabeculation.

HEART GROWTH: CELL NUMBER

It has been known for decades that organ growth closely parallels body growth. As body size doubles, so does the mass and output of the heart. How such precise, global regulation of organ growth is orchestrated is not known. We have discovered classes of mutations that cause the heart to either become too large, with increased cell number, or to remain diminutive.

The mutant *liebeskummer (lik)* is in the former category, those that result in hyperplastic hearts, and represents the first such global control gene (Fig. 4). *lik* encodes the zebrafish *reptin*, a protein that is part of chromatin remodeling complexes, with DNA-stimulated ATPase and helicase activities (Rottbauer et al. 2002). The orthologous gene in yeast regulates 5% of the genes in the genome. In vertebrates, *reptin* represses β-

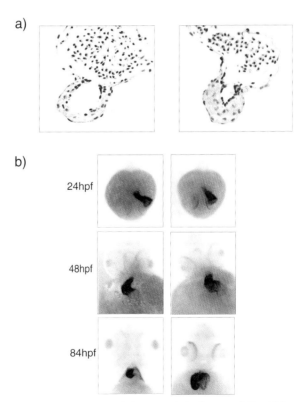

Figure 4. (*a*) The zebrafish mutant *liebeskummer (lik)* exhibits marked cardiac myocyte hyperplasia, seen here in hematoxylin and eosin stained sections of larval hearts. (*b*) The increase in myocyte number takes place relatively late in the course of cardiac development. In these whole-mount in situs, using the myosin light chain as the probe, it can be seen that the hearts are relatively normal in size at 24 or 48 hours postfertilization (hpf) but are substantially larger than wild-type controls by 84 hpf.

catenin/*TCF*-dependent transcription and is presumed to affect many other genes as well. In many circumstances the activity of *reptin* is antagonized by the protein *pontin* (Bauer et al. 2000). We find that the *lik* gene defect is an activating mutation of zebrafish *reptin*, and thus should have effects similar to those caused by diminishing *pontin*. Reduction of *pontin* expression using morpholinos does in fact cause a hypertrophic and hyperplastic heart, confirming that the mutual antagonism of these proteins is an important regulator of heart growth.

In contrast, the *island beat* (*isl*) mutation causes the embryonic heart to remain diminutive, with fewer cells. In the adult, where the heart is essentially postmitotic, increased use of the heart, whether in response to flow or pressure loads, leads to hypertrophy. Use is associated with calcium entry, and there is evidence that some aspects of hypertrophy involve calcium-induced activation of calcineurin-mediated signaling. *island beat* is the result of a null mutation in the zebrafish cardiac-specific L-type calcium channel (Rottbauer et al. 2001). The effect on cardiac growth is a result of abnormalities of calcium entry per se, rather than the electrical silence of the heart, as the heart grows to relatively normal size in other silent heart mutations. The exact mechanisms by which calcium entry remodels the heart during embryogenesis and adult life are unclear, but work from skeletal muscle suggests both frequency-dependent and frequency-independent pathways (Wu et al. 2000; McKinsey et al. 2002). Thus, zebrafish genetic analysis has provided among the first molecular handles on the regulation of organ size.

PHYSIOLOGY

Although form and function are tightly coupled, the extent to which vertebrate physiology might be amenable to genetic dissection was not known prior to the zebrafish genetic screens. The identification of a cadre of mutants with functional phenotypes, ranging from defects in heart rate regulation to abnormal chamber contractility, has confirmed the accessibility of function (Warren and Fishman 1998). These mutants also suggest that physiologic innovations have occurred in the same modular fashion as morphologic additions (Fishman and Olson 1997).

Many aspects of rhythmicity and contractility are affected. For example, heart rate is perturbed in the mutant *slow mo* (*smo*). *smo* mutant embryos exhibit bradycardia prior to the innervation of the heart. Cardiocytes from *smo* fish display a marked reduction in the amplitude of the fast component of the so-called pacemaking I_h current, proving the in vivo relevance of this current to pacemaking (Baker et al. 1997). This physiological defect persists through adulthood (Warren et al. 2001). Currently, positional cloning suggests that the mutation does not reside within the I_h channel genes themselves.

VESSEL GROWTH

Mutations isolated to date have shed light on three types of vessel growth: those that establish the first artery and vein or vasculogenesis; those that direct so-called sprouting angiogenesis of the intersomitic vessels of the trunk; and those that permit vascularization of an organ. These processes appear to be mechanistically quite distinct, although direct lineage mapping in the zebrafish, using laser-activated uncaging of fluorescent dextran, has demonstrated that angioblasts for all three types of vessels arise in the lateral posterior mesoderm (Zhong et al. 2001; Childs et al. 2002).

Vasculogenesis

The first vessels formed in the embryo are the major artery and vein of the trunk. These vessels alone accommodate the first embryonic blood flow. They assemble from angioblasts in the lateral posterior mesoderm. An individual progenitor cell provides progeny only to the artery or the vein, but not to both, implying that a cell fate decision is made prior to formation of the vessels. Thus, cells know whether they will become artery or vein (Zhong et al. 2001).

The *gridlock* mutation selectively interferes with the development of the artery. It causes dysmorphogenesis of the anterior branch point of the aorta, where it splits into left and right aortae. The *gridlock* (*grl*) gene encodes a divergent member of the hairy/enhancer of split (HES) family (Zhong et al. 2000). The HES proteins are transcription factors containing a bHLH domain and a carboxy-terminal groucho-binding domain. *gridlock* differs from HES proteins in the bHLH domain and at the carboxyl terminus. It is expressed in the artery but not the vein. Lowering of *grl* expression causes progressive truncation of the artery, interestingly with the most sensitive site being the anterior bifurcation. This suggests that the *grl* mutation may be hypomorphic and that the anterior aortic branch point is the arterial region most *grl*-dose-dependent. In parallel with morpholino-induced diminution of the artery, there is expansion of the adjacent vein. Overexpression of *grl* reduces the extent and size of the axial veins. Thus, as shown in Figure 5, it appears that there is a bimodal cell fate decision between arterial and venous cell fates, guided by *grl* suppression of the venous fate. This implies that additional factors are required to generate the artery. Like other HES proteins (Fig. 5), *grl* is downstream from *notch*, such that overexpressing *notch* increases *grl*. Blocking the *notch* intermediate, *Su(H)*, diminishes *grl* expression, concomitantly blocking artery formation in a dose-dependent manner identical to perturbation of *grl*. Downstream from *grl* are venous genes, such as *ephB4* (Zhong et al. 2000, 2001).

Sprouting Angiogenesis

Sprouting from preexisting vessels is believed to involve processes similar to those that drive tumor angiogenesis, with endothelial cell division providing the new cells for vessel elongation. Most such vessels are viewed as though they are homogeneous.

We find that the first vessel sprouting in the embryo, in the trunk from aorta to the caudal vein between the

a)

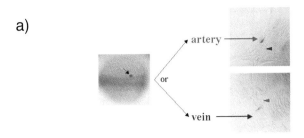

b)

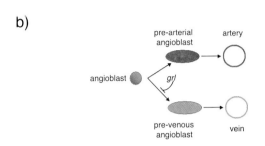

c)

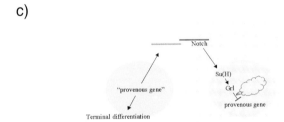

Figure 5. (*a*) Direct lineage mapping in the zebrafish, using laser-activated uncaging of fluorescent dextran to label cells in the lateral posterior mesoderm, has revealed that each progenitor cell provides progeny only to the artery or the vein, but not to both, implying that a cell fate decision is made prior to formation of the vessels. (*b*) A major implication of this result is that there are discrete populations of pre-arterial and pre-venous angioblasts which go on to form various components of the arterial and venous vessels. (*c*) During vasculogenesis there is a bimodal cell fate decision between arterial and venous cell fates which is guided by *gridlock* (*grl*) suppression of the venous fate. Additional factors are necessary for specification of the arterial fate. *grl* is downstream from *notch*, such that overexpressing *notch* increases *grl* and blocking the *notch* intermediate, *Su(H)*, diminishes *grl* expression, concomitantly blocking artery formation in a dose-dependent manner identical to perturbation of *grl*. The provenous genes downstream from *grl* have not yet been identified.

somites (intersomitic vessels), in fact proceeds by a carefully orchestrated migration of cells from the lateral posterior mesoderm, and not by local growth from preexisting vessels (Childs et al. 2002). The angioblasts migrate via the aorta and then assume one of three cell fates, the sum total of which is the new vessel. Thus, each sprout consists of three cell types. These distinctive fates are dictated by the local environment. The ventral half of each somite inhibits angioblast migration and growth, restricting the vessel to the intersomitic region. In the mutant known as *out of bounds*, this repressive signal is missing, leading to chaotic and misshapen intersomitic vessels (Childs et al. 2002). Thus, it appears that even small vessels are inhomogeneous along their length, and that apparent sprouting may rather reflect coalescence of angioblasts, a form of vasculogenesis.

Vascularization of the Kidney

The renal glomerulus is the site where blood is filtered to generate urine. The glomerulus is composed of a highly specialized capillary endothelium engulfed by renal podocytes. The combination of a fenestrated endothelium, a unique basement membrane, and the interdigitating podocyte foot processes results in the glomerular filtration barrier. The glomerulus of the fish pronephros is an analogous structure (Drummond et al. 1998). It forms just medial to the aorta in the trunk, as podocytes migrate medially from the intermediate cell mass (Serluca and Fishman 2001). A vessel sprouts from the aorta to ramify among these podocytes. Several mutations perturb this process, such that both endothelial and podocytic fates are present, but the glomerular structure does not form (Serluca et al. 2002). All of these mutations share the property of interfering with blood flow, suggesting that flow per se might somehow regulate formation of this vessel. Interference with blood flow by pharmacological means or by laser blockade of the aorta similarly prevents vessel formation specifically to the kidney. It seems that the mechanical hemodynamic forces themselves are responsible, rather than any humoral factor, as the vessel forms even when blood is replaced by saline. One important element in transducing the mechanical forces associated with blood flow into growth is matrix metalloproteinase 2 (*mmp2*). *mmp2* is expressed in endothelia in a stretch-responsive manner and diminished by genetic and embryological perturbations that block glomerulogenesis. Blockade of *mmp2* using *Timp2* prevents glomerulogenesis. Thus, this vessel is peculiar in manifesting a use-dependent regulation of growth (Serluca et al. 2002). It is interesting that glomerular capillaries are designed for filtration, so it may be sensible to prevent formation of this vessel until filtration forces are instituted.

DISEASE MODELS

The phenotypes of some of the zebrafish mutations have proved remarkably similar to human diseases (Shin and Fishman 2002). This clear phenotypic conservation provides an understanding of disease mechanisms, candidate genes for human disease, and targets for pharmaceutical development. This attribute of zebrafish biology was not entirely anticipated, as many human diseases are adult-onset dominant disorders, often due to haploinsufficiency, whereas the vast majority of mutations studied in zebrafish are recessive embryonic lethals. Of course, null mutations of orthologous genes in humans would be lethal during the first weeks of gestation and thus would not come to clinical attention. For example, mutation of

Tbx5 in humans causes the Holt-Oram syndrome, with skeletal malformation and cardiac defects, most prominent in the atria (Basson et al. 1997). Null mutation of *tbx5* in zebrafish causes a heart-fin syndrome, far more pronounced, with complete absence of pectoral fins and growth failure throughout both chambers (Garrity et al. 2002). However, partial diminution of *tbx5* by morpholino produces a syndrome more closely related to the human disease, with pectoral fin anomalies, often one-sided (as in the human), and less severe cardiac anomalies, although not atrial-restricted (Basson 2000).

The definition of the utility of the fish for prediction of human complex diseases must await the cloning of more of the mutations. However, initial findings are encouraging. One mutation with a phenotype resembling dilated cardiomyopathy (an enlarged heart with poor contraction) affects the *titin* gene (Xu et al. 2002). This same gene is mutated in some families with dilated cardiomyopathy (Gerull et al. 2002). The effects of other zebrafish mutations phenocopy a variety of cardiac disorders, including arrhythmias such as bradycardia, atrial fibrillation, and heart block. Their cloning will be important not only for dissection of disease but also for the understanding of the onset of physiological function in the embryo, a start on integrative physiology (Warren and Fishman 1998; Briggs 2002).

The zebrafish embryo is permeable to small molecules simply added to the aqueous environment. These can be screened in a high-throughput manner. Some provide remarkable phenocopies of genetic mutations and thus can be used to help to dissect the relevant pathways (Peterson et al. 2000). It will be of great interest to determine whether they also can suppress the effects of mutations, a first step toward pharmaceutical discovery based on phenotypic correction.

REFERENCES

Arap W., Kolonin M.G., Trepel M., Lahdenranta J., Cardo-Vila M., Giordano R.J., Mintz P.J., Ardelt P.U., Yao V.J., Vidal C.I., Chen L., Flamm A., and Valtanen H., et al. 2002. Steps toward mapping the human vasculature by phage display. *Nat. Med.* **8:** 121.

Baker K., Warren K.S., Yellen G., and Fishman M.C. 1997. Defective "pacemaker" current (Ih) in a zebrafish mutant with a slow heart rate. *Proc. Natl. Acad. Sci.* **94:** 4554.

Basson C.T. 2000. Holt-Oram syndrome vs heart-hand syndrome. *Circulation* **101:** E191.

Basson C.T., Bachinsky D.R., Lin R.C., Levi T., Elkins J.A., Soults J., Grayzel D., Kroumpouzou E., Traill T.A., Leblanc-Straceski J., Renault B., Kucherlapati R., Seidman J.G., and Seidman C.E. 1997. Mutations in human TBX5 [corrected] cause limb and cardiac malformation in Holt-Oram syndrome. *Nat. Genet.* **15:** 30.

Bauer A., Chauvet S., Huber O., Usseglio F., Rothbacher U., Aragnol D., Kemler R., and Pradel J. 2000. Pontin52 and reptin52 function as antagonistic regulators of beta-catenin signalling activity. *EMBO J.* **19:** 6121.

Briggs J.P. 2002. The zebrafish: A new model organism for integrative physiology. *Am. J. Physiol. Regul. Integr. Comp. Physiol.* **282:** R3.

Chen J.N. and Fishman M.C. 2000. Genetics of heart development. *Trends Genet.* **16:** 383.

Childs S., Chen J.N., Garrity D.M., and Fishman D.C. 2002. Patterning of angiogenesis in the zebrafish embryo. *Development* **129:** 973.

Driever W. and M.C. Fishman M.C. 1996. The zebrafish: Heritable disorders in transparent embryos. *J. Clin. Invest.* **97:** 1788.

Drummond I.A., Majumdar A., Hentschel H., Elger M., Solnica-Krezel L., Schier A.F., Neuhauss S.C., Stemple D.L., Zwartkruis F., Rangini Z., Driever W., and Fishman M.C. 1998. Early development of the zebrafish pronephros and analysis of mutations affecting pronephric function. *Development* **125:** 4655.

Fishman M.C. 2001. Genomics. Zebrafish—The canonical vertebrate. *Science* **294:** 1290.

Fishman M.C. and Chien K.R. 1997. Fashioning the vertebrate heart: Earliest embryonic decisions. *Development* **124:** 2099.

Fishman M.C. and Olson E.N. 1997. Parsing the heart: Genetic modules for organ assembly. *Cell* **91:** 153.

Gassmann M., Casagranda F., Orioli D., Simon H., Lai C., Klein R., and Lemke G. 1995. Aberrant neural and cardiac development in mice lacking the ErbB4 neuregulin receptor. *Nature* **378:** 390.

Gerull B., Gramlich M., Atherton J., McNabb M., Trombitas K., Sasse-Klaassen S., Seidman J.G., Seidman C., Granzier H., Labeit S., Frenneaux M., and Thierfelder L. 2002. Mutations of TTN, encoding the giant muscle filament titin, cause familial dilated cardiomyopathy. *Nat. Genet.* **30:** 201.

Horne-Badovinac S., Lin D., Waldron S., Schwarz M., Mbamalu G., Pawson T., Jan Y., Stainier D.Y., and Abdelilah-Seyfried S. 2001. Positional cloning of heart and soul reveals multiple roles for PKC lambda in zebrafish organogenesis. *Curr. Biol.* **11:** 1492.

Garrity D.N., Childs S., and Fishman M.C. 2002. The *heartstrings* mutation in zebrafish causes heart/fin Tbx5 deficiency syndrome. *Development* **129:** 4635.

Lee K.F., Simon H., Chen H., Bates B., Hung M.C., and Hauser C. 1995. Requirement for neuregulin receptor erbB2 in neural and cardiac development. *Nature* **378:** 394.

McKinsey T.A., Zhang C.L., and Olson E.N. 2001. MEF2: A calcium-dependent regulator of cell division, differentiation and death. *Trends Biochem. Sci.* **27:** 40.

Peterson R.T., Link B.A., Dowling J.E., and Schreiber S.L. 2000. Small molecule developmental screens reveal the logic and timing of vertebrate development. *Proc. Natl. Acad. Sci.* **97:** 12965.

Peterson R.T., Mably J.D., Chen J.N., and Fishman M.C. 2001. Convergence of distinct pathways to heart patterning revealed by the small molecule concentramide and the mutation heart-and-soul. *Curr. Biol.* **11:** 1481.

Rottbauer W., Baker K., Wo Z.G., Mohideen M.A., Cantiello H.F., and Fishman M.C. 2001. Growth and function of the embryonic heart depend upon the cardiac-specific L-type calcium channel alpha1 subunit. *Dev. Cell* **1:** 265.

Rottbauer W., Saurin A.J., Lickert H., Shen X., Burns C.G., Wo Z.G., Kemler R., Kingston R., Wu C., and Fishman M. 2002. Reptin and pontin antagonistically regulate heart growth in zebrafish embryos. *Cell* **111(5):** 661.

Serluca F.C. and Fishman M.C. 2001. Pre-pattern in the pronephric kidney field of zebrafish. *Development* **128:** 2233.

Serluca F.C., Drummond I.A., and Fishman M.C. 2002. Endothelial signaling in kidney morphogenesis. A role for hemodynamic forces. *Curr. Biol.* **12:** 492.

Shin J.T. and Fishman M.C. 2002. From zebrafish to human: Molecular medical models. *Annu. Rev. Genomics Hum. Genet.* **3:** 311.

Stainier D.Y., Lee R.K., and Fishman M.C. 1993. Cardiovascular development in the zebrafish. I. Myocardial fate map and heart tube formation. *Development* **119:** 31.

Stainier D.Y., Fouquet B., Chen J.N., Warren K.S., Weinstein B.M., Meiler S.E., Mohideen M.A., Neuhauss S.C., Solnica-Krezel L., Schier A.F., Zwartkruis F., Stemple D.L., Malicki J., Driever W., and Fishman M.C. 1996. Mutations affecting the formation and function of the cardiovascular system in the zebrafish embryo. *Development* **123:** 285.

Warren K.S. and Fishman M.C. 1998. "Physiological genomics": Mutant screens in zebrafish. *Am. J. Physiol.* **275:** H1.

Warren K.S., Baker K., and Fishman M.C. 2001. The slow mo mutation reduces pacemaker current and heart rate in adult zebrafish. *Am. J. Physiol. Heart Circ. Physiol.* **281:** H1711.

Warren K.S., Wu J.C., Pinet F., and Fishman M.C. 2000. The genetic basis of cardiac function: Dissection by zebrafish (*Danio rerio*) screens. *Philos. Trans. R. Soc. Lond. B Biol. Sci.* **355:** 939.

Wu H., Naya F.J., McKinsey T.A., Mercer B., Shelton J.M., Chin E.R., Simard A.R., Michel R.N., Bassel-Duby R., Olson E.N., and Williams R.S. 2000. MEF2 responds to multiple calcium-regulated signals in the control of skeletal muscle fiber type. *EMBO J.* **19:** 1963.

Xu X., Meiler S.E., Zhong T.P., Mohideen M., Crossley D.A., Burggren W.W., and Fishman M.C. 2002. Cardiomyopathy in zebrafish due to mutation in an alternatively spliced exon of titin. *Nat. Genet.* **30:** 205.

Zhong T.P., Childs S., Leu J.P., and Fishman M.C. 2001. Gridlock signalling pathway fashions the first embryonic artery. *Nature* **414:** 216.

Zhong T.P., Rosenberg M., Mohideen M.A., Weinstein B., and Fishman M.C. 2000. *gridlock*, an HLH gene required for assembly of the aorta in zebrafish. *Science* **287:** 1820.

Consomic Rats for the Identification of Genes and Pathways Underlying Cardiovascular Disease

R.J. Roman, A.W. Cowley, Jr., A. Greene, A.E. Kwitek, P.J. Tonellato, and H.J. Jacob

Department of Physiology, Medical College of Wisconsin, Milwaukee, Wisconsin 53005

The next step following completion of the first draft of the sequence of the human genome is to overlay this information with a complementary functional genomic map of the location of disease-causing genes and those involved in the regulation of normal function. To this end, our group has been developing two chromosomal substitution panels of 44 strains of consomic rats in which one chromosome at a time from a normal BN (Brown Norway) rat has been introgressed onto the genetic backgrounds of the SS/JrHsd/Mcw Dahl salt-sensitive (SS) and FHH/Eur Fawn Hooded Hypertensive rats. These strains were chosen because they exhibit >60% of all the genetic variation observed among 48 inbred strains of rats. In addition, the SS rat is a well-accepted model for salt-sensitive hypertension, insulin resistance, hyperlipidemia, endothelial dysfunction, vascular injury, cardiac hypertrophy, and glomerulosclerosis, whereas the FHH rat is a model of systolic hypertension, renal disease, pulmonary hypertension, a bleeding disorder, alcoholism, and depression. The chromosomal location of genes and pathways involved in the regulation of many cardiovascular functions can be identified by simply phenotyping 6–8 rats of each of the consomic strains and comparing the response to the parental strain and the other consomic strains. The region of the genome mediating the trait can be narrowed by phenotyping an F_2 population generated from the consomic and parental strains and isolated to a single locus model in congenic sublines in 6 months. The disease-causing or disease-resistance gene can then be identified by expression profiling and/or direct sequencing. The consomic strains are currently being screened for 203 traits related to renal, pulmonary, vascular, and cardiac function. We have already found that transfer of BN chromosome 13 or 18 attenuates the development of hypertension, proteinuria, and renal disease in SS.BN-13 and SS.BN-18 rats. These consomic panels will soon be commercially available and will provide investigators with a powerful tool to link functional data to the genome.

FUNCTIONAL GENOMICS/SYSTEMS BIOLOGY PLATFORMS

The sequencing of the human genome has presented mankind with the Rosetta stone of life and unprecedented opportunities for discovery of new genes, the genetic basis of complex human disease, and genomically validated targets for drug development. However, an examination of the publicly available genome databases, Locus link, Omim, HUGO, and GDB, indicates that about one-third of the 30,000 predicted genes have inferred function, and only about 5,000 of these have a confirmed action (Nadeau et al. 2001; Venter et al. 2001). The function of the other genes remains to be experimentally validated, since most of the current gene annotation is based on sequence homology with cDNAs and ESTs reported in other organisms. Thus, the enormous challenge is to attach function to the genome and identify the genes and pathways that underlie common human diseases that affect millions of people worldwide.

The challenge of attaching function to the genome has led to the rapid emergence of functional genomics/systems biology. A number of novel approaches have been developed to link sequence information with function, and the strengths and limitations of some of these model systems are summarized in Table 1. These techniques include mutagenesis screens in *Caenorhabditis elegans*, *Drosophila melanogaster* (Stanford et al. 2001; Adams and Sekelsky 2002; St Johnston 2002), zebrafish (Knapik 2000; Briggs 2002), and mice (Schimenti and Bucan 1998; Balling 2001); gene knockout approaches in mice (Merrill et al. 1997; James et al. 1998; Asa 2001); and chromosomal transfer techniques in consomic strains of rats and mice (Nadeau et al. 2000). The major advantage of mutagenesis screens in *C. elegans*, *Drosophila*, and zebrafish is that investigators can rapidly generate and phenotype large numbers of progeny. These models have proven to be especially useful to identify developmental defects (Justice et al.1999; Knapik 2000; Nolan et al. 2000; Wells and Brown 2000; Li 2001; Briggs 2002); changes in responses to various toxins, drugs (Law 2001; Lieschke 2001), or stressors (Bate 1999; Haddad 2000); and behavioral phenotypes (Nolan et al. 2000; Li 2001; Sokolowski 2001). They are also being used to look at the influence of mutated genes on transcriptional responses to various drugs and stimuli using expression profiling. However, because of anatomical differences and the small size of the organisms in lower orders, it is difficult to measure phenotypes directly related to complex human diseases, such as insulin resistance, blood pressure, changes in clinical chemistry, or pulmonary or renal function. Although the mouse offers tremendous promise, the limited physiological characterization of this organism,

Table 1. Strengths and Limitations of Various Systems Biology Platforms

Approach	Strengths	Limitations
ENU mutagenesis		
C. elegans	large number of progeny rapid screening tools complete genomic sequence	simple phenotypes lack of disease models lack of complex anatomy
Zebrafish	large number of progeny rapid screening tools complete genomic sequence	simple phenotypes lack of disease models lack of complex anatomy
Mice	sequence soon to be completed more complex phenotypes closely related anatomy to man	lack of known multigenic disease models limited number of progeny difficult phenotyping requires positional cloning
Gene knockout		
Mice	gene targeting techniques sequence soon to be completed more complex phenotypes closely related anatomy to man	limited disease models difficult phenotypic screens relatively slow process single gene disease phenotypes different from complex disease
Chromosomal transfer		
Rats	well-characterized disease models sequence soon to be completed complex phenotypes closely related anatomy to man	limited number of alleles requires positional cloning

the need to perform high-throughput but mechanistically simplistic physiological screens, and marked differences in responses observed in various strains of mice (genomic backgrounds) (Pearson 2002) are limiting the speed at which mouse mutants can be adopted and validated to be reflective of disease in man. Furthermore, the need to maintain huge numbers of strains or to rederive the strain from cryopreserved embryos or sperm is producing tremendous time, logistical, and expensive space problems for academic institutions and industry (Knight and Abbott 2002).

Another popular approach for genetic discovery is gene knockout or inactivation in mice. The idea is that one can inactivate a gene and then try to reproduce the symptoms of a human disease, such as an alteration in blood pressure. The knockout strategy has proven extremely successful to confirm hypotheses about the function of known genes or to identify genes that are bottlenecks within a developmental pathway. For example, knockouts of various components of the renin-angiotensin system have been shown to alter blood pressure (Nishimura and Ichikawa 1999; Oliverio and Coffman 2000). However, this general strategy has also run into problems with the failure to produce obvious phenotypes and genomic background effects (Nadeau 2001; Pearson 2002; Wolfer et al. 2002). For example, knockout of the epidermal growth factor receptor produces a lethal defect in the CF-1 strain of mice, but in the CD-1 strain, the knockout mice survive for several weeks (Threadgill et al. 1995). Similarly, knockout of the tumor-suppressor gene causes colon polyps in C57BL/6J strain of mice but has little effect in the AKR genetic background (Dietrich et al. 1993). These two problems have prevented this strategy from becoming routine for the discovery of genes with unknown function. An additional problem is that common diseases in man require multiple genes, each with a relatively small effect on the disease process, with the disease exhibiting a variable age of onset and in some cases requiring exposure to dietary challenges or other environmental stressors to be expressed. Thus, it is very difficult to design appropriate screening strategies that can capture the broadest range of disease-related outcomes with no preexisting information regarding the function of a gene. The gene knockout and phenotyping approach is very slow. One company proposes to create and profile 500–1000 knockouts a year. Even at this accelerated rate, it will take 25 years to complete the discovery process. Finally, pharmacologists have learned that surgically and pharmacologically induced disease models, although showing the same clinical phenotype, are often quite different from the human condition and are frequently of little utility in predicting which drugs and therapies will be efficacious in humans. It is highly likely that the same limitation will prove to be true for genetic modification models as well.

Moreover, all physiological functions are closed-loop systems that are influenced by many genes, physiological reflexes, and the environment. This poses a problem for interpretation of gene knockout studies, since inactivation of any gene will lead to compensating changes in the expression of many genes throughout the pathway. Thus, the disease phenotype could be caused by these compensatory changes rather than by loss of function of the gene of interest. The recognition of these problems has led to a growing concern for the need for additional model systems and approaches for systems biology.

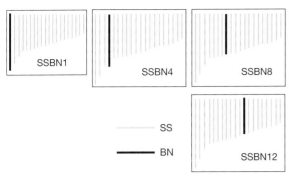

Figure 1. Genomic representation of a consomic strain. All genomic regions of the consomic strain are identical to the parental SS strain with the exception of a single chromosome transferred by backcross breeding from the BN strain. The bold dark line represents the chromosome from the BN rat and the gray lines represent unchanged chromosomes derived from the original SS background.

CONSOMIC STRAINS OF RATS FOR THE IDENTIFICATION OF DISEASE-CAUSING GENES

These concerns led our group to consider an alternate strategy for systems biology using chromosomal substitution panels in inbred strains of rats as a generalized tool for genetic discovery (Nadeau et al. 2000). Rather than mutating or knocking out single genes to create a model that mimics the symptoms of human disease, we propose to take advantage of the large number of well-accepted and validated inbred strains of rats. Presently there are 234 different inbred strains of rats that have been developed over the last 80 years for many complex human diseases such as hypertension (Rapp 2000; Stoll et al. 2000), obesity, hyperlipidemia, type I and II diabetes, and insulin resistance (Cole et al. 1983; Mordes et al. 1987; Gill et al. 1989; Kwitek-Black and Jacob 2001; Jacob and Kwitek 2002). A complete listing of the available inbred rat strains and model systems can be found at the rat genome database Web site (www.rgd.mcw.edu). These rat models of human complex disease have been functionally characterized extensively, and frequently have proven to be useful as model systems for drug discovery. However, the underlying genetic basis of the disease in most of these models remains unknown. What we are doing is to breed chromosomal substitution (consomic) panels of rats (Fig. 1) in which chromosomes from the disease-susceptible strain are systematically replaced, one at a time, with the corresponding chromosome from the disease-resistant strain. The chromosomal location of the genes and pathways contributing to the disease can be identified by simply phenotyping 8–10 rats of each of 22 (20 autosomes plus the X and Y chromosomes) consomic strains in a panel looking for those strains that were cured, or in which the symptoms of the disease are greatly diminished compared to the parental lines or other consomic strains. The advantage of this approach over traditional F_2 QTL mapping is that it allows investigators to rapidly identify the chromosomes harboring disease-modifying genes in a susceptible disease-causing genetic background without the need for genotyping.

Another advantage, as illustrated in Figure 2, is that since the consomic rats are isogenic with the parental strain on all other chromosomes, one can cross the consomic strain of interest with the parental strain to produce an F_2 population that only has recombination on the chromosome of interest. This population can then be geno-

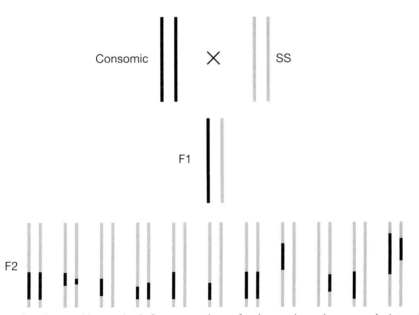

Figure 2. Method to produce F_2 recombinant animals from consomic rats for phenotyping and to narrow the interval around disease-causing genes. A consomic strain (SS.BN) is crossed with the parental strain (SS) to produce an F_1 generation with one copy of each gene on the chromosome of interest, and fixed for the original background SS strain on all other chromosomes. The F_1 rats are intercrossed to produce an F_2 population that exhibits recombination events only on the chromosome of interest, while the rest of the genome remains homogeneous and fixed for the disease-susceptible SS alleles.

typed and phenotyped looking for gain or loss of the disease-causing phenotype to narrow the interval around the potential region of interest to 1–2 centimorgans (cM) or 50–100 genes in only 6 months (Fig. 3), in contrast to the 3 or 4 years that has traditionally been required to generate a congenic strain. For example, generation of 1000 F_2 animals would yield an average of 10 overlapping congenic strains for every centimorgan, resulting in ~2 animals that would be homozygous at a 1-cM resolution, which is the desired size for positional cloning efforts. Therefore, consomic rats facilitate both the genetic mapping of a trait and the derivation of congenic strains to refine the location of the disease-causing genes. In our experience, it now only takes about 100–150 rats to narrow the interval to 1–2 cM. The F_2 rats derived from a cross of the consomic and parental strains also appear to have more power for genetic mapping than traditional mapping in F_2 populations derived from unrelated inbred strains. This is likely due to the fact that the F_2 rats derived from crosses of the consomic and parental strains are genetically identical on all the other chromosomes (same genetic background), in contrast to F_2 populations derived from unrelated strains, where each animal has a unique genomic mix of the two parental strains.

After narrowing the interval around the gene of interest, one can produce a congenic strain around the region of interest by breeding an appropriate pair of F_2 rats derived from the cross of the consomic and parental strains. Now that the sequencing of the rat genome is under way and scheduled for completion in the spring of 2003, it will also soon be practical to directly sequence all of the gene QTL regions of interest. In our studies, the genome of one of the strains (BN) used to construct the panels has already been sequenced. Thus, one only has to sequence corresponding regions from the SS or FHH strains to identify mutations. One can also use cDNA expression arrays to look for differentially expressed genes between the consomic/congenic and parental strains. In this case, the genomic background is controlled and the consomics and parental strains differ by at most 10% of the genome, in the case of the chromosome 1 consomic line. The genetic divergence is <0.5% when one is performing differential transcription analysis between the parental and congenic sublines. Genes that are differentially expressed and map to the region of interest are candidates for sequencing and further study. Those differentially expressed genes that map to other chromosomes likely are regulated by the gene of interest and may contribute to the pathogenesis of the disease or be involved in pathways that compensate for the disease process. Once a sequence variant or a differentially expressed gene has been identified, one can then perform genetic (transgenic) rescue; gene replacement or removal (RNAi, antisense, or adenovirus gene therapy); or pharmacologic rescue experiments in the congenic strain to confirm that the gene of interest plays a causal role in the development of the disease. Another advantage of this approach of looking for the genes that cause disease is that at the end of the discovery process following the creation of the congenic sublines, one creates single-gene models of complex human disease with both susceptible and resistant strains of rats that are ideal for screening compounds or performing interventional or mechanistic studies on the pathogenesis of the disease.

As a proof of concept of the utility of consomic rats derived from interesting inbred rat models of human disease, our group at MCW is in the process of creating a panel of 44 strains of consomic rats in which one chromosome at a time from a normotensive BN (Brown Norway) rat has been introgressed onto the genetic backgrounds of the SS/JrHsd/Mcw Dahl salt-sensitive (SS) or FHH/Eur Fawn Hooded Hypertensive rats (FHH) rats. These strains were chosen because they exhibit on average >60% of all the genetic variation observed among 48 inbred strains of rats studied with ~5000 genetic markers (Steen et al. 1999). In addition, the SS rat is a well-accepted model for salt-sensitive hypertension, insulin resistance, hyperlipidemia, endothelial dysfunction, vascular injury, cardiac hypertrophy, and glomerulosclerosis, whereas the FHH rat is a model of systolic hypertension,

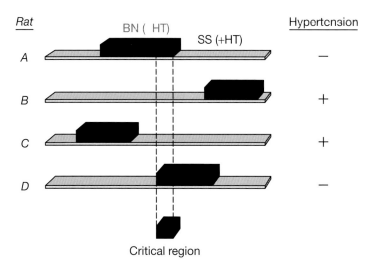

Figure 3. Recombinant animals, created from an F_2 cross of a consomic and parental strain, are genotyped and phenotyped for gain or loss of function of the phenotype of interest. Assessing the phenotypes of the congenic sublines allows the narrowing of the region containing the gene(s) of interest. In this example, the trait is hypertension. The darker shaded region represents the region of the chromosome obtained from the BN donor strain, and the light gray region represents the genome of the background SS strain. The vertical black bar represents the region that retains both the phenotype and the genomic region from the donor strain, and represents the narrowed critical QTL region. By selecting animals with the appropriate genotypes and phenotypes, one can rapidly produce a congenic line with a smallest interval of interest for positional cloning and expression arrays for gene hunting and for creating single-gene disease models for human disease for drug screening and mechanistic studies on the pathogenesis of the disease.

renal disease, pulmonary hypertension, a bleeding disorder, alcoholism, and depression.

HIGH-THROUGHPUT PHYSIOLOGICAL PHENOTYPING IN CONSOMIC RATS

As the value of the consomic rats is directly proportional to the number of traits assigned to each strain, we have developed a high-throughput physiology phenotyping protocol to study 203 heart, lung, blood, and renal function phenotypes in the consomic and parental strains. A flow diagram of the logistics of our breeding and phenotyping operation is presented in Figure 4. We produce over 3000 rats a year for study, with each rat microchipped and tracked for identification. The rats are preconditioned by exposure to stressors, either chronic hypoxia (12% O_2) or elevated salt intake (4% sodium chloride in the diet) for 3 weeks, and then studied using eight different protocols in which 203 different phenotypes are measured on a minimum of 8 male and 8 female rats per strain. The renal phenotypes include blood pressure, heart rate, blood pressure responses to i.v. administration of different doses of norepinephrine and angiotensin II, plasma renin activity on high-salt diet and after volume depletion with furosemide, and changes in blood pressure (salt sensitivity), heart rate, and plasma renin activity following volume depletion. In the pulmonary protocols, we measure changes in respiratory rate, tidal volume, ventilation, and plasma PCO_2 and PO_2 in response to acute hypoxia and hypercapnia and in response to exercise. We measure the vascular responses of aortic rings in vitro in response to phenylephrine and angiotensin II, direct and indirect endothelial-dependent vasodilators, and hypoxia and hyperoxia. In the cardiac phenotyping protocol, we measure heart rate, contractile function, and diastolic and systolic pressures in isolated perfused hearts under control conditions and after 20 minutes of complete ischemia. We collect urine and blood samples and perform a complete metabolic clinical chemistry profile, urinalysis, and CBC on each animal. Tissues are also collected from the heart, kidney, and vasculature and examined for histological changes, and RNA is extracted from the heart, liver, and kidney for expression profiling using 30,000 cDNAs per slide. A complete summary of all the phenotyping protocols and the phenotype data is publicly available on our Web site at www.pga.mcw.edu. This database currently houses over 100,000 compiled phenotypes from various consomic and inbred strains of rats that can be sorted by phenotype, gender, strain, and preconditioning protocol.

As an example of the utility of this genetic discovery approach, our group has recently reported (Cowley et al. 2001) that mean arterial blood pressure averages 170 ± 3 mm Hg in SS rats fed a high-salt diet for 4 weeks, whereas the BN salt-resistant strain remains normotensive with a blood pressure of 103 ± 1 mm Hg. Transfer of

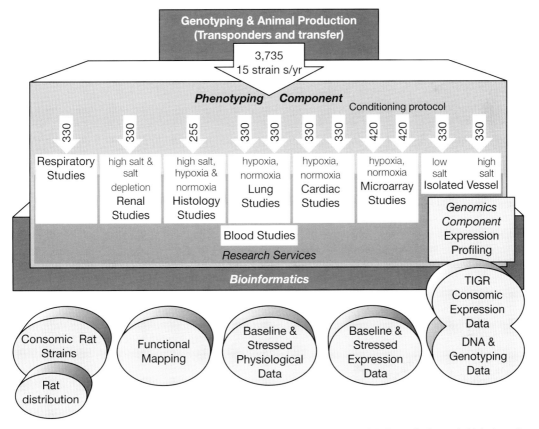

Figure 4. Overview of the "PhysGen" Program for Genomic Applications at the Medical College of Wisconsin high-throughput physiological phenotyping of the SS.BN and FHH.BN consomic panels of rats.

chromosome 13 from the BN rat largely prevents hypertension in SS.BN-13 rats. Blood pressure only averages 119 ± 2 mm Hg in SS.BN-13 rats fed a high-salt diet for 4 weeks. An examination of the data available on our Web site (www.pga.mcw.edu) indicates that transfer of chromosome 18 also lowers blood pressure in the SS.BN-18 consomic line but that transfer of other chromosomes (9 and 20) has little effect. Chromosome 18 is especially interesting, since we have recently reported that a large number of blood pressure and cardiovascular QTLs map to this chromosome in a large F_2 cross of SS and BN rats (Cowley et al. 2000; Stoll et al. 2001). The results in our consomic strains clearly indicate that there are important genes on chromosomes 13 and 18 that affect blood pressure. Positional cloning and identification of these genes may provide new therapeutic targets for the treatment of hypertension.

SS rats also develop severe glomerulosclerosis and proteinuria when fed a high-salt diet. We have reported that the excretion of protein averages 189 ± 30 mg/day in SS rats fed a high-salt diet versus 40 ± 6 mg/day in BN rats (Cowley et al. 2001). Transfer of chromosome 13 in SS.BN-13 rats onto the SS genetic background not only prevented hypertension, but also markedly reduced end organ damage in the kidney (Cowley et al. 2001). Accordingly, protein excretion fell to 63 ± 18 mg/day. The reduction in proteinuria was associated with an improvement in the histological appearance of the kidney compared to that seen in male SS rats, which exhibited extensive fibrosis of vasa recta capillaries in the outer medulla. This led to necrosis of the surrounding tubules and the formation of protein casts. The degree of fibrosis and tubular necrosis was markedly reduced in SS.BN-13 rats (Cowley et al. 2001). Examination of the data currently available on our Web site (www.pga.mcw.edu) clearly indicates that transfer of other chromosomes does not exhibit a similar renoprotective effect.

The consomic strains of rats are also useful for discovery of genes for resistance to cardiac ischemia. Baker et al. (2000) previously reported that $53 \pm 2\%$ of the left ventricle of SS rats is infarcted following 27 minutes of global ischemia followed by reperfusion, and that this is associated with a big deficit in contractile function and developed pressure. In contrast, the hearts of BN rats are remarkably resistant to the effects of ischemia, and only 20% of the left ventricle is infarcted. Examination of the preliminary data from the SS.BN consomic strains available on our Web site suggests that the resistance of the BN rat to cardiac ischemia likely involves genes on chromosomes 9, 13, and 20. It is also interesting that transfer of BN chromosomes 9 and 20 appears to provide cardiac protection without reducing blood pressure or the degree of cardiac hypertrophy.

USE OF cDNA MICROARRAYS TO IDENTIFY DIFFERENTIALLY EXPRESSED GENES IN CONSOMIC/ CONGENIC STRAINS OF RATS

As an example of this strategy, Liang et al. (2002) have recently completed a microarray study looking for differentially expressed genes in the kidney of SS.BN-13 and SS rats. In the renal medulla they identified 26 genes that were differentially expressed on a low-salt diet and 39 genes that were differentially expressed on a high-salt diet. Several of the genes have been previously implicated in the control of blood pressure, oxidative stress, apoptosis, and renal injury and fibrosis. One of these genes, kynurenine 3-hydroxylase, which is a mitochondrial enzyme involved in the synthesis of NAD and NADP, maps to chromosome 13 and is being sequenced and evaluated as a potential causal gene for the reduction in blood pressure in SS.BN-13 rats. All of the other differentially expressed genes identified in SS and SS.BN-13 rats in this study map to other chromosomes (Liang et al. 2002). As such, they must be considered as regulated genes and pathways that may contribute to the development of renal injury and hypertension in SS rats or are involved in compensatory pathways that counter the insult, but they are not the underlying genetic cause of the disease.

CONCLUSION

The present overview illustrates how chromosomal substitution panels of consomic rats created from well-accepted inbred rat models of complex human diseases such as asthma, diabetes, hypertension, and renal disease can provide an important platform for identification of genes and pathways underlying complex human disease. This strategy facilitates the ability to conduct high-throughput physiology, to annotate function onto the genome, and, when combined with the ability to conduct comparative genomics at the level of sequence in the context of systems biology, offers tremendous promise for advancing our knowledge of common, complex disease in man.

ACKNOWLEDGMENTS

This work was supported in part by grants HL-66579, HL-54998l, Hl-36279, and HL-29587 from the National Institutes of Health, Heart, Lung and Blood Institute. We thank the following personnel: John Baker, M.D.; Ph.D, Christopher Dawson, Ph.D., Hubert Forster, Ph.D., Julian Lombard, Ph.D, David Mattson, Ph.D., Kirkwood Pritchard, Ph.D., Abraham Provoost, Ph.D., Richard R. Roman, Ph.D., Becky Majewski, Sheri Jene, Lisa Philips, Jody Klingkammer, Jo Gullings-Handley, Kathleen Kennedy, Jessica Powlas, Mike Tschannen, Nadia Barreto, Audrey Siegrist, Rob Beauvais, Mike Bregantini, Beckie Bralich, Austin Brill, Candace Jones, Alison Kriegel, Jennifer Labecki, Erica Liss, Cynthia Maas, Jaime Peterson, Kathryn Privett, Lisa Gottschalk, Tracy Enslow, Angelo Piro, Greg McQuestion, Michael Kloehn, David Eick, Bonnie Pfau-Wentworth, Kevin Wollenzien, Glenn Slocum, Zhitao Wang, M.S., Nan Jiang, Ph.D., Zanchi Wang, Qunli Cheng, Weihong Jin, Sandy Grieger, Jane Brennan-Nelson, Diane Kelley, Julia Messinger, M.S. for their contributions to this paper. We also acknowledge the leadership of Mary Pat Kunert, RN, Ph.D., and Melinda Dwinell, Ph.D.

REFERENCES

Adams M.D. and Sekelsky J.J. 2002. From sequence to phenotype: Reverse genetics in *Drosophila melanogaster. Nat. Rev. Genet.* **2**: 189.

Asa S.L. 2001. Transgenic and knockout mouse models clarify pituitary development, function and disease. *Brain Pathol.* **11**: 371.

Baker J.E., Konorev E.A., Gross G.J., Chilian W.M., and Jacob H.J. 2000. Resistance to myocardial ischemia in five rat strains: Is there a genetic component of caridoprotection? *Am. J. Physiol.* **278**: H1395.

Balling R. 2001. ENU mutagenesis: Analyzing function in mice. *Annu. Rev. Genomics Hum. Genet.* **2**: 463.

Bate M. 1999. Development of motor behaviour. *Curr. Opin. Neurobiol.* **9**: 670.

Briggs J.P. 2002. The zebrafish: A new model organism for integrative physiology. *Am. J. Physiol.* **282**: R3.

Cole E., Guttman R.D., Seemayer T.A., and Michel F. 1983. Spontaneous diabetes mellitus syndrome in the rat. IV. Immunogenetic interactions of MHC and non-MHC components of the syndrome. *Metab. Clin. Exp.* **32**: 54.

Cowley A.W., Jr., Roman R.J., Kaldunski M.L., Dumas P., Dickhout J.G., Greene A.S., and Jacob H.J. 2001. Brown Norway chromosome 13 confers protection from high salt to consomic Dahl S rat. *Hypertension* **37**: 456.

Cowley A.W., Jr., Stoll M., Greene A.S., Kaldunski M.L., Roman R.J., Tonellato P.J., Shork N.J. Dumas P., and Jacob H.J. 2000. Genetically defined risk of salt sensitivity in an intercross of brown Norway and Dahl S rats. *Physiol. Genomics* **2**: 107.

Dietrich W.F., Lander E.S., Smith J.S., Moser A.R., Gould K.A., Lunongo C., Borenstein N., and Dove W. 1993. Genetic identification of Mom-1, a major modifier locus affecting Min-induced intestinal neoplasia in the mouse. *Cell* **75**: 631.

Gill T.J.D., Smith G.J., Wissler R.W., and Kunz H.W. 1989. The rat as an experimental animal. *Science* **245**: 269.

Haddad G.G. 2000. Enhancing our understanding of the molecular responses to hypoxia in mammals using *Drosophila melanogaster. J. Appl. Physiol.* **88**: 1481.

Jacob H.J. and Kwitek A.E. 2002. Rat genetics: Attaching physiology and pharmacology to the genome. *Nat. Rev. Genet.* **3**: 33.

James J.F., Hewett T.E., and Robbins J. 1998. Cardiac physiology in transgenic mice. *Circ. Res.* **82**: 407.

Justice M.J., Noveroske J.K., Weber J.S., Zheng B., and Bradley A. 1999. Mouse ENU mutagenesis. *Hum. Mol. Genet.* **8**: 1955.

Knapik E.W. 2000. ENU mutagenesis in zebrafish—From genes to complex diseases. *Mamm. Genome* **11**: 511.

Knight J. and Abbott A. 2002. Mouse genetics: Full house. *Nature* **417**: 785.

Kwitek-Black A.E. and Jacob H.J. 2001. The use of designer rats in the genetic dissection of hypertension. *Curr. Hypertens. Rep.* **3**: 12.

Law J.M. 2001. Mechanistic considerations in small fish carcinogenicity testing. *ILAR J.* **42**: 274.

Li L. 2001. Zebrafish mutants: Behavioral genetic studies of visual system defects. *Dev. Dyn.* **221**: 365.

Liang M., Yaun B., Rute E., Greene A., Zou A.P., Soares P., McQuestion G.D., Slocum G.R., Jacob H.J., and Cowley A.W., Jr. 2002. Renal medullary genes in salt-sensitive hypertension: A chromosomal substitution and cDNA microarray study. *Physiol. Genomics* **8**: 139.

Lieschke G.J. 2001. Zebrafish: An emerging genetic model for the study of cytokines and hematopoiesis in the era of functional genomics. *Int. J. Hematol.* **73**: 23.

Merrill D.C., Granwehr B.P., Davis D.R., and Sigmund C.D. 1997. Use of transgenic and gene-targeted mice to model the genetic basis of hypertensive disorders. *Proc. Assoc. Am. Physicians* **109**: 533.

Mordes J.P., Desemone J., and Rossini A.A. 1987. The BB rat. *Diabetes Metab. Rev.* **3**: 725.

Nadeau J.H. 2001. Modifier genes in mice and humans. *Nat. Rev. Genet.* **2**: 165.

Nadeau J.H., Singer J., Matin A., and Lander E. 2000. Analyzing complex genetic traits with chromosome substitution strains. *Nat. Genet.* **24**: 221.

Nadeau J.H., Balling R., Barsh G., Beir D., Brown S.D.M., Bucan M., Camper S., Carlson G., Copeland N., Eppig J., Fletcher C., Frankel W.N., Ganten D., Goldowitz D., Goodnow C., Guenet J.-L., Hicks G., Hrabe de Angelis M., Jackson I., Jacob H.J., Jenkins N., Jonson D., Justice M., Kay S., Kingsley D., Lehrach H., Magnuson T., Meisler M., Poustka A., Rossant J., Russell L.B., Schimenti J., Shiroishe T., Skarnes W.C., Soriano P., Stanford W., Takahashi J.S., Wurst W., and Zimmer A. et al. 2001. Functional annotation of mouse genome sequences. *Science* **291**: 1251.

Nishimura H. and Ichikawa I. 1999. What have we learned form gene targeting studies for the renin angiotensin system of the kidney? *Intern. Med.* **38**: 315.

Nolan P.M., Peters J., Vizor L., Strivens M., Washbourne R., Hough T., Wells C., Glenister P., Thornton C., Martin J., Fisher E., Rogers D., Hagan J., Reavill C., Gray I., Wood J., Spurr N., Browne M., Rastan S., Hunter J., and Brown S.D.M. 2000. Implementation of a large-scale ENU mutagenesis program: Towards increasing the mouse mutant resource. *Mamm. Genome* **11**: 500.

Oliverio M.I. and Coffman T.M. 2000. Angiotensin II receptor physiology using gene targeting. *News Physiol. Sci.* **15**: 171.

Pearson H. 2002. Surviving a knockout blow. *Nature* **415**: 8.

Rapp J. 2000. Genetic analysis of inherited hypertension in the rat. *Physiol. Rev.* **80**: 135.

Schimenti J. and Bucan M. 1998. Functional genomics in the mouse: Phenotype-based mutagenesis screens. *Genome Res.* **8**: 698.

Sokolowski M.B. 2001. *Drosophila:* Genetics meets behavior. *Nat. Rev. Genet.* **2**: 879.

Stanford W.L., Cohn J.B., and Cordes S.P. 2001. Gene-trap mutagenesis: Past, present and beyond. *Nat. Rev. Genet.* **2**: 756.

Steen R.G., Kwitek-Black A.E., Glenn C., Gullings-Handley J., Van Etten W., Atkinson O.S., Appel D., Twigger S., Muir W., Mull T., Granados M., Kissebah M., Russo K. Crane R., Popp M., Peden M., Matise T., Brown D.M., Lu J., Kingsmore S., Tonellato P.J., Rozen S., Slonim D., Young P., and Jacob H.J. et al. 1999. A high-density integrated genetic linkage and radiation hybrid map of the laboratory rat. *Genome Res.* **9**: 1.

St Johnston D. 2002. The art and design of genetic screens: *Drosophila melanogaster. Nat. Rev. Genet.* **3**: 176.

Stoll M., Cowley A.W., Jr., Tonellato P.J., Greene A.S., Kaldunski M.L., Roman R.J., Dumas P., Schork N.J., Wang Z., and Jacob H.J. 2001. A genomic-systems biology map for cardiovascular function. *Science* **294**: 1723.

Stoll M., Kwietek-Black A., Cowley A.W., Jr., Harris E.L., Harrap S.B., Krieger J.E., Printz M.P., Provoost A.P., Sassard J., and Jacob H.J. 2000. New target regions for human hypertension via comparative genomics. *Genome Res.* **10**: 473.

Threadgill D.W., Dlugosz A.A., Hansen L.A., Tennenbaum T., Lichti U., Yee D., LaMantia C. Mourton T., Herrup K., and Harris R.C. et al. 1995. Targeted disruption of mouse EGF receptor: Effect of genetic background on mutant phenotype. *Science* **269**: 230.

Wells C. and Brown S.D. 2000. Genomics meets genetics: Towards a mutant map of the mouse. *Mamm. Genome* **11**: 472.

Wolfer D.P., Crusio W.E., and Lipp H.P. 2002. Knockout mice: Simple solutions to the problems of genetic background and flanking genes. *Trends Neurosci.* **25**: 336.

Venter J.C., Adams M.D., Myers E.W., Li P.W., Mural R.J., Sutton G.G., Smith H.O., Yandell M., Evans C.A., Holt R.A., Gocayne J.D., Amanatides P., and Ballew R.M. et al. 2001. The sequence of the human genome. *Science* **291**: 1304.

A Mouse Model of Congenital Heart Disease: Cardiac Arrhythmias and Atrial Septal Defect Caused by Haploinsufficiency of the Cardiac Transcription Factor Csx/Nkx2.5

M. Tanaka,*# C.I. Berul,†# M. Ishii,‡# P.Y. Jay,*† H. Wakimoto,† P. Douglas,*
N. Yamasaki,* T. Kawamoto,¶ J. Gehrmann,† C.T. Maguire,† M. Schinke,*
C.E. Seidman,** J.G. Seidman,§ Y. Kurachi,‡ and S. Izumo*

*Cardiovascular Division, Beth Israel Deaconess Medical Center, and Department of Medicine,
†Department of Cardiology, Children's Hospital, §Department of Genetics, Howard Hughes Medical Institute;
**Cardiovascular Division, Howard Hughes Medical Institute, Brigham and Women's Hospital and
Department of Medicine, Harvard Medical School, Boston, Massachusetts 02215; ‡Department of Pharmacology,
Graduate School of Medicine, Osaka University, Osaka, Japan; ¶Department of Geriatric Medicine,
Graduate School of Medicine, Kyoto University, Kyoto, Japan

Congenital heart disease (CHD) is the most common birth defect in humans, affecting ~8 per 1000 live births (Friedman 1997). More children die from CHD than from all childhood malignancies combined. However, the molecular mechanisms underlying CHD are largely unknown.

Nkx2.5 (or Csx) (Komuro and Izumo 1993; Lints et al. 1993) is a vertebrate homolog of the *Drosophila* homeobox gene *tinman*, whose disruption causes loss of heart formation in the fly (Azpiazu and Frasch 1993; Bodmer 1993). In mice, homozygous loss-of-function mutation of Nkx2.5 causes the arrest of cardiac development at the looping stage, but heterozygous mice were initially deemed normal (Lyons et al. 1995; Tanaka et al. 1999).

Interestingly, heterozygous mutations of human NKX2.5 were found in families with CHD, predominantly secundum atrial septal defect (ASD). Moreover, cardiac arrhythmias such as atrioventricular (AV) conduction block, atrial fibrillation, and ventricular tachycardia were found in these patients (Schott et al. 1998; Benson et al. 1999).

Recently, cardiac morphological analysis of mice heterozygous for Nkx2.5 mutation was reported (Biben et al. 2000). Detailed morphological inspection showed increased frequencies of patent foramen ovale and septal aneurysm and decreased length of the septum primum flap valve in the heart of the heterozygous mutant mice (Biben at al. 2000). Moreover, the degree of septal dysmorphogenesis depended on genetic background (Biben et al. 2000).

In this study, we investigated the effects of Nkx2.5 haploinsufficiency on cardiac morphology and electrophysiology in the mouse. We found that Nkx2.5$^{+/-}$ mice had consistent AV conduction abnormalities, prolonged QRS intervals, and inducible atrial and ventricular arrhythmias, as well as ASD with reduced penetrance. A significantly decreased outward K$^+$ current and prolonged action potential duration were detected by patch clamp experiments. Furthermore, cardiac expression of Na$^+$/K$^+$ ATPase, connexin40, and Kv4.3 was down-regulated in Nkx2.5$^{+/-}$ mice. These results indicate a critical role of Nkx2.5 in the electrophysiological functions of the myocardium as well as in the morphogenesis of the heart.

Nkx2.5$^{+/-}$ MICE HAVE ASD WITH REDUCED PENETRANCE

A null mutation of Nkx2.5 was generated by deletion of the entire coding region as described previously (Tanaka et al. 1999). The Nkx2.5$^{+/-}$ mice were maintained on a hybrid 129 and C57BL/6 strain background. We first searched for structural defects in the hearts of Nkx2.5$^{+/-}$ mice. The heart was exposed and the left atrial appendage was removed. After careful examination under dissection microscope of the atrial septum from the left side, the right atrial appendage was removed and the atrial septum was inspected from the right side as well. When a defect in the atrial septum was found, 6-0 nylon thread was gently passed to confirm communication between the left and right atria. Hearts from 45 Nkx2.5$^{+/-}$ mice and 52 wild-type littermates were examined. Normally, the foramen ovale in the septum secundum is covered by the septum primum (Fig.1A). In 9 (20%) Nkx2.5$^{+/-}$ mice, the septum primum did not fully cover the foramen ovale with an open communication between the right and left atria, indicative of a secundum ASD (Fig.1B–E). The average size of ASD was 0.30 ± 0.03 mm × 0.20 ± 0.01 mm. There were no defects in the atrial septum of the other 36 Nkx2.5$^{+/-}$ mice or 52 wild-type littermates (Table 1). No other structural cardiac abnormalities were found in either Nkx2.5$^{+/-}$ or wild-type mice examined.

#These authors contributed equally to this work.

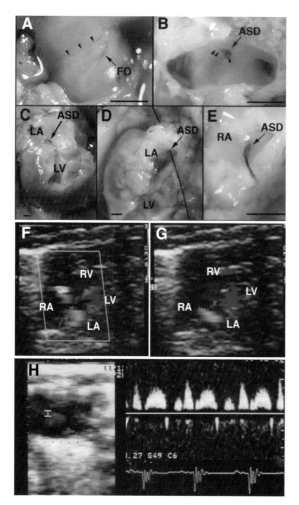

Figure 1. Morphology of the atrial septa of Nkx2.5$^{+/-}$ mice and wild-type mice. (*A*) The atrial septum of a 6-month-old wild-type mouse. The wall of the left atrium was removed and the atrial septum was examined from the left side. Note that the foramen ovale (FO) is completely covered by the septum primum. The arrowheads show the upper edge of the septum primum. (*B*) The atrial septum of a 6-month-old Nkx2.5$^{+/-}$ mouse. The wall of the left atrium was removed and the atrial septum was examined from the left side. Note a defect in the septum secundum that is not fully covered by the septum primum. The arrowheads show the upper edge of the septum primum. (*C–E*) ASD in an 8-month-old Nkx2.5$^{+/-}$ mouse. (LA) Left atrium, (LV) left ventricle. 6-0 nylon thread was placed through the defect, showing the connection between the left and right atria. Bars, 1 mm. (*F,G*) Color Doppler image of a Nkx2.5$^{+/-}$ mouse. Color Doppler performed in the subcostal view showing spontaneous left-to-right flow across the secundum atrial septal defect. (LA) Left atrium, (RA) right atrium, (SVC) superior vena cava, (AO) aorta. (*H*) Pulsed Doppler recording of a Nkx2.5$^{+/-}$ mouse, showing a biphasic change characteristic of ASD.

Table 1. Incidence of ASD

	Male	Female	Total
Csx/Nkx2.5$^{+/-}$	5/21	4/23	9/45 (20%)
Wild type	0/25	0/27	0/52 (0%)

The atrial septa of 45 Csx/Nkx2.5$^{+/-}$ mice and 52 wild-type mice were examined.

LEFT-TO-RIGHT SHUNT FLOW IS PRESENT IN VIVO

To determine cardiac function in vivo, we performed transthoracic echocardiography in 16 Nkx2.5$^{+/-}$ mice and 5 wild-type littermates. Images were obtained with a 12-MHz probe and Sonos 5500 (Hewlett Packard). Two-dimensionally guided M-mode analysis was obtained in the short axis to evaluate left ventricular chamber dimensions and contractile function. Color and pulsed Doppler was recorded in the short axis to evaluate left-to-right shunt flow between the atria using a 7.5-MHz probe and SSD-5500 (Aloka, Tokyo, Japan). Color Doppler echocardiography demonstrated an interatrial shunt in 5 of 16 Nkx2.5$^{+/-}$ mice (Fig.1F,G) and in none of the wild-type littermates. Pulsed Doppler interrogation distinguished the ASD shunt flow from the venous inflow, based on a characteristic phasic pattern (Porter et al. 1995). A positive Doppler signal began in late systole, diminished through mid-diastole, and was enhanced by atrial contraction (Fig.1H). We confirmed by gross dissection the presence of an ASD in the Nkx2.5$^{+/-}$ mice in which shunt flow was detected. No ASD was found in the remaining Nkx2.5$^{+/-}$ or wild-type mice. We did not observe dilatation or hypertrophy of the right atrium or ventricle in the mutant mice with ASD, suggesting that the shunt flow through the ASD was not large. Finally, there were no significant differences in LV chamber dimensions or systolic function between Nkx2.5$^{+/-}$ and wild-type mice (data not shown).

AV CONDUCTION DELAY AND QRS INTERVAL PROLONGATION IN Nkx2.5$^{+/-}$ MICE

We investigated whether Nkx2.5$^{+/-}$ mice have cardiac conduction abnormalities. A 6-lead surface electrocardiogram (ECG) was acquired under anesthesia in 17 Nkx2.5$^{+/-}$ mice and 18 littermate controls. A surface multi-lead ECG was obtained from subcutaneous electrodes placed in each extremity. Standard ECG intervals, including PR, QRS, and QT, were measured. To compensate for the rapid murine heart rate, ECG recording at 2000 Hz was required for precise measurement of ECG intervals.

The heart rates were similar in both groups, but the PR and QRS intervals were significantly longer in Nkx2.5$^{+/-}$ mice (Table 2, Fig. 2), indicating AV and intraventricular conduction defects. There were no significant differences in QT interval (Table 2). ECG telemetry was performed to exclude transient second- or third-degree AV block or ambient arrhythmias using PhysioTel radiotransmitters and receivers (Data Science International, St.Paul, Minnesota). Ambulatory ECG telemetry also showed that the average PR interval during a 24-hour recording was significantly prolonged in Nkx2.5$^{+/-}$ mice compared to wild type (36.6 ± 2.0 msec vs. 31.1 ± 1.2 msec, $p<0.001$, $n = 8$ in each group). No second- or third-degree AV block was recorded in any mouse. Since some human cases with NKX2.5 mutations showed progressive AV conduction abnormalities, we also examined 6 Nkx2.5$^{+/-}$ mice that were older than 1 year by ECG

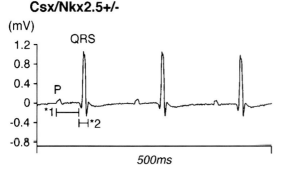

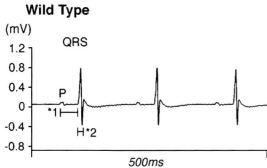

Figure 2. Slower atrioventricular and intraventricular conduction in Nkx2.5$^{+/-}$ mice. A surface electrocardiogram (ECG) was recorded at 2000 Hz in anesthetized mice. Lead I is shown. Nkx2.5$^{+/-}$ mice (*top*) displayed significantly longer PR and QRS intervals as compared to wild-type mice (*bottom*). In this example, the PR intervals (1*) were 38 and 48 msec in the wild-type and Nkx2.5$^{+/-}$ mouse, respectively. The QRS intervals (2*) were 13 and 19 msec, respectively. The sinus cycle length of each mouse is constant and measures 163 msec in the wild-type and 160 msec in the Nkx2.5$^{+/-}$ mouse.

telemetry. They maintained 1:1 AV conduction and a similar level of PR prolongation (data not shown).

INDUCTION OF ATRIAL AND VENTRICULAR TACHYARRHYTHMIAS IN Nkx2.5$^{+/-}$ MICE

We performed an in vivo electrophysiology (EP) study to examine arrhythmia vulnerability (Berul et al. 1997). The mouse EP study methodology has been described in detail previously (Berul et al. 1996). Because some patients with NKX2.5 mutations had arrhythmias and sudden death, we performed programmed electrical stimulation of the right atrium and right ventricle in the mouse to examine arrhythmia inducibility. Arrhythmia vulnerability (induction of atrial fibrillation and ventricular tachycardia) was also examined by a standard pacing protocol (Berul et al. 1996). Two independent observers, blinded to genotype, performed all measurements. Atrial fibrillation (Fig. 3), lasting 1–150 seconds, was easily and reproducibly inducible by atrial extra-stimuli in 13/23 Nkx2.5$^{+/-}$ mice, compared with 1/22 wild-type control mice (Table 3). Atrial fibrillation was also induced by ventricular extra-stimuli in 8/23 Nkx2.5$^{+/-}$ mice (Table 3). Ventricular tachycardia (Fig. 4), lasting 0.5–15 seconds, was induced by ventricular pacing in 6/23 Nkx2.5$^{+/-}$ mice but only in 1/22 wild-type controls (Table 3). These findings suggest that arrhythmia vulnerability is markedly increased in the atrium and ventricle of Nkx2.5$^{+/-}$ mice. Morphological and histological studies revealed no enlargement, hypertrophy, or fibrosis in the atrium or ventricle of Nkx2.5$^{+/-}$ mice (data not shown), suggesting that increased arrhythmia vulnerability was due not to gross cardiac structural changes but to functional abnormalities in cardiac myocytes.

DECREASED TRANSIENT OUTWARD K$^+$ CURRENT AND PROLONGED ACTION POTENTIAL DURATION IN VENTRICULAR CARDIAC MYOCYTES OF Nkx2.5$^{+/-}$ MICE

To examine ion channel function, we measured action potential and current density in ventricular myocytes by the whole-cell patch clamp technique in isolated ventricular myocytes. Left ventricular myocytes from male heterozygous mutant mice and wild-type littermates at 5–6 months of age were enzymatically isolated by the Langendorff coronary perfusion method. The action potential and K$^+$, Na$^+$, and Ca^{++} current densities were measured by whole-cell patch clamp techniques (Yamashita et al. 1996).

The action potential duration was significantly prolonged in cardiac myocytes of Nkx2.5$^{+/-}$ mice (Fig. 5A,B). There were no differences in the resting membrane potential (RMP) or maximum upstroke velocity (Vmax) (Fig. 5C). Moreover, early afterdepolarizations were observed in Nkx2.5$^{+/-}$ cardiac myocytes when stimulated at high frequencies (Fig. 5D).

Voltage clamp experiments revealed a decreased outward K$^+$ current in cardiac myocytes of Nkx2.5$^{+/-}$ mice (Fig. 6A,C). The outward K$^+$ currents in Nkx2.5$^{+/-}$ myocytes had a single slow phase of inactivation without a rapid initial phase (Fig. 6A), suggesting a substantial decrease in transient outward K$^+$ current (I_{to}), particularly a rapidly activating and inactivating component ($I_{to,f}$) (Nerbonne et al. 2001; Oudit et al. 2001). In agreement with this, the outward K$^+$ currents from Nkx2.5$^{+/-}$ myocytes were similar to those from wild-type myocytes in the presence of 4-aminopyridine (4-AP), an inhibitor of I_{to} (Fig. 6B, C). There were no differences in Na$^+$, L- or T-

Table 2. Summary of Surface ECG Data

	HR	PR	QRS	QTc
Nkx2.5$^{+/-}$	367.4 ± 69.7	47.8 ± 6.4*	17.1 ± 1.6*	26.5 ± 2.1
Wild type	390.7 ± 47.1	44.0 ± 3.9	12.8 ± 1.1	25.9 ± 1.7

The surface ECG data are summarized for 23 Csx/Nkx2.5$^{+/-}$ mice and 23 wild-type littermates. (HR) Heart rate/minute. All other values in milliseconds. *$p<0.05$.

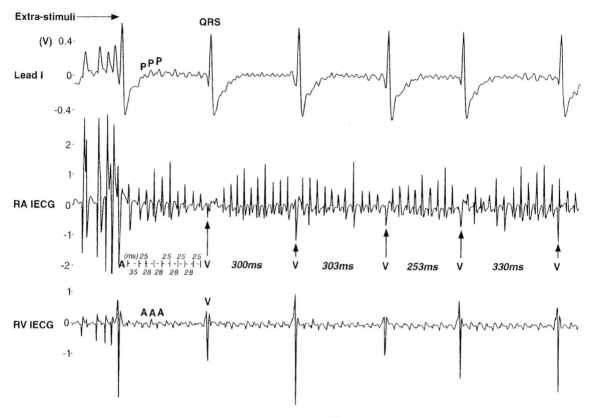

Figure 3. Atrial fibrillation induced by atrial extra-stimuli in Csx/Nkx2.5$^{+/-}$ mice. The atrial rhythm is irregular and conducts to the ventricles in an unorganized pattern. (Lead I) Lead I of a surface electrocardiogram (ECG), (RA IECG) right atrial intracardiac ECG, (RV IECG) right ventricular intracardiac ECG.

type Ca$^+$ channel currents between Nkx2.5$^{+/-}$ and wild-type cardiac myocytes (data not shown).

Cell lysates and membrane preparations were prepared from the atrium and ventricle for western blot analysis (Barry et al. 1995). Cell lysates containing 50–100 μg of protein or 40 μg of membrane proteins were separated by 10% SDS-PAGE, transferred onto nitrocellulose membrane, and probed with the following antibodies: polyclonal anti-α1 or anti-β1 subunit of Na$^+$/K$^+$ ATPase (Upstate Biotechnology), anti-connexin40 (Chemicon International), anti-connexin43, anti-Kv4.2, anti-Kv4.3, or anti-KChIP2 antibodies (Santa Cruz Biotechnology).

$I_{to,f}$ channels are assembled from the pore-forming α-subunits, Kv4.2 and Kv4.3, and the cytoplasmic regulatory protein KChIP2 (Guo et al. 2002). Consistent with the patch clamp results, the expression of Kv4.3 protein was down-regulated in the ventricle of Nkx2.5$^{+/-}$ mice (Fig. 7, 0.58 ± 0.03 folds, $p<0.01$ n = 4 each). Kv4.2 ex-pression was up-regulated (Fig. 7, 1.50 ± 0.11 folds, $p <0.01$ n = 4 each), and the expression level of KChIP2 proteins was unaffected (Fig. 7).

We also examined the expression of the α and β subunits of Na$^+$/K$^+$ ATPase, whose reduction could raise arrhythmia vulnerability (Hauptman and Kelly 1999; Ravens and Himmel 1999). The expression of the α subunit (the catalytic unit) was down-regulated in the atrium and ventricle of Nkx2.5$^{+/-}$ mice. The protein level of the β subunit (a regulatory unit) was similar to that of wild type (Fig. 7).

Gap junctions are responsible for electrical coupling of cardiac myocytes and consist of a hexameric assembly of connexins (Cx). Because altered Cx expression can cause cardiac arrhythmias (Kirchhoff et al. 1998; Simon et al. 1998; Hagendorff et al. 1999; Bevilacqua et al. 2000), we examined atrial and ventricular expression of Cx40 and 43. Western blot analysis showed reduced expression of

Table 3. Inducible Arrhythmias with Electrical Programmed Stimulation

	Atrial pacing protocols		Ventricular pacing protocol	
	Atrial fib.	Non-sust. VT	Atrial fib.	Non-sust. VT
Nkx2.5 $^{+/-}$	13/23 (57%)	1/23 (4%)	8/23 (35%)	6/23 (26%)
Wild type	1/22 (5%)	0/22 (0%)	0/22 (0%)	1/22 (5%)

Twenty-three Nkx2.5 $^{+/-}$ mice and 22 wild-type littermates were examined. (Atrial fib.) Atrial fibrillation, (Non-sust. VT) non-sustained ventricular tachycardia.

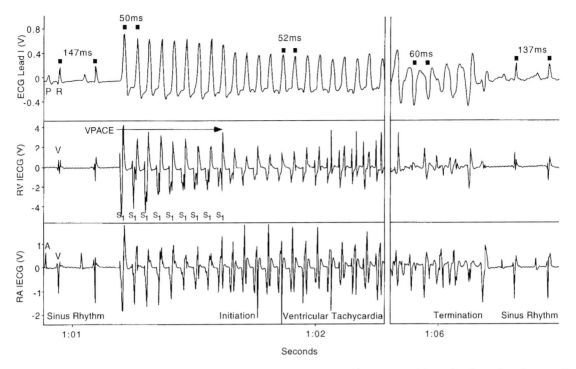

Figure 4. Ventricular tachycardia induced by ventricular pacing in Csx/Nkx2.5$^{+/-}$ mice. (Lead I) Lead I of a surface electrocardiogram (ECG), (RV IECG) right ventricular intracardiac ECG, (RA IECG) right atrial intracardiac ECG.

Cx40 in the atrium of Nkx2.5$^{-/-}$ mice (Fig. 7). Cx40 expression was undetectable by western blot in both Nkx2.5$^{+/-}$ and wild-type mice, most likely because Cx40

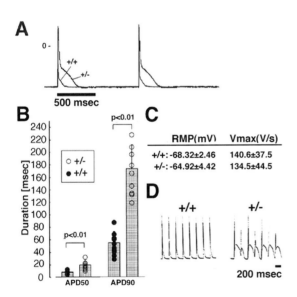

Figure 5. (*A*) Representative action potentials recorded from ventricular myocytes isolated from Nkx2.5$^{+/-}$ (+/–) and wild-type (+/+) mice. (*B*) Scattergram comparing action potential durations at 50% (APD50) and 90% (APD90) repolarization from Nkx2.5$^{+/-}$ and wild type cells. The differences in mean APD50 and APD90 were statistically significant. $n = 10$ for each. (*C*) Values for RMP (resting membrane potential) and Vmax. Significant differences were not detectable between Nkx2.5$^{+/-}$ and wild-type myocytes. (*D*) Early afterdepolarization was observed only in cardiac myocytes from Nkx2.5$^{+/-}$ mice when stimulated at 300 / min.

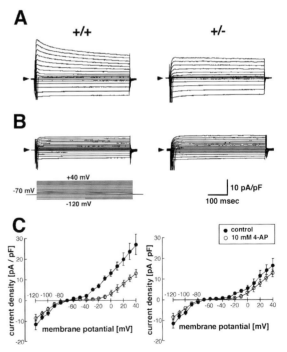

Figure 6. (*A*) Representative families of membrane currents from voltage-clamped ventricular myocytes from Nkx2.5$^{+/-}$ (+/–) and wild-type (+/+) mice. Currents were elicited by 500-msec voltage steps ranging between –120 and +40 mV in 10-mV increments from the holding potential of –70 mV. (*B*) Representative families of membrane currents from voltage-clamped ventricular myocytes from Nkx2.5$^{+/-}$ and wild-type mice in the presence of 4-AP. (*C*) Comparison of current–voltage relationships for ventricular myocytes from Nkx2.5$^{+/-}$ and wild-type mice in the absence (*left*) or presence (*right*) of 4-AP. $n = 8$ for each.

was specifically expressed in the conduction system in the adult ventricle. No difference was observed in the expression level of Cx43 (Fig. 7).

DISCUSSION

In patients with heterozygous mutations of NKX2.5, a variety of arrhythmias and cardiac structural defects have been observed (Schott et al. 1998; Benson et al. 1999). This murine model of Nkx2.5 haploinsufficiency recapitulated a limited number of cardiac defects seen in humans. The differences in the phenotype between humans and mice may be due to the presence of other factors with overlapping functions or differences in genetic background (Biben et al. 2000). A related gene Nkx2.6 has not been found in humans. In the mouse, Nkx2.6 is transiently expressed in the developing heart but is unlikely to have compensated for the loss of Nkx2.5 because more severe morphological abnormalities were not observed in the hearts of double mutant, Nkx2.5$^{+/-}$/Nkx2.6$^{-/-}$ mice (Tanaka et al. 2001).

Alternatively, the difference may be due to a selection bias in human studies, as previously suggested (Biben et al. 2000). It is plausible that only families with a high penetrance of severe phenotypes have been analyzed so far. Thus, there may be unidentified NKX2.5 mutations in families with milder phenotypes or lower penetrance. Most cases of congenital heart disease do not exhibit a typical monogenic (Mendelian) type inheritance. Heterozygous mutation of NKX2.5 may be responsible for some CHD cases that appear sporadic because of low disease penetrance. Indeed, the incidence of coding sequence mutations of NKX2.5 was reported to be 3% in randomly selected CHD patients (D. McElhinney and E. Goldmuntz, unpubl.). As for arrhythmias, AV conduction block is the major phenotype in the families so far reported. Further studies are necessary to determine why the mouse model does not develop second- or third-degree AV block.

It has been noted that some patients with ASD are more prone to atrial arrhythmias despite similarly sized defects (Craig and Selzer 1968; Campbell 1970; Gatzoulis et al. 1999; Oliver et al. 2002). It is also known that patients who undergo successful closure of an ASD or VSD may develop atrial fibrillation or ventricular arrhythmias many years after the repair (Brandenburg et al. 1983; Pastorek et al. 1994; Shah et al. 1994; Konstantinides et al. 1995; Gatzoulis et al. 1999; Oliver et al. 2002). It is therefore interesting to speculate that some of these patients may have transcription factor mutations that cause both the anatomic defect and physiologic abnormalities even after a successful operation. However, this hypothesis remains to be tested.

A detailed examination of the atrial septa of Nkx2.5 mutant mice in several genetic backgrounds by Biben et al. (2000) strongly suggested the presence of genetic modifiers of the atrial septal phenotype. The discrepancy in the frequency of ASD between their report and this study could be due to the fact that their mutant mice were genetically closer to C57BL/6J than our mutant mice because their mice were bred to C57BL/6J for two to six generations before the analysis. Indeed, in their analysis, septal dysmorphogenesis was most severe in the 129/Sv strain; 17% of the mutant hearts in this background showed borderline ASD, which was rare in the C57BL/6 strain (Biben et al. 2000). In the current study, Doppler echocardiography revealed a left-to-right shunt, indicating that the defects were not the potential communication of a patent foramen ovale but a true ASD. Nevertheless, the small ASDs did not seem to impose a significant hemodynamic burden, suggesting that the ASD observed in our model may be similar to borderline ASDs as defined by Biben et al. (2000).

Nkx2.5$^{+/-}$ mice exhibited PR and QRS prolongation on ECG and AV conduction abnormalities in EP studies (C.I. Berul et al., unpubl.). The discrepancies in these parameters between our study and the study by Biben et al. may be in part due to differences in the sensitivity of the method. In this study, we adopted a high signal sampling frequency (2000 Hz) and small decrements (5 msec) in measuring ECG intervals and testing AV node parameters. In fact, when we used a lower signal sampling frequency (400 Hz) and larger decrements (10 msec), we could not detect significant differences in AV conduction, either on ECG or EP studies.

What is the molecular basis for AV conduction abnormalities in the mutant mice? Patch clamp revealed no changes in the upstroke velocity or any depolarizing currents. We then examined expression of connexins, the

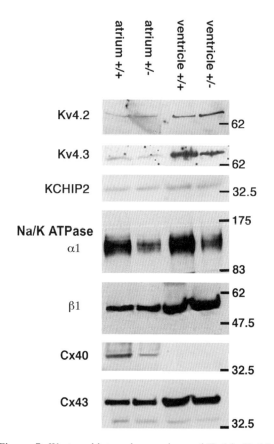

Figure 7. Western blot analyses using anti-Kv4.2, Kv4.3, KChIP2, α1 or β1 Na$^+$/K$^+$ ATPase, Cx40, or Cx43 antibody. Molecular standards are indicated in kilodaltons.

components of gap junctions, and found down-regulation of Cx40 in the atrium. Recently, it was shown that the heterozygous Tbx5 null mice have ASDs and AV conduction block (Bruneau et al. 2001). Cx40 expression is down-regulated in the hearts of heterozygous Tbx5 mutant mice. The Cx40 promoter contains binding sites for Tbx5 and Nkx2.5. A transient co-transfection analysis showed that Tbx5 and Nkx2.5 synergistically activate the Cx40 promoter (Bruneau et al. 2001). Since AV conduction abnormalities were reported in Cx40$^{-/-}$ mice (Kirchhoff et al. 1998; Simon et al. 1998; Hagendorff et al. 1999; Bevilacqua et al. 2000), reduction of Cx40 may be at least partly responsible for AV conduction abnormalities observed in Nkx2.5$^{+/-}$ mice and Tbx5$^{+/-}$ mice.

Interestingly, Nkx2.5$^{+/-}$ mice exhibited arrhythmia vulnerability in both the atrium and ventricle, although there were no detectable histological abnormalities in the tissue sections. Patch clamp revealed prolonged action potential duration in ventricular myocytes from Nkx2.5$^{+/-}$ mice. Longer action potential duration can be associated with repolarization abnormalities, which can predispose to triggered arrhythmias. Indeed, early afterdepolarizations were observed in myocytes from Nkx2.5$^{+/-}$ mice (Fig. 5D). Voltage clamp experiments suggested that a significant reduction in I_{to} is responsible, at least in part, for the prolonged action potentials.

In fetal mouse ventricular myocytes, I_{Kr} is the dominant repolarizing K$^+$ current, whereas I_{to} becomes dominant in adult mouse ventricular myocytes (Nuss and Marban 1994; Wang et al. 1996; Wang and Duff 1997). Because of its rapid activation kinetics, the developmental increase in I_{to} results in progressive decrease in action potential duration, suggesting that the alteration of I_{to} may be required for adaptation to rapid heart rate in the adult heart of the rodent (Nuss and Marban 1994; Wang et al. 1996; Wang and Duff 1997). Assembly of four Kv α subunits into a tetrameric structure creates a functional I_{to} channel (Jan and Jan 1992; Oudit et al. 2001). Kv1.4, Kv1.7, Kv3.4, Kv4.2, and Kv4.3 genes are expressed in the myocardium (Kalman et al. 1998; Coetzee et al. 1999; Oudit et al. 2001). Moreover, cytoplasmic proteins, such as Kv β subunits, Kv channel interacting proteins, and Kv channel-associated protein, modulate the expression and biophysical properties of voltage-gated K$^+$ channels (Oudit et al. 2001). Recently, it was reported that functional mouse ventricular $I_{to,f}$ channels are heteromeric, consisting of Kv4.2/Kv4.3 α subunits and KChIP2 (Guo et al. 2002). Interestingly, Kv4.3 was significantly down-regulated in the heart of Nkx2.5$^{+/-}$ mice. Significance of the up-regulation of Kv4.2 remains to be elucidated. It may be compensation for decreased expression of Kv4.3. Alternatively, the ratio of Kv4.2 and Kv4.3 may be critical for the proper formation of I_{to} (Guo et al. 2002).

The α subunit of Na$^+$/K$^+$ ATPase was down-regulated in the hearts of Nkx2.5$^{+/-}$ mice. Na$^+$/K$^+$ ATPase, the cellular sodium pump, is a heterodimer, the α subunit of which contains the Na$^+$, K$^+$, and ATP-binding sites (Geering 1997). Digitalis binds to the α subunit and inhibits its activity (Hauptman and Kelly 1999). Digitalis toxicity causes AV block and atrial and ventricular arrhythmias. It is not clear at present whether a similar mechanism may be partly responsible for AV conduction delay and arrhythmia vulnerability observed in the mutant mice and those seen in digitalis intoxication.

It should be pointed out that some of the data presented in this paper seem paradoxical. First, surface ECG did not reveal significant differences in QT interval between the groups, although patch clamp experiments clearly demonstrated prolonged action potential duration in ventricular myocytes of the mutant mice. This inconsistency was also observed in KChIP2$^{-/-}$ mice (Kuo et al. 2001) and is likely due to difficulty in precisely defining T wave and measuring QT interval offset on the mouse surface ECG. Second, surface ECG revealed prolongation both in PR and QRS intervals, whereas only longer PR was observed in ECG telemetry. It is possible that anesthetics used for surface ECG acquisition may unmask or exacerbate the phenotype. Alternatively, it may be due to low temporal resolution of the ECG telemetry measurement. To detect a 3-msec difference, the ECG telemetry would have to record at over 600 Hz.

CONCLUSIONS

Nkx2.5 (or Csx) is a homeobox transcription factor that plays a critical role in multiple aspects of cardiac development. Heterozygous mutations of the human NKX2.5 gene are associated with familial congenital heart disease (CHD) and arrhythmias. In this study, we performed morphological and electrophysiological analyses on the heterozygous null mutant mice for Nkx2.5. Small atrial septal defects (secundum type) were found in 20% of Nkx2.5$^{+/-}$ mice. They also exhibited prolonged PR and QRS intervals on electrocardiography, and atrioventricular (AV) conduction abnormalities on electrophysiological studies, indicating AV and intraventricular conduction defects. Moreover, atrial fibrillation and ventricular tachycardia were frequently induced in heterozygous mutant mice. Patch clamp experiments showed a significant decrease in the transient outward K$^+$ current (I_{to}), resulting in prolongation of the action potential duration in ventricular myocytes of the heterozygous mutant mice. Furthermore, the α1 subunit of Na/K ATPase, connexin40, and Kv4.3 were down-regulated in the heart of the heterozygous mutant mice. These results demonstrate that the heterozygous deletion of Nkx2.5 not only causes a defect in the atrial septum, but also affects the electrophysiological functions of the atrial and ventricular myocardium. At least some congenital heart disease and cardiac arrhythmias, if associated with a transcription factor mutation, may have a common genetic origin.

ACKNOWLEDGMENT

We thank Penny Thomas for her help in ASD detection in the mouse, Kimber Converso for echocardiography, James O. Mudd and Michael J. Healey for ECG telemetry, and D. Woodrow Benson for helpful comments. This work was supported by National Institutes of Health SCOR grant in congenital heart disease (S.I., C.E.S., C.B.). M.T. was a Paul Dudley White Fellow of American Heart Association Massachusetts Affiliate.

REFERENCES

Azpiazu N. and Frasch M. 1993. Tinman and bagpipe: Two homeo box genes that determine cell fates in the dorsal mesoderm of *Drosophila*. *Genes Dev.* **7:** 1325.

Barry D.M., Trimmer J.S., Merlic J.P., and Nerbonne J.M. 1995. Differential expression of voltage-gated K+ channel subunits in adult rat heart. Relation to functional K+ channels? *Circ. Res.* **77:** 361.

Benson D.W., Silberbach G.M., Kavanaugh-McHugh A., Cottrill C., Zhang Y., Riggs S., Smalls O., Johnson M.C., Watson M.S., Seidman J.G., Seidman C.E., Plowden J., and Kugler J.D. 1999. Mutations in the cardiac transcription factor NKX2.5 affect diverse cardiac developmental pathways. *J. Clin. Invest.* **104:** 1567.

Berul C.I., Aronovitz M.J., Wang P.J., and Mendelsohn M.E. 1996. In vivo cardiac electrophysiology studies in the mouse. *Circulation* **94:** 2641.

Berul C.I., Christe M.E., Aronovitz M.J., Seidman C.E., Seidman J.G., and Mendelsohn M.E. 1997. Electrophysiological abnormalities and arrhythmias in alpha MHC mutant familial hypertrophic cardiomyopathy mice. *J. Clin. Invest.* **99:** 570.

Bevilacqua L.M., Simon A.M., Maguire C.T., Gehrmann J., Wakimoto H., Paul D.L., and Berul C.I. 2000. A targeted disruption in connexin40 leads to distinct atrioventricular conduction defects. *J. Interv. Card. Electrophysiol.* **4:** 459.

Biben C., Weber R., Kesteven S., Stanley E., McDonald L., Elliott D.A., Barnett L., Koentgen F., Robb L., Feneley M., and Harvey R.P. 2000. Cardiac septal and valvular dysmorphogenesis in mice heterozygous for mutations in the homeobox gene Nkx2-5. *Circ. Res.* **87:** 888.

Bodmer R. 1993. The gene *tinman* is required for specification of the heart and visceral muscles in *Drosophila*. *Development* **118:** 719.

Brandenburg R.O., Jr., Holmes D.R., Jr., Brandenburg R.O., and McGoon D.C. 1983. Clinical follow-up study of paroxysmal supraventricular tachyarrhythmias after operative repair of a secundum type atrial septal defect in adults. *Am. J. Cardiol.* **51:** 273.

Bruneau B.G., Nemer G., Schmitt J.P., Charron F., Robitaille L., Caron S., Conner D.A., Gessler M., Nemer M., Seidman C.E., and Seidman J.G. 2001. A murine model of Holt-Oram syndrome defines roles of the T-box transcription factor Tbx5 in cardiogenesis and disease. *Cell* **106:** 709.

Campbell M. 1970. Natural history of atrial septal defect. *Br. Heart J.* **32:** 820.

Coetzee W.A., Amarillo Y., Chiu J., Chow A., Lau D., McCormack T., Moreno H., Nadal M.S., Ozaita A., Pountney D., Saganich M., Vega-Saenz de Miera E., and Rudy B. 1999. Molecular diversity of K+ channels. *Ann. N.Y. Acad. Sci.* **868:** 233.

Craig R.J. and Selzer A. 1968. Natural history and prognosis of atrial septal defect. *Circulation* **37:** 805.

Friedman W.F. 1997. Congenital heart disease in infancy and childhood. In *Heart disease* (ed. E. Braunwald), p. 877. W.B.Saunders, Philadelphia, Pennsylvania.

Gatzoulis M.A., Freeman M.A., Siu S.C., Webb G.D., and Harris L. 1999. Atrial arrhythmia after surgical closure of atrial septal defects in adults. *N. Engl. J. Med.* **340:** 839.

Geering K. 1997. Na,K-ATPase. *Curr. Opin. Nephrol. Hypertens.* **6:** 434.

Guo W., Li H., Aimond F., Johns D.C., Rhodes K.J., Trimmer J.S., and Nerbonne J.M. 2002. Role of heteromultimers in the generation of myocardial transient outward K+ currents. *Circ. Res.* **90:** 586.

Hagendorff A., Schumacher B., Kirchhoff S., Luderitz B., and Willecke K. 1999. Conduction disturbances and increased atrial vulnerability in connexin40-deficient mice analyzed by transesophageal stimulation. *Circulation* **99:** 1508.

Hauptman P.J. and Kelly R.A. 1999. Digitalis. *Circulation* **99:** 1265.

Jan L.Y. and Jan Y.N. 1992. Structural elements involved in specific K+ channel functions. *Annu. Rev. Physiol.* **54:** 537.

Kalman K., Nguyen A., Tseng-Crank J., Dukes I.D., Chandy G., Hustad C.M., Copeland N.G., Jenkins N.A., Mohrenweiser H., Brandriff B., Cahalan M., Gutman G.A., and Chandy K.G. 1998. Genomic organization, chromosomal localization, tissue distribution, and biophysical characterization of a novel mammalian Shaker-related voltage-gated potassium channel, Kv1.7. *J. Biol. Chem.* **273:** 5851.

Kirchhoff S., Nelles E., Hagendorff A., Kruger O., Traub O., and Willecke K. 1998. Reduced cardiac conduction velocity and predisposition to arrhythmias in connexin40-deficient mice. *Curr. Biol.* **8:** 299.

Komuro I. and Izumo S. 1993. Csx: A murine homeobox-containing gene specifically expressed in the developing heart. *Proc. Natl. Acad. Sci.* **90:** 8145.

Konstantinides S., Geibel A., Olschewski M., Gornandt L., Roskamm H., Spillner G., Just H., and Kasper W. 1995. A comparison of surgical and medical therapy for atrial septal defect in adults. *N. Engl. J. Med.* **333:** 469.

Kuo H.C., Cheng C.F., Clark R.B., Lin J.J., Lin J.L., Hoshijima M., Nguyen-Tran V.T., Gu Y., Ikeda Y., Chu P.H., Ross J., Giles W.R., and Chien K.R. 2001. A defect in the Kv channel-interacting protein 2 (KChIP2) gene leads to a complete loss of I(to) and confers susceptibility to ventricular tachycardia. *Cell* **107:** 801.

Lints T.J., Parsons L.M., Hartley L., Lyons I., and Harvey R.P. 1993. Nkx2.5: A novel murine homeobox gene expressed in early heart progenitor cells and their myogenic descendants. *Development* **119:** 419.

Lyons I., Parsons L.M., Hartley L., Li R., Andrews J.E., Robb L., and Harvey R.P. 1995. Myogenic and morphogenetic defects in the heart tubes of murine embryos lacking the homeo box gene Nkx2-5. *Genes Dev.* **9:** 1654.

Nerbonne J.M., Nichols C.G., Schwarz T.L., and Escande D. 2001. Genetic manipulation of cardiac K(+) channel function in mice: What have we learned, and where do we go from here? *Circ. Res.* **89:** 944.

Nuss H.B. and Marban E. 1994. Electrophysiological properties of neonatal mouse cardiac myocytes in primary culture. *J. Physiol.* **479:** 265.

Oliver J.M., Gallego P., Gonzalez A., Benito F., Mesa J.M., and Sobrino J.A. 2002. Predisposing conditions for atrial fibrillation in atrial septal defect with and without operative closure. *Am. J. Cardiol.* **89:** 39.

Oudit G.Y., Kassiri Z., Sah R., Ramirez R.J., Zobel C., and Backx P.H. 2001. The molecular physiology of the cardiac transient outward potassium current (I(to)) in normal and diseased myocardium. *J. Mol. Cell Cardiol.* **33:** 851.

Pastorek J.S., Allen H.D., and Davis J.T. 1994. Current outcomes of surgical closure of secundum atrial septal defect. *Am. J. Cardiol.* **74:** 75.

Porter C.J., Feldt R.H., Edwards W.D., Seward J.B., and Schaff H.V. 1995. Atrial septal defect. In *Heart disease in infants, children, and adolescents* (ed. G.C. Emmanouilides et al.), p. 687. Williams and Wilkins, Baltimore, Maryland.

Ravens U. and Himmel H.M. 1999. Drugs preventing Na+ and Ca2+ overload. *Pharmacol. Res.* **39:** 167.

Schott J.J., Benson D.W., Basson C.T., Pease W., Silberbach G.M., Moak J.P., Maron B.J., Seidman C.E., and Seidman J.G. 1998. Congenital heart disease caused by mutations in the transcription factor NKX2-5. *Science* **281:** 108.

Shah D., Azhar M., Oakley C.M., Cleland J.G., and Nihoyannopoulos P. 1994. Natural history of secundum atrial septal defect in adults after medical or surgical treatment: A historical prospective study. *Br. Heart J.* **71:** 224.

Simon A.M., Goodenough D.A., and Paul D.L. 1998. Mice lacking connexin40 have cardiac conduction abnormalities characteristic of atrioventricular block and bundle branch block. *Curr. Biol.* **8:** 295.

Tanaka M., Chen Z., Bartunkova S., Yamasaki N., and Izumo S. 1999. The cardiac homeobox gene Csx/Nkx2.5 lies genetically upstream of multiple genes essential for heart development. *Development* **126:** 1269.

Tanaka M., Schinke M., Liao H.S., Yamasaki N., and Izumo S.

2001. Nkx2.5 and Nkx2.6, homologs of *Drosophila tinman*, are required for development of the pharynx. *Mol. Cell. Biol.* **13:** 4391.

Turbay D., Wechsler S.B., Blanchard K.M., and Izumo S. 1996. Molecular cloning, chromosomal mapping, and characterization of the human cardiac-specific homeobox gene hCsx. *Mol. Med.* **2:** 86.

Wang L. and Duff H.J. 1997. Developmental changes in transient outward current in mouse ventricle. *Circ. Res.* **81:** 120.

Wang L., Feng Z.-P., Kondo C.S., Sheldon R.S., and Duff H.J. 1996. Developmental changes in the delayed rectifier K+ channels in mouse heart. *Circ. Res.* **79:** 79.

Yamashita T., Horio Y., Yamada M., Takahashi N., Kondo C., and Kurachi Y. 1996. Competition between Mg2+ and spermine for a cloned IRK2 channel expressed in a human cell line. *J. Physiol.* **493:** 143.

Genetic Dissection of the DiGeorge Syndrome Phenotype

F. VITELLI, E.A. LINDSAY, AND A. BALDINI
Department of Pediatrics (Cardiology), Baylor College of Medicine, Houston, Texas 77030

DiGeorge syndrome (DGS), velocardiofacial syndrome, and conotruncal anomaly face are phenotypically overlapping conditions associated with heterozygous deletion of chromosome 22 (Lindsay 2001). The deletion, generally referred to as *del22q11*, encompasses 1.5–3 Mbp of DNA encoding 20–30 genes. This is the most common chromosomal deletion associated with birth defects in humans, estimated to occur in 1 of 4000 live births (Emanuel and Shaikh 2001; McDermid and Morrow 2002).

DiGeorge syndrome is a typical pharyngeal arch/pouch disorder, as the clinical presentation includes developmental defects of most of the pharyngeal arch and pouch derivatives. Understanding the genetics of this syndrome should provide important insights into the developmental genetics of the pharyngeal apparatus, including the development of the pharyngeal arch arteries and of the outflow tract of the heart.

Generally, multi-gene deletion syndromes present special problems for genetic analysis because it is difficult to establish whether and how the deletion of individual genes contributes to the abnormal phenotype. Chromosome manipulation in mice has made it possible to physically dissect a chromosomal region and test the phenotypic consequences of deletions (or duplications) in vivo (Yu and Bradley 2001). This technology can be exploited for genetic analysis of deletion syndromes, or segmental aneuploidy syndromes, in general, provided that the genes deleted in the human syndrome are contiguous in the mouse genome. Using this technology, we generated the first mouse model of *del22q11*, which carries a deletion including most of the mouse homologs of genes deleted in *del22q11* (Lindsay et al. 1999). These mice, referred to as *Df1/+* mice, present with a subtle phenotype that includes some of the clinical features observed in *del22q11* patients. These are cardiovascular, thymic, parathyroid, and neurobehavioral abnormalities (Lindsay et al. 1999; Paylor et al. 2001; Taddei et al. 2001).

Through the generation and analysis of nested deletions, single-gene targeted mutations, and rescuing transgenic lines, we and other workers have identified the haploinsufficient gene, *Tbx1*, to be important for the development of pharyngeal arch and pouch derivatives (Jerome and Papaioannou 2001; Lindsay et al. 2001; Merscher et al. 2001). The analysis of *Tbx1* mutants, and the correlation between gene expression and phenotype, are providing insights into the role of this gene in the development of the pharyngeal apparatus. First of all, in contrast to the prevailing hypothesis about the pathogenesis of the *del22q11* phenotype, *Tbx1* is not expressed in neural crest cells.

Virtually all the chromosomal deletions within the 22q11.2 region and associated with a DiGeorge syndrome-like cardiovascular phenotype include *TBX1*. There are some exceptions, however, that we address.

Another important aspect of modeling and dissecting the DiGeorge syndrome phenotype is the incomplete penetrance of the cardiovascular defects. We discuss this issue in the context of strategies aimed at controlling genetically the penetrance of aortic arch abnormalities. Finally, we discuss new hypotheses and models of pharyngeal arch and pouch development based on the data derived from the analysis of the *Tbx1* mutant phenotype.

NONOVERLAPPING DELETIONS IN THE HUMAN SYNDROME

The great majority of patients with a typical DiGeorge syndrome phenotype, which includes cardiovascular, craniofacial, parathyroid, and thymic abnormalities, have a heterozygous chromosome deletion within 22q11.2. Of the patients with deletion, about 85% have a 3-Mbp deletion and about 10% have a smaller, 1.5-Mbp deletion (Emanuel and Shaikh 2001; McDermid and Morrow 2002). A small percentage of patients carry atypical deletions, and some of these deletions do not overlap with each other but do have phenotypic overlap (Scambler 2000; Lindsay 2001). Hypotheses have been put forward to explain these observations: e.g., long-range effects of the deletions on neighboring genes, and/or existence of multiple haploinsufficiency loci, each important for cardiovascular and pharyngeal development. Mouse genetics has provided evidence that at least two genes within the deleted region, *Tbx1* and *Crkol*, are important for cardiovascular and pharyngeal development, although apparently through different mechanisms. However, there is at least one human deletion that does not involve *TBX1* or *CRKOL* (Yamagishi et al. 1999). The patient with this deletion presents with the typical cardiovascular abnormalities involving both outflow tract and aortic arch, thymic abnormalities, and craniofacial defects. It was determined that the patient carries a small ~20-kbp deletion disrupting two genes, *Cdc45* and *Ufd1*. Both genes have been targeted in mice and have been shown not to be haploinsufficient, and homozygous deletion of either of these two genes is early embryonic lethal (Lindsay et al. 1999; Yoshida et al. 2001). We decided to establish the following: (1) whether combined deletion of *Cdc45* and *Ufd1* can cause phenotypic abnormalities in mice and (2) whether a chromosome rearrangement in this region can affect the nearby gene *Tbx1*.

To this end, we generated a deletion spanning ~150 kbp of the mouse region homologous to the *del22q11* region (Fig. 1). To generate this deletion, we used previously reported chromosome engineering techniques in mouse embryonic stem cells, using Cre-mediated loxP recombination (Su et al. 2000). Briefly, one loxP site was inserted into the *Hira* gene by homologous recombination as described previously (Lindsay et al. 2001), the second loxP site was inserted, using a retrovirus, in the region between the *Gbp1bb* and *Cld5* genes, 150 kbp centromeric to the other loxP site. The recombinase Cre was transiently expressed in the targeted ES cells to delete the region between the two loxP sites. To confirm the deletion, we used fluorescence in situ hybridization with a probe encompassing most of the region between *Cld5* and *Hira*, BAC256F19. Results show that the region had been deleted (Fig. 2). We named this deletion del(16)(*Cdl5-Hira*), hereafter referred to as *Df5*. The deletion includes genes *Cdl5*, *Cdc45*, *Ufd1*, *Nlvcf*, and *Hira* but does not include *Tbx1*, which is located ~150 kbp proximal to it (Fig. 3). ES cells carrying *Df5* were injected into mouse blastocysts to obtain chimeric mice. Chimeric mice were crossed with C57BL/6 mice and the mutation was transmitted through the germ line. Heterozygous (*Df5/+*) mice were viable and fertile. Crosses between *Df5/+* mice did not yield any homozygously deleted mice, consistent with the previous finding of embryonic lethality of three of the genes included in the deletion, *Cdc45*, *Ufd1*, and *Hira* (Lindsay et al. 1999; Yoshida et al. 2001; Roberts et al. 2002).

To understand whether the deletion could affect cardiovascular development, we have analyzed *Df5/+* embryos at E10.5 and at E18.5. At E10.5, virtually all the *Df1/+* and *Tbx1+/−* embryos have hypoplastic 4th PAAs. In contrast, none of the 13 *Df5/+* embryos tested presented with PAA abnormalities. In addition, we have ex-

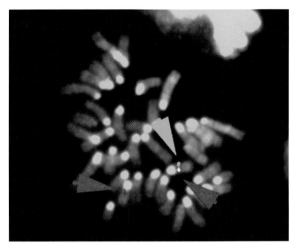

Figure 2. Photomicrograph of fluorescence in situ hybridization on mouse chromosomes from the embryonic stem cell line carrying *Df5*. The red signals are produced by a BAC320E18 (control probe), which includes the *Es2el* locus (located centromeric to *Df5*, see map in Fig. 3). The green signal is produced by BAC256F19, which includes the *Ufd1* and part of the *Hira* genes. This signal is present on only one of the two chromosome 16s, indicating heterozygous deletion of the locus. These cells were injected into mouse blastocysts to establish *Df5* in mice.

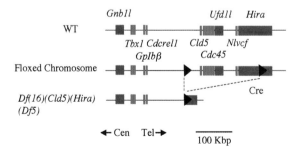

Figure 1. Generation of the chromosomal deficiency *Df5* on mouse chromosome 16 using a Cre-loxP strategy. LoxP sites are indicated as black triangles. Cre was expressed transiently in mouse embryonic stem cells. The map is approximately to scale, according to mouse genome sequence information available from the Wellcome Trust Sanger Institute data server http://www.ensembl.org/Mus_musculus/. Genes indicated in red are transcribed from centromere to telomere (gene names on top); genes indicated in green are transcribed from the opposite strand.

amined the cardiovascular morphology of 37 *Df5/+* embryos at E18.5 and found no abnormalities.

Df5 is the second chromosomal deletion that we have generated which is not associated with a cardiovascular phenotype, the other being *Df2* (Lindsay et al. 2001). The locations of these two deletions and their relationship to *Tbx1* are shown in Figure 3. *Df2* and *Df5* are ~150 kbp proximal and distal, respectively, to *Tbx1*, indicating that, in the mouse, there are no distant (>150 kbp) regulatory elements essential for the *Tbx1* function in cardiovascular development. Large (140–150 kbp) genomic fragments containing the mouse or human *Tbx1* genes have been successfully used in transgenic experiments to rescue the aortic arch defects due to chromosomal deletions including the *Tbx1* gene (Lindsay et al. 2001; Merscher et al. 2001). These results indicate that essential regulatory elements of the human or mouse *Tbx1* genes are contained within 150 kbp of genomic DNA. Overall, these genetic experiments do not support the hypothesis that patient deletions which do not include *Tbx1* may cause phenotypic abnormalities by removing *cis*-regulatory elements of this gene.

There are three additional hypotheses. (1) The association of phenotypic abnormalities with the very few cases of "atypical" deletions is by chance. (2) Chromosomal rearrangements in humans, but not in mice, affect chromatin structure and reduce the expression of genes over a long range. The relative vicinity of the 22q11.2 band to the centromere lends some ground to this view, but addressing this hypothesis is problematic because of the unavailability of relevant (embryonic) tissues from affected individuals. Perhaps in the near future, it will be possible to engineer chromosomal deletions in human stem cells and monitor gene expression during in vitro differentia-

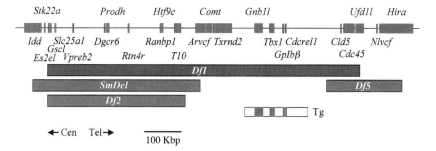

Figure 3. Map of the *del22q11*-homologous region showing the deletions proximal and distal to *Tbx1* that are not associated with a cardiovascular phenotype. The deletion *Df1* is also indicated for reference. *SmDel* was described in Puech et al. (2000). *Df2* was described in Lindsay et al. (2001). Tg indicates the transgene that rescues the *Tbx1* haploinsufficiency phenotype (Lindsay et al. 2001; Merscher et al. 2001). Genes indicated in red are transcribed from centromere to telomere (gene names on top); genes indicated in green are transcribed from the opposite strand.

tion. (3) The deletion of genes within 22q11.2 other than *Tbx1* may be sufficient to cause a DGS-like phenotype. An obvious candidate is *Crkol*, which encodes a SH2-SH3-SH3 adapter protein and is located within the 3-Mbp *del22q11* interval. Loss of function of this gene in mice causes, among other defects, aortic arch defects, thymic, parathyroid, and craniofacial abnormalities (Guris et al. 2001). In contrast to *Tbx1*, *Crkol* is not haploinsufficient in mice, at least not in the genetic background reported. Of course, it could be haploinsufficient in humans and could explain some, but not all, the "atypical" deletions. *Crkol* deletion appears not to be necessary for the expression of the DGS phenotype because the gene is not deleted in a substantial portion of patients (~10%).

PHENOTYPIC VARIABILITY AND *Tbx1* FUNCTIONS

Most patients have the same 3-Mbp deletion, but the clinical presentation is variable in terms of expressivity and penetrance of individual clinical findings. Although a degree of clinical variability is common among haploinsufficiency syndromes, the clinical presentation of *del22q11* syndrome is particularly variable, ranging from life-threatening cardiovascular defects or T-cell immunodeficiency to a virtually normal phenotype. Inconsistent phenotypes have also been reported in identical twins, suggesting that the clinical variability may not depend exclusively on the genetic makeup of individuals (Goodship et al. 1995; Fryer 1996; Hatchwell 1996; Yamagishi et al. 1998; Vincent et al. 1999; Lu et al. 2001). The use of mouse models is facilitating the understanding of phenotypic variability in DGS. The multigene deletion *Df1* has been bred into different strains of mice to obtain inbred *Df1*/+ 129SvEv and congenic *Df1*/+ C57BL/6 (Taddei et al. 2001). Phenotypic analysis revealed that the penetrance of aortic arch defects is substantially higher in the C57BL/6 strain than in the 129SvEv strain. In addition, thymic abnormalities were observed in the congenic and inbred strains but not in the mixed (129SvEv × C57BL/6) strain (Taddei et al. 2001). These results demonstrate that genes affect the haploinsufficiency phenotype, but, as in human patients, genetically identical individuals may have inconsistent phenotypes. This suggests that stochastic events may also play a role alongside genetic factors.

Whereas all the *Df1*/+ E10.5 embryos have at least one hypoplastic 4th PAA, most arteries tend to "normalize" later in development, so that many mutant embryos at term have a normal aortic arch (Lindsay and Baldini 2001). The reason that some arteries "recover" and some do not is unknown. The genetic background affects the recovery from, rather than the severity of, the 4th PAA hypoplasia (Taddei et al. 2001), at least in the strains tested. As in *Df1*/+ animals, the 4th PAA hypoplasia in *Tbx1*$^{+/-}$ animals also tends to recover (our unpublished data) and its penetrance is also strain-dependent (Jerome and Papaioannou 2001). However, extensive breeding in pure genetic backgrounds has not been accomplished yet. Because *Tbx1* is not expressed in the structural components of the 4th PAA, but rather in the pharyngeal endoderm surrounding it, we hypothesize that *Tbx1* triggers the expression of diffusible signals that directly or indirectly support the growth of the artery. One such signaling molecule could be Fgf8, because the gene that encodes it is coexpressed with *Tbx1* in the pharyngeal endoderm, and loss of *Tbx1* causes loss of *Fgf8* expression in this tissue (Vitelli et al. 2002b). Confirmation of a genetic interaction between *Tbx1* and *Fgf8* came from the finding that double heterozygous animals *Tbx1*$^{+/-}$;*Fgf8*$^{+/-}$ have a significantly higher penetrance of cardiovascular defects than *Tbx1*$^{+/-}$;*Fgf8*$^{+/+}$ animals (Vitelli et al. 2002b). *Fgf8* heterozygous mutation increases the penetrance of cardiovascular defects in *Tbx1*$^{+/-}$ mice by making the 4th PAA hypoplasia more severe. Hence, there are at least two ways for genes to affect the haploinsufficiency phenotype of *Df1*/+ or *Tbx1*$^{+/-}$ mice: by affecting the recovery process or by affecting the severity of artery hypoplasia.

The severity (penetrance and/or expressivity) of a haploinsufficiency phenotype may depend on the level of expression or structural variation of the remaining functional allele. Allelic variation across strains might account for the penetrance differences. However, this issue has been addressed, and it has been found that in *Df1*/+ mice the C57BL/6 *Tbx1* allele per se is not sufficient to cause the high penetrance of cardiovascular defects observed in C57BL/6 congenic mice, indicating that there are genetic modifiers elsewhere in the genome (Taddei et al. 2001). Modifiers may affect the expression of the functional copy of *Tbx1* (upstream modifiers), or the function of genes targeted by Tbx1 (downstream modifiers), or the functionality of potential alternative pathways that cells or tissues may utilize in the absence of *Tbx1* ("parallel" modifiers). *Fgf8*, for example, is most likely to be downstream from *Tbx1* and is perhaps a direct target of Tbx1 transcriptional activity. The strain-dependent differences, of course, are more difficult to dissect because they may be due to multiple loci, all yet to be identified. If strain differences were to exist in $Tbx1^{-/-}$ mice, then modifiers could not be upstream of *Tbx1* because they would not have any phenotypic effect. To test whether the null phenotype can be modified by different genetic backgrounds, we generated 129SvEv inbred $Tbx1^{-/-}$ embryos and compared them with mixed background C57BL/6 × 129SvEv $Tbx1^{-/-}$ embryos. The strain 129SvEv was selected because it confers the mildest phenotype in Df1/+ embryos. We have analyzed 7 $Tbx1^{-/-}$ embryos at E18.5 in the mixed genetic background and 14 in the inbred background. The overall morphological appearance of the two groups of embryos was not distinguishable, and cardiovascular defects were of the same type. We did find, however, that 129SvEv mutants have a milder palate phenotype. Although all the mixed genetic background mutants completely lack fusion of the shelves of the secondary palate, only 57% of the inbred 129SvEv embryos showed this severe defect, whereas the palate in the others was closed (21%) or almost completely closed (21%). These results show that the 129SvEv genetic background has a mild modifying effect on the phenotypic abnormalities of the pharyngeal apparatus in $Tbx1^{-/-}$ animals. Hence, modifiers in this strain may act downstream from *Tbx1* or facilitate compensatory pathways.

Tbx1 IS A MAJOR PLAYER IN THE DEVELOPMENT OF THE PHARYNGEAL APPARATUS

The pharyngeal apparatus is a transient and vertebrate-specific structure with a "modular" organization composed of pharyngeal arches (which include the pharyngeal arch arteries) and pouches. These modules develop progressively, in a cranial–caudal direction. The structure provides some important evolutionary "improvements" to organisms, e.g., efficient separation between the pulmonary and systemic circulation, T-cell-mediated immunity (thymus), and an improved system for controlling calcium metabolism (parathyroids) (Graham 2001). Ironically, developmental defects of the pharyngeal apparatus are also a rich source of birth defects. These include cardiovascular defects (of the outflow tract and aortic arch), craniofacial defects (cleft palate, micrognathia), velopharyngeal insufficiency, and probably some developmental defects of the ear.

The segmentation of the pharyngeal apparatus, i.e., its division into arch-pouch modules, is thought to be driven mainly by the pharyngeal endoderm. This hypothesis fits well with the *Tbx1* mutant phenotype. $Tbx1^{-/-}$ embryos lack proper segmentation of the pharyngeal apparatus because the pouches, which are invaginations of the pharyngeal endoderm, do not develop. Consequently, the embryonic pharynx is hypoplastic and has a tubular morphology (Jerome and Papaioannou 2001; Vitelli et al. 2002a). This leads to phenotypic abnormalities affecting the entire pharyngeal apparatus. We hypothesize that the initial function of *Tbx1*, which is expressed in the pharyngeal endoderm, is to cell-autonomously initiate and sustain the development of the pouches.

Neural crest cells migrate abnormally in $Tbx1^{-/-}$ mutants (Vitelli et al. 2002a). Because the pharynx abnormalities precede neural crest migration into the pharyngeal apparatus, and because neural crest cells are not required for pharyngeal segmentation (Veitch et al. 1999), we think that the migration defect is a consequence of the pharyngeal abnormalities caused by loss of *Tbx1*. We hypothesize that *Tbx1* in the endoderm triggers the expression of signals directed toward neural-crest-derived cells. This hypothesis is supported by the results of the genetic experiments with *Fgf8* mentioned earlier (Vitelli et al. 2002b). *Tbx1* is also expressed in the core mesoderm of the pharyngeal arches 1, 2, and 3. Hence, it is possible that *Tbx1* also triggers signals from this tissue that are directed toward the surrounding neural crest cells. This is consistent with the loss of *Fgf10* in the pharyngeal core mesoderm of $Tbx1^{-/-}$ embryos (Vitelli et al. 2002a). Hence, as supported by the analysis of the haploinsufficiency phenotype, *Tbx1* also has a cell nonautonomous function in pharyngeal development.

The pharyngeal apparatus also has a direct role in the development of the outflow tract of the heart. This portion of the heart is contributed by pharyngeal and splanchnic mesoderm cells (Kelly et al. 2001; Mjaatvedt et al. 2001; Waldo et al. 2001). These cells contribute to the muscle (outer) wall of the outflow tract. Neural-crest-derived cells, migrating through the pharyngeal apparatus, also contribute to the outflow tract by populating the outflow cushions. $Tbx1^{-/-}$ mice have severe outflow tract defects, some of which can be attributed to reduced contribution from the anterior heart field (hypoplasia of the outflow), whereas others are more likely to be secondary to reduced neural crest cell contribution (hypoplasia of the cushions and lack of septation) (Vitelli et al. 2002a, b).

CONCLUSIONS

The genetics of DGS is still puzzling investigators because excellent candidate genes, as assessed by mouse genetics experiments, have not yet been proven to be haploinsufficient in humans. This may be because the DGS phenotype, as assessed clinically, may require haploinsufficiency of more than one gene. A more likely situation is that patients with a single gene mutation are very rare. We show here that chromosomal rearrangements close to, but not including *Tbx1*, in the mouse do not affect *Tbx1* function. These results do not support the hypothesis of long-range effects of rearrangements in this locus.

The puzzling clinical variability is finding a parallel in mouse genetics. In mice, both the haploinsufficiency and the null phenotypes (as shown here) can be affected by the genetic background. It is predicted that multiple loci and multiple developmental mechanisms can contribute to phenotypic variability. In addition, nongenetic factors also appear to have a substantial impact on phenotypic presentation.

Finally, mouse genetics of DGS is gradually changing the way we view the development of the pharyngeal apparatus. Neural-crest-centered hypotheses to explain the pathogenesis of DGS should be revised in favor of hypotheses that consider all the possible sources of molecular instructions required to build the pharyngeal apparatus. Further studies are required in order to dissect the genetic networks controlling the expression of signaling molecules and signal transduction molecules important for communications between endoderm, neural crest cells, and pharyngeal mesoderm. Different genes may be important for different stages of the development of the pharyngeal apparatus, e.g., initial segmentation, growth, differentiation of derivatives, and the eventual partial regression of the system. However, the analysis of the mutant phenotype suggests that *Tbx1* may be involved in different phases of development of these structures, perhaps through interaction with different downstream targets.

ACKNOWLEDGMENTS

We thank Hedda Sobotka, Tom Huyhn, Ilaria Taddei, and Vesna Jurecic for technical contribution to this paper. We thank Allan Bradley for providing the ES line AB2.2 and chromosome engineering targeting cassettes. Research in the laboratories of A.B. and E.A.L. is funded by grants from the National Institutes of Health and American Heart Association Texas Affiliate. F.V. is the recipient of a fellowship from the Italian Telethon.

REFERENCES

Emanuel B.S. and Shaikh T.H. 2001. Segmental duplications: An 'expanding' role in genomic instability and disease. *Nat. Rev. Genet.* **2**: 791.

Fryer A. 1996. Monozygotic twins with 22q11 deletion and discordant phenotypes. *J. Med. Genet.* **33**: 173.

Goodship J., Cross I., Scambler P., and Burn J. 1995. Monozygotic twins with chromosome 22q11 deletion and discordant phenotype. *J. Med. Genet.* **32**: 746.

Graham A. 2001. The development and evolution of the pharyngeal arches. *J. Anat.* **199**: 133.

Guris D.L., Fantes J., Tara D., Druker B.J., and Imamoto A. 2001. Mice lacking the homologue of the human 22q11.2 gene CRKL phenocopy neurocristopathies of DiGeorge syndrome. *Nat. Genet.* **27**: 293.

Hatchwell E. 1996. Monozygotic twins with chromosome 22q11 deletion and discordant phenotype. *J. Med. Genet.* **33**: 261.

Jerome L.A. and Papaioannou V.E. 2001. DiGeorge syndrome phenotype in mice mutant for the T-box gene, *Tbx1*. *Nat. Genet.* **27**: 286.

Kelly R.G., Brown N.A., and Buckingham M.E. 2001. The arterial pole of the mouse heart forms from Fgf10-expressing cells in pharyngeal mesoderm. *Dev. Cell* **1**: 435.

Lindsay E.A. 2001. Chromosomal microdeletions: Dissecting del22q11 syndrome. *Nat. Rev. Genet.* **2**: 858.

Lindsay E.A. and Baldini A. 2001. Recovery from arterial growth delay reduces penetrance of cardiovascular defects in mice deleted for the DiGeorge syndrome region. *Hum. Mol. Genet.* **10**: 997.

Lindsay E.A., Botta A., Jurecic V., Carattini-Rivera S., Cheah Y.-C., Rosenblatt H.M., Bradley A., and Baldini A. 1999. Congenital heart disease in mice deficient for the DiGeorge syndrome region. *Nature* **401**: 379.

Lindsay E.A., Vitelli F., Su H., Morishima M., Huynh T., Pramparo T., Jurecic V., Ogunrinu G., Sutherland H.F., Scambler P.J., Bradley A., and Baldini A. 2001. Tbx1 haploinsufficieny in the DiGeorge syndrome region causes aortic arch defects in mice. *Nature* **410**: 97.

Lu J.H., Chung M.Y., Hwang B., and Chien H.P. 2001. Monozygotic twins with chromosome 22q11 microdeletion and discordant phenotypes in cardiovascular patterning. *Pediatr. Cardiol.* **22**: 260.

McDermid H.E. and Morrow B.E. 2002. Genomic disorders on 22q11. *Am. J. Hum. Genet.* **70**: 1077.

Merscher S., Funke B., Epstein J.A., Heyer J., Puech A., Lu M.M., Xavier R.J., Demay M.B., Russell R.G., Factor S., Tokooya K., Jore B.S., Lopez M., Pandita R.K., Lia M., Carrion D., Xu H., Schorle H., Kobler J.B., Scambler P.J., Wynshaw-Boris A., Skoultchi A.I., Morrow B.E., and Kucherlapati R. 2001. TBX1 is responsible for cardiovascular defects in velo-cardio-facial/DiGeorge syndrome. *Cell* **104**: 619.

Mjaatvedt C.H., Nakaoka T., Moreno-Rodriguez R., Norris R.A., Kern M.J., Eisenberg C.A., Turner D., and Markwald R.R. 2001. The outflow tract of the heart is recruited from a novel heart-forming field. *Dev. Biol.* **238**: 97.

Paylor R., McIlwain K.L., McAninch R., Nellis A., Yuva-Paylor L.A., Baldini A., and Lindsay E.A. 2001. Mice deleted for the DiGeorge/velocardiofacial syndrome region show abnormal sensorimotor gating and learning and memory impairments. *Hum. Mol. Genet.* **10**: 2645.

Puech A., Saint-Jore B., Merscher S., Russell R.G., Cherif D., Sirotkin H., Xu H., Factor S., Kucherlapati R., and Skoultchi A.I. 2000. Normal cardiovascular development in mice deficient for 16 genes in 550 kb of the velocardiofacial/DiGeorge syndrome region. *Proc. Natl. Acad. Sci.* **97**: 10090.

Roberts C., Sutherland H.F., Farmer H., Kimber W., Halford S., Carey A., Brickman J.M., Wynshaw-Boris A., and Scambler P.J. 2002. Targeted mutagenesis of the Hira gene results in gastrulation defects and patterning abnormalities of mesoendodermal derivatives prior to early embryonic lethality. *Mol. Cell. Biol.* **22**: 2318.

Scambler P.J. 2000. The 22q11 deletion syndromes. *Hum. Mol. Genet.* **9**: 2421.

Su H., Wang X., and Bradley A. 2000. Nested chromosomal deletions induced with retroviral vectors in mice. *Nat. Genet.* **24**: 92.

Taddei I., Morishima M., Huynh T., and Lindsay E.A. 2001. Genetic factors are major determinants of phenotypic variability in a mouse model of the DiGeorge/del22q11 syndromes. *Proc. Natl. Acad. Sci.* **98**: 11428.

Veitch E., Begbie J., Schilling T.F., Smith M.M., and Graham A. 1999. Pharyngeal arch patterning in the absence of neural crest. *Curr. Biol.* **9**: 1481.

Vincent M.C., Heitz F., Tricoire J., Bourrouillou G., Kuhlein E., Rolland M., and Calvas P. 1999. 22q11 deletion in DGS/VCFS monozygotic twins with discordant phenotypes. *Genet. Couns.* **10**: 43.

Vitelli F., Morishima M., Taddei I., Lindsay E.A., and Baldini A. 2002a. Tbx1 mutation causes multiple cardiovascular defects and disrupts neural crest and cranial nerve migratory pathways. *Hum. Mol. Genet.* **11**: 915.

Vitelli F., Taddei I., Morishima M., Meyers E.N., Lindsay E.A., and Baldini A. 2002b. A genetic link between Tbx1 and fibroblast growth factor signaling. *Development* **129:** 4605.

Waldo K.L., Kumiski D.H., Wallis K.T., Stadt H.A., Hutson M.R., Platt D.H., and Kirby M.L. 2001. Conotruncal myocardium arises from a secondary heart field. *Development* **128:** 3179.

Yamagishi H., Garg V., Matsuoka R., Thomas T., and Srivastava D. 1999. A molecular pathway revealing a genetic basis for human cardiac and craniofacial defects (comments). *Science* **283:** 1158.

Yamagishi H., Ishii C., Maeda J., Kojima Y., Matsuoka R., Kimura M., Takao A., Momma K., and Matsuo N. 1998. Phenotypic discordance in monozygotic twins with 22q11.2 deletion. *Am. J. Med. Genet.* **78:** 319.

Yoshida K., Kuo F., George E.L., Sharpe A.H., and Dutta A. 2001. Requirement of CDC45 for postimplantation mouse development. *Mol. Cell. Biol.* **21:** 4598.

Yu Y. and Bradley A. 2001. Engineering chromosomal rearrangements in mice. *Nat. Rev. Genet.* **2:** 780.

A Missense Mutation in a Highly Conserved Region of CASQ2 Is Associated with Autosomal Recessive Catecholamine-induced Polymorphic Ventricular Tachycardia in Bedouin Families from Israel

M. ELDAR,* E. PRAS,† AND H. LAHAT*
*Heart Institute, †Danek Gartner Institute of Human Genetics, Sheba Medical Center, 52621 Tel Hashomer, Israel

Sudden death due to cardiac arrhythmia is a devastating event, especially when it occurs in children. In the absence of structural heart defects, five major arrhythmogenic disorders manifesting as polymorphic ventricular tachycardia (PVT) and/or ventricular fibrillation have been described: the long QT syndrome, right bundle branch block and persistent ST elevations (Brugada syndrome), the "short coupled variant of torsade de pointes," idiopathic ventricular fibrillation with normal electrocardiogram, and PVT induced by catecholamines (Viskin and Belhassen 1998). The last entity was initially described as a case report in 1975 (Reid et al. 1975) and as a distinct clinical entity in 1995 (Leenhardt et al. 1995). Catecholamine-induced PVT is a rare disease that occurs in subjects without obvious organic heart disease, characterized by episodes of syncope, seizures, or sudden death in response to physiological or emotional stress (Viskin and Belhassen 1998). Leenhardt et al. described 21 children (mean ± S.D. age, 9.9 ± 4 years) with no organic heart disease and normal baseline ECG (except for sinus bradycardia), who suffered from episodes of syncope and seizures usually related to physiological or emotional stress. The mean age at which the first syncope occurred was 7.8 ± 4 years. They had no structural heart disease and a normal QT interval. In 30% of the patients there was a family history of syncope or sudden death. Interestingly, the ventricular tachycardia (VT) could be reproduced in these patients either by a treadmill exercise test or by infusing small amounts of isoproterenol at a rate sufficient to accelerate the cardiac rate to 150–160 beats per minute. Following treatment with beta-blockers, the patients' symptoms underwent almost complete resolution, and no episodes of VT could be found on 24-hour holter monitoring. PVT affects mainly children, but infants (Shaw 1981) as well as adults (Eisenberg et al. 1995) have also been reported as affected.

In 1999 Swan et al. from Finland described two unrelated families with typical catecholamine-induced PVT segregating in an autosomal dominant mode. However, symptoms in these two families began at a later age (mean ± S.D., 21 ± 10 years). Among affected family members, the cumulative cardiac mortality by the age of 30 years was 31%. Ventricular arrhythmias had a distinctive pattern of bidirectional VT, somewhat different from that described by Leenhardt et al. (1995), which was more chaotic in appearance. Affected members in these families were successfully treated with beta-blockers. Swan et al. (1999), using linkage analysis, mapped the disease-causing gene to the long arm of chromosome 1 at 1q42-1q43. In 2001, Priori et al. (2001) and Laitinen et al. (2001) identified several missense mutations in the ryanodine receptor 2 gene (RYR2) in four patients and three families harboring the disease. The ryanodine receptor is a sarcoplasmic reticulum (SR) channel, responsible for the release of Ca^{++} ions from the SR in response to Ca^{++} ingress to the cytoplasm through the sarcolemmal Ca^{++} channels ("calcium-induced calcium release") (Stokes and Wagenknecht 2000). Recently, the spectrum of RYR2-related disease was expanded when Tiso et al. (2001) described its mutations in patients suffering from arrhythmogenic right ventricular cardiomyopathy type 2 (ARVD2).

A RECESSIVE FORM OF CATECHOLAMINE-INDUCED PVT IN BEDOUIN FAMILIES FROM NORTH ISRAEL

Our group identified seven families belonging to a Bedouin tribe in the north of Israel, afflicted by catecholamine-induced PVT (Lahat et al. 2001a). Parents in these families are always consanguineously related (1st, 2nd, and 3rd degree cousins) and asymptomatic. Families in which one of the parents does not belong to the tribe are never afflicted by the disease. These data are consistent with an autosomal recessive form of catecholamine-induced PVT. During the last decade, nine children belonging to the afflicted families have died suddenly. Seven deaths occurred during vigorous exercise and two during excitement. Of the nine children who died in these families, five were males and four were females. The average age at the onset of symptoms was 5 ± 2 years, and average age at death was 7 ± 4 years. None was treated.

An additional twelve children have had recurrent syncope and seizures, always following exercise or excitement. Approximately 70% of these episodes appeared during vigorous physical activity, and 30% followed sudden excitement. The patients' clinical features are pre-

Table 1. Clinical Data on the Patients with Catecholamine-induced PVT

Patient	Gender	Age of 1st syncope (years)	Age at diagnosis (years)	Current age (years)	Resting heart rate (beats/min)	QTc[a] (sec)	VPB/PVT threshold (beats/min)	Symptoms
1	F	8	16	22	47	0.43	90	syncope
2	F	8	8	12	49	0.42	104	syncope
3	M	5	5	10	60	0.43	123	syncope
4	M	—	7	8	50	0.41	114	asympt
5	M	4	0.25	6	55	0.38	115	syncope
6	F	6	10	19	66	0.41	111	syncope
7	M	5	5	8	65	0.42	105	syncope
8	F	4	6	9	55	0.43	100	syncope
9	F	4	4	6	85	0.37	110	syncope
10	F	3.5	3.5	8	70	0.42	116	syncope
11	F	6	6	14	63	0.40	100	syncope
12	F	10	25	25	73	0.37	120	syncope
13	F	12	23	23	90	0.38	120	syncope
Mean ± S.D.		6 ± 3	9 ± 8	13 ± 7	64 ± 13	0.4 ± 0.02	110 ± 10	

[a]Bazzett's formula.

sented in Table 1. Syncopal episodes began at a mean age of 6 ± 3 years (range 3–12). One 7-year-old asymptomatic child was diagnosed during an exercise test.

Patients were of either gender (9 females, 4 males) and each had an unrevealing physical examination. ECG was normal, including the QT and QTc intervals, except for resting sinus bradycardia (63 ± 13 bpm), and echocardiogram did not disclose any pathology. All patients had PVT induced by exercise or isoproterenol infusion at a mean heart rate of 110 ± 10 bpm (Fig. 1). The duration and number of sequential beats of the tachycardia varied considerably. The tests were stopped once arrhythmia was detected, and therefore the significance of the number and duration of VT is unclear. In some patients, only short runs of polymorphic VT were noted, whereas in others continuous PVT lasted for >60 seconds. The VTs detected in these patients were polymorphic and similar to those described by Leenhardt et al. (1995). Patients' average QTc (before initiation of treatment) was normal, 0.4 ± 0.02 seconds (range: 0.37–0.43) compared to 0.37 ± 0.016 seconds (range: 0.36–0.41) in the unaffected sibs ($p<0.002$). Patients' mean QT interval, although in the normal range, was significantly longer than that of the unaffected siblings. The significance of these findings is unclear, but they may reflect a "mild" disorder in repolarization as the underlying pathology.

TREATMENT

All patients are treated with a beta-blocking agent, propranolol, 120 ± 40 mg/day. A 22-year-old female patient, who was treated with 300 mg/day, but continued to suffer from episodes of syncope, was treated with an implantable defibrillator.

Eleven patients reported complete resolution of symptoms, and in two patients (one of whom is mentioned above) with questionable treatment compliance, the episodes of syncope continued, although their frequency decreased. Poor compliance seems to be the reason for failure in at least one of them, but lack of efficacy of beta-blocker treatment, due to insufficient dose or inherent incomplete efficacy, cannot be ruled out. Since the initiation of this study (30 months) one 16-year-old girl (who was treated with beta-blockers) has died suddenly. We do not know whether she took the treatment on the day she died.

Comparison of the Israeli Bedouin families with those described by Leenhardt et al. (1995) shows striking phenotypic similarity, especially in age of onset and resting heart rate. Lack of family history in 2/3 of Leenhardt's patients may suggest a recessive inheritance, increasing the likelihood that at least some of Leenhardt's patients and the Bedouin patients described here harbor mutations in a common gene. In contrast, the patients described by

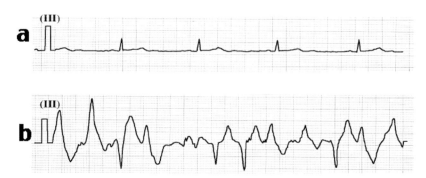

Figure 1. (*a*) Baseline ECG strip (lead 3) in a patient showing normal sinus rhythm. (*b*) An ECG strip (lead 3) in the same patient during an exercise test showing sustained polymorphic ventricular tachycardia. (Reprinted, with permission, from Lahat et al. 2001b [copyright University of Chicago Press].)

Swan et al. (1999) differ from the Bedouin patients in an older age of onset, a lower penetrance, an older age at death, and a dominant mode of inheritance.

THE GENETIC ANALYSIS OF THE BEDOUIN FAMILIES

Linkage to Chromosome 1

A genome-wide search using 300 polymorphic markers on DNA samples from the Bedouin families mapped the disease-causing gene locus (linkage interval) to the short arm of chromosome 1 (1p13-1p21). The utilization of historical recombinants in the Bedouin families allowed us to further narrow the linkage interval region to 16 Mb, between the polymorphic markers D1S187 and D1S534. A maximal LOD score of 8.24 was obtained with the marker D1S189 at $\theta = 0.00$ (Lahat et al. 2001a). Recombination and haplotype analysis with two additional markers refined the linkage interval to an 8-Mb interval (Lahat et al. 2001b). According to the current genomic databases, 46 known transcripts map to this interval.

The Gene: Calsequestrin 2

The recent implication of RYR2 in the autosomal dominant form of catecholamine-induced PVT (Laitinen et al. 2001; Priori et al. 2001) focused our attention on another gene located in the middle of the linkage interval and involved in the same pathway of SR calcium control, calsequestrin 2 (CASQ2). The gene *CASQ2* is composed of 11 exons and encodes a 399-amino acid protein. This gene is known to be the only calsequestrin expressed in cardiac muscle, whereas skeletal muscle expresses both CASQ2 and its homolog (91% identity) calsequestrin 1 (CASQ1) (Yano and Zarain-Hertzberg 1994).

Direct sequencing of the *CASQ2* exons revealed a G to C, nonsynonymous substitution (Fig. 2a) at nucleotide 1038 (exon 9), resulting in an aspartic acid to histidine change at position 307 of the protein, in an amino acid highly conserved throughout evolution.

No other sequence variations were noted in the coding region of this gene. The D307H mutation creates a *Bam*HI restriction site in the mutated sequence. A full segregation of the mutation was observed in all of the families, and an example of one family is shown in Figure 2b. Screening of 350 control subjects, including 250 Jews, 50 Bedouins, and 50 Israeli Arabs, identified only the wild-type form.

THE CALSEQUESTRIN 2 PROTEIN

The calsequestrin 2 protein serves as the major Ca^{++} reservoir within the SR of cardiac myocytes. Calsequestrin is localized to laminal spaces of the terminal cisternae of the SR of muscle cells and binds Ca^{++} cations with high capacity (40–50 mole Ca^{++} mole^{-1} calsequestrin) with moderate affinity ($K_d = 1$ mM). Ca^{++} is found in Ca^{++}/calsequestrin complexes in the terminal

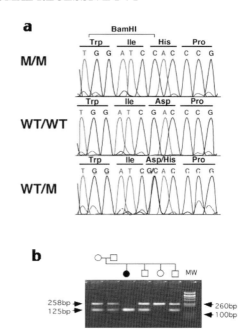

Figure 2. (*a*) Sequence chromatograms from a patient (M/M), a non-carrier (WT/WT), and a carrier (WT/M). G to C substitution at nucleotide 1038 in exon 9, converting aspartic acid to histidine at codon 307. (*b*) *Bam*HI restriction analysis of the D307H mutation in one of the Bedouin families. The 258-bp PCR product of exon 9 is cleaved in the carrier chromosome to yield 131-bp and 127-bp products that comigrate on the agarose gel. (Reprinted, with permission, from Lahat et al. 2001b [copyright University of Chicago Press].)

cisternae during muscle relaxation and is lost from those regions during tetanus (Yano and Zarain-Herzberg 1994). These properties suggest that calsequestrin acts as a Ca^{++} buffer and a Ca^{++} storage reservoir inside the SR, lowering free Ca^{++} concentrations and thereby facilitating further uptake by the Ca^{++}-ATPases. Calsequestrin may also be actively involved in the regulation of SR Ca^{++} release through protein–protein interactions with the ryanodine receptor, junctin, and triadin (Wang et al. 1998). The two additional proteins involved in the Ca^{++} cascade release, junctin and triadin, interact directly in the junctional SR membrane and stabilize a quaternary protein complex that anchors calsequestrin to the ryanodine receptor and may be required for normal operation of Ca^{++} release (Zhang et al. 1997).

THREE-DIMENSIONAL MODEL OF HUMAN CALSEQUESTRIN 2

For further insight, we constructed a three-dimensional model of human CASQ2 (Fig. 3a). The model was calculated using "homology" and "discovery" modules of Biosym/MSI (San Diego, California), and rabbit skeletal muscle calsequestrin structure as a template (Wang et al. 1998). The predicted structure of the protein is composed of three thioredoxin-like domains that contain a high aromatic amino acid composition and have a high net negative charge (MacLennan and Reithmeier 1998; Wang et al. 1998). The three domains enclose an interdomain

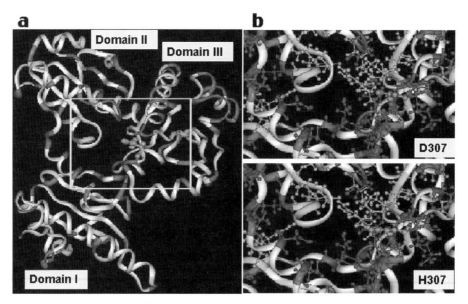

Figure 3. (*a*) Ribbon drawing of human CASQ2 three-dimensional homology model. The D307 amino acid is shown in a box. The model was calculated using "homology" and "discovery" modules of Biosym/MSI. The protein termini were omitted from the modeling calculation, leaving a G22-to-L366 amino acid core. The default parameters were used to generate a loop (E345 to P353). Energy minimization was performed using steepest-descent (100 iterations) and conjugate-gradient (1,000 iterations) algorithms. (*b*) A closer look at the mutant position domain. The amino acids are shown as "balls and sticks." The top model is of the wild-type protein (D307), and the bottom model is of the mutant protein (H307). The histidine in the mutant form appears to disrupt a band of negative amino acids and could thus upset the protein's charge balance. (Reprinted, with permission, from Lahat et al. 2001b [copyright University of Chicago Press].)

space with a strong net negative charge and a highly hydrophilic surface lining. It has been proposed that this acidic core is stabilized by the binding of Ca^{++} cations (Wang et al. 1998) and is likely to play an important role in the chelating and sequestration of Ca^{++} ions. According to the three-dimensional model of human CASQ2, residue 307, which harbors the mutation, protrudes into the negatively charged interdomain space (Fig. 3b). Its undergoing a change of two unit charges, from a negatively charged aspartic acid to a positively charged histidine, may upset a delicate charge balance and disrupt the normal chelating function of the protein.

DISCUSSION

In cardiac excitation–contraction coupling, the SR plays an essential role in the regulation of the cytosolic free Ca^{++} concentrations. Five major stages have been identified: (1) Membrane depolarization opens the L-type Ca^{++} channel allowing Ca^{++} influx. (2) Ca^{++} entering via the Ca^{++} channel binds to and opens the SR ryanodine receptors. (3) Ca^{++} released via the ryanodine receptor activates the contractile filaments and in part inactivates the Ca^{++} channel. (4) A fraction of Ca^{++} is then sequestered from the myoplasmic space into the SR by Ca^{++}-ATPase, while another fraction is extruded from the cell by Na^{+}-Ca^{++} exchanger and sarcolemmal Ca^{++}-ATPase. (5) Ca^{++} entering the SR is bound primarily to the Ca^{++}-binding protein of the SR, calsequestrin, and is primed for rerelease (Morad and Suzuki 1997).

To probe the physiological role of calsequestrin in excitation–contraction coupling, transgenic mice overexpressing cardiac calsequestrin were developed. Transgenic mice exhibited tenfold higher levels of calsequestrin in the myocardium and survived into adulthood, but had severe cardiac hypertrophy with a twofold increase in heart mass and cell size. In addition to resting bradycardia, they had a prolonged electrocardiographic QTc interval and an increased incidence of sudden death attributed to arrhythmia (Jones et al. 1998; Knollmann et al. 2000). The last three features are similar to those found in our patients, whereas cardiac hypertrophy and heart failure are absent.

Electrophysiological studies performed in these mice suggested that impairment in the Ca^{++} release process was the cause of part of the pathological manifestations. The detailed mechanism whereby the D307H mutation induces PVT remains to be elucidated. Calcium release from the SR is regulated by at least three factors: the magnitude of the triggering Ca^{++} influx, the state of the ryanodine receptors, and the Ca^{++} content within the SR (Volders et al. 2000). Increased intracellular Ca^{++} ("Ca^{++} overload") can trigger both early and delayed afterdepolarizations, which are oscillations of the membrane potential that occur during the plateau/repolarization phase of the action potential or after its completion, respectively (Rubart and Zipes 2001). Afterdepolarizations have been implicated as the pathophysiological basis for different clinical arrhythmias, including bidirectional ventricular tachycardia and PVT (Priori and Corr 1990). Thus, these experiments suggest that an intracellular Ca^{++} overload may be the triggering force initiating the arrhythmias.

The mutated CASQ2 may directly increase the Ca^{++} content within the SR, may alter the function of the ryanodine receptor to which it is connected, or, analogous to the transgenic mice model, may impair the Ca^{++} release process. Catecholamines induce intracellular overload of Ca^{++} by a variety of mechanisms (Rona 1985; Marban and O'Rourke 1995). In PVT patients, catecholamines may pathologically further increase intracellular Ca^{++} levels and trigger PVT.

CONCLUSION

The mechanism responsible for PVT in these Bedouin families is still not known. A definite link between the mutation and the clinical manifestations in this tribe awaits further investigation. Future identification of more mutations in CASQ2 in other families with this disorder, as well as the creation of a mouse knock-in model, will provide a definite link between the two.

Catecholamine-induced PVT may extend beyond the small number of patients described in the literature to date. It may also include sporadic unexplained cases of exercise- or stress-induced sudden death in children, adolescents, and adults, as well as some cases of the infant sudden death syndrome (SIDS).

REFERENCES

Eisenberg S.J., Scheinman M.M., Dullet N.K., Finkbeiner W.E., Griffin J.C., Eldar M., Franz M.R., Gonzalez R., Kadish A.H., and Lesh M.D. 1995. Sudden cardiac death and polymorphous ventricular tachycardia in patients with normal QT intervals and normal systolic cardiac function. *Am. J. Cardiol.* **75:** 687.

Jones L.R., Suzuki Y.J., Wang W., Kobayashi Y.M., Ramesh V., Franzini-Armstrong C., Cleemann L., and Morad M. 1998. Regulation of Ca^{2+} signaling in transgenic mouse cardiac myocytes overexpressing calsequestrin. *J. Clin. Invest.* **101:** 1385.

Knollmann B.C., Knollmann-Ritschel B.E., Weissman N.J., Jones L.R., and Morad M. 2000. Remodeling of ionic currents in hypertrophied and failing hearts of transgenic mice overexpressing calsequestrin. *J. Physiol.* **525:** 483.

Lahat H., Eldar M., Levy-Nissenbaum E., Bahan T., Friedman E., Khoury A., Lorber A., Kastner D.L., Goldman B., and Pras E. 2001a. Autosomal recessive catecholamine- or exercise-induced polymorphic ventricular tachycardia: Clinical features and assignment of the disease gene to chromosome 1p13-21. *Circulation* **103:** 2822.

Lahat H., Pras E., Olender T., Avidan N., Ben-Asher E., Man O., Levy-Nissenbaum E., Khoury A., Lorber A., Goldman B., Lancet D., and Eldar M. 2001b. A missense mutation in a highly conserved region of CASQ2 is associated with autosomal recessive catecholamine-induced polymorphic ventricular tachycardia in Bedouin families from Israel. *Am. J. Hum. Genet.* **69:** 1378.

Laitinen P.J., Brown K.M., Piippo K., Swan H., Devaney J.M., Brahmbhatt B., Donarum E.A., Marino M., Tiso N., Viitasalo M., Toivonen L., Stephan D.A., and Kontula K. 2001. Mutations of the cardiac ryanodine receptor (RyR2) gene in familial polymorphic ventricular tachycardia. *Circulation* **103:** 485.

Leenhardt A., Lucet V., Denjoy I., Grau F., Ngoc D.D., and Coumel P. 1995. Catecholaminergic polymorphic ventricular tachycardia in children: A 7 year follow up of 21 patients. *Circulation* **91:** 1512.

MacLennan D.H. and Reithmeier R.A. 1998. Ion tamers. *Nat. Struct. Biol.* **5:** 409.

Marban E. and O'Rourke B. 1995. Calcium channels: Structure, function, and regulation. In *Cardiac electrophysiology: From cell to bedside* (ed. D.P. Zipes and J. Jalife), p. 11. W.B. Saunders, Philadelphia, Pennsylvania.

Morad M. and Suzuki Y.J. 1997. Ca(2+)-signaling in cardiac myocytes: Evidence from evolutionary and transgenic models. *Adv. Exp. Med. Biol.* **430:** 3.

Priori S.G. and Corr P.B. 1990. Mechanisms underlying early and delayed afterdepolarizations induced by catecholamines. *Am. J. Physiol.* **258:** H1796.

Priori S.G., Napolitano C., Tiso N., Memmi M., Vignati G., Bloise R., Sorrentino V.V., and Danieli G.A. 2001. Mutations in the cardiac ryanodine receptor gene (hRyR2) underlie catecholaminergic polymorphic ventricular tachycardia. *Circulation* **103:** 196.

Reid D.S., Tynan M., Braidwood L., and Fitzgerald G.R. 1975. Bidirectional tachycardia in a child: A study using His bundle electrography. *Br. Heart J.* **37:** 339.

Rona G. 1985. Catecholamine cardiotoxicity. *J. Mol. Cell. Cardiol.* **17:** 291.

Rubart M. and Zipes D.P. 2001. Genesis of cardiac arrhythmias: Electrophysiological considerations. In *Heart disease: A textbook of cardiovascular medicine* (ed. E. Braunwald et al.), p. 680. W.B. Saunders, Philadelphia, Pennsylvania.

Shaw T. 1981. Recurrent ventricular fibrillation associated with normal QT intervals. *Q. J. Med.* **50:** 451.

Stokes D.L. and Wagenknecht T. 2000. Calcium transport across the sarcoplasmic reticulum: Structure and function of Ca^{2+}-ATPase and the ryanodine receptor. *Eur. J. Biochem.* **267:** 5274.

Swan H., Piippo K., Viitasalo M., Heikkila P., Paavonen T., Kainulainen K., Kere J., Keto P., Kontula K., and Toivonen L. 1999. Arrhythmic disorder mapped to chromosome 1q42-q43 causes malignant polymorphic ventricular tachycardia in structurally normal hearts. *J. Am. Coll. Cardiol.* **34:** 2035.

Tiso N., Stephan D.A., Nava A., Bagattin A., Devaney J.M., Stanchi F., and Larderet G. 2001. Identification of mutations in the cardiac ryanodine receptor gene in families affected with arrhythmogenic right ventricular cardiomyopathy type 2 (ARVD2). *Hum. Mol. Genet.* **10:** 189.

Viskin S. and Belhassen B. 1998. Polymorphic ventricular tachyarrhythmias in the absence of organic heart disease: Classification, differential diagnosis, and implication for therapy. *Prog. Cardiovasc. Dis.* **41:** 17.

Volders P.G.A., Vox M.A., Szabo B., Sipido K.R., de Groot S.H.M., Gorgels A.P.M., Wellens H.J.J., and Lazzara R. 2000. Progress in the understanding of cardiac early afterdepolarizations and torsades de pointes: Time to revise current concepts. *Cardiovasc. Res.* **46:** 376.

Wang S., Trumble W.R., Liao H., Wesson C.R., Dunker A.K, and Kang C.H. 1998. Crystal structure of calsequestrin from rabbit skeletal muscle sarcoplasmic reticulum. *Nat. Struct. Biol.* **5:** 476.

Yano K. and Zarain-Herzberg A. 1994. Sarcoplasmic reticulum calsequestrins: Structural and functional properties. *Mol. Cell. Biochem.* **135:** 61.

Zhang L., Kelley J., Schmeisser G., Kobayashi Y.M., and Jones L.R. 1997. Complex formation between junctin, triadin, calsequestrin, and the ryanodine receptor. Proteins of the cardiac junctional sarcoplasmic reticulum membrane. *J. Biol. Chem.* **272:** 23389.

Calcium-dependent Gene Regulation in Myocyte Hypertrophy and Remodeling

R.S. WILLIAMS AND P. ROSENBERG

Departments of Medicine and Pharmacology, Duke University Medical Center, Durham, North Carolina 27701

Striated myocytes from skeletal and cardiac muscle tissues are excitable cells that utilize calcium to trigger actomyosin cross-bridge formation in the generation of contractile force. Myocytes respond to different temporal patterns of activation and changing workloads by altering programs of gene expression that adjust cellular mass, kinetic properties of contractile proteins, and metabolic capacity to match muscle phenotypes to different physiological demands. In disease states, modulation of gene expression in myocytes as a function of contractile workload may have maladaptive consequences. We have considered the general hypothesis that changes in intracellular calcium resulting from different patterns of contractile activity not only serve to drive muscle contractions, but also provide a primary stimulus to activity-dependent changes in gene expression and muscle phenotype. Accordingly, we have investigated the role of calcium-regulated signaling molecules in controlling transcription of genes that are subject to activity-dependent regulation. Using cultured myocytes and transgenic mouse models, we have defined features of signaling cascades that modulate transcription of specific contractile protein isoforms, mitochondrial biogenesis, and myocyte mass. These pathways involve calmodulin-dependent protein kinases, the calcium-calmodulin-regulated protein phosphatase calcineurin, transcription factors of the MEF-2, NF-AT, and PGC-1 families, and proteins of the MCIP (DSCR1) gene family. Calcium released from discrete intracellular and extracellular pools exerts different effects on the kinetics of activation of specific transcription factors in striated myocytes. These results are reviewed and presented in the context of a conceptual model for activity-dependent gene regulation in myocytes.

Cardiomyocytes and myofibers of skeletal muscles are large, multinucleated cells specialized for generation of contractile force by energy-dependent sliding of highly organized filaments called sarcomeres that are composed of actin, myosin, and associated proteins (Huxley 1973). Neurotransmitters or hormones that bind to cell surface receptors modulate the rate and force of contraction by regulating the flux of calcium from intracellular stores (sarcoplasmic reticulum) or across the cell membrane. Calcium binding to proteins of the sarcomere initiates myofilament sliding and muscle contraction, and relaxation occurs when calcium is removed. Both classes of striated myocytes are capable of a spectrum of profound remodeling responses that serve to match the capabilities of muscle tissues to physiological demands for contractile work (Olson and Williams 2000a,b). Such remodeling responses include major changes in myocyte mass and volume, in the absolute and/or relative abundance of different organelles within the cell (sarcomeres, sarcoplasmic reticulum, mitochondria), in specialized programs of gene expression that involve hundreds of genes, and in tissue architecture (spatial relationships among myocytes, blood vessels, and fibrous elements). Various disease states, or advancing age, promotes pathological forms of remodeling responses in heart or skeletal muscles that initially may serve to preserve muscle function, but ultimately have deleterious or even lethal consequences.

In the heart, increases in contractile load provoked by hypertension or valvular disease promote increases in myocardial mass based on hypertrophy of myocytes. Injury or death to cardiomyocytes in a portion of the heart places a greater load on cells within uninjured areas and likewise stimulates cellular hypertrophy. Defects in genes encoding proteins of the sarcomere lead to familial forms of hypertrophic cardiomyopathy.

Hypertrophic hearts are susceptible to abnormalities of cardiac rhythm, and have impaired relaxation (diastolic dysfunction), although contractile performance can be preserved. However, when hypertrophic stimuli are unrelieved, the remodeling phenomenon progresses to dilated cardiomyopathy and increasing degrees of replacement of cardiomyocytes with fibrotic scar, and circulatory failure ensues. Morphological hypertrophy occurring in response to pathological stimuli is associated with changes in gene expression, prominently including the reactivation of genes expressed normally in the fetal heart but silenced in the adult heart (Hill et al. 2002; Williams 2002).

In skeletal muscles, the mass of individual myofibers and of muscle tissues also can be regulated by changes in contractile load. Unlike the heart, myofiber hypertrophy in skeletal muscles does not progress to contractile failure, but generally is associated with enhanced strength. However, loss of muscle mass is a prominent feature of many chronic diseases, complicates recovery from traumatic injury, and is a major cause of morbidity in the elderly. Skeletal myofibers also undergo profound remodeling responses as a function of changing patterns of neural activation that control specialized programs of gene expression that establish several different subtypes of myocytes (Vrbova 1963; Pette et al. 1997). Fast, glycolytic fibers are adapted for rapid generation of contractile force, but they fatigue rapidly, whereas slow, oxidative myofibers are capable of longer periods of repeated contraction without fatigue. A variety of intermediate forms exist. This diversification among skeletal myofibers is based on differential expression of hundreds of genes that encode different isoforms of sarcomeric proteins, enzymes of intermediary metabolism, signaling proteins, and cytoskeletal elements (Neufer et al. 1996).

What are the molecular signals that permit striated myocytes to sense and respond to changes in work activity by changes in gene expression that drive the various forms of activity-dependent muscle plasticity or remodeling? At least four general categories of signals have been considered: mechanical stresses sensed by cytoskeletal elements to activate signaling pathways; release of extracellular signaling molecules that act in a paracrine or autocrine manner through receptors at the cell surface; changes in intracellular metabolites sensed by cognate signaling proteins; and calcium itself, thereby serving not only as the proximate trigger for muscle contraction, but as the first messenger for signaling pathways that evoke remodeling responses (Neufer et al. 1996). All four categories of signaling mechanisms are known to exist in myocytes and likely contribute to remodeling responses. However, recent evidence suggests a central role for calcium-dependent signaling molecules in activity-dependent muscle remodeling and supports the working model shown in Figure 1 (Olson and Williams 2000a,b).

It is well established that the frequency and amplitude of intracellular calcium transients are altered by neural or hormonal inputs to cardiac or skeletal myocytes to modify the rate and/or the force of muscle contraction. It is not intuitively obvious, however, how these patterns could serve simultaneously as the proximate signal to control muscle mass, specialized programs of gene expression, and the contractile and metabolic capacities of muscle cells. Calcium-dependent signaling events activated by increased contractile load somehow must be insensitive to the large fluctuations in cytosolic calcium that occur as a function of ambient activity. Recent advances in cell imaging have suggested that calcium signals established by different microdomains within the cell are able to differentially activate target molecules. This consideration may explain why, although calcium-dependent signal transduction pathways that modulate gene expression have been well defined in other cell types for many years, only recently have similar signaling mechanisms been implicated in muscle plasticity (Dolmetsch et al. 1997; Crabtree 1999, 2001).

Substantial evidence based on biochemical and genetic experiments performed in cultured cells and intact animal models supports the conceptual model shown in Figure 1. Manipulations which promote gain of function in all of the signaling molecules that are illustrated drive phenotypic changes that recapitulate important features of work-dependent muscle plasticity. Likewise, loss-of-function strategies applied to these signaling proteins block activity-dependent changes in gene expression and muscle phenotype. In addition, reporter genes engineered to read out selectively the *trans*-activator function of NFAT or MEF2 transcription factors reveal a dose-dependent relationship between contractile work and *trans*-activator function (Chin et al. 1998; Molkentin et al. 1998; McKinsey et al. 2000; Naya et al. 2000; Wu et al. 2001).

Important lines of evidence supporting this model are based on studies of the MCIP (modulatory calcineurin-inhibitory proteins) family of proteins (Fig. 2), which were first identified as potentially relevant to this pathway on the basis of a conserved SP domain similar to the region of NFAT transcription factors that contains the phosphoamino acid residues that are the substrates for calcineurin. Nomenclature for this gene family has not yet been standardized, and other groups have proposed different terminology (DSCR1 or calcipressin proteins), but we favor the MCIP designation and use it here (Rothermel et al. 2000, 2001).

Sequencing of the human genome revealed three MCIP genes that each encode proteins of similar structure and function. Closely related orthologs of MCIP genes are found in both vertebrate and invertebrate species, and in unicellular eukaryotes, suggesting an ancient evolutionary origin. All of the MCIPs expressed in mice or humans are indeed substrates for calcineurin, and preliminary data suggest that the phosphorylation state of MCIPs reg-

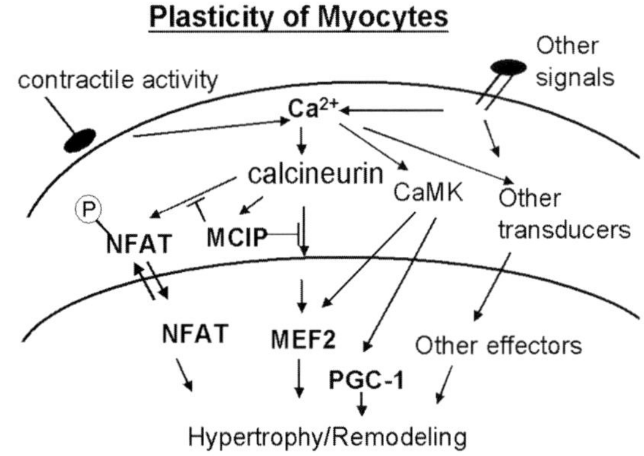

Figure 1. Signaling model for myocyte hypertrophy and remodeling as a function of changing patterns of contractile activity.

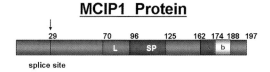

Figure 2. The MCIP gene family.

ulates their stability. Binding of MCIPs potently inhibits calcineurin activity in vitro and in vivo, through both competitive and noncompetitive (allosteric) inhibitory mechanisms. Although the functional properties of different MCIPs appear to be similar, the three different MCIP genes are subject to markedly different patterns of expression and transcriptional regulation. MCIP1 and MCIP2 are highly expressed in myocytes and neurons, whereas MCIP3 is expressed ubiquitously. MCIP2 expression is induced by thyroid hormone, but MCIP1 and MCIP2 do not manifest this response. Uniquely, the MCIP1 gene is potently up-regulated by calcineurin activity, a response that is based on the presence of a potent calcineurin response element comprising 14 NFAT-binding sites adjacent to an internal promoter site, transcriptional initiation from which produces a splice variant distinct from the form expressed under basal conditions. Increased expression of MCIP1 thereby establishes a negative feedback loop to restrain calcineurin-dependent signaling events (Fig. 3) (Rothermel et al. 2000, 2001; Yang et al. 2000).

The characterization of MCIPs as endogenous inhibitors of calcineurin has provided new and powerful tools to assess the role of calcineurin in myocyte remodeling phenomena. Previous efforts to utilize immunosuppressive drugs (cyclosporin and FK-506) for this purpose had produced a body of inconsistent results, probably because of systemic toxicities that preclude consistent achievement of inhibitory concentrations of these drugs in muscle tissues. Forced expression of MCIP1 in hearts of transgenic mice has no apparent deleterious consequences in unstressed animals but prevents cardiac hypertrophy, myocardial fibrosis, heart failure, and death in genetic, pharmacologic, and work-overload models (Rothermel et al. 2001; Hill et al. 2002). The relevance of these findings to common forms of cardiac hypertrophy and heart failure that afflict human populations remains to be established, but these data suggest that pharmacologic strategies that would enhance the expression, stability, or inhibitory potency of MCIPs may find clinical application (Williams 2002).

A recent experiment to promote gain of function of CaMK revealed a previously unexpected link to the gene encoding PGC-1, which was discovered as a coactivator of PPARγ target genes and found to promote mitochondrial biogenesis in brown adipose tissue. Many, if not all, nuclear genes encoding mitochondrial proteins are induced by the action of PGC-1, which appears to function as a master regulator of this large class of proteins (Puigserver et al. 1999; Wu et al. 1999). Forced expression of a constitutively active form of CaMKIV in skeletal muscles produced a variety of phenotypes, including mild hypertrophy, a shift in myosin isotype from fast to slower forms, and a prominent increase in the fractional volume of mitochondria within the myofibers. These effects closely resemble the morphological and biochemical remodeling responses evoked by endurance exercise, and muscles isolated from CaMKIV transgenic mice exhibited resistance to fatigue during repeated stimulation, a prominent consequence of endurance training. Increased mitochondrial biogenesis in this model was accompanied by a marked induction in expression of PGC-1 in transgenic animals, and the PGC-1 promoter is activated by CaMKIV in cultured myocytes (Fig. 4) (Wu et al. 2002).

Forced augmentation of calcineurin in skeletal muscles also promotes hypertrophy and myosin isoform switching, but much lesser effects on mitochondrial biogenesis and PGC-1 expression than those promoted by CaMK (Naya et al. 2000; Wu et al. 2001). The manner in which CaMK signals to the PGC-1 gene remains to be determined, since the CaMK-responsive promoter region examined in these experiments lacks apparent binding sites

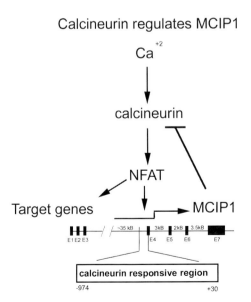

Figure 3. MCIP1 gene regulation by calcineurin establishes a negative feedback circuit.

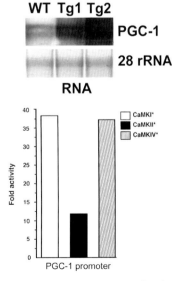

Figure 4. PGC-1 expression and promoter function are induced by CaMK gain of function in skeletal myocytes. (Reprinted, with permission, from Wu et al. 2002 [copyright American Association for the Advancement of Science].)

for NFAT or MEF2 transcription factors. In the heart, mitochondrial biogenesis and cardiac mass must be coordinated tightly to maintain energy homeostasis. Since energetic defects are implicated in the transition from compensated cardiac hypertrophy to dilated cardiomyopathy and circulatory failure, it will be interesting in future studies to determine whether impairment of the CaMK:PGC-1 signaling pathway participates in the pathobiology of heart failure.

As a different test of the signaling model shown in Figure 1, transgenic mice have been engineered to express a lacZ reporter gene under the control of multimerized, high-affinity binding sites for either the MEF2 or NFAT family of transcription factors. Unlike endogenous promoter/enhancer regions that comprise dense clusters of binding sites for multiple transcription factors, these indicator transgenes were designed to read out, selectively, the *trans*-activator function of these specific transcription factors. During embryonic and fetal life, both the MEF2 and NFAT indicator genes are expressed in spatial and temporal patterns consistent with the known developmental functions of these families of transcription factors. Remarkably, however, expression of both the MEF2 and NFAT indicator genes is largely extinguished in skeletal and cardiac muscles of adult mice housed under standard vivarium conditions. Both classes of proteins are abundant in muscle tissues and can be shown to bind DNA in vitro, but little or no *trans*-activator function is detectable at the limits of sensitivity of this assay. However, both the MEF2 and NFAT indicator genes are activated by an increased frequency of muscle contractions, promoted either by spontaneous treadmill running or by electrical pacing of the motor nerve (Fig. 5). The effects of muscle activity to augment the *trans*-activator function of MEF2 and NFAT transcription factors are limited to those muscles engaged in the locomotor activity, and are dependent on calcineurin, as shown by pharmacologic (cyclosporin) or genetic (MCIP) blockade (Wu et al. 2001).

Thus, multiple lines of evidence support the notion that calcineurin and CaMK play important roles in the remodeling responses that are induced by changing patterns of contractile work in cardiac and skeletal myocytes. Missing from the analysis so far, however, is a critical examination of the spatial and temporal relationships between distinct pools of calcium and the activation of these, and potentially other, signaling proteins germane to muscle plasticity. We have begun to explore these relationships by focusing on the kinetics of NFAT translocation from cytoplasm to nucleus in cultured skeletal myocytes. Detailed results of these investigations are soon to be presented elsewhere, but the principle that is emerging can be summarized here. Current evidence reveals a novel calcium signaling cascade that integrates pools of calcium involved in excitation–contraction coupling and store-operated calcium influx to establish a microdomain of calcium signaling that promotes NFAT-dependent transcriptional events. In this model, muscle exposed to enhanced contractile activity activates the store-operated calcium influx pathway, which is then capable of sustaining calcium-dependent transcriptional activity. Under basal conditions, calcium released during excitation–contraction coupling may permit NFAT access to the nucleus, but activation of this microdomain of calcium signaling alone is not sufficient to support calcineurin-dependent transcriptional activity. This principle is illustrated in Figure 6.

Another recent and intriguing conclusion has emerged from transcriptome analyses of different subtypes of skeletal muscles. Among the set of genes expressed differentially in slow, oxidative myofibers compared to fast, glycolytic fibers, the most prominent differences are observed in proteins of calcium metabolism and in calcium-regulated signaling molecules (Table 1). Since sustained patterns of tonic muscle contraction will convert fast glycolytic fibers into slow oxidative fibers, this observation implies that the signaling components themselves are modified as a response to the stimulus. Changes in the abundance of proteins that transduce contraction-dependent signals to the genome may alter the gain by which different patterns of muscle activity promote remodeling responses and may constitute a form of cellular memory that stabilizes the remodeled state.

The hypothesis that calcineurin plays a central role in cardiac hypertrophy and in remodeling phenomena in skeletal myofibers was first raised only a few years ago (Chin et al. 1998; Molkentin et al. 1998; Molkentin and Dorn 2001). Many experiments since that time have buttressed the notion that calcineurin is a nodal point in signaling pathways that govern muscle plasticity, and have revealed a parallel and synergistic pathway involving CaMK (Wu et al. 2002). Important roles for negative regulators in both calcineurin-dependent (MCIP) and CaMK-dependent (HDAC) signaling events have been established, and new and surprising conclusions have emerged from detailed studies of transcription factors

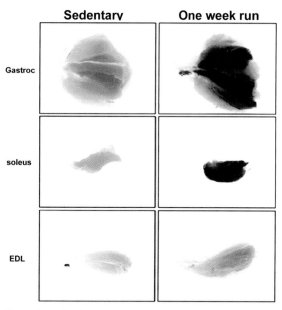

Figure 5. Activation of an NFAT-dependent indicator transgene by spontaneous wheel running in skeletal muscles of transgenic mice.

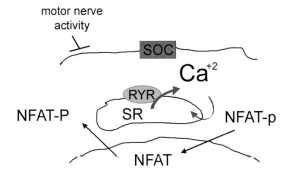

A. Basal contractile activity

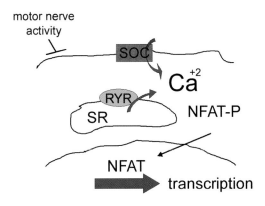

B. Augmented contractile activity

Figure 6. Discrete pools of calcium have differential effects in NFAT signaling. Basal contractile activity associated with excitation contraction coupling is necessary, but not sufficient for NFAT signaling. Augmented contractile activity results in activation of store-operated channels, which maintains NFAT in a transcriptionally active form.

(MEF2, NFAT, and PGC-1) controlled by these pathways. However, many other signaling proteins also have been implicated in the generation of cardiac or skeletal muscle hypertrophy, and in the control of specialized programs of gene expression. It will be of paramount importance to define the manner in which other signaling pathways intersect with the calcium/calmodulin-dependent signaling mechanisms that have been the focus of our recent work.

ACKNOWLEDGMENTS

This work is supported by National Institutes of Health grant R01-AR-40849. The authors also express their gratitude to their many colleagues who have contributed to work cited in this review.

REFERENCES

Chin E.R., Olson E.N., Richardson J.A., Yang Q., Humphries C., Shelton J.M., Wu H., Zhu W., Bassel-Duby R., and Williams R.S. 1998. A calcineurin-dependent transcriptional pathway controls skeletal muscle fiber type. *Genes Dev.* **12:** 2499.

Crabtree G.R. 1999. Generic signals and specific outcomes: Signaling through Ca^{++}, calcineurin, and NF-AT. *Cell* **96:** 611.

———. 2001. Calcium, calcineurin, and the control of transcription. *J. Biol. Chem.* **276:** 2313.

Dolmetsch R.E., Lewis R.S., Goodnow C.C., and Healy J.I. 1997. Differential activation of transcription factors induced by Ca^{2+} response amplitude and duration. *Nature* **386:** 855.

Hill J.A., Rothermel B., Yoo K.D., Cabuay B., Demetroulis E., Weiss R.M., Kutschke W., Bassel-Duby R., and Williams R.S. 2002. Targeted inhibition of calcineurin in pressure-overload cardiac hypertrophy. Preservation of systolic function. *J. Biol. Chem.* **277:** 10251.

Huxley H.E. 1973. Muscle contraction and cell motility. *Nature* **243:** 445.

Molkentin J.D. and Dorn I.G. 2001. Cytoplasmic signaling pathways that regulate cardiac hypertrophy. *Annu. Rev. Physiol.* **63:** 391.

Molkentin J.D., Lu J.R., Antos C.L., Markham B., Richardson J., Robbins J., Grant S.R., and Olson E.N. 1998. A calcineurin-dependent transcriptional pathway for cardiac hypertrophy. *Cell* **93:** 215.

McKinsey T.A., Zhang C.L., Olson E.N. 2000. Activation of the myocyte enhancer factor-2 transcription factor by calcium/calmodulin-dependent protein kinase-stimulated binding of 14-3-3 to histone deacetylase 5. *Proc. Natl. Acad. Sci.* **97:** 14400.

Naya F.J., Mercer B., Shelton J., Richardson J.A., Williams R.S., and Olson E.N. 2000. Stimulation of slow skeletal muscle fiber gene expression by calcineurin in vivo. *J. Biol. Chem.* **275:** 4545.

Neufer P.D., Ordway G.A., Hand G.A., Shelton J.M., Richardson J.A., Benjamin I.J., and Williams R.S. 1996. Continuous contractile activity induces fiber type specific expression of HSP70 in skeletal muscle. *Am. J. Physiol.* **271:** C1828.

Olson E.N. and Williams R.S. 2000a. Calcineurin signaling and muscle remodeling. *Cell* **101:** 689.

———. 2000b. Remodeling muscles with calcineurin. *Bioessays* **22:** 510.

Pette D. and Staron R.S., 1997. Mammalian skeletal muscle fiber type transitions. *Int. Rev. Cytol.* **170:** 143.

Puigserver P., Adelmant G., Wu Z., Fan M., Xu J., O'Malley B., and Spiegelman B.M. 1999. Activation of PPARgamma coactivator-1 through transcription factor docking. *Science* **286:** 1368.

Rothermel B., Vega R.B., Yang J., Wu H., Bassel-Duby R., and Williams R.S. 2000. A protein encoded within the Down syndrome critical region is enriched in striated muscles and inhibits calcineurin signaling. *J. Biol. Chem.* **275:** 8719.

Rothermel B.A., McKinsey T.A., Vega R.B., Nicol R.L., Mammen P., Yang J., Antos C.L., Shelton J.M., Bassel-Duby R., Olson E.N., and Williams R.S. 2001. Myocyte-enriched calcineurin-interacting protein, MCIP1, inhibits cardiac hyper-

Table 1. Genes Expressed Differentially in Murine Soleus Versus EDL Skeletal Muscles

Rank	Fold	Gene name	GenBank ID
1	8.5	sarcolipin	W.11587.1
2	8.3	calcsarcin	AA4333070.1
3	6.2	SERCA2	AA222567.1
4	4.4	Homer	AA013888.1
5	4.3	PP1	AA5431036.1
6	3.6	LIM protein	AA047966.1
7	3.5	NFAT2	AA451395.1
8	3.1	cbl-associated protein	W82192.1
9	2.9	slow twitch myosin	W59215.1
10	2.9	phospholamban	AA470249.1
11	2.5	calsequestrin	AA033488.1
12	2.2	MCIP1	AA200984.1

Total RNA isolated from individual skeletal muscles of 6 mice was subjected to microarray analysis and RT-PCR confirmation.

trophy in vivo. *Proc. Natl. Acad. Sci.* **98:** 3328.

Vrbova G. 1963. The effect of motorneurone activity on the speed of contraction of skeletal muscle. *J. Physiol.* **169:** 513.

Williams R.S. 2002. Calcineurin signaling in human cardiac hypertrophy. *Circulation* **105:** 2242.

Wu H., Kanatous S.B., Thurmond F.A., Gallardo T., Isotani E., Bassel-Duby R., and Williams R.S. 2002. Regulation of mitochondrial biogenesis in skeletal muscle by CaMK. *Science* **296:** 349.

Wu H., Rothermel B., Kanatous S., Rosenberg P., Naya F.J., Shelton J.M., Hutcheson K.A., DiMaio J.M., Olson E.N., Bassel-Duby R., and Williams R.S. 2001. Activation of MEF2 by muscle activity is mediated through a calcineurin-dependent pathway. *EMBO J.* **20:** 6414.

Wu Z., Puigserver P., Andersson U., Zhang C., Adelmant G., Mootha V., Troy A., Cinti S., Lowell B., Scarpulla R.C., and Spiegelman B.M. 1999. Mechanisms controlling mitochondrial biogenesis and respiration through the thermogenic coactivator PGC-1. *Cell* **98:** 115.

Yang J., Rothermel B., Vega R.B., Frey N., McKinsey T.A., Olson E.N., Bassel-Duby R., and Williams R.S. 2000. Independent signals control expression of the calcineurin inhibitory proteins MCIP1 and MCIP2 in striated muscles. *Circ. Res.* **87:** E61.

A Gradient of Myosin Regulatory Light-chain Phosphorylation across the Ventricular Wall Supports Cardiac Torsion

J.S. Davis,* S. Hassanzadeh,* S. Winitsky,* H. Wen,† A. Aletras,† and N.D. Epstein*

*Molecular Physiology Section, Cardiology Branch, and †Laboratory of Cardiac Energetics,
National Heart, Lung, and Blood Institute, National Institutes of Health, Bethesda, Maryland 20892

In large measure, medical researchers have undertaken disease gene identification through gene mapping and linkage analysis in order to provide increased knowledge of diseases. This approach promises improved diagnosis and prognosis together with the possibility of better treatments or cure through increased understanding of disease etiology and pathology. An additional consequence of studying diseases at the molecular level has been new insights into the concomitant normal proteins and their role in normal physiology. There are multiple instances in which discovery of a disease gene has revealed a previously unknown protein and web of associated proteins, such as the discovery of dystrophin and its associated glycoproteins through the mapping of the disease genes for muscular dystrophy.

The relatively recent history of the molecular biology of hypertrophic cardiomyopathy (HCM) has followed a similar path in our laboratory. The discovery by the Seidmans of the first HCM family with a molecular defect initiated a search for additional mutations in the identified β-myosin heavy-chain molecule. In our laboratory, 34 distinct mutations in 62 unrelated families from a total of 320 kindreds were identified. Presently, there are >70 known HCM mutations in this molecule alone, as well as mutations in another eight sarcomeric proteins, including cardiac troponin-T, troponin-I, α-tropomyosin, myosin-binding protein-C, the myosin essential and regulatory light chains, cardiac actin, and titin (Seidman and Seidman 2001 and references therein). Initially, our laboratory focused on myosin mutations and genotype–phenotype correlations in order to learn more about the disease. In the course of this study, we placed the mutations on the 3D structure of the homologous chicken skeletal muscle myosin soon after its publication by Ivan Rayment in 1993 (Rayment et al. 1993).

CARDIAC MYOSIN LIGHT-CHAIN MUTATIONS ARE ASSOCIATED WITH A RARE HCM PHENOTYPE

The distribution of the β-myosin heavy-chain mutations is not uniform, but is clustered in five regions of the myosin head: (1) in the actin-binding region, (2) surrounding the catalytic site where ATP is bound and cleaved, (3) in the proximity of two reactive cysteines known to be cross-linkable during the actomyosin crossbridge cycle, (4) at the base of the myosin neck adjacent to the essential light chain (ELC) in a region known as the converter domain, and (5) beyond the visualized crystal structure in the proximal portion of the S2 region (Rayment et al. 1995).

The identified mutations provided a marker allowing RT-PCR analysis of heart and skeletal muscle to demonstrate that both human slow skeletal muscle and cardiac myosin are transcribed off the same cardiac β-myosin gene (Cuda et al. 1993). This allowed the isolation, from skeletal muscle biopsies, of mutant and normal β-myosin for functional fiber studies (Lankford et al. 1995). Mutant myosins from six distinct mutations distributed around the globular portion of the myosin molecule, and one mutation in the proximal portion of the myosin rod, uniformly yielded a decreased velocity of actin translocation in this assay (Cuda et al. 1997). In contrast, in the same assay, a mutation at the base of the neck of the β-myosin heavy chain showed an increase in actin translocation velocity (Poetter et al. 1996).

This distinguishing functional effect highlighted a cluster of mutations at the base of the neck, adjacent to the ELC-binding region. The function of this portion of the myosin molecule had been controversial since the discovery that a truncated portion of the molecule lacking this region is sufficient for in vitro movement (Kurzawa et al. 1997). The cluster of mutations in this region among patients with HCM made it clear that this portion of the motor was important in the production of cardiac power and prompted a screen of the HCM population for mutations in the ELC itself. It was only after screening 320 unrelated HCM patients that we identified any mutation in the ELC. The mutation was a M149V substitution and was linked to cardiac hypertrophy in the kindred by a LOD score of >6. This HCM kindred was unusual in that 6 of 12 affected individuals with hypertrophic hearts had a rare phenotype involving massive hypertrophy of the papillary muscles and the adjacent myocardium.

In general, although morphologic changes associated with HCM are variable both within and between families, papillary muscle does not typically thicken to the extent of the septal and/or free wall hypertrophy. Since mid-cavity HCM is partly the result of an unusually large amount of papillary muscle hypertrophy, only a small fraction of HCM is characterized by mid-cavity obstruction. As rare as this phenotype is, it is even more rare in children.

Therefore, we were struck by the occurrence of the same mid-cavity phenotype in a child whom we found to have an adjacent ELC R154H mutation. The variant functional studies and cardiac morphology suggested a unique role for this portion of the molecule that might be consistent with a lever arm model of myosin motor activity. The subsequent search for mutations in the adjacent myosin regulatory light chain (RLC) was an obvious step and disclosed 2 unrelated individuals with an E22K RLC mutation showing the same rare mid-cavity cardiac morphology. Subsequently, additional individuals with HCM were found with either a P95R or an A13T RLC mutation (Poetter et al. 1996).

THE STRETCH-ACTIVATION RESPONSE IS MODULATED DIFFERENTIALLY IN THE HEART BY MYOSIN RLC SER-15 PHOSPHORYLATION

Inspection of the light-chain mutations on the 3D structure of homologous chicken S1 showed that the 3 RLC mutations were in close proximity to Ser-15, a phosphorylatable site that in the homologous smooth and nonmuscle myosins acts as a switch to activate myosin ATPase activity. However, previous studies have shown a lack of effect on the ATPase of cardiac/slow muscle myosin by Ser-15 phosphorylation. In fact, the ATPase remains active in the absence of light chains. Ser-15 phosphorylation does provide a staircase potentiation of isometric tension and a modest increase of calcium sensitivity at less than maximum activation levels (Sweeney et al. 1993). Despite this, a long-standing query has remained as to the function of Ser-15 phosphorylation and the reason for its evolutionary conservation in cardiac and skeletal myosins.

Toward this end, a telling experiment evaluated the role of the two homologous serines in the context of striated *Drosophila* flight muscle (Tohtong et al. 1995). An embryonic lethal transgenic fly, missing the RLC, was rescued with an RLC in which the homologous serines were substituted with alanines, prohibiting phosphorylation. The flies were born with morphologically normal flight muscle capable of producing normal isometric tension, yet the flies were incapable of flight. The property of stretch-activation was missing from the flight muscle of the rescued flies. The stretch-activation response is intrinsic to most muscle, exaggerated in insect flight muscle, and critical to the production of oscillatory power necessary for flight. The finding that the homologous phosphorylatable serines are necessary for the stretch-activation response drew our attention to the three mutations in the proximity of RLC Ser-15 that cause HCM. The frequent papillary muscle hypertrophy associated with the RLC E22K mutation, as well as the ELC mutations, highlighted a similarity between insect flight muscle and papillary muscle: First, they are both longitudinally aligned; second, they both produce oscillatory power; and third, they both have prominent stretch-activation responses (Steiger 1977).

Previous studies of the ability of the various myosin head mutations to translocate actin in an in vitro motility assay had uniformly shown a decreased velocity, whereas the ELC mutation and an adjacent heavy-chain mutation had yielded an increased velocity. This assay of unloaded velocity suggested an interesting difference in the two regions of the myosin molecule. Since both light-chain mutations were associated with a rare mid-cavity variety of cardiac hypertrophy, it seemed that this part of the myosin motor might be critical to the mechanics of the papillary muscle and adjacent ventricular tissue. The property of stretch-activation could in fact provide this mechanical contribution, as it fits well with the papillary muscle anatomy and physiology.

When Pringle first noted the stretch-activation property in a physiology experiment using a foreshortened wing preparation, he observed the ability of the flight muscles to oscillate independent of nervous innervation (Pringle 1978). The stretch-activation response is characterized by the tension transient following an imposed length change on an activated muscle fiber. After a stretch, tension increases immediately. A subsequent decay in tension is followed by a second rise in tension. The time between the stretch and the second rise in tension determines a resonant frequency of two muscle groups alternately contracting and stretching each other, similar to the biceps and triceps as the upper arm is alternately flexed and extended. Pringle found the insect flight muscles are asynchronous; that is, the rate of electrical stimulation is dissociated from the frequency of the wing beating. Note that the fact that we use the word "beating" to describe both wings and heart is a colloquial recognition that both produce oscillatory power. In contrast, the heart is clearly dependent on electrical innervation, but as Pringle noted, the matching of an intrinsic property of muscle to the rate of electrical innervation could be of great use in the production of efficient oscillatory power. The papillary muscle can take advantage of this matching. As one of the first portions of the ventricle to be activated, it is initially stretched by the chordae tendinae (that tether them to the mitral valve leaflets) when the valve closes during early systole. This stretch, initiated during active contraction of the papillary muscles, produces a brief increase in tension followed by a decay and then the second rise in tension characteristic of the stretch-activation response. If timed correctly, this delayed pulse of tension will help the papillary muscles keep the valves from prolapsing into the atria during end systole.

To test the importance of stretch-activation to the heart, we generated 10 lines of mouse transgenics expressing the human M149V ELC mutation and 6 lines expressing the normal human control on a B6SJLF1/J background. All of the mutant founders and many progeny developed a mid-cavity phenotype by one year, whereas the control transgenics had normal ventricular morphology (Vemuri et al. 1999). This abnormal morphology has subsequently been shown to be dependent on the mouse strain background and has been lost during backcross into pure C57BL/6J (N.D. Epstein, unpubl.). Although the mid-cavity phenotype was uniform by one year, none of the mice showed abnormal morphology at 5 months. The stretch-activation response was evaluated in papillary muscles from 3-month-old normal, mutant, and control human ELC mice. Despite the absence of any gross or histologic hypertrophy or disarray in these hearts, the rate constant of the stretch-activation response was much higher in the mutant papillary muscles compared to the normal or human control ELC mouse papillary muscles. The increased rate constant of the stretch-activation response in the mutant mice was consistent with a heart rate between 1700 and 3400 beats/minute (bpm), whereas the adjusted rate for the normal and human control ELC mice was between 600 and 1000 bpm, within the normal mouse heart rate (Vemuri et al. 1999). Thus, the ELC mutation did not destroy the stretch-activation response, but rather, increased the rate to a range above the physiologic murine heart rate, thus depriving the papillary muscle of its contribution. Lacking the stretch-activation response, these muscles and the adjacent ventricular tissue then presumably hypertrophy to compensate for the deficit in oscillatory power output.

THERE ARE MECHANICAL CONSEQUENCES OF A SPATIAL GRADIENT OF RLC PHOSPHORYLATION IN THE HEART

The light-chain mutations are rare and most interesting in that they suggest a novel mechanism of cardiac contraction. The possibility that phosphorylation of the RLC Ser-15 could regulate the stretch-activation response of the heart prompted us to study the pattern and consequences of this process. After multiple attempts, we were able to raise a polyclonal antibody (Fig. 1) that recognizes only the phosphorylated form of the RLC (RLCp). When this is put against longitudinal sections of normal mouse hearts, it is clear that there is a spatial gradient of RLCp that is highest at the epicardium and diminishes through the extent of the ventricular wall to the endocardium and the papillary muscles (Fig. 2). The apex is notably hyperphosphorylated while the middle of the ventricle shows RLC phosphorylation only at the epicardial surface. Similar findings were observed in western blots of human and rabbit cardiac tissue.

To evaluate the effect of cardiac myosin RLCp on cross-bridge cycle kinetics, we studied single rabbit slow muscle fibers. Our method of isolation and chemical skin-

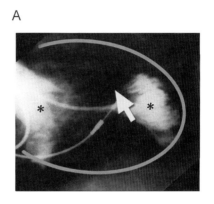

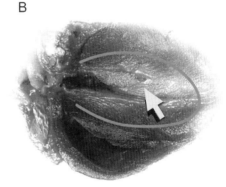

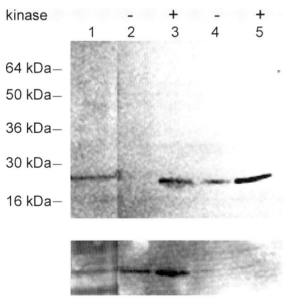

Figure 1. Western blot showing specificities of antibody against RLCp. (*Lane 1*) Mouse heart homogenate. Expressed human RLC without (–) (*lane 2*) and with (+) (*lane 3*) kinase. Extracted human cardiac myosin without (*lane 4*) and with (*lane 5*) kinase showing a baseline level of phosphorylation in extracted human cardiac myosin that is increased upon treatment with kinase. (Reprinted, with permission, from Davis et al. 2001 [copyright Cell Press].)

Figure 2. Comparison of (*A*) ventriculogram showing a patient's (M149V ELC) mid-cavity cardiac hypertrophy, (*B*) adult transgenic mouse heart expressing the M149V ELC, and (*C*) normal mouse heart stained with the antibody against RLCp. Arrow identifies hypertrophied muscle within the outline of a normal ventricular cavity in *A* and *B*. *C* shows this region is maintained in the normal heart, in a state of RLC hypophosphorylation (Reprinted, with permission, from Davis et al. 2001 [copyright Cell Press].)

ning produces single fibers without any RLC phosphorylation. With the use of a human myosin light-chain kinase (MLCK) that we cloned from heart and expressed in baculovirus, we are able to evaluate the mechanics of fully phosphorylated and dephosphorylated fibers (Davis et al. 2001). By extracting the Huxley-Simmons (H-S) phases from the mechanical transients of length stretches of 0.1–1.0% of rabbit slow muscle fibers, we made two major improvements to our original mechanical studies of murine transgenic papillary muscle bundles expressing mutant ELC (Vemuri et al. 1999). First, the rabbit slow muscle fibers express the cardiac β-myosin heavy-chain gene rather than the murine α-myosin heavy-chain gene. Second, by evaluating the four rates and five amplitudes of the H-S phases of the mechanical transient of an imposed length stretch, a unique and probably a complete "kinetic signature" of the dynamics of the muscle fiber function is obtained (Davis 2000). The most striking effect of β-myosin RLC phosphorylation is the dramatic increase in tension compared to fast muscle myosin RLC phosphorylation. At 20% activation, the former shows a 200% increase compared to the 15% increase in the latter (Davis et al. 2002). Analysis of the normalized amplitudes and rates shows that the major change is a decline in the amplitude of H-S phase 3. This H-S phase is the equivalent of the stretch-activation response. Thus, phosphorylation of the slow myosin RLC produces a dramatic increase in isometric tension while it produces a decrease in the amount of stretch-activation; i.e., delayed tension (Fig. 3).

Our studies of the distribution of RLCp in the normal heart had shown that a spatial gradient of cardiac myosin RLCp extends from high levels in the epicardium to low levels in the endocardium and papillary muscle (Davis et al. 2001). Therefore, the very region of the heart, i.e., papillary muscle and adjacent endocardium, that hypertrophies when the stretch-activation response is distorted by a mutant ELC is normally maintained as hypophosphorylated in the murine heart. Because the ELC mutations distort the frequency of the stretch-activation response, and RLC phosphorylation decreases the amplitude of the response, it is understandable how mutations in portions of either light chain can produce the same mid-cavity compensatory hypertrophy (Poetter et al. 1996). That is, distortions of either rate or amplitude of the stretch-activation response will impair the match between physiologic heart rate and the stretch-activation response that otherwise contributes to oscillatory power. The maintenance of both hyper- and hypophosphorylated regions is also likely to be important, since it has been recently shown that the E22K RLC mutation inhibits the phosphorylation of RLC Ser-15 (Szczesna et al. 2001).

The spatial gradient of RLCp from high to low, in cross-section across the ventricular wall, is accompanied by a gradient of high levels at the apex and lower levels at mid-ventricle when viewed in a longitudinal section. This pattern is consistent with the wringing motion of the heart when viewed from the feet in an opened chest. During systole, blood is wrung as well as squeezed out as the apex twists counterclockwise and the base rotates in the opposite direction. This torsion is generally believed to be due to the helical orientation of the muscle fibers as

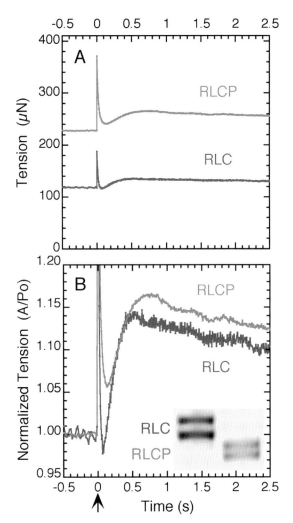

Figure 3. Changes in the kinetics of stretch-activation and fiber tension due to RLC phosphorylation. A large step-stretch of 12 nm was applied to a single isometrically contracting fiber at time zero (*arrow*) at 20% activation and 20°C. Note the large increase in baseline isometric tension before stretch at time zero from RLC phosphorylation in *A*. Note the drop in tension below the level of isometric tension in the nonphosphorylated fiber in *B*. (Reprinted, with permission, from Davis et al. 2001 [copyright Cell Press].)

they run in a left-handed helix at the epicardium, gradually rotating to a circumferential orientation in the midwall, ultimately rotating to a right-handed helix at the endocardial surface.

The spatial gradient of RLCp fits nicely with these anatomical and mechanical features and especially with recent findings using phase-tagged MRI to study motion across the ventricular wall (Fig. 4). These studies show that despite the counterhelical orientation of the endocardial fibers (right-handed) relative to the epicardial fibers, the latter dominate so that the endocardial fibers are apparently forced to rotate against their orientation. When this is modeled so that the stronger epicardial fibers dominate the endocardial fibers, the image simulates a short-axis view of the cardiac cycle (Fig. 5). As the epicardial fibers contract, the coil shortens and the number of helical turns decreases as the ends move closer to each

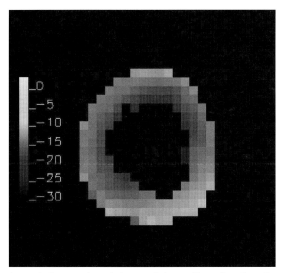

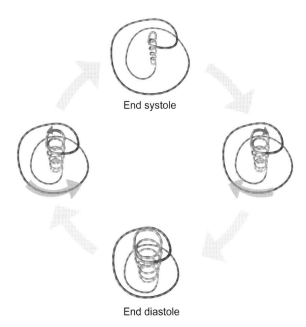

Figure 4. Gray-scale coded map of the gradient of angles of rotation around the left ventricle (LV) center. The slice is perpendicular to the LV at a distance of one-fourth the LV length from the apex. The scale ranges from 0° to −30°. The overall rotation is in the counterclockwise, left-handed helical direction. The darker scale of the endocardium represents higher rotation angles compared to the endocardium. (Reprinted, with permission, from Davis et al. 2001 [copyright Cell Press].)

Figure 5. Model of a dominant left-handed helix driving a right-handed helix in its center, drawn to represent the epicardial and endocardial surfaces. See text for details.

other and the apex rotates in a counterclockwise direction. Because these fibers produce more tension, by virtue of their phosphorylation, than the endocardial fibers, the latter will be moved against their orientation. This produces the thickening of the ventricular wall at the same time as the ventricular cavity is obliterated, thus mimicking systole as seen in a short-axis echocardiogram. The reverse occurs in diastole.

The mechanical consequences of cardiac myosin RLC phosphorylation fit very well with this scheme in that they explain how the epicardial fibers can dominate the endocardial fibers by virtue of the increased tension production of the former relative to the latter. At the same time, a small stretch imposed on the endocardial fibers as they are moved by the epicardial fibers is sufficient to produce a dramatic drop in endocardial tension, even to a level below isometric tension (Fig. 3). This drop in tension will reduce the resistance of the endocardial fibers to the epicardial fibers, decreasing shear, increasing efficiency, and thus decreasing the oxygen demand of both sets of fibers. This decrease in tension may help explain how the endocardial fibers move counter to their right-handed orientation and at a higher angular velocity than do the dominant left-handed epicardial fibers. Following the drop in tension in the endocardial fibers, the stretch-activation response, or delayed tension, will be initiated in the endocardial fibers. The rate constants for H-S phase 3 that have been calculated in rabbit slow muscle fibers fit the physiologic heart rate of rabbit (120–300 bpm) (Davis et al. 2002).

This delayed tension can be of great use in muscle groups such as the papillary muscles, which are among the first to be activated and must continue to contract throughout all of systole. Cine-MRI can be used to visualize the papillary muscles throughout the cardiac cycle.

Figure 6 shows that despite the fact that much of the heart is contracting and shortening during systole, the papillary muscles are being stretched by their attachments to the mitral leaflets. This stretch, which is occurring while the papillary muscles are activated, produces a delayed tension that is matched to the physiologic heart rate. Steiger (1977) showed a linear relationship between the rate of papillary muscle stretch-activation and physiologic heart rate in a variety of vertebrate hearts. In a similar manner, other portions of the heart are being stretched by other contracting parts of the heart. The counterhelical model of the epicardial and endocardial fibers predicts a stretching of the endocardial fibers by the epicardial fibers in the early stages of systole. This, following an immediate spike in tension, would lead to an abrupt drop in tension followed by a second rise in tension sometime later in systole. If the stretch-induced second rise in tension occurs in the endocardial fibers while calcium levels are still elevated at late systole, diastole might be initiated by this delayed pulse of tension. This delayed pulse of tension in the endocardial fibers might serve to reverse the systolic contraction. This active process would add to the passive recoil of myocellular elements such as titin, together with the restoring forces of the extracellular matrix. If, in fact, diastole is partly active, new approaches to diastolic failure can be imagined based on energetic manipulations of the cardiac myosin molecules.

THERE IS CLINICAL RELEVANCE TO THE DIFFERENCES IN THE CROSS-BRIDGE CYCLE KINETICS OF FAST AND CARDIAC MYOSIN

In the course of studying the cross-bridge cycle kinetics of cardiac myosin, we observed a dramatic difference,

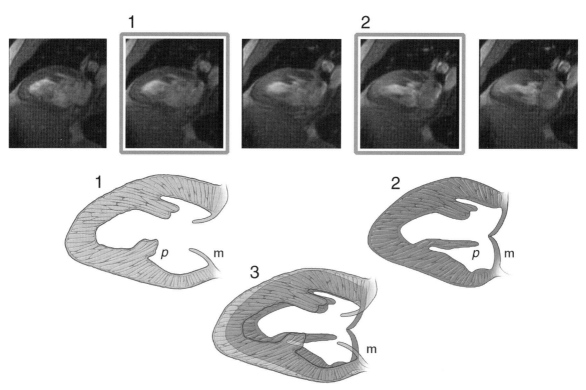

Figure 6. Series of cine-MRI images of the left ventricle taken every 25 msec, showing the stretching of the papillary muscles during early systole (posterior papillary muscle is marked with a "p"). Frames *1* and *2* are traced below to highlight the papillary muscles, and an overlay (*3*) provides comparison. A mitral valve leaflet is marked with an "m." (Reprinted, with permission, from Davis et al. 2001 [copyright Cell Press].)

with significant clinical implications, between slow and fast myosins. It has already been demonstrated that slow skeletal muscle is identical to and in fact transcribed off the cardiac myosin gene. Below the three graphs of Figure 7 is a simplified schematic of the actomyosin cross-bridge cycle. Above it are temperature-dependent plots of the rates of the force-producing state (i.e., the H-S phase 2_{slow}). The Q_{10} values for each of these rates, included in each graph, represent the temperature sensitivity of the force-producing state for the three fiber types: fast (IIb), less fast (IIa), and slow (type 1). A Q_{10} of less than 2 for fast muscle fibers is consistent with a single isolated step

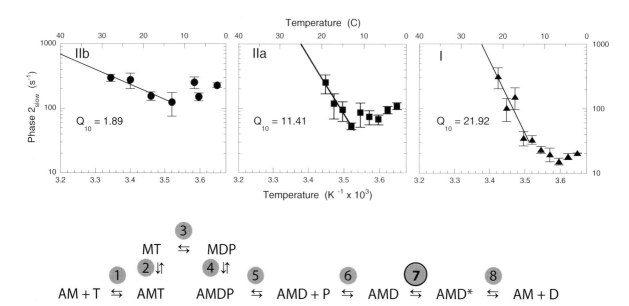

Figure 7. Temperature-dependence plots of H-S, phase 2_{slow}, the force-producing step of the cross-bridge cycle (depicted as step 7) in the scheme at the bottom of the figure. Fiber type progresses from fast to slow in panels *1–3*; respectively. (A) Actin, (T) ATP, (D) ADP, and (P) P_i (inorganic phosphate). See text for details.

of force production at step 7 as depicted in the cross-bridge scheme when actomyosin with bound ADP, having previously released P_i, moves from a pre-force-producing state (AMD) to a force-producing state (AMD*). A Q_{10} of ~2 is indicative of a single step-2-state structural change, as has been put forward in studies of cross-bridge cycle kinetics of fast skeletal muscle fibers (Davis and Harrington 1993; Davis 1998). The dramatic increase in the Q_{10} and other observations of the force-producing step 7, as fiber type progresses from fast to slow, suggest that the force-producing step of cardiac muscle myosin is no longer isolated from the earlier and subsequent cross-bridge cycle steps, as it is in fast muscle myosin. This stretch-induced reversibility of the cross-bridge cycle back from tension generation suggests that cardiac muscle myosin cross-bridges can readily detach from and reattach to actin without loss of energy, thereby conserving ATP (Davis and Epstein 2003).

The clinical relevance of this is rooted in the fact that during systole, activated portions of the heart that shorten with the onset of contraction pull against and thus lengthen other activated portions of the heart. The stretching of activated muscle fibers by opposing forces is called eccentric contraction and generally produces greater levels of tension than does active shortening (concentric contraction). These are the conditions under which the stretch-activation response occurs. Examples of eccentric contraction are evident throughout vertebrate locomotion. Quadriceps, for example, undergo eccentric contraction during a hike downhill when they are stretched by the force of gravity, while continuing to contract actively, preventing collapse. Because of the complicated fiber orientations of the heart, eccentric contraction is not easily imaged. We have hypothesized that the counterhelical orientations of the epicardial and endocardial fibers, together with the differences in muscle mechanics mediated by the spatial gradient of myosin RLC phosphorylation, promote a stretch of the inner fibers by the outer fibers in early systole. This effect depends on the reciprocal gradients of increased tension and decreased stretch-activation from the epicardial to the endocardial surface of the left ventricle. As shown in Figure 5, the counterclockwise uncoiling of the outer helix can wind and stretch the inner helix at the beginning of systole. This is only a model, obtained by superimposing known cardiac fiber orientations onto the observed overall pattern of cardiac contraction; i.e., counterclockwise torsion viewed from the cardiac apex. On the other hand, the linear arrangement of the papillary muscles allows the imaging of eccentric contraction. As shown in Figure 6, this stretch initiates a delayed pulse of tension that reinforces the papillary muscles as they prevent prolapse of the closed mitral leaflets in late systole.

Portions of the heart are, thus, likely to be stretched during their active contraction, either directly or indirectly, by the shortening of other regions of the heart. The reversibility typical of cardiac β-myosin suggests that the cross-bridge cycle can use this mechanical stretch of activated muscle fibers to drive the cycle in reverse and that the β-myosin heads will be more easily pulled off actin in the midst of the tension-producing step as the kinetic cycle is reversed. Since RLC phosphorylation slows the cross-bridge cycle, this effect will be accentuated in the regions of the heart that are hypophosphorylated, as is evident in the increased drop in tension following stretch of the phosphorylated slow fibers (Fig. 3). This process, characteristic of cardiac muscle, can be thought of as a method of energy conservation available to the heart by virtue of the co-functioning of the unique kinetic properties of cardiac β-myosin and cyclic eccentric contraction.

CONCLUSION

The study of a relatively rare cardiac disease (HCM) at the molecular level, and the distribution of the β-myosin mutations on the 3D structure, have yielded valuable insights into the function of the normal molecule. This, together with study of a rare subtype of the disease, led to the identification of myosin light-chain mutations. Functional assays of intact mutant myosins and insights gleaned from flightless transgenic flies with homologous light-chain mutations led to the hypothesis that the stretch-activation response is important to cardiac function. A diminishing spatial gradient of phosphorylated myosin regulatory light chain from the epicardium to the endocardium changes the mechanical properties of the muscle fibers across the ventricular wall in a fashion that supports cardiac torsion produced by the counterhelical orientations of muscle fibers in the epicardium and endocardium. Kinetic analysis of slow and fast muscle fibers with and without myosin RLC phosphorylation has suggested a unique ability of cardiac muscle to conserve energy. This property occurs in systole when some shortening portions of the heart, either directly or indirectly, produce eccentric contraction of other regions.

REFERENCES

Cuda G., Fananapazir L., Epstein N.D., and Sellers J.R. 1997. The in vitro motility activity of beta-cardiac myosin depends on the nature of the beta-myosin heavy chain gene mutation in hypertrophic cardiomyopathy. *J. Muscle Res. Cell. Motil.* **18:** 275.

Cuda G., Fananapazir L., Zhu W.S., Sellers J.R., and Epstein N.D. 1993. Skeletal muscle expression and abnormal function of beta-myosin in hypertrophic cardiomyopathy. *J. Clin. Invest.* **91:** 2861.

Davis J.S. 1998. Force generation simplified. Insights from laser temperature-jump experiments on contracting muscle fibers. *Adv. Exp. Med. Biol.* **453:** 343.

Davis J.S. 2000. Kinetic analysis of dynamics of muscle function. *Methods Enzymol.* **321:** 23.

Davis J.S. and Harrington W.F. 1993. A single order-disorder transition generates tension during the Huxley-Simmons phase 2 in muscle. *Biophys. J.* **65:** 1886.

Davis J.S., Hassanzadeh S., Winitsky S., Lin H., Satorius C., Vemuri R., Aletras A.H., Wen H., and Epstein N.D. 2001. The overall pattern of cardiac contraction depends on a spatial gradient of myosin regulatory light chain phosphorylation. *Cell* **107:** 631.

Davis J.S., Satorius C.L., and Epstein N.D. 2002. Kinetic effects of myosin regulatory light chain phosphorylation on skeletal muscle contraction. *Biophys. J.* **83:** 359.

Davis J.S. and Epstein N.D. 2003. Crossbridge cycle reversibility back from tension generation to phosphate release is marked in slow, detectable in medium, and absent in fast muscle fibers contracting isometrically. *Biophys. J.* **84:** 140a.

Kurzawa S.E., Manstein D.J., and Geeves M.A. 1997. Dictyostelium discoideum myosin II: characterization of functional myosin motor fragments. *Biochemistry* **36:** 317.

Lankford E.B., Epstein N.D., Fananapazir L., and Sweeney H.L. 1995. Abnormal contractile properties of muscle fibers expressing beta-myosin heavy chain gene mutations in patients with hypertrophic cardiomyopathy. *J. Clin. Invest.* **95:** 1409.

Poetter K., Jiang H., Hassanzadeh S., Master S.R., Chang A., Dalakas M.C., Rayment I., Sellers J.R., Fananapazir L., and Epstein N.D. 1996. Mutations in either the essential or regulatory light chains of myosin are associated with a rare myopathy in human heart and skeletal muscle. *Nat. Genet.* **13:** 63.

Pringle J.W. 1978. The Croonian Lecture, 1977. Stretch activation of muscle: Function and mechanism. [Review]. *Proc. R. Soc. Lond. Biol.* **201:** 107.

Rayment I., Holden H.M., Sellers J.R., Fananapazir L., and Epstein N.D. 1995. Structural interpretation of the mutations in the beta-cardiac myosin that have been implicated in familial hypertrophic cardiomyopathy. *Proc. Natl. Acad. Sci.* **92:** 3864.

Rayment I., Holden H.M., Whittaker M., Yohn C.B., Lorenz M., Holmes K.C., and Milligan R.A. 1993. Structure of the actin-myosin complex and its implications for muscle contraction. *Science* **261:** 58.

Seidman J.G. and Seidman C. 2001. The genetic basis for cardiomyopathy: From mutation identification to mechanistic paradigms. *Cell* **104:** 557.

Steiger G.J. 1977. Stretch activation and tension transients in cardiac, skeletal, and insect flight muscle. In *Insect Flight Muscle* (ed. R.T. Tregear), p. 221. North-Holland, Amsterdam.

Sweeney H.L., Bowman B.F., and Stull J.T. 1993. Myosin light chain phosphorylation in vertebrate striated muscle: regulation and function. *Am. J. Physiol.* **264:** C1085.

Szczesna D., Ghosh D., Li Q., Gomes A.V., Guzman G., Arana C., Zhi G., Stull J.T., and Potter J.D. 2001. Familial hypertrophic cardiomyopathy mutations in the regulatory light chains of myosin affect their structure, Ca2+ binding, and phosphorylation. *J. Biol. Chem.* **276:** 7086.

Tohtong R., Yamashita H., Graham M., Haeberle J., Simcox A., and Maughan D. 1995. Impairment of muscle function caused by mutations of phosphorylation sites in myosin regulatory light chain [see comments]. *Nature* **374:** 650.

Vemuri R., Lankford E.B., Poetter K., Hassanzadeh S., Takeda K., Yu Z.X., Ferrans V.J., and Epstein N.D. 1999. The stretch-activation response may be critical to the proper functioning of the mammalian heart. *Proc. Natl. Acad. Sci.* **96:** 1048.

Molecular and Functional Maturation of the Murine Cardiac Conduction System

S. RENTSCHLER,* G.E. MORLEY,† AND G.I. FISHMAN†

*Department of Biochemistry and Molecular Biology, Mount Sinai School of Medicine, New York, New York 10029; †Division of Cardiology, Department of Medicine, New York University School of Medicine, New York, New York 10016

Evolution of mammals from the ancestral sea squirt was associated with a significant change in the structure and function of the cardiovascular system (Fishman and Chien 1997). In particular, the development of a systemic circulation with unidirectional high-pressure blood flow necessitated the formation of valvular structures and highly synchronized cardiac electrical activation and mechanical contraction. Moreover, the ability of the heart to respond to the ever-changing metabolic needs of the organism by altering its cardiac output also imposed a need for regulation of the rate of electrical activation. The specialized cardiac conduction system (CCS) is responsible for meeting these critically important functional requirements. Comprising a diverse, highly heterogeneous assembly of specialized myocytes with distinct electrical and mechanical properties, the CCS is responsible for setting the heart's intrinsic beat rate, for transducing neurohumoral signals and modulating the heart rate commensurate for its immediate needs, and for synchronizing the temporal pattern of excitation of the millions of individual contractile elements to assure highly efficient pump function.

Understanding the genetic circuitry responsible for directing the formation of the CCS is of great interest for both developmental biologists and physician-scientists. Abnormalities of cardiac impulse generation or propagation, i.e., cardiac arrhythmias, affect several million individuals in the United States and lead to more than 400,000 deaths each year.

In this paper, we briefly review the current understanding of CCS development, based largely on studies in the chick. We also describe work from our own laboratory exploring the relevance of these findings to the mammalian heart, focusing on recent studies using murine systems. Areas of uncertainty and controversy are highlighted, which we hope will serve to stimulate additional experimentation by the various laboratories around the world focusing on this interesting and medically relevant topic.

FUNCTIONAL DEVELOPMENT OF THE CCS

Cardiomyocytes can generally be separated into two different functional classes—"working" cardiomyocytes primarily involved in contraction, and "conductive" cardiomyocytes that comprise the components of the CCS. Cells within the sinoatrial (SA) node have a higher rate of automaticity than the other myocytes, and consequently function as the normal pacemaker site within the heart. Electrical impulses originating within the SA node, residing within the right atrium, travel through atrial tissue to the atrioventricular (AV) node. An important function of the AV node is to provide a short delay in impulse propagation to allow for contraction of the atria before initiation of ventricular contraction occurs. From the AV node, impulses travel into the ventricles through the rapidly conducting AV (or His) bundle located at the crest of the interventricular septum. The AV bundle splits into a left bundle branch and right bundle branch, each of which further ramifies into the most distal component of the conduction system—the Purkinje fibers located within the right and left ventricles. These rapidly conducting components are required for efficient delivery of the impulse to the apex of the heart, which causes activation of the ventricles from apex to base and is responsible for efficient ejection of blood from the heart.

On the basis of studies in the chick heart, the pacemaker activity in the tubular heart of the embryo is thought to reside within the right sinus horn at the inflow region of the heart. From this site, activation proceeds in a peristaltic-like wave, with slow and sequential activation of the atrium, ventricle, and outflow tract (Kamino 1991). As maturation proceeds and the need for forward flow of blood exists, this smooth peristaltic-like activation pattern is eventually replaced by sequential, brisk contractions of the atrium and ventricle that rely on rapid propagation within each chamber and a pause at the AV junction. The temporal delay between atrium and ventricle at this stage is due to a region of slowly conducting tissue between the chambers, and in later-stage embryonic and adult hearts the delay is achieved by the specialized function of the AV node (de Jong et al. 1992). During the stages of development subsequent to the septation of the ventricular chamber and outflow tract, activation of both ventricles from apex to base is necessary to ensure the efficient ejection of blood.

EVOLUTIONARY DIVERSITY

There is significant conservation of the general circuitry of the CCS throughout vertebrate evolution, with the most primordial system described in the fish heart. Nonetheless, there are important differences even among higher vertebrates, such as the insulating connective tissue sheaths found in the sheep (Ansari et al. 1999). The nature of the Purkinje fibers differs significantly among vertebrates; in birds and hoofed animals the Purkinje fibers are histologically distinct from the surrounding working myocardium, whereas in humans and dogs the distinction is less clear. Importantly, in the mouse heart the Purkinje fibers are histologically indistinguishable from the contractile myocytes. In addition, the avian CCS appears unique in that it comprises not only the commonly observed subendocardial Purkinje fiber network, but also a distinct periarterial network (discussed further below).

CCS LINEAGE TRACING

Cells within the CCS express a variety of genes normally associated with neuronal structures or skeletal muscle tissue, but absent from the working cardiomyocytes. However, in addition to differences in histological appearance, patterns of gene expression within the CCS vary significantly among species, and thus it has not been possible to assign a universal "molecular phenotype" to the conduction system. Consequently, each marker must be evaluated on a species-by-species basis (Moorman et al. 1998). Because of this complexity, it was unclear for a number of years whether the conduction system is of neural crest or myocyte origin. Lineage tracing studies using replication-defective retroviruses to infect, and thus "tag," cardiomyocyte progenitor cells in E3 embryonic chick hearts resulted in clones of cardiomyocytes at later developmental stages containing both working myocytes and components of the CCS (Gourdie et al. 1995; Cheng et al. 1999). However, similar labeling of neural crest progenitor cells did not demonstrate a contribution of these cells to the CCS. These experiments have thereby demonstrated that components of the avian CCS, including the AV node, AV bundle, bundle branches, and Purkinje fibers, arise through differentiation of the cardiomyocyte lineage. Cell fate studies to determine which lineage contributes to the CCS have not been performed in a mammalian heart. However, indirect evidence using Cre/lox technology to label neural crest derivatives in the murine heart did not detect a contribution of this lineage to the CCS (Epstein et al. 2000; Jiang et al. 2000).

MODELS OF CCS FORMATION

Studies carried out primarily in the avian heart have led to one model for CCS development that involves the specification of an initial conduction framework within the looped tubular heart, with elaboration of the network through progressive recruitment of additional cardiomyocytes, rather than proliferation of cells within the initially specified pool. In fact, it has been proposed that this process of continual recruitment persists throughout in ovo development almost until hatching (Cheng et al. 1999). This idea came about during the course of the lineage-tracing studies described above, where tagging of cardiomyocytes at E3 was performed and the proportion of clones containing periarterial Purkinje fibers increased between E14 and E18. On the basis of this observation, it was proposed that Purkinje fiber recruitment remains ongoing between E14 and E18.

To further support the continual recruitment hypothesis, Gourdie and colleagues performed cell birthdating studies (Cheng et al. 1999). These studies demonstrated that conduction cells become quiescent before working myocytes, and the timing of cell quiescence was shown to be much earlier in some CCS components, such as the His bundle, than in the bundle branches and Purkinje fibers. This differential timing of cellular quiescence in various components of the avian CCS was cited to support a paradigm of continual recruitment. Additionally, it was argued that continual recruitment to the CCS must be occurring, otherwise progenitor cells in the His bundle would need to continue to proliferate to account for the increase in cell number associated with the addition of the Purkinje fiber network observed at later stages.

The molecular signals controlling differentiation of the CCS are rather poorly understood. However, some progress has been made in elucidating the signals involved in Purkinje fiber development in the chick heart, where, as noted earlier, the Purkinje fiber network consists of both subendocardial and periarterial populations. The close proximity of a subset of chick Purkinje fibers to the arterial system led to the hypothesis that a secreted factor originating in the arterial system may be responsible for converting working myocytes into Purkinje fibers (Gourdie et al. 1995; Mikawa and Fischman 1996). A substantial body of data has accumulated in support of this hypothesis, including studies modulating coronary arterial branching during avian development and examining the effects on periarterial Purkinje fiber differentiation (Hyer et al. 1999). These studies, all performed in the chick, suggest that endothelin-1 (ET-1) is the major arterially derived factor responsible for conversion of embryonic myocytes to a Purkinje-like phenotype (Gourdie et al. 1998, 1999; Takebayashi-Suzuki et al. 2000; Kanzawa et al. 2002).

INSIGHTS FROM NEW MURINE MODELS

Although much has been learned regarding CCS development in the chick, it is not clear whether these findings are generally applicable to mammalian CCS formation. Recently, we reported the discovery of a line of transgenic mice harboring a lacZ reporter gene that unexpectedly expressed β-galactosidase activity in a pattern suggestive of the CCS (Rentschler et al. 2001). Comprising regulatory elements from the *Engrailed-2* locus (Logan et al. 1993), expression in the CCS appears to result from the site of integration, rather than the En-2 regulatory elements per se. This strain of mice has allowed us to examine directly some of the issues related to mammalian CCS formation.

Visualization of the Mammalian CCS

As reported recently (Rentschler et al. 2001), expression of the reporter gene in the hearts of CCS-lacZ transgenic mice is first detectable at ~8.5 dpc, with β-galac-

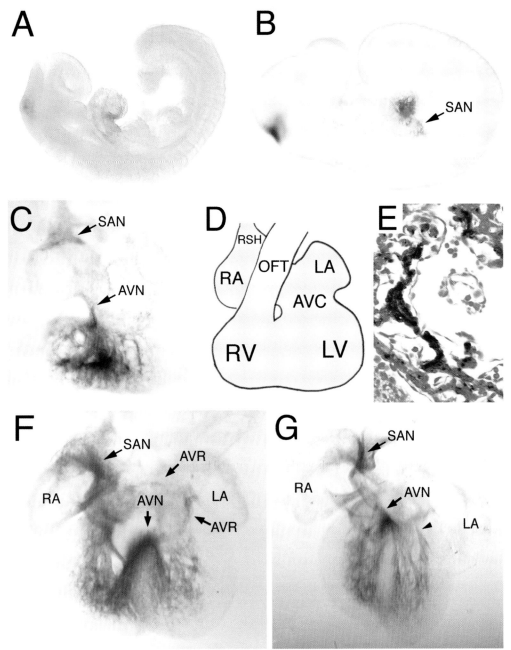

Figure 1. Maturation of the CCS during murine heart development. (*A*) Initiation of expression within the heart occurred at 8.5 dpc predominantly in CCS precursors along the dorsal wall of the AV canal. (*B*) By 9.5 dpc, expression could be seen in precursors of the SA node (SAN) within the right sinus horn (RSH) as well as in precursors of the ventricular CCS located in the region between the undivided left and right ventricles. The labeled illustration in *D* corresponds to the 10.5-dpc heart shown in *C*, where the transparent OFT cannot be visualized. At 10.5 dpc, the location of the CCS is similar to 9.5 dpc; however, discrete fibers within the ventricles can clearly be seen, as well as a group of cells along the right AV canal (AVC), where the developing AV node (AVN) is located. (*E*) An eosin-stained section through the ventricular region of a 10.5-dpc embryo revealed that the location of trabecular CCS cells is predominantly subendocardial. (*F*) In the 12.5-dpc heart, the SA node and the presumptive SA ring, a bundle in the posterior right atrial wall leading toward the AV node, were detected. The AV node, located in the posterior AV canal, was continuous with the His bundle and bundle branches developing on top and astride the budding IVS. Components of the AV ring (AVR) were also detected. (*G*) The CCS of the 13.5-dpc heart appears nearly mature, including the distal Purkinje networks within right and left ventricles. Fibers coursing from LA to LV are indicated (*arrowhead*). (RA) Right atrium; (LA) left atrium. (Reprinted from Rentschler et al. 2001 [copyright Company of Biologists].)

tosidase activity strongest along the dorsal wall, consistent with the location of the first specialized cells of the AV conduction system identified in the classic morphological studies of Viragh and Challice (1977), as shown in Figure 1. Within the next 24 hours, a number of key components of the CCS are identifiable, including cells within the right sinus horn that comprise the sino-atrial node, as well as the "primary interventricular ring,"

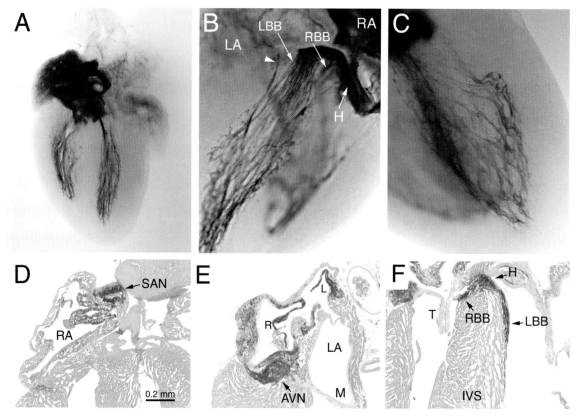

Figure 2. *LacZ* expression in the CCS of neonatal hearts. (*A*) Low magnification view of *lacZ* expression within the heart, which delineates components of the entire cardiac conduction system. (*B*) Higher magnification of the region of the His bundle (H) and bundle branches. Several fibers split off from the His bundle and travel along the right side of the IVS, giving rise to the RBB, which is out of the plane of focus. The termination of the His bundle gives rise to the fibers of the LBB, which has a characteristic fan-shaped appearance. Fibers coursing directly from left atrium (LA) to left ventricle (LV) are indicated (*arrowhead*). (*C*) Higher magnification of the extensive Purkinje fiber network within the LV. (*D–F*) Analysis of eosin-stained sections demonstrates preferential transgene expression within specific regions of the right atrium (RA), including the SA node (SAN), the right (R) and left (L) venous valves, and the AV node (AVN). Ventricular transgene expression delineated the His bundle, located beneath the tricuspid annulus, and the bundle branches. (M) Mitral valve; (T) tricuspid valve. Scale bar for *D–F* is shown. (Reprinted from Rentschler et al. 2001 [copyright Company of Biologists].)

which is thought to give rise to the His bundle and bundle branches (Moorman et al. 1998). Soon thereafter, a near-continuous pathway from SA node to AV node and His bundle is seen, as well as transgene expression along the right side of the AV canal, the presumptive site of the forming AV node. In the 12.5-dpc heart, CCS-*lacZ* transgene expression delineated a near-continuous pathway from SA node to AV node and bundle, which was in turn contiguous with the developing bundle branches located straddling the interventricular septum. In addition, the trabecular component of both right and left ventricles from which the peripheral Purkinje fibers are derived also expressed the transgene. One day later, at 13.5 dpc, the IVS was almost fully developed, and it separated the networks of *lacZ*-positive cells within the two ventricular chambers, forming a pattern generally similar to the appearance within the neonatal heart.

Within neonatal hearts, as shown in Figure 2, we noted robust β-galactosidase activity within the right atrium, with the most discrete expression within the sino-atrial (SA) node, atrioventricular (AV) node and ring, and the left and right venous valves (Lev and Thaemert 1973; Viragh and Challice 1980). We also noted fibers originating within the right atrium and coursing toward the left atrium which are likely to be components of Bachmann's bundle. Within the ventricular conduction system of the neonatal heart, there was transgene expression demarcating the AV (or His) bundle, the left (LBB) and right (RBB) bundle branches, and the extensive ramifications of the Purkinje fiber network along the ventricular free walls.

Induction of the Murine CCS

Inasmuch as lacZ expression in the CCS-lacZ transgenic mouse strain serves as a readout of the CCS, we tested the ability of a number of paracrine growth and/or differentiation factors known to play a role in embryonic heart development to convert contractile cardiomyocytes into cells of the CCS. In these studies, embryonic CCS-*lacZ* hearts were excised from 9.5-dpc embryos, and the beating hearts were subsequently cultured in the presence of paracrine factors for 48 hours (Rentschler et al. 2002). Conversion of contractile cardiomyocytes to a conduc-

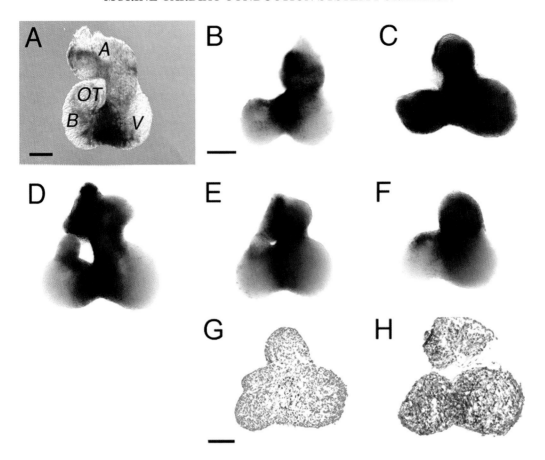

Figure 3. NRG-1 converts 9.5-dpc embryonic cardiomyocytes to a CCS phenotype. X-gal staining of a 9.5-dpc CCS-*lacZ* transgenic heart (*A*). Highest *lacZ* expression can be seen in the area surrounding the bulboventricular junction in the presumptive "interventricular ring," as well as in an area located on the right side of the common atrial chamber, which may correspond to the developing sinoatrial node. CCS-*lacZ* hearts were cultured for 48 hours in (*B*) media alone, or in the presence of paracrine factors including (*C*) a biologically active NRG-1 peptide, (*D*) ET-1, (*E*) AT-II, or (*F*) IGF-1, followed by X-gal staining. Panels *G* and *H* show a representative eosin-stained section from a control and NRG-1-treated heart, respectively. (A) Common atrium; (V) ventricle; (B) bulbus cordis; (OT) outflow tract. Scale bar in *A* is 0.2 mm, in *B–F* is 0.2 mm, and in *G* and *H* is 0.1 mm. (Reprinted from Rentschler et al. 2002 [copyright National Academy of Sciences].)

tion phenotype was assessed by expression of *lacZ* in ectopic sites within the cultured hearts, as shown in Figure 3. Angiotensin II (AT-II) and insulin-like growth factor-I (IGF-I) had no effect on *lacZ* expression, whereas ET-1, which appears to have a potent role in avian CCS development, as discussed earlier (Gourdie et al. 1998; Takebayashi-Suzuki et al. 2000; Kanzawa et al. 2002), resulted in only a slight induction of ectopic expression. We next tested the effects of neuregulin-1 (NRG-1) as a candidate promoting ventricular conduction system induction. NRG-1 is structurally related to epidermal growth factor (EGF) and known to play a role in diverse cellular processes including differentiation, proliferation, apoptosis, survival, migration, and fate (Bacus et al. 1992; Shah et al. 1994; Carraway et al. 1995; Syroid et al. 1996; Daly et al. 1997; Zhao et al. 1998). Homozygous NRG-1 mutant mice die around 10.5 dpc due to cardiac defects (Meyer and Birchmeier 1995). Moreover, they have an irregular heartbeat and a complete absence of ventricular trabeculae, the fingerlike projections where the embryonic ventricular CCS forms.

Indeed, as shown in Figure 3, we found that a biologically active recombinant peptide containing the β variant of the neuregulin-1 (NRG-1) epidermal growth factor (EGF)-like domain induced dramatic ectopic *lacZ* expression in the myocytes of the cultured hearts.

Using a combination of TUNEL and BrdU staining to examine the effects of NRG-1 on cell death and proliferation, respectively (Rentschler et al. 2002), we concluded that NRG-1 must directly convert embryonic cardiomyocytes to a conduction phenotype, rather than acting through a mechanism involving selective proliferation or cell death signals. Interestingly, we found that NRG-1 induced ectopic *lacZ* expression in nearly all myocytes of 8.5- and 9.5-dpc embryonic hearts, but by 10.5 dpc there were areas within the heart, including the ventricular free walls, that no longer responded to the NRG-1 signal, and at later stages, the response continued to diminish, as shown in Figure 4.

Significantly, we found that NRG-1 treatment of 9.5-dpc hearts not only led to molecular evidence of conversion of myocytes to a CCS phenotype, but addition of the

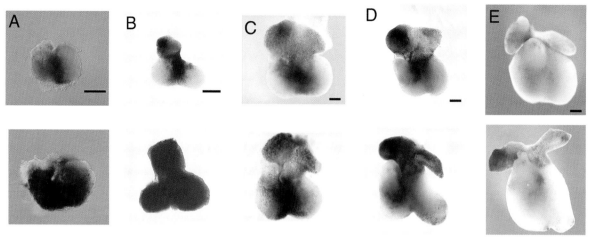

Figure 4. Response to NRG-1 is developmentally regulated. CCS-*lacZ* hearts from (*A*) 8.5-dpc, (*B*) 9.5-dpc, (*C*) 10.5-dpc, (*D*) 11.5-dpc, and (*E*) 12.5-dpc embryos were cultured for 48 hours in the absence (*upper panel*) or presence (*lower panel*) of NRG-1 EGF-like domain followed by X-gal staining. Hearts were oriented with the atria located at the top of the picture for each stage except 8.5 dpc, where the atria are not visible. Scale bars represent 0.2 mm. (Reprinted from Rentschler et al. 2002 [copyright National Academy of Sciences].)

ligand also resulted in measurable effects on cardiac electrical properties (Rentschler et al. 2002). In control hearts, electrical activation proceeded dorsally from the AV canal into the ventricle and then traveled toward the ventral surface. In contrast, treatment with NRG-1 led to near-simultaneous activation from the AV canal along both the dorsal and ventral surfaces, in concert with conversion of many working ventricular myocytes to a CCS phenotype as assayed by *lacZ* expression. These data, along with a proposed model explaining the effects of NRG-1 on CCS development and cardiac conduction are presented in Figure 5.

CONCLUSIONS

Our recent studies raise some interesting questions regarding the process of CCS formation. Although there is now general agreement that the CCS derives from the

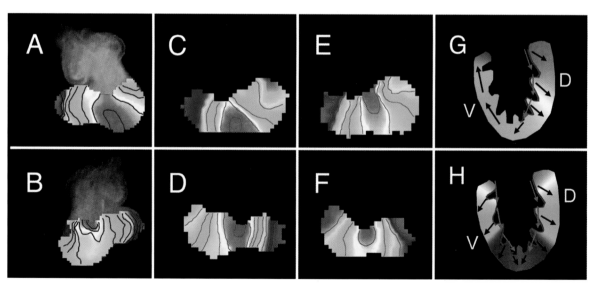

Figure 5. NRG-1 induces electrophysiological changes within embryonic hearts. Activation maps of 9.5-dpc wild-type hearts from either the ventral (*A, C, E*) or dorsal (*B, D, F*) surfaces are shown. Acutely dissociated hearts (*A, B*); hearts cultured for 24 hours in the absence (*C, D*) or presence (*E, F*) of NRG-1. Activation progresses from red to purple, with isochrone lines drawn every 1 msec. Model depicting CCS-dependent activation of the ventricles in control (*G*) and NRG-1-treated (*H*) hearts. A sagittal view is shown with ventral (V) and dorsal (D) surfaces indicated. The CCS is indicated by the blue coloring along the subendocardial surface; activation sequence is indicated by colors progressing from red to purple. Red arrows indicate activation of the CCS; black arrows indicate activation of the working myocardium. (Reprinted from Rentschler et al. 2002 [copyright National Academy of Sciences].)

myocyte lineage, there is significant uncertainty as to the timing of cardiomyocyte recruitment to the CCS. In the chick it is thought that only a CCS framework is initially specified, and that elaboration of the conduction network through progressive recruitment of cardiomyocytes occurs throughout in ovo development almost until hatching (Cheng et al. 1999). In contrast, studies of morphology and gene expression in mammalian embryos support the idea that cell fates become fixed early within the embryonic heart (Wenink 1976; Moorman et al. 1998; MacNeill et al. 2000). Specific domains within the heart are thought to contain the precursor cells from which the CCS subsequently develops. However, in part due to limited tools for experimentally manipulating mammalian embryos and subsequently maintaining the capacity for normal development, it remains unclear whether the process of recruitment to the CCS and/or timing of commitment to the CCS is fundamentally different between avian and mammalian species.

Our analysis of cellular proliferation in the CCS-lacZ mice demonstrates that BrdU incorporation into $lacZ^+$ ventricular CCS cells was less than incorporation into $lacZ^-$ working myocytes. Nonetheless, there was measurable DNA synthesis, and presumably, proliferation of $lacZ^+$ CCS cells (Rentschler et al. 2002). In addition, we performed BrdU labeling of CCS-lacZ embryos in vivo during later stages of development when a functional conduction system has been shown to exist (S. Rentschler and G.I. Fishman, unpubl.). The results of these experiments were consistent with the cultured heart data, in that $lacZ^+$ cells continue to proliferate, albeit at a slower rate than $lacZ^-$ cells. Thus, we conclude that $lacZ^+$ CCS cells are indeed still synthesizing DNA and are most likely proliferating at these later stages after specification. On the basis of these findings, as well as the timing of NRG-1-induced CCS specification, we favor the model that murine CCS cell fate is specified early, likely between 8.5 and 10 dpc. This initial pool of specified cells proliferates at a slower rate than the surrounding working myocytes. In accordance with this model, the embryonic conduction system is a much larger structure in relation to the total size of the embryonic heart than the adult conduction system is in relation to the total size of the adult heart. However, definite proof for either early commitment or continued recruitment of the murine CCS awaits in-depth lineage-tracing studies

What accounts for these apparent differences in avian and mammalian CCS development? In the avian heart, connexin isoform diversity serves as the earliest known marker for avian Purkinje fibers. However, we are able to detect specialization of conduction cells both with the lacZ molecular marker and functionally using optical mapping techniques *before* connexin specialization within the heart is known to occur. Therefore, it is possible that subendocardial Purkinje fiber fate is specified early in the avian heart, but the available markers may not detect these cells at the first stages of differentiation.

In addition to uncertainties as to the timing of CCS formation, the characterization of molecular cues that initiate and/or maintain a CCS phenotype are incompletely known. On the basis of our functional studies using optical mapping techniques, which demonstrated preferential conduction pathways in the heart as early as 10.5 dpc, we hypothesized that potential inducers of the murine ventricular conduction system should exert its effect prior to this stage of development. Taking advantage of the CCS-lacZ mice, we examined likely candidates to determine whether any of these had the ability to convert 9.5-dpc myocytes into cells of the CCS. As noted previously, there is substantial evidence establishing an association between the development of the coronary arterial network, expression of arterially derived ET-1 and the periarterial Purkinje fibers within the avian heart. However, there is no comparable periarterial Purkinje network in the mammalian heart, and even in the avian heart this network develops at relatively late stages of cardiac development. Therefore, whereas ET-1 was a plausible candidate inducing factor in the murine heart, it was perhaps not surprising that treatment of embryonic hearts showed only minimal induction of the lacZ marker. Our data are consistent with gene inactivation studies of various components of the murine ET-1 pathway. Although these knockout mice have revealed an important role for the ET signaling pathway in murine heart development, to date CCS defects have not been observed in any of these genetically altered murine strains, although redundancy does exist (Kurihara et al. 1995; Clouthier et al. 1998; Yanagisawa et al. 1998, 2000). Thus, it is conceivable that unique mechanisms may have evolved to account for elaboration of the avian periarterial network, and one should not generalize these observations to induction and patterning of the mammalian subendocardial Purkinje fiber network.

In mammalian species the ventricular conduction system develops in close spatial association with the endocardium. Moreover, co-culture experiments (with Dr. Takashi Mikawa, Cornell University) demonstrate a requirement for endocardial cells in the induction and/or maintenance of *lacZ* expression in cultured 9.5-dpc CCS-*lacZ* myocytes (Pennisi et al. 2002). On the basis of these observations, we hypothesized that endocardial cells are the source of a factor required for CCS formation. High levels of NRG-1 Type I are expressed by ventricular endocardial cells, and to a lesser extent by atrial endocardial cells as early as 8.5 dpc. Hence, NRG-1 is found in the right place, at the right time, for a potential role in ventricular CCS induction. Additionally, the NRG-1 receptors ErbB2 and ErbB4 are expressed in cardiomyocytes during this same time frame. Moreover, targeted inactivation of NRG-1, ErbB2, and ErbB4 in mice results in a nearly identical phenotype of embryonic lethality due to an absence of ventricular trabeculae, the fingerlike projections where the ventricular CCS forms.

Treatment of embryonic hearts with NRG-1 led to a rather dramatic induction of lacZ expression, as well as functional changes in cardiac conduction patterns. Interestingly, the conversion of cardiomyocytes to a CCS phenotype in response to NRG-1 was developmentally regulated, with sensitivity to the ligand declining sharply between 9.5 and 10.5 dpc, and minimal conversion after

10.5 dpc. Since NRG-1 continues to be expressed at later stages in vivo and may play a role in subsequent cardiac developmental processes, this decline in cardiomyocyte potential may be critically important for preventing further recruitment to the CCS at these stages. It remains to be seen whether continued production of NRG-1 is required at later stages to maintain a CCS phenotype within already differentiated cells.

Our results do not prove that NRG-1 acts directly on myocytes, as opposed to a more complex paracrine or other signaling system. However, treatment at 8.5 dpc results in ectopic *lacZ* expression, a stage when the heart consists only of endocardial and myocardial cells, thus the effect is most likely directly on the myocytes. Experiments to determine which ErbB receptors on the myocytes are involved in NRG-1-induced ectopic *lacZ* expression are being pursued.

A number of outstanding questions must still be resolved. For instance, can the processes of NRG-1-dependent CCS specification and myocardial trabeculation be dissociated? Initial studies with the embryo culture technique at 11.5–12.5 dpc suggest that NRG-1 can exert an effect on trabeculation at stages after CCS specification no longer occurs (Hertig et al. 1999). Additionally, treatment of neonatal and adult myocytes with NRG-1 has been shown to cause cellular hypertrophy (Zhao et al. 1998). Since many of the hypertrophic molecular markers are also markers of the CCS, it will be interesting to see whether the processes of CCS specification, trabeculation, and hypertrophy involve similar or distinct downstream effector pathways.

NRG-1 Type 1 appears to be expressed within all ventricular endocardial cells; however, the ventricular CCS at 9.5 dpc is only specified within a specific region of the heart that includes the primary interventricular ring. What factors are responsible for the spatially restricted induction of the CCS in the early embryonic heart in vivo? We hypothesize that myocytes from different regions of the developing heart may have differential sensitivity to the inductive effects of NRG-1. This may reflect a gradient in the expression of ErbB receptors or downstream signaling factors.

After the initial specification of the CCS occurs, likely before 10 dpc, what processes are responsible for patterning these cells into the intricate subendocardial network seen at later developmental stages? Expression of *lacZ* in 9.5-dpc ventricles delineates a broad interventricular ring; however, by 10.5 dpc, organized "fibers" are already seen. One possibility is that the organization of CCS cells into fibers may involve differential rates of proliferation. Once CCS cells are specialized, they proliferate at a slower rate. The presence of cells with a faster rate of proliferation, such as working myocytes, or the migration of cells such as the epicardial-derived cells, into specific subendocardial regions may push the original CCS lining apart and thereby, in a sense, pattern this intricate network. The biology of cardiac conduction system induction and patterning is interesting and very complex, and a great deal of work still needs to be done to understand the mechanisms involved. Future studies will not only uncover exciting aspects of cardiac developmental biology, but may also provide greater insight into the pathogenesis of cardiovascular disease and provide novel targets for therapy.

ACKNOWLEDGMENTS

The work presented in this chapter was supported by a Burroughs Wellcome Fund Clinical Scientist Award in Translational Research and by grant HL-64757 from the National Institutes of Health to G.I.F. and a Scientist Development Award from the American Heart Association to G.E.M. Portions of this work were submitted by S.R. in partial fulfillment of the Ph.D. requirements of the Graduate School of Biological Sciences of Mount Sinai School of Medicine.

REFERENCES

Ansari A., Ho S.Y., and Anderson R.H. 1999. Distribution of the Purkinje fibres in the sheep heart. *Anat. Rec.* **254:** 92.

Bacus S.S., Huberman E., Chin D., Kiguchi K., Simpson S., Lippman M., and Lupu R. 1992. A ligand for the erbB-2 oncogene product (gp30) induces differentiation of human breast cancer cells. *Cell Growth Differ.* **3:** 401.

Carraway K.L., III, Soltoff S.P., Diamonti A.J., and Cantley L.C. 1995. Heregulin stimulates mitogenesis and phosphatidylinositol 3-kinase in mouse fibroblasts transfected with erbB2/neu and erbB3. *J. Biol. Chem.* **270:** 7111.

Cheng G., Litchenberg W.H., Cole G.J., Mikawa T., Thompson R.P., and Gourdie R.G. 1999. Development of the cardiac conduction system involves recruitment within a multipotent cardiomyogenic lineage. *Development* **126:** 5041.

Clouthier D.E., Hosoda K., Richardson J.A., Williams S.C., Yanagisawa H., Kuwaki T., Kumada M., Hammer R.E., and Yanagisawa M. 1998. Cranial and cardiac neural crest defects in endothelin-A receptor-deficient mice. *Development* **125:** 813.

Daly J.M., Jannot C.B., Beerli R.R., Graus-Porta D., Maurer F.G., and Hynes N.E. 1997. Neu differentiation factor induces ErbB2 down-regulation and apoptosis of ErbB2-overexpressing breast tumor cells. *Cancer Res.* **57:** 3804.

de Jong F., Opthof T., Wilde A.A., Janse M.J., Charles R., Lamers W.H., and Moorman A.F. 1992. Persisting zones of slow impulse conduction in developing chicken hearts. *Circ. Res.* **71:** 240.

Epstein J.A., Li J., Lang D., Chen F., Brown C.B., Jin F., Lu M.M., Thomas M., Liu E., Wessels A., and Lo C.W. 2000. Migration of cardiac neural crest cells in Splotch embryos. *Development* **127:** 1869.

Fishman M.C. and Chien K.R. 1997. Fashioning the vertebrate heart: Earliest embryonic decisions. *Development* **124:** 2099.

Gourdie R.G., Kubalak S., and Mikawa T. 1999. Conducting the embryonic heart: Orchestrating development of specialized cardiac tissues. *Trends Cardiovasc. Med.* **9:** 18.

Gourdie R.G., Mima T., Thompson R.P., and Mikawa T. 1995. Terminal diversification of the myocyte lineage generates Purkinje fibers of the cardiac conduction system. *Development* **121:** 1423.

Gourdie R.G., Wei Y., Kim D., Klatt S.C., and Mikawa T. 1998. Endothelin-induced conversion of embryonic heart muscle cells into impulse-conducting Purkinje fibers. *Proc. Natl. Acad. Sci.* **95:** 6815.

Hertig C.M., Kubalak S.W., Wang Y., and Chien K.R. 1999. Synergistic roles of neuregulin-1 and insulin-like growth factor-I in activation of the phosphatidylinositol 3-kinase pathway and cardiac chamber morphogenesis. *J. Biol. Chem.* **274:** 37362.

Hyer J., Johansen M., Prasad A., Wessels A., Kirby M.L., Gourdie R.G., and Mikawa T. 1999. Induction of Purkinje fiber differentiation by coronary arterialization. *Proc. Natl. Acad. Sci.* **96:** 13214.

Jiang X., Rowitch D.H., Soriano P., McMahon A.P., and Sucov H.M. 2000. Fate of the mammalian cardiac neural crest. *Development* **127:** 1607.

Kamino K. 1991. Optical approaches to ontogeny of electrical activity and related functional organization during early heart development. *Physiol. Rev.* **71:** 53.

Kanzawa N., Poma C.P., Takebayashi-Suzuki K., Diaz K.G., Layliev J., and Mikawa T. 2002. Competency of embryonic cardiomyocytes to undergo Purkinje fiber differentiation is regulated by endothelin receptor expression. *Development* **129:** 3185.

Kurihara Y., Kurihara H., Oda H., Maemura K., Nagai R., Ishikawa T., and Yazaki Y. 1995. Aortic arch malformations and ventricular septal defect in mice deficient in endothelin-1. *J. Clin. Invest.* **96:** 293.

Lev M. and Thaemert J.C. 1973. The conduction system of the mouse heart. *Acta Anat.* **85:** 342.

Logan C., Khoo W.K., Cado D., and Joyner A.L. 1993. Two enhancer regions in the mouse En-2 locus direct expression to the mid/hindbrain region and mandibular myoblasts. *Development* **117:** 905.

MacNeill C., French R., Evans T., Wessels A., and Burch J.B. 2000. Modular regulation of cGATA-5 gene expression in the developing heart and gut. *Dev. Biol.* **217:** 62.

Meyer D. and Birchmeier C. 1995. Multiple essential functions of neuregulin in development. *Nature* **378:** 386.

Mikawa T. and Fischman D.A. 1996. The polyclonal origin of myocyte lineages. *Annu. Rev. Physiol.* **58:** 509.

Moorman A.F., de Jong F., Denyn M.M., and Lamers W.H. 1998. Development of the cardiac conduction system. *Circ. Res.* **82:** 629.

Pennisi D.J., Rentschler S., Gourdie R.G., Fishman G.I., and Mikawa T. 2002. Induction and patterning of the cardiac conduction system. *Int. J. Dev. Biol.* **46:** 765.

Rentschler S., Vaidya D.M., Tamaddon H., Degenhardt K., Sassoon D., Morley G.E., Jalife J., and Fishman G.I. 2001. Visualization and functional characterization of the developing murine cardiac conduction system. *Development* **128:** 1785.

Rentschler S., Zander J., Meyers K., France D., Levine R., Porter G., Rivkees S.A., Morley G.E., and Fishman G.I. 2002. Neuregulin-1 promotes formation of the murine cardiac conduction system. *Proc. Natl. Acad. Sci.* **99:** 10464.

Shah N.M., Marchionni M.A., Isaacs I., Stroobant P., and Anderson D.J. 1994. Glial growth factor restricts mammalian neural crest stem cells to a glial fate. *Cell* **77:** 349.

Syroid D.E., Maycox P.R., Burrola P.G., Liu N., Wen D., Lee K.F., Lemke G., and Kilpatrick T.J. 1996. Cell death in the Schwann cell lineage and its regulation by neuregulin. *Proc. Natl. Acad. Sci.* **93:** 9229.

Takebayashi-Suzuki K., Yanagisawa M., Gourdie R.G., Kanzawa N., and Mikawa T. 2000. In vivo induction of cardiac Purkinje fiber differentiation by coexpression of preproendothelin-1 and endothelin converting enzyme-1. *Development* **127:** 3523.

Viragh S. and Challice C.E. 1977. The development of the conduction system in the mouse embryo heart. I. The first embryonic A-V conduction pathway. *Dev. Biol.* **56:** 382.

———. 1980. The development of the conduction system in the mouse embryo heart. *Dev. Biol.* **80:** 28.

Wenink A.C.G. 1976. Development of the human cardiac conducting system. *J. Anat.* **121:** 617.

Yanagisawa H., Hammer R.E., Richardson J.A., Emoto N., Williams S.C., Takeda S., Clouthier D.E., and Yanagisawa M. 2000. Disruption of ECE-1 and ECE-2 reveals a role for endothelin-converting enzyme-2 in murine cardiac development. *J. Clin. Invest.* **105:** 1373.

Yanagisawa H., Yanagisawa M., Kapur R.P., Richardson J.A., Williams S.C., Clouthier D.E., de Wit D., Emoto N., and Hammer R.E. 1998. Dual genetic pathways of endothelin-mediated intercellular signaling revealed by targeted disruption of endothelin converting enzyme-1 gene. *Development* **125:** 825.

Zhao Y.Y., Sawyer D.R., Baliga R.R., Opel D.J., Han X., Marchionni M.A., and Kelly R.A. 1998. Neuregulins promote survival and growth of cardiac myocytes. Persistence of ErbB2 and ErbB4 expression in neonatal and adult ventricular myocytes. *J. Biol. Chem.* **273:** 10261.

The Yin/Yang of Innate Stress Responses in the Heart

D.L. MANN

Winters Center for Heart Failure Research, Department of Medicine, Baylor College of Medicine, and Houston Veterans Administration Medical Center, Houston, Texas 77030

The ability of the myocardium to successfully adapt to superimposed environmental stress, whether it results from hemodynamic overloading or myocardial ischemia and/or infarction, ultimately determines whether the heart will decompensate and fail, or whether instead it will maintain preserved function. Three of the critical mechanisms that allow the heart to withstand environmental stress are depicted in Figure 1: cardiac hypertrophy, cardiac myocyte cytoprotective responses, and cardiac repair. Despite the long-standing recognition that cardiac hypertrophy, cytoprotection, and repair are critical for maintaining myocardial homeostasis, the molecules that mediate and integrate the myocardial response to environmental stress, both at the level of the intact myocyte and for the ventricle as a whole, remain poorly understood. In this review, we focus on the emerging role that proinflammatory cytokines play as autocrine/paracrine factors that are responsible for initiating, integrating, and maintaining the myocardial response to environmental stress. The theme that emerges from this discussion is that the short-term expression of proinflammatory cytokines may provide the heart with an adaptive response to environmental injury, whereas long-term expression of these cytokines may be frankly maladaptive by producing overt cardiac decompensation.

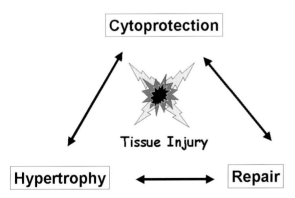

Figure 1. Myocardial homeostasis. Three of the important mechanisms that allow the heart to adapt to a superimposed environmental stress include cardiac hypertrophy, cardiac myocyte cytoprotective responses, and cardiac repair. The integration and coordination of these responses is critical to maintaining myocardial homeostasis.

OVERVIEW OF MYOCARDIAL RESPONSES TO STRESS

The supposition that resident cell types within the mammalian myocardium adapt to environmental stress by synthesizing, as well as responding to, a variety of endogenous stress-induced protein factors was first suggested by a series of thoughtful studies in hemodynamically overloaded canine hearts (Hammond et al. 1979, 1982). Although these sentinel studies did not succeed in identifying the nature of these stress-induced proteins (Hammond et al. 1984), subsequent reports have identified the biochemical nature of at least some of these myocardial-derived protein factors, such as acidic and basic fibroblast growth factor, platelet-derived growth factor, transforming growth factor-β (TGFβ), vascular endothelial-derived growth factor, and angiotensin II (Eghbali 1989; Weiner and Swain 1989; Casscells et al. 1990; Sadoshima et al. 1993; Schorb et al. 1993; Li et al. 1997; Yue et al. 1998). Although the precise role that these endogenous stress-induced proteins play is not known, it has been suggested that these proteins may contribute to the myocardial growth/remodeling and angiogenesis that occurs in response to tissue injury. Moreover, recent studies have shown that the myocardium is capable of synthesizing a variety of cytoprotective factors in response to ischemic injury, including adenosine, bradykinin, and nitric oxide (Downey et al. 1993; Goto et al. 1995; Bolli et al. 1997). Taken together, the extant literature suggests that the myocardial response to environmental stress comprises at least two interdependent homeostatic mechanisms: one that allows this tissue to delimit cell injury through up-regulation of cytoprotective factors and a second that facilitates tissue repair when and if these cytoprotective responses are insufficient to prevent cell death. However, the mechanisms that are responsible for orchestrating these different stress responses within the myocardium are not known.

ADAPTIVE EFFECTS OF CYTOKINES IN THE HEART

The portfolio of cytokines that comprise the focus of much of this review includes tumor necrosis factor (TNF), interleukin-1 (IL-1), and the interleukin-6 (IL-6) family of cytokines. These molecules have been referred to as "proinflammatory cytokines," insofar as they were traditionally thought to be derived exclusively from the immune system and were therefore considered to be primarily responsible for initiating inflammatory responses in tissues. However, these molecules are now known to be expressed by all nucleated cell types residing in the myocardium, including the cardiac myocyte itself (Kapadia et al. 1995, 1997). This latter observation has raised a number of important questions regarding the biological role that proinflammatory cytokines play in the heart.

Two important themes have emerged from recent studies of proinflammatory cytokine gene regulation in the heart thus far. The first is that proinflammatory cytokines are not constitutively expressed in the heart (Kapadia et al. 1995, 1997). The second theme is that these molecules are consistently and rapidly expressed in response to a variety of different forms of myocardial injury (see Table 1). The observation that proinflammatory cytokine gene expression is not coupled to a specific form of cardiac injury, but is instead observed in *all* forms of cardiac injury, suggests that these molecules constitute part of an intrinsic or "innate" stress response system in the heart. Thus, analogous to the role that proinflammatory cytokines play as effector molecules in the innate immune system, which is intended to act as an "early warning system" that allows the host to rapidly discriminate self from non-self (Hoffmann et al. 1999), the expression of proinflammatory cytokines in the heart may permit the myocardium to rapidly respond to tissue injury as part of an early warning system that coordinates and integrates a panoply of homeostatic responses within the heart following tissue injury. Indeed, as discussed below, there is a growing body of evidence which supports the point of view that short-term expression of proinflammatory cytokines is beneficial in the heart. However, it bears emphasis that the family of proinflammatory molecules which comprise this innate stress response system are phylogenetically ancient, and thus likely evolved in organisms with relatively short life spans (weeks to months). Thus, activation of the innate stress response system was never intended to provide long-term adaptive responses to the host organism. As discussed toward the end of this review, sustained and/or dysregulated expression of proinflammatory cytokines is sufficient to produce tissue injury and provoke overt cardiac decompensation.

The first line of evidence in support of a beneficial role for cytokines in the heart is implicit in phylogenetic studies of so-called "primitive cytokines" such as TNF and IL-1. TNF protostome vertebrates (annelids) and IL-1-like activity have been identified in both protostome vertebrates (annelids) and deuterostome invertebrates (echinoderms) (Raftos et al. 1991; Beck and Habicht 1991), thus suggesting that these molecules came into existence during the onset of the Cambrian period, before the split of the major animal phyla into vertebrate and invertebrate species. The evolutionary development of cytokines was probably necessary for the development of large multicellular organisms that required intercellular messengers such as cytokines to coordinate complex biological cellular responses. The observation that primitive cytokines such as TNF and IL-1 have been conserved by nature throughout the animal kingdom for nearly 600,000 million years, coupled with the observation that these same cytokines are expressed in virtually all forms of cardiac injury (see Table 1), suggests that these molecules may in some way confer a survival benefit in the host organism. Nonetheless, this argument is based on teleological evidence and must therefore be regarded as indirect proof in support of the point of view that cytokines confer beneficial responses in the heart.

The second line of evidence in support of a beneficial role for proinflammatory cytokines in the heart comes from a series of "gain-of-function" studies which have shown that proinflammatory cytokines confer cytoprotective responses in the heart. The first study to demonstrate the potential beneficial effects of cytokines showed that pretreating rats with TNF protected the heart from ischemic reperfusion injury ex vivo (Eddy et al. 1992). Following ischemia reperfusion injury, the hearts from TNF-pretreated animals had an approximately threefold reduction in the amount of lactate dehydrogenase release (Eddy et al. 1992) and showed an increase in the percent of recovery of developed left ventricular pressure when compared to control hearts (Nelson et al. 1995). IL-1 has also been shown to protect rat hearts against ischemia reperfusion injury in vitro (Terracio et al. 1988; Maulik et al. 1993). Subsequent in vitro studies have demonstrated that "physiological" levels of TNF are sufficient to protect cardiac myocytes against either hypoxic or ischemic injury, respectively (Nakano et al. 1998). Moreover, the cytoprotective effects of TNF could be mimicked by stimulating either the type 1 (p55, TNFR1) or the type 2 (p75, TNFR2) receptor, thus suggesting the cytoprotective effects of TNF were mediated by activation of TNFR1 or TNFR2. Although the above studies did not clearly identify the mechanism for these findings, proinflammatory cytokines are known to up-regulate the expression of at least two sets of protective proteins in the heart: the free radical scavenger manganese superoxide dismutase (MnSOD) (Wong and Goeddel 1988; Eddy et al. 1992) and the cytoprotective heat shock proteins (HSPs) (Low-Friedrich et al. 1992; Nakano et al. 1996). Relevant to this discussion is the finding that TNF-induced MnSOD induction is very rapid (< 1 hour) and requires very low levels of TNF (0.1 ng/ml^{-1}), consistent with the proposed homeostatic role for these proteins (Wong and Goeddel 1988). Given that contracting my-

Table 1. Cardiac Pathophysiological Conditions Associated with Proinflammatory Cytokines

Acute viral myocarditis
Cardiac allograft rejection
Myocardial infarction
Unstable angina
Myocardial reperfusion injury
Hypertrophic cardiomyopathy*
Heart failure*
Cardiopulmonary bypass*
Magnesium deficiency*
Pressure overload*

*Indicates conditions not traditionally associated with immunologically mediated inflammation.

ocardial cells are continually susceptible to oxygen-derived free radicals, TNF and IL-1 may play important roles in protecting the heart against oxidative stress, particularly during ischemia and reperfusion injury. TNF has also been shown recently to up-regulate the expression of heat shock protein 72 (HSP 72) (Nakano et al. 1996), a protein that is thought to protect the heart against ischemia reperfusion injury (Marber et al. 1995; Plumier et al. 1995). Finally, proinflammatory cytokines such as TNF and IL-1β have been shown to activate the transcription factor nuclear factor-kappa B (NF-κB), which has been shown to be cytoprotective under certain circumstances, presumably through up-regulation of one or more cytoprotective genes, including MnSOD, the cellular inhibitors of apoptosis 1 and 2 (c-IAP1 and cIAP2), and the members of the Bcl-2 family, including Bcl-2, Bfl-1, and Bcl-xL (Stehlik et al. 1998; Wang et al. 1998; Erl et al. 1999; Lee et al. 1999; Narula et al. 1999).

Similar findings have been obtained in gain-of-function studies for the so-called IL-6 family of cytokines that include interleukin-6 (IL-6), leukemia inhibitory factor (LIF), cardiotrophin-1 (CT-1), ciliary neurotrophic factor (CNTF), interleukin-11 (IL-11), and oncostatin M (OSM). This family of cytokines triggers downstream signaling pathways in multiple cell types, including cardiac myocytes, either through the homodimerization of the gp130 receptor or through the heterodimerization of gp130 with a related transmembrane receptor. Studies with CT-1 have shown that CT-1 blunts serum-deprivation-induced apoptosis in isolated neonatal cardiac myocytes through a pathway that is dependent on activation of the mitogen-activated protein kinase (MAPK). In these studies, transfection of a MAP kinase kinase 1 (MEK1) dominant negative mutant cDNA into myocardial cells or treatment with a MEK-specific inhibitor (PD098059) blocked the anti-apoptotic effects of CT-1, indicating a requirement of the MAP kinase pathway for the survival effect of CT-1. Similarly, studies have shown that LIF confers cytoprotective responses in isolated myocytes, as well as intact myocardial tissue (Nelson et al. 1995; Fujio et al. 1997). However, the mechanisms for LIF-mediated cytoprotective effects appear to be more complex than those reported for CT-1. That is, whereas studies in isolated neonatal myocytes suggest an important role for the Janus kinase (JAK) and the signal transducer and activator of transcription (STAT)-mediated signaling pathways (Fujio et al. 1997), more recent studies in adult myocytes suggest that the cytoprotective effects of LIF are mediated through activation of the MAPK pathway, consistent with what has been reported for CT-1. One explanation for these apparent differences in the cytoprotective mechanism for LIF is that there may be functionally significant cross-talk between the MAPK and the JAK/STAT pathways, as has been suggested recently (Fujio et al. 1997). Thus, the cytoprotective signaling pathways that are downstream from gp130-mediated signaling may involve both JAK/STAT- and MAPK-mediated signaling pathways.

The third, and perhaps most striking, line of evidence in support of a beneficial role for proinflammatory cytokines in the heart comes from a series of "loss-of-function" studies in mice deficient in proinflammatory cytokine-receptor-mediated signaling. Mice with targeted disruption of gp130 have been developed recently (Yoshida et al. 1996). Mice homozygous for the gp130 knockout (gp130$^{-/-}$) died between 12.5 days postcoitum (dpc) and term. The ventricular myocardium in these mice developed normally until 14.5 dpc; however, beyond 16.5 dpc, the gp130$^{-/-}$ mice demonstrated a markedly hypoplastic ventricle, with an abnormally thin ventricular wall that had a minimum thickness of one cell. Studies employing mice that harbor a ventricular-restricted knockout of the gp130 cytokine receptor via Cre-LoxP-mediated recombination showed that these mice have normal embryonic viability and no evidence of cardiac morphological abnormalities that were observed in the gp130 knockout (gp130$^{-/-}$) (Hirota et al. 1999). Moreover, these mice had normal cardiac structure (Fig. 2B) and function under basal conditions. Thus, the most likely explanation for the findings in the gp130$^{-/-}$ mice is that the cardiac developmental defects and embryonic lethality in these mice were the result of hematopoietic abnormalities and associated oxygen deprivation, as opposed to a primary gp130-mediated defect in myocytes. Interestingly, mice harboring the ventricular-restricted knockout of the gp130 cytokine receptor demonstrated a critical role for a gp130-dependent myocyte survival pathway following aortic banding (Hirota et al. 1999). That is, following hemodynamic overloading by transaortic constriction, these mice displayed a decrease in survival (Fig. 2A), ventricular enlargement involving both the right and left ventricular chambers (Fig. 2B), and a striking increase in the prevalence of cardiac myocyte apoptosis (Fig. 2E) when compared to control mice that exhibited normal compensatory cardiac hypertrophy (Hirota et al. 1999). These studies suggest that the gp130 pathway is an essential stress-activated myocyte survival pathway.

More recently, studies in mice that are doubly deficient for the type 1 and type 2 TNF receptors (TNFR1$^{-/-}$/TNFR2$^{-/-}$) have been shown to have an increase in infarct size in response to ischemic injury (Kurrelmeyer et al. 2000). Unlike gp130 mice, TNFR1$^{-/-}$/TNFR2$^{-/-}$ mice displayed a normal cardiac phenotype under nonstressed conditions. However, following acute coronary artery ligation, there was a striking increase in infarct size in TNFR1$^{-/-}$/TNFR2$^{-/-}$ mice (Fig. 3B) compared to littermate controls (Fig. 3B). As shown by the group data summarized in Figure 3C, infarct size was the same in wild-type, TNFR1$^{-/-}$ mice, and TNFR2$^{-/-}$ mice; however, infarct size was ~40% greater in the TNFR1$^{-/-}$/TNFR2$^{-/-}$ mice. Interestingly, the increase in infarct size in the TNFR1$^{-/-}$/TNFR2$^{-/-}$ mice was shown to be secondary to accelerated apoptosis in the TNFR1$^{-/-}$/TNFR2$^{-/-}$ mice, as opposed to increased myocyte necrosis. Although this study did not identify the biological mechanisms that were responsible for the cytoprotective effects of TNF, the observation that deletion of both TNFR1 and TNFR2 was necessary to provoke increased tissue injury suggested that TNFR1 and TNFR2 activated redundant cytoprotective signaling pathways in

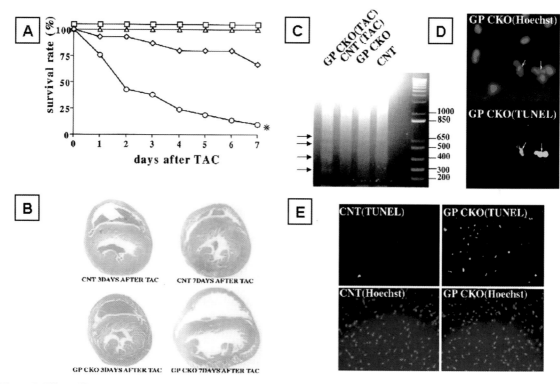

Figure 2. Effect of hemodynamic overload in mice with a ventricular-restricted knockout of the gp130 signaling pathway. (*A*) Analysis of survival of gp130 conditional knockout mice after transaortic constriction (TAC). Control mice (CNT) were subjected to a sham operation (*squares*) or transaortic constriction (*diamonds*). Mice with conditional ventricular-restricted knockout of the gp130 signaling pathway (CKO) were subjected to a sham operation (*triangles*) or transaortic constriction (*circles*). (*B*) Pathological analysis of the gp130 conditional knockout hearts. (*C*) A DNA laddering assay (TUNEL) revealed evidence of increased apoptosis in the gp130 conditional knockout mice after. (*D*) Low-power (100x) and (*E*) high-power (1000x) images showing DNA labeling visualized by fluoresence (*green*) and couterstained with Hoechst dye (*blue*). (Modified, with permission, from Hirota et al. 1999 [copyright Elsevier Science].)

the heart. Although the complete portfolio of cytoprotective signaling pathways that are common to both TNFR1 and TNFR2 is not known, it is interesting to note that NF-κB activation is common to both TNF receptors (Rothe et al. 1995). As noted above, NF-κB activation has been shown to be cytoprotective in certain settings.

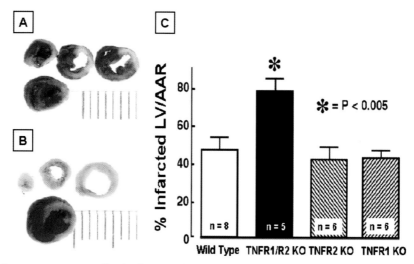

Figure 3. Effect of acute coronary artery ligation in TNFR1/TNFR2 knockout mice. The TTC (triphenyltetrazolium chloride) staining deficit, a marker of infarct size, was significantly greater in mice lacking both TNF receptors (*B*) when compared to littermate control mice (*A*). (TNFR1/TNFR2 KO) TNFR1/TNFR2 knockout mice; (TNFR1 KO) TNFR1 knockout mice; (TNFR2 KO) TNFR2 knockout mice. (Modified, with permission, from Kurrelmeyer et al. 2000 [copyright National Academy of Sciences].)

Thus, taken together, the above gain-of-function and loss-of-function studies for gp130- and TNF-mediated signaling suggest that proinflammatory cytokines may play an important role in the orchestration and the timing of the myocardial stress response, both by providing early antiapoptotic cytoprotective signals that are responsible for delimiting tissue injury, and also by providing delayed signals that facilitate tissue repair and/or tissue remodeling once myocardial tissue damage has supervened. In keeping with this latter point of view, previous studies have shown that CT 1, LIF, and TNF are all sufficient to provoke modest hypertrophic growth response in cardiac myocytes (Yokoyama et al. 1997), and that TNF is sufficient to lead to degradation and remodeling of the extracellular matrix in the heart (Bozkurt et al. 1998).

MALADAPTIVE EFFECTS OF CYTOKINES IN THE HEART

Although the short-term, self-limited expression of stress-activated cytokines may provide the heart with an adaptive response to environmental injury, this protective response may occur at the cost of unwanted deleterious effects that occur either when cytokines are elaborated for sustained periods of time, or when cytokines are expressed at supraphysiological, pathological levels. As shown in Table 2, there is a substantial body of evidence which suggests that the sustained expression of cytokines may produce frank maladaptive effects in the heart. Although it is clear from a large body of studies that proinflammatory cytokine expression becomes dysregulated (i.e., sustained) in the heart in a variety of pathophysiological contexts, the mechanisms that are responsible for the sustained expression of proinflammatory cytokines in the heart are not known. Accordingly, in the following section, we focus on the maladaptive downstream consequences of sustained proinflammatory cytokine expression in the heart, with an emphasis on the adverse consequences of cytokines on cardiac remodeling.

Effects of Cytokines on Left Ventricular Remodeling

The term "left ventricular remodeling" has been used to describe the multitude of changes that occur in cardiac shape, size, and composition in the failing heart that are not related to a preload-mediated increase in sarcomere length (Cohn 1995). Cytokines have a number of important effects on myocardial biology that may directly contribute to the process of LV remodeling, including myocyte hypertrophy (Yokoyama et al. 1997), alterations in fetal gene expression (Thaik et al. 1995; Kubota et al. 1997), contractile defects, and progressive myocyte loss through myocyte apoptosis (Krown et al. 1996). In addition to the above effects, several recent lines of evidence suggest that TNF may promote left ventricular remodeling through alterations in the extracellular matrix component of the myocardium. First, when human volunteers were administered endotoxin (a potent stimulus for TNF

Table 2. Maladaptive Cardiovascular Effects of Stress-activated Cytokines

Produce left ventricular dysfunction
Produce pulmonary edema in humans
Produce cardiomyopathy in humans
Promote left ventricular remodeling experimentally
Produce abnormalities in myocardial metabolism experimentally
Produce β-receptor uncoupling from adenylate cyclase experimentally
Produce abnormalities of mitochondrial energetics
Produce activation of the fetal gene program experimentally

production) intravenously, there was ~20% increase in left ventricular end-diastolic volume within 5 hours (Suffredini et al. 1989). Second, a recent experimental study shows that when concentrations of TNF that are observed in patients with heart failure were infused continuously in rats, there was a time-dependent change in LV dimension that was accompanied by progressive degradation of the extracelluar matrix (Bozkurt et al. 1998). Third, lines of transgenic mice that overexpress TNF in the cardiac compartment develop progressive LV dilation that is accompanied by activation of matrix metalloproteinases (MMP), as well as increased denaturation of collagen (Li et al. 2000; Sivasubramanian et al. 2001). As shown in Figures 4 and 5, respectively, there was progressive loss of fibrillar collagen and increased MMP activation in the hearts of the transgenic mice overexpressing TNF in the cardiac compartment. The dissolution of the fibrillar collagen weave that surrounds the individual cardiac myocytes and links the myocytes together would be expected to allow for rearrangement (slippage) of myofibrillar bundles within the ventricular wall (Weber 1989). However, Figure 4 shows that long-term stimulation (i.e., 8–12 weeks) with TNF resulted in an increase in fibrillar collagen content that was accompanied by decreased MMP activity (Fig. 5A, B) and increased expression of the tissue inhibitors of matrix metalloproteinases (TIMPs [Fig. 5C]). Taken together, these observations suggest that sustained myocardial inflammation provokes time-dependent changes in the balance between MMP activity and TIMP activity. That is, during the early stages of inflammation there is an increase in the ratio of MMP activity to TIMP levels that fosters LV dilation. However, with chronic inflammatory signaling there is a time-dependent increase in TIMP levels, with a resultant decrease in the ratio of MMP activity to TIMP activity, and a subsequent increase in myocardial fibrillar collagen content. Although the molecular mechanisms that are responsible for the transition between excessive degradation and excessive synthesis of the extracellular matrix are not known, studies in experimental models of chronic injury/inflammation in an array of different organs, including liver, lung, and kidney, wherein an initial increase in MMP expression is superseded by increased TIMP expression and progressive tissue fibrosis, have implicated the increased expression of a number of fibrogenic cytokines, most notably TGFβ (Sime et al. 1998; Knittel et al. 2000). Thus, excessive activation of proin-

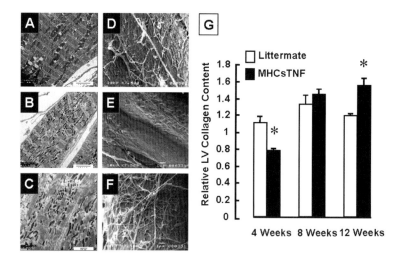

Figure 4. Effects of sustained proinflammatory cytokine expression on myocardial ultrastructure and collagen content. Panels *A–C* show representative transmission electron micrographs in littermate controls (*A*) and the TNF transgenic mice at 4 (*B*) and 8 (*C*) weeks of age. Panels *D–F* show representative scanning electron micrographs in littermate controls (*D*) and the TNF transgenic mice at 4 (*E*) and 8 (*F*) weeks of age. Panel *G* summarizes the results of group data for collagen content, as determined by the picrosirius red technique. (Reprinted, with permission, from Sivasubramanian et al. 2001 [copyright Lippincott Williams and Wilkins].)

flammatory cytokines may contribute to LV remodeling through a variety of different mechanisms that involve both the myocyte and non-myocyte components of the heart.

CONCLUSION

Yin/Yang reflects the Chinese philosophy that for every positive there is a negative. Under yang are the principles of maleness: the sun, creation, heat, light, and heaven, whereas under yin are the principles of femaleness: the moon, completion, cold, darkness, and material forms. The principle of yin/yang suggests that each of these opposites produces the other. The production of yin from yang and yang from yin occurs cyclically and constantly, so that no one principle continually dominates the other or determines the other. In this review, we have summarized experimental material which suggests that activation of cytokines within the heart following cardiac injury may have beneficial, or alternatively, detrimental consequences for the host, depending on the duration and degree of cytokine exposure. That is, short-term expres-

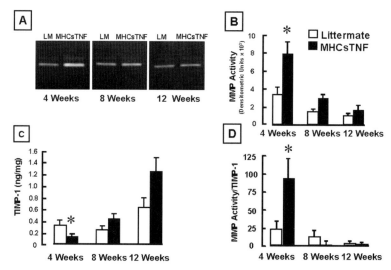

Figure 5. Effects of sustained proinflammatory cytokine expression on MMP activity and TIMP levels. Panel *A* shows a zymogram of total MMP activity in the TNF transgenic mice (TNF-TG) and littermate (LM) control mice at 4, 8, and 12 weeks of age, whereas panel *B* summarizes the results of group data for total MMP zymographic activity. Panel *C* depicts the time-dependent changes in TIMP levels at 4, 8, and 12 weeks in the TNF transgenic and littermate control mice. Panel *D* depicts the time-dependent changes in the ratio of MMP activity/TIMP levels in the TNF transgenic and littermate control mice. (Reprinted, with permission, from Sivasubramanian et al. 2001 [copyright Lippincott Williams and Wilkins].)

sion of these stress-activated cytokines may be beneficial by up-regulating the expression of families of so-called protective proteins in the heart, as well as by integrating the various components of the myocardial stress response, namely cardiac hypertrophy, cardiac remodeling, and cardiac repair. This statement notwithstanding, the short-term beneficial effects of stress-activated cytokines may be lost if myocardial expression of these molecules becomes either sustained and/or excessive, in which case the salutary effects of these proteins may be contravened by their known deleterious effects. Thus, one of the challenges that faces investigators in this field will be to delineate the nature of the cell signaling pathways that are responsible for the adaptive and maladaptive aspects of cytokine in order to develop effective strategies that maximize the potential spectrum of beneficial responses conferred by stress-activated cytokines in the heart, without simultaneously activating their known potential deleterious effects.

ACKNOWLEDGMENTS

The author gratefully acknowledges Ms. Mary Helen Soliz for secretarial assistance and Dr. Andrew I. Schafer for his past and present guidance and support. This research was supported by research funds from the Veterans Administration and the National Institutes of Health (P50 HL-O6H and RO1 HL-58081-01, RO1 HL-61543-01, HL-42250-10/10).

REFERENCES

Beck G. and Habicht G.S. 1991. Primitive cytokines: Harbingers of vertebrate defense. *Immunol. Today* **12**: 180.

Bolli R., Bhatti Z.A., Tang X.L., Qiu Y., Zhang Q., Guo Y., and Jadoon A.K. 1997. Evidence that late preconditioning against myocardial stunning in conscious rabbits is triggered by the generation of nitric oxide. *Circ. Res.* **81**: 42.

Bozkurt B., Kribbs S.B., Clubb F.J., Jr., Michael L.H., Didenko V.V., Hornsby P.J., Seta Y., Oral H., Spinale F.G., and Mann D.L. 1998. Pathophysiologically relevant concentrations of tumor necrosis factor-α promote progressive left ventricular dysfunction and remodeling in rats. *Circulation* **97**: 1382.

Casscells W., Speir E., Sasse J., Klagsbrun M., Allen P., Lee M., Calvo B., Chiba M., Haggroth L., Folkman J., et al. 1990. Isolation, characterization, and localization of heparin-binding growth factors in the heart. *J. Clin. Invest.* **85**: 433.

Cohn J.N. 1995. Structural basis for heart failure: Ventricular remodeling and its pharmacological inhibition. *Circulation* **91**: 2504.

Downey J.M., Liu G.S., and Thornton J.D. 1993. Adenosine and the anti-infarct effects of preconditioning. *Cardiovasc. Res.* **27**: 3.

Eddy L.J., Goeddel D.V., and Wong G.H.W. 1992. Tumor necrosis factor-α pretreatment is protective in a rat model of myocardial ischemia-reperfusion injury. *Biochem. Biophys. Res. Commun.* **184**: 1056.

Eghbali M. 1989. Cellular origin and distribution of transforming growth factor-β_1 in the normal rat myocardium. *Tissue Res.* **256**: 553.

Erl W., Hansson G.K., de Martin R., Draude G., Weber K.S., and Weber C. 1999. Nuclear factor-kappa B regulates induction of apoptosis and inhibitor of apoptosis protein-1 expression in vascular smooth muscle cells. *Circ. Res.* **84**: 668.

Fujio Y., Kunisada K., Hirota H., Yamauchi-Takihara K., and Kishimoto T. 1997. Signals through gp130 upregulate *bcl-x* gene expression via STAT1-binding *cis*-element in cardiac myocytes. *J. Clin. Invest.* **99**: 2898.

Goto M., Liu Y., Yang X.M., Ardell J.L., Cohen M.V., and Downey J.M. 1995. Role of bradykinin in protection of ischemic preconditioning in rabbit hearts. *Circ. Res.* **77**: 611.

Hammond G.L., Lai Y.K., and Markert C.L. 1982. The molecules that initiate cardiac hypertrophy are not species specific. *Science* **216**: 529.

———. 1984. Preliminary characterization of molecules that increase cell free translational activity of cardiac cytoplasmic RNA. *Eur. Heart J.* (suppl. F) **5**: 225.

Hammond G.L., Wieben E., and Markert C.L. 1979. Molecular signals for initiating protein synthesis in organ hypertrophy. *Proc. Natl. Acad. Sci.* **76**: 2455.

Hirota H., Chen J., Betz U.A., Rajewsky K., Gu Y., Ross J., Jr., Muller W., and Chien K.R. 1999. Loss of a gp130 cardiac muscle cell survival pathway is a critical event in the onset of heart failure during biomechanical stress. *Cell* **97**: 189.

Hoffmann J.A., Kafatos F.C., Janeway C.A., Jr., and Ezekowitz R.A. 1999. Phylogenetic perspectives in innate immunity. *Science* **284**: 1313.

Kapadia S., Lee J.R., Torre-Amione G., Birdsall H.H., Ma T.S., and Mann D.L. 1995. Tumor necrosis factor gene and protein expression in adult feline myocardium after endotoxin administration. *J. Clin. Invest.* **96**: 1042.

Kapadia S., Oral H., Lee J., Nakano M., Taffet G.E., and Mann D.L. 1997. Hemodynamic regulation of tumor necrosis factor-α gene and protein expression in adult feline myocardium. *Circ. Res.* **81**: 187.

Knittel T., Mehde M., Grundmann A., Saile B., Scharf J.G., and Ramadori G. 2000. Expression of matrix metalloproteinases and their inhibitors during hepatic tissue repair in the rat. *Histochem. Cell Biol.* **113**: 443.

Krown K.A., Page M.T., Nguyen C., Zechner D., Gutierrez V., Comstock K.L., Glembotski C.C., Quintana P.J., and Sabbadini R.A. 1996. Tumor necrosis factor alpha-induced apoptosis in cardiac myocytes: Involvement of the sphingolipid signaling cascade in cardiac cell death. *J. Clin. Invest.* **98**: 2854.

Kubota T., McTiernan C.F., Frye C.S., Slawson S.E., Lemster B.H., Koretsky A.P., Demetris A.J., and Feldman A.M. 1997. Dilated cardiomyopathy in transgenic mice with cardiac-specific overexpression of tumor necrosis factor-alpha. *Circ. Res.* **81**: 627.

Kurrelmeyer K., Michael L., Baumgarten G., Taffet G.E., Peschon J.J., Sivasubramanian N., Entman M.L., and Mann D.L. 2000. Endogenous myocardial tumor necrosis factor protects the adult cardiac myocyte against ischemic-induced apoptosis in a murine model of acute myocardial infarction. *Proc. Natl. Acad. Sci.* **97**: 5456.

Lee J.P., Palfrey H.C., Bindokas V.P., Ghadge G.D., Ma L., Miller R.J., and Roos R.P. 1999. The role of immunophilins in mutant superoxide dismutase-1linked familial amyotrophic lateral sclerosis. *Proc. Natl. Acad. Sci.* **96**: 3251.

Li J., Hampton T., Morgan J.P., and Simons M. 1997. Stretch-induced VEGF expression in the heart. *J. Clin. Invest.* **100**: 18.

Li Y.Y., Feng Y.Q., Kadokami T., McTiernan C.F., Draviam R., Watkins S.C., and Feldman A.M. 2000. Myocardial extracellular matrix remodeling in transgenic mice overexpressing tumor necrosis factor alpha can be modulated by anti-tumor necrosis factor alpha therapy. *Proc. Natl. Acad. Sci.* **97**: 12746.

Low-Friedrich I., Weisensee D., Mitrou P., and Schoeppe W. 1992. Cytokines induce stress protein formation in cultured cardiac myocytes. *Basic Res. Cardiol.* **87**: 12.

Marber M.S., Mestril R., Chi S.H., Sayen R., Yellon D.M., and Dillmann W.H. 1995. Overexpression of the rat inducible 70-kd heat stress protein in a transgenic mouse increases the resistance of the heart to ischemic injury. *J. Clin. Invest.* **95**: 1446.

Maulik N., Engelman R.M., Wei Z.J., Lu D., Rousou J.A., and Das D.K. 1993. Interleukin-1 alpha preconditioning reduces myocardial ischemia reperfusion injury. *Circulation* **88**: 387.

Nakano M., Knowlton A.A., Dibbs Z., and Mann D.L. 1998. Tumor necrosis factor-α confers resistance to injury induced by hypoxic injury in the adult mammalian cardiac myocyte. *Circulation* **97:** 1392.

Nakano M., Knowlton A.A., Yokoyama T., Lesslauer W., and Mann D.L. 1996. Tumor necrosis factor-α induced expression of heat shock protein 72 in adult feline cardiac myocytes. *Am. J. Physiol.* **270:** H1231.

Narula J., Pandey P., Arbustini E., Haider N., Narula N., Kolodgie F.D., Dal Bello B., Semigran M.J., Bielsa-Masdeu A., Dec G.W., Israels S., Ballester M., Virmani R., Saxena S., and Kharbanda S. 1999. Apoptosis in heart failure: Release of cytochrome c from mitochondria and activation of caspase-3 in human cardiomyopathy. *Proc. Natl. Acad. Sci.* **96:** 8144.

Nelson S.K., Wong G.H.W., and McCord J.M. 1995. Leukemia inhibitory factor and tumor necrosis factor induce manganese superoxide dismutase and protect rabbit hearts from reperfusion injury. *J. Mol. Cell Cardiol.* **27:** 223.

Plumier J.C.L., Ross B.M., Currie R.W., Angelidis C.E., Kazlaris H., Kollias G., and Pagoulatos G.N. 1995. Transgenic mice expressing the human heat shock protein 70 have improved post-ischemic myocardial recovery. *J. Clin. Invest.* **95:** 1854.

Raftos D.A., Cooper E.L., Habicht G.S., and Beck G. 1991. Invertebrate cytokines: Tunicate cell proliferation stimulated by an interleukin 1-like molecule. *Proc. Natl. Acad. Sci.* **88:** 9518.

Rothe M., Sarma V., Dixit V.M., and Goeddel D.V. 1995. TRAF2-mediated activation of NF-κB by TNF receptor 2 and CD40. *Science* **269:** 1424.

Sadoshima J.I., Xu Y., Slayter H.S., and Izumo S. 1993. Autocrine release of angiotensin II mediates stretch-induced hypertrophy of cardiac myocytes in vitro. *Cell* **75:** 977.

Schorb W., Booz G.W., Dostal D.E., Conrad K.M., Chang K.C., and Baker K.M. 1993. Angiotensin II is mitogenic in neonatal rat cardiac fibroblasts. *Circ. Res.* **72:** 1245.

Sime P.J., Marr R.A., Gauldie D., Xing Z., Hewlett B.R., Graham F.L., and Gauldie J. 1998. Transfer of tumor necrosis factor-alpha to rat lung induces severe pulmonary inflammation and patchy interstitial fibrogenesis with induction of transforming growth factor-beta1 and myofibroblasts. *Am. J. Pathol.* **153:** 825.

Sivasubramanian N., Coker M.L., Kurrelmeyer K., MacLellan W.R., DeMayo F.J., Spinale F.G., and Mann D.L. 2001. Left ventricular remodeling in transgenic mice with cardiac restricted overexpression of tumor necrosis factor. *Circulation* **104:** 826.

Stehlik C., de Martin R., Binder B.R., and Lipp J. 1998. Cytokine induced expression of porcine inhibitor of apoptosis protein (iap) family member is regulated by NF-kappa B. *Biochem. Biophys. Res. Commun.* **243:** 827.

Suffredini A.F., Fromm R.E., Parker M.M., Brenner M., Kovacs J.A., Wesley R.A., and Parrillo J.E. 1989. The cardiovascular response of normal humans to the administration of endotoxin. *N. Engl. J. Med.* **321:** 280.

Terracio L., Miller B., and Borg T.K. 1988. Effects of cyclic mechanical stimulation of the cellular components of the heart: In vitro. *In Vitro Cell. Dev. Biol.* **24:** 53.

Thaik C.M., Calderone A., Takahashi N., and Colucci W.S. 1995. Interleukin-1β modulates the growth and phenotype of neonatal rat cardiac myocytes. *J. Clin. Invest.* **96:** 1093.

Wang C.Y., Mayo M.W., Korneluk R.G., Goeddel D.V., and Baldwin A.S.J. 1998. NF-kappaB antiapoptosis: Induction of TRAF1 and TRAF2 and c-IAP1 and c-IAP2 to suppress caspase-8 activation. *Science* **281:** 1680.

Weber K.T. 1989. Cardiac interistium in health and disease: The fibrillar collagen network. *J. Am. Coll. Cardiol.* **13:** 1637.

Weiner H.L. and Swain J.L. 1989. Acidic fibroblast growth factor mRNA is expressed by cardiac myocytes in culture and the protein is localized to the extracellular matrix. *Proc. Natl. Acad. Sci.* **86:** 2683.

Wong G.H.W. and Goeddel D.V. 1988. Induction of manganous superoxide dismutase by tumor necrosis factor: Possible protective mechanism. *Science* **242:** 941.

Yokoyama T., Nakano M., Bednarczyk J.L., McIntyre B.W., Entman M.L., and Mann D.L. 1997. Tumor necrosis factor-α provokes a hypertrophic growth response in adult cardiac myocytes. *Circulation* **95:** 1247.

Yoshida K., Taga T., Saito M., Suematsu S., Kumanogoh A., Tanaka T., Fujiwara H., Hirata M., Yamagami T., Nakahata T., Hirabayashi T., Yoneda Y., Tanaka K., Wang W.Z., Mori C., Shiota K., Yoshida N., and Kishimoto T. 1996. Targeted disruption of gp130, a common signal transducer for the interleukin 6 family of cytokines, leads to myocardial and hematological disorders. *Proc. Natl. Acad. Sci.* **93:** 407.

Yue P., Long C.S., Austin R., Chang K.C., Simpson P.C., and Massie B.M. 1998. Post-infarction heart failure in the rat is associated with distinct alterations in cardiac myocyte molecular phenotype. *J. Mol. Cell Cardiol.* **30:** 1615.

Regulatory Networks Controlling Mitochondrial Energy Production in the Developing, Hypertrophied, and Diabetic Heart

B.N. FINCK,* J.J. LEHMAN,* P.M. BARGER,† AND D.P. KELLY*

Center for Cardiovascular Research, Department of Medicine; Departments of Molecular Biology & Pharmacology and Pediatrics, Washington University School of Medicine, St. Louis, Missouri 63110; †Department of Medicine, Baylor College of Medicine, Houston, Texas 77030

The adult mammalian heart is highly specialized for efficient and high-capacity energy production to meet the diverse physiologic demands of the postnatal environment. In contrast to the fetal heart, which relies largely on glucose, the adult heart is programmed to rely on multiple energy sources. Although the postnatal heart continues to utilize glucose, the primary source of ATP is the mitochondrial fatty acid oxidation (FAO) pathway (Bing 1955; Neely et al. 1972; Schulz 1991). The expression of FAO pathway enzymes is induced following birth via developmentally programmed nuclear gene regulatory events. The relative importance of glucose and fatty acids as fuel sources for the adult heart is a function of developmental, physiologic, and dietary contexts (Lockwood and Bailey 1970; Neely et al. 1972; Schulz 1991). For example, energy production in the fetal heart is primarily via glycolysis because mitochondrial oxidative capacity is limited. Furthermore, although the energy demands of the fully developed heart are mainly met by the oxidation of fats, glucose utilization is important in postprandial states and with sudden increases in hemodynamic load. Thus, the normal adult heart exhibits "plasticity" in its energy substrate choices.

Evidence has emerged that the normal balance of myocardial energy substrate utilization is compromised in several common cardiovascular disease states (Fig. 1). For example, the extraordinary capacity of the heart to catabolize fatty acids within mitochondria is diminished in pathologic forms of cardiac hypertrophy (Bishop and Altschuld 1970; Taegtmeyer and Overturf 1988; Christe and Rodgers 1994) and in the hypoxic or ischemic heart (Fig. 1) (Abdel-aleem et al. 1998; Rumsey et al. 1999). Conversely, glucose utilization is severely reduced in the diabetic heart such that it relies almost exclusively on FAO (Rodrigues et al. 1995; Stanley et al. 1997; Belke et al. 2000). Do these metabolic "switches" serve an adaptive function? For example, does the shift from mitochondrial FAO to glucose utilization in the hypoxic or hypertrophied heart reduce oxygen consumption costs as an adaptive response? Do such changes become maladaptive, leading to heart failure? Similarly, does the reduction in glucose utilization by the diabetic heart lead to the dysfunction and increased cardiovascular morbidity of the diabetic population? Recent studies in the field of cardiac metabolism have begun to address these and related questions. The answers should lead to the development of novel therapeutic strategies aimed at modulating cardiac metabolism in common diseases of the myocardium.

This review covers three topics related to the control of cardiac mitochondrial function in the normal and diseased heart. First, current knowledge of a gene regulatory program involved in the developmental maturation of the cardiac mitochondrial FAO pathway is reviewed. The control of the cardiac mitochondrial FAO by the nuclear receptor peroxisome proliferator-activated receptor α (PPARα) and the retinoid X receptor (RXR) will serve as the starting point, followed by a description of recently discovered links to the program of mitochondrial biogenesis. Second, mechanisms leading to derangements in the PPARα-mediated control of cardiac energy metabolism in the hypertrophied, hypoxic, and failing heart are summarized. Third, the molecular regulatory events responsible for altered energy metabolism in the diabetic heart are reviewed, and a mouse model of diabetic cardiomyopathy is described.

DEVELOPMENTAL MATURATION OF CARDIAC ENERGY PRODUCTION PATHWAYS

The Fetal-to-Adult Energy Metabolic "Switch": Postnatal Induction of Cellular Mitochondrial Energy Transduction/Production Pathways

Following birth, the heart undergoes a remarkable metabolic maturation such that the normal adult myocardium is capable of high-capacity utilization of both glucose and fatty acids for ATP production (Bing 1955; Neely et al. 1972; Warshaw 1972). During the fetal stages, the heart relies mainly on glucose for energy. Following birth, the increased hemodynamic demands imposed on the left ventricle, which serves as a constant pump throughout the life of the organism, mandates a reliable, high-capacity energy production system. The increased postnatal energy demands are met largely through the rapid proliferation of mitochondria and coor-

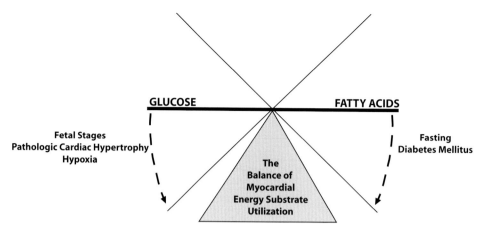

Figure 1. The balance of myocardial energy substrate utilization in the normal and diseased heart. The normal adult heart derives ~50–70% of its energy from the oxidation of FA. However, during fetal stages and in pathologic cardiac hypertrophy and hypoxic conditions, the heart relies mainly on glycolysis to produce ATP. Conversely, following a fast or in the uncontrolled diabetic state, over 90% of energy produced in the heart is derived from FAO.

dinate induction of the expression of nuclear genes encoding a variety of mitochondrial enzymes, including those of the FAO pathway (Kelly et al. 1989; Nagao et al. 1993). The postnatal mammalian diet, which is rich in milk fat, provides a new source of fatty acid substrate. In addition, the expression of proteins involved in fatty acid import such as fatty acid transport protein and FAT/CD36 are induced in heart following birth. Collectively, these developmentally programmed events lead to a switch from glucose to fatty acids as the chief myocardial energy substrate.

The perinatal switch in myocardial energy substrate preference is considered to be adaptive for several reasons. First, per mole of substrate, fatty acids provide a much greater yield of ATP compared to glucose, albeit at a higher oxygen consumption cost. Second, the mammalian diet provides an ample quantity of fatty acids to serve as a substrate source. Third, as noted above, the energy metabolic maturation of the postnatal heart provides flexibility in substrate preference. Specifically, in postprandial states and certain pathologic conditions, including ischemia or hypertrophy, glucose re-emerges as the chief substrate for energy production. The balanced use of fatty acids and glucose provides an energy substrate reserve that protects the heart from periods of mismatch between energy demands and production.

The Mitochondrial FAO Pathway as a Focus for the Characterization of Gene Regulatory Pathways Involved in the Cardiac Metabolic Maturation Program

The postnatal myocardial metabolic switch is associated with a dramatic induction in the expression of nuclear genes involved in the uptake and oxidation of fatty acids (Fig. 2A) (Kelly et al. 1989; Nagao et al. 1993). Following birth, the increase in FAO enzyme gene expression occurs in parallel with a marked increase in cellular mitochondrial volume density (mitochondrial biogenesis) (Fig. 2B). Two mitochondrial FAO enzyme genes have been used as a starting point to delineate the upstream transcriptional regulatory events involved in the cardiac energy metabolic maturation program. The first, muscle carnitine palmitoyltransferase I (M-CPT I), catalyzes a tightly regulated, rate-limiting step in the mitochondrial import of long-chain fatty acids (see Fig. 2A). The second, medium-chain acyl-CoA dehydrogenase (MCAD), catalyzes the initial step within the FAO spiral (Fig. 2A) and is the most commonly deficient enzyme among the human inborn errors of FAO. The levels of mRNA encoding MCAD (Fig. 2B) and M-CPT I parallel the developmental switch in energy substrate utilization in the rodent heart (Kelly et al. 1989; Lehman et al. 2000). The expression of the MCAD and M-CPT I genes is also coordinately increased in dietary and physiologic conditions known to increase cardiac and skeletal muscle FAO rates such as short-term fasting (Nagao et al. 1993; Leone et al. 1999) and chronic stimulation of muscle (Cresci et al. 1996).

The postnatal induction of MCAD and M-CPT I gene expression in the developing heart provided proof of concept that these genes were targets for upstream regulatory pathways involved in the transcriptional control of myocardial energy metabolism. The initial approach involved mapping the *cis*-acting regulatory elements within the promoter regions of these genes. Two complementary experimental strategies were employed. First, gene promoter regulatory regions involved in the cardiac developmental control of MCAD gene expression were mapped in vivo in promoter-reporter transgenic mice (Disch et al. 1996). Second, relevant M-CPT I and MCAD gene promoter elements were defined in rat neonatal cardiac myocytes in culture (Disch et al. 1996; Brandt et al. 1998). The latter approach revealed that fatty acid substrate activated the transcription of the FAO enzyme genes. Both strategies demonstrated that DNA sequences containing recognition sites for members of the nuclear hormone receptor superfamily were necessary for developmental and fatty acid-mediated control of MCAD and M-CPT I gene

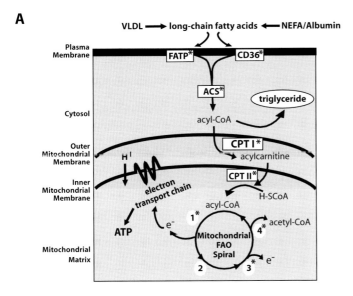

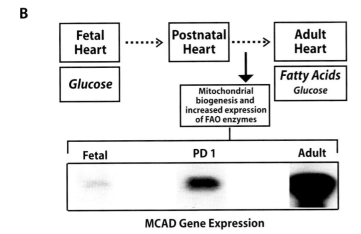

Figure 2. (*A*) The cellular fatty acid oxidation (FAO) pathway. The diagram depicts the major routes of fatty acid uptake and oxidation in the cardiac myocyte. Abbreviations: (VLDL) very low density lipoproteins; (NEFA) non-esterified fatty acids; (FATP) fatty acid transport protein; (ACS) acyl-CoA synthetase; (CPT I) carnitine palmitoyltransferase I; (CPT II) carnitine palmitoyltransferase II; (1) very long-chain (VLCAD), long-chain (LCAD), and medium-chain (MCAD) acyl-CoA dehydrogenases; (2) enoyl-CoA hydratase; (3) 3-hydroxyacyl-CoA dehydrogenase; (4) 3-ketoacyl-CoA thiolase. The asterisks denote known PPARα target genes. (*B*) Developmental shifts in cardiac energy substrate preference and FAO enzyme gene expression. (*Top*) The fetal heart relies primarily on anaerobic glucose utilization pathways. Following birth, the heart increases its capacity for and dependence on the oxidation of fatty acids to produce ATP. The normal adult mammalian heart relies principally on FA for energy production. The changes in myocardial substrate preference following birth are associated with robust mitochondrial biogenesis and increased expression of genes encoding FAO enzymes. (*Bottom*) A representative autoradiograph of a Northern blot analysis demonstrating the developmental regulation of the nuclear gene encoding the mitochondrial FAO enzyme MCAD (PD1 = postnatal day 1).

expression. These observations led to the discovery that the fatty acid-activated nuclear receptor transcription factor, PPARα, and its heterodimeric partner, RXR, serve as critical components of the postnatal control of mitochondrial FAO enzyme gene expression in heart.

PPARα: A Critical Transcriptional Regulator of Cardiac Lipid Utilization Pathways

PPARα was originally identified by its involvement in the hepatic peroxisomal proliferative response to fibrates (Issemann and Green 1990). Subsequently, two other members of the PPAR family were identified (PPARβ/δ and PPARγ) (for review, see Desvergne and Wahli 1999). The results of studies by a large number of laboratories have shown that PPARα and γ serve distinct but critical roles in the control of lipid metabolism and other biologic processes. The functional role of PPARβ remains unclear. PPARs regulate the transcription of target genes by binding their target DNA regulatory elements as a heterodimer with the retinoid X receptor (RXR). PPARα, a focus of this review, is enriched in tissues with high capacity for FAO, including liver and heart, and activates cellular *lipid utilization* pathways. PPARγ, which is adipose-enriched, plays a critical role in the differentiation and function of the adipocyte; a *lipid storage* cell. Importantly, the activity of the PPARs is ligand-dependent. A variety of activating ligands for the PPARs have been identified, most of which are long-chain fatty acids or their derivatives, although certain prostaglandin derivatives can also serve to activate the PPARs (Xu et al. 1999). The endogenous ligands for the PPARs have not been identified with certainty. Of interest is that certain activators are PPAR-specific. For example, thiazolidinediones, a new class of insulin-sensitizing drugs, are PPARγ-specific activators, whereas fibrates (hypolipidemic drugs) are more specific for PPARα (Xu et al. 1999). The existence of PPAR-specific activators has generated an intense interest in the development of drugs that have customized metabolic effects by virtue of the fact that PPARα and γ have distinct tissue expression patterns and generally opposing effects on lipid metabolism.

It is now known that PPARα activates the expression of cardiac genes involved in multiple FA utilization pathways including lipid uptake, thioesterification, and peroxisomal and mitochondrial FAO (see Fig. 2A) (for review, see Barger and Kelly 2000). Studies of mice null for PPARα (PPARα$^{-/-}$) have shown that the basal expression of mitochondrial and peroxisomal FAO enzymes in liver and heart is reduced (Lee et al. 1995; Aoyama et al. 1998; Djouadi et al. 1998). The PPARα$^{-/-}$ mice, which were produced by the Gonzalez laboratory (Lee et al. 1995), appear normal under basal physiologic conditions. Studies of the isolated working PPARα$^{-/-}$ heart have shown that FAO rates are markedly diminished, whereas glucose utilization is increased similar to that of the fetal heart (Campbell et al. 2002). PPARα$^{-/-}$ mice are unable to appropriately increase the expression of target genes involved in myocardial or hepatic FA utilization in response to physiologic or dietary demands known to increase FAO rates, such as fasting (Kersten et al. 1999; Leone et al. 1999). Gain-of-function studies performed by our group have also demonstrated the importance of PPARα in the control of cellular FA utilization pathways. Cardiac-specific overexpression of PPARα results in increased myocardial FAO enzyme expression and palmitate oxidation rates (Finck et al. 2002). Collectively, these studies have shown that PPARα plays a critical role in the developmental and physiologic control of cardiac FA utilization pathways.

PGC-1α: An Inducible, Cardiac-enriched PPARα Coactivator and Master Regulator of Cardiac Mitochondrial Functional Capacity

The developmental and physiologic control of the cardiac FAO pathway is linked to the regulation of mitochondrial biogenesis. Mitochondrial number and respiratory capacity increase markedly in parallel with the induction of genes involved in FAO and other energy transduction pathways such as the tricarboxylic acid cycle during the postnatal period in heart and many other tissues. Recent evidence indicates that FAO and mitochondrial functional capacity are coordinately deranged in a variety of myocardial disease states including hypertrophy, ischemic heart disease, certain cardiomyopathies, and in the aging heart (Kelly and Strauss 1994; Wallace 1999; Barger et al. 2000).

What is the link between the transcriptional control of the FAO pathway and mitochondrial biogenesis? The answer to this question came, in part, from a discovery made by studies focused on adipose tissue. Spiegelman and coworkers (Puigserver et al. 1998) sought to identify transcriptional regulatory factors that distinguished mitochondria-rich brown adipose tissue (BAT) from mitochondria-poor white adipose tissue (WAT). For these studies, the adipose tissue-enriched relative of PPARα, PPARγ, was used as a "bait" in a two-hybrid screen of a BAT cDNA library. Earlier studies had shown that PPARγ served a critical role in the early steps of adipocyte differentiation. The screen led to the identification of a transcriptional coactivator termed PPARγ coactivator-1 (now known as PGC-1α) (Puigserver et al. 1998). PGC-1α was shown to coactivate PPARγ as well as a number of other nuclear receptors including RXR, thyroid receptor, and the estrogen receptor. Subsequently, PGC-1α was shown to activate components of the mitochondrial biogenesis program in BAT and C_2C_{12} myotubes through a distinct transcription factor, the nuclear respiratory factor 1 (NRF-1) (Wu et al. 1999). The identification of PGC-1α as a key regulator of mitochondrial function in BAT led to the hypothesis that this transcriptional coactivator may play a similar role in heart via PPARα. This hypothesis was supported by the observation that PGC-1α expression is cardiac-enriched and induced in rodent heart following birth (Lehman et al. 2000). Indeed, studies in several mammalian cell types in culture demonstrated that PGC-1α coactivates PPARα to increase the expression of mitochondrial FAO enzyme genes including MCAD and M-CPT I (Lehman et al. 2000; Vega et al. 2000). More importantly, forced expression of PGC-1 in cardiac myocytes in culture using an adenoviral expression system and in vivo in transgenic mice demonstrated that PGC-1 induced mitochondrial biogenesis and increased respiration rates (Lehman et al. 2000). Interestingly, PGC-1α was shown to induce *coupled* respiration in cardiac myocytes in culture in contrast to uncoupled respiration in noncardiac tissue such as BAT (Lehman et al. 2000). These latter results suggest that cell context (and presumably the availability of downstream transcription factor partners) dictates the metabolic phenotype of mitochondria induced by the PGC-1α pathway. Last, the expression of the PGC-1α gene is induced in physiologic scenarios related to increased mitochondrial FAO such as fasting (Lehman et al. 2000), diabetes (Finck et al. 2002), and exercise (Goto et al. 2000). Collectively, these results have identified PGC-1α as an inducible, cardiac-enriched coactivator of the PPARα pathway and a critical regulator of mitochondrial capacity in the postnatal heart (Fig. 3).

DERANGEMENTS IN PPARα SIGNALING IN THE HYPERTROPHIED, FAILING, AND HYPOXIC HEART

Differential Regulation of the Mitochondrial FAO Pathway in Physiologic Versus Pathologic Forms of Cardiac Hypertrophic Growth

It is well known that cardiac hypertrophic growth may occur as a purely adaptive form ("physiologic" hypertrophy) or as a response to a pathophysiologic stimulus such as chronic pressure or volume overload ("pathologic" hypertrophy). Physiologic hypertrophy is known to occur during normal postnatal developmental growth of the mammalian heart and in response to exercise training (Blomquist and Saltin 1983; Crisman and Tomanek 1985; White et al. 1987). Physiologic hypertrophic growth is purely adaptive, providing appropriate structural and metabolic adaptation to increased hemodynamic demands. Pathologic forms of cardiac hypertrophic growth, such as occur in response to longstanding pressure overload (e.g., hypertension in humans), also likely

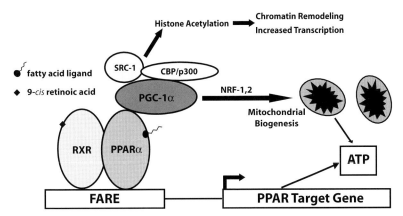

Figure 3. The PPARα/PGC-1α transcriptional regulatory complex. The schematic depicts key components of the PPARα regulatory complex that play a critical role in the activation of FAO enzyme expression and mitochondrial function in the postnatal heart. The nuclear receptor PPARα binds to cognate promoter DNA response elements (FARE; fatty acid response element) with its obligate partner, the retinoid X receptor (RXR). The cardiac-enriched, inducible PPARα coactivator, PGC-1α, interacts with PPARα and recruits other cofactors (including SRC-1 and CBP/P300) necessary to initiate target gene transcription via histone acetylation leading to chromatin remodeling. Formation of the PPARα/RXR dimer, DNA binding, and recruitment of coactivator are influenced by the presence of ligands for PPARα (fatty acids and their metabolites) or RXR (9-cis retinoic acid). PGC-1 also serves to activate other energy metabolic processes including mitochondrial biogenesis via distinct transcription factors, including nuclear respiratory factor (NRF)-1 and 2. Thus, the PPARα regulatory complex serves as a "master" regulator of cardiac ATP production.

occur as an adaptive response. However, the chronic pathologically hypertrophied heart often undergoes remodeling that leads to the development of contractile dysfunction, dilatation, and cardiac rhythm disturbances. Evidence from epidemiologic studies in humans suggests that pathologic hypertrophic growth is associated with increased remodeling, morbidity, and mortality, and therefore can be maladaptive (Casale et al. 1986; Levy et al. 1989). Although significant progress has been made in identifying extracelluar agonists and linked signal transduction events involved in the hypertrophic growth response, little is known about the specific regulatory events that distinguish pathologic from physiologic cardiac hypertrophic growth leading to the remodeling phase.

One approach to delineating the regulatory circuitry distinguishing the two broad forms of cardiac hypertrophic growth is to identify a downstream response that differs between the growth programs. This approach avoids the potential pitfall of identifying common upstream signaling events. Once a distinguishing downstream response is identified, regulatory pathways can be dissected in reverse fashion. The mitochondrial FAO pathway has proven to be a useful starting point for such a strategy. Two initial observations indicate that cardiac metabolism, FAO in particular, is differentially regulated in different forms of cardiac hypertrophic growth. First, inborn errors in mitochondrial FAO enzymes are known to cause a pathologic hypertrophic cardiomyopathic state in humans (Kelly and Strauss 1994). Second, metabolic studies performed in a variety of mammalian species including humans have demonstrated that in acquired forms of pathologic cardiac hypertrophy due to pressure overload, the heart reverts to a reliance on glucose rather than FAO as its chief energy substrate; a reversion to the "fetal" energy substrate utilization pattern (Wittels and Spann 1968; Bishop and Altschuld 1970; Taegtmeyer and Overturf 1988; Christe and Rodgers 1994; Massie et al. 1995). In contrast, physiologic cardiac hypertrophy due to postnatal development or training is not associated with a fetal metabolic shift (Blomquist and Saltin 1983; Crisman and Tomanek 1985; White et al. 1987).

The expression of nuclear genes encoding FAO enzymes has been characterized in the pathologically hypertrophied heart. The results of studies performed in a variety of animal models of ventricular pressure overload have shown that the expression of genes encoding M-CPT I, MCAD, and other β-oxidation spiral enzymes are down-regulated in the pathologically hypertrophied heart in parallel with the shift toward glucose utilization (Sack et al. 1996, 1997; Depre et al. 1998; Barger et al. 2000). In addition, studies of samples from the ventricle of humans with end-stage cardiomyopathy have shown that FAO enzyme gene expression is down-regulated in the failing human heart (Fig. 4A) (Sack et al. 1996).

Deactivation of the PPARα/PGC-1α Complex during Pathologic Cardiac Hypertrophic Growth

The next step was to determine whether, as predicted by the alteration in FAO enzyme expression, the activity of PPARα and its coactivator PGC-1α is altered in the hypertrophied heart in parallel with the metabolic changes. The results of several studies of rodent models of pressure overload hypertrophy have now demonstrated that the cardiac expression of the genes encoding PPARα and PGC-1α are down-regulated within 7 days of the onset of ventricular pressure overload (Fig. 4B) (Sack et al. 1997; Barger et al. 2000). Studies have shown that steady-state nuclear levels of PPARα protein are decreased in the pathologically hypertrophied heart, whereas the levels of its transcriptional antagonist,

COUP-TF, are increased (Sack et al. 1997). In addition, studies performed in cardiac myocytes in culture revealed that the deactivation of PPARα-mediated control of M-CPT I gene transcription following exposure to a hypertrophic agonist occurs within 24 hours, a surprisingly rapid response (Barger et al. 2000). This latter observation suggested that posttranslational mechanisms also served to rapidly deactivate PPARα during hypertrophic growth. Indeed, further studies demonstrated that activation of extracellular regulated kinase-mitogen-activated protein kinase (ERK-MAPK) leads to a rapid reduction in PPARα trans-activation following exposure to hypertrophic agonists such as the α_1-adrenergic agonist phenylephrine. Last, recent studies have shown that in contrast to pressure overload-induced cardiac hypertrophy, the expression of the PPARα and PGC-1α genes is maintained at high levels in physiologic cardiac hypertrophy due to exercise training (A.R. Wende and D.P. Kelly, unpubl.). In summary, the PPARα/PGC-1α complex is deactivated at the level of gene expression and posttranslationally following the onset of pressure overload but not in physiologic forms of hypertrophic growth (Fig. 4C).

Do PPARα-mediated Alterations in Cardiac Energy Metabolism Influence the Cardiac Hypertrophic Phenotype?

Several lines of evidence suggest that alterations in cardiac energy substrate utilization are linked directly to hypertrophic growth programs. First, as noted above, children with inborn errors in the FAO pathway develop cardiac hypertrophy. Second, pharmacologic inhibition of CPT I and other FAO enzymes causes cardiac hypertrophy in cell culture and in vivo (Litwin et al. 1990; Vetter et al. 1995). Several recent studies have added further support to the notion that derangements in myocyte lipid metabolism serve as primary triggers of cardiac hypertrophy. A recent study by Schaffer and coworkers (Chiu et al. 2001) has provided evidence that abnormalities in myocardial lipid homeostasis, such as might occur when the capacity for FAO is diminished, lead to hypertrophic growth. Mice with cardiac-specific overexpression of acyl-CoA synthetase (ACS) develop cardiac hypertrophy associated with neutral lipid accumulation within myocytes. These results suggest that lipid-mediated signaling pathways may trigger a growth response. A second study in humans has suggested that the activity of PPARα may modify the cardiac hypertrophic response. Jamshidi and colleagues (Jamshidi et al. 2002; Kelly 2002) found that a single nucleotide polymorphism within the PPARα gene is associated with the degree of left ventricular hypertrophy due to exercise training in British Army volunteers. In addition, this same study found that PPARα genotype was a determinant of the degree of left ventricular hypertrophy caused by hypertension in a large cohort. Interestingly, the latter association was only observed in males. Taken together, these results suggest that PPARα-mediated alterations in FAO influence the hypertrophic response.

Deactivation of the PPARα/RXRα Complex in the Hypoxic Cardiac Myocyte

A critical cellular adaptive response to conditions of reduced oxygen availability involves the suppression of cellular energy consumption and production (Fahey and Lister 1989; Hochachka et al. 1996). Under hypoxic conditions, decreased oxygen consumption is achieved in part by increasing glycolysis while down-regulating mitochondrial FAO flux (Abdel-aleem et al. 1998; Rumsey et al. 1999). This metabolic switch, which is similar to that of the pathologically hypertrophied heart, would also reduce the generation of potentially toxic reactive species within the mitochondrion in hypoxic conditions. Recently, studies performed with neonatal cardiac myocytes in culture have shown that the PPARα-mediated activation of M-CPT I is diminished following exposure to hypoxic conditions (Huss et al. 2001). Gel mobility shift studies demonstrated that exposure to hypoxia leads to a reduction in PPARα/RXR DNA-binding activity. However, the acute reduction in PPARα/RXR activity is not caused by altered levels of PPARα as occurs in the hypertrophied heart. Rather, the levels of RXRα are reduced by hypoxic exposure. These results indicate that short-term (24 hours) exposure to hypoxic conditions leads to a rapid fall in the availability of RXRα, the obligate PPARα partner, effectively deactivating the PPARα-mediated control of its target genes involved in FAO. In a separate study performed in vivo, longer periods of hypoxia (days) were shown to cause a reduction in PPARα gene expression (Razeghi et al. 2001), identifying a second mechanism whereby hypoxia alters PPARα signaling.

DERANGEMENTS IN MYOCARDIAL ENERGY METABOLISM IN THE DIABETIC HEART: THE ROLE OF PATHOLOGIC ACTIVATION OF THE PPARα/PGC-1α REGULATORY PATHWAY

The PPARα Gene Regulatory Pathway Is Activated in the Diabetic Heart

Diabetes mellitus is associated with increased cardiac morbidity and mortality (Kannel et al. 1974). Cardiomyopathy commonly occurs in diabetics independent of known risk factors such as coronary disease or hypertension (Rubler et al. 1972). Although little is known about the pathogenesis of diabetic cardiomyopathy, evidence is emerging that functional abnormalities are directly related to derangements in myocardial energy metabolism (Rodrigues et al. 1995; Stanley et al. 1997). In diabetes, the capacity of the heart to switch between utilization of fatty acids and glucose is severely constrained because the uptake and utilization of glucose is dependent on an intact insulin signaling pathway (Fig. 1). Accordingly, in the uncontrolled diabetic state, the heart relies almost exclusively on FAO to fulfill its ATP requirements (Rodrigues et al. 1995; Stanley et al. 1997; Belke et al. 2000). The chronic dependence of the diabetic heart on mitochondrial FAO could have detrimental consequences, in-

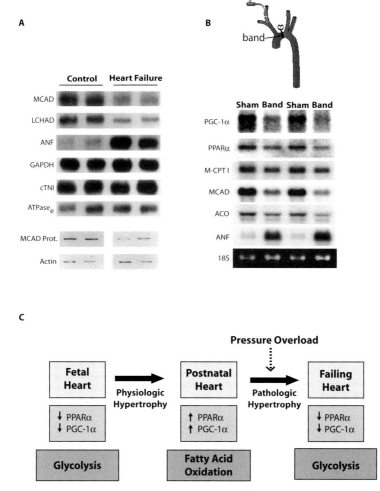

Figure 4. Down-regulation of PPARα and FAO enzyme gene expression in the hypertrophied and failing heart. (*A*) FAO enzyme (MCAD and LCHAD) levels in LV from failing human hearts. Representative autoradiographs of northern (*top*) and western (*bottom two rows*) blot analyses performed with total RNA or protein prepared from LV of (normal LV function) controls and age-matched subjects with heart failure. Northern blot analyses were performed using radiolabeled cDNA probes listed at left. Western blot analyses were performed with a polyclonal anti-MCAD antibody and actin control antibody. (*B*) The PPARα pathway is deactivated in pressure overload-induced hypertrophied mouse heart. Representative autoradiographs of northern blot analyses performed with RNA from the LV of mice 7 days after placement of a constricting band around the transverse aortic arch (Band) or sham operation (Sham). A schematic of the aortic arch banding procedure is shown at top. Abbreviations: (MCAD) medium-chain acyl-CoA dehydrogenase; (LCHAD) 3-OH long-chain acyl-CoA dehydrogenase; (ANF) atrial natriuretic factor; (cTNI) cardiac troponin I; (ATPase$_e$) ATPase subunit e; (M-CPT I) muscle-type carnitine palmitoyltransferase I; (ACO) acyl-CoA oxidase. (*C*) Schematic of the energy substrate and transcriptional switches known to occur during normal postnatal physiologic cardiac hypertrophic growth compared to the pathologic form caused by pressure overload. The boxes at the bottom denote chief source of energy. The deactivation of PPARα and PGC-1 occurs at both transcriptional and posttranscriptional levels as described in the text.

cluding increased myocardial oxygen demand and accumulation of toxic intermediates derived from increased cellular fatty acid uptake and catabolism.

We have recently shown that the myocardial expression of PPARα, PGC-1α, and multiple downstream PPARα target genes involved in mitochondrial FAO is abnormally elevated in the hearts of mice with insulin-deficient or insulin-resistant forms of diabetes (Finck et al. 2002). Interestingly, the expression of the downstream PPARα target genes is not increased in the hearts of diabetic PPARα-null mice (B.N. Finck and D.P. Kelly, unpubl.). These results suggest that the increased capacity of the diabetic heart for fatty acid utilization is mediated, at least in part, by chronic activation of the PPARα gene regulatory pathway. To evaluate the potential role of chronic activation of the PPARα regulatory pathway in the development of the metabolic and functional disturbances of the diabetic heart, a transgenic mouse model with cardiac-restricted overexpression of PPARα was established. This transgenic model allowed the evaluation of the effects of PPARα on cardiac metabolism and function in the absence of systemic abnormalities related to the diabetic state. Mice transgenic for a PPARα cDNA downstream from the cardiac α-myosin heavy-chain promoter (MHC-PPAR mice) were generated (Fig. 5). As predicted, MHC-PPAR mice exhibited increased expression of PPARα target genes involved in the cellular uptake and oxidation of fatty acids (Finck et al. 2002). Interestingly, the expression of genes involved in glucose uptake (GLUT4), glycolysis, and glucose oxidation were coordinately down-regulated. These results provide evidence that primary activation of the fatty acid utilization pathway via PPARα leads to a counter-regulatory reduction in the expression of genes involved in glucose utilization. The mechanism for the transcriptional repression of glucose utilization enzymes by PPARα overexpression is unknown but is likely mediated through indirect pathways, given that the majority of these genes are not known to be direct PPARα targets.

Several lines of evidence demonstrated that, as predicted by the derangements in gene expression, myocardial metabolism is altered in the MHC-PPAR heart. First, micro-positron emission tomographic (micro-PET) studies of cardiac substrate uptake were performed with intact MHC-PPAR mice. The micro-PET studies revealed that the myocardial uptake of the tracer [^{11}C]palmitate was increased and [^{18}F]fluorodeoxyglucose (FDG) import was reduced in MHC-PPAR mouse heart (Fig. 6A). Second, analyses of substrate utilization rates using isolated working hearts from MHC-PPAR and nontransgenic littermate mice demonstrated that palmitate oxidation rates were increased ~65%, whereas glucose oxidation was reduced by over 60% in MHC-PPAR hearts compared to controls (Fig. 6B). Taken together with the gene expression data, these results indicate that the metabolic phenotype of the MHC-PPAR heart is remarkably similar to that of the diabetic heart.

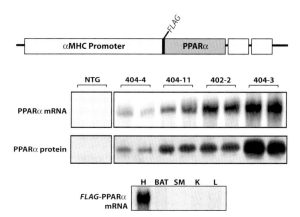

Figure 5. Generation of transgenic mouse lines that overexpress PPARα in a cardiac-restricted manner (MHC-PPAR mice). The schematic at the top depicts the MHC-PPAR construct used to generate transgenic mice with cardiac-specific overexpression of PPARα. A *FLAG*-tagged PPARα cDNA was inserted downstream from the α-cardiac myosin heavy-chain promoter. Four independent lines of mice transgenic for the MHC-PPAR construct were established (*404-4, 404-11, 402-2, 404-3*). Representative autoradiographs of northern (*top*) and western (*middle*) blot studies performed with heart samples from 6-week-old mice from each transgenic line are displayed. At the exposure shown, endogenous PPARα could not be detected in nontransgenic (NTG) samples. *FLAG*-PPARα protein was detected in the transgenic mice using an antibody directed against PPARα. Cardiac-specific expression of FLAG-PPARα mRNA was confirmed by northern blot analysis performed with multiple tissues. Abbreviations: (NTG) nontransgenic; (H) heart; (BAT) brown adipose tissue; (SM) skeletal muscle; (K) kidney; (L) liver.

MHC-PPAR Mice Provide Evidence for a Link between Metabolic and Functional Derangements in the Diabetic Heart

The MHC-PPAR mice afforded the opportunity to determine whether altered myocardial metabolism due to chronic activation of the PPARα pathway leads to functional abnormalities known to occur in the diabetic heart (diabetic cardiomyopathy). The initial stages of diabetic cardiomyopathy are characterized by ventricular hypertrophy and diastolic dysfunction. In severe cases, cardiac abnormalities can progress to systolic ventricular dysfunction and overt congestive heart failure. Indeed, MHC-PPAR mice exhibit an increased biventricular weight/body weight ratio in a pattern dependent on the level of transgene expression (Finck et al. 2002). In addition, MHC-PPAR mice exhibit reduced ventricular systolic function and chamber dilatation as determined by echocardiography (Fig. 7). The degree of ventricular dysfunction in MHC-PPAR mice was worsened when the mice were rendered insulin-deficient by administration of the pancreatic islet cell toxin streptozotocin (B.N. Finck and D.P. Kelly, unpubl.). In summary, the MHC-PPAR mice develop both metabolic and functional characteristics of the diabetic heart, indicating that primary derangements in mitochondrial substrate utilization can lead to functional consequences. Although the heightened rate of mitochondrial FAO in the diabetic heart likely represents an adaptive response in the short term, the dysfunctional phenotype of the MHC-PPAR heart suggests that this metabolic alteration may become maladaptive, contributing to the development of diabetic cardiomyopathy.

The Potential Role of "Lipotoxicity" in the Development of Cardiomyopathy in MHC-PPAR Mice: Relevance to the Diabetic State

How do the metabolic derangements of the MHC-PPAR heart or the diabetic heart lead to cardiac hypertrophy and dysfunction? A clue was provided by the observation that in hearts of diabetic animals, histologic analyses often reveal evidence of lipid droplet accumulation within myocytes (Murthy and Shipp 1977; Paulson and Crass 1982; Zhou et al. 2000). Indeed, recent evidence indicates that lipid accretion in the myocyte may trigger hypertrophic growth and apoptosis (Chiu et al. 2001). It is possible that the increased rate of myocardial fatty acid uptake, especially in the context of elevated circulating plasma lipids as seen in the uncontrolled diabetic, outpaces the increased oxidative capacity of the diabetic cardiomyocyte. This possibility was evaluated in MHC-PPAR mice by subjecting them to a short-term fast to acutely elevate circulating free fatty acid levels. Histologic studies using oil red O staining and quantitative

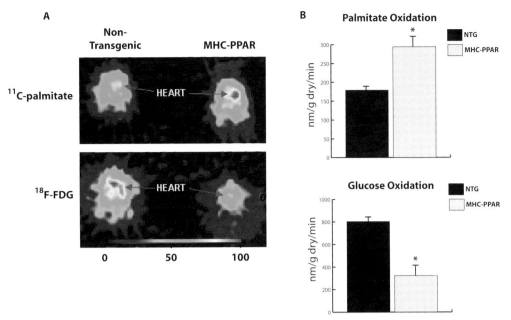

Figure 6. Myocardial palmitate utilization is increased and glucose utilization reduced in MHC-PPAR mice. (*A*) Panels at left contain representative images of ^{11}C-palmitate and ^{18}F-fluorodeoxyglucose (FDG) uptake into myocardium as determined by micropositron emission tomography (microPET) in nontransgenic and MHC-PPAR littermate mice. The relative amount of tracer uptake into the mouse heart 15 seconds after bolus injection of ^{11}C-palmitate or ^{18}F-FDG into the jugular vein is indicated by the color scale (0–100). The arrow indicates the cardiac field. As shown in the upper panel, the color field is increased into the red scale in the hearts of MHC-PPAR mice infused with ^{11}C-palmitate, which is indicative of enhanced myocardial uptake of FA. Conversely, uptake of ^{18}F-FDG is substantially lower in hearts of MHC-PPAR mice compared to NTG littermates. (*B*) Myocardial palmitate oxidation is increased and glucose oxidation reduced in MHC-PPAR mice. The oxidation of [9,10-^{3}H]palmitate and [U-^{14}C]glucose was assessed in isolated working hearts of MHC-PPAR or nontransgenic littermate mice. Bars represent mean (± S.E.M.) oxidation rates expressed as nmole of substrate oxidized/min/g dry mass. *p<0.05 versus nontransgenic littermate mice. (*A*, Reprinted, with permission, from Finck et al. 2002.)

analyses of myocardial triacylglyceride (TAG) using electrospray ionization mass spectrometry (ESIMS) revealed markedly increased TAG levels in fasted MHC-PPAR mice (Fig. 8) (Finck et al. 2002). A similar pattern of myocardial lipid accumulation was observed in diabetic wild-type mice, albeit of lesser magnitude. Strikingly, the combination of insulin deficiency and the MHC-PPAR genotype leads to a massive increase in myocardial TAG content (B.N. Finck and D.P. Kelly, unpubl.). These results suggest that abnormalities in myocardial lipid balance lead to ventricular hypertrophy and dysfunction in the diabetic heart.

CONCLUSIONS

The adult mammalian heart meets the high energy demands of the postnatal environment via several high-capacity ATP-generating pathways. As described herein, the maturation of the cardiac mitochondrial FAO system occurs during the postnatal period. Under normal postnatal conditions, the mitochondria provide the majority of ATP through the oxidation of fatty acids. The expression of nuclear genes encoding cellular enzymes and proteins involved in the uptake, thioesterification, and mitochondrial oxidation of fatty acids is coordinately controlled by

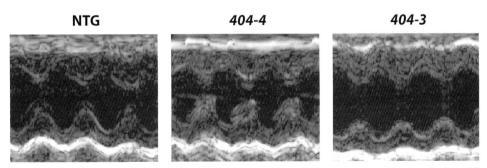

Figure 7. Cardiac dysfunction and myocardial lipid accumulation in MHC-PPAR mice. Ventricular dysfunction in MHC-PPAR mice is related to transgene expression level. Representative two-dimensional guided M-mode echocardiographic images of the left ventricle obtained from the parasternal view at the midventricular level of nontransgenic (NTG) and low- (*404-4*) or high- (*404-3*) expressing transgenic lines of female mice at 2 months of age. Ventricular dysfunction is apparent in the *404-3* line.

the PPARα transcriptional regulatory complex. This complex is activated by increased dietary fatty acids that serve as ligands for PPARα coincident with increased expression of PPARα and its coactivator PGC-1α following birth. The PPARα complex not only controls the high basal expression of FAO target genes in heart, but also serves to dynamically regulate the capacity for mitochondrial and peroxisomal FAO in response to dietary and physiologic stimuli. In summary, the PPARα/RXR/PGC-1α transcriptional regulatory complex serves to transduce energy demands and substrate availability to the expression of genes involved in the generation of ATP via oxidation of fatty acids.

The discovery of PGC-1α has also provided new insight into the link between transcriptional control of the FAO pathway and the broad program of mitochondrial biogenesis. PGC-1α is now recognized as a global regulator of cellular energy metabolism via its ability to coactivate a variety of transcription factors. PGC-1α promotes an increase in mitochondrial number and function in heart during cardiac postnatal development via coactivation of nuclear respiratory factor-1, PPARα, and probably additional factors. The inducibility of PGC-1α allows it to respond rapidly to physiologic and dietary conditions that dictate increased cellular energy demands.

The balance of substrate choices available for the normal adult heart becomes constrained in certain disease states. In pathologic forms of cardiac hypertrophy due to chronic pressure overload and in the hypoxic or ischemic heart, the activity of the PPARα complex is down-regulated, leading to a fall in FAO rates and the re-emergence of glucose as the chief energy substrate. This shift away from mitochondrial FAO likely serves to reduce oxygen consumption and the potential for generation of reactive intermediates within the mitochondrion. This notion is supported by the results of a recent study demonstrating that treatment of rats with PPARα activators during pressure overload hypertrophy leads to ventricular dysfunction (Young et al. 2001). It is possible, however, that long-term reduction in mitochondrial FAO capacity becomes deleterious due to energy starvation or alterations in myocardial lipid balance. Future studies with murine gain-of-function and loss-of-function models should prove useful to further test this hypothesis.

In contrast to the pathologically hypertrophied and hypoxic heart, the diabetic myocardium exhibits derangements in glucose utilization leading to increased fatty acid utilization mediated at least in part by chronic activation of the PPARα pathway. Activation of PPARα likely comprises an initial adaptive response in the diabetic heart to provide adequate ATP production via a single substrate. However, recent evidence stemming from studies of the MHC-PPAR mouse model of diabetic cardiomyopathy suggests that over the long term, excessive activation of the PPARα/PGC-1α complex leads to ventricular hypertrophy and dysfunction.

FUTURE DIRECTIONS

Recent progress in our understanding of the molecular regulatory mechanisms involved in the control of cardiac mitochondrial energy production has opened new avenues for future investigative efforts. First, the precise mechanisms whereby the transcriptional coactivator PGC-1α and related family members control the broad program of mitochondrial function should be further characterized. It is likely that additional, as-yet-unidentified PGC-1α interacting partners exist and contribute to the control of mitochondrial biogenesis and function in the developing and adult heart. Second, it will be important to identify and characterize additional upstream signaling pathways that converge on PGC-1α, PPARα, and RXR in response to diverse physiologic conditions. Third, given that PPARα and its heterodimeric partner RXR are ligand-activated transcription factors, each represents potential targets for the development of novel therapeutics aimed at the prevention or treatment of heart failure. However, as indicated above, additional studies are necessary to determine whether the alterations in PPARα signaling known to occur in various myocardial diseases such as cardiac hypertrophy are adaptive or mal-

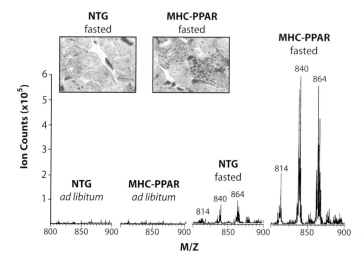

Figure 8. Myocardial TAG accumulation in MHC-PPAR mice. Photomicrographs depict the histologic appearance of ventricular tissue sections (stained with oil red O) from NTG and MHC-PPAR mice following a 24-hour fast. The stained droplets represent neutral lipid deposits. The spectra depict the lipid profile of mouse ventricle samples prepared from MHC-PPAR or NTG mice given ad libitum access to food or fasted for 24 hours. Lipid species were separated and analyzed using electrospray ionization mass spectrometry (ESIMS). Mass/charge (m/z) ratios of 814, 840, and 864 denote TAGs containing fatty acyl groups containing chain lengths of 16:0/16:0/16:0, 16:0/16:0/18:1, and 16:0/18:1/18:2, respectively.

adaptive. The most compelling data for a PPARα-mediated maladaptive response is in the diabetic heart, where chronic activation appears to contribute to the cardiomyopathy. Future studies using loss-of-function and gain-of-function strategies in genetically engineered mice should prove useful in determining whether the myocardial switches in energy substrate utilization are beneficial or contribute to pathologic remodeling, and ultimately, heart failure. Last, recent evidence suggests that components of the PPAR regulatory complex and its downstream target genes are candidates for genomic modifiers of the cardiac hypertrophic phenotype. Useful genomic markers predictive of cardiac remodeling or the response to therapeutic approaches is a major goal of current cardiovascular research. The genes described herein should be added to the list of candidate modifier genes relevant to cardiovascular disease.

ACKNOWLEDGMENTS

This manuscript is dedicated to the current and previous members of the Kelly laboratory involved in the studies described herein. Special thanks to Mary Wingate for assistance with the preparation of this manuscript. This work was supported by the following National Institutes of Health grants: RO1 DK-45416, RO1 HL-58493, P50 HL-61006, PO1 HL-57278, P30 DK-52574, and P30 DK-56341.

REFERENCES

Abdel-aleem S., St. Louis J., Hendrickson S.C., El-Shewy H.M., El-Dawy K., Taylor D.A., and Lowe J.E. 1998. Regulation of carbohydrate and fatty acid utilization by L-carnitine during cardiac development and hypoxia. *Mol. Cell. Biol.* **180:** 95.

Aoyama T., Peters J.M., Iritani N., Nakajima T., Furihata K., Hashimoto T., and Gonzalez F.J. 1998. Altered constitutive expression of fatty acid-metabolizing enzymes in mice lacking the peroxisome proliferator-activated receptor α (PPARα). *J. Biol. Chem.* **273:** 5678.

Barger P.M. and Kelly D.P. 2000. PPAR signaling in the control of cardiac energy metabolism. *Trends Cardiovasc. Med.* **10:** 238.

Barger P.M., Brandt J.M., Leone T.C., Weinheimer C.J., and Kelly D.P. 2000. Deactivation of peroxisome proliferator-activated receptor-α during cardiac hypertrophic growth. *J. Clin. Invest.* **105:** 1723.

Belke D.D., Larsen T.S., Gibbs E.M., and Severson D.L. 2000. Altered metabolism causes cardiac dysfunction in perfused hearts from diabetic (db/db) mice. *Am. J. Physiol.* **279:** E1104.

Bing R.J. 1955. The metabolism of the heart. *Harvey Lect.* **50:** 27.

Bishop S.P. and Altschuld R.A. 1970. Increased glycolytic metabolism in cardiac hypertrophy and congestive failure. *Am. J. Physiol.* **218:** 153.

Blomquist C.G. and Saltin B. 1983. Cardiovascular adaptations to physical training. *Annu. Rev. Physiol.* **45:** 169.

Brandt J., Djouadi F., and Kelly D.P. 1998. Fatty acids activate transcription of the muscle carnitine palmitoyltransferase I gene in cardiac myocytes via the peroxisome proliferator-activated receptor α. *J. Biol. Chem.* **273:** 23786.

Campbell F.M., Kozak R., Wagner A., Altarejos J.Y., Dyck J.R.B., Belke D.D., Severson D.L., Kelly D.P., and Lopaschuk G.D. 2002. A role for PPARα in the control of cardiac malonyl-CoA levels: Reduced fatty acid oxidation rates and increased glucose oxidation rates in the hearts of mice lacking PPARα are associated with higher concentrations of malonyl-CoA and reduced expression of malonyl-CoA decarboxylase. *J. Biol. Chem.* **277:** 4098.

Casale P.N., Devereux R.B., Milner M., Zullo G., Harshfield G.A., Pickering T.G., and Laragh J.H. 1986. Value of echocardiographic measurement of left ventricular mass in predicting cardiovascular morbid events in hypertensive men. *Ann. Int. Med.* **105:** 173.

Chiu H.-C., Kovacs A., Ford D.A., Hsu F.-F., Garcia R., Herrero P., Saffitz J.E., and Schaffer J.E. 2001. A novel mouse model of lipotoxic cardiomyopathy. *J. Clin. Invest.* **107:** 813.

Christe M.D. and Rodgers R.L. 1994. Altered glucose and fatty acid oxidation in hearts of the spontaneously hypertensive rat. *J. Mol. Cell. Cardiol.* **26:** 1371.

Cresci S., Wright L.D., Spratt J.A., Briggs F.N., and Kelly D.P. 1996. Activation of a novel metabolic gene regulatory pathway by chronic stimulation of skeletal muscle. *Am. J. Physiol.* **270:** C1413.

Crisman R.P. and Tomanek R.J. 1985. Exercise training modifies myocardial mitochondria and myofibril growth in spontaneously hypertensive rats. *Am. J. Physiol.* **248:** H8.

Depre C., Shipley G.L., Chen W., Han Q., Doenst T., Moore M.L., Stepkowski S., Davies P.J.A., and Taegtmeyer H. 1998. Unloaded heart *in vivo* replicates fetal gene expression of cardiac hypertrophy. *Nat. Med.* **4:** 1269.

Desvergne B. and Wahli W. 1999. Peroxisome proliferator-activated receptors: Nuclear control of metabolism. *Endocr. Rev.* **20:** 649.

Disch D.L., Rader T.A., Cresci S., Leone T.C., Barger P.M., Vega R., Wood P.A., and Kelly D.P. 1996. Transcriptional control of a nuclear gene encoding a mitochondrial fatty acid oxidation enzyme in transgenic mice: Role for nuclear receptors in cardiac and brown adipose expression. *Mol. Cell. Biol.* **16:** 4043.

Djouadi F., Weinheimer C.J., Saffitz J.E., Pitchford C., Bastin J., Gonzalez F.J., and Kelly D.P. 1998. A gender-related defect in lipid metabolism and glucose homeostasis in peroxisome proliferator-activated receptor α-deficient mice. *J. Clin. Invest.* **102:** 1083.

Fahey J.T. and Lister G. 1989. Response to low cardiac output: Developmental differences in metabolism during oxygen deficit and recovery in lambs. *Pediatr. Res.* **26:** 180.

Finck B., Lehman J.J., Leone T.C., Welch M.J., Bennett M.J., Kovacs A., Han X., Gross R.W., Kozak R., Lopaschuk G.D., and Kelly D.P. 2002. The cardiac phenotype induced by PPARα overexpression mimics that caused by diabetes mellitus. *J. Clin. Invest.* **109:** 121.

Goto M., Terada S., Kato M., Katoh M., Yokozeki T., Tabata I., and Shimokawa T. 2000. cDNA cloning and mRNA analysis of PGC-1 in epitrochlearis muscle in swimming-exercised rats. *Biochem. Biophys. Res. Commun.* **274:** 350.

Hochachka P.W., Buck L.T., Doll C.J., and Land S.C. 1996. Unifying theory of hypoxia tolerance: Molecular/metabolic defense and rescue mechanisms for surviving oxygen lack. *Proc. Natl. Acad. Sci.* **93:** 9493.

Huss J.M., Levy F.H., and Kelly D.P. 2001. Hypoxia inhibits the PPARα/RXR gene regulatory pathway in cardiac myocytes. *J. Biol. Chem.* **276:** 27605.

Issemann I. and Green S. 1990. Activation of a member of the steroid hormone receptor superfamily by peroxisome proliferators. *Nature* **347:** 645.

Jamshidi Y., Montgomery H.E., Hense H.-W., Myerson S.G., Torra I.P., Staels B., World M.J., Doering A., Erdmann J., Hengstenberg C., Humphries S.E., Schunkert H., and Flavell D.M. 2002. The PPARα gene regulates left ventricular growth in response to exercise and hypertension. *Circulation* **105:** 950.

Kannel W.B., Hjortland M., and Castelli W.P. 1974. Role of diabetes in congestive heart failure: The Framingham study. *Am. J. Cardiol.* **34:** 29.

Kelly D.P. 2002. Peroxisome proliferator-activated receptor α as a genetic determinant of cardiac hypertrophic growth. Culprit or innocent bystander? *Circulation* **105:** 1025.

Kelly D.P. and Strauss A.W. 1994. Inherited cardiomyopathies. *N. Engl. J. Med.* **330:** 913.

Kelly D.P., Gordon J.I., Alpers R., and Strauss A.W. 1989. The tissue-specific expression and developmental regulation of the two nuclear genes encoding rat mitochondrial proteins: Medium-chain acyl-CoA dehydrogenase and mitochondrial malate dehydrogenase. *J. Biol. Chem.* **264:** 18921.

Kersten S., Seydoux J., Peters J.M., Gonzalez F.J., Desvergne B., and Wahli W. 1999. Peroxisome proliferator-activated receptor α mediates the adaptive response to fasting. *J. Clin. Invest.* **103:** 1489.

Lee S.S.T., Pineau T., Drago J., Lee E.J., Owens J.W., Kroetz D.L., Fernandez-Salguero P.M., Westphal H., and Gonzalez F.J. 1995. Targeted disruption of the α isoform of the peroxisome proliferator-activated receptor gene in mice results in abolishment of the pleiotropic effects of peroxisome proliferators. *Mol. Cell. Biol.* **15:** 3012.

Lehman J.J., Barger P.M., Kovacs A., Saffitz J.E., Medeiros D., and Kelly D.P. 2000. PPARγ coactivator-1 (PGC-1) promotes cardiac mitochondrial biogenesis. *J. Clin. Invest.* **106:** 847.

Leone T.C., Weinheimer C.J., and Kelly D.P. 1999. A critical role for the peroxisome proliferator-activated receptor alpha (PPARα) in the cellular fasting response: The PPARα-null mouse as a model of fatty acid oxidation disorders. *Proc. Natl. Acad. Sci.* **96:** 7473.

Levy D., Garrison R.J., Savage D.D., Kannel W.B., and Castelli W.P. 1989. Left ventricular mass and incidence of coronary heart disease in an elderly cohort. The Framingham heart study. *Ann. Int. Med.* **110:** 101.

Litwin S.E., Raya T.E., Gay R.G., Bedotto J.B., Bahl J.J., Anderson P.G., Goldman S., and Bressler R. 1990. Chronic inhibition of fatty acid oxidation: New model of diastolic dysfunction. *Am. J. Physiol.* **258:** H51.

Lockwood E.A. and Bailey E. 1970. Fatty acid utilization during development of the rat. *Biochem. J.* **120:** 49.

Massie B.M., Schaefer S., Garcia J., McKirnan M.D., Schwartz G.G., Wisneski J.A., Weiner M.W., and White F.C. 1995. Myocardial high-energy phosphate and substrate metabolism in swine with moderate left ventricular hypertrophy. *Circulation* **91:** 1814.

Murthy V.K. and Shipp J.C. 1977. Accumulation of myocardial triacylglycerols in ketotic diabetes. *Diabetes* **26:** 222.

Nagao M., Parimoo B., and Tanaka K. 1993. Developmental, nutritional, and hormonal regulation of tissue-specific expression of the genes encoding various acyl-CoA dehydrogenases and α-subunit of electron transfer flavoprotein in rat. *J. Biol. Chem.* **268:** 24114.

Neely J.R., Rovetto M.J., and Oram J.F. 1972. Myocardial utilization of carbohydrate and lipids. *Prog. Cardiovasc. Dis.* **15:** 289.

Paulson D.J. and Crass M.F. 1982. Endogenous triacylglycerol metabolism in diabetic heart. *Am. J. Physiol.* **242:** 1084.

Puigserver P., Wu Z., Park C.W., Graves R., Wright M., and Spiegelman B.M. 1998. A cold-inducible coactivator of nuclear receptors linked to adaptive thermogenesis. *Cell* **92:** 829.

Razeghi P., Young M.E., Abbasi S., and Taegtmeyer H. 2001. Hypoxia in vivo decreases peroxisome proliferator-activated receptor alpha-regulated gene expression in rat heart. *Biochem. Biophys. Res. Commun.* **287:** 5.

Rodrigues B., Cam M.C., and McNeill J.H. 1995. Myocardial substrate metabolism: Implications for diabetic cardiomyopathy. *J. Mol. Cell. Cardiol.* **27:** 169.

Rubler S., Dlugash J., Yuceoglu Y.Z., Kumral T., Branwood A.W., and Grishman A. 1972. New type of cardiomyopathy associated with glomerulosclerosis. *Am. J. Cardiol.* **30:** 595.

Rumsey W.L., Abbott B., Bertelsen D., Mallamaci M., Hagan K., Nelson D., and Erecinska M. 1999. Adaptation to hypoxia alters energy metabolism in rat heart. *Am. J. Physiol.* **276:** H71.

Sack M.N., Disch D.L., Rockman H.A., and Kelly D.P. 1997. A role for Sp and nuclear receptor transcription factors in a cardiac hypertrophic growth program. *Proc. Natl. Acad. Sci.* **94:** 6438.

Sack M.N., Rader T.A., Park S., Bastin J., McCune S.A., and Kelly D.P. 1996. Fatty acid oxidation enzyme gene expression is downregulated in the failing heart. *Circulation* **94:** 2837.

Schulz H. 1991. Beta oxidation of fatty acids. *Biochim. Biophys. Acta* **1081:** 109.

Stanley W.C., Lopaschuk G.D., and McCormack J.G. 1997. Regulation of energy substrate metabolism in the diabetic heart. *Cardiovasc. Res.* **34:** 25.

Taegtmeyer H. and Overturf M.L. 1988. Effects of moderate hypertension on cardiac function and metabolism in the rabbit. *Hypertension* **11:** 416.

Vega R.B., Huss J.M., and Kelly D.P. 2000. The coactivator PGC-1 cooperates with peroxisome proliferator-activated receptor α in transcriptional control of nuclear genes encoding mitochondrial fatty acid oxidation enzymes. *Mol. Cell. Biol.* **20:** 1868.

Vetter R., Kott M., and Rupp H. 1995. Differential influences of carnitine palmitoyltransferase-1 inhibition and hyperthyroidism on cardiac growth and sarcoplasmic reticulum phosphorylation. *Eur. Heart J.* **16:** 15.

Wallace D.C. 1999. Mitochondrial diseases in man and mouse. *Science* **283:** 1482.

Warshaw J.B. 1972. Cellular energy metabolism during fetal development. IV. Fatty acid activation, acyl transfer and fatty acid oxidation during development of the chick and rat. *Dev. Biol.* **28:** 537.

White F.C., McKirnan M.D., Breisch E.A., Guth B.D., Liu Y.M., and Bloor C.M. 1987. Adaptation of the left ventricle to exercise-induced hypertrophy. *J. Appl. Physiol.* **62:** 1097.

Wittels B. and Spann J.F., Jr. 1968. Defective lipid metabolism in the failing heart. *J. Clin. Invest.* **47:** 1787.

Wu Z., Puigserver P., Andersson U., Zhang C., Adelmant G., Mootha V., Troy A., Cinti S., Lowell B., Scarpulla R.C., and Spiegelman B.M. 1999. Mechanisms controlling mitochondrial biogenesis and respiration through the thermogenic coactivator PGC-1. *Cell* **98:** 115.

Xu H.E., Lambert M.H., Montana V.G., Parks D.J., Blanchard S.G., Brown P.J., Sternbach D.D., Lehmann J.M., Wisely G.B., Willson T.M., Kliewer S.A., and Milburn M.V. 1999. Molecular recognition of fatty acids by peroxisome proliferator-activated receptors. *Mol. Cell* **3:** 397.

Young M.E., Laws F.A., Goodwin G.W., and Taegtmeyer H. 2001. Reactivation of peroxisome proliferator-activated receptor alpha is associated with contractile dysfunction in hypertrophied rat heart. *J. Biol. Chem.* **276:** 44390.

Zhou Y.T., Grayburn P., Karim A., Shimabukuro M., Higa M., Baetens D., Orci L., and Unger R.H. 2000. Lipotoxic heart disease in obese rats: Implications for human disease. *Proc. Natl. Acad. Sci.* **97:** 1784.

Molecular Epidemiology of Hypertrophic Cardiomyopathy

H. MORITA,* S.R. DEPALMA,* M. ARAD,* B. MCDONOUGH,* S. BARR,* C. DUFFY,*
B.J. MARON,† C.E. SEIDMAN,*‡ AND J.G. SEIDMAN*

*Department of Genetics, Harvard Medical School and Howard Hughes Medical Institute, Boston,
Massachusetts 02115; †Minneapolis Heart Institute Foundation, Minneapolis, Minnesota 55407;
‡Cardiovascular Division, Brigham and Women's Hospital, Boston, Massachusetts 02115*

Hypertrophic cardiomyopathy (HCM) is a genetically heterogeneous autosomal dominant trait, in which affected individuals develop left ventricular hypertrophy or thickening of the heart wall (Fig. 1) (for reviews, see Fatkin et al. 2000; Towbin 2000; Seidman and Seidman 2001). Although affected individuals may live for many years without obvious clinical signs of their disease, many of these individuals go on to develop chest pain, dyspnea, palpitations, and even sudden death. Associated with cardiac hypertrophy, heart muscle from affected individuals demonstrates myocyte disarray and fibrosis (Fig. 2). Eleven different disease genes have been identified, including β-cardiac myosin (MHC) (Geisterfer-Lowrance et al. 1990), cardiac actin (Olson et al. 2000), α-tropomyosin (Thierfelder et al. 1994), cardiac troponin T (TNT; Thierfelder et al. 1994), cardiac myosin-binding protein C (MyBP-C; Watkins et al. 1995), essential myosin light chain (MLC; Poetter et al. 1996), regulatory myosin light chain (Poetter et al. 1996), cardiac troponin I (TNI; Kimura et al. 1997), α-cardiac myosin (Niimura et al. 2002), and titin (Gerull et al. 2002). Each of these disease genes encodes a component of the cardiac sarcomere (Fig. 3). Multiple different mutations have been identified in each gene (Table 1). That is, HCM displays both intergenic and intragenic heterogeneity. The distribution of mutations (Table 1) found in HCM patients is poorly defined, because relatively small numbers of HCM patient genomes have been characterized. Each mutated gene encodes a defective polypeptide, or poison polypeptide, that becomes incorporated into the growing sarcomere thereby creating a defective sarcomere, and these defective sarcomeres initiate the hypertrophic response (Seidman and Seidman 2001). We imagine that only a finite number of mutations can create poison polypeptides which can induce a hypertrophic response; we are attempting to define the full catalog of mutations that can cause HCM. We have also demonstrated that different HCM populations have different distributions of HCM-causing mutations (Niimura et al. 1998, 2002). For example, the distribution of disease-causing mutations found in individuals with early-onset disease is different from the distribution found in individuals with elderly-onset disease. Neither the number of HCM disease genes nor the number of HCM-causing mutations that can occur in each gene is known.

The identification of disease-causing mutations in large numbers of HCM patients is technically challenging. Because the mutated genes encode poison polypeptides, we and other workers hypothesize that only the coding portions of disease genes need to be screened for disease-causing mutations. Nevertheless, more than 100,000 base pairs of DNA, which code for more than 35,000 amino acids, need to be screened from each indi-

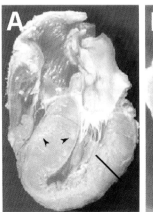
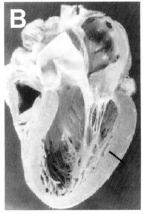

Figure 1. Hearts from an HCM patient (*A*) and an unaffected individual (*B*). Note the thickness of the left ventricular free wall (*line*) as well as fibrotic material in the interventricular septum and free wall of heart from an HCM patient (*arrow*).

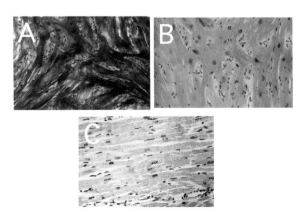

Figure 2. Light micrographs of sections taken from hearts of an HCM patient (*A,B*) and an unaffected individual (*C*). Sections were stained with Masson's Tricombe (*A*) and hematoxylin and eosin (*B,C*) and magnified 200-fold.

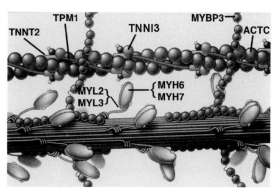

Figure 3. Model of the cardiac sarcomere. Mutations in the genes encoding thin filament proteins (*top*) cardiac troponin I (TNNI3), α-tropomyosin (TPM1), cardiac actin (ACTC), and thick filament proteins (*bottom*) β-cardiac myosin heavy chain (βMHC), α-cardiac myosin heavy chain (αMHC), regulatory myosin light chain (MYL2), and essential myosin light chain (MYL3), cause HCM. Cardiac myosin binding protein C (MYBP3) associates with both thick and thin filament. Mutations in titin and myosin light-chain kinase (not shown) can also cause HCM.

vidual in order to search all known HCM disease genes for mutations (Table 1). A variety of methods, including single-strand conformation polymorphism, denaturing gradient gel electrophoresis, and denaturing high-pressure liquid chromatography, have been proposed to identify mutations that cause autosomal dominant disease (for a discussion of these various methods, see Dracopoli et al. 2002). The concern with these techniques is that none of them detects all genetic variants with absolute certainty. That is, investigators using each of these methods have demonstrated that some variant sequences cannot be distinguished from normal sequences. High-throughput DNA sequencing detects almost all sequence variants. Unfortunately, the cost of screening large numbers of DNA samples is high both because DNA sample preparation requires expensive reagents and because reading DNA sequence traces is labor-intensive. Nevertheless, we have elected to use high-throughput DNA sequence analyses to search for HCM-causing mutations because of the high mutation detection rate of this method.

The extreme genetic heterogeneity of HCM is presumed to occur because affected individuals have a reduced life expectancy, hence over the course of many years such mutations are lost from the population (for review, see Towbin 2000; Watkins 2000; Seidman and Seidman 2001). No strong founder effect has been demonstrated for any known HCM-causing mutation. Rather, instances of de novo mutations arising in affected families have been documented. Poison polypeptides can be encoded by two different types of mutationally altered genes. One class of poison polypeptides are caused by missense mutations that alter only a single nucleotide. To date, mutations in β-cardiac MHC and α-tropomyosin all fall in this category. The other class of poison polypeptides are encoded by mutations that cause protein truncation. Many cardiac MyBP-C and cardiac troponin T mutations are of this type. These mutations frequently alter sequences that control RNA splicing. Because we hypothesize that the number of different mutational alterations that can create poison polypeptides by creating missense peptides is finite, we hypothesize that if we sequence the β-cardiac MHC and α-tropomyosin genes of enough affected individuals we will define the complete set of simple genetic alterations that can generate functional dominant-acting mutationally altered sarcomere proteins. The complete set of mutations in these sarcomere protein genes that can cause HCM may eventually provide clues as to the mechanism by which these mutant proteins cause disease, as well as allow the production of useful diagnostic tools for affected individuals.

Most previous searches of HCM-causing mutations have been in patients from tertiary referral centers (for discussion, see Spirito et al. 1997; Maron 2002). These individuals, and often one or more family members, demonstrate clinical symptoms associated with HCM. Here we have attempted to search for HCM-causing mutations among individuals who are either self-referred or referred by local cardiologists and may thus constitute a different population of HCM patients. Because these individuals may have milder disease than individuals who are in tertiary care centers and may have life-threatening clinical features, the distribution of HCM-causing mutations in this population may be different from the distribution of mutations found in previous studies. We may also generate a rough estimate of the total number of HCM-causing mutations and the fraction of these mutations which have already been discovered.

Table 1. Distribution of HCM-causing Mutations

Gene name	Protein	Mut.[a]	Percent	Coding (AA)	RNA (NT)
MYLK	myosin light-chain kinase	2	<2	1910	5926
ACTC	cardiac actin	5	<5	376	1549
MYBP3	cardiac MyBP C	53	15–20	1275	4200
MYH6	α-cardiac MHC	1	<1	1938	5820
MYH7	β-cardiac MHC	84	25–30	1934	6008
MYL2	regulatory MLC	8	<1	165	813
MYL3	ventricular MLC	4	<1	196	872
TNNI3	cardiac troponin I	12	~10	209	2073
TNNT2	cardiac troponin T	14	10–15	287	1184
TPM1	α-tropomyosin	7	<5	283	1633
TTN	titin	2	<1	26927	82027

[a]The number of mutations (Mut.) found in each HCM-causing disease gene. The percentage (Percent) of mutations found in the disease gene based on earlier studies (Watkins et al. 1992; Fatkin et al. 2000), as well as the size of the encoded protein and RNA transcript are indicated.

RESULTS

Sixty-two unrelated HCM probands were selected for study. DNA samples and clinical information, including echocardiogram and EKG, were obtained from these individuals. HCM was generally diagnosed in these individuals by echocardiography; these individuals have maximum left ventricular wall thickness (LVWT) > 1.3 cm. They also did not have evidence of other conditions (such as untreated hypertension, congenital heart disease, or known coronary artery disease) that might have caused their ventricular hypertrophy.

The complete coding portions of eight HCM disease genes—cardiac actin, regulatory MLC, essential MLC, cardiac MyBP-C, cardiac troponin T, cardiac troponin I, α-tropomyosin, and β-cardiac MHC (head region only)—have been searched for disease-causing mutations in these 62 probands' DNA. Only exons encoding the head or head–rod junction of β-cardiac MHC were sequenced because, until recently (Blair et al. 2002), other studies (MacRae et al. 1994) have failed to find disease-causing mutations in exons encoding the rod portion. DNA corresponding to each of the coding exons of these genes was PCR-amplified and subjected to DNA sequence analyses (Table 2). From each proband DNA, 108 separate exons and flanking sequence were amplified and subjected to sequence analyses. About 22,500 base pairs were analyzed from these 108 amplified DNA fragments for each individual, or roughly 1.4 million base pairs in all. Multiple sequence variants, both disease-causing and non-disease-causing, were identified by this approach. Sequence variants identified in these studies were all confirmed by restriction enzyme digestion of PCR-amplified DNA fragments.

Because sequence variants are found among most genes in any collection of individuals, and because the collection of HCM patients enrolled in this study was relatively unique, we needed to carefully define our criteria for characterizing sequence variants in HCM disease genes as disease-causing mutations. The criteria for identifying a nucleotide sequence variant as a disease-causing mutation normally include the following: (1) The sequence variant must alter the structure or amount of the encoded protein. (2) The sequence variant occurs in a highly conserved residue that is relatively invariant during the course of evolution. (3) The sequence variant is not found among more than 100 chromosomes from unaffected individuals. (4) The sequence variant is found only among affected family members and not in unaffected family members. (5) Sequence variants are found segregating in other families in which the disease is inherited that affect the same gene. Since the eight genes selected for this study are recognized as HCM disease genes, we have modified the criteria for designating a sequence variant as a disease-causing mutation. All sequence variants that met the following two criteria were considered HCM-causing mutations. First, the sequence variant must alter the structure of the encoded protein. Second, the sequence variant did not correspond to a previously recognized polymorphism. Because the nucleotide sequences of the eight sarcomere protein genes selected for study here have been evaluated in previous studies of the genomes of more than 400 affected individuals, we assumed that all sequence variants that did not correspond to known polymorphisms were disease-causing mutations.

To determine the severity of the HCM caused by each mutation, we have attempted to identify other family members bearing the mutation. In many instances, other family members with the same mutation have been identified. The clinical features of these individuals are reported. Unfortunately, although more than 50 genetically affected relatives of probands have been identified, no single mutation is sufficiently represented in this population to establish the clinical consequences of the mutation. As described in previous studies, the range of clinical signs and symptoms of HCM caused by a particular mutation are quite diverse (for review, see Seidman and Seidman 2001; for specific examples, see Watkins et al. 1993; Gruver et al. 1999). Even in a single family, where all affected individuals bear the same disease-causing mutation, the range of LVWT can be from 1.3 cm to 4 cm, and the hypertrophy can be in the septal or free wall. Nevertheless, patterns involving symptoms such as survival of patients do emerge. Eventually, after more mutation-carrying individuals are clinically characterized, we anticipate that we will be able to provide more specific prognostic information to individuals diagnosed with a particular mutation.

Considerable information can be gained from defining the effects of mutations on protein structure and function. Among the sarcomere proteins that are mutated in HCM, cardiac MHC is the best characterized. That is, the three-dimensional structure of the protein has been determined, as have the structures of its interacting molecular partners, myosin light chain and actin. Several groups (for review, see Towbin 2000) have hypothesized that from these structures and from the location of the mutations in the protein one might be able to suggest how the mutant proteins lead to the morphological changes associated with HCM.

β-cardiac MHC has multiple functions—movement along the actin polymer or thin filament, light-chain binding, and Ca^{++}-dependent ATPase activity (for review, see Weiss and Leinwand 1996). Structural features of the βMHC head, including an actin-binding domain and an ATP-binding domain, that carry out these functions have been identified through a combination of crystallographic

Table 2. Description of HCM Disease Genes Selected for DNA Sequence Analysis

Gene	Symbol	Exons screened[a]	Exons
ACTC	cardiac actin	6	6
MYBP3	cardiac MYBP-C	34	37
TNNI3	cardiac troponin I	8	8
TNNT2	cardiac troponin T	15	16
MYL2	regulatory MLC	7	7
MYL3	essential MLC	6	7
TPM1	α-tropomyosin	9	9
MYH7	β-cardiac MHC	23	40

[a]The entire coding region of each disease gene, except β-cardiac MHC, was screened by PCR amplification and nucleotide sequence analysis using an ABI 3100 DNA sequencer.

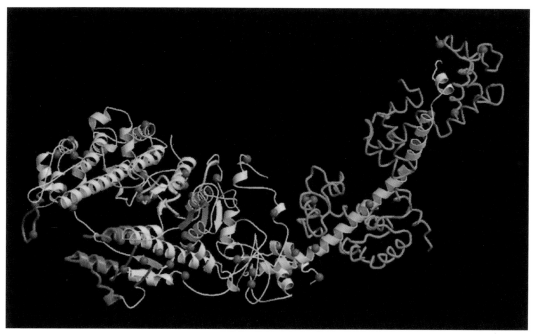

Figure 4. The three-dimensional structure of the head and head–rod junction of MHC showing the locations of mutated residues (*blue* spheres) that cause HCM. The MHC structure is of chicken skeletal muscle MHC which has a 69% identity over 841 amino acid residues (Genbank). The ATP-binding site (*green*), actin-binding site (*yellow*), and the myosin light chains (*orange* and *purple*) define active regions in the myosin head; there does not appear to be a clustering of mutations in these regions.

and biochemical studies. Because the first HCM-causing mutations in βMHC appeared to fall in regions of the myosin head that bind actin or ATP, one hypothesis (Rayment et al. 1995) was that disruption of these functions might be required to cause disease. This model has not been supported by the identification of further β-cardiac MHC mutations that cause HCM. That is, the 84 missense mutations in the βMHC head are scattered throughout the head and head–rod junction region of the molecule (Fig. 4). Mutations are located in virtually every α-helix in the protein. Apparently any mutation that alters βMHC function can cause disease. Presumably, all of these mutations affect a common function of the protein. Studies of mutant protein isolated from patients and from mice have suggested that this function may be related to the velocity with which the myosin head can move along the actin thin filament. The analyses of more mutant proteins will further test this hypothesis. Regardless, the pattern of mutations recognized to cause HCM is consistent with the model that changes in a variety of βMHC functions can induce the hypertrophic response.

We have previously suggested that mutant sarcomere proteins initiate the hypertrophic response by creating defective sarcomeres that function as a "Ca^{++} trap"(Semsarian et al. 2002). That is, the defective sarcomere requires a higher Ca^{++} concentration to generate the same amount of force. We imagine that this Ca^{++} trap reduces the concentration of Ca^{++} in the sarcoplasmic reticulum. Furthermore, we imagine that this Ca^{++} deficiency in the sarcoplasmic reticulum sends a signal that causes the myocyte to undergo the morphologic changes associated with left ventricular hypertrophy. We assume that there is a finite set of missense changes that can alter a sarcomere protein in a way that initiates the hypertrophic response. Other mutations might result in death or a different cardiac phenotype. The current study gives clues as to the numbers of sarcomere protein gene mutations that can cause HCM. Here we have screened 62 DNAs for mutations, and we have found mutations in 34 probands. We do not know whether the disease found in the remaining 28 individuals is due to mutations in other parts of the same genes, in other sarcomere protein genes (such as titin or α-cardiac MHC), or in other as-yet-unidentified genes. Further studies of the genomes of these individuals should provide answers to this question.

The 34 mutations identified in the present study were found in six genes. The distribution of mutations is similar to the distribution in previous studies. However, there are two significant differences. The fractions of individuals with mutations in βMHC and cardiac troponin T in this cohort are reduced compared to the fraction of individuals with mutations in these genes from previous studies (Table 3). Because βMHC and cardiac troponin T cause more severe forms of HCM, we hypothesize that fewer of these mutations were found in the present study because the patients in the study cohort were not identified at a referral center but rather were either self-identified or were identified by their cardiologist. Previous studies have addressed patient populations from tertiary care centers (for discussion, see Spirito et al. 1997).

How many HCM-causing mutations are there? Two-thirds of the mutations identified in this study were new mutations. These numbers suggest ~600 HCM-causing mutations, of which only about 200 have been identified. More than 3000 genomes of HCM patients must be screened to have a 90% confidence of finding these 400

Table 3. Distribution of HCM-causing Mutations among a Cohort of 62 Patients

Gene	Protein	No. of probands[a]	Percent	Novel mutations	Known mutations	Total mutations	Previous studies (%)
ACTC	cardiac actin	0	0	0	0	5	<5
MYBP3	cardiac MyBP-C	14	22	9	5	53	15–20
MYH7	α-cardiac MHC	11	18	7	4	84	25–30
MYL2	regulatory MLC	1	2	0	1	8	<1
MYL3	essential MLC	0	0	0	0	4	<1
TNNI3	cardiac troponin I	4	6	3	0	12	~10
TNNT2	cardiac troponin T	2	3	2	0	14	10–15
TPM1	α-tropomyosin	2	3	1	1	7	<5

[a]The number of probands with mutations, and the percentage (Percent) of the 62 study probands with mutations in this gene, as well as the number of the mutations found in that gene which were novel (not determined in previous studies) and the number of mutations detected previously are reported. The total number of mutations (current) as well as the distribution of mutations among HCM patients from previous studies are compared to results obtained in this study.

new mutations. Eventually a complete catalog of HCM-causing mutations should facilitate clinical diagnosis of HCM as well as provide further insights into the mechanism by which these mutations alter sarcomere function and induce cardiac hypertrophy.

ACKNOWLEDGMENTS

We thank all of the patients, physicians, and family members who made this work possible through the generous contributions of their time and DNA. This work was supported by a National Institutes of Health Program in Genomics grant (J.G.S. and C.E.S.) and by the Howard Hughes Medical Foundation (J.G.S., C.E.S., S.B., S.R.D., and B.M.).

REFERENCES

Blair E., Redwood C., de Jesus Oliveira M., Moolman-Smook J.C., Brink P., Corfield V.A., Ostman-Smith I., and Watkins H. 2002. Mutations of the light meromyosin domain of the beta-myosin heavy chain rod in hypertrophic cardiomyopathy. *Circ. Res.* **90:** 263.

Dracopoli N., Haines J.L., Korf B.R., Moir D.T., Morton C.C., Seidman C.E., Seidman J.G., and Smith D.R., Eds. 2002. Searching candidate genes for mutations. *Current Protocols in human genetics, Chap. 7.* Current Protocols. Wiley and Sons, New York.

Fatkin D., Seidman J.G., and Seidman C.E. 2000. Hypertrophic cardiomyopathy. In *Cardiovascular medicine, 2nd edition* (ed. J.T. Willerson and J.N. Cohn), p. 1055. Churchill Livingstone, New York.

Geisterfer-Lowrance A.A., Kass S., Tanigawa G., Vosberg H.P., McKenna W., Seidman C.E., and Seidman J.G. 1990. A molecular basis for familial hypertrophic cardiomyopathy: A beta cardiac myosin heavy chain gene missense mutation. *Cell* **62:** 999.

Gerull B., Gramlich M., Atherton J., McNabb M., Trombitas K., Sasse-Klaassen S., Seidman J.G., Seidman C., Granzier H., Labeit S., Frenneaux M., and Thierfelder L. 2002. Mutations of TTN, encoding the giant muscle filament titin, cause familial dilated cardiomyopathy. *Nat. Genet.* **30:** 201.

Gruver E.J., Fatkin D., Dodds G.A., Kisslo J., Maron B.J., Seidman J.G., and Seidman C.E. 1999. Familial hypertrophic cardiomyopathy and atrial fibrillation caused by Arg663His beta-cardiac myosin heavy chain mutation. *Am. J. Cardiol.* **83:** 13H.

Kimura A., Harada H., Park J.E., Nishi H., Satoh M., Takahashi M., Hiroi S., Sasaoka T., Ohbuchi N., Nakamura T., Koyanagi T., Hwang T.H., Choo J.A., Chung K.S., Hasegawa A., Nagai R., Okazaki O., Nakamura H., Matsuzaki M., Sakamoto T., Toshima H., Koga Y., Imaizumi T., and Sasazuki T. 1997. Mutations in the cardiac troponin I gene associated with hypertrophic cardiomyopathy. *Nat. Genet.* **16:** 379.

MacRae C.A., Watkins H.C., Jarcho J.A., Thierfelder L., McKenna W.J., Seidman J.G., and Seidman C.E. 1994. An evaluation of ribonuclease protection assays for the detection of beta-cardiac myosin heavy chain gene mutations. *Circulation* **89:** 33.

Maron B.J. 2002. Hypertrophic cardiomyopathy: A systematic review. *J. Am. Med. Assoc.* **287:** 1308.

Niimura H., Patton K.K., McKenna W.J., Soults J., Maron B.J., Seidman J.G., and Seidman C.E. 2002. Sarcomere protein gene mutations in hypertrophic cardiomyopathy of the elderly. *Circulation* **105:** 446.

Niimura H., Bachinski L.L., Sangwatanaroj S., Watkins H., Chudley A.E., McKenna W., Kristinsson A., Roberts R., Sole M., Maron B.J., Seidman J.G., and Seidman C.E. 1998. Mutations in the gene for cardiac myosin-binding protein C and late-onset familial hypertrophic cardiomyopathy. *N. Engl. J. Med.* **338:** 1248.

Olson T.M., Doan T.P., Kishimoto N.Y., Whitby F.G., Ackerman M.J., and Fananapazir L. 2000. Inherited and de novo mutations in the cardiac actin gene cause hypertrophic cardiomyopathy. *J. Mol. Cell. Cardiol.* **32:** 1687.

Poetter K., Jiang H., Hassanzadeh S., Master S.R., Chang A., Dalakas M.C., Rayment I., Sellers J.R., Fananapazir L., and Epstein N.D. 1996. Mutations in either the essential or regulatory light chains of myosin are associated with a rare myopathy in human heart and skeletal muscle. *Nat. Genet.* **13:** 63.

Rayment I., Holden H.M., Sellers J.R., Fananapazir L., and Epstein N.D. 1995. Structural interpretation of the mutations in the beta-cardiac myosin that have been implicated in familial hypertrophic cardiomyopathy. *Proc. Natl. Acad. Sci.* **92:** 3864.

Seidman C.E. and Seidman J.G. 2001. Hypertrophic cardiomyopathy. In *The metabolic and molecular bases of inherited disease, 8th edition* (ed. C.R. Scriver et al.), vol. IV, p. 5433. McGraw-Hill Medical, New York, New York.

Semsarian C., Ahmad I., Giewat M., Georgakopoulos D., Schmitt J.P., McConnell B.K., Reiken S., Mende U., Marks A.R., Kass D.A., Seidman C.E., and Seidman J.G. 2002. The L-type calcium channel inhibitor diltiazem prevents cardiomyopathy in a mouse model. *J. Clin. Invest.* **109:** 1013.

Spirito P., Seidman C.E., McKenna W.J., and Maron B.J. 1997. The management of hypertrophic cardiomyopathy. *N. Engl. J. Med.* **336:** 775.

Thierfelder L., Watkins H., MacRae C., Lamas R., McKenna W., Vosberg H.P., Seidman J.G., and Seidman C.E. 1994. Alpha-tropomyosin and cardiac troponin T mutations cause familial hypertrophic cardiomyopathy: A disease of the sarcomere. *Cell* **77:** 701.

Towbin J.A. 2000. Molecular genetics of hypertrophic cardiomyopathy. *Curr. Cardiol. Rep.* **2:** 134.

Watkins H. 2000. Sudden death in hypertrophic cardiomyopathy. *N. Engl. J. Med.* **342:** 422.

Watkins H., Rosenzweig A., Hwang D.S., Levi T., McKenna W., Seidman C.E., and Seidman J.G. 1992. Characteristics and prognostic implications of myosin missense mutations in familial hypertrophic cardiomyopathy. *N. Engl. J. Med.* **326:** 1108.

Watkins H., MacRae C., Thierfelder L., Chou Y.H., Frenneaux M., McKenna W., Seidman J.G., and Seidman C.E. 1993. A disease locus for familial hypertrophic cardiomyopathy maps to chromosome 1q3. *Nat. Genet.* **3:** 333.

Watkins H., Conner D., Thierfelder L., Jarcho J.A., MacRae C., McKenna W.J., Maron B.J., Seidman J.G., and Seidman C.E. 1995. Mutations in the cardiac myosin binding protein-C gene on chromosome 11 cause familial hypertrophic cardiomyopathy. *Nat. Genet.* **11:** 434.

Weiss A. and Leinwand L.A. 1996. The mammalian myosin heavy chain gene family. *Annu. Rev. Cell Dev. Biol.* **12:** 417.

The Sarcoglycan Complex in Striated and Vascular Smooth Muscle

M.T. Wheeler,* M.J. Allikian,† A. Heydemann,† and E.M. McNally†‡

*Department of Molecular Genetics and Cell Biology, †Department of Medicine, and ‡Department of Human Genetics, The University of Chicago, Chicago, Illinois 60637

Congestive heart failure (CHF), a leading cause of hospitalization and death, arises when cardiac output is insufficient to meet demand. Cardiomyopathy, defined as dysfunction of cardiac muscle, frequently results in CHF and may arise as a consequence of ischemic, toxic, immunologic, or genetic insult. Decompensated cardiomyopathy may occur as a result of progressive ventricular dysfunction and is often aggravated by cardiac arrhythmias. Understanding the molecular and physiological mechanisms that underlie cardiomyopathy progression is central to improving treatment options.

Inherited forms of cardiomyopathy are relatively rare (Codd et al. 1989) but have led to considerable insight on the pathogenesis of cardiac muscle dysfunction (Seidman and Seidman 2001; Towbin and Bowles 2001). Many of the same genetic defects that produce cardiac muscle dysfunction also produce skeletal muscle dysfunction (Heydemann et al. 2001), reflecting the overlapping program of gene expression in cardiac and skeletal muscle. Skeletal muscle disease most commonly manifests as muscular dystrophy, a progressive degenerative process of skeletal muscle. In the muscular dystrophies, ongoing muscle degeneration is accompanied by regeneration. Over time, the regenerative capacity of muscle becomes exhausted and the muscle is replaced by fibrofatty infiltrate. In cardiac muscle, degeneration occurs at a slower pace but is unaccompanied by regeneration, leading to increased cardiac fibrosis and cardiomyopathy. In many cases, the clinical constellation of simultaneous skeletal and cardiac muscle weakness may limit the degree to which each is detected. In this paper, we discuss genetic mechanisms of cardiomyopathy and muscular dystrophy with particular emphasis on the role of the membrane-associated dystrophin glycoprotein complex (DGC) and the role of vasospasm as a mechanism for cardiomyopathy progression.

DYSTROPHIN AND ITS ASSOCIATED PROTEINS

The DGC is a specialized complex found at the plasma membrane of striated muscle that mediates interactions between the cytoskeleton, the membrane, and the extracellular matrix (Fig. 1) (Cohn and Campbell 2000; Hack et al. 2000a). The DGC copurifies as a macromolecular assembly from the membranous fraction and is composed of both cytoskeletal and transmembrane elements (Yoshida and Ozawa 1990; Ervasti and Campbell 1991). The cytoskeletal proteins of the DGC include dystrophin, the protein product of the Duchenne muscular dystrophy (DMD) locus, dystrobrevin, and the syntrophins (Rafael and Brown 2000). The DGC transmembrane components include dystroglycan and sarcoglycan. Dystroglycan is a broadly expressed protein that is essential for normal development. Dystroglycan appears to order the extracellular matrix, and specifically, the extracellular matrix protein laminin (Henry and Campbell 1998). The expression of the sarcoglycan subcomplex is restricted to muscle (Hack et al. 2000a). In striated muscle, the sarcoglycan complex includes at least four sarcoglycan proteins, α, β, γ, and δ, and likely includes additional components such as ε and ζ sarcoglycan (Ettinger et al. 1997; McNally et al. 1998; Straub et al. 1999; Hack et al. 2000a; Wheeler et al. 2002).

Mutations in dystrophin lead to muscular dystrophy and, importantly, to disruption of the entire DGC (Ervasti et al. 1990). In the absence of dystrophin, sarcoglycan, dystroglycan, dystrobrevin, and syntrophin are reduced at the plasma membrane leading to abnormal membrane permeability and membrane instability (Matsuda et al. 1995; Straub et al. 1997). It is this membrane instability that causes skeletal and cardiac myocyte loss, leading to muscular dystrophy and cardiomyopathy. Dystrophin is thought to provide mechanical stability to the membrane to protect it from contraction-induced damage (Petrof et al. 1993; Danialou et al. 2001). However, it is also clear that dystrophin plays an essential role to scaffold the remaining DGC proteins to the membrane and submembrane locales (Chamberlain et al. 1997).

Mutations in the genes encoding the sarcoglycan proteins also cause muscular dystrophy and cardiomyopathy (Bonnemann et al. 1996; Cohn and Campbell 2000). Autosomal recessive mutations in α-, β-, γ-, and δ-sarcoglycan produce limb girdle muscular dystrophy (LGMD). In many instances, cardiomyopathy accompanies both DMD and the LGMDs, leading to reduced cardiac output, congestive heart failure, and arrhythmias (Perloff et al. 1966; Melacini et al. 1999; Calvo et al. 2000; Emery 2002). In the case of primary mutations in sarcoglycan genes, there is instability of the sarcoglycan complex and secondary reduction of the remaining sarcoglycan proteins (Bonnemann et al. 1995; Lim et al. 1995; Noguchi

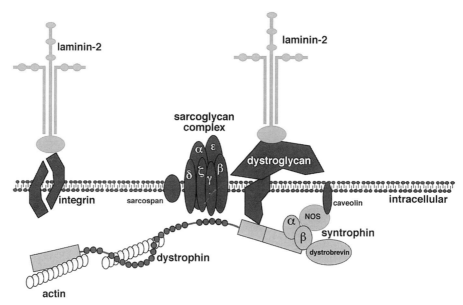

Figure 1. Schematic of the dystrophin glycoprotein complex (DGC). The DGC is composed of transmembrane elements that include the sarcoglycan complex, sarcospan, and dystroglycan. Caveolin-3 copurifies with the DGC and interacts with the PDZ domain of syntrophin. Integrins and the DGC both bind the extracellular matrix protein laminin. The cytoskeletal proteins that constitute the DGC include dystrophin, cytoplasmic actin, dystrobrevin, syntrophins, and nitric oxide synthase (NOS). The DGC is a mechanosignaling complex.

et al. 1995). However, the expression of dystrophin and dystroglycan is normal (Bonnemann et al. 1995; Noguchi et al. 1995). Sarcoglycan loss is a common molecular feature in genetically diverse forms of muscular dystrophy and cardiomyopathy.

Each of the sarcoglycan genes encodes a transmembrane protein with a single transmembrane-spanning region. The short cytoplasmic domains may be serine/threonine- and tyrosine-phosphorylated (Yoshida et al. 1998). β-, γ-, and δ-sarcoglycan each have conserved extracellular cysteine residues resembling epidermal growth factor (McNally et al. 1996b). Mutations in these cysteines result in reduced protein stability (McNally et al. 1996b; Nigro et al. 1996a; Piccolo et al. 1996); these residues may serve as a ligand for proteins in the extracellular matrix. A new sarcoglycan gene, ζ-sarcoglycan, was recently described (Wheeler et al. 2002). The genes encoding γ-, δ-, and ζ-sarcoglycan have an identical intron–exon structure, suggesting these genes arose from gene duplication events (McNally et al. 1996a; Nigro et al. 1996b; Wheeler et al. 2002). α- and ε-sarcoglycan similarly appear to have arisen from a gene duplication event (McNally et al. 1998). The *Drosophila* and *Caenorhabditis elegans* databases have three sarcoglycan genes each, suggesting that vertebrates acquired additional sarcoglycan genes, potentially to support the diversity of muscle types.

Mutations in the ε-sarcoglycan gene are associated with the autosomal dominant movement disorder, myoclonus dystonia (Zimprich et al. 2001). Because the ε-sarcoglycan gene is imprinted (Piras et al. 2000), mutations in the ε-sarcoglycan gene show parent-of-origin effects. The contribution of ε-sarcoglycan in skeletal muscle is not known at this time and will likely require tissue-specific conditionally targeted ablations of the murine ε-sarcoglycan gene to elucidate fully the mechanism of ε-sarcoglycan gene mutations.

ANIMAL MODELS: MOLECULAR MECHANISMS OF MUSCLE DEGENERATION

The *mdx* mouse is a naturally occurring animal model of dystrophin mutations (Coulton et al. 1988). A point mutation in exon 23 leads to premature termination and the absence of full-length dystrophin (Sicinski et al. 1989). In the *mdx* mouse, smaller dystrophin forms produced from internally active promoters remain present but are insufficient to protect from myocyte damage (Cox et al. 1993). Cardiomyopathy is a common source of later-stage morbidity and mortality in DMD patients, yet in the *mdx* mouse, the cardiomyopathy remains comparatively mild (Perloff et al. 1966). The difference in cardiovascular disease severity between human DMD patients and the *mdx* mouse may arise from protective genetic modifier loci in *mdx*. It has been suggested that up-regulation of the chromosome-6-encoded dystrophin homolog utrophin ameliorates the dystrophic and cardiac phenotype in the *mdx* mouse (Tinsley et al. 1996). Ablation of the utrophin gene in the *mdx* background produces more profound skeletal and cardiac muscle phenotypes, arguing that utrophin can partially compensate for dystrophin (Deconinck et al. 1997; Grady et al. 1997).

THE DGC AND SARCOGLYCAN FUNCTION AS A MECHANOSIGNALING COMPLEX

Mouse models of sarcoglycan gene disruptions have helped to elucidate functions of this subcomplex (Allamand and Campbell 2000; Heydemann et al. 2001). Mice with a targeted deletion of γ-sarcoglycan were generated first and revealed that sarcoglycan loss is independent of dystrophin (Hack et al. 1998). Moreover, γ-sarcoglycan mutant muscle displayed a similar phenotype to dystrophin mutant muscle, although γ-sarcoglycan mutant muscle was more severely affected. The characteristic finding in dystrophin or sarcoglycan mutant muscle is focal degeneration or focal necrosis (Fig. 2) that is characterized in its earliest stages by regional uptake of the vital tracer Evans blue dye (EBD) (Hack et al. 1998). Normal muscle is impermeable to EBD, whereas sarcoglycan mutant muscle readily takes up EBD focally. EBD uptake is also seen in dystrophin mutant muscle where sarcoglycan is secondarily reduced (Matsuda et al. 1995; Straub et al. 1997). The mechanism responsible for membrane permeability defects in sarcoglycan-deficient muscle is not well understood. Electron microscopy of sarcoglycan mutant muscle does not reveal membrane disruptions of sufficient size to visualize (Hack et al. 1998). Because of abnormal membrane permeability, muscle-specific proteins are increased in the serum. Both human patients and mice mutant for γ-sarcoglycan have elevated serum creatine kinase (Ben Hamida et al. 1996; Hack et al. 1998).

TUNEL-positive nuclei are found in and around regions of focal necrosis, suggesting that programmed cell death is increased in DGC-mediated degeneration (Hack et al. 1998). However, although TUNEL nuclei are increased in DGC-mediated dystrophy, the number of TUNEL-positive nuclei is not sufficient to fully account for the degree of muscle degeneration. A mononuclear cell infiltrate also is found near regions of degeneration and is likely to promote further degeneration. In cardiac muscle, focal degeneration is accompanied by EBD uptake as an early marker and replacement by fibrosis as a late histologic equivalent. The stimulus that triggers focal degeneration in sarcoglycan and dystrophin mutations remains a central question for understanding the pathophysiology of these disorders.

It has been suggested that contraction-associated mechanical forces produce focal degeneration. Eccentric contraction studies performed on *mdx* muscle have shown that contraction induces damage to dystrophin-deficient myofibers, but only minimally does so to normal muscle fibers (Petrof et al. 1993). Eccentric contraction mimics in vivo physiological exercise since contraction is stimulated on a lengthened muscle. In the heart, pressure overload hypertrophy creates additional force. Consistent with this, *mdx* hearts are abnormally sensitive to pressure overload hypertrophy (Danialou et al. 2001).

In γ-sarcoglycan mutant muscle, dystrophin, dystroglycan, and laminin are normally localized (Hack et al. 1998). Dystrophin, dystroglycan, and laminin are the components of the mechanical linkage thought to be most relevant for stabilizing the plasma membrane against the contraction-associated force. To test whether this linkage is functional, eccentric contraction experiments were performed on γ-sarcoglycan mutant muscle. Surprisingly, γ-sarcoglycan mutant muscle showed no increase in damage when subjected to eccentric contraction experiments

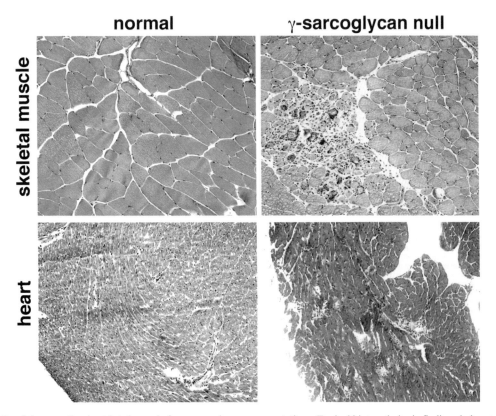

Figure 2. Focal degeneration in striated muscle from sarcoglycan gene mutations. Typical histopathologic findings in heart and skeletal muscle are shown and include focal degeneration and fibrosis. Similar findings are seen in dystrophin mutant mice.

(Hack et al. 1999). Because sarcoglycan is secondarily reduced when dystrophin is absent, this argues that nonmechanical functions may be relevant for the pathophysiology of muscular dystrophy and cardiomyopathy in genetically diverse forms of disease. Furthermore, these findings suggest that sarcoglycan, and the entire DGC, have both mechanical-stabilizing functions and nonmechanical functions. Targeted disruption of α-dystrobrevin produces muscular dystrophy independent of dystrophin (Grady et al. 1999). In this form of muscular dystrophy, the process by which striated muscle degeneration occurs is thought to involve a nonmechanical process. Recently, the interaction between the sarcoglycans and dystrobrevins was more fully elucidated (Yoshida et al. 2000).

SKELETAL MUSCLE DISEASE

In DGC-mediated muscular dystrophies, skeletal muscle weakness predominates early in the course of disease. The primary muscles affected by the dystrophic process include those of the limbs and trunk. In addition, the respiratory musculature is often severely affected, leading to hypoventilation and hypoxemia (Melacini et al. 1996). Cardiomyopathy develops as a relatively late consequence, suggesting that cardiomyopathy may develop as a consequence of severe skeletal muscle disease. In this case, hypoventilation/hypoxemia may cause pulmonary hypertension and right ventricular failure or cor pulmonale. Supporting this, the severe cardiomyopathy seen in mdx/MyoD mutant mice is thought to arise as a secondary consequence of severe muscle disease (Megeney et al. 1999).

To assess whether severe skeletal muscle disease can contribute to cardiomyopathy, we used a tissue-specific transgenic approach to rescue skeletal muscle expression of γ-sarcoglycan in the background of the null γ-sarcoglycan allele (Zhu et al. 2002). The full-length murine γ-sarcoglycan was expressed under the control of the myosin light-chain 1-3 gene promoter (Donoghue et al. 1988) to drive expression in skeletal but not cardiac muscle. This transgene, MLCgsg, was bred into the background of the γ-sarcoglycan null allele. The skeletal muscle dystrophy was fully corrected by the presence of the MLCgsg transgene, yet the cardiomyopathic process remained evident. Focal necrosis and EBD uptake persisted in the hearts of MLCgsg transgene-rescued mice. These findings demonstrate that cardiac and skeletal muscle diseases are independent processes. Therefore, gene therapy for these disorders should target both cardiac and skeletal muscle.

For gene replacement therapy, the level and tissue specificity of sarcoglycan expression must also be considered (Cordier et al. 2000, 2001). In a transgenic approach, we expressed high levels of γ-sarcoglycan using the muscle creatine kinase (MCK) promoter (Zhu et al. 2001). The MCK promoter drives expression in both skeletal and cardiac muscle, although it typically leads to greater levels of expression in skeletal muscle (Gossett et al. 1989; Johnson et al. 1989). High-level overexpression of γ-sarcoglycan in a wild-type background resulted in severe muscle wasting and weakness. Interestingly, the pathological mechanism of γ-sarcoglycan overexpression appeared to involve abnormal cytoplasmic accumulation of γ-sarcoglycan as well as abnormal accumulation in the secretory system. Thus, high-level (50-fold) overexpression of γ-sarcoglycan is toxic to myocytes.

SMOOTH MUSCLE SARCOGLYCAN AND CORONARY ARTERY VASOSPASM

Focal necrosis also characterizes the degenerative pattern seen in heart and skeletal muscle of the BIO 14.6 Syrian hamster model of cardiomyopathy (Factor et al. 1982). The genetic defect in the BIO 14.6 hamster is a large deletion in the promoter region of δ-sarcoglycan (Nigro et al. 1997; Sakamoto et al. 1997). It was believed that regional degeneration arose from arterial vasospasm (Factor et al. 1982) since focal "infarct-like" regions are commonly seen in the myocardium and skeletal muscle of the BIO 14.6 hamster. The appearance and location of these lesions, sometimes near vessels, suggested that intermittent arterial spasm caused ischemia and the infarct appearance.

A targeted disruption of the murine δ-sarcoglycan gene produced a phenotype similar to that of the BIO 14.6 hamster and identical to that seen in mice lacking γ-sarcoglycan (Coral-Vazquez et al. 1999; Hack et al. 2000b). Interestingly, disruption of the smooth muscle sarcoglycan complex was noted in δ-sarcoglycan mutant mice (Coral-Vazquez et al. 1999), and it was suggested that disruption of smooth muscle sarcoglycan was responsible for vasospasm and cardiomyopathy. Smooth muscle sarcoglycan differs from striated muscle sarcoglycan in that it is composed of β-, δ-, and ε-sarcoglycan (Straub et al. 1999). There has been debate whether γ-sarcoglycan is expressed in smooth muscle (Barresi et al. 2000). This confusion may be explained by the identification of ζ-sarcoglycan, a highly related sarcoglycan sequence that is homologous to both γ- and δ-sarcoglycan (Wheeler et al. 2002).

In γ-sarcoglycan mutant mice, the smooth muscle sarcoglycan complex is intact (M.T. Wheeler and E.M. McNally, unpubl.). Yet, evidence for focal degeneration and, importantly, evidence for vasospasm are present (Hack et al. 1998). This indicates that vasospasm is not a cell-intrinsic defect of smooth muscle. Instead, primary myocyte degeneration may lead to cytokine release and regional vasospasm. Foci of damage are common in sarcoglycan mutant mice; thus, the degree of degeneration can easily support a secondary vasospasm model. In this model, secondary vasospasm would further potentiate damage by worsening ischemia. This mechanism, where primary cardiac damage produces secondary vasospasm, may not be limited to these rare models of cardiomyopathy and may accompany any form of myocardial damage including myocardial infarction, myocarditis, and other inherited forms of cardiomyopathy. Vasospasm may also occur as a response to intrinsic skeletal muscle damage. Supporting this, defective contraction-induced vasodilation has been noted in DMD patients as well as in the *mdx* mouse (Thomas et al. 1998; Sander et al. 2000).

Secondary vasospasm may prove an important target for therapeutic intervention. In both δ-sarcoglycan mutant mice and the BIO 14.6 Syrian hamster model, the calcium channel antagonist verapamil reduced cardiomyopathy progression (Sonnenblick et al. 1985; Cohn et al. 2001). Mice lacking γ-sarcoglycan were similarly treated with verapamil to determine whether reduction of secondary vasospasm is effective (M.T. Wheeler and E.M. McNally, unpubl.). A significant improvement in cardiac output was achieved with 5 months of verapamil treatment. Microfil, a liquid plastic, was used to define the coronary anatomy and demonstrate anatomic evidence for reduced coronary artery vasospasm. Verapamil may have a multifold effect that includes a reduction of vasospasm, blood pressure, and cardiac contractility.

To more fully address the role of the cell-intrinsic nature of cardiomyocyte damage, a transgenic cardiomyocyte rescue of δ-sarcoglycan was achieved (M. Allikian and E.M. McNally, unpubl.). Using the α-myosin heavy-chain promoter to drive δ-sarcoglycan expression exclusively in cardiomyocytes, δ-sarcoglycan expression was restored in cardiomyocytes in the background of the δ-sarcoglycan null mouse. Cardiomyocyte expression of δ-sarcoglycan was sufficient to reduce vasospasm, confirming that arterial vasospasm arises non-cell-autonomously.

The cellular mechanisms that give rise to vasospasm may include nitric oxide (NO). NO is a mediator of vasomotor tone, and neuronal nitric oxide synthase (nNOS) directly associates with the DGC by way of its interaction with the PDZ domain containing protein α-syntrophin (Adams et al. 2001). nNOS mislocalization from the plasma membrane is insufficient to mediate muscular dystrophy (Crosbie et al. 1998). However, regional mislocalization of NO may play a role in the progression of cardiac and skeletal muscle dysfunction.

PRIMARY VASOSPASM

The mechanisms that underlie primary arterial vasospasm are not well understood. Prinzmetal variant angina arises from coronary artery vasospasm in the absence of atherosclerotic burden, and therefore represents a "pure" form of vascular spasm. Clinically, Prinzmetal patients develop repeated episodes of transient coronary artery vasospasm manifesting as chest pain, arrhythmias, and myocardial infarction. Infusion of vasoconstricting agents, such as ergonovine, into the coronary arteries is used to demonstrate angiographic and electrocardiographic evidence of vasospasm. Coronary artery vasospasm also frequently accompanies atherosclerotic disease because abnormal vasoconstriction may develop adjacent to areas of atherosclerotic plaque in response to acetylcholine (Ganz et al. 1991; Sueda et al. 1999).

K_{ATP} channels are known to modulate vasomotor tone (Quayle et al. 1997; Yokoshiki et al. 1998). K_{ATP} channels couple the intracellular energy state to membrane potential because K_{ATP} channels are responsive to intracellular ATP/ADP ratios (Ashcroft and Gribble 1998; Seino 1999). Changes in K_{ATP} activity produce membrane potential alterations that, in turn, activate a variety of cellu-

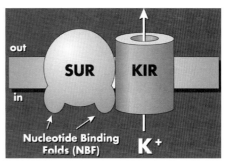

Figure 3. K_{ATP} channels are regulators of vasomotor tone. K_{ATP} channels are composed of a sulfonylurea subunit, either SUR1 or SUR2, and a pore-forming potassium channel, either Kir6.1 or 6.2. Targeted disruption of the Sur2 gene leads to episodic coronary artery vasospasm and hypertension, elucidating a molecular mechanism of primary vascular spasm.

lar responses. These responses include the activation of voltage-sensitive calcium channels or the stimulation of hormone release. K_{ATP} channels are composed of a regulatory subunit, the sulfonylurea receptor (SUR), and a pore-forming potassium channel Kir6.1 or Kir6.2 (Fig. 3) (Babenko et al. 1998). SUR1 and SUR2 are highly related regulatory subunits that are expressed in different tissues (Aguilar-Bryan et al. 1995; Chutkow et al. 1996). SUR1 is expressed primarily in the pancreas, whereas SUR2 is expressed in striated and smooth muscle. A targeted disruption of the SUR2 subunit was generated and found to stimulate an abnormal response to insulin that resulted from the skeletal muscle loss of SUR2 (Chutkow et al. 2001). SUR2 mutant mice also develop sudden cardiac death that is more pronounced in male versus female mice (Chutkow et al. 2002).

Telemetric monitoring of SUR2 mutant mice revealed in vivo evidence for transient, repeated episodes of coronary artery ischemia consistent with primary vasospasm. Episodes lasting 30–60 seconds were frequently seen and often resulted in arrhythmias. Microfil studies of SUR2 mutant mice showed evidence for anatomic narrowing of the coronary arteries consistent with vasospasm similar to what was seen in sarcoglycan mutant mice (Fig. 4). During these episodes, transient ischemia produced arrhythmias, and, at times, these arrhythmias were lethal. Vasospasm was suppressed by treatment with the calcium channel antagonist nifedipine, the agent of choice currently used for treating vasospasm in human patients. Interestingly, gene targeting of the vascular smooth muscle partner protein of SUR2, Kir6.1, also resulted in a similar picture of spontaneous vasospasm (Miki et al. 2002).

In contrast to the sarcoglycan-deficient model of secondary vasospasm, K_{ATP}-channel-related vasospasm did not produce cardiomyopathy. The absence of cardiomyopathy or focal cardiac damage in SUR2 mutant mice may relate to the relatively brief episodes of vasospasm and the high incidence of cardiac arrhythmias.

CONCLUSIONS

Disruption of the DGC produces membrane instability, and specifically, dissolution of the sarcoglycan complex

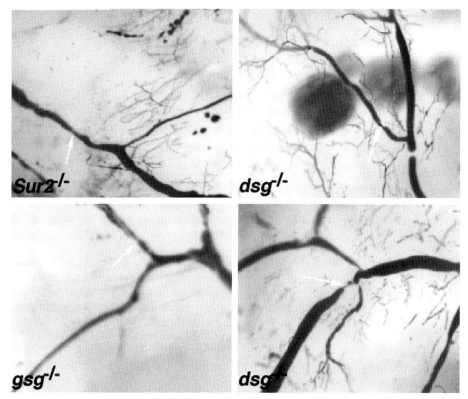

Figure 4. Anatomic evidence of coronary artery vasospasm from Microfil. Microfil infusions were used to outline the coronary arteries of δ-sarcoglycan mutant ($dsg^{-/-}$), γ-sarcolglycan mutant ($gsg^{-/-}$), or SUR2 mutant ($Sur2^{-/-}$) mice. Evidence for arterial vasospasm (*arrows*) was seen in each of these mutant mice. In the case of sarcoglycan mutant mice, vasospasm arises as a non-cell-intrinsic defect in the vasculature. Rescue of cardiomyocyte sarcoglycan expression corrects vasospasm. Thus, secondary vasospasm arises from primary cardiomyocyte degeneration.

leads to increased membrane permeability. Membrane disruption is thought to increase intracellular calcium pools and activate calcium-sensitive proteases. It is these mechanisms that are thought to result in myocyte damage and loss. The interplay between regions of focal damage in the myocardium, or within skeletal muscle, and the nearby vasculature have recently drawn attention. We present data that primary myocardial degeneration leads to secondary vascular spasm (Fig. 5). We have also found that inhibition of secondary vasospasm limits the pro-

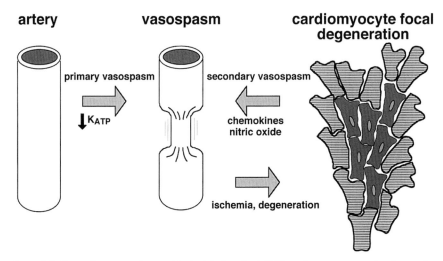

Figure 5. Schematic model of vascular spasm. A vessel is depicted on the left. Vascular spasm can arise from reduced Sur2-K_{ATP} activity. Alternatively, primary cardiomyocyte degeneration can lead to vascular spasm as shown on the right. In this case, vascular spasm may arise from regional changes in chemokines and nitric oxide (NO). Vasospasm can reduce blood supply and further potentiate cardiomyocyte damage. Inhibition of secondary vasospasm is an important therapeutic target that may apply to many different forms of cardiomyocyte damage.

gression of cardiomyopathy in an animal model of cardiomyopathy. Future studies will evaluate the degree to which vasospasm contributes to other forms of cardiac degeneration. Inhibition of vasospasm may be a more broadly applicable therapeutic option to limit cardiac degeneration.

As a contrast, we described a model of primary coronary vasospasm that derives from disruption of the sulfonylurea receptor, the regulatory subunit of the major K_{ATP} channel expressed in vascular smooth muscle (Fig. 5). Despite repeated episodes of ischemia, cardiomyopathy did not develop, in part owing to the severity of coronary artery vasospasm. Potassium channel openers that stimulate K_{ATP} channel activity may become more useful in the treatment of hypertension and vasospasm. An improved understanding of secondary and primary vascular spasm will improve therapy for cardiovascular disease, and this understanding may have applicability to human disease beyond the cardiovascular system.

ACKNOWLEDGMENTS

This work was supported by the National Institutes of Health (HL-61322 and HL-63783), the American Heart Association, the Muscular Dystrophy Association, and the Burroughs Wellcome Fund. M.T.W. is supported by the Medical Scientist Training Program (NIH-GM007281). M.A. is supported by the National Institutes of Health (HL-68472). A.H. is supported by the National Institutes of Health (HL-10432).

REFERENCES

Adams M.E., Mueller H.A., and Froehner S.C. 2001. In vivo requirement of the alpha-syntrophin PDZ domain for the sarcolemmal localization of nNOS and aquaporin-4. *J. Cell Biol.* **155:** 113.

Aguilar-Bryan L., Nichols C.G., Wechsler S.W., Clement J.P., IV, Boyd A.E., III, Gonzalez G., Herrera-Sosa H., Nguy K., Bryan J., and Nelson D.A. 1995. Cloning of the beta cell high-affinity sulfonylurea receptor: A regulator of insulin secretion. *Science* **268:** 423.

Allamand V. and Campbell K.P. 2000. Animal models for muscular dystrophy: Valuable tools for the development of therapies. *Hum. Mol. Genet.* **9:** 2459.

Ashcroft F.M. and Gribble F.M. 1998. Correlating structure and function in ATP-sensitive K+ channels. *Trends Neurosci.* **21:** 288.

Babenko A.P., Aguilar-Bryan L., and Bryan J. 1998. A view of sur/KIR6.X, KATP channels. *Annu. Rev. Physiol.* **60:** 667.

Barresi R., Moore S.A., Stolle C.A., Mendell J.R., and Campbell K.P. 2000. Expression of gamma-sarcoglycan in smooth muscle and its interaction with the smooth muscle sarcoglycan-sarcospan complex. *J. Biol. Chem.* **275:** 38554.

Ben Hamida M., Ben Hamida C., Zouari M., Belal S., and Hentati F. 1996. Limb-girdle muscular dystrophy 2C: Clinical aspects. *Neuromuscul. Disord.* **6:** 493.

Bonnemann C.G., McNally E.M., and Kunkel L.M. 1996. Beyond dystrophin: Current progress in the muscular dystrophies. *Curr. Opin. Pediatr.* **8:** 569.

Bonnemann C.G., Modi R., Noguchi S., Mizuno Y., Yoshida M., Gussoni E., McNally E.M., Duggan D.J., Angelini C., and Hoffman E.P. 1995. Beta-sarcoglycan (A3b) mutations cause autosomal recessive muscular dystrophy with loss of the sarcoglycan complex. *Nat. Genet.* **11:** 266.

Calvo F., Teijeira S., Fernandez J.M., Teijeiro A., Fernandez-Hojas R., Fernandez-Lopez X.A., Martin E., and Navarro C. 2000. Evaluation of heart involvement in gamma-sarcoglycanopathy (LGMD2C). A study of ten patients. *Neuromuscul. Disord.* **10:** 560.

Chamberlain J.S., Corrado K., Rafael J.A., Cox G.A., Hauser M., and Lumeng C. 1997. Interactions between dystrophin and the sarcolemma membrane. *Soc. Gen. Physiol. Ser.* **52:** 19.

Chutkow W.A., Simon M.C., Le Beau M.M., and Burant C.F. 1996. Cloning, tissue expression, and chromosomal localization of SUR2, the putative drug-binding subunit of cardiac, skeletal muscle, and vascular KATP channels. *Diabetes* **45:** 1439.

Chutkow W.A., Pu J., Wheeler M.T., Wada T., Makielski J., Burant C.F., and McNally E.M. 2002. Hypertension and episodic coronary artery vasospasm in mice deficient for SUR2 KATP channels. *J. Clin. Invest.* **110:** 203.

Chutkow W.A., Samuel V., Hansen P.A., Pu J., Valdivia C.R., Makielski J.C., and Burant C.F. 2001. Disruption of Sur2-containing K(ATP) channels enhances insulin-stimulated glucose uptake in skeletal muscle. *Proc. Natl. Acad. Sci.* **98:** 11760.

Codd M.B., Sugrue D.D., Gersh B.J., and Melton L.J., III. 1989. Epidemiology of idiopathic dilated and hypertrophic cardiomyopathy. A population-based study in Olmsted County, Minnesota, 1975–1984. *Circulation* **80:** 564.

Cohn R.D. and Campbell K.P. 2000. Molecular basis of muscular dystrophies. *Muscle Nerve* **23:** 1456.

Cohn R.D., Durbeej M., Moore S.A., Coral-Vazquez R., Prouty S., and Campbell K.P. 2001. Prevention of cardiomyopathy in mouse models lacking the smooth muscle sarcoglycan-sarcospan complex. *J. Clin. Invest.* **107:** R1.

Coral-Vazquez R., Cohn R.D., Moore S.A., Hill J.A., Weiss R.M., Davisson R.L., Straub V., Barresi R., Bansal D., Hrstka R.F., Williamson R., and Campbell K.P. 1999. Disruption of the sarcoglycan-sarcospan complex in vascular smooth muscle: A novel mechanism for cardiomyopathy and muscular dystrophy. *Cell* **98:** 465.

Cordier L., Gao G.P., Hack A.A., McNally E.M., Wilson J.M., Chirmule N., and Sweeney H.L. 2001. Muscle-specific promoters may be necessary for adeno-associated virus-mediated gene transfer in the treatment of muscular dystrophies. *Hum. Gene Ther.* **12:** 205.

Cordier L., Hack A.A., Scott M.O., Barton-Davis E.R., Gao G., Wilson J.M., McNally E.M., and Sweeney H.L. 2000. Rescue of skeletal muscles of gamma-sarcoglycan-deficient mice with adeno-associated virus-mediated gene transfer. *Mol. Ther.* **1:** 119.

Coulton G.R., Morgan J.E., Partridge T.A., and Sloper J.C. 1988. The mdx mouse skeletal muscle myopathy. I. A histological, morphometric and biochemical investigation. *Neuropathol. Appl. Neurobiol.* **14:** 53.

Cox G.A., Phelps S.F., Chapman V.M., and Chamberlain J.S. 1993. New mdx mutation disrupts expression of muscle and nonmuscle isoforms of dystrophin. *Nat. Genet.* **4:** 87.

Crosbie R.H., Straub V., Yun H.Y., Lee J.C., Rafael J.A., Chamberlain J.S., Dawson V.L., Dawson T.M., and Campbell K.P. 1998. mdx muscle pathology is independent of nNOS perturbation. *Hum. Mol. Genet.* **7:** 823.

Danialou G., Comtois A.S., Dudley R., Karpati G., Vincent G., Des Rosiers C., and Petrof B.J. 2001. Dystrophin-deficient cardiomyocytes are abnormally vulnerable to mechanical stress-induced contractile failure and injury. *FASEB J.* **15:** 1655.

Deconinck A.E., Rafael J.A., Skinner J.A., Brown S.C., Potter A.C., Metzinger L., Watt D.J., Dickson J.G., Tinsley J.M., and Davies K.E. 1997. Utrophin-dystrophin-deficient mice as a model for Duchenne muscular dystrophy. *Cell* **90:** 717.

Donoghue M., Ernst H., Wentworth B., Nadal-Ginard B., and Rosenthal N. 1988. A muscle-specific enhancer is located at the 3′ end of the myosin light-chain 1/3 gene locus. *Genes Dev.* **2:** 1779.

Emery A.E. 2002. The muscular dystrophies. *Lancet* **359:** 687.

Ervasti J.M. and Campbell K.P. 1991. Membrane organization

of the dystrophin-glycoprotein complex. *Cell* **66**: 1121.

Ervasti J.M., Ohlendieck K., Kahl S.D., Gaver M.G., and Campbell K.P. 1990. Deficiency of a glycoprotein component of the dystrophin complex in dystrophic muscle. *Nature* **345**: 315.

Ettinger A.J., Feng G., and Sanes J.R. 1997. Epsilon-sarcoglycan, a broadly expressed homologue of the gene mutated in limb-girdle muscular dystrophy 2D. *J. Biol. Chem.* **272**: 32534.

Factor S.M., Minase T., Cho S., Dominitz R., and Sonnenblick E.H. 1982. Microvascular spasm in the cardiomyopathic Syrian hamster: A preventable cause of focal myocardial necrosis. *Circulation* **66**: 342.

Ganz P., Weidinger F.F., Yeung A.C., Vekshtein V.I., Vita J.A., Ryan T.J., Jr., McLenachan J.M., and Selwyn A.P. 1991. Coronary vasospasm in humans: The role of atherosclerosis and of impaired endothelial vasodilator function. *Basic Res. Cardiol.* **86**: 215.

Gossett L.A., Kelvin D.J., Sternberg E.A., and Olson E.N. 1989. A new myocyte-specific enhancer-binding factor that recognizes a conserved element associated with multiple muscle-specific genes. *Mol. Cell. Biol.* **9**: 5022.

Grady R.M., Teng H., Nichol M.C., Cunningham J.C., Wilkinson R.S., and Sanes J.R. 1997. Skeletal and cardiac myopathies in mice lacking utrophin and dystrophin: A model for Duchenne muscular dystrophy. *Cell* **90**: 729.

Grady R.M., Grange R.W., Lau K.S., Maimone M.M., Nichol M.C., Stull J.T., and Sanes, J.R. 1999. Role for alpha-dystrobrevin in the pathogenesis of dystrophin-dependent muscular dystrophies. *Nat. Cell Biol.* **1**: 215.

Hack A.A., Groh M.E., and McNally E.M. 2000a. Sarcoglycans in muscular dystrophy. *Microsc. Res. Tech.* **48**: 167.

Hack A.A., Cordier L., Shoturma D.I., Lam M.Y., Sweeney H.L., and McNally E.M. 1999. Muscle degeneration without mechanical injury in sarcoglycan deficiency. *Proc. Natl. Acad. Sci.* **96**: 10723.

Hack A.A., Ly C.T., Jiang F., Clendenin C.J., Sigrist K.S., Wollmann R.L., and McNally E.M. 1998. Gamma-sarcoglycan deficiency leads to muscle membrane defects and apoptosis independent of dystrophin. *J. Cell Biol.* **142**: 1279.

Hack A.A., Lam M.Y., Cordier L., Shoturma D.I., Ly C.T., Hadhazy M.A., Hadhazy M.R., Sweeney H.L., and McNally E.M. 2000b. Differential requirement for individual sarcoglycans and dystrophin in the assembly and function of the dystrophin-glycoprotein complex. *J. Cell Sci.* **113**: 2535.

Henry M.D. and Campbell K.P. 1998. A role for dystroglycan in basement membrane assembly. *Cell* **95**: 859.

Heydemann A., Wheeler M.T., and McNally E.M. 2001. Cardiomyopathy in animal models of muscular dystrophy. *Curr. Opin. Cardiol.* **16**: 211.

Johnson J.E., Wold B.J., and Hauschka S.D. 1989. Muscle creatine kinase sequence elements regulating skeletal and cardiac muscle expression in transgenic mice. *Mol. Cell. Biol.* **9**: 3393.

Lim L.E., Duclos F., Broux O., Bourg N., Sunada Y., Allamand V., Meyer J., Richard I., Moomaw C., Slaughter C., et al. 1995. Beta-sarcoglycan: Characterization and role in limb-girdle muscular dystrophy linked to 4q12. *Nat. Genet.* **11**: 257.

Matsuda R., Nishikawa A., and Tanaka H. 1995. Visualization of dystrophic muscle fibers in mdx mouse by vital staining with Evans blue: Evidence of apoptosis in dystrophin-deficient muscle. *J. Biochem.* **118**: 959.

McNally E.M., Ly C.T., and Kunkel L.M. 1998. Human epsilon-sarcoglycan is highly related to alpha-sarcoglycan (adhalin), the limb girdle muscular dystrophy 2D gene. *FEBS Lett.* **422**: 27.

McNally E.M., Passos-Bueno M.R., Bonnemann C.G., Vainzof M., de Sa Moreira E., Lidov H.G., Othmane K.B., Denton P.H., Vance J.M., Zatz M., and Kunkel L.M. 1996a. Mild and severe muscular dystrophy caused by a single gamma-sarcoglycan mutation. *Am. J. Hum. Genet.* **59**: 1040.

McNally E.M., Duggan D., Gorospe J.R., Bonnemann C.G., Fanin M., Pegoraro E., Lidov H.G., Noguchi S., Ozawa E., Finkel R.S., Cruse R.P., Angelini C., Kunkel L.M., and Hoffman E.P. 1996b. Mutations that disrupt the carboxyl-terminus of gamma-sarcoglycan cause muscular dystrophy. *Hum. Mol. Genet.* **5**: 1841.

Megeney L.A., Kablar B., Perry R.L., Ying C., May L., and Rudnicki M.A. 1999. Severe cardiomyopathy in mice lacking dystrophin and MyoD. *Proc. Natl. Acad. Sci.* **96**: 220.

Melacini P., Vianello A., Villanova C., Fanin M., Miorin M., Angelini C., and Dalla Volta S. 1996. Cardiac and respiratory involvement in advanced stage Duchenne muscular dystrophy. *Neuromuscul. Disord.* **6**: 367.

Melacini P., Fanin M., Duggan D.J., Freda M.P., Berardinelli A., Danieli G.A., Barchitta A., Hoffman E.P., Dalla Volta S., and Angelini C. 1999. Heart involvement in muscular dystrophies due to sarcoglycan gene mutations. *Muscle Nerve* **22**: 473.

Miki T., Suzuki M., Shibasaki T., Uemura H., Sato T., Yamaguchi K., Koseki H., Iwanaga T., Nakaya H., and Seino S. 2002. Mouse model of Prinzmetal angina by disruption of the inward rectifier Kir6.1. *Nat. Med.* **8**: 466.

Nigro V., de Sa Moreira E., Piluso G., Vainzof M., Belsito A., Politano L., Puca A.A., Passos-Bueno M.R., and Zatz M. 1996a. Autosomal recessive limb-girdle muscular dystrophy, LGMD2F, is caused by a mutation in the delta-sarcoglycan gene. *Nat. Genet.* **14**: 195.

Nigro V., Okazaki Y., Belsito A., Piluso G., Matsuda Y., Politano L., Nigro G., Ventura C., Abbondanza C., Molinari A.M., Acampora D., Nishimura M., Hayashizaki Y., and Puca G.A. 1997. Identification of the Syrian hamster cardiomyopathy gene. *Hum. Mol. Genet.* **6**: 601.

Nigro V., Piluso G., Belsito A., Politano L., Puca A.A., Papparclla S., Rossi E., Viglietto G., Esposito M.G., Abbondanza C., Medici N., Molinari A.M., Nigro G., and Puca G.A. 1996b. Identification of a novel sarcoglycan gene at 5q33 encoding a sarcolemmal 35 kDa glycoprotein. *Hum. Mol. Genet.* **5**: 1179.

Noguchi S., McNally E.M., Ben Othmane K., Hagiwara Y., Mizuno Y., Yoshida M., Yamamoto H., Bonnemann C.G., Gussoni E., Denton P.H., et al. 1995. Mutations in the dystrophin-associated protein gamma-sarcoglycan in chromosome 13 muscular dystrophy. *Science* **270**: 819.

Perloff J.K., de Leon A.C., Jr., and O'Doherty D. 1966. The cardiomyopathy of progressive muscular dystrophy. *Circulation* **33**: 625.

Petrof B.J., Shrager J.B., Stedman H.H., Kelly A.M., and Sweeney H.L. 1993. Dystrophin protects the sarcolemma from stresses developed during muscle contraction. *Proc. Natl. Acad. Sci.* **90**: 3710.

Piccolo F., Jeanpierre M., Leturcq F., Dode C., Azibi K., Toutain A., Merlini L., Jarre L., Navarro C., Krishnamoorthy R., Tome F.M., Urtizberea J.A., Beckmann J.S., Campbell K.P., and Kaplan J.C. 1996. A founder mutation in the gamma-sarcoglycan gene of gypsies possibly predating their migration out of India. *Hum. Mol. Genet.* **5**: 2019.

Piras G., El Kharroubi A., Kozlov S., Escalante-Alcalde D., Hernandez L., Copeland N.G., Gilbert D.J., Jenkins N.A., and Stewart C.L. 2000. Zac1 (Lot1), a potential tumor suppressor gene, and the gene for epsilon-sarcoglycan are maternally imprinted genes: Identification by a subtractive screen of novel uniparental fibroblast lines. *Mol. Cell. Biol.* **20**: 3308.

Quayle J.M., Nelson M.T., and Standen N.B. 1997. ATP-sensitive and inwardly rectifying potassium channels in smooth muscle. *Physiol. Rev.* **77**: 1165.

Rafael J.A. and Brown S.C. 2000. Dystrophin and utrophin: Genetic analyses of their role in skeletal muscle. *Microsc. Res. Tech.* **48**: 155.

Sakamoto A., Ono K., Abe M., Jasmin G., Eki T., Murakami Y., Masaki T., Toyo-oka T., and Hanaoka F. 1997. Both hypertrophic and dilated cardiomyopathies are caused by mutation of the same gene, delta-sarcoglycan, in hamster: An animal model of disrupted dystrophin-associated glycoprotein complex. *Proc. Natl. Acad. Sci.* **94**: 13873.

Sander M., Chavoshan B., Harris S.A., Iannaccone S.T., Stull J.T., Thomas G.D., and Victor R.G. 2000. Functional muscle ischemia in neuronal nitric oxide synthase-deficient skeletal muscle of children with Duchenne muscular dystrophy. *Proc. Natl. Acad. Sci.* **97**: 13818.

Seidman J.G. and Seidman C. 2001. The genetic basis for cardiomyopathy: From mutation identification to mechanistic paradigms. *Cell* **104:** 557.

Seino S. 1999. ATP-sensitive potassium channels: A model of heteromultimeric potassium channel/receptor assemblies. *Annu. Rev. Physiol.* **61:** 337.

Sicinski P., Geng Y., Ryder-Cook A.S., Barnard E.A., Darlison M.G., and Barnard P.J. 1989. The molecular basis of muscular dystrophy in the mdx mouse: A point mutation. *Science* **244:** 1578.

Sonnenblick E.H., Fein F., Capasso J.M., and Factor S.M. 1985. Microvascular spasm as a cause of cardiomyopathies and the calcium-blocking agent verapamil as potential primary therapy. *Am. J. Cardiol.* **55:** 179B.

Straub V., Rafael J.A., Chamberlain J.S., and Campbell K.P. 1997. Animal models for muscular dystrophy show different patterns of sarcolemmal disruption. *J. Cell Biol.* **139:** 375.

Straub V., Ettinger A.J., Durbeej M., Venzke D.P., Cutshall S., Sanes J.R., and Campbell K.P. 1999. epsilon-sarcoglycan replaces alpha-sarcoglycan in smooth muscle to form a unique dystrophin-glycoprotein complex. *J. Biol. Chem.* **274:** 27989.

Sueda S., Ochi N., Kawada H., Matsuda S., Hayashi Y., Tsuruoka T., and Uraoka T. 1999. Frequency of provoked coronary vasospasm in patients undergoing coronary arteriography with spasm provocation test of acetylcholine. *Am. J. Cardiol.* **83:** 1186.

Thomas G.D., Sander M., Lau K.S., Huang P.L., Stull J.T., and Victor R.G. 1998. Impaired metabolic modulation of alpha-adrenergic vasoconstriction in dystrophin-deficient skeletal muscle. *Proc. Natl. Acad. Sci.* **95:** 15090.

Tinsley J.M., Potter A.C., Phelps S.R., Fisher R., Trickett J.I., and Davies K.E. 1996. Amelioration of the dystrophic phenotype of mdx mice using a truncated utrophin transgene. *Nature* **384:** 349.

Towbin J.A. and Bowles N.E. 2001. Molecular genetics of left ventricular dysfunction. *Curr. Mol. Med.* **1:** 81.

Wheeler M.T., Zarnegar S., and McNally E.M. 2002. ζ-sarcoglycan is a novel component of the dystrophin-glycoprotein complex. *Hum. Mol. Gen.* **11:** 2147.

Yokoshiki H., Sunagawa M., Seki T., and Sperelakis N. 1998. ATP-sensitive K+ channels in pancreatic, cardiac, and vascular smooth muscle cells. *Am. J. Physiol.* **274:** C25.

Yoshida M. and Ozawa E. 1990. Glycoprotein complex anchoring dystrophin to sarcolemma. *J. Biochem.* **108:** 748.

Yoshida M., Hama H., Ishikawa-Sakurai M., Imamura M., Mizuno Y., Araishi K., Wakabayashi-Takai E., Noguchi S., Sasaoka T., and Ozawa E. 2000. Biochemical evidence for association of dystrobrevin with the sarcoglycan-sarcospan complex as a basis for understanding sarcoglycanopathy. *Hum. Mol. Genet.* **9:** 1033.

Yoshida T., Pan Y., Hanada H., Iwata Y., and Shigekawa M. 1998. Bidirectional signaling between sarcoglycans and the integrin adhesion system in cultured L6 myocytes. *J. Biol. Chem.* **273:** 1583.

Zhu X., Wheeler M.T., Hadhazy M., Lam M.Y., and McNally E.M. 2002. Cardiomyopathy is independent of skeletal muscle disease in muscular dystrophy. *FASEB J.* **8:** 8.

Zhu X., Hadhazy M., Groh M.E., Wheeler M.T., Wollmann R., and McNally E.M. 2001. Overexpression of gamma-sarcoglycan induces severe muscular dystrophy. Implications for the regulation of sarcoglycan assembly. *J. Biol. Chem.* **276:** 21785.

Zimprich A., Grabowski M., Asmus F., Naumann M., Berg D., Bertram M., Scheidtmann K., Kern P., Winkelmann J., Muller-Myhsok B., Riedel L., Bauer M., Muller T., Castro M., Meitinger T., Strom T.M., and Gasser T. 2001. Mutations in the gene encoding epsilon-sarcoglycan cause myoclonus-dystonia syndrome. *Nat. Genet.* **29:** 66.

The MLP Family of Cytoskeletal Z Disc Proteins and Dilated Cardiomyopathy: A Stress Pathway Model for Heart Failure Progression

M. HOSHIJIMA, M. PASHMFOROUSH, R. KNÖLL, AND K.R. CHIEN
Institute of Molecular Medicine, University of California San Diego, La Jolla, California 92093

Heart failure is a major cause of human morbidity and mortality, predicted to reach epidemic proportions in the early period of the 21st century. This growing incidence reflects our current lack of understanding of the fundamental pathways that drive the progressive dilation of the cardiac chamber and the associated decreases in cardiac contractility that accompany the diverse environmental stimuli that can trigger the disease, including post-infarction injury, side effects of chemotherapeutic agents, viral infection, and chronic exposure to volume and pressure overload. In short, the disease is complex, the pathologic stimuli are diverse, genetic susceptibility is varied, and the mechanisms that lead to progression are unclear. However, an integrative approach, employing insights from genetic-based studies in mouse and humans, in vivo somatic gene transfer studies, bioinformatics, and computational biology (Chien 2000), is beginning to provide new insights into the disease process and suggesting new therapeutic targets and strategies for intervening in the disease. Recent studies are beginning to apply these tools to clinical studies, representing a major scientific and translational opportunity in this post-genome era.

As an approach to unraveling the pathways for heart failure, our laboratory and others have utilized mouse model systems to dissect pathways that lead to dilated cardiomyopathy (DCM). Growing evidence suggests that a major component of this process appears to be chronic increases in wall stress that trigger specific signaling pathways for critical cell responses, which include cell death pathways, survival cues, hypertrophic responses, and associated changes in the downstream pathways (Chien 1999; Hoshijima and Chien 2002). This paper highlights the initial framework for utilizing mouse models to study the physiology of heart failure, and the subsequent scientific and therapeutic advances that have arisen from the initial discovery of a genetic link between defects in cytoskeletal Z disc proteins and dilated cardiomyopathy in mice that harbor a mutation in a muscle-specific LIM protein (MLP).

MOUSE AS A GENETICALLY BASED MODEL SYSTEM FOR DCM AND HEART FAILURE

Traditionally, the small animal model system of choice for the study of heart failure has been based on the post-infarcted rat heart, which has been shown to bear a close resemblance to features of human heart failure (Yue et al. 1998; Swynghedauw 1999; Gomez et al. 2001; Loennechen et al. 2002). During the process of recovery from the initial stimulus of the irreversible loss of cardiac muscle, the heart initiates a hypertrophic response that ultimately leads to cardiac enlargement in the viable myocardium, and later undergoes a transition where there is progressive chamber dilation and wall thinning that is accompanied by marked increases in wall stress and decreases in cardiac contractile performance. These latter features are also found in genetic forms of human DCM, where an intrinsic defect in cardiac muscle leads to pathological changes. Over the past decade, our laboratory and others have developed a combination of miniaturized in vivo physiological assay systems for cardiac failure (Table 1), as well as genetic tools, such as Cre-loxP conditional mutagenesis in ventricular muscle cells (Hirota et al. 1999; Crone et al. 2002; Ozcelik et al. 2002), genetic complementation (Minamisawa et al. 1999), and transcoronary somatic gene transfer, to uncover new disease and therapeutic pathways for heart failure in mouse-based disease model systems. Mechanistic studies in these mouse models have identified candidate genes and pathways that have been further examined in larger animal model systems and in carefully enriched patient populations whose clinical phenotype matches that of the in vivo mouse cardiac phenotypes.

A GENETICALLY BASED MODEL OF DCM IN MLP-DEFICIENT MICE: INITIAL LINK BETWEEN CYTOSKELETAL DEFECTS AND DCM

One of the first genetic links between cardiac cytoskeletal defects and DCM arose from studies of mice that harbor a deficiency in MLP. MLP, also known as cysteine-rich protein3 (CRP3), is a member of the CRP gene family (Arber et al. 1994; Weiskirchen et al. 1995) and contains two highly conserved LIM domains that have been shown to serve as protein–protein interaction modules in a series of LIM proteins (Bach 2000). MLP is expressed in cardiac and skeletal muscle (Arber et al. 1994, 1997) and is predominantly localized in the cardiac cytoskeleton, where it binds to α-actinin at the Z disc

Table 1. In Vivo Cardiac Physiological Phenotyping and Biomechanical Stress Assays in Genetically Engineered Mice

	References
Non-invasive phenotyping	
Transthoracic echocardiography	Tanaka et al. (1996)
Transesophageal echocardiography	Scherrer-Crosbie et al. (1998)
Magnetic resonance imaging	Kubota et al. (1997); Williams et al. (2001)
Invasive phenotyping	
LV pressure analysis	Milano et al. (1994)
LV pressure-volume loop analysis	McConnell et al. (1999)
PA/RV pressure analysis	Zhao et al. (2002)
LV and RV angiography	Rockman et al. (1994); Pashmforoush et al. (2001)
Aortography	Nakamura et al. (2002)
Electrophysiological phenotyping	
Surface electrocardiography	Berul et al. (1996); Sah et al. (1999)
Telemetric electrocardiography	Kramer et al. (1993); Nguyen-Tran et al. (2000)
Transesophageal cardiac pacing	Scherrer-Crosbie et al. (1998)
Open-chest in vivo EP study	Berul et al. (1997); Nguyen-Tran et al. (2000)
Closed-chest in vivo endocardial EP study	Berul et al. (1998); Kuo et al. (2001)
Biomechanical stress assays	
In vivo	
Thoracic aortic banding	Rockman et al. (1991)
Abdominal aortic banding	Shimoyama et al. (1999)
PA banding	Rockman et al. (1994)
Arteriovenous fistula	Tanaka et al. (1996)
Coronary artery ligation	Michael et al. (1995)
In vitro	
Isolated perfused heart system	Grupp et al. (1993); Geisterfer-Lowrance et al. (1996)
Papillary muscle stretch	Knöll et al. (2002)
Cultured neonatal cardiomyocyte stretch	Kudoh et al. (1998); Knöll et al. (2002)

(LV) Left ventricle; (RV) right ventricle; (PA) pulmonary artery; (EP) electrophysiology.

structure (Clark et al. 2002). Interestingly, mice that lack MLP display a marked cardiac phenotype which completely reproduces the phenotype of human DCM, including progressive enlargement of all four cardiac chambers, ventricular wall thinning, decreases in a panel of physiological endpoints of cardiac contractility (Arber et al. 1997), defects in sarcoplasmic reticulum (SR) Ca^{++} cycling (Minamisawa et al. 1999; Esposito et al. 2000), increases in wall stress, loss of the transient outward (Ito) current (W. Giles and K. Chien, unpubl.), and the desensitization β-adrenergic signaling (Fig. 1) (Rockman et al. 1998). This remarkable recapitulation of the entire phenotypic spectrum of clinical DCM and associated heart failure has led to the establishment of the $MLP^{-/-}$ mouse system as a tool for unraveling pathways that lead to heart failure initiation and progression (Knöll et al. 2002), as well as the identification of new therapeutic strategies via genetic complementation (Minamisawa et al. 1999). In addition, this initial link between the cardiac cytoskeleton in general, and cardiac Z disc proteins in particular, and DCM has now been extended to include an expanding number of known and novel cytoskeletal/Z disc components, such as telethonin (titin-cap:T-cap), actinin-associated LIM protein (ALP), and ZASP/Cypher/Oracle (Chien and Olson 2002; Clark et al. 2002).

A MLP–TELETHONIN–TITIN COMPLEX IS AN ESSENTIAL COMPONENT OF THE CARDIAC STRETCH SENSOR MACHINERY

Since DCM can result from mutations in a diverse group of cytoskeletal proteins, the possibility exists that there may be discrete pathways by which each of these proteins is linked to the disease phenotype, thereby reflecting individual, distinct roles in the regulation of normal cardiac function that lie beyond their known role as a structural scaffold for cardiac sarcomeric organization and assembly. Utilizing the MLP-null mouse model of cardiomyopathy, a critical role for MLP as an essential component of the cardiac muscle stretch sensor machinery has recently been uncovered, identifying MLP as a molecular link between the Z disc and the titin complex in cardiac muscle cells (Fig. 2) (Knöll et al. 2002). Via a combination of yeast two-hybrid, GST pull-down, and immunohistological studies, it has been shown that MLP binds to T-cap, a titin-associated protein that is localized at Z disc, interacting with the amino-terminal end of the titin molecule. The T-cap interacting domain of MLP is localized in a short amino-terminal domain of MLP that precedes the first LIM domain. Titin serves a structural role in cardiac muscle in sarcomeric organization and assembly, as well as a crucial role in elastic recoil to restore cardiac sarcomeric length during each cardiac contraction (Gregorio et al. 1999; Granzier and Labeit 2002). Recent studies have also implicated titin function in passive stretch-mediated responses in cardiac muscle (Knöll et al. 2002). Interestingly, $MLP^{-/-}$ cardiac papillary muscle displays a selective defect in stretch-mediated responses, with an impairment in tension generation following the delivery of a 10% increase in passive stretch of the muscle, implying that there may be a primary defect in muscle stretch activation responses. MLP-deficient neonatal cardiac muscle cells display a nearly complete loss of stretch-dependent induction of a genetic marker of cardiac mechanical stress, BNP, but maintain a normal induction of the gene in response to hormonal stimulation, providing direct support for a primary defect in muscle stretch sensing in $MLP^{-/-}$ cardiac muscle. Taken together, these studies suggest that a MLP–T-cap–titin complex is an essential component of the cardiac muscle stretch sensor machinery, and that a loss of this function leads to DCM via the loss of the ability to activate stretch-mediated survival cues, such as gp130 survival pathways (Fig. 2).

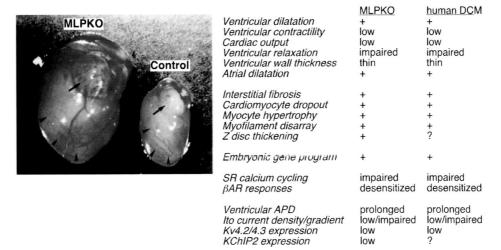

Figure 1. MLP-deficient (MLPKO) mice share a broad spectrum of pathological phenotypes with human DCM. (βAR) β-adrenergic receptor; (APD) action potential duration.

A FAMILY OF MUSCLE-RESTRICTED LIM DOMAIN PROTEINS: ALP LOCALIZED AT THE CARDIAC Z DISC AND ASSOCIATED WITH RIGHT VENTRICULAR CARDIOMYOPATHY

MLP is a member of a family of cytoskeletal LIM domain proteins that can be divided into distinct subfamilies according to the presence of accessory protein domains (Bach 2000). One of the largest classes of these include the PDZ-LIM domain proteins, several of which are noted to be expressed at the cardiac Z disc and display a restricted pattern of expression in cardiac and skeletal muscle. Recent studies have implicated two separate members of this family in cardiomyopathy. ALP is an actinin-associated PDZ-LIM domain protein that is localized at the cardiac Z disc and forms a complex with α-actinin, and possibly with γ-catenin, at the cardiac intercalated disc (Pashmforoush et al. 2001). Interestingly, mutations in γ-catenin have recently been shown to be the disease gene in a rare form of RV dysplasia that is characterized by a thin right ventricular wall and a form of predominant right ventricular cardiomyopathy (McKoy et al. 2000), although γ-catenin is expressed in both the right and left ventricular chambers. Intriguingly, ALP is predominantly expressed in the RV chamber during cardiac development, and recent studies in ALP-deficient mice have documented a form of RV dysplasia characterized by a predominant RV cardiomyopathy (Fig. 3) (Pashmforoush et al. 2001). These results again support a critical role for Z disc proteins in the pathogenesis of cardiomyopathy, in this case via contributing to the stability of the cardiac Z disc in the embryonic RV chamber, which supports greater stroke volume than the left ventricle in the embryonic circulation system (Sutton et al. 1991). Interestingly, another member of the muscle-restricted Z disc-associated PDZ-LIM domain family has been linked to congenital myopathy in a mouse model that harbors a deficiency in ZASP/Cypher/Oracle, a protein that also interacts with protein kinase C (Zhou et al. 2001). In addition, a growing number of cytoskeletal LIM proteins have also been uncovered (Fig. 4). The systematic mutations of these genes in mice and translational studies in enriched patient populations should be informative as to their potential role in the initiation and progression of DCM and heart failure.

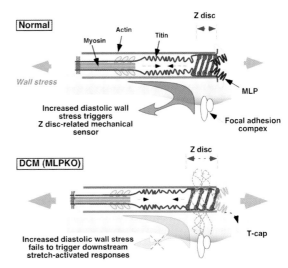

Figure 2. MLP–T-cap–titin complex and the cardiac stretch sensor machinery. At diastole, as ventricular filling pressure rises, titin elastic segments are reshaped and entropy elasticity generates diastolic wall stress. The change in intrinsic elasticity functions as the sensor of mechanical load. In MLP$^{-/-}$ heart, we speculate the molecular involvement of sensor dysfunction as follows. Ventricular chamber dilatation and wall thinning together enhance diastolic wall stress; T-cap, which is mechanically supported by MLP, gradually dislocates because of instability without MLP; MLP/T-cap absence causes Z disc and related titin abnormalities; the MLP$^{-/-}$ cardiomyocytes fail in building up titin-related entropy elasticity; translation of mechanical stress to the signaling to regulate nuclear gene program is not functionally triggered.

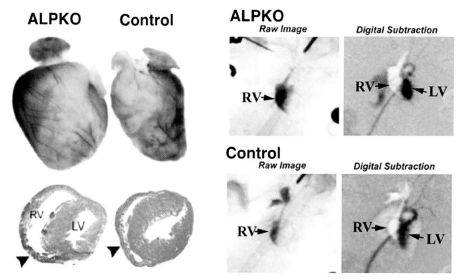

Figure 3. ALPKO mice display RV dominant DCM phenotypes. Note significantly enlarged RV chamber, which is documented in vivo by chamber-selective angiography. (Adapted, with permission, from Pashmforoush et al. 2001 [copyright Nature Publishing Group].)

A FOUNDER EFFECT W4RMLP MUTATION IS ASSOCIATED WITH DCM IN A SUBSET OF PATIENTS WITH DCM

Since the initial report of the link between defects in MLP in DCM, a growing body of evidence both in experimental model systems and in familial forms of the disease have implicated mutations in cytoskeletal proteins in general, and Z disc proteins in particular, in the pathogenesis of this cardiac disease (Seidman and Seidman 2001; Towbin and Bowles 2002). In this regard, the finding that MLP is part of an endogenous MLP–T-cap complex became of interest, as mutations in T-cap are linked to limb-girdle muscular dystrophy type 2G (Moreira et al. 2000), raising the question as to whether MLP mutations might be linked to forms of DCM in human populations. By screening over 1400 well-phenotyped control individuals and patients with DCM, we have recently reported a close association between a subset of human DCM and a mutation in a highly conserved amino acid residue in the amino-terminal domain of MLP (Knöll et al. 2002). The 10 patients harboring the heterozygous

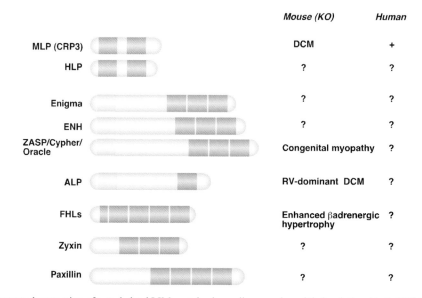

Figure 4. The segregated expression of cytoskeletal LIM proteins in cardiac muscle and their relationship to DCM. References: MLP (Arber et al. 1997); HLP, heart LIM protein (Yu et al. 2002); Enigma (Guy et al. 1999); ENH, enigma homolog (Nakagawa et al. 2000); ZASP, Z-band alternatively spliced PDZ motif protein/Cypher/Oracle (Faulkner et al. 1999; Zhou et al. 1999, 2001; Passier et al. 2000); ALP (Pashmforoush et al. 2001); CLIM1/Elfin (Kotaka et al. 1999); FHL, four and a half LIM-only protein (Chan et al. 1998; Chu et al. 2000; Kong et al. 2001); Zyxin (Beckerle 1997); Paxillin (Schaller 2001).

W4R mutant MLP gene display a clinical phenotype that is consistent with the MLP-deficient mouse heart. Although the formal linkage analysis in these families is limited by several factors, including the size of available pedigrees and age-dependent penetrance, the haplotype analysis provides strong evidence of a founder effect in this population. The strong association of the W4RMLP mutation with a neighboring polymorphic marker suggests that these patients share a distant common ancestor, and supports vertical transmission of the DCM phenotype related to W4RMLP over many generations in the European population (Fig. 5). Haploinsufficiency is the most likely mechanistic basis for the appearance of the disease phenotype in the W4R patients, as precise left ventricular pressure measurements reveal a partial defect in cardiac contractility and β-adrenergic agonist sensitivity in MLP$^{+/-}$ animals (J. Ross and K.R. Chien, unpubl.). Furthermore, the W4R mutation in MLP results in the complete loss of T-cap interaction, and myocardial biopsies from patients with this mutation show the same phenotype as the one found in the MLP$^{-/-}$ myocardium. Independent data in MLP-deficient cardiac muscle support the concept that T-cap binding to MLP is critical for maintenance of the cardiac Z disc, and that, in the absence of MLP, additional mechanical stress facilitates the mislocalization of T-cap from the titin complex, which is linked to the onset of Z disc defects and progressive DCM (Fig. 2).

GENETIC COMPLEMENTATION UNCOVERS A CRITICAL ROLE OF CALCIUM CYCLING DEFECTS IN DCM: PHOSPHOLAMBAN ABLATION RESCUES HEART FAILURE IN MLP-DEFICIENT MICE

Defects in SR Ca^{++} cycling are a conserved feature of all forms of human heart failure, including DCM. One of the most prominent defects relates to a decrease in the activity of the SR calcium ATPase 2 (SERCA2), which is responsible for transporting Ca^{++} from the cardiac cytoplasm into the SR lumen. The SR calcium level is directly proportional to the amount of Ca^{++} that is released via the ryanodine receptor molecule, a calcium release channel, during each cardiac contraction. In this manner, the activity of SERCA2 is critical for the maintenance of both normal cardiac contractility (systolic function) and relaxation (diastolic function). Cardiac muscle cells also have an endogenous inhibitor of SERCA2, namely PLN, which is a highly conserved 52-amino-acid peptide expressed almost exclusively in the heart and slow skeletal muscle. PLN is phosphorylated by AMP-dependent protein kinase (PKA), which results in releasing the PLN inhibitory effects on the SERCA2 activity, which represents an important mechanism by which β-adrenergic agonists activate cardiac contractility (Bers 2002). In many forms of human and experimental heart failure, the down-regulation of the SERCA2 expression level, impaired SERCA2/PLN ratio, and a marked decrease in the phosphorylated form of PLN have been noted (Hasenfuss 1998; Tomaselli and Marban 1999; Houser et al. 2000), raising the question as to whether this might represent a nodal point in the pathway of heart failure progression. To address this question, we have utilized a genetic complementation strategy to provide strong evidence that defects in SR Ca^{++} cycling play a pivotal role in heart failure progression. Accordingly, we have generated double knockout mice, which harbor the MLP cardiomyopathic mutation but also lack the PLN, and subsequently examined whether removing PLN inhibition alone would have a measurable effect on heart failure progression. Remarkably, the ablation of PLN completely prevented the spectrum of heart failure phenotypes found in a mouse model of DCM caused by MLP deficiency (Minamisawa et al. 1999).

Subsequently, recent studies have further supported the beneficial effect of PLN ablation in other transgenic cardiomyopathy mouse models (Freeman et al. 2001; Sato et al. 2001). In addition, independent studies employing short-term adenoviral-mediated somatic gene transfer of SERCA, or overexpression of upstream regulators of SERCA, such as a β-adrenergic receptor kinase inhibitory peptide, have been shown to delay the development of cardiac dysfunction (Miyamoto et al. 2000; White et al. 2000). Taken together, these studies suggest that the inhibition of a single inhibitory Ca^{++} cycling gene is sufficient to completely block heart failure progression. In fact, a 50% reduction of PLN significantly blocks heart failure progression in MLP-deficient mice (Minamisawa et al. 1999), thereby validating this as a potential therapeutic target for heart failure.

PLN INHIBITION RESCUES MULTIPLE FORMS OF GENETIC AND ACQUIRED FORMS OF HEART FAILURE: A NEW PARADIGM FOR THE DISCOVERY OF THERAPEUTIC STRATEGIES FOR HEART FAILURE PROGRESSION

As noted above, the genetic complementation of the DCM phenotype in MLP-deficient mice via interbreeding with mice harboring an additional ablation of the PLN gene results in a complete, chronic prevention of the on-

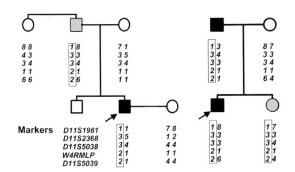

Figure 5. W4RMLP mutation is transmitted with DCM phenotype in the subset of European DCM population. The pedigrees of two index DCM patients (*arrows*) are illustrated. Four neighboring allelic markers are co-transmitted with the W4R mutation in MLP, suggesting the founder effect for the origin of this mutation. Filled, shaded, and open symbols denote clinically affected, mildly affected, and unaffected disease status.

set of heart failure (Minamisawa et al. 1999). Since the onset of the DCM phenotype is blocked in these mice, it remained unclear as to whether PLN inhibition would prove therapeutically efficacious after the onset of cardiomyopathy. Utilizing a new transcoronary gene delivery system via recombinant adeno-associated virus (rAAV) vectors (Ikeda et al. 2002), we have recently documented the effects of a pseudophosphorylated (S16E) mutant of PLN, designed to mimic the conformational change in PLN following phosphorylation by PKA, to constitutively activate SR Ca^{++} cycling in the myocardium of diverse species, including rats and hamsters. The well-characterized BIO14.6 cardiomyopathic (CM) hamster displays a progressive form of DCM and associated severe heart failure (Ikeda et al. 2000) that arises due to a mutation in the δ-sarcoglycan gene and corresponds to the genetic defect in patients with an autosomal recessive form of cardiomyopathy found in limb-girdle muscular dystrophy type F (Straub and Campbell 1997). In recent studies, we constructed the S16E mutant of PLN (S16EPLN) in a rAAV vector (rAAV/S16EPLN) and delivered it into BIO14.6 CM hamsters, providing clear evidence that chronic enhancement of SR Ca^{++} cycling promotes the improvement of progressive cardiac contractile dysfunction and assists in cardiac cytoprotection in this animal model of DCM and heart failure (Hoshijima et al. 2002). The chronic inhibition of PLN by the pseudophosphorylated PLN peptide, which leads to the constitutive activation of cardiac contractility and relaxation in the absence of adrenergic stimuli, was analogous to the null phenotype of the PLN mutant mice. Furthermore, the extended studies provide clear evidence that the in vivo rAAV/S16EPLN delivery blocks the pathological remodeling with left ventricular chamber dilation in the chronic progressive heart failure of post-myocardial infarction rats (Y. Iwanaga et al., unpubl.).

The molecular mechanism of S16EPLN action has been speculated as follows. The recent 2.6 Å resolution X-ray crystallographic analysis of SERCA (Toyoshima et al. 2000) and high-resolution NMR spectroscopy of PLN (Cornea et al. 1997; Pollesello et al. 1999) have provided a structural framework for understanding the molecular interaction of SERCA-PLN. In these models, positive charged arginines (R9[+], R13[+], and possibly R14[+]) are located in the proximity of the neutral S16T17 and interact with the charged amino acid cluster of D397(–)D(–)K(+)PV402 in SERCA1a. The phosphorylation of S16 and T17 (S16Pi[-] and T17Pi[-], respectively) neutralizes the positive charges of R9(+), R13(+), and R14(+). On the other hand, R25(+) can interact with phosphorylated side chains of S16Pi(–)T17Pi(–) to stabilize the oligomeric structure of PLN, as proposed earlier (Cornea et al. 1997). Interestingly, recent independent observations support the working model that a phosphorylated PLN remains associated with SERCA (Asahi et al. 2000; Negash et al. 2000). Thus, a slight conformational change induced by a side-chain charge shift at S16T17 most likely exerts its effects by dramatically modifying the functional interaction of PLN-SERCA (Fig. 6).

The unique specificity of the S16EPLN peptide to precisely target SR Ca^{++} uptake, and the inherent cardiac and slow skeletal muscle specificity of PLN, suggested the potential therapeutic value of in vivo rAAV/S16EPLN gene delivery. Previous experimental and clinical studies have documented that chronic increases in contractility mediated by β-adrenergic agonists or phosphodiesterase inhibitors can lead to the rapid progression of heart dysfunction in chronic heart failure (Stevenson 1998). Furthermore, administration of β-blockers may improve survival and the progression of clinical heart failure (Braunwald 2001). The mechanisms that underlie this

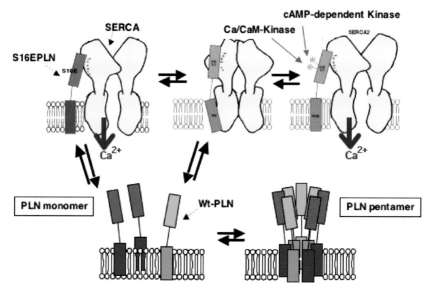

Figure 6. The working model of the pseudophosphorylated mutant of PLN (S16EPLN) to activate SERCA2. S16PLN replaces endogenous PLN (Wt-PLN) for SERCA interaction. S16PLN also segregates with Wt-PLN to increase the fraction of the oligomeric form of PLNs, which are inactive or less functional.

detrimental long-term effect have been ascribed to catecholamine toxicity, and have raised the query as to whether the chronic stimulation of cardiac performance by catecholamines is inherently driving heart failure progression. Consistent with this notion, a recent study has documented the development of DCM in cardiac transgenic mice overexpressing a constitutive active PKA, pointing out the substantial risk of modifying upstream signaling molecules in the β-adrenergic pathway (Antos et al. 2001). rAAV/S16EPLN gene therapy targets the most downstream point in this pathway, without chronic increases in cAMP or chronic PKA activation. In addition, the biological effects of S16EPLN are restricted to the cell types, which express SERCA1/2, avoiding any unexpected long-term toxic side effects in nontargeted tissues or cell types. Accordingly, chronic improvement of excitation-contraction coupling in a cAMP-independent manner with a high efficiency, long term, and cardiotropic rAAV/S16EPLN gene delivery represents a novel therapeutic strategy for slowing heart failure progression. Given that there is currently no therapy for end-stage heart failure aside from heart transplantation, future studies appear warranted to translate these findings into larger animal models and the clinical setting.

CONCLUSION

The MLP-associated defects in DCM and heart failure, based on studies in mice and humans, have led to the initial identification of a critical role of cytoskeletal proteins in general, and Z disc proteins in particular, in the pathways for disease development. In addition, MLP is an essential component of molecular machinery for cardiac muscle stretch sensing, and therefore represents one of the most proximal steps in the initiation of growth and survival cues that originate from heart muscle itself in response to increases in wall stress and the associated increase in passive stretch of cardiac muscle. Mutations in MLP in experimental models and in human patients implicate defects in the muscle stretch sensor function as a novel pathway for heart failure progression, and genetic studies indicate that the loss of this MLP-related function may account for the subset of European DCM patient populations due to a founder effect of W4R mutation in MLP. Furthermore, the recent finding that MLP protein levels are decreased in human idiopathic and ischemic cardiomyopathy patients (Zolk et al. 2000) suggests that defects in stabilization of the macromolecular titin–Z disc complex including MLP could potentially underlie other genetic and acquired forms of heart failure, as well. On the other hand, genetic complementation studies in the MLP mutant mouse model have uncovered a critical role for SR Ca^{++} cycling in heart failure progression and have led to the identification of new therapeutic targets and pathways for interrupting heart failure progression that will necessitate further clinical studies. Taken together, studies of MLP have provided a compelling case for a critical role of increases in wall stress as the one of the most important risk factors for heart failure progression, and may underlie the activation of multiple pathways for myocardial cell death and pathological remodeling that characterize end-stage heart failure. It will become of interest to determine precisely how the mechanical sensor function interacts with specific sets of signaling effectors, and to determine how the sensor function is precisely activated only during increases in passive stretch, which extend the sarcomere length changes during normal cardiac cycles, and how this sensor signaling cross-talks with other cardiac hypertrophic and cell survival signaling pathways (Fig. 7). The recent discovery of a growing number of components of the cardiac Z disc and their associated proteins forms an excellent starting point for future work in this field that is likely to be propelled by an integrative biology approach based on experimental studies in mouse models, somatic genetic transfer in acquired forms of heart failure in large animal models, and genetic and clinical studies in carefully phenotyped patient populations.

ACKNOWLEDGMENT

We thank the Jean Le Ducq Foundation for supporting these entire studies.

REFERENCES

Antos C.L., Frey N., Marx S.O., Reiken S., Gaburjakova M., Richardson J.A., Marks A.R., and Olson E.N. 2001. Dilated cardiomyopathy and sudden death resulting from constitutive activation of protein kinase a. *Circ. Res.* **89:** 997.

Arber S., Halder G., and Caroni P. 1994. Muscle LIM protein, a novel essential regulator of myogenesis, promotes myogenic differentiation. *Cell* **79:** 221.

Arber S., Hunter J.J., Ross J., Jr., Hongo M., Sansig G., Borg J., Perriard J.C., Chien K.R., and Caroni P. 1997. MLP-deficient mice exhibit a disruption of cardiac cytoarchitectural organization, dilated cardiomyopathy, and heart failure. *Cell* **88:** 393.

Asahi M., McKenna E., Kurzydlowski K., Tada M., and

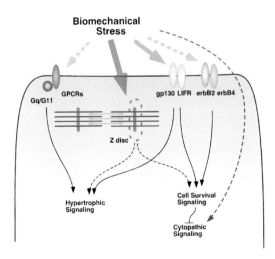

Figure 7. Biomechanical stress triggers multiple signaling of myocyte hypertrophy, survival, and death. Z disc-related mechanical sensor may cross-talk with other receptor-dependent signaling, including gp130-LIF receptor (LIFR) (Hirota et al. 1999; Yasukawa et al. 2001), erbB2-erbB4 (Crone et al. 2002), and G-protein coupling receptor (GPCR) dependent pathways (Wettschureck et al. 2001).

MacLennan D. H. 2000. Physical interactions between phospholamban and sarco(endo)plasmic reticulum Ca^{2+}-ATPases are dissociated by elevated Ca^{2+}, but not by phospholamban phosphorylation, vanadate, or thapsigargin, and are enhanced by ATP. *J. Biol. Chem.* **275:** 15034.

Bach I. 2000. The LIM domain: Regulation by association. *Mech. Dev.* **91:** 5.

Beckerle M.C. 1997. Zyxin: Zinc fingers at sites of cell adhesion. *Bioessays* **19:** 949.

Bers D.M. 2002. Cardiac excitation-contraction coupling. *Nature* **415:** 198.

Berul C.I., Aronovitz M.J., Wang P.J., and Mendelsohn M.E. 1996. In vivo cardiac electrophysiology studies in the mouse. *Circulation* **94:** 2641.

Berul C.I., Christe M.E., Aronovitz M.J., Seidman C.E., Seidman J.G., and Mendelsohn M.E. 1997. Electrophysiological abnormalities and arrhythmias in alpha MHC mutant familial hypertrophic cardiomyopathy mice. *J. Clin. Invest.* **99:** 570.

Berul C.I., Christe M.E., Aronovitz M.J., Maguire C.T., Seidman C.E., Seidman J.G., and Mendelsohn M.E. 1998. Familial hypertrophic cardiomyopathy mice display gender differences in electrophysiological abnormalities. *J. Interv. Card. Electrophysiol.* **2:** 7.

Braunwald E. 2001. Expanding indications for beta-blockers in heart failure. *N. Engl. J. Med.* **344:** 1711.

Chan K.K., Tsui S.K., Lee S.M., Luk S.C., Liew C.C., Fung K.P., Waye M.M., and Lee C.Y. 1998. Molecular cloning and characterization of FHL2, a novel LIM domain protein preferentially expressed in human heart. *Gene* **210:** 345.

Chien K.R. 1999. Stress pathways and heart failure. *Cell* **98:** 555.

———. 2000. Genomic circuits and the integrative biology of cardiac diseases. *Nature* **407:** 227.

Chien K.R. and Olson E.N. 2002. Converging pathways and principles in heart development and disease: CV@CSH. *Cell* **110:** 153.

Chu P.H., Bardwell W.M., Gu Y., Ross J., Jr., and Chen J. 2000. FHL2 (SLIM3) is not essential for cardiac development and function. *Mol. Cell. Biol.* **20:** 7460.

Clark K.A., McElhinny A.S., Beckerle M.C., and Gregorio C.C. 2002. Striated muscle cytoarchitecture: An intricate web of form and function. *Annu. Rev. Cell Dev. Biol.* **18:** 637.

Cornea R.L., Jones L.R., Autry J.M., and Thomas D.D. 1997. Mutation and phosphorylation change the oligomeric structure of phospholamban in lipid bilayers. *Biochemistry* **36:** 2960.

Crone S.A., Zhao Y.Y., Fan L., Gu Y., Minamisawa S., Liu Y., Peterson K.L., Chen J., Kahn R., Condorelli G., Ross J., Jr., Chien K.R., and Lee K.F. 2002. ErbB2 is essential in the prevention of dilated cardiomyopathy. *Nat. Med.* **8:** 459.

Esposito G., Santana L.F., Dilly K., Cruz J.D., Mao L., Lederer W.J., and Rockman H.A. 2000. Cellular and functional defects in a mouse model of heart failure. *Am. J. Physiol. Heart Circ. Physiol.* **279:** H3101.

Faulkner G., Pallavicini A., Formentin E., Comelli A., Ievolella C., Trevisan S., Bortoletto G., Scannapieco P., Salamon M., Mouly V., Valle G., and Lanfranchi G. 1999. ZASP: A new Z-band alternatively spliced PDZ-motif protein. *J. Cell Biol.* **146:** 465.

Freeman K., Lerman I., Kranias E.G., Bohlmeyer T., Bristow M.R., Lefkowitz R.J., Iaccarino G., Koch W.J., and Leinwand L.A. 2001. Alterations in cardiac adrenergic signaling and calcium cycling differentially affect the progression of cardiomyopathy. *J. Clin. Invest.* **107:** 967.

Geisterfer-Lowrance A.A., Christe M., Conner D.A., Ingwall J.S., Schoen F.J., Seidman C.E., and Seidman J.G. 1996. A mouse model of familial hypertrophic cardiomyopathy. *Science* **272:** 731.

Gomez A.M., Guatimosim S., Dilly K.W., Vassort G., and Lederer W.J. 2001. Heart failure after myocardial infarction: Altered excitation-contraction coupling. *Circulation* **104:** 688.

Granzier H. and Labeit D. 2002. Cardiac titin: An adjustable multi-functional spring. *J. Physiol.* (part 2) **541:** 335.

Gregorio C.C., Granzier H., Sorimachi H., and Labeit S. 1999. Muscle assembly: A titanic achievement? *Curr. Opin. Cell Biol.* **11:** 18.

Grupp I.L., Subramaniam A., Hewett T.E., Robbins J., and Grupp G. 1993. Comparison of normal, hypodynamic, and hyperdynamic mouse hearts using isolated work-performing heart preparations. *Am. J. Physiol.* **265:** H1401.

Guy P.M., Kenny D.A., and Gill G.N. 1999. The PDZ domain of the LIM protein enigma binds to beta-tropomyosin. *Mol. Biol. Cell* **10:** 1973.

Hasenfuss G. 1998. Alterations of calcium-regulatory proteins in heart failure. *Cardiovasc. Res.* **37:** 279.

Hirota H., Chen J., Betz U.A., Rajewsky K., Gu Y., Ross J., Jr., Muller W., and Chien K.R. 1999. Loss of a gp130 cardiac muscle cell survival pathway is a critical event in the onset of heart failure during biomechanical stress. *Cell* **97:** 189.

Hoshijima M. and Chien K.R. 2002. Mixed signals in heart failure: Cancer rules. *J. Clin. Invest.* **109:** 849.

Hoshijima M., Ikeda Y., Iwanaga Y., Minamisawa S., Date M.O., Gu Y., Iwatate M., Li M., Wang L., Wilson J.M., Wang Y., Ross J., Jr., and Chien K.R. 2002. Chronic suppression of heart-failure progression by a pseudophosphorylated mutant of phospholamban via in vivo cardiac rAAV gene delivery. *Nat. Med.* **8:** 864.

Houser S.R., Piacentino V., III, and Weisser J. 2000. Abnormalities of calcium cycling in the hypertrophied and failing heart. *J. Mol. Cell. Cardiol.* **32:** 1595.

Ikeda Y., Martone M., Gu Y., Hoshijima M., Thor A., Oh S.S., Peterson K.L., and Ross J. 2000. Altered membrane proteins and permeability correlate with cardiac dysfunction in cardiomyopathic hamsters. *Am. J. Physiol. Heart Circ. Physiol.* **278:** H1362.

Knöll R., Hoshijima M., Hoffman H.M., Person V., Lorenzen-Schmidt I., Bang M.-L., Hayashi T., Shiga N., Yasukawa H., Schaper W., McKenna W., Yokoyama M., Schork N.J., Omens J.H., McCulloch A.D., Kimura A., Gregorio C.G., Poller W., Schaper J., Schultheiss H.P., and Chien K.R. 2002. The cardiac mechanical stretch sensor machinery involves a Z disc complex that is defective in a subset of human dilated cardiomyopathy. *Cell* **111:** 943.

Ikeda Y., Gu Y., Iwanaga Y., Hoshijima M., Oh S.S., Giordano F.J., Chen J., Nigro V., Peterson K.L., Chien K.R., and Ross J., Jr. 2002. Restoration of deficient membrane proteins in the cardiomyopathic hamster by in vivo cardiac gene transfer. *Circulation* **105:** 502.

Kong Y., Shelton J.M., Rothermel B., Li X., Richardson J.A., Bassel-Duby R., and Williams R.S. 2001. Cardiac-specific LIM protein FHL2 modifies the hypertrophic response to beta-adrenergic stimulation. *Circulation* **103:** 2731.

Kotaka M., Ngai S.M., Garcia-Barcelo M., Tsui S.K., Fung K.P., Lee C.Y., and Waye M.M. 1999. Characterization of the human 36-kDa carboxyl terminal LIM domain protein (hCLIM1). *J. Cell. Biochem.* **72:** 279.

Kramer K., van Acker S.A., Voss H.P., Grimbergen J.A., van der Vijgh W.J., and Bast A. 1993. Use of telemetry to record electrocardiogram and heart rate in freely moving mice. *J. Pharmacol. Toxicol. Methods* **30:** 209.

Kubota T., McTiernan C.F., Frye C.S., Slawson S.E., Lemster B.H., Koretsky A.P., Demetris A.J., and Feldman A.M. 1997. Dilated cardiomyopathy in transgenic mice with cardiac-specific overexpression of tumor necrosis factor-alpha. *Circ. Res.* **81:** 627.

Kudoh S., Komuro I., Hiroi Y., Zou Y., Harada K., Sugaya T., Takekoshi N., Murakami K., Kadowaki T., and Yazaki Y. 1998. Mechanical stretch induces hypertrophic responses in cardiac myocytes of angiotensin II type 1a receptor knockout mice. *J. Biol. Chem.* **273:** 24037.

Kuo H.C., Cheng C.F., Clark R.B., Lin J.J., Lin J.L., Hoshijima M., Nguyen-Tran V.T., Gu Y., Ikeda Y., Chu P.H., Ross J., Giles W.R., and Chien K.R. 2001. A defect in the Kv channel-interacting protein 2 (KChIP2) gene leads to a complete loss of I(to) and confers susceptibility to ventricular tachycardia. *Cell* **107:** 801.

Loennechen J.P., Wisloff U., Falck G., and Ellingsen O. 2002. Cardiomyocyte contractility and calcium handling partially recover after early deterioration during post-infarction failure in rat. *Acta Physiol. Scand.* **176:** 17.

McConnell B.K., Jones K.A., Fatkin D., Arroyo L.H., Lee R.T., Aristizabal O., Turnbull D.H., Georgakopoulos D., Kass D., Bond M., Niimura H., Schoen F.J., Conner D., Fischman D.A., Seidman C.E., Seidman J.G., and Fischman D.H. 1999. Dilated cardiomyopathy in homozygous myosin-binding protein-C mutant mice. *J. Clin. Invest.* **104:** 1235.

McKoy G., Protonotarios N., Crosby A., Tsatsopoulou A., Anastasakis A., Coonar A., Norman M., Baboonian C., Jeffery S., and McKenna W.J. 2000. Identification of a deletion in plakoglobin in arrhythmogenic right ventricular cardiomyopathy with palmoplantar keratoderma and woolly hair (Naxos disease). *Lancet* **355:** 2119.

Michael L.H., Entman M.L., Hartley C.J., Youker K.A., Zhu J., Hall S.R., Hawkins H.K., Berens K., and Ballantyne C.M. 1995. Myocardial ischemia and reperfusion: A murine model. *Am. J. Physiol.* **269:** H2147.

Milano C.A., Allen L.F., Rockman H.A., Dolber P.C., McMinn T.R., Chien K.R., Johnson T.D., Bond R.A., and Lefkowitz R.J. 1994. Enhanced myocardial function in transgenic mice overexpressing the beta 2-adrenergic receptor. *Science* **264:** 582.

Minamisawa S., Hoshijima M., Chu G., Ward C.A., Frank K., Gu Y., Martone M.E., Wang Y., Ross J., Jr., Kranias E.G., Giles W.R., and Chien K.R. 1999. Chronic phospholamban-sarcoplasmic reticulum calcium ATPase interaction is the critical calcium cycling defect in dilated cardiomyopathy. *Cell* **99:** 313.

Miyamoto M.I., del Monte F., Schmidt U., DiSalvo T.S., Kang Z.B., Matsui T., Guerrero J.L., Gwathmey J.K., Rosenzweig A., and Hajjar R.J. 2000. Adenoviral gene transfer of SERCA2a improves left-ventricular function in aortic-banded rats in transition to heart failure. *Proc. Natl. Acad. Sci.* **97:** 793.

Moreira E.S., Wiltshire T.J., Faulkner G., Nilforoushan A., Vainzof M., Suzuki O.T., Valle G., Reeves R., Zatz M., Passos-Bueno M.R., and Jenne D.E. 2000. Limb-girdle muscular dystrophy type 2G is caused by mutations in the gene encoding the sarcomeric protein telethonin. *Nat. Genet.* **24:** 163.

Nakagawa N., Hoshijima M., Oyasu M., Saito N., Tanizawa K., and Kuroda S. 2000. ENH, containing PDZ and LIM domains, heart/skeletal muscle-specific protein, associates with cytoskeletal proteins through the PDZ domain. *Biochem. Biophys. Res. Commun.* **272:** 505.

Nakamura T., Lozano P.R., Ikeda Y., Iwanaga Y., Hinek A., Minamisawa S., Cheng C.F., Kobuke K., Dalton N., Takada Y., Tashiro K., Ross J., Jr., Honjo T., and Chien K.R. 2002. Fibulin-5/DANCE is essential for elastogenesis in vivo. *Nature* **415:** 171.

Negash S., Yao Q., Sun H., Li J., Bigelow D.J., and Squier T.C. 2000. Phospholamban remains associated with the Ca^{++}- and Mg^{++}-dependent ATPase following phosphorylation by cAMP-dependent protein kinase. *Biochem. J.* **351:** 195.

Nguyen-Tran V.T., Kubalak S.W., Minamisawa S., Fiset C., Wollert K.C., Brown A.B., Ruiz-Lozano P., Barrere-Lemaire S., Kondo R., Norman L.W., Gourdie R.G., Rahme M.M., Feld G.K., Clark R.B., Giles W.R., and Chien K.R. 2000. A novel genetic pathway for sudden cardiac death via defects in the transition between ventricular and conduction system cell lineages. *Cell* **102:** 671.

Ozcelik C., Erdmann B., Pilz B., Wettschureck N., Britsch S., Hubner N., Chien K.R., Birchmeier C., and Garratt A.N. 2002. Conditional mutation of the ErbB2 (HER2) receptor in cardiomyocytes leads to dilated cardiomyopathy. *Proc. Natl. Acad. Sci.* **99:** 8880.

Pashmforoush M., Pomies P., Peterson K.L., Kubalak S., Ross J., Jr., Hefti A., Aebi U., Beckerle M.C., and Chien K.R. 2001. Adult mice deficient in actinin-associated LIM-domain protein reveal a developmental pathway for right ventricular cardiomyopathy. *Nat. Med.* **7:** 591.

Passier R., Richardson J.A., and Olson E.N. 2000. Oracle, a novel PDZ-LIM domain protein expressed in heart and skeletal muscle. *Mech. Dev.* **92:** 277.

Pollesello P., Annila A., and Ovaska M. 1999. Structure of the 1-36 amino-terminal fragment of human phospholamban by nuclear magnetic resonance and modeling of the phospholamban pentamer. *Biophys. J.* **76:** 1784.

Rockman H.A., Chien K.R., Choi D.J., Iaccarino G., Hunter J.J., Ross J., Jr., Lefkowitz R.J., and Koch W.J. 1998. Expression of a beta-adrenergic receptor kinase 1 inhibitor prevents the development of myocardial failure in gene-targeted mice. *Proc. Natl. Acad. Sci.* **95:** 7000.

Rockman H.A., Ono S., Ross R.S., Jones L.R., Karimi M., Bhargava V., Ross J., Jr., and Chien K.R. 1994. Molecular and physiological alterations in murine ventricular dysfunction. *Proc. Natl. Acad. Sci.* **91:** 2694.

Rockman H.A., Ross R.S., Harris A.N., Knowlton K.U., Steinhelper M.E., Field L.J., Ross J., Jr., and Chien K.R. 1991. Segregation of atrial-specific and inducible expression of an atrial natriuretic factor transgene in an in vivo murine model of cardiac hypertrophy. *Proc. Natl. Acad. Sci.* **88:** 8277.

Sah V.P., Minamisawa S., Tam S.P., Wu T.H., Dorn G.W., II, Ross J., Jr., Chien K.R., and Brown J.H. 1999. Cardiac-specific overexpression of RhoA results in sinus and atrioventricular nodal dysfunction and contractile failure. *J. Clin. Invest.* **103:** 1627.

Sato Y., Kiriazis H., Yatani A., Schmidt A.G., Hahn H., Ferguson D.G., Sako H., Mitarai S., Honda R., Mesnard-Rouiller L., Frank K.F., Beyermann B., Wu G., Fujimori K., Dorn G.W., II, and Kranias E.G. 2001. Rescue of contractile parameters and myocyte hypertrophy in calsequestrin overexpressing myocardium by phospholamban ablation. *J. Biol. Chem.* **276:** 9392.

Schaller M.D. 2001. Paxillin: A focal adhesion-associated adaptor protein. *Oncogene* **20:** 6459.

Scherrer-Crosbie M., Steudel W., Hunziker P.R., Foster G.P., Garrido L., Liel-Cohen N., Zapol W.M., and Picard M.H. 1998. Determination of right ventricular structure and function in normoxic and hypoxic mice: A transesophageal echocardiographic study. *Circulation* **98:** 1015.

Seidman J.G. and Seidman C. 2001. The genetic basis for cardiomyopathy: From mutation identification to mechanistic paradigms. *Cell* **104:** 557.

Shimoyama M., Hayashi D., Takimoto E., Zou Y., Oka T., Uozumi H., Kudoh S., Shibasaki F., Yazaki Y., Nagai R., and Komuro I. 1999. Calcineurin plays a critical role in pressure overload-induced cardiac hypertrophy. *Circulation* **100:** 2449.

Stevenson L.W. 1998. Inotropic therapy for heart failure. *N. Engl. J. Med.* **339:** 1848.

Straub V. and Campbell K.P. 1997. Muscular dystrophies and the dystrophin-glycoprotein complex. *Curr. Opin. Neurol.* **10:** 168.

Sutton M.S., Gill T., Plappert T., Saltzman D.H., and Doubilet P. 1991. Assessment of right and left ventricular function in terms of force development with gestational age in the normal human fetus. *Br. Heart J.* **66:** 285.

Swynghedauw B. 1999. Molecular mechanisms of myocardial remodeling. *Physiol. Rev.* **79:** 215.

Tanaka N., Dalton N., Mao L., Rockman H.A., Peterson K.L., Gottshall K.R., Hunter J.J., Chien K.R., and Ross J., Jr. 1996. Transthoracic echocardiography in models of cardiac disease in the mouse. *Circulation* **94:** 1109.

Tomaselli G.F. and Marban E. 1999. Electrophysiological remodeling in hypertrophy and heart failure. *Cardiovasc. Res.* **42:** 270.

Towbin J.A. and Bowles N.E. 2002. The failing heart. *Nature* **415:** 227.

Toyoshima C., Nakasako M., Nomura H., and Ogawa H. 2000. Crystal structure of the calcium pump of sarcoplasmic reticulum at 2.6 Å resolution. *Nature* **405:** 647.

Weiskirchen R., Pino J.D., Macalma T., Bister K., and Beckerle M.C. 1995. The cysteine-rich protein family of highly related LIM domain proteins. *J. Biol. Chem.* **270:** 28946.

Wettschureck N., Rutten H., Zywietz A., Gehring D., Wilkie T.M., Chen J., Chien K.R., and Offermanns S. 2001. Absence of pressure overload induced myocardial hypertrophy after conditional inactivation of Galphaq/Galpha11 in cardiomyocytes. *Nat. Med.* **7:** 1236.

White D.C., Hata J.A., Shah A.S., Glower D.D., Lefkowitz R.J., and Koch W.J. 2000. Preservation of myocardial beta-adrenergic receptor signaling delays the development of heart failure after myocardial infarction. *Proc. Natl. Acad. Sci.* **97:** 5428.

Williams S.P., Gerber H.P., Giordano F.J., Peale F.V., Jr., Bernstein L.J., Bunting S., Chien K.R., Ferrara N., and van Bruggen N. 2001. Dobutamine stress cine-MRI of cardiac function in the hearts of adult cardiomyocyte-specific VEGF knockout mice. *J. Magn. Reson. Imaging* **14:** 374.

Yasukawa H., Hoshijima M., Gu Y., Nakamura T., Pradervand S., Hanada T., Hanakawa Y., Yoshimura A., Ross J., Jr., and Chien K.R. 2001. Suppressor of cytokine signaling-3 is a biomechanical stress-inducible gene that suppresses gp130-mediated cardiac myocyte hypertrophy and survival pathways. *J. Clin. Invest.* **108:** 1459.

Yu T.S., Moctezuma-Anaya M., Kubo A., Keller G., and Robertson S. 2002. The heart LIM protein gene (Hlp), expressed in the developing and adult heart, defines a new tissue-specific LIM-only protein family. *Mech. Dev.* **116:** 187.

Yue P., Long C.S., Austin R., Chang K.C., Simpson P.C., and Massie B.M. 1998. Post-infarction heart failure in the rat is associated with distinct alterations in cardiac myocyte molecular phenotype. *J. Mol. Cell. Cardiol.* **30:** 1615.

Zhao Y.Y., Liu Y., Stan R.V., Fan L., Gu Y., Dalton N., Chu P.H., Peterson K., Ross J., Jr., and Chien K.R. 2002. Defects in caveolin-1 cause dilated cardiomyopathy and pulmonary hypertension in knockout mice. *Proc. Natl. Acad. Sci.* **99:** 11375.

Zhou Q., Ruiz-Lozano P., Martone M.E., and Chen J. 1999. Cypher, a striated muscle-restricted PDZ and LIM domain-containing protein, binds to alpha-actinin-2 and protein kinase C. *J. Biol. Chem.* **274:** 19807.

Zhou Q., Chu P.H., Huang C., Cheng C.F., Martone M.E., Knoll G., Shelton G.D., Evans S., and Chen J. 2001. Ablation of Cypher, a PDZ-LIM domain Z-line protein, causes a severe form of congenital myopathy. *J. Cell Biol.* **155:** 605.

Zolk O., Caroni P., and Bohm M. 2000. Decreased expression of the cardiac LIM domain protein MLP in chronic human heart failure. *Circulation* **101:** 2674.

From Sarcomeric Mutations to Heart Disease: Understanding Familial Hypertrophic Cardiomyopathy

A. MAASS,* J.P. KONHILAS, B.L. STAUFFER, AND L.A. LEINWAND

Department of Molecular, Cellular, and Developmental Biology, University of Colorado, Boulder, Colorado 80309

Familial hypertrophic cardiomyopathy (FHC) is a heterogeneous disease with variable phenotypes caused by autosomal-dominant mutations in one of several contractile proteins (Roberts and Sigwart 2001a). The mutations typically lead to myocyte hypertrophy, myocellular/myofibrillar disarray, increased interstitial collagen, and hypertrophy in small arteries (Wigle et al. 1995). FHC is estimated to have a prevalence of about 0.2% in the population of young adults as recognized by echocardiography (Maron et al. 1995) and is the most common cause of sudden death in adolescents and young adults, especially in athletes. Clinical symptoms can include dyspnea, palpitations, chest pain, sudden death, or severe heart failure, whereas many other patients remain asymptomatic (Maron 1997). As of yet, no specific therapy is available for this disease, with only symptomatic treatments for aspects of the syndrome (Roberts and Sigwart 2001b).

The identification of a mutation in the β-myosin heavy-chain (MyHC) gene was the first evidence for a gene defect underlying any intrinsic heart-muscle disease (Geisterfer-Lowrance et al. 1990). Since then, mutations in nine genes, all of which encode proteins of the myofibrillar apparatus, have been found to cause FHC (Table 1). Although major advances have been made in the genetics of FHC, the molecular mechanisms by which sarcomeric mutations lead to hypertrophy and other features of this disease remain unresolved.

Even more difficult to explain is the high degree of phenotypic variability among patients with sarcomeric mutations, even among patients harboring the same or similar mutations in the same protein (Fananapazir and Epstein 1994; Klues et al. 1995). For example, a recent report documents a mutation in α-cardiac actin, a gene previously only thought to cause familial dilated cardiomyopathy (DCM) when mutated, that leads to FHC (Mogensen et al. 1999). Similarly, mutations in troponin T or β-MyHC genes previously thought to only cause FHC can also cause familial DCM, without any evidence of preceding cardiac hypertrophy (Kamisago et al. 2000). The existence of genetic modifiers and/or gene polymorphisms has been suggested to be an important contributor to these phenotypic variations (for review, see Bonne et al. 1998). Recently, mice with a missense mutation in α-MyHC from the 129 strain were crossed into the Black Swiss background. Only half of the animals developed hypertrophy, demonstrating that a polymorphic modifier gene can determine the hypertrophic response to a dominant-negative sarcomere protein mutation (Semsarian et al. 2001). Moreover, in our laboratory, we have recently detailed the phenotypic variability of inbred mouse strains, finding significant differences in parameters such as cardiac mass and function among the strains studied (Lerman et al. 2002). Additionally, gender-specific factors seem to play a role, since penetrance is significantly lower in females (Charron et al. 1997). Clearly, these genetic components cannot be ignored when evaluating this disease.

Table 1. Sarcomeric Mutations and Familial Hypertrophic Cardiomyopathy

Gene	Chromosome	Frequency (%)	Number of mutations
β-MyHC*	14q1	30–50	>60
Troponin T*	1q3	15–20	>20
MyBP-C	11q11	15–20	>15
α-Tropomyosin	15q2	<5	3
Troponin I	19q13	<1	3
MLC-1	3p	<1	2
MLC-2	12q	<1	2
α-Cardiac actin*	15q11	?	2
Titin*	2q31	?	?
Unknown	7q3	?	?

Listed above are the known mutations in the myofibrillar apparatus that have been found to cause FHC. Also indicated are the chromosome location, frequency of the disease, and number of mutations to date for each myofibrillar protein. The asterisk (*) denotes myofibrillar proteins in which mutations can lead to either hypertrophic cardiomyopathy or dilated cardiomyopathy.

TRANSGENIC MODELS OF FHC

To elucidate the mechanisms responsible for the phenotypic variability and to define both genetic modifiers and gender-specific differences in cardiac phenotype, an increasing number of transgenic animals have been created that model human FHC mutations and disease (Geisterfer-Lowrance et al. 1996; Vikstrom et al. 1996; Tardiff et al. 1998, 1999; McConnell et al. 2001). Because of the

---

*Present address: Department of Medicine, University of Wuerzburg, Josef-Schneider-Str. 2, 97080 Wuerzburg, Germany.

genetic identity of inbred mouse strains, they are well suited for this task. Our laboratory has focused on transgenic mouse lines mimicking human mutations in troponin T gene (Tardiff et al. 1998, 1999) or the α-myosin heavy-chain gene (Vikstrom et al. 1996). Mice harboring a missense mutation in the α-MyHC gene (R403Q) corresponding to a human mutation in β-MyHC causing FHC also carry an additional deletion of amino acids 468–527 in the actin-binding domain of the protein (Vikstrom et al. 1996). These mice show typical histologic features of FHC and, in addition, significant left- and right-ventricular hypertrophy at 4 months of age. Interestingly, at an older age (8 months), the male mice develop progressive left-ventricular dilation whereas female counterparts show increasing hypertrophy without dilation (Vikstrom et al. 1996). The hearts from female mice express hypertrophic markers in both RV and LV such as atrial natriuretic factor (ANF), skeletal α-actin, and β-MyHC (Vikstrom et al. 1996). ANF expression is localized to hypertrophic myocytes and regions around fibrosis, suggesting that this protein is a sensitive marker not only of hypertrophy, but also of cardiac pathogenesis. An additional finding is that ventricular hypertrophy can occur in the absence of induction of these markers and vice versa, demonstrating that there is not an obligate relationship between the two. Recent experiments carried out by our laboratory show an impaired exercise tolerance in a treadmill test in these mice. Furthermore, isoproterenol response is impaired, suggesting adrenergic desensitization. Both contractility and diastolic function are abnormal in a working heart preparation (Freeman et al. 2001). Therefore, these mice show features present in human FHC and heart failure, making the further characterization of this murine model important for the study of the pathogenesis of FHC and heart failure.

Transgenic mouse lines modeling human mutations in the TnT gene have also been described (Tardiff et al. 1998, 1999). The first model using the mouse TnT as the transgene mimics the product of a splice donor site mutation leading to a truncated TnT missing the last coding exon 16, containing a tropomyosin-binding domain. Mouse lines expressing 4% and 6% of their TnT protein as the truncated isoform are viable (Tardiff et al. 1998). These mice exhibit smaller left ventricles and severe diastolic and milder systolic dysfunction. These hearts demonstrate myocellular disarray but no fibrosis and no induction of hypertrophic markers. The decrease in cardiac mass is due to a reduced number of cardiac myocytes that are smaller in size. When these lines of mice are crossed to increase transgene expression, all *doubly transgenic* mice die within 8 hours of birth, demonstrating the severity of this allele.

A second mutant allele was analyzed in transgenic mouse lines expressing a mouse TnT missense mutation (R92Q) in the heart. Lines were established expressing 30%, 67%, and 92% of their total TnT protein as the missense allele (Tardiff et al. 1999). These mice also have smaller left ventricles, but in contrast to the mice with the truncated TnT, they have significant fibrosis and induction of the hypertrophic markers, ANF and β-MyHC, in the absence of myocyte hypertrophy. Isolated cardiac myocytes show sarcomeric activation, impaired relaxation, and shorter sarcomere length. Isolated working heart preparations demonstrate hypercontractility and diastolic dysfunction. The finding of hypercontractility is intriguing because this characteristic is seen in a subset of FHC patients.

PATHOGENESIS OF HYPERTROPHIC CARDIOMYOPATHY

Two mechanisms for the dominant inheritance of FHC have been proposed. First is the *poison peptide* theory. Mutant proteins are incorporated into the sarcomere and lead to dominant negative inhibition of contractile or diastolic function. The other theory is that the mutant protein acts as a null allele and leads to haploinsufficiency, ultimately causing altered stoichiometry of thick- or thin-filament components and thereby altering structure and function of the sarcomere (Bonne et al. 1998). Recent in vitro data and data from animal models (Tardiff et al. 1998, 1999) suggest, at least for the TnT alleles studied, that the truncated or mutant proteins are incorporated into the sarcomere and are, therefore, exhibiting a dominant-negative effect on the phenotype (Bonne et al. 1998). However, in human endocardial biopsies taken from a myosin-binding protein C (MyBP-C; a component of the thick filament) FHC patient, no mutated protein was detected (Rottbauer et al. 1997). Yet, two transgenic models mimicking the MyBP-C class of FHC demonstrated normal incorporation of the mutant MyBP-C transgene (Yang et al. 1998; McConnell et al. 1999). Only when a truncated mutant form of MyBP-C is expressed in a homozygous fashion do animals express reduced amounts of the mutant protein and still exhibit a marked phenotype (McConnell et al. 1999).

The dominant negative effects most likely result from a deficit in myocellular force production in the mutant sarcomeres initiating a remodeling process. The cellular signals leading to FHC phenotypes most certainly involve several cell types (fibroblasts, vascular smooth muscle cells, cardiomyocytes), even though the mutant protein is expressed exclusively in cardiomyocytes. For example, subsets of patients (Geisterfer-Lowrance et al. 1990) and certain animal models of FHC (Vikstrom et al. 1996; Welikson et al. 1999) exhibit significant small vessel disease and valvular pathology. This suggests signaling from the cardiomycyte to neighboring cells in an autocrine or paracrine fashion (Fig. 1). Nevertheless, because of the numerous signals and growth factors implicated in these processes, it has been difficult to ascribe a specific signal to any one of the many pathologies.

SIGNALING PATHWAYS IN HYPERTROPHY AND HEART FAILURE

Cardiac function is ultimately dependent on the ability of the myofilaments to generate force, which, in turn, is dependent on the amount of cytosolic Ca^{++}, phosphorylation state, and isoform content of the cardiac myocyte (Solaro and Rarick 1998; de Tombe and Solaro 2000). From the analysis of the above-mentioned transgenic

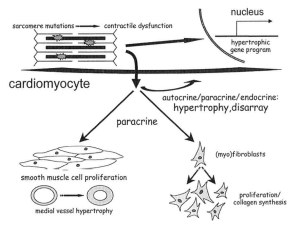

Figure 1. Cardiomyocyte response to mutant sarcomeric proteins. The contractile dysfunction that results from the mutant sarcomeric protein induces a number of paracrine, autocrine, endocrine, and intracrine signals. These signals lead to the phenotypes typically seen in FHC as indicated.

mouse models harboring mutations in troponin T and MyHC, it is already clear that there are allele-specific phenotypes and that mutations of each sarcomeric protein are likely to result in a distinct set of clinical characteristics. Although it is unclear how myofilament dysfunction resulting from these mutant proteins translates into the complex phenotypes associated with FHC, it is most certainly related to a functional impairment of the contractile apparatus and subsequent coordinated integration of the multiple hypertrophic signals to counteract the contractile dysfunction. As discussed below, emerging data are shedding light on signal transduction pathways that lead to cardiac hypertrophy.

Ca^{++}/Calmodulin-dependent Pathways

One of the primary signaling pathways for hypertrophy is the Ca^{++}/calmodulin-dependent system. In fact, overexpression of calmodulin in the hearts of transgenic mice induces hypertrophy (Gruver et al. 1993). When Ca^{++} binds calmodulin, this complex can activate Ca^{++}/calmodulin (CaM)-dependent protein kinase (Passier et al. 2000) and a CaM-dependent phosphatase, calcineurin (Molkentin et al. 1998; Wu et al. 2000; Molkentin and Dorn 2001). Over the last few years, there have been conflicting reports about calcineurin having a central rolie in myocyte hypertrophy. Overexpression of activated calcineurin in the murine heart resulted in profound cardiac hypertrophy that progressed to dilated cardiomyopathy and heart failure similar to human patients with decompensated hypertrophy (Molkentin et al. 1998), a response that could be inhibited by cyclosporin A (CsA) and FK506 (calcineurin inhibitors) in vitro and in vivo (Molkentin et al. 1998; De Windt et al. 2001). Moreover, it was shown that treatment with CsA could prevent hypertrophy and subsequent development of heart failure in mice expressing a mutant MLC2v that leads to typical features of hypertrophic cardiomyopathy (HCM) (Sussman et al. 1998). In a mouse model of HCM caused by a mutation in α-MyHC, however, treatment with CsA led to a worsening of the phenotype with augmented hypertrophy, worsened histopathology, and increase in sudden cardiac death (Fatkin et al. 2000). In this model, the L-type Ca^{++} channel blocker diltiazem, however, restores changes in sarcoplasmic reticular proteins and prevents cardiomyopathy, suggesting that disruption of sarcoplasmic reticulum Ca^{++} homeostasis is an important early step in the pathogenesis of FHC (Semsarian et al. 2002). An interesting recent report shows activation of calcineurin in human hearts with both aortic stenosis and hypertrophic obstructive cardiomyopathy (Ritter et al. 2002). As a new mechanism, the authors propose proteolytic cleavage of the autoinhibitory domain of calcineurin, rendering a constitutively active isoform of this protein.

In addition to targeting cellular proteins involved with Ca^{++} handling in the cardiac myocyte, CaM kinase (CaMK) has been implicated as an integral component in hypertrophic signaling. CaMK can induce expression of ANF (Sprenkle et al. 1995), an established indicator of cardiac hypertrophy. Moreover, CaMK activity is elevated in failing human hearts (Kirchhefer et al. 1999). Overexpression of activated CaMKIV in the heart results in increased ventricular mass and decreased fractional shortening (Passier et al. 2000). In addition, CaMK signaling activates transcription factors of the MEF2 family (Passier et al. 2000), which has long been known to establish myogenic cell lineages in the embryo (Naya and Olson 1999; Naya et al. 1999) and a putative marker of pathological hypertrophy (Molkentin and Markham 1993; Passier et al. 2000). This enzyme has recently been implicated in regulation of mitochondrial biogenesis (Wu et al. 2002). Since mitochondria are the critical organelles in cells of high-energy demand, such as cardiac myocytes, this pathway might be pivotal in the development of cardiac and skeletal muscle phenotypes.

Mitogen-activated Protein Kinases

A variety of hypertrophic stimuli have been shown to stimulate signaling cascades of the mitogen-activated protein kinases (MAPK). Phosphorylated ERK proteins augment ANF promoter activity and are required for agonist-induced hypertrophy and sarcomeric organization following hypertrophy (Bueno et al. 2000). Similarly, expression of activated MEK1 in the hearts of transgenic mice stimulates hypertrophy (Bueno et al. 2000). JNKs and p38-MAPKs, upon activation by specific MKKs, phosphorylate a number of transcription factors, such as c-Jun, ATF2, and Elk1 (Sugden and Clerk 1998), that have been shown to be involved in the hypertrophic response (Chien et al. 1991). The coordination and integration of these factors may be a result of selective activation by upstream regulators and specific stimuli. In support of this, targeted disruption of MEKK1 expression selectively inhibits JNK activation in cell culture (Yujiri et al. 1999) and in transgenic mice (Yujiri et al. 2000), whereas MEKK2 gene disruption prevents receptor-mediated JNK activation but has little effect on the stimulation of ERK or p38 pathways (Garrington et al. 2000). Moreover, the absence of MEKK1 in mice abolishes the increase in cardiac mass and ANF induction associated with pathologic hypertrophy of a transgenic mouse model overexpressing Gαq in the heart (Sadoshima et al. 2002). Interestingly, it has been demonstrated that subunits of the G-protein complex can directly enhance intermediates of several hypertrophic cascades including MAPK (Crespo et al. 1995), phosphatidylinositol 3-kinase (PI3K) (Naga Prasad et al. 2000), and the low-molecular-weight G proteins such as Ras signaling pathways (Pumiglia et al. 1995). Moreover, second messengers following G-protein-coupled receptor activation, for example PKC, can directly affect the activity of these intermediates (Wakasaki et al. 1997).

Mouse models with cardiac-specific overexpression of the proto-oncogenes c-myc (Jackson et al. 1991), c-H-ras (Hunter et al. 1995), and the enzyme-activator calmodulin (Gruver et al. 1993) have been created and mimic some of the phenotypic features of FHC such as cardiac hypertrophy and hypertrophic markers. For example, transgenic mice expressing Ras, upstream activator of the MAPK, under the control of the MLC2v promoter have a phenotype very similar to FHC with the typical histologic findings accompanied by diastolic dysfunction (Hunter et al. 1995). Similarly, human myocardium from patients with FHC mutations in β-MyHC or troponin T express the proto-oncogenes c-H-ras and c-myc (Kai et al. 1998). In these studies, expression levels correlated with myocyte size. Thus, Ras seems to be sufficient to induce an FHC-like disease, yet it still remains to be shown whether it is a necessary component of the signal transduction cascade and a possible target for therapy.

GSK3

Glycogen synthase kinase-3 (GSK-3β) has recently gained prominence in the literature as an important negative regulator of hypertrophic signaling (Antos et al. 2002; Hardt and Sadoshima 2002). GSK-3β is a ubiquitous serine/threonine kinase that phosphorylates amino-terminal regulatory regions of NFAT. Since the dephosphorylation of NFAT by calcineurin is a critical step in eliciting the hypertrophic response, it was hypothesized that activation of GSK-3β could suppress cardiac hypertrophy in vivo. It was shown that the hearts of transgenic mice expressing activated GSK-3β in addition to activated calcineurin are resistant to cardiac hypertrophy (Antos et al. 2002). Similarly, these mice do not exhibit extensive cardiac hypertrophy in response to adrenergic stimulation or pressure overload (Antos et al. 2002). Thus, the fact that GSK-3β is a potent negative regulator of cardiac hypertrophy suggests that a necessary component of the hypertrophic response is the inhibition of GSK-3β (Hardt and Sadoshima 2002).

TRANSCRIPTION FACTORS
GATA4

A variety of pathologic stimuli induce hypertrophic growth in the myocardium via many signaling pathways. Given such redundancy in cytosolic signal transduction, it is unclear whether these multiple signals converge on a single or on multiple nuclear effectors. Our hypothesis and that of other investigators (Liang et al. 2001) is that there might be only a few critically important transcription factors that regulate hypertrophy-specific gene expression. GATA4 DNA-binding sequences have been identified within promoters of numerous cardiac-specific genes, making it a logical potential target of numerous hypertrophic stimuli (Liang et al. 2001). Activation of GATA4 can occur via one of several cytosolic signals including Gαq-protein-coupled receptors and the various MAPKs (Liang and Molkentin 2002). Moreover, overexpression of GATA4 in cultured neonatal cardiomyocytes or in hearts of transgenic mice (Liang et al. 2001) can induce cardiomyocyte hypertrophy, providing support for a central role of GATA4 in this process.

MEF2

As mentioned above, MEF2 has been implicated as a nuclear target for Ca^{++}-regulated signals (Passier et al. 2000) and MAPK intermediates (Han and Molkentin 2000) in the hypertrophic response. It has been known for some time that the predominant transcription factor involved in cardiac differentiation is MEF2 (see Cripps and Olson 2002), making it a logical target for cardiac hypertrophy. MEF2 activity is stimulated by MAPK-dependent phosphorylation of the *trans*-activating domain (Han and Molkentin 2000) or by disruption of the histone deacetylase (HDAC)/MEF2 complex by a CaMK-dependent mechanism (Lu et al. 2000; McKinsey et al. 2000). Recently, a transgenic mouse line harboring a lacZ transgene under the transcriptional control of the consensus binding site for MEF2 was generated (Black and Olson 1998), allowing direct monitoring of MEF2 activity in vivo. Crossing the MEF2 indicator line with a mouse expressing activated CaMKIV in the heart shows robust activation of MEF2 accompanied with profound myocardial hypertrophy (Passier et al. 2000). Consistent with the role of MEF2 activation in pathological hypertrophy, we observe that transgenic mice with FHC mutations in α-MyHC and cardiac TnT on the MEF2-lacZ background demonstrate a substantial increase in MEF2 activity (L. Leinwand et al., unpubl.). Thus, MEF2 is strongly implicated in the hypertrophic response, yet the coordination and integration of the signaling pathways that ultimately converge on MEF2 remain elusive.

GENDER EFFECTS IN THE MYOCARDIUM

Many recent reports suggest gender differences in pressure-overload hypertrophy (Weinberg et al. 1999) and FHC (Maron et al. 1999). In addition, penetrance of FHC has been shown to be significantly lower in females (Charron et al. 1997). A recent report by Maron et al. (1999) describes that women diagnosed with HCM are older (postmenopausal) and present with a more severe form of the disease with higher mortality. Similarly, we have observed gender-specific differences in our MyHC mutation model of FHC (Olsson et al. 2001). In the FHC mice, hearts exhibit increased systolic contraction in an isolated working heart model at 4 months of age with decreased diastolic function, probably due to hypertrophy in both sexes. At 10 months of age, however, hearts of female mice show preserved systolic function with continued diastolic dysfunction, whereas hearts of male mice are dilated with systolic impairment and worsening diastolic function (Olsson et al. 2001). Evidence has been accumulating which suggests that this cardioprotective effect seen in females may be due to specific effects of estrogen. Indeed, estrogen has been shown to have direct effects in the vasculature (Mendelsohn 2002) and on activation of the MAPK intermediate, ERK1/2, in the heart (de Jager et al. 2001). Interestingly, a recent report shows that disruption of the FKBP12.6 gene in mice leads to development of cardiac hypertrophy only in male, but not female, mice (Xin et al. 2002). Treatment of female mice with the estrogen receptor antagonist tamoxifen leads to hypertrophy similar to male mice, establishing a protective role of estrogen at least for this model of cardiac hypertrophy (Xin et al. 2002). Yet, in postmenopausal women, the Women's Health Initiative reports that the combined hormone therapy with estrogen and progestin increased the risk for coronary heart disease, challenging the use of estrogens as a protective measure against heart disease (Women's Health Iniative 2002).

CONCLUSIONS

As detailed above, there are many signals that contribute to the phenotype of FHC. It is believed that sarcomeric mutations trigger a remodeling process to overcome a deficit created by the mutations. The ability to recapitulate many aspects of the disease state in transgenic models has hastened this learning process. One of the more confounding issues is the fact that the same mutations can induce a clinically heterogeneous phenotype in humans. Similarly, the same mutation inserted into the murine genome does not exactly mimic the human disease. Clearly, the context, including gender and physiological environment, plays a critical role in determining the phenotype. Nevertheless, transgenic mouse models are important in describing the molecular pathogenesis of the disease. The elucidation of the signaling pathways that participate in the remodeling process associated with these mutations will provide the necessary link between sarcomeric mutations, muscle dysfunction, and the manifestation of the disease and might reveal potential therapeutic targets.

ACKNOWLEDGMENTS

A. Maass was supported by a grant from the Deutsche Forschungsgemeinschaft (Ma 2185/1-1). L.A. Leinwand is supported by National Institutes of Health grant HL-50561. J. Konhilas is supported by a National Research Service Award from the National Institutes of Health grant F32 HL-70509-01. B. Stauffer is supported by a National Research Service Award from the National Institutes of Health grant HL-67543-01.

REFERENCES

Antos C.L., McKinsey T.A., Frey N., Kutschke W., McAnally J., Shelton J.M., Richardson J.A., Hill J.A., and Olson E.N. 2002. Activated glycogen synthase-3 beta suppresses cardiac hypertrophy in vivo. *Proc. Natl. Acad. Sci.* **99:** 907.

Black B.L. and Olson E.N. 1998. Transcriptional control of muscle development by myocyte enhancer factor-2 (MEF2) proteins. *Annu. Rev. Cell Dev. Biol.* **14:** 167.

Bonne G., Carrier L., Richard P., Hainque B., and Schwartz K. 1998. Familial hypertrophic cardiomyopathy: From mutations to functional defects. *Circ. Res.* **83:** 580.

Bueno O.F., De Windt L.J., Tymitz K.M., Witt S.A., Kimball T.R., Klevitsky R., Hewett T.E., Jones S.P., Lefer D.J., Peng C.F., Kitsis R.N., and Molkentin J.D. 2000. The MEK1-ERK1/2 signaling pathway promotes compensated cardiac hypertrophy in transgenic mice. *EMBO J.* **19:** 6341.

Charron P., Carrier L., Dubourg O., Tesson F., Desnos M., Richard P., Bonne G., Guicheney P., Hainque B., Bouhour J.B., Mallet A., Feingold J., Schwartz K., and Komajda M. 1997. Penetrance of familial hypertrophic cardiomyopathy. *Genet. Couns.* **8:** 107.

Chien K.R., Knowlton K.U., Zhu H., and Chien S. 1991. Regulation of cardiac gene expression during myocardial growth and hypertrophy: Molecular studies of an adaptive physiologic response. *FASEB J.* **5:** 3037.

Crespo P., Cachero T.G., Xu N., and Gutkind J.S. 1995. Dual effect of beta-adrenergic receptors on mitogen-activated protein kinase. Evidence for a beta gamma-dependent activation and a G alpha s-cAMP-mediated inhibition. *J. Biol. Chem.* **270:** 25259.

Cripps R.M. and Olson E.N. 2002. Control of cardiac development by an evolutionarily conserved transcriptional network. *Dev. Biol.* **246:** 14.

de Jager T., Pelzer T., Muller-Botz S., Imam A., Muck J., and Neyses L. 2001. Mechanisms of estrogen receptor action in the myocardium. Rapid gene activation via the ERK1/2 pathway and serum response elements. *J. Biol. Chem.* **276:** 27873.

de Tombe P.P. and Solaro R.J. 2000. Integration of cardiac myofilament activity and regulation with pathways signaling hypertrophy and failure. *Ann. Biomed. Eng.* **28:** 991.

De Windt L.J., Lim H.W., Bueno O.F., Liang Q., Delling U., Braz J.C., Glascock B.J., Kimball T.F., del Monte F., Hajjar R.J., and Molkentin J.D. 2001. Targeted inhibition of calcineurin attenuates cardiac hypertrophy in vivo. *Proc. Natl. Acad. Sci.* **98:** 3322.

Fananapazir L. and Epstein N.D. 1994. Genotype-phenotype correlations in hypertrophic cardiomyopathy. Insights provided by comparisons of kindreds with distinct and identical beta-myosin heavy chain gene mutations. *Circulation* **89:** 22.

Fatkin D., McConnell B.K., Mudd J.O., Semsarian C., Moskowitz I.G., Schoen F.J., Giewat M., Seidman C.E., and Seidman J.G. 2000. An abnormal Ca(2+) response in mutant sarcomere protein-mediated familial hypertrophic cardiomyopathy. *J. Clin. Invest.* **106:** 1351.

Freeman K., Colon-Rivera C., Olsson M.C., Moore R.L., Weinberger H.D., Grupp I.L., Vikstrom K.L., Iaccarino G., Koch W.J., and Leinwand L.A. 2001. Progression from hypertrophic to dilated cardiomyopathy in mice that express a mutant myosin transgene. *Am. J. Physiol. Heart Circ. Physiol.* **280:** H151.

Garrington T.P., Ishizuka T., Papst P.J., Chayama K., Webb S., Yujiri T., Sun W., Sather S., Russell D.M., Gibson S.B., Keller G., Gelfand E.W., and Johnson G.L. 2000. MEKK2 gene disruption causes loss of cytokine production in response to IgE and c-Kit ligand stimulation of ES cell-derived mast cells. *EMBO J.* **19:** 5387.

Geisterfer-Lowrance A.A., Christe M., Conner D.A., Ingwall J.S., Schoen F.J., Seidman C.E., and Seidman J.G. 1996. A mouse model of familial hypertrophic cardiomyopathy. *Science* **272:** 731.

Geisterfer-Lowrance A.A., Kass S., Tanigawa G., Vosberg H.P., McKenna W., Seidman C.E., and Seidman J.G. 1990. A molecular basis for familial hypertrophic cardiomyopathy: A beta cardiac myosin heavy chain gene missense mutation. *Cell* **62:** 999.

Gruver C.L., DeMayo F., Goldstein M.A., and Means A.R. 1993.

Targeted developmental overexpression of calmodulin induces proliferative and hypertrophic growth of cardiomyocytes in transgenic mice. *Endocrinology* **133**: 376.

Han J. and Molkentin J.D. 2000. Regulation of MEF2 by p38 MAPK and its implication in cardiomyocyte biology. *Trends Cardiovasc. Med.* **10**: 19.

Hardt S.E. and Sadoshima J. 2002. Glycogen synthase kinase-3beta: A novel regulator of cardiac hypertrophy and development. *Circ. Res.* **90**: 1055.

Hunter J.J., Tanaka N., Rockman H.A., Ross J., Jr., and Chien K.R. 1995. Ventricular expression of a MLC-2v-ras fusion gene induces cardiac hypertrophy and selective diastolic dysfunction in transgenic mice. *J. Biol. Chem.* **270**: 23173.

Jackson T., Allard M.F., Sreenan C.M., Doss L.K., Bishop S.P., and Swain J.L. 1991. Transgenic animals as a tool for studying the effect of the c-myc proto-oncogene on cardiac development. *Mol. Cell. Biochem.* **104**: 15.

Kai H., Muraishi A., Sugiu Y., Nishi H., Seki Y., Kuwahara F., Kimura A., Kato H., and Imaizumi T. 1998. Expression of proto-oncogenes and gene regulation of sarcomeric proteins in patients with hypertrophic cardiomyopathy. *Circ. Res.* **83**: 594.

Kamisago M., Sharma S.D., DePalma S.R., Solomon S., Sharma P., McDonough B., Smoot L., Mullen M.P., Woolf P.K., Wigle E.D., Seidman J.G., and Seidman C.E. 2000. Mutations in sarcomere protein genes as a cause of dilated cardiomyopathy. *N. Engl. J. Med.* **343**: 1688.

Kirchhefer U., Schmitz W., Scholz H., and Neumann J. 1999. Activity of cAMP-dependent protein kinase and Ca^{++}/calmodulin-dependent protein kinase in failing and nonfailing human hearts. *Cardiovasc. Res.* **42**: 254.

Klues H.G., Schiffers A., and Maron B.J. 1995. Phenotypic spectrum and patterns of left ventricular hypertrophy in hypertrophic cardiomyopathy: Morphologic observations and significance as assessed by two-dimensional echocardiography in 600 patients. *J. Am. Coll. Cardiol.* **26**: 1699.

Lerman I., Harrison B.C., Freeman K., Hewett T.E., Allen D.L., Robbins J., and Leinwand L.A. 2002. Genetic variability in forced and voluntary endurance exercise performance in seven inbred mouse strains. *J. Appl. Physiol.* **92**: 2245.

Liang Q. and Molkentin J.D. 2002. Divergent signaling pathways converge on GATA4 to regulate cardiac hypertrophic gene expression. *J. Mol. Cell. Cardiol.* **34**: 611.

Liang Q., De Windt L.J., Witt S.A., Kimball T.R., Markham B.E., and Molkentin J.D. 2001. The transcription factors GATA4 and GATA6 regulate cardiomyocyte hypertrophy in vitro and in vivo. *J. Biol. Chem.* **276**: 30245.

Lu J., McKinsey T.A., Nicol R.L., and Olson E.N. 2000. Signal-dependent activation of the MEF2 transcription factor by dissociation from histone deacetylases. *Proc. Natl. Acad. Sci.* **97**: 4070.

Maron B.J. 1997. Hypertrophic cardiomyopathy. *Lancet* **350**: 127.

Maron B.J., Casey S.A., Gohmann T.E., and Aeppli D.M. 1999. Impact of gender on the clinical and morphologic expression of hypertrophic cardiomyopathy. *Circulation* **100**: I.

Maron B.J., Gardin J.M., Flack J.M., Gidding S.S., Kurosaki T.T., and Bild D.E. 1995. Prevalence of hypertrophic cardiomyopathy in a general population of young adults. Echocardiographic analysis of 4111 subjects in the CARDIA Study (Coronary Artery Risk Development in [Young] Adults). *Circulation* **92**: 785.

McConnell B.K., Jones K.A., Fatkin D., Arroyo L.H., Lee R.T., Aristizabal O., Turnbull D.H., Georgakopoulos D., Kass D., Bond M., Niimura H., Schoen F.J., Conner D., Fischman D.A., Seidman C.E., and Seidman J.G. 1999. Dilated cardiomyopathy in homozygous myosin-binding protein-C mutant mice. *J. Clin. Invest.* **104**: 1235.

McConnell B.K., Fatkin D., Semsarian C., Jones K.A., Georgakopoulos D., Maguire C.T., Healey M.J., Mudd J.O., Moskowitz I.P., Conner D.A., Giewat M., Wakimoto H., Berul C.I., Schoen F.J., Kass D.A., Seidman C.E., and Seidman J.G. 2001. Comparison of two murine models of familial hypertrophic cardiomyopathy. *Circ. Res.* **88**: 383.

McKinsey T.A., Zhang C.L., Lu J., and Olson E.N. 2000. Signal-dependent nuclear export of a histone deacetylase regulates muscle differentiation. *Nature* **408**: 106.

Mendelsohn M.E. 2002. Genomic and nongenomic effects of estrogen in the vasculature. *Am. J. Cardiol.* **90**: 3F.

Mogensen J., Klausen I.C., Pedersen A.K., Egeblad H., Bross P., Kruse T.A., Gregersen N., Hansen P.S., Baandrup U., and Borglum A.D. 1999. Alpha-cardiac actin is a novel disease gene in familial hypertrophic cardiomyopathy. *J. Clin. Invest.* **103**: R39.

Molkentin J.D. and Dorn G.W., II. 2001. Cytoplasmic signaling pathways that regulate cardiac hypertrophy. *Annu. Rev. Physiol.* **63**: 391.

Molkentin J.D. and Markham B.E. 1993. Myocyte-specific enhancer-binding factor (MEF-2) regulates alpha-cardiac myosin heavy chain gene expression in vitro and in vivo. *J. Biol. Chem.* **268**: 19512.

Molkentin J.D., Lu J.R., Antos C.L., Markham B., Richardson J., Robbins J., Grant S.R., and Olson E.N. 1998. A calcineurin-dependent transcriptional pathway for cardiac hypertrophy. *Cell* **93**: 215.

Naga Prasad S.V., Esposito G., Mao L., Koch W.J., and Rockman H.A. 2000. Gbetagamma-dependent phosphoinositide 3-kinase activation in hearts with in vivo pressure overload hypertrophy. *J. Biol. Chem.* **275**: 4693.

Naya F.J., Wu C., Richardson J.A., Overbeek P., and Olson E.N. 1999. Transcriptional activity of MEF2 during mouse embryogenesis monitored with a MEF2-dependent transgene. *Development* **126**: 2045.

Naya F.S. and Olson E. 1999. MEF2: A transcriptional target for signaling pathways controlling skeletal muscle growth and differentiation. *Curr. Opin. Cell Biol.* **11**: 683.

Olsson M.C., Palmer B.M., Leinwand L.A., and Moore R.L. 2001. Gender and aging in a transgenic mouse model of hypertrophic cardiomyopathy. *Am. J. Physiol. Heart Circ. Physiol.* **280**: H1136.

Passier R., Zeng H., Frey N., Naya F.J., Nicol R.L., McKinsey T.A., Overbeek P., Richardson J.A., Grant S.R., and Olson E.N. 2000. CaM kinase signaling induces cardiac hypertrophy and activates the MEF2 transcription factor in vivo. *J. Clin. Invest.* **105**: 1395.

Pumiglia K.M., LeVine H., Haske T., Habib T., Jove R., and Decker S.J. 1995. A direct interaction between G-protein beta gamma subunits and the Raf-1 protein kinase. *J. Biol. Chem.* **270**: 14251.

Ritter O., Hack S., Schuh K., Rothlein N., Perrot A., Osterziel K.J., Schulte H.D., and Neyses L. 2002. Calcineurin in human heart hypertrophy. *Circulation* **105**: 2265.

Roberts R. and Sigwart U. 2001a. New concepts in hypertrophic cardiomyopathies, part I. *Circulation* **104**: 2113.

———. 2001b. New concepts in hypertrophic cardiomyopathies, part II. *Circulation* **104**: 2249.

Rottbauer W., Gautel M., Zehelein J., Labeit S., Franz W.M., Fischer C., Vollrath B., Mall G., Dietz R., Kubler W., and Katus H.A. 1997. Novel splice donor site mutation in the cardiac myosin-binding protein-C gene in familial hypertrophic cardiomyopathy. Characterization of cardiac transcript and protein. *J. Clin. Invest.* **100**: 475.

Sadoshima J., Montagne O., Wang Q., Yang G., Warden J., Liu J., Takagi G., Karoor V., Hong C., Johnson G.L., Vatner D.E., and Vatner S.F. 2002. The MEKK1-JNK pathway plays a protective role in pressure overload but does not mediate cardiac hypertrophy. *J. Clin. Invest.* **110**: 271.

Semsarian C., Healey M.J., Fatkin D., Giewat M., Duffy C., Seidman C.E., and Seidman J.G. 2001. A polymorphic modifier gene alters the hypertrophic response in a murine model of familial hypertrophic cardiomyopathy. *J. Mol. Cell. Cardiol.* **33**: 2055.

Semsarian C., Ahmad I., Giewat M., Georgakopoulos D., Schmitt J.P., McConnell B.K., Reiken S., Mende U., Marks A.R., Kass D.A., Seidman C.E., and Seidman J.G. 2002. The L-type calcium channel inhibitor diltiazem prevents cardiomyopathy in a mouse model. *J. Clin. Invest.* **109**: 1013.

Solaro R.J. and Rarick H.M. 1998. Troponin and tropomyosin: Proteins that switch on and tune in the activity of cardiac my-

ofilaments. *Circ. Res.* **83:** 471.
Sprenkle A.B., Murray S.F., and Glembotski C.C. 1995. Involvement of multiple *cis* elements in basal- and alpha-adrenergic agonist-inducible atrial natriuretic factor transcription. Roles for serum response elements and an SP-1-like element. *Circ. Res.* **77:** 1060.
Sugden P.H. and Clerk A. 1998. "Stress-responsive" mitogen-activated protein kinases (c-Jun N-terminal kinases and p38 mitogen-activated protein kinases) in the myocardium. *Circ. Res.* **83:** 345.
Sussman M.A., Lim H.W., Gude N., Taigen T., Olson E.N., Robbins J., Colbert M.C., Gualberto A., Wieczorek D.F., and Molkentin J.D. 1998. Prevention of cardiac hypertrophy in mice by calcineurin inhibition. *Science* **281:** 1690.
Tardiff J.C., Hewett T.E., Palmer B.M., Olsson C., Factor S.M., Moore R.L., Robbins J., and Leinwand L.A. 1999. Cardiac troponin T mutations result in allele-specific phenotypes in a mouse model for hypertrophic cardiomyopathy. *J. Clin. Invest.* **104:** 469.
Tardiff J.C., Factor S.M., Tompkins B.D., Hewett T.E., Palmer B.M., Moore R.L., Schwartz S., Robbins J., and Leinwand L.A. 1998. A truncated cardiac troponin T molecule in transgenic mice suggests multiple cellular mechanisms for familial hypertrophic cardiomyopathy. *J. Clin. Invest.* **101:** 2800.
Vikstrom K.L., Factor S.M., and Leinwand L.A. 1996. Mice expressing mutant myosin heavy chains are a model for familial hypertrophic cardiomyopathy. *Mol. Med.* **2:** 556.
Wakasaki H., Koya D., Schoen F.J., Jirousek M.R., Ways D.K., Hoit B.D., Walsh R.A., and King G.L. 1997. Targeted overexpression of protein kinase C beta2 isoform in myocardium causes cardiomyopathy. *Proc. Natl. Acad. Sci.* **94:** 9320.
Weinberg E.O., Thienelt C.D., Katz S.E., Bartunek J., Tajima M., Rohrbach S., Douglas P.S., and Lorell B.H. 1999. Gender differences in molecular remodeling in pressure overload hypertrophy. *J. Am. Coll. Cardiol.* **34:** 264.
Welikson R.E., Buck S.H., Patel J.R., Moss R.L., Vikstrom K.L., Factor S.M., Miyata S., Weinberger H.D., and Leinwand L.A. 1999. Cardiac myosin heavy chains lacking the light chain binding domain cause hypertrophic cardiomyopathy in mice. *Am. J. Physiol.* **276:** H2148.
Wigle E.D., Rakowski H., Kimball B.P., and Williams W.G. 1995. Hypertrophic cardiomyopathy. Clinical spectrum and treatment. *Circulation* **92:** 1680.
Women's Health Initiative. 2002. Risks and benefits of estrogen plus progestin in healthy postmenopausal women: Principal results from the Women's Health Initiative randomized controlled trial. *J. Am. Med. Assoc.* **288:** 321.
Wu H., Kanatous S.B., Thurmond F.A., Gallardo T., Isotani E., Bassel-Duby R., and Williams R.S. 2002. Regulation of mitochondrial biogenesis in skeletal muscle by CaMK. *Science* **296:** 349.
Wu H., Naya F.J., McKinsey T.A., Mercer B., Shelton J.M., Chin E.R., Simard A.R., Michel R.N., Bassel-Duby R., Olson E.N., and Williams R.S. 2000. MEF2 responds to multiple calcium-regulated signals in the control of skeletal muscle fiber type. *EMBO J.* **19:** 1963.
Xin H.B., Senbonmatsu T., Cheng D.S., Wang Y.X., Copello J.A., Ji G.J., Collier M.L., Deng K.Y., Jeyakumar L.H., Magnuson M.A., Inagami T., Kotlikoff M.I., and Fleischer S. 2002. Oestrogen protects FKBP12.6 null mice from cardiac hypertrophy. *Nature* **416:** 334.
Yang Q., Sanbe A., Osinska H., Hewett T.E., Klevitsky R., and Robbins J. 1998. A mouse model of myosin binding protein C human familial hypertrophic cardiomyopathy. *J. Clin. Invest.* **102:** 1292.
Yujiri T., Fanger G.R., Garrington T.P., Schlesinger T.K., Gibson S., and Johnson G.L. 1999. MEK kinase 1 (MEKK1) transduces c-Jun NH2-terminal kinase activation in response to changes in the microtubule cytoskeleton. *J. Biol. Chem.* **274:** 12605.
Yujiri T., Ware M., Widmann C., Oyer R., Russell D., Chan E., Zaitsu Y., Clarke P., Tyler K., Oka Y., Fanger G.R., Henson P., and Johnson G.L. 2000. MEK kinase 1 gene disruption alters cell migration and c-Jun NH2-terminal kinase regulation but does not cause a measurable defect in NF-kappa B activation. *Proc. Natl. Acad. Sci.* **97:** 7272.

Vascular Endothelium in Tissue Remodeling: Implications for Heart Failure

S.M. Dallabrida[†] and M.A. Rupnick[*†‡]

[*]Brigham and Women's Hospital, Division of Cardiovascular Medicine, Boston, Massachusetts 02115;
[†]Children's Hospital, Division of Surgical Research, Boston, Massachusetts 02115; [‡]Massachusetts
Institute of Technology, Department of Chemical Engineering, Cambridge, Massachusetts 02139;
Affiliates of Harvard Medical School, Boston, Massachusetts 02115

Cardiac remodeling markedly contributes to cardiovascular disease progression, emerging as a primary therapeutic target for heart failure of all causes. It is a crisis management response of an injured heart to counterbalance stress and salvage function through changes in composition and architecture. Although initially stabilizing, the changes cannot offset the impairment. Progressive remodeling results in ventricular hypertrophy, dilation, and a spherical geometry. The evolving cardiac phenotype is maladaptive and inevitably decompensates, manifesting as heart failure. Cardiac remodeling is a determinant of clinical course and indicates an ominous prognosis.

Cardiac remodeling is prominently involved in the progression of most cardiovascular disorders (Florea et al. 1999). The inciting injury may be ischemia (infarct), infection (myocarditis), pressure (hypertension) or volume overload (valvular disease), or idiopathic (cardiomyopathy). This establishes cardiac remodeling as a final common pathway in the development of heart failure. Thus, a major objective of heart failure therapy is to prevent, slow, or reverse remodeling.

Advances in the management of hypertension and coronary artery and valvular diseases have considerably improved survival. This has increased the prevalence and the life expectancy of patients with diminished cardiac function. Unfortunately, abating injury from the primary disease has proven necessary but not sufficient to prevent the development of heart failure in this population. Once triggered, remodeling begets remodeling, regardless of whether the inciting pathology was acute (infarction) or chronic (hypertension) (Katz 1990). A prolonged course of progressive deterioration ensues, culminating in heart failure.

The United States averages nearly 550,000 new cases of heart failure annually, with 287,200 deaths in 1999 (Levy et al. 2002). The incidence is particularly high in persons age 65 or older, a segment of the population projected to significantly increase in the United States (25.5 million in 1980 vs. 34.8 million in 2000 vs. 53.7 million in 2020) (National Population Projections 2002). Not surprisingly, it is predicted that "for the first time in human history cardiovascular disease will become the most common cause of death worldwide"(Braunwald 1997).

Once cardiac reserve is exhausted, mortality rates surpass those of cancer (Levy et al. 2002). However, decline in cardiac function associated with remodeling can progress insidiously for years before becoming evident. The ongoing nature of pathologic remodeling provides an opportunity to intervene before symptoms occur. Furthermore, the extent of the structural changes correlates with subsequent loss of ventricular function and increased mortality (Cohn 1995). This offers means of both identifying patients and gauging treatment efficacy early in the course. Therapeutic objectives have thus graduated from alleviating symptoms to altering the natural history of remodeling in an effort to avoid them all together.

A noteworthy challenge in therapeutic development is that symptoms and mortality are not conjoined endpoints. For example, flosequinan (hydralazine plus isorsorbide dinitrate) improves hemodynamics and exercise capacity as much or more than ACE inhibitors, but the survival benefit is less (Cohn et al. 1991; Massie et al. 1993). In contrast, β-adrenergic blockers improve survival, despite the aggravated symptoms acutely experienced by many patients (Heidenreich et al. 1997). Digoxin provides clinical relief without significantly affecting overall mortality (Digitalis Investigation Group 1997). Changes occur on systemic, organ, and cellular levels. An intervention that improves ventricular performance, such as certain inotropic agents (e.g., phosphodiesterase inhibitors) (Colucci et al. 1986; Om and Hess 1993), may harm cellular energetics resulting in better exercise capacity, but reduced survival. Appreciating the interdependent adaptive and maladaptive dynamics underlying the clinical sequelae has led to significant therapeutic advances.

Heart failure regimens have expanded over the past three decades as growing understanding of the pathophysiology has been translated to clinical practice. Early interventions aiming to correct volume status (diuretics) and stimulate failing myocardium (digoxin) improved symptoms but not outcome. Survival benefits were later achieved by optimizing hemodynamic measures (vasodilators) (Stevenson 1999, 2002). However, disease progression was still not deterred, and mortality rates remained high. This related to the compensatory structural and neurohormonal responses following cardiac injury. Remodeling does not restore ventricular performance, and sustained neurohormonal activation further depresses pump function. Over time, the defense mechanisms themselves are deleterious and propagate the disease. Counteracting remodeling by antagonizing neurohormonal pathways, such as renin-angiotensin-aldosterone (RAS)

(SOLVD Investigators 1991) and adrenergic systems (Bristow et al. 1996; Packer et al. 1996), was then added to established regimens and prolonged survival. Recently, spironolactone has been included with further reduction in heart failure mortality, presumably in part by reducing aldosterone-induced ventricular hypertrophy and fibrosis (Pitt et al. 1999).

A growing number of other therapeutic candidates or targets are also under investigation. Neutral endopeptidase (NEP) degrades atrial and brain natriuretic peptides (ANP, BNP) and bradykinin (vasodilators), functioning as an endogenous RAS system inhibitor (Corti et al. 2001). Vasopeptidase inhibitors inactivate NEP and ACE, offering a new therapeutic target (Chen et al. 2002). Strategies that improve the myocardial microenvironment may also confer benefit. Molecules classically termed "neurohormones" (angiotensin II, norepinephrine, TNF-α, endothelin) may contribute to heart failure in novel capacities, including autocrine and paracrine interactions within the myocardium (for review, see Mann et al. 2002). Antioxidants may counter the increased oxidative stress in heart failure (Serdar et al. 2001) as ejection fraction correlates with vitamin A and E levels (Polidori et al. 2002).

As the knowledge of heart failure advances from consequences to mechanisms, clinical approaches evolve from managing symptoms to extending survival. Despite these strides, mortality remains high, with over a 50% mortality rate within 5 years of diagnosis (Levy et al. 2002). Moreover, the survival benefit gained in the first year is gradually lost thereafter (Shepherd et al. 1995; Mann et al. 2002), illustrating the incompleteness of current approaches and the need for novel targets, which may include remodeling itself.

Chronic neurohormonal activation and hemodynamic stresses impel the remodeling process, but do not entirely account for it. Evidence suggests that remodeling may contribute to heart failure progression independently, since antagonizing factors that propagate cardiac remodeling using a combinatorial approach are efficacious. However, progress has been incremental, and clinical deterioration is not yet preventable. Halting remodeling will likely require a more comprehensive inventory of the begetting factors. An adjuvant approach would be to prevent the tissue from responding to these factors altogether by blocking essential steps in the remodeling process. This is predicated upon greater understanding of the molecular and cellular mechanisms underlying the structural changes in the heart.

Cardiac remodeling reflects changes in myocyte morphology (Samuel et al. 1984, 1986) and function (Katz 1990; Swynghedauw 1999), including hypertrophy (for review, see Cohn et al. 2000), apoptosis (Sharov et al. 1996; Teiger et al. 1996; Olivetti et al. 1997), shifts in contractility (for review, see Swynghedauw 1999), and energy metabolism (Katz 1990). Changes in expression of myosin heavy chain to slower isoforms (Lompre et al. 1979; Lowes et al. 1997) and less able calcium exchanger proteins, and up-regulation of atrial and brain natriuretic peptides (Takahashi et al. 1992; Yoshimura et al. 1993), reflect a reversion to a fetal phenotype. Extracellular matrix changes also accompany remodeling (Lee and Libby 2000; Libby and Lee 2000), including increased matrix metalloproteases (MMP) (Spinale et al. 2000; Maytin and Colucci 2002), and changes in interstitial collagen content cause fibrosis and cardiac rigidity (Weber and Brilla 1991).

In contrast to the myocyte and matrix components of cardiac remodeling, little is known about the role of the microvascular endothelium, in spite of its prominence. The adult heart has nearly 2,249 capillaries/mm^2 (Rakusan et al. 1992), and the microvasculature holds 35% of the coronary blood volume (Weber et al. 1987). Myocytes reside within 3–5 μm of cardiac microvascular endothelial cells (CMEC) (Shah et al. 1996), and communication between these cell types affects a number of cardiac functions. In response to hemodynamic forces, CMEC release substances that regulate inotropic activity, apoptosis, hypertrophic response, and extracellular matrix content of subjacent myocardium (Li et al. 1993; Shah et al. 1996). On the basis of the intimate working relationship between CMEC and myocytes, one would expect that the endothelium is also involved in the pathophysiology of cardiac remodeling.

Causes of heart failure as diverse as heredity (Factor et al. 1982), infection (Factor et al. 1985; Sole and Liu 1993), hypertension (Factor et al. 1984), infarction (Ren et al. 2002), or idiopathic forces (Chang et al. 1998) are associated with microvascular abnormalities such as spasm, aneurysm, focal constriction, and tortuosity. Endothelial cell permeability and proliferation increase and pericyte coverage lessens, consistent with vascular remodeling. Treatment of heart failure with ACE inhibitors (Kobayashi et al. 2000) or calcium channel blockers (Morris et al. 1989; Dong et al. 1992) reduces microvascular abnormalities, suggesting CMEC may be a target for intervention.

We propose that interventions directed at improving the health of the microvasculature may favorably affect cardiac remodeling and sustain or restore cardiac function. This may involve maintaining the viability of the existing endothelium or enhancing the population with additional endothelial cells. Recent evidence supports this concept. Endothelial progenitor cells can localize to the heart, particularly after a myocardial infarction (Dimmeler et al. 2002), and improve cardiac function (Morgan et al. 2002). Transepicardial injection of autologous mature endothelial cells post-ischemic cardiomyopathy results in myocardial angiogenesis and improved left ventricular function (Dangas et al. 2002). Furthermore, the number of circulating apoptotic endothelial cells (CAEC), and the ratio of CAEC to total circulating endothelial cells, increase in coronary artery disease (Zeiher et al. 2002). The ramification of these studies is that endothelial cell health may be important in the treatment of cardiac pathologies that can progress to heart failure.

Extensive CMEC/myocyte cross-talk suggests that coordinated interactions enable proper cardiac function. During development, the cell types grow in parallel and proportionate with increases in cardiac mass (Anversa et al. 1986). However, injury such as infarction provokes disproportionate remodeling. The initial myocardial response to ischemia has an angiogenic period that is critical for survival and recovery of function. However, microvascular expansion is inadequate, and the hypertrophic ventricle becomes more ischemic. Myocyte cell volume (28%), cross-sectional area (12%), and length (14%) increase within 3 days. Hypertrophy also precipitates increases in the distance between capillaries and myocytes (Motz et al. 1995), which is exacerbated by thickened basement membrane (Tanganelli et al. 1990) and interstitial collagen (Eghbali and Weber 1990). Thus, cross-talk is spatially uncoupled, and growth curves temporally diverge. Approaches bridging such gaps via endothelial recuperation (quantity or quality) or reducing cell–cell separation enhance cardiac performance. Angioblasts injected into infarcted myocardium avert myocyte apoptosis and collagen deposition (Kocher et al. 2001). Bolstering cardiac VEGF levels promotes endothelial cell survival and angiogenesis, and improves cardiac function (Gustafsson et al. 2001; Leotta et al. 2002). Animal (Gallego et al. 2001; Bauersachs et al. 2002) and human (Pitt et al. 1999) studies show that spironolactone, which prevents collagen accumulation, reduces heart failure mortality (Zannad et al. 2001).

Many strategies for treating heart failure that demonstrate survival benefits regulate vascular remodeling and angiogenesis. They include ACE inhibitors, ATR_1 blockers, β-blockers, calcium-channel blockers, and aspirin. ACE inhibitors and ATR_1 blockers inhibit tumor (Volpert et al. 1996; Fujita et al. 2002; Miyajima et al. 2002), ischemia-induced (Sasaki et al. 2002), and possibly, proliferative diabetic retinopathy-mediated (Chaturvedi et al. 1998) angiogenesis. Dual therapy in which both modalities have antiangiogenic actions (spironolactone [Doggrell and Brown 2001] and an ACE inhibitor) markedly reduces heart failure mortality (Pitt et al. 1999). β-Blockers inhibit norepinephrine-stimulated up-regulations in VEGF (Fredriksson et al. 2000) and bFGF (Yamashita et al. 1995). Benidipine, a calcium-channel blocker, reduces cardiac expressions of VEGF, bFGF, VEGF-R2, and TGF-β1, and decreases capillary density to baseline, whereas hydrazaline, which similarly reduced blood pressure, had no effect (Jesmin et al. 2002). Cyclooxygenase inhibitors, such as aspirin, inhibit endothelial cell migration and sprouting and reduce the production of angiogenic factors (Tsujii et al. 1998). Overall, these findings support the concept of targeting vascular remodeling as a means of affecting the pathophysiology of heart failure.

A SYSTEM FOR STUDYING VASCULAR REMODELING

Remodeling is limited in healthy adult myocardium but is prominent in cardiac pathologies. Therefore, much of our understanding is derived from studies of diseased hearts. However, post-insult cardiac remodeling may be aberrant, since the consequences are maladaptive and ultimately progress to heart failure. Greater understanding of the processes that result in balanced growth would be needed to address this question. During development, vascular and cardiac remodeling are coordinated and commensurate events. However, there is essentially no other condition in which the result is comparably balanced cardiac remodeling. Although myocyte growth accompanying exercise is adaptive, the capillaries do not expand in proportion to the muscle mass (Anversa et al. 1986), increasing vulnerability to ischemia.

An alternative approach would be to first study an adult tissue that normally undergoes physiological remodeling to identify key regulatory pathways, and then examine these pathways in the heart. Adipose tissue provides such an opportunity. It has a number of distinct and relevant features conducive to these investigations. Adipose tissue, like cardiac tissue, is highly vascularized (Crandall et al. 1997; Sierra-Honigmann et al. 1998) with capillary networks around each adipocyte (Gersh and Still 1945). Unlike cardiac tissue, adipose tissue preserves the capacity to grow and regress, rapidly and substantially, throughout adulthood. Adipocyte and vascular remodeling are tightly coordinated and maintain balance during both tissue growth and regression. Sizable and repeated shifts in mass can be induced without surgical manipulation and are well tolerated. One would expect remodeling of such magnitude to provide an amplified view of the genomic, cellular, and molecular dynamics involved. These pathways may not be as readily detectable in tissues that undergo less extensive physiological remodeling, such as the heart. Thus, our approach has been to define the mechanisms that regulate adipose tissue growth, remodeling, and vascular plasticity, and to determine whether they are functional in cardiac remodeling.

ADIPOSE TISSUE GROWTH AND REGRESSION REQUIRE COMMENSURATE VASCULAR REMODELING

Studies were conducted using a monogenetic strain of mice (C57BL/6J, *ob/ob*), which develops morbid obesity because of the inability to synthesize leptin (Zhang et al. 1994), a hormone that regulates appetite and metabolism (Pelleymounter et al. 1995; Campfield et al. 1996; Spiegelman and Flier 1996; Halaas et al. 1997). Leptin-deficient (*ob/ob*) mice can gain (or lose, if dieted) nearly 1.0 g/d of weight, principally in the form of fat tissue. This far exceeds the average weight changes of 0.1 g/d, which are typical of wild-type mice. The substantial shifts in fat mass in *ob/ob* mice provide an opportunity to study the mechanisms coordinating tissue and vascular remodeling.

Through a series of studies, we have shown that adipose tissue remodeling requires commensurate and coordinated remodeling of its vascular supply (Rupnick et al. 2002). Adipose tissue growth depends on new blood vessel formation and can be prevented by angiogenesis inhibitors. When *ob/ob* mice were treated with distinct antiangiogenic agents, such as TNP-470 (Morita et al. 1994; Wang et al. 2000), endostatin (O'Reilly et al. 1997), angiostatin (O'Reilly et al. 1994), thalidomide (D'Amato et al. 1994), and Bay-129566 (Gatto et al. 1999), they gained less or lost weight compared to controls (Fig. 1A). This was associated with significant and selective loss of adipose tissue, while the animals remained healthy (Rupnick et al. 2002). Mice from other obesity models (A^y, Cpe^{fat}) and wild-type C57BL/6 mice fed a high-fat diet also lost weight with TNP-470 (Fig. 1B). Thus, adipose tissue growth depends on concomitant blood vessel growth.

Vascular regulation of adipose tissue growth was further illustrated by intermittently treating *ob/ob* mice with TNP-470 for a total of four cycles (Fig. 2A). Mice similarly and repeatedly lost weight while receiving the angiogenesis inhibitor and regained when the drug was discontinued. The magnitude of the weight changes can be appreciated by comparing representative mice from the bottom of the 4th cycle (Fig. 2B). The control *ob/ob* mouse is morbidly obese, whereas the treated one is similar in weight to the age-matched wild-type animal. Of note, the pliability of adipose tissue far exceeds that of the cardiac tissue, perhaps offering an amplified view of remodeling. This may assist in identifying governing mechanisms, which can then be examined in the heart.

Treated *ob/ob* mice nearly doubled and halved their body weights repeatedly. One would expect the vasculature to have an equal capacity to remodel. This was confirmed by the observations that endothelial cell proliferation occurred with adipose tissue growth, whereas endothelial cell apoptosis became abundant during adi-

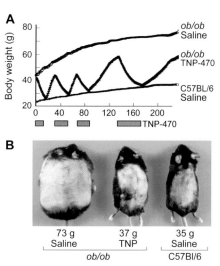

Figure 2. Intermittent treatment of *ob/ob* mice with TNP-470. Body weights of obese mice were cycled with intermittent administration of TNP-470, demonstrating vascular regulation of adipose tissue growth. (*A*) Obese mice were treated with vehicle or TNP-470 (10 mg/kg/d) ($n = 15$) until they reduced to the weight of age-matched C57BL/6 mice. Treatment was then discontinued and the mice were permitted to regain to approximately their starting weights. The drug was then restarted and the cycle was repeated three times. (*B*) Photograph of representative mice from each group on day 173 (bottom of cycle 4). Except for superficial scarring, TNP-470 treatment was well tolerated. The marked remodeling capacity of adipose tissue is notable. (Reprinted, with permission, from Rupnick et al. 2002 [copyright National Academy of Sciences].)

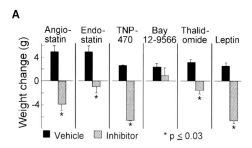

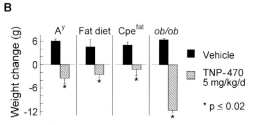

Figure 1. Body weight responses to angiogenesis inhibitors. Obese mice treated with various angiogenesis inhibitors gain less or lose weight compared with controls. (*A*) Eight-week-old *ob/ob* mice were treated with various angiogenesis inhibitors (*hatched bars*) for 7 days and compared to control mice receiving the corresponding vehicles (*solid bars*). The response to leptin is shown for comparison. (*B*) Twelve-week-old mice from various obesity models (agouti mice, C57BL/6J-A^y [A^y]; fat mice, C57BlKS-Cpe^{fat}/J [Cpe^{fat}]; *ob/ob* mice, C57BL/6J-Lep^{ob}; and wild-type C57BL/6J mice fed a diet consisting of 40% fat by calorie) were treated with vehicle (*solid bars*) or TNP-470 (*hatched bars*) for 7 days. $n = 3$/group. (Reprinted, with permission, from Rupnick et al. 2002 [copyright National Academy of Sciences].)

pose tissue loss (Fig. 3). Previous studies showed similar findings (Crandall et al. 1997; Cohen et al. 2001). We further found that the shift from endothelial proliferation to apoptosis occurred under a variety of weight-loss conditions, such as angiogenesis inhibitors, diet restriction, or leptin. This suggests that vascular remodeling is a critical and common event during adipose tissue growth or regression. Whether weight loss leads to vascular regression or vice versa is likely a function of the inciting mechanism. In either event, these results demonstrate that vascular and adipose tissue remodeling are tightly coordinated and may be orchestrated, in part, by paracrine interactions between endothelial cells and adipocytes. Adipocytes produce endothelial cell mitogens and angiogenic factors (Castellot et al. 1980; Crandall et al. 1997), such as monobutyrin (Dobson et al. 1990), vascular endothelial growth factor (Zhang et al. 1997), and leptin (Sierra-Honigmann et al. 1998). Endothelial cells also produce preadipocyte mitogens (Varzaneh et al. 1994; Lau et al. 1996). Elucidating these cell–cell interactions may reveal pathways that regulate vascular responses to tissue remodeling, which may also be functional in the heart.

VASCULAR MATURATION REGULATES ADIPOSE TISSUE PLASTICITY

We have shown thus far that, relative to most adult organs, adipose tissue is distinct in pliability, capacity for

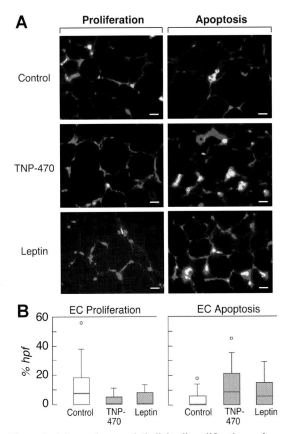

Figure 3. Adipose tissue endothelial cell proliferation and apoptosis in *ob/ob* mice. Vasculature remodeling accompanies adipose tissue growth or regression. (*A*) Endothelial cell proliferation and apoptosis were evaluated in epididymal adipose tissue sections from *ob/ob* mice following 7 days of TNP-470 or leptin treatment. Endothelial cells (*red*) were identified using vWF. BrdU was used to identify proliferating cells (*green*; left images) and the TUNEL assay was used to identify apoptotic cells (*green*; right images). Double-stained cells are yellow. Adipose tissue from control *ob/ob* mice contained numerous proliferating endothelial cells (*yellow, top left*), occasional proliferating nonendothelial cells (*green*), and few apoptotic endothelial cells (*yellow, top right*). In contrast, adipose tissue from TNP-470- and leptin-treated mice contained abundant apoptotic endothelial cells (*yellow, middle* and *bottom right*). Bar, 25 µm. (*B*) Median percentages of endothelial cell proliferation and apoptosis were quantified from 30 high-power fields per section (*n* = 5/group). Box and whisker plots illustrate the median (*red*) and range of the measurements. A shift from endothelial cell proliferation in growing adipose tissue to apoptosis in regressing tissue is seen. (Reprinted, with permission, from Rupnick et al. 2002 [copyright National Academy of Sciences].)

markers of vessel maturation, the angiopoietin/Tie system.

Angiopoietins bind to the Tie2 receptor (Davis et al. 1996; Maisonpierre et al. 1997), which is largely restricted to endothelium (Dumont et al. 1992; Korhonen et al. 1992). Angiopoietin-1 (Ang1) and angiopoietin-2 (Ang2) have similar binding affinities but opposing effects (Fig. 4) (Davis et al. 1996; Maisonpierre et al. 1997). Ang1 binding activates Tie2 in a paracrine manner and facilitates vessel maturation. Specifically, Ang1 promotes vascular contact with smooth muscle cells, pericytes (Holash et al. 1999; Zagzag et al. 2000), and extracellular matrix (Dumont et al. 1994), reducing vascular permeability and fostering endothelial survival (Thurston et al. 2000). Ang2 is a competitive antagonist that inactivates Tie2 in an autocrine fashion (Maisonpierre et al. 1997). It destabilizes vessels by reducing contact with support cells and matrix. Vessels are more immature and thus able to grow or regress, depending on the local signals (Asahara et al. 1998; Goede et al. 1998).

We measured Ang1 and Ang2 mRNA levels in the adipose tissue from wild-type mice undergoing different changes in body weight to examine the role of vascular maturation in fat tissue remodeling. Ang1 mRNA (Fig. 5A) and protein (data not shown) levels were lower in animals undergoing changes in weight compared to those that were weight-stable (Fig. 5A). Ang2 levels were similar. Reduced Ang1 favors a more immature vasculature during adipose tissue remodeling. Further studies using both wild-type and *ob/ob* mice revealed that Ang1 levels decreased as the rate of remodeling increased, regardless of the etiology (TNP-470, diet, leptin) or direction (gain, loss) of the weight change (Fig. 5B).

These findings suggest that the degree of vessel maturation in adipose tissue falls along a continuum and may be described in terms of Ang1 expression. Low Ang1 levels would favor an immature vascular bed, which is able to remodel and respond to angiogenesis regulators (Fig. 6). Conversely, high Ang1 levels promote vascular matu-

vascular remodeling, and susceptibility to angiogenesis inhibitors. Mature vascular beds do not typically exhibit these characteristics. However, immature vessels, which lack an established structure, may be more readily mobilized for growth or regression. Furthermore, they would also be sensitive to antiangiogenic agents that target growing or newly formed vessels, whereas established vascular beds are unresponsive (Folkman 1995a,b). This rationale suggests that vascular maturation may be an important variable governing the remodeling capacity of adipose tissue. We tested this concept by examining

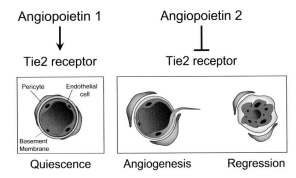

Figure 4. The role of angiopoietins in vascular remodeling. Ang1 binding to the Tie2 receptor increases endothelial cell associations with support cells, such as pericytes and matrix, thereby promoting vascular maturation and quiescence. Ang2 competes with Ang1 for binding to Tie2 and has opposing effects. Vessels have an immature phenotype and are more readily able to grow in the presence of angiogenesis stimulators or regress in the presence of angiogenesis inhibitors.

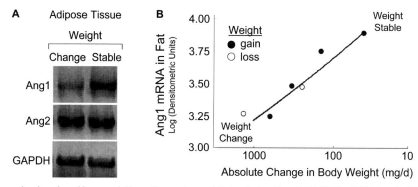

Figure 5. Ang1 expression is reduced in remodeling adipose tissue. (*A*) Ang1, Ang2, and GAPDH mRNA levels (RT-PCR) were measured in the epididymal adipose tissues from control and diet-restricted C57BL/6 and *ob/ob* mice on day 12 ($n = 4$/group). Representative expressions for C57BL/6 mice show that during weight loss (Change), Ang1 levels are low, whereas weight-stable (Stable) control mice have relatively high Ang1. (*B*) Ang1 mRNA levels in epididymal adipose tissue are plotted against the absolute rates of change in body weight for C57BL/6 and *ob/ob* mice. Representative points are shown. Ang1 levels inversely correlate with the rate of change in body weight, regardless of the direction of the weight change (gain or loss).

ration, reducing the capacity to remodel and conferring resistance to angiogenesis regulators. Thus, we propose that vascular maturation is a determinant of adipose tissue plasticity. The implication is that manipulating the extent of vascular maturation may afford a novel means of regulating the remodeling capacity and responsiveness of the tissue. Furthermore, this may represent a common pathway coordinating vascular and tissue remodeling that applies to non-adipose tissues as well, such as the heart. Reports that Ang1 overexpression reduces VEGF-induced vascular permeability (Thurston et al. 1999) and tumor growth (Hayes et al. 2000; Hawighorst et al. 2002; Tian et al. 2002) support this supposition.

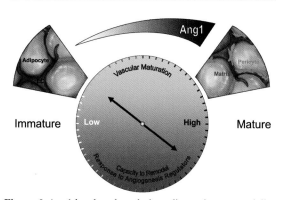

Figure 6. Ang1 levels reduce during adipose tissue remodeling. Schematic representation of the proposed relationship between vascular maturation and adipose tissue pliability. An immature vasculature, as indicated by low Ang1 mRNA expression, increases adipose tissue remodeling capacity and responsiveness to angiogenesis regulators, whereas a mature vasculature with high Ang1 expression would impede tissue remodeling and reduce susceptibility to angiogenesis regulators. Adipose tissue rendering (Advanced Medical Graphics, Boston, Massachusetts).

SIMILARITIES BETWEEN CARDIAC AND ADIPOSE TISSUE REMODELING

During development, the heart undergoes considerable growth and remodeling. This reflects an increase in myocyte volume from hypertrophy and hyperplasia (Anversa et al. 1980) and a greater increase in capillary volume from angiogenesis (Olivetti et al. 1980). In the rat, growth is maximal in the first week and tapers thereafter, largely ceasing by the time the animal reaches adulthood at 2 months (Riva and Hearse 1991). If Ang1 expression in the heart mimics that in adipose tissue, we would expect the levels to be lowest in the postnatal growing heart and highest in the adult quiescent heart.

Ang mRNA expression was measured in the left ventricles of neonatal and adult C57BL/6 mice. Using RT-PCR, we found that Ang1 expression was relatively low in the neonatal left ventricle and significantly higher in the adult (Fig. 7A). Ang2 levels were comparable. Examining Ang1 expression throughout development, when the left ventricular increases in weight more than fivefold, revealed a correlation with remodeling similar to that seen in adipose tissue (Fig. 7B). During the postnatal period, the rate of angiogenesis and ventricular growth is greatest, corresponding to the lowest Ang1 levels. As the heart increases in size in the immature animal, the growth rate slows and Ang1 expression increases. By adulthood, growth has all but ceased, and the Ang1 levels are highest and stable.

We determined the cellular identity of Ang1 expressions in both adipose and cardiac tissues. Fresh cell isolates of adipocytes and cardiac myocytes express Ang1 mRNAs (data not shown). Thus, these distinct tissues share a similar mechanism of paracrine signaling to the endothelium when undergoing remodeling.

Angiopoietins are emerging as important mediators in physiological and pathological cardiovascular remodeling (Carmeliet 1999). Embryonic expression of Ang1 is initially greatest in myocardium, where it is critical to myocardial development. Knockout studies revealed the requisite and temporal role of angiopoietins in vascular

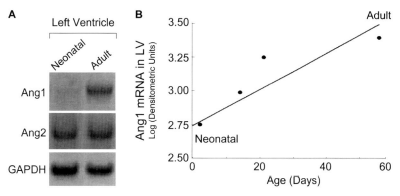

Figure 7. Ang1 expression is reduced in developing myocardium. (*A*) Ang1, Ang2, and GAPDH mRNA expressions (RT-PCR) were measured in the left ventricles of neonatal and adult C57BL/6 mice. (*B*) Analysis of Ang1 mRNA levels in the left ventricle at various times during development shows that Ang1 levels are highest in weight-stable adult cardiac tissue and lowest in growing neonatal myocardium. This pattern is similar to that seen in adipose tissue. $n = 6$/group.

development and integrity (for review, see Beck and D'Amore 1997). Ventricular and atrial defects are prominent in Ang1 knockout mice (Suri et al. 1996). Endocardial complexity and maturation are decreased, with little or no trabecula formation, predisposing the endocardium to retraction and collapse. Contact with matrix and pericytes is also markedly reduced. The Tie2 receptor is constitutively activated in wild-type mice (Wong et al. 1997), suggesting a role for Ang1 in maintaining cardiac endothelium in a mature, quiescent state. However, angiopoietins also respond during adult cardiac remodeling. Ang1 and Ang2 levels shift during hypoxia and reoxygenation of the rat myocardium (Ray et al. 2000). This raises the question of whether cardiovascular remodeling may be regulated via the angiopoietin system as a novel means of impeding the development of heart failure.

CONCLUSIONS

Considerable progress has been made in our understanding of the pathophysiology of cardiac remodeling and heart failure, especially in the past few decades (Fig. 8). This has generated a number of new therapies that not only alleviate symptoms, but also affect the disease process. These include angiotensin-converting-enzyme inhibitors, β-adrenergic antagonists, and spironolactone, which, when prescribed in a combinatorial fashion, result in incremental improvements in survival. However, the mortality rate from heart failure remains high, and the search for further advancements continues. In this regard, we propose that the microvascular endothelium may play a regulatory role in cardiac tissue remodeling and, therefore, serve as a therapeutic target to prevent the development of heart failure.

Toward this goal, our efforts have been directed at elucidating the mechanisms governing cardiac remodeling from a vascular perspective. Using adipose tissue as an experimental system of remodeling, we identified vascular maturation as a key determinant of tissue plasticity. Adipose tissue remodeling was associated with reduced Ang1 expression favoring a more immature vasculature, able to grow or regress in accordance with the tissue.

Analysis of Ang1 levels in developing myocardium revealed a similar expression pattern. Thus, adipose and cardiac tissues both utilize paracrine signaling via Ang1 to synchronize with the vascular endothelium during remodeling. This suggests that balanced vascular and tissue remodeling may be coordinated through this common pathway in a variety of different organs. The implication is that the capacity of a tissue to remodel or to respond to

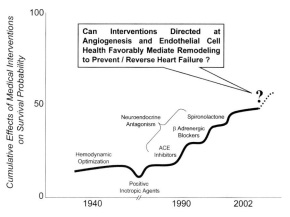

Figure 8. The evolution of heart failure therapy. Considerable advancements in our understanding of cardiac remodeling have spawned a number of new treatments over the past few decades that relieve symptoms and also affect disease progression, resulting in incremental improvements in survival. However, the mortality rate from heart failure remains high, and new advancements are continually being sought. In this regard, we propose that microvascular endothelium may play a regulatory role in cardiac tissue remodeling and, therefore, potentially serve as a therapeutic target to favorably affect the phenotype and prevent the development of heart failure. If so, therapeutic modalities directed at enhancing vascularization and maintaining microvascular endothelial cell health may contribute survival benefit to that achieved by current medical regimens.

angioactive agents may be regulated by altering the maturation state of its vasculature, perhaps via the Ang/Tie system.

In addition to facilitating angiogenesis, regulating vascular maturation may also favorably affect cardiac remodeling by maintaining the health of the existing endothelium. For example, Ang1 blocks endothelial cell apoptosis (Fujikawa et al. 1999; Kwak et al. 1999, 2000; Kim et al. 2000; Papapetropoulos et al. 2000), potentiates the antiapoptotic actions of VEGF and aFGF, and facilitates vascular stability (Papapetropoulos et al. 1999). Thus Ang1, and other factors such as VEGF, bFGF, endothelin-1, natriuretic peptides, and cytokine antagonists, which preserve the health and viability of the cardiac endothelium, may be beneficial as adjuvant therapies for cardiac remodeling and heart failure.

ACKNOWLEDGMENTS

Generous gifts were provided by Takeda Chemical Company (TNP-470), Bayer Corporation (Bay 12-9566), and Entremed (endostatin, thalidomide). Technical assistances of Michelle Euloth and Bin Shi are gratefully recognized. Sincere appreciation is extended to Joan Ebert for her guidance during the preparation of this manuscript. This study was supported, in part, by National Institutes of Health grant K08 HL-03698 to M.A.R.

REFERENCES

Anversa P., Olivetti G., and Loud A.V. 1980. Morphometric study of early postnatal development in the left and right ventricular myocardium of the rat. I. Hypertrophy, hyperplasia, and binucleation of myocytes. *Circ. Res.* **46:** 495.

Anversa P., Ricci R., and Olivetti G. 1986. Quantitative structural analysis of the myocardium during physiologic growth and induced cardiac hypertrophy: A review. *J. Am. Coll. Cardiol.* **7:** 1140.

Asahara T., Chen D., Takahashi T., Fujikawa K., Kearney M., Magner M., Yancopoulos G.D., and Isner J.M. 1998. Tie2 receptor ligands, angiopoietin-1 and angiopoietin-2, modulate VEGF-induced postnatal neovascularization. *Circ. Res.* **83:** 233.

Bauersachs J., Heck M., Fraccarollo D., Hildemann S.K., Ertl G., Wehling M., and Christ M. 2002. Addition of spironolactone to angiotensin-converting enzyme inhibition in heart failure improves endothelial vasomotor dysfunction: Role of vascular superoxide anion formation and endothelial nitric oxide synthase expression. *J. Am. Coll. Cardiol.* **39:** 351.

Beck L., Jr. and D'Amore P.A. 1997. Vascular development: Cellular and molecular regulation. *FASEB J.* **11:** 365.

Braunwald E. 1997. Cardiovascular medicine at the turn of the millennium: Triumphs, concerns, and opportunities. *N. Engl. J. Med.* **337:** 1360.

Bristow M.R., Gilbert E.M., Abraham W.T., Adams K.F., Fowler M.B., Hershberger R.E., Kubo S.H., Narahara K.A., Ingersoll H., Krueger S., Young S., and Shusterman N. 1996. Carvedilol produces dose-related improvements in left ventricular function and survival in subjects with chronic heart failure. *Circulation* **94:** 2807.

Campfield L.A., Smith F.J., and Burn P. 1996. The OB protein (leptin) pathway—A link between adipose tissue mass and central neural networks. *Horm. Metab. Res.* **28:** 619.

Carmeliet P. 1999. Basic concepts of (myocardial) angiogenesis: Role of vascular endothelial growth factor and angiopoietin. *Curr. Interv. Cardiol. Rep.* **1:** 322.

Castellot J.J., Jr., Karnovsky M.J., and Spiegelman B.M. 1980. Potent stimulation of vascular endothelial cell growth by differentiated 3T3 adipocytes. *Proc. Natl. Acad. Sci.* **77:** 6007.

Chang C.J., Chou Y.Y., and Lee Y.S. 1998. Electron microscopic studies of microvasculature and sympathetic nerve fibers in dilated cardiomyopathy. *Chin. Med. J.* **111:** 929.

Chaturvedi N., Sjolie A.K., Stephenson J.M., Abrahamian H., Keipes M., Castellarin A., Rogulja-Pepeonik Z., and Fuller J.H. 1998. Effect of lisinopril on progression of retinopathy in normotensive people with type 1 diabetes. The EUCLID Study Group. EURODIAB controlled trial of lisinopril in insulin-dependent diabetes mellitus. *Lancet* **351:** 28.

Chen H.H., Lainchbury J.G., Harty G.J., and Burnett J.C., Jr. 2002. Maximizing the natriuretic peptide system in experimental heart failure: Subcutaneous brain natriuretic peptide and acute vasopeptidase inhibition. *Circulation* **105:** 999.

Cohen B., Barkan D., Levy Y., Goldberg I., Fridman E., Kopolovic J., and Rubinstein M. 2001. Leptin induces angiopoietin-2 expression in adipose tissues. *J. Biol. Chem.* **276:** 7697.

Cohn J.N. 1995. Structural basis for heart failure. Ventricular remodeling and its pharmacological inhibition. *Circulation* **91:** 2504.

Cohn J.N., Ferrari R., and Sharpe N. 2000. Cardiac remodeling—Concepts and clinical implications: A consensus paper from an international forum on cardiac remodeling. Behalf of an International Forum on Cardiac Remodeling. *J. Am. Coll. Cardiol.* **35:** 569.

Cohn J.N., Johnson G., Ziesche S., Cobb F., Francis G., Tristani F., Smith R., Dunkman W.B., Loeb H., and Wong M., et al. 1991. A comparison of enalapril with hydralazine-isosorbide dinitrate in the treatment of chronic congestive heart failure. *N. Engl. J. Med.* **325:** 303.

Colucci W.S., Wright R.F., and Braunwald E. 1986. New positive inotropic agents in the treatment of congestive heart failure. Mechanisms of action and recent clinical developments. 2. *N. Engl. J. Med.* **314:** 349.

Corti R., Burnett, Jr., J.C., Rouleau J.L., Ruschitzka F., and Luscher T.F. 2001. Vasopeptidase inhibitors: A new therapeutic concept in cardiovascular disease? *Circulation* **104:** 1856.

Crandall D.L., Hausman G.J., and Kral J.G. 1997. A review of the microcirculation of adipose tissue: Anatomic, metabolic, and angiogenic perspectives. *Microcirculation* **4:** 211.

D'Amato R.J., Loughnan M.S., Flynn E., and Folkman J. 1994. Thalidomide is an inhibitor of angiogenesis. *Proc. Natl. Acad. Sci.* **91:** 4082.

Dangas G., Chekanov V., Roubin G.S., Chekanov G., and Kipshidze N. 2002. Transplantation of autologous endothelial cells induces neoangiogenesis and improves LV function in an animal model of ischemic cardiomyopathy (*Abstract*). American Heart Association, Dallas, Texas.

Davis S., Aldrich T.H., Jones P.F., Acheson A., Compton D.L., Jain V., Ryan T.E., Bruno J., Radziejewski C., P.C. Maisonpierre, and Yancopoulos G.D. 1996. Isolation of angiopoietin-1, a ligand for the TIE2 receptor, by secretion-trap expression cloning. *Cell* **87:** 1161.

Digitalis Investigation Group. 1997. The effect of digoxin on mortality and morbidity in patients with heart failure. The Digitalis Investigation Group. *N. Engl. J. Med.* **336:** 525.

Dimmeler S., Henze E., Assmus B., Mssoudi S., Badorff C., Zuhayra M., Brenner W., and Aicher A. 2002. Monitering the fate and tissue distribution of transplanted endothelial progenitor cells by radiolabeling in a rat model of myocardial infarction (*Abstract*). American Heart Association, Dallas, Texas.

Dobson, D.E., A. Kambe, E. Block, T. Dion, H. Lu, J.J. Castellot, Jr., and B.M. Spiegelman. 1990. 1-Butyryl-glycerol: A novel angiogenesis factor secreted by differentiating adipocytes. *Cell* **61:** 223.

Doggrell S.A. and Brown L. 2001. The spironolactone renaissance. *Expert Opin. Investig. Drugs* **10:** 943.

Dong R., Liu P., Wee L., Butany J., and Sole M.J. 1992. Verapamil ameliorates the clinical and pathological course of murine myocarditis. *J. Clin. Invest.* **90:** 2022.

Dumont D.J., Yamaguchi T.P., Conlon R.A., Rossant J., and

Breitman M.L. 1992. tek, a novel tyrosine kinase gene located on mouse chromosome 4, is expressed in endothelial cells and their presumptive precursors. *Oncogene* **7:** 1471.

Dumont D.J., Gradwohl G., Fong G.H., Puri M.C., Gertsenstein M., Auerbach A., and Breitman M.L. 1994. Dominant-negative and targeted null mutations in the endothelial receptor tyrosine kinase, tek, reveal a critical role in vasculogenesis of the embryo. *Genes Dev.* **8:** 1897.

Eghbali M. and Weber K.T. 1990. Collagen and the myocardium: Fibrillar structure, biosynthesis and degradation in relation to hypertrophy and its regression. *Mol. Cell. Biochem.* **96:** 1.

Factor S.M., Cho S., Wittner M., and Tanowitz H. 1985. Abnormalities of the coronary microcirculation in acute murine Chagas' disease. *Am. J. Trop. Med. Hyg.* **34:** 246.

Factor S.M., Minase T., Cho S., Dominitz R., and Sonnenblick E.H. 1982. Microvascular spasm in the cardiomyopathic Syrian hamster: A preventable cause of focal myocardial necrosis. *Circulation* **66:** 342.

Factor S.M., Minase T., Cho S., Fein F., Capasso J.M., and Sonnenblick E.H. 1984. Coronary microvascular abnormalities in the hypertensive-diabetic rat. A primary cause of cardiomyopathy? *Am. J. Pathol.* **116:** 9.

Florea, V.G., Mareyev V.Y., Samko A.N., Orlova I.A., Coats A.J., and Belenkov Y.N. 1999. Left ventricular remodelling: Common process in patients with different primary myocardial disorders. *Int. J. Cardiol.* **68:** 281.

Folkman J. 1995a. Angiogenesis in cancer, vascular, rheumatoid and other disease. *Nat. Med.* **1:** 27.

———. 1995b. Seminars in Medicine of the Beth Israel Hospital, Boston. Clinical applications of research on angiogenesis. *N. Engl. J. Med.* **333:** 1757.

Fredriksson J.M., Lindquist J.M., Bronnikov G.E., and Nedergaard J. 2000. Norepinephrine induces vascular endothelial growth factor gene expression in brown adipocytes through a beta-adrenoreceptor/cAMP/protein kinase A pathway involving Src but independently of Erk1/2. *J. Biol. Chem.* **275:** 13802.

Fujikawa K., de Aos Scherpenseel I., Jain S.K., Presman E., Christensen R.A., and Varticovski L. 1999. Role of PI 3-kinase in angiopoietin-1-mediated migration and attachment-dependent survival of endothelial cells. *Exp. Cell Res.* **253:** 663.

Fujita M., Hayashi I., Yamashina S., Itoman M., and Majima M. 2002. Blockade of angiotensin AT1a receptor signaling reduces tumor growth, angiogenesis, and metastasis. *Biochem. Biophys. Res. Commun.* **294:** 441.

Gallego M., Espina L., Vegas L., Echevarria E., Iriarte M.M., and Casis O. 2001. Spironolactone and captopril attenuates isoproterenol-induced cardiac remodelling in rats. *Pharmacol. Res.* **44:** 311.

Gatto C., Rieppi M., Borsotti P., Innocenti S., Ceruti R., Drudis T., Scanziani E., Casazza A.M., Taraboletti G., and Giavazzi R. 1999. BAY 12-9566, a novel inhibitor of matrix metalloproteinases with antiangiogenic activity. *Clin. Cancer Res.* **5:** 3603.

Gersh I. and Still M.A. 1945. Blood vessels in fat tissue: Relation to problems of gas exchange. *J. Exp. Med.* **81:** 219.

Goede V., Schmidt T., Kimmina S., Kozian D., and Augustin H.G. 1998. Analysis of blood vessel maturation processes during cyclic ovarian angiogenesis. *Lab. Invest.* **78:** 1385.

Gustafsson T., Bodin K., Sylven C., Gordon A., Tyni-Lenne R., and Jansson E. 2001. Increased expression of VEGF following exercise training in patients with heart failure. *Eur. J. Clin. Invest.* **31:** 362.

Halaas J.L., Boozer C., Blair-West J., Fidahusein N., Denton D.A., and Friedman J.M. 1997. Physiological response to long-term peripheral and central leptin infusion in lean and obese mice. *Proc. Natl. Acad. Sci.* **94:** 8878.

Hawighorst T., Skobe M., Streit M., Hong Y.K., Velasco P., Brown L.F., Riccardi L., Lange-Asschenfeldt B., and Detmar M. 2002. Activation of the tie2 receptor by angiopoietin-1 enhances tumor vessel maturation and impairs squamous cell carcinoma growth. *Am. J. Pathol.* **160:** 1381.

Hayes A.J., Huang W.Q., Yu J., Maisonpierre P.C., Liu A., Kern F.G., Lippman M.E., McLeskey S.W., and Li L.Y. 2000. Expression and function of angiopoietin-1 in breast cancer. *Br. J. Cancer* **83:** 1154.

Heidenreich P.A., Lee T.T., and Massie B.M. 1997. Effect of beta-blockade on mortality in patients with heart failure: A meta-analysis of randomized clinical trials. *J. Am. Coll. Cardiol.* **30:** 27.

Holash J., Maisonpierre P.C., Compton D., Boland P., Alexander C.R., Zagzag D., Yancopoulos G.D., and Wiegand S.J. 1999. Vessel cooption, regression, and growth in tumors mediated by angiopoietins and VEGF. *Science* **284:** 1994.

Jesmin S., Sakuma I., Hattori Y., Fujii S., and Kitabatake A. 2002. Long-acting calcium channel blocker benidipine suppresses expression of angiogenic growth factors and prevents cardiac remodelling in a type II diabetic rat model. *Diabetologia* **45:** 402.

Katz A.M. 1990. Cardiomyopathy of overload. A major determinant of prognosis in congestive heart failure. *N. Engl. J. Med.* **322:** 100.

Kim I., Kim H.G., So J.N., Kim J.H., Kwak H.J., and Koh G.Y. 2000. Angiopoietin-1 regulates endothelial cell survival through the phosphatidylinositol 3´-kinase/Akt signal transduction pathway. *Circ. Res.* **86:** 24.

Kobayashi N., Hara K., Watanabe S., Higashi T., and Matsuoka H. 2000. Effect of imidapril on myocardial remodeling in L-NAME-induced hypertensive rats is associated with gene expression of NOS and ACE mRNA. *Am. J. Hypertens.* **13:** 199.

Kocher A.A., Schuster M.D., Szabolcs M.J., Takuma S., Burkhoff D., Wang J., Homma S., Edwards N.M., and Itescu S. 2001. Neovascularization of ischemic myocardium by human bone-marrow-derived angioblasts prevents cardiomyocyte apoptosis, reduces remodeling and improves cardiac function. *Nat. Med.* **7:** 430.

Korhonen J., Partanen J., Armstrong E., Vaahtokari A., Elenius K., Jalkanen M., and Alitalo K. 1992. Enhanced expression of the tie receptor tyrosine kinase in endothelial cells during neovascularization. *Blood* **80:** 2548.

Kwak H.J., Lee S.J., Lee Y.H., Ryu C.H., Koh K.N., Choi H.Y., and Koh G.Y. 2000. Angiopoietin-1 inhibits irradiation- and mannitol-induced apoptosis in endothelial cells. *Circulation* **101:** 2317.

Kwak H.J., So J.N., Lee S.J., Kim I., and Koh G.Y. 1999. Angiopoietin-1 is an apoptosis survival factor for endothelial cells. *FEBS Lett.* **448:** 249.

Lau D.C., Schillabeer G., Li Z.H., Wong K.L., Varzaneh F.E., and Tough S.C. 1996. Paracrine interactions in adipose tissue development and growth. *Int. J. Obes. Relat. Metab. Disord.* (suppl. 3) **20:** S16.

Lee R.T. and Libby P. 2000. Matrix metalloproteinases: Not-so-innocent bystanders in heart failure. *J. Clin. Invest.* **106:** 827.

Leotta E., Patejunas G., Murphy G., Szokol J., McGregor L., Carbray J., Hamawy A., Winchester D., Hackett N., Crystal R., and Rosengart T. 2002. Gene therapy with adenovirus-mediated myocardial transfer of vascular endothelial growth factor 121 improves cardiac performance in a pacing model of congestive heart failure. *J. Thorac. Cardiovasc. Surg.* **123:** 1101.

Levy D., Kenchaiah S., Larson M.G., Benjamin E.J., Kupka M.J., Ho K.K.L., Murabito J.M., and Vasan R.S. 2002. Long-term trends in the incidence of and survival with heart failure. *N. Engl. J. Med.* **347:** 1397.

Li K., Rouleau J.L., Andries L.J., and Brutsaert D.L. 1993. Effect of dysfunctional vascular endothelium on myocardial performance in isolated papillary muscles. *Circ. Res.* **72:** 768.

Libby P. and Lee R.T. 2000. Matrix matters. *Circulation* **102:** 1874.

Lompre A.M., Schwartz K., d'Albis A., Lacombe G., Van Thiem N., and Swynghedauw B. 1979. Myosin isoenzyme redistribution in chronic heart overload. *Nature* **282:** 105.

Lowes B.D., Minobe W., Abraham W.T., Rizeq M.N., Bohlmeyer T.J., Quaife R.A., Roden R.L., Dutcher D.L., Robertson A.D., Voelkel N.F., Badesch D.B., Groves B.M., Gilbert E.M., and Bristow M.R. 1997. Changes in gene expression in the intact human heart. Downregulation of alpha-myosin heavy chain in hypertrophied, failing ventricular myocardium. *J. Clin. Invest.* **100:** 2315.

Maisonpierre P.C., Suri C., Jones P.F., Bartunkova S., Wiegand S.J., Radziejewski C., Compton D., McClain J., Aldrich T.H., Papadopoulos N., Daly T.J., Davis S., T.N. Sato T.N., and Yancopoulos G.D. 1997. Angiopoietin-2, a natural antagonist for Tie2 that disrupts in vivo angiogenesis. *Science* **277:** 55.

Mann D.L., Deswal A., Bozkurt B., and Torre-Amione G. 2002. New therapeutics for chronic heart failure. *Annu. Rev. Med.* **53:** 59.

Massie B., Berk M., Brozena S., Elkayam U., Plehn J., Kukin M., Packer M., Murphy B., Neuberg G., and Steingart R. 1993. Can further benefit be achieved by adding flosequinan to patients with congestive heart failure who remain symptomatic on diuretic, digoxin, and an angiotensin converting enzyme inhibitor? Results of the flosequinan-ACE inhibitor trial (FACET). *Circulation* **88:** 492.

Maytin M. and Colucci W.S. 2002. Molecular and cellular mechanisms of myocardial remodeling. *J. Nucl. Cardiol.* **9:** 319.

Miyajima A., Kosaka T., Asano T., Seta K., Kawai T., and Hayakawa M. 2002. Angiotensin II type I antagonist prevents pulmonary metastasis of murine renal cancer by inhibiting tumor angiogenesis. *Cancer Res.* **62:** 4176.

Morgan J.P., Min J.-Y., Huang X., and Oettgen P. 2002. Transplantation of embryonic endothelial progenitor cells promote neovascularization and myocardial tissue regeneration after myocardial infarction (*Abstract*). American Heart Association, Dallas, Texas.

Morita T., Shinohara N., and Tokue A. 1994. Antitumour effect of a synthetic analogue of fumagillin on murine renal carcinoma. *Br. J. Urol.* **74:** 416.

Morris S.A., Weiss L.M., Factor S., Bilezikian J.P., Tanowitz H., and Wittner M. 1989. Verapamil ameliorates clinical, pathologic and biochemical manifestations of experimental chagasic cardiomyopathy in mice. *J. Am. Coll. Cardiol.* **14:** 782.

Motz W., Schwartzkopff B., and Vogt M. 1995. Hypertensive heart disease: Cardioreparation by reversal of interstitial collagen in patients. *Eur. Heart J.* **16:** 69.

National Population Projections. 2002. Census Bureau, Washington, D.C.: http://www.census.gov/population/www/projections/natproj.html.

Olivetti G., Anversa P., and Loud A.V. 1980. Morphometric study of early postnatal development in the left and right ventricular myocardium of the rat. II. Tissue composition, capillary growth, and sarcoplasmic alterations. *Circ. Res.* **46:** 503.

Olivetti G., Abbi R., Quaini F., Kajstura J., Cheng W., Nitahara J.A., Quaini E., Di Loreto C., Beltrami C.A., Krajewski S., Reed J.C., and Anversa P. 1997. Apoptosis in the failing human heart. *N. Engl. J. Med.* **336:** 1131.

Om A. and Hess M.L. 1993. Inotropic therapy of the failing myocardium. *Clin. Cardiol.* **16:** 5.

O'Reilly M.S., Boehm T., Shing Y., Fukai N., Vasios G., Lane W.S., Flynn E., Birkhead J.R., Olsen B.R., and Folkman J. 1997. Endostatin: An endogenous inhibitor of angiogenesis and tumor growth. *Cell* **88:** 277.

O'Reilly M.S., Holmgren L., Shing Y., Chen C., Rosenthal R.A., Moses M., Lane W.S., Cao Y., Sage E.H., and Folkman J. 1994. Angiostatin: A novel angiogenesis inhibitor that mediates the suppression of metastases by a Lewis lung carcinoma. *Cell* **79:** 315.

Packer M., Bristow M.R., Cohn J.N., Colucci W.S., Fowler M.B., Gilbert E.M., Shusterman N.H., and the U.S. Carvedilol Heart Failure Study Group. 1996. The effect of carvedilol on morbidity and mortality in patients with chronic heart failure. *N. Engl. J. Med.* **334:** 1349.

Papapetropoulos A., Fulton D., Mahboubi K., Kalb R.G., O'Connor D.S., Li F., Altieri D.C., and Sessa W.C. 2000. Angiopoietin-1 inhibits endothelial cell apoptosis via the Akt/survivin pathway. *J. Biol. Chem.* **275:** 9102.

Papapetropoulos A., Garcia-Cardena G., Dengler T.J., Maisonpierre P.C., Yancopoulos G.D., and Sessa W.C. 1999. Direct action of angiopoietin-1 on human endothelium: Evidence for network stabilization, cell survival, and interaction with other angiogenic growth factors. *Lab. Invest.* **79:** 213.

Pelleymounter M.A., Cullen M.J., Baker M.B., Hecht R., Winters D., Boone T., and Collins F. 1995. Effects of the obese gene product on body weight regulation in ob/ob mice. *Science* **269:** 540.

Pitt B., Zannad F., Remme W.J., Cody R., Castaigne A., Perez A., Palensky J., and Wittes J. 1999. The effect of spironolactone on morbidity and mortality in patients with severe heart failure. Randomized Aldactone Evaluation Study Investigators. *N. Engl. J. Med.* **341:** 709.

Polidori M.C., Savino K., Alunni G., Freddio M., Senin U., Sies H., Stahl W., and Mecocci P. 2002. Plasma lipophilic antioxidants and malondialdehyde in congestive heart failure patients: Relationship to disease severity. *Free Radic. Biol. Med.* **32:** 148.

Rakusan K., Flanagan M.F., Geva T., Southern J., and Van Praagh R. 1992. Morphometry of human coronary capillaries during normal growth and the effect of age in left ventricular pressure-overload hypertrophy. *Circulation* **86:** 38.

Ray P.S., Estrada-Hernandez T., Sasaki H., Zhu L., and Maulik N. 2000. Early effects of hypoxia/reoxygenation on VEGF, ang-1, ang-2 and their receptors in the rat myocardium: Implications for myocardial angiogenesis. *Mol. Cell. Biochem.* **213:** 145.

Ren G., Michael L.H., Entman M.L., and Frangogiannis N.G. 2002. Morphological characteristics of the microvasculature in healing myocardial infarcts. *J. Histochem. Cytochem.* **50:** 71.

Riva E. and Hearse D.J. 1991. *The developing myocardium*. Futura Publishing, Mt. Kisco, New York.

Rupnick M.A., Panigrahy D., Zhang C.Y., Dallabrida S.M., Lowell B.B., Langer R., and Folkman M.J. 2002. Adipose tissue mass can be regulated through the vasculature (issue cover). *Proc. Natl. Acad. Sci.* **99:** 10730.

Samuel J.L., Marotte F., Delcayre C., and Rappaport L. 1986. Microtubule reorganization is related to rate of heart myocyte hypertrophy in rat. *Am. J. Physiol.* **251:** H1118.

Samuel J.L., Bertier B., Bugaisky L., Marotte F., Swynghedauw B., Schwartz K., and Rappaport L. 1984. Different distributions of microtubules, desmin filaments and isomyosins during the onset of cardiac hypertrophy in the rat. *Eur. J. Cell Biol.* **34:** 300.

Sasaki K., Murohara T., Ikeda H., Sugaya T., Shimada T., Shintani S., and Imaizumi T. 2002. Evidence for the importance of angiotensin II type 1 receptor in ischemia-induced angiogenesis. *J. Clin. Invest.* **109:** 603.

Serdar A., Yesilbursa D., Serdar Z., Dirican M., Turel B., and Cordan J. 2001. Relation of functional capacity with the oxidative stress and antioxidants in chronic heart failure. *Congest. Heart Fail.* **7:** 309.

Shah A.M., Grocott-Mason R.M., Pepper C.B., Mebazaa A., Henderson A.H., Lewis M.J., and Paulus W.J. 1996. The cardiac endothelium: Cardioactive mediators. *Prog. Cardiovasc. Dis.* **39:** 263.

Sharov V.G., Sabbah H.N., Shimoyama H., Goussev A.V., Lesch M., and Goldstein S. 1996. Evidence of cardiocyte apoptosis in myocardium of dogs with chronic heart failure. *Am. J. Pathol.* **148:** 141.

Shepherd J., Cobbe S.M., Ford I., Isles C.G., Lorimer A.R., MacFarlane P.W., McKillop J.H., and Packard C.J. 1995. Prevention of coronary heart disease with pravastatin in men with hypercholesterolemia. West of Scotland Coronary Prevention Study Group. *N. Engl. J. Med.* **333:** 1301.

Sierra-Honigmann M.R., Nath A.K., Murakami C., Garcia-Cardena G., Papetropoulos A., Sessa W.C., Madge L.A., Schechner J.S., Schwabb M.B., Polverini P.J., and Flores-Riveros J.R. 1998. Biological action of leptin as an angiogenic factor. *Science* **281:** 1683.

Sole M.J. and Liu P. 1993. Viral myocarditis: A paradigm for understanding the pathogenesis and treatment of dilated cardiomyopathy. *J. Am. Coll. Cardiol.* **22:** 99A.

SOLVD Investigators. 1991. Effect of enalapril on survival in patients with reduced left ventricular ejection fractions and congestive heart failure. The SOLVD Investigators. *N. Engl. J. Med.* **325:** 293.

Spiegelman B.M. and Flier J.S. 1996. Adipogenesis and obesity: Rounding out the big picture. *Cell* **87:** 377.

Spinale F.G., Coker M.L., Heung L.J., Bond B.R., Gunasinghe H.R., Etoh T., Goldberg A.T., Zellner J.L., and Crumbley A.J. 2000. A matrix metalloproteinase induction/activation system exists in the human left ventricular myocardium and is upregulated in heart failure. *Circulation* **102:** 1944.

Stevenson L.W. 1999. Tailored therapy to hemodynamic goals for advanced heart failure. *Eur. J. Heart Fail.* **1:** 251.

———. 2002. Treatment of congestive heart failure. *J. Am. Med. Assoc.* **287:** 2209.

Suri C., Jones P.F., Patan S., Bartunkova S., Maisonpierre P.C., Davis S., Sato T.N., and Yancopoulos G.D. 1996. Requisite role of angiopoietin-1, a ligand for the TIE2 receptor, during embryonic angiogenesis. *Cell* **87:** 1171.

Swynghedauw B. 1999. Molecular mechanisms of myocardial remodeling. *Physiol. Rev.* **79:** 215.

Takashi T., Allen P.D., and Izumo S. 1992. Expression of A-, B-, and C-type natriuretic peptide genes in failing and developing human ventricles. Correlation with expression of the Ca(2+)-ATPase gene. *Circ. Res.* **71:** 9.

Tanganelli P., Pierli C., Bravi A., Del Sordo M., Salvi A., Bussani R., Silvestri F., and Camerini F. 1990. Small vessel disease (SVD) in patients with unexplained ventricular arrhythmia and dilated congestive cardiomyopathy. *Am. J. Cardiovasc. Pathol.* **3:** 13.

Teiger E., Than V.D., Richard L., Wisnewsky C., Tea B.S., Gaboury L., Tremblay J., Schwartz K., and Hamet P. 1996. Apoptosis in pressure overload-induced heart hypertrophy in the rat. *J. Clin. Invest.* **97:** 2891.

Thurston G., Suri C., Smith K., McClain J., Sato T.N., Yancopoulos G.D., and McDonald D.M. 1999. Leakage-resistant blood vessels in mice transgenically overexpressing angiopoietin-1. *Science* **286:** 2511.

Thurston G., Rudge J.S., Ioffe E., Zhou H., Ross L., Croll S.D., Glazer N., Holash J., McDonald D.M., and Yancopoulos G.D. 2000. Angiopoietin-1 protects the adult vasculature against plasma leakage. *Nat. Med.* **6:** 460.

Tian S., Hayes A.J., Metheny-Barlow L.J., and Li L.Y. 2002. Stabilization of breast cancer xenograft tumour neovasculature by angiopoietin-1. *Br. J. Cancer* **86:** 645.

Tsujii M., Kawano S., Tsuji S., Sawaoka H., Hori M., and DuBois R.N. 1998. Cyclooxygenase regulates angiogenesis induced by colon cancer cells. *Cell* **93:** 705.

Varzaneh F.E., Shillabeer G., Wong K.L., and Lau D.C. 1994. Extracellular matrix components secreted by microvascular endothelial cells stimulate preadipocyte differentiation in vitro. *Metabolism* **43:** 906.

Volpert O.V., Ward W.F., Lingen M.W., Chesler L., Solt D.B., Johnson M.D., Molteni A., Polverini P.J., and Bouck N.P. 1996. Captopril inhibits angiogenesis and slows the growth of experimental tumors in rats. *J. Clin. Invest.* **98:** 671.

Wang J., Lou P., and Henkin J. 2000. Selective inhibition of endothelial cell proliferation by fumagillin is not due to differential expression of methionine aminopeptidases. *J. Cell. Biochem.* **77:** 465.

Weber K.T. and Brilla C.G. 1991. Pathological hypertrophy and cardiac interstitium. Fibrosis and renin-angiotensin-aldosterone system. *Circulation* **83:** 1849.

Weber K.T., Clark W.A., Janicki J.S., and Shroff S.G. 1987. Physiologic versus pathologic hypertrophy and the pressure-overloaded myocardium. *J. Cardiovasc. Pharmacol.* **10:** S37.

Wong A.L., Haroon Z.A., Werner S., Dewhirst M.W., Greenberg C.S., and Peters K.G. 1997. Tie2 expression and phosphorylation in angiogenic and quiescent adult tissues. *Circ. Res.* **81:** 567.

Yamashita H., Sato N., Kizaki T., Oh-ishi S., Segawa M., Saitoh D., Ohira Y., and Ohno H. 1995. Norepinephrine stimulates the expression of fibroblast growth factor 2 in rat brown adipocyte primary culture. *Cell Growth Differ.* **6:** 1457.

Yoshimura M., Yasue H., Okumura K., Ogawa H., Jourasaki M., Mukoyama M., Nakao K., and Imura H. 1993. Different secretion patterns of atrial natriuretic peptide and brain natriuretic peptide in patients with congestive heart failure. *Circulation* **87:** 464.

Zagzag D., Amirnovin R., Greco M.A., Yee H., Holash J., Wiegand S.J., Zabski S., Yancopoulos G.D., and Grumet M. 2000. Vascular apoptosis and involution in gliomas precede neovascularization: A novel concept for glioma growth and angiogenesis. *Lab. Invest.* **80:** 837.

Zannad F., Dousset B., and Alla F. 2001. Treatment of congestive heart failure: Interfering the aldosterone-cardiac extracellular matrix relationship. *Hypertension* **38:** 1227.

Zeiher A.M., Dimmeler S., Fichtlscherer S., Kamper U., and Rossig L. 2002. Levels of circulating apoptotic endothelial cells reflect disease activity in patients with coronary artery disease (*Abstract*). American Heart Association, Dallas, Texas.

Zhang Q.X., Magovern C.J., Mack C.A., Budenbender K.T., Ko W., and Rosengart T.K. 1997. Vascular endothelial growth factor is the major angiogenic factor in omentum: Mechanism of the omentum-mediated angiogenesis. *J. Surg. Res.* **67:** 147.

Zhang Y., Proenca R., Maffei M., Barone M., Leopold L., and Friedman J.M. 1994. Positional cloning of the mouse obese gene and its human homologue. *Nature* **372:** 425.

Using a Gene-switch Transgenic Approach to Dissect Distinct Roles of MAP Kinases in Heart Failure

B.G. Petrich, P. Liao,* and Y. Wang

Department of Physiology, University of Maryland School of Medicine, Baltimore, Maryland 21201

Heart failure is a leading cause of death and medical needs in most developed countries, and is particularly prevalent in the aging population of the United States (Cohn et al. 1997; American Heart Association 2002). An estimate of more than 4,000,000 people have already been affected by this disease, ~75% of them over the age of 65. Still, more than 400,000 new cases of heart failure are being registered each year. Although heart failure shares common manifestations in impaired ventricular contractile function and other aspects of cardiac remodeling, the origins of the disease are highly heterogeneous, ranging from coronary atherosclerosis, hypertension, myocarditis, and valvular diseases, to myocardial infarction and congenital cardiomyopathy (Chien and Grace 1997). A commonly accepted paradigm for the development of heart failure divides the pathological process into two distinct stages; an initial compensatory hypertrophy in response to excess hemodynamic loading, followed by a critical transition to decompensated failure under persistent stress (Chien 1999; Sugden 1999; Molkentin and Dorn 2001; Wang 2001). Therefore, studies to uncover the underlying mechanisms for the stress-induced hypertrophic response and the critical transition from compensatory hypertrophy to failing heart should hold great promise to better treatments for this devastating disease.

Stress-induced cardiac hypertrophy and pathological remodeling involve changes of different aspects in addition to depressed contractility, including gene expression profile from adult to a "fetal-like" program, increase in myocyte size and protein contents, induction of sarcomere disorganization and interstitial fibrosis, loss of intercellular conduction, and loss of myocyte viability (Sadoshima and Izumo 1997; Ruwhof and van der Laarse 2000; Kyriakis and Avruch 2001). At the cellular level, multitudes of potential mechanisms can contribute to the loss of contractility, including abnormal calcium regulation and changes in sarcomere calcium sensitivity (Houser et al. 2000). Dissecting the molecular components involved in each aspect of the remodeling processes and establishing the causal/effect relationship between different intracellular signaling pathways and specific pathology should provide significant inroads toward a better understanding of the disease mechanism. Recent progress in molecular genetics and cellular/organ physiology has provided powerful tools to dissect different signaling pathways and link genes with specific pathological process in the heart (Izumo and Shioi 1998; Wang 2001). Efforts from our laboratory focusing on the mitogen-activated protein kinases (MAPK) have yielded exciting new insights from both in vitro and in vivo model systems to suggest that individual MAPK cascades can lead to pathological changes in the heart that may contribute to the specific aspects of cardiac pathology in the stressed heart.

MAPKs

MAPKs are a family of protein kinases ubiquitously expressed in different tissues and are highly conserved across species with major roles in intracellular signal transduction (Widmann et al. 1999; Chang and Karin 2001). Upstream MAP kinase kinase kinases (MAP3K), MAP kinase kinase (MKKs, serine-threonine- and tyrosine-specific), and downstream MAP kinases (serine-threonine-specific) form distinct phosphorylation cascades (Marshall 1994). The prototypic MAPK pathway involves extracellular signal-regulated protein kinases (ERKs) that are phosphorylated by Ras-Raf-MEK-1/2 in response to growth stimulation (Chang and Karin 2001). The more recently identified two MAPK cascades involving c-jun amino-terminal kinases (JNK) and p38 kinases are collectively termed stress-activation protein kinases (SAPK), because they both can be activated by stress-related stimuli (Karin 1995; Ono and Han 2000). In mammalian cells, JNK activity is dramatically induced by UV irradiation (Derijard et al. 1994; Kyriakis et al. 1994, 1995), whereas p38 induction is sensitive to hyperosmotic shock and LPS (Han et al. 1994; Lee and Young 1996), and both can be activated by inflammatory cytokines and a variety of cellular stressors. Certain upstream kinases (MEKK-1 and JNKK1/MKK4) can activate both JNK and p38 (Derijard et al. 1995; Lin et al. 1995). On the other hand, MKK7/JNKK2 can specifically activate JNK without affecting the p38 pathway (Lu et al. 1997; Tournier et al. 1997; Wang et al. 1998a), and MKK3 and MKK6 are specific activators for p38, but not for JNK (Fig. 1) (Derijard et al. 1995; Han et al. 1996; Raingeaud et al. 1996). The specific role of MAPKs in heart failure has generated intense interest because they have been implicated as part of the downstream signaling effectors for G-protein-coupled receptors (α/β-adrener-

---

*Present address: Heart and Stroke/Richard Lewar Centre of Excellence, University of Toronto, Ontario, Canada.

gic receptors, ET-1/angiotensinII), tyrosine kinase receptors (IGF-1/gp130/FGF), TGFβ signaling, and mechanical, chemical, and biological stresses (Sugden and Clerk 1998a; McKinsey and Olson 1999; Kyriakis and Avruch 2001). In heart, activation of ERK, JNK, and p38 is associated with the onset of cardiac hypertrophy and progression of heart failure in a variety of animal models (McKinsey and Olson 1999; Molkentin and Dorn 2001). In human, activation of MAPKs is also reported in diseased hearts with hypertrophic, dilated, and ischemic cardiomyopathy (Cook et al. 1999). Although the linkage between MAPK induction and cardiac pathology is compelling, the specific roles of each MAPK pathway in the development of heart failure and the causal/effect relationship between individual MAPK cascades and the specific aspects of cardiac remodeling have not yet been fully established from in vivo studies (Fig. 1).

GENETIC MANIPULATION IN TRANSGENIC ANIMALS AND CULTURED MYOCYTES

The 5.5-kilobase murine α-myosin heavy chain (αMHC) promoter, first characterized by Dr. Robbins, is able to confer highly specific and uniform expression of a transgene in cardiomyocytes in vivo (Subramaniam et al. 1991). Using such cardiac-specific promoters, targeted expression of transgenes in intact mouse heart is routinely achieved (Izumo and Shioi 1998). However, several potential limitations may hinder its in vivo application, particularly for studies of signaling molecules. (1) When the transgene expression has a deleterious effect on cardiac function, the viability of the founder animals can be affected so that stable transgenic lines are difficult to establish. (2) The αMHC-directed transgene expression coincides with embryonic and postnatal development processes of the heart; thus, the pathological phenotype revealed in adult heart may be complicated by potential effects on cardiac development at earlier stages. (3) The transgene expression is constitutively turned on throughout the postnatal process, making it difficult to dissect the primary effects from the secondary, compensatory processes in the transgenic heart. In light of these problems, we have developed a unique conditional transgenesis system based on Cre-loxP-mediated DNA recombination to achieve cardiac-specific and temporally regulated gene expression in cardiomyocytes. As shown in Figure 2A, the transgenic construct used an αMHC promoter to drive a marker gene flanked by two tandem repeats of the loxP sequence (Orban et al. 1992), followed by an additional expression cassette for the target transgene. The floxed transgene construct should only express the marker gene product, not the transgene. While in the presence of Cre protein (by breeding with Cre-expressing transgenic mice; Agah et al. 1997; Chen et al. 1998; Sohal et al. 2001), the DNA recombination event mediated by the loxP elements will delete the intervening marker gene expression cassette and initiate the transgene expression. This Cre-loxP-mediated gene-switch system has certain advantages over conventional transgenesis. One is to facilitate the establishment of stable transgenic lines with potential early lethal phenotypes. Second is to target the transgene expression in the myocytes of the ventricle instead of the whole heart by breeding the floxed allele with MLC-2v/cre mice which target the cre expression to ventricular myocytes in heart by a knock-in strategy in the myosin light-chain 2-v gene (Chen et al. 1998). The other is to allow further temporal specific regulation by breeding the floxed allele with a transgenic line expressing tamoxifen-inducible Cre protein (Mer-Cre-Mer), developed by Dr. Molkentin (Fig. 3A) (Sohal et al. 2001). By analyzing the expression level of both the marker and the transgene, we demonstrated that the Cre-loxP-mediated gene switch effectively induced the transgene expression in the double transgenic mouse hearts carrying both the floxed alleles and the MLC-2v/cre (Fig. 2B) (Liao et al. 2001). In double transgenic animals harboring both the floxed alleles and Mer-Cre-Mer transgene (Sohal et al. 2001), no recombination event or transgene expression can be detected at adult age prior to tamoxifen treatment (B. Petrich et al., unpubl.). However, within 5 days after tamoxifen administration, gene switch was induced in cardiac myocytes with very high efficiency (Fig. 3B). Taking this strategy, we have generated a number of transgenic lines with activation of different MAPK pathways either specifically targeted in ventricular myocytes, or temporally regulated in adult myocytes. These animal models, as described below, provided an array of new information regarding the functional roles of different MAPK cascades in cardiac hypertrophy, myocardial remodeling, and modulation of contractility and cell–cell communication in intact hearts.

Ras-MEDIATED HYPERTROPHY CARDIOMYOPATHY AND DIASTOLIC DYSFUNCTION

Small GTP-binding protein Ras is a potent upstream activator of MAPKs that mainly activates ERK and, to a much lesser extent, also activates JNK and p38 in cardiomyocytes (Y. Wang, unpubl.). Expression of Ras was reported to be correlated with the development of hypertrophy in hypertrophic cardiomyopathy patients and pressure-overloaded rats (Kai et al. 1998). In cultured my-

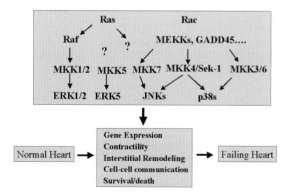

Figure 1. Simplified version of the MAPK cascades and their potential roles in the development of heart failure under stress stimulation.

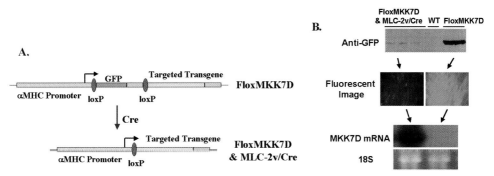

Figure 2. (*A*) Schematics of Cre-loxP-mediated gene-switch transgenic strategy to achieve targeted gene expression in cardiomyocytes in vivo. (αMHC) 5.5-kb murine α-myosin heavy-chain promoter in purple; (loxP) tandem repeat of two loxP sequences; (GFP) green fluorescent protein in green, SV40 polyA sequence in yellow. Targeted transgene, cDNA in blue and polyA sequence in yellow. (*B*) Efficient gene-switch and targeted expression of MKK7D in mouse ventricle. GFP expression was determined by immunoblot using anti-GFP antibody and protein samples prepared from ventricles of the FloxMKK7D, wild-type, and FloxMKK7D/MLC-2v/Cre double transgenic hearts (*top* panel) and direct fluorescent imaging from FloxMKK7D and FloxMKK7D/MLC-2v/Cre transgenic hearts (*middle* panel). MKK7D expression was detected by northern blot (*bottom* panel) with 18S rRNA bands as loading control visualized by ethidium bromide staining. (Modified, with permission, from Petrich et al. 2002 [copyright Lippincott Williams and Wilkins].)

ocytes, Ras-mediated signaling was shown to be critical in α-adrenergic-induced hypertrophy (LaMorte et al. 1994; Sah et al. 1996), and activation of the Ras pathway is sufficient to induce ERK activity and myocyte hypertrophy (Thorburn et al. 1993; Abdellatif et al. 1998; Hoshijima et al. 1998). In early transgenic studies, Ras expression was also associated with hypertrophy and diastolic dysfunction, but the phenotype was not consistent and was evident only after selected breeding due to very low level expression of the transgene (Hunter et al. 1995). Using the gene-switch approach, we achieved substantial expression of the activated H-Ras-v12 mutant in ventricular cardiac myocytes. Marked ventricular hypertrophy developed in the transgenic hearts as determined by an almost twofold increase in heart weight and left ventricular weight relative to the body weight, as well as histology and 2D echocardiographic analysis (Fig. 4A). The contractile dysfunction as determined by direct hemodynamic measurement includes elevated end-diastolic pressure (Y. Wang et al., unpubl.), and significantly prolonged relaxation time constant (τ) in Ras transgenic animals (15.9 ± 4.0 vs. 10.6 ± 2.1 in controls, $p<0.01$), suggesting a profound state of diastolic failure. The abnormal relaxation in Ras transgenic hearts is most likely contributed by induced interstitial fibrosis and impaired SR Ca^{++} uptake (Y. Wang, unpubl.). Furthermore, the Ras-expressing transgenic animals died prematurely between 6 and 8 weeks of age from cardiac sudden death associated with prevalent ventricular arrhythmia (Fig. 4B). Recent genetic evidence suggests that defects in SR Ca^{++} uptake play a significant role in Ras-mediated diastolic dysfunction and induction in arrhythmia (S. Minimisawa et al., unpubl.). Therefore, Ras transgenic animals established by gene-switch strategy developed a characteristic form of hypertrophic cardiomyopathy with morphological remodeling, contractile dysfunction, and arrhythmic sudden death, resembling many of the clinical symptoms observed in hypertrophic cardiomyopathy patients. Given the correlation of Ras activation in hypertrophic cardiomyopathy heart and its well-established function in mediating signal transduction from both G-protein-coupled receptor and tyrosine kinase receptor, it is plausible

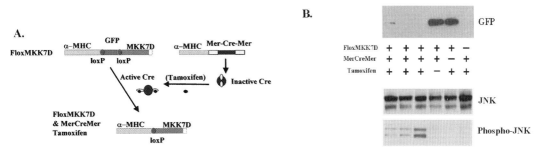

Figure 3. (*A*) Schematics of temporally regulated gene expression in transgenic heart via an inducible gene-switch approach. (MCM) Mer-Cre-Mer transgenic line as described previously. (FloxMKK7D) Same as described in Fig. 2. (*B*) Tamoxifen-induced gene-switch in the ventricles of double transgenic animals carrying both FloxMKK7D and MCM alleles. Tamoxifen treatment in adult hearts as indicated switched off the GFP expression as determined by immunoblot for GFP (*top*) and turned on the target gene MKK7D that led to induction in JNK activity as determined by immunoblotting for phospho-JNK (*bottom*) (B. Petrich et al., unpubl.).

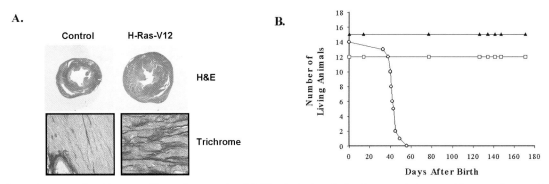

Figure 4. Ras-mediated hypertrophic cardiomyopathy. (*A*) Histological analysis of transgenic hearts with targeted expression of H-Ras-V12 in ventricle. (*Top* panels) Cross-section of H&E staining from control and H-Ras-V12 transgenic heart. (*Bottom* panels) Higher magnification of trichrome staining of the ventricles from the same hearts. (*B*) Survival curve of H-Ras-V12 (FloxRas & MLC-2v/Cre) double transgenic animals (*open circle*) vs. wild type (*open square*) and FloxRas (*closed triangle*) controls.

that the Ras pathway plays a critical role in the development of cardiac hypertrophy and its related pathology, including interstitial fibrosis, diastolic dysfunction, and arrhythmic sudden death. Although the ERK pathway is one of the main downstream targets of Ras, and the Raf-MEK1-ERK signaling cascade has been implicated in the regulation of SR calcium regulation in cultured myocytes (Ho et al. 1998, 2001), activation of ERK by MEK1 in intact heart did not lead to any contractile defects other than a mild form of cardiac hypertrophy (Bueno et al. 2000). Combining this effect with our in vivo observation in Ras transgenic heart, we would speculate that other branches of the downstream signaling, including PI3K and other MAPK pathways, may also contribute to overall pathology in the hypertrophic cardiomyopathy heart.

p38-MEDIATED RESTRICTIVE CARDIOMYOPATHY AND CONTRACTILE REGULATION

p38 MAPKs are implicated in various stress responses, particularly by endotoxin, in regulating gene expression by phosphorylating a number of nuclear transcription factors, including ATF-2 and MEF-2 (Ono and Han 2000). Genetic evidence also supports a critical role for p38 in the early developmental process (Adams et al. 2000; Allen et al. 2000; Mudgett et al. 2000; Tamura et al. 2000). There are conflicting reports regarding the potential function of p38 in cardiac hypertrophy, apoptosis, or ischemia protection/injury (Sugden amd Clerk 1998b; Ping and Murphy 2000). However, most of the observations were made in either neonatal myocytes or intact animals using transient gene transfer vectors or pharmacological inhibitors (Zechner et al. 1997, 1998; Nemoto et al. 1998; Wang et al. 1998b; Craig et al. 2000; Hoover et al. 2000; Mocanu et al. 2000; Saurin et al. 2000; Au-Yeung et al. 2001; Behr et al. 2001; Martin et al. 2001; Schneider et al. 2001). To gain more insights into this important signaling pathway, we established transgenic models to achieve targeted activation of p38 in heart using the Cre-loxP-mediated-gene switch approach (Liao et al. 2001). Expression of constitutively activated mutants of p38 upstream activators, MKK3b(E) and MKK6b(E), was targeted to ventricular myocytes and led to significant activation of p38 activities (Liao et al. 2001). Contrary to earlier observations in vitro, both MKK3bE and MKK6bE transgenic hearts showed no hypertrophy at the organ level, where heart weight/body weight stayed constant, but a marked increase in interstitial fibrosis and premature death suggested a status of severe pathological remodeling (Fig. 5A). By direct hemodynamic measurement, the hearts from both transgenic lines displayed profound contractile dysfunction (Fig. 5B). Compared to controls, MKK3bE and MKK6bE hearts had 21% and 28% lower end-systolic pressure, but 3.2- and 3.5-fold induction in end-diastolic pressure, 44% and 42% lower dP/dT_{max}, and 52% and 55% lower dP/dT_{min}, respectively. As a result, cardiac outputs from MKK3bE and MKK6bE transgenic hearts were only 47.9% and 39.1% of the controls. A distinct feature of the contractile defects in both transgenic models was the loss of end-systolic elastance and over 20-fold induction in diastolic stiffness. However, there are also significant differences in myocardial morphology between the two models, where the MKK3bE transgenic heart developed a thinner myocardial wall with evidence of heterogeneous myocyte loss, while the MKK6bE transgenic heart developed a thicker myocardial wall. As a result, MKK3bE hearts had dilated chambers and increased end-systolic wall stress, while MKK6bE transgenic hearts had a small chamber at systole and near-normal level of wall stress (Fig. 5B). In neither transgenic heart was a significant level of apoptotic activity identified while a fetal gene expression program was induced as measured at mRNA level. All these observations suggested that, in the intact heart, the most significant effects of p38 activation are loss of contractile function and myocardial remodeling rather than myocyte hypertrophy and apoptosis.

The cellular mechanism underlying the loss of contractile function in the p38-activated heart is further studied in adult cardiomyocytes at the single-myocyte level (Liao et al. 2002). In isolated adult cardiomyocytes, activation

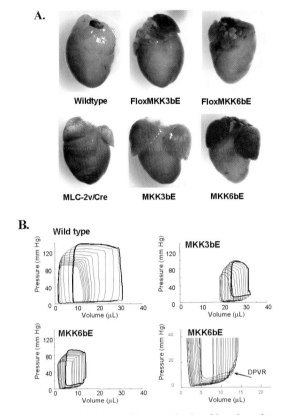

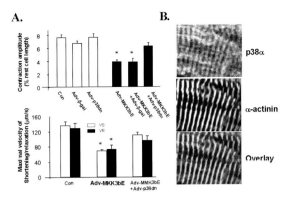

Figure 5. Restrictive cardiomyopathy in p38-activated transgenic hearts. (*A*) Morphological remodeling in MKK3bE and MKK6bE transgenic hearts comparing wild type and single transgenic controls as indicated. (*B*) Pressure-volume loops in MKK3bE and MKK6bE transgenic hearts comparing to wild-type controls, showing basal (*thick-line loop*) and reduction of preload by vena cava occlusion (*thin-line loops*). The rightward shift of end systolic p-v relation in MKK3bE indicates systolic chamber dilatation; in contrast, the leftward shift of diastolic p-v relation in MKK6bE indicates chamber thickening. Chamber passive stiffness is increased in both transgenic hearts as determined by DPVR (end diastolic pressure-volume relationship) shown in an expanded scale at right bottom panel. (Reprinted, with permission, from Liao et al. 2001 [copyright National Academy of Sciences].)

Figure 6. Reduction of myocyte contractility mediated specifically by p38 MAPK activation. (*A*) (*Top*) Cell contraction amplitude is markedly decreased in myocytes expressing MKK3bE compared with uninfected myocytes or those infected by Adv-β-gal. Coexpression of a p38 dominant-negative mutant (p38dn) reverses the MKK3bE-induced reduction of contraction amplitude. *$p<0.01$ vs. cells in the absence of Adv-MKK3bE or those coinfected by Adv-MKK3bE and Adv-p38dn. (*B*) Activation of p38 MAPK significantly reduces the maximal shortening velocity (VS) and relaxation velocity (VR); this effect is also largely prevented by coexpression of the p38 dominant-negative mutant. *$p<0.01$ vs. the control group and the group coinfected by Adv-MKK3bE and Adv-p38dn. (*B*) Immunohistochemistry showing p38α protein colocalized with α-actinin at sarcomeric structures in ventricular myocytes. (*A*, Modified, with permission, from Liao et al. 2002 [copyright Lippincott Williams and Wilkins].)

of p38 was achieved by adenovirus-mediated gene transfer of the upstream activator, MKK3bE, and the cellular contractile function was analyzed by measuring twitch amplitude and kinetics using an edge-detection system (Fig. 6A). Contractility was reduced significantly by p38 activation, whereas basal contractile function was increased significantly by p38 inhibition. Interestingly, increased contractility by p38 inhibitor was not associated with changes in peak calcium transient or membrane calcium currents, suggesting a potential role of altered calcium sensitivity in p38-mediated contraction modulation (Liao et al. 2002). This observation is in good agreement with our in vivo study that showed no changes in cardiac relaxation time constant (τ) in both MKK3bE and MKK6bE transgenic hearts (Liao et al. 2001). Colocalization of p38 protein with sarcomere structure as determined by immunohistochemistry (Fig. 6B) further supports a potential role of p38 kinase in interacting with sarcomeric proteins and modulating contractile function without affecting intracellular calcium cycling.

Although p38 activation is highly associated with the development of heart failure, its specific role in any specific aspects of cardiac pathology and the underlying mechanisms are yet to be uncovered. Current studies from transgenic animals with temporally regulated and targeted activation of p38 activity in the heart should provide valuable insights to the signaling mechanisms involved in the loss of cardiac contractility and myocardial remodeling under stressed conditions. A better understanding of the physiological role and molecular mechanisms of the p38 function in heart will also help to determine whether the p38-mediated signaling pathway can serve as a novel therapeutic target for treating heart failure.

JNK-MEDIATED CARDIAC REMODELING AND REGULATION OF CELL–CELL COMMUNICATION

JNK is a stress-activated MAPK pathway with a distinct activation profile under stress stimulation as compared to p38 (Paul et al. 1997; Widmann et al. 1999). JNK activity has been implicated in the process of pressure-overload-induced cardiac hypertrophy and ischemia/reperfusion injury (Bogoyevitch et al. 1996; Knight and Buxton 1996; Xu et al. 1996; Kudoh et al. 1997; Ramirez et al. 1997; Clerk et al. 1998; Li et al. 1998; Nemoto et al. 1998; Wang et al. 1998a, Cook et al. 1999; Hreniuk et al. 2001). However, the physiological effect of JNK activation in intact hearts has not been fully established. Targeted JNK activation in heart was achieved by expressing an activated mutant of upstream MKK7(D) via the same Cre-loxP-mediated gene-switch strategy as described earlier by breeding with the MLC-2v/cre mouse line (Figs. 2 and 3) (B. Petrich et al., unpubl.) Targeted expression of MKK7D led to a significant increase in JNK activity in the transgenic ventricular myocardium (Fig. 3) (Petrich et al. 2002). Induction of fetal marker genes was detected in the MKK7D transgenic hearts, but ventricular morphology as determined by histology and echocardiography was not significantly affected, whereas the atrium was markedly enlarged (Fig. 7A). MKK7D transgenic hearts had an elevated end-diastolic pressure as directly measured in the left ventricle using a miniaturized catheter (B. Petrich et al., unpubl.) and decreased time constant for acceleration/deceleration of blood flow at the mitral valve as determined by Doppler echocardiography (B. Petrich et al., unpubl.), suggesting an abnormal compliance and an increase in myocardial stiffening. The stiffening of the transgenic myocardium was associated with a significant increase in interstitial deposition of fibronectin but no induction of collagen (Fig. 7B). In contrast with early in vitro studies, JNK activation in intact hearts did not induce apoptosis or signs of myocyte dropout (B. Petrich et al., unpubl.). Therefore, the major defects in JNK-activated hearts appeared to be related to changes in gene expression and interstitial remodeling.

Although the ventricular morphology and fraction shortening in JNK-activated transgenic hearts were largely unaffected, the transgenic animals died prematurely between 6 and 8 weeks of age (Petrich et al. 2002), suggesting other underlying detrimental defects in the MKK7D animals. Cx43 is the major gap junction protein in the ventricle with a short half-life of 1.5 hours in heart (Beardslee et al. 1998). Rapid loss of Cx43 is one of the first pathological remodeling events in response to myocardial stress and injury, and loss of Cx43 is implicated in conduction defects, arrhythmia, and contractile dysfunction (Saffitz et al. 2000). We found that Cx43 expression was dramatically reduced at both mRNA and protein levels in the MKK7D heart (Fig. 7C), while expression of Cx40 and Cx45 were largely not affected (Petrich et al. 2002). The loss of Cx43 protein was associated with disrupted gap junction structure as observed under the electron microscope, and an ~30% reduction in conduction velocity as determined using a combination of voltage-sensitive fluorescent dye and optic mapping technique (Petrich et al. 2002 and unpubl.). Loss of Cx43 and impaired cell–cell communication have been widely reported in diseased hearts originated from different stressors that can contribute to contractile defects and increased incidence of arrhythmia (Fernandez-Cobo et al. 1999; Huang et al. 1999; Lerner et al. 2000; Daleau et al. 2001; Gutstein et al. 2001a, b). Linking a major stress-inducible signaling pathway, JNK, with this specific aspect of cardiac remodeling provided a potential molecular mechanism for the loss of Cx43 and impaired cell–cell communication in failing hearts. To demonstrate that JNK-mediated down-regulation of Cx43 is a specific consequence of JNK activation rather than a nonspecific secondary event, we further studied the role of JNK in the regulation of Cx43 expression in cultured myocytes. Indeed, stress stimulation in cultured myocytes also led to Cx43 down-regulation coinciding with JNK activation, and the loss of Cx43 was partially reversed by pretreating myocytes with a JNK-specific inhibitor (Petrich et al. 2002). On the other hand, specific activation of JNK via adenovirus-vector-mediated gene transfer of MKK7D also resulted in a dramatic loss of Cx43 at both mRNA and protein levels as observed in transgenic hearts (Fig. 7D). All this evidence suggests that JNK-mediated stress-signaling may play an important and specific role in the pathological remodeling of myocardium, particularly in modulating extracellular matrix and cell–cell communication.

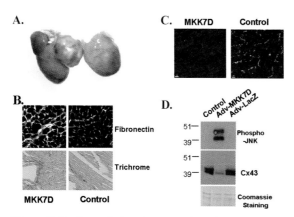

Figure 7. Cardiac pathology in JNK-activated hearts. (*A*) Cardiac remodeling in MKK7D transgenic hearts as observed at whole-heart level. No ventricular hypertrophy was observed, but both atria were significantly enlarged and filled with thrombi. (*B*) Interstitial remodeling detected by immunohistological staining for fibronectin and trichrome staining for collagen. (*C*) Loss of Cx43 expression in MKK7D transgenic heart as shown in the representative immunohistochemical staining of Cx43 protein in the ventricle (Petrich et al. 2002). (*D*) Down-regulation of Cx43 by JNK activation in cultured myocytes. Adenovirus-mediated MKK7D expression in cultured neonatal myocytes led to JNK activation as determined by immunoblotting for phospho-JNK and down-regulation of Cx43 protein. (*C,* Reprinted from Petrich et al. 2002; *D,* modified from Petrich et al. 2002; both with permission [copyright Lippincott Williams and Wilkins].)

SUMMARY

We have demonstrated that Cre-loxP-mediated gene-switch transgenesis is an effective approach to achieve targeted and temporally regulated gene manipulation in the heart. Using this approach, we have established animal models with targeted activation of different MAPK pathways. From these animal models, we identified distinct features of cardiac pathology associated with individual MAPK branches (summarized in Fig. 8). Specifically, Ras activation appears to promote cardiac hypertrophy, whereas p38 and JNK activation does not. Whereas Ras activation leads to depressed diastolic func-

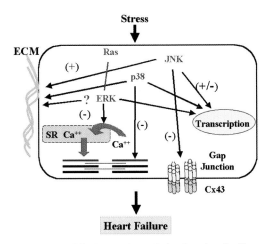

Figure 8. Simplified overview of the functional effects of MAPK activation as determined from transgenic studies. Arrows indicate specific cellular process potentially targeted by corresponding MAPK cascades. (SR) Sarcoplasmic reticulum, (ECM) extracellular matrix proteins.

tion associated with suppressed calcium transients and SR calcium uptake, p38 activity seems to modulate cellular contractility without affecting intracellular calcium cycling. Although all three models displayed extensive remodeling in the myocardium, the extent and the composition of interstitial fibrosis are different among them, with Ras- and p38-activated hearts promoting collagen-based fibrosis, and JNK activation leading to induction in fibronectin-based reticular fiber. In addition, JNK activation leads to loss of Cx43 expression and abnormal cell–cell communication. Therefore, ERK, p38, and JNK are three distinct intracellular signaling pathways that contribute to different aspects of cardiac pathology during heart failure. Combining sophisticated genetic manipulation with comprehensive analysis at physiological, molecular, and genomic levels, the transgenic animals established in these studies should serve as valuable model systems to identify and dissect the underlying mechanisms for different aspects of cardiac pathology such as hypertrophy, contractile dysfunction, and abnormal cell–cell communication. The insights learned from these investigations may help to develop novel therapeutic approaches to confront this devastating disease.

ACKNOWLEDGMENTS

The authors acknowledge Ms. Haiying Pu for her excellent technical support. We also thank collaborators and their associates for their kind support; specifically, Drs. Ju Chen and Kenneth R. Chien from the University of California, San Diego; Drs. Jeffrey E. Saffitz and Deborah Lerner from Washington University; Dr. Jeffery D. Molkentin at Children's Hospital, Cincinnati; Dr. David Kass from Johns Hopkins University, Drs. Rui-ping Xiao and Peace Cheng from the National Institute of Aging. The work presented here is supported in part by grants from the National Institutes of Health Heart, Lung and Blood Institute, a Grant-in-Aid from the American Heart Association Mid-Atlantic Affiliate (Y.W.), and a National Institutes of Health predoctoral fellowship from the Interdisciplinary Training Program in Muscle Biology (B.G.P.).

REFERENCES

Abdellatif M., Packer S.E., Michael L.H., Zhang D., Charng M.J., and Schneider M.D. 1998. A Ras-dependent pathway regulates RNA polymerase II phosphorylation in cardiac myocytes: Implications for cardiac hypertrophy. *Mol. Cell. Biol.* **18:** 6729.

Adams R.H., Porras A., Alonso G., Jones M., Vintersten K., Panelli S., Valladares A., Perez L., Klein R., and Nebreda A.R. 2000. Essential role of p38alpha MAP kinase in placental but not embryonic cardiovascular development. *Mol. Cell* **6:** 109.

Agah R., Frenkel P.A., French B.A., Michael L.H., Overbeek P.A., and Schneider M.D. 1997. Gene recombination in postmitotic cells. Targeted expression of Cre recombinase provokes cardiac-restricted, site-specific rearrangement in adult ventricular muscle in vivo. *J. Clin. Invest.* **100:** 169.

Allen M., Svensson L., Roach M., Hambor J., McNeish J., and Gabel C.A. 2000. Deficiency of the stress kinase p38alpha results in embryonic lethality: Characterization of the kinase dependence of stress responses of enzyme-deficient embryonic stem cells. *J. Exp. Med.* **191:** 859.

American Heart Association. 2002. *2002 Heart and stroke statistical update* (PDF). American Heart Association, Dallas, Texas, 2001.

Au-Yeung K.K., Zhu D.Y., O K., and Siow Y.L. 2001. Inhibition of stress-activated protein kinase in the ischemic/reperfused heart: Role of magnesium tanshinoate B in preventing apoptosis. *Biochem. Pharmacol.* **62:** 483.

Beardslee M.A., Laing J.G., Beyer E.C., and Saffitz J.E. 1998. Rapid turnover of connexin43 in the adult rat heart. *Circ. Res.* **83:** 629.

Behr T.M., Nerurkar S.S., Nelson A.H., Coatney R.W., Woods T.N., Sulpizio A., Chandra S., Brooks D.P., Kumar S., Lee J.C., Ohlstein E.H., Angermann C.E., Adams J.L., Sisko J., Sackner-Bernstein J.D., and Willette R.N. 2001. Hypertensive end-organ damage and premature mortality are p38 mitogen-activated protein kinase-dependent in a rat model of cardiac hypertrophy and dysfunction. *Circulation* **104:** 1292.

Bogoyevitch M.A., Gillespie B.J., Ketterman A.J., Fuller S.J., Ben L.R., Ashworth A., Marshall C.J., and Sugden P.H. 1996. Stimulation of the stress-activated mitogen-activated protein kinase subfamilies in perfused heart. p38/RK mitogen-activated protein kinases and c-Jun N-terminal kinases are activated by ischemia/reperfusion. *Circ. Res.* **79:** 162.

Bueno O.F., De Windt L.J., Tymitz K.M., Witt S.A., Kimball T.R., Klevitsky R., Hewett T.E., Jones S.P., Lefer D.J., Peng C.F., Kitsis R.N., and Molkentin J.D. 2000. The MEK1-ERK1/2 signaling pathway promotes compensated cardiac hypertrophy in transgenic mice. *EMBO J.* **19:** 6341.

Chang L. and Karin M. 2001. Mammalian MAP kinase signalling cascades. *Nature* **410:** 37.

Chen J., Kubalak S.W., and Chien K.R. 1998. Ventricular muscle-restricted targeting of the RXRalpha gene reveals a non-cell-autonomous requirement in cardiac chamber morphogenesis. *Development* **125:** 1943.

Chien K.R. 1999. Stress pathways and heart failure. *Cell* **98:** 555.

Chien K.R. and Grace A.A. 1997. Principles of cardiovascular molecular biology and cellular biology. In *Heart disease: A textbook of cardiovascular medicine* (ed. E. Braunwald), p. 1626. W.B. Saunders, Philadelphia, Pennsylvania.

Clerk A., Fuller S.J., Michael A., and Sugden P.H. 1998. Stimulation of "stress-regulated" mitogen-activated protein kinases (stress-activated protein kinases/c-Jun N-terminal kinases and p38-mitogen-activated protein kinases) in perfused rat hearts by oxidative and other stresses. *J. Biol. Chem.* **273:** 7228.

Cohn J.N., Bristow M.R., Chien K.R., Colucci W.S., Frazier O.H., Leinwand L.A., Lorell B.H., Moss A.J., Sonnenblick E.H., Walsh R.A., Mockrin S.C., and Reinlib L. 1997. Report of the National Heart, Lung, and Blood Institute Special Emphasis Panel on Heart Failure Research (comments). *Circulation* **95:** 766.

Cook S.A., Sugden P.H., and Clerk A. 1999. Activation of c-Jun N-terminal kinases and p38-mitogen-activated protein kinases in human heart failure secondary to ischaemic heart disease. *J. Mol. Cell. Cardiol.* **31:** 1429.

Craig R., Larkin A., Mingo A.M., Thuerauf D.J., Andrews C., McDonough P.M., and Glembotski C.C. 2000. p38 MAPK and NF-kappa B collaborate to induce interleukin-6 gene expression and release. Evidence for a cytoprotective autocrine signaling pathway in a cardiac myocyte model system. *J. Biol. Chem.* **275:** 23814.

Daleau P., Boudriau S., Michaud M., Jolicoeur C., and Kingma J.G., Jr. 2001. Preconditioning in the absence or presence of sustained ischemia modulates myocardial Cx43 protein levels and gap junction distribution. *Can. J. Physiol. Pharmacol.* **79:** 371.

Derijard B., Raingeaud J., Barrett T., Wu I.H., Han J., Ulevitch R.J., and Davis R.J. 1995. Independent human MAP-kinase signal transduction pathways defined by MEK and MKK isoforms (erratum in *Science* [1995] **269:** 17). *Science* **267:** 682.

Derijard B., Hibi M., Wu I.H., Barrett T., Su B., Deng T., Karin M., and Davis R.J. 1994. JNK1: A protein kinase stimulated by UV light and Ha-Ras that binds and phosphorylates the c-Jun activation domain. *Cell* **76:** 1025.

Fernandez-Cobo M., Gingalewski C., Drujan D., and De Maio A. 1999. Downregulation of connexin 43 gene expression in rat heart during inflammation. The role of tumour necrosis factor. *Cytokine* **11:** 216.

Gutstein D.E., Morley G.E., Vaidya D., Liu F., Chen F.L., Stuhlmann H., and Fishman G.I. 2001a. Heterogeneous expression of Gap junction channels in the heart leads to conduction defects and ventricular dysfunction. *Circulation* **104:** 1194.

Gutstein D.E., Morley G.E., Tamaddon H., Vaidya D., Schneider M.D., Chen J., Chien K.R., Stuhlmann H., and Fishman G.I. 2001b. Conduction slowing and sudden arrhythmic death in mice with cardiac-restricted inactivation of connexin43. *Circ. Res.* **88:** 333.

Han J., Lee J.D., Bibbs L., and Ulevitch R.J. 1994. A MAP kinase targeted by endotoxin and hyperosmolarity in mammalian cells. *Science* **265:** 808.

Han J., Lee J.D., Jiang Y., Li Z., Feng L., and Ulevitch R.J. 1996. Characterization of the structure and function of a novel MAP kinase kinase (MKK6). *J. Biol. Chem.* **271:** 2886.

Ho P.D., Zechner D.K., He H., Dillmann W.H., Glembotski C.C., and McDonough P.M. 1998. The Raf-MEK-ERK cascade represents a common pathway for alteration of intracellular calcium by Ras and protein kinase C in cardiac myocytes. *J. Biol. Chem.* **273:** 21730.

Ho P.D., Fan J.S., Hayes N.L., Saada N., Palade P.T., Glembotski C.C., and McDonough P.M. 2001. Ras reduces L-type calcium channel current in cardiac myocytes. Corrective effects of L-channels and SERCA2 on [Ca(2+)](i) regulation and cell morphology. *Circ. Res.* **88:** 63.

Hoover H.E., Thuerauf D.J., Martindale J.J., and Glembotski C.C. 2000. alpha B-crystallin gene induction and phosphorylation by MKK6-activated p38. A potential role for alpha B-crystallin as a target of the p38 branch of the cardiac stress response. *J. Biol. Chem.* **275:** 23825.

Hoshijima M., Sah V.P., Wang Y., Chien K.R., and Brown J.H. 1998. The low molecular weight GTPase Rho regulates myofibril formation and organization in neonatal rat ventricular myocytes. Involvement of Rho kinase. *J. Biol. Chem.* **273:** 7725.

Houser S.R., Piacentino V., III, and Weisser J. 2000. Abnormalities of calcium cycling in the hypertrophied and failing heart. *J. Mol. Cell. Cardiol.* **32:** 1595.

Hreniuk D., Garay M., Gaarde W., Monia B.P., McKay R.A., and Cioffi C.L. 2001. Inhibition of c-Jun N-terminal kinase 1, but not c-Jun N-terminal kinase 2, suppresses apoptosis induced by ischemia/reoxygenation in rat cardiac myocytes. *Mol. Pharmacol.* **59:** 867.

Huang X.D., Sandusky G.E., and Zipes D.P. 1999. Heterogeneous loss of connexin43 protein in ischemic dog hearts. *J. Cardiovasc. Electrophysiol.* **10:** 79.

Hunter J.J., Tanaka N., Rockman H.A., Ross J., Jr., and Chien K.R. 1995. Ventricular expression of a MLC-2v-ras fusion gene induces cardiac hypertrophy and selective diastolic dysfunction in transgenic mice. *J. Biol. Chem.* **270:** 23173.

Izumo S. and Shioi T. 1998. Cardiac transgenic and gene-targeted mice as models of cardiac hypertrophy and failure: A problem of (new) riches (editorial; comment). *J. Card. Fail.* **4:** 263.

Kai H., Muraishi A., Sugiu Y., Nishi H., Seki Y., Kuwahara F., Kimura A., Kato H., and Imaizumi T. 1998. Expression of proto-oncogenes and gene mutation of sarcomeric proteins in patients with hypertrophic cardiomyopathy. *Circ. Res.* **83:** 594.

Karin M. 1995. The regulation of AP-1 activity by mitogen-activated protein kinases. *J. Biol. Chem.* **270:** 16483.

Knight R.J. and Buxton D.B. 1996. Stimulation of c-Jun kinase and mitogen-activated protein kinase by ischemia and reperfusion in the perfused rat heart. *Biochem. Biophys. Res. Commun.* **218:** 83.

Kudoh S., Komuro I., Mizuno T., Yamazaki T., Zou Y., Shiojima I., Takekoshi N., and Yazaki Y. 1997. Angiotensin II stimulates c-Jun NH2-terminal kinase in cultured cardiac myocytes of neonatal rats. *Circ. Res.* **80:** 139.

Kyriakis J.M. and Avruch J. 2001. Mammalian mitogen-activated protein kinase signal transduction pathways activated by stress and inflammation. *Physiol. Rev.* **81:** 807.

Kyriakis J.M., Woodgett J.R., and Avruch J. 1995. The stress-activated protein kinases. A novel ERK subfamily responsive to cellular stress and inflammatory cytokines. *Ann. N.Y. Acad. Sci.* **766:** 303.

Kyriakis J.M., Banerjee P., Nikolakaki E., Dai T., Rubie E.A., Ahmad M.F., Avruch J., and Woodgett J.R. 1994. The stress-activated protein kinase subfamily of c-Jun kinases. *Nature* **369:** 156.

LaMorte V.J., Thorburn J., Absher D., Spiegel A., Brown J.H., Chien K.R., Feramisco J.R., and Knowlton K.U. 1994. Gq- and ras-dependent pathways mediate hypertrophy of neonatal rat ventricular myocytes following alpha 1-adrenergic stimulation. *J. Biol. Chem.* **269:** 13490.

Lee J.C. and Young P.R. 1996. Role of CSB/p38/RK stress response kinase in LPS and cytokine signaling mechanisms. *J. Leukoc. Biol.* **59:** 152.

Lerner D.L., Yamada K.A., Schuessler R.B., and Saffitz J.E. 2000. Accelerated onset and increased incidence of ventricular arrhythmias induced by ischemia in Cx43-deficient mice. *Circulation* **101:** 547.

Li W.G., Zaheer A., Coppey L., and Oskarsson H.J. 1998. Activation of JNK in the remote myocardium after large myocardial infarction in rats. *Biochem. Biophys. Res. Commun.* **246:** 816.

Liao P., Wang S.Q., Wang S., Zheng M., Zhang S.J., Cheng H., Wang Y., and Xiao R.P. 2002. p38 Mitogen-activated protein kinase mediates a negative inotropic effect in cardiac myocytes. *Circ. Res.* **90:** 190.

Liao P., Georgakopoulos D., Kovacs A., Zheng M., Lerner D., Pu H., Saffitz J., Chien K., Xiao R.P., Kass D.A., and Wang Y. 2001. The in vivo role of p38 MAP kinases in cardiac remodeling and restrictive cardiomyopathy. *Proc. Natl. Acad. Sci.* **98:** 12283.

Lin A., Minden A., Martinetto H., Claret F.X., Lange-Carter C., Mercurio F., Johnson G.L., and Karin M. 1995. Identification of a dual specificity kinase that activates the Jun kinases and p38-Mpk2. *Science* **268:** 286.

Lu X., Nemoto S., and Lin A. 1997. Identification of c-Jun NH2-terminal protein kinase (JNK)-activating kinase 2 as an activator of JNK but not p38. *J. Biol. Chem.* **272:** 24751.

Marshall C.J. 1994. MAP kinase kinase kinase, MAP kinase kinase and MAP kinase. *Curr. Opin. Genet. Dev.* **4:** 82.

Martin J.L., Avkiran M., Quinlan R.A., Cohen P., and Marber M.S. 2001. Antiischemic effects of SB203580 are mediated through the inhibition of p38alpha mitogen-activated protein kinase: Evidence from ectopic expression of an inhibition-resistant kinase. *Circ. Res.* **89:** 750.

McKinsey T.A. and Olson E.N. 1999. Cardiac hypertrophy: Sorting out the circuitry. *Curr. Opin. Genet. Dev.* **9:** 267.

Mocanu M.M., Baxter G.F., Yue Y., Critz S.D., and Yellon D.M. 2000. The p38 MAPK inhibitor, SB203580, abrogates ischaemic preconditioning in rat heart but timing of administration is critical. *Basic Res. Cardiol.* **95:** 472.

Molkentin J.D. and Dorn G.W., II. 2001. Cytoplasmic signaling pathways that regulate cardiac hypertrophy. *Annu. Rev. Physiol.* **63:** 391.

Mudgett J.S., Ding J., Guh-Siesel L., Chartrain N.A., Yang L., Gopal S., and Shen M.M. 2000. Essential role for p38alpha mitogen-activated protein kinase in placental angiogenesis. *Proc. Natl. Acad. Sci.* **97:** 10454.

Nemoto S., Sheng Z., and Lin A. 1998. Opposing effects of Jun kinase and p38 mitogen-activated protein kinases on cardiomyocyte hypertrophy. *Mol. Cell. Biol.* **18:** 3518.

Ono K. and Han J. 2000. The p38 signal transduction pathway: Activation and function. *Cell Signal.* **12:** 1.

Orban P.C., Chui D., and Marth J.D. 1992. Tissue- and site-specific DNA recombination in transgenic mice. *Proc. Natl. Acad. Sci.* **89:** 6861.

Paul A., Wilson S., Belham C.M., Robinson C.J., Scott P.H., Gould G.W., and Plevin R. 1997. Stress-activated protein kinases: Activation, regulation and function. *Cell Signal.* **9:** 403.

Petrich B.G., Gong X., Lerner D.L., Wang X., Brown J.H., Saffitz J.E., and Wang Y. 2002. c-Jun N-terminal kinase activation mediates downregulation of connexin43 in cardiomyocytes. *Circ. Res.* **91:** 640.

Ping P. and Murphy E. 2000. Role of p38 mitogen-activated protein kinases in preconditioning: A detrimental factor or a protective kinase? *Circ. Res.* **86:** 921.

Raingeaud J., Whitmarsh A.J., Barrett T., Derijard B., and Davis R.J. 1996. MKK3- and MKK6-regulated gene expression is mediated by the p38 mitogen-activated protein kinase signal transduction pathway. *Mol. Cell. Biol.* **16:** 1247.

Ramirez M.T., Sah V.P., Zhao X.L., Hunter J.J., Chien K.R., and Brown J.H. 1997. The MEKK-JNK pathway is stimulated by alpha1-adrenergic receptor and ras activation and is associated with in vitro and in vivo cardiac hypertrophy. *J. Biol. Chem.* **272:** 14057.

Ruwhof C. and van der Laarse A. 2000. Mechanical stress-induced cardiac hypertrophy: Mechanisms and signal transduction pathways. *Cardiovasc. Res.* **47:** 23.

Sadoshima J. and Izumo S. 1997. The cellular and molecular response of cardiac myocytes to mechanical stress. *Annu. Rev. Physiol.* **59:** 551.

Saffitz J.E., Green K.G., Kraft W.J., Schechtman K.B., and Yamada K.A. 2000. Effects of diminished expression of connexin43 on gap junction number and size in ventricular myocardium. *Am. J. Physiol. Heart Circ. Physiol.* **278:** H1662.

Sah V.P., Hoshijima M., Chien K.R., and Brown J.H. 1996. Rho is required for Galphaq and alpha1-adrenergic receptor signaling in cardiomyocytes. Dissociation of Ras and Rho pathways. *J. Biol. Chem.* **271:** 31185.

Saurin A.T., Martin J.L., Heads R.J., Foley C., Mockridge J.W., Wright M.J., Wang Y., and Marber M.S. 2000. The role of differential activation of p38-mitogen-activated protein kinase in preconditioned ventricular myocytes. *FASEB J.* **14:** 2237.

Schneider S., Chen W., Hou J., Steenbergen C., and Murphy E. 2001. Inhibition of p38 MAPK alpha/beta reduces ischemic injury and does not block protective effects of preconditioning. *Am. J. Physiol. Heart Circ. Physiol.* **280:** H499.

Sohal D.S., Nghiem M., Crackower M.A., Witt S.A., Kimball T.R., Tymitz K.M., Penninger J.M., and Molkentin J.D. 2001. Temporally regulated and tissue-specific gene manipulations in the adult and embryonic heart using a tamoxifen-inducible Cre protein. *Circ. Res.* **89:** 20.

Subramaniam A., Jones W.K., Gulick J., Wert S., Neumann J., and Robbins J. 1991. Tissue-specific regulation of the alpha-myosin heavy chain gene promoter in transgenic mice. *J. Biol. Chem.* **266:** 24613.

Sugden P.H. 1999. Signaling in myocardial hypertrophy: Life after calcineurin (comments)? *Circ. Res.* **84:** 633.

Sugden P.H. and Clerk A. 1998a. Regulation of mitogen-activated protein kinase cascades in the heart. *Adv. Enzyme Regul.* **38:** 87.

———. 1998b. "Stress-responsive" mitogen-activated protein kinases (c-Jun N-terminal kinases and p38 mitogen-activated protein kinases) in the myocardium. *Circ. Res.* **83:** 345.

Tamura K., Sudo T., Senftleben U., Dadak A.M., Johnson R., and Karin M. 2000. Requirement for p38alpha in erythropoietin expression: A role for stress kinases in erythropoiesis. *Cell* **102:** 221.

Thorburn A., Thorburn J., Chen S.Y., Powers S., Shubeita H.E., Feramisco J.R., and Chien K.R. 1993. HRas-dependent pathways can activate morphological and genetic markers of cardiac muscle cell hypertrophy. *J. Biol. Chem.* **268:** 2244.

Tournier C., Whitmarsh A.J., Cavanagh J., Barrett T., and Davis R.J. 1997. Mitogen-activated protein kinase kinase 7 is an activator of the c-Jun NH2-terminal kinase. *Proc. Natl. Acad. Sci.* **94:** 7337.

Wang Y. 2001. Signal transduction in cardiac hypertrophy-dissecting compensatory versus pathological pathways utilizing a transgenic approach. *Curr. Opin. Pharmacol.* **1:** 134.

Wang Y., Su B., Sah V.P., Brown J.H., Han J., and Chien K.R. 1998a. Cardiac hypertrophy induced by mitogen-activated protein kinase kinase 7, a specific activator for c-Jun NH2-terminal kinase in ventricular muscle cells. *J. Biol. Chem.* **273:** 5423.

Wang Y., Huang S., Sah V.P., Ross J., Jr., Brown J.H., Han J., and Chien K.R. 1998b. Cardiac muscle cell hypertrophy and apoptosis induced by distinct members of the p38 mitogen-activated protein kinase family. *J. Biol. Chem.* **273:** 2161.

Widmann C., Gibson S., Jarpe M.B., and Johnson G.L. 1999. Mitogen-activated protein kinase: Conservation of a three-kinase module from yeast to human. *Physiol. Rev.* **79:** 143.

Xu Q., Liu Y., Gorospe M., Udelsman R., and Holbrook N.J. 1996. Acute hypertension activates mitogen-activated protein kinases in arterial wall. *J. Clin. Invest.* **97:** 508.

Zechner D., Thuerauf D.J., Hanford D.S., McDonough P.M., and Glembotski C.C. 1997. A role for the p38 mitogen-activated protein kinase pathway in myocardial cell growth, sarcomeric organization, and cardiac-specific gene expression. *J. Cell Biol.* **139:** 115.

Zechner D., Craig R., Hanford D.S., McDonough P.M., Sabbadini R.A., and Glembotski C.C. 1998. MKK6 activates myocardial cell NF-kappaB and inhibits apoptosis in a p38 mitogen-activated protein kinase-dependent manner. *J. Biol. Chem.* **273:** 8232.

G-Protein-coupled Receptor Function in Heart Failure

S.V. NAGA PRASAD, J. NIENABER, AND H.A. ROCKMAN

Departments of Medicine, Cell Biology, and Genetics, Duke University Medical Center, Durham, North Carolina 27710

G-protein-coupled receptors (GPCRs) are a family of receptors that contain a conserved structure of seven transmembrane α-helices (Rockman et al. 2002). Of these, the adrenergic receptors are of particular importance for the heart because they are involved in the homeostatic regulation of the cardiovascular system (Rockman et al. 2002). Although GPCRs lack catalytic activity, interaction of the receptor with agonist promotes the dissociation of heterotrimeric guanine-nucleotide-binding regulatory proteins (G proteins) into Gα and Gβγ subunits. The G-protein subunits (both Gα and Gβγ) then amplify and propagate signals within the cell by modulating the activity of one or more effector molecules such as the adenylyl cyclases, phospholipases, and ion channels (Clapham and Neer 1997). The activity of these effector enzymes and ion channels in turn regulates the production of second messenger molecules, which elicit cellular responses by the activation of an array of signaling pathways. Examples of second messengers are adenosine 3′,5′ monophosphate (cAMP) produced by the stimulation of β-adrenergic receptors (βARs), and inositol (1,4,5)-triphosphate (IP_3) and diacylglycerol (DAG) by stimulation of $α_1$-adrenoceptors. Different G-protein complexes are involved in transmitting signals to the variety of effector molecules that exist on cells. For the effector adenylyl cyclase, Gαs is stimulatory as a consequence of βAR activation, whereas Gαi is inhibitory following muscarinic receptor stimulation. The signaling molecules IP_3 and DAG are generated by the effector phospholipase C, which is activated by the G protein, Gαq, in response to $α_1$-adrenoceptor, angiotensin II, and endothelin receptor stimulation.

β-ADRENERGIC RECEPTOR FUNCTION

In the heart, the $β_1$AR is the most predominant subtype, comprising 75–80% of total βARs (Hoffman and Lefkowitz 1996). In the heart, ligand activation of βARs not only leads to the stimulation of adenylyl cyclase and a physiological response, but also promotes the rapid desensitization and targeting of the activated receptor to clathrin-coated pits for internalization (Ferguson et al. 1996; Goodman et al. 1996). An early step in this process involves the rapid phosphorylation of agonist-occupied receptors by a G-protein-coupled receptor kinase (GRK, commonly known as the β-adrenergic receptor kinase or βARK) (Lefkowitz 1998). The phosphorylation of activated βARs by βARK1 requires translocation of the primarily cytosolic βARK to the plasma membrane, a process facilitated by the liberated Gβγ subunits and membrane phospholipids (Pitcher et al. 1998). The second step involves binding of the protein β-arrestin to the phosphorylated receptor, resulting in termination of the signal (Lefkowitz 1998). The binding of β-arrestin serves to target phosphorylated βARs for internalization, through the recruitment of the AP-2 adapter protein and clathrin to the receptor complex (Laporte et al. 2000).

ROLE OF PHOSPHOINOSITIDE 3-KINASES

There is increasing evidence that phosphatidylinositol (PtdIns) phospholipids are important molecules in endocytosis of membrane proteins (Czech 2000). Since a key enzyme for the generation of phospholipids is phosphoinositide 3-kinase (PI3K), it suggests a possible role for PI3K in the internalization of GPCRs. Thus, although the endocytic process is a multistep event involving the coordinate interaction between proteins as well as the control of lipid modification, the precise molecular mechanisms for this interaction are not well understood. PI3Ks are a conserved family of lipid kinases that catalyze the addition of phosphate on the third position of the inositol ring (Rameh and Cantley 1999). Stimulation of a variety of receptor tyrosine kinases and GPCRs results in the activation of PI3K and leads to an increase in the level of D-3 PtdIns, which in turn are potent signaling molecules that modulate a number of diverse cellular effects including cell proliferation, cell survival, cytoskeletal rearrangements, and receptor endocytosis (Martin 1998; Sato et al. 2001). In this context, GPCR stimulation leads to the activation of the I_B subclass of PI3Ks mediated by the Gβγ subunits of G proteins (Stoyanov et al. 1995). Studies have suggested a role of phosphoinositides in the process of receptor internalization. For example, deletion of the phosphoinositide-binding site from β-arrestin impairs GPCR endocytosis, and the binding of PtdIns (3,4,5) P_3 and PtdIns (4,5) P_2 to AP-2 promotes targeting of the receptor–arrestin complex to clathrin-coated pits (Gaidarov and Keen 1999; Gaidarov et al. 1999).

Previous studies have shown that in the pathological condition of in vivo pressure overload hypertrophy, both βARK1 and PI3Kγ activities are increased (Choi et al. 1997; Naga Prasad et al. 2000), suggesting their involvement in the physiological state. Indeed, recent studies have shown the involvement of PI3K in the regulation of βAR internalization (Naga Prasad et al. 2001) and have demonstrated a direct protein–protein interaction between PI3K and βARK1 (Naga Prasad et al. 2002). The region of the PI3K molecule that provides the necessary structure for this interaction is known as the PIK domain (Fig. 1). Importantly, the interaction between PI3K and βARK1 is not dependent on Gβγ, and overexpression of PIK domain competitively displaces PI3K from the βARK1/PI3K complex, leading to a loss in βARK1-associated PI3K activity. Using the PIK peptide as a reagent to disrupt the βARK1/PI3K interaction results in the inhibition of both agonist-stimulated AP-2 adapter protein recruitment to βArs and βAR endocytosis (Fig. 1) (Naga Prasad et al. 2002). This suggests that impairing the local production of PtdIns-3,4,5-P_3 lipid molecules within the agonist-occupied receptor complex affects the recruitment of critical molecules necessary for efficient receptor endocytosis. These studies demonstrate a role for the localized generation of D-3 phosphoinositides in regulating the recruitment of the receptor/cargo to clathrin-coated pits.

In addition to the role β-arrestins play in turning off βAR-mediated second messenger production, it has become increasingly clear that β-arrestins serve other important functions, such as mediating receptor internalization (Laporte et al. 2000) and acting as a scaffold to bring together various elements of complex receptor-stimulated signaling pathways (Claing et al. 2002). For example, β-arrestins are able to serve as adapters that bind members of the Src family and, in a ligand-dependent fashion, bring these tyrosine kinases into complex with heptahelical receptors (Luttrell et al. 1999, 2001; DeFea et al. 2000). This often serves as a means of linking the GPCRs with elements of the MAP kinase cascades, such as, for example, the extracellular regulated kinase (ERK) cascade. Moreover, β-arrestins can serve to actually scaffold the various members of several MAP kinase cascades (i.e., a MAPKKK, MAPKK, and MAPK). The scaffolding function has been demonstrated both for the c-Jun amino-terminal kinase3 (JNK3) (McDonald et al. 2000) and ERK (DeFea et al. 2000; Luttrell et al. 2001) modules and serves to bring their activation and subcellular localization under the direct control of the stimulated receptor.

MYOCARDIAL HYPERTROPHY

Cardiac hypertrophy is the physiological response of the heart to an increased work demand (Rockman et al. 2002). It has been central dogma that the development of myocardial hypertrophy is a compensatory mechanism in response to increased cardiac workload that acts to normalize wall stress, thereby preventing the development of heart failure (Grossman et al. 1975). Although this remodeling may indeed normalize wall stress, results from the Framingham Heart Study show an association between ventricular hypertrophy and increased cardiac mortality, casting doubt on the validity of the wall stress hypothesis (Levy et al. 1990). To determine whether cardiac hypertrophy is adaptive or maladaptive, two genetically altered mouse models that have an attenuated hypertrophic response to pressure overload were studied after long-term mechanical overload (Esposito et al. 2002). The models included transgenic mice with overexpression of the Gαq inhibitor peptide (Akhter et al. 1998), and knockout mice that lack endogenous norepinephrine and epinephrine (Rapacciuolo et al. 2001). In vivo end-systolic wall stress was measured by sonomicrometry in sham and pressure- overloaded mice

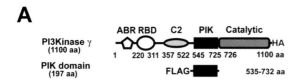

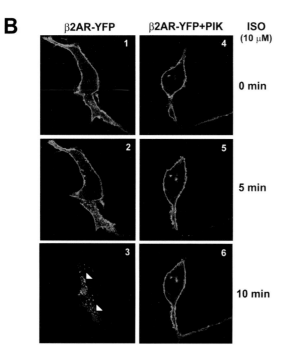

Figure 1. βARK1 directly interacts with the PIK domain of PI3K to inhibit βAR endocytosis. (A) Schematic representation of full-length PI3Kp110γ. (ABR) Adapter-binding region, (RBD) ras-binding domain, (C2) similar to PLCd, which is involved in Ca^{++}-dependent or -independent phospholipid binding, (PIK) domain thought to be involved in protein–protein interactions, (HA)-hemagglutinin tag, (FLAG) flag peptide tag. (B) Endocytosis of $β_2$AR-YFP in live cells following isoproterenol (10 mM) stimulation in the absence or presence of PIK domain protein coexpression using laser scanning confocal microscopy. Panels on the left represent cells transfected with the $β_2$AR-YFP alone (panels 1, 2, and 3 show the same cell monitored at 0, 5, and 10 minutes following stimulation). Panels on the right represent cells transfected with $β_2$AR-YFP and PIK (panels 4, 5, and 6 show the same cell monitored as above). In the absence of PIK, isoproterenol causes the internalization of $β_2$ARs as shown by the formation of distinct cytoplasmic aggregates (arrowheads) and complete loss of membrane fluorescence. In contrast, in the presence of PIK domain protein, there is no redistribution of $β_2$AR-YFP following agonist stimulation, indicating that the process of receptor endocytosis is completely inhibited. (Adapted, with permission, from Naga Prasad et al. 2002 [copyright Rockefeller University Press].)

(Takaoka et al. 2002). One week after banding there was complete normalization of end-systolic wall stress in pressure-overloaded wild-type mice, whereas, as expected, transgenic mice with a blunted hypertrophic response showed nearly a twofold increase in wall stress (Esposito et al. 2002). Remarkably, despite the lack of adequate hypertrophy to normalize wall stress in the genetically altered mice, there was less deterioration in heart function compared to control mice with a "normal" hypertrophic response (Fig. 2). Furthermore, analysis of downstream signaling pathways in the late stages of pressure overload suggested that activation of PI3K may play a pivotal role in the transition from hypertrophy to heart failure (Esposito et al. 2002). These data suggest that, contrary to previous hypotheses, the development of cardiac hypertrophy and normalization of wall stress may not be necessary to preserve cardiac function (Esposito et al. 2002; Sano and Schneider 2002).

Taken together, the above experiments demonstrate a critical role for Gq and catecholamine signaling in the activation of the hypertrophic process. In addition, contrary to previous hypotheses, hypertrophy itself, not elevated wall stress, appears to be a trigger for cardiac decompensation. These data point toward a potential novel strategy for preventing pathophysiological signaling that occurs under conditions of pressure overload by modeling a strategy as used in the transgenic experiments through design of novel therapeutic agents that target the GPCR-Gq interface.

β-ADRENERGIC RECEPTOR SIGNALING IN THE FAILING HEART

Heart failure is characterized by left ventricular (LV) dysfunction associated with activation of the sympathetic nervous system leading to multiple abnormalities in the βAR system. Indeed, whereas $β_1$ARs have been shown to be selectively down-regulated, both $β_1$ and $β_2$ARs are markedly uncoupled from G proteins in failing hearts (Bristow et al. 1982; Bristow 1998), a process attributed to increased levels of myocardial βARK1 (Ungerer et al. 1993, 1994) and myocardial Gαi (Feldman et al. 1988). This not only results in the attenuation of βAR signaling, but may also enhance the coupling of receptors to other signaling pathways that play a role in ventricular remodeling, such as the MAPK and PI3K cascades (Daaka et al. 1997).

To determine whether the level and activity of βARK1 (the primary GRK expressed in the heart) influence cardiac contractile responsiveness to catecholamines, transgenic mice have been generated that overexpress the full-length protein or a peptide inhibitor of βARK (βARKct) (Koch et al. 1995). The βARKct is of particular interest because it is a 194-amino-acid peptide containing the Gβγ-binding domain of βARK1, which competes with endogenous βARK1 for Gβγ-mediated membrane translocation and activation. These studies have shown that in the normal heart in vivo, the level of contractile function and the heart's response to catecholamine stimulation are primarily determined by the level of βARK1 activity (Koch et al. 1995; Rockman et al. 1998b). These data have important implications in disease states, characterized by increased βARK1 activity, since even partial inhibition of βARK1 activity may lead to improved functional catecholamine responsiveness.

To test whether βARK1 plays an important role in the pathophysiology of cardiac failure, transgenic mice with cardiac overexpression of βARKct (resulting in βARK inhibition) have been mated with several of these mouse models. These have included the MLP knockout (Rockman et al. 1998a; Esposito et al. 2000), cardiac overexpression of calsequestrin (Harding et al. 2001), and transgenic overexpression of a mutant cardiac α-myosin

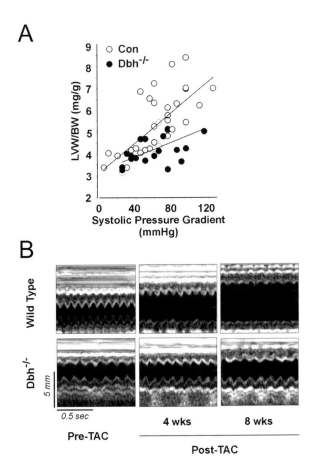

Figure 2. Attenuation of cardiac hypertrophy preserves cardiac function: Reappraisal of current dogma. Pressure overload cardiac hypertrophy was induced by surgical constriction of the transverse aortic in genetically altered mice that lack endogenous norepinephrine and epinephrine created by disruption of the dopamine β-hydroxylase gene ($Dbh^{-/-}$), the essential enzyme in the biosynthetic pathway converting dopamine to norepinephrine. (A) Pressure-overloaded $Dbh^{-/-}$ mice develop less cardiac hypertrophy (shown as a reduction in left ventricular weight to body weight ratio, LVW/BW) compared to similarly pressure-overloaded control mice. (B) Representative serial M-mode echocardiographic tracings before, 4 weeks after, and 8 weeks after creation of mechanical pressure overload. Remarkably, even though cardiac hypertrophy in $Dbh^{-/-}$ mice was significantly inhibited, there was no deterioration of cardiac function. Thus, blunting of the hypertrophic response, and the resultant increase in wall stress, are associated with less deterioration in cardiac function despite the continued presence of a pressure load on the heart. (Adapted, with permission, from Esposito et al. 2002 [copyright Lippincott Williams and Wilkins].)

heavy chain (Freeman et al. 2001). Remarkably, cardiac overexpression of the βARKct resulted in prevention of progressive deterioration in cardiac dysfunction, improved exercise tolerance, correction in the classic biochemical alterations of βAR function, and when assessed, markedly improved survival. Moreover, effects on survival were synergistic with those achieved by βAR blocker treatment (Harding et al. 2001), an interesting finding given the recent clinical data showing improved survival in patients with chronic heart failure who were treated with β-blockers (Packer et al. 2001).

The nature and extent of heart failure associated with the MLP$^{-/-}$ mouse and that rescued by the βARKct have been examined in vivo using a pressure-volume analysis and $[Ca^{++}]_i$ signaling using patch-clamp methods and confocal Ca^{++} imaging (Esposito et al. 2000). In order to examine the murine myocardial contractile state in vivo, a method was developed to measure the intrinsic myocardial contractility in mice (Fig. 3). Using two pairs of endocardial implanted piezoelectric crystals (Fig. 3) and a high-fidelity micromanometer in the LV, in vivo pressure-volume (PV) relationships were obtained. These PV relationships were obtained in the presence and absence of βAR stimulation and showed that in the MLP$^{-/-}$/βARKct animals, end-diastolic and end-systolic cardiac volumes became near normal, and the slope of the end-systolic pressure volume relation (indicating contractile state) was similar to that observed in wild-type animals (Esposito et al. 2000). Furthermore, there was functional restoration of βAR responsiveness in the MLP$^{-/-}$/βARKct animals. These findings were associated with normalization of the EC coupling gain function in cells from the MLP$^{-/-}$/βARKct animals (Esposito et al. 2000).

CONCLUSIONS

Heart failure is characterized by defects in βAR signaling that include down-regulation of β$_1$ARs, increased βARK1 activity, and uncoupling of both β$_1$ and β$_2$AR subtypes from their effector, adenylyl cyclase. Chronic exposure to agonist promotes the rapid desensitization of βARs, which not only leads to the attenuation of signaling, but also targets the activated receptor to clathrin-coated pits for internalization, thereby further activating signaling pathways. The overwhelming evidence supports the hypothesis that the loss of normal βAR function in chronic heart failure contributes to the pathogenesis of this syndrome, since correcting these defects has been associated with improved cardiac function in experimental models of heart failure. The discovery of new molecules and interacting pathways has important implications with regard to the molecular mechanisms of heart failure. In this regard, the identification of a previously unappreciated protein–protein interaction between βARK1 and PI3K has revealed how the disruption of this interaction can prevent agonist-induced βAR endocytosis. Taken together, restoration of normal βAR signaling in the failing heart appears to be salutary and represents a new therapeutic approach to the treatment of chronic human heart failure.

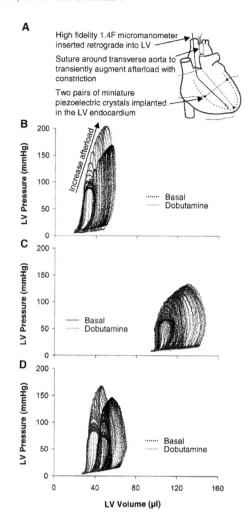

Figure 3. Pressure-volume (PV) loops in wild-type, MLP$^{-/-}$, and MLP$^{-/-}$/βARKct mice. (*A*) Schematic of the instrumented mouse heart. (*B*) Representative PV loops are displayed during transient constriction of the transverse aorta to augment afterload under basal conditions and following dobutamine (2 mg/kg/min) infusion for wild-type heart. (*C*) MLP$^{-/-}$ heart. (*D*) MLP$^{-/-}$/βARKct heart. At baseline the end diastolic volume (EDV) is 32 μl, which increases to 55 μl as afterload is increased by aortic constriction as shown by the shift to the right and upward as expected for a normal heart. With the addition of the βAR agonist dobutamine, there is an increase in the slope of the relation describing end-systolic pressure and volume as afterload increases. The MLP$^{-/-}$ heart is dilated and does not respond to βAR stimulation, which is corrected by overexpression of the βARKct. (Adapted, with permission, from Esposito et al. 2000 [copyright American Physiological Society].)

REFERENCES

Akhter S.A., Luttrell L.M., Rockman H.A., Iaccarino G., Lefkowitz R.J., and Koch W.J. 1998. Targeting the receptor-Gq interface to inhibit in vivo pressure overload myocardial hypertrophy. *Science* **280**: 574.

Bristow M.R. 1998. Why does the myocardium fail? Insights from basic science. *Lancet* (suppl. 1) **352**: 8.

Bristow M.R., Ginsburg R., Minobe W., Cubicciotti R.S., Sageman W.S., Lurie K., Billingham M.E., Harrison D.C., and Stinson E.B. 1982. Decreased catecholamine sensitivity and beta-adrenergic-receptor density in failing human hearts. *N.*

Engl. J. Med. **307:** 205.

Choi D.J., Koch W.J., Hunter J.J., and Rockman H.A. 1997. Mechanism of beta-adrenergic receptor desensitization in cardiac hypertrophy is increased beta-adrenergic receptor kinase. *J. Biol. Chem.* **272:** 17223.

Claing A., Laporte S.A., Caron M.G., and Lefkowitz R.J. 2002. Endocytosis of G protein-coupled receptors: Roles of G protein-coupled receptor kinases and beta-arrestin proteins. *Prog. Neurobiol.* **66:** 61.

Clapham D.E. and Neer E.J. 1997. G protein beta gamma subunits. *Annu. Rev. Pharmacol. Toxicol.* **37:** 167.

Czech M.P. 2000. PIP2 and PIP3: Complex roles at the cell surface. *Cell* **100:** 603.

Daaka Y., Luttrell L.M., and Lefkowitz R.J. 1997. Switching of the coupling of the beta2-adrenergic receptor to different G proteins by protein kinase A. *Nature* **390:** 88.

DeFea K.A., Zalevsky J., Thoma M.S., Dery O., Mullins R.D., and Bunnett N.W. 2000. β-Arrestin-dependent endocytosis of proteinase-activated receptor 2 is required for intracellular targeting of activated ERK1/2. *J. Cell Biol.* **148:** 1267.

Esposito G., Rapacciuolo A., Naga Prasad S.V., Takaoka H., Thomas S.A., Koch W.J., and Rockman H.A. 2002. Genetic alterations that inhibit in vivo pressure-overload hypertrophy prevent cardiac dysfunction despite increased wall stress. *Circulation* **105:** 85.

Esposito G., Santana L.F., Dilly K., Cruz J.D., Mao L., Lederer W.J., and Rockman H.A. 2000. Cellular and functional defects in a mouse model of heart failure. *Am. J. Physiol. Heart Circ. Physiol.* **279:** H3101.

Feldman A.M., Cates A.E., Veazey W.B., Hershberger R.E., Bristow M.R., Baughman K.L., Baumgartner W.A., and Van Dop C. 1988. Increase of the 40,000-mol wt pertussis toxin substrate (G protein) in the failing human heart. *J. Clin. Invest.* **82:** 189.

Ferguson S.S., Downey W.E., III, Colapietro A.M., Barak L.S., Menard L., and Caron M.G. 1996. Role of beta-arrestin in mediating agonist-promoted G protein-coupled receptor internalization. *Science* **271:** 363.

Freeman K., Lerman I., Kranias E.G., Bohlmeyer T., Bristow M.R., Lefkowitz R.J., Iaccarino G., Koch W.J., and Leinwand L.A. 2001. Alterations in cardiac adrenergic signaling and calcium cycling differentially affect the progression of cardiomyopathy. *J. Clin. Invest.* **107:** 967.

Gaidarov I. and Keen J.H. 1999. Phosphoinositide-AP-2 interactions required for targeting to plasma membrane clathrin-coated pits. *J. Cell Biol.* **146:** 755.

Gaidarov I., Krupnick J.G., Falck J.R., Benovic J.L., and Keen J.H. 1999. Arrestin function in G protein-coupled receptor endocytosis requires phosphoinositide binding. *EMBO J.* **18:** 871.

Goodman O.B., Jr., Krupnick J.G., Santini F., Gurevich V.V., Penn R.B., Gagnon A.W., Keen J.H., and Benovic J.L. 1996. Beta-arrestin acts as a clathrin adaptor in endocytosis of the beta2-adrenergic receptor. *Nature* **383:** 447.

Grossman W., Jones D., and McLaurin L.P. 1975. Wall stress and patterns of hypertrophy in the human left ventricle. *J. Clin. Invest.* **56:** 56.

Harding V.B., Jones L.R., Lefkowitz R.J., Koch W.J., and Rockman H.A. 2001. Cardiac beta ARK1 inhibition prolongs survival and augments beta blocker therapy in a mouse model of severe heart failure. *Proc. Natl. Acad. Sci.* **98:** 5809.

Hoffman B.B. and Lefkowitz R.J. 1996. Catecholamines, sympathetic drugs, and adrenergic receptor antagonists. In *Goodman and Gilman's the pharmacological basis of therapeutics* (ed. J.G. Hardman et al.), p. 199. McGraw-Hill, New York.

Koch W.J., Rockman H.A., Samama P., Hamilton R.A., Bond R.A., Milano C.A., and Lefkowitz R.J. 1995. Cardiac function in mice overexpressing the beta-adrenergic receptor kinase or a beta ARK inhibitor. *Science* **268:** 1350.

Laporte S.A., Oakley R.H., Holt J.A., Barak L.S., and Caron M.G. 2000. The interaction of beta-arrestin with the AP-2 adaptor is required for the clustering of beta 2-adrenergic receptor into clathrin-coated pits. *J. Biol. Chem.* **275:** 23120.

Lefkowitz R.J. 1998. G protein-coupled receptors. III. New roles for receptor kinases and beta-arrestins in receptor signaling and desensitization. *J. Biol. Chem.* **273:** 18677.

Levy D., Garrison R.J., Savage D.D., Kannel W.B., and Castelli W.P. 1990. Prognostic implications of echocardiographically determined left ventricular mass in the Framingham Heart Study. *N. Engl. J. Med.* **322:** 1561.

Luttrell L.M., Roudabush F.L., Choy E.W., Miller W.E., Field M.E., Pierce K.L., and Lefkowitz R.J. 2001. Activation and targeting of extracellular signal-regulated kinases by beta-arrestin scaffolds. *Proc. Natl. Acad. Sci.* **98:** 2449.

Luttrell L.M., Ferguson S.S., Daaka Y., Miller W.E., Maudsley S., Della Rocca G.J., Lin F., Kawakatsu H., Owada K., Luttrell D., Caron M.G., and Lefkowitz R.J. 1999. Beta-arrestin-dependent formation of beta2 adrenergic receptor-Src protein kinase complexes (comments). *Science* **283:** 655.

Martin T.F. 1998. Phosphoinositide lipids as signaling molecules: Common themes for signal transduction, cytoskeletal regulation, and membrane trafficking. *Annu. Rev. Cell Dev. Biol.* **14:** 231.

McDonald P.H., Chow C.W., Miller W.E., Laporte S.A., Field M.E., Lin F.T., Davis R.J., and Lefkowitz R.J. 2000. Beta-arrestin 2: A receptor-regulated MAPK scaffold for the activation of JNK3. *Science* **290:** 1574.

Naga Prasad S.V., Barak L.S., Rapacciuolo A., Caron M.G., and Rockman H.A. 2001. Agonist-dependent recruitment of phosphoinositide 3-kinase to the membrane by beta-adrenergic receptor kinase 1. A role in receptor sequestration. *J. Biol. Chem.* **276:** 18953.

Naga Prasad S.V., Esposito G., Mao L., Koch W.J., and Rockman H.A. 2000. Gβγ-dependent phosphoinositide 3-kinase activation in hearts with in vivo pressure overload hypertrophy. *J. Biol. Chem.* **275:** 4693.

Naga Prasad S.V., Laporte S.A., Chamberlain D., Caron M.G., Barak L.S., and Rockman H.A. 2002. Phosphoinositide 3-kinase regulates β2-adrenergic receptor endocytosis by AP-2 recruitment to the receptor/β-arrestin complex. *J. Cell Biol.* **158:** 563.

Packer M., Coats A.J., Fowler M.B., Katus H.A., Krum H., Mohacsi P., Rouleau J.L., Tendera M., Castaigne A., Roecker E.B., Schultz M.K., DeMets D.L., and the Carvedilol Prospective Randomized Cumulative Survival Study Group. 2001. Effect of carvedilol on survival in severe chronic heart failure. *N. Engl. J. Med.* **344:** 1651.

Pitcher J.A., Freedman N.J., and Lefkowitz R.J. 1998. G protein-coupled receptor kinases. *Annu. Rev. Biochem.* **67:** 653.

Rameh L.E. and Cantley L.C. 1999. The role of phosphoinositide 3-kinase lipid products in cell function. *J. Biol. Chem.* **274:** 8347.

Rapacciuolo A., Esposito G., Caron K., Mao L., Thomas S.A., and Rockman H.A. 2001. Important role of endogenous norepinephrine and epinephrine in the development of in vivo pressure-overload cardiac hypertrophy. *J. Am. Coll. Cardiol.* **38:** 876.

Rockman H.A., Koch W.J., and Lefkowitz R.J. 2002. Seven-transmembrane-spanning receptors and heart function. *Nature* **415:** 206.

Rockman H.A., Chien K.R., Choi D.J., Iaccarino G., Hunter J.J., Ross J., Jr., Lefkowitz R.J., and Koch W.J. 1998a. Expression of a beta-adrenergic receptor kinase 1 inhibitor prevents the development of myocardial failure in gene-targeted mice. *Proc. Natl. Acad. Sci.* **95:** 7000.

Rockman H.A., Choi D.J., Akhter S.A., Jaber M., Giros B., Lefkowitz R.J., Caron M.G., and Koch W.J. 1998b. Control of myocardial contractile function by the level of beta-adrenergic receptor kinase 1 in gene-targeted mice. *J. Biol. Chem.* **273:** 18180.

Sano M. and Schneider M.D. 2002. Still stressed out but doing fine: Normalization of wall stress is superfluous to maintaining cardiac function in chronic pressure overload. *Circulation* **105:** 8.

Sato T.K., Overduin M., and Emr S.D. 2001. Location, location, location: Membrane targeting directed by PX domains. *Science* **294:** 1881.

Stoyanov B., Volinia S., Hanck T., Rubio I., Loubtchenkov M.,

Malek D., Stoyanova S., Vanhaesebroeck B., Dhand R., and Nurnberg B., et al. 1995. Cloning and characterization of a G protein-activated human phosphoinositide-3 kinase. *Science* **269:** 690.

Takaoka H., Esposito G., Mao L., Suga H., and Rockman H.A. 2002. Heart size-independent analysis of myocardial function in murine pressure overload hypertrophy. *Am. J. Physiol. Heart Circ. Physiol.* **282:** H2190.

Ungerer M., Bohm M., Elce J.S., Erdmann E., and Lohse M.J. 1993. Altered expression of beta-adrenergic receptor kinase and beta 1-adrenergic receptors in the failing human heart. *Circulation* **87:** 454.

Ungerer M., Parruti G., Bohm M., Puzicha M., DeBlasi A., Erdmann E., and Lohse M.J. 1994. Expression of beta-arrestins and beta-adrenergic receptor kinases in the failing human heart. *Circ. Res.* **74:** 206.

Salt and Blood Pressure: New Insight from Human Genetic Studies

R.P. Lifton, F.H. Wilson, K.A. Choate, and D.S. Geller

Howard Hughes Medical Institute, Department of Genetics, Yale University School of Medicine, New Haven, Connecticut 06510

Heart disease and stroke are the number one and number three leading causes of death in the United States, accounting for a third of all deaths annually. Epidemiologic studies have established a number of risk factors for these diseases, including hypertension, high cholesterol, diabetes mellitus, and smoking. Prospective randomized trials of blood pressure and cholesterol reduction and smoking cessation have established the causal relationship of these risk factors to disease, because modifying these parameters prevents adverse clinical outcomes, including death.

In the case of cholesterol, initial therapeutic agents had modest cholesterol-lowering effects that reduced risk of heart attack but did not reduce overall mortality. Understanding the cholesterol biosynthetic pathway ultimately led to identification of rate-limiting steps in the pathway and development of highly potent cholesterol-lowering agents, the HMG-CoA reductase inhibitors. These agents lower cholesterol far more than their predecessors and markedly reduce both morbidity and overall mortality.

The case of cholesterol provides a useful paradigm for hypertension. Hypertension is perhaps the most common disease of the industrialized world, affecting more than 20% of the adult population, with the occurrence in the elderly rising to over 70%. Treatment of hypertension has a clear benefit to reduce the incidence of stroke and heart attack (Multiple Risk Factor Intervention Trial Research Group 1982; Medical Research Council Working Party 1985). Nonetheless, the blood pressure reduction achieved with current medications is very modest in most treated subjects; importantly, only a small minority of hypertensive individuals achieve the goals of blood pressure reduction, leaving considerable room for improvement in therapy. Efforts to improve therapy, however, have been complicated by a lack of understanding of the main pathways that determine long-term blood pressure homeostasis and the key points in this pathway that might represent improved targets for therapy. This uncertainty arises from the complex physiologic regulation of blood pressure, with inputs from the brain, heart, adrenal, kidney, vasculature, and endocrine systems. Consequently, the field has been plagued with many ideas as to where primary abnormalities might lie with little data to substantiate any.

To address this problem, our group has taken a genetic approach to the understanding of hypertension in humans, using the paradigm of positional cloning to identify genes underlying blood pressure variation. We anticipate that identification of specific genetic abnormalities that alter blood pressure will firmly establish pathways involved in long-term blood pressure regulation and identify points in the pathway(s) that might be exploited for therapeutic intervention. At the outset, we have focused on the extremes of the blood pressure distribution, searching for individuals and families with early onset of severely high or low blood pressure, attempting to identify families in which single genes impart very large effects on blood pressure. Our expectation is that success with this effort should identify pathways that underlie variation in long-term blood pressure and identify new targets for development of novel antihypertensive agents.

This program has succeeded thus far in identifying six genes that raise blood pressure and another eight genes in which mutation lowers blood pressure. The most striking finding of these studies is that all of these genes fall into a final common pathway that regulates renal salt handling, with all mutations that increase net renal salt reabsorption raising blood pressure and mutations that reduce salt reabsorption lowering blood pressure.

To understand this relationship, an understanding of renal salt homeostasis is required. Every day the kidneys of a normal adult filter 170 liters of plasma containing 1500 grams of salt. Consequently, on a normal western 5-gram salt diet, the kidneys must reabsorb all but about 0.5% of the filtered load of salt to maintain homeostasis. The amount of water reabsorbed is regulated to achieve a serum sodium concentration of 140 mM. As a consequence, increased salt reabsorption results in higher plasma volume, and reduced salt reabsorption results in lower plasma volume. The kidneys achieve salt homeostasis by an integrated set of exchangers, cotransporters, and channels that act in distinct nephron segments. Bulk reabsorption occurs proximally, with progressively smaller fractions in more distal nephron segments. Sixty percent of the filtered load of salt is reabsorbed by Na^+/H^+ exchange in the proximal tubule, 30% by the action of a $Na^+-K^+-2Cl^-$ cotransporter in the thick ascending limb of Henle, 7% by the action of a Na–Cl cotransporter in the distal convoluted tubule, and the last 2% by the action of an epithelial sodium channel. The activity of this final channel, ENaC, is regulated by the renin-angiotensin system, which is activated when the kidney senses inadequate chloride delivery to the thick ascending limb of

Henle, resulting in secretion of the aspartyl protease renin by cells of the juxtaglomerular apparatus. This ultimately results in formation of the active peptide hormone angiotensin II, which increases secretion of aldosterone from the adrenal glomerulosa, ultimately resulting in induction of a transcriptional program that leads to increased activity of the epithelial sodium channel. Thus, although ENaC activity accounts for only a small minority of all sodium reabsorption, its activity is normally the major determinant of net salt reabsorption, since it is the main regulated step.

The mutations our group has identified all underlie Mendelian diseases; in each case, multiple independent mutations have been identified that show specificity for the disease and co-segregate with the disease in families. In nearly every case, biochemical and clinical studies have established the specific mechanisms by which mutations increase or decrease renal salt reabsorption. Below is a description of the diseases and mutations that account for them.

MUTATIONS THAT RAISE BLOOD PRESSURE

Glucocorticoid-remediable aldosteronism (GRA) is an autosomal dominant trait featuring early onset of severe hypertension caused by a novel gene duplication produced by unequal crossing over between two genes involved in adrenal steroid biosynthesis (Lifton et al. 1992a,b). One of these genes encodes aldosterone synthase, whose enzymatic activity is the rate-limiting step in aldosterone biosynthesis in the adrenal glomerulosa and is normally regulated there by signaling via angiotensin II. The other encodes steroid 11-β hydroxylase, which is employed in cortisol biosynthesis in adrenal fasciculate and whose expression is regulated by adrenocorticotropic hormone (ACTH). These two genes have recently evolved from a common ancestor and are 95% identical in DNA sequence; they are tightly linked on chromosome 8, arranged in a head-to-tail orientation. The chimeric gene duplications that cause GRA fuse ACTH-responsive regulatory elements from 11-hydroxylase onto coding sequences that result in aldosterone synthase enzymatic activity. This results in ectopic expression of the rate-limiting enzyme for aldosterone secretion in the adrenal fasciculate under control of ACTH rather than the normal secretagogue, angiotensin II. At the expense of maintaining ACTH levels to support normal cortisol secretion, affected subjects have sustained aldosterone secretion.

This results in hypertension by the following sequence: Chronic aldosterone secretion leads to increased sodium reabsorption via ENaC; water follows to maintain isotonicity of plasma, leading to increased intravascular volume; this increases venous blood return to the heart, and the cardiac stroke volume consequently increases, thereby increasing cardiac output which by Ohm's law results in elevated blood pressure.

The second Mendelian form of hypertension we have solved is Liddle's syndrome, which is characterized by early-onset hypertension despite suppression of the secretion of renin and aldosterone. This disease is caused by mutations in either the β or γ subunits of the epithelial sodium channel ENaC (Shimkets et al. 1994; Hansson et al. 1995). Disease-causing mutations eliminate or alter single amino acids in the cytoplasmic carboxyl termini of these proteins. The target sequence for these mutations is a PPPXY segment shared by these two subunits (Schild et al. 1996). This common sequence motif is required for the normal removal of ENaC from the cell surface by endocytosis via clathrin-coated pits (Shimkets et al. 1997). This sequence is recognized by WW domains of the Nedd-4 protein, a ubiquitin ligase (Staub et al. 1996). These mutations thus result in increased numbers of active ENaC at the cell surface, owing to a markedly prolonged half-life (Shimkets et al. 1997). These channels are active and mediate increased sodium reabsorption, resulting in the same final common pathway for hypertension as GRA.

A third Mendelian form of hypertension was identified by investigation of a young male with severe hypertension with suppressed renin and aldosterone secretion (Geller et al. 2000). Resequencing of genes of the salt-retaining pathway identified a missense mutation in the ligand-binding domain of the mineralocorticoid receptor. This mutation was found to co-segregate with early hypertension in the extended family. Biochemical studies have demonstrated the molecular mechanism of this mutation. The mutation results in a partial activation of the receptor in the absence of added steroid; more impressively, progesterone, which normally binds but fails to activate the normal receptor, is now a very potent agonist for the mutant receptor. Molecular modeling followed by site-directed mutagenesis has shown that the disease-causing mutation creates a key van der Waals interaction between helix 5 and helix 3 of the ligand-binding domain. This interaction eliminates the requirement for an interaction between the 21-hydroxyl group of aldosterone and helix 3 of the receptor, allowing steroids lacking this moiety, such as progesterone, to become potent receptor agonists. An important in vivo correlate of this finding is that when women harboring this mutation become pregnant, and serum progesterone levels rise 100-fold, they develop extremely severe pregnancy-induced hypertension. This is the first demonstration of a molecular mechanism underlying this common complication of pregnancy and shows that it can arise from the abnormal action of a normal hormone.

Pseudohypoaldosteronism type II is another autosomal dominant trait featuring hypertension, hyperkalemia, and renal tubular acidosis despite normal glomerular filtration. It is caused by mutations in two closely related serine-threonine kinases, WNK1 and WNK4 (Wilson et al. 2001). Disease-causing mutations in WNK1 are large deletions in the first intron of this gene, whereas WNK4 mutations cluster within a highly conserved 10-amino-acid segment distal to the kinase domain. These kinases are expressed in the distal convoluted tubule and collecting duct of the nephron. WNK1 is distributed in the cytoplasm of these nephron segments, whereas WNK4 is localized predominantly to the tight junctions of these segments. The precise mechanism by which these mutations alter blood pressure is uncertain; two possibilities are that the mutations result in increased activity of the thiazide-sensitive Na–Cl cotransporter or that they increase permeability of the paracellular path for Cl⁻ flux in the distal nephron. These findings establish the role of a new signaling pathway in blood pressure regulation; identification of the upstream regulators and downstream targets of these kinases will clearly be of interest.

MUTATIONS THAT LOWER BLOOD PRESSURE

A host of mutations that lower blood pressure have also been identified and act to lower net renal salt reabsorption. Homozygous loss-of-function mutations in ENaC subunits result in impaired renal salt handling, causing the recessive form of pseudohypoaldosteronism type I, an often fatal disease of newborns featuring profound intravascular volume depletion with impaired ability to secrete K^+ and H^+ (Chang et al. 1996). In these children, salt depletion activates the renin-angiotensin system; however, this fails to increase salt reabsorption because the key target, ENaC, is missing. Thus, gain-of-function mutations in ENaC cause hypertension, whereas loss-of-function mutations in this same channel cause life-threatening hypotension. There is also an autosomal dominant form of this disease, which is caused by heterozygous loss-of-function mutations in the mineralocorticoid receptor (Geller et al. 1998). Interestingly, although patients with both forms of this disease can be severely ill in the neonatal period, those with mutations in the mineralocorticoid receptor become asymptomatic after about 2 years of age, and as adults are only recognizable by a markedly elevated aldosterone level. In contrast, patients with ENaC mutations remain dependent on massive salt supplementation throughout life. These observations indicate the critical dependence on ENaC for salt and blood pressure homeostasis and have implications for the development of new antihypertensive agents (see below).

Homozygous loss-of-function mutations in the thiazide-sensitive Na–Cl cotransporter of the renal distal convoluted tubule cause Gitelman's syndrome, which features renal salt wasting with hypokalemia, metabolic alkalosis, and so-called normal blood pressure (Simon et al. 1996c). We hypothesized that the relatively mild salt wasting seen in this disease should reduce blood pressure. This was tested by comparing the blood pressures of family members identified as harboring zero, one, or two mutant copies of the gene. The results demonstrated about 8 mm Hg reduction in blood pressure among homozygous mutant subjects, with no reduction in blood pressure of heterozygotes. Interestingly, however, both the heterozygous and homozygous mutant subjects consume markedly more salt than their wild-type relatives, indicating dietary compensation for the inherited renal defect (Cruz et al. 2001). It is noteworthy that the individuals with the lowest blood pressure in such families are those who eat the most salt, confounding the presumed positive relationship between salt intake and blood pressure in the general population.

The related disease, Bartter's syndrome, features a typically far more severe form of salt wasting with hypokalemic alkalosis that has high morbidity and mortality in the neonatal period. Affected subjects have hypercalciuria, and a fraction have nephrocalcinosis and renal failure. This disease proves to result from a failure of salt reabsorption in the renal thick ascending limb (TAL) of Henle. To date, four genes have been identified that result in related phenotypes. One of these, the Na–K–2Cl cotransporter encoded by NKCC2, mediates the entry of salt from the lumen into the epithelium of this nephron segment; homozygous loss-of-function mutations in this gene are one cause of Bartter's syndrome (Simon et al. 1996a). Potassium entering epithelial cells of the TAL via this cotransporter is recycled back into the lumen by the K^+ channel ROMK, and mutation in this channel results in a related phenotype (Simon et al. 1996b). The explanation for this phenocopy is that fluid in the TAL is high in Na and Cl but low in K. Consequently, without ROMK, there is not sufficient K to permit efficient extraction of Na and Cl. Similarly, NaCl entering the epithelial cell must return to the bloodstream by traversing the basolateral membrane. Mutation in the Cl channel encoded by CLCNKB in another subset of patients with Bartter's syndrome identifies this channel as a required component of this exit step (Simon et al. 1997). Recently, Birkenhager and colleagues (2001) have identified an accessory subunit of this channel in which mutation causes a similar phenotype. Genotype–phenotype correlations have demonstrated that much of the clinical variability seen among Bartter's patients is explained by the different genes that are mutated to produce this phenotype. For example, patients with mutations in CLCNKB or Barttin have more modest hypercalciuria and less often have nephrocalcinosis, a feature which is nearly always found among patients with mutations in NKCC2 and ROMK. Similarly, patients with mutations in ROMK typically present in the neonatal period with paradoxical hyperkalemia and evolve over time to hypokalemia, although they rarely require K supplementation to the extent that Bartter's patients with mutations in other genes do.

IMPLICATIONS

These studies have unequivocally established that alteration of net renal salt reabsorption alters blood pressure in humans, with mutations that increase salt balance raising blood pressure, and mutations that reduce it lowering blood pressure. Moreover, they have also identified specific genes whose perturbed function either increases or decreases this balance. There are a host of implications of these findings. First, although physiologic studies have identified many mediators of salt handling, it has been difficult to conclusively establish their essential roles in normal physiology in humans. Moreover, although one can estimate the fraction of salt reabsorption each pathway mediates in vivo under normal circumstances, it is very difficult to predict what the quantitative consequences of loss of each might be, owing to the potential for compensation from other mediators or redundant functions. These molecular genetic studies have firmly established the essential roles played by many elements in normal salt homeostasis. For example, the finding that mutation in CLCKB results in severe salt wasting indicates that this Cl^- channel is not merely one of many redundant Cl^- channels in the kidney, but rather that it plays a required role for normal salt homeostasis that cannot be compensated by other gene products.

In addition, these molecular genetic studies provide an improved understanding of integrated physiology. For example, although it has long been known that salt reabsorption in the TAL, DCT, and distal nephron, respectively, accounts for reabsorption of approximately 30%, 7%, and 2% of the filtered load, the capacity of these various systems to compensate for increased or decreased activities at other sites has not been clear. Thus, although ENaC normally mediates the reabsorption of only about 2% of the filtered sodium load, the loss of this function results in severely impaired salt homeostasis that cannot be adequately compensated by other salt-retaining pathways. Conversely, whereas substantially more salt reabsorption normally occurs via the Na–Cl cotransporter of the DCT, loss of this activity does not severely impair intravascular volume, as ENaC activity, augmented via aldosterone signaling, can substantially compensate for loss of cotransporter function, albeit at the expense of increased loss of K^+ and H^+. This ability to increase salt reabsorption in the distal nephron is not infinite, however, as illustrated by patients with loss of salt reabsorption in the TAL. These patients' salt wasting from the TAL vastly exceeds what can be compensated by the Na–Cl cotransporter and ENaC, with consequent severe salt wasting.

Finally, these studies also provide further insight into integrative physiology by identifying secondary effects of mutations in these salt-handling pathways. For example, the marked effects of loss-of-function Na–Cl cotransporter mutations in producing renal Mg^{++} loss and Ca^{++} retention reveal unanticipated interrelationships between handling of salt in the DCT and these divalent cations.

However, these findings are of particular relevance to understanding the pathogenesis and treatment of hypertension. These findings firmly establish the central role of the kidney and renal salt balance in the long-term determination of blood pressure in humans, and strongly validate the general approach of lowering net salt balance in the treatment of hypertension.

Moreover, this relationship between salt and blood pressure in inherited forms of hypertension prompts one to consider the known causes of acquired forms of hypertension. These, too, can in whole or in part be attributed to increased net salt balance. Aldosterone-producing adenoma is one obvious example. Another is renal artery stenosis, which leads to increased plasma renin activity, ultimately increasing salt balance via increased aldosterone secretion. The excess catechols in pheochromocytoma increase proximal renal tubular salt reabsorption and also decrease renal blood flow. Impaired salt clearance in end-stage renal disease likely accounts for the very high prevalence of hypertension among patients on dialysis. At present, one can make the case that all of the known causes of hypertension act by increasing net salt balance, either by altering renal salt handling directly, or as a secondary effect resulting either from impaired blood flow to the kidney or from impaired glomerular filtration.

These findings are not surprising, as they fit nicely into our understanding of the physiology of blood pressure regulation, with the increased intravascular volume resulting in increased cardiac output due to increased stroke volume and a rise in blood pressure according to Ohm's law. One surprising aspect, however, is that this hemodynamic pattern (i.e., increased intravascular volume with increased cardiac output) is absent among nearly all patients with long-standing hypertension. This is true even for models of pure volume overload, such as animal models of aldosterone excess. Guyton (1991) has argued persuasively that this is explained by local tissue beds matching their blood flow to their metabolic need—thus a primary increase in flow beyond metabolic need is countered by an increase in vascular resistance to flow, directing blood elsewhere. This results in increased renal perfusion and natriuresis, ultimately resulting in a hemodynamic profile of elevated blood pressure with high systemic vascular resistance and normal cardiac output.

These findings have particular impact in several clinical arenas. For example, hypertension has a prevalence that approaches 100% in the more than 300,000 patients in the U.S. on dialysis for end-stage renal disease. It is logical to propose that this high prevalence is attributable to inadequate removal of salt and water on dialysis, a consequence of efforts to reduce the number and time of dialysis sessions. Recent studies that employ more frequent dialysis permit more net salt and water removal, with dramatic reduction in the prevalence of hypertension.

These findings also have implications for new therapeutic development for the treatment of hypertension. The phenotype resulting from loss of a channel is highly predictive of the phenotype that would result from a selective antagonist of that same channel. The finding that loss-of-function mutations in a number of ion channels and cotransporters result in marked salt wasting and reduced blood pressure indicates that these channels are poised at key points in integrated physiology and are not of redundant function; these genetic findings suggest that pharmacologic antagonists of these same channels and transporters would have potent antihypertensive diuretic effects. New targets thus include the potassium channel ROMK, the chloride channel CLCNKB, and the WNK kinases. Importantly, the absence of clinical phenotypes in other organs among patients lacking these channels indicates that a specific antagonist would not have adverse effects in other organs. There are specific features of these targets that might give their antagonists particular utility. For example, a limiting aspect of treatment with antagonists of the Na–K–2Cl cotransporter is the development of marked hypokalemia. The potassium channel ROMK is used both for recycling in the TAL and for potassium secretion in the collecting duct. As a result, individuals deficient for ROMK have only mild potassium wasting. These findings suggest that ROMK antagonists might have diuretic effects like furosemide without the hypokalemic side effects.

Why is it that diuretic agents have not had better efficacy in the treatment of hypertension? One answer is that all agents that act in the TAL or the DCT activate the rennin-angiotensin system and increase salt reabsorption by the sodium channel ENaC in the collecting duct, limiting the efficacy of these agents and also producing untoward side effects that are dose limiting. ENaC itself has been believed to be largely inactive on a typical high-salt diet, owing to its regulation by aldosterone. This notion largely derives from the low efficacy of amiloride, an antagonist of this channel. The massive and sustained salt wasting among patients genetically missing this channel belies this notion, and indicates that ENaC is required for normal salt balance, even on a high-salt diet. The reason amiloride is a poor antihypertensive drug is that it acts as a competitive antagonist with sodium, and, consequently, on a high-salt diet luminal levels of drug are too low to effectively antagonize the channel. The considerations strongly suggest high efficacy of a potent ENaC antagonist; in this case, the resulting volume depletion would not result in a compensatory increase in sodium reabsorption, since ENaC is itself the means by which this compensation is achieved. Such an agent might have far greater efficacy than other antihypertensive agents, just as the HMG-CoA reductase inhibitors, which target a key step in cholesterol biosynthesis, have far greater efficacy than earlier agents that targeted other mechanisms. It is important to point out that the use of such agents need not be restricted to individuals with inherited mutations in these genes. Just as HMG-CoA reductase inhibitors are efficacious in patients who do not have mutations in the LDL receptor, one can anticipate that such agents would have efficacy in the general hypertensive population.

CONCLUSIONS

Genetic studies of humans with blood pressures at the extremes of the distribution have established that primary alteration of net renal salt reabsorption alters blood pressure in humans. The finding that the same pathway is involved in many different diseases that feature altered blood pressure suggests that altered salt handling may underlie not only rare Mendelian forms of hypertension, but essential hypertension as well. These findings have implications for the selection of medications to lower blood pressure. In addition, a number of new potential therapeutic targets have been identified for development of antihypertensive agents; these targets are validated in that the effects of selective antagonists of specific channels and cotransporters can be predicted with high likelihood from the phenotypes of individuals lacking these proteins. Several targets have been identified that may have greater efficacy and reduced side effects compared with existing agents. Such agents have the potential to provide substantial benefit in the treatment of hypertension.

ACKNOWLEDGMENTS

R.P.L. is an investigator of the Howard Hughes Medical institute. K.A.C. is supported by the National Institutes of Health Medical Scientist Training Program. D.S.G. is the recipient of a KO8 award from the National Institute of Diabetes and Digestive Kidney Diseases. This work was supported in part by the National Institutes of Health Specialized Center of Research in Hypertension.

REFERENCES

Birkenhager R., Otto E., Schurmann M.J., Vollmer M., Ruf E.M., Maier-Lutz I., Beekmann F., Fekete A., Omran H., Feldmann D., Milford D.V., Jeck N., Konrad M., Landau D., Knoers N.V., Antignac C., Sudbrak R., Kispert A., and Hildebrandt F. 2001. Mutation of BSND causes Bartter syndrome with sensorineural deafness and kidney failure. *Nat. Genet.* **29:** 310.

Chang S.S., Grunder S., Hanukoglu A., Rvsler A., Mathew P.M., Hanukoglu I., Schild L., Lu Y., Shimkets R.A., Nelson-Williams C., Rossier B.C., and Lifton R.P. 1996. Mutations in subunits of the epithelial sodium channel cause salt wasting with hyperkalemic acidosis, pseudohypoaldosteronism type 1. *Nat. Genet.* **12:** 248.

Cruz D.N., Simon D.B., Nelson-Williams C., Farhi A., Finberg K., Burleson L., Gill J.R., and Lifton R.P. 2001. Mutations in the Na-Cl cotransporter reduce blood pressure in humans. *Hypertension* **37:** 1458.

Geller D.S., Rodriguez-Soriano J., Vallo Boado A., Schifter S., Bayer M., Chang S.S., and Lifton R.P. 1998. Mutations in the mineralocorticoid receptor gene cause autosomal dominant pseudohypoaldosteronism type I. *Nat. Genet.* **19:** 279.

Geller D.S., Farhi A., Pinkerton N., Fradley M., Moritz M., Spitzer A., Meinke G., Tsai F.T., Sigler P.B., and Lifton R.P. 2000. Activating mineralocorticoid receptor mutation in hypertension exacerbated by pregnancy. *Science* **289:** 119.

Guyton A.C. 1991. Blood pressure control: Special role of the kidneys and body fluids. *Science* **252:** 1813.

Hansson J.H., Nelson-Williams C., Suzuki H., Schild L., Shimkets R., Lu Y., Canessa C., Iwasaki T., Rossier B., and Lifton R.P. 1995. Hypertension caused by a truncated epithelial sodium channel gamma subunit: Genetic heterogeneity of Liddle syndrome. *Nat. Genet.* **11:** 76.

Lifton R.P., Dluhy R.G., Powers M., Rich G.M., Cook S., Ulick S., and Lalouel J.M. 1992a. A chimaeric 11 beta-hydroxylase/aldosterone synthase gene causes glucocorticoid-remediable aldosteronism and human hypertension. *Nature* **355:** 262.

Lifton R.P., Dluhy R.G., Powers M., Rich G.M., Gutkin M., Fallo F., Gill J.R., Jr., Feld L., Ganguly A., Laidlaw J.C., et al. 1992b. Hereditary hypertension caused by chimaeric gene duplications and ectopic expression of aldosterone synthase. *Nat. Genet.* **2:** 66.

Multiple Risk Factor Intervention Trial Research Group. 1982. Multiple risk factor intervention trial. Risk factor changes and mortality results. *J. Am. Med. Assoc.* **248:** 1465.

Schild L., Lu Y., Gautschi I., Schneeberger E., Lifton R.P., and Rossier B.C. 1996. Identification of a PY motif in the epithelial Na channel subunits as a target sequence for mutations causing channel activation found in Liddle syndrome. *EMBO J.* **15:** 2381.

Shimkets R.A., Lifton R.P., and Canessa C.M. 1997. The activity of the epithelial sodium channel is regulated by clathrin-mediated endocytosis. *J. Biol. Chem.* **272:** 25537.

Shimkets R.A., Warnock D.G., Bositis C.M., Nelson-Williams C., Hansson J.H., Schambelan M., Gill J.R., Jr., Ulick S., Milora R.V., Findling J.W., Canessa C.M., Rossier B.C., and Lifton R.P. 1994. Liddle's syndrome: Heritable human hypertension caused by mutations in the beta subunit of the epithelial sodium channel. *Cell* **79:** 407.

Simon D.B., Karet F.E., Hamdan J.M., DiPietro A., Sanjad S.A., and Lifton R.P. 1996a. Bartter's syndrome, hypokalaemic alkalosis with hypercalciuria, is caused by mutations in the Na-K-2Cl cotransporter NKCC2. *Nat. Genet.* **13:** 183.

Simon D.B., Karet F.E., Rodriguez-Soriano J., Hamdan J.H., DiPietro A., Trachtman H., Sanjad S.A., and Lifton R.P. 1996b. Genetic heterogeneity of Bartter's syndrome revealed by mutations in the K+ channel, ROMK. *Nat. Genet.* **14:** 152.

Simon D.B., Nelson-Williams C., Bia M.J., Ellison D., Karet F.E., Molina A.M., Vaara I., Iwata F., Cushner H.M., Koolen M,, Gainza F.J., Gitleman H.J., and Lifton R. 1996c. Gitelman's variant of Bartter's syndrome, inherited hypokalaemic alkalosis, is caused by mutations in the thiazide-sensitive Na-Cl cotransporter. *Nat. Genet.* **12:** 24.

Simon D.B., Bindra R.S., Mansfield T.A., Nelson-Williams C., Mendonca E., Stone R., Schurman S., Nayir A., Alpay H., Bakkaloglu A., Rodriguez-Soriano J., Morales J.M., Sanjad S.A., Taylor C.M., Pilz D., Brem A., Trachtman H., Griswold W., Richard G.A., John E., and Lifton R.P. 1997. Mutations in the chloride channel gene, CLCNKB, cause Bartter's syndrome type III. *Nat. Genet.* **17:** 171.

Staub O., Dho S., Henry P., Correa J., Ishikawa T., McGlade J., and Rotin D. 1996. WWdomains of Nedd4 bind to the proline-rich PY motifs in the epithelial Na^+ channel deleted in Liddle's syndrome. *EMBO J.* **15:** 2371.

Wilson F.H., Disse-Nicodème S., Choate K.A., Ishikawa K., Nelson-Williams C., Desitter I., Gunel M., Milford D.V., Lipkin G.W., Achard J.M., Feely M.P., Dussol B., Berland Y., Unwin R.J., Mayan H., Simon D.B., Farfel Z., Jeunemaitre X., and Lifton R.P. 2001. Human hypertension caused by mutations in WNK kinases. *Science* **293:**1107.

Gene-targeting Studies of the Renin–Angiotensin System: Mechanisms of Hypertension and Cardiovascular Disease

S.B. GURLEY, T.H. LE, AND T.M. COFFMAN

Division of Nephrology, Department of Medicine, Duke University and Durham VA Medical Centers, Durham, North Carolina 27705

As a major regulator of blood pressure in humans, the renin–angiotensin system has been the subject of extensive biochemical, pharmacological, and physiological investigation. The development of efficient approaches for gene targeting in mice has allowed detailed examination of each component of this highly complex endocrine system. In addition to confirming previous concepts about the functions of this system, genetic manipulation of the renin–angiotensin system has uncovered novel pathways in development, hypertension, renal and cardiac function, and immunology. Homologous recombination in embryonic stem cells (Bronson and Smithies 1994) and pronuclear injection of transgenes (Gordon and Ruddle 1981) have both been used to create transgenic mice in which genes are "knocked-out" or genetically altered. All major components of the renin–angiotensin system have been genetically manipulated in mice, and these animals have been used to study the functions of specific candidate genes, both in physiological and pathophysiological conditions. In this paper, we review the results of experiments using these knockout mice, highlighting information revealed by these studies in understanding the physiological functions of the renin–angiotensin system.

The renin–angiotensin system is depicted in Figure 1 (Timmermans et al. 1993). Angiotensinogen is the major substrate of this enzymatic cascade and is primarily made in the liver. After its release into the circulation, angiotensinogen is cleaved by renin, which is synthesized within the juxtaglomerular apparatus of the kidney, to form the inactive decapeptide, angiotensin I. Renin is encoded by a single gene in humans, but in some mouse strains there are two renin genes. Angiotensin-converting enzyme (ACE) converts angiotensin I to angiotensin II, an octapeptide that acts as the major effector molecule of the renin–angiotensin system. Angiotensin II exerts its biological effects through its cell surface receptors, members of the G-protein-coupled family of receptors. Two classes of angiotensin II receptors have been identified: AT_1 and AT_2, with AT_1 receptors mediating most of the classically recognized effects of angiotensin II. Two subtypes of AT_1 receptors have been cloned in rodents (AT_{1A} and AT_{1B}); however, only a single isoform is believed to exist in humans. A homolog of ACE, known as ACE2, was recently identified (Donoghue et al. 2000; Tipnis et al. 2000), which metabolizes both angiotensin I and angiotensin II in vitro. ACE2 expression is more limited than conventional ACE, with highest levels of expression in kidney, heart, and testis. Gene-targeting strategies have suggested physiological functions for ACE2 in the cardiovascular system. Using genomic methods, a novel transmembrane glycoprotein, collectrin, with significant homology to ACE2, was recently discovered (Zhang et al. 2001). Collectrin is expressed in the collecting duct of the kidney, and its expression is up-regulated following partial nephrectomy. The physiological function of collectrin has not yet been identified.

ANGIOTENSINOGEN

Because it encodes the major substrate for the renin–angiotensin system, the angiotensinogen gene has been targeted for study to examine the role of this system in several areas, including development, renal function, and blood pressure regulation. A role for the renin–angiotensin system in development had been suggested by observations that renin–angiotensin system genes are

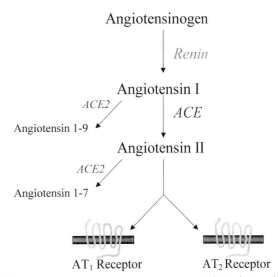

Figure 1. The renin–angiotensin system. Angiotensinogen is cleaved by renin to form angiotensin I, which is then cleaved by angiotensin converting enzyme (ACE) to form angiotensin II. The effects of angiotensin II are mediated by specific cell-surface receptors, AT_1 and AT_2, which exist in the kidney, adrenal glands, heart, and vascular smooth muscle. ACE2 is a recently identified homolog of ACE that cleaves both angiotensin I and angiotensin II to form angiotensin 1-9 and angiotensin 1-7, respectively.

highly expressed during development (Gomez et al. 1988a,b; Millan et al. 1989; Jones et al. 1990; Grady et al. 1991; Kakuchi et al. 1995) and that inhibition of ACE during pregnancy is associated with teratogenic effects in humans (Pryde et al. 1993). Surprisingly, mice completely lacking expression of the angiotensinogen gene appear normal at birth and have normal organs, including kidneys (Kim et al. 1995; Niimura et al. 1995); however, their survival in the perinatal period is reduced. Moreover, surviving adult $Agt^{-/-}$ mice have marked renal structural and vascular abnormalities (Table 1). These vascular changes are not seen in other organs (Niimura et al. 1995; Yababa et al. 1996). Davisson et al. (1997) demonstrated rescue of this abnormal phenotype in transgenic mice expressing human renin and angiotensinogen genes.

In $Agt^{-/-}$ mice, the absence of angiotensinogen leads to a significant reduction in blood pressure in conscious mice (Kim et al. 1995), substantiating the important role of the renin–angiotensin system in the normal regulation of blood pressure. To examine the effects of increased angiotensinogen levels, Kim et al. used gene targeting followed by selective breeding to generate duplication of the angiotensinogen gene locus in mice (Smithies and Kim 1994). Mice with one to four copies of Agt genes were produced, and there was a linear relationship between the number of gene copies and blood pressure. The dependence of blood pressure on angiotensinogen levels is analogous to what has been observed in humans with elevated plasma angiotensinogen levels (Jeunemaitre et al. 1992). Taken together, these studies have provided direct evidence for the role of angiotensinogen in renal development and maintenance of blood pressure.

In addition to its contribution to renal and vascular physiology as a circulating system, the renin–angiotensin system has been demonstrated to exist locally within specific tissues. One example of the importance of a local renin–angiotensin system is seen in $Agt^{-/-}$ mice (Yanai et al. 2000), where the density of the granular layer cells of the hippocampus and the blood-brain barrier are abnormal. Again, genetically modified animals have provided a novel construct for studying specific components of the renin–angiotensin system, leading to the discovery of new functions for previously identified molecules.

RENIN

In contrast to most other species which only have a single renin gene, certain strains of wild and laboratory mice, including 129Sv, have two closely linked genes that encode renin (Piccini et al. 1982; Dickinson et al. 1984). Loss of either one of these genes, $Ren-1$ or $Ren-2$, does not cause an abnormal phenotype (Sharp et al. 1996), perhaps due to compensation by the other remaining renin gene. Mice completely lacking renin, however, have low blood pressures and marked renal structural abnormalities, similar to those seen in $Agt^{-/-}$ mice (Yanai et al. 2000), indicating a dependence of these phenotypes on the generation of angiotensin II (Table 1). Additionally, physiological studies of these animals reveal that these renin-deficient mice are polydipsic and polyuric, which reflects their impaired urinary concentrating function. The renin-deficient mice do not have the abnormal granular cell layers in the hippocampus seen in $Agt^{-/-}$ mice, which suggests a distinct function for angiotensinogen in the central nervous system.

ANGIOTENSIN-CONVERTING ENZYME

ACE enzyme exists as two forms in humans: a somatic form that is expressed on cellular membranes in vascular endothelium and renal proximal tubular epithelium and a testicular form that is important for fertility. The tissue-bound form of the enzyme can be cleaved to produce a soluble form of the enzyme that circulates in plasma (Beldent et al. 1993). The key role of ACE in generating the vasoactive peptide angiotensin II has made it an attractive target for pharmacological manipulation. The importance of this pathway in the pathogenesis of disease is apparent from the numerous clinical trials showing that inhibition of ACE is beneficial in several human diseases (CONSENSUS Trial Study Group 1987; Pfeffer et al. 1988; SOLVD Invesigators 1991; Lewis et al. 1993; Viberti et al. 1994; Yusuf et al. 2000).

The phenotype of $Ace^{-/-}$ mice is similar to that of $Agt^{-/-}$ mice, with prominent hypotension, vascular hyperplasia, atrophy of the renal papillae, and impaired renal function (Table 1) (Krege et al. 1995; Esther et al. 1996; Tian et al. 1997). These similarities further underscore the importance of angiotensin II deficiency in this phenotype. Fertility is also impaired in male ACE-deficient mice, despite otherwise normal mating behavior, testis histology, sperm count, and morphology. Because infertility has not been observed in other renin–angiotensin system mutant lines, this suggests that the contribution of ACE to male fertility is not related to metabolism of angiotensin I.

In 1997, Bernstein et al. employed gene-targeting experiments to further investigate the physiological role of endothelial expression of ACE (Esther et al. 1997). Mice were generated that expressed the circulating without the endothelial form of tissue ACE. These mice that were de-

Table 1. Phenotypes of Mice with Targeted Mutations of Renin–Angiotension System Genes

Mutation	Postnatal mortality	Blood pressure	Renal vascular lesions	Atrophy of the renal papilla
$Agt^{-/-}$	+++	decreased	+++	+++
$Ren^{-/-}$	+++	decreased	+++	+++
$Ace^{-/-}$	+++	decreased	+++	+++
$Agtr1a^{-/-}/Agtr1b^{-/-}$	+++	decreased	+++	+++
$Agtr1a^{-/-}$	+	decreased	+	+
$Agtr1b^{-/-}$	no	normal	no	no
$Agtr2^{-/-}$	no	increased?	no	no

ficient in tissue ACE had hypotension and abnormally thickened renal arteriolar walls, similar to mice that were globally deficient in ACE. However, they did not develop characteristic atrophy of the renal papilla. This study suggested that the tissue and plasma forms of ACE differ in their contribution to the regulation of blood pressure and kidney structure. To further investigate the role of tissue-specific expression of ACE, a mouse line was engineered in which the transcriptional control of the *Ace* gene was placed under control of the albumin promoter (Cole et al. 2002). These mice express ACE only in hepatocytes. Endothelial or tissue ACE is completely absent in these animals, but plasma levels of ACE are 87 times normal. Despite the absence of tissue ACE, these mice were indistinguishable from their wild-type littermates, suggesting that elevated levels of plasma ACE can compensate for the absence of the endothelial enzyme.

Molecular variation in the expression of ACE may provide an explanation for the predisposition to hypertension and cardiovascular disease in certain populations. Insertion and deletion polymorphisms of the ACE gene have been described in humans. Homozygotes for the deletion allele have the highest level of serum ACE, homozygotes for the insertion allele have low levels of the enzyme, and heterozygotes have intermediate levels (Rigat et al. 1990). A number of studies have suggested association between increased risk for cardiovascular disease and the deletion allele (Doria et al. 1994; Ludwig et al. 1995; Markus et al. 1995; Missouris et al. 1996; Singer et al. 1996; Agerholm-Larsen et al. 1997; Kario et al. 1997). To determine the direct effect of ACE on blood pressure, Krege et al. (1997) used gene duplication to create mice with one, two, or three functional copies of the enzyme. Despite a linear increase in enzyme activity with gene copy number, there was no significant difference in blood pressure in these animals. These experiments in mice indicate that unlike angiotensinogen, increased levels of ACE activity do not affect resting blood pressure. Compensation by other components of the renin–angiotensin system could account for this lack of difference. Increased levels of ACE did lead to decreased heart rates, cardiac weights, and renal tubulointerstitial volume in these mice, suggesting that other physiological factors are indeed affected by altered gene expression.

HOMOLOGS OF ACE

On the basis of genomic strategies, homologs of ACE have recently been described (Donoghue et al. 2000; Tipnis et al. 2000; Zhang et al. 2001). One of these, ACE2 (also referred to as ACEH in the literature) shares more than 40% homology at the protein level with the catalytic domains of ACE, and its expression appears to be limited to heart, kidney, and testis. Similar to ACE, ACE2 is expressed on endothelial surfaces but also exists as a circulating form. The catalytic activity of ACE2, however, differs from that of ACE. Whereas ACE cleaves the dipeptide histidine-leucine from angiotensin I to form angiotensin II, ACE2 metabolizes angiotensin I to form a nonapeptide (angiotensin 1-9) by cleaving the carboxy-terminal leucine residue. Moreover, the formation of angiotensin 1-9 was not inhibited by ACE inhibitors (Donoghue et al. 2000). ACE2, through its carboxypeptidase actions, also generates angiotensin 1-7 from angiotensin II (Fig. 1) (Tipnis et al. 2000). ACE2 also possesses enzymatic activity against neurotensin, kinetensin, and des-Arg bradykinin but not against bradykinin or the enkephalins (Turner and Hooper 2002).

Because it is expressed in the heart and was originally cloned from a cDNA library derived from ventricular tissue from a patient with heart failure, it was hypothesized that ACE2 may play a role in cardiovascular function. To further investigate the physiological roles of ACE2, lines of mice with targeted disruption of the *Ace2* gene have been generated (Crackower et al. 2002; T.M. Coffman et al., in prep.). The phenotypes of these animals are divergent. One of the lines of ACE2-deficient mice develops left ventricular failure. These animals were found, by echocardiography under anesthesia, to have reduced cardiac contractility. The defects in cardiac function worsened with age and were more pronounced in male than female null mice. These animals had lower blood pressures and elevated angiotensin II levels, presumably due to their impaired cardiac function. The congestive heart failure phenotype was rescued in double mutant mice that are deficient in both ACE and ACE2.

In a distinct line of ACE2-deficient mice, we have reported a very different phenotype. Using echocardiography in conscious animals, we found no evidence of left ventricular failure in male mice more than one year old. In contrast, we found that these animals have elevated blood pressures and increased susceptibility to angiotensin II-induced hypertension. The phenotypic differences observed in these two mouse lines are difficult to explain. It is possible that the use of anesthesia affected the assessments of cardiac function in the mice described by Crackower et al. (2002). Genetic background differences may also contribute to the dramatic difference in phenotypes observed. Although both lines were created from embryonic stem cells derived from 129 strain mice, and chimeras were back-crossed onto C57BL/6, the relative composition of background genes might differ between the two mutant lines and account for some of the differences if true modifier genes exist.

A homolog of ACE2, collectrin, was identified and cloned from mouse kidneys following partial nephrectomy. The protein is a novel transmembrane glycoprotein expressed in renal collecting tubules that shares considerable homology with ACE2 but not ACE (Zhang et al. 2001). Given its presence in injured renal tissue, collectrin may have a role in renal hypertrophy, perhaps as part of a local up-regulation of a more extensive biochemical cascade that includes this ACE2 homolog. Gene-targeting studies may help determine the role of this molecule in normal physiology and in the pathogenesis of renal disease.

ANGIOTENSIN II RECEPTORS

The actions of the major biologically active peptide generated by the renin–angiotensin system, angiotensin

II, are mediated via specific cell-surface receptors. These receptors can be divided into two pharmacologically distinct classes: AT_1 and AT_2. AT_1 receptors are thought to mediate most of the traditional functions of angiotensin II. Although in humans there appears to be only a single AT_1 gene, rodents possess two subtypes of the AT_1 receptor: AT_{1A} and AT_{1B}. In most tissues, AT_{1A} receptors are more highly expressed than AT_{1B} (Timmermans et al. 1992; Burson et al. 1994; Gasc et al. 1994; Llorens-Cortes et al. 1994). These two isoforms are the products of two distinct genes (Yoshida et al. 1992).

Homologous recombination has been used to generate mice deficient in the type 1_A receptor for angiotensin II (Ito et al. 1995; Sugaya et al. 1995; Matsusaka et al. 1996; Oliverio et al. 1998a). Unexpectedly, absence of the major AT_1 receptor subtype did not recapitulate the phenotype seen with angiotensinogen, ACE, or renin deficiencies (Table 1). Their renal structure was intact except for hypertrophy of the juxtaglomerular apparatus and modest mesangial expansion. They lacked the atrophy of the renal inner medulla that was seen in $Agt^{-/-}$ and $Ace^{-/-}$ mice. However, these animals did have a renal concentration defect and a stepwise reduction in resting blood pressure associated with loss of $Agtr1a$ expression, confirming an important role for the AT_{1A} receptor in the regulation of blood pressure (Ito et al. 1995; Matsusaka et al. 1996; Oliverio et al. 1998a).

Less was known about the role of the AT_{1B} receptor type that exists in mice. AT_{1B}-deficient mice have phenotypes indistinguishable from those of their wild-type controls (Benetos and Safar 1996; Tsuchida et al. 1998). Studies of the $Agtr1a^{-/-}$ mice, however, provided evidence for a role of the AT_{1B} receptor type in blood pressure control and vascular response to angiotensin II (Oliverio 1997). After pretreatment with enalapril to reduce endogenous angiotensin II levels, $Agtr1a^{-/-}$ mice displayed a modest pressor response to angiotensin II infusion that was subsequently ameliorated by AT_1 receptor antagonists. The response to angiotensin II was not affected by sympatholytic agents and suggests a direct effect of AT_{1B} receptors on vascular tone. Chronic administration of losartan caused a further lowering of systolic blood pressures in AT_{1A}-deficient mice, implying that in the absence of the dominant angiotensin II receptor, AT_{1B} receptors contribute to the regulation of blood pressure. A non-redundant role for the two subtypes of AT_1 receptors in the brain was defined by examining the differential response to angiotensin II in both $Agtr1a^{-/-}$ and $Agtr1b^{-/-}$ mice (Davisson et al. 2000). Angiotensin II mediates its CNS actions on blood pressure regulation primarily via the AT_{1A} receptors, whereas the dipsogenic actions of angiotensin II are mediated by AT_{1B} receptors. AT_1 blockade by losartan eliminated the drinking response in wild-type, $Agtr1a^{-/-}$, and $Agtr1b^{-/-}$ mice, indicating that non-AT_1 receptors are not involved in the dipsogenic actions of angiotensin II.

Because the genes are on separate chromosomes, mice completely deficient in AT_1 receptors could be generated by intercrossing the $Agtr1a^{-/-}$ and $Agtr1b^{-/-}$ mice (Oliverio et al. 1998b; Tsuchida et al. 1998). The phenotype of these double knockouts is virtually identical to the angiotensinogen- and ACE-deficient mice; they have significantly reduced postnatal survival, very low baseline blood pressure, typical renal vascular changes, and marked hypoplasia of the renal papilla (Table 1). Studies with mice completely lacking in receptors for angiotensin II have demonstrated the importance of both the AT_1 receptor and its agonist in renal development and blood pressure regulation.

To study the effect of genetic variants that increase AT_1 receptor expression on blood pressure, transgenic mice with targeted duplication of the $Agtr1a$ gene were generated by our laboratory (T.H. Le et al., in prep.). On an inbred 129/Sv background, mice with three and four copies of the $Agtr1a$ locus were generated and had increased AT_{1A} mRNA expression and binding of ^{125}I-labeled angiotensin II according to gene copy number. Survival and weights were similar to those of wild-type animals. However, pressor response to angiotensin II was blunted in the four-copy mice. There were also gender differences in the effects of gene duplication on blood pressure. In male mice, there was no correlation between blood pressure and $Agtr1a$ gene copy number. However, in females, there was a highly significant positive correlation between blood pressure and $Agtr1a$ gene expression. Thus, the impact of increased AT_{1A} receptor expression on the phenotypes of these mice is complex.

Gene-targeting has also been useful in clarifying the functions of the type 2 angiotensin II receptor. AT_2 receptors are abundantly expressed during fetal development, but their expression decreases in most mature tissues (Millan et al 1989; Grady et al. 1991; Szombathy et al. 1998). However, it has been suggested that AT_2 receptors may be up-regulated in human failing hearts in association with a concomitant down-regulation of AT_1 receptors (Asano et al. 1997; Haywood et al. 1997). Mice lacking expression of AT_2 receptors have behavioral changes with decreased spontaneous movements, rearing activity, and impaired drinking responses (Hein et al. 1995; Ichiki et al. 1995). These mice also manifest increased sensitivity to the vascular actions of angiotensin II, and one of these AT_2-deficient mouse lines had elevated baseline blood pressure and heart rate.

To further investigate the actions of the AT_2 receptor in cardiac myocytes, mice overexpressing the AT_2 receptor under the control of a cardiac-specific promoter were generated (Masaki et al. 1998). These mice had decreased sensitivity to AT_1-mediated pressor and chronotropic actions. Pressor responses to angiotensin II were also attenuated in these AT_2 transgenic mice, and this attenuation was completely reversed after pretreatment with a specific AT_2 receptor blocker. Functions for the AT_2 receptor in blood pressure regulation were also suggested in studies of AT_{1A}/AT_{1B} knockouts. In these animals, administration of an ACE inhibitor caused a paradoxical increase in blood pressure (Oliverio et al. 1998b). This effect may be due to initiation of signaling through the AT_2 receptor. It appears that AT_2 receptors can form heterodimers with AT_1 and as a result cause inhibition of AT_1 stimulation. This antagonism of the AT_1 receptor was reversed by decreased AT_2 expression or heterodimerization inhibition (AbdAlla et al. 2001). Taken together, these studies using transgenic animals support a role for the AT_2 receptor in negatively modulating the actions of the AT_1 receptor. The mechanism for this modulation may occur through nitric oxide production, as basal levels of bradykinin and cGMP are lower in the null mice. Alternatively, ligand and/or receptor interactions may also play a role.

CONCLUSIONS

The physiological importance of individual genes in the renin–angiotensin system has been examined through genetic manipulation of mice. As described, these studies demonstrate the utility of transgenic approaches in physiological experiments. Further refinement and broader application of these techniques should allow additional insights into the functions of the renin–angiotensin system. These discoveries may lead to novel approaches to therapies for kidney and heart disease.

ACKNOWLEDGMENT

The authors acknowledge support from National Institutes of Health grants HL-56122 and HL-49277.

REFERENCES

AbdAlla S., Lother H., Abdel-tawab A.M., and Quitterer U. 2001. The angiotensin II AT_2 receptor is an AT_1 receptor antagonist. *J. Biol. Chem.* **276:** 39721.

Agerholm-Larsen B., Nordestgaard B.G., Steffensen R., Sorensen T.I., Jensen G., and Tybjaerg-Hansen A. 1997. ACE gene polymorphism: Ischemic heart disease and longevity in 10,150 individuals. A case-reference and restrospective cohort study based on the Copenhagen City Heart Study. *Circulation* **95:** 2358.

Asano K., Dutcher D.L., Port J.D., Minobe W.A., Tremmel K.D., Roden R.L., Bohlmeyer T.J., Bush E.W., Jenkin M.J., Abraham W.T., Raynolds M.V., Zisman L.S., Perryman M.B., and Bristow M.R. 1997. Selective downregulation of the angiotensin II AT1-receptor subtype in failing human ventricular myocardium (see comments) (erratum in *Circulation* [1997] **96:** 4435). *Circulation* **95:** 1193.

Beldent V., Michaud A., Wei L., Chauvet M.T., and Corvol P. 1993. Proteolytic release of human angiotensin-converting enzyme. Localization of the cleavage site. *J. Biol. Chem.* **268:** 26428.

Benetos A. and Safar M.E. 1996. Aortic collagen, aortic stiffness, and AT1 receptors in experimental and human hypertension. *Can. J. Physiol. Pharmacol.* **74:** 862.

Bonnardeaux A., Davies E., Jeunemaitre X., Fery I., Charru A., Clauser E., Tiret L., Cambien F., Corvol P., and Soubrier F. 1994. Angiotensin II type 1 receptor gene polymorphisms in human essential hypertension. *Hypertension* **24:** 63.

Bronson S. and Smithies O. 1994. Altering mice by homologous recombination using embryonic stem cells. *J. Biol. Chem.* **269:** 27155.

Burson J.M., Aguilera G., Gross K.W., and Sigmund C.D. 1994. Differential expression of angiotensin receptor 1A and 1B in mouse. *Am. J. Physiol.* **267:** E260.

Cole J., Quach D.L., Sundaram K., Corvol P., Capecchi M.R., and Bernstein K.E. 2002. Mice lacking endothelial angiotensin-converting enzyme have a normal blood pressure. *Circ. Res.* **90:** 87.

CONSENSUS Trial Study Group. 1987. Effects of enalapril on mortality in severe congestive heart failure. Results of the Cooperative North Scandinavian Enalapril Survival Study (CONSENSUS). The CONSENSUS Trial Study Group. *N. Engl. J. Med.* **316:** 429.

Crackower M.A., Sarao R., Oudit G.Y., Yagil C., Kozieradzki I., Scanga S.E., Oliveira-dos-Santos A.J., da Costa J., Zhang L., Pei Y., Scholey J., Ferrario C.M., Manoukian A.S., Chappell M.C., Backx P.H., Yagil Y., and Penninger J.M. 2002. Angiotensin-converting enzyme 2 is an essential regulator of heart function. *Nature* **417:** 822.

Davisson R.L., Oliverio M.I., Coffman T.M., and Sigmund C.D. 2000. Divergent functions of angiotensin II receptor isoforms in the brain. *J. Clin. Invest.* **106:** 103.

Davisson R.L., Kim H.S., Krege J.H., Lager D.J., Smithies O., and Sigmund C.D. 1997. Complementation of reduced survival, hypotension, and renal abnormalities in angiotensinogen-deficient mice by the human renin and human angiotensinogen genes. *J. Clin. Invest.* **99:** 1258.

Dickinson D.P., Gross K.W., Piccini N., and Wilson C.M. 1984. Evolution and variation of renin genes in mice. *Genetics* **108:** 651.

Donoghue M., Hsieh F., Baronas E., Godbout K., Gosselin M., Stagliano N., Donovan M., Woolf B., Robison K., Jeyaseelan R., Breitbart R.E., and Acton S. 2000. A novel angiotensin-converting enzyme-related carboxypeptidase (ACE2) converts angiotensin I to angiotensin 1-9. *Circ. Res.* **87:** E1.

Doria A., Warram J.H., and Krolewshi A.S. 1994. Genetic predisposition to diabetic nephropathy. Evidence for a role of the angiotensin I-converting enzyme gene. *Diabetes* **43:** 690.

Esther C.R., Howard T.E., Marino E.M., Goddard J.M., Capecchi M.R., and Bernstein K.E. 1996. Mice lacking angiotensin-converting enzyme have low blood pressure, renal pathology, and reduced male fertility. *Lab. Invest.* **74:** 953.

Esther C.R., Marino E.M., Howard T.E., Machada A., Corvol P., Capecchi M.R., and Bernstein K.E. 1997. The critical role of tissue angiotensin-converting enzyme as revealed by gene targeting in mice. *J. Clin. Invest.* **99:** 2375.

Gasc J.M., Shanmugam S., Sibony M., and Corvol P. 1994. Tissue-specific expression of type 1 angiotensin II receptor subtypes. An in situ hybridization study. *Hypertension* **24:** 531.

Gomez R.A., Lynch K.R., Chevalier R.L., Wilfong N., Everett A., Carey R.M., and Peach M.J. 1988a. Renin and angiotensinogen gene expression in maturing rat kidney. *Am. J. Physiol.* **254:** F582.

Gomez R.A., Lynch K.R., Chevalier R.L., Everett A.D., Johns D.W., Wilfong N., Peach M.J., and Carey R.M. 1988b. Renin and angiotensinogen gene expression and intrarenal renin during ACE inhibition. *Am. J. Physiol.* **254:** F900.

Gordon J.W. and Ruddle F.H. 1981. Integration and stable germ line transmission of genes injected into mouse pronuclei. *Science* **214:** 1244.

Grady E.F., Sechi L.A., Griffin C.A., Schambelan M., and Kalinyak J.E. 1991. Expression of AT2 receptors in the developing rat fetus. *J. Clin. Invest.* **88:** 921.

Haywood G.A., Gullestad L., Katsuya T., Hutchinson H.G., Pratt R.E., Horiuchi M., and Fowler M.B. 1997. AT1 and AT2 angiotensin receptor gene expression in human heart failure. *Circulation* **95:** 1201.

Hein L., Barsh G.S., Pratt R.E., Dzau V.J., and Kobilka B.K. 1995. Behavioural and cardiovascular effects of disrupting the angiotensin II type-2 receptor in mice (erratum in *Nature* [1996] **380:** 366). *Nature* **377:** 744.

Ichiki T., Labosky P.A., Shiota C., Okuyama S., Imagawa Y., Fogo A., Niimura F., Ichikawa I., Hogan B.L., and Inagami T. 1995. Effects on blood pressure and exploratory behaviour of mice lacking angiotensin II type-2 receptor. *Nature* **377:** 748.

Ito M., Oliverio M.I., Mannon P.J., Best F., Maeda N., Smithies O., and Coffman T.M. 1995. Regulation of blood pressure by the type 1A angiotensin II receptor gene. *Proc. Natl. Acad. Sci.* **92:** 3521.

Jeunemaitre X., Soubrier F., Kotelevtsev Y.V., Lifton R.P., Williams C.S., Charru A., Hunt S.C., Hopkins P.N., Williams R.R., and Lalouel J.M., et al. 1992. Molecular basis of human hypertension: Role of angiotensinogen. *Cell* **71:** 169.

Jones C.A., Sigmund C.D., McGowan R.A., Kane-Haas C.M., and Gross K.W. 1990. Expression of murine renin genes during fetal development. *Mol. Endocrinol.* **4:** 375.

Kakuchi J., Ichiki T., Kiyama S., Hogan B.L., Fogo A., Inagami T., and Ichikawa I. 1995. Developmental expression of renal angiotensin II receptor genes in the mouse. *Kidney Int.* **47:** 140.

Kario K., Kanai N., Nishiuma S., Fujii T., Saito K., Matsuo T., Matsuo M., and Shimada K. 1997. Hypertensive nephropathy and the gene for angiotensin-converting enzyme. *Arterioscler. Thromb. Biol.* **17:** 252.

Kim H.S., Krege J.H., Kluckman K.D., Hagaman J.R., Hodgin J.B., Best C.F., Jennette J.C., Coffman T.M., Maeda N., and Smithies O. 1995. Genetic control of blood pressure and the angiotensinogen locus. *Proc. Natl. Acad. Sci.* **92:** 2735.

Krege J.H., Kim H.S., Moyer J.S., Jennette J.C., Peng L., Hiller S.K., and Smithies O. 1997. Angiotensin-converting enzyme gene mutations, blood pressures, and cardiovascular homeostasis. *Hypertension* **29:** 150.

Krege J.H., John S.W., Langenbach L.L., Hodgin J.B., Hagaman J.R., Bachman E.S., Jennette J.C., O'Brien D.A., and Smithies O. 1995. Male-female differences in fertility and blood pressure in ACE-deficient mice. *Nature* **375:** 146.

Lewis E.J., Hunsicker L.G., Bain R.P., and Rohde R.D. 1993. The effect of angiotensin-converting enzyme inhibitor on diabetic nephropathy. *N. Engl. J. Med.* **329:** 1456.

Llorens-Cortes C., Greenberg B., Huang H., and Corvol P. 1994. Tissular expression and regulation of type 1 angiotensin II receptor subtypes by quantitative reverse transcriptase-polymerase chain reaction analysis. *Hypertension* **24:** 538.

Ludwig E., Corneli P.S., Anderson J.L., Marshall H.W., Lalouel J.M., and Ward R.H. 1995. Angiotensin-converting enzyme gene polymorphism is associated with myocardial infarction but not with development of coronary stenosis (see comments). *Circulation* **91:** 2120.

Markus H.S., Barley J., Lunt R., Bland J.M., Jeffrey S., Carter N.D., and Brown M.M. 1995. Angiotensin-converting enzyme gene deletion polymorphism. A new risk factor for lacunar stroke but not carotid atheroma (see comments). *Stroke* **26:** 1329.

Masaki H., Kurihara T., Yamaki A., Inomata N., Nozawa Y., Mori Y., Murasawa S., Kizima K., Maruyama K., Horiuchi M., Dzau V.J., Takahashi H., Iwasaka T., Inada M., and Matsubara H. 1998. Cardiac-specific overexpression of angiotensin II AT_2 receptor causes attenuated response to AT_1 receptor-mediated pressor and chronotropic effects. *J. Clin. Invest.* **101:** 527.

Matsusaka T., Nishimura H., Utsunomiya H., Kakuchi J., Nimura F., Inagami T., Fogo A., and Ichikawa I. 1996. Chimeric mice carrying 'regional' targeted deletion of the angiotensin type 1A receptor gene. Evidence against the role for local angiotensin in the in vivo feedback regulation of renin synthesis in juxtaglomerular cells. *J. Clin. Invest.* **98:** 1867.

Millan M.A., Carvallo P., Izumi S., Zemel S., Catt K.J., and Aguilera G. 1989. Novel sites of expression of functional angiotensin II receptors in the late gestation fetus. *Science* **244:** 1340.

Missouris C.G., Barley J., Jeffery S., Carter N.D., Singer D.R., and MacGregor G.A. 1996. Genetic risk for renal artery stenosis: Association with deletion polymorphism in angiotensin I-converting enzyme gene. *Kidney Int.* **49:** 534.

Niimura F., Labosky P.A., Kakuchi J., Okubo S., Yoshida H., Oikawa T., Ichiki T., Naftilan A.J., Fogo A., and Inagami T., et al. 1995. Gene targeting in mice reveals a requirement for angiotensin in the development and maintenance of kidney morphology and growth factor regulation. *J. Clin. Invest.* **96:** 2947.

Oliverio M.I., Best C.F., Kim H-S., Arendshorst W.J., Smithies O., and Coffman T.M. 1997. Angiotension II responses in AT1A receptor-deficient mice: A role for AT1B receptors in blood pressure regulation. *Am. J. Physiol.* **272:** f515.

Oliverio M.I., Madsen K., Best C.F., Ito M., Maeda N., Smithies O., and Coffman T.M. 1998a. Renal growth and development in mice lacking AT_{1A} receptors for angiotensin II. *Am. J. Physiol.* **274:** F43.

Oliverio M.I., Kim H., Ito M., Le T., Audoly L., Best C.F., Hiller S., Kluckman K., Maeda N., Smithies O., and Coffman T.M. 1998b. Reduced growth, abnormal kidney structure, and type 2 (AT2) angiotensin receptor-mediated blood pressure regulation in mice lacking both AT1A and AT1B receptors for angiotensin II. *Proc. Natl. Acad. Sci.* **95:** 15496.

Pfeffer M.A., Lamas G.A., Vaughn D.E., Parisi A.F., and Braunwald E. 1988. Effect of captopril on progressive ventricular dilatation after anterior myocardial infarction. *N. Engl. J. Med.* **319:** 80.

Piccini N., Knopf J.L., and Gross K.W. 1982. A DNA polymorphism, consistent with gene duplication, correlates with high renin levels in the mouse submaxillary gland. *Cell* **30:** 205.

Pryde P.G., Sedman A.B., Nugent C.E., and Barr M., Jr. 1993. Angiotensin-converting enzyme inhibitor fetopathy. *J. Am. Soc. Nephrol.* **3:** 1575.

Rigat B., Hubert C., Alhenc-Gelas F., Cambien F., Corvol P., and Soubrier F. 1990. An insertion/deletion polymorphism in the angiotensin I-converting enzyme gene accounting for half the variance of serum enzyme levels. *J. Clin. Invest.* **86:** 1343.

Sharp M.G., Fettes D., Brooker G., Clark A.F., Peters J., Fleming S., and Mullins J.J. 1996. Targeted inactivation of the Ren-2 gene in mice. *Hypertension* **28:** 1126.

Singer D.R., Missouris C.G., and Jeffery S. 1996. Angiotensin-converting enzyme gene polymorphism. What to do about all the confusion (editorial) (see comments). *Circulation* **94:** 236.

Smithies O. and Kim H.S. 1994. Targeted gene duplication and disruption for analyzing quantitative genetic traits in mice. *Proc. Natl. Acad. Sci.* **91:** 3612.

SOLVD Investigators. 1991. Effect of enalapril on survival in patients with reduced left ventricular ejection fraction and congestive heart failure. The SOLVD Investigators. *N. Engl. J. Med.* **325:** 303.

Sugaya T., Nishimatsu S., Tanimoto K., Takimoto E., Yamagishi T., Imamura K., Goto S., Imaizumi K., Hisada Y., and Otsuka A., et al. 1995. Angiotensin II type 1a receptor-deficient mice with hypotension and hyperreninemia. *J. Biol. Chem.* **270:** 18719.

Szombathy T., Szalai C., Katalin B., Palicz T., Romics L., and Csaszar A. 1998. Association of angiotensin II type 1 receptor polymorphism with resistant essential hypertension. *Clin. Chim. Acta* **269:** 91.

Tian B., Meng Q.C., Chen Y.F., Krege J.H., Smithies O., and Oparil S. 1997. Blood pressures and cardiovascular homeostasis in mice having reduced or absent angiotensin-converting enzyme gene function. *Hypertension* **30:** 128.

Timmermans P., Chiu A., Herblin W., Wong P., and Smith R. 1992. Angiotensin II receptor subtypes. *Am. J. Hypertens.* **5:** 406.

Timmermans P.B., Wong P.C., Chiu A.T., Herblin W.F., Benfield P., Carini D.J., Lee R.J., Wexler R.R., Saye J.A., and Smith R.D. 1993. Angiotensin II receptors and angiotensin II receptor antagonists. *Pharmacol. Rev.* **45:** 205.

Tipnis S.R., Hooper N.M., Hyde R., Karran E., Christie G., and Turner A.J. 2000. A human homolog of angiotensin-converting enzyme. Cloning and functional expression as a captopril-insensitive carboxypeptidase. *J. Biol. Chem.* **275:** 33238.

Tsuchida S., Matsusaka T., Chen X., Okubo S., Niimura F., Nishimura H., Fogo A., Utsunomiya H., Inagami T., and Ichikawa I. 1998. Murine double nullizygotes of the angiotensin type 1A and 1B receptor genes duplicate severe abnormal phenotypes of angiotensinogen nullizygotes. *J. Clin. Invest.* **101:** 755.

Turner A.J. and Hooper N.M. 2002. The angiotensin-converting enzyme gene family: Genomics and pharmacology. *Trends Pharmacol. Sci.* **23:** 177.

Viberti G., Mogensen C.E., Groop L.C., and Pauls J.F. 1994. Effect of captopril on progression to clinical proteinuria in patients with insulin-dependent diabetes mellitus and microalbuminuria. *J. Am. Med. Assoc.* **271:** 275.

Yababa M., Umemura S., Kihara M., Tamara K., Nyui N., Ishigami T., Ishii M., Kiuchi Y., Yagami K., Tanimoto K.,

Sugiyama F., Fukamizu A., and Murakami K. 1996. Altered development of vasculature in the kidney but not in the other main organs in angiotensinogen-deficient mice. *Hypertension* **28:** 543A.

Yanai K., Saito T., Kakinuma Y., Kon Y., Hirota K., Taniguchi-Yanai K., Nishijo N., Shigematsu Y., Horiguchi H., Kasuya Y., Sugiyama F., Yagami K., Murakami K., and Fukamizu A. 2000. Renin-dependent cardiovascular functions and renin-independent blood-brain barrier functions revealed by renin-deficient mice. *J. Biol. Chem.* **275:** 5.

Yoshida H., Kakuchi J., Guo D.F., Furuta H., Iwai N., van der Meer-de Jong R., Inagami T., and Ichikawa I. 1992. Analysis of the evolution of angiotensin II type 1 receptor gene in mammals (mouse, rat, bovine and human). *Biochem. Biophys. Res. Commun.* **186:** 1042.

Yusuf S., Sleight P., Pogue J., Bosch J., Davies R., and Dagenais G. 2000. Effects of an angiotensin-converting enzyme inhibitor, ramipril, on cardiovascular events in high-risk patients. The Heart Outcomes Prevention Evaluation Study Investigators. *N. Engl. J. Med.* **342:** 145.

Zhang H., Wada J., Hida K., Tsuchiyama Y., Hiragushi K., Shikata K., Wang H., Lin S., Kanwar Y.S., and Makino H. 2001. Collectrin, a collecting duct-specific transmembrane glycoprotein, is a novel homolog of ACE2 and is developmentally regulated in embryonic kidneys. *J. Biol. Chem.* **276:** 17132.

Modulation of Endothelial NO Production by High-density Lipoprotein

C. MINEO AND P.W. SHAUL

Department of Pediatrics, University of Texas Southwestern Medical Center, Dallas, Texas 75390-9063

Atherosclerosis is the primary cause of cardiovascular morbidity and mortality, and hypercholesterolemia is frequently a key contributing factor. During the early stages of hypercholesterolemia-induced vascular disease there is a dramatic decrease in bioavailable endothelium-derived nitric oxide (NO), which is a potent vasodilator with multiple additional cardiovascular functions (Vane et al. 1990). The NO is generated by the endothelial isoform of nitric oxide synthase (eNOS) upon the conversion of L-arginine to L-citrulline, and eNOS activity is modulated by agonists of diverse G-protein-coupled cell surface receptors and by physical stimuli such as hemodynamic shear stress (Shaul 2002). In the initial phase of hypercholesterolemia, there is exclusive impaired responsiveness to receptor-dependent stimuli, such as acetylcholine (ACh), whereas responsiveness to receptor-independent stimuli such as the calcium ionophore A23187 is not altered. As the disease progresses, there is nonspecific inhibition of NO bioavailability that is at least partly due to enhanced inactivation of NO by superoxide anions (Flavahan 1992; Harrison 1994; Cohen 1995). These processes result in increased neutrophil adherence to the endothelium, thereby comprising key components of the pathogenesis of atherosclerosis (Lefer and Ma 1993). It is also known that NO deficiency enhances smooth muscle cell proliferation and platelet aggregation and adhesion (Shaul 2002). In vivo evidence of these mechanisms includes studies in rabbit models of hypercholesterolemia in which the chronic inhibition of NO synthesis causes marked acceleration of the development of vascular dysfunction and intimal lesions (Cayatte et al. 1994; Naruse et al. 1994). In addition, mice lacking apoE, which develop spontaneous atherosclerosis and display attenuated NO-mediated vasodilation, have more rapid progression of atherosclerosis when subjected to either long-term NOS antagonism or genetic eNOS deficiency (Yang et al. 1999; Kauser et al. 2000; Kuhlencordt et al. 2001). Thus, multiple lines of investigation indicate that NO is atheroprotective, and that NO deficiency is critically involved in the pathogenesis of hypercholesterolemia-induced vascular disease.

From a clinical standpoint, many epidemiologic studies have evaluated the relative risk of atherosclerosis and coronary artery disease, and the overall consensus is that circulating levels of total high-density lipoprotein (HDL) and its principal protein component, apolipoprotein A-I (apoA-I), are major negative predictors of disease (Gordon and Rifkind 1989; Zambon and Hokanson 1998; Fidge 1999). HDL is one of the primary lipoproteins involved in cholesterol metabolism, and it classically serves as a shuttle removing free cholesterol from peripheral tissues for transfer to target cells or to other lipoproteins. The principal organ to which HDL delivers cholesterol is the liver, and this occurs through a process known as reverse cholesterol transport (Krieger 1999). Importantly, although HDL and apoA-I are involved in reverse cholesterol transport, serum concentrations of HDL and apoA-I do not control the degree of reverse cholesterol transport (Jolley et al. 1998), yet the levels are directly and unequivocally related to atheroprotection. Thus, there is strong evidence that HDL and apoA-I are antiatherogenic, but the fundamental basis for the atheroprotection is not well understood.

In the past few years, our laboratory has discovered novel mechanisms of HDL action that provide a biological link between circulating levels of the lipoprotein and the capacity for endothelial NO production. The processes that we have uncovered may therefore underlie the marked antiatherogenic characteristics of HDL. In this review, we summarize our recent studies indicating that HDL binding to the class B scavenger receptor SR-BI modulates both the subcellular localization of eNOS within endothelial cells and the level of activity of the enzyme. First, the mechanisms underlying eNOS trafficking to plasma membrane domains known as caveolae are reviewed. The impact of lipoproteins on eNOS localization in caveolae is then examined, and the existence of an eNOS signaling module in caveolae is evaluated. Finally, the additional ability of HDL binding to SR-BI to cause robust activation of eNOS is reviewed. It is anticipated that issues related to HDL and SR-BI actions in endothelium will continue to warrant strong consideration in our ongoing efforts to understand the mechanisms underlying the potent atheroprotective features of HDL.

eNOS TRAFFICKING AND LOCALIZATION IN CAVEOLAE

Caveolae are specialized, lipid-ordered plasma membrane microdomains that are enriched in cholesterol, glycosphingolipids, sphingomyelin, and lipid-anchored membrane proteins, and are characterized by a light buoyant density and resistance to solubilization by Triton X-100 at 4°C. Caveolae are most likely a subset of lipid rafts. Following the identification of the caveola coat protein caveolin, it became possible to purify this specialized membrane domain. It was then discovered that caveolae also contain a variety of signal transduction molecules including G-protein-coupled receptors such as the muscarinic acetylcholine receptor, G-proteins, and molecules involved in the regulation of intracellular calcium homeostasis (Shaul and Anderson 1998). Knowledge that eNOS activity is acutely regulated by extracellular factors and that the protein is primarily associated with the plasma membrane prompted the initial examinations of eNOS in caveolae.

Using cultured endothelial cells, eNOS localization was first assessed in subcellular fractions including the caveolae and noncaveolae fractions of the plasma membrane (Fig. 1a) (Shaul et al. 1996). Within the plasma membrane, caveolin was detected specifically in the caveolae fraction, and eNOS protein was also highly enriched in caveolae. The enrichment of eNOS protein in caveolae correlated with the distribution of NOS enzymatic activity (Fig. 1b), which was 7-fold greater in plasma membrane versus cytosol. Within the plasma membrane, NOS activity was not detected in noncaveolae membranes, whereas it was 9- to 10-fold more prevalent in the caveolae fraction compared to whole plasma membrane. Over a range of experiments, 51–86% of the total enzymatic activity in the postnuclear supernatant was recovered in plasma membrane, and 57–100% of the activity in plasma membrane was recovered in caveolae, revealing that the majority of functional enzyme in quiescent endothelial cells is localized to caveolae. eNOS was also localized to caveolae by immunoelectron microscopy (Fig. 1c), indicating that the enrichment of eNOS protein and enzymatic activity in isolated endothelial caveolae membranes is not related to nonspecific association of the protein with the caveolae during purification; alternatively, it accurately reflects enzyme localization in the microdomain.

The basis for eNOS localization to caveolae was then explored. The process was faithfully reconstituted by transient transfection of wild-type eNOS cDNA into COS-7 cells, which possess caveolae but do not express NOS constitutively. The distribution of wild-type eNOS in transfected COS-7 cells mimicked that in endothelial cells, with NOS activity in caveolae enriched 7- to 12-fold versus whole plasma membrane. The mechanisms underlying trafficking to the microdomain were then determined using the knowledge that eNOS is amino-terminally myristoylated at the glycine residue in position 2 and palmitoylated at the cysteine residues in positions 15 and 26 (Lamas et al. 1992; Robinson and Michel 1995). In distinct contrast to wild-type eNOS, myristoylation-deficient mutant eNOS, which is incapable of both myristoylation and palmitoylation, was not enriched in either whole plasma membranes or the caveolae fraction of plasma membranes. This finding indicated that acylation targets eNOS to caveolae. Experiments were then performed with a palmitoylation-deficient mutant of eNOS that is capable of myristoylation. The palmitoylation-deficient form of eNOS was partially enriched in plasma membrane versus cytosol, and it was partially enriched in caveolae versus noncaveolae fractions of the plasma membrane, indicating a modest targeting to caveolae mediated by myristoylation alone. Repeated studies indicated that there is approximately a 10-fold enhancement in eNOS targeting to caveolae due to myristoylation alone, and the targeting is augmented another 10-fold by palmitoylation for a combined enhancement of 100-fold. Thus, both acylation processes are necessary for optimal targeting of eNOS to caveolae. Importantly, the results of studies of caveolae obtained from rat lung agreed with those obtained in cultured endothelial cells (Garcia-Cardena et al. 1996), indicating that eNOS localization to caveolae occurs in intact endothelium. Further studies of the localization of the palmitoylation-deficient form of eNOS showed that the mutant protein is found principally in Golgi apparatus, whereas the myristoylation-deficient mutant is ubiquitously distributed, and that normal dual acylation is necessary for optimal function of eNOS in a physiological context (Liu et al. 1996, 1997; Sowa et al. 1999).

LDL, HDL, AND eNOS LOCALIZATION IN CAVEOLAE

In addition to the modifications of the eNOS protein described above, the specialized lipid environment within caveolae is critical to the targeting and regulation of the enzyme within the microdomain. Since membrane cholesterol is essential for normal caveolae function (Chang et al. 1992), and the initiating events in atherogenesis are characterized by diminished endothelial NO production in response to extracellular stimuli, the effects of oxLDL on eNOS subcellular localization and function were investigated (Blair et al. 1999). In cultured endothelial cells briefly exposed to control conditions including lipoprotein-deficient serum (LPDS), HDL, or native LDL (nLDL), eNOS and caveolin both remained highly enriched in caveolae (Fig. 2a, panels 1 and 2). In contrast,

Figure 1. eNOS is localized to endothelial cell caveolae. (*a*) Immunoblot analysis for eNOS, caveolin-1, and calmodulin in subcellular fractions from endothelial cells. Samples of postnuclear supernatant (PNS), cytosol, plasma membrane (PM), noncaveolae membrane (NCM), and caveolae membrane (CM) were evaluated. (*b*) NOS enzymatic activity in subcellular fractions from endothelial cells. [^{3}H]-L-arginine conversion to [^{3}H]-L-citrulline was measured in the presence of excess substrate, cofactors, calcium, and calmodulin. Enzymatic activity was undetectable in NCM. Values are mean ± S.E.M., $n = 4$–6, *$p < 0.05$ versus plasma membrane. (*c*) Localization of caveolin and eNOS in plasma membranes by immunoelectron microscopy. Immunogold labeling was performed using antibody to caveolin-1 in fibroblasts (panel *1*) and endothelial cells (panel *2*), and antibody to eNOS in fibroblasts (panel *3*) and endothelial cells (panel *4*). Caveolae (*arrows*) are evident in both cell types. Bar, 0.45 μm. (Reprinted, with permission, from Shaul et al. 1996 [copyright American Society for Biochemistry and Molecular Biology].)

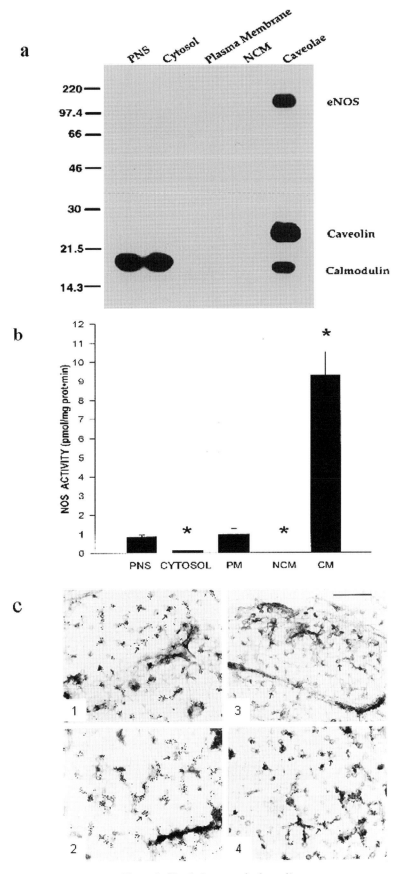

Figure 1. (*See facing page for legend.*)

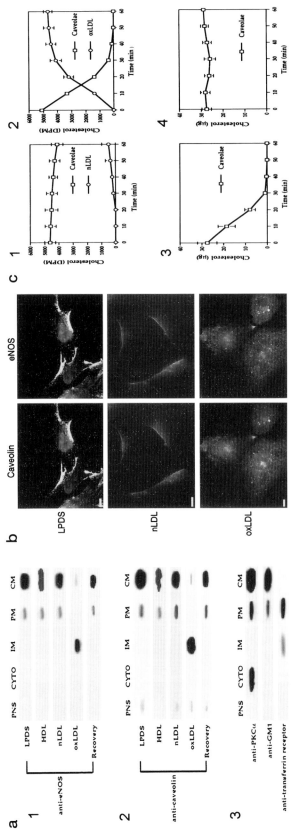

Figure 2. Oxidized LDL displaces eNOS from endothelial cell caveolae due to depletion of caveolae cholesterol. (*a*) Oxidized LDL (oxLDL) but not lipoprotein-deficient serum (LPDS), HDL, or nLDL exposure (60 minutes) alters the subcellular distribution of eNOS and caveolin. Additional studies were performed in oxLDL-treated cells that were washed and incubated for an additional 120 minutes in LPDS only (Recovery). Samples of postnuclear supernatant (PNS), cytosol (CYTO), intracellular membranes (IM), plasma membrane (PM), and caveolae membrane (CM) were isolated, and immunoblot analysis was performed for eNOS (panel *1*) and caveolin-1 (panel *2*). Control experiments in oxLDL-treated cells included immunoblot analyses for the caveolae resident proteins PKCα and GM1, and for transferrin receptors that are found in coated pits (panel *3*). (*b*) OxLDL but not LPDS or nLDL induces eNOS and caveolin to colocalize in an internal membrane compartment. Cells were incubated for 60 minutes, fixed, and processed for double-label immunofluorescence. (*c*) oxLDL depletes caveolae of cholesterol. For panels *1* and *2*, endothelial cells were labeled with [^{3}H]acetate for 18 hours at 37°C, washed and incubated with nLDL (panel *1*) or oxLDL (panel *2*) for 0–60 minutes at 37°C. The medium was collected, and the nLDL or oxLDL was isolated by centrifugation. The cells were washed, processed to isolate caveolae, and [^{3}H]cholesterol was extracted and measured. For panels *3* and *4*, unlabeled cells were incubated with oxLDL (panel *3*) or HDL (panel *4*) for 0–60 minutes at 37°C. The cells were washed, caveolae were isolated, and the amount of total cholesterol associated with caveolae was determined. Values are mean ± S.E.M., $n = 8$. (Reprinted, with permission, from Blair et al. 1999 [copyright American Society for Biochemistry and Molecular Biology]).

oxLDL exposure induced eNOS and caveolin to move from caveolae and plasma membrane to an internal membrane fraction containing endoplasmic reticulum, Golgi apparatus, mitochondria, and other intracellular organelles. However, PKCα and GM_1, two other resident proteins in caveolae, did not translocate from the caveolae fraction upon oxLDL exposure (Fig. 2a, panel 3). Comparable findings were obtained in cells treated with the protein synthesis inhibitor cycloheximide, and the removal of oxLDL and continued treatment with cycloheximide (recovery) enabled eNOS and caveolin to return to caveolae membranes (Fig. 2a, panels 1 and 2). Studies employing indirect immunofluorescence were confirmatory (Fig. 2b).

Having demonstrated that oxLDL causes eNOS redistribution in endothelial cells, the effects of oxLDL on the activation of the enzyme by acetylcholine were evaluated. oxLDL attenuated eNOS stimulation at all concentrations of acetylcholine tested, yielding a shift in the dose-response curve to acetylcholine to the right by 100-fold. Potential effects of oxLDL on eNOS modification were assessed, and both the palmitoylation and myristoylation of the enzyme were unperturbed. In addition, eNOS phosphorylation was unaffected by oxLDL (Blair et al. 1999). Thus, oxLDL provokes dramatic subcellular redistribution of eNOS, and this process is not due to changes in the co- and posttranslation modifications of the enzyme.

The next series of studies determined whether modifications in the caveolae lipid environment underlie the effects of oxLDL on eNOS distribution. Endothelial cells were labeled with [^{3}H]acetate and were exposed to nLDL or oxLDL, the media were collected and caveolae were isolated, and the amounts of sterol in the extracellular media and caveolae were determined (Fig. 2c). nLDL did not alter the amount of radiolabeled sterol in caveolae, and negligible amounts accumulated extracellularly (Fig. 2c, panel 1). However, radiolabeled sterol was readily transferred from caveolae to oxLDL in the media, such that the caveolae were essentially devoid of labeled sterol by 30 minutes (Fig. 2c, panel 2). In parallel, the total cholesterol content of the caveolae fraction was depleted by oxLDL, whereas HDL chosen as a control had no effect (Fig. 2c, panels 3 and 4). Thus, by acting as an acceptor of cholesterol, oxLDL causes marked depletion of caveolae cholesterol. To determine whether this process mediates eNOS displacement, the effect of cyclodextrin was determined. Similar to oxLDL, cyclodextrin caused redistribution of eNOS and caveolin from caveolae to the intracellular membrane fraction. Thus, oxLDL causes a depletion of caveolae cholesterol, and the alteration in the lipid environment results in eNOS redistribution and an attenuated capacity to activate the enzyme. Later experiments employing antibody blockade revealed that oxLDL-induced inhibition of eNOS localization and activation is mediated by the class B receptor CD36 (Uittenbogaard et al. 2000). Recent further studies by the Smart laboratory have shown that apoE$^{-/-}$ mice on a high-fat diet do not have a normal acute fall in blood pressure with Ach infusion. In addition, caveolae isolated in vivo from apoE$^{-/-}$ vessels do not contain eNOS and are depleted of cholesterol. However, the Ach response and eNOS localization to caveolae are conserved in apoE/CD36 double knockout mice. Thus, the findings in cell culture have been effectively extrapolated to intact vasculature by Smart and colleagues (Kincer et al. 2002). These cumulative observations indicate that oxLDL causes depletion of caveolae cholesterol in endothelial cells via CD36, leading to eNOS redistribution and an attenuated capacity to activate the enzyme. This process may play a critical role in the early pathogenesis of hypercholesterolemia-induced vascular disease and atherosclerosis.

Since there is a strong negative correlation between HDL levels and the risk for atherosclerosis (Grundy 1986; Tall 1990) and HDL mediates cholesterol trafficking (Krieger 1998; Fidge 1999), the possibility was then tested that HDL modifies the effects of oxLDL on eNOS localization and activation in caveolae (Uittenbogaard et al. 2000). The addition of HDL to medium containing oxLDL prevented eNOS and caveolin displacement from caveolae (Fig. 3a), and it additionally restored acetylcholine-induced stimulation of the enzyme (Fig. 3b). To determine the mechanisms underlying the effects of HDL on eNOS targeting, further studies of caveolae cholesterol homeostasis were performed (Fig. 3c). Whereas oxLDL alone caused a fall in caveolae sterol content (Fig. 3c, panel 1), cotreatment with HDL prevented the decline (Fig. 3c, panel 3). Further studies showed that the ability of HDL to maintain the concentration of caveolae-associated cholesterol in the face of oxLDL is not related to the inhibition of cholesterol transport from caveolae by oxLDL (Fig. 3c, panel 4); instead, it is due to the provision of cholesterol esters by HDL (Fig. 3c, panel 5). Thus, in the presence of oxLDL, HDL preserves the unique lipid environment within caveolae, thereby maintaining the normal subcellular localization and function of eNOS and possibly explaining at least a portion of the antiatherogenic properties of HDL.

Studies of the principal HDL receptor, scavenger receptor B-I, were done to determine whether the receptor is expressed in endothelial cells, and SR-BI protein was indeed detected and found to be highly enriched in endothelial cell caveolae. Antibody blockade experiments were performed to determine whether HDL binding to SR-BI underlies the reversal of the impact of oxLDL on eNOS localization and function. Endothelial cells were treated with oxLDL or oxLDL plus HDL in the presence of nonspecific IgG or SR-BI IgG, and eNOS and caveolin-1 distribution was evaluated (Fig. 4a). In the presence of nonspecific IgG, eNOS and caveolin-1 localization to caveolae was restored by HDL. However, SR-BI blocking antibody prevented the restoration of eNOS and caveolin-1 localization by HDL. In parallel, SR-BI blocking antibody prevented HDL-mediated restoration of eNOS stimulation by acetylcholine (Fig. 4b). These cumulative observations indicate that SR-BI mediates the capacity of HDL to preserve eNOS localization and function in the presence of oxLDL.

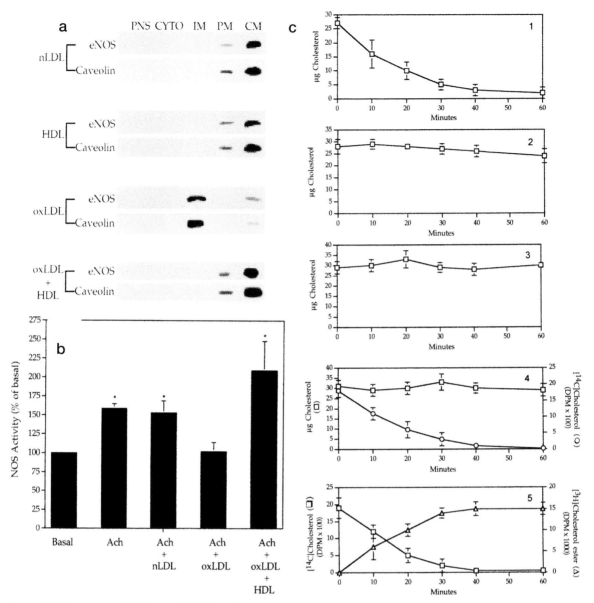

Figure 3. HDL prevents oxLDL-induced inhibition of eNOS localization and activation in caveolae. (*a*) Endothelial cells were treated for 60 minutes with nLDL, HDL, or oxLDL, or with oxLDL followed by HDL for an additional 15 minutes. Samples of postnuclear supernatant (PNS), cytosol (CYTO), intracellular membranes (IM), plasma membrane (PM), and caveolae membrane (CM) were isolated, and immunoblot analysis was performed for eNOS and caveolin-1. (*b*) After treating endothelial cells as described above, NOS activity was evaluated in intact cells by measuring [^{3}H]-L-arginine conversion to [^{3}H]-L-citrulline over 15 minutes in the absence of exogenous stimulation (Basal) or in the presence of acetylcholine (Ach, 10^{-6}M). Values are mean ± S.E.M., $n = 4$, * $p < 0.05$ versus basal. (*c*) HDL maintains the sterol content of caveolae. Endothelial cells were incubated with oxLDL (panel *1*) or HDL (panel *2*) for 0–60 minutes at 37°C. For panel *3*, cells were pretreated with oxLDL and then HDL was added for an additional 15 minutes. The cells were fractionated to isolate caveolae and the mass of cholesterol associated with caveolae was determined. In panel *4*, cells were radiolabeled with [^{14}C]acetate for 18 hours, and HDL and oxLDL were added simultaneously for 0–60 minutes. The cells were processed to measure the amount of [^{14}C]cholesterol associated with caveolae and the mass of cholesterol associated with caveolae. In panel *5*, cellular cholesterol pools were radiolabeled as described in panel *4*, and HDL was labeled with [^{3}H]cholesterol ester. [^{3}H]HDL and oxLDL were incubated with cells as described in panel *4*, and the amount of [^{14}C]cholesterol and [^{3}H]cholesterol associated with caveolae was determined. Values are mean ± S.E.M., $n = 3$. (Reprinted, with permission, from Uittenbogaard et al. 2000 [copyright American Society for Biochemistry and Molecular Biology].)

eNOS SIGNALING MODULE IN CAVEOLAE

There has been an accumulation of observations over the past several years indicating that many of the signal transduction molecules described above are colocalized with eNOS in caveolae (Shaul and Anderson 1998). Such colocalization has implied the existence of a functional signaling module that serves to compartmentalize the multiple events regulating the degree of NO production. However, direct evidence of such a module was unavailable until recent studies were done to further understand how eNOS is stimulated by estradiol via nongenomic ac-

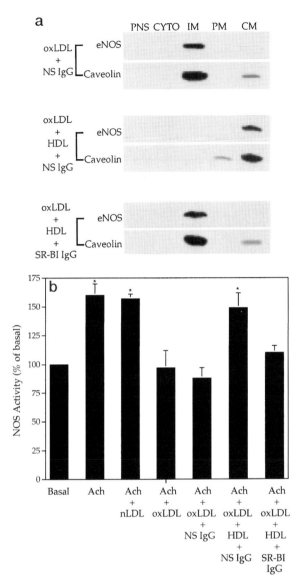

Figure 4. SR-BI mediates the effects of HDL on eNOS localization and activation in caveolae. (*a*) Endothelial cells were treated for 1 hour with oxLDL, or oxLDL and HDL in the presence of nonspecific (NS) IgG or SR-BI IgG. Samples of postnuclear supernatant (PNS), cytosol (CYTO), intracellular membranes (IM), plasma membrane (PM), and caveolae membrane (CM) were isolated, and immunoblot analysis was performed for eNOS and caveolin-1. Representative data from four independent experiments are shown. (*b*) After treating endothelial cells as described above, NOS activity was evaluated in intact cells by measuring [^{3}H]-L-arginine conversion to [^{3}H]-L-citrulline over 15 minutes in the absence of exogenous stimulation (Basal) or in the presence of acetylcholine (Ach, 10^{-6} M). Values are mean ± S.E.M., $n = 4$, * $p < 0.05$ versus basal. (Reprinted, with permission, from Uittenbogaard et al. 2000 [copyright American Society for Biochemistry and Molecular Biology].)

cumstances, affording atheroprotection to premenopausal women compared to men and to postmenopausal women receiving certain forms of estrogen replacement, and also modifying responses to vascular injury (Mendelsohn and Karas 1999).

The subcellular site of interaction between ERα and eNOS was first explored in experiments employing isolated endothelial cell plasma membranes (Chambliss et al. 2000). Estradiol caused an increase in eNOS activity in the isolated membranes in the absence of added calcium, calmodulin, or eNOS cofactors; activation by estradiol was blocked by the ERα antagonist ICI 182,780 and also by ERα antibody; and immunoidentification experiments detected ERα protein in endothelial cell plasma membranes. Similarly, plasma membranes from COS-7 cells coexpressing eNOS and ERα displayed ER-mediated eNOS stimulation, whereas membranes from cells expressing eNOS alone or ERα plus myristoylation-deficient mutant eNOS were insensitive. Further fractionation of endothelial cell plasma membranes revealed ERα protein in caveolae, and estradiol caused eNOS stimulation in isolated caveolae that was ER-dependent; in contrast, noncaveolae membranes were insensitive. Responses to estradiol and the more classic eNOS agonists acetylcholine and bradykinin were also compared in isolated caveolae fractions, and equivalent, robust eNOS activation was stimulated by all three agents (Fig. 5). Furthermore, calcium chelation prevented estradiol-stimulated eNOS activity in either isolated whole plasma membranes or caveolae. These cumulative findings indicate that a subpopulation of ERα and multiple other receptors for eNOS agonists are compartmentalized in endothelial cell caveolae where they are coupled to the enzyme in a functional signaling module which regulates the local calcium environment. Processes occurring within this signaling module are most likely critical to the atheropro-

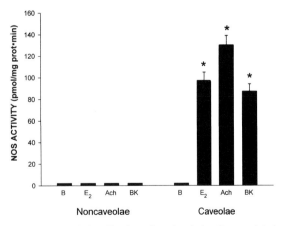

Figure 5. eNOS is localized to a functional signaling module in endothelial cell caveolae. [^{3}H]-L-arginine conversion to [^{3}H]-L-citrulline was measured in isolated noncaveolae and caveolae fractions of endothelial cell plasma membranes, in the absence (basal, B) or presence of 10^{-8} M estradiol (E2), 10^{-6} M acetylcholine (Ach), or 10^{-6} M bradykinin (BK). Values are mean ± S.E.M., $n = 4–6$, *$p < 0.05$ versus basal. (Reprinted, with permission, from Chambliss et al. 2000 [copyright Lippincott Williams and Wilkins].)

tions of estrogen receptor α (ERα), which was previously known to serve exclusively as a nuclear transcription factor (Chen et al. 1999). These mechanisms were pursued because estradiol is an important agonist for endothelial eNOS in both physiological and pathophysiological cir-

tective properties of estradiol (Mendelsohn and Karas 1999).

HDL-MEDIATED ACTIVATION OF eNOS

More recent studies have determined whether the dramatic antiatherogenic features of HDL are further explained by caveolae-related mechanisms in addition to those involving changes in membrane cholesterol content. Studies of the impact of HDL on eNOS function in the absence of oxLDL were undertaken (Yuhanna et al. 2001), and they demonstrated that HDL causes rapid and dramatic stimulation of the enzyme in cultured endothelial cells (Fig. 6a). HDL stimulation of eNOS was concentration-dependent (Fig. 6b), it was detected within 3 minutes of exposure to HDL, the maximal effect was seen by 6 minutes, and stimulation persisted for at least 60 minutes. The HDL fraction of human serum caused eNOS activation, whereas the non-HDL fraction did not, and whole serum yielded an effect comparable to HDL, but lipoprotein-deficient serum did not (Fig. 6c). In contrast to HDL, nLDL did not stimulate eNOS, and simultaneous exposure to excess nLDL prevented the HDL response (Fig. 6d,e). In contrast to native plasma HDL, lipid-free apoA-I or apoA-II did not cause eNOS activation (Fig. 6f).

To examine the role of SR-BI in eNOS activation, heterologous expression experiments were performed in Chinese hamster ovary (CHO) cells. Bovine eNOS cDNA was transiently transfected into SR-BI-deficient CHO cells (ldlA7) or CHO cells stably transformed with murine SR-BI (ldlA7-SR-BI). Whereas HDL had no effect in SR-BI-deficient cells, it caused a 4-fold increase in eNOS activity in SR-BI-transfected cells (Fig. 7a). Confirming the demonstration of SR-BI protein expression in endothelial cells, SR-BI mRNA was demonstrable in endothelium by RT-PCR (Fig. 7b). To determine whether functional HDL-SR-BI coupling to eNOS occurs in cave-

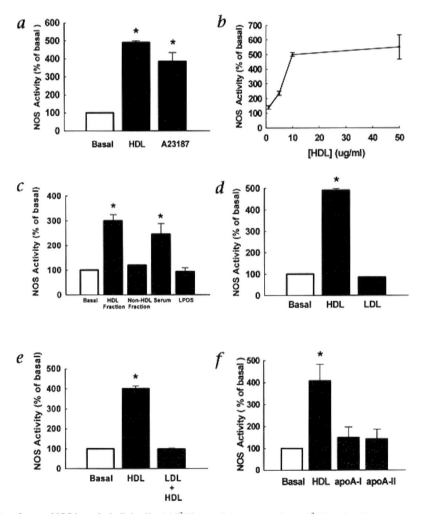

Figure 6. HDL activates eNOS in endothelial cells. (*a*) [^{3}H]-L-arginine conversion to [^{3}H]-L-citrulline was measured under basal conditions or in the presence of HDL or A23187. (*b*) The dose response to HDL was also assessed. (*c*) eNOS activation was evaluated under basal conditions or in the presence of the HDL fraction or the non-HDL fraction of human serum, or whole serum, or lipoprotein-deficient serum (LPDS) added at equal volumes. (*d*) The responses to HDL and LDL were compared. (*e*) eNOS activation by HDL was assessed in the absence or presence of excess LDL. (*f*) eNOS activation was determined in the presence of HDL or recombinant apoA-I or lipid-free apoA-II purified from human plasma. Values are mean ± S.E.M., $n = 3–6$, *$p<0.05$ versus basal. (Reprinted, with permission, from Yuhanna et al. 2001 [copyright Nature Publishing Group].)

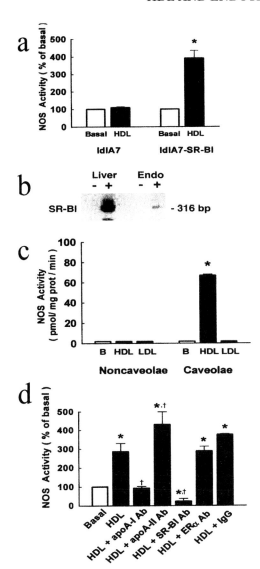

Figure 7. eNOS activation by HDL requires SR-BI and occurs in endothelial caveolae. (*a*) Human eNOS cDNA was transiently transfected into the CHO cell line ldlA7, which expresses minimal levels of SR-BI, or ldlA7 stably transformed with SR-BI (ldlA7-SR-BI). After 72 hours, eNOS activation was evaluated under basal conditions or in the presence of HDL. (*b*) RT-PCR products from SR-BI mRNA in murine liver and ovine endothelial cells (Endo). RT was performed in the absence (–) or presence (+) of reverse transcriptase. Findings were replicated in three independent experiments. (*c*) [^{3}H]-L-arginine conversion to [^{3}H]-L-citrulline was measured in noncaveolae and caveolae membranes from endothelial cells during incubations performed in the absence (basal, B) or presence of HDL or LDL. NOS activity was undetectable in noncaveolae fractions in all groups, and it was also not detected in caveolae under basal conditions or with LDL added. (*d*) [^{3}H]-L-arginine conversion to [^{3}H]-L-citrulline was measured in isolated endothelial cell plasma membranes during incubations done in the absence (basal) or presence of HDL, with or without the addition of polyclonal antibodies to apoA-I, apoA-II, SR-BI, or estrogen receptor α (ERα), or unrelated IgG. Values are mean ± S.E.M., n=3–6, *p<0.05 versus basal, †p<0.05 versus HDL alone. (Reprinted, with permission, from Yuhanna et al. 2001 [copyright Nature Publishing Group].)

olae, NOS activation was examined in noncaveolae and caveolae membranes from endothelial cells (Fig. 7c). In studies done in the absence of exogenous calcium, calmodulin, or eNOS cofactors, NOS activity was undetectable in noncaveolae membranes under any conditions. In addition, basal NOS activity was not measurable in caveolae membranes. However, HDL caused robust activation of eNOS in isolated caveolae, whereas LDL did not. In antibody blockade studies in isolated endothelial plasma membranes (Fig. 7d), HDL stimulation of eNOS was fully attenuated by antibodies to apoA-I or to the carboxy-terminal cytoplasmic tail of SR-BI. In contrast, eNOS activation by HDL was enhanced by antibody to apoA-II, and it was unaltered by antibody to ERα, which as mentioned above is colocalized and functionally linked to eNOS in caveolae, or unrelated IgG. These findings indicate that SR-BI mediates eNOS activation by HDL, that the receptor and enzyme are coupled in endothelial cell caveolae, and that the process entails apoA-I binding to the receptor.

The effect of HDL on endothelial NO production was also evaluated in a physiological context by assessing the impact of the lipoprotein on endothelium-dependent vascular relaxation in isolated mouse aortae (Yuhanna et al. 2001). HDL caused direct relaxation of phenylephrine-precontracted rings of thoracic aortae from wild-type CD-1 mice, the response was blocked by pretreatment with NOS antagonist, and relaxation was not observed in endothelium-denuded rings (Fig. 8a,b). The ability of HDL to alter acetylcholine-mediated vasorelaxation was also examined. Relaxation with intermediate concentrations of acetylcholine was greater with coexposure to HDL, whereas responses with higher levels of acetylcholine were not (Fig. 8c). The role of SR-BI in endothelial NO production was also determined by assessing both direct responses to HDL and acetylcholine-stimulated relaxation in the absence or presence of HDL using aortae from homozygous null SR-BI knockout mice (Rigotti et al. 1997). Aortae from wild-type 129/C57BL/6 mice exhibited direct relaxation to HDL, mimicking the observations in the CD-1-derived vascular rings. In contrast, HDL did not cause discernable relaxation in aortae from homozygous null mutants (Fig. 8d). In addition, relaxation to acetylcholine was increased by HDL in aortae from wild-type mice, whereas HDL did not modify the acetylcholine response in aortae from SR-BI-null mice (Fig. 8e). Thus, HDL enhances endothelial NO production in intact arteries, and this process is mediated by SR-BI. This process may be critical to the atheroprotective features of HDL.

CONCLUSIONS

The severity of atherosclerotic lesion formation in both humans and animal models is inversely related to serum concentrations of HDL cholesterol (Fidge 1999). In addition, studies of SR-BI/apolipoprotein E double homozygous knockout mice indicate that SR-BI protects against early-onset atherosclerosis and myocardial infarction (Trigatti et al. 1999; Braun et al. 2002). However, the mechanisms by which HDL and SR-BI are atheroprotective are complex and poorly understood (Gordon and Rifkind 1989; Krieger 1998). Although HDL is involved in reverse cholesterol transport from peripheral tissues to

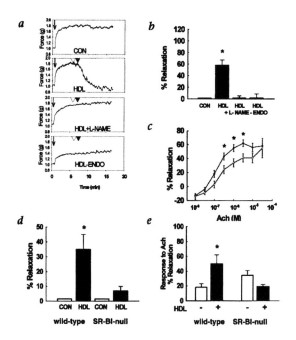

Figure 8. HDL enhances endothelial NO production in aortae from wild-type mice, but not in aortae from homozygous SR-BI null mutant mice. (*a*) Following precontraction of rings of thoracic aortae from wild-type CD-1 mice with phenylephrine (*arrow*), direct relaxation responses to control buffer (CON), HDL at 10 μg/ml (*open arrowhead*) or 25 μg/ml (*closed arrowhead*), HDL with prior L-NAME treatment, or HDL with endothelium-denuded (– endo) rings were evaluated. (*b*) Cumulative findings (mean ± S.E.M.) for maximal relaxation to 25 μg/ml HDL in $n = 10, 10, 3$, and 5 studies, respectively, which were performed as in *a*. (*c*) HDL augments the relaxation response to acetylcholine (Ach) in aortae from wild-type CD-1 mice. Studies were performed in control rings (*open circles*) and rings exposed to HDL (*closed circles*). (*d*) The direct effects of control buffer (CON) or HDL were tested on phenylephrine contracted rings of thoracic aortae from wild-type 129/C57BL/6 mice or homozygous null mutant mice (SR-BI-null) mice. (*e*) Rings of thoracic aortae from wild-type or SR-BI-null mice were precontracted with phenylephrine, and the response to Ach was determined before and after exposure to HDL. In *c*, *d* and *e*, values are mean ± S.E.M., and a minimum of 4 rings from 4 different mice were studied in each group. *$p<0.05$ versus no HDL. (Reprinted, with permission, from Yuhanna et al. 2001 [copyright Nature Publishing Group].)

the liver, serum concentrations of HDL do not control the degree of reverse cholesterol transport (Jolley et al. 1998). Thus, it is quite likely that other processes are critical to the capacities of HDL to attenuate hypercholesterolemia-induced vascular disease. Our recent work has provided a biological link between HDL and endothelial cell signaling. We have demonstrated that HDL has important direct effects on endothelial cells via SR-BI to both maintain eNOS localization in caveolae and to activate the enzyme, thereby increasing NO availability. NO is a critical mediator of vasomotor tone and an important atheroprotective signaling molecule that attenuates both cell adherence to the lumenal surface and vascular cell growth (Harrison 1994; Cohen 1995). However, it is yet to be directly determined whether HDL modulation of eNOS contributes to blood pressure regulation or to atheroprotection. The molecular basis by which HDL-SR-BI binding modifies endothelial cell membrane cholesterol content also warrants clarification. In addition, the processes by which HDL-SR-BI binding is coupled to eNOS activation are yet to be elucidated. Other downstream targets of HDL-SR-BI binding in endothelial cells, both of nongenomic and genomic nature, should also be considered. Further studies along these lines of investigation will increase our knowledge of the role of HDL in vascular health and disease. In addition, the regulation of endothelial NO production by HDL and SR-BI may provide an entirely new therapeutic target for the design of strategies to prevent or treat atherosclerosis.

ACKNOWLEDGMENTS

The authors express their appreciation to numerous colleagues involved in this work, and to Ms. DiAna Randle and Ms. Marilyn Dixon for preparing this manuscript. This work was supported by National Institutes of Health grants HL-58888, HL-53546, and HD-30276; the Lowe Foundation; and the Crystal Charity Ball.

REFERENCES

Blair A., Shaul P.W., Yuhanna I.S., Conrad P.A., and Smart E.J. 1999. Oxidized low density lipoprotein displaces endothelial nitric-oxide synthase (eNOS) from plasmalemmal caveolae and impairs eNOS activation. *J. Biol. Chem.* **274:** 32512.

Braun A., Trigatti B.L., Post M.J., Sato K., Simons M., Edelberg J.M., Rosenberg R.D., Schrenzel M., and Krieger M. 2002. Loss of SR-BI expression leads to the early onset of occlusive atherosclerotic coronary artery disease, spontaneous myocardial infarctions, severe cardiac dysfunction, and premature death in apolipoprotein E-deficient mice. *Circ. Res.* **90:** 270.

Cayatte A.J., Palacino J.J., Horten K., and Cohen R.A. 1994. Chronic inhibition of nitric oxide production accelerates neointima formation and impairs endothelial function in hypercholesterolemic rabbits. *Arterioscler. Thromb.* **14:** 753.

Chambliss K.L., Yuhanna I.S., Mineo C., Liu P., German Z., Sherman T.S., Mendelsohn M.E., Anderson R.G., and Shaul P.W. 2000. Estrogen receptor alpha and endothelial nitric oxide synthase are organized into a functional signaling module in caveolae. *Circ. Res.* **87:** E44.

Chang W.J., Rothberg K.G., Kamen B.A., and Anderson R.G. 1992. Lowering the cholesterol content of MA104 cells inhibits receptor-mediated transport of folate. *J. Cell Biol.* **118:** 63.

Chen Z., Yuhanna I.S., Galcheva-Gargova Z., Karas R.H., Mendelsohn M.E., and Shaul P.W. 1999. Estrogen receptor alpha mediates the nongenomic activation of endothelial nitric oxide synthase by estrogen. *J. Clin. Invest.* **103:** 401.

Cohen R.A. 1995. The role of nitric oxide and other endothelium-derived vasoactive substances in vascular disease. *Prog. Cardiovasc. Dis.* **38:** 105.

Fidge N.H. 1999. High density lipoprotein receptors, binding proteins, and ligands. *J. Lipid Res.* **40:** 187.

Flavahan N.A. 1992. Atherosclerosis or lipoprotein-induced endothelial dysfunction. Potential mechanisms underlying reduction in EDRF/nitric oxide activity. *Circulation* **85:** 1927.

Garcia-Cardena G., Oh P., Liu J., Schnitzer J.E., and Sessa W.C. 1996. Targeting of nitric oxide synthase to endothelial cell caveolae via palmitoylation: Implications for nitric oxide signaling. *Proc. Natl. Acad. Sci.* **93:** 6448.

Gordon D.J. and Rifkind B.M. 1989. High-density lipoprotein—The clinical implications of recent studies. *N. Engl. J. Med.* **321:** 1311.

Grundy S.M. 1986. Cholesterol and coronary heart disease. A new era. *J. Am. Med. Assoc.* **256:** 2849.

Harrison D.G. 1994. Endothelial dysfunction in atherosclerosis. *Basic Res. Cardiol.* (suppl. 1) **89:** 87.

Jolley C.D., Woollett L.A., Turley S.D., and Dietschy J.M. 1998. Centripetal cholesterol flux to the liver is dictated by events in the peripheral organs and not by the plasma high density lipoprotein or apolipoprotein A-I concentration. *J. Lipid Res.* **39:** 2143.

Kauser K., da Cunha V., Fitch R., Mallari C., and Rubanyi G.M. 2000. Role of endogenous nitric oxide in progression of atherosclerosis in apolipoprotein E-deficient mice. *Am. J. Physiol. Heart Circ. Physiol.* **278:** H1679.

Kincer J.F., Uittenbogaard A., Dressman J., Guerin T.M., Febbraio M., Guo L., and Smart E.J. 2002. Hypercholesterolemia promotes a CD36-dependent and endothelial nitric oxide synthase-mediated vascular dysfunction. *J. Biol. Chem.* **277:** 23525.

Krieger M. 1998. The "best" of cholesterols, the "worst" of cholesterols: A tale of two receptors. *Proc. Natl. Acad. Sci.* **95:** 4077.

———. 1999. Charting the fate of the "good cholesterol": Identification and characterization of the high-density lipoprotein receptor SR-BI. *Annu. Rev. Biochem.* **68:** 523.

Kuhlencordt P.J., Gyurko R., Han F., Scherrer-Crosbie M., Aretz T.H., Hajjar R., Picard M.H., and Huang P.L. 2001. Accelerated atherosclerosis, aortic aneurysm formation, and ischemic heart disease in apolipoprotein E/endothelial nitric oxide synthase double-knockout mice. *Circulation* **104:** 448.

Lamas S., Marsden P.A., Li G.K., Tempst P., and Michel T. 1992. Endothelial nitric oxide synthase: Molecular cloning and characterization of a distinct constitutive enzyme isoform. *Proc. Natl. Acad. Sci.* **89:** 6348.

Lefer A.M. and Ma X.L. 1993. Decreased basal nitric oxide release in hypercholesterolemia increases neutrophil adherence to rabbit coronary artery endothelium. *Arterioscler. Thromb.* **13:** 771.

Liu J., Garcia-Cardena G., and Sessa W.C. 1996. Palmitoylation of endothelial nitric oxide synthase is necessary for optimal stimulated release of nitric oxide: Implications for caveolae localization. *Biochemistry* **35:** 13277.

Liu J., Hughes T.E., and Sessa W.C. 1997. The first 35 amino acids and fatty acylation sites determine the molecular targeting of endothelial nitric oxide synthase into the Golgi region of cells: A green fluorescent protein study. *J. Cell Biol.* **137:** 1525.

Mendelsohn M.E. and Karas R.H. 1999. The protective effects of estrogen on the cardiovascular system. *N. Engl. J. Med.* **340:** 1801.

Naruse K., Shimizu K., Muramatsu M., Toki Y., Miyazaki Y., Okumura K., Hashimoto H., and Ito T. 1994. Long-term inhibition of NO synthesis promotes atherosclerosis in the hypercholesterolemic rabbit thoracic aorta. PGH2 does not contribute to impaired endothelium-dependent relaxation. *Arterioscler. Thromb.* **14:** 746.

Rigotti A., Trigatti B.L., Penman M., Rayburn H., Herz J., and Krieger M. 1997. A targeted mutation in the murine gene encoding the high density lipoprotein (HDL) receptor scavenger receptor class B type I reveals its key role in HDL metabolism. *Proc. Natl. Acad. Sci.* **94:** 12610.

Robinson L.J. and Michel T. 1995. Mutagenesis of palmitoylation sites in endothelial nitric oxide synthase identifies a novel motif for dual acylation and subcellular targeting. *Proc. Natl. Acad. Sci.* **92:** 11776.

Shaul P.W. 2002. Regulation of endothelial nitric oxide synthase: Location, location, location. *Annu. Rev. Physiol.* **64:** 749.

Shaul P.W. and Anderson R.G.W. 1998. Role of plasmalemmal caveolae in signal transduction. *Am. J. Physiol.* **275:** L843.

Shaul P.W., Smart E.J., Robinson L.J., German Z., Yuhanna I.S., Ying Y., Anderson R.G., and Michel T. 1996. Acylation targets emdothelial nitric-oxide synthase to plasmalemmal caveolae. *J. Biol. Chem.* **271:** 6518.

Sowa G., Liu J., Papapetropoulos A., Rex-Haffner M., Hughes T.E., and Sessa W.C. 1999. Trafficking of endothelial nitric-oxide synthase in living cells. Quantitative evidence supporting the role of palmitoylation as a kinetic trapping mechanism limiting membrane diffusion. *J. Biol. Chem.* **274:** 22524.

Tall A.R. 1990. Plasma high density lipoproteins. Metabolism and relationship to atherogenesis. *J. Clin. Invest.* **86:** 379.

Trigatti B., Rayburn H., Vinals M., Braun A., Miettinen H., Penman M., Hertz M., Schrenzel M., Amigo L., Rigotti A., and Krieger M. 1999. Influence of the high density lipoprotein receptor SR-BI on reproductive and cardiovascular pathophysiology. *Proc. Natl. Acad. Sci.* **96:** 9322.

Uittenbogaard A., Shaul P.W., Yuhanna I.S., Blair A., and Smart E.J. 2000. High density lipoprotein prevents oxidized low density lipoprotein-induced inhibition of endothelial nitric-oxide synthase localization and activation in caveolae. *J. Biol. Chem.* **275:** 11278.

Vane J.R., Anggard E.E., and Botting R.M. 1990. Regulatory functions of the vascular endothelium. *N. Engl. J. Med.* **323:** 27.

Yang R., Powell-Braxton L., Ogaoawara A.K., Dybdal N., Bunting S., Ohneda O., and Jin H. 1999. Hypertension and endothelial dysfunction in apolipoprotein E knockout mice. *Arterioscler. Thromb. Vasc. Biol.* **19:** 2762.

Yuhanna I.S., Zhu Y., Cox B.E., Hahner L.D., Osborne-Lawrence S., Lu P., Marcel Y.L., Anderson R.G., Mendelsohn M.E., Hobbs H.H., and Shaul P.W. 2001. High-density lipoprotein binding to scavenger receptor-BI activates endothelial nitric oxide synthase. *Nat. Med.* **7:** 853.

Zambon A. and Hokanson J.E. 1998. Lipoprotein classes and coronary disease regression. *Curr. Opin. Lipidol.* **9:** 329.

Plaque Angiogenesis:
Its Functions and Regulation

K.S. MOULTON

Cardiovascular Division, Brigham and Women's Hospital, Vascular Biology, Children's Hospital, Boston, Massachusetts 02115

Therapeutic "angiogenesis" trials refer to the stimulation of collateral arterioles or larger-caliber blood vessels that perfuse ischemic myocardium and tissues caused by impaired blood flow in occluded or stenotic native arteries. Angiogenesis, or the formation of small-caliber capillary-like blood vessels, occurs within the atherosclerotic lesions that are responsible for the vascular occlusions themselves. Plaque neovascularization comprises a network of capillaries that arise from the adventitial vasa vasorum and extend into the thickened intimal layer associated with atherosclerosis and other conditions of vascular injury and inflammation. These plaque capillaries are conduits for the exchange of inflammatory cells in atherosclerotic lesions and therefore may affect their growth rate or susceptibility to rupture. Thus, the development of agents that are either positive or negative regulators of arteriogenesis and angiogenesis may have dual therapeutic applications in cardiovascular diseases—to ameliorate impaired tissue perfusion in the setting of established flow-limiting vascular stenosis, or to delay the progression of atherosclerotic lesions that cause ischemia, thereby obviating the need for collaterals. Although strategies to promote arteriogenesis have advanced into clinical investigations, this paper focuses on the functions and potential regulators of plaque angiogenesis in atherosclerosis.

CARDIOVASCULAR APPLICATION OF ANGIOGENESIS RESEARCH

The outcomes of research in the fields of tumor angiogenesis and developmental biology have resulted in emerging applications for cardiovascular medicine and other diseases that manifest pathologic angiogenesis such as diabetic retinopathy, macular degeneration, and rheumatoid arthritis. Angiogenesis and vascular remodeling are manifested in several cardiovascular conditions including atherosclerosis, myocardial infarction, myocardial hypertrophy, pulmonary hypertension, vasculitis, and vascular malformations. The elucidation of molecular pathways that regulate vasculogenesis, angiogenesis, and the assembly of larger arteries may ultimately lead to therapeutic strategies that manipulate the formation or regression of blood vessels in order to modify some of these diseases. Understanding how angiogenesis is regulated in these angiogenic disease states also provides important insights into their pathogenic or repair mechanisms.

For cardiovascular diseases, there may be applications for both positive and negative regulators of angiogenesis. Clinical trials are currently testing the therapeutic effects of endothelial cell growth factors to augment the formation of collaterals in order to relieve chronic ischemia in the heart and peripheral limbs (Baumgartner et al. 1998; Losordo et al. 1998; Schumacher et al. 1998; Laham et al. 1999). Ischemia is a prerequisite for the formation of collateral vessels that can develop as a native response to impaired perfusion of distal tissues. The perfusion capacity provided by collaterals is less than that of the native artery, yet in many cases it can be sufficient to preserve the viability of tissues and ameliorate the effects of ischemia. Unfortunately, the native collateral response of patients with significant vascular disease is often inadequate, thereby requiring interventions to revascularize ischemic tissue or providing the clinical need to augment mechanisms of collateral formation.

Although these clinical trials are considered as examples of "therapeutic angiogenesis," the process of collateral formation is more precisely called "arteriogenesis" because collaterals have a larger caliber and more complex structure compared to small capillaries, which consist of an endothelial cell layer with basement membrane and few mural cells. This distinction between the mechanisms of arteriogenesis and angiogenesis may be important in refining strategies to selectively regulate collateral growth or capillary proliferation at sites of chronic inflammation and in tumors.

MECHANISMS OF ARTERIOGENESIS VERSUS PLAQUE ANGIOGENESIS

It is conceivable that applications of both angiogenesis stimulators and inhibitors for the same disease may be feasible when one considers the potentially different mechanisms that regulate the formation of collaterals compared to the formation of a capillary network within atherosclerotic lesions. Both arteriogenesis (formation of collaterals) and angiogenesis (formation of capillaries) may develop in response to an endothelial cell growth factor, but additional and potentially unique factors may be further required to complete either of these respective processes. Collateral vessels are larger-caliber vessels with smooth muscle cell layer components that are of sufficient size to be visible on an angiogram (Fig. 1). In contrast, plaque capillaries are microscopic (<25–50 μm)

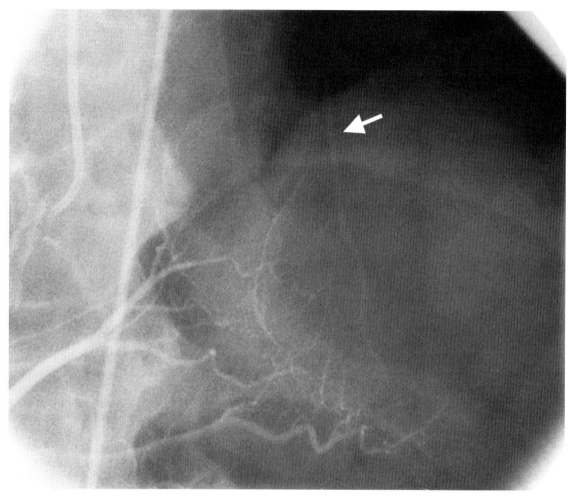

Figure 1. Coronary artery collaterals. Collateral vessels are larger-caliber channels shown here on the right coronary artery angiogram. Distal branches of the left coronary artery fill via numerous collateral vessels derived from the distal right coronary artery. (Courtesy of Brigham and Women's Hospital catheterization laboratory.)

thin-walled endothelial cell channels with a basement membrane and few pericytes (Fig. 2). These structural differences imply that these two processes may activate distinct molecular pathways. For example, collateral vessels may have a greater dependence on the Tie2 receptor agonist angiopoietin-1, a factor that is important in vessel maturation and the recruitment of smooth muscle cells to develop the wall of larger-caliber blood vessels. Interestingly, animal studies have shown that both angiopoietin-1 and angiopoietin-2 augment vascular endothelial growth factor (VEGF)-stimulated angiogenesis, but only angiopoietin-1 augments collateral development (Asahara et al. 1998; Shyu et al. 1998). Recently, placental growth factor acting via the VEGF receptor 1 on endothelial cells and monocytes showed enhanced collateral formation in a hindlimb model of ischemia. Importantly, the caliber, the extent of smooth muscle cell recruitment, perfusion capacity, and other markers of vessel maturation were improved by VEGF receptor 1 activation (Luttun et al. 2002). The enhanced collaterals induced by placental growth factor may be due to its effect on both monocytes and endothelial cells that cooperate to develop collaterals. Future distinctions between the mechanisms of collateral formation and plaque angiogenesis could lead to the design of refined therapeutic methods that augment collaterals or inhibit plaque angiogenesis selectively.

Some investigators have proposed that larger collateral vessels form via the remodeling of existing vessels, which can nonetheless produce dramatic enlargement of the vessel caliber and greatly increase blood flow to ischemic tissues. Although the remodeling of existing vessels may be a common mechanism for the development of collaterals, generation of arteries de novo can be observed in some conditions. New arteries, complete with an internal elastic membrane and medial layer, can form within an artery occluded with thrombus (Roberts and Virmani 1984). Second, large-caliber arteries and veins often develop in association with an enlarging mass such as a neoplasm. These examples of new artery formation suggest that the complex biologic program to grow new arteries can be activated postnatally.

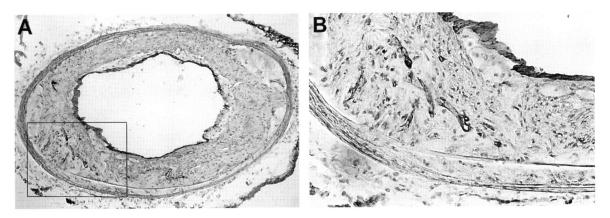

Figure 2. Plaque neovascularization. Plaque neovascularization comprises microscopic capillary-like endothelial channels that arise from the adventitial network of vasa vasorum and extend into the base of lesions. (A) The concentric lesion harvested from an ApoE$^{-/-}$ mouse fed a cholesterol diet for 6 months shows a focal area of neovascularization in the boxed frame. Higher magnification of this inset area shows plaque capillaries detected by positive endothelial cell staining with an anti-CD31 antibody. Magnification 250x.

The strategies employed in early therapeutic angiogenesis trials tested the delivery of endothelial cell growth factors as single agents to augment collateral growth. These studies have been based on benefits shown in animal studies with porcine, rabbit, or canine models of myocardial and limb ischemia (Ware and Simons 1997). The structure of collateral vessels comprises several cell types (smooth muscle cells and endothelial cells), and growth of collaterals requires remodeling of the medial layer. The ability of a single endothelial cell growth factor to achieve collateral growth implies that this single agent activates or complements a cascade of cellular events in order to remodel a complex vascular structure with normal arterial and venous connections. Although single agents have shown promising results in preclinical studies on young adult animals, results in clinical trials have been less favorable in patients with significant atherosclerosis. A reduced arteriogenesis response perhaps may also be due to the presence of clinical conditions such as hypercholesterolemia, endothelial cell dysfunction, diabetes, and increased-age-related factors that impair optimal collateral responses (Van Belle et al. 1997; Baumgartner et al. 1998; Rivard et al. 1999a,b). Current and future arteriogenesis trials aim to identify and compensate for the deficiencies associated with a suboptimal collateral response. Further improvements may be seen by the use of angiogenic factors in combination and by targeting non-endothelial cells such as monocytes and smooth muscle cells that cooperate with endothelial cells in arteriogenesis (Arras et al. 1998).

ANGIOGENESIS IN VASCULAR DISEASE

For over a century, it has been recognized that angiogenesis develops within the inner layer of atherosclerotic lesions which themselves can cause occlusion and flow-limiting stenosis of native large vessels. The normal network of small capillaries that perfuse the outer wall of large blood vessels is called the vasa vasorum. In diseases of chronic inflammation or injury of large vessels, the vasa vasorum proliferate and then extend into the expanding intimal layer of atheromas and other types of intimal hyperplasia (Fig. 3). These pathologic conditions include atherosclerosis, vasculitis, transplant-associated vascular disease, and thrombosis of large veins (Barger et al. 1984; Tanaka et al. 1994; Kaiser et al. 1999). The functions of the vasa vasorum in these diseases are not fully understood, but it is hypothesized that the inward growth of neovascularization into the innermost layer of the artery wall may promote plaque growth, intraplaque hemorrhage, or lesion instability that results in myocardial infarction and stroke. If this hypothesis is true, then negative regulators of angiogenesis in the artery wall may be beneficial to delay disease progression or reduce the incidence of acute ischemic complications that are the major causes of cardiovascular morbidity and mortality.

PATHOLOGIC DESCRIPTION OF PLAQUE ANGIOGENESIS

The observation of intimal neovascularization within atherosclerotic plaques has been well described by Virchow, Winternitz, Barger, and others (Koester 1876; Barger et al. 1984). In the normal artery, the adventitial first-order vasa vasorum run parallel to the artery and give rise to secondary branches that extend around the artery wall. In regions of lesion development, the proliferation of vasa vasorum also disrupts their native orientation and anatomy (Kwon et al. 1998). Casting studies and confocal microscopy demonstrate that these intimal vessels more frequently arise from the dense network of vessels in the adventitia adjacent to a plaque rather than from the main artery lumen (Zhang et al. 1993). Although these

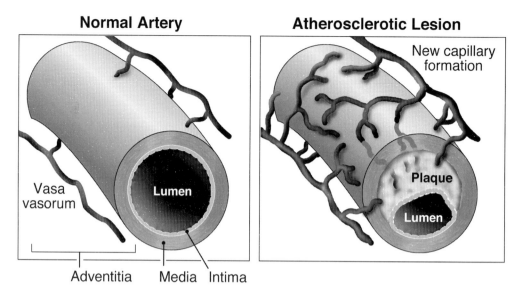

Figure 3. Diagram of vasa vasorum in normal and diseased arteries. The vasa vasorum in regions without lesion involvement are less abundant and are confined to the outer adventitial and medial layer. In regions with atherosclerotic lesions, the network of vasa vasorum is more extensive and extends into the intimal layer of the plaque. (Illustration by Steven Moskowitz.)

anatomical studies have shown a clear spatial association between the proliferation of vasa vasorum and the presence of atherosclerotic lesions, the function of these capillaries in the progression and manifestations of atherosclerosis have not been fully determined.

FUNCTION TO SUSTAIN GROWTH BEYOND LIMITS OF DIFFUSION FROM THE VESSEL LUMEN

It has been speculated that plaque neovascularization can promote plaque growth by sustaining the growth of tissue beyond critical limits of diffusion from the artery lumen. Several observations support this function. Vasa vasorum are more developed in the normal vessels of large animal species and humans that have increased numbers of lamellar units or cell layers in the vessel wall (Wolinsky and Glagov 1967). Surgical interruption or compression of vasa vasorum can result in medial necrosis (Werber et al. 1987; Williams et al. 1988a). Microsphere-injection techniques in nonhuman primate models of atherosclerosis have shown that the increase in blood flow to the intima and media relative to blood flow in the adventitia is due to proliferation of the vasa vasorum (Williams et al. 1988b). The dependence of the artery wall for diffusional support from the vasa vasorum network has also been demonstrated in diffusion studies; however, this perfusion is often heterogeneous, and residual hypoxic tissues can be detected in advanced atheromas even in the presence of intimal neovascularization (Werber et al. 1987; Bjornheden et al. 1999). Plaque capillaries are observed more frequently in human atherosclerotic lesions with wall thicknesses that exceed 0.5 mm, which suggests that wall ischemia may be one determinant for the development of intimal neovascularization (Geiringer 1951). Similarly, despite the limited dimensions of the thin-walled mouse aorta, the incidence of plaque neovascularization was increased ninefold in atherosclerotic lesions of apolipoprotein E-deficient (ApoE$^{-/-}$) mice with a thickened intima that exceeded 250 µm (Moulton et al. 1999). These dimensions suggest there may be critical limits of perfusion beyond which the development of neovascularization is a precondition for further growth. This situation may be comparable to the physiologic constraints regulating tumor mass and growth, where, for example, intravital measurements detect low oxygen tensions ($pO_2 < 50$) and acidosis (pH <6.9) when the distance from a capillary exceeds even 100 µm (Torres Filho et al. 1994).

If the function of plaque neovascularization is limited to providing passive perfusion to the expanding atheroma, then the extent of neovascularization would be correlated with plaque size, and the distribution of plaque capillaries would be more prevalent in regions of the plaque distant from adventitial capillaries or blood flow in the main artery. A more detailed evaluation of plaque size and intimal neovascularization, however, does not demonstrate a strong linear correlation between both parameters (Zhang et al. 1993). Withdrawal of an atherogenic diet from primates with established atherosclerosis resulted in a sixfold reduction of plaque blood flow but only a minimal regression of plaque size (Williams et al. 1988b). In murine lesions that contain intimal capillaries, the frequency of microvessels does not show a linear correlation with plaque thickness (Moulton et al. 2003). Instead, intimal neovascularization often occurs in "hot spots" in the shoulders of atheromas and is more abundant in restenotic compared to similar-sized primary lesions (O'Brien et al. 1994). Temporal fluctuations in the development and regression of intimal vessels during the evolution of atheromas also make the relationship of neovascularization and plaque size more complex. Thus, additional factors independent of plaque size, such as the inflammatory cell and matrix composition of lesions, may be more important predictors of plaque angiogenesis.

INTIMAL CAPILLARIES: POTENTIAL ROUTES FOR INFLAMMATORY CELL ENTRY INTO LESIONS

Decades of research have led to the recognition of atherosclerosis as an inflammatory disease (Ross 1999). Angiogenesis and inflammation are independent processes but are closely related in biologic processes such as wound healing and response to injury. Pathologic neovascularization accompanies chronic inflammation in psoriasis, synovial pannus formation in rheumatoid arthritis, and chronic granulomatous diseases (Folkman 1995). In acute inflammation, microvessels dilate and increase their permeability. In chronic inflammation, neovascularization is the prominent vascular response and may function to sustain it (Folkman and Brem 1992). Chronic inflammation of large blood vessels affected by vasculitis or atherosclerosis show intimal neovascularization in regions that are rich with inflammatory cells (O'Brien et al. 1996; Kaiser et al. 1999). Several types of leukocytes, including macrophages, T cells, and mast cells, have been shown to activate angiogenesis (Polverini et al. 1977; Auerbach and Sidky 1979; Kaartinen et al. 1996). In experimental models, the pro-angiogenic activity of these leukocytes is due to their elaboration of endothelial cell growth factors or their release of metalloproteases that mobilize matrix-bound growth factors and facilitate binding to receptors on endothelial cells (Coussens et al. 1999; Bergers et al. 2000). The close proximity to inflammatory infiltrates and the expression of adhesion molecules, such as VCAM-1, ICAM-1, and E-selectin, on the endothelium of plaque vessels suggest that these vessels may recruit inflammatory cells into lesions, which can in turn stimulate further angiogenesis (O'Brien et al. 1996). Although the initial inflammatory infiltrate may initiate angiogenesis, once established, a positive feedback cycle may operate in chronic inflammation and atherosclerosis whereby inflammatory cells stimulate angiogenesis in the plaque and new vessels further recruit leukocytes. Recent observations in an experimental animal model of atherosclerosis support this function of plaque neovascularization as a conduit for leukocyte entry in atherosclerosis. The densities of vasa vasorum show a strong linear and spatial correlation with focal collections of inflammatory cells, but not the intimal thickness of murine atheromas. In addition, treatments of established advanced atheromas with an angiogenesis inhibitor secondarily reduced macrophage accumulation in plaques but had no apparent direct effect on circulating and resident inflammatory cells or direct effect on monocyte chemotaxis and early recruitment (Moulton et al. 2003). These results suggest that the effect of angiogenesis inhibitors on the progression of atherosclerosis may be linked to their reciprocal effects on chronic inflammation.

The concept that plaques can grow not just from the inside out via the arterial surface endothelium, but also from the outside in via the adventitial vasa vasorum, is also supported by studies which showed that labeled-plasma constituents were deposited into plaque tissues adjacent to intimal vessels (Zhang et al. 1993). The endothelium of the plaque microvasculature may therefore provide an important route for the infiltration of inflammatory cells and the deposition of plasma constituents into advanced lesions. Some of the molecular mechanisms that regulate inflammatory cell entry via the plaque microvasculature may be distinct from those of the arterial endothelium on the surface of the plaque. In early-stage fatty streak lesions, the endothelial area of exchange on the surface of the atheroma is large relative to the volume of the atheroma. As plaque tissue expands outward in advanced lesions, the ratio of endothelial surface area to the volume of the atheroma decreases. Thus, an advanced atheroma that acquires plaque neovascularization may have an expanded area of endothelium and hence a different rate of growth compared to atheromas that lack neovascularization. This difference is relevant if the extent of neovascularization in atheromas can be modified.

INTIMAL VESSELS AND CELL PROLIFERATION IN ATHEROSCLEROTIC LESIONS

Active angiogenesis was first demonstrated in human atherosclerotic lesions that showed dual staining for an endothelial cell marker and proliferating cell nuclear antigens (O'Brien et al. 1994). Analysis of atherectomy specimens from primary and restenotic human atherosclerotic lesions demonstrated plaque vessels in 42% of primary lesions and 63% of restenotic lesions, but only a minority of plaque capillaries contained replicating endothelial cells. Interestingly, the presence of intraplaque vessels in primary lesions was associated with proliferation of other cell types in the atheroma. These findings suggest that intimal capillaries may be markers of proliferating lesions or may in fact contribute to plaque growth. This correlation between proliferating cells and active plaque angiogenesis, however, was not significant for restenotic lesions.

PLAQUE ANGIOGENESIS PROMOTES PLAQUE GROWTH

In summary, the relationships of plaque angiogenesis with inflammatory cells and proliferating cells in lesions provide correlative evidence that these capillaries promote the progression of atherosclerosis. The experimental demonstration of a functional contribution of plaque angiogenesis in the development of atherosclerosis was previously impeded by the limited availability of specific agents that can modulate angiogenesis in an animal model of atherosclerosis. Prior studies demonstrated that advanced atheromas of ApoE$^{-/-}$ mice develop plaque neovascularization and chronic treatments with endothelial cell inhibitors, TNP-470 or endostatin, significantly reduced plaque neovascularization and growth (Moulton et al. 1999). Conversely, stimulators of angiogenesis, such as VEGF or nicotine, have been shown to induce lesion progression (Celletti et al. 2001; Heeschen et al. 2001). Although these positive and negative endothelial cell regulators may alter plaque growth via diverse mechanisms, together these results support the hypothesis that plaque neovascularization promotes the progression of atherosclerosis. If studies in other models confirm these observations, then new treatments directed at plaque angiogenesis might be considered to reduce the progression of atherosclerosis.

ASSOCIATION OF NEOVASCULARIZATION WITH PLAQUE INSTABILITY

Angiographic evaluations of atherosclerotic lesions to determine their effect on impaired blood flow do not accurately predict their tendency to cause acute ischemic syndromes. Early angiographic examinations of coronary lesions at the time of thrombolytic therapy have demonstrated that the majority of "culprit lesions" were not hemodynamically significant prior to plaque rupture, erosion, hemorrhage, and thrombosis (Falk et al. 1995). Consequently, the tightest stenosis does not necessarily confer the most risk for myocardial infarction. This is illustrated by the series of angiograms which show a small non-flow-restricting lesion in the right coronary artery that develops an acute occlusion and myocardial infarction only 6 months later (Fig. 4). Although angiographic evidence of coronary disease may stratify patients into different risk groups, the limited detection of unstable lesions by this and other screening modalities has prompted new investigations to identify and detect vulnerable features of plaque that render them more prone to rupture. The benefits of these screening methods can be realized only if a potential therapy exists to modify these vulnerable features and reduce the incidence of acute ischemic complications.

The clinical importance of intimal vessels in vulnerable plaques is suggested by their higher incidence in culprit lesions that result in unstable angina, myocardial infarction, and stroke (Paterson 1938; Tenaglia et al. 1998). Acute ischemic complications constitute the major causes of morbidity and mortality from atherosclerosis. Coronary artery plaques harvested from patients with coronary artery disease who died suddenly have shown two plaque morphologies associated with plaque disruption (Burke et al. 1997). The more common morphology consists of a dense infiltration of macrophages within a fibrous cap that overlies a relatively hypocellular collection of lipid. The occlusive thrombus makes a direct connection with the lipid core material through a broken thin fibrous cap in this type of culprit lesion. The second pattern of vascular occlusion reported in these pathologic studies is an acute thrombus that overlies an eroded area of plaque tissue rich in smooth muscle cells and proteoglycans. Although plaque neovascularization has not been evaluated within these two "vulnerable" plaque morphologies, plaque hemorrhage has been previously reported as a potential mechanism for plaque instability, particularly in the carotid arteries (Paterson 1938; Davies and Thomas 1985; Barger and Beeuwkes 1990).

Plaque neovascularization may directly promote plaque rupture if these microvessels are a source of intraplaque hemorrhage (van der Wal et al. 1994; Libby 1995). Similar to new vessels in diabetic retinopathy, the plaque capillaries in the intima may be mechanically weak and more prone to bleed as they are exposed to arterial hemodynamic forces. The colocalization of hemosiderin-containing macrophages and intimal vessels and the occurrence of intraplaque hemorrhage with no connection to the arterial lumen suggest that these vessels

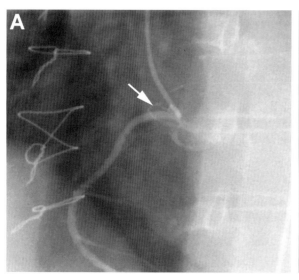
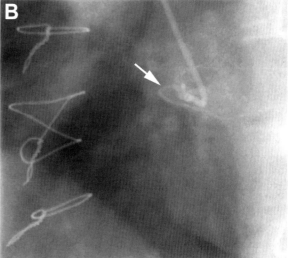

Figure 4. Serial coronary artery angiograms in the same patient. Initial angiogram of the right coronary artery shows a patient artery with a small non-flow-limiting stenosis at the site of a coronary artery lesion (*A*, *arrow*). About 6 months later, a subsequent angiogram performed due to the onset of rest angina showed an occlusion of the right coronary artery at the site of the previously small narrowing. (Courtesy of Brigham and Women's Hospital catheterization laboratory.)

may be involved in disruption events of some lesions (Virmani and Roberts 1983; Brogi et al. 1993; Kumamoto et al. 1995). Hemorrhage within the plaque may secondarily initiate other inflammatory and biochemical responses that lead to plaque rupture and arterial thrombosis.

Intimal neovascularization may indirectly promote plaque rupture due to protease activation that occurs during angiogenesis and due to its correlation with focal collections of inflammatory cells. First, angiogenesis requires proteolysis of the existing basement membrane before endothelial cells can develop vascular sprouts, which then invade the adjacent tissue (Brooks et al. 1996). Endothelial cells in sprouting capillaries express matrix metalloprotease-1 (MMP-1, or interstitial collagenase-1) and membrane type-1 matrix metalloproteinases (MT1-MMP), which are capable of degrading collagen and fibrin matrix (Gross et al. 1983; Hiraoka et al. 1998). In sites of angiogenesis, sprouting capillaries digest their way through tissues, which may weaken the mechanical properties of the plaque. The convergence of mechanical forces and the higher prevalence of intimal neovascularization in the shoulder regions of plaques may render these sites more susceptible for plaque disruption. Atherosclerotic lesions show increased activation of the matrix metalloproteases MMP-2 and MMP-9 in these vulnerable regions (Galis et al. 1994). Thus, factors that stimulate plaque angiogenesis may also activate tissue remodeling and alter plaque stability.

Second, plaque neovascularization can indirectly promote plaque disruption through its correlation with inflammatory cells. Vulnerable plaque morphologies have shown that the disrupted areas in the fibrous cap, which covers the lipid pool in a vulnerable plaque, are frequently infiltrated with macrophages. A shift toward the predominance of macrophages over smooth muscle cells in the fibrous cap is a condition that promotes plaque rupture by both increasing the cells that degrade matrix and reducing the cells that synthesize matrix (Davies et al. 1993; Libby 1995).

POTENTIAL REGULATORS OF PLAQUE ANGIOGENESIS

Although neovascularization may have significant functions in the progression and acute manifestations of atherosclerosis, little is known about the mechanisms that regulate angiogenesis within the plaque. Several endothelial cell growth factors and angiogenic stimulators, including VEGF, FGF family members, chemokines, platelet-derived endothelial cell growth factor, placental growth factor, granulocyte monocyte colony stimulation factor GM-CSF, and others, have been demonstrated in human atherosclerotic lesions (Brogi et al. 1993; Couffinhal et al. 1997; Inoue et al. 1998). Atherosclerotic tissues induce angiogenesis on the chorioallantoic membrane assay and in the cornea micropocket assay of angiogenesis (Moulton et al. 2003). Some of the pro-angiogenic activities associated with atherosclerotic lesions have been subsequently identified as VEGF (Bo et al. 1992; Kuzuya et al. 1995). Blocking antibodies to the VEGF receptor 2 on endothelial cells, however, did not impair plaque growth and angiogenesis during a relatively short period of treatment in vivo (Luttun et al. 2002).

Although endothelial cell growth factors are abundant in atherosclerotic lesions (even in nonvascularized early-stage fatty streak lesions), these factors are present well before the onset of plaque angiogenesis. Furthermore, their expression patterns do not spatially coincide with the distribution of plaque neovascularization (Brogi et al. 1993; Couffinhal et al. 1997). Thus, endothelial cell growth factors may be necessary, but not sufficient, for angiogenesis to extend into the intimal layer of atherosclerotic lesions. Angiogenesis may depend on the net balance of positive and negative regulators in the plaque tissue. Either activation of stimulatory signals to overcome endogenous inhibitors, or down-regulation of inhibitors, could initiate the onset of plaque angiogenesis.

Mechanisms of the angiogenic switch have been studied in transgenic mouse models of tumor progression where the progression of non-angiogenic dormant tumors to angiogenic growing tumors can be monitored temporally. The non-angiogenic tumor nodules express high levels of pro-angiogenic factors, but the onset of tumor angiogenesis correlated with mobilization of sequestered growth factors or down-regulation of an angiogenesis inhibitor (Hanahan and Folkman 1996; Bergers et al. 2000). It is possible that progressive loss of an inhibitor as the atheroma becomes more advanced may be an additional condition to stimulate plaque angiogenesis.

Little information is currently available about the levels of known endogenous endothelial cell inhibitory molecules in atherosclerotic lesions. Thrombospondin-1, previously identified as an inhibitor of angiogenesis regulated by the p53 tumor suppressor gene, is prevalent in atherosclerotic lesions and may inhibit plaque angiogenesis while promoting smooth muscle cell proliferation (Good et al. 1990; Roth et al. 1998; Chen et al. 1999). Interestingly, single-nucleotide polymorphisms of thrombospondin 1, 2, and 4 were associated with reduced plasma levels of thrombospondin and increased risk for the premature onset of myocardial infarction (Topol et al. 2001). However, it remains to be determined whether these gene variants affect angiogenesis, platelet function, matrix interactions, or other functions reported for these large complex pluripotent molecules.

Additional studies have shown that another endothelial cell inhibitor, endostatin, a 20-kD protein derived from collagen XVIII, is a highly abundant protein in the aorta wall. The endostatin domain of collagen XVIII binds to laminin, fibulin, and nidogen proteins that are associated with elastin fibers in the medial layer (Miosge et al. 1999; Javaherian et al. 2002). These studies have not identified the production source of endostatin in the aorta, nor have they determined whether the deposited proteins are functionally active. If the endostatin deposits in the artery wall are biologically active, they may impede the inward growth of vasa vasorum toward the intima layer. In addition, the abundance of endostatin in the artery wall may be altered in the setting of atherosclerosis and other conditions of vascular inflammation. Studies are in progress to evaluate the effect of collagen XVIII deficiency on plaque angiogenesis and the progression of atherosclerosis in genetically susceptible mice. Since capillary growth within a vessel wall is a pathologic process, these future results may demonstrate a paradigm for the presence of additional factors in the wall of native large blood vessels that inhibit the proliferation of vasa vasorum and antagonize the formation of abnormal fistulous connections between blood vessels that develop in parallel.

EVIDENCE THAT BLOCKING PLAQUE ANGIOGENESIS WILL STABILIZE LESIONS

It is not yet determined whether therapies that directly inhibit or regress intimal vessels will result in lesion stabilization. The experimental verifications of these questions are partly limited by animal models of atherosclerosis that are widely accepted to represent plaque rupture and hemorrhage events in human disease, although some murine strains can manifest specific features (Palinski and Napoli 2002). In several animal models of atherosclerosis, lesion regression after withdrawal of an atherogenic diet is often accompanied by regression of intimal vessels, decreased macrophages, reduced metalloprotease activities, and reduced intraplaque hemorrhage (Williams et al. 1988b; Aikawa et al. 1998). Non-cholesterol-lowering doses of HMG-CoA reductase agents have anti-angiogenic effects and reduce plaque angiogenesis in vivo; however, these properties may not account for the mechanisms of plaque stabilization observed in widespread clinical use (Park et al. 2002; Wilson et al. 2002). The available data therefore show a correlation between lesion stabilization and reduced plaque angiogenesis, but it is not sufficient to conclude a causal relationship.

POTENTIAL INDICATIONS FOR ANGIOGENESIS INHIBITOR TREATMENTS

The treatment of ischemia symptoms due to flow-limiting stenosis in arteries is the premise for coronary artery and vascular bypass surgeries, angioplasty, and stent procedures and is also the goal in clinical therapeutic angiogenesis trials. Since collaterals rarely develop in the absence of ischemia or vascular occlusion, collateral trials are a palliative treatment for established vascular disease. The potential use of angiogenesis inhibitors as treatments for cardiovascular diseases is at an early stage of investigation. These agents are mostly used as experimental tools in animal studies to characterize the basic functions and mechanisms of plaque angiogenesis. Considerable research and time are required to validate the feasibility and utility of regulating plaque neovascularization as a putative treatment goal in atherosclerosis. Future studies are also required to determine whether angiogenesis inhibitors are able to regress established intimal vessels. Endothelial cell inhibitors that inhibit endothelial cell proliferation or migration may in principle have few effects on quiescent nonproliferating endothelium. Therefore, regression of established vessels versus inhibition of new vessel formation may require different biologic activities.

The experimental treatments of Apo $E^{-/-}$ mice with angiogenesis inhibitors utilized a secondary prevention strategy, which resulted in reduced intimal vessels and the growth of atheromas during the treatment period. A preventative strategy that inhibits plaque growth via inhibition of plaque angiogenesis could in principle obviate the need for collaterals by controlling the causes of impaired blood flow in diseased arteries. Several important hurdles must be addressed before these strategies could become practice. First, atherosclerosis is highly prevalent, and the costs of providing chronic treatments are substantial. Second, the selection criterion used to identify appropriate candidates must be established. For example, what portion of the large population with coronary artery disease would most benefit from this type of treatment? Is it possible to screen and identify those patients that have more severe plaque neovascularization? Third, once a target group of patients is identified, it will be necessary to have reliable clinical and biologic parameters to assess the effect of these interventions on plaque angiogenesis. A modality to monitor angiogenesis activity in vulnerable lesions may be a useful clinical tool. Ultimately, the practical applications of angiogenesis inhibitors to block plaque angiogenesis may also depend on the cost and clinical availability of well-tolerated oral agents.

If plaque angiogenesis is a feature of unstable lesions, it may be possible to provide short-term treatments to inhibit neovascularization and inflammation in order to stabilize lesions. Some patients present with unstable angina but have no clear culprit lesion on angiogram, yet remain at high risk to develop future cardiovascular events. Autopsy studies of patients presenting with an acute myocardial infarction from a known culprit lesion have a high chance of having several other coincidental unstable lesions (Burke et al. 1997; Goldstein et al. 2000). The presenting symptoms of angina may be produced from one unstable plaque, but the local culprit lesion may be the index marker of more widespread disease. These patients have a high risk for recurrent symptoms and cardiovascular mortality, even when the culprit lesion is managed and blood flow is restored. For this population of unstable patients, systemic treatments to stabilize lesions may be useful adjuvants to catheter-directed treatments of the culprit lesion because they work independently of lesion location.

USE OF ANGIOGENESIS STIMULATORS VERSUS INHIBITORS

The clinical uses of endothelial cell growth factors and inhibitors for patients with cardiovascular disease raise a number of relevant questions. Will exogenous growth factor treatments designed to promote collaterals have an untoward effect to enhance intimal angiogenesis or destabilize lesions? In the corneal micropocket assay, the amount of angiogenesis induced by a pellet that contains bFGF or VEGF is not enhanced when additional systemic growth factors are administered. Plaque tissue already has a significant local supply of VEGF and acidic FGF, so it is not clear that systemic doses of these growth factors will activate angiogenesis in the plaque. On the other hand, ischemic myocardium and limbs also have increased expression of VEGF and bFGF, yet systemic delivery of these growth factors results in angiogenesis and the development of collateral vessels. Recent studies have shown atherosclerosis is enhanced after short-term exposure to endothelial cell stimulators (Celletti et al. 2001; Heeschen et al. 2001). Some of these effects of growth factors could be related to other actions of these agents and not be restricted to their effects on plaque angiogenesis alone. Many angiogenic factors have effects on other cell types found within atherosclerotic lesions. For example, bFGF may promote smooth muscle cell proliferation and the growth of lesions independent of an effect on plaque capillaries. VEGF has effects on monocyte cells via the VEGF receptor 1 and may promote the release of endothelial or hematopoietic progenitor cells from the bone marrow that can migrate into atherosclerotic lesions (Barleon et al. 1996; Lyden et al. 2001; Sata et al. 2002). In the setting of hypercholesterolemia, factors such as VEGF or angiopoietin-2 may promote vascular permeability that can increase delivery of plasma constituents and lipids into atherosclerotic lesions (Thurston et al. 2000). Given the relatively short period of time required for collaterals to grow, the change in plaque size, inflammation, and angiogenesis could prove to be self-limited. The small number of patients treated in these trials may not be high enough to detect changes in cardiovascular events that are significantly different compared to patients with similar extents of severe atherosclerosis.

The converse question also arises. Will angiogenesis inhibitors block collateral development? Angiogenesis inhibitors such as thrombospondin and platelet factor 4 can inhibit collateral development when provided at the time of vascular occlusion but are not effective if the inhibitor treatments are delayed (Couffinhal et al. 1998). Long-term evaluations of collaterals have shown that once formed, collateral vessels are stable and remain potent even after endothelial cell growth factor levels reduce, which may occur after successful revascularization.

MAKING ARTERIES GROW AND BLOCKING PLAQUE ANGIOGENESIS

The use of both endothelial cell growth factors and inhibitors to treat the same disease may initially seem contradictory or mutually exclusive applications of these regulatory factors. These alternative strategies have different objectives appropriate for different stages and different manifestations of atherosclerosis. Strategies to promote collateral vessels relieve ischemia caused by flow-limiting obstructions in native vessels. Thus, these treatments have palliative treatment goals similar to coronary artery bypass grafting (CABG), stents, and angioplasty. The obvious treatment goals are relief of angina or claudication and avoidance of amputation. Ultimately, it is hoped that the stimulation of coronary collaterals may correct ischemia sufficiently to preserve left ventricular function or reduce the area of infarction that results after acute occlusion of a native artery. It remains to be shown, however, whether collateral treatments will alter cardiovascular mortality. Despite the marked enhancement of coronary perfusion produced by coronary artery bypass grafts, angioplasty, or stent interventions, these modalities relieve chest pain events and improve the functional capacity of patients, but only affect cardiovascular mortality and morbidity in specific subsets of patients (BARI Investigators 1996). Therapeutic attempts to promote collaterals may also have limitations for the treatment of proximal left main coronary lesions where the large ischemic myocardial territory has high clinical risk and where collaterals that arise directly from the aorta are less frequently seen.

In contrast, potential strategies to inhibit plaque angiogenesis are based on their function to promote plaque growth and their putative role in plaque instability. In this regard, the potential therapeutic goals would be preventative and designed to delay progression of the disease or reduce the incidence of acute ischemic complications. Pertinent to these discussions, it is important to recognize that the potential vascular applications of angiogenesis inhibitors to block plaque angiogenesis are still in an early stage of basic investigation compared to the animal and clinical studies of collateral development. Although there are no current trials of angiogenesis inhibitors for cardiovascular disease, clinical applications of angiogenesis inhibitors for cancer, retinopathy, and other conditions may have concurrent effects on vascular diseases, which are prevalent in these patient populations. Ironically, such treatments have the potential to affect plaque progression and stability while ameliorating another disease.

ANGIOGENESIS INHIBITORS IN CLINICAL USE

Clinical trials of angiogenesis inhibitors are currently under investigation for the treatment of cancer, macular degeneration, hemangiomas, and arthritis. Angiogenesis itself is a complex biologic process involved principally with the endothelial cell but also with other cell types and biologic processes involved in tissue remodeling. Agents that inhibit this process are diverse because they can be directed at multiple targets and cellular events, such as the endothelial cell itself (proliferation, migration, apoptosis, endothelial cell receptors), the availability of endothelial cell growth factors, downstream signaling responses of growth factor receptors, the activity of metalloprotease enzymes, and the recruitment of pericytes to promote the maturation and stabilization of the newly formed capillary. In the conduct of clinical trials to test the efficacy of angiogenesis regulators for other disease, there may be coincidental treatment-related effects on the cardiovascular system that can be monitored. In some cases, however, the validity of these uncontrolled observations may be confounded by the presence of significant co-morbid diseases. Nevertheless, these studies may produce important observations that will improve our understanding about the cardiovascular effects of these agents. Due to the diverse mechanisms of action for different angiogenesis inhibitors, it is not expected that the effects of separate agents of plaque neovascularization will be similar. Although VEGF may be an important angiogenic molecule in inflammation-related angiogenesis, this factor may also have important properties to maintain normal endothelial cell function and permeability in specific capillary beds like the kidneys and lungs (Gerber et al. 1999; Ng et al. 2001). Different angiogenesis inhibitors may have additional effects (either beneficial or undesirable) on different cell types and biochemical pathways that are also present in atherosclerotic lesions. Similar to the experience with other oncology drugs, there may be various classes of angiogenesis regulators that have the potential to modify the cardiovascular system, and it will be important for cardiologists to appreciate both the potential and complexity of these interactions. The use of angiogenesis inhibitors to treat cancer requires chronic treatment. Thus, understanding the functions and regulation of plaque angiogenesis will be important and relevant information as the clinical and experimental use of these agents expands.

FUTURE DIRECTIONS AND CONCLUSION

The practical application of strategies to inhibit plaque angiogenesis will require the testing and development of effective oral agents. Oral agents provide the advantages of easy administration and chronic systemic treatment. Local delivery of angiogenesis inhibitors via catheters, gene transfer techniques, and endovascular stents or polymers may also be used to control localized disease. However, because atherosclerosis affects multiple vessels and the risk that a plaque might rupture cannot be predicted, the rationale to select certain lesions for local delivery of angiogenesis inhibitors needs to be defined. Justification for these strategies will require an improved understanding of the positive and negative regulators that control arteriogenesis or plaque angiogenesis, and a further characterization of the functions that these vessels impart in the progression of atherosclerosis.

ACKNOWLEDGMENT

The author's studies summarized in this review were supported by a grant from the National Heart, Lung, and Blood Institute (R01 HL-67255).

REFERENCES

Aikawa M., Rabkin E., Okada Y., Voglic S., Clinton S., Brinckerhoff C., Sukhova G., and Libby P. 1998. Lipid lowering by diet reduces matrix metalloproteinase activity and increases collagen content of rabbit atheroma. A potential mechanism of lesion stabilization. *Circulation* **97:** 2433.

Arras M., Ito W.D., Scholz D., Winkler B., Schaper J., and Schaper W. 1998. Monocyte activation in angiogenesis and collateral growth in the rabbit hindlimb. *J. Clin. Invest.* **101:** 40.

Asahara T., Chen D., Takahashi T., Fujikawa K., Kearney M., Magner M., Yancopoulos G.D., and Isner J.M. 1998. Tie2 receptor ligands, angiopoietin-1 and angiopoietin-2, modulate VEGF-induced postnatal neovascularization. *Circ. Res.* **83:** 233.

Auerbach R. and Sidky Y.A. 1979. Nature of the stimulus leading to lymphocyte-induced angiogenesis. *J. Immunol.* **123:** 751.

Barger A.C. and Beeuwkes R., III. 1990. Rupture of coronary vasa vasorum as a trigger of acute myocardial infarction. *Am. J. Cardiol.* **66:** 41G.

Barger A.C., Beeuwkes R., III, Lainey L.L., and Silverman K.J. 1984. Hypothesis: Vasa vasorum and neovascularization of human coronary arteries. A possible role in the pathophysiology of atherosclerosis. *N. Engl. J. Med.* **310:** 175.

BARI (Bypass Angioplasty Revascularization Investigation) Investigators. 1996. Comparison of coronary bypass surgery with angioplasty in patients with multivessel disease (erratum in *N. Engl. J. Med.* [1997] **336:**147). *N. Engl. J. Med.* **335:** 217.

Barleon B., Sozzani S., Zhou D., Weich H.A., Mantovani A., and Marme D. 1996. Migration of human monocytes in response to vascular endothelial growth factor (VEGF) is mediated via the VEGF receptor flt-1. *Blood* **87:** 3336.

Baumgartner I., Pieczek A., Manor O., Blair R., Kearney M., Walsh K., and Isner J.M. 1998. Constitutive expression of phVEGF165 after intramuscular gene transfer promotes collateral vessel development in patients with critical limb ischemia. *Circulation* **97:** 1114.

Bergers G., Brekken R., McMahon G., Vu T.H., Itoh T., Tamaki K., Tanzawa K., Thorpe P., Itohara S., Werb Z., and Hanahan D. 2000. Matrix metalloproteinase-9 triggers the angiogenic switch during carcinogenesis. *Nat. Cell Biol.* **2:** 737.

Bjornheden T., Levin M., Evaldsson M., and Wiklund O. 1999. Evidence of hypoxic areas within the arterial wall in vivo. *Arterioscler. Thromb. Vasc. Biol.* **19:** 870.

Bo W.J., Mercuri M., Tucker R., and Bond M.G. 1992. The human carotid atherosclerotic plaque stimulates angiogenesis on the chick chorioallantoic membrane. *Atherosclerosis* **94:** 71.

Brogi E., Winkles J.A., Underwood R., Clinton S.K., Alberts G.F., and Libby P. 1993. Distinct patterns of expression of fibroblast growth factors and their receptors in human atheroma and nonatherosclerotic arteries. Association of acidic FGF with plaque microvessels and macrophages. *J. Clin. Invest.* **92:** 2408.

Brooks P.C., Stromblad S., Sanders L.C., von Schalscha T.L., Aimes R.T., Stetler-Stevenson W.G., Quigley J.P., and Cheresh D.A. 1996. Localization of matrix metalloproteinase MMP-2 to the surface of invasive cells by interaction with integrin alpha v beta 3. *Cell* **85:** 683.

Burke A.P., Farb A., Malcom G.T., Liang Y.H., Smialek J., and Virmani R. 1997. Coronary risk factors and plaque morphology in men with coronary disease who died suddenly. *N. Engl. J. Med.* **336:** 1276.

Celletti F.L., Waugh J.M., Amabile P.G., Brendolan A., Hilfiker P.R., and Dake M.D. 2001. Vascular endothelial growth factor enhances atherosclerotic plaque progression. *Nat. Med.* **7:** 425.

Chen D., Asahara T., Krasinski K., Witzenbichler B., Yang J., Magner M., Kearney M., Frazier W.A., Isner J.M., and Andres V. 1999. Antibody blockade of thrombospondin accelerates reendothelialization and reduces neointima formation in balloon-injured rat carotid artery. *Circulation* **100:** 849.

Couffinhal T., Silver M., Zheng L.P., Kearney M., Witzenbichler B., and Isner J.M. 1998. Mouse model of angiogenesis. *Am. J. Pathol.* **152:** 1667.

Couffinhal T., Kearney M., Witzenbichler B., Chen D., Murohara T., Losordo D.W., Symes J., and Isner J.M. 1997. Vascular endothelial growth factor/vascular permeability factor (VEGF/VPF) in normal and atherosclerotic human arteries. *Am. J. Pathol.* **150:** 1673.

Coussens L.M., Raymond W.W., Bergers G., Laig-Webster M., Behrendtsen O., Werb Z., Caughey G.H., and Hanahan D. 1999. Inflammatory mast cells up-regulate angiogenesis during squamous epithelial carcinogenesis. *Genes Dev.* **13:** 1382.

Davies M.J. and Thomas A.C. 1985. Plaque fissuring: The cause of acute myocardial infarction, sudden ischaemic death, and crescendo angina. *Br. Heart J.* **53:** 363.

Davies M.J., Richardson P.D., Woolf N., Katz D.R., and Mann

J. 1993. Risk of thrombosis in human atherosclerotic plaques: Role of extracellular lipid, macrophage, and smooth muscle cell content. *Br. Heart J.* **69:** 377.

Falk E., Shah P.K., and Fuster V. 1995. Coronary plaque disruption. *Circulation* **92:** 657.

Folkman J. 1995. Angiogenesis in cancer, vascular, rheumatoid and other disease. *Nat. Med.* **1:** 27.

Folkman J. and Brem H. 1992. Angiogenesis and inflammation. In *Inflammation: Basic principles and clinical correlates* (ed. J.I. Gallin et al.), p. 821. Raven Press, New York.

Galis Z.S., Sukhova G.K., Lark M.W., and Libby P. 1994. Increased expression of matrix metalloproteinases and matrix degrading activity in vulnerable regions of human atherosclerotic plaques. *J. Clin. Invest.* **94:** 2493.

Geiringer E. 1951. Intimal vascularization and atherosclerosis. *J. Pathol. Bacteriol.* **63:** 210.

Gerber H.P., Hillan K.J., Ryan A.M., Kowalski J., Keller G.A., Rangell L., Wright B.D., Radtke F., Aguet M., and Ferrara N. 1999. VEGF is required for growth and survival in neonatal mice. *Development* **126:** 1149.

Goldstein J.A., Demetriou D., Grines C.L., Pica M., Shoukfeh M., and O'Neill W.W. 2000. Multiple complex coronary plaques in patients with acute myocardial infarction. *N. Engl. J. Med.* **343:** 915.

Good D.J., Polverini P.J., Rastinejad F., Le Beau M.M., Lemons R.S., Frazier W.A., and Bouck N.P. 1990. A tumor suppressor-dependent inhibitor of angiogenesis is immunologically and functionally indistinguishable from a fragment of thrombospondin. *Proc. Natl. Acad. Sci.* **87:** 6624.

Gross J.L., Moscatelli D., and Rifkin D.B. 1983. Increased capillary endothelial cell protease activity in response to angiogenic stimuli in vitro. *Proc. Natl. Acad. Sci.* **80:** 2623.

Hanahan D. and Folkman J. 1996. Patterns and emerging mechanisms of the angiogenic switch during tumorigenesis. *Cell* **86:** 353.

Heeschen C., Jang J.J., Weis M., Pathak A., Kaji S., Hu R.S., Tsao P.S., Johnson F.L., and Cooke J.P. 2001. Nicotine stimulates angiogenesis and promotes tumor growth and atherosclerosis. *Nat. Med.* **7:** 833.

Hiraoka N., Allen E., Apel I., Gyetko M., and Weiss S. 1998. Matrix metalloproteinases regulate neovascularization by acting as pericellular fibrinolysins. *Cell* **95:** 365.

Inoue M., Itoh H., Ueda M., Naruko T., Kojima A., Komatsu R., Doi K., Ogawa Y., Tamura N., Takaya K., Igaki T., Yamashita J., Chun T.H., Masatsugu K., Becker A.E., and Nakao K. 1998. Vascular endothelial growth factor (VEGF) expression in human coronary atherosclerotic lesions : Possible pathophysiological significance of VEGF in progression of atherosclerosis. *Circulation* **98:** 2108.

Javaherian K., Park S.Y., Pickl W.F., LaMontagne K.R., Sjin R.T., Gillies S., and Lo K.M. 2002. Laminin modulates morphogenic properties of the collagen XVIII endostatin domain. *J. Biol. Chem.* **277:** 45211.

Kaartinen M., Penttila A., and Kovanen P.T. 1996. Mast cells accompany microvessels in human coronary atheromas: Implications for intimal neovascularization and hemorrhage. *Atherosclerosis* **123:** 123.

Kaiser M., Younge B., Bjornsson J., Goronzy J.J., and Weyand C.M. 1999. Formation of new vasa vasorum in vasculitis. Production of angiogenic cytokines by multinucleated giant cells. *Am. J. Pathol.* **155:** 765.

Koester W. 1876. Endarteritis and arteritis. *Berl. Klin. Wochenschr.* **13:** 454.

Kumamoto M., Nakashima Y., and Sueishi K. 1995. Intimal neovascularization in human coronary atherosclerosis: Its origin and pathophysiological significance. *Hum. Pathol.* **26:** 450.

Kuzuya M., Satake S., Esaki T., Yamada K., Hayashi T., Naito M., Asai K., and Iguchi A. 1995. Induction of angiogenesis by smooth muscle cell-derived factor: Possible role in neovascularization in atherosclerotic plaque. *J. Cell. Physiol.* **164:** 658.

Kwon H.M., Sangiorgi G., Ritman E.L., McKenna C., Holmes D.R., Jr., Schwartz R.S., and Lerman A. 1998. Enhanced coronary vasa vasorum neovascularization in experimental hypercholesterolemia. *J. Clin. Invest.* **101:** 1551.

Laham R.J., Sellke F.W., Edelman E.R., Pearlman J.D., Ware J.A., Brown D.L., Gold J.P., and Simons M. 1999. Local perivascular delivery of basic fibroblast growth factor in patients undergoing coronary bypass surgery: Results of a phase I randomized, double-blind, placebo-controlled trial. *Circulation* **100:** 1865.

Libby P. 1995. Molecular bases of the acute coronary syndromes. *Circulation* **91:** 2844.

Losordo D., Vale P., Symes J., Dunnington C., Esakof D., Maysky M., Ashare A., Lathi K., and Isner J. 1998. Gene therapy for myocardia angiogenesis. Initial clinical results with direct myocardial injection of phVEGF165 as sole therapy for myocardial ischemia. *Circulation* **98:** 2800.

Luttun A., Tjwa M., Moons L., Wu Y., Angelillo-Scherrer A., Liao F., Nagy J.A., Hooper A., Priller J., De Klerck B., Compernolle V., Daci E., Bohlen P., Dewerchin M., Herbert J.M., Fava R., Matthys P., Carmeliet G., Collen D., Dvorak H.F., Hicklin D.J., and Carmeliet P. 2002. Revascularization of ischemic tissues by PlGF treatment, and inhibition of tumor angiogenesis, arthritis and atherosclerosis by anti-Flt1. *Nat. Med.* **8:** 831.

Lyden D., Hattori K., Dias S., Costa C., Blaikie P., Butros L., Chadburn A., Heissig B., Marks W., Witte L., Wu Y., Hicklin D., Zhu Z., Hackett N.R., Crystal R.G., Moore M.A., Hajjar K.A., Manova K., Benezra R., and Rafii S. 2001. Impaired recruitment of bone-marrow-derived endothelial and hematopoietic precursor cells blocks tumor angiogenesis and growth. *Nat. Med.* **7:** 1194.

Miosge N., Sasaki T., and Timpl R. 1999. Angiogenesis inhibitor endostatin is a distinct component of elastic fibers in vessel walls. *FASEB J.* **13:** 1743.

Moulton K.S., Heller E., Konerding M.A., Flynn E., Palinski W., and Folkman J. 1999. Angiogenesis inhibitors endostatin or TNP-470 reduce intimal neovascularization and plaque growth in apolipoprotein E-deficient mice. *Circulation* **99:** 1726.

Moulton K.S., Vakili K., Zurakowski D., Soliman M., Butterfiels C., Sylvin E., Lo K., Gillies S., Javaherian K., and Folkman J. 2003. Inhibition of plaque neovascularization reduces macrophage accumulation and progression of advanced atherosclerosis. *Proc. Natl. Acad. Sci.* **100:** 4736.

Ng Y.S., Rohan R., Sunday M.E., Demello D.E., and D'Amore P.A. 2001. Differential expression of VEGF isoforms in mouse during development and in the adult. *Dev. Dyn.* **220:** 112.

O'Brien E.R., Garvin M.R., Dev R., Stewart D.K., Hinohara T., Simpson J.B., and Schwartz S.M. 1994. Angiogenesis in human coronary atherosclerotic plaques. *Am. J. Pathol.* **145:** 883.

O'Brien K.D., McDonald T.O., Chait A., Allen M.D., and Alpers C.E. 1996. Neovascular expression of E-selectin, intercellular adhesion molecule-1, and vascular cell adhesion molecule-1 in human atherosclerosis and their relation to intimal leukocyte content. *Circulation* **93:** 672.

Palinski W. and Napoli C. 2002. Unraveling pleiotropic effects of statins on plaque rupture. *Arterioscler. Thromb. Vasc. Biol.* **22:** 1745.

Park H.J., Kong D., Iruela-Arispe L., Begley U., Tang D., and Galper J.B. 2002. 3-hydroxy-3-methylglutaryl coenzyme A reductase inhibitors interfere with angiogenesis by inhibiting the geranylgeranylation of RhoA. *Circ. Res.* **91:** 143.

Paterson J.C. 1938. Capillary rupture with intimal hemorrhage as a causative factor in coronary thrombosis. *Arch. Pathol.* **25:** 474.

Polverini P.J., Cotran R.S., Gimbrone M.A., Jr., and Unanue E.R. 1977. Activated macrophages induce vascular proliferation. *Nature* **269:** 804.

Rivard A., Silver M., Chen D., Kearney M., Magner M., Annex B., Peters K., and Isner J.M. 1999a. Rescue of diabetes-related impairment of angiogenesis by intramuscular gene therapy with adeno-VEGF. *Am. J. Pathol.* **154:** 355.

Rivard A., Fabre J.E., Silver M., Chen D., Murohara T., Kearney M., Magner M., Asahara T., and Isner J.M. 1999b. Age-dependent impairment of angiogenesis. *Circulation* **99:** 111.

Roberts W.C. and Virmani R. 1984. Formation of new coronary arteries within a previously obstructed epicardial coronary artery (intraarterial arteries): A mechanism for occurrence of angiographically normal coronary arteries after healing of acute myocardial infarction. *Am. J. Cardiol.* **54:** 1361.

Ross R. 1999. Atherosclerosis: An inflammatory disease. *N. Engl. J. Med.* **340:** 115.

Roth J.J., Gahtan V., Brown J.L., Gerhard C., Swami V.K., Rothman V.L., Tulenko T.N., and Tuszynski G.P. 1998. Thrombospondin-1 is elevated with both intimal hyperplasia and hypercholesterolemia. *J. Surg. Res.* **74:** 11.

Sata M., Saiura A., Kunisato A., Tojo A., Okada S., Tokuhisa T., Hirai H., Makuuchi M., Hirata Y., and Nagai R. 2002. Hematopoietic stem cells differentiate into vascular cells that participate in the pathogenesis of atherosclerosis. *Nat. Med.* **8:** 403.

Schumacher B., Pecher P., von Specht B.U., and Stegmann T. 1998. Induction of neoangiogenesis in ischemic myocardium by human growth factors: First clinical results of a new treatment of coronary heart disease. *Circulation* **97:** 645.

Shyu K.G., Manor O., Magner M., Yancopoulos G.D., and Isner J.M. 1998. Direct intramuscular injection of plasmid DNA encoding angiopoietin-1 but not angiopoietin-2 augments revascularization in the rabbit ischemic hindlimb. *Circulation* **98:** 2081.

Tanaka H., Sukhova G.K., and Libby P. 1994. Interaction of the allogeneic state and hypercholesterolemia in arterial lesion formation in experimental cardiac allografts. *Arterioscler. Thromb.* **14:** 734.

Tenaglia A.N., Peters K.G., Sketch M.H., Jr., and Annex B.H. 1998. Neovascularization in atherectomy specimens from patients with unstable angina: Implications for pathogenesis of unstable angina. *Am. Heart J.* **135:** 10.

Thurston G., Rudge J.S., Ioffe E., Zhou H., Ross L., Croll S.D., Glazer N., Holash J., McDonald D.M., and Yancopoulos G.D. 2000. Angiopoietin-1 protects the adult vasculature against plasma leakage. *Nat. Med.* **6:** 460.

Topol E.J., McCarthy J., Gabriel S., Moliterno D.J., Rogers W.J., Newby L.K., Freedman M., Metivier J., Cannata R., O'Donnell C.J., Kottke-Marchant K., Murugesan G., Plow E.F., Stenina O., and Daley G.Q. 2001. Single nucleotide polymorphisms in multiple novel thrombospondin genes may be associated with familial premature myocardial infarction. *Circulation* **104:** 2641.

Torres Filho I., Leunig M., Yuan F., Intaglietta M., and Jain R. 1994. Noninvasive measurement of microvascular and interstitial oxygen profiles in a human tumor in SCID mice. *Proc. Natl. Acad. Sci.* **91:** 2081.

Van Belle E., Rivard A., Chen D., Silver M., Bunting S., Ferrara N., Symes J.F., Bauters C., and Isner J.M. 1997. Hypercholesterolemia attenuates angiogenesis but does not preclude augmentation by angiogenic cytokines. *Circulation* **96:** 2667.

van der Wal A.C., Becker A.E., van der Loos C.M., and Das P.K. 1994. Site of intimal rupture or erosion of thrombosed coronary atherosclerotic plaques is characterized by an inflammatory process irrespective of the dominant plaque morphology. *Circulation* **89:** 36.

Virmani R. and Roberts W.C. 1983. Extravasated erythrocytes, iron, and fibrin in atherosclerotic plaques of coronary arteries in fatal coronary heart disease and their relation to luminal thrombus: Frequency and significance in 57 necropsy patients and in 2958 five mm segments of 224 major epicardial coronary arteries. *Am. Heart J.* **105:** 788.

Ware J.A. and Simons M. 1997. Angiogenesis in ischemic heart disease. *Nat. Med.* **3:** 158.

Werber A.H., Armstrong M.L., and Heistad D.D. 1987. Diffusional support of the thoracic aorta in atherosclerotic monkeys. *Atherosclerosis* **68:** 123.

Williams J.K., Armstrong M.L., and Heistad D.D. 1988a. Blood flow through new microvessels: Factors that affect regrowth of vasa vasorum. *Am. J. Physiol.* **254:** H126.

———. 1988b. Vasa vasorum in atherosclerotic coronary arteries: Responses to vasoactive stimuli and regression of atherosclerosis. *Circ. Res.* **62:** 515.

Wilson S.H., Herrmann J., Lerman L.O., Holmes D.R., Jr., Napoli C., Ritman E.L., and Lerman A. 2002. Simvastatin preserves the structure of coronary adventitial vasa vasorum in experimental hypercholesterolemia independent of lipid lowering. *Circulation* **105:** 415.

Wolinsky H. and Glagov S. 1967. A lamellar unit of aortic medial structure and function in mammals. *Circ. Res.* **20:** 99.

Zhang Y., Cliff W.J., Schoefl G.I., and Higgins G. 1993. Immunohistochemical study of intimal microvessels in coronary atherosclerosis. *Am. J. Pathol.* **143:** 164.

Extracellular Superoxide Dismutase, Uric Acid, and Atherosclerosis

H.U. HINK*† AND T. FUKAI*

*Division of Cardiology, Department of Medicine, Emory University School of Medicine and the Atlanta Veterans Administration Hospital, Atlanta, Georgia 30322; †University Hospital Hamburg-Eppendorf, Department of Cardiology, Hamburg 20246, Germany

Accumulating evidence suggests that atherosclerosis is associated with increased production of reactive oxygen species, especially superoxide anion, in the vessel wall, both from vascular cells and from macrophages that accumulate within the atherosclerotic lesion (Griendling et al. 2000). Indeed, many of the manifestations of atherosclerosis likely derive from the actions of reactive oxygen species: modification of lipids (Steinberg 1997), induction of proinflammatory genes (Marui et al. 1993; Cominacini et al. 1997; Wung et al. 1997), increased cellular migration and proliferation (Rao and Berk 1992), and endothelial dysfunction (Kojda and Harrison 1999).

One of the major cellular defenses against the superoxide anion ($O_2^{\cdot -}$) and formation of peroxynitrite is the superoxide dismutases (Beyer et al. 1991; Fridovich 1997). These include cytosolic Cu/ZnSOD (Cu/ZnSOD or SOD1), manganese SOD (MnSOD or SOD2), and extracellular Cu/ZnSOD (ecSOD or SOD3). In the vessel wall, ecSOD is highly expressed (Strålin et al. 1995). Although cytosolic Cu/ZnSOD functions as an antioxidant by catalyzing dismutation of $O_2^{\cdot -}$ to H_2O_2, some studies showed that Cu/ZnSOD has pro-oxidative properties, such as neurologic disorders (Wiedau-Pazos et al. 1996) or rather pro-atherogenic effects (Tribble et al. 1997). Several mechanisms have been proposed to account for such pro-oxidative effects: overproduction of H_2O_2 (Scott et al. 1987), peroxidative activity of SOD, and opposing roles played by $O_2^{\cdot -}$ in both initiation and termination of radical chain reactions (Nelson et al. 1994). Peroxidase properties of cytosolic Cu/ZnSOD have been studied extensively in vitro (Hodgson and Fridovich 1975; Singh et al. 1998; Sankarapandi and Zweier 1999). In vitro, the peroxidase substrate H_2O_2 leads to inhibition of Cu/ZnSOD (Hodgson and Fridovich 1975). However, evidence is lacking to prove that SOD activity is affected by H_2O_2 in vivo.

Interestingly, we and other investigators have found that ecSOD also has peroxidase activity similar to the cytosolic Cu/ZnSOD (Marklund 1984; Hink et al. 2002). In our recent studies, we have provided evidence that both cytosolic Cu/ZnSOD and ecSOD are partially inactivated in aortas of ApoE$^{-/-}$ mice, in which superoxide production is increased, and that the activity of these enzymes can be restored by elevating endogenous levels of uric acid, a potent inhibitor of peroxidase activity of both cytosolic Cu/ZnSOD and ecSOD (Hink et al. 2002). Thus, levels of uric acid commonly encountered in vivo may play an important role in regulating vascular redox state by preserving the activity of these SODs. In this paper, based on our recent findings about peroxidase activity of ecSOD, the role of ecSOD and uric acid in atherosclerosis is discussed.

EXTRACELLULAR SUPEROXIDE DISMUTASES

Three forms of superoxide dismutase are present in mammalian tissues (Beyer et al. 1991): cytosolic Cu/ZnSOD, MnSOD, and ecSOD. These isoforms differ in their tissue distribution and location: Cytosolic Cu/ZnSOD is localized in cytosol, MnSOD in mitochondria, and ecSOD in extracellular space. Importantly, ecSOD activity is ~10-fold higher in the vessel wall than in other tissues, where cytosolic Cu/ZnSOD and MnSOD constitute the majority of SOD activity (Strålin et al. 1995; Oury et al. 1996a). Thus, the relatively high levels of ecSOD in the vessel wall suggest that it may play a critical role in regulating the extracellular vascular redox state.

The ecSOD is a secretory, glycosylated copper- and zinc-containing SOD that was first described and characterized by Marklund and coworkers (Marklund 1982). Furthermore, it is the major SOD isoform in extracellular fluids, including plasma, synovial fluid (Marklund et al. 1986), and lymph (Marklund et al. 1982). In most species, ecSOD is a tetramer composed of two disulfide-linked dimers (Oury et al. 1996b). Each subunit has a molecular weight of ~25,000–34,000. The enzyme is composed of an amino-terminal signal peptide that permits secretion from the cell, an N-linked glycosylation site at Asn-89, an active domain that binds copper and zinc, and a carboxy-terminal region that is involved in heparin binding (Fig. 1). This affinity to heparin, as well as its affinity to other sulfated glycosaminoglycans, derives from its positively charged carboxy-terminal region (Sandstrom et al. 1992). Nonproteolyzed subunits with high heparin affinity are classified as type C, whereas the proteolyzed subunits lacking heparin affinity are classified as type A (Oury et al. 1996a). In vivo, both circulating (type A) and tissue-bound (type C) ecSOD are present, with ~99% of the total ecSOD being tissue-bound (Karlsson and Marklund 1987). The middle portion of the ecSOD includes the catalytic region containing both copper and zinc. This region

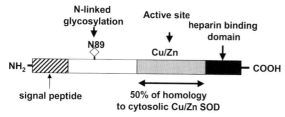

Figure 1. ecSOD protein structure. The ecSOD protein is composed of four domains. First, an amino-terminal signal peptide permits secretion from the cell. Second, an N-linked glycosylation site at Asn-89 is useful in the separation of ecSOD from cytosolic Cu/ZnSOD and greatly increases the solubility of the protein. Third, a domain containing active site (amino acid residues 96–193) shows about 50% homology with Cu/ZnSOD. All the ligands to Cu and Zn and the arginine in the entrance to the active site in Cu/ZnSOD can be identified in this domain of ecSOD. Finally, a carboxy-terminal region corresponding to heparin-binding domain has a cluster of positively charged residues. This region is critical for binding to heparan sulfate proteoglycans.

shares about 50% homology with the respective region of the cytosolic Cu/ZnSOD (Hjalmarsson et al. 1987). As compared to the x-ray crystallography structure data for cytosolic Cu/ZnSOD (Tainer et al. 1983), all ligands to the Cu and Zn atoms can be identified in ecSOD, as can the cysteines forming the intramolecular disulfide bond (Hjalmarsson et al. 1987). Thus, although the three-dimensional structure of ecSOD is not currently available, these findings suggest that the conformation of the copper-binding site of ecSOD and cytosolic Cu/ZnSOD might be similar. Indeed, the rate constants for dismutation of $O_2^{\cdot-}$ are similar, and both enzymes are inhibited similarly by cyanide, azide, phenylglyoxal, diethyldithiocarbamate, and hydrogen peroxide (Marklund 1984).

The predominant site of production of ecSOD in the healthy vessel is vascular smooth muscle cells (Strålin et al. 1995), whereas in the atherosclerotic vessel the enzyme is produced by both vascular smooth muscle cells and macrophages (Fukai et al. 1998; Luoma et al. 1998). Of note, studies from other laboratories and our own indicate that endothelial cells do not produce ecSOD (Marklund 1990; Strålin et al. 1995). However, ecSOD binds to the heparan sulfates on the endothelial cell surface and has been found to be internalized by endothelial cells (Ohta et al. 1994). Thus, this potent antioxidant enzyme can be made by vascular smooth muscle cells and end up inside adjacent endothelial cells.

Despite the fact that cytosolic Cu/ZnSOD and ecSOD are considered to be constitutively expressed, they seem to be substantially regulated by various stimuli (Fukai et al. 2002; Zelko et al. 2002). Details regarding the regulation of their expression have been reviewed recently (Fukai et al. 2002; Zelko et al. 2002).

ecSOD AND ATHEROSCLEROSIS

In macrophage-rich atherosclerotic lesions, enzymatic activity and protein expression of ecSOD are markedly increased (Fukai et al. 1998; Luoma et al. 1998). Interestingly, lipid-laden macrophages produce a large amount of ecSOD, and the ecSOD is colocalized with epitopes for oxidized LDL in atherosclerotic lesions. In cell culture studies, ecSOD markedly reduced LDL oxidation by endothelial cells (Laukkanen et al. 2000; Takatsu et al. 2001). In contrast, in vascular smooth muscle cells from atherosclerotic vessels, ecSOD expression is decreased (Fukai et al. 1998).

There are at least two mechanisms by which ecSOD is regulated in atherosclerotic vessels (for review, see Fukai et al. 2002). First, nitric oxide (NO) up-regulates ecSOD expression in vascular smooth muscle cells (Fukai et al. 2000). Therefore, the decrease in ecSOD expression in human atherosclerotic lesions might be due to a decline in the production and/or biological activity of endothelium-derived NO. Indeed, in patients with coronary artery disease, the activity of endothelium-bound ecSOD, released by heparin bolus injection, is positively correlated with flow-dependent, endothelium-mediated dilation (Landmesser et al. 2000). Second, ecSOD activity may be reduced by H_2O_2 produced by lipid-laden macrophages. Lipid-laden macrophages produce a large amount of both superoxide anion and ecSOD, resulting in the formation of H_2O_2. H_2O_2 in turn may inactivate the ecSOD via peroxidase-like reaction, as discussed below. Indeed, Sentman et al. (2001) recently observed that genetic deletion of ecSOD paradoxically caused a slight decrease in atherosclerotic lesions in ApoE$^{-/-}$ mice after 1 month on an atherogenic diet, but had no effect on the amount of atherosclerosis in these animals after 3 months on the atherogenic diet or after 8 months on standard chow. Although the mechanisms for proatherogenic effects of ecSOD remain unclear, peroxidase activity of ecSOD might contribute.

PEROXIDASE ACTIVITY OF ecSOD

The peroxidase activity of cytosolic Cu/ZnSOD has been extensively studied since 1975. In addition to its known capability to catalyze the dismutation of $O_2^{\cdot-}$ to H_2O_2, the cytosolic Cu/ZnSOD can use H_2O_2 as a substrate, forming a copper-bound hydroxyl ($^{\cdot}OH$)-like radical (Hodgson and Fridovich 1975). After the reduction of copper (II) upon reaction with H_2O_2 and the release of dioxygen, this resulting radical attacks one or more adjacent histidine residues that bind copper, leading inactivation of the enzyme (Fig. 2) (for review, see Jewett et al. 1999). This inactivation of cytosolic Cu/ZnSOD by H_2O_2 may be prevented by co-incubation with small anions or other reductants such as formate, tocopherol, nitrite, and uric acid, resulting in the formation of their respective radicals (Hodgson and Fridovich 1975; Singh et al. 1998; Goss et al. 1999; Zhang et al. 2000). Further oxidative damage as a consequence of this peroxidase-like activity, i.e., oxidation of lipids, strongly depends on the oxidation/reduction potential of the respective radical formed. Recently, it has been recognized that HCO_3^- augments consequent oxidative reactions, either by facilitating the reaction of H_2O_2 with Cu^{++} in the enzyme's active site (Sankarapandi and Zweier 1999) or, most probably, by formation of a putative bicarbonate radical (Goss et al. 1999; Liochev and Fridovich 1999; Zhang et al. 2000).

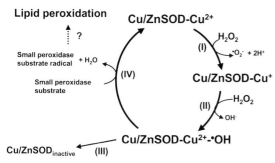

Figure 2. Peroxidase activity of cytosolic Cu/ZnSOD. In a first step, H_2O_2 donates an electron to reduce the copper ion of the active site of the enzyme to its cupric state (I). In a Fenton-type reaction, the Cu^+ ion reacts with a second molecule of H_2O_2 to form a SOD-copper-bound hydroxyl radical (II). This radical is very reactive and is able to oxidize a histidine residue in the active site of the enzyme. This reaction is followed by the release of copper from the active site and the inactivation of the enzyme (III). Small peroxidase substrates such as nitrite, however, will prevent inactivation of the enzyme by becoming oxidized by the hydroxyl radical to form their respective radicals (IV). Depending on the oxidative potential of the resulting small-molecule radical, they may further react with other molecules, such as lipids.

This bicarbonate radical enhances oxidative damage by mediating hydroxylation, e.g., of the spin trap DMPO, oxidation and nitration of tyrosine, and oxidation of the peroxidizable substrate ABTS to the ABTS cation radical (Zhang et al. 2000).

Interestingly, we and other workers have demonstrated that ecSOD is also inactivated by H_2O_2 in a similar fashion to cytosolic Cu/ZnSOD in the absence of small peroxidase substrates (Fig. 3A). Our recent studies extend these observations by showing that this involves a peroxidase-like activity with formation of a hydroxyl-like radical in the presence of HCO_3^-. For example, the reaction of ecSOD and H_2O_2 catalyzed hydroxylation of DEPMPO only in the presence of HCO_3^- to DEPMPO-OH as detected by electron spin resonance spectroscopy (Fig. 3B), consistent with previous work with cytosolic Cu/ZnSOD (Goss et al. 1999; Liochev and Fridovich 1999; Sankarapandi and Zweier 1999; Zhang et al. 2000). These findings are in line with the concept that HCO_3^-, because of its negative charge and small size, readily enters the active site of cytosolic Cu/ZnSOD, where it is oxidized to the bicarbonate radical ($CO_3^{\cdot -}$) by the hydroxyl radical formed in the catalytic region of cytosolic Cu/ZnSOD upon reaction with H_2O_2 (Goss et al. 1999; Zhang et al. 2000). The bicarbonate radical can readily diffuse out of the active site of cytosolic Cu/ZnSOD to react with other small molecules. Thus, ecSOD undergoes similar oxidative changes as compared to cytosolic Cu/ZnSOD.

ROLE OF URIC ACID IN MODULATING ACTIVITY OF ecSOD

In Vitro Experiments

Inactivation of cytosolic Cu/ZnSOD by H_2O_2 can be prevented by the presence of small peroxidase substrates

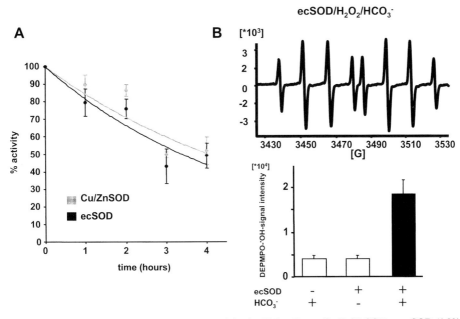

Figure 3. (*A*) Inhibition of cytosolic Cu/ZnSOD and ecSOD activity by H_2O_2. Cytosolic Cu/ZnSOD or ecSOD (1 U/ml) were incubated in the presence or absence of H_2O_2 (100 μM) for up to 4 hours at 37°C. SOD activity was determined with the cytochrome *c* reduction assay at indicated time points in the presence of catalase (1000 U/ml) and DTPA (100 μM). The data are expressed as the relative inhibition by H_2O_2 as mean ±S.E.M. from three independent experiments performed in duplicates. (*B*) Formation of DEPMPO-OH adduct in the presence of ecSOD and H_2O_2. The spin-adduct from DEPMPO (10 mM) resulting from the reaction between ecSOD (20 U/ml each) and H_2O_2 (10 mM) was measured by electron spin resonance (ESR) spectroscopy. Representative DEPMPO-·OH tracing (*upper panel*): ESR spectrometer settings are as given in the text. The spectra shown are the average of ten sequential scans. Quantification of the obtained ESR signals are shown in arbitrary units as the mean of 5 independent experiments ±S.E.M (*lower panel*).

(Fig. 2). Because of the favorable electrostatic interaction, anionic radical scavengers such as formate, urate, and azide can protect against this inactivation (Hodgson and Fridovich 1975). To determine whether ecSOD shared this property, we examined the effect of various reductants on inactivation of both SODs by H_2O_2 by using recombinant ecSOD protein overexpressed in *Pichia pastoris*. In these experiments, we focused on molecules previously reported to prevent inactivation of cytosolic Cu/ZnSOD, and in particular those present in vivo. Supraphysiological levels of nitrite (1 mM) prevented inactivation of cytosolic Cu/ZnSOD, but only partially prevented inactivation of ecSOD (Fig. 4). Of note, the plasma concentrations of nitrite range from 4.2 to 6.1 μM (Moshage et al. 1995). In contrast, physiological concentrations of urate encountered in humans (≤500 μM) almost completely prevented inactivation of both cytosolic Cu/ZnSOD and ecSOD by H_2O_2 (Fig. 4). High concentration of other potential substrates (1 mM), including formate, glutamate, tyrosine, lactate, γ-tocopherol, histidine, arginine, and uracil, had minimal effect on inactivation of either enzyme by H_2O_2. Thus, physiological levels of uric acid could effectively prevent inactivation of both cytosolic Cu/ZnSOD and ecSOD in vitro.

Although small peroxidase substrates prevent inactivation of enzymes, they will be oxidized by the copper-bound hydroxyl radical produced by the reaction between H_2O_2 and Cu/ZnSODs. Therefore, if the resulting respective radical is highly reactive, it will oxidize other substrates, such as lipids. Importantly, the resultant urate radical is relatively stable (redox potential = 0.59 V at pH 7) and does not react with oxygen, which contributes to the radical chain-breaking potential of urate. Furthermore, in human plasma, urate can be regenerated from the urate radical by its reaction with ascorbate. The resulting ascorbate radical (Asc$^•$) with a redox potential of only 0.28 V is even more innocuous (Maples and Mason 1988). These findings indicate a new and potentially important antioxidant role of urate, aside from its known ability to be an antagonist of potent oxidants such as peroxynitrite (Whiteman and Halliwell 1996).

In Vivo Experiments

To determine whether endogenous uric acid plays an important role in preventing inactivation of Cu/ZnSODs by H_2O_2 in vivo, we examined the effect of increased uric acid on SOD activity in aortas from atherosclerotic vessels in ApoE$^{-/-}$ mice, in which superoxide production is increased. To increase endogenous levels of uric acid in these mice, oxonic acid was infused. Oxonic acid is a potent inhibitor of uricase and is the enzyme responsible for catalyzing the reaction of uric acid to allantoin in mice (Fridovich 1965). A 7-day infusion of oxonic acid increased plasma levels of uric acid by 3-fold in both control and ApoE$^{-/-}$ mice. In ApoE$^{-/-}$ mice, oxonic acid increased activity of both ecSOD (2.5-fold) and cytosolic Cu/ZnSOD (1.7-fold) (Fig. 5A) while having no effect on SOD activity of control animals. As shown previously (Fukai et al. 1998), the protein level of ecSOD, but not cytosolic Cu/ZnSOD, was higher in aortas from ApoE$^{-/-}$ mice as compared to those of control mice, but changes in uric acid levels had no effect on expression of either protein (Fig. 5B). The effects of oxonic acid on SOD activity are most likely caused by increases in uric acid levels, because oxonic acid had no effect on inactivation of ecSOD by H_2O_2 in vitro or on the $O_2^{•-}$ production by cultured mouse aortic vascular smooth muscle cells.

In contrast to the 3-fold increase in the protein expression of ecSOD in atherosclerotic vessels in comparison to control vessels, the activity of ecSOD in atherosclerotic vessels was increased by only 2-fold. Our current findings provide some possible insight into this apparent discrepancy and suggest that cytosolic Cu/ZnSOD and ecSOD are partially inactivated in the aortas of ApoE$^{-/-}$ mice. Thus, in the presence of a pro-oxidant state such as

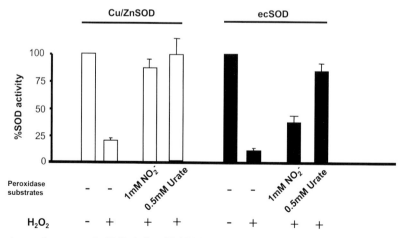

Figure 4. Effect of various reductants on the H_2O_2-induced inhibition of cytosolic Cu/ZnSOD and ecSOD: 2.5 U/ml of either enzyme was incubated with H_2O_2 (100 μM) in the presence or absence of the indicated reductants for 3.5 hours at 37°C. Thereafter, SOD activity was measured by cytochrome *c* reduction assay in the presence of catalase (1000 U/ml) and DTPA (100 μM). Data are expressed as the mean ±S.E.M. from at least three independent experiments performed in duplicates.

atherosclerosis, uric acid plays an important role in regulating the vascular redox state by preventing the inactivation of both SOD isoforms (Fig. 6).

ROLE OF URIC ACID IN ATHEROSCLEROSIS

Although a possible correlation between uric acid and cardiovascular disease has been suggested for more than half a century (Gertler et al. 1951), the role of uric acid in atherosclerosis remains unclear. There have been several epidemiological, prospective studies reporting a positive correlation between hyperuricemia and cardiovascular morbidity (Brand et al. 1985; Freedman et al. 1995; Lehto et al. 1998; Alderman et al. 1999; Liese et al. 1999); however, others, including the data of the Framingham Heart Study cohort (117,376 person years observational interval) (Culleton et al. 1999), suggested that this association was confounded by the coexistence of known independent risk factors for atherosclerosis, such as hypertension, obesity, hyperlipidemia, and diabetes mellitus (Wannamethee et al. 1997; Culleton et al. 1999; Moriarity et al. 2000). In contrast to most mammals that further metabolize uric acid to allantoin by the enzyme uricase, uric acid is the end product of purine metabolism in humans and higher primates (Wu et al. 1989). It has been postulated that uric acid might be an evolutionary antioxidant substitute for loss of the ability to synthesize ascorbate in higher primates (Ames et al. 1981). Indeed, in vitro experiments have demonstrated that uric acid is an effective scavenger of free radicals (Ames et al. 1981) and can chelate transition metal ions (Davies et al. 1986). On the other hand, uric acid also has been reported to enhance platelet adhesiveness (Ginsberg et al. 1977) and to stimulate vascular smooth muscle cell growth (Rao et al. 1991). Hence, the overall functional importance of uric acid in vivo remains unknown. Our data suggest that one potential benefit of uric acid may relate to its ability to prevent the inactivation of both cytosolic Cu/ZnSOD and ecSOD by H_2O_2.

CONCLUSIONS

It remains unclear whether in atherosclerotic vessels ecSOD highly expressed in lipid-laden macrophages, which also produce a large amount of superoxide anion, represents a compensatory adaptive function, or participates in lipid peroxidation through its peroxidase activities. Indeed, there are conflicting results about the functional role of ecSOD in atherosclerosis. In cell culture studies, ecSOD markedly reduced LDL oxidation by endothelial cells (Laukkanen et al. 2000; Takatsu et al. 2001). In contrast, genetic deletion of ecSOD paradoxically caused a slight decrease in atherosclerotic lesions in ApoE$^{-/-}$ mice after 1 month on an atherogenic diet (Sentman et al. 2001).

Our recent studies have demonstrated that in atherosclerotic vessels, cytosolic Cu/ZnSOD and ecSOD are partially inactivated, and that this inactivation can be prevented by increasing circulating levels of uric acid (Hink et al. 2002). Thus, the inactivation of these SODs by H_2O_2 may represent a new mechanism by which oxidative stress contributes to atherosclerosis, and show an important role of uric acid in modulating this phenomenon (Fig. 6).

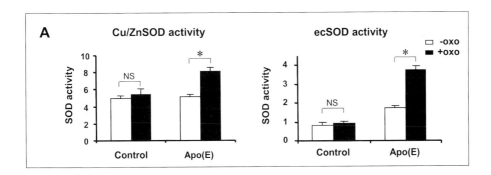

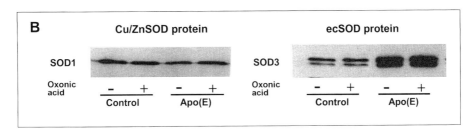

Figure 5. Effect of oxonic acid on cytosolic Cu/ZnSOD and ecSOD activity and protein expression. Oxonic acid was infused in control (C57BL/6) and ApoE$^{-/-}$ mice for 7 days by using the osmotic minipumps. (*A*) Aortas were homogenized, and SOD activity was assayed by examining inhibition of cytochrome *c* reduction by xanthine/xanthine oxidase at pH 7.8. The activity of ecSOD was determined after separation with ConA-Sepharose. Experiments were performed on four pooled aortas on three separate occasions. The results were presented as mean ± S.E.M., and the values were expressed as units per milligram of total protein. *p < 0.05 vs. control. (*B*) Protein levels of ecSOD and cytosolic Cu/ZnSOD were determined by western analysis. Representative blots from $n = 4$ individual experiments.

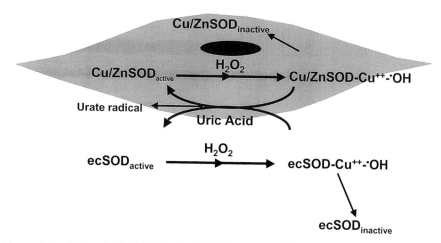

Figure 6. Peroxidase activity of cytosolic Cu/ZnSOD and ecSOD. The active ecSOD (cytosolic Cu/ZnSOD) reacts with H_2O_2 to form an intermediate (ecSOD-OH radical) that can interact with the bicarbonate to ultimately lead to formation of a DEPMPO-OH adduct. In the absence of a co-substrate, ecSOD (cytosolic Cu/ZnSOD) is inactivated by H_2O_2. In the presence of physiological levels of uric acid, however, the inactivation of ecSOD (cytosolic Cu/ZnSOD) by H_2O_2 is prevented. The urate radical formed has low oxidative potential and may react with ascorbic acid to regenerate uric acid.

Although abundant data from epidemiological observational studies have shown an inverse association between antioxidant intake and the risk of cardiovascular diseases, recent clinical trials, such as the recent GISSI and HOPE trials, have found no benefits of vitamin E supplementation on cardiovascular disease risk (Marchioli 1999; Witztum and Steinberg 2001). It is clear that additional studies of ecSOD and other vascular antioxidant defense systems are needed. For example, further analysis of ecSOD enzymatic properties, and studies of mice with ecSOD conditionally "knocked-out," will be very revealing. Overall, understanding the role of ecSOD in atherosclerosis would provide a new insight into the vascular redox state and the antioxidant therapy.

ACKNOWLEDGMENTS

This research was supported by National Institutes of Health grants RO-1 HL-39006, RO-1 HL-59248, and R01 HL-70187; Program Project grant HL-5800, a VA Merit grant, the Deutsche Forschungs Gemeinschaft HI 712/1-1, and an American Heart Association National Scientist Development Grant 0030180N.

REFERENCES

Alderman M.H., Cohen H., Madhavan S., and Kivlighn S. 1999. Serum uric acid and cardiovascular events in successfully treated hypertensive patients (comments). *Hypertension* **34:** 144.

Ames B.N., Cathcart R., Schwiers E., and Hochstein P. 1981. Uric acid provides an antioxidant defense in humans against oxidant- and radical-caused aging and cancer: A hypothesis. *Proc. Natl. Acad. Sci.* **78:** 6858.

Beyer W., Imlay J., and Fridovich I. 1991. Superoxide dismutases. *Prog. Nucleic Acid Res. Mol. Biol.* **40:** 221.

Brand F.N., McGee D.L., Kannel W.B., Stokes J., III, and Castelli W.P. 1985. Hyperuricemia as a risk factor of coronary heart disease: The Framingham Study. *Am. J. Epidemiol.* **121:** 11.

Cominacini L., Garbin U., Pasini A.F., Davoli A., Campagnola M., Contessi G.B., Pastorino A.M., and Lo Cascio V. 1997. Antioxidants inhibit the expression of intercellular cell adhesion molecule-1 and vascular cell adhesion molecule-1 induced by oxidized LDL on human umbilical vein endothelial cells. *Free Radic. Biol. Med.* **22:** 117.

Culleton B.F., Larson M.G., Kannel W.B., and Levy D. 1999. Serum uric acid and risk for cardiovascular disease and death: The Framingham Heart Study. *Ann. Intern. Med.* **131:** 7.

Davies K.J., Sevanian A., Muakkassah-Kelly S.F., and Hochstein P. 1986. Uric acid-iron ion complexes. A new aspect of the antioxidant functions of uric acid. *Biochem. J.* **235:** 747.

Freedman D.S., Williamson D.F., Gunter E.W., and Byers T. 1995. Relation of serum uric acid to mortality and ischemic heart disease. The NHANES I Epidemiologic Follow-up Study. *Am. J. Epidemiol.* **141:** 637.

Fridovich I. 1965. The competitive inhibition of uricase by oxonate and related derivatives of s-triazines. *J. Biol. Chem.* **240:** 2491.

———. 1997. Superoxide anion radical (O2-.), superoxide dismutases, and related matters. *J. Biol. Chem.* **272:** 18515.

Fukai T., Folz R.J., Landmesser U., and Harrison D.G. 2002. Extracellular superoxide dismutase and cardiovascular disease. *Cardiovasc. Res.* **55:** 239.

Fukai T., Galis Z.S., Meng X.P., Parthasarathy S., and Harrison D.G. 1998. Vascular expression of extracellular superoxide dismutase in atherosclerosis. *J. Clin. Invest.* **101:** 2101.

Fukai T., Siegfried M.R., Ushio-Fukai M., Cheng Y., Kojda G., and Harrison D.G. 2000. Regulation of the vascular extracellular superoxide dismutase by nitric oxide and exercise training. *J. Clin. Invest.* **105:** 1631.

Gertler M.M., Garn S.M., and Levine S.A. 1951. Serum uric acid in relation to age and physique in health and in coronary artery disease. *Ann. Intern. Med.* **34:** 1421.

Ginsberg M.H., Kozin F., O'Malley M., and McCarty D.J. 1977. Release of platelet constituents by monosodium urate crystals. *J. Clin. Invest.* **60:** 999.

Goss S.P., Singh R.J., and Kalyanaraman B. 1999. Bicarbonate enhances the peroxidase activity of Cu,Zn-superoxide dismutase. Role of carbonate anion radical. *J. Biol. Chem.* **274:** 28233.

Griendling K.K., Sorescu D., and Ushio-Fukai M. 2000. NAD(P)H oxidase: Role in cardiovascular biology and disease. *Circ. Res.* **86:** 494.

Hink H.U., Santanam N., Dikalov S., McCann L., Nguyen A.D., Parthasarathy S., Harrison D.G., and Fukai T. 2002. Peroxidase properties of extracellular superoxide dismutase. Role of uric acid in modulating in vivo activity. *Arterioscler. Thromb. Vasc. Biol.* **22:** 1402.

Hjalmarsson K., Marklund S.L., Engstrom A., and Edlund T. 1987. Isolation and sequence of complementary DNA encoding human extracellular superoxide dismutase. *Proc. Natl. Acad. Sci.* **84:** 6340.

Hodgson E.K. and Fridovich I. 1975. The interaction of bovine erythrocyte superoxide dismutase with hydrogen peroxide: Inactivation of the enzyme. *Biochemistry* **14:** 5294.

Jewett S.L., Rocklin A.M., Ghanevati M., Abel J.M., and Marach J.A. 1999. A new look at a time-worn system: Oxidation of CuZn-SOD by H_2O_2. *Free Radic. Biol. Med.* **26:** 905.

Karlsson K. and Marklund S.L. 1987. Heparin-induced release of extracellular superoxide dismutase to human blood plasma. *Biochem. J.* **242:** 55.

Kojda G. and Harrison D. 1999. Interactions between NO and reactive oxygen species: Pathophysiological importance in atherosclerosis, hypertension, diabetes and heart failure. *Cardiovasc. Res.* **43:** 562.

Landmesser U., Merten R., Spiekermann S., Buttner K., Drexler H., and Hornig B. 2000. Vascular extracellular superoxide dismutase activity in patients with coronary artery disease: Relation to endothelium-dependent vasodilation. *Circulation* **101:** 2264.

Laukkanen M.O., Lehtolainen P., Turunen P., Aittomaki S., Oikari P., Marklund S.L., and Yla-Herttuala S. 2000. Rabbit extracellular superoxide dismutase: Expression and effect on LDL oxidation. *Gene* **254:** 173.

Lehto S., Niskanen L., Ronnemaa T., and Laakso M. 1998. Serum uric acid is a strong predictor of stroke in patients with non-insulin-dependent diabetes mellitus. *Stroke* **29:** 635.

Liese A.D., Hense H.W., Lowel H., Doring A., Tietze M., and Keil U. 1999. Association of serum uric acid with all-cause and cardiovascular disease mortality and incident myocardial infarction in the MONICA Augsburg cohort. World Health Organization Monitoring Trends and Determinants in Cardiovascular Diseases. *Epidemiology* **10:** 391.

Liochev S.I. and Fridovich I. 1999. On the role of bicarbonate in peroxidations catalyzed by Cu,Zn superoxide dismutase. *Free Radic. Biol. Med.* **27:** 1444.

Luoma J.S., Stralin P., Marklund S.L., Hiltunen T.P., Sarkioja T., and Yla-Herttuala S. 1998. Expression of extracellular SOD and iNOS in macrophages and smooth muscle cells in human and rabbit atherosclerotic lesions: Colocalization with epitopes characteristic of oxidized LDL and peroxynitrite-modified proteins. *Arterioscler. Thromb. Vasc. Biol.* **18:** 157.

Maples K.R. and Mason R.P. 1988. Free radical metabolite of uric acid. *J. Biol. Chem.* **263:** 1709.

Marchioli R. 1999. Antioxidant vitamins and prevention of cardiovascular disease: Laboratory, epidemiological and clinical trial data. *Pharmacol. Res.* **40:** 227.

Marklund S.L. 1982. Human copper-containing superoxide dismutase of high molecular weight. *Proc. Natl. Acad. Sci.* **79:** 7634.

———. 1984. Properties of extracellular superoxide dismutase from human lung. *Biochem. J.* **220:** 269.

———. 1990. Expression of extracellular superoxide dismutase by human cell lines. *Biochem. J.* **266:** 213.

Marklund S.L., Bjelle A., and Elmqvist L.G. 1986. Superoxide dismutase isoenzymes of the synovial fluid in rheumatoid arthritis and in reactive arthritides. *Ann. Rheum. Dis.* **45:** 847.

Marklund S.L., Holme E., and Hellner L. 1982. Superoxide dismutase in extracellular fluids. *Clin. Chim. Acta* **126:** 41.

Marui N., Offermann M.K., Swerlick R., Kunsch C., Rosen C.A., Ahmad M., Alexander R.W., and Medford R.M. 1993. Vascular cell adhesion molecule-1 (VCAM-1) gene transcription and expression are regulated through an antioxidant-sensitive mechanism in human vascular endothelial cells. *J. Clin. Invest.* **92:** 1866.

Moriarity J.T., Folsom A.R., Iribarren C., Nieto F.J., and Rosamond W.D. 2000. Serum uric acid and risk of coronary heart disease: Atherosclerosis Risk in Communities (ARIC) Study. *Ann. Epidemiol.* **10:** 136.

Moshage H., Kok B., Huizenga J.R., and Jansen P.L. 1995. Nitrite and nitrate determinations in plasma: A critical evaluation. *Clin. Chem.* **41:** 892.

Nelson S.K., Bose S.K., and McCord J.M. 1994. The toxicity of high-dose superoxide dismutase suggests that superoxide can both initiate and terminate lipid peroxidation in the reperfused heart. *Free Radic. Biol. Med.* **16:** 195.

Ohta H., Adachi T., and Hirano K. 1994. Internalization of human extracellular-superoxide dismutase by bovine aortic endothelial cells. *Free Radic. Biol. Med.* **16:** 501.

Oury T.D., Day B.J., and Crapo J.D. 1996a. Extracellular superoxide dismutase: A regulator of nitric oxide bioavailability. *Lab. Invest.* **75:** 617.

Oury T.D., Crapo J.D., Valnickova Z., and Enghild J.J. 1996b. Human extracellular superoxide dismutase is a tetramer composed of two disulphide-linked dimers: A simplified, high-yield purification of extracellular superoxide dismutase. *Biochem. J.* **317:** 51.

Rao G.N. and Berk B.C. 1992. Active oxygen species stimulate vascular smooth muscle cell growth and proto-oncogene expression. *Circ. Res.* **70:** 593.

Rao G.N., Corson M.A., and Berk B.C. 1991. Uric acid stimulates vascular smooth muscle cell proliferation by increasing platelet-derived growth factor A-chain expression. *J. Biol. Chem.* **266:** 8604.

Sandstrom J., Carlsson L., Marklund S.L., and Edlund T. 1992. The heparin-binding domain of extracellular superoxide dismutase C and formation of variants with reduced heparin affinity. *J. Biol. Chem.* **267:** 18205.

Sankarapandi S. and Zweier J.L. 1999. Bicarbonate is required for the peroxidase function of Cu, Zn-superoxide dismutase at physiological pH. *J. Biol. Chem.* **274:** 1226.

Scott M.D., Meshnick S.R., and Eaton J.W. 1987. Superoxide dismutase-rich bacteria. Paradoxical increase in oxidant toxicity. *J. Biol. Chem.* **262:** 3640.

Sentman M.L., Brannstrom T., Westerlund S., Laukkanen M.O., Yla-Herttuala S., Basu S., and Marklund S.L. 2001. Extracellular superoxide dismutase deficiency and atherosclerosis in mice. *Arterioscler. Thromb. Vasc. Biol.* **21:** 1477.

Singh R.J., Goss S.P., Joseph J., and Kalyanaraman B. 1998. Nitration of gamma-tocopherol and oxidation of alpha-tocopherol by copper-zinc superoxide dismutase/H_2O_2/NO_2-: Role of nitrogen dioxide free radical. *Proc. Natl. Acad. Sci.* **95:** 12912.

Steinberg D. 1997. Low density lipoprotein oxidation and its pathobiological significance. *J. Biol. Chem.* **272:** 20963.

Strålin P., Karlsson K., Johansson B.O., and Markland S.L. 1995. The interstitium of the human arterial wall contains very large amounts of extracellular superoxide dismutase. *Arterioscler. Thromb. Vasc. Biol.* **15:** 2032.

Tainer J.A., Getzoff E.D., Richardson J.S., and Richardson D.C. 1983. Structure and mechanism of copper, zinc superoxide dismutase. *Nature* **306:** 284.

Takatsu H., Tasaki H., Kim H.N., Ueda S., Tsutsui M., Yamashita K., Toyokawa T., Morimoto Y., Nakashima Y., and Adachi T. 2001. Overexpression of EC-SOD suppresses endothelial-cell-mediated LDL oxidation. *Biochem. Biophys. Res. Commun.* **285:** 84.

Tribble D.L., Gong E.L., Leeuwenburgh C., Heinecke J.W., Carlson E.L., Verstuyft J.G., and Epstein C.J. 1997. Fatty streak formation in fat-fed mice expressing human copper-zinc superoxide dismutase (erratum in *Arterioscler. Thromb. Vasc. Biol.* [1997] **17:** 3363). *Arterioscler. Thromb. Vasc. Biol.* **17:** 1734.

Wannamethee S.G., Shaper A.G., and Whincup P.H. 1997. Serum urate and the risk of major coronary heart disease events. *Heart* **78:** 147.

Whiteman M. and Halliwell B. 1996. Protection against peroxynitrite-dependent tyrosine nitration and alpha 1-antiproteinase inactivation by ascorbic acid. A comparison with other biological antioxidants. *Free Radic. Res.* **25:** 275.

Wiedau-Pazos M., Goto J.J., Rabizadeh S., Gralla E.B., Roe J.A., Lee M.K., Valentine J.S., and Bredesen D.E. 1996. Altered reactivity of superoxide dismutase in familial amyotrophic lateral sclerosis (comments). *Science* **271:** 515.

Witztum J.L. and Steinberg D. 2001. The oxidative modification hypothesis of atherosclerosis: Does it hold for humans? *Trends Cardiovasc. Med.* **11:** 93.

Wu X.W., Lee C.C., Muzny D.M., and Caskey C.T. 1989. Urate oxidase: Primary structure and evolutionary implications. *Proc. Natl. Acad. Sci.* **86:** 9412.

Wung B.S., Cheng J.J., Hsieh H.J., Shyy Y.J., and Wang D.L. 1997. Cyclic strain-induced monocyte chemotactic protein-1 gene expression in endothelial cells involves reactive oxygen species activation of activator protein 1. *Circ. Res.* **81:** 1.

Zelko I.N., Mariani T.J., and Folz R.J. 2002. Superoxide dismutase multigene family: A comparison of the CuZn-SOD (SOD1), Mn-SOD (SOD2), and EC-SOD (SOD3) gene structures, evolution, and expression. *Free Radic. Biol. Med.* **33:** 337.

Zhang H., Joseph J., Felix C., and Kalyanaraman B. 2000. Bicarbonate enhances the hydroxylation, nitration, and peroxidation reactions catalyzed by copper, zinc superoxide dismutase. Intermediacy of carbonate anion radical. *J. Biol. Chem.* **275:** 14038.

SREBPs: Transcriptional Mediators of Lipid Homeostasis

J.D. HORTON,*† J.L. GOLDSTEIN,* AND M.S. BROWN*

*Departments of *Molecular Genetics and †Internal Medicine, University of Texas Southwestern Medical Center, Dallas, Texas 75390*

Sterol regulatory element-binding proteins (SREBPs) are a family of membrane-bound transcription factors that regulate cellular lipid synthesis and the clearance of atherogenic lipoproteins from the blood. They were initially isolated as proteins that transcriptionally regulated two genes involved in cholesterol homeostasis, the low-density lipoprotein (LDL) receptor, a transmembrane protein responsible for binding and internalizing LDL cholesterol from plasma, and 3-hydroxy-3-methylglutaryl-coenzyme A (HMG-CoA) synthase, a key enzyme in the cholesterol biosynthesis pathway (Goldstein et al. 2002). Subsequent studies have revealed that SREBPs directly activate the expression of more than 30 genes dedicated to the synthesis and uptake of cholesterol, fatty acids, triglycerides, and phospholipids, as well as the NADPH cofactor required to synthesize these molecules (Horton et al. 2002). In this paper, we will focus on the transcriptional activating properties of each SREBP family member, and how the regulation of each SREBP isoform provides a mechanism for cells to coordinately and independently regulate the lipid biosynthetic pathways.

SREBP STRUCTURE AND ACTIVATION

SREBPs belong to the large basic-helix-loop-helix-leucine zipper (bHLH-Zip) family of transcription factors, but differ from other bHLH-Zip proteins in that they are synthesized as inactive precursors bound to the endoplasmic reticulum (ER) (Brown and Goldstein 1997; Goldstein et al. 2002). SREBP precursor proteins are ~1150 amino acids in length, and all have three domains. The first domain is the amino-terminal ~480 amino acids that contains the bHLH-Zip region responsible for binding DNA and protein dimerization. The second domain contains two hydrophobic transmembrane-spanning segments separated by a ~30-amino-acid loop that projects into the lumen of the ER. The third domain is the carboxy-terminal segment of ~590 amino acids that performs an essential regulatory function through its interaction with the SREBP cleavage-activating protein (SCAP) described below.

The amino-terminal segment of each SREBP, designated nuclear SREBP (nSREBP), must be released from the membrane proteolytically to gain access to the nucleus for DNA binding and to activate transcription (Fig. 1). Three additional proteins are required for SREBP processing: SCAP, Site-1 protease (S1P), and Site-2 protease (S2P). SCAP, S1P, and S2P were all identified and characterized in cultured cells using tools of somatic cell genetics (Goldstein et al. 2002). SCAP serves as both a chaperone of SREBPs and a sensor of sterols. SREBP is first inserted into the membranes of the ER, where its carboxy-terminal regulatory domain binds to the carboxy-terminal domain of SCAP (Fig. 1). When cells require cholesterol, SCAP senses this need and escorts SREBPs from the ER to the Golgi apparatus, where the S1P and S2P reside. In the Golgi, S1P, a membrane-bound serine protease, cleaves SREBPs in the luminal loop that separates the two membrane-spanning segments (Fig. 1). Following the first cleavage, a membrane-bound zinc metal-

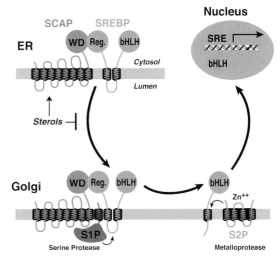

Figure 1. Model for the sterol-mediated proteolytic release of SREBPs from membranes. SCAP functions as a sensor of sterols and a chaperone of SREBPs. When cells are depleted of sterols, SCAP transports SREBPs from the ER to the Golgi apparatus, where the Site-1 protease (S1P) and Site-2 protease (S2P) act to release the amino-terminal bHLH-Zip domain from the membrane. The bHLH-Zip domain enters the nucleus and binds to a sterol response element (SRE) in the enhancer/promoter region of target genes, activating their transcription. When cellular cholesterol rises, the SCAP/SREBP complex is no longer transported to the Golgi apparatus, and the bHLH-Zip domain cannot be released from the membrane. (Reprinted, with permission, from Goldstein et al. 2002 [copyright Academic Press].)

loproteinase, S2P, cleaves the amino-terminal half of SREBP in the first transmembrane domain, releasing the bHLH-Zip domain (nSREBP) from the membrane. nSREBP then translocates to the nucleus where it binds to sterol response elements (SREs) in the promoter/enhancer regions of target genes to activate transcription.

If cells accumulate cholesterol, the sterol-sensing domain of SCAP senses the excess cholesterol and changes its conformation in such a way that the SCAP/SREBP complex is no longer incorporated into ER transport vesicles for transport to the Golgi. Therefore, SREBPs no longer gain access to S1P or S2P, and the transcriptionally active bHLH-Zip domains cannot be released from the ER membrane. This results in decreased transcription of SREBP target genes and lowers the rates of cholesterol and fatty acid formation (Brown and Goldstein 1997; Goldstein et al. 2002). The mechanism by which SCAP senses sterol levels and regulates movement of the SCAP/SREBP complex from the ER to the Golgi is currently not known.

SREBP ISOFORMS

Three SREBP isoforms, designated SREBP-1a, SREBP-1c, and SREBP-2, are present in mammalian cells. SREBP-1a and SREBP-1c are derived from a single gene through the use of two alternate forms of exon 1, designated 1a and 1c. SREBP-2 is encoded by a separate gene, and the protein is ~45% similar to SREBP-1 (Brown and Goldstein 1997). In most cultured cells, SREBP-1a and SREBP-2 are the predominant SREBP isoforms, whereas SREBP-1c and SREBP-2 are the predominant isoforms in most animal tissues (Shimomura et al. 1997a). SREBP-1a is a potent activator of all SREBP-responsive genes. The potent transcriptional activation properties of the SREBP-1a isoform are dependent on the longer acidic *trans*-activation segment encoded by the 1a exon. The 1c exon encodes only six amino acids, which shortens the *trans*-activation segment and reduces its transcriptional activation strength (Shimano et al. 1997). As a result, SREBP-1c preferentially enhances the transcription of genes required for fatty acid synthesis but not cholesterol synthesis (Horton et al. 2002). SREBP-2, like SREBP-1a, has a long transcriptional activation domain, but it preferentially activates cholesterol synthesis (Horton et al. 2002).

Figure 2 shows the cholesterol and fatty acid biosynthetic pathways with the SREBP-regulated genes denoted in italics (Horton et al. 2002). Common to both lipid biosynthetic pathways is acetyl-CoA, the 2-carbon precursor for cholesterol and fatty acids. Acetyl-CoA is generated in the cytosol by ATP citrate lyase and acetyl-CoA synthetase. ATP citrate lyase produces acetyl-CoA from citrate, and acetyl-CoA synthetase generates acetyl-CoA from acetate. Both enzymes are transcriptionally activated by all SREBPs (Horton et al. 2002).

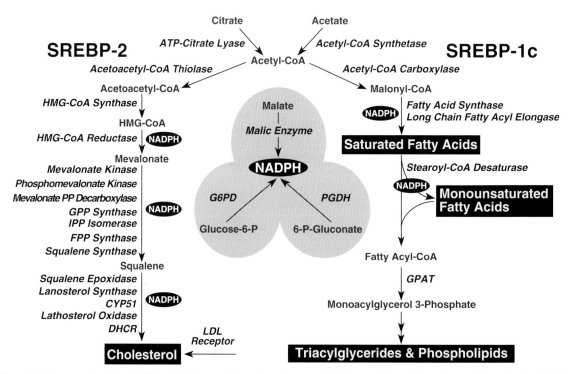

Figure 2. Genes regulated by SREBPs. The diagram shows the major metabolic intermediates in the pathways for synthesis of cholesterol, fatty acids, and triglycerides. In vivo, SREBP-2 preferentially activates genes of cholesterol metabolism, whereas SREBP-1c preferentially activates genes of fatty acid and triglyceride metabolism. (DHCR) 7-Dehydrocholesterol reductase; (FPP) farnesyl diphosphate; (GPP) geranylgeranyl pyrophosphate synthase; (CYP51) lanosterol 14α-demethylase; (G6PD) glucose-6-phosphate dehydrogenase; (PGDH) 6-phosphogluconate dehydrogenase; (GPAT) glycerol-3-phosphate acyltransferase. (Reprinted, with permission, from Horton et al. 2002 [copyright American Society for Clinical Investigation].)

The SREBP-2-responsive genes include all known enzymes in the cholesterol biosynthesis pathway as well as the LDL receptor (Fig. 2). The SREBP-1c-responsive genes include acetyl-CoA carboxylase and fatty acid synthase, which together produce palmitate (C16:0). SREBP-1c preferentially activates the long-chain fatty acyl elongase, a microsomal enzyme that carries out the rate-limiting condensation reaction required for elongating palmitate (16:0) to stearate (18:0), and stearoyl-CoA desaturase-1, a $\Delta9$-desaturase that generates monounsaturated fatty acids. Glycerol-3-phosphate acyltransferase, the enzyme responsible for the first committed step in triglyceride and phospholipid synthesis, is also selectively increased by the SREBP-1 isoforms (Edwards et al. 2000). All SREBP isoforms activate the three cytosolic enzymes that generate NADPH, a required cofactor for multiple reduction reactions in both lipid biosynthetic pathways (Fig. 2) (Horton et al. 2002).

SREBP-1c also contributes to the regulation of polyunsaturated fatty acid synthesis. Polyunsaturated fatty acids are produced in mammals from the essential fatty acids linoleic acid (18:2n-6) and α-linolenic acid (18:3n-3) through a series of elongation and desaturation reactions. The Δ^5-desaturase and Δ^6-desaturase enzymes carry out the desaturation steps, and both genes are transcriptionally activated by SREBP-1 isoforms (Matsuzaka et al. 2002). Interestingly, polyunsaturated fatty acids are also potent negative regulators of SREBP-1c expression through multiple mechanisms discussed in the following sections.

FUNCTION OF SREBPS IN VIVO

A series of genetically manipulated mouse models that either lack or overexpress a single component of the SREBP pathway have been generated to delineate the function of the SREBPs in vivo (for review, see Horton et al. 2002). Cultured cells that lack either SCAP or S1P have no nuclear SREBPs and require exogenous cholesterol and monounsaturated fatty acid supplementation for survival (Goldstein et al. 2002). Therefore, it was predicted, and subsequently confirmed, that mice carrying a homozygous deletion of either *Scap* or *S1p* die during embryonic development. This is illustrated by the germ-line deletion of *S1p*, which results in death before embryonic day 4 (Mitchell et al. 2001; Yang et al. 2001). Embryos that lack S1P are unable to process any SREBPs, which presumably lowers the rates of cholesterol and fatty acid synthesis to levels that are not compatible with normal cellular growth and development.

Embryonic lethality also occurred as a result of the disruption of *Srebp-1* (isoforms 1a and 1c) or *Srebp-2*. Deletion of the *Srebp-2* gene resulted in 100% lethality, but at a later stage of embryonic development than deletion of *S1p* (~p.c.7–8). The germ-line deletion of both SREBP-1a and SREBP-1c transcripts resulted in a partially lethal phenotype in which ~60–85% of the *Srebp-1*$^{-/-}$ mice died in utero (Horton et al. 2002). The surviving *Srebp-1*$^{-/-}$ mice had elevated levels of SREBP-2 in liver, presumably to compensate for the loss of SREBP-1a and -1c. In contrast, selectively deleting the SREBP-1c transcript resulted in no embryonic lethality, suggesting that the partial embryonic lethality in the *Srebp-1*$^{-/-}$ mice is due to the loss of the SREBP-1a transcript (Liang et al. 2002). The differences in embryonic lethality observed between SREBP-1 and SREBP-2 knockout mice may be explained by the relative contribution of each SREBP isoform in maintaining the absolute rates of cholesterol or fatty acid synthesis. In mutant Chinese hamster ovarian (CHO) cells that lack the S2P, and thus all active SREBP isoforms, the rate of fatty acid synthesis is only reduced 30%; however, essentially no cholesterol biosynthesis can be measured (Pai et al. 1998). This suggests that SREBPs are essential for maintaining cholesterol synthesis but that other factors may sustain basal fatty acid synthesis at levels necessary for cellular survival.

To bypass the embryonic lethality described above, tissue-specific knockout mice that lack all SREBPs in liver were produced using Cre/*LoxP* technology. To generate these mice, *loxP* recombination sites were inserted into introns that flank key exons in the *Scap* or *S1p* genes (floxed alleles). Mice with the integrated floxed alleles were bred with transgenic mice expressing Cre recombinase under control of an interferon-inducible promoter (MX1-Cre). These mice were then injected with polyinosinic acid–polycytidylic acid to induce the Cre recombinase production in liver, which disrupts the floxed gene by recombination between the *loxP* sites. Cre-mediated deletion of *Scap* or *S1p* in the adult mouse liver resulted in dramatic reductions in the amounts of nSREBP-1 and nSREBP-2, and their associated target genes in both cholesterol and fatty acid synthetic pathways. As a result, the rates of cholesterol and fatty acid synthesis were reduced by 70–80% in livers of both the *Scap*$^{-/-}$ and *S1p*$^{-/-}$ mice. The net effect on plasma lipids was a 20–40% reduction in plasma cholesterol and a 40–60% reduction in plasma triglycerides (Horton et al. 2002). These studies showed that SCAP and S1P were absolutely required for SREBP-1 and SREBP-2 processing (Goldstein et al. 2002), and that SREBPs are required to maintain basal levels of cholesterol and fatty acid biosynthesis in liver.

The germ-line deletion of the SREBP-1 isoforms further defined the in vivo function of SREBP-1a and SREBP-1c in mice (Horton et al. 2002; Liang et al. 2002). Most mice that lacked both SREBP-1a and SREBP-1c isoforms (*Srebp-1*$^{-/-}$) died in utero; however, those that survived had ~50% reduction in hepatic fatty acid synthesis, owing to reduced expression of mRNAs for all fatty acid synthetic enzymes (Horton et al. 2002). The nSREBP-2 levels in livers of the *Srebp-1*$^{-/-}$ knockout mice were increased, presumably to compensate for the loss of nSREBP-1. Increased nSREBP-2 expression stimulated the transcription of cholesterol biosynthetic genes and produced a 3-fold increase in hepatic cholesterol synthesis. Mice with the homozygous deletion of only SREBP-1c had no associated embryonic lethality but had a phenotype that was indistinguishable from the surviving mice that lack both 1a and 1c isoforms (Liang et al. 2002). These results suggest that SREBP-1a is important during embryonic development and that it has a relatively minor role in lipid metabolism in adult liver.

Transgenic mice that overexpress transcriptionally active nSREBPs in liver were produced to study the transcriptional activation properties of each SREBP isoform in vivo. For each isoform, the transgene protein terminated prior to the membrane attachment domain, which bypasses the regulated proteolytic cleavage steps, permitting direct access to the nucleus (Fig. 3). By studying the effects of overexpressing each nSREBP isoform separately, the distinct activating properties of each SREBP isoform could be determined.

Hepatic overexpression of nSREBP-1c, the principal SREBP-1 isoform in the liver, produced a triglyceride-enriched fatty liver with no increase in cholesterol (Fig. 3). The fatty acid synthetic enzyme mRNA levels and the rates of fatty acid synthesis were similarly increased 4-fold, whereas the mRNAs for cholesterol synthetic enzymes and the rate of cholesterol synthesis were unchanged (Horton et al. 2002). The ability of SREBP-1c to preferentially activate lipogenic genes provides a mechanism for the liver to increase fatty acid synthesis without altering cholesterol metabolism. Despite the increased rates of hepatic fatty acid synthesis, the plasma triglyceride levels were ~40% lower in SREBP-1c transgenic mice compared to those in wild-type mice.

Although the SREBP-1a isoform is normally expressed at low levels in the livers of adult animals, the consequences of its overexpression were also studied in mice. nSREBP-1a transgenic mice developed a massive fatty liver owing to the 6-fold increase in cholesterol content and the 22-fold increase in triglyceride content (Fig. 3). The hepatic mRNAs for genes controlling fatty acid synthesis were increased 17- to 20-fold, whereas a less dramatic 5- to 6-fold activation was measured in cholesterol biosynthetic enzyme mRNAs. The increased gene expression led to a preferential activation of fatty acid synthesis (26-fold increase) relative to cholesterol synthesis (5-fold increase) as measured in vivo by the incorporation of tritiated water into newly formed cholesterol and fatty acids. The larger increase in fatty acid synthesis explained the greater accumulation of triglycerides in the SREBP-1a transgenic livers. The fatty acids that accumulated in these livers were dramatically enriched in oleate (C18:1) as a result of the induction of the long-chain fatty acid elongase and stearoyl-CoA desaturase-1 genes. The transgenic nSREBP-1a livers contained ~65% oleate, which is markedly different from normal livers, which typically contain only 15–20% oleate. Again, despite the massive increase in hepatic cholesterol and fatty acid biosynthesis, plasma cholesterol and triglycerides were ~50% lower than levels in control mice (Horton et al. 2002).

Overexpression of nSREBP-2 in the liver increased the mRNAs encoding all cholesterol biosynthetic enzymes, the most dramatic of which was a 75-fold increase in HMG-CoA reductase. There was a lesser effect on the mRNAs for genes required for fatty acid synthesis. The in vivo rate of cholesterol synthesis was increased 28-fold in the transgenic nSREBP-2 livers, whereas fatty acid synthesis was only increased 4-fold. The plasma cholesterol levels were not consistently altered by SREBP-2 overexpression; however, the plasma triglyceride levels were reduced by 50% (Horton et al. 2002).

The following conclusions regarding the function of SREBPs can be drawn from the studies in knockout and transgenic mice: (1) Normal SREBP processing in liver requires SCAP and S1P; (2) SREBP-1c is the predominant SREBP-1 isoform in adult liver and it preferentially activates genes required for fatty acid synthesis; (3) SREBP-2 preferentially activates genes required for cholesterol synthesis; (4) SREBP-1a is a strong activator of cholesterol and fatty acid synthesis, but its function in adult liver is uncertain; and (5) SREBP-1a and SREBP-2, but not SREBP-1c, are required for embryogenesis.

SREBP EXPRESSION AND PLASMA LIPID LEVELS

SREBPs clearly stimulate lipid synthesis, but they also enhance LDL receptor expression. These properties have opposing effects on plasma lipid levels, inasmuch as the final plasma lipid concentration ultimately depends on the balance between production and clearance (Brown and Goldstein 1997). Cholesterol and triglycerides are secreted from the liver in very-low-density lipoprotein (VLDL) particles. In most hepatocyte cell lines, VLDL production and secretion are tightly linked and positively associated with rates of cellular lipid synthesis. Transgenic mice that overexpress nSREBPs in liver have elevated rates of lipid synthesis; however, as discussed in the preceding section, their plasma cholesterol and triglycerides are generally lower than those of wild-type mice (Horton et al. 2002).

To determine how altered SREBP expression levels altered key parameters determining plasma lipid levels, we studied isolated hepatocytes from mice overexpressing

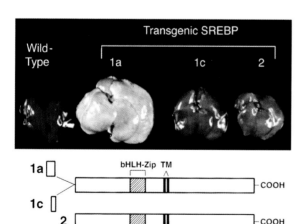

Figure 3. Photograph of livers from 12-week-old wild-type and transgenic mice expressing the dominant-positive SREBP-1a, -1c, and -2. The lower panel is a schematic diagram of the domain structure of SREBP-1a, -1c, and -2, illustrating the site of truncation used for construction of the transgenes. (Reprinted, with permission, from Horton et al. 1998b [copyright American Society for Clinical Investigation].)

nSREBP-1a and demonstrated that they secrete ~10-fold more cholesterol and triglycerides in the form of VLDL than do hepatocytes isolated from wild-type mice (Horton et al. 1999). The VLDL particles do not accumulate in the plasma because the VLDLs are rapidly removed through the action of LDL receptors, whose expression is also enhanced by the increased nSREBP-1a expression (Horton et al. 1999). The overall importance of the LDL receptor in preventing VLDL accumulation in plasma was revealed by experiments in which nSREBP-1a transgenic mice were bred with LDL receptor knockout mice. The plasma cholesterol and triglyceride levels rose by 10-fold in the doubly mutant mice principally due to the accumulation of VLDL-size particles (Fig. 4) (Horton et al. 1999). These studies demonstrated that increased clearance of plasma lipids through the action of the LDL receptor reduced plasma lipid levels in transgenic mice and that this action more than compensated for the massive increase in VLDL production and secretion.

The beneficial effects of SREBP-induced increased LDL receptor expression on plasma lipid levels raised the possibility that reduced SREBP levels could have deleterious effects on plasma lipid profiles. However, mice that lack all SREBPs in liver as a result of disrupting either *Scap* or *S1p* also have ~40% lower plasma cholesterol and triglyceride levels, presumably owing to the ~70–80% reduction in hepatic cholesterol and fatty acid synthesis. In these mice, the reduced cholesterol and fatty acid synthesis is associated with a ~90% reduction in VLDL secretion from hepatocytes. Interestingly, the LDL receptor mRNA and LDL clearance from plasma are also significantly lower; however, the reduced LDL clearance is less than the overall reduction in VLDL secretion. Therefore, the combined effect of reduced SREBP expression in liver is a net decrease in plasma lipid levels (Horton et al. 2002).

It is well established that lipoprotein metabolism in humans differs substantially from that in mice. Unlike human plasma, mouse plasma contains very little LDL, and mouse liver seems to have high basal LDL receptor activity. Humans seem to produce fewer LDL receptors and, thus, have much higher LDL levels in the basal state. Therefore, how plasma lipids carried in VLDL and LDL change when the SREBP pathway is blocked or activated will ultimately require demonstration in humans.

REGULATION OF SREBP ISOFORM EXPRESSION

SREBP expression is determined by posttranscriptional regulatory mechanisms that alter SREBP cleavage and by mechanisms that regulate gene transcription. Posttranscriptional regulation involves sterol-mediated suppression of SREBP cleavage, which occurs by inhibiting the movement of the SCAP/SREBP complex from the ER to the Golgi apparatus (Fig. 1). This form of regulation blocks all SREBP isoform precursor cleavage in cells cultured in the presence of sterols (Goldstein et al. 2002) and in the livers of rodents fed cholesterol-enriched diets (Shimomura et al. 1997b).

The transcriptional regulation of the SREBPs is more complex, but it provides a mechanism for the cell to independently regulate SREBP isoform expression. SREBP-1c and SREBP-2 are subject to distinct forms of transcriptional regulation, whereas SREBP-1a appears to be constitutively expressed at low levels in most tissues of adult animals (Shimomura et al. 1997a). One mechanism of regulation shared by SREBP-1c and SREBP-2 involves a feed-forward regulation mediated by SREs present in the enhancer/promoters of each gene (Sato et al. 1996; Amemiya-Kudo et al. 2000). Through this feed-forward loop, nSREBPs regulate the transcription of their own genes. In SREBP transgenic mice, the mRNAs for SREBP-1c and SREBP-2 are increased in liver. Conversely, when nSREBP levels are low, as in livers of $Scap^{-/-}$ or $S1p^{-/-}$ mice, SREBP-1c and SREBP-2 mRNA levels are reduced (Horton et al. 2002).

The transcription of SREBP-1c is regulated by at least three additional factors: (1) liver X-activated receptors (LXRs); (2) insulin; and (3) glucagon. LXRs are nuclear hormone receptors that act by forming heterodimers with retinoid X receptors (RXR). They are activated by interacting with a variety of sterols, including intermediates formed during the synthesis of cholesterol (Janowski et al. 1999). The *SREBP-1c* promoter contains an LXR-binding site that activates SREBP-1c transcription in cells cultured in the presence of an LXR agonist or in mice fed an LXR agonist (Repa et al. 2000). The importance of LXR-mediated SREBP-1c regulation was confirmed in LXR knockout mice that lack both isoforms of LXR (α and β) (Repa et al. 2000). The LXR knockout mice have markedly reduced SREBP-1c expression, and the mRNAs for

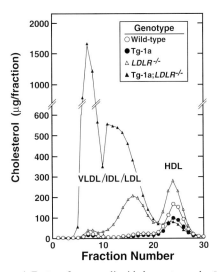

Figure 4. Fast performance liquid chromatography (FPLC) profiles of plasma lipoproteins from wild-type, SREBP-1a transgenic (TgBP-1a), LDL receptor knockout ($LDLR^{-/-}$), or double mutant (TgBP-1a;$LDLR^{-/-}$) mice. The plasma lipoprotein fractions (d = 1.215 g/ml) were subjected to gel filtration by FPLC, and the cholesterol content of each fraction was measured. (Reprinted, with permission, from Horton et al. 1999 [copyright American Society for Clinical Investigation].)

SREBP-1c lipogenic target genes are suppressed in liver. LXR knockout mice administered a synthetic LXR agonist exhibit a blunted increase in SREBP-1c and SREBP-1c-controlled lipogenic gene mRNAs compared to normal mice (Repa et al. 2000). A similar blunted response to a synthetic LXR agonist was also found in mice that lack SREBP-1c, confirming that LXR increases fatty acid synthesis largely by inducing SREBP-1c (Liang et al. 2002).

Previously identified LXR-regulated genes such as ABCA1, ABCG5, and ABCG8 are involved in mediating cholesterol efflux from the cell (Lawn et al. 1999; Repa et al. 2002). Currently, no known SREBP-1c-regulated gene has been directly linked to cholesterol efflux. However, LXR-mediated activation of SREBP-1c transcription does provide a mechanism to induce the synthesis of oleate when sterols are in excess (Repa et al. 2000). Oleate is the preferred fatty acid substrate for the synthesis of cholesteryl esters, which are necessary for both the transport and the storage of cholesterol.

As discussed above, SREBP-1c stimulates two enzymes required for the synthesis of polyunsaturated fatty acids. Polyunsaturated fatty acids also mediate a negative-feedback loop that suppresses SREBP-1c expression and lipogenesis. LXR-mediated regulation of *SREBP-1c* appears to be one mechanism by which *SREBP-1c* transcription and fatty acid synthesis are suppressed by polyunsaturated fatty acids. In vitro, unsaturated fatty acids competitively block LXR activation of SREBP-1c transcription by antagonizing the activation of LXR by its endogenous ligands (Ou et al. 2001). In vivo, rodents fed diets containing polyunsaturated fatty acids have reduced SREBP-1c levels and low rates of lipogenesis in liver (Xu et al. 1999). In addition to LXR-mediated transcriptional inhibition, polyunsaturated fatty acids lower SREBP-1c expression by accelerating degradation of its mRNA (Xu et al. 2001). These combined effects result in decreased SREBP-1c expression, which may contribute to the ability of polyunsaturated fatty acids to suppress fatty acid synthesis and lower plasma triglycerides.

Transcriptional regulation of SREBP-1c expression is also mediated by the hormones insulin and glucagon. A classic action of insulin is to stimulate fatty acid synthesis in liver during times of carbohydrate excess. This action is normally opposed by the action of glucagon, which acts by raising cyclic AMP. In vitro and in vivo evidence suggests that insulin's ability to stimulate fatty acid synthesis in liver is mediated by SREBP-1c. In primary hepatocytes isolated from rat or mouse, insulin treatment increases the amount of mRNA for SREBP-1c in parallel with the mRNAs of its target genes (Foretz et al. 1999a; Shimomura et al. 1999c). Insulin-mediated induction of SREBP-1c target genes can be blocked by the expression of a dominant-negative form of SREBP-1c (Foretz et al. 1999b). Conversely, incubating primary hepatocytes with glucagon or dibutyryl cAMP decreases the expression of SREBP-1c and its associated lipogenic target genes (Foretz et al. 1999b; Shimomura et al. 2000).

In vivo, the total amount of SREBP-1c mRNA and protein in liver is reduced by fasting (low insulin/high glucagon) and elevated by refeeding (high insulin/low glucagon) (Horton et al. 1998a). The SREBP-1c target gene mRNA levels parallel the changes in SREBP-1c expression. Plasma insulin levels can also be lowered by administering streptozotocin, a drug that destroys the insulin-producing β cells of the pancreas. In rats treated with streptozotocin, SREBP-1c mRNA levels fall in liver but are restored if the animal is administered exogenous insulin (Shimomura et al. 1999c). Overexpression of nSREBP-1c in livers of transgenic mice prevents the reduction in lipogenic mRNAs that normally follows a fall in plasma insulin levels. Conversely, if SREBP-1c levels are reduced by the genetic disruption of *Srebp-1c*, *Srebp-1a* and *-1c*, or *Scap$^{-/-}$*, there is a marked decrease in the insulin-induced stimulation of lipogenic gene expression that normally occurs after fasting/refeeding (Shimano et al. 1999; Horton et al. 2002).

Taken together, the above evidence suggests that SREBP-1c mediates insulin's lipogenic actions in liver. This observation led to the discovery that SREBP-1c contributes to the development of hepatic steatosis or "fatty liver" associated with diabetes, obesity, and the metabolic syndrome. Hepatic steatosis is the most commonly encountered liver abnormality in the United States, owing to its strong association with obesity and insulin-resistant diabetes mellitus (Mokdad et al. 2001). Conservative estimates indicate that 40–60% of individuals with obesity or diabetes develop fatty livers. A subset of patients with fatty liver will subsequently develop fibrosis, cirrhosis, and liver failure. Data initially obtained in mice indicate that the fatty liver associated with insulin resistance is caused, in part, by increased SREBP-1c expression. The increased SREBP-1c expression occurs in response to the high insulin levels present in insulin-resistant states. Thus, SREBP-1c levels are elevated in the fatty livers of obese (*ob/ob*) mice with insulin resistance and hyperinsulinemia caused by leptin deficiency (Shimomura et al. 1999a). Despite the presence of insulin resistance in peripheral tissues, insulin continues to activate SREBP-1c transcription and cleavage in the livers of these insulin-resistant mice. Elevated nSREBP-1c increases lipogenic gene expression, enhances fatty acid synthesis, and accelerates hepatic triglyceride accumulation.

The metabolic abnormalities observed in the diabetic mice can be reversed by three different modalities, all of which decrease SREBP-1c expression in liver. First, leptin administration corrects the insulin resistance and lowers plasma insulin levels. This reduces SREBP-1c expression and ameliorates the fatty liver in insulin-resistant mice (Shimomura et al. 1999b). Second, the administration of Metformin, a biguanide drug used to treat insulin-resistant diabetes, reduces hepatic nSREBP-1 levels and dramatically lowers the lipid accumulation in livers of insulin-resistant *ob/ob* mice (Lin et al. 2000). Metformin stimulates AMP-activated protein kinase, an enzyme that inhibits lipid synthesis through phosphorylation and inactivation of key lipogenic enzymes, but the mechanism by which AMP-activated protein kinase lowers SREBP-1c mRNA in liver is not understood (Zhou et al. 2001). Finally, if SREBP-1c is eliminated in livers of diabetic mice by mating SREBP-1 knockout mice with *ob/ob* mice, the double mutant mice have a significant reduction in the amount of fat that accumulates in the liver (Yahagi et al. 2002). Therefore, all lines of evidence suggest that increased SREBP-1c expression and subsequent activation of its target genes is responsible for the fatty livers observed in diabetic mice.

CONCLUSION

SREBPs are transcriptional regulators of lipid metabolism within the cell. To date, in vivo and in vitro studies suggest that SREBP-2 is absolutely required for the normal cholesterol homeostasis. SREBP-1c is required to maintain normal levels of fatty acid synthesis, and it mediates the insulin- and LXR-induced transcriptional activation of fatty acid synthesis. However, unlike cholesterol biosynthesis, other factors in addition to SREBP-1c contribute to the regulation of fatty acid synthesis. This has been shown most clearly in cultured CHO cells that lack all nSREBPs as a result of a defective Site-2 protease. In these cells, the levels of mRNAs encoding cholesterol biosynthetic enzymes and the rates of cholesterol synthesis decline nearly to undetectable levels, whereas the rate of fatty acid synthesis is reduced by only 30% (Pai et al. 1998). Other factors such as Upstream stimulatory factors 1 and 2, and LXR, have been identified as independent regulators of fatty acid synthesis, and it is likely that they contribute to the basal lipogenic gene expression when SREBP-1c levels are reduced (Kersten 2001; Joseph et al. 2002). Therefore, SREBP-1c is one important component of a more complex system that regulates the overall rate of lipogenesis in cells.

The transcriptional activating properties of the three SREBP isoforms provide a mechanism for the cell to both coordinately and independently regulate cholesterol and fatty acid synthesis depending on the relative amount of each SREBP isoform expressed. Transcriptional regulation of each SREBP isoform permits independent regulation of SREBP isoform expression. However, all SREBP isoforms apparently use the same SCAP-regulated cleavage pathway to become active. Understanding the mechanism of how SCAP independently regulates SREBP-1 and SREBP-2 cleavage will likely provide key insights into the mechanisms by which lipid levels are sensed by the cell.

ACKNOWLEDGMENTS

Support for the research cited from the authors' laboratories was provided by grants from the National Institutes of Health (HL-20948), Moss Heart Foundation, Keck Foundation, and Perot Family Foundation. J.D.H. is a Pew Scholar in the Biomedical Sciences and is the recipient of an Established Investigator Grant from the American Heart Association and a Research Scholar Award from the American Digestive Health Industry.

REFERENCES

Amemiya-Kudo M., Shimano H., Yoshikawa T., Yahagi N., Hasty A.H., Okazaki H., Tamura Y., Shionoiri F., Iizuka Y., Ohashi K., Osuga J.-I., Harada K., Gotoda T., Sato R., Kimura S., Ishibashi S., and Yamada N. 2000. Promoter analysis of the mouse sterol regulatory element-binding protein-1c gene. *J. Biol. Chem.* **275:** 31078.

Brown M.S. and Goldstein J.L. 1997. The SREBP pathway: Regulation of cholesterol metabolism by proteolysis of a membrane-bound transcription factor. *Cell* **89:** 331.

Edwards P.A., Tabor D., Kast H.R., and Venkateswaran A. 2000. Regulation of gene expression by SREBP and SCAP. *Biochim. Biophys. Acta* **1529:** 103.

Foretz M., Guichard C., Ferre P., and Foufelle F. 1999a. Sterol regulatory element binding protein-1c is a major mediator of insulin action on the hepatic expression of glucokinase and lipogenesis-related genes. *Proc. Natl. Acad. Sci.* **96:** 12737.

Foretz M., Pacot C., Dugail I., Lemarchand P., Guichard C., Le Liepvre X., Berthelier-Lubrano C., Spiegelman B., Kim J.B., Ferre P., and Foufelle F. 1999b. ADD1/SREBP-1c is required in the activation of hepatic lipogenic gene expression by glucose. *Mol. Cell. Biol.* **19:** 3760.

Goldstein J.L., Rawson R.B., and Brown M.S. 2002. Mutant mammalian cells as tools to delineate the sterol regulatory element-binding protein pathway for feedback regulation of lipid synthesis. *Arch. Biochem. Biophys.* **397:** 139.

Horton J.D., Goldstein J.L., and Brown M.S. 2002. SREBPs: Activators of the complete program of cholesterol and fatty acid synthesis in the liver. *J. Clin. Invest.* **109:** 1125.

Horton J.D., Bashmakov Y., Shimomura I., and Shimano H. 1998a. Regulation of sterol regulatory element binding proteins in livers of fasted and refed mice. *Proc. Natl. Acad. Sci.* **95:** 5987.

Horton J.D., Shimano H., Hamilton R.L., Brown M.S., and Goldstein J.L. 1999. Disruption of LDL receptor gene in transgenic SREBP-1a mice unmasks hyperlipidemia resulting from production of lipid-rich VLDL. *J. Clin. Invest.* **103:** 1067.

Horton J.D., Shimomura I., Brown M.S., Hammer R.E., Goldstein J.L., and Shimano H. 1998b. Activation of cholesterol synthesis in preference to fatty acid synthesis in liver and adipose tissue of transgenic mice overproducing sterol regulatory element-binding protein-2. *J. Clin. Invest.* **101:** 2331.

Janowski B.A., Grogan M.J., Jones S.A., Wisely G.B., Kliewer S.A., Corey E.J., and Mangelsdorf D.J. 1999. Structural requirements of ligands for the oxysterol liver X receptors LXRa and LXRb. *Proc. Natl. Acad. Sci.* **96:** 266.

Joseph S.B., Laffitte B.A., Patel P.H., Watson M.A., Matsukuma K.E., Walczak R., Collins J.L., Osborne T.F., and Tontonoz P. 2002. Direct and indirect mechanisms for regulation of fatty acid synthase gene expression by liver X receptors. *J. Biol. Chem.* **277:** 11019.

Kersten S. 2001. Mechanisms of nutritional and hormonal regulation of lipogenesis. *EMBO Rep.* **2:** 282.

Lawn R.M., Wade D.P., Garvin M.R., Wang X., Schwartz K., Porter J.G., Seilhamer J.J., Vaughan A.M., and Oram J.F. 1999. The Tangier disease gene product ABC1 controls the cellular apolipoprotein-mediated lipid removal pathway. *J. Clin. Invest.* **104:** R25.

Liang G., Yang J., Horton J.D., Hammer R.E., Goldstein J.L., and Brown M.S. 2002. Diminished hepatic response to fasting/refeeding and LXR agonists in mice with selective deficiency of SREBP-1c. *J. Biol. Chem.* **277:** 9520.

Lin H.Z., Yang S.Q., Chuckaree C., Kuhajda F., Ronnet G., and Diehl A.M. 2000. Metformin reverses fatty liver disease in obese, leptin-deficient mice. *Nat. Med.* **6:** 998.

Matsuzaka T., Shimano H., Yahagi N., Amemiya-Kudo M., Yoshikawa T., Hasty A.H., Tamura Y., Osuga J.-I., Okazaki H., Iizuka Y., Takahashi A., Sone H., Gotoda T., Ishibashi S., and Yamada N. 2002. Dual regulation of mouse $\Delta 5$- and $\Delta 6$-desaturase gene expression by SREBP-1 and PPARα. *J. Lipid Res.* **43:** 107.

Mitchell K.J., Pinson K.I., Kelly O.G., Brennan J., Zupicich J., Scherz P., Leighton P.A., Goodrich L.V., Lu X., Avery B.J., Tate P., Dill K., Pangilinan E., Wakenight P., Tessier-Lavigne M., and Skarnes W.C. 2001. Functional analysis of secreted and transmembrane proteins critical to mouse development. *Nat. Genet.* **28:** 241.

Mokdad A.H., Bowman B.A., Ford E.S., Vinicor F., Marks J.S., and Koplan J.P. 2001. The continuing epidemics of obesity and diabetes in the United States. *J. Am. Med. Assoc.* **286:** 1195.

Ou J., Tu H., Shan B., Luk A., DeBose-Boyd R.A., Bashmakov Y., Goldstein J.L., and Brown M.S. 2001. Unsaturated fatty acids inhibit transcription of the sterol regulatory element-binding protein-1c (SREBP-1c) gene by antagonizing ligand-dependent activation of the LXR. *Proc. Natl. Acad. Sci.* **98:** 6027.

Pai J.-T., Guryev O., Brown M.S., and Goldstein J.L. 1998. Differential stimulation of cholesterol and unsaturated fatty acid biosynthesis in cells expressing individual nuclear sterol regulatory element-binding proteins. *J. Biol. Chem.* **273:** 26138.

Repa J.J., Berge K.E., Pomajzl C., Richardson J.A., Hobbs H., and Mangelsdorf D.J. 2002. Regulation of ATP-binding cassette sterol transporters ABCG5 and ABCG8 by the liver X receptors alpha and beta *J. Biol. Chem.* **277:** 18793.

Repa J.J., Liang G., Ou J., Bashmakov Y., Lobaccaro J.M., Shimomura I., Shan B., Brown M.S., Goldstein J.L., and Mangelsdorf D.J. 2000. Regulation of mouse sterol regulatory element-binding protein-1c gene (SREBP-1c) by oxysterol receptors, LXRα and LXRβ. *Genes Dev.* **14:** 2819.

Sato R., Inoue J., Kawabe Y., Kodama T., Takano T., and Maeda M. 1996. Sterol-dependent transcriptional regulation of sterol regulatory element-binding protein-2. *J. Biol. Chem.* **271:** 26461.

Shimano H., Horton J.D., Shimomura I., Hammer R.E., Brown M.S., and Goldstein J.L. 1997. Isoform 1c of sterol regulatory element binding protein is less active than isoform 1a in livers of transgenic mice and in cultured cells. *J. Clin. Invest.* **99:** 846.

Shimano H., Yahagi N., Amemiya-Kudo M., Hasty A.H., Osuga J., Tamura Y., Shionoiri F., Iizuka Y., Ohashi K., Harada K., Gotoda T., Ishibashi S., and Yamada N. 1999. Sterol regulatory element-binding protein-1 as a key transcription factor for nutritional induction of lipogenic enzyme genes. *J. Biol. Chem.* **274:** 35832.

Shimomura I., Bashmakov Y., and Horton J.D. 1999a. Increased levels of nuclear SREBP-1c associated with fatty livers in two mouse models of diabetes mellitus. *J. Biol. Chem.* **274:** 30028.

Shimomura I., Hammer R.E., Ikemoto S., Brown M.S., and Goldstein J.L. 1999b. Leptin reverses insulin resistance and diabetes mellitus in mice with congenital lipodystrophy. *Nature* **401:** 73.

Shimomura I., Shimano H., Horton J.D., Goldstein J.L., and Brown M.S. 1997a. Differential expression of exons 1a and 1c in mRNAs for sterol regulatory element binding protein-1 in human and mouse organs and cultured cells. *J. Clin. Invest.* **99:** 838.

Shimomura I., Bashmakov Y., Ikemoto S., Horton J.D., Brown M.S., and Goldstein J.L. 1999c. Insulin selectively increases SREBP-1c mRNA in livers of rats with streptozotocin-induced diabetes. *Proc. Natl. Acad. Sci.* **96:** 13656.

Shimomura I., Bashmakov Y., Shimano H., Horton J.D., Goldstein J.L., and Brown M.S. 1997b. Cholesterol feeding reduces nuclear forms of sterol regulatory element binding proteins in the hamster liver. *Proc. Natl. Acad. Sci.* **94:** 12354.

Shimomura I., Matsuda M., Hammer R.E., Bashmakov Y., Brown M.S., and Goldstein J.L. 2000. Decreased IRS-2 and increased SREBP-1c lead to mixed insulin resistance and sensitivity in livers of lipodystrophic and *ob/ob* mice. *Mol. Cell* **6:** 77.

Xu J., Nakamura M.T., Cho H.P., and Clarke S.D. 1999. Sterol regulatory element binding protein-1 expression is suppressed by dietary polyunsaturated fatty acids. A mechanism for the coordinate suppression of lipogenic genes by polyunsaturated fats. *J. Biol. Chem.* **274:** 23577.

Xu J., Teran-Garcia M., Park J.H.Y., Nakamura M.T., and Clarke S.D. 2001. Polyunsaturated fatty acids suppress hepatic sterol regulatory element-binding protein-1 expression by accelerating transcript decay. *J. Biol. Chem.* **276:** 9800.

Yahagi N., Shimano H., Hasty A.H., Matsuzaka T., Ide T., Yoshikawa T., Amemiya-Kudo M., Tomita S., Okazaki H., Tamura Y., Iizuka Y., Ohashi K., Osuga J.-I., Harada K., Gotoda T., Nagai R., Ishibashi S., and Yamada N. 2002. Absence of sterol regulatory element-binding protein-1 (SREBP-1) ameliorates fatty livers but not obesity or insulin resistance in Lepob/Lepob mice. *J. Biol. Chem.* **277:** 19353.

Yang J., Goldstein J.L., Hammer R.E., Moon Y.-A., Brown M.S., and Horton J.D. 2001. Decreased lipid synthesis in livers of mice with disrupted site-1 protease gene. *Proc. Natl. Acad. Sci.* **98:** 13607.

Zhou G., Myers R., Li Y., Chen Y., Shen X., Fenyk-Melody J., Wu M., Ventre J., Doebber T., Fujii N., Musi N., Hirshman M.F., Goodyear L.J., and Moller D.E. 2001. Role of AMP-activated protein kinase in mechanism of metformin action. *J. Clin. Invest.* **108:** 1167.

Genetic Defenses against Hypercholesterolemia

H.H. Hobbs,*‡§ G.A. Graf,*‡§ L. Yu,*‡ K.R. Wilund,*‡ and J.C. Cohen*†§
*McDermott Center for Human Growth and Development, †Center for Human Nutrition,
Department of ‡Molecular Genetics and §Internal Medicine, University of Texas
Southwestern Medical Center, Dallas, Texas 75390

An elevated plasma cholesterol level is the primary risk factor for coronary atherosclerosis (NCEP Report Summary 2001). In humans, most circulating cholesterol is transported in low-density lipoproteins (LDL). Between 60% and 70% of circulating LDL is removed from the circulation by LDL receptor (LDLR)-mediated endocytosis in the liver (Bilheimer et al. 1984; Goldstein et al. 2001). Reductions in LDLR activity result in elevations in plasma LDL-C and deposition of cholesterol in body tissues (Goldstein et al. 2001). The critical importance of the LDLR pathway in preventing the accumulation of LDL is illustrated by the very high plasma levels of LDL and severe, premature atherosclerosis that characterizes the two autosomal dominant forms of Mendelian hypercholesterolemia: familial hypercholesterolemia (FH) and familial defective apolipoprotein B-100 (FDB).

FH is the most common and most severe of the single-gene disorders of LDL metabolism (Goldstein et al. 2001). FH is caused by mutations in the *LDLR* gene. Approximately 1 in 500 individuals in the general population are FH heterozygotes and have one mutant *LDLR* allele. FH heterozygotes have two- to fourfold elevations in plasma LDL-cholesterol levels. If these patients are not effectively treated with lipid-lowering agents, ~50% of the men will have a myocardial infarction before age 60 (Stone et al. 1974). Individuals with two defective *LDLR* alleles have a markedly delayed rate of clearance of LDL from the circulation, resulting in a sixfold increase in plasma LDL-C levels frequently accompanied by myocardial infarction in the first or second decade of life. Cultured skin fibroblasts from FH homozygotes do not internalize LDL and have provided an invaluable model system for studies of LDLR-mediated endocytosis (Goldstein et al. 2001). FDB patients have a clinical picture very similar to that of FH and have delayed clearance of circulating LDL (Myant 1993). The defect in this disorder is in apolipoprotein B-100, the ligand that mediates binding of LDL to the LDLR (Innerarity et al. 1990). In contrast to FH fibroblasts, LDL uptake is normal in fibroblasts from FDB patients. Thus, the LDLR constitutes a critical defense against the accumulation of atherogenic lipoproteins in the circulation.

AUTOSOMAL RECESSIVE HYPERCHOLESTEROLEMIA IS DUE TO DEFECTS IN LDLR-MEDIATED ENDOCYTOSIS

More recently, the molecular defects responsible for two autosomal recessive forms of severe hypercholesterolemia have been identified. Khachadurian described the first of these disorders, known as autosomal recessive hypercholesterolemia (ARH), which is a phenocopy of FH (Khachadurian and Uthman 1973). The index cases were four Lebanese siblings who had the clinical features of homozygous FH, including severe hypercholesterolemia (mean plasma cholesterol level of 728 mg/dl), huge tendon xanthomas, and premature CAD, but only a modest reduction in LDLR activity in cultured fibroblasts. The parents of the four siblings were normocholesterolemic (Khachadurian and Uthman 1973). Subsequently, additional subjects with an autosomal recessive form of hypercholesterolemia who had normal LDLR function in their fibroblasts were described (Harada-Shiba et al. 1992; Zuliani et al. 1995, 1999; Schmidt et al. 1998), including five families from Sardinia (Zuliani et al. 1995, 1999). LDL turnover studies performed in two of the Sardinian probands revealed a fourfold reduction in the rate of clearance of LDL compared to control subjects, which was similar to the rate observed in a FH homozygote studied simultaneously (Zuliani et al. 1999).

To elucidate the molecular basis of ARH, we performed a whole-genome linkage study in four families with the following characteristics:

1. The probands of the four families were offspring of normolipidemic first cousins, and all other hypercholesterolemic family members were siblings or first cousins.
2. The affected family members were all severely hypercholesterolemic (plasma LDL-cholesterol levels ranging from 400 to 600 mg/dl) and had normal triglyceride and HDL-cholesterol levels.
3. The affected family members all had xanthomas and many had xanthelasmas, aortic stenosis, and premature CAD.
4. LDLR function studies in cultured fibroblasts from representative affected family members were normal or only moderately reduced, thus ruling out a diagnosis of FH.

ARH IS CAUSED BY MUTATIONS IN AN ADAPTER PROTEIN

Multipoint linkage analysis revealed linkage to a 5.7-cM interval on the short arm of chromosome 1 (1p34-1p35), and DNA sequencing revealed mutations in the coding sequence of an EST (DKFZp586D0624) that mapped to this region (Garcia et al. 2001). The cDNA corresponding to this EST encodes a 308-amino-acid protein that was designated ARH. The gene encoding ARH spans ~20 kb and has nine exons and eight introns. The predicted amino acid sequence of ARH contains a ~130-amino-acid motif that shares significant sequence similarity to the phosphotyrosine binding (PTB) domains of several adapter proteins (Margolis et al. 1999). The PTB domains of adapter proteins, such as SHC, IRS-1, X11, Dab-1, and Dab-2, bind a canonical sequence motif (NPXY) in the cytoplasmic tails of various cell-surface receptors (Forman-Kay and Pawson 1999). The cytoplasmic tail of the LDLR contains an NPVY sequence that is required for LDLR-mediated endocytosis (Davis et al. 1987; Chen et al. 1990). Accordingly, we proposed that ARH functions as an adapter protein that binds to the NPVY sequence in LDLR and couples the receptor to the endocytic machinery (Garcia et al. 2001).

Seven mutations in *ARH* have been reported to date. We described six mutations associated with ARH in patients from Lebanon, Sardinia, Iran, Italy, and the United States (Garcia et al. 2001). Five of these mutations introduced premature termination codons, and one changed codon 202 from proline to histidine. Subsequent studies (K.R. Wilund et al., unpubl.) have revealed that the P202H allele also contains a single-nucleotide substitution in intron 7 that creates a splice donor site. This mutation leads to the splicing of an additional exon between exon 7 and exon 8, introducing irrelevant codons and a premature termination signal downstream of exon 7. Al-Kateb et al. (2002) have recently identified a G to C mutation in the splice acceptor sequence of exon 1 in three Syrian siblings with ARH. Thus, all known ARH mutations preclude the expression of a full-length protein, and almost all have been identified in individuals of Mediterranean or Middle Eastern origin.

ARH IS REQUIRED FOR LDLR INTERNALIZATION

The specific role of ARH in LDLR endocytosis has not been fully defined. Norman et al. (1999) reported that lymphocytes from two patients with a clinical phenotype consistent with ARH had increased cell-surface binding but markedly reduced internalization of LDL. Treatment of the cells with pronase indicated that almost all of the LDLRs were on the plasma membrane. These data suggest that ARH is not required for trafficking of the LDLR to the cell surface, nor for the binding of LDL to the receptor. Rather, ARH plays a critical role in the internalization of the receptor. LDLRs are internalized via clathrin-coated pits (Brown and Goldstein 1986). ARH may be required for the translocation of LDLRs to clathrin-coated pits, or it may couple the receptor to the endocytic machinery in the pit.

ARH appears to be a perfect phenocopy of homozygous FH (Arca et al. 2002), which is consistent with all clinical sequelae of ARH mutations being attributable to defective LDLR activity. ARH may bind specifically to the LDLR. Both ARH and LDLR are ubiquitously expressed, although LDLR expression is relatively low in some tissues that express high levels of ARH (kidney, placenta) (Garcia et al. 2001), raising the possibility that this protein may be involved in other receptor pathways. To date, no other phenotypes that would indicate defective function of other NPXY-containing proteins have been reported in ARH patients. More detailed clinical phenotyping of these probands is now in progress to explore the possibility of defects in other pathways.

Analysis of a large collection of Sardinian patients with ARH indicated that the mean plasma levels of LDL cholesterol are lower in these patients than in patients with homozygous FH (Arca et al. 2002). Approximately half of the Sardinian ARH patients had evidence of coronary artery disease at the time of diagnosis, and six of these patients were below the age of 40. Thus, the elevated plasma cholesterol levels in ARH patients are clearly associated with premature coronary artery disease. However, the onset of disease tends to be later in ARH patients than in patients with homozygous FH. Although patients with ARH have somewhat lower plasma cholesterol levels and delayed onset of coronary atherosclerosis compared to FH homozygotes, they tend to have large xanthomas. The LDLR is not thought to play a direct role in the formation of cholesterol-rich foam cells, since the LDLR is expressed at very low levels in cholesterol-loaded macrophages (Goldstein et al. 2001). However, studies in which bone marrow from normal or LDLR-deficient mice was transplanted into C57BL/6 mice suggested that LDLR expression may promote cholesterol accumulation in macrophages. The transplanted mice that received LDLR-deficient bone marrow had reduced lipid accumulation in the aorta despite having increased plasma levels of cholesterol (Linton et al. 1999), suggesting that the expression of the LDLR in macrophages promotes lipid accumulation. Maintenance of LDLR activity in the macrophages of patients with ARH may promote increased cellular uptake and accumulation of cholesterol, resulting in accelerated xanthoma formation.

Although ARH patients do not reach an optimal plasma cholesterol level on lipid-lowering medications and are maintained on LDL apheresis (Arca et al. 2002), these patients appear to respond to lipid-lowering medications, especially statins, with more substantial plasma cholesterol reductions than do patients with homozygous FH. Whereas reductase inhibitors lead to modest reductions in plasma cholesterol levels of FH homozygotes even when given at very high doses (Raal et al. 2000), these agents produce striking reductions in plasma cholesterol in some ARH patients (Arca et al. 2002). The increased level of expression of the LDLR associated with statins (Bilheimer et al. 1983) may compensate to some degree for the defective hepatic LDLR function in ARH.

SITOSTEROLEMIA IS A DISORDER OF NEUTRAL STEROL TRAFFICKING

Defects in LDLR function are the primary cause of three of the four Mendelian forms of severe hypercholesterolemia. The fourth form, sitosterolemia, is a more generalized disorder of neutral sterol transport (Bjorkhem et al. 2001) that is distinguished from the other monogenic forms of hypercholesterolemia in three important aspects. First, patients with sitosterolemia have markedly increased concentrations of plant sterols, as well as animal-derived sterols, in their blood and body tissues (Bhattacharyya and Connor 1974). Whereas the plasma concentrations of sitosterol and campesterol, the major plant sterols, almost invariably constitute less than 1% of the circulating sterols in normal individuals, they can exceed 20% of the plasma sterols in sitosterolemic patients (Salen et al. 1992). Neutral sterols also accumulate in the tissues of sitosterolemic patients in proportion to their concentrations in the circulation (Salen et al. 1985). Second, the plasma cholesterol concentrations of sitosterolemic individuals are remarkably sensitive to dietary cholesterol intakes. For example, Belamarich et al. (1990) reported that the plasma cholesterol levels of an 11-year-old sitosterolemic patient decreased from 444 mg/dl to 177 mg/dl after 3 months on a low-cholesterol diet. The responses of normal individuals are far more modest: On average, a 100 mg/day decrease in the dietary cholesterol content results in a 1–2 mg/dl decrease in plasma cholesterol levels. In contrast to patients with homozygous FH or ARH, diet and drug therapy can reduce plasma cholesterol concentrations to the lower end of the normal range in some sitosterolemic patients. Third, the plasma cholesterol concentrations of sitosterolemic patients show little response to HMG-CoA reductase inhibitors (Nguyen et al. 1991; Moghadasian and Frohlich 1999), but are extremely sensitive to bile-acid-binding resins (Moghadasian and Frohlich 1999). This finding is consistent with the observation that cholesterol biosynthesis is markedly reduced in sitosterolemic individuals, whereas bile acid synthesis is largely preserved.

The accumulation of exogenous sterols is prevented at two levels. First, the intestine provides a barrier to the entry of dietary sterols into the body. Each day, the average American synthesizes ~800 mg of cholesterol (von Bergmann et al. 1979) and consumes ~400 mg of dietary cholesterol and 200 mg of dietary plant sterols (primarily sitosterol and campesterol) (Weihrauch and Gardner 1978; Nair et al. 1984). However, less than 5% of dietary sitosterol and less than 50% of the dietary cholesterol consumed enters the circulation (Salen et al. 1970; Heinemann et al. 1993). Second, dietary plant sterols transported from the gut to the liver are efficiently excreted into the bile (Salen et al. 1970). Cholesterol can be excreted directly into bile or converted to bile acids prior to excretion. The molecular mechanisms underlying the intestinal and hepatic barriers to cholesterol accumulation remain poorly understood. Cholesterol entering the proximal small intestine is emulsified in mixed micelles containing bile acids and phospholipids. The solubilized sterol is shuttled across the unstirred water layer abutting the absorptive surfaces of the intestine and transferred to brush border membranes of enterocytes. It is not known whether sterol import into the enterocyte is a passive or an active process. Both in vitro and in vivo evidence suggest that cholesterol and non-cholesterol sterols are taken up by enteroctyes at similar rates (Compassi et al. 1997). The sterols are then transported from the apical surface by a poorly characterized process to the endoplasmic reticulum (ER), where the cholesterol is esterified by acyl-CoA acyltransferase-2 (ACAT-2) (Cases et al. 1998), incorporated into lipoproteins, and then secreted from the basolateral surface. Plant sterols are a poor substrate for ACAT (Tavani et al. 1982), and the small amounts of dietary plant sterols absorbed appear as free sterol in the lymph (Bjorkhem et al. 2001). The neutral sterols are absorbed (absorption being defined as the transfer of substrate from the diet into the blood) with varying efficiencies. The fractional absorption of cholesterol (35–50%) is significantly greater than that of campesterol (10–15%), and dietary sitosterol is very poorly absorbed (<5%) (Lutjohann et al. 1995). The mechanisms by which the intestine discriminates among these sterols have not been defined.

The major pathway for the elimination of neutral sterols is via secretion into the bile. Cholesterol can either be secreted directly into the bile or converted to bile acids and then secreted. Plant sterols such as sitosterol cannot be converted into bile acids but are efficiently secreted into bile as free sterols (Bjorkhem et al. 2001). Free sterols are secreted into bile together with phospholipids and bile acids, each of which has a specific ATP-dependent transporter in the bile canalicular (apical) membrane (ABCB11 for bile acids and ABCB4 for phospholipids). These apical transporters are either delivered to the biliary membrane from the basolateral (sinusoidal) surface (Sztul et al. 1991) or are transported from the *trans*-Golgi directly to the subapical region (Sai et al. 1999). Subapical vesicles containing ABC transporters fuse with the apical membrane in response to various stimuli (Kipp et al. 2001). A critical unanswered question is how cholesterol and non-cholesterol sterols are secreted into the bile. Is there a specific canalicular membrane transporter for sterols?

SITOSTEROLEMIC PATIENTS HAVE INCREASED FRACTIONAL ABSORPTION AND DECREASED BILIARY EXCRETION OF NEUTRAL STEROLS

Elegant metabolic studies have revealed that sitosterolemia is caused by the combination of increased absorption of sterols from the intestine (Bhattacharyya and Connor 1974; Lutjohann et al. 1995) and decreased excretion of sterols into bile (Miettinen 1980; Gregg et al. 1986; Salen et al. 1989). Cholesterol absorption is elevated by ~15% in these patients, and sitosterol absorption is increased about threefold compared to normal individuals (Lutjohann et al. 1995). Biliary excretion of cholesterol and other neutral sterols is reduced by more than twofold (Miettinen 1980). Interestingly, the metabolic defect in sitosterolemia appears to be specific for neutral sterols, as the synthesis and excretion of phospholipids and bile acids are essentially normal in affected individuals.

SITOSTEROLEMIA IS CAUSED BY MUTATIONS IN THE ATP-BINDING CASSETTE PROTEINS ABCG5 AND ABCG8

A clue to the molecular basis of sitosterolemia emerged from studies of the orphan nuclear hormone receptor, LXR, a transcription factor that coordinates the hepatic and intestinal responses to dietary cholesterol (Janowski et al. 1996). Mice treated with an LXR agonist showed markedly reduced cholesterol absorption and an increase in the mRNA levels of ABCA1, an ATP-binding cassette transporter implicated in the transport of cholesterol (Repa et al. 2000). To identify LXR-regulated genes that may limit cholesterol absorption, we used DNA microarrays to identify mRNAs induced by an LXR agonist in mouse liver and intestine (Berge et al. 2000). One transcript, designated ABCG5, was induced ~2.5-fold in the liver and intestine by the LXR agonist. ABCG5 resembles the *Drosophila* ABC half-transporters *brown*, *scarlet*, and *white* (Bingham et al. 1981) and was located within the genomic region to which Patel had previously mapped the gene defective in the autosomal recessive disorder, sitosterolemia (Patel et al. 1998). Analysis of several sitosterolemic patients revealed mutations in either *ABCG5* or in a previously unidentified ABC half-transporter, *ABCG8*, that resides adjacent to *ABCG5* (Berge et al. 2000; Hubacek et al. 2001; Lu et al. 2001). Mutations in either *ABCG5* or *ABCG8* cause an identical clinical picture, whereas heterozygotes for mutations in either gene have no clinical phenotype.

ABCG5 AND *ABCG8* PROMOTE NEUTRAL STEROL EXCRETION INTO BILE

The specific defects in sterol transport observed in sitosterolemic patients indicate that ABCG5 and ABCG8 facilitate the excretion of sterols from the liver and intestine. To test this hypothesis, we developed P1 transgenic mice expressing human *ABCG5* and *ABCG8* (Yu et al. 2002). The transgenes were expressed primarily in the liver and small intestine, mirroring the tissue distribution of expression of the endogenous genes. Transgene expression had only modest effects on plasma and liver cholesterol levels, but biliary cholesterol levels were increased more than fivefold compared to wild-type mice. This finding demonstrates that biliary secretion of cholesterol can be driven by increased expression of ABCG5 and ABCG8, leading to a compensatory increase in cholesterol synthesis and a selective increase in fecal neutral sterol excretion.

The fractional absorption of dietary cholesterol was reduced ~50% in the transgenic mice (Yu et al. 2002). Biliary phospholipid concentrations were modestly increased in the transgenic mice, and no significant changes were detected in the pool size, composition, or excretion of bile acids. The reduction in fractional cholesterol absorption in the transgenic mice may be due in part to increased efflux of sterols from enteroctyes by ABCG5 and ABCG8. In humans, the absence of ABCG5 or ABCG8 is associated with increased fractional cholesterol absorption, suggesting that ABCG5 and ABCG8 mediate the efflux of sterols across the apical membrane of the enteroctye and back into the gut lumen. Increased expression of these two proteins in the enterocytes likely limits the amount of cholesterol delivered to ACAT-2 for esterification.

Alternatively, the reduction in cholesterol absorption in transgenic mice may be secondary to the increased delivery of cholesterol to the intestinal lumen from the liver. The fractional absorption of dietary cholesterol is inversely related to the rate of biliary cholesterol secretion, presumably because the biliary cholesterol competes with the dietary cholesterol for micellar solubilization and uptake. The development of mice in which ABCG5 and ABCG8 are expressed exclusively in the intestine or in the liver will be required to determine the role of each tissue in the reduction in sterol absorption in these mice.

The amino acid sequences of ABCG5 and ABCG8 provide some insights into the specific role of the two proteins in sterol trafficking. Both ABCG5 and ABCG8 are members of a superfamily of ATP-driven proteins that actively transport a wide variety of substances across membranes (Higgins 1992). All ABC transporters share a common molecular architecture, which includes a pair of nucleotide-binding folds (NBFs) consisting of ATP-binding motifs (Walker A and B motifs) and an ABC-transporter signature motif (C motif), and two transmembrane (TM) domains each containing 6–11 membrane-spanning α helices. In eukaryotes, the genes encoding ABC transporters are organized either as full transporters containing both TM domains and both NBFs, or as half-transporters that must form homo- or heterodimers to be functionally active. The 48 ABC transporters identified in the human genome include both full and half-transports that can be further classified into seven groups (A–G), based on similarities in gene structure, domain order, and sequence homology. In members of the G subfamily (which includes ABCG5 and ABCG8), the NBF is located in the amino-terminal portion of the protein, and the six transmembrane domains are located in the carboxy-terminal portion.

To elucidate the basic mechanisms by which ABCG5 and ABCG8 facilitate sterol transport, we have undertaken a series of studies to address three critical questions:

1. *Where are ABCG5 and ABCG8 located in the cell?* Although the flux of neutral sterols across the bile canalicular membrane is clearly impaired in sitosterolemia, it is not known at what step in the pathway ABCG5 and ABCG8 promote sterol efflux. An intracellular location for these proteins would suggest a role in the partitioning of sterols among intracellular membranes or vesicles, whereas a plasma membrane location would be more consistent with ABCG5 and ABCG8 playing a direct role in sterol excretion from cells. Until recently, functional dimers of mammalian half-transporters were all thought to be localized to internal membranes (Dean et al. 2001). Among the best-characterized pairs of ABC half-transporters, TAP1 (ABCB2) and TAP2 (ABCB3) reside in the ER (Abele and Tampe 1999), and PMP70 (ABCD3) and ALD (ABCD1) reside in the peroxisomal membrane (Gartner et al. 1992; Contreras et al. 1994). In *Drosophila*, the orthologs of ABCG5 and ABCG8 are located on the delimiting membrane of the pigmentary granules in pigmentary eye cells (Mackenzie et al. 2000). However, ABCG2, which transports the chemotherapeutic agent mitozantrone, has been localized to the bile canalicular membrane in liver (Maliepaard et al. 2001).

To determine the subcellular localization of ABCG5 and ABCG8, we expressed epitope-tagged versions of mouse ABCG5 and ABCG8 in Chinese hamster ovary (CHO)-K1 cells, cultured rat hepatocytes (CRL-1601), and polarized hepatocyte-derived cells (WIF-B cells) (Graf et al. 2002). Both proteins were N-glycosylated. When either protein was expressed individually, the oligosaccharide chains remained sensitive to endoglycosidase H (Endo-H). When ABCG5 and ABCG8 were expressed together in cells, the attached sugars became resistant to Endo-H and sensitive to neuraminidase, indicating that both proteins were transported from the ER to the *trans*-Golgi. Immunofluorescence microscopy indicated that ABCG5 and ABCG8 were located on the plasma membrane in a rat hepatocyte cell line (CRL-1601). Immunoelectron microscopy confirmed that the proteins were located in the membrane itself, rather than in a subplasmalemmal compartment, in these cells. Since CRL-1601 cells are not polarized, these cells cannot be used to determine whether ABCG5 and ABCG8 are specifically targeted to the bile canalicular (apical) surface. Therefore, additional studies were performed in a polarized cell line (WIF-B) derived from a fusion between a rat hepatoma and human fibroblast cells (Ihrke et al. 1993). These cells form canaliculi and sort proteins in a manner similar to that observed in vivo (Song et al. 1994), and they have been used previously to define the subcellular localization of biliary ABC transporters (Sai et al. 1999). In polarized hepatocytes, recombinant ABCG5 colocalized with an apical membrane marker protein (aminopeptidase N) when coexpressed with ABCG8 but not when expressed alone. These data indicate that ABCG5 and ABCG8 are located on the bile canalicular surfaces of hepatocytes and that trafficking to the cell membrane requires the expression of both proteins. Furthermore, the localization of ABCG5 and ABCG8 at the cell surface is consistent with the notion that the two proteins are directly and specifically involved in the translocation of neutral sterols across the plasma membranes of hepatocytes and enterocytes.

2. *What is the functional form of ABCG5 and ABCG8?*
ABC half-transporters must form either homodimers or heterodimers to constitute a functional transporter (Dean and Allikmets 1995). The phenotype resulting from deficiency of ABCG5 (i.e., sitosterolemia) is identical to that caused by deficiency of ABCG8, indicating that the two proteins participate in the same transport process, presumably as a heterodimer. The close juxtaposition of *ABCG5* and *ABCG8* in the human genome, the similarity in tissue distribution, and coordinate regulation of the two mRNA transcripts (Berge et al. 2000) are also consistent with these two proteins forming a functional complex. Co-immunoprecipitation experiments using myc- and HA-tagged versions of ABCG5 and ABCG8 indicate that the two proteins are associated in cells, and suggest that they heterodimerize (Graf et al. 2002). Mixing of lysates from cells transfected with either ABCG5 or ABCG8 indicated that the interaction between the two proteins was specific and did not occur after cell lysis. Co-immunoprecipitation of the Endo-H-resistant forms of ABCG5-myc and ABCG8-HA was almost quantitative (>95%). Taken together, these data indicate that ABCG5/ABCG8 function as a complex, and that formation of the complex permits the two proteins to exit the endoplasmic reticulum.

3. *What substrate(s) does ABCG5 and ABCG8 transport and what are the components of the transport system?*
The most fundamental characteristic of a transporter is the nature of its substrate. Although previous studies have clearly indicated the involvement of ABCG5 and ABCG8 in sterol transport and metabolism, the direct substrate(s) for this transport system has not been identified. The observation that overexpression of ABCG5 and ABCG8 in mice leads to a specific increase in biliary cholesterol concentrations suggests that neutral sterols are the primary transport substrates of the complex, but further studies will be required to directly demonstrate this. In addition, very little is known about the enzymatic properties and other biochemical characteristics of the ABCG5/ABCG8 transport system, including whether or not other protein components are required and whether the lipid environment of the transporter influences its functional characteristics. Elucidation of the questions must await the purification and reconstitution of the ABCG5/ABCG8 transport system.

CONCLUSIONS

Atherogenic lipoproteins are generated in the circulation as a by-product of triglyceride transport. The LDLR system provides a mechanism for clearing these atherogenic lipoproteins from the circulation into the liver. Mutations in the receptor, its ligand, or the accessory protein(s) required for internalization of the receptor lead to severe hypercholesterolemia and atherosclerotic vascular disease. Once cholesterol is delivered to the liver, it can be excreted from the body via the bile. The ABCG5/ABCG8 transport complex appears to be a specific transporter that facilitates the excretion of cholesterol and other neutral sterols into the bile.

ACKNOWLEDGMENTS

We thank Scott Grundy, David Russell, Richard Anderson, Jay Horton, Michael Brown, and Joseph Goldstein for helpful discussion. Our laboratory is supported by grants from the National Institutes of Health (HL-20948 and HL-53917), the W.M. Keck Foundation, the Perot Family Fund, and the Donald W. Reynolds Cardiovascular Clinical Research Center at Dallas.

REFERENCES

Abele R. and Tampe R. 1999. Function of the transport complex TAP in cellular immune recognition. *Biochim. Biophys. Acta* **1461**: 405.

Al-Kateb H., Bahring S., Hoffmann K., Strauch K., Busjahn A., Nurnberg G., Jouma M., Bautz E. K., Dresel H.A., and Luft F.C. 2002. Mutation in the ARH gene and a chromosome 13q locus influence cholesterol levels in a new form of digenic-recessive familial hypercholesterolemia. *Circ. Res.* **90:** 951.

Arca M., Zuliani G., Wilund K., Campagna F., Fellin R., Bertolini S., Calandra S., Ricci G., Glorioso N., Maioli M., Pintus P., Carru C., Cossu F., Cohen J., and Hobbs H.H. 2002. Autosomal recessive hypercholesterolaemia in Sardinia, Italy, and mutations in ARH: A clinical and molecular genetic analysis. *Lancet* **359:** 841.

Belamarich P.F., Deckelbaum R.J., Starc T.J., Dobrin B.E., Tint G.S., and Salen G. 1990. Response to diet and cholestyramine in a patient with sitosterolemia. *Pediatrics* **86:** 977.

Berge K.E., Tian H., Graf G.A., Yu L., Grishin N.V., Schultz J., Kwiterovich P., Shan B., Barnes R., and Hobbs H.H. 2000. Accumulation of dietary cholesterol in sitosterolemia caused by mutations in adjacent ABC transporters. *Science* **290:** 1771.

Bhattacharyya A.K. and Connor W.E. 1974. Beta-sitosterolemia and xanthomatosis. A newly described lipid storage disease in two sisters. *J. Clin. Invest.* **53:** 1033.

Bilheimer D.W., Grundy S.M., Brown M.S., and Goldstein J.L. 1983. Mevinolin and colestipol stimulate receptor-mediated clearance of low density lipoprotein from plasma in familial hypercholesterolemia heterozygotes. *Proc. Natl. Acad. Sci.* **80:** 4124.

Bilheimer D.W., Goldstein J.L., Grundy S.M., Starzl T.E., and Brown M.S. 1984. Liver transplantation to provide low-density-lipoprotein receptors and lower plasma cholesterol in a child with homozygous familial hypercholesterolemia. *N. Engl. J. Med.* **311:** 1658.

Bingham P.M., Levis R., and Rubin G.M. 1981. Cloning of DNA sequences from the white locus of *D. melanogaster* by a novel and general method. *Cell* **25:** 693.

Bjorkhem I., Boberg K., and Leitersdorf E. 2001. Inborn errors in bile acid biosynthesis and storage of sterols other than cholesterol. In *The metabolic and molecular bases of inherited disease* (ed. C. Scriver et al.), vol. 2, p. 2961. McGraw-Hill, New York.

Brown M.S. and Goldstein J.L. 1986. A receptor-mediated pathway for cholesterol homeostasis. *Science* **232:** 34.

Cases S., Novak S., Zheng Y.W., Myers H.M., Lear S.R., Sande E., Welch C.B., Lusis A.J., Spencer T.A., Krause B.R., Erickson S.K., and Farese R.V., Jr. 1998. ACAT-2, a second mammalian acyl-CoA:cholesterol acyltransferase: Its cloning, expression, and characterization. *J. Biol. Chem.* **273:** 26755.

Chen W.-J., Goldstein J.L., and Brown M.S. 1990. NPXY, a sequence often found in cytoplasmic tails, is required for coated pit-mediated internalization of the low density lipoprotein receptor. *J. Biol. Chem.* **265:** 3116.

Compassi S., Werder M., Weber F.E., Boffelli D., Hauser H., and Schulthess G. 1997. Comparison of cholesterol and sitosterol uptake in different brush border membrane models. *Biochemistry* **36:** 6643.

Contreras M., Mosser J., Mandel J.L., Aubourg P., and Singh I. 1994. The protein coded by the X-adrenoleukodystrophy gene is a peroxisomal integral membrane protein. *FEBS Lett.* **344:** 211.

Davis C.G., van Driel I.R., Russell D.W., Brown M.S., and Goldstein J.L. 1987. The low density lipoprotein receptor. Identification of amino acids in cytoplasmic domain required for rapid endocytosis. *J. Biol. Chem.* **262:** 4075.

Dean M. and Allikmets R. 1995. Evolution of ATP-binding cassette transporter genes. *Curr. Opin. Genet. Dev.* **5:** 779.

Dean M., Hamon Y., and Chimini G. 2001. The human ATP-binding cassette (ABC) transporter superfamily. *J. Lipid Res.* **42:** 1007.

Forman-Kay J.D. and Pawson T. 1999. Diversity in protein recognition by PTB domains. *Curr. Opin. Struct. Biol.* **9:** 690.

Garcia C.K., Wilund K., Arca M., Zuliani G., Fellin R., Maioli M., Calandra S., Bertolini S., Cossu F., Grishin N., Barnes R., Cohen J.C., and Hobbs H.H. 2001. Autosomal recessive hypercholesterolemia caused by mutations in a putative LDL receptor adaptor protein. *Science* **292:** 1394.

Gartner J., Moser H., and Valle D. 1992. Mutations in the 70K peroxisomal membrane protein gene in Zellweger syndrome. *Nat. Genet.* **1:** 16.

Goldstein J., Hobbs H., and Brown M. 2001. Familial hypercholesterolemia. In *The metabolic and molecular bases of inherited disease* (ed. C. Scriver et al.), vol. 2, p. 2863. McGraw Hill, New York.

Graf G.A., Li W.-P., Gerard R.D., Gelissen I., White A.L., Cohen J.C., and Hobbs H.H. 2002. Coexpression of ATP-binding cassette proteins ABCG5 and ABCG8 permits their transport to the apical surface. *J. Clin. Invest.* (in press).

Gregg R.E., Connor W.E., Lin D.S., and Brewer H.B., Jr. 1986. Abnormal metabolism of shellfish sterols in a patient with sitosterolemia and xanthomatosis. *J. Clin. Invest.* **77:** 1864.

Harada-Shiba M., Tajima S., Yokoyama S., Miyake Y., Kojima S., Tsushima M., Kawakami M., and Yamamoto A. 1992. Siblings with normal LDL receptor activity and severe hypercholesterolemia. *Arterioscler. Thromb.* **12:** 1071.

Heinemann T., Axtmann G., and von Bergmann K. 1993. Comparison of intestinal absorption of cholesterol with different plant sterols in man. *Eur. J. Clin. Invest.* **23:** 827.

Higgins C. 1992. ABC transporters: From microorganisms to man. *Annu. Rev. Cell Biol.* **8:** 67.

Hubacek J.A., Berge K.E., Cohen J.C., and Hobbs H.H. 2001. Mutations in ATP-cassette binding proteins G5 (*ABCG5*) and G8 (*ABCG8*) causing sitosterolemia. *Hum. Mutat.* **18:** 359.

Ihrke G., Neufeld E.B., Meads T., Shanks M.R., Cassio D., Laurent M., Schroer T.A., Pagano R.E., and Hubbard A.L. 1993. WIF-B cells: An *in vitro* model for studies of hepatocyte polarity. *J. Cell Biol.* **123:** 1761.

Innerarity T.L., Mahley R.W., Weisgraber K.H., Bersot T.P., Krauss R.M., Vega G.L., Grundy S.M., Friedl W., Davignon J., and McCarthy B.J. 1990. Familial defective apolipoprotein B-100: A mutation of apolipoprotein B that causes hypercholesterolemia. *J. Lipid Res.* **31:** 1337.

Janowski B.A., Willy P.J., Devi T.R., Falck J.R., and Mangelsdorf D.J. 1996. An oxysterol signalling pathway mediated by the nuclear receptor LXR alpha. *Nature* **383:** 728.

Khachadurian A.K. and Uthman S.M. 1973. Experiences with the homozygous cases of familial hypercholesterolemia. A report of 52 patients. *Nutr. Metab.* **15:** 132.

Kipp H., Pichetshote N., and Arias I.M. 2001. Transporters on demand; intrahepatic pools of canalicular ATP binding cassette transporters in rat liver. *J. Biol. Chem.* **276:** 7218.

Linton M.F., Babaev V.R., Gleaves L.A., and Fazio S. 1999. A direct role for the macrophage low density lipoprotein receptor in atherosclerotic lesion formation. *J. Biol. Chem.* **274:** 19204.

Lu K., Lee M.-H., Hazard S., Brooks-Wilson A., Hidaka H., Kojima H., Ose L., Stalenhoef A.F.H., Mietinnen T., Bjorkhem I., Bruckert E., Pandya A., Brewer H.B., Jr., Salen G., Dean M., Srivastava A., and Patel S.B. 2001. Two genes that map to the *STSL* locus cause sitosterolemia: Genomic structure and spectrum of mutations involving sterolin-1 and sterolin-2, encoded by *ABCG5* and *ABCG8*, respectively. *Am. J. Hum. Genet.* **69:** 278.

Lutjohann D., Bjorkhem I., Beil U.F., and von Bergmann K. 1995. Sterol absorption and sterol balance in phytosterolemia evaluated by deuterium-labeled sterols: Effect of sitostanol treatment. *J. Lipid Res.* **36:** 1763.

Mackenzie S.M., Howells A.J., Cox G.B., and Ewart G.D. 2000. Sub-cellular localisation of the white/scarlet ABC transporter to pigment granule membranes within the compound eye of *Drosophila melanogaster*. *Genetica* **108:** 239.

Maliepaard M., Scheffer G.L., Faneyte I.F., van Gastelen M.A., Pijnenborg A.C.L.M., Schinkel A.H., van De Vijver M.J., Scheper R.J., Schellens J.H.M. 2001. Subcellular localization and distribution of the breast cancer resistance protein transporter in normal human tissues. *Cancer Res.* **61:** 3458.

Margolis B., Borg J.P., Straight S., and Meyer D. 1999. The function of PTB domain proteins. *Kidney Int.* **56:** 1230.

Miettinen T.A. 1980. Phytosterolaemia, xanthomatosis and premature atherosclerotic arterial disease: A case with high plant sterol absorption, impaired sterol elimination and low cholesterol synthesis. *Eur. J. Clin. Invest.* **10:** 27.

Moghadasian M.H. and Frohlich J.J. 1999. Effects of dietary phytosterols on cholesterol metabolism and atherosclerosis: Clinical and experimental evidence. *Am. J. Med.* **107:** 588.

Myant N.B. 1993. Familial defective apolipoprotein B-100: A review, including some comparisons with familial hypercholesterolaemia. *Atherosclerosis* **104:** 1.

Nair P.P., Turjman N., Kessie G., Calkins B., Goodman G.T., Davidovitz H., and Nimmagadda G. 1984. Diet, nutrition intake, and metabolism in populations at high and low risk for colon cancer. Dietary cholesterol, beta-sitosterol, and stigmasterol. *Am. J. Clin. Nutr.* **40:** 927.

NCEP Report Summary. 2001. Executive Summary of the Third Report of the National Cholesterol Education Program (NCEP) Expert Panel on Detection, Evaluation, and Treatment of High Blood Cholesterol in Adults (Adult Treatment Panel III). *J. Am. Med. Assoc.* **285:** 2486.

Nguyen L.B., Cobb M., Shefer S., Salen G., Ness G.C., and Tint G.S. 1991. Regulation of cholesterol biosynthesis in sitosterolemia: Effects of lovastatin, cholestyramine, and dietary sterol restriction. *J. Lipid Res.* **32:** 1941.

Norman D., Sun X.M., Bourbon M., Knight B.L., Naoumova R.P., and Soutar A.K. 1999. Characterization of a novel cellular defect in patients with phenotypic homozygous familial hypercholesterolemia. *J. Clin. Invest.* **104:** 619.

Patel S.B., Salen G., Hidaka H., Kwiterovich P.O., Stalenhoef A.F., Miettinen T.A., Grundy S.M., Lee M.H., Rubenstein J.S., Polymeropoulos M.H., and Brownstein M.J. 1998. Mapping a gene involved in regulating dietary cholesterol absorption. The sitosterolemia locus is found at chromosome 2p21. *J. Clin. Invest.* **102:** 1041.

Raal F.J., Pappu A.S., Illingworth D.R., Pilcher G.J., Marais A.D., Firth J.C., Kotze M.J., Heinonen T.M., and Black D.M. 2000. Inhibition of cholesterol synthesis by atorvastatin in homozygous familial hypercholesterolaemia. *Atherosclerosis* **150:** 421.

Repa J.J., Turley S.D., Lobaccaro J.A., Medina J., Li L., Lustig K., Shan B., Heyman R.A., Dietschy J.M., and Mangelsdorf D.J. 2000. Regulation of absorption and ABC1-mediated efflux of cholesterol by RXR heterodimers (comments). *Science* **289:** 1524.

Sai Y., Nies A.T., and Arias I.M. 1999. Bile acid secretion and direct targeting of mdr1-green fluorescent protein from Golgi to the canalicular membrane in polarized WIF-B cells. *J. Cell Sci.* **112:** 4535.

Salen G., Ahrens E.H., Jr., and Grundy S.M. 1970. Metabolism of *beta*-sitosterol in man. *J. Clin. Invest.* **49:** 952.

Salen G., Shefer S., Nguyen L., Ness G.C., Tint G.S., and Shore V. 1992. Sitosterolemia. *J. Lipid Res.* **33:** 945.

Salen G., Horak I., Rothkopf M., Cohen J.L., Speck J., Tint G.S., Shore V., Dayal B., Chen T., and Shefer S. 1985. Lethal atherosclerosis associated with abnormal plasma and tissue sterol composition in sitosterolemia with xanthomatosis. *J. Lipid Res.* **26:** 1126.

Salen G., Shore V., Tint G.S., Forte T., Shefer S., Horak I., Horak E., Dayal B., Nguyen L., and Batta A.K., et al. 1989. Increased sitosterol absorption, decreased removal, and expanded body pools compensate for reduced cholesterol synthesis in sitosterolemia with xanthomatosis. *J. Lipid. Res.* **30:** 1319.

Schmidt H.H., Stuhrmann M., Shamburek R., Schewe C.K., Ebhardt M., Zech L.A., Buttner C., Wendt M., Beisiegel U., Brewer H.B., Jr., and Manns M.P. 1998. Delayed low density lipoprotein (LDL) catabolism despite a functional intact LDL-apolipoprotein B particle and LDL-receptor in a subject with clinical homozygous familial hypercholesterolemia. *J. Clin. Endocrinol. Metab.* **83:** 2167.

Song W., Apodaca G., and Mostov K. 1994. Transcytosis of the polymeric immunoglobulin receptor is regulated in multiple intracellular compartments. *J. Biol. Chem.* **269:** 29474.

Stone N.J., Levy R.I., Fredrickson D.S., and Verter J. 1974. Coronary artery disease in 116 kindred with familial type II hyperlipoproteinemia. *Circulation* **49:** 476.

Sztul E., Kaplin A., Saucan L., and Palade G. 1991. Protein traffic between distinct plasma membrane domains: Isolation and characterization of vesicular carries involved in transcytosis. *Cell* **64:** 81.

Tavani D.M., Nes W.R., and Billheimer J.T. 1982. The sterol substrate specificity of acyl CoA:cholesterol acyltransferase from rat liver. *J. Lipid Res.* **23:** 774.

von Bergmann K., Mok H.Y., Hardison W.G., and Grundy S.M. 1979. Cholesterol and bile acid metabolism in moderately advanced, stable cirrhosis of the liver. *Gastroenterology* **77:** 1183.

Weihrauch J.L. and Gardner J.M. 1978. Sterol content of foods of plant origin. *J. Am. Diet. Assoc.* **73:** 39.

Yu L., Li-Hawkins J., Hammer R.E., Berge K.E., Horton J.D., Cohen J.C., and Hobbs H.H. 2002. Overexpression of ABCG5/ABCG8 promotes biliary cholesterol secretion and reduces fractional absorption of dietary cholesterol. *J. Clin. Invest.* (in press).

Zuliani G., Vigna G.B., Corsini A., Maioli M., Romagnoni F., and Fellin R. 1995. Severe hypercholesterolaemia: Unusual inheritance in an Italian pedigree. *Eur. J. Clin. Invest.* **25:** 322.

Zuliani G., Arca M., Signore A., Bader G., Fazio S., Chianelli M., Bellosta S., Campagna F., Montali A., Maioli M., Pacifico A., Ricci A., and Fellin R. 1999. Characterization of a new form of inherited hypercholesterolemia: Familial recessive hypercholesterolemia. *Arterioscler. Thromb. Vasc. Biol.* **19:** 802.

Insulin-like Growth Factor Isoforms in Skeletal Muscle Aging, Regeneration, and Disease

N. WINN,* A. PAUL,* A. MUSARÓ,† AND N. ROSENTHAL*

*EMBL Mouse Biology Programme, 32 -00016 Monterotondo (Rome), Italy; †Department of Histology and Medical Embryology, University of Rome "La Sapienza," 14 -00161 Rome, Italy

Recent studies on the role of insulin-like growth factor-1 (IGF-1) isoforms in skeletal muscle growth and differentiation have implicated this factor as an important mediator of anabolic pathways and have provided new insights into their function in muscle homeostasis. These studies suggest promising new avenues for systemic as well as local intervention in the defects associated with many muscle pathologies. Since IGF-1 levels decline with age in both humans and rodents, this growth factor has also been considered a promising therapeutic agent in staving off advancing muscle weakness during aging. Multiple transcripts of the IGF-1 gene encode different isoforms generated by alternate promoter usage, differential splicing, and posttranslational modification. Although IGF-1 has long been implicated as a key player in the physiology of growth, the complexity of IGF-1 gene expression and its protein products has not been systematically documented.

COMPLEXITIES OF IGF-1 TRANSCRIPTION

As its name implies, IGF-1 is similar to insulin in structure. However, unlike the insulin gene, the single-copy IGF-1 gene locus encodes multiple proteins with variable amino- and carboxy-terminal amino acid sequences (Fig. 1). Although the IGF-1 gene is highly conserved in numerous species, its relatively large size (>70 kb), and its complex transcriptional and splicing pattern, have complicated its analysis. The mature IGF-1 is a single-chain protein of 70 amino acids and differs from insulin by retention of the C-domain, by a short extension of the A-domain to include a novel domain D, and by the presence of variable carboxy-terminal extension peptides (E-peptides). The A- and B-domains of IGF-1 are 48% identical to the A- and B-chains of insulin (Rinderknecht and Humbel 1978).

The rodent IGF-1 gene contains six exons, separated by five introns (Fig. 1B) (Shimatsu and Rotwein 1987a; Bucci et al. 1989). Exons 1 and 2 encode distinct 5´UTRs, as well as different parts of the signal peptide, and are therefore termed leader exons. Exon 3 encodes 27 amino acids that are part of the signal peptide and common to all isoforms, as well as part of the mature IGF-1 peptide. Exon 4 encodes the rest of the mature peptide and 16 amino acids of the amino-terminal region of the E-peptide, which is also common to all IGF-1 mRNAs. Exons 5 and 6 encode two distinct carboxy-terminal E-peptides and the 3´UTR (Fig. 1D).

Separate promoters drive exon 1 and exon 2 transcription (Fig. 1C) (for review, see Adamo 1995), giving rise to transcripts containing exon 1 but lacking exon 2 (class 1) or to transcripts that are initiated by exon 2 and contain no exon-1-derived sequences (class 2). Exon 1 transcription initiation can occur from at least four distinct sites (Fig. 1C), located between –380 to –30 bases upstream of the 3´ end of exon 1, with start sites 2 and 3 representing major start sites. Additional complexity is given by the fact that in some class 1 mRNAs, a 186-base region, including the major start site 3 (–245 bp) is spliced out (Shimatsu and Rotwein 1987b). Compared to the dispersed pattern of exon 1 start sites, transcription initiation for exon 2 is more localized, occurring from a major cluster of start sites at ~70 bp and a minor cluster of start sites ~50 bp upstream of the exon-2 3´ end (Fig. 1C).

By the usage of alternate promoters and transcription start sites, three distinct signal peptides can thus be generated from the IGF-1 gene. The generation of different signal peptides has been reported to affect the precise amino-terminal cleavage site of the signal peptide (LeRoith and Roberts 1991), as well as the glycosylation of the E-peptides in vitro (Simmons et al. 1993).

EVOLVING THEORIES OF IGF-1 ACTION

Early investigations into the biological actions of IGF-1 led to the formulation of the somatomedin hypothesis (Daughaday et al. 1972), according to which growth is determined by GH acting primarily on the liver, which was thought to be the only source of IGF-1, where it stimulates the synthesis of IGF-1. Acting in an endocrine mode, it is then released into the circulation and transported to the main target organs, such as cartilage, bone, and muscle. This hypothesis turned out to be an oversimplification of a complex biological problem, since GH has direct growth effects of its own (for review, see Le Roith et al. 2001), and in addition it turned out that IGF-1 was also expressed in extrahepatic tissues (D'Ercole et al. 1980). This evidence for local IGF-1 production suggested that IGF-1 had an additional autocrine/paracrine effect, which was supported by subsequent investigations into the tissue distribution of IGF-1 mRNAs and their response to GH. These investigations led to the hypothesis

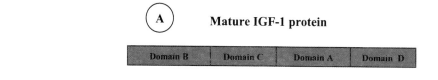
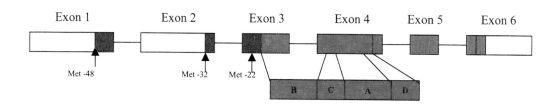
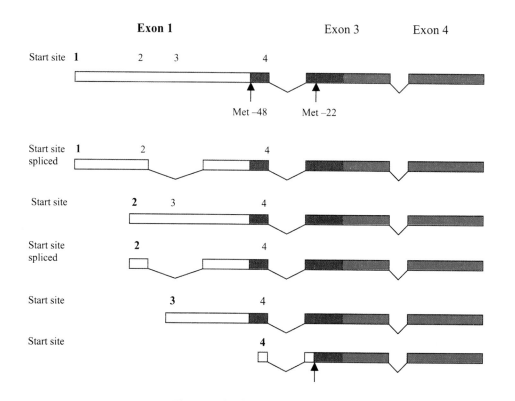

Figure 1. (*See facing page for legend.*)

that class 1 transcripts, together with the Ea form of the E-peptide, might encode the autocrine/paracrine form of IGF-1, as they are expressed in all tissues and are less GH responsive. Class 2 transcripts, together with the Eb-peptide, in contrast, are regulated by GH and are primarily found in liver, which is thought of as the main source of circulating IGF-1 and therefore might encode the endocrine form of IGF-1 (Hoyt et al. 1988; Lowe et al. 1988). The Class 1-Eb isoform has also been shown to be up-regulated in skeletal muscle that has been subjected to stretch (Yang et al. 1996; McKoy et al. 1999), and therefore has been named mechano growth factor (MGF).

DIVERSE FUNCTIONS OF IGF-1 IN VIVO

The complicated structure and regulation of the IGF-1 gene (Fig. 1), together with the fact that the encoded proteins can either circulate as hormones or act as local growth factors, have also complicated the analysis of IGF-1 functions, which vary widely (Table 1). Unlike insulin, which primarily acts as a hormone of intermediary metabolism, IGF-1 functions primarily as a growth, survival, and differentiation factor. IGF-1 shares many anabolic functions with insulin (for review, see Florini and Ewton 1990), but differs in its metabolic actions, as IGF-1 stimulates whole-body protein metabolism by increas-

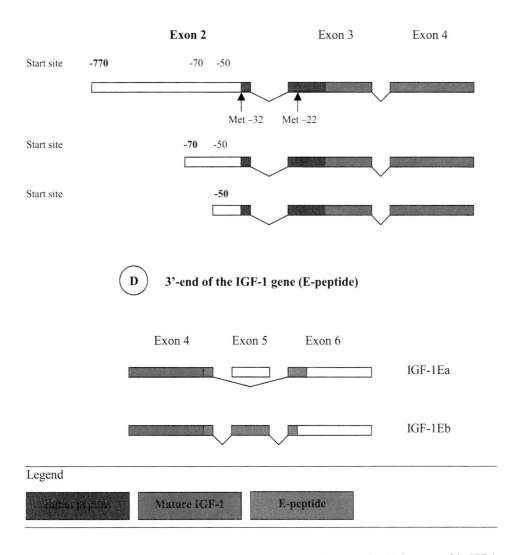

Figure 1. Organization of the IGF-1 gene and protein. (*A*) Domains of the mature IGF-1 protein; (*B*) Structure of the IGF-1 gene. Regions that code for the mature IGF-1 protein are shown in red. Contributions of exons 3 and 4 to the mature peptide are indicated by lines connecting the mature peptide shown again below the gene. Parts of exons 1 and 2, which are present in the different signal peptides, are shown in dark red. Positions of the translation initiation codons are indicated by arrows. Parts of exons 4, 5, and 6, which contribute to the distinct E-peptides, are blue. (*C*) 5′ end of the IGF-1 gene, showing the different possible 5′UTRs and signal peptides, resulting from transcription initiation in either exon 1 or exon 2. Methionine translation initiation codons are indicated with arrows. Resulting signal peptides are colored in dark red. (*D*) 3′end of the IGF-1 gene. The two possible IGF-1 E-peptides are colored blue.

Table 1. Effects of IGF-1 on Tissue Homeostasis

Skeletal muscle	Cardiac muscle	Nervous system	Skin	Reproductive system
Induction of hypertrophy (Coleman et al. 1995; Musarò et al. 2001)	Protection from apoptosis, limitation of ventricular dilation, myocardial loading, and cardiac hypertrophy (Reiss et al. 1996; Li et al. 1997; Leri et al. 1999; Li et al. 1999)	Action as a neurotrophic factor (Carro et al. 2001)	Induction of hyperplasia, dermal abnormalities, and spontaneous tumor formation in tg mice (Bol et al. 1997)	Promotion of ovarian cyst formation (Dyck et al. 2001)
Increase of DHPR number in adult tg mice (Renganathan et al. 1997)	Inhibition of diabetic cardiomyopathy (Kajstura et al. 2001)	Stimulation of postnatal brain growth (Gao et al. 1999; Lee et al. 1999)		Induction of IGFBP localization in testis with no obvious alterations in offspring generation (Dyck et al. 1999)
Prevention of age-related decline in DHPR number (Renganathan and Delbono 1998) No prevention of unloading-induced atrophy (Criswell et al. 1998)	Induction of physiologic and pathologic cardiac hypertrophy (Delaughter et al. 1999) Attenuation of cardiovascular diseases in fetal programming (Vickers et al. 2001)	Promotion of neuron survival and growth (Dentremont et al. 1999; O'Kusky et al. 2000) Increase in the rate of axon regeneration in crush-injured sciatic nerve (Tiangco et al. 2001) Protection of myelination from undernutrition damage during development (Ye et al. 2000)		
Extension of in vitro replicative life span of satellite cells (Barton-Davis et al. 1999) Long-term expression attenuates proliferative ability of satellite cells (Chakravarthy et al. 2001) Increase of muscle strength (Musarò et al. 2001) Preservation of muscle phenotype and function during aging (Musarò et al. 2001) Promotion of muscle regeneration in old mice (Musarò et al. 2001)				

(Reprinted, with permission, from Musarò and Rosenthal 2002.)

ing protein synthesis, as well as by inhibition of protein proteolysis (Fryburg 1994). Unlike other growth factors, such as FGF, EGF, or PDGF, which act as mitogens while inhibiting differentiation (for review, see Florini et al. 1991), IGF-1 promoted proliferation as well as differentiation of skeletal muscle cells, by first acting upon myoblast replication and subsequently promoting myogenic differentiation (Engert et al. 1996).

Distinct components of the IGF system in vivo have been studied by gene targeting experiments (for review, see D'Ercole 1999), showing that IGF-1 (Liu et al. 1993; Powell-Braxton et al. 1993), as well as the IGF-1 receptor (Liu et al. 1993), is necessary for normal embryonic and postnatal growth, and that the development of many tissues and organs is regulated to some degree by components of the IGF system. Although obviously also affecting other tissues, like brain, skin, and bone (for review, see Stewart and Rotwein 1996), as well as the immune and reproductive system (for review, see Clark 1997; van Buul-Offers Kooijman 1998; Bartke 1999), IGF-1 plays a central role in the control of muscle development, growth, and differentiation, as well as in maintenance of skeletal muscle, both in culture and in intact animals. However, the results of liver-specific IGF-1 gene deletion (Sjogren et al. 1999) further emphasize the critical role of locally produced autocrine/paracrine IGF-1 and bring the function of circulating IGF-1 under renewed scrutiny.

IGF-1 TRANSGENE ACTION IN SKELETAL MUSCLE

The role of IGF-1 signaling in the physiology of striated muscle has been studied with supplemental expression of various IGF-1 isoforms in transgenic mice (Coleman et al. 1995; Reiss et al. 1996; Li et al. 1997; Musarò et al. 2001). The discrepancies in cellular responses to overexpression of different IGF-1 isoforms underscores the complexity of IGF-1 function and highlights the need for a comprehensive characterization of the different isoforms in various physiological contexts. Whereas knockout mice lacking IGF-1, IGF-II, or the IGF-1R exhibit marked hypoplasia, mice overexpressing various muscle-specific IGF-1 transgenes showed enlarged myofibers (hypertrophy) (Coleman et al. 1995; Musarò et al. 2001).

Increases in fiber size diameter, protein contents per myotube, and increased nuclei within myofibers of transgenic mice expressing the class 1-Ea isoform (mIGF-1) were accompanied by activation of GATA-2 (Musarò et al. 2001), a novel marker of myocyte hypertrophy (Musarò and Rosenthal 1999a). Further studies of these mice revealed that mIGF-1 plays an important role in conferring muscle mass and strength during aging, in neuromuscular diseases, and after injury (Musarò et al. 2001; Barton et al. 2002). Similar beneficial effects have been observed with an mIGF-1 transgene expressed in cardiac muscle, which results in increases in heart size due to a combination of hyperplasia and adaptive hypertrophy (L. Tsao et al., unpubl.).

In contrast, a circulating class2-Ea transgene expressed in both heart and skeletal muscle (Coleman et al. 1995) initially induced an analog of physiological cardiac hypertrophy, but later progressed to a pathological condition (Delaughter et al. 1999). In the muscles this IGF-1 isoform increased the number of dihydropyridine receptors and prevented their age-related decline (Renganathan et al. 1997; Renganathan and Delbono 1998) and extended the in vitro replicative life span of skeletal muscle satellite cells, via PI3-kinase signal transduction pathway; however, there is no further beneficial effect on enhancing satellite cell proliferation ability with persistent long-term transgene expression (Chakravarthy et al. 2001). The disparity of these observations highlights the complexity of IGF-1 action and underscores for more extensive comparative analysis of IGF-1 isoform distribution and function under different physiological conditions.

IGF-1 ISOFORM DIVERSITY IN SKELETAL MUSCLE

To gain further insight into the differential role of IGF-1 isoforms in skeletal muscle, we undertook a comprehensive characterization of the distribution of IGF-1 isoform expression in diverse physiological conditions. The analyses in Figure 2a confirmed the expression of class 1 (initiated by exon 1) and Ea-containing IGF-1 transcripts in all tissues studied, whereas the expression of class 2 (initiated by exon 2) Eb-containing transcripts was mainly detected in the liver (Hoyt et al. 1988; Lowe et al. 1988; Adamo et al. 1989). Interestingly, muscle-specific overexpression of the mIGF-1 transgene, which contains the spliced form of exon 1 and the Ea-sequence, strongly up-regulated endogenous IGF-1 isoforms containing either exon 2, the unspliced version of exon 1, or the Eb-sequence exclusively in skeletal muscle. An unexpected fragment of ~500 bp and as-yet-unidentified origin was exclusively amplified in the MLC/mIGF-1 tissues. This was identified as a nonsignificant rearrangement in the 5´UTR of the transgene. This obviously needs to be further investigated, as additional bands were also seen with exon-2-specific amplification.

As seen in the Northern analysis in Figure 2b, the extent to which the endogenous isoforms were up-regulated in the different muscles correlates with the amount of mIGF-1 transgene expressed (Musarò et al. 2001). Thus, the transgenic IGF-1 product initiates a cascade of events that ultimately lead to up-regulation of endogenous IGF-1 gene products. The present findings also raise the possibility that the phenotype seen in the transgenic mIGF-1 mice is due not only to the overexpression of the exogenous mIGF-1 isoform, but also to increases in all endogenous IGF-1 isoforms, which then act synergistically to stimulate muscle hypertrophy, to maintain muscle functionality, and to promote regeneration in these animals.

IGF-1 ISOFORM SHIFTS IN AGING SKELETAL MUSCLE

Skeletal muscle aging involves an alteration in the heterogeneity of muscle fibers and therefore in muscle plasticity (Musarò et al. 1999b). Hallmarks of aging skeletal muscle include restriction of adaptability, and decline in muscle mass and function, as well as quality (innervation patterns, contractility, capillary density, and endurance), leading to a decrease in muscle strength and force output. Both GH and IGF-1 decline with age (Velasco et al. 1998), which might contribute to the progress of muscle atrophy in senescence and limit the reparative and regenerative abilities of skeletal muscle. This assumption was confirmed by muscle-specific overexpression of IGF-1 in a transgenic mouse model, which overexpresses IGF-1 exclusively in skeletal muscle (MLC/mIGF-1) (Musarò et al. 2001). The study showed that muscle mass and fiber type distribution in old animals were maintained at levels similar to those of younger adults. Even in aged muscle, the mIGF-1 transgenic mice retained the proliferative response to muscle injury characteristic of younger animals.

To characterize the distribution of IGF-1 isoforms in skeletal muscle during aging, IGF-1 transcripts from muscles of wild-type mice from 4 to 17 months old were analyzed for inclusion of full-length and spliced exon 1 or exon 2 sequences, as well as the Ea- and the Eb-peptide (Fig. 3a). These isoforms were present at all ages studied. Northern analysis (Fig. 3b) confirmed the presence of all isoforms at 4 months of age, with highest levels in the diaphragm. Transcripts encoding all isoforms progressively declined thereafter, but were re-activated in the diaphragm at 17 months (Fig. 3b). Since the diaphragm is the one muscle that is most extensively used during the life of an organism, it is more susceptible to work-induced muscle damage and age-related degeneration. Thus, the up-regulation of IGF-1 in the diaphragm may be a compensatory mechanism to promote the maintenance of continuous muscle function.

This possibility is supported by studies showing that MyoD and myogenin, as well as muscle-specific genes that are *trans*-activated by these MRFs, were expressed at high levels in the skeletal muscle of senile mice (Musarò et al. 1995). It is likely that the up-regulation of IGF-1 isoforms is connected with MRF activation, since IGF-1 has been shown to stimulate myogenin expression (for review, see Florini et al. 1996). Thus, factors that are involved in the proper development and maturation of

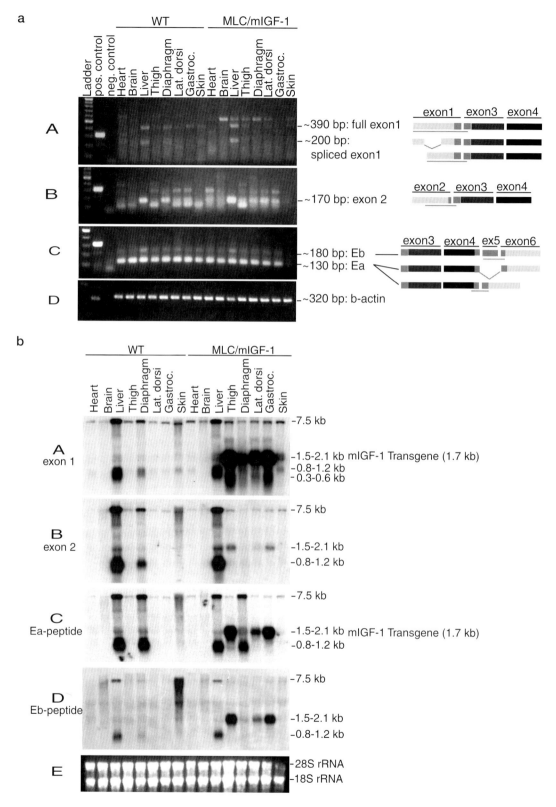

Figure 2. IGF-1 isoform diversity in skeletal muscle. Experiments were performed with total RNA from heart, brain, liver, thigh, diaphragm, latissimus dorsi, gastrocnemius, and skin of 6-month-old wild-type and 6-month-old MLC/mIGF-1 mice. (*a*) RT-PCR with 1 μg of total RNA was performed with specific primer pairs amplifying the sequences coding for the Ea- and the Eb-peptide (*A*), the full-length and spliced exon 1 (*B*), and exon 2 (*C*). The samples were normalized by specific amplification of β-actin (*D*). The sizes of the expected fragments are indicated. (*b*) Northern blots were performed with 10 μg of total RNA and probes specific for the sequences derived from the full-length exon 1 (*A*), exon 2 (*B*), and the Ea- (*C*) and Eb-peptide (*D*) sequences. Ethidium bromide staining was used to verify equal loading of RNA samples and RNA integrity (*E*). The probes hybridized to different previously described size classes of IGF-1 mRNAs, which are due to the usage of different polyadenylation signals in the 3′UTR and therefore differ in the length of their 3′UTRs.

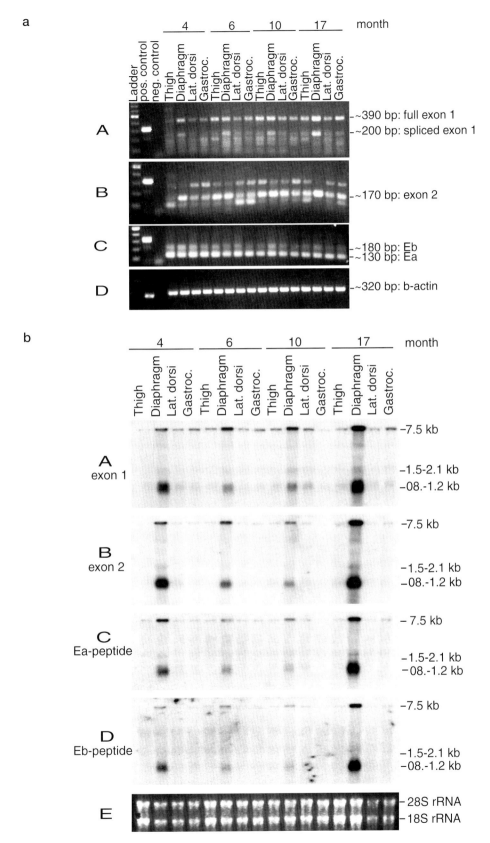

Figure 3. IGF-1 isoform shifts in aging skeletal muscle. Experiments were performed with total RNA from thigh, diaphragm, latissimus dorsi, and gastrocnemius of 4-month-, 6-month-, 10-month-, and 17-month-old mice. (*a*) RT-PCR with 1 μg of total RNA and specific primer pairs amplifying the sequences coding for the full-length and spliced exon 1 (*A*), exon 2 (*B*) and the Ea- and the Eb-peptide (*C*). The samples were normalized by specific amplification of β-actin (*D*). (*b*) Northern blots were performed with 10 μg of total RNA and probes specific for the sequences derived from the full-length exon 1 (*A*), exon 2 (*B*), and the Ea- (*C*) and Eb-peptide (*D*) sequences. Ethidium bromide staining was used to verify equal loading of RNA samples and RNA integrity (*E*).

skeletal muscle early in life also seem to be of importance later in life, but then to maintain and not to establish muscle functionality.

IGF-1 ISOFORMS IN MUSCLE DISEASE

Two mouse models of muscle diseases with different origins were investigated for IGF-1 isoform distribution. Amyotrophic lateral sclerosis (ALS) is one of the most common (5 per 100,000 individuals) adult-onset neurodegenerative diseases in humans, where the progressive decline in muscular function is a secondary effect, caused by the selective degeneration and loss of motor neurons that control muscle movement. The loss of these motor neurons results in severe atrophy of skeletal muscle. To date, genetically characterized causes of ALS include missense mutations in the gene encoding Cu/Zn superoxide dismutase 1 (SOD1) (for review, see Rowland and Shneider 2001), an ubiquitously expressed metalloenzyme that catalyzes the conversion of superoxide anions to hydrogen peroxide and molecular oxygen. Although the precise molecular pathways leading to motor neuron death remain unknown (Cleveland 1999), possible primary mechanisms are toxic gain-of-function effects of mutant SOD1 (for review, see Julien 2001; Rowland and Shneider 2001).

The SOD1 transgenic line (Gurney et al. 1994) expresses high levels of the human SOD1 gene carrying a substitution of glycine for alanine at position 93. This mutation does not affect the activity of the enzyme (Borchelt et al. 1994) but nevertheless was found in patients that were seriously affected by familial ALS. In the transgenic SOD1 mice, overexpression of the mutated human SOD1 gene causes muscle paralysis in one or more limbs as a result of motor neuron loss from the spinal cord, leading to death by 5 to 6 months of age (Gurney et al. 1994). Transcript analysis of 5-month SOD1 limb muscles compared to C57/6 wild-type controls, using both RT-PCR (Fig. 4a) and northern analysis (data not shown), revealed that despite the extensive paralysis following motor neuronal degeneration, the disease does not induce any changes in IGF-1 isoform expression or levels in the affected skeletal muscles, which are incapable of sustaining mass and function in the absence of innervation.

In contrast to ALS, in which skeletal muscle is indirectly affected, Duchenne muscular dystrophy (DMD) is an X-linked progressive skeletal and cardiac muscle-wasting disease in which deficiency of the dystrophin protein in muscle cells leads to progressive paralysis of limb and trunk muscles. Dystrophin is part of a complex (dystrophin-associated protein complex, DPC) that links the cytoskeleton to the extracellular matrix (Ervasti and Campbell 1993) and is necessary for the maintenance of the integrity of skeletal muscle surface membrane. In dystrophic muscle, muscle fibers develop normally but are easily damaged. Muscles undergo successive rounds of degeneration and insufficient regeneration, leading to gradual replacement of muscle by connective tissue until the functional capacity of muscle diminishes to a point below required force output (for review, see McArdle et al. 1995). The *mdx* mouse (Bulfield et al. 1984; Sicinski et al. 1989) is a well-characterized model for DMD in which the dystrophin protein is absent due to a spontaneously occurring point mutation. Although the progression of the disease in most of the *mdx* skeletal muscle groups is much milder than in humans (for review, see McArdle et al. 1995), the diaphragm exhibits a pattern of degeneration that is similar to that seen in patients with DMD (Stedman et al. 1991).

To investigate the distribution of IGF-1 isoforms in muscular dystrophy, transcripts from 12-month-old *mdx* mouse muscle were compared to age-matched C57/6 wild-type controls. RT-PCR analysis confirmed the presence of all IGF-1 isoforms in the *mdx* liver, brain, and all skeletal muscle groups (Fig. 4a). In northern analysis, *mdx* muscles contained higher levels of IGF-1 transcripts encoding exon 1 and either Ea- or Eb-peptide sequences (Fig. 4b). These latter results were unexpected, as an earlier report failed to detect IGF-1 transcripts containing the Eb sequence in dystrophic *mdx* muscle (Goldspink et al. 1996).

IGF-1 AND MUSCLE REGENERATION

A role for IGF-1 in dystrophic muscle regeneration has been suggested by previous studies (De Luca et al. 1999), which demonstrated that during the period of spontaneous regeneration seen after 8–10 weeks in the hind limb muscles of the *mdx* mouse, IGF-1 levels were markedly up-regulated in the muscle itself, as well as in the plasma. The *mdx* mouse shows a multistage pathogenesis, where the muscle appears histologically normal in the immediate postnatal period, but then an acute phase of muscle necrosis occurs around weaning with visible muscle weakness. This period is followed by apparent stability of the myopathy with increasing age, where skeletal muscle recovery is seen. Muscle of *mdx* mice from 15 months atrophy progressively, and some fibrosis is evident (Pastoret and Sebille 1993). It is therefore possible that the periods of muscle stability seen in the *mdx* mice are due to the synergistic action of IGF-1 isoforms that maintain the muscle function for a limited period of time.

This argument is supported by a recent study, in which mIGF-1 transgenic mice, which show pronounced muscle hypertrophy, escape age-related muscle atrophy, and retain the regenerative capacity in response to injury characteristic for younger animals (Musaró et al. 2001), were crossed with *mdx* mice (Barton et al. 2002). This study demonstrated the complete rescue of the dystrophic phenotype, suggesting that supplementary levels of the mIGF-1 isoform are sufficient to stimulate a regenerative response, which overcomes the degeneration of muscle fibers and is capable of maintaining the functionality of the muscle in these animals. Although up-regulation of endogenous IGF-1 isoforms in *mdx* muscles is clearly not sufficient to maintain the hypertrophic muscle phenotype seen in the *mdx*/mIGF-1 muscles, these results highlight the potential therapeutic value of at least the mIGF-1 isoform in the maintenance of muscle function and stimulation of regeneration.

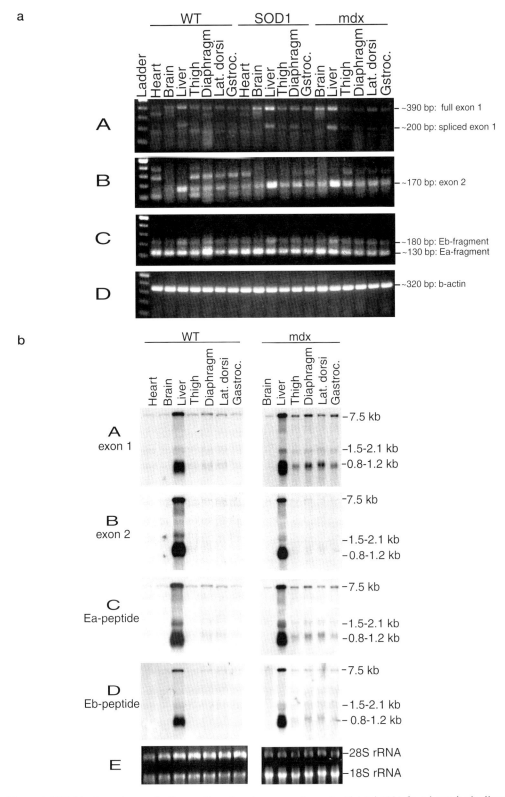

Figure 4. IGF-1 isoforms in muscle disease. Experiments were performed with total RNA from heart, brain, liver, thigh, diaphragm, latissimus dorsi, and gastrocnemius of a 12-month-old WT, a 5-month-old SOD1, and a 12-month-old *mdx* mouse. (*a*) RT-PCR with 1 μg of total RNA and specific primer pairs amplifying the sequences coding for the full-length and spliced exon 1 (*A*), exon 2 (*B*), and the Ea- and the Eb-peptide (*C*). The samples were normalized by specific amplification of β-actin (*D*). (*b*) Northern blots were performed with 10 μg of total RNA and probes specific for the sequences derived from the full-length exon 1 (*A*), exon 2 (*B*), and the Ea- (*C*) and Eb-peptide (*D*) sequences. Ethidium bromide staining was used to verify equal loading of RNA samples and RNA integrity (*E*).

CONCLUSIONS AND FUTURE VIEW

The studies summarized in this review implicate IGF-1 in development, postnatal growth, and maintenance of skeletal muscle. Due to the complex structure and regulation of the IGF-1 gene, and the fact that IGF-1 can either circulate as a hormone or act as a local growth factor, it is still not clear which of the various isoforms generated from the IGF-1 gene is involved in promoting its respective functions. The systemic documentation of isoform function is therefore a fundamental prerequisite for understanding and interpreting IGF-1 actions, and for effective exploitation of its beneficial effects in eventual therapeutic treatments. This is especially critical in light of disparate transgenic phenotypes obtained with different IGF-1 isoforms (Coleman et al. 1995; Reiss et al. 1996; Li et al. 1997; Delaughter et al. 1999; Musaró et al. 2001).

Our preliminary survey of IGF-1 isoform changes in response to muscle aging and dystrophy suggests that differential expression of these proteins is important for their potentially disparate action in the organism. This is a particularly relevant consideration for eventual IGF-1-based therapies, as is the increasing evidence that besides its beneficial effects on various tissues and organ systems, high levels of IGF-1 might also be connected to several progressive diseases including prostate, colon, and breast cancer (Table 1). Further delineation of IGF-1 isoform function, and subsequent choice of the appropriate isoform for gene-based muscle therapies, might circumvent these undesirable side effects and permit the beneficial use of IGF-1 in more widespread clinical applications.

REFERENCES

Adamo M.L. 1995. Regulation of insulin-like growth factor I gene expression. *Diabetes Rev.* **3:** 2.

Adamo M., Lowe W.L., Jr., LeRoith D., and Roberts C.T., Jr. 1989. Insulin-like growth factor I messenger ribonucleic acids with alternative 5′-untranslated regions are differentially expressed during development of the rat. *Endocrinology* **124:** 2737.

Bartke A. 1999. Role of growth hormone and prolactin in the control of reproduction: What are we learning from transgenic and knock-out animals? *Steroids* **64:** 598.

Barton E.R., Morris L., Musaró A., Rosenthal N., and Sweeney H.L. 2002. Muscle-specific expression of insulin-like growth factor I counters muscle decline in mdx mice. *J. Cell Biol.* **157:** 137.

Barton-Davis E.R., Shoturma D.I., and Sweeney H.L. 1999. Contribution of satellite cells to IGF-I induced hypertrophy of skeletal muscle. *Acta Physiol. Scand.* **167:** 301.

Bol D.K., Kiguchi K., Gimenez-Conti I., Rupp T., and DiGiovanni J. 1997. Overexpression of insulin-like growth factor-1 induces hyperplasia, dermal abnormalities, and spontaneous tumor formation in transgenic mice. *Oncogene* **14:** 1725.

Borchelt D.R., Lee M.K., Slunt H.S., Guarnieri M., Xu Z.S., Wong P.C., Brown R.H., Jr., Price D.L., Sisodia S.S., and Cleveland D.W. 1994. Superoxide dismutase 1 with mutations linked to familial amyotrophic lateral sclerosis possesses significant activity. *Proc. Natl. Acad. Sci.* **91:** 8292.

Bucci C., Mallucci P., Roberts C.T., Frunzio R., and Bruni C.B. 1989. Nucleotide sequence of a genomic fragment of the rat IGF-I gene spanning an alternate 5′ non coding exon. *Nucleic Acids Res.* **17:** 3596.

Bulfield G., Siller W.G., Wight P.A., and Moore K.J. 1984. X chromosome-linked muscular dystrophy (mdx) in the mouse. *Proc. Natl. Acad. Sci.* **81:** 1189.

Carro E., Trejo J.L., Busiguina S., and Torres-Aleman I. 2001. Circulating insulin-like growth factor I mediates the protective effects of physical exercise against brain insults of different etiology and anatomy. *J. Neurosci.* **21:** 5678.

Chakravarthy M.V., Fiorotto M.L., Schwartz R.J., and Booth F.W. 2001. Long-term insulin-like growth factor-I expression in skeletal muscles attenuates the enhanced in vitro proliferation ability of the resident satellite cells in transgenic mice. *Mech. Ageing Dev.* **122:** 1303.

Clark R. 1997. The somatogenic hormones and insulin-like growth factor-1: Stimulators of lymphopoiesis and immune function. *Endocr. Rev.* **18:** 157.

Cleveland D.W. 1999. From Charcot to SOD1: Mechanisms of selective motor neuron death in ALS. *Neuron* **24:** 515.

Coleman M., DeMayo F., Yin K.C., Lee H.M., Geske R., Montgomery C., and Schwartz R.J. 1995. Myogenic vector expression of insulin-like growth factor I stimulate myocyte differentiation and myofiber hypertrophy in transgenic mice. *J. Biol. Chem.* **270:** 12109.

Criswell D.S., Booth F.W., DeMayo F., Schwartz R.J., Gordon S.E., and Fiorotto M.L. 1998. Overexpression of IGF-I in skeletal muscle of transgenic mice does not prevent unloading-induced atrophy. *Am. J. Physiol.* **275:** E373.

Daughaday W.H., Hall K., Raben M.S., Salmon W.D., Jr., van den Brande J.L., and Van Wyk J.J. 1972. Somatomedin: Proposed designation for sulphation factor. *Nature* **235:** 107.

Delaughter M.C., Taffet G.E., Fiorotto M.L., Entman M.L., and Schwartz R.J. 1999. Local insulin-like growth factor I expression induces physiologic, then pathologic cardiac hypertrophy in transgenic mice. *FASEB J.* **13:** 1923.

De Luca A., Pierno S., Camerino C., Cocchi D., and Camerino D.C. 1999. Higher content of insulin-like growth factor-I in dystrophic mdx mouse: Potential role in the spontaneous regeneration through an electrophysiological investigation of muscle function. *Neuromuscul. Disord.* **9:** 11.

Dentremont K.D., Ye P., D'Ercole A.J., and O'Kusky J.R. 1999. Increased insulin-like growth factor-I (IGF-I) expression during early postnatal development differentially increases neuron number and growth in medullary nuclei of the mouse. *Brain Res. Dev. Brain Res.* **114:** 135.

D'Ercole A.J. 1999. Actions of IGF system proteins from studies of transgenic and gene knockout models. In *The IGF system* (ed. R. Rosenfeld and C.T. Roberts, Jr.), p. 545. Humana Press, Totowa, New Jersey.

D'Ercole A.J., Applewhite G., and Underwood L. 1980. Evidence that sometomedin is synthesized by multiple tissues in the fetus. *Dev. Biol.* **75:** 315.

Dyck M.K., Parlow A.F., Senechal J.F., Sirard M.A., and Pothier F. 2001. Ovarian expression of human insulin-like growth factor-I in transgenic mice results in cyst formation. *Mol. Reprod. Dev.* **59:** 178.

Dyck M.K., Ouellet M., Gagn M., Petitclerc D., Sirard M.A., and Pothier F. 1999. Testes-specific transgene expression in insulin-like growth factor-I transgenic mice. *Mol. Reprod. Dev.* **54:** 32.

Engert J., Berglund E.B., and Rosenthal N. 1996. Proliferation precedes differentiation in IGF-I stimulated myogenesis. *J. Cell Biol.* **135:** 431.

Ervasti J.M. and Campbell K.P. 1993. A role for the dystrophin-glycoprotein complex as a transmembrane linker between laminin and actin. *J. Cell Biol.* **122:** 809.

Florini J.R. and Ewton D.Z. 1990. Highly specific inhibition of IGF-I-stimulated differentiation by an antisense oligodeoxyribonucleotide to myogenin mRNA. No effects on other actions of IGF-T. *J. Biol. Chem.* **265:** 13435.

Florini J.R., Ewton D.Z., and Coolican S.A. 1996. Growth hormone and the insulin-like growth factor system in myogenesis. *Endocr. Rev.* **17:** 481.

Florini J.R., Ewton D.Z., and Magri K.A. 1991. Hormones, growth factors, and myogenic differentiation. *Annu. Rev. Physiol.* **53:** 201.

Fryburg D.A. 1994. Insulin-like growth factor I exerts growth hormone- and insulin-like actions on human muscle protein

metabolism. *Am. J. Physiol.* **267:** E331.
Gao W.Q., Shinsky N., Ingle G., Beck K., Elias K.A., and Powell-Braxton L. 1999. IGF-I deficient mice show reduced peripheral nerve conduction velocities and decreased axonal diameters and respond to exogenous IGF-I treatment. *J. Neurobiol.* **39:** 142.
Goldspink G., Yang S.Y., Skarli M., and Vrbova G. 1996. Local growth regulation is associated with an isoform of IGF-1 that is expressed in normal but not in dystrophic muscle when subjected to stretch. *J. Physiol.* **495P:** 162.
Gurney M.E., Pu H., Chiu A.Y., Dal Canto M.C., Polchow C.Y., Alexander D.D., Caliendo J., Hentati A., Kwon Y.W., and Deng H.X., et al. 1994. Motor neuron degeneration in mice that express a human Cu,Zn superoxide dismutase mutation. *Science* **264:** 1772.
Hoyt E., Van Wyk J.J., and Lund P.K. 1988. Tissue and development specific regulation of a complex family of rat insulin-like growth factor I messenger ribonucleic acids. *Mol. Endocrinol.* **2:** 1077.
Julien J.P. 2001. Amyotrophic lateral sclerosis. Unfolding the toxicity of the misfolded. *Cell* **104:** 581.
Kajstura J., Fiordaliso F., Andreoli A.M., Li B., Chimenti S., Medow M.S., Limana F., Nadal-Ginard B., Leri A., and Anversa P. 2001. IGF-1 overexpression inhibits the development of diabetic cardiomyopathy and angiotensin II-mediated oxidative stress. *Diabetes* **50:** 1414.
Lee K.H., Calikoglu A.S., Ye P., and D'Ercole A.J. 1999. Insulin-like growth factor-I (IGF-I) ameliorates and IGF binding protein-1 (IGFBP-1) exacerbates the effects of undernutrition on brain growth during early postnatal life: Studies in IGF-I and IGFBP-1 transgenic mice. *Pediatr. Res.* **45:** 331.
Leri A., Liu Y., Wang X., Kajstura J., Malhotra A., Meggs L.G., and Anversa P. 1999. Overexpression of insulin-like growth factor-1 attenuates the myocyte renin-angiotensin system in transgenic mice. *Circ. Res.* **84:** 752.
LeRoith D. and Roberts C.T., Jr. 1991. Insulin-like growth factor I (IGF-1): A molecular basis for endocrine versus local action. *Mol. Cell. Endocrinol.* **77:** C57.
LeRoith D., Bondy C., Yakar S., Liu J.L., and Butler A. 2001. The somatomedin hypothesis: 2001. *Endocr. Rev.* **22:** 53.
Li B., Setoguchi M., Wang X., Andreoli A.M., Leri A., Malhotra A., Kajstura J., and Anversa P. 1999. Insulin-like growth factor-1 attenuates the detrimental impact of nonocclusive coronary artery constriction on the heart. *Circ. Res.* **84:** 1007.
Li Q., Li B., Wang X., Leri A., Jana K.P., Liu Y., Kajstura J., Baserga R., and Anversa P. 1997. Overexpression of insulin-like growth factor-I in mice protects from myocyte death after infarction attenuating ventricular dilation, wall stress, and cardiac hypertrophy. *J. Clin. Invest.* **100:** 1991.
Liu J., Baker J., Perkins A.S., Robertson E.J., and Efstratiadis A. 1993. Mice carrying null mutations of the genes encoding insulin-like growth factor I (IGF-I) and type I IGF receptor (IGF-Ir). *Cell* **75:** 59.
Lowe W.L., Jr., Lasky S.R., LeRoith D., and Roberts C.T., Jr. 1988. Distribution and regulation of rat insulin-like growth factor I messenger ribonucleic acids encoding alternative carboxyterminal E- peptides: Evidence for differential processing and regulation in liver. *Mol. Endocrinol.* **2:** 528.
McArdle A., Edwards R.H., and Jackson M.J. 1995. How does dystrophin deficiency lead to muscle degeneration? Evidence from the mdx mouse. *Neuromuscul. Disord.* **5:** 445.
McKoy G., Ashley W., Mander J., Yang S.Y., Williams N., Russell B., and Goldspink G. 1999. Expression of insulin growth factor-1 splice variants and structural genes in rabbit skeletal muscle induced by stretch and stimulation. *J. Physiol.* **516:** 583.
Musaró A. and Rosenthal N. 1999a. Maturation of the myogenic program is induced by post-mitotic expression of IGF-I. *Mol. Cell. Biol.* **19:** 3115.
———. 1999b. Transgenic mouse models of muscle aging. *Exp. Gerontol.* **34:** 147.
———. 2002. The role of local insulin-like growth factor-1 isoforms in the pathophysiology of skeletal muscle. *Curr. Genomics* **3:** 149.
Musaró A., Cusella De Angelis M.G., Germani A., Ciccarelli C., Molinaro M., and Zani B.M. 1995. Enhanced expression of myogenic regulatory factors in aging skeletal muscle. *Exp. Cell Res.* **221:** 241.
Musaró A., McCullagh K., Paul A., Houghton L., Dobrowolny G., Molinaro M., Barton E.R., Sweeney H.L., and Rosenthal N. 2001. Localized Igf-1 transgene expression sustains hypertrophy and regeneration in senescent skeletal muscle. *Nat. Genet.* **27:** 195.
O'Kusky J.R., Ye P., and D'Ercole A.J. 2000. Insulin-like growth factor-I promotes neurogenesis and synaptogenesis in the hippocampal dentate gyrus during postnatal development. *J. Neurosci.* **20:** 8435.
Pastoret C. and Sebille A. 1993. Further aspects of muscular dystrophy in mdx mice. *Neuromuscul. Disord.* **3:** 471.
Powell-Braxton L., P. Hollingshead P., Warburton C., Dowd M., Pitts-Meek S., Dalton D., Gillett N., and Stewart T.A. 1993. IGF-I is required for normal embryonic growth in mice. *Genes Dev.* **7:** 2609.
Reiss K., Cheng W., Ferber A., Kajstura J., Li P., Li B., Olivetti G., Homcy C.J., Baserga R., and Anversa P. 1996. Overexpression of insulin-like growth factor-1 in the heart is coupled with myocyte proliferation in transgenic mice. *Proc. Natl. Acad. Sci.* **93:** 8630.
Renganathan M. and Delbono O. 1998. Caloric restriction prevents age-related decline in skeletal muscle dihydropyridine receptor and ryanodine receptor expression. *FEBS Lett.* **434:** 346.
Renganathan M., Messi M.L., Schwartz R., and Delbono O. 1997. Overexpression of hIGF-1 exclusively in skeletal muscle increases the number of dihydropyridine receptors in adult transgenic mice. *FEBS Lett.* **417:** 13.
Rinderknecht E. and Humbel R.E. 1978. The amino acid sequence of human insulin-like growth factor I and its structural homology with proinsulin. *J. Biol. Chem.* **253:** 2769.
Rowland L.P. and Shneider N.A. 2001. Amyotrophic lateral sclerosis. *N. Engl. J. Med.* **344:** 1688.
Shimatsu A. and Rotwein P. 1987a. Mosaic evolution of the insulin-like growth factors: Organization, sequence, and expression of the rat insulin-like growth factor I gene. *J. Biol. Chem.* **262:** 7894.
———. 1987b. Sequence of two rat insulin-like growth factor I mRNAs differing within the 5´ untranslated region. *Nucleic Acids Res.* **15:** 7196.
Sicinski P., Geng Y., Ryder-Cook A.S., Barnard E.A., Darlison M.G., and Barnard P.J. 1989. The molecular basis of muscular dystrophy in the mdx mouse: A point mutation. *Science* **244:** 1578.
Simmons J.G., Van Wyk J.J., Hoyt E.C., and Lund P.K. 1993. Multiple transcription start sites in the rat insulin-like growth factor-I gene give rise to IGF-I mRNAs that encode different IGF-I precursors and are processed differently in vitro. *Growth Factors* **9:** 205.
Sjogren K., Liu J.L., Blad K., Skrtic S., Vidal O., Wallenius V., LeRoith D., Tornell J., Isaksson O.G., Jansson J.O., and Ohlsson C. 1999. Liver-derived IGF-I is the principal source of IGF-I in blood but is not required for postnatal body growth in mice. *Proc. Natl. Acad. Sci.* **96:** 7088.
Stedman H.H., Sweeney H.L., Shrager J.B., Maguire H.C., Panettieri R.A., Petrof B., Narusawa M., Leferovich J.M., Sladky J.T., and Kelly A.M. 1991. The mdx mouse diaphragm reproduces the degenerative changes of Duchenne muscular dystrophy. *Nature* **352:** 536.
Stewart C.E. and Rotwein P. 1996. Growth, differentation, and survival: Multiple physiological functions for insulin-like growth factors. *Physiol. Rev.* **76:** 1005.
Tiangco D.A., Papakonstantinou K.C., Mullinax K.A., and Terzis J.K. 2001. IGF-I and end-to-side nerve repair: A dose-response study. *J. Reconstr. Microsurg.* **17:** 247.
van Buul-Offers S.C. and Kooijman R. 1998. The role of growth hormone and insulin-like growth factors in the immune system. *Cell Mol. Life Sci.* **54:** 1083.
Velasco B., Cacicedo L., Escalada J., Lopez-Fernandez J., and Sanchez-Franco F. 1998. Growth hormone gene expression

and secretion in aging rats is age dependent and not age-associated weight increase related. *Endocrinology* **139:** 1314.

Vickers M.H., Ikenasio B.A., and Breier B.H. 2001. IGF-I treatment reduces hyperphagia, obesity, and hypertension in metabolic disorders induced by fetal programming. *Endocrinology* **142:** 3964.

Yang S., Alnaqeeb M., Simpson H., and Goldspink G. 1996. Cloning and characterization of an IGF-1 isoform expressed in skeletal muscle subjected to stretch. *J. Muscle Res. Cell Motil.* **17:** 487.

Ye P., Lee K.H., and D'Ercole A.J. 2000. Insulin-like growth factor-I (IGF-I) protects myelination from undernutritional insult: Studies of transgenic mice overexpressing IGF-I in brain. *J. Neurosci. Res.* **62:** 700.

Cellular Therapies for Myocardial Infarct Repair

C.E. MURRY,*† M.L. WHITNEY,*† M.A. LAFLAMME,* H. REINECKE,* AND L.J. FIELD‡
*Departments of *Pathology and †Bioengineering, University of Washington, Seattle, Washington 98195;
‡Wells Center for Pediatric Research, Indiana University School of Medicine, Indianapolis, Indiana 46202*

Despite significant advances in prevention, diagnosis, and treatment, myocardial infarction remains the number one cause of death in the United States and other industrialized nations. Myocardial infarctions almost always result from complications of atherosclerosis, typically arising when an atherosclerotic plaque ruptures, leading to thrombotic occlusion of a major coronary artery. Myocardium has one of the highest metabolic rates of any tissue in the body, making it very susceptible to injury by ischemia (deprivation of blood flow). Irreversible ischemic injury (defined as cell death occurring despite reperfusion) begins within 20 minutes of severe ischemia in vivo in all species studied. A wavefront of cell death then progresses from subendocardium to subepicardium (Reimer et al. 1977; Reimer and Jennings 1979). The time to completion of irreversible injury varies with species, from 1–2 hours in the mouse (Michael et al. 1999) to 3–6 hours in the dog (Reimer and Jennings 1979). Although cardiac myocytes are the most ischemia-sensitive cell population, prolonged ischemia also kills most of the fibroblasts and vascular cells in the tissue.

MYOCARDIAL INFARCT REPAIR

The heart has very little intrinsic regenerative capacity. Numerous studies have shown that surviving cardiomyocytes do not re-enter the cell cycle to any significant extent (for review, see Soonpaa and Field 1998), although there are isolated reports in the literature of relatively high proliferation rates (Beltrami et al. 2001). Additionally, unlike skeletal muscle, it is not at all clear whether there is a myogenic stem cell population in the heart. As a result, infarcted myocardium is removed by phagocytes and replaced by a fibroblast-rich, provisional wound repair tissue termed granulation tissue. Granulation tissue remodels over time to form a relatively hypocellular, collagen-rich scar. Although the scar tissue has high tensile strength and serves to bridge the muscle defect in the heart, it does not contract. As a result, many patients suffer from heart failure after a myocardial infarct.

The heart undergoes significant changes in its global anatomy post-infarction (for review, see Yousef et al. 2000). The infarcted wall thins dramatically, with the final scar often being only 25% the thickness of the original wall. Wall thinning results in marked dilation of the ventricular cavity, which in turn significantly increases the mechanical stress on the non-infarcted wall. Although they do not divide significantly, cardiomyocytes have a tremendous capacity to undergo hypertrophy after infarction. Indeed, Beltrami et al. (1994) reported that human hearts in end-stage failure actually have more muscle mass than normal hearts (although the mass is in the wrong location). Finally, infarction results in activation of fibroblasts in the non-infarcted region of the heart, resulting in excess collagen deposition and eventual interstitial fibrosis. Interstitial fibrosis is thought to interfere with mechanical function of the heart by impeding relaxation and increasing internal work with each beat. This constellation of left ventricular dilation, hypertrophy, and interstitial fibrosis has been termed *left ventricular remodeling*. Remodeling is clearly detrimental to cardiac function. Clinical studies have shown that left ventricular dilation is the strongest predictor of heart failure and death following a myocardial infarct (White et al. 1987).

CARDIOMYOCYTE GRAFTING

Since myocardial infarction results in cardiomyocyte deficiency, a reasonable hypothesis is that replacing cardiomyocytes through cellular grafting would form new myocardium and improve left ventricular function. Initial studies with cardiomyocyte grafting showed that fetal or neonatal cardiomyocytes could form new myocardium when implanted into the normal hearts of mice (Soonpaa et al. 1994; Koh et al. 1995), rats (Leor et al. 1996; Connold et al. 1997; Reinecke et al. 1999; Scorsin et al. 2000; Muller-Ehmsen et al. 2002), and dogs (Koh et al. 1995). In contrast, adult cardiomyocytes did not form viable grafts (Reinecke et al. 1999). Furthermore, Soonpaa et al. (1994) showed that the transplanted cardiomyocytes formed intercalated disks (specialized junctions for electromechanical coupling) with host cardiomyocytes, suggesting they would beat synchronously with the host. Subsequent studies showed that fetal or neonatal cardiomyocyte transplantation could form new myocardium in injured hearts (Leor et al. 1996; Reinecke et al. 1999; Scorsin et al. 2000), although graft success rates were much lower in injured than in normal myocardium (Watanabe et al. 1998). Although formation of intercalated disks between graft and host cardiomyocytes was occasionally observed in the injured heart, more commonly the grafts were insulated from the host myocardium by a barrier of scar tissue (Reinecke et al. 1999).

These initial studies led to much excitement and suggested that it might be possible to have complete infarct repair via cardiomyocyte grafting. To this end, our group performed dose-escalation studies in cryoinjured rat hearts, where the number of engrafted cardiomyocytes ranged from 3 million to 24 million (Zhang et al. 2001.) (For comparison, the normal rat left ventricle contains ~20 million cardiomyocytes [Olivetti et al. 1990].) The results of this analysis were sobering: The amount of myocardium that formed 1 week after grafting did not increase over this dose range, and furthermore, graft size never exceeded 2% of the left ventricle. Analysis of cell death via TUNEL staining showed that extensive cell death occurred in the initial days after transplantation, peaking at 24 hours and continuing for at least 4 days (Fig. 1). Estimates of survival based on TUNEL staining indicated that only 1–10% of the grafted cardiomyocytes survived to form long-term grafts. Furthermore, increasing the number of transplanted cells simply resulted in more cell death, not increased graft size. Our initial data suggest that graft cells are dying from ischemic injury, and that death can be blocked in part by activating antiapoptotic, Akt-dependent signaling pathways or by heat shocking the cells prior to transplantation (Zhang et al. 2001).

Thus, the problem of extensive graft cell death, coupled with very little cell proliferation after grafting (Reinecke et al. 1999), makes the final amount of new myocardium that forms in the injured heart quite small. Despite this, some investigators have reported improved global ventricular function following cardiomyocyte grafting (Scorcin et al. 1995, 2000; Li et al. 1996; Etzion et al. 2001). The improved function appears to result from an impact of cell grafting on ventricular remodeling, rather than by a direct contractile effect. This view is supported by the observation that transplantation of non-contractile cells can effect a similar functional improvement in some laboratories (Li et al. 1999). These data indicate that left ventricular remodeling may be a more tractable problem to target for initial cell transplantation studies than restoration of contractile function in the injured region.

SKELETAL MUSCLE GRAFTING

Numerous studies have explored the effects of skeletal muscle grafting on cardiac repair. Skeletal muscle cells are quite ischemia-resistant and survive grafting into the injured heart much better than do cardiomyocytes (Murry et al. 1996). Furthermore, unlike cardiomyocytes, primary isolates of skeletal myoblasts can proliferate for several days after grafting into the heart. The myoblasts then withdraw from the cell cycle, fuse to form multinucleated myotubes, and differentiate further to form mature skeletal muscle fibers. In other words, they undergo a normal developmental program despite their unusual location in the injured heart. The combination of enhanced survival and proliferation results in larger grafts using skeletal muscle rather than cardiac muscle. Some investigators have hypothesized that skeletal muscle transdifferentiates into cardiomyocytes after intracardiac grafting (Chiu et al. 1995; Taylor et al. 1998). We re-

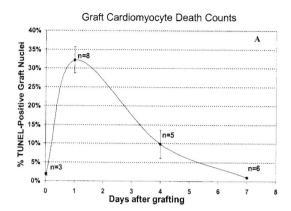

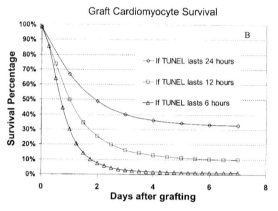

Figure 1. Graft cardiomyocyte death. Neonatal rat cardiomyocytes were labeled with Orange Cell Tracker and engrafted into acutely cryoinjured syngeneic rat hearts (5 million cells/heart). TUNEL staining was performed to detect DNA fragmentation at the indicated time after grafting. (*A*) Cells appeared viable immediately after grafting, but 1 day later, 33% of graft cells were TUNEL-positive. Significant cell death continued for at least 4 days after grafting, and by 7 days, death appeared largely complete. (*B*) Graft survival was estimated from the TUNEL data, using indicated assumptions for the duration of TUNEL-positivity. The most likely estimates of graft survival are between 1% and 10% at 7 days. (Reprinted, with permission, from Zhang et al. 2001 [copyright Academic Press].)

cently completed a study to test this hypothesis (Reinecke et al. 2002). A rigorous, double immunolabeling protocol showed clearly that graft cells (identified by BrdU prelabeling) expressed skeletal muscle markers such as fast myosin heavy chain and failed to up-regulate cardiac-specific genes such as α-myosin heavy chain, cardiac troponin I, and atrial natriuretic factor. Thus, myoblasts and satellite cells appear to be committed to the skeletal muscle phenotype irrespective of their location.

A critical question that needs to be addressed with regard to myoblast grafting studies is, What are the functional consequences of grafting skeletal muscle cells into the injured heart? Although the muscle grafts will contract when exogenously stimulated (Murry et al. 1996), skeletal muscle grafts do not form electromechanical junctions with host cardiomyocytes, due to absence of gap junction and mechanical adhesion proteins (Reinecke et al. 2000). Thus, skeletal muscle grafts are unlikely to be contracting synchronously with the host myocardium. Despite this shortcoming, multiple studies indicate that skeletal muscle engraftment improves global function in the injured heart. Taylor et al. have published several studies where global function of cryoinjured rabbit hearts was improved by satellite cell grafting (Taylor et al. 1998; Atkins et al. 1999). Jain et al. (2001) recently reported a careful functional study, showing skeletal muscle cell grafting one week after infarction enhanced exercise tolerance in rats compared to vehicle injections. Furthermore, skeletal muscle grafts attenuated ventricular dilation and enhanced infarct wall thickness, shifting the in vitro pressure–volume curve toward normal. Similar work from Menasche's group has shown clear benefits to global ventricular function in vivo from skeletal muscle grafting (Pouzet et al. 2001). *Importantly, they showed a dose dependency of functional change, with a linear improvement in ejection fraction as satellite cell dose increased.*

Taken together, these studies indicate that although skeletal muscle grafts may not contract, they can attenuate deleterious effects of post-infarct ventricular remodeling and improve cardiac function. The dose dependency observed by Menasche et al. indicates that graft size is a critical determinant of functional improvement. Based on the above functional data, three clinical trials of autologous satellite cell grafting for myocardial infarct repair are currently under way (Menasche et al. 2001).

"MOLECULAR PHARMACOLOGY" FOR CONTROL OF GRAFT SIZE

To date, most studies have not generated an appreciable amount of new muscle in the injured heart. It seems likely that, in order to improve the function of the infarcted heart, a large fraction of the lost myocardium must be replaced. Although this remains a significant challenge for cardiomyocyte grafting, skeletal muscle grafting presents some unique opportunities. The simplest approach to increasing myoblast cell number would be increasing the dose of injected cells. We tested this strategy using primary cultures of myoblasts in the rat heart and found that the relation between myoblast dose and graft size was quite variable (Reinecke and Murry 2000). At the low end (<3 million cells), most grafts were small. At the higher end (≥6 million cells), some grafts were small, some had optimal replacement of the injured region, and still others had overgrown the heart. This overgrowth pattern was the most concerning for a possible therapy, because these grafts distended the epi- and endocardial borders and could well have interfered with proper ventricular function. (As a side note, variable graft size likely results in large part from variable seeding efficiency at the time of injection. Muller-Ehmsen et al. [2002], working with an experienced cell grafting group, recently showed that initial cell seeding varies from 10% to 90%.)

These studies illustrate the need for a better way to control graft size in the heart. An alternative strategy is to inject a smaller number of myoblasts, have them proliferate to a desired size in vivo, and then induce their differentiation into mature muscle. FGF is a potent mitogen for myoblasts, but it also stimulates proliferation in multiple other cell types and might cause fibrous overgrowth of the graft. We took advantage of a genetic system, originally developed by Schreiber and Crabtree (Spencer et al. 1993; Belshaw et al. 1996) and marketed by ARIAD Pharmaceuticals, to stimulate FGF signaling selectively in transfected myoblasts and not other cells. Our initial results have recently been published (Whitney et al. 2001). The system consists of the cytoplasmic domain of a growth factor receptor (FGFR1 in our case) fused in-frame to a drug-binding domain (a modified version of the FK506-binding protein termed F36V). The chimeric receptor is targeted to the cytoplasmic membrane with a src myristoylation domain. We hypothesized that administration of a bifunctional F36V ligand (a small molecule having an internal plane of symmetry, capable of binding to two chimeric receptors) termed AP20187, would dimerize the receptor domains and mimic the effects of bFGF. The FGF-dependent mouse myoblast line MM14 was stably transfected with a bicistronic retrovirus encoding eGFP and the chimeric receptor, and nonclonal populations of transfected myoblasts were selected by FACS.

Both the transfected and wild-type myoblasts proliferated in response to bFGF and ceased proliferating upon mitogen withdrawal (Fig. 2). When exposed to the dimerizing compound AP20187, proliferation of the transfected myoblasts was comparable to that following bFGF treatment. The wild-type cells, however, showed no proliferation in response to dimerizer treatment. Immunostaining for expression of sarcomeric myosin heavy chain revealed that both wild-type and transfected cells remained mononucleated and myosin-negative when treated with bFGF (Fig. 3). Each cell population fused to form normal-appearing, myosin-positive myotubes when mitogens were withdrawn. Treatment with the dimerizer blocked myotube formation and myosin expression in the

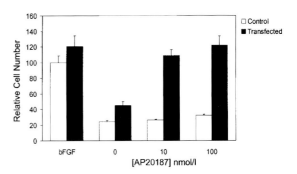

Figure 2. Cell proliferation in wild-type and chimeric receptor-transfected skeletal myoblasts. The vertical axis shows relative cell number, normalized to the bFGF-treated control cells. Both control and transfected cells proliferated in response to 6 ng/ml bFGF. The transfected cells proliferated in response to dimerizer treatment, similar to bFGF treatment. The control cells showed no proliferative response to dimerizer. (Reprinted, with permission, from Whitney et al. 2001 [copyright American Society for Biochemistry and Molecular Biology].)

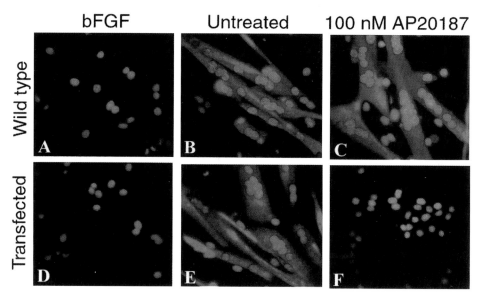

Figure 3. Differentiation of wild-type and chimeric receptor-transfected skeletal myoblasts. Both cell types remained mononucleated and did not express myosin heavy chain with bFGF treatment. Both cell types formed myotubes and expressed myosin heavy chain (*red*) upon growth factor withdrawal. Dimerizer treatment blocked myotube formation and myosin expression in transfected cells, similar to bFGF, whereas wild-type cells differentiated normally in the presence of dimerizer (Whitney et al. 2001). (Reprinted, with permission, from Whitney et al. 2001 [copyright American Society for Biochemistry and Molecular Biology].)

transfected cells, whereas wild-type cells differentiated normally in the presence of dimerizer. We were interested to learn whether the MAP kinase pathway was activated comparably by bFGF and dimerizer treatment. Immunoblotting with a phospho-ERK 1/2 antibody revealed that both wild-type and transfected cells phosphorylated the p44/p42 MAP kinases in response to bFGF treatment (Fig. 4). When dimerizer was administered, the transfected cells phosphorylated p44/p42, whereas the wild-type cells did not activate this pathway.

As a final experiment, we wanted to determine whether prolonged stimulation with the dimerizer might induce a hyperproliferative state where the cells could be tumorigenic in vivo. To this end, transfected cells were stimulated for 30 days in the continuous presence of dimerizer and compared to "naïve" transfected cells, which had not previously been exposed to dimerizer. BrdU labeling showed that both the 30d and naïve cells proliferated in response to the dimerizer. Importantly, both populations differentiated comparably upon dimerizer removal, withdrawing from the cell cycle and forming myosin-positive, multinucleated myotubes (data not shown). Thus, forced dimerization of FGFR1 recapitulates critical aspects of bFGF-induced signaling in MM14 cells, including activation of proliferation, inhibition of the differentiation program, and activation of the MAP kinase pathway. These effects were reversible even after prolonged stimulation of the pathway.

The results from these experiments indicate that FGFR1 signaling can be faithfully reproduced by dimerizing the receptor's cytoplasmic domain with a small molecule. Our next goals for this system are to extend this system to primary (vs. immortalized) skeletal muscle cells, and to determine whether administration of the dimerizing compound in vivo will control skeletal muscle graft size in injured hearts.

HEMATOPOIETIC STEM CELL GRAFTING

Stem cell plasticity has generated an enormous amount of interest in the last few years. There are numerous reports in the literature of "transdifferentiation" events, where cells appear to adopt unexpected phenotypes after transplantation, coculture, or other interventions (for review, see Weissman et al. 2001). One study of particular relevance was performed by Jackson et al. (2001), in which genetically tagged, Hoechst dye-excluding "side population" marrow stem cells were transplanted into irradiated congenic mice. Myocardial infarcts were induced 2 months after transplantation, and the hearts were studied 2 and 4 weeks post-infarction. They reported ex-

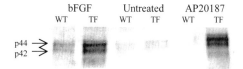

Figure 4. Immunoblot for phosphorylated forms of p44/p42 MAP kinase in wild-type and transfected myoblasts. bFGF treatment caused MAP kinase phosphorylation in both cell types. Untreated cells showed little MAP kinase phosphorylation. Dimerizer treatment resulted in robust phosphorylation of MAP kinase in transfected cells, similar to bFGF treatment, whereas the wild-type cells were unaffected. (Reprinted, with permission, from Whitney et al. 2001 [copyright American Society for Biochemistry and Molecular Biology].)

pression of the LacZ genetic tag in endothelial cells, smooth muscle cells, and cardiomyocytes around the infarcted region, albeit at low frequencies (~2% for endothelial cells; 0.02% for cardiomyocytes).

A remarkable study from Orlic et al. (2001a) reported extensive regeneration of myocardial infarcts 9 days after direct injection of bone-marrow-derived stem cells. Marrow was isolated from mice constitutively expressing GFP under control of the CMV promoter. The marrow was depleted of differentiated cells using antibodies to surface markers coupled to magnetic beads. These "lineage-negative" cells then were sorted to isolate the subpopulation expressing the stem cell antigen c-kit. They reported that 68% of the infarct was replaced by cells coexpressing myosin heavy chain and GFP, and that this regeneration was associated with enhanced ventricular function determined by echocardiography. This group subsequently reported almost complete regeneration of infarcts by mobilizing endogenous bone marrow stem cells with stem cell factor and granulocyte colony stimulating factor. In this study, however, the bone marrow was not tagged, so it was not possible to trace the origin of cells in the infarct (Orlic et al. 2001b).

Because of the obvious importance of these reports, we sought to test whether direct injection of hematopoietic stem cells could form cardiac myocytes in infarcted hearts, using a more rigorous genetic test of cell lineage and phenotype. These studies have been published in abstract form (Murry et al. 2001). We utilized transgenic mice in which the cardiac-specific α-MHC promoter drove expression of nuclear localized β-galactosidase as the donor animals. Lineage-negative, c-kit$^+$ cells were isolated from these animals and injected into acute infarcts of congenic mice. Hearts were harvested 1–2 weeks after engraftment and sectioned in their entirety into 300-micron slices, and whole-mount β-gal histochemistry was performed. The readout for this experiment was activation of nuclear-localized β-galactosidase, which would provide a test of both the cells' lineage (from the transgenic donor) and their phenotype as cardiomyocytes. It should be noted that this assay is highly sensitive, capable of detecting a single transgenic cell in cardiomyocyte grafting experiments. Despite the high degree of sensitivity, not one cell with a blue nucleus was detected in over 50 grafting experiments involving infarcted hearts (Table 1). Furthermore, immunostaining for myosin heavy chain showed that the infarcts did not contain regenerating cardiomyocytes, but rather the usual content of surviving and necrotic cardiomyocytes. Experiments performed using donor cells from GFP-transgenic mice showed consistent engraftment, indicating that the cell injections were successful and that the cells persisted throughout the duration of the experiment. In separate studies, coculture of lineage-negative, c-kit-positive cells with fetal cardiomyocytes, or formation of chimeric embryoid bodies with embryonic stem cells, did not result in activation of the cardiac reporter gene.

These studies clearly do not support the hypothesis that lineage-negative, c-kit-positive bone marrow cells form new myocardium after grafting into infarcted hearts. The

Table 1. Analysis of Cardiac Potential of Bone Marrow Stem Cells

Model	Cells	Result
Normal heart	Lin-/c-kit$^+$	0/7 hearts
Infarct	Lin-/c-kit$^+$	0/50 hearts
Cautery injury	Lin-/c-kit$^+$/Sca$^+$	0/8 hearts
Isoproterenol infusion	Lin-/c-kit$^+$/Sca$^+$	0/3 hearts
Coculture with fetal cardiomyocytes	Lin-/c-kit$^+$	0/10^6 cells
Coculture with embryonic stem cells	Lin-/c-kit$^+$	0/1000 embryoid bodies

In the results column, the numerator is the number of β-galactosidase-positive nuclei observed (an indicator of cardiac transdifferentiation), and the denominator is the number of hearts or chimeric embryoid bodies studied. See text for additional details.

basis for the discrepancy with the above studies is not clear at present. It is possible that the transgene is inactivated after passing through a hematopoietic lineage, although it clearly is not inactivated after passing through the germ line for multiple generations. Nevertheless, inactivation of the transgene could not explain failure to detect the new myocardium by myosin staining. There may be technical problems with the thresholds used to define cells as myosin-positive or myosin-negative in the different laboratories. Immunostaining requires a judgment as to whether a given staining intensity is significantly above background. Importantly, however, we were unable to detect a difference in myosin staining patterns between bone-marrow-engrafted hearts and sham-injected hearts. Another possibility is that the cell-sorting parameters may have differed from those previously published, although considerable care was taken to reproduce them.

CARDIAC PROGENITOR CELLS IN HUMANS

Given the results of the above studies, we wanted to ask in a more open-ended way whether any cells in the adult could give rise to cardiomyocytes. We therefore took advantage of a "naturally occurring," therapeutic experiment, where men received heart transplants from female donors. We reasoned that the combination of immune rejection, ischemia from graft coronary disease, and frequent mechanical injury from surveillance biopsies might be stimuli for recruitment of progenitor cells. The presence of a Y chromosome in a cardiomyocyte or other cell type would provide clear evidence for its origin from the host, rather than the donor heart. Five autopsy cases were studied by a combination of Y chromosome in situ hybridization and sarcomeric myosin immunostaining. Very strict criteria were used to define a nucleus as belonging to a cardiomyocyte, thereby minimizing the chance of mistaking a Y-positive leukocyte for a transdifferentiation event. In all patients, a small but readily detectable population of Y-positive cardiomyocytes was detected (Fig. 5A,B), whereas in five female control hearts no Y-positive myocytes were identified. A mean of 0.04% of the cardiomyocytes in the transplanted hearts was donor-derived, based on evaluation of 5.4 × 10^5 car-

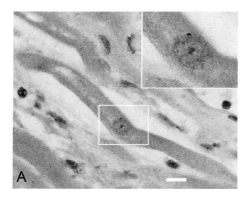
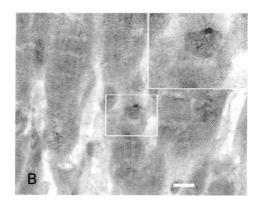
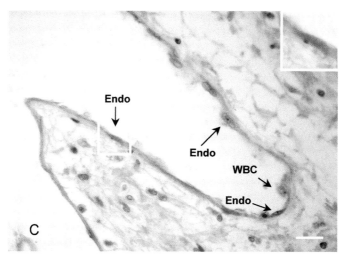

Figure 5. Identification of donor-derived cardiomyocytes and endothelial cells in transplanted human hearts. Female hearts transplanted into male recipients were studied. Y chromosome in situ hybridization (*brown nuclear dot*) was used to track cells of recipient lineage, and sarcomeric myosin heavy-chain immunostaining (*red*) was used to identify cardiomyocytes. CD31 staining was used to identify endothelial cells. Panels *A* and *B* each show one cardiomyocyte containing an eccentrically located Y chromosome signal, indicating they arose from the donor after transplantation. The boxed regions are shown at higher magnification in the upper right. Bar, 10 microns. Panel *C* shows a sinusoidal epicardial vessel containing multiple Y-positive endothelial cells (Endo) and a Y-positive white blood cell (WBC). The boxed region is shown at higher magnification at upper right. Bar, 20 microns. (Reprinted, with permission, from Laflamme et al. 2002 [copyright Lippincott Williams and Wilkins].)

diomyocyte nuclei. In one patient, however, there were two "hot spots" of 1 mm², having 22% and 29% donor-derived cardiomyocytes. Interestingly, this patient was the only one who died of acute cellular rejection, raising the possibility that transdifferentiation requires an injury event.

More recently, we have undertaken an analysis of other cell types in these hearts. We have found that chimerism is much more common in microvascular endothelial cells, averaging ~24% from extracardiac origins (Fig. 5C). Smooth muscle cells and perineural Schwann cells were also found that contain Y chromosomes, although the quantitative data are not yet available for these cell types. Thus, it appears that cardiomyocytes can be derived from extracardiac progenitors in transplanted human hearts (albeit at low levels) and that other cell types can be derived at significantly greater frequencies.

Our data contrast somewhat with a recently published study by Quaini et al. (2002). This group also studied male patients who received heart transplants from female donors. They reported very high frequencies of cardiomyocytes arising from outside the donor organ, with a mean of 18% being host-derived. Counterintuitively, they reported the highest rates of chimerism in patients having the shortest posttransplant intervals, with a mean of 30% host-derived cardiomyocytes (corresponding to ~100 g of new myocardium) in patients who died at 4 and 28 days posttransplantation. Although we cannot definitively rule out differences in the clinical situations of these patients, we believe technical differences most likely underlie this 500-fold difference. We observed comparable rates of false-negative in situ hybridization in control sections of male heart (~50%), so this cannot explain the difference. A likely culprit is inflammatory cells. Shortly after transplantation, the donor heart is significantly infiltrated by host leukocytes, principally T cells and macrophages. In a female-to-male transplant, all of these leukocytes will, of course, contain Y chromosomes. Leukocytes are very plastic cells, capable of closely abutting myocytes or invading their cytoplasm after injury. In addition to employing the aforementioned stringent criteria, we deliberately avoided the early posttransplant interval due to concerns about mistaking leukocyte nuclei for cardiomyocyte nuclei. Thus, we believe the high degree of cardiomyocyte chimerism reported by Quaini et al. (2002) reflects misidentification of leukocyte nuclei. On the other hand, they reported endothelial chimerism to a degree comparable to what we observed, indicating the tissues and assays were comparable in this regard. It should be noted that two other groups (Hruban et al. 1993; Glaser et al. 2002) failed to find any Y chromosome-positive cardiomyocytes in comparably designed experiments, although their sample sizes were much smaller, and that one other group (Muller et al. 2002) identified a much lower frequency of Y-positive myocytes than did Quaini et al.

These data indicate that many of the components of human myocardium can be derived from extracardiac progenitor cells. The frequency of cell replacement, however, appears to vary tremendously by cell type. Endothelial cells, smooth muscle cells, and perineural Schwann cells arise at relatively high frequencies, whereas cardiomyocytes appear to arise very infrequently from progenitor cells. The high frequency of endothelial cells arising from the host may have implications for immune tolerance of grafts, given the importance of endothelial cells in antigen presentation in solid organ transplantation. These findings raise multiple questions, such as the location and identity of the progenitor cell, mechanisms of progenitor mobilization and homing, and what the signals are that mediate transdifferentiation. As future studies identify answers to these questions, it may be possible to exploit these pathways to enhance repair of the injured heart or to induce better tolerance of the transplanted heart.

ACKNOWLEDGMENTS

This work was supported by National Institutes of Health grants R01 HL-61553, PO1 HL-03174, and R24 HL-64387.

REFERENCES

Atkins B.Z., Lewis C.W., Kraus W.E., Hutcheson K.A., Glower D.D., and Taylor D.A. 1999. Intracardiac transplantation of skeletal myoblasts yields two populations of striated cells in situ. *Ann. Thorac. Surg.* **67:** 124.

Belshaw P.J., Ho S.N., Crabtree G.R., and Schreiber S.L. 1996. Controlling protein association and subcellular localization with a synthetic ligand that induces heterodimerization of proteins. *Proc. Natl. Acad. Sci.* **93:** 4604.

Beltrami C.A., Finato N., Rocco M., Feruglio G.A., Puricelli C., Cigola E., Quaini F., Sonnenblick E.H., Olivetti G., and Anversa P. 1994. Structural basis of end-stage failure in ischemic cardiomyopathy in humans. *Circulation* **89:** 151.

Beltrami A.P., Urbanek K., Kajstura J., Yan S.M., Finato N., Bussani R., Nadal-Ginard B., Silvestri F., Leri A., Beltrami C.A., and Anversa P. 2001. Evidence that human cardiac myocytes divide after myocardial infarction. *N. Engl. J. Med.* **344:** 1750.

Chiu R.C., Zibaitis A., and Kao R.L. 1995. Cellular cardiomyoplasty: Myocardial regeneration with satellite cell implantation. *Ann. Thorac. Surg.* **60:** 12.

Connold A.L., Frischknecht R., Dimitrakos M., and Vrbov'a G. 1997. The survival of embryonic cardiomyocytes transplanted into damaged host rat myocardium. *J. Muscle Res. Cell Motil.* **18:** 63.

Etzion S., Battler A., Barbash I.M., Cagnano E., Zarin P., Granot Y., Kedes L.H., Kloner R.A., and Leor J. 2001. Influence of embryonic cardiomyocyte transplantation on the progression of heart failure in a rat model of extensive myocardial infarction. *J. Mol. Cell. Cardiol.* **33:** 1321.

Glaser R., Lu M.M., Narula N., and Epstein J.A. 2002. Smooth muscle cells, but not myocytes, of host origin in transplanted human hearts. *Circulation* **106:** 17.

Hruban R.H., Long P.P., Perlman E.J., Hutchins G.M., Baumgartner W.A., Baughman K.L., and Griffin C.A. 1993. Fluorescence in situ hybridization for the Y-chromosome can be used to detect cells of recipient origin in allografted hearts following cardiac transplantation. *Am. J. Pathol.* **142:** 975.

Jackson K.A., Majka S.M., Wang H., Pocius J., Hartley C.J., Majesky M.W., Entman M.L., Michael L.H., Hirschi K.K., and Goodell M.A. 2001. Regeneration of ischemic cardiac muscle and vascular endothelium by adult stem cells. *J. Clin. Invest.* **107:** 1395.

Jain M., DerSimonian H., Brenner D.A., Ngoy S., Teller P., Edge A.S., Zawadzka A., Wetzel K., Sawyer D.B., Colucci W.S., et al. 2001. Cell therapy attenuates deleterious ventricular remodeling and improves cardiac performance after myocardial infarction. *Circulation* **103:** 1920.

Koh G.Y., Soonpaa M.H., Klug M.G., Pride H.P., Cooper B.J., Zipes D.P., and Field L.J. 1995. Stable fetal cardiomyocyte grafts in the hearts of dystrophic mice and dogs. *J. Clin. Invest.* **96:** 2034.

Laflamme M.A., Myerson D., Saffitz J.E., and Murry C.E. 2002. Evidence for cardiomycyte repopulation by extracardiac progenitors in transplanted human hearts. *Circ. Res.* **90:** 634.

Leor J., Patterson M., Quinones M.J., Kedes L.H., and Kloner R.A. 1996. Transplantation of fetal myocardial tissue into the infarcted myocardium of rat. A potential method for repair of infarcted myocardium? *Circulation* **94:** II332.

Li R.K., Jia Z.Q., Weisel R.D., Merante F., and Mickle D.A. 1999. Smooth muscle cell transplantation into myocardial scar tissue improves heart function. *J. Mol. Cell. Cardiol.* **31:** 513.

Li R.K., Jia Z.Q., Weisel R.D., Mickle D.A., Zhang J., Mohabeer M.K., Rao V., and Ivanov J. 1996. Cardiomyocyte transplantation improves heart function. *Ann. Thorac. Surg.* **62:** 654.

Menasche P., Hagege A.A., Scorsin M., Pouzet B., Desnos M., Duboc D., Schwartz K., Vilquin J.T., and Marolleau J.P. 2001. Myoblast transplantation for heart failure. *Lancet* **357:** 279.

Michael L.H., Ballantyne C.M., Zachariah J.P., Gould K.E., Pocius J.S., Taffet G.E., Hartley C.J., Pham T.T., Daniel S.L., Funk E., and Entman M.L. 1999. Myocardial infarction and remodeling in mice: Effect of reperfusion. *Am. J. Physiol.* **277:** H660.

Muller P., Pfeiffer P., Koglin J., Schafers H.J., Seeland U., Janzen I., Urbschat S., and Bohm M. 2002. Cardiomyocytes of noncardiac origin in myocardial biopsies of human transplanted hearts. *Circulation* **106:** 31.

Muller-Ehmsen J., Whittaker P., Kloner R.A., Dow J.S., Sakoda T., Long T.I., Laird P.W., and Kedes L. 2002. Survival and development of neonatal rat cardiomyocytes transplanted into adult myocardium. *J. Mol. Cell. Cardiol.* **34:** 107.

Murry C.E., Wiseman R.W., Schwartz S.M., and Hauschka S.D. 1996. Skeletal myoblast transplantation for repair of myocardial necrosis. *J. Clin. Invest.* **98:** 2512.

Murry C.E., Rupart M.J., Soonpaa M.H., Nakajima H., Nakajima H., and Field L.J. 2001. Absence of cardiac differentiation in hematopoietic stem cells transplanted into normal and injured hearts. *Circulation* (suppl. II) **104:** II-599.

Olivetti G., Capasso J.M., Sonnenblick E.H., and Anversa P. 1990. Side-to-side slippage of myocytes participates in ventricular wall remodeling acutely after myocardial infarction in rats. *Circ. Res.* **67:** 23.

Orlic D., Kajstura J., Chimenti S., Bodine D.M., Leri A., and

Anversa P. 2001a. Transplanted adult bone marrow cells repair myocardial infarcts in mice. *Ann. N.Y. Acad. Sci.* **938:** 221.

Orlic D., Kajstura J., Chimenti S., Limana F., Jakoniuk I., Quaini F., Nadal-Ginard B., Bodine D.M., Leri A., and Anversa P. 2001b. Mobilized bone marrow cells repair the infarcted heart, improving function and survival. *Proc. Natl. Acad. Sci.* **98:** 10344.

Pouzet B., Vilquin J.T., Hagege A.A., Scorsin M., Messas E., Fiszman M., Schwartz K., and Menasche P. 2001. Factors affecting functional outcome after autologous skeletal myoblast transplantation. *Ann. Thorac. Surg.* **71:** 844.

Quaini F., Urbanek K., Beltrami A.P., Finato N., Beltrami C.A., Nadal-Ginard B., Kajstura J., Leri A., and Anversa P. 2002. Chimerism of the transplanted heart. *N. Engl. J. Med.* **346:** 5.

Reimer K.A. and Jennings R.B. 1979. The "wavefront phenomenon" of myocardial ischemic cell death. II. Transmural progression of necrosis within the framework of ischemic bed size (myocardium at risk) and collateral blood flow. *Lab. Invest.* **40:** 633.

Reimer K.A., Lowe J.E., Rasmussen M.M., and Jennings R.B. 1977. The wavefront phenomenon of ischemic cell death. 1. Myocardial infarct size vs duration of coronary occlusion in dogs. *Circulation* **56:** 786.

Reinecke H. and Murry C.E. 2000. Transmural replacement of myocardium after skeletal myoblast grafting into the heart. Too much of a good thing? *Cardiovasc. Pathol.* **9:** 337.

Reinecke H., Poppa V., and Murry C.E. 2002. Skeletal muscle stem cells do not transdifferentiate into cardiomyocytes after cardiac grafting. *J. Mol. Cell. Cardiol.* **34:** 241.

Reinecke H., MacDonald G.H., Hauschka S.D., and Murry C.E. 2000. Electromechanical coupling between skeletal and cardiac muscle. Implications for infarct repair. *J. Cell Biol.* **149:** 731.

Reinecke H., Zhang M., Bartosek T., and Murry C.E. 1999. Survival, integration, and differentiation of cardiomyocyte grafts: A study in normal and injured rat hearts. *Circulation* **100:** 193.

Scorcin M., Marotte F., Sabri A., Le Dref O., Demirag M., Samuel J.-L., Rappaport L., and Menasche P. 1995. Can grafted cardiomyocytes colonize peri-infarction myocardial areas? *Circulation* (suppl.) **92:** I50.

Scorsin M., Hagege A., Vilquin J.T., Fiszman M., Marotte F., Samuel J.L., Rappaport L., Schwartz K., and Menasche P. 2000. Comparison of the effects of fetal cardiomyocyte and skeletal myoblast transplantation on postinfarction left ventricular function. *J. Thorac. Cardiovasc. Surg.* **119:** 1169.

Soonpaa M.H. and Field L.J. 1998. Survey of studies examining mammalian cardiomyocyte DNA synthesis. *Circ. Res.* **83:** 15.

Soonpaa M.H., Koh G.Y., Klug M.G., and Field L.J. 1994. Formation of nascent intercalated disks between grafted fetal cardiomyocytes and host myocardium. *Science* **264:** 98.

Spencer D.M., Wandless T.J., Schreiber S.L., and Crabtree G.R. 1993. Controlling signal transduction with synthetic ligands. *Science* **262:** 1019.

Taylor D.A., Atkins B.Z., Hungspreugs P., Jones T.R., Reedy M.C., Hutcheson K.A., Glower D.D., and Kraus W.E. 1998. Regenerating functional myocardium: Improved performance after skeletal myoblast transplantation. *Nat. Med.* **4:** 929.

Watanabe E., Smith D.M., Jr., Delcarpio J.B., Sun J., Smart F.W., Van M.C., Jr., and Claycomb W.C. 1998. Cardiomyocyte transplantation in a porcine myocardial infarction model. *Cell Transplant.* **7:** 239.

Weissman I.L., Anderson D.J., and Gage F. 2001. Stem and progenitor cells: Origins, phenotypes, lineage commitments, and transdifferentiations. *Annu. Rev. Cell Dev. Biol.* **17:** 387.

White H.D., Norris R.M., Brown M.A., Brandt P.W., Whitlock R.M., and Wild C.J. 1987. Left ventricular end-systolic volume as the major determinant of survival after recovery from myocardial infarction. *Circulation* **76:** 44.

Whitney M.L., Otto K.G., Blau C.A., Reinecke H., and Murry C.E. 2001. Control of myoblast proliferation with a synthetic ligand. *J. Biol. Chem.* **276:** 41191.

Yousef Z.R., Redwood S.R., and Marber M.S. 2000. Postinfarction left ventricular remodeling: A pathophysiological and therapeutic review. *Cardiovasc. Drugs Ther.* **14:** 243.

Zhang M., Methot D., Poppa V., Fujio Y., Walsh K., and Murry C.E. 2001. Cardiomyocyte grafting for cardiac repair: Graft cell death and anti-death strategies. *J. Mol. Cell. Cardiol.* **33:** 907.

Gene Therapy for Cardiac Arrhythmias

E. MARBÁN, H.B. NUSS, AND J.K. DONAHUE
Institute for Molecular Cardiobiology, The Johns Hopkins University, Baltimore, Maryland 21205

The heart requires a steady rhythm and rate in order to fulfill its physiological role as the pump for the circulation. An excessively rapid heart rate (tachycardia) allows insufficient time for the mechanical events of ventricular emptying and filling. Cardiac output drops, the lungs become congested, and, in the extreme, the circulation collapses. An equally morbid chain of events ensues if the heart beats too slowly (bradycardia). Serious disturbances of cardiac rhythm, known as arrhythmias, afflict more than 3 million Americans and account for > 400,000 deaths annually (American Heart Association 2001). Current therapy has serious limitations: Antiarrhythmic drugs can sometimes be effective, but their utility is limited by their propensity to create new, potentially fatal, arrhythmias while suppressing others (Coplen et al. 1990; Echt et al. 1991; Siebels et al. 1993; Waldo et al. 1996). Ablation of targeted tissue can readily cure simple wiring errors, but it is less effective in treating more complex and common arrhythmias, such as atrial fibrillation or ventricular tachycardia (Richardson and Josephson 1999; Falk 2001). Implantable devices can serve as surrogate pacemakers to sustain heart rate, or as defibrillators to treat excessively rapid rhythms. Such devices are expensive, and implantation involves a number of acute and chronic risks (pulmonary collapse, bacterial infection, lead or generator failure; Bernstein and Parsonnet 2001). In short, arrhythmias are a serious threat of public health proportions, and current treatment is inadequate. Given these limitations, we have begun to develop gene therapy as an alternative to conventional treatment.

The most obvious application of gene therapy is to correct monogenic deficiency disorders such as hemophilia or adenosine deaminase deficiency. Indeed, the latter is the only disease to have been cured (in a few infants) by gene therapy (Blaese et al. 1995). Gene therapy for cardiovascular disorders, as it is most commonly being developed today, focuses not on correcting deficiency disorders but rather on attempts to foster angiogenesis in ischemic myocardium (Losordo et al. 1998; Rosengart et al. 1999) or to suppress vascular stenosis in a variety of iatrogenic settings (Ohno et al. 1994; Mann et al. 1999). The concept of gene therapy for cardiac arrhythmias differs conceptually from conventional applications. We seek to achieve functional re-engineering of cardiac tissue, so as to alter a specific electrical property of the tissue in a salutary manner. For example, genes are introduced to alter the velocity of electrical conduction in a defined region of the heart, or to create a spontaneously active biological pacemaker from normally quiescent myocardium. A relevant analogy is the use of off-the-shelf or customized parts to improve the performance of a lackluster automobile engine. Our "parts" are wild-type or mutant genes; our engine is the heart.

Here we review our progress in two areas: the treatment of atrial fibrillation, and the creation of biological pacemakers. The reader is referred to the original articles describing the work for more details (Donahue et al. 2000; Miake et al. 2002). We then conclude by considering future directions of this type of gene therapy.

ATRIAL FIBRILLATION

Atrial fibrillation (AF) affects more than 2 million people in the United States, including 5–10% of people over the age of 65 and 10–35% of the 5 million patients with congestive heart failure (Chugh et al. 2001). Accepted therapy for AF includes either antiarrhythmic drugs to maintain sinus rhythm, or drugs that suppress conduction in the atrioventricular (AV) node to control the ventricular rate. Although appealing, the maintenance of sinus rhythm is often unsuccessful. Within 1 year of conversion to sinus rhythm, 25–50% of patients revert to AF despite antiarrhythmic drug treatment (Khand et al. 2000). The usual long-term clinical scenario, then, is to accept the inevitability of chronic AF and to focus on ventricular rate control. The equivalence of rate and rhythm control strategies in asymptomatic patients was recently confirmed in AFFIRM and RACE, two multicenter clinical trials (Williams and Miller 2002).

Medical therapies to control heart rate during AF have targeted the conduction properties of the AV node by suppressing the calcium current (calcium channel blockers) or by affecting the balance of adrenergic and cholinergic tone (β-blockers and digoxin). For patients with normal ventricular function, AV nodal suppressing drugs reduce the heart rate by 15–30% (Khand et al. 2000), but the frequent systemic effects of these drugs include bronchospasm, hypotension, depression, fatigue, and gastrointestinal side effects (Prystowsky et al. 1996). When AV nodal blocking drugs are not tolerated or not effective, the only remaining option is radiofrequency ablation of the AV node and implantation of a pacemaker. We recently evaluated the possibility of gene transfer-mediated control of the ventricular rate during AF (Donahue et al. 2000). In a porcine model of acute AF, we delivered recombinant adenoviruses containing a gene for the α subunit of the inhibitory G protein ($G\alpha_{i2}$) into the AV nodal artery. This transgene was chosen following the premise that overexpression of $G\alpha_{i2}$ would suppress basal adenylate cyclase activity and thereby indirectly suppress calcium channel activity in the AV node. The efficiency of gene transfer was optimized by pretreatment with nitroglycerin, vascular endothelial growth factor, and sildenafil (Donahue et al. 1998; Nagata et al. 2001). Slightly less than half of all cells in the AV node showed evidence of gene transfer, and western blot analysis documented a sixfold overexpression of $G\alpha_{i2}$. Gene transfer occurred rarely (<1% of cells) throughout the rest of the body. The transgene had obvious phenotypic consequences that were evident by measuring various indices of cardiac conduction. During sinus rhythm, the treatment group had evidence of conduction slowing and increased refractoriness in the AV node. In AF, the heart rate was reduced by 20% when compared to controls (Fig. 1). The relative decrease in heart rate persisted after the application of a β-adrenergic agonist. Since these results were obtained using clinically available catheters and equipment, translation to human therapy would be relatively straightforward. Major steps between the current state of development and clinical applicability include the use of longer-lasting, noninflammatory vectors (such as adeno-associated virus), and tests of safety and efficacy in animal models of chronic AF.

BIOLOGICAL PACEMAKER

The pacemaker of the heart is normally encompassed within a small region known as the sinoatrial (SA) node. The SA node initiates the heartbeat, sustains the circulation, and sets the rate and rhythm of cardiac contraction (Brooks and Lu 1972). The working muscle of the heart (myocardium), comprising the pumping chambers known as the atria and the ventricles, is normally excited by pacemaker activity originating in the SA node. However, in the absence of such activity, the myocardium lacks spontaneous activity. Therefore, loss of specialized pacemaker cells in the SA node, as occurs in a variety of common diseases, results in circulatory collapse, necessitating the implantation of an electronic pacemaker (Kusumoto and Goldschlager 1996). To create an alternative to electronic pacemakers, we sought to render electrically quiescent myocardium spontaneously active.

Our strategy to effect such a conversion was based on the premise that ventricular myocardium contains all it requires to pace, but that pacing is normally suppressed by an expressed gene. The reasoning is as follows. In the early embryonic heart, each cell possesses intrinsic pacemaker activity. The mechanism of spontaneous beating in the early embryo is remarkably simple (Wobus et al. 1995). The opening of L-type calcium channels produces depolarization; the subsequent voltage-dependent opening of transient outward potassium channels leads to repolarization. With further development, the heart differentiates into specialized functional regions, each with its own distinctive electrical signature. The atria and ventricles become electrically quiescent; only a small number of pacemaker cells, within compact "nodes," set the overall rate and rhythm. Nevertheless, there is reason to wonder whether pacemaker activity may be latent within adult ventricular myocytes and masked by the differential expression of many other ionic currents. Among these, the inward rectifier potassium current (I_{K1}) is notable for its intense expression in electrically quiescent atria and ventricle, but not in nodal pacemaker cells. I_{K1}, encoded by the Kir2 gene family (Kubo et al. 1993), stabilizes a strongly negative resting potential and thereby would be expected to suppress excitability. We thus explored the possibility that dominant-negative suppression of Kir2-encoded inward rectifier potassium channels in the ventricle would suffice to produce spontaneous, rhythmic electrical activity.

Replacement of three critical residues in the pore region of Kir2.1 by alanines ($GYG_{144-146} \rightarrow AAA$, or Kir2.1AAA) creates a dominant-negative construct (Herskowitz 1987). The GYG motif plays a key role in ion selectivity and pore function (Slesinger et al. 1996). Kir2.1AAA and GFP were packaged into a bicistronic adenoviral vector (AdEGI-Kir2.1AAA) and injected into the left ventricular cavity of guinea pigs during transient cross-clamp of the great vessels (Miake et al. 2002). This

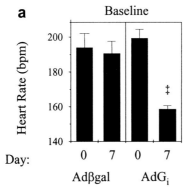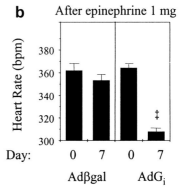

Figure 1. Ventricular rate after induction of atrial fibrillation in a porcine model. The heart rate was measured during acutely induced AF at baseline and 7 days after AV nodal gene transfer of β-galactosidase (control) or $G\alpha_{i2}$. (*a*) Drug-free state. (*b*) After 1 mg of epinephrine IV. (Reprinted, with permission, from Donahue et al. 2000 [copyright Nature Publishing Group].)

method of delivery sufficed to achieve transduction of ~20% of ventricular myocytes. Myocytes isolated 3–4 days after in vivo transduction with Kir2.1AAA exhibited ~80% suppression of I_{K1}, but the L-type calcium current was unaffected.

Non-transduced (non-green) left ventricular myocytes isolated from AdEGI-Kir2.1AAA-injected animals, as well as green cells from AdEGI-injected hearts, exhibited no spontaneous activity, but fired single action potentials in response to depolarizing external stimuli (Fig. 2a). In contrast, Kir2.1AAA myocytes exhibited either of two phenotypes: a stable resting potential from which prolonged action potentials could be elicited by external stimuli (7 of 22 cells, not shown) or spontaneous activity (Fig. 2b). The spontaneous activity, which was seen in all cells in which I_{K1} was suppressed below 0.4 pA/pF (at –50 mV; cf. >1.5 pA/pF in controls, or 0.4–1.5 pA/pF in non-pacing Kir2.1AAA cells), resembles that of genuine pacemaker cells; the maximum diastolic potential (-60.7 ± 2.1 mV, $n = 15$ of 22 Kir2.1AAA cells, $p<0.05$ t-test) is relatively depolarized, with repetitive, regular, and incessant electrical activity initiated by gradual "phase 4" depolarization and a slow upstroke (Brooks and Lu 1972; Irisawa et al. 1993). Kir2.1AAA pacemaker cells responded to β-adrenergic stimulation (isoproterenol) just as SA nodal cells do, increasing their pacing rate (Brown et al. 1975; Irisawa et al. 1993).

Electrocardiography revealed two phenotypes in vivo. What we most often observed was simple prolongation of the QT interval (not shown). Nevertheless, 40% of the animals exhibited an altered cardiac rhythm indicative of spontaneous ventricular foci. In normal sinus rhythm, every P wave is succeeded by a QRS complex (Fig. 2c). In two of five animals after transduction with Kir2.1AAA, premature beats of ventricular origin can be distinguished by their broad amplitude, and can be seen to "march through" to a beat independent of, and more rapid than, that of the physiological sinus pacemaker (Fig. 2d). In these proof-of-concept experiments, the punctate transduction required for pacing occurred by chance rather than by design, in that the distribution of the transgene throughout the ventricles was not controlled. Nevertheless, ectopic beats, arising from foci of induced pacemakers, cause the entire heart to be paced from the ventricle.

Our findings provide new insights into the biological basis of pacemaker activity. The conventional wisdom postulates that pacemaker activity requires the highly localized expression in nodal cells of "pacemaker genes," such as those of the HCN family (Santoro and Tibbs 1999), although an important role for I_{K1} has also been recognized (Irisawa et al. 1993). Exposure to barium induces automaticity in ventricular muscle and myocytes because of its time- and voltage-dependent block of I_{K1} (Imoto et al. 1987; Hirano and Hiraoka 1988). However,

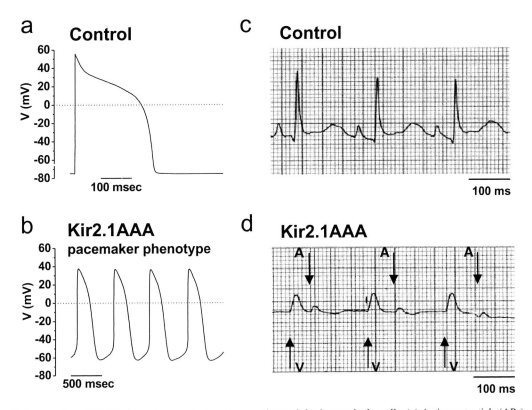

Figure 2. Suppression of Kir2.1 channels unmasks latent pacemaker activity in ventricular cells. (*a*) Action potentials (APs) evoked by depolarizing external stimuli in control ventricular myocytes. (*b*) Spontaneous APs in Kir2.1AAA-transduced myocytes with depressed I_{K1}. (*c*) Baseline electrocardiograms in normal sinus rhythm. (*d*) Ventricular rhythms 72 hours after gene transfer of Kir2.1AAA. P waves (*A* and arrow) and wide QRS complexes (*V* and arrow) march through to their own rhythm. (Reprinted, with permission, from Miake et al. 2002 [copyright Nature Publishing Group].)

barium also permeates L-type calcium channels in mixed solutions of Ca^{++} (4 mM) and Ba^{++} (1 mM) (Rodriguez-Contreras et al. 2002) and slows their inactivation (Campbell et al. 1988), effects which make it difficult to interpret barium effects strictly in terms of I_{K1}. Our dominant-negative approach is durable and regionally specific; the barium effect is not.

Thus, the specific suppression of Kir2 channels suffices to unleash pacemaker activity in ventricular myocytes. The crucial factor for pacing is the absence of the strongly polarizing I_{K1}, rather than the presence of special genes (although such genes may play an important modulatory role in genuine pacemaker cells) (Brown et al. 1984). In addition to the conceptual insight into the genesis of pacing, our work implies that localized delivery of constructs such as Kir2.1AAA to the myocardium may be useful in the creation of biological pacemakers for therapeutic purposes. Focal injection into a focal area of the ventricle, possibly via an endocardial injection catheter, would be a logical means of trying to reduce this concept to practice in a larger animal.

FUTURE DIRECTIONS

The most tractable targets for near-term development are arrhythmias in which very local modifications of electrical properties suffice for effective treatment. Here, the singular advantages of gene therapy include:

- Highly localized gene delivery suffices to treat the problem. The amount of gene delivered can be correspondingly reduced, and potential problems due to widespread dissemination can be more readily averted.
- Treated cells can remain responsive to endogenous nerves and hormones. Such was the case with Gi overexpression in the AV node: AV conduction remained responsive to β-adrenergic stimulation. Likewise, the induced pacemakers appropriately boosted their firing rate in response to β-adrenergic stimulation.
- Implantable hardware is avoided, obviating long-term risks and decreasing the expense and morbidity associated with battery and lead replacements.
- The localized coronary circulation may allow isolated delivery, as in the case of the AV node.
- The proximity to the inner lining of the heart, the endocardium, allows access by intracardiac injection, providing a potential alternative delivery route.
- The therapeutic effects can be readily detected by physical examination or by electrocardiography.
- Gene transfer-induced changes can be rescued by conventional electrophysiological methods (focal ablation and pacemaker implantation).

The concepts are generalizable to ventricular arrhythmias such as those associated with heart failure or heritable long QT syndrome. In heart failure, for example, overexpression of K channels can be used to antagonize the acquired long QT syndrome (Nuss et al. 1996, 1999); the attendant loss of contractility may be amenable to co-administration of a second gene to augment calcium cycling, in a dual gene therapy strategy. Although such work is conceptually attractive, widespread delivery with long-term expression will be required before human trials can be anticipated.

ACKNOWLEDGMENTS

This work was supported by the Richard S. Ross Clinician-Scientist Award, Johns Hopkins University, the National Institutes of Health (P50 HL-52307, R01 HL-67148, R01 HL-66381), and the American Heart Association (SDG 130350N). E.M. holds the Michel Mirowski, M.D. Professorship of Cardiology at the Johns Hopkins University. The authors dedicate this paper to the memory of Dr. Mirowski.

REFERENCES

American Heart Association. 2001. *2002 Heart and stroke statistical update*. American Heart Association, Dallas, Texas.

Bernstein A.D. and Parsonnet V. 2001. Survey of cardiac pacing and implanted defibrillator practice patterns in the United States in 1997. *Pacing Clin. Electrophysiol.* **24:** 842.

Blaese R.M., Culver K.W., Miller A.D., Carter C.S., Fleisher T., Clerici M., Shearer G., Chang L., Chiang Y., and Tolstoshev P. et al. 1995. T lymphocyte-directed gene therapy for ADA-SCID: Initial trial results after 4 years. *Science* **270:** 475.

Brooks C.M. and Lu H.-H. 1972. *The sinoatrial pacemaker of the heart*. Charles C. Thomas, Springfield, Illinois.

Brown H.F., McNaughton P.A., Noble D., and Noble S.J. 1975. Adrenergic control of cardiac pacemaker currents. *Philos. Trans. R. Soc. Lond. B Biol. Sci.* **270:** 527.

Brown H.F., Kimura J., Noble D., Noble S.J., and Taupignon A. 1984. The slow inward current, Isi, in the rabbit sino-atrial node investigated by voltage clamp and computer simulation. *Proc. R. Soc. Lond. B Biol. Sci.* **222:** 305.

Campbell D.L., Giles W.R., and Shibata E.F. 1988. Ion transfer characteristics of the calcium current in bull-frog atrial myocytes. *J. Physiol.* **403:** 239.

Chugh S.S., Blackshear J.L., Shen W.K., Hammill S.C., and Gersh B.J. 2001. Epidemiology and natural history of atrial fibrillation: Clinical implications. *J. Am. Coll. Cardiol.* **37:** 371.

Coplen S.E., Antman E.M., Berlin J.A., Hewitt P., and Chalmers T.C. 1990. Efficacy and safety of quinidine therapy for maintenance of sinus rhythm after cardioversion: A meta-analysis of randomized controlled trials. *Circulation* **82:** 1106.

Donahue J.K., Kikkawa K., Thomas A.D., Marbán E., and Lawrence J.H. 1998. Acceleration of widespread adenoviral gene transfer to intact rabbit hearts by coronary perfusion with low calcium and serotonin. *Gene Ther.* **5:** 630.

Donahue J.K., Heldman A.W., Fraser H., McDonald A.D., Miller J.M., Rade J.J., Eschenhagen T., and Marbán E. 2000. Focal modification of electrical conduction in the heart by viral gene transfer. *Nat. Med.* **6:** 1395.

Echt D.S., Liebson P.R., Mitchell L.B., Peters R.W., Obias-Manno D., Barker A.H., Arensberg D., Baker A., Friedman L., and Greene H.L. et al. 1991. Mortality and morbidity in patients receiving encainide, flecainide, or placebo. The Cardiac Arrhythmia Suppression Trial. *N. Engl. J. Med.* **324:** 781.

Falk R.H. 2001. Atrial fibrillation. *N. Engl. J. Med.* **344:** 1067.

Herskowitz I. 1987. Functional inactivation of genes by dominant negative mutations. *Nature* **329:** 219.

Hirano Y. and Hiraoka M. 1988. Barium-induced automatic activity in isolated ventricular myocytes from guinea pig hearts. *J. Physiol.* **395:** 455.

Imoto Y., Ehara T., and Matsuura H. 1987. Voltage- and time-dependent block of Ik1 underlying Ba2+-induced ventricular automaticity. *Am. J. Physiol.* **252:** H325.

Irisawa H., Brown H.F., and Giles W. 1993. Cardiac pacemaking in the sinoatrial node. *Physiol. Rev.* **73:** 197.

Khand A.U., Rankin A.C., Kaye G.C., and Cleland J.G. 2000. Systematic review of the management of atrial fibrillation in patients with heart failure. *Eur. Heart J.* **21:** 614.

Kubo Y., Baldwin T.J., Jan Y.N., and Jan L.Y. 1993. Primary structure and functional expression of a mouse inward rectifier potassium channel. *Nature* **362:** 127.

Kusumoto F.M. and Goldschlager N. 1996. Cardiac pacing. *N. Engl. J. Med.* **334:** 89.

Losordo D.W., Vale P.R., Symes J.F., Dunnington C.H., Esakof D.D., Maysky M., Ashare A.B., Lathi K., and Isner J.M. 1998. Gene therapy for myocardial angiogenesis: Initial clincial results with direct myocardial injection of phVEGF165 as sole therapy for myocardial ischemia. *Circulation* **98:** 2800.

Mann M.J., Whittemore A.D., Donaldson M.C., Belkin M., Conte M.S., Polak J.F., Orav E.J., Ehsan A., Dell'Acqua G., and Dzau V.J. 1999. Ex-vivo gene therapy of human vascular bypass grafts with E2F decoy: The PREVENT single-centre, randomised, controlled trial. *Lancet* **354:** 1493.

Miake J., Marbán E., and Nuss H.B. 2002. Creation of a biological pacemaker by gene transfer. *Nature* **419:** 132.

Nagata K., Marbán E., Lawrence J.H., and Donahue J.K. 2001. Phosphodiesterase inhibitor-mediated potentiation of adenovirus delivery to myocardium. *J. Mol. Cell. Cardiol.* **33:** 575.

Nuss H.B., Marbán E., and Johns D.C. 1999. Overexpression of a human potassium channel suppresses cardiac hyperexcitability in rabbit ventricular myocytes. *J. Clin. Invest.* **103:** 889.

Nuss H.B., Johns D.C., Kaab S., Tomaselli G.F., Kass D., Lawrence J.H., and Marbán E. 1996. Reversal of potassium channel deficiency in cells from failing hearts by adenoviral gene transfer: A prototype for gene therapy for disorders of cardiac excitability and contractility. *Gene Ther.* **3:** 900.

Ohno T., Gordon D., San H., Pompili V.J., Imperiale M.J., Nabel G.J., and Nabel E.G. 1994. Gene therapy for vascular smooth muscle cell proliferation after arterial injury. *Science* **265:** 781.

Prystowsky E.N., Benson D.W., Jr., Fuster V., Hart R.G., Kay G.N., Myerberg R.J., Naccarelli G.V., and Wyse D.G. 1996. Management of patients with atrial fibrillation: A statement for healthcare professionals. From the Subcommittee on Electrocardiography and Electrophysiology, American Heart Association. *Circulation* **93:** 1262.

Richardson A.W. and Josephson M.E. 1999. Ablation of ventricular tachycardia in the setting of coronary artery disease. *Curr. Cardiol. Rep.* **1:** 157.

Rodriguez-Contreras A., Nonner W., and Yamoah E.N. 2002. Ca^{2+} transport properties and determinants of anomalous mole fraction effects of single voltage-gated Ca^{2+} channels in hair cells from bullfrog saccule. *J. Physiol.* **538:** 729.

Rosengart T.K., Lee L.Y., Patel S.R., Kligfield P.D., Okin P.M., Hackett N.R., Isom O.W., and Crystal R.G. 1999. Six-month assessment of a phase I trial of angiogenic gene therapy for the treatment of coronary artery disease using direct intramyocardial administration of an adenovirus vector expressing the VEGF121 cDNA. *Ann. Surg.* **230:** 466.

Santoro B. and Tibbs G.R. 1999. The HCN gene family: Molecular basis of the hyperpolarization-activated pacemaker channels. *Ann. N.Y. Acad. Sci.* **868:** 741.

Siebels J., Cappato R., Ruppel R., Schneider M.A., and Kuck K.H. 1993. Preliminary results of the Cardiac Arrest Study Hamburg (CASH). CASH Investigators. *Am. J. Cardiol.* **72:** 109F.

Slesinger P.A., Patil N., Liao Y.J., Jan Y.N., Jan L.Y., and Cox D.R. 1996. Functional effects of the mouse weaver mutation on G protein-gated inwardly rectifying K+ channels. *Neuron* **16:** 321.

Waldo A.L., Camm A.J., deRuyter H., Friedman P.L., MacNeil D.J., Pauls J.F., Pitt B., Pratt C.M., Schwartz P.J., and Veltri E.P. 1996. Effect of d-sotalol on mortality in patients with left ventricular dysfunction after recent and remote myocardial infarction. The SWORD Investigators. Survival with oral d-sotalol. *Lancet* **348:** 7.

Williams E.S. and Miller J.M. 2002. Results from late-breaking clinical trial sessions at the American College of Cardiology 51st Annual Scientific Session. *J. Am. Coll. Cardiol.* **40:** 1.

Wobus A.M., Rohwedel J., Maltsev V., and Hescheler J. 1995. Development of cardiomyocytes expressing cardiac-specific genes, action potentials, and ionic channels during embryonic stem cell-derived cardiogenesis. *Ann. N.Y. Acad. Sci.* **752:** 460.

Myocardial Disease in Failing Hearts: Defective Excitation–Contraction Coupling

X.H.T. WEHRENS AND A.R. MARKS

Center for Molecular Cardiology, Departments of Medicine and Pharmacology, Columbia University College of Physicians and Surgeons, New York, New York 10032

Congestive heart failure is a major public health problem (Franz et al. 2001). Each year in the U.S. alone, approximately 5 million patients are treated for heart failure and 1 million are admitted to hospitals, with about 550,000 new cases of heart failure being diagnosed (Givertz 2000). Despite advances in treatment, mortality associated with heart failure remains high. In 1998, heart failure contributed to 240,000 deaths in the U.S. (Franz et al. 2001).

Patients with heart failure suffer from a progressive deterioration of cardiac function. Although loss of contractility can be attributed to the loss of cardiomyocytes due to the major cause of heart failure, myocardial infarction, it is likely that additional factors reduce contractility of the remaining ventricular cardiomyocytes, although the mechanism underlying this important contributing factor to heart failure remains largely unknown. This review focuses on the contribution of defective excitation–contraction (EC) coupling to the deterioration of cardiomyocyte function in heart failure and, in particular, on the role of the ryanodine receptor (RyR2)/calcium (Ca^{++}) release channel on the cardiac sarcoplasmic reticulum (SR).

MOLECULAR BASIS OF EC COUPLING

Cardiac EC coupling is the process of electrical excitation of the myocyte, leading to contraction of the heart. During the cardiac action potential, Ca^{++} enters the cell through depolarization-activated L-type Ca^{++} channels, which conduct inward Ca^{++} current during the action potential plateau phase ($I_{Ca,L}$) (Fig. 1). Additional Ca^{++} may enter via the T-type Ca^{++} channel ($I_{Ca,T}$) and via reverse mode Na^+/Ca^{++} exchange (Sipido et al. 1997, 1998). Ca^{++} entry triggers Ca^{++} release from the SR, a phenomenon referred to as Ca^{++}-induced Ca^{++} release (CICR) (Fabiato 1983; Nabauer et al. 1989). The combination of Ca^{++} influx and release from intracellular Ca^{++} release channels (ryanodine receptors) on the SR raises the free intracellular Ca^{++} concentration ($[Ca^{++}]_i$) approximately tenfold. Ca^{++} binds to the myofilament protein troponin C, which undergoes a conformational change allowing actin–myosin cross-bridging to occur. Relaxation only occurs after $[Ca^{++}]_i$ declines, allowing Ca^{++} to dissociate from troponin C. Relaxation requires Ca^{++} removal from the cytosol by four pathways: SR Ca^{++}-ATPase (SERCA2a), sarcolemmal Na^+/Ca^{++} exchange, sarcolemmal Ca^{++}-ATPase, and mitochondrial Ca^{++} uniport. Of these pathways, the dominant one is SERCA2a-mediated Ca^{++} re-uptake into the SR.

It is widely accepted that the amplitude of the Ca^{++} transient generated by SR Ca^{++} release determines the contractile force in cardiomyocytes. Systems that regulate SR Ca^{++} release include (1) the triggers (predominantly Ca^{++} influx through the voltage-gated L-type Ca^{++} channel on the plasma membrane and secondary Ca^{++} influx via the Na^+/Ca^{++} exchanger); (2) the SR Ca^{++} release channel or type 2 ryanodine receptor (RyR2); (3) the SR Ca^{++} re-uptake pump (SERCA2a) and its regulator, phospholamban. All three systems are modulated by signaling pathways including the β-adrenergic (βAR) signaling

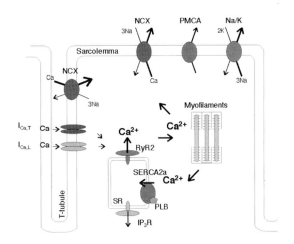

Figure 1. Molecular basis of EC coupling and Ca^{++} transport in ventricular myocytes. EC coupling involves depolarization of the transverse tubule (T-tubule) that activates L-type Ca^{++} channels (dihydropyridine receptors, DHPR) in the plasma membrane. Ca^{++} influx via DHPR triggers Ca^{++} release from the sarcoplasmic reticulum (SR) via RyR2 (SR Ca^{++} release channel). Free intracellular Ca^{++} concentration increases tenfold from ~100 nM to ~1 μM, such that Ca^{++} binds to troponin C, inducing a conformational change that permits actin–myosin cross-bridging and muscle contraction. In order to enable relaxation, intracellular Ca^{++} is pumped out of the cytoplasm by the SR Ca^{++}-ATPase (SERCA2a), which is inhibited by phospholamban (PLB). Other Ca^{++} regulatory molecules include the sarcolemmal Ca^{++}-ATPase (PMCA) and the sarcolemmal Na^+/Ca^{++} exchanger (NCX). The Na-K ATPase (Na/K) indirectly regulates $[Ca^{++}]$ by controlling intracellular $[Na^+]$.

pathway (i.e., phosphorylation by cAMP-dependent protein kinase A [PKA]).

STRUCTURE AND FUNCTION OF RYANODINE RECEPTORS

Ca^{++} release channel proteins are the largest ion channels yet described. Indeed, these channels can be visualized using electron microscopy as the "foot" processes that span the gap between the SR and surface membranes. Three different isoforms of ryanodine receptors, each encoded by different genes, have been identified. RyR1 is the major SR Ca^{++} release channel required for skeletal muscle EC coupling; RyR2 is required for cardiac muscle EC coupling; and RyR3, although present in striated muscles, does not appear to play a major role in EC coupling. In addition to muscle, RyRs are widely expressed in almost all non-muscle tissues.

The SR Ca^{++} release channels have been cloned from skeletal and cardiac muscles as well as brain (Marks et al. 1989; Takeshima et al. 1989; Nakai et al. 1990; Otsu et al. 1990; Hakamata et al. 1992). RyR2 is a protein of 4,969 amino acids that is 66% homologous to RyR1. Four RyR subunits, each with a molecular mass of ~560 kD, assemble to form a functional Ca^{++} release channel. RyR channels are characterized by enormous amino-terminal domains protruding into the cytosol, which contain binding sites for channel modulators (that regulate the channel pore located in the carboxyl terminus).

RyR2 MACROMOLECULAR COMPLEX

A growing list of regulatory molecules have been shown to bind to the cytoplasmic scaffold domain of the RyR channels. These include calmodulin (which contributes to Ca^{++}-dependent regulation of RyR) (Fruen et al. 2000), FK-506 binding protein (FKBP12.6; which stabilizes RyR gating and is required for coupled gating of groups of RyR channels that form Ca^{++} release units, [CRUs]) (Marks et al. 1989; Jayaraman et al. 1992; Marx et al. 2001a), PKA (which phosphorylates RyR1 and RyR2, regulates the binding of FKBP12/12.6 to the channels, and increases the sensitivity of the channels to Ca^{++}-dependent activation) (Marx et al. 1998, 2000, 2001a,b), phosphatases 1 and 2A (PP1/PP2A) (Marx et al. 2000, 2001b), and sorcin (which binds to RyR and DHPR) (Meyers et al. 1998). Ryanodine receptors are also bound to proteins at the luminal SR surface (triadin, junctin, and calsequestrin) (Zhang et al. 1997).

Highly conserved leucine/isoleucine zippers (LIZ) in RyR1 and RyR2 form binding sites for LIZs present in adapter/targeting proteins for kinases (e.g., PKA) and phosphatases (e.g., PP1 and PP2A) that regulate RyR function (Fig. 2) (Marx et al. 2000, 2001b). LIZs are heptad repeats of hydrophobic residues (Leu/Ile) that line up on one side of the helix (3.5 residues/turn) and oligomerize with other coiled-coil helices. Hydrophobic residues occupy the "*a*" as well as the "*d*" (Leu/Ile) positions of the repeating heptad of *a–g* residues that form the helix. Ile/Leu residues can be substituted by Val and can include "skips" and "hiccups" (Leung and Lassam 1998). The amino acids in the other positions in the α-helix determine the specificity of each LIZ-binding site. PP1 is targeted to RyR2 by the adapter protein spinophilin; PP2A is targeted to RyR2 via the adapter protein PR130, and mAKAP (muscle A-kinase anchoring protein) is the adapter protein that targets PKA/RII to RyR2 (Fig. 2) (Marx et al. 2000, 2001b).

FKBP12.6 IS REQUIRED FOR NORMAL RyR FUNCTION

The immunophilin FKBP12 (FK506-binding protein 12 kD) is a *cis-trans* peptidyl-prolyl isomerase that was originally identified as a peptide KC7 (Marks et al. 1989, 1990) that copurifies with RyR1 on the basis of binding to the channel with a stoichiometry of 1 (RyR protomer): 1 (FKBP12) or 4 FKBPs per single functional channel. FKBP12 is the cytosolic receptor for the immunosuppressant drugs FK506 and rapamycin (Schreiber 1991); FK506 and rapamycin compete FKBP12 off from RyR1 and FKBP12.6 from cardiac RyR2 (Brillantes et al. 1994; Timerman et al. 1994). FKBP12.6-devoid RyRs reconstituted in lipid bilayers exhibit subconductance states and increased open probability (Kaftan et al. 1996; Xiao et al. 1997).

In addition to stabilizing individual RyR channels, FKBP12 and FKBP12.6 are required for coupled gating between RyRs (Marx et al. 1998, 2001a). Coupled gating between RyR channels enables groups of channels to open and close (gate) simultaneously such that they function as Ca^{++} release units (Marx et al. 1998, 2001a). Dissociation of FKBP12.6 from coupled RyR2 channels results in functional but not physical uncoupling of the channels (Marx et al. 2001a). Functional uncoupling of RyR2 channels is thought to contribute to the decrease in EC coupling gain that is observed in heart failure.

Mice deficient in FKBP12.6 displayed a marked increase in CICR gain, suggesting that FKBP12.6-deficient myocytes released substantially more Ca^{++} from the SR (Xin et al. 2002). Ca^{++} sparks were increased in amplitude and size, and were markedly longer in duration due to an increase in decay time. These results are consistent with the RyR2s' remaining open for longer periods of time in mice lacking FKBP12.6 and support the concept that the RyR2–FKBP12.6 interaction plays a role in modulating EC coupling in cardiac muscle.

REGULATION OF THE RYANODINE RECEPTOR BY PKA PHOSPHORYLATION

βAR signaling plays a fundamental role in regulating cardiac performance, and abnormalities in this pathway have been implicated as important determinants of reduced contractility of the failing heart (Bristow et al. 1982, 1992; Ungerer et al. 1993). The β1- and β2-adrenergic receptors, which bind circulating catecholamines, activate adenylyl cyclase via a stimulatory G protein, $G\alpha_s$. Adenylyl cyclase increases cAMP levels. cAMP activates the cAMP-dependent protein kinase (PKA) by

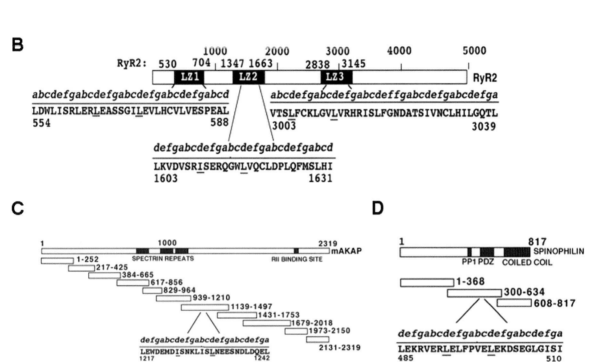

Figure 2. (*A*) Location of the three leucine-isoleucine zipper (LIZ) motifs in RyR2 that bind to LIZ motifs in adapter/targeting proteins for PKA, PP1, and PP2A. The numbers below RyR2 refer to the amino acid residues (aa). (*B*) LZ1 (aa 555–604) binds to spinophilin, LZ2 (aa 1603–1631) binds to PR130, and LZ3 (aa 3003–3039) binds to mAKAP. (*C*) The PKA/RII/mAKAP complex binds to RyR2 via a LIZ motif in mAKAP (aa 1217–1242). (*D*) PP1/spinophilin binds to RyR2 via a LIZ motif (aa 485–510) in spinophilin. These studies provided the first description of a role of LIZ in the formation of ion channel macromolecular complexes. Elucidation of this role for LIZ sequences in ion channel complexes has provided a road map for subsequent studies demonstrating the essential role of these motifs in other ion channels including voltage-gated channels (Marx et al. 2002). (Reprinted, with permission, from Marx et al. 2001b [copyright Rockefeller University Press].)

causing dissociation of the catalytic subunit from the regulatory subunit (RII). Among the many PKA substrates in cardiomyocytes are several that influence contractility in response to βAR signaling including the voltage-gated (L-type) Ca^{++} channel in the sarcolemmal transverse tubules, the Ca^{++} release channel (RyR2), and phospholamban, which in its unphosphorylated state inhibits Ca^{++} uptake via the Ca^{++}-ATPase (SERCA2a) in the SR.

PKA phosphorylation of RyR2 activates the channel (Hain et al. 1995; Valdivia et al. 1995; Marx et al. 2000). We have shown that the mechanism by which PKA phosphorylation of RyR2 activates the channel involves dissociation of FKBP12.6 from the channel, resulting in increased open probability (Po) due to increased sensitivity of RyR2 to Ca^{++}-dependent activation (Marx et al. 2000). PKA phosphorylation of RyR2 is an important physiological pathway, part of the "fight or flight" response, that regulates cardiac EC coupling and specifically enhances EC coupling gain by increasing the amount of Ca^{++} released for a given trigger (Marks 2000). This signaling pathway provides a mechanism whereby activation of the sympathetic nervous system in response to stress results in increased cardiac output required to meet the metabolic demands of the stress responses (Fig. 3).

HEART FAILURE

Chronic heart failure is a hyperadrenergic state in which chronic sympathetic nervous system stimulation is associated with a significant increase in circulating catecholamines (Chidsey et al. 1962). The chronic hyperadrenergic state of heart failure is maladaptive and results in several pathologic consequences, including PKA hyperphosphorylation of RyR2 (Fig. 3) (Marx et al. 2000). This maladaptive response to stress, PKA hyperphosphorylation of RyR2, results in depletion of FKBP12.6 from the channel macromolecular complex and leads to a shift to the left in the sensitivity of RyR2 to Ca^{++}-induced

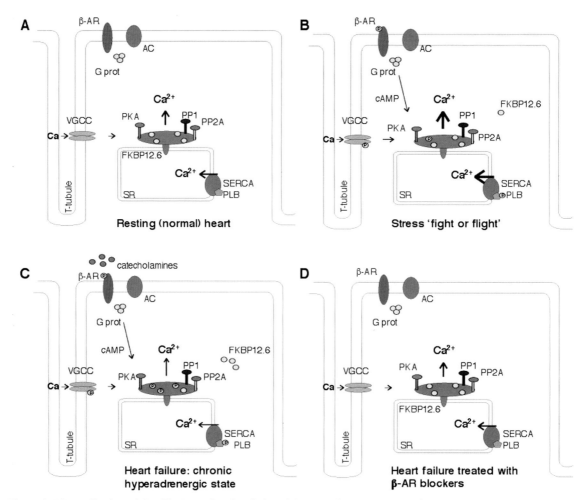

Figure 3. EC coupling is modulated by PKA phosphorylation of the ryanodine receptor. (*A*) The RyR2 macromolecular complex includes four identical RyR2 subunits, each with a binding site for one FKBP12.6, as well as the PKA catalytic and regulatory subunits (RII), mAKAP, PP2A and its targeting protein PR130, and PP1 and its targeting protein sphinophilin (only one of each is shown except for the four FKBP12.6 molecules). βAR signaling molecules include βAR in the plasma membrane, which activates adenylyl cyclase (AC) via G proteins (G prot). (*B*) The stress-activated "fight or flight" response involves sympathetic nervous system-mediated catecholamine release that leads to activation of PKA. PKA phosphorylates and activates voltage-gated Ca^{++} channels (VGCC) increasing the Ca^{++} influx that activates RyR2; RyR2, increasing Ca^{++}-dependent activation and EC coupling gain; and PLB, releasing inhibition of SERCA and increasing SR Ca^{++} uptake. (*C*) In failing hearts, impaired cardiac function leads to chronic sympathetic nervous system activation, resulting in a chronic hyperadrenergic state. PKA hyperphosphorylation, possibly due in part to reduced PP1 and PP2A levels in the RyR2 macromolecular complex, depletes FKBP12.6 from RyR2 complex, pathologically increasing Ca^{++}-dependent activation of RyR2 and resulting in depletion of SR Ca^{++} stores, uncoupling of RyR2 from each other (reducing EC coupling gain). These alterations may cause diastolic SR Ca^{++} release that can activate depolarizations and trigger fatal ventricular cardiac arrhythmias. (*D*) In failing hearts, βAR blockade restores FKBP12.6, PP1, and PP2A levels and RyR2 function to normal.

Ca^{++} release, resulting in "leaky channels" (channels with pathologically increased sensitivity to Ca^{++}-induced activation) (Fig. 4). Over time, the increased "leak" through RyR2 results in resetting of the SR Ca^{++} content to a lower level, which in turn reduces EC coupling gain and contributes to impaired systolic contractility. In addition, a subpopulation of RyR2 that are particularly leaky can release SR Ca^{++} during the resting phase of the cardiac cycle, diastole, resulting in depolarizations of the cardiomyocyte membrane called delayed after depolarizations (DADs) that are known to trigger fatal ventricular cardiac arrhythmias (Fig. 4) (Marks et al. 2000, 2002b).

As discussed above, we have shown that PKA phosphorylation of RyR2 is regulated by kinases and phosphatases that are bound to the channel via targeting proteins (Marx et al. 2000). The amount of the phosphatases PP1 and PP2A in the RyR2 macromolecular complex in failing hearts is reduced (Marx et al. 2000). This reduction in phosphatase levels in the RyR2 macromolecular complex may contribute to PKA hyperphosphorylation of RyR2 and defective channel function (Marx et al. 2000). PKA hyperphosphorylated RyR2 channels in failing hearts are depleted of FKBP12.6 and exhibit single-channel properties similar to those observed in RyR1 in the absence of FKBP12 (Brillantes et al. 1994; Marx et al. 2000) and in RyR2 pharmacologically depleted of FKBP12.6 (Kaftan et al. 1996). Moreover, PKA hyperphosphorylation of RyR2 in failing hearts reduces coupled gating between RyR2 (Marx et al. 2001b). Loss of coupled gating between RyR2 channels can impair EC

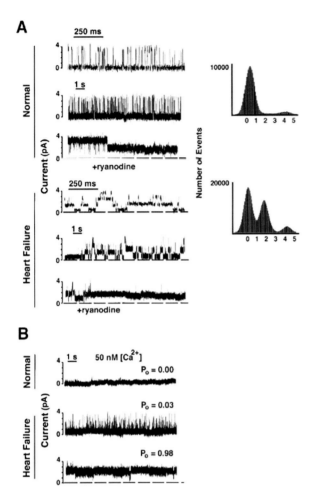

Figure 4. Altered RyR2 channel function in heart failure. (*A*) Single-channel tracings from normal and failing canine hearts, with corresponding amplitude histograms at the right. Bottom tracing in each set of three shows the characteristic modification of the RyR2 channels by ryanodine (1 μM). (*B*) Canine heart failure (pacing induced) RyR2 channels have increased sensitivity to Ca^{++}-dependent activation compared to channels from normal hearts, which are inactive at <50 nM free Ca^{++} in the *cis* (cytosolic) chamber (*top* trace). RyR2 channels from failing hearts can exhibit two types of Ca^{++}-dependent activation at <50 nM Ca^{++}, some are active with a low Po (*second* tracing), and others are extremely active at <50 nM Ca^{++}, remaining open in a subconductance state (*bottom* tracing). (Reprinted, with permission, from Marx et al. 2000 [copyright Elsevier Science].)

coupling gain, as we have shown that the open probability of coupled channels is greater than that of uncoupled channels when both are measured at the exact same cytosolic [Ca^{++}] (Marx et al. 2001a). In addition, reduced coupled gating between RyR2 channels could cause incomplete closure of RyR2 during diastole (Marks 2001; Marx et al. 2001b), resulting in a diastolic leak of SR Ca^{++} that can activate triggers of potentially lethal cardiac arrhythmias (DADs). The findings from our laboratory regarding PKA hyperphosphorylation of RyR2 and depletion of FKBP12.6 from the RyR2 macromolecular complex have been confirmed by other workers (Yano et al. 2000; Doi et al. 2002).

LEAKY RyR2 MAY CAUSE TRIGGERED ARRHYTHMIAS

One of the major causes of mortality in heart failure (HF) is the onset of life-threatening ventricular arrhythmias. Ventricular tachycardia (VT) is an immediate precursor of ventricular fibrillation (VF), which is a major cause of sudden cardiac death in HF. Three-dimensional mapping studies in patients with HF indicate that most VT initiate by non-reentrant mechanisms, especially in nonischemic HF (Pogwizd et al. 1998). Triggered arrhythmias caused by delayed after depolarizations (DADs) and early after depolarizations (EADs) are major initiators of VT.

EADs are secondary depolarizations that occur during the action potential plateau phase or later during phase 3 of repolarization (Volders et al. 2000). EADs are more common with long action potential (AP) durations, during slow cycle lengths, and in patients with long QT syndrome (Priori et al. 1999; Wehrens et al. 2000). The smaller Ca^{++} transients that occur in HF may cause incomplete $I_{Ca,L}$ inactivation during the early phases of the AP, which increases the likelihood of reactivation of inward $I_{Ca,L}$ late in the AP (Zeng and Rudy 1995).

DADs are defined as oscillations in membrane potential that occur after repolarization of an action potential (Volders et al. 2000). Spontaneous Ca^{++} release is usually observed when the Ca^{++} content of the SR is very high (a condition referred to as Ca^{++} overload). Spontaneous release could result from an increased sensitivity of the SR Ca^{++} release channel to a small increase in diastolic [Ca^{++}], possibly from a small Ca^{++} influx across the sarcolemma. DADs are more common at normal or shorter cycle lengths, especially in association with βAR activation. βAR activation increases $I_{Ca,L}$ and SR Ca^{++} uptake, which tends to increase the SR Ca^{++} load, overcoming the low SR Ca^{++} content typically seen in HF (which is responsible for the poor contractile function). We propose that leaky ryanodine receptors in failing hearts predispose to the development of DADs. The leaky phenotype of the channel could be due to increased Ca^{++}-induced activation of the channel such as is seen when RyR2 is PKA hyperphosphorylated and depleted of FKBP12.6.

MOLECULAR MECHANISMS UNDERLYING βAR BLOCKADE THERAPY FOR HEART FAILURE

Down-regulation of βARs in failing heart muscle and desensitization of these receptors attributable to uncoupling from their downstream signaling molecules, G proteins, have been documented previously (Bristow et al. 1982). A common finding in patients with heart failure is that a hyperadrenergic state and elevated levels of circulating catecholamines are markers for increased risk of mortality (Cohn et al. 1984). Clinical trials have shown that treatment with βAR blockers can restore cardiac function and reduce mortality rates in patients with heart failure (CIBIS-II, 1999; MERIT-HF, 1999). It is not intuitively obvious, however, how βAR blockers can improve cardiac function in failing hearts, because they are known to decrease contractility in normal hearts (Epstein et al. 1965; Parmley and Braunwald 1967; Higgins et al. 1973).

We have recently shown in a well-characterized canine model of pacing-induced heart failure that systemic oral administration of a βAR blocker reverses PKA hyperphosphorylation of RyR2, restores the stoichiometry of the RyR2 macromolecular complex, and normalizes single-channel function (Reiken et al. 2001). In a similar dog model of heart failure, Doi et al. (2002) reported that propranolol reversed the hyperphosphorylation of RyR2 in conjunction with a reassociation of FKBP12.6. Jeyakumar et al. (2001) have demonstrated that the canine heart is exceptional with respect to the interaction of FKBP with RyR. In contrast to seven other mammals studied, including human, in canine hearts only FKBP12.6 is bound to RyR2, which could potentially make a translation from the dog data to the human troublesome. However, we have recently demonstrated that βAR blocker treatment also corrects the stoichiometry of RyR in failing human hearts, which is associated with normalization of RyR2 single-channel gating (S. Reiken et al., in prep.). In conclusion, the restoration of normal RyR2 structure and function contributes to a mechanistic understanding that may in part explain the improved cardiac function observed in heart-failure patients treated with βAR blockers.

CELLULAR EFFECTS OF PKA HYPERPHOSPHORYLATION OF RyR2 IN HEART FAILURE

Other investigators have reported that short-term treatment with caffeine, which activates RyR2, fails to alter cardiac EC coupling (Trafford et al. 2000), and that short-term administration of isoproterenol decreases Ca^{++} spark heterogeneity (Litwin et al. 2000; Song et al. 2001) or improves EC coupling in failing cardiomyocytes (Gomez et al. 1997). However, these findings are consistent with the RyR2 PKA hyperphosphorylation, FKBP12.6 depletion from the RyR2 macromolecular complex, and increased Ca^{++} sensitivity of RyR2 in failing cardiomyocytes that we have reported (Marx et al. 2000). Heart failure is a chronic disease and the functional consequences of alterations in the RyR2 macromolecular complex (including depletion of FKBP12.6, PP1, and PP2A) in failing hearts persist for months or years (Marks 2000) and are not the equivalent of acute short-term administration of agonists such as caffeine or isoproterenol. These drugs can acutely modulate RyR2 function, allowing other Ca^{++}-handling molecules in the cell to restore homeostasis when RyR2 function returns to normal. In failing hearts, the chronic alteration of RyR2 structure and function can reset SR Ca^{++} content to a lower level, in part because of increased leak through PKA-hyperphosphorylated RyR2. Diminished Ca^{++} transients and slowed contractions have been reported in cardiomyocytes from rats with heart failure (Gomez et al. 2001), although the L-type Ca^{++} channel remained unchanged. EC coupling gain measured as $\Delta[Ca^{++}]_i/I_{Ca}$ was also significantly decreased. Reduced SR Ca^{++} content can also contribute to a reduced EC coupling gain (Marks 2000; Marx et al. 2000). In heart failure, decreased SERCA2a expression and function and increased levels of Na^+/Ca^{++} exchanger can exacerbate SR Ca^{++} depletion by reducing Ca^{++} re-uptake.

CONGENITAL RyR2 MUTATIONS LINKED TO SUDDEN CARDIAC DEATH

RyR2 mutations are linked to two forms of sudden cardiac death (SCD): (1) catecholaminergic polymorphic ventricular tachycardia (CPVT) or familial polymorphic ventricular tachycardia (FPVT) (Laitinen et al. 2001; Priori et al. 2001) and (2) arrhythmogenic right ventricular dysplasia type 2 (ARVD2) (Tiso et al. 2001). An autosomal-dominant disorder, CPVT/FPVT is characterized by syncope and stress-related, bidirectional, or polymorphic VTs without structural heart disease or prolonged QT interval (Leenhardt et al. 1995; Swan et al. 1999). Mortality is 30–50% by age 30 (Fisher et al. 1999). The CPVT/FPVT disease gene locus was localized to chromosome 1q42-q43 (Swan et al. 1999).

Previously, ARVD2 had been mapped to the same locus (Rampazzo et al. 1995). ARVD includes degeneration and progressive fibro-fatty replacement of the right ventricular myocardium (Corrado et al. 1997; Runge et al. 2000) and arrhythmias of right ventricular origin that can result in sudden cardiac death. Of particular note, ARVD2 is distinct from other subtypes because it is associated with exercise-induced ventricular tachycardia and SCD (Nava et al. 1988). The association with exercise and sympathetic nervous system stimulation is notable because mutations in RyR2 are also linked to exercise and stress-induced SCD in the absence of any structural heart disease (Laitinen et al. 2001; Priori et al. 2001; Tiso et al. 2001). To date, all of the 21 reported RyR2 mutations linked to SCD cluster into three regions of the channel that correspond to three malignant hyperthermia (MH)/central core disease (CCD) regions found in the skeletal muscle ryanodine receptor (RyR1) (Fig. 5). On this basis we have proposed that RyR2 mutations linked to exercise-induced SCD likely share functional abnormalities with RyR1 mutations linked to MH and CCD. MH and CCD mutations may alter the Ca^{++}-dependent regulation of RyR1 (Lynch et al. 1997; Richter et al. 1997; Tong et al. 1999). Moreover, the link between RyR2 mutations and stress-induced VT suggests a role for PKA phosphorylation of the channel (which is increased by activation of the sympathetic nervous system) and altered channel function that predisposes to SCD. Indeed, PKA hyperphosphorylation of RyR2 may exacerbate functional defects in mutant RyR2 channels (Marks et al. 2002b). The ARVD2 RyR2 mutations N2368I and T2504M occur in a region known to be important for interaction with FKBP12.6, the regulatory subunit that stabilizes the RyR2 channel (Marx et al. 2000). The net effect of RyR2 mutations linked to SCD on the biophysical properties of the channels could be to make them more active in vivo, thus contributing to SR Ca^{++} leak that could explain the pathophysiology of MH and CCD, as well as triggers for arrhythmias in CPVT/FPVT and ARVD2 (Marks et al. 2002b).

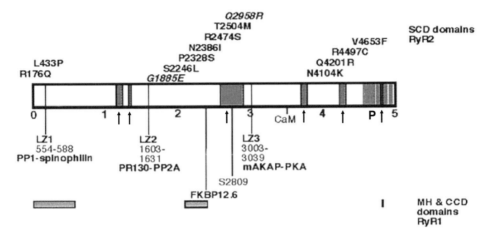

Figure 5. Locations of RyR2 SCD-linked mutations compared to MH and CCD regions in RyR1. Eleven of the SCD-linked RyR2 mutations (Laitinen et al. 2001; Priori et al. 2001; Tiso et al. 2001) cluster in three regions homologous to three MH/CCD regions (Jurkat-Rott et al. 2000; MacLennan 2000). Two common polymorphisms are indicated by italics. Three RyR2 leucine/isoleucine zippers (LZ) that target PP1, PP2A, and PKA to RyR2 are shown (Marx et al. 2000, 2001b), as is the FKBP12.6-binding region (Gaburjakova et al. 2001), and the CaM-binding site (Yamaguchi et al. 2001). Arrows denote six surface exposed regions of RyR1 (Marks et al. 1990), and the pore region is at the carboxyl terminus. (Reprinted, with permission, from Marks et al. 2002b [copyright Wiley-Liss].).

CONCLUSIONS

Maladaptive regulation of RyR2 that results in leaky SR Ca^{++} release channels likely contributes to impaired contractility and SCD in heart failure. Our studies have shown that PKA hyperphosphorylation of the channel is associated with altered stoichiometry of the channel macromolecular complex that includes depletion of FKBP12.6. These RyR2 defects in heart failure can be reversed by therapy with oral βAR blockers. Future studies will be aimed at designing more specifically targeted therapy to overcome the unacceptable side effects of systemic β-blocker therapy in heart failure patients.

ACKNOWLEDGMENTS

This work was supported by grants to A.R.M. from the National Institutes of Health and the American Heart Association. A.R.M. is a Doris Duke Charitable Foundation Distinguished Clinical Scientist. Correspondence and requests for materials should be addressed to A.R.M. (arm42@columbia.edu).

REFERENCES

Brillantes A.B., Ondrias K., Scott A., Kobrinsky E., Ondriasova E., Moschella M.C., Jayaraman T., Landers M., Ehrlich B.E., and Marks A.R. 1994. Stabilization of calcium release channel (ryanodine receptor) function by FK506-binding protein. *Cell* **77:** 513.

Bristow M.R., Ginsburg R., Minobe W., Cubicciotti R.S., Sageman W.S., Lurie K., Billingham M.E., Harrison D.C., and Stinson E.B. 1982. Decreased catecholamine sensitivity and beta-adrenergic-receptor density in failing human hearts. *N. Engl. J. Med.* **307:** 205.

Bristow M.R., Minobe W., Rasmussen R., Larrabee P., Skerl L., Klein J.W., Anderson F.L., Murray J., Mestroni L., Karwande S.V., et al. 1992. Beta-adrenergic neuroeffector abnormalities in the failing human heart are produced by local rather than systemic mechanisms. *J. Clin. Invest.* **89:** 803.

Chidsey C.A., Harrison D.C., and Braunwald E. 1962. Augmentation of plasma norepinephrine response to exercise in patients with congestive heart failure. *N. Engl. J. Med.* **267:** 650.

Cohn J.N., Levine T.B., Olivari M.T., Garberg V., Lura D., Francis G.S., Simon A.B., and Rector T. 1984. Plasma norepinephrine as a guide to prognosis in patients with chronic congestive heart failure. *N. Engl. J. Med.* **311:** 819.

Corrado D., Basso C., Thiene G., McKenna W.J., Davies M.J., Fontaliran F., Nava A., Silvestri F., Blomstrom-Lundqvist C., Wlodarska E.K., Fontaine G., and Camerini F. 1997. Spectrum of clinicopathologic manifestations of arrhythmogenic right ventricular cardiomyopathy/dysplasia: A multicenter study. *J. Am. Coll. Cardiol.* **30:** 1512.

Doi M., Yano M., Kobayashi S., Kohno M., Tokuhisa T., Okuda S., Suetsugu M., Hisamatsu Y., Ohkusa T., and Matsuzaki M. 2002. Propranolol prevents the development of heart failure by restoring FKBP12.6-mediated stabilization of ryanodine receptor. *Circulation* **105:** 1374.

Epstein S., Robinson B.F., Kahler R.L., and Braunwald E. 1965. Effects of beta-adrenergic blockade on the cardiac response to maximal and submaximal exercise in man. *J. Clin. Invest.* **44:** 1745.

Fabiato A. 1983. Calcium-induced release of calcium from the cardiac sarcoplasmic reticulum. *Am. J. Physiol.* **245:** C1.

Fisher J.D., Krikler D., and Hallidie-Smith K.A. 1999. Familial polymorphic ventricular arrhythmias: A quarter century of successful medical treatment based on serial exercise-pharmacologic testing. *J. Am. Coll. Cardiol.* **34:** 2015.

Franz W.M., Muller O.J., and Katus H.A. 2001. Cardiomyopathies: From genetics to the prospect of treatment. *Lancet* **358:** 1627.

Fruen B.R., Bardy J.M., Byrem T.M., Strasburg G.M., and Louis C.F. 2000. Differential Ca(2+) sensitivity of skeletal and cardiac muscle ryanodine receptors in the presence of calmodulin. *Am. J. Physiol. Cell Physiol.* **279:** C724.

Gaburjakova M., Gaburjakova J., Reiken S., Huang F., Marx S.O., Rosemblit N., and Marks A.R. 2001. FKBP12 binding modulates ryanodine receptor channel gating. *J. Biol. Chem.* **276:** 16931.

Givertz M.M. 2000. Underlying causes and survival in patients with heart failure. *N. Engl. J. Med.* **342:** 1120.

Gomez A.M., Guatimosim S., Dilly K.W., Vassort G., and Lederer W.J. 2001. Heart failure after myocardial infarction: Altered excitation-contraction coupling. *Circulation* **104:** 688.

Gomez A.M., Valdivia H.H., Cheng H., Lederer M.R., Santana L.F., Cannell M.B., McCune S.A., Altschuld R.A., and Lederer W.J. 1997. Defective excitation-contraction coupling in experimental cardiac hypertrophy and heart failure. *Science* **276:** 800.

Hain J., Onoue H., Mayrleitner M., Fleischer S., and Schindler H. 1995. Phosphorylation modulates the function of the calcium release channel of sarcoplasmic reticulum from cardiac muscle. *J. Biol. Chem.* **270:** 2074.

Hakamata Y., Nakai J., Takeshima H., and Imoto K. 1992. Primary structure and distribution of a novel ryanodine receptor/calcium release channel from rabbit brain. *FEBS Lett.* **312:** 229.

Higgins C.B., Vatner S.F., Franklin D., and Braunwald E. 1973. Extent of regulation of the heart's contractile state in the conscious dog by alteration in the frequency of contraction. *J. Clin. Invest.* **52:** 1187.

Jayaraman T., Brillantes A.-M.B., Timerman A.P., Erdjument-Bromage H., Fleischer S., Tempst P., and Marks A.R. 1992. FK506 binding protein associated with the calcium release channel (ryanodine receptor). *J. Biol. Chem.* **267:** 9474.

Jeyakumar L.H., Ballester L., Cheng D.S., McIntyre J.O., Chang P., Olivey H.E., Rollins-Smith L., Barnett J.V., Murray K., Xin H.B., and Fleischer S. 2001. FKBP binding characteristics of cardiac microsomes from diverse vertebrates. *Biochem. Biophys. Res. Commun.* **281:** 979.

Jurkat-Rott K., McCarthy T., and Lehmann-Horn F. 2000. Genetics and pathogenesis of malignant hyperthermia. *Muscle Nerve* **23:** 4.

Kaftan E., Marks A.R., and Ehrlich B.E. 1996. Effects of rapamycin on ryanodine receptor/Ca(2+)-release channels from cardiac muscle. *Circ. Res.* **78:** 990.

Laitinen P.J., Brown K.M., Piippo K., Swan H., Devaney J.M., Brahmbhatt B., Donarum E.A., Marino M., Tiso N., Viitasalo M., Toivonen L., Stephan D.A., and Kontula K. 2001. Mutations of the cardiac ryanodine receptor (RyR2) gene in familial polymorphic ventricular tachycardia. *Circulation* **103:** 485.

Leenhardt A., Lucet V., Denjoy I., Grau F., Ngoc D.D., and Coumel P. 1995. Catecholaminergic polymorphic ventricular tachycardia in children. A 7-year follow-up of 21 patients. *Circulation* **91:** 1512.

Leung I.W. and Lassam N. 1998. Dimerization via tandem leucine zippers is essential for the activation of the mitogen-activated protein kinase kinase kinase, MLK-3. *J. Biol. Chem.* **273:** 32408.

Litwin S.E., Zhang D., and Bridge J.H. 2000. Dyssynchronous Ca(2+) sparks in myocytes from infarcted hearts. *Circ. Res.* **87:** 1040.

Lynch P.J., Krivosic-Horber R., Reyford H., Monnier N., Quane K., Adnet P., Haudecoeur G., Krivosic I., McCarthy T., and Lunardi J. 1997. Identification of heterozygous and homozygous individuals with the novel RYR1 mutation Cys35Arg in a large kindred. *Anesthesiology* **86:** 620.

MacLennan D.H. 2000. Ca^{2+} signalling and muscle disease. *Eur. J. Biochem.* **267:** 5291.

Marks A.R. 2000. Cardiac intracellular calcium release channels: Role in heart failure. *Circ. Res.* **87:** 8.

———. 2001. Ryanodine receptors/calcium release channels in heart failure and sudden cardiac death. *J. Mol. Cell. Cardiol.* **33:** 615.

Marks A.R., Fleischer S., and Tempst P. 1990. Surface topography analysis of the ryanodine receptor/junctional channel complex based on proteolysis sensitivity mapping. *J. Biol. Chem.* **265:** 13143.

Marks A.R., Reiken S., and Marx S.O. 2002a. Progression of heart failure: Is protein kinase a hyperphosphorylation of the ryanodine receptor a contributing factor? *Circulation* **105:** 272.

Marks A.R., Priori S., Memmi M., Kontula K., and Laitinen P.J. 2002b. Involvement of the cardiac ryanodine receptor/calcium release channel in catecholaminergic polymorphic ventricular tachycardia. *J. Cell. Physiol.* **190:** 1.

Marks A.R., Tempst P., Hwang K.S., Taubman M.B., Inui M., Chadwick C., Fleischer S., and Nadal-Ginard B. 1989. Molecular cloning and characterization of the ryanodine receptor/junctional channel complex cDNA from skeletal muscle sarcoplasmic reticulum. *Proc. Natl. Acad. Sci.* **86:** 8683.

Marx S.O., Ondrias K., and Marks A.R. 1998. Coupled gating between individual skeletal muscle Ca^{2+} release channels (ryanodine receptors). *Science* **281:** 818.

Marx S.O., Gaburjakova J., Gaburjakova M., Henrikson C., Ondrias K., and Marks A.R. 2001a. Coupled gating between cardiac calcium release channels (ryanodine receptors). *Circ. Res.* **88:** 1151.

Marx S.O., Kurokawa J., Reiken S., Motoike H., D'Armiento J., Marks A.R., and Kass R.S. 2002. Requirement of a macromolecular signaling complex for beta adrenergic receptor modulation of the KCNQ1-KCNE1 potassium channel. *Science* **295:** 496.

Marx S.O., Reiken S., Hisamatsu Y., Jayaraman T., Burkhoff D., Rosemblit N., and Marks A.R. 2000. PKA phosphorylation dissociates FKBP12.6 from the calcium release channel (ryanodine receptor): Defective regulation in failing hearts. *Cell* **101:** 365.

Marx S.O., Reiken S., Hisamatsu Y., Gaburjakova M., Gaburjakova J., Yang Y.M., Rosemblit N., and Marks A.R. 2001b. Phosphorylation-dependent regulation of ryanodine receptors. A novel role for leucine/isoleucine zippers. *J. Cell Biol.* **153:** 699.

Meyers M.B., Puri T.S., Chien A.J., Gao T., Hsu P.H., Hosey M.M., and Fishman G.I. 1998. Sorcin associates with the pore-forming subunit of voltage-dependent L-type Ca^{2+} channels. *J. Biol. Chem.* **273:** 18930.

Nabauer M., Callewart G., Cleeman L., and Morad M. 1989. Regulation of calcium release is gated by calcium current, not gating charge, in cardiac, myocytes. *Science* **244:** 800.

Nakai J., Imagawa T., Hakamata Y., Shigekawa M., Takeshima H., and Numa S. 1990. Primary structure and functional expression from cDNA of the cardiac ryanodine receptor/calcium release channel. *FEBS Lett.* **271:** 169.

Nava A., Canciani B., Daliento L., Miraglia G., Buja G., Fasoli G., Martini B., Scognamiglio R., and Thiene G. 1988. Juvenile sudden death and effort ventricular tachycardias in a family with right ventricular cardiomyopathy. *Int. J. Cardiol.* **21:** 111.

Otsu K., Willard H.F., Khanna V.K., Zorzato F., Green N.M., and MacLennan D.H. 1990. Molecular cloning of cDNA encoding the Ca^{2+} release channel (ryanodine receptor) of rabbit cardiac muscle sarcoplasmic reticulum. *J. Biol. Chem.* **265:** 13472.

Parmley W.W. and Braunwald E. 1967. Comparative myocardial depressant and anti-arrhythmic properties of d-propranolol, dl-propranolol and quinidine. *J. Pharmacol. Exp. Ther.* **158:** 11.

Pogwizd S.M., McKenzie J.P., and Cain M.E. 1998. Mechanisms underlying spontaneous and induced ventricular arrhythmias in patients with idiopathic dilated cardiomyopathy. *Circulation* **98:** 2404.

Priori S.G., Napolitano C., Tiso N., Memmi M., Vignati G., Bloise R., Sorrentino V.V., and Danieli G.A. 2001. Mutations in the cardiac ryanodine receptor gene (hRyR2) underlie catecholaminergic polymorphic ventricular tachycardia. *Circulation* **103:** 196.

Priori S.G., Barhanin J., Hauer R.N., Haverkamp W., Jongsma H.J., Kleber A.G., McKenna W.J., Roden D.M., Rudy Y., Schwartz K., Schwartz P.J., Towbin J.A., and Wilde A.M. 1999. Genetic and molecular basis of cardiac arrhythmias: Impact on clinical management parts I and II. *Circulation* **99:** 518.

Rampazzo A., Nava A., Erne P., Eberhard M., Vian E., Slomp P., Tiso N., Thiene G., and Danieli G.A. 1995. A new locus for arrhythmogenic right ventricular cardiomyopathy (ARVD2) maps to chromosome 1q42-q43. *Hum. Mol. Genet.* **4:** 2151.

Reiken S., Gaburjakova M., Gaburjakova J., He Kl K.L., Prieto A., Becker E., Yi Gh G.H., Wang J., Burkhoff D., and Marks A.R. 2001. β-Adrenergic receptor blockers restore cardiac calcium release channel (ryanodine receptor) structure and function in heart failure. *Circulation* **104:** 2843.

Richter M., Schleithoff L., Deufel T., Lehmann-Horn F., and Herrmann-Frank A. 1997. Functional characterization of a distinct ryanodine receptor mutation in human malignant hyperthermia-susceptible muscle. *J. Biol. Chem.* **272:** 5256.

Runge M.S., Stouffer G.A., Sheahan R.G., Yamamoto S., Tsyplenkova V.G., and James T.N. 2000. Morphological patterns of death by myocytes in arrhythmogenic right ventricular dysplasia. *Am. J. Med. Sci.* **320:** 310.

Schreiber S. 1991. Chemistry and biology of the immunophilins and their immunosuppressive ligands. *Science* **251:** 283.

Sipido K.R., Carmeliet E., and Van de Werf F. 1998. T-type Ca^{2+} current as a trigger for Ca^{2+} release from the sarcoplasmic reticulum in guinea-pig ventricular myocytes. *J. Physiol.* **508:** 439.

Sipido K.R., Maes M., and Van de Werf F. 1997. Low efficiency of Ca^{2+} entry through the Na(+)-Ca^{2+} exchanger as trigger for Ca^{2+} release from the sarcoplasmic reticulum. A comparison between L-type Ca^{2+} current and reverse-mode Na(+)-Ca^{2+} exchange. *Circ. Res.* **81:** 1034.

Song L.S., Wang S.Q., Xiao R.P., Spurgeon H., Lakatta E.G., and Cheng H. 2001. β-adrenergic stimulation synchronizes intracellular Ca(2+) release during excitation-contraction coupling in cardiac myocytes. *Circ. Res.* **88:** 794.

Swan H., Piippo K., Viitasalo M., Heikkila P., Paavonen T., Kainulainen K., Kere J., Keto P., Kontula K., and Toivonen L. 1999. Arrhythmic disorder mapped to chromosome 1q42-q43 causes malignant polymorphic ventricular tachycardia in structurally normal hearts. *J. Am. Coll. Cardiol.* **34:** 2035.

Takeshima H., Nishimura S., Matsumoto T., Ishida H., Kangawa K., Minamino N., Matsuo H., Ueda M., Hanaoka M., Hirose T., and Numa S. 1989. Primary structure and expression from complementary DNA of skeletal muscle ryanodine receptor. *Nature* **339:** 439.

Timerman A.P., Jayaraman T., Wiederrecht G., Onoue H., Marks A.R., and Fleischer S. 1994. The ryanodine receptor from canine heart sarcoplasmic reticulum is associated with a novel FK-506 binding protein. *Biochem. Biophys. Res. Commun.* **198:** 701.

Tiso N., Stephan D.A., Nava A., Bagattin A., Devaney J.M., Stanchi F., Larderet G., Brahmbhatt B., Brown K., Bauce B., Muriago M., Basso C., Thiene G., Danieli G.A., and Rampazzo A. 2001. Identification of mutations in the cardiac ryanodine receptor gene in families affected with arrhythmogenic right ventricular cardiomyopathy type 2 (ARVD2). *Hum. Mol. Genet.* **10:** 189.

Tong J., McCarthy T.V., and MacLennan D.H. 1999. Measurement of resting cytosolic Ca^{2+} concentrations and Ca^{2+} store size in HEK-293 cells transfected with malignant hyperthermia or central core disease mutant Ca^{2+} release channels. *J. Biol. Chem.* **274:** 693.

Trafford A.W., Diaz M.E., Sibbring G.C., and Eisner D.A. 2000. Modulation of CICR has no maintained effect on systolic Ca^{2+}: Simultaneous measurements of sarcoplasmic reticulum and sarcolemmal Ca^{2+} fluxes in rat ventricular myocytes. *J. Physiol.* (pt. 2) **522:** 259.

Ungerer M., Bohm M., Elce J.S., Erdmann E., and Lohse M.J. 1993. Altered expression of β-adrenergic receptor kinase and $β_1$-adrenergic receptors in failing human heart. *Circulation* **87:** 454.

Valdivia H.H., Kaplan J.H., Ellis-Davies G.C., and Lederer W.J. 1995. Rapid adaptation of cardiac ryanodine receptors: Modulation by Mg^{2+} and phosphorylation. *Science* **267:** 1997.

Volders P.G., Vos M.A., Szabo B., Sipido K.R., de Groot S.H., Gorgels A.P., Wellens H.J., and Lazzara R. 2000. Progress in the understanding of cardiac early afterdepolarizations and torsades de pointes: Time to revise current concepts. *Cardiovasc. Res.* **46:** 376.

Wehrens X.H., Abriel H., Cabo C., Benhorin J., and Kass R.S. 2000. Arrhythmogenic mechanism of an LQT-3 mutation of the human heart Na(+) channel alpha-subunit: A computational analysis. *Circulation* **102:** 584.

Xiao R.P., Valdivia H.H., Bogdanov K., Valdivia C., Lakatta E.G., and Cheng H. 1997. The immunophilin FK506-binding protein modulates Ca^{2+} release channel closure in rat heart. *J. Physiol.* **500:** 343.

Xin H.B., Senbonmatsu T., Cheng D.S., Wang Y.X., Copello J.A., Ji G.J., Collier M.L., Deng K.Y., Jeyakumar L.H., Magnuson M.A., Inagami T., Kotlikoff M.I., and Fleischer S. 2002. Oestrogen protects FKBP12.6 null mice from cardiac hypertrophy. *Nature* **416:** 334.

Yamaguchi N., Xin C., and Meissner G. 2001. Identification of apocalmodulin and Ca^{2+}-calmodulin regulatory domain in skeletal muscle Ca^{2+} release channel, ryanodine receptor. *J. Biol. Chem.* **276:** 22579.

Yano M., Ono K., Ohkusa T., Suetsugu M., Kohno M., Hisaoka T., Kobayashi S., Hisamatsu Y., Yamamoto T., Noguchi N., Takasawa S., Okamoto H., and Matsuzaki M. 2000. Altered stoichiometry of FKBP12.6 versus ryanodine receptor as a cause of abnormal Ca(2+) leak through ryanodine receptor in heart failure. *Circulation* **102:** 2131.

Zeng J. and Rudy Y. 1995. Early afterdepolarizations in cardiac myocytes: Mechanism and rate dependence. *Biophys. J.* **68:** 949.

Zhang L., Kelley J., Schmeisser G., Kobayashi Y.M., and Jones L.R. 1997. Complex formation between junctin, triadin, calsequestrin, and the ryanodine receptor. Proteins of the cardiac junctional sarcoplasmic reticulum membrane. *J. Biol. Chem.* **272:** 23389.

The Coming of Age of Cardiovascular Science

C.E. SEIDMAN AND J.G. SEIDMAN

Department of Medicine and Genetics, Harvard Medical School, Howard Hughes Medical Institutes, Boston, Massachusetts 02115

THE CLINICAL CHALLENGES

The socioeconomic and human burden of cardiovascular disease is staggering. One in 5 Americans, or 61,800,000 individuals, live with high blood pressure, coronary heart disease, stroke, congenital cardiovascular disease, or congestive heart failure. Cardiovascular disease is the leading cause of death in the U.S., accounting for 1 of every 2.5 deaths, and is the primary or contributing cause in 65% of all deaths. Each day 2,600 Americans die of cardiovascular disease, and in a third of these, death is premature, occurring below the current average U.S. life expectancy of 75 years.

Costs associated with cardiovascular disease are enormous. Medicare expenses for cardiovascular problems that required hospitalization equaled $26.4 billion in 1998. Almost 60 million physician office visits for treatment of heart and blood vessel disorders occurred in 1999. The American Heart Association (2002 Heart and Stroke Statistical Update. Dallas, TX 2001) database estimates that direct medical costs of cardiovascular disease and indirect costs attributed to lost productivity from cardiovascular disease would reach $329 billion in 2002.

Cardiovascular disease affects people of all ages, sex, race, and ethnicity. Despite widespread occurrence, cardiovascular disease prevalence varies in particular populations. The age-adjusted prevalence for cardiovascular disease is 26.9% and 27.7%, respectively, for non-Hispanic white and Hispanic Americans, but 40.0% for non-Hispanic black Americans. Although comparable data are unavailable for Native Americans, surveys suggest that 60% of this population have important risk factors (tobacco abuse, diabetes, and hypertension) for ischemic heart disease. For obese Americans (in 2000, 19.8% of Americans had a body mass index above 30 kg/m^2), risk factors for cardiovascular disease have reached epidemic proportions. Dramatic increase in the prevalence of diabetes has already been recognized (33% increase between 1990 and 1998); obesity in children has changed both the incidence and type of diabetes diagnosed in the young. The accomplishments of cardiovascular medicine further contribute to the changing clinical profiles of these pathologies. With the enormous improvements in medical therapies for diabetes, hypertension, dyslipidemias, and surgical interventions for congenital heart malformations and coronary artery disease, patients are living longer, but are at increased risk for heart failure, a diagnosis that has surged to 550,000 new cases annually.

THE RESEARCH RESPONSE

The sheer magnitude of this biomedical problem has prompted considerable research to understand normal cardiovascular biology, to discover causes of cardiovascular pathology, and to define evidence-based therapy. Yet despite vast numbers of cardiovascular studies and publications over the past century, 2002 heralded the first-ever Cold Spring Harbor Symposium on heart and vessels. Why has it taken 60 years? Undoubtedly the answers to this question are many, but we suspect the root answer has to do with much of the past orientation of cardiovascular science. Unlike immunology, oncology, and neurobiology, biomedical disciplines that have each been the subject of multiple Cold Spring Harbor Symposia, until recently cardiovascular science has been firmly grounded in pathophysiologic studies, not molecular biology. Although multiple reasons likely account for the organ-based, integrative focus of cardiovascular research over the past half-century, the absence of immortalized cell lines has certainly contributed to the orientation of the field. Without these critical reagents, cell and molecular approaches that have fueled decades of progress in virtually every other medical discipline have until recently been almost absent from cardiovascular science. Instead, the community has relied on primary cultures of myocardial or vascular cells, heterogeneous populations that require technically rigorous isolation procedures, are unable to divide, and have short survival, which have been poor substitutes for well-characterized immortal cardiac cell lines. Although the problem of immortalized cardiovascular cell lines remains unresolved, cardiovascular science has finally bypassed this obstacle by embracing genetic manipulation of model organisms as a means of probing development, cell biology, and the molecular basis of pathophysiology. Indeed, the preponderance of science reported in this first Symposium on the cardiovascular system comes from discoveries made by manipulating and characterizing the genomes of frogs, flies, fish, and rodents. Although certainly the ability to manipulate genomes and study cell biology has fostered science in every discipline, these methodologies have quite simply revolutionized cardiovascular science.

A second factor contributing to the exponential growth of cardiovascular discoveries may actually be attributable to clinical care efforts. Given the enormous burden of human cardiovascular disease, physicians have invested heavily in developing diagnostics and interventions. Two unanticipated benefits have come from their efforts. First, the precision with which human cardiovascular phenotypes are routinely characterized has propelled the application of human molecular genetics to discover cardiovascular disease genes. Second, the same powerful diagnostic human tools, once miniaturized, have proven to be enormously effective for characterizing cardiovascular phenotypes in genetically manipulated model organisms. Despite the broad schism created by different language and cultures, cross-fertilization between clinical medicine and cardiovascular research bodes well for ultimately changing the burden of cardiovascular disease.

Convening the first-ever Cold Spring Harbor Symposium on the heart and vessels in 2002 dates the coming of age of cardiovascular science. Discoveries reported here unmask only some of the many molecular signals required for heart and vessel formation, hint at the molecular interplay between cardiovascular pathophysiology and gene transcription, and provide a framework for potential treatment of selected cardiovascular disorders. But as cardiovascular science matures, many more issues need to be considered. This synopsis cannot hope to address all of these, nor to recite the entire array of impressive data presented at Symposium LXVII, but will instead attempt to integrate individual results into broader themes and to provide a personal perspective on questions that cardiovascular science is now posed to address. A second and related disclaimer relates to referencing of this summary; the citations provided herein are only intended to refer the reader to the relevant papers included in this volume, where extensive and definitive references of primary materials can be found.

DEVELOPMENT OF VASCULAR BEDS

Arterial, venous, and lymphatic vessels make up a continuous and integrated circulatory system that both supplies nutrients and transports wastes in complex organisms. Subservient to the shared functions of these vessels are endothelial, smooth muscle, and connective tissue cells fashioned into exquisitely arborized lumens that extend into all tissues, while maintaining full integration of the entire system. Morphologic distinctions (vascular smooth muscle content, presence or absence of valves) distinguish arteries, veins, and lymphatics to accommodate the distinct and important physiologic functions of these vessels. Efforts from multiple laboratories evidence a rapidly growing knowledge base of the genetic programs that characterize each vascular tree.

Considerable evidence supports the critical role of two classes of ligands: vascular endothelial growth factors (VEGFs) and angiopoietins, and their relevant receptors in development of the vasculature. The birth of a new vessel begins when angioblasts derived from mesodermal precursor cells form a primitive endothelial network. VEGFs are critical to induce endothelial cell proliferation and migration, and VEGF activity is specifically targeted to the developing vasculature through the highly selective expression of three receptors (VEGFR), members of the tyrosine kinase receptor superfamily (Gale et al.). VEGF functions are highly integrated with the activity of angiopoietins, paracrine growth factors secreted by perivascular cells in vessel walls. Four distinct angiopoietins (Sato et al.) have been identified that bind Tie receptors, also tyrosine kinase receptors that are selectively expressed on the vasculature.

VEGF haploinsufficiency is lethal in mice and produces severe deficits of vasculogenesis and angiogenesis, a phenotype recapitulated in VEGFR-2-deficient mice. Mice lacking angiopoietin-1 or Tie receptor (Sato et al.) develop an immature vasculature that fails to undergo remodeling into an adult vessel. Overexpression of VEGF leads primarily to increased numbers of vessels, suggesting a primary role in angiogenic sprouting. Overexpression of angiopoietin increases vessel size, implying a role in circumferential growth, particularly in veins. Vascular permeability appears to be reciprocally modulated by these two growth factors: Immature vessels resulting from VEGF overexpression are leaky, whereas vessels produced under excessive angiopoietin-1 show tight endothelial-matrix cell interactions that resist leak.

Definition of molecular distinctions between arteries, veins, and lymphatics has also fostered insights into the developmental processes specific for each circulatory system. Coordinated expression of particular VEGFs and angiopoietins appear critical for selection, development, and maturation of these distinct vascular trees. Angiogenesis induced by VEGF results primarily in arterial vessels. An important downstream target of VEGF is Notch signaling (Weinstein and Lawson), which further promotes arterial over venous differentiation, and repression or activation of Notch signaling in zebrafish caused reciprocal expansion or suppression of venous markers. Notch signaling also results in the induction of markers of arterial identity, including ephrin B2 expression, and the comparable phenotypes of knockout mice deficient in Notch or ephrin B2 genes imply that ephrin B2 has a major role in establishing arterial identity.

Angiopoietin-1 and -2 are needed for the promotion of venous identity: Expression of Angiopoietin-1 is enhanced in the sinus venosus from which the venous tree develops and is preferentially expressed around veins (Sato et al.; Gale et al.). Angiopoietin-2 can suppress the markers of arterial identity, but does not appear to enhance venous development.

A temporal sequence for establishing arterial–venous vascular networks has also been identified (Nagy et al.). Capillaries that sprout in response to VEGF signaling are arterial, whereas venous capillaries appear to be patterned from this arterial template, either from local endothelial progenitors, or perhaps from a vasculature of unspecified type. Expansion of the system requires continued growth and remodeling, so that an initially symmetric vascular tree evolves into the asymmetric mature network. Asymmetric expression of angiopoietin-1 in concert with Tie1 (Sato et al.) may be an important signal for the handedness of the venous system, but comparable clues that establish an asymmetric arterial system are unknown.

The lymphatic tree develops after blood vessel formation and is initiated by sprouting from the cardinal vein.

Not surprisingly, both VEGFs and angiopoietins appear essential for establishing this vascular tree (Mäkinen and Alitalo) in addition to blood vessels. VEGF C and D induce the proliferation and migration of lymphatic endothelial cells through interactions with the VEGFR-3. Whereas VEGFR-3 is expressed early in developing blood vessels, the receptor becomes almost exclusively restricted to lymphatic vessels. However, lymphatic vessels are distinguished from blood vessels, by expression of endothelial-specific molecules. Prox-1 appears to establish the lymphatic endothelial phenotype, because when this transcription factor is absent, endothelial cells assume blood vessel phenotypes. Angiopoietin-2 appears particularly important for maturation of the lymphatics, as targeted disruption in mice results in widespread lymphatic malfunction (Gale et al.), with disseminated tissue lymphedema. Because angiopoietin-1 replacement in angiopoietin-2-deficient mice restores normal lymphatic integrity, maintenance of tight endothelial and smooth muscle interactions of lymphatics may, like blood vessels, be predicated on Tie2 signaling.

VEGF and angiopoietin-1 and -2 are regulated in part by metabolic clues (Upalakalin et al.), particularly oxygen levels. The hypoxia inducible factor (HIF) transcriptional complex is made up of three vertebrate HIF proteins that heterodimerize with ARNT, aryl hydrocarbon receptor nuclear translocator (Simon et al.). Genetic manipulation of ARNT and HIF genes in the mouse evidences the primary role of these molecules in vasculogenesis and vascular remodeling so as to influence vascular density.

All immature vessels acquire layers of pericytes and smooth muscle cells. As noted above, there are important distinctions in the quantity of smooth muscle found in arteries versus veins and the endothelial integrity (Mukhopadhyay and Zeng) of lymphatic versus either blood vessel. Several lines of evidence indicate angiopoietin-1 activation of Tie2 receptors in endothelial cells (Anghelina et al.), and endothelial-derived platelet-derived growth factors (PDGFs) mature nascent vessels by stimulating pericyte coverage and recruitment of smooth muscle cells (Upalakalin et al.). Intriguing new data (Coughlin) suggest that components of the coagulation cascade may participate in these processes. Protease-activated receptors (PARs) are G-protein-coupled receptors that, when activated by thrombin, increase the permeability of endothelium to allow platelet and leukocyte recruitment to damaged tissues. As a consequence, homeostasis and/or inflammation results. PARs may play a parallel role in stabilizing leaks in developing vessels, in that embryonic mice deficient in PAR-1, an abundant molecule in endothelial cells, exsanguinate not from coagulation deficits but apparently from an immature vasculature. Additional studies indicate a role for other inflammatory molecules, such as type IV collagen (Kalluri) fragments, which through their interactions with endothelial integrins (Hynes et al.) regulate angiogenesis during development. Phenotypes associated with selective inactivation of integrins in mice suggest these molecules are critical for adhesion of the vessel pericytes to tissue parenchymal cells and may in particular also modulate tissue-specific vascular development.

Molecules that direct vasculogenesis also play critical roles in vascular remodeling, and a considerable knowledge base about these processes comes from detailed study of tumor angiogenesis (Hood and Cheresh; Benezra et al.). Angiopoietin-2 expression is increased at sites of vascular remodeling and, in the presence of VEGF, facilitates vascular growth. Tissue oxygen excess reduces VEGF-A expression, prompting endothelial apoptosis and vessel regression as do negative regulators such as angiostatin, endostatin, and thrombostatin (Kalluri).

Analyses of the complex developmental interactions between vasculature beds and surrounding tissues (LeCouter et al.) are just beginning. Evidence from zebrafish mutants suggests vascular development is shaped by multiple extrinsic factors. The *out of bound* mutant (MacRae and Fishman) appears to result from the lack of tissue-derived repressive signals for angioblast migration and growth, such that vessels are not restricted to normal intersomitic domains, but assume chaotic patterns. Other mutants hint that mechanical factors contribute to vessel structures; reduced blood flow mutants cause abnormal glomerular capillary networks in the developing kidney, but these can be rescued when flow is restored by saline replacement.

Further discovery of the similarities and differences between arteries, veins, and lymphatics has great potential for treating human vascular diseases. For example, longitudinal follow-up of the more than 500,000 coronary artery bypass procedures annually performed in the U.S. demonstrate that a saphenous vein grafted onto an artery bed prematurely becomes affected by atherosclerosis. Although these data have modified surgical procedures to employ arterial grafts (internal mammary or radial artery), appropriate and dispensable arteries are often too few. Understanding the anatomic and biologic distinctions between arteries and veins may suggest opportunities to manipulate processes that contribute to venous demise. Discovery that molecules involved in inflammation also participate in developmental vascular remodeling suggests a link between cardiac pro-inflammatory states such as transplantation and the associated coronary arteriopathy, which most often accounts for transplant demise and death. There is also great medicinal potential in deciphering the critical genetic determinants of the lymphatic vasculature. At present there are exceedingly few and only paltry mechanical treatments for lymphedema, the clinical manifestation of acquired and intrinsic lymphatic system leak. The possibility to develop rational therapeutics for these disorders and for the many metastasizing tumors that hone to lymphatics must be predicated on a better understanding of the biology of the poorly understood vascular tree.

In addition to the gaps in our knowledge about distinctions between vascular beds, there are many unknowns about how vascular segments become specialized for unique physiologic functions. How is the coronary tree of the heart adapted to accommodate high systolic ventricular pressures? What factors acclimatize the pulmonary bed for the dramatic shifts in oxygen tension? What factors derived from tissues participate in remodeling the developing vasculature?

LOCAL AND SYSTEMIC VASCULAR DISEASE

Extrapolation of insights derived from developmental studies, tumor biology, and model systems so as define novel approaches for therapeutic angiogenesis remains complex. VEGF-A emerges as the critical signaling molecule to initiate angiogenesis (Nagy et al.; Mäkinen and Alitalo; Gale et al.), but responses to this cytokine differ by cell type and model system, therein suggesting the active participation of targeted tissue in angiogenesis. For example, induction of VEGF expression in skeletal muscle (Banfi et al.) resulted in co-expression of angiopoietin-2 but not other angiogenic molecules, a finding that may indicate myocytes and presumably other cells can become actively engaged in stimulating angiogenesis. Yet triggering angiogenesis in ischemic myocytes of the heart has been an exceedingly difficult task.

Cross-talk between angiogenic signals and tissue remodeling pathways, although poorly understood, also suggest that there are critical differences in whether expression of angiogenic factors has salutary or detrimental effects on organ function (Dallabrida and Rupnick). For example, bolstering VEGF levels in the ischemic myocardium can promote endothelial cell survival, angiogenesis, and improved cardiac function, whereas in heart failure patients and model systems, drugs with antiangiogenic actions (angiotensin converting enzyme [ACE] inhibitors, β-adrenergic receptor inhibitors, and spironolactone), almost universally result in improved myocardial performance. Understanding these apparent incongruencies is critical for devising angiogenic strategies to revascularize target tissues.

Hypertension and atherosclerosis are undoubtedly the two most prevalent human vascular disorders. Although there are likely to be multiple genetic loci predisposing to these complex disorders, most contemporary insights into the molecular causes of hypertension or atherosclerosis stem from analyses of single-gene mutations that cause these conditions.

Human molecular genetic dissection of disorders of blood pressure regulation indicates the critical role of salt reabsorption in hereditary hypertension. All monogeneic causes of blood pressure (Lifton et al.), whereas mutations associated with hypotension produce salt wasting. Further evidence for the critical role of salt and water homeostasis in blood pressure regulation and vascular biology comes from genetic manipulations of the renin angiotensin system in mice (Gurley et al.). Increasing copy numbers of the angiotensinogen genes causes a linear increase in blood pressure, whereas renin or angiotensinogen deficiency produces low blood pressure. Ongoing genomic approaches indicate a great likelihood that other critical modulators of blood pressure will soon be defined. Studies of consomic rat strains (Roman et al.) suggest genes encoded on chromosome 13 and chromosome 18 from normotensive rats attenuate blood pressure and kidney disease associated with two hypertensive rat strains.

In addition to the role of the kidney in global regulation of circulatory pressure, analyses of model systems have defined regulators that modulate local vascular tone (Wheeler et al.). K_{ATP} channels, multimeric structures composed of a pore-forming potassium channel and a regulatory subunit, the sulfonylurea receptor, respond to local ATP/ADP levels. Genetic ablation of either the regulatory subunit or the pore-forming potassium channel in mice causes striking coronary vasospasm, therein defining K_{ATP} channels and presumably ATP/ADP levels as important local modulators of vasomotor tone. Myocardium surrounding the microvasculature (Wheeler et al.) also appears capable of inducing coronary vasospasm. Degenerating myocytes that lack δ sarcoglycan have vasospasm and ischemia, which could be extinguished by selective restoration of δ sarcoglycan to myocytes. Nitric oxide (NO) may be the important signaling molecule enabling this cross-talk between myocytes and vascular endothelium (Mineo and Shaul). Endothelial-derived NO is a highly potent vasodilator that also attenuates the adherence of monocytes and macrophages and reduces smooth muscle growth in response to vascular diseases such as atherosclerosis. Evidence suggests that some of the protective effect of high-density lipoprotein (HDL), a molecule that shuttles cholesterol from peripheral tissues to the liver, may also result from HDL-stimulated synthesis of endothelial NO (Mineo and Shaul) and vasorelaxation.

Signaling molecules produced by vascular smooth muscle cells (Nabel et al.) also influence vasomotor tone. In healthy arteries, smooth muscle cells are quiescent and constitutively express the cyclin-dependent kinase (CDK) inhibitor $p27^{Kip1}$. Following vascular injury, $p27^{Kip1}$ is down-regulated, and smooth muscle cell proliferation ensues with collagen and extracellular matrix synthesis. Genetic ablation of $p27^{Kip1}$ of the related CDK inhibitor $p21^{Cip1}$ prevents wound repair in arteries and accelerates atherogenesis in apolipoprotein E mice, implying that these molecules are critical regulators of vascular smooth muscle proliferation.

Ongoing dissection of the genes that regulate lipid metabolism continues to suggest avenues to treat atherosclerosis (Hink and Fukai). In addition to previously identified human mutations in low-density lipoprotein receptors (LDLR) that clear atherogenic lipoproteins, mutations in an adapter protein which plays a critical role for internalizing LDLR have been identified to cause recessive familial hypercholesterolemia. In addition, a newly described ABCG5/ABCG8 transport system (Hobbs et al.) provides complementary data about the cellular machinery for handling neutral sterols. Recessive human mutations in the genes encoding these transporters cause sitosterolemia, characterized by marked increases in serum concentrations of plant sterols. Mouse and cell studies indicate ABCG5/ABCG8 function to translocate sterols across plasma membranes of hepatocytes and enterocytes so as to increase biliary and fecal excretion of neutral sterols.

Recent information about transcriptional regulators of LDL receptor suggests new potential targets for treating lipid metabolism disorders. Sterol regulatory element-binding proteins (SREBPs) are membrane-bound transcription factors that activate genes involved in fatty acid and cholesterol synthesis and enhance the clearance of atherogenic lipoproteins from the blood by increasing LDL receptor expression (Horton et al.). Genetic manipulations of three SREBP isoforms in mice define essential roles for SREBP-1a and SREBP-2 in cholesterol biosynthesis during development and further indicate that SREBP-1c is involved in the promotion of diabetic fatty livers.

Discovering the changes that hypertension, atherosclerosis, and other systemic disorders cause in vascular biology remains a critical challenge. Although ongoing discovery of the mechanisms that shape vascular structure in development will undoubtedly suggest clues, a more precise understanding of integrated vascular and tissue biology seems essential for understanding vascular disease. For example, is it merely coincident that angiogenesis is part and parcel of solid tumor growth, but rarely occurs in nondividing and ischemic myocardial cells? What factors trigger microvascular sprouting in the diabetic retina that are absent in the ischemic diabetic limb? The considerable intellectual excitement, in concert with an impressive array of cell and molecular tools and models available to address these and other puzzles, indicates that answers will soon be forthcoming.

CARDIAC DEVELOPMENT

Mammalian cardiac development is often divided into three distinct stages: recruitment and specification of cells to a cardiomyocyte lineage, formation of a linear heart tube, and looping and septation of this tube to shape a mature four-chambered heart. Progression through these stages has been previously recorded through detailed microscopic investigations of developing embryos. With the application of molecular approaches to study cardiac development, the initiating stimuli and cellular responses that underlie these remarkable morphologic events are increasingly being discovered. A primary enabling factor in all studies that utilize genetics, biochemistry, and cell physiology has been the dramatic advances in genome sequences. The availability of the complete mouse and *Drosophila* genome sequences and significant portions of the zebrafish genome have made these powerful model systems to analyze cardiogenesis. Major advances have proceeded in two directions, the development and enhancement of vertebrate and non-vertebrate models of failed heart development, and the identification and characterization of transcription factors that play critical roles in the maturing heart. Although a complete compendium of the molecular steps in cardiac embryogenesis remains a work in progress, new technologies and model systems recruited to this study have added considerable new data and hold the promise to more fully define key components of these steps in the near future.

Genetic dissection of cardiac development began in *Drosophila* with the demonstration that the heartless mutant, tinman, resulted from deletion of *Nkx2.5*. Confirmation that the murine *Nkx2.5* homolog played a comparable essential function in the developing mammalian heart (Zaffran et al.; Tanaka et al.) validated the appropriateness of deducing general principles about cardiac embryogenesis from the studying of different species. The two-chamber zebrafish heart has proven to be a particularly fruitful source of genetic variants in heart development. Early heart fields (Yelon et al.) are specified by a genetic cascade (*Nodal/Oep* and *Swirl/bmp2* induces *Faust/Gata5* induces *Nkx2.5*). From these fields, definitive myocardial precursors are selected through the expression of *Nkx2.5*, *Hand2*, and other unknown transcription factors. Zebrafish mutants also demonstrate the critical role of genes involved in general regulation of transcription (such as transcription elongation factors *Spt 5/6*) in myocardial differentiation. *Nodal* signaling pathways establish asymmetric heart development; analyses of zebrafish mutants with altered heart jogging and looping led to the discovery of the southpaw gene as an early organizer of downstream left–right patterning genes, including *Cyclops*, *Lefty1*, *Left2*, and *Pitx2* (Long et al.). Axis definition is further defined by PKCλ, which, when mutated in zebrafish *Heart and Soul*, disturbs cardiac growth along an anterior–posterior axis, such that there is anterior displacement of the primitive heart tube and subsequent development of the ventricle within the atria (MacRae and Fishman). In contrast, cardiac concentric growth is altered by mutation of the novel gene, *Heg* (zebrafish mutants *Heart of Glass*, *Santa*, and *Valentine*). Definition of atrioventricular boundaries and the molecular establishment of cardiac valves is growing from analyses of zebrafish mutant *Jekyll* (encoding uridine 5′ diphosphate-glucose dehydrogenase, an enzyme involved in proteoglycan biosynthesis) and *Cardiofunk* (unknown gene) and 20 novel mutants (Stainier et al.). The considerable excitement in this field comes from the virtual certainty that many more cardiac malformations will arise from extensive mutagenesis screens. Pathways can then be readily classified by both complementation analyses and powerful transgenesis and morpholino oligonucleotide strategies. With further positional cloning of zebrafish mutants, one can expect dramatic strides in the identification of the genetic program for cardiovascular development.

There are important limitations in zebrafish as a model organism for cardiac development. Pathways specific for development of the right-sided chambers of the mammalian heart will not be found by mutations of a two-chambered heart, and depend on studies of higher animal systems. Molecular dissection of four-chamber heart development poses considerable challenges because of greater structural complexity, longer generation times, and limited approaches for saturation mutagenesis. Creative approaches to these issues continue to emerge. For example, cell fate studies suggested that specialized heart cells such as the cardiac conduction system evolve from contractile myocardial precursors. The fortuitous insertion of a lacZ reporter transgene in an unknown gene locus of the mouse now allows meticulous dissection of development of this system (Rentschler et al.). Evidence from studies of this reporter mouse indicates that a cardiac conduction system fate is specified early (day 8.5–10 dpc) and likely involves *neuregulin-1* signaling. A novel heart slice culture system (Stuckmann and Lassar) has revealed some of the molecules (retinoic acid and erythropoietin) derived from the epicardium that are necessary and sufficient for myocardial cell proliferation. Discovery of molecules using this system and proteins disrupted by the zebrafish *Heg* mutants has the potential for elucidating whether cell cycle exit by cardiac cells is reversible, data with enormous therapeutic implications.

A complementary approach to understanding cardiogenesis has been the identification and characterization of transcription factors expressed throughout cardiac development. Evidence demonstrates that induction of an essential cardiogenic homeobox transcription factor, *Nkx2.5*, is dependent both on *Dpp*, the *Drosophila* homolog of *BMP* (Zaffran et al.) and on auto-activation of *Nkx2.5*. Combinatorial binding of *Nkx2.5* and MADS box (*Mad* and *Medea* as well as a high mobility group protein [*HMG-D*] results in a fully active transcriptional complex. *Myocardin* is another early marker of cardiomyocyte lineages, and also interacts with MADS box transcription factors (particularly serum response factor) to activate downstream promoters that contain a minimum of two CarG boxes. A novel inhibitory gene, *HOP* (homeodomain-only protein, also termed *Cameo*) may function to limit cardiogenesis by causing myocytes to exit from the cell cycle (Gitler et al.; Wang et al.); *HOP*-deficient mice have excess cardiomyocytes. Examination of the promoters of cardiac-specific genes including *ANF*, *myosin light chain* (Small and Krieg), and *cardiac actin* (Mohun et al.) indicate the particular importance of combinatorial expression of *Nkx2.5* and MADS proteins *SRF*, *MEF2*, and *GATA 4/5* in establishing a myocyte cell program.

Multiple transcription factors have also been defined that direct cardiac morphogenesis. *PitX* families of transcription factors play a critical role in determining left/right asymmetry in the early embryo and play important roles in axis definition of the developing heart. *Pitx2*, initially identified as the cause of Rieger syndrome, a multisystem human disorder with craniofacial and cardiac malformations, has emerged as a marker of left cardiac lineages. Expressed in atrial myocardium associated with the left but not the right vena cava, and pulmonary vein (Campione et al.), Pitx2 may also be required for endothelin A receptor signaling of cardiac neural crest cells into the developing outflow tract (Kioussi et al.). However, other factors that participate in recruiting neural crest cell to the heart remain largely unknown; migration of these cells in the developing mouse was not prevented by ablation of genes encoding *Pax3*, *Splotch*, *type I neurofibromatosis*, or *Tbx1* (Gitler et al.).

Specification of cardiac chamber identities may depend on other transcription factors that exhibit restricted patterns of expression of factors including *dHand* (right ventricle and outflow tract) and *eHand* (left ventricle and outflow tract) and *Irx4* (ventricular specific). But as was recognized through the study of *Spt 5/6* mutations in zebrafish, mutations in genes with more general roles in transcription, such as *mBop*, which promotes heterochromatin condensation to silence transcription (Svrivastava et al.), also subserve chamber-specific development. Ablation of *mBop* in mice caused right ventricular hypoplasia, immature ventricular myocytes, but normal atrial development.

Deducing the role of some transcription factors in particular developmental programs of the heart can be difficult when compiling data from different species. Three mouse Hey genes, a subfamily of the *Hairy/enhancer of split*-related genes, demonstrate overlapping patterns of expression (Fischer et al.) in cardiovascular compartments (*Hey1* in sinus venosus, atria, and aorta; *Hey 2* in ventricular and vascular precursors; *HeyL* in vascular smooth muscle cells), whereas expression of *Hey* genes is more restricted in zebrafish. The functional significance of seemingly subtle differences is expanded by analyses of *Hey* mutations in these species: gridlock, a hypomorphic allele of the zebrafish *Hey2* homolog, grl, produces aortic coarctation, whereas *Hey2* ablation in mice caused cardiac enlargement and dysfunction with septal defects and markedly immature myocytes containing reduced numbers of disorganized sarcomeres. Since *Hey* genes are known direct targets of *Notch* signaling, both mutant studies implicate *Notch* signaling in cardiac and vascular development. Human genetics data further support *Notch* signaling in mammalian heart development, since mutation of *JAG1*, a Notch ligand, causes congenital cardiac malformations in Alagille syndrome. However, since human mutations in *Notch-1* cause leukemia, there must be important tissue-specific aspects of Notch signaling still unknown.

DISORDERS OF CARDIAC STRUCTURE AND FUNCTION

The powerful dissection of cardiac development in model organisms would be expected to provide information that is critical to deciphering the molecular basis for congenital heart disease. Remarkably, this has not yet been the case. Discovery that *NKX2.5* (Tanaka et al.), *TBX1* (Vitelli et al.), and *TBX5* (Huang et al.) cause structural heart malformations came directly from human molecular genetics. The demonstration that *NKX2.5*, *TBX1*, and *TBX5* caused defects in cardiac septation, development of the outflow tract, and maturation of the proximal conduction system indicates the power of integrating information from complex and simple model organisms. Unlike the many mutations engineered in model systems, human cardiac malformations are produced by heterozygous mutation in each of these transcription factor genes, data that indicate cardiac development is dependent on appropriate dosage of the molecules. The importance of transcription factor dose is not unique to humans, in that heterozygous ablation of mouse *Nkx2.5* (Tanaka et al.) or *Tbx1* (Vitelli et al.) is also recognized to have structural malformations and electrophysiologic defects comparable to those found in humans with these mutations. Furthermore, mouse and humans exhibit clinical variation with transcription factor haploinsufficiency (Huang et al.). Evidence indicates that background genes, environmental factors, and differences in mechanical force due to cardiovascular hemodynamics each contribute to phenotype variability. Discovery of critical factors that modulate phenotype may also help to define target genes regulated by these transcription factors.

The phenomenal wealth of clinical techniques developed to care for heart patients, and the adaptation of this equipment to enable comparable investigations in mice, has focused functional studies of the heart in mouse and man. Molecular genetic analyses of inherited human disorders continue to expand the compendium of disease genes that cause arrhythmias, cardiomyopathy, and heart failure. On a background of considerable evidence that mutation of genes encoding ion channels expressed in the heart cause long QT syndrome, new data expand knowledge about other arrhythmias. Mutations in the ryanodine receptor and calsequestrin-2 (Eldar et al.) are demonstrated to cause, respectively, dominant or recessive polymorphic ventricular tachycardia. Localized within the sarcoplasmic reticulum of myocytes, calsequestrin acts as a calcium buffer and storage reservoir, but may also regulate calcium release through ryanidine interactions. Mutations in ryanodine receptor or calsequestrin-2 may overload cells with calcium and thereby trigger arrhythmias. Understanding the molecular basis for cardiac electrophysiology also shows promise for manipulation of critical regulators of heart rhythm.

Expression of exogenous genes in cardiac tissues can selectively alter fundamental electrophysiologic properties and thereby indicates the potential to functionally change working myocardial tissues into pacemaker cells (Marbán et al.). Alternatives to gene therapy of cardiovascular disease include cell replacement strategies to restore myocardial function. Engraftment of cardiac and skeletal myoblasts has some promise (Murry et al.), but definitive evidence for the ability of hematopoietic stem cells to populate the myocardium and differentiate into myocytes remains limited.

Calcium imbalance has also been implicated in hypertrophic cardiac remodeling. Human mutations that cause inherited hypertrophic cardiomyopathy arise most commonly in genes encoding sarcomere proteins (Morita et al.). Mouse models developed from these human studies indicate that, as a consequence of sarcomere mutation, myocyte calcium homeostasis is perturbed. The signaling cascade that results in cardiac hypertrophy remains an area of active investigation (Maass et al.; Wang et al.). Studies in cells and genetically engineered mice indicate the central role of calcium in triggering potential CaM kinase, calcineurin, and *NFAT* signaling cascades, which in turn activate transcription factors such as *MEF2* and promote expression of hypertrophy genes. Identification of inhibitory proteins such as MCIP (Williams and Rosenberg), which decreases calcineurin-induced phosphorylation of NFAT, indicates there are molecular checks to the hypertrophic program that may be important therapeutic targets. Two other signaling pathways, MAP kinases and PPARs, can also trigger cardiac hypertrophy. Targeted activation of selected MAP kinase pathways implicates Ras activation, but not p38 or JNK activation in hypertrophic signals (Petrich et al.). Ras-activated MAP kinase altered calcium homeostasis, again suggesting that critical signals emanate from fluxes in cation concentrations. Cardiac metabolism changes in hypertrophy, due to altered expression of PPARs (fatty acid-activated nuclear receptor transcription factor) and the heterodimeric partner RXR (Finck et al.). Chronic activation of PPAR in transgenic mice caused heightened rates of fatty acid oxidation and cardiac hypertrophy, indicating this as another transcriptional pathway in hypertrophy that may be particularly relevant to diabetic cardiomyopathies.

Dissection of the pathways that cause the heart to fail indicates that this may be even more complicated than hypertrophic signaling. Human mutations in a disparate set of genes, including sarcomere protein, cytoskeletal and intermediate filament proteins, and nuclear membrane proteins, cause cardiac dilation and dysfunction. Mouse models engineered to contain selected mutations are beginning to define some common and several unique pathways that are triggered by distinct gene mutations. As noted above, δ sarcoglycan ablation (Wheeler et al.) disrupts the myocyte cytoskeleton, but cardiac dysfunction comes in part from the superimposed ischemic insult due to coronary vasospasm. Intrinsic myocyte dysfunction has been studied by targeted ablation of *MLP*, a member of the *CRP* gene family with two highly conserved LIM domains, which binds α actinin and titin at Z-discs. *MLP*-deficient mice may impair transmission of sarcomere forces throughout the myocyte by uncoupling the stretch sensor machinery contained at Z-discs (Hoshijima et al.). As a consequence, dilated cardiomyopathy and heart failure ensue. The phenotype of *MLP*-deficient mice (Naga Prasad et al.) recapitulates many morphologic features of human heart failure. Genetic complementation studies performed to dissect the mechanisms underpinning cardiac dysfunction demonstrated that ablation of phospholamban, an endogenous inhibitor of the SR calcium ATPase pump, could improve cardiac function. The logical conclusion from these data is that tweaking calcium cycling between the SR and sarcomere is a potential target for ameliorating poor myocyte function. Analyses of excitation–contraction coupling in normal and failing myocardium (Wehrens and Marks) resulted in a similar conclusion; heart failure is characterized by maladaptive regulation of proteins involved in calcium cycling. PKA-mediated hyperphosphorylation of the ryanidine receptor causes a depletion of FKBP12.6 from the channel complex, thereby increasing the open probability of the channel. As a consequence, Ca^{++} leaks from the SR, even during the resting phase of the cardiac cycle; excitation–contraction-coupling gain is reduced; and contractile function is impaired. A noted complexity in this elegant scheme comes from analyses of β-adrenergic receptors, important determinants of PKA signaling. In human and model systems with heart failure, β-adrenergic receptors are selectively down-regulated and uncoupled from G-proteins (Naga Prasad et al.), implying that PKA signaling should be attenuated. Whether these findings are late and inadequate compensatory events, or whether uncoupled β-adrenergic receptors signal through other pathways that adversely affect Ca^{++} regulation in heart failure, remains unknown, but restoration of normal adrenergic signaling appears quite important. Prevention of β-adrenergic receptor down-regulation in MLP-deficient mice restored cardiac performance. (Naga Prasad et al.).

Pro-inflammatory molecules may also contribute to cardiac dysfunction (Mann). TNF and IL-1, although not constitutively expressed in the heart, are rapidly induced in response to myocardial injury, as a possible protective response to oxygen-derived free radicals during states of ischemia and reperfusion. However, continued expression of pro-inflammatory molecules is ultimately maladaptive. Transgenic mice that overexpress TNF develop progressive ventricular remodeling with increased activation of matrix metalloproteinases and collagen denaturation, which fosters ventricular dilation and dysfunction.

The considerable evidence for calcium signaling in remodeling the heart along a hypertrophic and dilated pathway raises several questions. Despite considerable differences in anatomic and hemodynamic manifestations, does a common pathway lead to both pathologies? Is there a temporal relationship (i.e., hypertrophy precedes dilation)? What are the critical nodal points in the common pathway that fosters hypertrophy with sustained contractile performance or dilation with heart failure? Or is the attraction of a single signaling pathway the artifact of too great a focus on too few models?

FINAL THOUGHTS

The accomplishments evidenced by the collections in this volume attest to the emergence of cardiovascular science. A highly efficient and complex system of pump and tubes, the structure and biology of the cardiovascular system are remarkable. Unraveling the mysteries of its development, understanding the regulation of hemodynamic flow, and discovering the mechanisms for cardiovascular disease pose impressive challenges that will likely be met only through continued integration of molecular and cell biology with classical disciplines of embryology and physiology. Meeting these challenges holds great promise for continued maturation of the discipline of cardiovascular science and great opportunities for cardiovascular medicine.

Author Index

A

Acosta L., 89
Ahmad N., 27
Akyurek L.M., 163
Aletras A., 345
Alitalo K., 189
Allikian M.J., 389
Amann K., 63
Anghelina M., 209
Aránega A., 89
Arad M., 383
Arap W., 223
Arbeit J.M., 133

B

Baek S.H., 81
Baldini A., 327
Barger P.M., 371
Barr S., 383
Bartman T., 49
Basson C., 115
Beis D., 49
Benezra R., 249
Benjamin L.E., 181
Bergers G., 293
Berul C.I., 317
Biben C., 107
Boehm M., 163
Briata P., 81
Brown C.B., 57
Brown L.F., 227
Brown M.S., 491

C

Campione M., 89
Cheresh D.A., 285
Chien K.R., 399
Choate K.A., 445
Coffman T.M., 451
Cohen J.C., 499
Costa M., 107
Coughlin S.R., 197
Cowley A.W., Jr., 309
Crook M.F., 163

D

Dallabrida S.M., 417
Davis E.C., 171
Davis S., 267
Davis J.S., 345
de Candia P., 249
Dehio C., 181
DePalma S.R., 383
Douglas P., 317
Donahue J.K., 527
Dvorak A.M., 227
Dvorak H.F., 227
Duffy C., 383

E

Eldar M., 333
Elliott D., 107
Elmaari H., 63
Epstein J.A., 57
Epstein N.D., 345

F

Feldman J.L., 19
Feng D., 227
Ferrara N., 217
Field L.J., 519
Finck B.N., 371
Fischer A., 63
Fishman G.I., 353
Fishman M.C., 301
Frasch M., 1
Francis S.E., 143
Franco D., 89
Furtado M., 107
Fukai T., 483

G

Gale N.W., 267
Gehrmann J., 317
Geller D.S., 445
Gessler M., 63
Gitler A.D., 57
Goldstein J.L., 491
Gottlieb P.D., 121
Graf G.A., 499
Greene A., 309
Groves N., 107
Gurley S.B., 451

H

Hanahan D., 293
Harvey R.P., 107
Hassanzadeh S., 345
Helisch A., 63
Hemo I., 181
Heydemann A., 389
Hink H.U., 483
Hobbs H.H., 499
Hodivala-Dilke K., 143
Holash J., 267
Hood J.D., 285
Horton J.D., 491
Hoshijima M., 399
Huang T., 115
Hu C.-J., 127
Hynes R.O., 143
Hyun C., 107

I

Icardo J.M., 89
Ishii M., 317
Izumo S., 317

J

Jacob H.J., 309
Jay P.Y., 317
Jain R.K., 239
Jungblut B., 49

K

Kalluri R., 255
Kawamoto T., 317
Keegan B.R., 19
Kelly D.P., 371
Keshet E., 181
Kioussi C., 81
Kirk E., 107
Klamt B., 63
Knobeloch K.-P., 63
Knöll R., 399
Kochilas L., 57
Konhilas J.P., 409
Kotecha S., 13
Kramer K.L., 37
Krieg P.A., 71, 97
Krishnan P., 209
Kurachi Y., 317
Kwitek A.E., 309

L

Laflamme M.A., 519
Lahat H., 333
Lai D., 107
Lassar A.B., 45
Latinkic B., 13
Lavulo L., 107
Lawson N.D., 155
LeCouter J., 217
Le T.H., 451
Lee H.-H., 1
Lehman J.J., 371
Leimeister C., 63
Leinwand L.A., 409
Li H., 249
Li J., 57
Li S., 97
Liao P., 429
Lifton R.P., 445
Lin R., 217
Lindsay E.A., 327
Liu Z.-P., 97
Lively J.C., 143
Lo P.C.H., 1
Lock J.E., 115
Long S., 27
Loughna S., 171
Lyden D., 249

M

Maass A., 409
Mack F., 127
MacRae C.A., 301

Maguire C.T., 317
Maier M., 63
Mäkinen T., 189
Mann D.L., 363
Manseau E.J., 227
Mansfield K., 127
Marbán E., 527
Marks A.R., 533
Maron B.J., 383
Marshall A.C., 115
Martínez S., 89
McCarty J.H., 143
McDonough B., 383
McNally E.M., 389
Mineo C., 459
Mohun T., 13
Moldovan L., 209
Moldovan N.I., 209
Morley G.E., 353
Morita H., 383
Moulton K.S., 471
Mukhopadhyay D., 275
Murry C.E., 519
Musarò A., 507

N

Nabel E.G., 163
Naga Prasad S.V., 439
Nagy J.A., 227
Nienaber J., 439
Nuss H.B., 527

O

Olive M., 163
Olson E.N., 97, 121

P

Pan Y., 127
Pashmforoush M., 399
Pasqualini R., 223
Passier R., 97
Paul A., 507
Petrich B.G., 429
Prall O., 107
Pras E., 333

Q

Qu X., 163

R

Rafii S., 249
Ramirez-Bergeron D., 127
Rebagliati M., 27

551

Reinecke H., 519
Rentschler S., 353
Richardson C.D., 171
Rockman H.A., 439
Roman R.J., 309
Romero E., 249
Rose D.W., 81
Rosenberg P., 339
Rosenfeld M.G., 81
Rosenthal N., 507
Rudge J.S., 267
Rupnick M.A., 417
Ruzinova M., 249

S

San H., 163
Sato T.N., 171
Schindeler A., 107
Schinke M., 317
Schumacher N., 63
Schmeisser A., 209
Sendtner M., 63
Seidman C.E., 115, 317, 383, 543

Seidman J.G., 115, 317, 383, 543
Shaul P.W., 459
Shin C.H., 97
Simon M.C., 127
Small E., 97
Small E.M., 71
Solloway M., 107
Srivastava D., 121
Stainier D.Y.R., 49
Stauffer B.L., 409
Steidl C., 63
Stennard F., 107
Strasser R.H., 209
Stuckmann I., 45
Sutherland L.B., 97
Sundberg C., 227

T

Tanaka M., 317
Taverna D., 143
Thurston G., 267
Tonellato P.J., 309
Towers N., 13

U

Upalakalin J.N., 181

V

Vasile E., 227
Visconti R.P., 171
Vitelli F., 327

W

Wakimoto H., 317
Wang D., 97
Wang Y., 429
Wang Z., 97
Wehrens X.H.T., 533
Weinstein B.M., 155
Wen H., 345
Wheeler M.T., 389
Whitney M.L., 519
Wiegand S.J., 267
Williams R.S., 339
Wilson F.H., 445
Wilund K.R., 499

Winitsky S., 345
Winkler C., 63
Winn N., 507
Wynshaw-Boris A., 81

X

Xiao Q., 143
Xu X., 1

Y

Yamasaki N., 317
Yancopoulos G.D., 267
Yelon D., 19
Yeoh T., 107
Yost H.J., 37
Yoshimoto T., 163
Yu L., 499

Z

Zaffran S., 1
Zeng H., 275

Subject Index

A

α-Linolenic fatty acid, 493
α1 and α2 isoforms, 257–258
α1β1 and α2β1, 144
α5β1 and fibronectin, 143–144
ABCG5 and ABCG genes, 501–503
Acardiac structure and function, disorders of, 548–550
ACE (angiotensin-converting enzyme), 451–453
    ACE inhibitors, 417–418
    ACE2 (ACEH), 453
    angiotensin I to angiotensin II conversion, 451–453
ACTH (adrenocorticotropic hormone), 446
Acute coronary artery ligation, 366f
Ad-VEGF-A
    angiogenic response induced by, 232f
    arteriogenesis induced by, 232f
Adaptive cardiac hypertrophy, 440
Adenosine deaminase deficiency, 527
Adenoviral vectors, 227–228
Adhesion and migration of cancer cells, 239
Adipocytes, 373
Adipose tissue
    growth and regression, 419–420
    plasticity, 420–422
Adult vasculogenesis, 134
Alcoholism, 309
ALS (amyotrophic lateral sclerosis), 514
αMHC gene, 74
ANF (atrial natriuretic factor)
    ANF mRNA, 74
    expression, atrial restriction of, 75–77
    promoter, transgenic analysis of, 71–79
Ang-1/TIE1 null mutant, 176f
Ang1 and Ang2, 421–423
Angiogenesis
    antiangiogenic therapies, 150–151, 293
    and basement membranes (BMs), 255–257
    EG-VEGF regulation of, 217–221
    enhancement of, 134
    induced by VEGF-A, 227–237
    inhibitors, 295
    inhibitors and body weight, 420f
    negative regulation by αv integrins, 148–150
    pathological, 146
    plaque angiogenesis, 471–482
        arterial growth and blocking of, 479
        vs. arteriogenesis, 471–473
        blocking to stabilize lesions, 478
        cardiovascular application of research, 471
        cell proliferation in atherosclerotic lesions, 475
        in clinical use, 479–480
        function to sustain growth, 474
        inhibitor treatments, 478
        intimal capillaries, 475
        neovascularization and plaque instability, 476–477
        pathologic description of, 473–474
        plaque growth promotion, 475–476
        potential regulators of, 477–478
        stimulators vs. inhibitors, 478–479
        in vascular disease, 473
    regulation by integrins, 285–286
    regulation by VEGF and angiopoietins, 177f
    response to Ad-VEGF-A, 228–231
    roles of integrins and ligands in, 143–153
    sprouting (zebrafish), 304–305
    and survival of endothelial cells, 181
    therapeutic, 133–134
    in tumors, 271
        antiangiogenic therapy, 244–246
        host-tumor interactions, 240–242
        intravital microscopy, 239–240
        lymphangiogenesis and, 239–248
        lymphatic role in metastasis, 243–244
        VEGFs and angiopoietins, 249–254
    tunneling and, 209–215
    Type IV collagen and, 255–266
Angiogenic cytokines
    adenoviral vectors and, 227–228
    tumor secretion of, 275
Angiogenic factors, selectivity of, 175–180
Angiogenic therapies, 128
Angiopoietins
    angiopoietin-1 (Ang-1), 133–134, 268–269
    angiopoietin-2, 269–270
    angiopoietin/Tie system, 421
    in cardiovascular remodeling, 422–423
    in vasculature development, 544–546
    in vascular remodeling, 421f
Angiopoietins and VEGFs
    effects on blood vessels, 171–180
        Angiopoietin-1 as venous-specific regulator, 173
        Angiopoietin-1, expression of, 171–172
        arterial and venous selectivity, 173–175
        selectivity of angiogenic factors, 175–180
    in vascular formation, 267–273
Angiostatin, 148
Angiotensin II and receptors, 446, 451, 453–455
Angiotensinogen, 451–452
Anomalous pulmonary venous return, 115
Anterior heart fields, 91
Anti-epo receptor antibody, 47
Anti-melanoma activity, 263–264
Anti-VEGF therapy, 246
Antiangiogenic agents, 293–300
Antiangiogenic therapies, 150–151, 244–246
Antiarrhythmic drugs, 527
Aortic arch
    abnormalities, 327
    arteries, 57
Aortic coarctation, 68
Aortic stenosis, 115
Apoptosis, 181, 184
Arnt embryonic stem (ES) cells, 128–130
Arresten, human (26 kD), 262
Arrhythmias, cardiac. See also Cardiac arrhythmias
    cardiac conduction system (CCS), 353
    induced with electrical stimulation, 320t
    leaky RyR2 and triggered, 537
    polymorphic ventricular tachycardia, 333
    ventricular, 536

Page numbers followed by f indicate a figure and by t indicate a table.

Arteries and veins
  arterial and venous vascular endothelia, 155–156
  arterial growth, 479
  arterial-venous networks, assembly of, 159–160
  arterial vs. venous network development, 176–177
  arterio-venous malformations, 231–232, 234f
  and lymphatics comparison, 545–546
  notch and VEGF, 155–162
Arteriogenesis
  induced by VEGF-A, 232, 236
  vs. plaque angiogenesis, 471–473
Arthritis, 479
Aryl hydrocarbon receptor nuclear translocator (ARNT), 545
Asymmetrical expression of Angiopoietin-1, 171–172
Asymmetries/monocilia in *Xenopus* embryos, 39–40
Asymmetry, left-right in zebrafish, 27–36
AT_{1A} and AT_{1B} receptors, 454
Atherogenic lipoproteins, 503
Atherosclerosis, 459
  atherosclerotic lesions
    blocking plaque angiogenesis to stabilize, 478
    cell proliferation in, 475
    ecSOD and uric acid in, 483–490
    as inflammatory disease, 475
    MC/MPH and plaque formation, 212–213
    myocardial infarction and, 519
    and single-gene mutations, 546
ATP production, 371–372
ATPµ-Raf, 287–288
Atrial and ventricular septum defects, 69
Atrial and ventricular tachyarrhythmias in Nkx2.5 mice, 319
Atrial fibrillation (AF), 320f, 527–528
Atrial natriuretic factor. *See* ANF (atrial natriuretic factor)
Atrial restriction of ANF expression, 75–77
Atrial septal defects (ASDs)
  caused by Csx/Nkx2.5, 317–325
  in zebrafish, 49
Atrioventricular (AV) boundary formation, 51f, 52f
Atrioventricular canal, 115
Autosomal recessive hypercholesterolemia (ARH), 499–500
AV canal defects, 49

AV conduction abnormalities, 318–319, 323
AV (His) bundle, 353
αv integrins, 144–147
αvβ3 integrins
  αvβ3-NP/Raf(-), 289
  and αvβ5, 286–287
  effects on cell survival, 150f
Avian CCS, 354

B

β-Adrenergic blockers, 417–418
β-Adrenergic receptor (BAR) function, 439, 441–442
b3 tubulin, 72–73
βAR blockade therapy for heart failure, 537–538
Bartter's syndrome, 447
Basement membranes (BMs)
  and angiogenesis, 255–257
  described, 143
  early studies of, 258–260
BAY 12-9566 angiogenesis inhibitor, 295–297
BB-94 angiogenesis inhibitor, 295–297
BECs (blood vascular endothelial cells), 193–195
Bedouin families, Israeli, 333–337
BHLH-PAS proteins, 127
BHLH proteins, 249–250
Bilateral cardiac fields, 3
Binding and multimerization motifs (angiopoietins), 268
Biological pacemaker, 528–530
Blood pressure, salt and, 445–450, 546
Blood vessels
  development of, 190f
  effects of angiopoietins and VEGF on, 171–180
  molecular diversity of, 223–225
Bmp4 gene mutations, 50
Bone marrow transplantation, 252f
Bone morphogenic proteins (BMP), 32
Bradycardia, 527
Brain and cardiac L-R asymmetry, 27–36
Brain edema, 269
Brains, mutant, 146
Breast cancer, 242

C

Ca^{++}. *See* Calcium
Ca^{++}/calmodulin-dependent pathways, 411
CACGTG binding site, 65
CAEC (circulating apoptotic endothelial cells), 418–419
Caenorhabditis elegans, 255

Calcineurin, 339–342
Calcium
  calcium (Ca^{++}) release channel, 533–537
  calcium-dependent gene regulation, 339–344
  channel blockers, 527
  chelation, 465
  cycling defects in DCM, 403
  imbalance and signaling, 549–550
Calsequestrin 2 (CASQ2) protein, 335–336
CaMK, 341–342, 411
CAMP, 534–535
Canstatin, human (24 kD), 262–263
Capillaries
  intimal, 475
  vs. larger vessels, 179
  morphology, 174–175
  plaque, 471
Carcinogenesis, 294
Cardiac actin promoters, 15, 73
Cardiac actin.GFP transgene, 15
Cardiac and adipose tissue remodeling, 422–423
Cardiac and central nervous system asymmetries (zebrafish), 33–34
Cardiac arrhythmias
  cardiac conduction system (CCS) and, 353
  caused by Csx/Nkx2.5, 317–325
  gene therapy for, 527–531
    atrial fibrillation (AF), 527–528
    biological pacemaker, 528–530
    polymorphic ventricular tachycardia (PVT), 333
Cardiac asymmetry, left/right, 90–91
Cardiac β-myosin, 351
Cardiac conduction abnormalities, 318–319
Cardiac conduction system (CCS), 353–361
  evolutionary diversity, 353–354
  functional development of, 353
  lineage tracing, 354
  mammalian, 354–356
  models of CCS formation, 354
  murine models, 354–358
Cardiac development
  mammalian, 547–548
  Pitx2 and, 89–95
Cardiac energy production pathways, 371–374
Cardiac growth, regulation by SRF, 97–105
Cardiac hypertrophic growth, 374–376
Cardiac hypertrophy, 309, 345, 363, 374–376, 440–441

SUBJECT INDEX

Cardiac induction, wingless signals in, 9–11
Cardiac left-right development
  asymmetries and monocilia in *Xenopus* embryos, 39–40
  conservation of early steps, 37–43
  early development in vertebrates, 42–43
  fertilization and L-R development, 40–41
  gap junctions, 41
  H+/K+-ATPase in ectoderm cells, 41–42
  L-R pathways among vertebrates, 37–38
  monociliated node cells in vertebrates, 38–39
  nodal signaling in, 27–36
  syndecans in ectoderm, 40
Cardiac metabolic maturation program, 372–373
Cardiac morphogenesis, 121f
Cardiac muscle differentiation in *Xenopus laevis* embryos, 13–18
  cardiac actin promoter, 15
  cardiac-specific transgenic lines, 14–15
  induction of cardiac tissue, 16–18
  myosin light-chain 2a promoter, 15–16
  transgenesis in amphibians, 13–14
Cardiac myocyte proliferation, 45–48
Cardiac myosin RLCp, 347–348
Cardiac neural crest, 57, 82–84
Cardiac outflow tract development, 86
Cardiac pathophysiological conditions, 364t
Cardiac patterning in zebrafish, 19–25
Cardiac physiological phenotyping, 400t
Cardiac progenitor cells, 523–525
Cardiac progenitors in *Drosophila*, 1–12
Cardiac remodeling
  and cardiovascular disease progression, 417–419
  JNK-mediated, 433–434
Cardiac sarcomere model, 384f
Cardiac-specific transgenic lines, 14–15
Cardiac stretch sensor machinery, 400–401
Cardiac tissue
  induction of in animal pole explants, 17
  induction of in embryonic tissue, 16–18
Cardiac torsion, 345–352

Cardiac transcription factor pathways, 110
Cardiac valves, 49
cardiofunk mutants, 52–53
Cardiogenesis in *Drosophila* model, 1–12
  bilateral cardiac fields, 3
  dorsal vessel morphology/patterning, 1–3
  Dpp and wingless signals, 9–11
  Dpp signals, transmitting to *tinman* gene, 7–9
  HMG-D association with Tinman protein, 7
  methods, 4
  protein/DNA and protein/protein complexes, 7–9
  results, 4–7
  tinman induction by Dpp signals, 3–4
Cardiomyocytes
  contractile force of, 339
  grafting, 519–520
  transplantation, 519
Cardiomyopathy
  as consequence of skeletal muscle disease, 392
  defined, 389
  and diastolic dysfunction, 430–432
  dilated (DCM), 399–408
    calcium cycling defects in, 403
    cardiac stretch sensor machinery, 400–401
    cytoskeletal defects and, 399–400
    LIM domain proteins, 401–402
    MLP–telethonin–titin complex, 400–401
    mouse model for, 399
    phospholamban ablation, 403
    PLN inhibition, 405
    W4RMLP mutation, 402–403
  lipotoxicity in development of, 378–379
  restrictive, 432–433
Cardiotrophin-1 (CT-1), 365–367
Cardiovascular compendium, zebrafish, 301–307
Cardiovascular development
  Hey bHLH factors in, 63–70
  hypoxia inducible factors (HIFs) and, 127–132
  Pitx genes during, 81–87
Cardiovascular disease
  angiogenesis research and, 471
  cell cycle signaling, 163–170
  costs of, 543
  demographics of, 543
  genes and pathways underlying, 309–315
  history of research, 543–544

  renin-angiotensin system, 451–457
  uric acid and, 487
Cardiovascular science, 543–550
  cardiac development, 547–548
  clinical challenges, 543
  disorders of cardiac structure and function, 548–550
  research response, 543–544
  vascular beds development, 544–546
  vascular disease, local/systemic, 546–547
Cardiovascular system
  protease-activated receptors in, 197–208
  zebrafish, assembly and growth, 301–302
CASQ2, 333–337
Catecholamine-induced PVT, 333–337
  calsequestrin 2 (CASQ2) protein, 335
  genetic analysis of Bedouins, 335
  human CASQ2 model, 335–336
  recessive form in Bedouins, 333–334
  treatment, 334–335
Catecholaminergic polymorphic ventricular tachycardia (CPVT), 538
Caveolae, 459–466
Cdc45 gene, 327–328
CDK inhibitors, 546
CDKs, cyclins, and CKIs, 163–164
cDNA microarrays, 314
Cell cycle signaling and cardiovascular disease, 163–170
  cell cycle pathways, 163f
  CIP/CIK structure and function, 164–165
  CKI expression in vascular disease, 165
  cyclins, CDKs, and CKIs, 163–164
  Kip1 loci, deletion of, 165–167
  p27Kip1 protein, 168
Cell growth/muscle differentiation by SRF, 97–98
Cell-targeting, endothelium-derived, 224
Cellular fatty acid oxidation (FAO) pathway, 373f
Cellular lesion development, 165–167
Cellular mitochondrial energy transduction, 371–374
Cellular therapies for myocardial infarction, 519–526
Central nervous system and cardiac asymmetries (zebrafish), 33–34
Cerebral hemorrhage in mice, 145f

SUBJECT INDEX

Chambers
    chamber-specific myocardial gene expression, 71–79
    form and pattern, 302–303
    formation, vertebrate, 73–75
    myocardium regulators, 111
Chemotherapy, metronomic, 293–300
CHF1/2 gene, 63
Cholesterol. *See also* Hypercholesterolemia, 445, 491
Chromatin
    immunoprecipitation assays, 82
    structure, 123
Ciliary neurotrophic factor (CNTF), 365–367
CIP/CIK structure and function, 164–165
Cip1 gene, 165–166
cis-binding sites, 15–16
CKIs (cyclin-dependent kinase inhibitors)
    cyclins and CDKs, 163–164
    expression in vascular disease, 165–166
    to treat vascular diseases, 167f
Classical angiogenesis, 134
Classical heart fields, 91–92
CMEC (cardiac microvascular endothelial cells), 418–419
Coagulation
    factors and targets, 204f
    and inflammation links, 203f
    protease thrombin, 197
Coarctation of aorta, 115
Collateral vessels, 472
Collectrin, 453
Common atrioventricular canal (CAVC), 91, 93
Concentric contraction, 351
Conduction
    abnormalities, cardiac, 318–319
    conductive cardiomyocytes, 353
    system, cardiac (CCS). *See* Cardiac conduction system (CCS)
Congenital heart disease (CHD)
    AV canal defects and, 49
    Left/right signaling, 89–95
    mouse models of, 57–62, 317–325
    Nkx2-5, 111–112
    Pitx2, 93–94
Congenital heart malformations, 118f, 119
Congenital RyR2 mutations, 538–539
Congestive heart failure, 533
Connexin 45-deficient embryos, 50
Connexin RNA, 41
Consomic rats
    cDNA microarrays to identify genes in, 314
    for identifying disease-causing genes, 311–313

    physiological phenotyping in, 313–314
Coronary artery
    bypass procedures, 545
    collaterals, 472f
    vasospasm, 392–393
    vasospasm from Microfil, 394f
CPEC (circulating precursor endothelial cells), 211–212
Craniofacial defects, 330
Cre-lox
    approaches, 58
    Cre-loxP-mediated gene-switch system, 430–431
    Cre/LoxP technology, 493
Crkol gene, 327, 329
Cross-bridge cycle kinetics, 347, 349–351
CRP3. *See* MLP-deficient mice and DCM
Csx/Nkx2.5 cardiac transcription factor, 317–325
CTX (cyclophosphamide), 297–299
Cu/ZnSOD, 483–488
Culprit lesions, 476
Cx40 gene, 74
Cx43 gap junction protein, 434
Cyclins, CDKs, and CKIs, 163–164
Cyclophosphamide (CTX), 297–299
cyclops gene, 29–32
Cytokines
    effects in heart
        adaptive, 363–367
        maladaptive, 367–368
        primitive, 364
        proinflammatory, 363–368
Cytoprotection, 363
Cytoskeletal Z disc proteins, 399–408
Cytotoxic and antiangiogenic dosing, 297t
Cytotoxic drugs, 293

D

Dahl salt-sensitive (SS) rats, 309, 312–314
Daughter capillaries and glomeruloid bodies, 231
Defibrillators, 527
del22q11-homologous region map, 329f
Delayed after depolarizations (DADs), 536–537
Delta ligands, 66
deltaC Notch signaling gene, 157
Depression, 309
Df5 chromosomal deficiency, 328f
DGC (dystrophin glycoprotein complex)
    and associated proteins, 389–390
    and sarcoglycan as mechanosignaling complex, 390–392

dHand protein, 122–123, 124f
dHey gene, 63
Diabetes mellitus, 376–378
    diabetic heart, 371, 376–379
    diabetic retinopathy, 269
Diastolic dysfunction and cardiomyopathy, 430–432
DiGeorge syndrome (DGS)
    AV canal defects and, 49
    phenotype
        genetic dissection of, 327–332
        nonoverlapping deletions in human syndrome, 327–329
        pharyngeal apparatus and *Tbx1*, 330
    Tbx1 functions, 329–330
    Tbx1 and, 60–61
Digoxin, 417
Dimerizer treatment, 521–522
Disease-causing genes, identification of, 311–313
Disease models in zebrafish, 304–305
Distal enhancer (DE) of cardiac actin promoter, 16
Dmef2, 72–73
DNA
    -binding transcription factors, ventricular-specific, 122–123
    high-throughput sequencing, 384
Dorsal vessel
    Drosophila model, 71–73
    morphology and patterning of, 1–3
Double-inlet left ventricle (DILV), 91, 94
Double-outlet right ventricle (DORV), 59–60, 91, 93
Down syndrome, 49
Dpp signals
    transmitting to *tinman* gene, 7–9
    and wingless signals, 9–11
Drosophila
    dorsal vessel model, 71–73
    Enhancer-of-split proteins, 65
    history of research, 547–548
    model, cardiogenesis in, 1–12
DSIF (DRB sensitivity inducing factor) complex, 23
Duchenne muscular dystrophy (DMD), 514
Dystrophin. *See also* DGC (dystrophin glycoprotein complex)
    sarcoglycan complex and, 389–390
    skeletal muscles and, 514

E

E-box motif, 65t
ε-Sarcoglycan gene, 390

Early after depolarizations (EADs), 537
EC coupling, 536f
Eccentric contraction, 351
ecSOD. *See* Extracellular superoxide dismutases (ecSOD)
EG-VEGF regulation of endocrine angiogenesis, 217–221
  EG-VEGF as tissue-specific factor, 218–219
  organ-specific regulation of endothelial function, 217–218
  role in pathological conditions, 219
  signaling in epithelial cells, 219
EGDR and EGLT chimeric receptors, 276f
eHand protein, 122–123
Ellis-van Creveld syndrome, 49
Eluting stents, 167f
Embryogenesis
  and lymphatic vasculature, 189–190
  and organogenesis, 127
  Pitx1/Pitx2 genes during, 82
Embryonic development
  of mammalian heart, 107, 109f
  thrombin signaling in, 203–205
Embryonic stem cells (ES), 241
Embryonic tissue, induction of cardiac tissue in, 16–18
ENaC, 445–447, 449
Endocardial and epicardial fibers, 349
Endocardial cushion formation in zebrafish, 49–56
  development time line, 51
  endocardial cushion mutants, 53f
  mutations affecting development, 51–54
Endocardial layer, 111
Endocrine glands, EG-VEGF regulation of angiogenesis in, 217–221
Endogenous Syndecan-2, 40f
Endostatin, 148
Endothelial cells
  antiangiogenic therapy and, 245
  arterial and venous, 155–156
  endothelial activation, PARS in, 202–203
  endothelial dysfunction, 309
  lymphatic markers, 193t
  migration and proliferation of, 240
  organ-specific regulation of function, 217–218
  regulation of, 285
Endothelial nitric oxide (NO) production, 459–469
  eNOS signaling module in caveolae, 464–466

eNOS trafficking and localization in caveolae, 459–460
HDL-mediated activation of eNOS, 466–467
LDL, HDL, and eNOS, 460–464
Endothelin (ET) receptor as *Pitx1/Pitx2* target, 84–86
Endothelium-derived human cells, targeting, 224
Endurance exercise, 341
Engineered transposons, 14
Enhancer-of-split genes in *Drosophila*, 63
eNOS (endothelial isoform of nitric oxide synthase)
  HDL-mediated activation of eNOS, 466–467
  LDL, HDL, and, 460–464
  signaling module in caveolae, 464–466
  trafficking and localization in caveolae, 459–460
EphrinB2 and EphB4 genes, 155–158
Epicardium, erythropoietin and retinoic signaling in, 45–48
Epidermal growth factors (EGFs), 267, 357
Epigenetic factors in ventricular development, 123–124
Epithelial to mesenchymal transformation (EMT), 49–50
ERKs (extracellular signal-regulated protein kinases), 429, 432, 440
Erythropoietin (EPO)/retinoic acid (RA) signaling, 45–48
eve gene, 9–10
Excitation-contraction (EC) coupling, 533–541
  βAR blockade therapy for heart failure, 537–538
  congenital RyR2 mutations, 538–539
  FKBP12.6 for normal RyR function, 534
  heart failure, 535–537
  leaky RyR2 and triggered arrhythmias, 537
  molecular basis of, 533–534
  PKA hyperphosphorylation of RyR2, 538
  PKA phosphorylation regulation of ryanodine receptors, 534–535
  ryanodine receptors (RyR2), 534
  RyR2 macromolecular complex, 534
Exercise, endurance, 341
Extra-cardiac lineages and secondary heart field, 108
Extracellular matrix (ECM)
  antiangiogenic fragments of, 149

basement membranes and, 256
protein fragments, 148
structural modifications of, 209
transition during angiogenesis, 261t
Extracellular superoxide dismutases (ecSOD), 483–490
  and atherosclerosis, 484
  forms of SOD, 483–484
  peroxidase activity of, 483–485
  uric acid and modulating activity of, 485–487
  uric acid role in atherosclerosis, 487

F

Familial defective apolipoprotein B-100 (FDB), 499
Familial hypercholesterolemia (FH), 499
Familial hypertrophic cardiomyopathy (FHC), 409–415
  GATA4 transcription factor, 412
  GSK3 (glycogen synthase kinase-3β), 412
  MEF2 transcription factor, 412
  myocardium, gender effects in, 412–413
  pathogenesis of, 410
  signaling pathways in, 410–412
  transgenic models of, 409–410
Familial polymorphic ventricular tachycardia (FPVT), 538
Fatty acid oxidation (FAO) pathway, 371–373
Fatty acids, 491–494
Fertilization and L-R development, 40–41
Fetal-to-adult energy metabolic switch, 371–374
FHH rats, 309
Fibroblast growth factors (FGFs)
  FGF signaling, 521
  inducing angiogenesis, 255
  receptor tyrosine kinase family, 267
Fibronectin, 143–144, 256–258
FK-506 binding protein (FKBP12.6), 534
FKBP12.6 for normal RyR function, 534–539
Fli.GFP reporter, 15
Flight or fight response, 535
Flosequinan, 417
Flt-1 and KDR signaling pathways, 275–277, 279–281
Flt4 marker, 157
foggy and *pandora* mutations, 22–23
Framingham Heart Study, 440
Frt or Cre recombinases, 14
Functional genomics, 309–311

G

G-Protein-coupled receptor (GPCR) function, 439–444
  β-adrenergic receptor (BAR) function, 439
    signaling, 441–442
    myocardial hypertrophy, 440–441
    phosphoinositide 3-kinases (P13K), 439–440
G proteins in VPF/VEGF signaling, 275–283
  heterotrimeric G proteins, 277–278
  KDR and Flt-1 signaling pathways, 275–277
  PLC-β3 impairment, 278–279
  PLC involvement, 277
GAL4/UAS binary expression system, 54f
Gap junctions, 41, 434
GATA- and Nkx2-5-binding sites, 75–77
GATA transcription factors, 16–17, 412
GATA4-GR fusion protein, 17f
Gβγ minigene, 279–280
Gene expression profiling, 194–195
Gene knockout technique in mice, 310
Gene regulatory pathways, 372–373
Gene-switch system, Cre-loxP-mediated, 430
Gene-targeting, 451–455
Gene therapy for arrhythmias, 527–531
Genentech anti-VEGF antibody, 246
Genes
  expressed differentially, 343t
  regulated by SREBPs, 492f
Genetic dissection of DiGeorge syndrome (DGS) phenotype, 327–332
Genetic manipulations (*Pitx* genes), 81
Genetic screening pyramid, 301f
Genetic (transgenic) rescue, 312
Genomics, 246, 309–311
GFP reporter, 14
Glomerulogenesis, 305
Glomeruloid bodies (GB), 231, 233f
Glomerulosclerosis, 309
Glucagon, 496
Glycogen synthase kinase-3. *See* GSK3 (glycogen synthase kinase-3β)
Gq family proteins, 277–279, 281
GRA (glucocorticoid-remediable aldosteronism), 446
Grafting
  cardiomyocyte, 519–520
  hematopoietic stem cell, 522–523
  skeletal muscle, 520–521
gridlock (*grl*) mutation, zebrafish, 67–68, 124, 157, 301, 304
Groucho proteins, 64
Growth factors
  endocrine-gland derived vascular endothelial. *See* EG-VEGF regulation of endocrine angiogenesis
  fibroblast (FGF). *See* Fibroblast growth factors (FGFs)
  VEGF (vascular endothelial growth factor). *See* VEGF (vascular endothelial growth factor)
GSK3 (glycogen synthase kinase-3β), 412
GST-*Hey1*-binding site, 65

H

H+/K+-ATPase in ectoderm cells, 41–42
H3 and H4 histones, 123
hairy/Enhancer-of-split-related genes, 63–64, 157
hairy-related bHLH factors, 63–64
Hand1 and *Hand2* genes, 20t, 111
hands off (*han*) mutation, 20t, 21f, 22
Haploinsufficiency
  of Csx/Nkx2.5, 317–325
  TBX5, 115–116, 119
has (*heart and soul*) mutant embryos, 50, 302–303
HβARK1(495), 279, 281
HCM (hypertrophic cardiomyopathy)
  HCM-causing mutations, 383–387
  phenotype, 345–346
hdf (*heart defect*) mutation, 50
HDL (high-density lipoprotein). *See* High-density lipoprotein (HDL)
Heart
  congenital defects of, 27
  growth
    early, 302f
    zebrafish, 303–304
  morphogenesis overview, 107–108
  and placenta development, 129
  repair of, 519
  transplantation, 523–524
Heart development
  control mechanisms of, 1
  and disease
    heart building blocks, 108–110
    Nkx2-5 in, 107–114
Heart failure
  defective EC coupling and, 535–538
  evolution of therapy, 423f
  G-protein-coupled receptor (GPCR) function, 439–444
    β-adrenergic receptor (BAR) function, 439
    β-adrenergic receptor (BAR) signaling, 441–442
    myocardial hypertrophy, 440–441
    phosphoinositide 3-kinases (P13K), 439–440
  MAP kinases in, 429–437
  progression of, 399–408
  vascular endothelium in tissue remodeling, 417–427
Heart fields
  anterior or secondary, 91
  classical, 91–92
  posterior, 91–92
  zebrafish, 19f, 20–22
heg (*heart of glass*) mutants, 303
Hemangiomas, 479
Hematopoiesis and HIF, 129
Hematopoietic precursors, 134
Hematopoietic stem cell grafting, 522–523
Hematopoietic therapies, 128
Hemodynamic overload, 366f
Hemophilia, 527
Heparan sulfate proteoglycans (perlecans), 256–258
her2/neu, 242, 246
Herceptin treatment, 242, 246
Hes family of helix-loop-helix(bHLH) proteins, 63
HES (hairy/enhancer of split) proteins, 304
Hesr1/2 gene, 63
Heterotaxia, 92
Heterotrimeric G proteins, 277–278
Heterozygous mutations of Nkx2.5, 322
Hey bHLH factors in cardiovascular development, 63–70
  DNA binding of Hey proteins, 65
  expression patterns in mice, 65–66
  hairy-related bHLH factors, 63–64
  Hey genes, 63–64
  Hey proteins, 64–65
  Hey1 gene, 63
  Hey1 loss in mice, 69
  Hey2 loss in mice, 68–69
  mutual compensation in mammals, 69
  Notch signals, 66–67
  zebrafish expression of *Hey* genes, 66
  zebrafish *grl* (*Hey2*) to build aorta, 67–68
HIFs (hypoxia inducible factors). *See also* Hypoxia

HIF-1 target genes, 135f
HIF-1α
and angiogenesis, 137–139
in disease, 137
function in quiescent hypervascularity, 133–142
HIFα
and HIFβ subunits, 129–130
molecules, regulation of, 136–137
High-density lipoprotein (HDL), 459–469
HDL, LDL, and eNOS, 460–464
HDL-mediated activation of eNOS, 466–467
High-throughput DNA sequencing, 384
Histones, 123
HMG-D association with Tinman protein, 1, 7–8
Holt-Oram syndrome (HOS), 115–116
Homeodomain factor Nkx2-5. *See* Nkx2-5
HOP (homeodomain-only protein)
abnormalities in HOP mutant mice, 102
antagonism of SRF activity by, 102–103
antithetical phenotypes in HOP mutant mice, 103
feedback loop between Nkx2.5 and, 103–104
inhibition of SRF by, 101–102
SRF gene expression in HOP mutant hearts, 102
Hormone-dependent human tumors, 245–246
Host-tumor interactions, 240–242, 246
Hrt1/2/3 gene, 63
Hrt1, 2, and 3 transcription factors, 123–124
HRT2 gene, 157
hsp70 gene, 23, 24f
Human arresten (26 kD). *See* Arresten, human (26 kD)
Human canstatin (24 kD). *See* Canstatin, human (24 kD)
Human CASQ2 homology model, 336f
Hypercalciuria, 447
Hypercholesterolemia, 499–505
ABCG5 and ABCG8, 502–503
ARH, cause of, 499–500
ARH requirement for LDLR internalization, 500
autosomal recessive hypercholesterolemia (ARH), 499
sitosterolemia, 500–502
Hyperlipidemia, 309

Hypertension
inherited forms of, 448
potassium channel openers and, 395
renin-angiotensin system and, 451–457
risk factors, 445
and single-gene mutations, 546
Hypertrophic cardiomyopathy (HCM), 383–388
Hypertrophy
familial hypertrophic cardiomyopathy (FHC), 409
myocardial, 440–441
myocyte, 339–344
PPARα signaling, 374–376, 379–381
pressure overload hypertrophy, 412
signaling pathways in, 410
stress-induced cardiac, 429
Hypervascularity, quiescent. *See* Quiescent hypervascularity
Hypoplasia, ventricular, 121–125
Hypoplastic left heart syndrome, 115
Hypoxia
angiogenesis and HIF, 134–136
HIFs and, 127–132
hematopoiesis and HIF, 129
HIF and oxygen homeostasis in embryos, 128–129
HIFα and HIFβ subunits, 129–130
hypoxia-mediated signal transduction, 127–128
placenta and heart, development of, 129
von Hippel-Lindau (VHL) disease, 130–131
hypoxia inducible factors (HIFs), 47, 545
tissue, 139
in tumors, 246
Hypoxic heart, 374–376, 379–381

I

Id proteins
biology of, 249–250
for mobilization of endothelial precursors, 252–253
role in angiogenesis within tumors, 250–252
targeting, 253–254
IGF-1 (insulin-like growth factor) isoforms, 507–518
complexities of IGF-1 transcription, 507
diversity in skeletal muscle, 511, 512f, 513f
effects on tissue homeostasis, 510t
functions of IGF-1 in vivo, 509–510

in muscle disease, 514, 515f
and muscle regeneration, 514–516
shifts in aging skeletal muscle, 511–514
theories of IGF-1 action, 507–509
transgene action in skeletal muscle, 510–511
Implantable devices, 527
In-stent restenosis, 167
Inflammation, 197
Insulin. *See also* IGF-1 (insulin-like growth factor) isoforms
lipogenic actions in liver, 496
resistance, 309
Integrin αvβ3, 286–287
Integrins
regulation of angiogenesis by, 285–286
and their ligands in angiogenesis, 143–153
α1β1 and α2β1, 144
α5β1 and fibronectin, 143–144
angiogenesis assays in mice, 146f
antiangiogenic therapies, 150–151
αv integrins, 144–147
extracellular matrix protein fragments, 148
negative regulation by αv integrins, 148–150
thrombospondins (TSPs), 147–148
Intercellular cooperation, 213
Interleukin-1 (IL-1) and interleukin-6 (IL-6), 363–365
Interstitial diffusion, convection, and binding, 239
Interstitial fibrosis, 519
Interstitial hypertension, 243, 246
Intimal capillaries, 475
Intravital microscopy (IVM), 239–240
Irx4 gene
ANF promoters and, 15
AV boundary formation, 49
role in ventricular differentiation, 74
isl (*island beat*) mutants, 304
Isoforms, SREBP, 492–493, 495–497
Isomerism
atrial and ventricular molecular, 92
syndromes, 49

J

Jagged ligands, 66
jekyll mutants, 52

JNK (c-jun amino-terminal kinases)
  JNK-activated hearts, 434f
  JNK-mediated cardiac remodeling, 433–434
  SAPK (stress-activation protein kinases), 429

K

Kaplan-Meyer survival curves, 299f
KDR and Flt-1 signaling pathways, 275–277, 279–281
Kidneys
  role in regulation of circulatory pressure. See also Salt and blood pressure, 546
  vascularization of (zebrafish), 305
Kip1 gene, 164–167
Kir2 channels, 528–530
Knockout mice, 451
Kynurenine 3-hydroxylase, 314

L

L-R pathways, conservation in vertebrates, 37–38
LacZ
  activities, yeast assays for, 4
  expression, 356–360
  transgenes, 58
ladybird (*lb*) gene, 73
Laminin, 256–258
LDL (low-density lipoprotein)
  HDL and eNOS, 460–464
  hypercholesterolemia and, 499–500
  receptors, 491, 494–495, 546
LDLR (low-density lipoprotein receptor)
  internalization, 500
  LDLR-mediated endocytosis, 499
Leaky channels, 536
Leaky RyR2 and triggered arrhythmias, 537
Leaky vessels, 269
LECs (lymphatic endothelial cells), 193–195
Left–right asymmetries (zebrafish), 27–30
Left–right development, cardiac, 37–43
Left–right dynein expressions, 38–39
Left–right signaling
  and congenital heart disease, 89–95
  defect, 92–93
Left-to-right shunt flow, 318
Left ventricular remodeling, 367–368, 519–520
lefty genes
  expressions, 37–38, 40–41
  lefty-1 and *-2*, 89
  and *pitx2* genes, 29–30

Leucine/isoleucine zippers (LIZ), 534–535
Leukemia inhibitory factor (LIF), 365–367
Liddle's syndrome, 446
Ligands. See Integrins
lik (*liebeskummer*) mutants, 303
LIM domain proteins, 401–402
Linoleic fatty acid, 493
Lipid homeostasis, 491–498
Lipid utilization pathways, 373
Lipotoxicity, 378–379
LIZ (leucine/isoleucine zippers), 534–535
Local vascular disease, 546–547
LXR hormone receptor, 501–502
Lymphangioblasts, 190
Lymphangiogenesis
  induced by VEGF-A, 232, 236
  molecular mechanisms of, 189–196
    lymphatic vasculature development, 189–190
    lymphatic vs. blood vascular endothelial cells, 192–194
    lymphedema, 192
    molecular regulation of, 190–191
    Prox-1 and lymphatic endothelial cells, 194
    tumors and lymphatic metastasis, 191–192
  in tumors, 243–244
Lymphatics
  compared to veins and arteries, 545–546
  endothelial cell markers, 193t
  endothelial phenotype, 545
  lymphatic hyperplasia, 244
  lymphatic vs. blood vascular endothelial cells, 192–194
  phenotypes in mouse mutants, 193t
  role in metastasis, 243–244
  tree development, 544–545
  vessels
    angiopoietin-2 and development of, 270–271
    definition and history, 189–190
    development of, 190f
Lymphedema, 192
LYVE-1
  receptor, 193–194
  tumors stained with, 244

M

M-CPT I gene, 372
Macrophages/monocytes (MC/MPH), migration of, 209–215
  Matrigel plugs, 211
  MC/MPH and CPEC, 211

tunneling
  in clinical setting, 212–213
  morphology of, 209–210
  in vitro, 210–211
  in vivo, 211–212
Macular degeneration, 479
MADS family proteins, 1, 5–6, 16–17, 97
Maladaptive cardiac hypertrophy, 440
Mammalian cardiac development, 547–548
Mammalian CCS, 354–356
MAPK (mitogen-activated protein kinases)
  in heart failure, 429–437
    cardiomyopathy and diastolic dysfunction, 430–432
    genetic manipulation in transgenic animals, 430
    JNK-mediated cardiac remodeling, 433–434
    MAPK fundamentals, 429–430
    Ras-mediated hypertrophy, 430–432
    restrictive cardiomyopathy, 432–433
  signaling cascades, 411–412
Matrigel plugs, 210–211
Matrix degrading enzymes, 209
Matrix metalloproteases. See MMPs (matrix metalloproteases)
mBop factor, 123
MCAD gene, 372
MCIP protein family, 340–341
mdx mouse muscles, 390–391, 514
Mechano growth factor, 509
MEF2 factors, 16, 97, 123, 192, 412
Melanoma
  M21-L implantation in mice, 288
  tumstatin and anti-melanoma activity, 263–264
Membrane-bound transcription factors, 491
Mendelian diseases, 446
Mendelian hypercholesterolemia, 499
Mesenchymal cells, 108
Metastasis
  arresting, 239
  lymphatic, 191–192, 243–244
Metronomic chemotherapy, 293–300, 298f
MHC-PPAR mice and diabetic hearts, 378–379
Microscopy
  intravital (IVM), 239–240
  multiphoton, 241
Microvascular remodeling, 181–187
Microvasculature, survival mechanisms in developing, 182–183
Microvessel architecture, 133–134
Migration signaling, 277–278

mindbomb (*mib*) mutant, 157–158
Missense mutations, 384
Mitochondria
    mitochondrial biogenesis, 341–342
    mitochondrial energy production, 371–382
        pathologic activation of PPARα/PGC-1α regulatory pathway, 376–379
        postnatal induction of pathways, 371–374
        PPARα signaling, derangements in, 374–376
    mitochondrial function, 128
Mitogen-activated protein kinases (MAPK). *See* MAPK (mitogen-activated protein kinases)
Mitral valve prolapse, 115
MLC-2v gene, 74
MLC2a transgene, 14
MLP-deficient mice and DCM, 399–401
MLP family of cytoskeletal Z disc proteins, 399–408
MLP-telethonin-titin complex, 400–401
MMP12, 209–215, 305
MMPIs in last-stage disease, 295–297, 300
MMPs (matrix metalloproteases), 149–150, 209, 211, 367, 477
Molecular pharmacology, 521–522
Monociliated node cells in vertebrates, 38–40
Monocytes/macrophages, migration of. *See* Macrophages/monocytes (MC/MPH), migration of
Morphogenesis
    cardiac, 121f
    heart overview, 107–108
Morphology of dorsal vessel, 2f
Mother lymphatics, enlarged, 235f
Mother vessels
    daughter capillaries and glomeruloid bodies, 231
    formation of, 229–231
Mouse models
    of congenital heart disease, 57–62
    for DCM and heart failure, 399
MRTF-A and -B genes, 99–101
MTD (maximal tolerated doses) therapies, 293
Multi-gene deletion syndromes, 327
Multimerization and binding motifs (angiopoietins), 268
Multiphoton microscopy, 241
Mural cells, 143
Murine cardiac conduction system (CCS), 353–361

Muscles
    degeneration, molecular mechanisms of, 390
    differentiation and cell growth by SRF, 97–98
    diseases and IGF-1 isoforms, 514
    muscle myosin, 72
    regeneration and IGF-1 isoforms, 514–516
    skeletal, 339–340
Muscular dystrophy, 389, 392, 514
Mutagenesis screens, 309
Mutant brains, 146
Mutant Raf kinase, 285–291
Mutations
    in adapter proteins, 499–500
    affecting endocardial cushion formation, 51–54
    in ATP-binding proteins ABCG5 and ABCG8, 501–502
    blood pressure lowering, 447
    blood pressure raising, 446–447
    Bmp4 gene, 50
    disruptive cardiac patterning, 20t
    in dystrophin, 389
    effects on vascularization of embryoid bodies, 144t
    Has2, 50
    HCM-causing, 383–387
    hdf (*heart defect*), 50
    heterozygous, 59
    homozygous, 59
    out of bound mutants, 545
    Vinculin, 50
    zebrafish, 303–306
Myoblasts, skeletal, 520–521
Myocardial differentiation (zebrafish), 22–23
Myocardial disease, 533–541
Myocardial energy substrate utilization, 372f
Myocardial gene expression, 71–79
    atrial restriction of ANF expression, 75–77
    Drosophila dorsal vessel model, 71–73
    vertebrate chamber formation, 73–75
Myocardial homeostasis, 363f
Myocardial hypertrophy, 440–441
Myocardial infarction, 519–526
    cardiac progenitor cells, 523–525
    cardiomyocyte grafting, 519–520
    graft size and molecular pharmacology, 521–522
    hematopoietic stem cell grafting, 522–523
    repair of, 519
    skeletal muscle grafting, 520–521
Myocardial precursors, 19
Myocardial stress response, 363

Myocardin
    definition and fundamentals, 98–99
    frog embryo gene expression by, 100–101
    smooth muscle-specific expression of, 99
    SRF-dependent transcription by, 99–100
Myocardium
    chamber regulators, 111
    gender effects ln, 412–413
Myocyte hypertrophy, 339–344, 405
Myocyte proliferation. *See* Cardiac myocyte proliferation
Myofiber hypertrophy, 339–340
Myosin
    light-chain 2a promoter, 15–16
    regulatory light chain (RLC) phosphorylation, 345–352
        cross-bridge cycle kinetics, 349–351
        HCM phenotype, 345–346
        Ser-15, 346–347
        spatial gradient consequences of, 347–349
        stretch-activation response, 346–347, 348f

N

NC1 domains inhibiting angiogenesis, 259–262
Neovascularization
    monocyte/macrophage migration and, 209–215
    neovascular apoptosis and tumor regression, 287–289
    plaque, 473f
    and plaque instability, 476–477
    VEGF domination over, 220
NEP (neutral endopeptidase), 418
Nephrocalcinosis, 447
Networks controlling mitochondrial energy production, 371–382
Neural crest cells, 59f, 108, 330
Neural crest migration, 57–62
    cardiac neural crest, 57
    Cre-lox approaches to fate-map neural crest, 58
    DiGeorge syndrome and Tbx1, 60–61
    double-outlet right ventricle, 59–60
    neurofibromatosis gene, type 1 (NF1), 59–60
    Pax3 function in Splotch mice, 57–58
    Sema3C deficiency, 58–59
Neural left-right asymmetry, 27–36
Neuregulin molecule, 111, 303

Neurofibromatosis gene, type 1 (*NF1*), 59–60
Neurohormones, 418
Neuronal nitric oxide synthase (nNOS), 393
Neuropilin, 58
Neutral sterol trafficking, 500–501
NFAT transcription factors, 342
Nidogen, 256–258
Nifedipine, 393
Nitric oxide (NO). *See also* Endothelial nitric oxide (NO) production, 393
Nkx2-5
  cardiac transcription factor, 317–325
  feedback loop between HOP and, 103–104
  and GATA-binding sites, 75–77
  in heart development and disease, 107–114
    cardiac transcription factor pathways, 110
    chamber myocardium regulators, 111
    congenital heart disease (CHD) and, 111–112
    genetic dissection in mice, 110–111
    heart building blocks, 108–110
    morphogenesis overview, 107–108
    secondary heart field and extracardiac lineages, 108
  transcription factor gene, 21–22
Nodal expression/regulation in vertebrates, 32–33, 37–38, 40–41
Nodal-Lefty-Pitx2 cassette (zebrafish), 28–29
Nodal-related genes, 28–29, 33f, 89–90
Nodal signaling in zebrafish, 27–36
Notch
  notch5 signaling gene, 157
  pathways, 66–67
  signal transduction, 63
  signaling, 66–67, 156–159, 544
NRG-1 (neuregulin-1), 357–360

O

ob/ob mice, 419–420
Obesity risk factor, 543
ODD (oxygen-dependent degradation domain), 136–138
Oncostatin M (OSM), 365
one-eyed-pinhead mutation (zebrafish), 30
Organ fields, zebrafish, 19
Organ selectivity of vascular system, 177–179
Outflow tract, heart, 327

out of bound mutants, 305, 545
Oxygen homeostasis in embryos, 128–129

P

$p21^{Cip1}$ protein, 165, 169
$p27^{Kip1}$ protein, 164–165, 168
p38 kinases, 429, 432–433
Pacemakers
  biological, 528–530
  fundamentals, 527–528
  site and activity, 353
Pancreatic islet carcinomas, 293–300
pandora and *foggy* mutations, 22–23
pannier (*pnr*) gene, 73
PAR1 activation
  of intracellular signaling pathways, 200f
  irreversibility of, 199
PARs (protease-activated receptors)
  in cardiovascular system, 197–208
  cellular responses triggered by thrombin, 198–199
  definition and basics, 545
  in endothelial activation, 202–203
  family of, 199–201
  platelet activation and, 201–202
  thrombin action on cells, 197–198
  thrombin signaling in embryonic development, 203–205
Patent ductus arteriosis, 115, 118
Pathogenesis of FHC, 410
Pathological angiogenesis, 146
Pathological conditions, EG-VEGF role in, 219
Pathophysiological conditions, cardiac, 364t
Pax3 function in Splotch mice, 57–58
Peptide library, 223–224
Peptides. *See also* PARs (protease-activated receptors), 144, 147, 223–224
Pericytes, 143, 183, 545
Peroxidase activity of ecSOD, 483–485
PGC-1 expression, 341f
PGC-1α gene, 374
Phage display libraries, 223
Pharmacologic rescue experiments, 312
Pharyngeal apparatus, 330
Pharyngeal arch arteries, 327, 330
Phenotypic variability in DGS, 329–330
Phenotyping, physiological, 313–314
Phosphatases 1 and 2A (PP1/PP2A), 534
Phosphoinositide 3-kinases (P13K), 439–440
Phospholamban ablation, 403

Phosphorylation, myosin regulatory light chain (RLC), 345–352
PhysGen program for genomic applications, 313f, 314
Physiological phenotyping, 313–314
Physiomic approach, 246
Pitx genes during cardiovascular development
  antibodies, immunohistochemistry, and in situ hybridization, 82
  chromatin immunoprecipitation assays, 82
  endothelin receptor as *Pitx1/Pitx2* target, 84–86
  genetic manipulations, 81
  Pitx1/Pitx2
    in cardiac neural crest, 82–84
    in cardiac outflow tract development, 86
    during embryogenesis, 82
  Pitx2 and cardiac development, 89–95
    abnormal venous return, 92
    congenital heart disease, 93–94
    developing heart, 90
    domains during development, 91–92
    isomerism, atrial and ventricular, 92
    left/right cardiac asymmetry, 90–91
    left/right signaling defect, 92–93
  pitx2 and *lefty* genes, 29–30
  pitx2 expressions, 37–38, 40–41
  single-cell nuclear microinjection assays, 82
PKA phosphorylation
  hyperphosphorylation of RyR2, 538
  of RyR2, 534–535
Placenta and heart development, 129
Plaque angiogenesis, 471–482
  angiogenesis in clinical use, 479–480
  angiogenesis inhibitor treatments, 478
  angiogenesis stimulators vs. inhibitors, 478–479
  arterial growth and blocking of, 479
  vs. arteriogenesis, 471–473
  blocking to stabilize lesions, 478
  cardiovascular application of research, 471

cell proliferation in atherosclerotic lesions, 475
function to sustain growth, 474
intimal capillaries, 475
neovascularization and plaque instability, 476–477
pathologic description of, 473–474
plaque growth promotion, 475–476
potential regulators of, 477–478
in vascular disease, 473
Plaque neovascularization, 473f
Plasma lipid levels, 494–495
Platelet activation, PARS and, 201–202
Platelet derived growth factors (PDGFs), 267, 545
Platelet factor 4, 479
PLC-β3 activation, impairment of, 278–279
PLC in VPF/VEGF signaling, 277
Plexin subunits, 58
PlGF (placenta growth factor)
survival during microvascular remodeling, 181–187
and VEGF-A signaling mechanisms, 184
PLN gene, 403–404
PLN inhibition for heart failure, 405
Podoplanin mucoprotein, 193
Poison peptide theory, 410
Poison polypeptides, 384
Polarized expression of angiopoietin-1, 171–172
Polycystic ovary syndrome (PCOS), 219
Polymorphic ventricular tachycardia (PVT), 333–337
Polypeptides, poison, 384
Polyunsaturated fatty acids, 493, 496
Pontin protein, 304
Potassium channel openers, 395
PP1/PP2A (phosphatases 1 and 2A), 534
PPARα
mediating alterations in energy metabolism, 376
PPARα/PGC-1α complex, 375f
PPARα/PGC-1α regulatory pathway, 376–379
PPARα/RXRα complex, 376
regulating cardiac lipid utilization, 373–374
signaling, derangements in, 374–376
Pressure overload hypertrophy, 412
Primary vasospasm, 393
Primitive cytokines, 364
Proinflammatory cytokines, 363–368
Prokineticin-1, 218
Proliferation of cardiac myocytes, 45–48
Protease-activated receptors (PARs).
See PARs (protease-activated receptors)
Protein A (VORA), 218
Protein-binding assay, in vitro, 4
Protein/DNA and protein/protein complexes, 7–9
Proteolytic mechanism of PAR1 activation, 199f
Proteomics, 224, 246
Prox-1
and lymphatic endothelial cells, 194
tumors stained with, 244
Pseudohypoaldosteronism type II, 446
Pulmonary hypertension, 309
Purkinje fibers, 111, 353–354, 359
Putative lymphatic markers, 244
pVHL protein, 130

Q

QRS interval prolongation in Nkx2.5 mice, 318–319
Quiescent hypervascularity and HIF-1α function, 133–142
enhancement of angiogenesis, 134
gain of HIF-1α and angiogenesis, 137–139
HIF-1α in disease, 137
HIF-α molecules, regulation of, 136–137
hypoxia and HIF, 134–136
microvessel architecture, 133–134
therapeutic angiogenesis, 133–134
tissue hypoxia, 139

R

Raf kinases, mutant, 285–291
angiogenesis regulation by integrins, 285–286
neovascular apoptosis and tumor regression, 287–289
Raf as antiangiogenic target, 286
targeted delivery of, 287–289
RAS (renin-angiotensin-aldosterone)
Ras-mediated hypertrophy, 430–432
Ras-Raf-MEK-ERK cascade, 286
RBPJ-k binding sites, 68f
RBPJ-k protein, 66
Receptor tyrosine kinases, 267–268
Remodeling and sprouting, 181–182
Renal cell carcinoma xenograft tumors, 262
Renal disease, 309
Renal failure, 447
Renal glomerulus (zebrafish), 305
Renal hypertrophy, 453
Renal salt homeostasis, 445
Renin-angiotensin system, 451–457
angiotensin-converting enzyme (ACE), 452–453
angiotensin II receptors, 453–455
angiotensinogen, 451–452
homologs of ACE, 453
renin gene, 452
Reptin protein, 303–304
Restrictive cardiomyopathy, 432–433
Retinoic acid (RA) and erythropoietin (EPO) signaling, 45–48
Retinopathy, 146, 479
RGD peptides, 147
Right ventricular cardiomyopathy, 401
Rip1Tag2 model of multistage carcinogenesis, 294
RLC (regulatory light chain) phosphorylation, 345–352
RNA, synthetic, 16
Ro-415253, 47
ROMK potassium channel, 448–449
RyR2 (ryanodine receptors)
macromolecular complex, 534
mutations, congenital, 538–539
regulation by PKA phosphorylation, 534–535
triggered arrhythmias and leaky, 537–539

S

Salt and blood pressure, 445–450
blood pressure-lowering mutations, 447
blood pressure-raising mutations, 446–447
human genetic studies, 447–449
Salt-sensitive hypertension, 309
Santa mutants, 303
SAP domain proteins, 99–100
SAPK (stress-activation protein kinases), 429
Sarcoglycan complex in muscles, 389–397
coronary artery vasospasm, 392–393
DGC and sarcoglycan as mechanosignaling complex, 390–392
dystrophin and associated proteins, 389–390
muscle degeneration, 390
primary vasospasm, 393
skeletal muscle disease, 392
smooth muscle sarcoglycan, 392–393
Sarcomere protein genes, 384
Sarcomeric mutations, 409–415
Sarcoplasmic reticulum (SR), 533
schmalspur mutation (zebrafish), 30
Schwann cells, 158
Secondary heart field and extra-cardiac lineages, 108
Secretory Trap expression cloning, 268

Semaphorins, 58–59
Septal defects, 49
Ser-15 phosphorylation, 346–347
SERCA2 (SR calcium ATPase 2), 403–404
Serrate ligands, 66
seven-up (*svp*) gene, 73
Signal transduction, hypoxia-mediated, 127–128
Signaling
    cell cycle, 163–170
    EG-VEGF in endocrine epithelial cells, 219
    pathways, 10f, 410–412, 429, 435
    during sprouting and remodeling, 181–182
    survival, 181–182
Single-cell nuclear microinjection assays, 82
Sinoatrial (SA) node, 528
Sitosterolemia, 500–502
Skeletal muscles
    disease of, 392
    grafting, 520–521
    IGF-1 isoforms and disease of, 507–518
    myofiber hypertrophy, 339–340
Skeletal myoblasts, 520–521
slow MyHC3 gene, 74–76
Smad proteins, 1, 4–7
smo (*slow mo*) mutants, 304
Smooth muscle cells, 143
Smooth muscle sarcoglycan, 392–393
SOD1 gene, 514
SOD (superoxide dismutase). *See* Extracellular superoxide dismutases (ecSOD)
southpaw gene, 30–31, 33
Spatial gradient of RLC phosphorylation, 347–349
Splotch embryos, 58
Sprouting
    angiogenesis (zebrafish), 304–305
    and remodeling, 181–182
spt5 and *spt6* genes, 20t, 23
squint gene, 30, 32f
SR-BI, 463, 467–468
SREBPs (sterol regulatory element-binding proteins), 491–498
    definition and fundamentals, 547
    isoforms, 492–493, 495–497
    SREBP expression and plasma lipid levels, 494–495
    structure and activation, 491–492
    in vivo function of, 493–494
SRF (serum response factor)
    affinity, 15
    in cardiac development, 97–105
    abnormalities in HOP mutant mice, 102
    antagonism of SRF by HOP, 102–103
    antithetical phenotypes of cardiomyocytes in HOP mutants, 103
    cell growth and muscle differentiation, 97–98
    feedback loop between Nkx2.5 and HOP, 103–104
    frog embryo gene expression by myocardin, 100–101
    inhibition of SRF by HOP, 101–102
    MADS box proteins, 97
    myocardin, 98–99
    smooth muscle-specific expression of myocardin, 99
    SRF-dependent transcription by myocardin, 99–100
    SRF gene expression in HOP mutant hearts, 102
    and MEF2 factors, 16
Stem cells
    bone marrow, 523t
    enhancement of angiogenesis using, 134
    grafting, hematopoietic, 522–523
Steroidogenic glands, 219–220
Stress responses, heart, 363–370
    cytokines
        adaptive effects of, 363–367
        maladaptive effects of, 367–368
    myocardial stress response overview, 363
Stretch-activation response, 346–347, 348f
Stromal cells, 241–242
SU5416 angiogenesis inhibitor, 295–297
Sudden cardiac death (SCD), 538–539
suppressor of hairless (su(H)) protein, 66, 157
Survival mechanisms of VEGF and PlGF, 181–187
    in developing microvasculature, 182–183
    pericytes and, 183
    signaling during sprouting/remodeling, 181–182
    in tumor angiogenesis, 183–184
    VEGF-A and PlGF signaling, 184
Syndecans in ectoderm, 40
Synthetic RNA, 16
Systemic vascular disease, 546–547
Systems biology platforms, 309–311
Systolic hypertension, 309

T

T-Box transcription factors, 115
Tachycardia. *See also* Catecholaminergic polymorphic ventricular tachycardia (CPVT); Polymorphic ventricular tachycardia (PVT); Ventricular tachycardia, 527
Targeted therapies, 224
TATFGp peptide, 277–279
Tbx1
    and DiGeorge syndrome, 60–61
    functions, 327–330
TBX5 mutations
    effects in humans and zebrafish, 306
    human, 115–120
        Holt-Oram syndrome (HOS), 115–116
        modifying factors of, 116–119
    Nkx2-5 and, 112
Tetralogy of Fallot, 115
TGFβ
    suppression of angiogenesis by, 242
    TGFβ1, 2, and 3, 50
Therapeutic angiogenesis, 133–134
Thrombin
    action on cells, 197–198
    cellular responses triggered by, 198–199
    generation of, 197
    receptors in human and mouse platelets, 201–202
    signaling in embryonic development, 203–205
Thrombosis, 197, 202
Thrombospondins (TSPs), 147–148, 479
Tie receptors, cloning, 267–268
TIMPs (tissue inhibitors of matrix metalloproteases), 367
Tin derivatives, 6f
tinman gene
    expression and transcripts, 73
    induction by Dpp signals, 3–4
    induction of, 1
    Nkx2-5 and, 110
    protein interaction with HMG-D, 8f
    and Smad proteins interaction, 5f
Tissue factor, 197
Tissue homeostasis, 510t
Tissue hypoxia, 139
Tissue remodeling, 417–427
titin gene, 306
Torsion, cardiac, 345–352
Trabecular myocytes, 111
Transcription factors
    cardiac pathways, 110
    membrane-bound, 491

SUBJECT INDEX

TBX5. *See* TBX5 mutations, human
Transdifferentiation events, 522
Transdominant inhibition of integrins, 149f
Transgene reporters, 15
Transgenesis
 in amphibians, 13–14
 transgenic analysis of ANF promoter, 71–79
 transgenic animals, genetic manipulation in, 430
 transgenic mouse lines, 410
 transgenic studies
  of cardiac actin promoter, 15
  of myosin light-chain 2a promoter, 15–16
 transgenic VEGF, 269
Tricuspid artesia, 115
Truncus arteriosis, 115
TSP-1 and TSP-2, 147–148
Tumor angiogenesis
 genetic approach to understanding, 249–254
 Id as target, 253–254
 Id proteins
  biology of, 249–250
  and endothelial precursors, 252–253
  role within tumors, 250–252
 negative regulators of, 146–147
 roles of angiopoietins and VEGFs, 271
 survival mechanisms in, 183–184
 tumor growth and, 144
Tumors. *See also* Pancreatic islet carcinomas
 angiogenesis and lymphangiogenesis in, 239–248
 delivery of mutant Raf to tumor vessels, 289f
 growth of, 239
 hormone-dependent, 245f
 human gall bladder, 242
 hypoxia in, 246
 inhibition of growth in Id knockout mice, 251f
 and lymphatic metastasis, 191–192
 regression of, 285–291
 secretion of angiogenic cytokines, 275
 tumor necrosis factor (TNF), 363–365
 VEGF inhibition to treat, 217
Tumstatin, 148, 263–264
Tunnels
 in clinical setting, 212–213
 definition and fundamentals, 209
 morphology of, 209–210
 in vitro, 210–211
 in vivo, 211–212
Type IV collagen, 255–266
 angiogenesis and basement membranes (BMs), 255–257
 arresten, human (26 kD), 262
 basement membranes, early studies with, 258–260
 canstatin, human (24 kD), 262–263
 gene organization, structure, and function, 257–258
 NC1 domains inhibiting angiogenesis, 260–262
 tumstatin, 263–264

U

UAS-tinΔ42-301 and UAS-tinΔ42-124 transformants, 4
Ufd1 gene, 327–328
Uric acid, 485–487

V

valentine mutants, 303
Valves
 cardiac, 49
 valve morphogenesis, 50
 valvular lesions, 115
Vasa vasorum, 474
Vascular basement membrane (VBM), 255–256, 260f, 261f
Vascular beds, development of, 544–546
Vascular development, handedness of, 175–176
Vascular disease
 local and systemic, 546–547
 plaque angiogenesis in, 473
Vascular domains concept, 179–180
Vascular endothelia
 arterial and venous, 155–156
 KDR and Flt-1 signaling pathways in, 275–277
 in tissue remodeling, 417–427
  adipose tissue growth and regression, 419–420
  adipose tissue plasticity, 420–422
  cardiac and adipose remodeling similarities, 422–423
  vascular remodeling, studying, 419
Vascular formations
 malformations induced by VEGF-A, 227–237
 normal and pathologic, 267–273
Vascular injury, 309
Vascular network, polarity of, 176
Vascular permeability
 Ang1 overexpression and, 422
 angiopoietin-1 and, 269
 effect of host-tumor interaction on, 242f
 insights from intravital microscopy, 239–240
 in mice, 139f
 VPF (vascular permeability factor), 275–283
Vascular precursors, 134
Vascular receptors, 223
Vascular remodeling, studying, 419
Vascular spasm schematic, 394f
Vascular stenosis, 527
Vascularization in tumors, 241f
Vasculature
 lymphatic development, 189–190
 mapping human, 223–224
Vasculogenesis
 defined, 217
 zebrafish, 304
Vasopeptidase inhibitors, 418
Vasospasms
 coronary artery, 392–393
 from Microfil, 394f
 primary, 393
VEGD-A, 271
VEGF (vascular endothelial growth factor)
 acting upstream of Notch, 158–159
 and angiopoietins, 171–180
 effects on blood vessels, 171–180
 EG-VEGF and, 218–219
 G proteins and, 275–283
 inducer of angiogenesis, 255, 258
 and lymphatic vessels, 189
 in mutant embryos, 128–129
 production and release of, 240–242
 and retinal angiogenesis, 146
 role in vasculature development, 544–546
 survival during microvascular remodeling, 181–187
VEGF-A
 angiogenesis, arteriogenesis, and lymphangiogenesis induced by, 133–134, 227–237
 biological properties of, 138
 and PlGF signaling mechanisms, 184
 regulator of vascular development, 158
 role in survival of new blood vessels, 182–183
 vascular response to, 228
VEGF-C/-D signaling, 244
VEGF-C growth factor, 189, 270
VEGF-D, 190, 270
VEGF-R2 cells, 253
VEGF/VEGFR inhibitors, 295
VEGF/VEGFR2 signaling, 149
VEGFR-3 growth factor, 189–192
VEGFs and angiopoietins in vascular formation, 267–273

VEGFs and angiopoietins in vascular formation (*continued*)
    angiopoietin-1, 268–269
    angiopoietin-2, 269–270
    angiopoietin-2 and lymphatic vessel development, 270–271
    binding and multimerization motifs, 268
    Tie receptors, cloning of, 267–268
    tumor angiogenesis, 271
Veins
    arteries, and lymphatics compared, 545–546
    arteries, VEGF, and Notch, 155–162
    venous return, abnormal, 92
    venous vs. arterial network development, 176–177
Ventricular and atrial septum defects, 69
Ventricular cardiac myocytes of Nkx2.5 mice, 319–322
Ventricular fibrillation, 537
Ventricular hypoplasia, 121–125
    DNA-binding transcription factors, 122–123
    epigenetic factors in ventricular development, 123–124
Ventricular morphogenesis, 122
Ventricular septal defects (VSDs), 49, 116
Ventricular tachycardia (VT)
    ablation and, 527
    induced by ventricular pacing, 321f
    initiated by DADs, 537
Verapamil treatment, 393
Vertebrates
    chamber formation, 73–75
    early cardiac development in, 42–43
    myocardial layer of vertebrate heart, 71
Vinblastine, 297

Vinculin mutations, 50
Vitamin E supplementation, 488
VLDL (very-low-density lipoprotein), 494–495
Von Hippel-Lindau (VHL) disease, 130–131
VPF (vascular permeability factor)/VEGD signaling, 275–283
VVOs (vesiculo-vacuolar organelles), 228–231

W

W4RMLP mutation, 402–403
Wingless signals in cardiac induction, 9–11
Working cardiomyocytes, 353

X

Xenopus embryos
    asymmetries and monocilia in, 39–40
    cardiac muscle differentiation in, 13–18
    early steps in, 42
    ectopic gene expression in, 38–39
    gastrula stages of development, 40f
    myocardin injected, 100–101
    nodal signaling pathway, 29–30
Xenopus transgenesis procedure, 75
XMLC2a promoter, 16f
XNkx2-5.GFP transgene, 14

Y

Yeast screens and assays, 4
Yin/yang of innate stress responses, 363–370

Z

Z disc proteins, cytoskeletal, 399–408
Zebrafish
    and arterial-venous identity, 156
    cardiovascular compendium, 301–307
    chamber form and pattern, 302–303
    disease models, 304–305
    heart growth, 303–304
    system assembly and growth, 301–302
    vessel growth, 304–305
    endocardial cushion formation in, 49–56
    expression of *hey* genes in, 66
    genetic regulation of cardiac patterning in, 19–25
    heart fields, 20–22
    as model organism, 20
    myocardial differentiation, 22–23
    organ fields, 19
    pandora and *foggy* mutations, 22–23
    gridlock (grl) mutation, 67–68
    grl (*Hey2*) to build aorta, 67–68
    hearts and minds, 27–36
        BMP antagonism by *cyclops*, 31–32
        cardiac/central nervous system asymmetries, 33–34
        left-right asymmetries, 27–30
        nodal expression in vertebrates, 32–33
        Nodal-Lefty-Pitx2 cassette, 28–29
        southpaw gene, 30–31
    history and limitations of, 547
    monociliated cells in, 39
    and Notch signaling, 156
Zygotic mutants, *pan* and *fog*, 23